YOUR GUIDE TO
Developmental Biology

9

The Genetics of Axis Specification in *Drosophila*

THANKS LARGELY TO STUDIES spearheaded by Thomas Hunt Morgan's laboratory during the first two decades of the twentieth century, we know more about the genetics of *Drosophila melanogaster* than that of any other multicellular organism. The reasons have to do with both the flies themselves and with the people who first studied them. *Drosophila* is easy to breed, hardy, prolific, and tolerant of diverse conditions. Moreover, in some larval cells, the DNA replicates several times without separating. This leaves hundreds of strands of DNA adjacent to each other, forming polytene (Greek, "many strands") chromosomes (**FIGURE 9.1**). The unused DNA is more condensed and stains darker than the regions of active DNA. The banding patterns were used to indicate the physical location of the genes on the chromosomes. Morgan's laboratory established a database of mutant strains, as well as an exchange network whereby any laboratory could obtain them.

Historian Robert Kohler noted in 1994 that "The chief advantage of *Drosophila* initially was one that historians have overlooked: it was an excellent organism for student projects." Indeed, undergraduates (starting with Calvin Bridges and Alfred Sturtevant) played important roles in *Drosophila* research. The *Drosophila* genetics program, says Kohler, was "designed by young persons to be a young person's game," and the students set the rules for *Drosophila* research: "No trade secrets, no monopolies, no p

Jack Schultz (originally in Morgan's laboratory) and other burgeoning supply of data on the genetics of *Drosophila* to its dev was a difficult organism on which to study embryology. Fly e and intractable, being neither large enough to manipulate experi

What changes in development caused this fly to have four wings instead of two?

The Punchline

The development of the fruit fly is extremely rapid, and its body axes are specified by factors in the maternal cytoplasm even before the sperm enters the egg. The anterior-posterior axis is specified by proteins and mRNAs made in maternal nurse cells and transported into the oocyte, such that each region of the egg contains different ratios of anterior- and posterior-promoting proteins. Eventually, gradients of these proteins control a set of transcription factors—the homeotic proteins—that specify the structures to be formed by each segment of the adult fly. The dorsal-ventral axis is also initiated in the egg, which sends a signal to its surrounding follicle cells. The follicle cells respond by initiating a molecular cascade that leads both to cell-type specification and to gastrulation. Specific organs form at the intersection of the anterior-posterior axis and the dorsal-ventral axis.

Scientists Speak

In these interviews, emerging topics in developmental biology are discussed by leading experts in the field.

(A) Antenna   (B)

 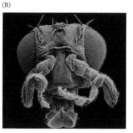

FIGURE 9.26 (A) Head of a wild-type fruit fly. (B) Head of a fly containing the *Antennapedia* mutation that converts antennae into legs. (A © Eye of Science/Science Source; B © Science VU/Dr. F. Rudolph Turner/Visuals Unlimited, Inc.)

can be detected in specific regions of the embryo (see Figure 9.24B) and are especially prominent in the central nervous system.

SCIENTISTS SPEAK 9.3 Listen to this interview with Dr. Walter Gehring, who spearheaded investigations that unified genetics, development, and evolution, leading to the discovery of the homeobox and its ubiquity throughout the animal kingdom.

Web Topic

Here you are provided with more information about cutting-edge topics, as well as historical, philosophical, and ethical perspectives, in addition to links to online resources.

WEB TOPIC 9.5 **INITIATION AND MAINTENANCE OF HOMEOTIC GENE EXPRESSION** Homeotic genes make specific boundaries in the *Drosophila* embryo. Moreover, the protein products of the homeotic genes activate batteries of other genes, specifying the segment.

Generating the Dorsal-Ventral Axis

Dorsal-ventral patterning in the oocyte

As oocyte volume increases, the oocyte nucleus is pushed by the growin_ to an anterior dorsal position (Zhao et al. 2012). Here the *gurken* mess_ been critical in establishing the anterior-posterior axis, initiates the f_

Developing Questions

Homeobox genes specify the anterior-posterior body axis in both *Drosophila* and humans. How come we do not see homeotic mutations that result in extra sets of limbs in humans, as can happen in flies?

Developing Questions

These questions are an entry-way for independent research, empowering you to expand your knowledge and enhance your participation in class discussion.

Next Step Investigation

This feature provides insights into some of the field's greatest challenges, inspiring curiosity and further exploration.

Next Step Investigation

The precision of *Drosophila* transcription patterning is remarkable, and a transcription factor may specify whole regions or small parts. Some of the most important regulatory genes in *Drosophila*, such as the gap genes, have been found to have "shadow enhancers," secondary enhancers that may be quite distant from the gene. These shadow enhancers seem to be critical for the fine-tuning of gene expression, and they may cooperate or compete with the main enhancer. Some of these shadow enhancers may work under particular physiological stresses. New studies are showing that the robust phenotypes of flies may result from an entire series of secondary enhancers that are able to improvise for different conditions (Bothma et al. 2015).

Closing Thoughts on the Opening Photo

In the fruit fly, inherited genes produce proteins that interact to specify the normal orientation of the body, with the head at one end and the tail at the other. As you studied this chapter, you should have observed how these interactions result in the specification of entire blocks of the fly's body as modular units. A patterned array of homeotic proteins specifies the structures to be formed in each segment of the adult fly. Mutations in the genes for these proteins, called homeotic mutations, can change the structure specified, resulting in wings where there should have been halteres, or legs where there should have been antennae (see pp. 242–243). Remarkably, the proximal-distal orientation of the mutant appendages corresponds to the original appendage's proximal-distal axis, indicating that the appendages follow similar rules for their extension. We now know that many mutations affecting segmentation of the adult fly in fact work on the embryonic modular unit, the parasegment (see pp. 234 and 240). You should keep in mind that, in both invertebrates and vertebrates, the units of embryonic construction often are not the same units we see in the adult organism. (Photograph courtesy of Nipam Patel.)

Closing Thoughts on the Opening Photo

Coming full circle, this feature relates chapter concepts back to the Opening Question and Photo.

Snapshot Summary

This closing feature provides you with a step-by-step breakdown of the chapter text.

9 Snapshot Summary
Drosophila Development and Axis Specification

1. *Drosophila* cleavage is superficial. The nuclei divide 13 times before forming cells. Before cell formation, the nuclei reside in a syncytial blastoderm. Each nucleus is surrounded by actin-filled cytoplasm.

2. When the cells form, the *Drosophila* embryo undergoes a mid-blastula transition, wherein the cleavages become asynchronous and new mRNA is made. At this time, there is a transfer from maternal to zygotic control of development.

3. Gastrulation begins with the invagination of the most ventral region (the presumptive mesoderm), which causes the formation of a ventral furrow. The germ band expands such that the future posterior segments curl just

• There is a *temporal order* _ genes are transcribed, and _ often regulate the expressi_

• *Boundaries* of gene expre_ the interaction between transcription factors and their gene targets. Here, the transcription factors transcribed earlier regulate the expression of the next set of genes.

• *Translational control* is extremely important in the early embryo, and localized mRNAs are critical in patterning the embryo.

• *Individual cell fates* are not defined immediately. Rather, there is a stepwise specification wherein

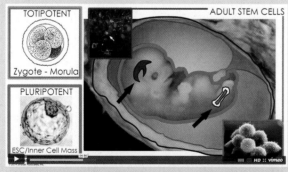

ties at high frequency (Huet et al. 2007; del Valle Rodríguez et al. 2011). Researchers are now able to identify developmental interactions taking place in very small regions of the embryo, to identify enhancers and their transcription factors, and to mathematically model the interactions to a remarkable degree of precision (Hengenius et al. 2014).

Early *Drosophila* Development

We have already discussed the specification of early embryonic cells by cytoplasmic determinants stored in the oocyte. The cell membranes that form during cleavage establish the region of cytoplasm incorporated into each new blastomere, and the morphogenetic determinants in the incorporated cytoplasm then direct differential gene expression in each cell. But in *Drosophila* development, cell membranes do not form until after the thirteenth nuclear division. Prior to this time, the dividing nuclei all share a common cytoplasm and material can diffuse throughout the whole embryo. The specification of cell types along the anterior-posterior and dorsal-ventral axes is accomplished by the interactions of components *within* the single multinucleated cell. Moreover, these axial differences are initiated at an earlier developmental stage by the position of the egg within the mother's egg chamber. Whereas the sperm entry site may fix the axes in nematodes and tunicates, the fly's anterior-posterior and dorsal-ventral axes are specified by interactions between the egg and its surrounding follicle cells prior to fertilization.

The Stem Cell Concept

Division and self-renewal

A cell is a stem cell if it can divide and in doing so produce a replica of itself (a process called **self-renewal**) as well as a daughter cell that can undergo further development. Stem cells are often referred to as undifferentiated due to this maintenance of proliferative properties[1]. Upon division, a stem cell may also produce a daughter cell that can mature into a terminally differentiated cell type. Cell division can occur either symmetrically or asymmetrically. If a stem cell divides symmetrically, it could produce two self-renewing stem cells or two daughter cells that are committed to differentiate, resulting in either the expansion or reduction of the resident stem cell population, respectively. In contrast, if the stem cell divides asymmetrically, it could stabilize the stem cell pool as well as generate a daughter cell that goes on to differentiate. This strategy, in which two types of cells (a stem cell and a developmentally committed cell) are produced at each division, is called the *single stem cell asymmetry* mode and is seen in many types of stem cells (**FIGURE 5.1A**). An alternative (but not mutually exclusive) mode of retaining cell homeostasis is the *population asymmetry* mode of stem cell division. Here, some stem cells are more prone to produce differentiated progeny, and this is compensated for by another set of stem cells that divide symmetrically to maintain the stem cell pool within this population (**FIGURE 5.1B**; Watt and Hogan 2000; Simons and Clevers 2011).

 DEV TUTORIAL
Stem Cell Basics

[1]There are many different stem cells and so their status as "undifferentiated" really only pertains to the retained ability to divide, but they are in fact a defined cell type.

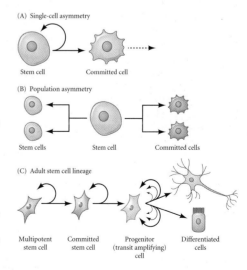

(A) Single-cell asymmetry

Stem cell   Committed cell

(B) Population asymmetry

Stem cells   Stem cell   Committed cells

(C) Adult stem cell lineage

Multipotent stem cell   Committed stem cell   Progenitor (transit amplifying) cell   Differentiated cells

m cell concept. (A) The fundamental characteristic of a stem cell is that it can make more of itself while also producing cells committed to differentiation. This is called asymmetric division. A population of stem cells may be maintained through population asymmetry, where a stem cell is shown to have the ability to divide symmetrically to produce either two stem cells (increasing the stem cell pool by 1) or to produce two committed cells (thus decreasing the pool). These divisions are termed symmetrical self-renewal and symmetrical differentiating. (C) In many adult tissues, lineages pass from a multipotent stem cell (capable of forming numerous types of cells) to a committed stem cell that makes one or very few cell types, to a progenitor (transit amplifying) cell that can undergo multiple rounds of divisions but whose fate is committed to becoming a particular differentiated cell.

Developmental Biology ■ Eleventh Edition

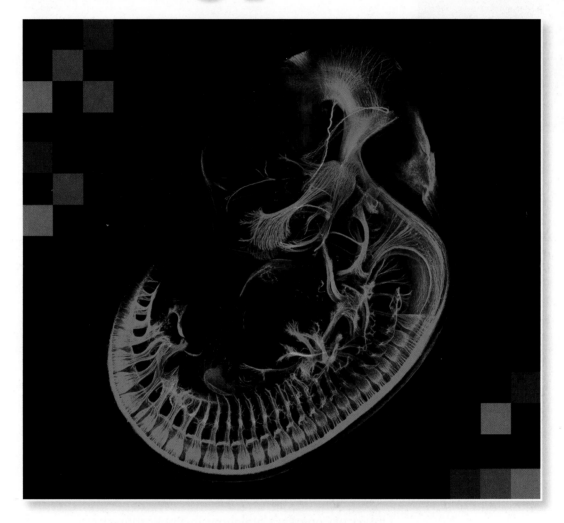

Scott F. Gilbert
Swarthmore College and the University of Helsinki

Michael J. F. Barresi
Smith College

 Sinauer Associates, Inc., Publishers ■ Sunderland, Massachusetts U.S.A.

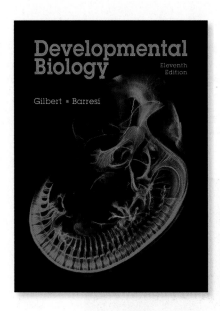

The Cover

The axons of the developing peripheral nervous system are stained red in this confocal micrograph of a whole mount mouse embryo at day 11.5 of development. The growth and specific targeting of axons during vertebrate development are discussed in Chapter 15. Photograph courtesy of Zhong Hua and Jeremy Nathans, Johns Hopkins University.

For more information, address:
Sinauer Associates, Inc.
P.O. Box 407
Sunderland, MA 01375 USA
FAX 413-549-1118
publish@sinauer.com
www.sinauer.com

Library of Congress Cataloging-in-Publication Data

Names: Gilbert, Scott F., 1949- | Barresi, Michael J. F., 1974-
Title: Developmental biology / Scott F. Gilbert, Swarthmore College and
  the University of Helsinki, Michael J.F. Barresi, Smith College.
Description: Eleventh edition. | Sunderland, Massachusetts :
  Sinauer Associates, Inc., [2016] | Includes bibliographical references and index.
Identifiers: LCCN 2016012601 | ISBN 9781605354705 (casebound)
Subjects: LCSH: Embryology--Textbooks. | Developmental biology--Textbooks.
Classification: LCC QL955 .G48 2016 | DDC 571.8--dc23
LC record available at http://lccn.loc.gov/2016012601

Printed in U.S.A.

5 4 3 2

Brief Contents

Contents

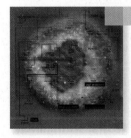

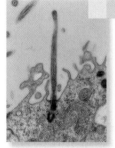

CHAPTER 4
Cell-to-Cell Communication: Mechanisms of Morphogenesis **95**

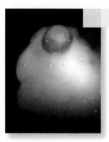

CHAPTER 5
Stem Cells: Their Potential and Their Niches **143**

PART II ■ Gametogenesis and Fertilization: The Circle of Sex

CHAPTER 6
Sex Determination and Gametogenesis 181

CHAPTER 7
Fertilization: Beginning a New Organism 217

PART III ■ Early Development: Cleavage, Gastrulation, and Axis Formation

CHAPTER 11
Amphibians and Fish 333

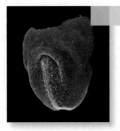

CHAPTER 12
Birds and Mammals 379

PART IV ■ Building with Ectoderm:
The Vertebrate Nervous System and Epidermis

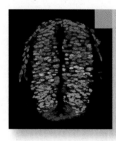
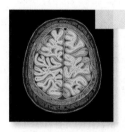
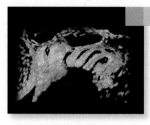

CHAPTER 16
Ectodermal Placodes and the Epidermis 517

PART V ■ Building with Mesoderm and Endoderm: Organogenesis

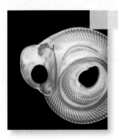

CHAPTER 17
Paraxial Mesoderm:
The Somites and Their Derivatives **539**

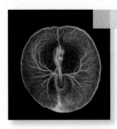

CHAPTER 18
Intermediate and Lateral Plate Mesoderm: Heart, Blood, and Kidneys **581**

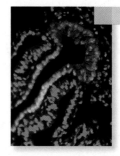

PART VI ■ Postembryonic Development

CHAPTER 22
Regeneration 693

CHAPTER 23
Aging and Senescence 723

PART VII ■ Development in Wider Contexts

CHAPTER 24
Development in Health and Disease: Birth Defects, Endocrine Disruptors, and Cancer 735

CHAPTER 25
Development and the Environment: Biotic, Abiotic, and Symbiotic Regulation of Development 763

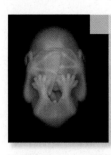

From the Authors

From Scott Gilbert

A BIOLOGIST, A PHILOSOPHER, AND A THEOLOGIAN WALK INTO A BAR. Yes, it actually happened, in the chill of a winter night in Finland! A group of enthusiastic people listened as the moderator asked what each of them considered to be the most important story people need to know. The Christian theologian said that the most important story was salvation through God's grace. The analytic philosopher disagreed, saying that the most important story for mankind was that of the Enlightenment. The developmental biologist knew that he was supposed to say "evolution." But evolution is the consequence of another, more fundamental story. So the biologist claimed that most inspiring and meaningful story was how the embryo constructs itself. You pass from unformed zygote to the adult organism with its heart, brain, limbs, and gut all properly differentiated and organized. It is a story of how newness is created, how one keeps one's identity while building oneself, and how global forces and local forces work together to generate a functional entity. This is the story we tell in this book.

In the Ninth and Tenth editions of *Developmental Biology*, we speculated that the study of animal development was undergoing metamorphosis. The field has not reached the climax phase yet, but certain differences between the previous edition and one in your hands (or on your screen) are definitely apparent. The first can be seen on the cover. Developmental biology has been charged with a huge undertaking—nothing less than discovering the anatomical and genetic bases of neural organization and behaviors. This task was part of developmental biology when it was reformulated in the early 1900s (especially by the American C. O. Whitman), but it had dropped out of the portfolio as being "too complicated" and not amenable for study. Today, however, developmental neurobiology is an increasingly large part of developmental biology. Among many other things, developmental biology is becoming necessary for cognitive science.

The second difference between this and previous editions is the prominence of stem cells. From being a small area of developmental biology, stem cell research has grown so fast as to have its own scientific societies. Not only do stem cells provide explanations for organ development, they also hold the tantalizing possibility of organ regeneration. Recent work, detailed in this book, shows how knowledge of developmental biology has been critical in turning adult cells into stem cells that can functionally replace missing and damaged tissue in laboratory animals.

A third difference is the incredible revolution in lineage studies made possible by in vivo labeling. We can look at each cell developing in an early, living, embryo and discern which adult cells are its descendants. The techniques of computer-enhanced visualization have given scientists amazing new technologies to see embryonic development.

A fourth difference is the idea that animal development, even that of mammals, is significantly influenced by the environment. The data that have accumulated for developmental plasticity and the roles of microbes in normal development have increased remarkably over the past several years.

Finally, a fifth difference concerns the way science is taught. The "sage on the stage" model, where lectures generate the flow of information down a gradient from higher concentration to lower, has been supplemented by the "guide on the side." Here, the professor becomes a facilitator or capacitator of discussion while the students are encouraged to discover the information for themselves.

Indeed, education is sometimes referred to as "development," and there are many similarities between education and embryology. The two fields have exchanged metaphors constantly for the past two centuries, and two German words that have been used for both development and education—*Bildung* and *Entwicklung*—connote education by experience and education by instruction, respectively. Both work in different situations. So in this edition of *Developmental Biology*, we have tried to facilitate those professors who wish to experiment with different teaching methods. As in embryology, we don't expect one method to be best for all occasions.

To all these ends, this book has metamorphosed to embrace a co-author. Michael J. F. Barresi is expert in all these areas of stem cells, developmental neurobiology, and new techniques of learning and teaching. It's been 30 years since the first edition of this book was published, and I wanted a young professor to reconfigure this book into a learning tool that a new generation of teachers could use to inspire a new generation of students. Enter Michael. Michael did not want just cosmetic changes in the book. He proposed a radical re-envisioning of its mission: to educate students to appreciate and participate in developmental biology.

Michael convinced us that we needed to rearrange the order of the chapters, add some chapters and shorten others, alter the ways that the material is presented within the chapters, and give all chapters more supplemental material for "flipped" classes, case studies, and other means of learning. The extra thought and effort that went into incorporating Michael's new approaches have clearly been worth it.

One other thing that has changed in the past decade is the realization of how much our understanding of biology depends on our knowledge of development. If "nothing in biology makes sense except in the light of evolution," we now find that "nothing in morphological evolution makes sense without knowledge of development." Changes in adult anatomy and physiology are predicated on changes in morphogenesis and differentiation during development.

This is also true of the history of biology, where developmental biology can be seen to play the unique role of "the stem cell of biological disciplines," constantly regenerating its own identity while simultaneously producing lineages that can differentiate in new directions. As Fred Churchill noted, *cell biology* "derived from descriptive embryology." The founders of cell biology were each trying to explain development, and their new conception of the cell helped them do it. The original theories of *evolution* concerned themselves with how new variants arose from the altered development of ancestors. Charles Darwin's friend and champion Thomas Huxley, expanded on this idea, which would eventually flourish into the field of *evolutionary developmental biology*.

Also during this Victorian age, a variant of developmental biology grew to become the field of *immunology*. Elie Metchnikoff (who showed the pole cells of flies to be germ cell precursors and who studied gastrulation throughout the animal kingdom) proposed a new cell theory of immunology in his attempt to find universal characters of the embryonic and larval mesoderm. Similarly, but with more anguish, *genetics* directly descended from a generation of embryologists who dealt with whether the nucleus or cytoplasm contained the determinants of embryonic development. Before his association with *Drosophila*, Thomas Hunt Morgan was a well-known embryologist who worked on sea urchin embryos, wrote a textbook on frog development, and was an authority on regeneration. Many of the first geneticists were originally embryologists, and it was only in the 1920s that Morgan formally separated the two fields. And *regeneration* is still intimately linked with development, for regeneration often is a recapitulation of embryonic processes. Ross Granville Harrison and Santiago Ramon y Cajal founded the science of *neurobiology* by showing how the brain and axons develop. To this day, neurology requires an understanding of the developmental origins of the central and peripheral nervous systems.

Several medical disciplines descend from embryology. *Teratology* (the study of congenital anomalies) has always studied altered or disrupted development, but other medical disciplines also trace back to embryology. *Cancer biology*—oncology— derives from developmental biology, as cancers have long been perceived and studied as a cell's reversion to an embryonic state. Although this view was at one time eclipsed by a strictly genetic view of cancer, today it is being revived and revised by the discoveries of cancer stem cells, paracrine factor regulation of tumor initiation, and embryonic modes of cell migration used by tumor cells. Medical disciplines such as cardiology and diabetes research are being invigorated by new developmental perspectives. And the new fields of *endocrine disruption* and the *developmental origin of health and disease*, looking at how environmental factors experienced during pregnancy can alter adult phenotypes, have emerged from developmental biology with their own paradigms and rules of evidence.

The developmental biology stem cell produces new disciplines even as it keeps its own identity. The field of *stem cell biology* is directly linked to its parent discipline, and new studies (many of them documented in this text) show how directing stem cells to differentiate in particular ways demands knowledge of their normal development.

Developmental biology interacts with other disciplines to induce new ways of thinking. Ecological developmental biology, for instance, looks at the interactions between developing organisms and their abiotic and biotic environments. Even the field of paleontology has been revolutionized by developmental perspectives that allow new and often surprising phylogenies to be constructed.

In short, this is an exciting time for this textbook to promote an interactive way of perceiving and studying the natural world. Pascal wrote that science is like a balloon expanding into the unknown. The more that we know, the greater the area in contact with the unknown. Developmental biology is a discipline where the unknown contains important questions yet to be answered, with new techniques and ideas for those ready to try.

Acknowledgments

It is becoming increasingly difficult to distinguish between an author, a curator, and a "nexus" in a node-link diagram. This book is a developing and symbiotic organism whose acknowledgments must either be confined to an inner layer or else expand throughout the world. First and foremost, I sincerely acknowledge that without Michael Barresi's enthusiasm, expertise, and passion for this project, this edition of the book would not exist.

The Sinauer Associates team, headed by Andy Sinauer and Rachel Meyers, has been remarkable. I have been incredibly privileged over the years to work with Sinauer Associates. I am also lucky to have had my words, sentences, and paragraphs, rearranged, reordered, and realigned by Carol Wigg, who has worked with me on all eleven editions to communicate the wonder of developmental biology in prose that is as clear, accessible, and enjoyable for students as we can possibly make it.

This is a beautiful book, and I can say that because it is not my doing. It is due to talent of Chris Small and his production staff; to Jefferson Johnson and his artistic mastery of Adobe InDesign; to the expertise of the artists at Dragonfly Media Graphics; and to photo editor extraordinaire David McIntyre, who manages to find incredible photographs to complement the many wonderful images my colleagues have so generously supplied for each edition.

I have been blessed with remarkable students who have never been shy about asking me questions. Even today they continue to send me "Did you see this?" emails that make sure I'm keeping current. I also thank all those people who continue to send me emails of encouragement or who come up to me at meetings to pass on good words about the book and provide me with even more information. This book is, and always has been, a community endeavor.

My wife, Anne Raunio, has put up with my textbook writing for most of our married life, and I know she'll be glad that this edition is finished. Indeed, just as this book goes to press our lives have shifted greatly with our move away from Swarthmore. I would certainly be remiss if I didn't acknowledge the many years of support I have enjoyed at Swarthmore College, a wonderful academic institution that deems textbook writing a service to the scientific community and that encourages interdisciplinary ventures.

—S.F.G.

From Michael Barresi

A NEUROSCIENTIST, AN ECO-EVO-DEVO BIOLOGIST, AND A DEVELOPMENTAL BIOLOGIST WALK INTO A POOL. Yes, it actually happened, on a scorching hot summer day in Cancun, Mexico! It was at the first Pan American Society for Developmental Biology Conference when Scott Gilbert mentioned to Kathryn Tosney and me that he was considering a co-author for the upcoming Eleventh Edition of *Developmental Biology*. While I waded in the water next to two of my heroes, Scott asked whether I might be interested in such an opportunity.

A combination of shock, excitement, and fear set in, pretty much in that order. *Shock*, because I was in wonderment of how I could be considered; after all, I had neither published a dozen papers a year nor had the historical perspective and cultural scope that Scott has woven so intricately and uniquely through each edition. *Excitement,* because this textbook has had such a great impact on my life. The chance to be part of a book that has been with me throughout my entire science education would be a true honor. Then *fear* set in because, as it does to me, this book means so much to so many in this field. The undertaking required to maintain the standard that Scott Gilbert has set for this work was daunting. However, if there is one thing I have learned in 11 years as a college professor, it is that fear can be the most significant barrier to innovative teaching and learning.

I agreed to be Scott's co-author because it presented an opportunity to influence how this subject is taught around the world. My enthusiasm for all aspects of the book is limitless, and I am passionately committed to improving the learning experience for all students. There is certainly no replacing Scott Gilbert, and I do not pretend to be Scott's equivalent. What I can offer to this and future editions of *Developmental Biology* is a complementary approach that builds upon Scott's accuracy and style with increased creativity and an overarching philosophy of student empowerment to learn about developmental biology.

The textbook and the classroom have something in common. Neither can survive this digital age as a mere vessel for information: pages of dense content paired with even denser lectures are not effective methods for "deep" learning. There is overwhelming evidence that true active learning pedagogies provide the most effective gains in conceptual understanding, longer retention of material, better problem-solving abilities, and greater persistence in STEM majors, particularly for underprepared students (Waldrop 2015; Freeman et al. 2014; Chi and Wylie 2014). I want my students and yours to learn the core concepts in developmental biology not by simply memorizing the text or stressfully scribing bullets off of a PowerPoint, but by *experiencing* how these concepts can explain known and unknown phenomena of development. How can a textbook adapt to (1) support teachers in implementing effective active learning approaches, and (2) encourage students to become active learners?

Carrying out effective active learning exercises in class that target concept acquisition and problem solving skill development is challenging. Potential challenges include a lack of activities to offer students and a lack of training on behalf of the instructor to administer those exercises, a shortage of available class time (real or perceived), student reluctance to participate in novel and challenging activities, uneven preparedness by students, and a whole range of associated fears.

We have transformed the Eleventh Edition of *Developmental Biology* to support a movement in pedagogy toward an active experience for both the professor and student. For many of the chapters, Scott Gilbert and I have written and produced "Dev Tutorials," short (10–20 minutes) video recordings of us explaining some of the basic principles of development. These professionally produced videos are designed to deliver some basic amount of content outside of class, thus providing instructors with a mechanism to conduct a "flipped" classroom (see Seery 2015).

To satisfy the in-class half of the "flipped" classroom, we wrote a set of case study problems that accompany the "Dev Tutorials" to encourage team-based learning approaches. Prior to conducting a case study activity, consider asking students to read the "Punchline" for a specific chapter as well as watching the related "Dev Tutorial." Completing this won't take students very long, so instructors can expect that each student will walk into class with a baseline of content exposure sufficient to *actively* engage in solving the case study. We intend to add more "Dev Tutorials" and "Case Studies" in the future, as user interest demands. We are excited to see how "Dev Tutorials" and case study problems can be tailored to meet the learning objectives of your own courses, and I, in particular, welcome the chance to work with faculty to help support their implementation of these new active learning resources.

Traditionally, the role of a textbook has been to introduce students to the core concepts of a given field; however, I don't feel this should be its only role. Textbooks can take advantage of the fact that, usually, the student is reading about the subject for the first time. This is the moment to capture a student's inquisitive spirit, build their confidence in discussing and asking questions about the subject, and fuel their future learning through a determined ownership of their place in the field. Gaining a sense of identity in a particular field of science often begins with the ability to engage in a dialogue. Unfortunately, for a student learning "the facts" for the first time, one of the most difficult barriers is being able to articulate the questions that would open up a substantive conversation.

Several unique mechanisms in the Eleventh Edition are intended to empower students to engage actively with the field of developmental biology. The "Developing Questions" found throughout each chapter function as suggested extensions and potential areas of future research on the topics being covered, and indirectly provide a model for the type of thinking and questions that developmental biologists might ask. These questions would be a huge success if students repeated them in class as a sort of ice-breaker to begin or further the discussion, or used them as entry points for supplemental literature research on their own. Most of these questions do not have definitive answers. Sorry, but they are designed to spur interaction in the classroom and engage students with the actual research. The potential of the thrill of discovery to motivate student interest cannot be underestimated. And students know the difference between quiz questions and life questions. To that end, each chapter ends with a "Next Step Investigation"; these play a similar role to the "Developing Questions," except they attempt to present a broader view of the directions the field may be moving in. The hope is that students can use "Next Step Investigations" as logical entry points for their own research.

One other objective for the Eleventh Edition has been to introduce the actual voices of the biologists working today. "Scientists Speak" is a new resource linked throughout the textbook to provide students (and faculty) with direct access to recorded conversations with leading developmental biologists. Many of these discussions took place between the lead investigators from current and seminal papers and my own students at Smith College through web conferencing technology. For students, the unique benefit of this type of resource is a highly approachable dialogue with the scientists combined with a fantastic array of questions asked by their peers—often the only individuals students really trust.

I sincerely hope these many new resources help to increase student engagement, improve their confidence to communicate, and truly invite everyone to become a significant participant in this most amazing science of development.

Acknowledgments

I wish to express my special and sincere thanks to Mary Tyler, who played a pivotal role as a content editor for my chapters. Mary has held a great love for this textbook over the years, and her perspectives helped me achieve a perfect balance between the past and present in this new edition. Thank you, Mary, for all of your support and focused, substantive input.

The field of developmental biology is ever-expanding and the pace of research seems like it is increasing exponentially. This comprehensive edition was only possible with the keen oversight of the expert reviewers listed on the following page. Thanks to Johannah Walkowicz for her unique balance of persistence and kindness in organizing all of the reviewers. I extend a special acknowledgmentf to Willy Lensch and Bill Anderson, who spent significant time with me discussing the field of stem cells, which directly influenced the organization of the new stem cell chapter.

I have been continually amazed by the stellar team at Sinauer Associates, Publishers. I have been humbled by Andy Sinauer's complete acceptance of me into this family. His open-minded consideration of all of my ideas was a critical factor in my acceptance to co-author this great book; thank you Andy for your support, and for compiling the most amazing staff! First Azelie Aquadro Fortier and then Rachel Meyers oversaw the entire production of this edition, and both provided this new co-author nothing but genuine encouragement and support at all times. Carol Wigg, Sydney Carroll, and Laura Green worked together to provide the precise editorial eyes needed, especially for this tired, father-of-four, first-time author. Your determination and equally long hours on this project produced a new edition that I know I can only be proud of because of your contributions.

I sincerely appreciate the vast amount of energy and time Sinauer's art director Chris Small and the entire group at Dragonfly Media Graphics took to produce such a beautiful art program. They also had to deal with me, an overprotective visual artist who was likely too critical to changes to his original drawings! Thanks for your patience. I'd also like to thank Chris again, as well as Joanne Delphia and Jefferson Johnson, for their excellent design and layout of the book. David McIntyre, thank you for your help in researching and obtaining the many new photographs.

A new book can only reach the hands of the students with the help of strategic marketing, and Dean Scudder, Marie Scavotto, and Susan McGlew have been remarkable in highlighting all of the new features. I thank you for always managing to present this new author in the best light. Jason Dirks and all of the people working in Sinauer's Media and Supplements department deserve a special thanks for designing an appealing website and brainstorming with me about the best ways to present all of our new interactive features.

The support of Smith College cannot go unrecognized. Smith has allowed me to produce and disseminate my "Web Conferences," "Developmental Documentaries," and the "Dev Tutorials" used in this text. The commitment and talent of Kate Lee and the overall support by Smith's education technology services department have also made the production of these features possible. I would be remiss if I did not thank all of the scientists who over the years have volunteered their time to speak with my students about their research. Hopefully your shared insights will now reach many more students.

To my students at Smith College, both in my courses and in my research lab, I thank you for being my collaborators and the best teachers I have ever had. Your enthusiasm, hard work, and crazy ideas make all that I do worth it.

There are many things we do in our lives that could not be possible without the support of family. However, in my experience, I have never had to rely on my family quite as much as was required for this endeavor. True sacrifices were made by all in my family to meet the demands of this work. In my book, you are all my co-authors! I thank you for your unconditional love and support.

—M.J.F.B.

Reviewers of the Eleventh Edition

William Anderson, *Harvard University*

David Angelini, *Colby College*

Robert Angerer, *University of Rochester* and *NIH*

John Belote, *Syracuse University*

James Briscoe, *The Francis Crick Institute*

Frank Costantini, *Columbia University*

Gregory Davis, *Bryn Mawr College*

Stephen Devoto, *Wesleyan University*

Richard Dorsky, *University of Utah*

Gregg Duester, *Sanford Burnham Prebys Medical Discovery Institute*

Miguel Turrero Garcia, *Harvard Medical School*

Laura Grabel, *Wesleyan University*

Erik Griffin, *Dartmouth College*

Corey Harwell, *Harvard Medical School*

Jason Hodin, *Stanford University*

Nathalia Holtzman, *Queens College, City University of New York*

Lara Hutson, *University at Buffalo*

Rebecca Ihrie, *Vanderbilt University*

Dan Kessler, *University of Pennsylvania*

Rebecca Landsberg, *The College of Saint Rose*

Kersti Linask, *University of South Florida*

Barbara Lom, *Davidson College*

Frank Lovicu, *University of Sydney*

Laura Anne Lowery, *Boston College*

Deirdre Lyons, *Duke University*

Francesca Mariani, *University of Southern California*

Marja Mikkola, *University of Helsinki*

Lee Niswander, *University of Colorado, Denver*

Isabelle Peter, *California Institute of Technology*

Dominic Poccia, *Amherst College*

Olivier Pourquié, *Harvard Medical School*

Jodi Schottenfeld-Roames, *Swarthmore College*

Gerhard Schlosser, *NUI Galway*

Claudio Stern, *University College London*

Nicole Theodosiou, *Union College*

Mary Tyler, *University of Maine*

Andrea Ward, *Adelphi University*

Media and Supplements

to accompany Developmental Biology, *Eleventh Edition*

For the Student

Companion Website

devbio.com

Significantly enhanced for the Eleventh Edition, and referenced throughout the textbook, the *Developmental Biology* Companion Website provides students with a range of engaging resources to help them learn the material presented in the textbook. The companion site is available free of charge and includes resources in the following categories:

- **Dev Tutorials**: Professionally-produced video tutorials, presented by the textbook's authors, reinforce key concepts.

- **Watch Development**: Putting concepts into action, these informative videos show real-life developmental biology processes.

- **Web Topics**: These extensive topics provide more information for advanced students, historical, philosophical, and ethical perspectives on issues in developmental biology, and links to additional online resources.

- **Scientists Speak**: In these question-and-answer interviews, developmental biology topics are explored by leading experts in the field.

- **Flashcards**: Per-chapter flashcard sets help students learn and review the many new terms introduced in the textbook.

- **Bibliography**: Full citations are provided for all of the literature cited in the textbook (most linked to their PubMed citations).

DevBio Laboratory: Vade Mecum³: An Interactive Guide to Developmental Biology

labs.devbio.com

MARY S. TYLER and RONALD N. KOZLOWSKI

Included with each new copy of the textbook, *Vade Mecum³* is an interactive website that helps students understand the organisms discussed in the course, and prepare them for the lab. The site includes videos of developmental processes and laboratory techniques, and has chapters on the following organisms: slime mold (*Dictyostelium discoideum*), planarian, sea urchin, fruit fly (*Drosophila*), chick, and amphibian. (Also available for purchase separately.)

Developmental Biology: A Guide for Experimental Study, Third Edition

(Included in *DevBio Laboratory*: *Vade Mecum³*)

MARY S. TYLER

This lab manual teaches students to work as independent investigators on problems in development and provides extensive background information and instructions for each experiment. It emphasizes the study of living material, intermixing developmental anatomy in an enjoyable balance, and allows students to make choices in their work.

For the Instructor

(Available to qualified adopters)

Instructor's Resource Library

The *Developmental Biology*, Eleventh Edition Instructor's Resource Library includes the following resources:

- **Case Studies in Dev Bio**: This new collection of case study problems accompanies the Dev Tutorials and provides instructors with ready-to-use in-class active learning exercises. The case studies foster deep learning in developmental biology by providing students an opportunity to apply course content to the critical analysis of data, to generate hypotheses, and to solve novel problems in the field. Each case study includes a PowerPoint presentation and a student handout with accompanying questions.

- **Developing Questions**: Answers, references, and recommendations for further reading are provided so that you and your students can explore the Developing Questions that are posed throughout each chapter.

- **Textbook Figures & Tables**: All of the textbook's figures, photos, and tables are provided both in JPEG (high- and low-resolution) and PowerPoint formats. All images have been optimized for excellent legibility when projected in the classroom.

- **Video Collection**: Includes video segments depicting a wide range of developmental processes, plus segments from *DevBio Laboratory: Vade Mecum³*, and *Differential Expressions²*. For each video collection, an instructor's guide is provided.

- **Vade Mecum³ PowerPoints**: Chick serial sections and whole mounts, provided in both labeled and unlabeled versions, for use in creating quizzes, exams, or in-class exercises.

- **Developmental Biology: A Guide for Experimental Study, Third Edition** (by Mary S. Tyler): The complete lab manual, in PDF format.

Value Options

eBook

Developmental Biology is available as an eBook, in several different formats, including VitalSource, RedShelf, and BryteWave. The eBook can be purchased as either a 180-day rental or a permanent (non-expiring) subscription. All major mobile devices are supported. For details on the eBook platforms offered, please visit www.sinauer.com/ebooks.

Looseleaf Textbook

(ISBN 978-1-60535-604-4)

Developmental Biology is also available in a three-hole punched, looseleaf format. Students can take just the sections they need to class and can easily integrate instructor material with the text.

Making New Bodies
Mechanisms of Developmental Organization

What stays the same when a tadpole becomes a frog, and what changes?

BETWEEN FERTILIZATION AND BIRTH, the developing organism is known as an embryo. The concept of an embryo is a staggering one. As an embryo, you had to build yourself from a single cell. You had to respire before you had lungs, digest before you had a gut, build bones when you were pulpy, and form orderly arrays of neurons before you knew how to think. One of the critical differences between you and a machine is that a machine is never required to function until after it is built. Every multicellular organism has to function even as it builds itself. Most human embryos die before being born. You survived.

Multicellular organisms do not spring forth fully formed. Rather, they arise by a relatively slow process of progressive change that we call **development**. In nearly all cases, the development of a multicellular organism begins with a single cell—the fertilized egg, or **zygote**, which divides mitotically to produce all the cells of the body. The study of animal development has traditionally been called **embryology**, after that phase of an organism that exists between fertilization and birth. But development does not stop at birth, or even at adulthood. Most organisms never stop developing. Each day we replace more than a gram of skin cells (the older cells being sloughed off as we move), and our bone marrow sustains the development of millions of new red blood cells every minute of our lives. Some animals can regenerate severed parts, and many species undergo metamorphosis (such as the transformation of a tadpole into a frog, or a caterpillar into a butterfly).

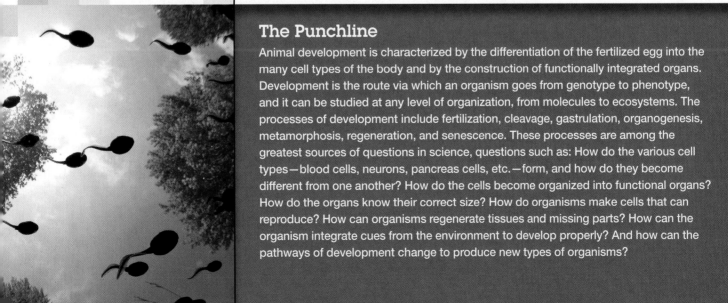

The Punchline

Animal development is characterized by the differentiation of the fertilized egg into the many cell types of the body and by the construction of functionally integrated organs. Development is the route via which an organism goes from genotype to phenotype, and it can be studied at any level of organization, from molecules to ecosystems. The processes of development include fertilization, cleavage, gastrulation, organogenesis, metamorphosis, regeneration, and senescence. These processes are among the greatest sources of questions in science, questions such as: How do the various cell types—blood cells, neurons, pancreas cells, etc.—form, and how do they become different from one another? How do the cells become organized into functional organs? How do the organs know their correct size? How do organisms make cells that can reproduce? How can organisms regenerate tissues and missing parts? How can the organism integrate cues from the environment to develop properly? And how can the pathways of development change to produce new types of organisms?

Therefore, in recent years it has become customary to speak of **developmental biology** as the discipline that studies embryonic and other developmental processes.

"How Are You?" The Questions of Developmental Biology

Aristotle, the first known embryologist, said that wonder was the source of knowledge, and animal development, as Aristotle knew well, is a remarkable source of wonder. This development, this formation of an orderly body from relatively homogeneous material, provokes profound and fundamental questions that *Homo sapiens* have asked since the dawn of self-awareness: How does the body form with its head always above its shoulders? How come the heart is on the left side of our body? How does a simple tube become the complex structures of the brain and spinal cord that generate both thought and movement? Why can't we grow back new limbs? How do the sexes develop their different anatomies?

Our answers to these questions must respect the complexity of the inquiry and must explain a coherent causal network from gene through functional organ. To say that mammals with two X chromosomes are usually females and those with XY chromosomes are usually males does not explain sex determination to a developmental biologist, who wants to know *how* the XX genotype produces a female and *how* the XY genotype produces a male. Similarly, a geneticist might ask how globin genes are transmitted from one generation to the next, and a physiologist might ask about the function of globin proteins in the body. But the developmental biologist asks how it is that the globin genes come to be expressed only in red blood cells and how these genes become active only at specific times in development. (We don't have all the answers yet.) The particular set of questions asked defines the field of biology, as we, too, become defined (at least in part) by the questions we ask. *Welcome to a wonderful and important set of questions!*

Development accomplishes two major objectives. First, it generates cellular diversity and order within the individual organism; second, it ensures the continuity of life from one generation to the next. Put another way, there are two fundamental questions in developmental biology. How does the fertilized egg give rise to the adult body? And, how does that adult body produce yet another body? These huge questions can be subdivided into several categories of questions scrutinized by developmental biologists:

- **The question of differentiation**   A single cell, the fertilized egg, gives rise to hundreds of different cell types—muscle cells, epidermal cells, neurons, lens cells, lymphocytes, blood cells, fat cells, and so on. This generation of cellular diversity is called **differentiation**. Since every cell of the body (with very few exceptions) contains the same set of genes, how can this identical set of genetic instructions produce different types of cells? How can a single fertilized egg cell generate so many different cell types?[1]

- **The question of morphogenesis**   How can the cells in our body organize into functional structures? Our differentiated cells are not randomly distributed. Rather, they are organized into intricate tissues and organs. During development, cells divide, migrate, and die; tissues fold and separate. Our

---

[1] More than 210 different cell types are recognized in the *adult* human, but this number tells us little about how many cell types a human body produces over the course of development. A particular cell may play many roles during development, going through stages that are no longer seen in adulthood. In addition, the role of some cell types is to activate specific genes in neighboring cells, and once this function is accomplished, the activating cell type dies. The primary notochord cells, for example, are not even listed in medical histology texts. Once this task is done, most of them undergo programmed cell death so as not to disturb further neural development. Because such a cell type is not seen in the adult, it and its importance are known mainly by developmental biologists.

fingers are always at the tips of our hands, never in the middle; our eyes are always in our heads, not in our toes or gut. This creation of ordered form is called **morphogenesis**, and it involves coordinating cell growth, cell migration, and cell death.

- **The question of growth**   If each cell in our face were to undergo just one more cell division, we would be considered horribly malformed. If each cell in our arms underwent just one more round of cell division, we could tie our shoelaces without bending over. How do our cells know when to stop dividing? Our arms are generally the same size on both sides of the body. How is cell division so tightly regulated?

- **The question of reproduction**   The sperm and egg are highly specialized cells, and only they can transmit the instructions for making an organism from one generation to the next. How are these germ cells set apart, and what are the instructions in the nucleus and cytoplasm that allow them to form the next generation?

- **The question of regeneration**   Some organisms can regenerate every part of their bodies. Some salamanders regenerate their eyes and their legs, while many reptiles can regenerate their tails. While mammals are generally poor at regeneration, there are some cells in our bodies—**stem cells**—that are able to form new structures even in adults. How do stem cells retain this capacity, and can we harness it to cure debilitating diseases?

- **The question of environmental integration**   The development of many (perhaps all) organisms is influenced by cues from the environment that surrounds the embryo or larva. The sex of many species of turtles, for instance, depends on the temperature the embryo experiences while in the egg. The formation of the reproductive system in some insects depends on bacteria that are transmitted inside the egg. Moreover, certain chemicals in the environment can disrupt normal development, causing malformations in the adult. How is the development of an organism integrated into the larger context of its habitat?

- **The question of evolution**   Evolution involves inherited changes of development. When we say that today's one-toed horse had a five-toed ancestor, we are saying that changes in the development of cartilage and muscles occurred over many generations in the embryos of the horse's ancestors. How do changes in development create new body forms? Which heritable changes are possible, given the constraints imposed by the necessity of the organism to survive as it develops?

The questions asked by developmental biologists have become critical in molecular biology, physiology, cell biology, genetics, anatomy, cancer research, neurobiology, immunology, ecology, and evolutionary biology. The study of development has become essential for understanding all other areas of biology. In turn, the many advances of molecular biology, along with new techniques of cell imaging, have finally made these questions answerable. This is exciting; for, as the Nobel-prize winning developmental biologist Hans Spemann stated in 1927, "We stand in the presence of riddles, but not without the hope of solving them. And riddles with the hope of solution—what more can a scientist desire?"

So, we come bearing questions. They are questions bequeathed to us by earlier generations of biologists, philosophers, and parents. They are questions with their own histories, questions discussed on an anatomical level by people such as Aristotle, William Harvey, St. Albertus Magnus, and Charles Darwin. More recently, these questions have been addressed on the cellular and molecular levels by men and women throughout the world, each of whom brings to the laboratory his or her own perspectives and training. For there is no one way to become a developmental biologist, and the field has benefitted by having researchers trained in cell biology, genetics, biochemistry, immunology, and even anthropology, engineering, physics, and art.

The Cycle of Life

For animals, fungi, and plants, the sole way of getting from egg to adult is by developing an embryo. The embryo is where genotype is translated into phenotype, where inherited genes are expressed to form the adult. The developmental biologist usually finds the transient stages leading up to the adult to be the most interesting. Developmental biology studies the building of organisms. It is a science of becoming, a science of process.

One of the major triumphs of descriptive embryology was the idea of a generalizable animal life cycle. Modern developmental biology investigates the temporal changes of gene expression and anatomical organization along this life cycle. Each animal, whether earthworm or eagle, termite or beagle, passes through similar stages of development: fertilization, cleavage, gastrulation, organogenesis, birth, metamorphosis, and gametogenesis. The stages of development between fertilization and hatching (or birth) are collectively called **embryogenesis**.

1. **Fertilization** involves the fusion of the mature sex cells, the sperm and egg, which are collectively called the **gametes**. The fusion of the gamete cells stimulates the egg to begin development and initiates a new individual. The subsequent fusion of the gamete nuclei (the male and female **pronuclei**, each of which has only half the normal number of chromosomes characteristic for the species) gives the embryo its **genome**, the collection of genes that helps instruct the embryo to develop in a manner very similar to that of its parents.

2. **Cleavage** is a series of extremely rapid mitotic divisions that immediately follow fertilization. During cleavage, the enormous volume of zygote cytoplasm is divided into numerous smaller cells called **blastomeres**. By the end of cleavage, the blastomeres have usually formed a sphere, known as a **blastula**.

3. After the rate of mitotic division slows down, the blastomeres undergo dramatic movements and change their positions relative to one another. This series of extensive cell rearrangements is called **gastrulation**, and the embryo is said to be in the **gastrula** stage. As a result of gastrulation, the embryo contains three **germ layers** (**endoderm**, **ectoderm**, and **mesoderm**) that will interact to generate the organs of the body.

4. Once the germ layers are established, the cells interact with one another and rearrange themselves to produce tissues and organs. This process is called **organogenesis**. Chemical signals are exchanged between the cells of the germ layers, resulting in the formation of specific organs at specific sites. Certain cells will undergo long migrations from their place of origin to their final location. These migrating cells include the precursors of blood cells, lymph cells, pigment cells, and gametes (eggs and sperm).

5. In many species, the organism that hatches from the egg or is born into the world is not sexually mature. Rather, the organism needs to undergo **metamorphosis** to become a sexually mature adult. In most animals, the young organism is a called a **larva**, and it may look significantly different from the adult. In many species, the larval stage is the one that lasts the longest, and is used for feeding or dispersal. In such species, the adult is a brief stage whose sole purpose is to reproduce. In silkworm moths, for instance, the adults do not have mouthparts and cannot feed; the larva must eat enough so that the adult has the stored energy to survive and mate. Indeed, most female moths mate as soon as they eclose from the pupa, and they fly only once—to lay their eggs. Then they die.

6. In many species, a group of cells is set aside to produce the next generation (rather than forming the current embryo). These cells are the precursors of the gametes. The gametes and their precursor cells are collectively called **germ cells**, and they are set aside for reproductive function. All other cells of the body are called **somatic cells**. This separation of somatic cells (which give rise to the individual body) and germ cells (which contribute to the formation of a new generation) is often

one of the first differentiations to occur during animal development. The germ cells eventually migrate to the gonads, where they differentiate into gametes. The development of gametes, called **gametogenesis**, is usually not completed until the organism has become physically mature. At maturity, the gametes may be released and participate in fertilization to begin a new embryo. The adult organism eventually undergoes senescence and dies, its nutrients often supporting the early embryogenesis of its offspring and its absence allowing less competition. Thus, the cycle of life is renewed.

▶ **DEV TUTORIAL**   *Personhood*   **Scott Gilbert discusses the human life cycle and the question of when in this cycle the embryo may be said to achieve "personhood."**

WEB TOPIC 1.1   **WHEN DOES A HUMAN BECOME A PERSON?**   Scientists have proposed different answers to this question. Fertilization, gastrulation, the first signs of brain function, and the time around birth—each of these stages has its supporters as the starting point of human personhood.

An Example: A Frog's Life

All animal life cycles are modifications of the generalized one described above. Here we will present a concrete example, the development of the leopard frog *Rana pipiens* (**FIGURE 1.1**).

VADE MECUM

As seen in the segment on amphibians, frogs display some of the most dramatic of vertebrate life cycles.

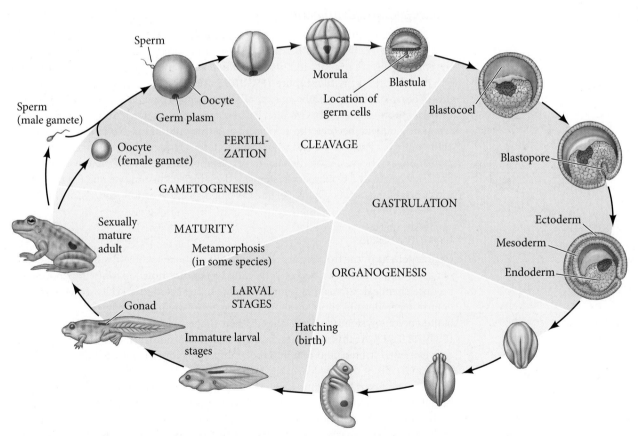

FIGURE 1.1 Developmental history of the leopard frog, *Rana pipiens*. The stages from fertilization through hatching (birth) are known collectively as embryogenesis. The region set aside for producing germ cells is shown in purple. Gametogenesis, which is completed in the sexually mature adult, begins at different times during development, depending on the species. (The sizes of the varicolored wedges shown here are arbitrary and do not correspond to the proportion of the life cycle spent in each stage.)

Gametogenesis and fertilization

The end of one life cycle and the beginning of the next are often intricately intertwined. Life cycles are often controlled by environmental factors (tadpoles wouldn't survive if they hatched in the fall, when their food is dying), so in most frogs, gametogenesis and fertilization are seasonal events. A combination of photoperiod (hours of daylight) and temperature informs the pituitary gland of the mature female frog that it is spring. The pituitary secretions cause the eggs and sperm to mature.

In most species of frogs, fertilization is external (**FIGURE 1.2A**). The male frog grabs the female's back and fertilizes the eggs as the female releases them (**FIGURE 1.2B**). Some species lay their eggs in pond vegetation, and the egg jelly adheres to the plants and anchors the eggs. The eggs of other species float into the center of the pond without any support. So an important thing to remember about life cycles is that they are intimately involved with environmental factors.

Fertilization accomplishes both sex (genetic recombination) and reproduction (the generation of a new individual). The genomes of the haploid male and female pronuclei merge and recombine to form the diploid zygote nucleus. In addition, the entry of the sperm facilitates the movement of cytoplasm inside the newly fertilized egg. This migration will be critical in determining the three body axes of the frog: anterior-posterior (head-tail), dorsal-ventral (back-belly), and right-left. And, importantly, fertilization activates those molecules necessary to begin cell cleavage and gastrulation (Rugh 1950).

Cleavage and gastrulation

During cleavage, the volume of the frog egg stays the same, but it is divided into tens of thousands of cells (**FIGURE 1.2C,D**). Gastrulation in the frog begins at a point on the embryo surface roughly 180° opposite the point of sperm entry with the formation of a dimple called the **blastopore** (**FIGURE 1.2E**). The blastopore, which marks the future dorsal side of the embryo, expands to become a ring. Cells migrating through the blastopore to the embryo's interior become the mesoderm and endoderm; cells remaining outside become the ectoderm, and this outer layer expands to enclose the entire embryo. Thus, at the end of gastrulation, the ectoderm (precursor of the epidermis, brain, and nerves) is on the outside of the embryo, the endoderm (precursor of the lining of the gut and respiratory systems) is deep inside the embryo, and the mesoderm (precursor of the connective tissue, muscle, blood, heart, skeleton, gonads, and kidneys) is between them.

Organogenesis

Organogenesis in the frog begins when the cells of the most dorsal region of the mesoderm condense to form a rod of cells called the **notochord**.[2] These notochord cells produce chemical signals that redirect the fate of the ectodermal cells above it. Instead of forming epidermis, the cells above the notochord are instructed to become the cells of the nervous system. The cells change their shapes and rise up from the round body (**FIGURE 1.2F**). At this stage, the embryo is called a **neurula**. The neural precursor cells elongate, stretch, and fold into the embryo, forming the **neural tube**. The future epidermal cells of the back cover the neural tube.

Once the neural tube has formed, it and the notochord induce changes in the neighboring regions, and organogenesis continues. The mesodermal tissue adjacent to the neural tube and notochord becomes segmented into **somites**—the precursors of the frog's back muscles, spinal vertebrae, and dermis (the inner portion of the skin). The embryo develops a mouth and an anus, and it elongates into the familiar tadpole structure (**FIGURE 1.2G**). The neurons make connections to the muscles and to other

---

[2] Although adult vertebrates do not have notochords, this embryonic organ is critical for establishing the fates of the ectodermal cells above it, as we shall see in Chapter 13.

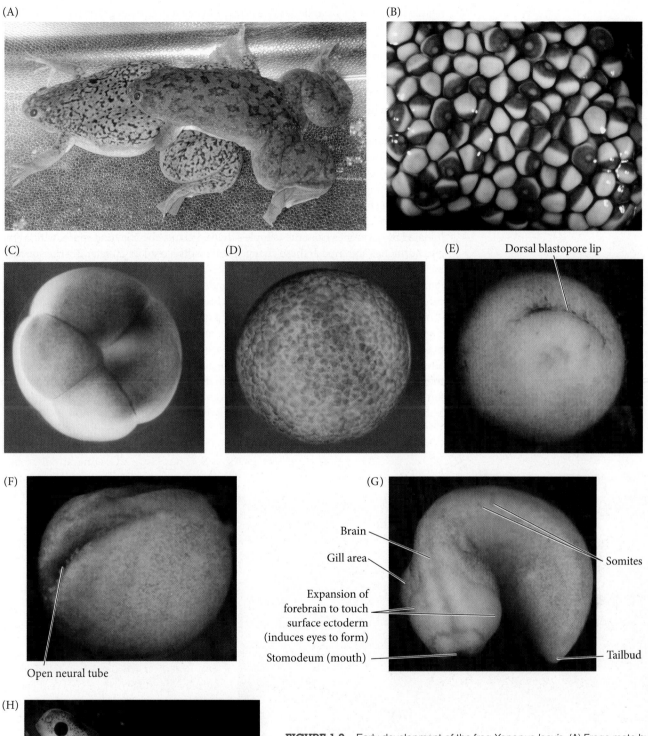

(A)

(B)

(C)

(D)

(E)

Dorsal blastopore lip

(F)

Open neural tube

(G)

Brain

Gill area

Expansion of
forebrain to touch
surface ectoderm
(induces eyes to form)

Stomodeum (mouth)

Somites

Tailbud

(H)

FIGURE 1.2 Early development of the frog *Xenopus laevis*. (A) Frogs mate by amplexus, the male grasping the female around the belly and fertilizing the eggs as they are released. (B) A newly laid clutch of eggs. The cytoplasm has rotated such that the darker pigment is where the nucleus resides. (C) An 8-cell embryo. (D) A late blastula, containing thousands of cells. (E) An early gastrula, showing the blastopore lip through which the mesodermal and some endoderm cells migrate. (F) A neurula, where the neural folds come together at the dorsal midline, creating a neural tube. (G) A pre-hatching tadpole, as the protrusions of the forebrain begin to induce eyes to form. (H) A mature tadpole, having swum away from the egg mass and feeding independently. (Courtesy of Michael Danilchik and Kimberly Ray.)

neurons, the gills form, and the larva is ready to hatch from its egg. The hatched tadpole will feed for itself as soon as the yolk supplied by its mother is exhausted.

Metamorphosis and gametogenesis

Metamorphosis of the fully aquatic tadpole larva into an adult frog that can live on land is one of the most striking transformations in all of biology. Almost every organ is subject to modification, and the resulting changes in form are striking and very obvious (**FIGURE 1.3**). The hindlimbs and forelimbs the adult will use for locomotion differentiate as the tadpole's paddle tail recedes. The cartilaginous tadpole skull is replaced by the predominantly bony skull of the young frog. The horny teeth the tadpole uses to tear up pond plants disappear as the mouth and jaw take a new shape, and the fly-catching tongue muscle of the frog develops. Meanwhile, the tadpole's lengthy intestine—a characteristic of herbivores—shortens to suit the more carnivorous diet of the adult frog. The gills regress and the lungs enlarge. Amphibian metamorphosis is initiated by hormones from the tadpole's thyroid gland; the mechanisms by which thyroid hormones accomplish these changes will be discussed in Chapter 21. The speed of metamorphosis is keyed to environmental pressures. In temperate regions, for instance, *Rana* metamorphosis must occur before ponds freeze in winter. An adult leopard frog can burrow into the mud and survive the winter; its tadpole cannot.

As metamorphosis ends, the development of the germ cells (sperm and egg) begins. Gametogenesis can take a long time. In *Rana pipiens*, it takes 3 years for the eggs to mature in the female's ovaries. Sperm take less time; *Rana* males are often fertile soon after metamorphosis. To become mature, the germ cells must be competent to complete **meiosis**. Having undergone meiosis, the mature sperm and egg nuclei can unite in fertilization, restoring the diploid chromosome number and initiating the events that lead to development and the continuation of the circle of life.

(A)

(B)

(C)

(D)

(E)

(F)

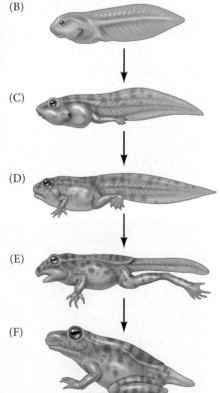

FIGURE 1.3 Metamorphosis of the frog. (A) Huge changes are obvious when one contrasts the tadpole and the adult bullfrog. Note especially the differences in jaw structure and limbs. (B) Premetamorphic tadpole. (C) Prometamorphic tadpole, showing hindlimb growth. (D) Onset of metamorphic climax as forelimbs emerge. (E,F) Climax stages. (A © Patrice Ceisel/Visuals Unlimited.)

Comparative Embryology

The fertilized egg has no heart. Where does the heart come from? Does it form the same way in both insects and vertebrates? How is heart development in these two groups similar and how is it different? How do the tissues that form a bird's wing relate to the tissues that form a fish fin or a human hand? Many of the questions in developmental biology are of this type, and they stem from the field's embryological heritage. The first known study of comparative developmental anatomy was undertaken by Aristotle. In *The Generation of Animals* (ca. 350 BCE), he noted some of the variations on the life cycle themes: some animals are born from eggs (**oviparity**, as in birds, frogs, and most invertebrates); some by live birth (**viviparity**, as in placental mammals); and some by producing an egg that hatches inside the body (**ovoviviparity**, as in certain reptiles and sharks). Aristotle also identified the two major cell division patterns by which embryos are formed: the **holoblastic** pattern of cleavage (in which the entire egg is divided into successively smaller cells, as it is in frogs and mammals) and the **meroblastic** pattern of cleavage (as in chicks, wherein only part of the egg is destined to become the embryo while the other portion—the yolk—serves as nutrition for the embryo). And should anyone want to know who first figured out the functions of the mammalian placenta and umbilical cord, it was Aristotle.

There was remarkably little progress in embryology for the two thousand years following Aristotle. It was only in 1651 that William Harvey concluded that all animals—even mammals—originate from eggs. *Ex ovo omnia* ("All from the egg") was the motto on the frontispiece of Harvey's *On the Generation of Living Creatures*, and this precluded the spontaneous generation of animals from mud or excrement.[3] Harvey also was the first to see the blastoderm of the chick embryo (the small region of the egg containing the yolk-free cytoplasm that gives rise to the embryo), and he was the first to notice that "islands" of blood tissue form before the heart does. Harvey also suggested that the amniotic fluid might function as a "shock absorber" for the embryo.

As might be expected, embryology remained little but speculation until the invention of the microscope allowed detailed observations (**FIGURE 1.4**). Marcello Malpighi published the first microscopic account of chick development in 1672. Here, for the first time, the neural groove (precursor of the neural tube), the muscle-forming somites, and the first circulation of the arteries and veins—to and from the yolk—were identified.

Epigenesis and preformationism

With Malpighi began one of the great debates in embryology: the controversy over whether the organs of the embryo are formed de novo ("from scratch") at each generation, or whether the organs are already present, in miniature form, within the egg or sperm. The first view, **epigenesis**, was supported by Aristotle and Harvey. The second view, **preformationism**, was reinvigorated with Malpighi's support. Malpighi showed that the unincubated[4] chick egg already had a great deal of structure, and this observation provided him with reasons to question epigenesis and advocate the preformationist view, according to which all the organs of the adult were prefigured in miniature within the sperm or (more usually) the egg. Organisms were not seen to be "constructed" but rather "unrolled" or "unfurled."

The preformationist view had the backing of eighteenth-century science, religion, and philosophy (Gould 1977; Roe 1981; Churchill 1991; Pinto-Correia 1997). First, if all organs were prefigured, embryonic development merely required the growth of existing structures, not the formation of new ones. No extra mysterious force was needed for

---

[3] Harvey did not make this statement lightly, for he knew that it contradicted the views of Aristotle, whom Harvey venerated. Aristotle had proposed that menstrual fluid formed the substance of the embryo, while the semen gave it form and animation.

[4] As pointed out by Maître-Jan in 1722, the eggs Malpighi examined may technically be called "unincubated," but as they were left sitting in the Bolognese sun in August, they were not unheated. Such eggs would be expected to have developed into chicks.

(A)

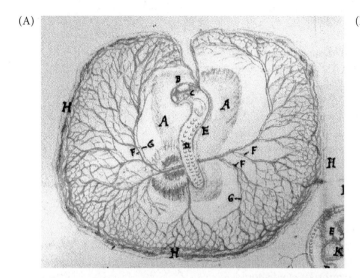

(B)

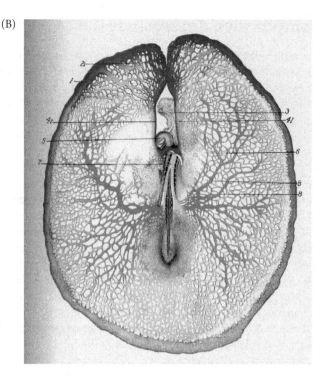

(C)

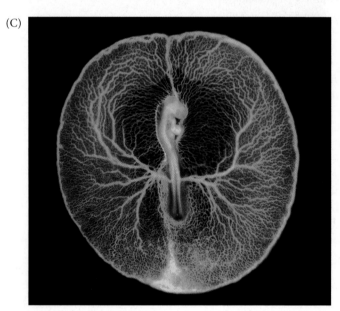

FIGURE 1.4 Depictions of chick developmental anatomy. (A) Dorsal view (looking "down" at what will become the back) of a 2-day chick embryo, as depicted by Marcello Malpighi in 1672. (B) Ventral view (looking "up" at the prospective belly) of a chick embryo at a similar stage, seen through a dissecting microscope and rendered by F. R. Lillie in 1908. (C) Dorsal view of a late 2-day chick embryo, about 45 hours after the egg was laid. The heart starts beating during day 2. The vascular system of this embryo was revealed by injecting fluorescent beads into the circulatory system. The three-dimensionality is achieved by superimposing two separate images. (A from Malpighi 1672; B from Lillie 1908; C © Vincent Pasque, Wellcome Images.)

embryonic development. Second, just as the adult organism was prefigured in the germ cells, another generation already existed in a prefigured state within the germ cells of the first prefigured generation. The preformationists had no cell theory to provide a lower limit to the size of their preformed organisms (the cell theory did not arise until the mid-1800s).

Preformationism's principal failure was its inability to account for the intergenerational variations revealed by even the limited genetic evidence of the time. It was known, for instance, that the children of a white and a black parent would have intermediate skin color—an impossibility if inheritance and development were solely through either the sperm or the egg. In more scientific studies, the German botanist Joseph Kölreuter (1766) produced hybrid tobacco plants with characteristics of both species.

The embryological case for epigenesis was revived at the same time by Kaspar Friedrich Wolff. By carefully observing the development of chick embryos, Wolff demonstrated that the embryonic parts develop from tissues that have no counterpart in the adult organism. The heart, intestine, and blood vessels (which, according to preformationism, must be present from the beginning) could be seen to develop anew in each embryo. So Wolff (1767) was able to state, "when the formation of the intestine in this manner has been duly weighed, almost no doubt can remain, I believe, of the truth of epigenesis." To explain how an organism is created anew each generation, however, Wolff had to postulate an unknown force—the *vis essentialis* ("essential force")—which, acting according to natural laws analogous to those such as gravity or magnetism, would organize embryonic development.

A reconciliation between preformationism and epigenesis was attempted by the German philosopher Immanuel Kant (1724–1804) and his colleague, biologist Johann Friedrich Blumenbach (1752–1840). Blumenbach postulated a mechanical, goal-directed force he called *Bildungstrieb* ("developmental force"). Such a force, he said, was not theoretical, but could be shown to exist by experimentation. A hydra, when cut, regenerates its amputated parts by rearranging existing elements (as we will see Chapter 22). Some purposeful organizing force could be observed in operation, and it was thought to be inherited through the germ cells. Thus, development could proceed through a predetermined force inherent in the matter of the embryo (Cassirer 1950; Lenoir 1980). In this hypothesis, wherein epigenetic development is directed by preformed instructions, we are not far from the view held by modern biologists that most (but by no means all) of the instructions for forming the organism are already present in the fertilized egg.

An Overview of Early Development

Patterns of cleavage

E. B. Wilson, one of the pioneers in applying cell biology to embryology, noted in 1923, "To our limited intelligence, it would seem a simple task to divide a nucleus into equal parts. The cell, manifestly, entertains a very different opinion." Indeed, different organisms undergo cleavage in distinctly different ways, and the mechanisms for these differences remain at the frontier of cell and developmental biology. Cells in the cleavage-stage cells are called **blastomeres**.[5] In most species (mammals being the chief exception), both the initial rate of cell division and the placement of the blastomeres with respect to one another are under the control of proteins and mRNAs stored in the oocyte. Only later do the rates of cell division and the placement of cells come under the control of the newly formed organism's own genome. During the initial phase of development, when cleavage rhythms are controlled by maternal factors, cytoplasmic volume does not increase. Rather, the zygote cytoplasm is divided into increasingly smaller cells—first in half, then quarters, then eighths, and so forth. Cleavage occurs very rapidly in most invertebrates, probably as an adaptation to generate a large number of cells quickly and to restore the somatic ratio of nuclear volume to cytoplasmic volume. The embryo often accomplishes this by abolishing the gap periods of the cell cycle (the G1 and G2 phases), when growth can occur. A frog egg, for example, can divide into 37,000 cells in just 43 hours. Mitosis in cleavage-stage *Drosophila* embryos occurs every 10 minutes for more than 2 hours, forming some 50,000 cells in just 12 hours.

The pattern of embryonic cleavage peculiar to a species is determined by two major parameters: (1) the amount and distribution of yolk protein within the cytoplasm, which determine where cleavage can occur and the relative sizes of the blastomeres; and (2) factors in the egg cytoplasm that influence the angle of the mitotic spindle and the timing of its formation.

In general, yolk inhibits cleavage. When one pole of the egg is relatively yolk-free, cellular divisions occur there at a faster rate than at the opposite pole. The yolk-rich pole is referred to as the **vegetal pole**; the yolk concentration in the **animal pole** is relatively low. The zygote nucleus is frequently displaced toward the animal pole. **FIGURE 1.5** provides a classification of cleavage types and shows the influence of yolk on cleavage symmetry and pattern. At one extreme are the eggs of sea urchins, mammals, and snails. These eggs have sparse, equally distributed yolk and are thus **isolecithal** (Greek, "equal yolk"). In these species, cleavage is **holoblastic** (Greek *holos*, "complete"), meaning that the cleavage furrow extends through the entire egg. With little yolk, these embryos must have some other way of obtaining food. Most will generate a voracious larval form, while mammals will obtain their nutrition from the maternal placenta.

At the other extreme are the eggs of insects, fish, reptiles, and birds. Most of their cell volumes are made up of yolk. The yolk must be sufficient to nourish these animals throughout embryonic development. Zygotes containing large accumulations of yolk

---

[5]We will be using an entire "blast" vocabulary in this book. A *blastomere* is a cell derived from cleavage in an early embryo. A *blastula* is an embryonic stage composed of blastomeres; a mammalian blastula is called a *blastocyst* (see Chapter 12). The cavity within the blastula is the *blastocoel*. A blastula that lacks a blastocoel is called a *stereoblastula*. The invagination where gastrulation begins is the *blastopore*.

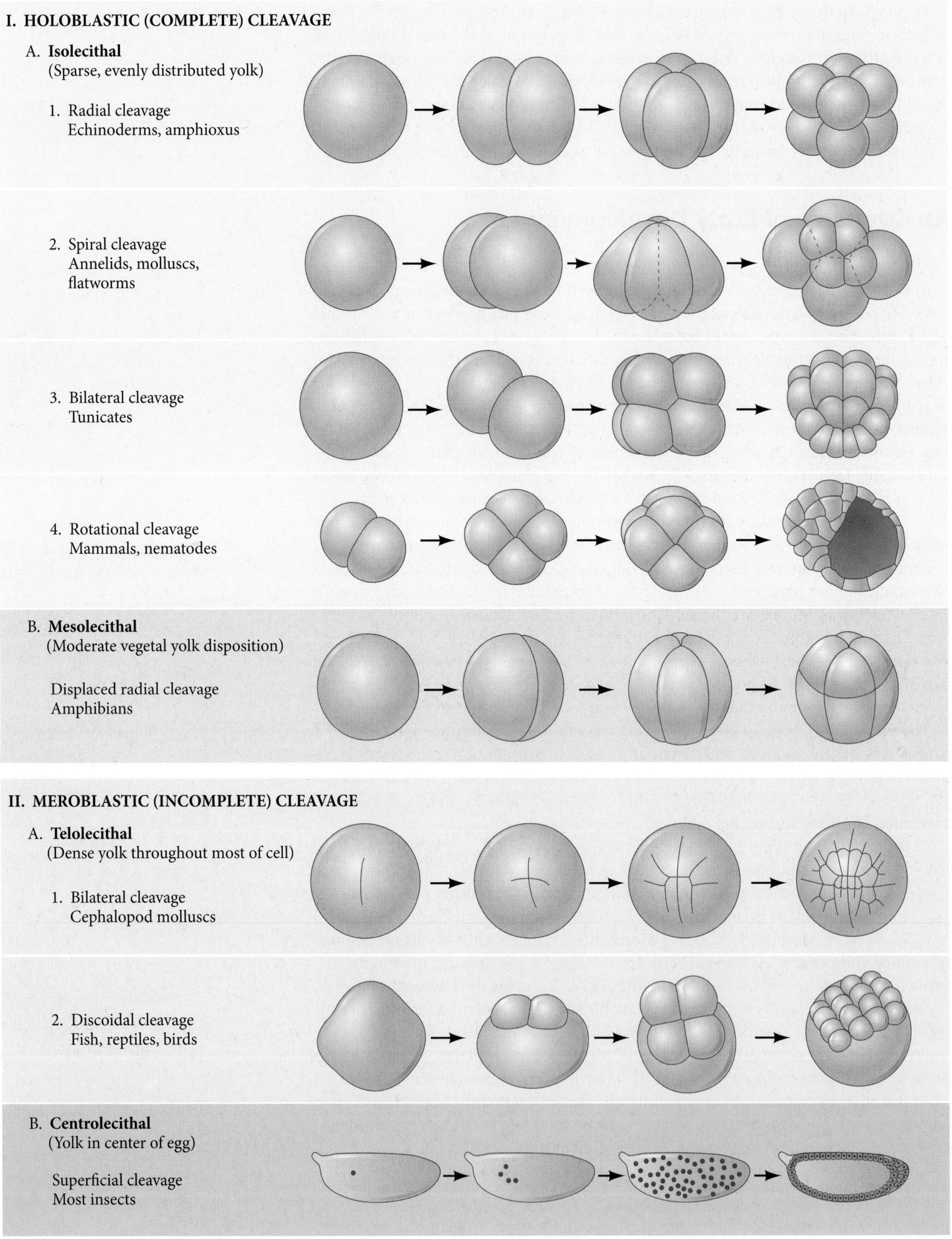

I. HOLOBLASTIC (COMPLETE) CLEAVAGE

A. Isolecithal
(Sparse, evenly distributed yolk)

1. Radial cleavage
   Echinoderms, amphioxus

2. Spiral cleavage
   Annelids, molluscs,
   flatworms

3. Bilateral cleavage
   Tunicates

4. Rotational cleavage
   Mammals, nematodes

B. Mesolecithal
(Moderate vegetal yolk disposition)

Displaced radial cleavage
Amphibians

II. MEROBLASTIC (INCOMPLETE) CLEAVAGE

A. Telolecithal
(Dense yolk throughout most of cell)

1. Bilateral cleavage
   Cephalopod molluscs

2. Discoidal cleavage
   Fish, reptiles, birds

B. Centrolecithal
(Yolk in center of egg)

Superficial cleavage
Most insects

FIGURE 1.5 Summary of the main patterns of cleavage.

undergo **meroblastic cleavage** (Greek *meros*, "part"), wherein only a portion of the cytoplasm is cleaved. The cleavage furrow does not penetrate the yolky portion of the cytoplasm because the yolk platelets impede membrane formation there. Insect eggs have yolk in the center (i.e., they are **centrolecithal**), and the divisions of the cytoplasm occur only in the rim of cytoplasm, around the periphery of the cell (i.e., **superficial cleavage**). The eggs of birds and fish have only one small area of the egg that is free of yolk (**telolecithal** eggs), and therefore the cell divisions occur only in this small disc of cytoplasm, giving rise to **discoidal cleavage**. These are general rules, however, and even closely related species have evolved different patterns of cleavage in different environments.

Yolk is just one factor influencing a species' pattern of cleavage. There are also, as Conklin had intuited, inherited patterns of cell division superimposed on the constraints of the yolk. The importance of this inheritance can readily be seen in isolecithal eggs. In the absence of a large concentration of yolk, **holoblastic cleavage** takes place. Four major patterns of this cleavage type can be described: *radial, spiral, bilateral,* and *rotational* holoblastic cleavage.

WEB TOPIC 1.2   **THE CELL BIOLOGY OF EMBRYONIC CLEAVAGE**   Cell cleavage is accomplished by a remarkable coordination between the cytoskeleton and the chromosomes. This integration of part and whole is becoming better understood as better imaging technologies become available.

TABLE 1.1   Types of cell movement during gastrulation[a]

Type of movement	Description	Illustration	Example
Invagination	Infolding of a sheet (epithelium) of cells, much like the indention of a soft rubber ball when it is poked.		Sea urchin endoderm
Involution	Inward movement of an expanding outer layer so that it spreads over the internal surface of the remaining external cells.		Amphibian mesoderm
Ingression	Migration of individual cells from the surface into the embryo's interior. Individual cells become mesenchymal (i.e., separate from one another) and migrate independently.		Sea urchin mesoderm, *Drosophila* neuroblasts
Delamination	Splitting of one cellular sheet into two more or less parallel sheets. While on a cellular basis it resembles ingression, the result is the formation of a new (additional) epithelial sheet of cells.		Hypoblast formation in birds and mammals
Epiboly	Movement of epithelial sheets (usually ectodermal cells) spreading as a unit (rather than individually) to enclose deeper layers of the embryo. Can occur by cells dividing, by cells changing their shape, or by several layers of cells intercalating into fewer layers; often, all three mechanisms are used.		Ectoderm formation in sea urchins, tunicates, and amphibians

[a] The gastrulation of any particular organism is an ensemble of several of these movements.

Gastrulation: "The most important time in your life"

According to embryologist Lewis Wolpert (1986), "It is not birth, marriage, or death, but gastrulation which is truly the most important time in your life." This is not an overstatement. **Gastrulation** is what makes animals animals. (Animals gastrulate; plants and fungi do not.) During gastrulation, the cells of the blastula are given new positions and new neighbors and the multilayered body plan of the organism is established. The cells that will form the endodermal and mesodermal organs are brought to the inside of the embryo, while the cells that will form the skin and nervous system are spread over its outside surface. Thus, the three germ layers—outer ectoderm, inner endoderm, and interstitial mesoderm—are first produced during gastrulation. In addition, the stage is set for the interactions of these newly positioned tissues.

Gastrulation usually proceeds by some combination of several types of movements. These movements involve the entire embryo, and cell migrations in one part of the gastrulating embryo must be intimately coordinated with other movements that are taking place simultaneously. Although patterns of gastrulation vary enormously throughout the animal kingdom, all of the patterns are different combinations of the five basic types of cell movements—**invagination, involution, ingression, delamination,** and **epiboly**—described in **TABLE 1.1** on the previous page.

In addition to establishing which cells will be in which germ layer, embryos must develop three crucial axes that are the foundation of the body: the anterior-posterior axis, the dorsal-ventral axis, and the right-left axis (**FIGURE 1.6**). The **anterior-posterior (AP** or **anteroposterior) axis** is the line extending from head to tail (or mouth to anus in those organisms that lack a head and tail). The **dorsal-ventral (DV** or **dorsoventral) axis** is the line extending from back (dorsum) to belly (ventrum). The **right-left axis** separates the two lateral sides of the body. Although humans (for example) may look symmetrical, recall that in most of us, the heart is in the left half of the body, while the liver is on the right. Somehow, the embryo knows that some organs belong on one side and other organs go on the other.

Naming the parts: The primary germ layers and early organs

The end of preformationism did not come until the 1820s, when a combination of new staining techniques, improved microscopes, and institutional reforms in German universities created a revolution in descriptive embryology. The new techniques enabled microscopists to document the epigenesis of anatomical structures, and the institutional reforms provided audiences for these reports and students to carry on the work of their teachers. Among the most talented of this new group of microscopically inclined investigators were three friends, born within a year of each other, all of whom came from the Baltic region and studied in northern Germany. The work of Christian Pander, Heinrich Rathke, and Karl Ernst von Baer transformed embryology into a specialized branch of science.

FIGURE 1.6 Axes of a bilaterally symmetrical animal. (A) A single plane, the midsagittal plane, divides the animal into left and right halves. (B) Cross sections bisecting the anterior-posterior axis.

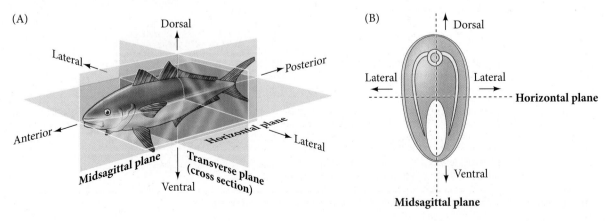

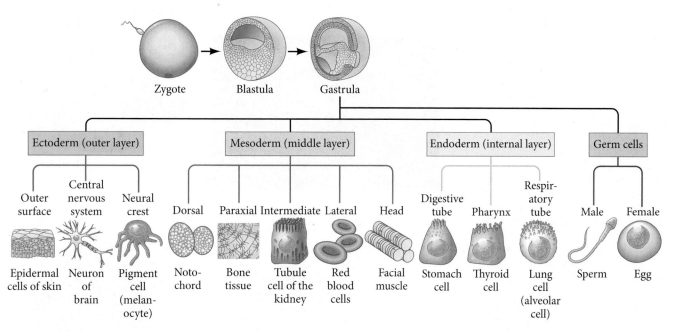

FIGURE 1.7   The dividing cells of the fertilized egg form three distinct embryonic germ layers. Each of the germ layers gives rise to myriad differentiated cell types (only a few representatives are shown here) and distinct organ systems. The germ cells (precursors of the sperm and egg) are set aside early in development and do not arise from any particular germ layer.

Studying the chick embryo, Pander discovered that the embryo was organized into **germ layers**[6]—three distinct regions of the embryo that give rise through epigenesis to the differentiated cells types and specific organ systems (**FIGURE 1.7**). These three layers are found in the embryos of most animal phyla:

- The **ectoderm** generates the outer layer of the embryo. It produces the surface layer (epidermis) of the skin and forms the brain and nervous system.

- The **endoderm** becomes the innermost layer of the embryo and produces the epithelium of the digestive tube and its associated organs (including the lungs).

- The **mesoderm** becomes sandwiched between the ectoderm and endoderm. It generates the blood, heart, kidney, gonads, bones, muscles, and connective tissues.

Pander also demonstrated that the germ layers did not form their respective organs autonomously (Pander 1817). Rather, each germ layer "is not yet independent enough to indicate what it truly is; it still needs the help of its sister travelers, and therefore, although already designated for different ends, all three influence each other collectively until each has reached an appropriate level." Pander had discovered the tissue interactions that we now call induction. No vertebrate tissue is able to construct organs by itself; it must interact with other tissues, as we will describe in Chapter 4.

Meanwhile, Rathke followed the intricate development of the vertebrate skull, excretory systems, and respiratory systems, showing that these became increasingly complex. He also showed that their complexity took on different trajectories in different classes of vertebrates. For instance, Rathke was the first to identify the **pharyngeal arches** (**FIGURE 1.8**). He showed that these same embryonic structures became gill supports in fish and the jaws and ears (among other things) in mammals.

The four principles of Karl Ernst von Baer

Karl Ernst von Baer extended Pander's studies of the chick embryo. He recognized that there is a common pattern to all vertebrate development—that each of the three germ

---

[6] From the same root as "germination," the Latin *germen* means "sprout" or "bud." The names of the three germ layers are from the Greek: ectoderm from *ektos* ("outside") plus *derma* ("skin"); mesoderm from *mesos* ("middle"); and endoderm from *endon* ("within").

FIGURE 1.8 Evolution of pharyngeal arch structures in the vertebrate head. (A) Pharyngeal arches (also called branchial arches) in the embryo of the salamander *Ambystoma mexicanum.* The surface ectoderm has been removed to permit visualization of the arches (highlighted in color) as they form. (B) In adult fish, pharyngeal arch cells form the hyo-mandibular jaws and gill arches. (C) In amphibians, birds, and reptiles (a croco-dile is shown here), these same cells form the quadrate bone of the upper jaw and the articular bone of the lower jaw. (D) In mammals, the quadrate has become internalized and forms the incus of the middle ear. The articular bone retains its contact with the quadrate, becoming the malleus of the middle ear. Thus, the cells that form gill supports in fish form the middle ear bones in mammals. (A courtesy of P. Falck and L. Olsson; B–D after Zangerl and Williams 1975.)

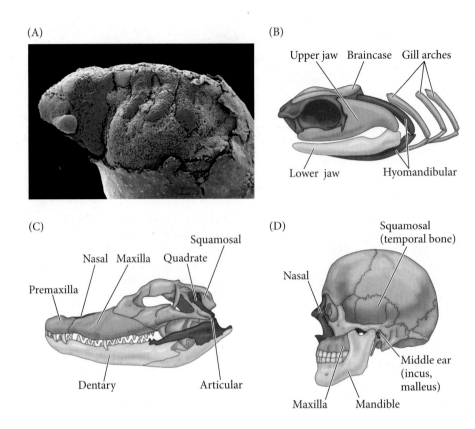

layers generally gives rise to the same organs, whether the organism is a fish, a frog, or a chick. He discovered the notochord, the rod of mesoderm that separates the embryo into right and left halves and instructs the ectoderm above it to become the nervous system (**FIGURE 1.9**). He also discovered the mammalian egg, that minuscule, long-sought cell that everyone believed existed but no one before von Baer had ever seen.

In 1828, von Baer reported, "I have two small embryos preserved in alcohol, that I forgot to label. At present I am unable to determine the genus to which they belong. They may be lizards, small birds, or even mammals." Drawings of such early-stage embryos allow us to appreciate his quandary (**FIGURE 1.10**). From his detailed study of chick development and his comparison of chick embryos with the embryos of other vertebrates, von Baer derived four generalizations. Now often referred to as "von Baer's laws," they are stated here with some vertebrate examples.

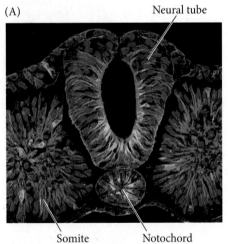

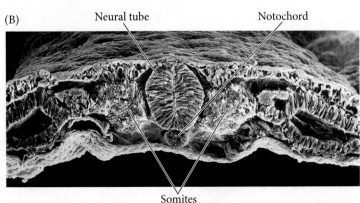

FIGURE 1.9 Two types of microscopy are used to visualize the notochord and its separation of vertebrate embryos (in this case a chick) into right and left halves. The notochord instructs the ectoderm above it to become the nervous system (the neural tube at this stage of development). To either side of the notochord and the neural tube are the mesodermal masses called somites, which will form vertebrae, ribs, and skeletal muscles. (A) Fluorescence micrograph stained with different dyes to highlight nuclear DNA (blue), cytoskeletal microtubules (red, yellow), and the extracellular matrix (green). (B) Scanning electron micrograph of the same stage, highlighting the three-dimensional relationship of the structures. (A courtesy of M. Angeles Rabadán and E. Martí Gorostiza; B courtesy of K. Tosney and G. Schoenwolf.)

1. *The general features of a large group of animals appear earlier in development than do the specialized features of a smaller group.* Although each vertebrate group may start off with different patterns of cleavage and gastrulation, they converge at a very similar structure when they begin forming their neural tube. All developing vertebrates appear very similar right after gastrulation. All vertebrate embryos have gill arches, a notochord, a spinal cord, and primitive kidneys. The structure in Figure 1.9—a notochord below a neural tube, flanked by somites—is seen in every vertebrate embryo. It is only later in development that the distinctive features of class, order, and finally species emerge.

2. *Less general characters develop from the more general, until finally the most specialized appear.* All vertebrates initially have the same type of skin. Only later does the skin develop fish scales, reptilian scales, bird feathers, or the hair, claws, and nails of mammals. Similarly, the early development of limbs is essentially the same in all vertebrates. Only later do the differences between legs, wings, and arms become apparent.

3. *The embryo of a given species, instead of passing through the adult stages of lower animals, departs more and more from them.*[7] For example, as seen in Figure 1.8, the pharyngeal arches start off the same in all vertebrates. But the arch that becomes the jaw support in fish becomes part of the skull of reptiles and becomes part of the middle ear bones of mammals. Mammals never go through a fishlike stage (Riechert 1837; Rieppel 2011).

4. *Therefore, the early embryo of a higher animal is never like a lower animal, but only like its early embryo.* Human embryos never pass through a stage equivalent to an adult fish or bird. Rather, human embryos initially share characteristics in common with fish and avian embryos. Later in development, the mammalian and other embryos diverge, none of them passing through the stages of the others.

Recent research has confirmed von Baer's view that there is a "phylotypic stage" at which the embryos of the different phyla of vertebrates all have a similar physical structure, such as the stage seen in Figure 1.10. At this same stage there appears to be the least amount of difference among the *genes* expressed by the different groups within the same vertebrate phylum (Irie and Kuratani 2011).[8]

Lizard   Human

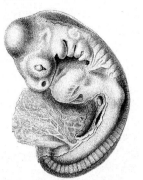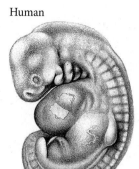

FIGURE 1.10   The vertebrates—fish, amphibians, reptiles, birds, and mammals—all start development very differently because of the enormous differences in the sizes of their eggs. By the beginning of neurulation, however, all vertebrate embryos have converged on a common structure. Here, a lizard embryo is shown next to a human embryo at a similar stage. As they develop beyond the neurula stage, the embryos of the different vertebrate groups become less and less like each other. (From Keibel 1904, 1908; see Galis and Sinervo 2002.)

> **VADE MECUM**
>
> The compound microscope has been the critical tool of developmental anatomists. Mastery of microscopic techniques allows one to enter an entire world of form and pattern.

Keeping Track of Moving Cells: Fate Maps and Cell Lineages

By the late 1800s, it had been conclusively demonstrated that the cell is the basic unit of all anatomy and physiology. Embryologists, too, began to base their field on the cell. But, unlike those who studied the adult organism, developmental anatomists found that cells in the embryo do not "stay put." Indeed, one of the most important conclusions of developmental anatomists is that embryonic cells do not remain in one place, nor do they keep the same shape (Larsen and McLaughlin 1987).

There are two major types of cells in the embryo: **epithelial cells**, which are tightly connected to one another in sheets or tubes; and **mesenchymal cells**, which are unconnected or loosely connected to one another and can operate as independent units.

---

[7] Von Baer formulated these generalizations prior to Darwin's theory of evolution. "Lower animals" would be those having simpler anatomies.

[8] Indeed, one definition of a phylum is that it is a collection of species whose gene expression at the phylotypic stage is highly conserved among them, yet different from that of other species (see Levin et al. 2016).

Within these two types of arrangements, morphogenesis is brought about through a limited repertoire of variations in cellular processes:

- *Direction and number of cell divisions.* Think of the faces of two dog breeds—say, a German shepherd and a poodle. The faces are made from the same cell types, but the number and orientation of the cell divisions are different (Schoenebeck et al. 2012). Think also of the legs of a German shepherd compared with those of a dachshund. The skeleton-forming cells of the dachshund have undergone fewer cell divisions than those of taller dogs.

- *Cell shape changes.* Cell shape change is a critical feature of development. Changing the shapes of epithelial cells often creates tubes out of sheets (as when the neural tube forms), and a shape change from epithelial to mesenchymal is critical when individual cells migrate away from the epithelial sheet (as when muscle cells are formed). (As we will see in Chapter 24, this same type of epithelial-to-mesenchymal change operates in cancer, allowing cancer cells to migrate and spread from the primary tumor to new sites.)

- *Cell migration.* Cells have to move in order to get to their appropriate locations. The germ cells have to migrate into the developing gonad, and the primordial heart cells meet in the middle of the vertebrate neck and then migrate to the left part of the chest.

- *Cell growth.* Cells can change in size. This is most apparent in the germ cells: the sperm eliminates most of its cytoplasm and becomes smaller, whereas the developing egg conserves and adds cytoplasm, becoming comparatively huge. Many cells undergo an "asymmetric" cell division that produces one big cell and one small cell, each of which may have a completely different fate.

- *Cell death.* Death is a critical part of life. The embryonic cells that constitute the webbing between our toes and fingers die before we are born. So do the cells of our tails. The orifices of our mouth, anus, and reproductive glands all form through **apoptosis**—the programmed death of certain cells at particular times and places.

- *Changes in the composition of the cell membrane or secreted products.* Cell membranes and secreted cell products influence the behavior of neighboring cells. For instance, extracellular matrices secreted by one set of cells will allow the migration of their neighboring cells. Extracellular matrices made by other cell types will *prohibit* the migration of the same set of cells. In this way, "paths and guiderails" are established for migrating cells.

Fate maps

Given such a dynamic situation, one of the most important programs of descriptive embryology became the tracing of **cell lineages**: following individual cells to see what those cells become. In many organisms, resolution of individual cells is not possible, but one can label groups of embryonic cells to see what that area becomes in the adult organism. By bringing such studies together, one can construct a **fate map**. These diagrams "map" larval or adult structures onto the region of the embryo from which they arose. Fate maps constitute an important foundation for experimental embryology, providing researchers with information on which portions of the embryo normally become which larval or adult structures. **FIGURE 1.11** shows fate maps of some vertebrate embryos at the early gastrula stage.

Fate maps can be generated in several ways, and the technology has changed greatly over the past few years. The ability to follow cells with molecular dyes and computer imaging has altered our understanding of the origins of several cell types. Even our views of where heart cells originate has been changed (Lane and Sheets 2006; Camp et al. 2012). Mammalian embryos are among the most difficult to map (since they develop inside another organism), and researchers are actively constructing, refining, and arguing about the fate maps of mammalian embryos.

FIGURE 1.11 Fate maps of vertebrates at the early gastrula stage. All are dorsal surface views (looking "down" on the embryo at what will become its back). Despite the different appearances of the adult animals, fate maps of these four vertebrates show numerous similarities among the embryos. The cells that will form the notochord occupy a central dorsal position, while the precursors of the neural system lie immediately anterior to it. The neural ectoderm is surrounded by less dorsal ectoderm, which will form the epidermis of the skin. A indicates the anterior end of the embryo, P the posterior end. The dashed green lines indicate the site of ingression—the path cells will follow as they migrate from the exterior to the interior of the embryo.

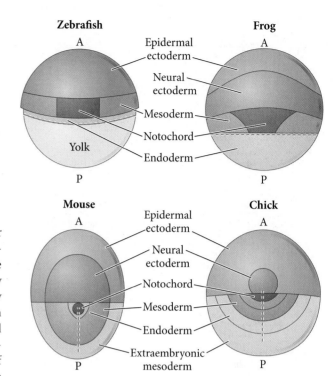

Direct observation of living embryos

Some embryos have relatively few cells, and the cytoplasms of their early blastomeres have differently colored pigments. In such fortunate cases, it is actually possible to look through the microscope and trace the descendants of a particular cell into the organs they generate. E. G. Conklin patiently followed the fates of each early cell of the tunicate (sea squirt) *Styela partita* (**FIGURE 1.12**; Conklin 1905). The muscle-forming cells of the *Styela* embryo always had a yellow color, derived from a region of cytoplasm found one particular pair of blastomeres at the 8-cell sage. Removal of this pair of blastomeres (which according to Conklin's fate map should produce the tail musculature) in fact resulted in larvae with no tail muscles, thus confirming Conklin's map (Reverberi and Minganti 1946).

WEB TOPIC 1.3   **CONKLIN'S ART AND SCIENCE**   The plates from Conklin's remarkable 1905 paper are online. Looking at them, one can see the precision of his observations and how he constructed his fate map of the tunicate.

Dye marking

Most embryos are not so accommodating as to have cells of different colors. In the early years of the twentieth century, Vogt (1929) traced the fates of different areas of amphibian eggs by applying **vital dyes** to the region of interest. Vital dyes stain cells but do not kill them. Vogt mixed such dyes with agar and spread the agar on a microscope

(A)   (B)   (C)

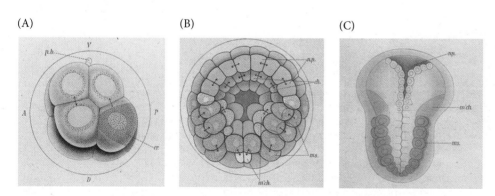

FIGURE 1.12   The fates of individual cells. Edwin Conklin mapped the fates of early cells of the tunicate *Styela partita*, using the fact that in embryos of this species many of the cells can be identified by their different colored cytoplasms. Yellow cytoplasm marks the cells that form the trunk muscles. (A) At the 8-cell stage, two of the eight blastomeres contain this yellow cytoplasm. (B) Early gastrula stage, showing the yellow cytoplasm in the precursors of the trunk musculature. (C) Early larval stage, showing the yellow cytoplasm in the newly formed trunk muscles. (From Conklin 1905.)

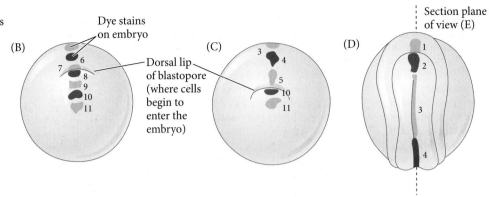

FIGURE 1.13 Vital dye staining of amphibian embryos. (A) Vogt's method for marking specific cells of the embryonic surface with vital dyes. (B–D) Dorsal surface views of stain on successively later embryos. (E) Newt embryo dissected in a medial sagittal section to show the stained cells in the interior. (After Vogt 1929.)

slide to dry. The ends of the dyed agar were very thin. Vogt cut chips from these ends and placed them on a frog embryo. After the dye stained the cells, he removed the agar chips and could follow the stained cells' movements within the embryo (**FIGURE 1.13**).

One problem with vital dyes is that as they become more diluted with each cell division, they become difficult to detect. One way around this is to use **fluorescent dyes** that are so intense that once injected into individual cells, they can still be detected in the progeny of these cells many divisions later. Fluorescein-conjugated dextran, for example, can be injected into a single cell of an early embryo, and the descendants of that cell can be seen by examining the embryo under ultraviolet light (**FIGURE 1.14**).

Genetic labeling

One way of permanently marking cells and following their fates is to create embryos in which the same organism contains cells with different genetic constitutions. One of the best examples of this technique is the construction of **chimeric embryos**—embryos made from tissues of more than one genetic source. Chick-quail chimeras, for example, are made by grafting embryonic quail cells inside a chick embryo while the chick is still

---

VADE MECUM

Most cells must be stained in order to see them; different dyes stain different types of molecules. A segment on histotechniques offers instructions on staining cells to observe particular structures (such as the nucleus).

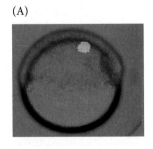

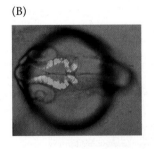

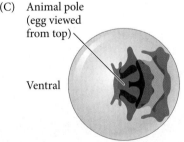

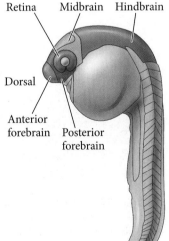

FIGURE 1.14 Fate mapping using a fluorescent dye. (A) Specific cells of a zebrafish embryo were injected with a fluorescent dye that will not diffuse from the cells. The dye was then activated by laser in a small region (about 5 cells) of the late cleavage stage embryo. (B) After formation of the central nervous system had begun, cells that expressed the active dye were visualized by fluorescent light. The fluorescent dye is seen in particular cells that generate the forebrain and midbrain. (C) Fate map of the zebrafish central nervous system. Fluorescent dye was injected into cells 6 hours after fertilization (left), and the results are color-coded onto the hatched fish (right). Overlapping colors indicate that cells from these regions of the 6-hour embryo contribute to two or more regions. (A,B from Kozlowski et al. 1998, photographs courtesy of E. Weinberg; C after Woo and Fraser 1995.)

in the egg. Chicks and quail embryos develop in a similar manner (especially during the early stages), and the grafted quail cells become integrated into the chick embryo and participate in the construction of the various organs (**FIGURE 1.15A**). The chick that hatches will have quail cells in particular sites, depending on where the graft was placed. Quail cells also *differ* from chick cells in several important ways, including the species-specific proteins that form the immune system. There are quail-specific proteins can be used to find individual quail cells, even when they are "hidden" within a large population of chick cells (**FIGURE 1.15B**). By seeing where these cells migrate, fine-structure maps of the chick brain and skeletal system have been produced (Le Douarin 1969; Le Douarin and Teillet 1973).

Chimeras dramatically confirmed the extensive migrations of the neural crest cells during vertebrate development. Mary Rawles (1940) showed that the pigment cells (melanocytes) of the chick originate in the **neural crest**, a transient band of cells that joins the neural tube to the epidermis. When she transplanted small regions of neural crest-containing tissue from a pigmented strain of chickens into a similar position in an embryo from an unpigmented strain of chickens, the migrating pigment cells entered the epidermis and later entered the feathers (**FIGURE 1.15C**). Ris (1941) used similar techniques to show that, although almost all of the external pigment of the chick embryo came from the migrating neural crest cells, the pigment of the retina formed in the retina itself and was not dependent on migrating neural crest cells. This pattern was confirmed in the chick-quail chimeras, in which the quail neural crest cells produced their own pigment and pattern in the chick feathers.

Transgenic DNA chimeras

In most animals, it is difficult to meld a chimera from two species. One way of circumventing this problem is to transplant cells from a genetically modified organism. In such a technique, the genetic modification can then be traced only to those cells that express it. One version is to infect the cells of an embryo with a virus whose genes have

FIGURE 1.15 Genetic markers as cell lineage tracers. (A) Experiment in which cells from a particular region of a 1-day quail embryo have been grafted into a similar region of a 1-day chick embryo. After several days, the quail cells can be seen by using an antibody to quail-specific proteins (photograph below). This region produces cells that populate the neural tube. (B) Chick and quail cells can also be distinguished by the heterochromatin of their nuclei. Quail cells have a single large nucleus (dense purple), distinguishing them from the diffuse nuclei of the chick. (C) Chick resulting from transplantation of a trunk neural crest region from an embryo of a pigmented strain of chickens into the same region of an embryo of an unpigmented strain. The neural crest cells that gave rise to the pigment migrated into the wing epidermis and feathers. (A,B from Darnell and Schoenwolf 1997, courtesy of the authors; C from the archives of B. H. Willier.)

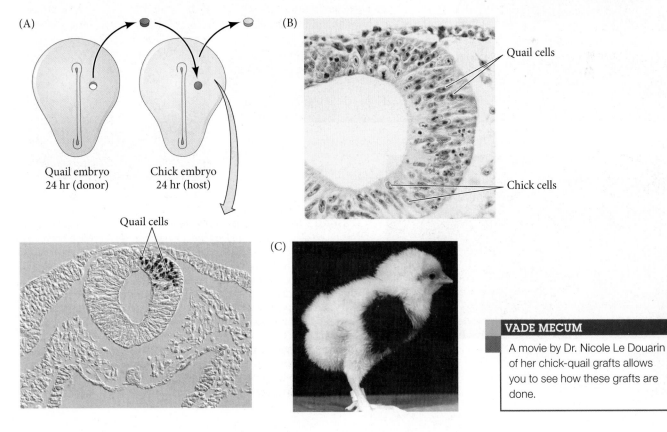

(A)

Quail embryo
24 hr (donor)   Chick embryo
                     24 hr (host)

Quail cells

(B)

Quail cells

Chick cells

(C)

VADE MECUM

A movie by Dr. Nicole Le Douarin of her chick-quail grafts allows you to see how these grafts are done.

been altered such that they express the gene for a fluorescently active protein such as **green fluorescent protein**, or **GFP**.[9] A gene altered in this way is called a **transgene**, because it contains DNA from another species. When the infected embryonic cells are transplanted into a wild-type host, only the donor cells will express GFP; these emit a visible green glow (see Affolter 2016; Papaioannou 2016). Variations on transgenic labeling can give us a remarkably precise map of the developing body.

For example, Freem and colleagues (2012) used transgenic techniques to study the migration of neural crest cells to the gut of chick embryos, where they form the neurons that coordinate peristalsis—the muscular contractions of the gut necessary to eliminate solid waste. The parents of the GFP-labeled chick embryo were infected with a replication-deficient virus that carried an active gene for GFP. This virus was inherited by the chick embryo and expressed in every cell. In this way, Freem and colleagues generated embryos in which every cell glowed green when placed under ultraviolet light (**FIGURE 1.16A**). They then transplanted the neural tube and neural crest of a GFP-transgenic embryo into a similar region of a normal chick embryo (**FIGURE 1.16B**). A day later, they could see GFP-labeled cells migrating into the stomach region (**FIGURE1.16C**), and by 7 days, the entire gut showed GFP staining up to the anterior region of the hindgut (**FIGURE 1.16D**).

[9] Green fluorescent protein occurs naturally in certain jellyfish. It emits bright green fluorescence when exposed to ultraviolet light and is widely used as a transgenic label. GFP labeling will be seen in many photographs throughout this book.

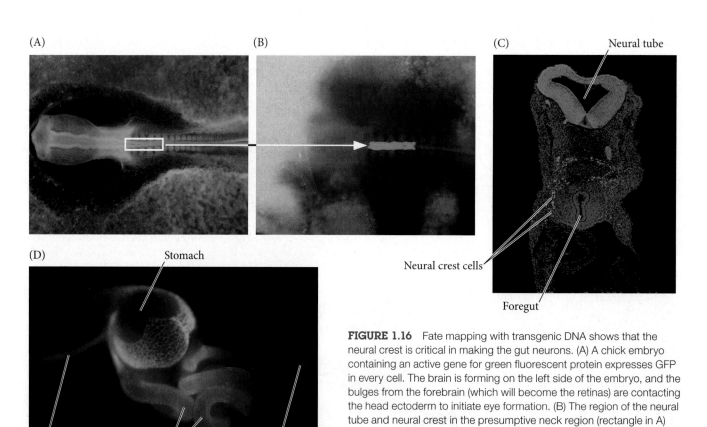

FIGURE 1.16 Fate mapping with transgenic DNA shows that the neural crest is critical in making the gut neurons. (A) A chick embryo containing an active gene for green fluorescent protein expresses GFP in every cell. The brain is forming on the left side of the embryo, and the bulges from the forebrain (which will become the retinas) are contacting the head ectoderm to initiate eye formation. (B) The region of the neural tube and neural crest in the presumptive neck region (rectangle in A) is excised and transplanted into a similar position in an unlabeled wild-type embryo. One can see it by its green fluorescence. (C) A day later, one can see the neural crest cells migrating from the neural tube to the stomach region. (D) In 4 more days, the neural crest cells have spread in the gut from the esophagus to the anterior end of the hindgut. (From Freem et al. 2012; photographs courtesy of A. Burns.)

Evolutionary Embryology

Charles Darwin's theory of evolution restructured comparative embryology and gave it a new focus. After reading Johannes Müller's summary of von Baer's laws in 1842, Darwin saw that embryonic resemblances would be a strong argument in favor of the genetic connectedness of different animal groups. "Community of embryonic structure reveals community of descent," he would conclude in *On the Origin of Species* in 1859. Darwin's evolutionary interpretation of von Baer's laws established a paradigm that was to be followed for many decades—namely, that relationships between groups can be established by finding common embryonic or larval forms.

Even before Darwin, larval forms were used in taxonomic classification. In the 1830s, for instance, J. V. Thompson demonstrated that larval barnacles were almost identical to larval shrimp, and therefore he (correctly) counted barnacles as arthropods rather than mollusks (**FIGURE 1.17**; Winsor 1969). Darwin, himself an expert on barnacle taxonomy, celebrated this finding: "Even the illustrious Cuvier did not perceive that a barnacle is a crustacean, but a glance at the larva shows this in an unmistakable manner." Alexander Kowalevsky (1871) made the similar discovery that tunicate larvae had a notochord and pharyngeal pouches, and that these came from the same germ layers as those same structures in fish and chicks. Thus, Kowalevsky reasoned, the invertebrate tunicate is related to the vertebrates, and the two great domains of the animal kingdom—invertebrates and vertebrates— are thereby united through larval structures (see Chapter 10). As he had endorsed Thompson, Darwin also applauded Kowalevsky's finding, writing in *The Descent of Man* (1874) that "if we may rely on embryology, ever

(A) Barnacle

(B) Shrimp

FIGURE 1.17 Larval stages reveal the common ancestry of two crustacean arthropods. (A) Barnacle. (B) Shrimp. Barnacles and shrimp both exhibit a distinctive larval stage (the nauplius) that underscores their common ancestry as crustacean arthropods, even though adult barnacles—once classified as mollusks—are sedentary, differing in body form and lifestyle from the free-swimming adult shrimp. (A © Wim van Egmond/Visuals Unlimited and © Barrie Watts/OSF/Getty; B courtesy of U.S. National Oceanic and Atmospheric Administration and © Kim Taylor/Naturepl.com.)

the safest guide in classification, it seems that we have at last gained a clue to the source whence the Vertebrata were derived." Darwin further noted that embryonic organisms sometimes form structures that are inappropriate for their adult form, but demonstrate their relatedness to other animals. He pointed out the existence of eyes in embryonic moles, pelvic bone rudiments in embryonic snakes, and teeth in baleen whale embryos.

Darwin also argued that adaptations that depart from the "type" and allow an organism to survive in its particular environment develop late in the embryo.[10] He noted that the differences among species within genera become greater as development persists, as predicted by von Baer's laws. Thus, Darwin recognized two ways of looking at "descent with modification." One could emphasize *common descent* by pointing out embryonic similarities between two or more groups of animals, or one could emphasize the *modifications* to show how development has been altered to produce structures that enable animals to adapt to particular conditions.

Embryonic homologies

One of the most important distinctions made by evolutionary embryologists was the difference between analogy and homology. Both terms refer to structures that appear to be similar. **Homologous** structures are those organs whose underlying similarity arises from their being derived from a common ancestral structure. For example, the wing of a bird and the arm of a human are homologous, both having evolved from the forelimb bones of a common ancestor. Moreover, their respective parts are homologous (**FIGURE 1.18**).

Analogous structures are those whose similarity comes from their performing a similar function rather than their arising from a common ancestor. For example, the wing of a butterfly and the wing of a bird are analogous; the two share a common function (and thus both are called wings), but the bird wing and insect wing did not arise from a common ancestral structure that became modified through evolution into bird wings and butterfly wings. Homologies must always refer to the level of organization being compared. For instance, bird and bat wings are homologous as forelimbs but not as wings. In other words, they share an underlying structure of forelimb bones because birds and mammals share a common ancestor that possessed such bones. Bats, however, descended from a long line of non-winged mammals, whereas bird wings evolved independently, from the forelimbs of ancestral reptiles. As we will see, the structure of a bat's wing is markedly different from that of a bird's wing.

As we will see in Chapter 26, evolutionary change is based on developmental change. The bat wing, for example, is made in part by (1) maintaining a rapid growth rate in the cartilage that forms the fingers and (2) preventing the cell death that

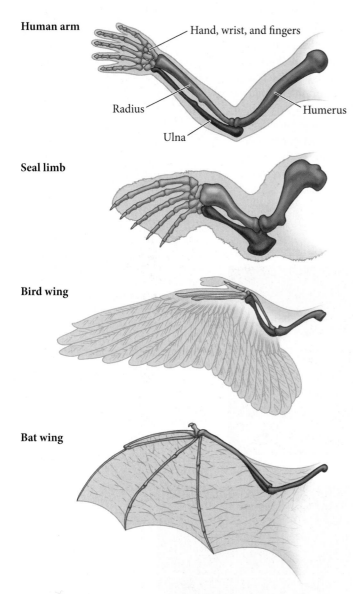

FIGURE 1.18 Homologies of structure among a human arm, a seal forelimb, a bird wing, and a bat wing; homologous supporting structures are shown in the same color. All four were derived from a common tetrapod ancestor and thus are homologous as forelimbs. The adaptations of bird and bat forelimbs to flight, however, evolved independently of each other, long after the two lineages diverged from their common ancestor. Therefore, as wings they are not homologous, but analogous.

_____

[10]As first noted by Weismann (1875), larvae must have their own adaptations. The adult viceroy butterfly mimics the monarch butterfly, but the viceroy caterpillar does not resemble the beautiful larva of the monarch. Rather, the viceroy larva escapes detection by resembling bird droppings (Begon et al. 1986).

(A)

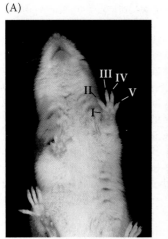

(B)

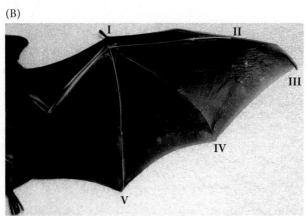

FIGURE 1.19 Development of bat and mouse forelimbs. (A,B) Mouse and bat torsos, showing the mouse forelimb and the elongated fingers and prominent webbing in the bat wing. The digits are numbered on both animals (I, thumb; V, "pinky"). (C) Comparison of mouse and bat forelimb morphogenesis. Both limbs start as webbed appendages, but the webbing between the mouse's digits dies at embryonic day 14 (arrow). The webbing in the bat fore-limb does not die and is sustained as the fingers grow. (A courtesy of D. McIntyre; B,C from Cretekos et al. 2008, courtesy of C. J. Cretekos.)

(C)

Bat

Mouse

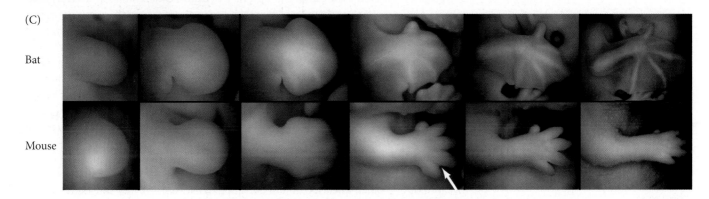

normally occurs in the webbing between the fingers. As seen in **FIGURE 1.19**, mice start off with webbing between their digits (as do humans and most other mammals). This webbing is important for creating the anatomical distinctions between the fingers. Once the webbing has served that function, genetic signals cause its cells to die, leaving free digits that can grasp and manipulate. Bats, however, use their fingers for flight, a feat accomplished by changing the genes that are active in the webbing. The genes activated in embryonic bat webbing encode proteins that *prevent* cell death, as well as proteins that accelerate finger elongation (Cretekos et al. 2005; Sears et al. 2006; Weatherbee et al. 2006). Thus, homologous anatomical structures can differentiate by altering development, and such changes in development provide the variation needed for evolutionary change.

Charles Darwin observed artificial selection in pigeon and dog breeds, and these examples remain valuable resources for studying selectable variation. For instance, the short legs of dachshunds were selected by breeders who wanted to use these dogs to hunt badgers (German *Dachs*, "badger" + *Hund*, "dog") in their underground burrows. The mutation that causes the dachshund's short legs involves an extra copy of the *Fgf4* gene, which makes a protein that informs the cartilage precursor cells that they have divided enough and can start differentiating. With this extra copy of *Fgf4*, cartilage cells are told that they should stop dividing earlier than in most other dogs, so the legs stop growing (Parker et al. 2009). Similarly, long-haired dachshunds differ from their short-haired relatives in having a mutation in the *Fgf5* gene (Cadieu et al. 2009). This gene is involved in hair production and allows each follicle to make a longer hair shaft (Ota et al. 2002; see Chapter 16). Thus, mutations in genes controlling developmental processes can generate selectable variation.

Medical Embryology and Teratology

While embryologists could look at embryos to describe the evolution of life and how different animals form their organs, physicians became interested in embryos for more practical reasons. Between 2% and 5% of human infants are born with a readily observable anatomical abnormality (Winter 1996; Thorogood 1997). These abnormalities may include missing limbs, missing or extra digits, cleft palate, eyes that lack certain parts, hearts that lack valves, and so forth. Some birth defects are produced by mutant genes or chromosomes, and some are produced by environmental factors that impede development. The study of birth defects can tell us how the human body is normally formed. In the absence of experimental data on human embryos, nature's "experiments" sometimes offer important insights into how the human body becomes organized.

Genetic malformations and syndromes

Abnormalities caused by genetic events (gene mutations, chromosomal aneuploidies, and translocations) are called **malformations**, and a **syndrome** is a condition in which two or more malformations are expressed together. For instance, a hereditary disease called Holt-Oram syndrome is inherited as an autosomal dominant condition. Children born with this syndrome usually have a malformed heart (the septum separating the right and left sides fails to grow normally; see Chapter 18) and absent wrist or thumb bones. Holt-Oram syndrome was found to be caused by mutations in the *TBX5* gene (Li et al 1997; Basson et al 1997). The TBX5 protein is expressed in the developing heart and the developing hand and is important for normal growth and differentiation in both locations.

Disruptions and teratogens

Developmental abnormalities caused by exogenous agents (certain chemicals or viruses, radiation, or hyperthermia) are called **disruptions**. The agents responsible for these disruptions are called **teratogens** (Greek, "monster-formers"), and the study of how environmental agents disrupt normal development is called teratology. Substances that can cause birth defects include relatively common substances such as alcohol and retinoic acid (often used to treat acne), as well many chemicals used in manufacturing and released into the environment. Heavy metals (e.g., mercury, lead, selenium) can alter brain development.

Teratogens were brought to the attention of the public in the early 1960s. In 1961, Lenz and McBride independently accumulated evidence that the drug thalidomide, prescribed as a mild sedative to many pregnant women, caused an enormous increase in a previously rare syndrome of congenital anomalies. The most noticeable of these anomalies was phocomelia, a condition in which the long bones of the limbs are deficient or absent (**FIGURE 1.20A**). More than 7,000 affected infants were born to women who took thalidomide, and a woman need only have taken one tablet for her child to be born with all four limbs deformed (Lenz 1962, 1966; Toms 1962). Other abnormalities induced by ingesting this drug included heart defects, absence of the external ears, and malformed intestines. Nowack (1965) documented the period of susceptibility during which thalidomide caused these abnormalities (**FIGURE 1.20B**). The drug was found to be teratogenic only during days 34–50 after the last menstruation (i.e., 20–36 days postconception). From days 34 to 38, no limb abnormalities are seen, but during this period thalidomide can cause the absence or deficiency of ear components. Malformations of the upper limbs are seen before those of the lower limbs because the developing arms form slightly before the legs. These and other teratogens will be discussed extensively in Chapter 24.

The integration of anatomical information about congenital malformations with our new knowledge of the genes responsible for development has resulted in an ongoing restructuring of medicine. This integrated information is allowing us to discover the genes responsible for inherited malformations, and to identify exactly which steps in development are disrupted by specific teratogens. We will see examples of this integration throughout this text.

(A)

(B)

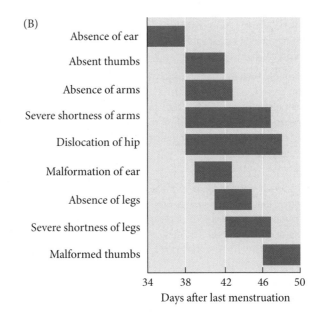

Days after last menstruation

FIGURE 1.20   A developmental anomaly caused by an environmental agent. (A) Phocomelia, the lack of proper limb development, was the most visible of the birth defects that occurred in many children born in the early 1960s whose mothers took the drug thalidomide during pregnancy. These children are now middle-aged adults; this photograph is of Grammy-nominated German singer Thomas Quasthoff. (B) Thalidomide disrupts different structures at different times of human development. (A © dpa picture alliance archive/Alamy Stock Photo; B after Nowack 1965.)

Closing Thoughts on the Opening Photo

For many animal species, larvae are a critical part of normal development. Larvae (such as these tadpoles) are often the food-gathering and dispersal stage of the organism. They also often have different habitats than the adults. The frog genome has two sets of genes, one set for the larval stage and another for the adult; which set of genes is expressed is regulated by a cascade of hormones, as we will describe in Chapter 21. Metamorphosis means getting rid of some organs, making new organs, and re-purposing other organs. The tadpole keeps its same three body axes as it becomes an adult frog. However, its retinal pigments, blood hemoglobins, urea cycle enzymes, and skin transform from those of an aquatic animal to those characteristic of a terrestrial animal. Its eyes change location and the digestive system changes from that of a herbivore to that of a carnivore. (Photograph by Bert Willaert © Nature Picture Library/Alamy Stock Photo.)

 Snapshot Summary
Mechanisms of Developmental Organization

1. The life cycle can be considered a central unit in biology; the adult form need not be paramount. The basic animal life cycle consists of fertilization, cleavage, gastrulation, germ layer formation, organogenesis, metamorphosis, adulthood, and senescence.

2. In gametogenesis, the germ cells (i.e., those cells that will become sperm or eggs) undergo meiosis. Eventually, usually after adulthood is reached, the mature gametes

are released to unite during fertilization. The resulting new generation then begins development.

3. Epigenesis happens. Organisms are created de novo each generation from the relatively disordered cytoplasm of the egg.

4. Preformation is not found in the anatomical structures themselves, but in the genetic instructions that instruct their formation. The inheritance of the fertilized egg

includes the genetic potentials of the organism. These preformed nuclear instructions include the ability to respond to environmental stimuli in specific ways.

5. The three germ layers give rise to specific organ systems. The ectoderm gives rise to the epidermis, nervous system, and pigment cells; the mesoderm generates the kidneys, gonads, muscles, bones, heart, and blood cells; and the endoderm forms the lining of the digestive tube and the respiratory system.

6. Karl von Baer's principles state that the general features of a large group of animals appear earlier in the embryo than do the specialized features of a smaller group. As each embryo of a given species develops, it diverges from the adult forms of other species. The early embryo of a "higher" animal species is not like the adult of a "lower" animal.

7. Labeling cells with dyes shows that some cells differentiate where they form, whereas others migrate from their original sites and differentiate in their new locations. Migratory cells include neural crest cells and the precursors of germ cells and blood cells.

8. "Community of embryonic structure reveals community of descent" (Charles Darwin, *On the Origin of Species*).

9. Homologous structures in different species are those organs whose similarity is due to sharing a common ancestral structure. Analogous structures are those organs whose similarity comes from serving a similar function (but which are not derived from a common ancestral structure).

10. Congenital anomalies can be caused by genetic factors (mutations, aneuploidies, translocations) or by environmental agents (certain chemicals, certain viruses, radiation).

11. Teratogens—environmental compounds that can alter development—act at specific times when certain organs are being formed. Similar genetic malformations can occur when communication between cells is interrupted or eliminated. The molecular signal and its receptor on the responding cell are both critical.

Further Reading

Affolter, M. 2016. Seeing is believing, or how GFP changed my approach to science. *Curr. Top. Dev. Biol.* 116: 1–16.

Cadieu, E. and 19 others. 2009. Coat variation in the domestic dog is governed by variants in three genes. *Science* 326: 150–153.

Larsen, E. and H. McLaughlin. 1987. The morphogenetic alphabet: Lessons for simple-minded genes. *BioEssays* 7: 130–132.

Le Douarin, N. M. and M.-A. Teillet. 1973. The migration of neural crest cells to the wall of the digestive tract in the avian embryo. *J. Embryol. Exp. Morphol.* 30: 31–48.

Nishida, H. 1987. Cell lineage analysis in ascidian embryos by intracellular injection of a tracer enzyme. III. Up to the tissue-restricted stage. *Dev. Biol.* 121: 526–541.

Papaioannou, V. E. 2016. Concepts of cell lineage in mammalian embryos. *Curr. Top. Dev. Biol.* 117: 185–198.

Pinto-Correia, C. 1997. *The Ovary of Eve: Egg and Sperm and Preformation*. University of Chicago Press, Chicago.

Weatherbee, S. D., R. R. Behringer, J. J. Rasweiler 4th and L. A. Niswander. 2006. Interdigital webbing retention in bat wings illustrates genetic changes underlying amniote limb diversification. *Proc. Natl. Acad. Sci. USA* 103: 15103–15107.

Winter, R. M. 1996. Analyzing human developmental abnormalities. *BioEssays* 18: 965–971.

Woo, K. and S. E. Fraser. 1995. Order and coherence in the fate map of the zebrafish embryo. *Development* 121: 2595–2609.

GO TO WWW.DEVBIO.COM…

…for Web Topics, Scientists Speak interviews, Watch Development videos, Dev Tutorials, and complete bibliographic information for all literature cited in this chapter.

Specifying Identity
Mechanisms of Developmental Patterning

IN 1883, ONE OF AMERICA'S FIRST EMBRYOLOGISTS, William Keith Brooks, reflected on "the greatest of all wonders of the material universe: the existence, in a simple, unorganized egg, of a power to produce a definite adult animal." He noted that the process is so complex that "we may fairly ask what hope there is of discovering its solution, of reaching its true meaning, its hidden laws and causes." Indeed, how to get from "a simple, unorganized egg" to an exquisitely ordered body is the fundamental mystery of development. Biologists today have come a long way along the road to discovering the solution to this mystery, piecing together its "hidden laws and causes." They include how the unorganized egg becomes organized, how different cells interpret the same genome differently, and the many modes of communication by which cells signal one another and thus orchestrate the unique patterns of their differentiation.

In this chapter, we will introduce the concept of *cell specification*—how cells become specified to a specific fate—and explore how cells of different organisms use different mechanisms for determining cell fate. In Chapters 3 and 4, we will delve deeper into the genetic mechanisms underlying cell differentiation and the cell signaling involved. Chapter 5, the final chapter of this unit, focuses on development of stem cells, which exemplifies all the principles defined in this first unit.

A crowd of individuals or a gang of clones?

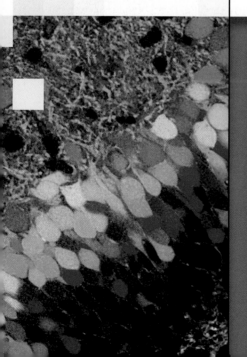

The Punchline

Undifferentiated cells go through a process of maturation that begins when they become committed to a specific cell lineage, progresses through a stage when cell fate is determined to become that of a specific cell type, and ends in differentiation as cells acquire the gene expression pattern characteristic of a specific cell type. In some organisms, cell fate is determined very early by the specific molecules present in the cytoplasm apportioned to each cell as the fertilized egg divides. In other organisms, cell fate remains plastic or changeable in the early embryo and becomes restricted over time through cell-cell interactions. In some species (notably the fruit fly), initially only the nuclei divide, creating a syncytium of many nuclei within a single undivided cytoplasm. In these embryos, anterior-posterior gradients of informational molecules in the cytoplasm determine which genes will be expressed in the different nuclei once they become separated into individual cells. Using powerful new imaging techniques such as Brainbow, researchers today are able to map the fates of individual cells from zygote to adult.

FIGURE 2.1   From sand grains to an organized octopus sculpture. (Photograph courtesy of Michael J. F. Barresi, 2014.)

FIGURE 2.1   From sand grains to an organized octopus sculpture. (Photograph courtesy of Michael J. F. Barresi, 2014.)

Levels of Commitment

To the naked eye, individual grains of sand on an expansive beach look unorganized, yet the grains can be molded together to create complex structures, as illustrated by a sand sculpture of an octopus holding children in its tentacles (**FIGURE 2.1**). How can disordered units become ordered, a pile of sand become a structured creation, or a collection of cells become a highly complex embryo? Did the sand grains that became the octopus's eye *know* they were going to become an eye as they washed up on the beach earlier that morning? Obviously, significant energy had to be applied to the inanimate and inorganic sand grains to get them to become the sculpture's eye. What about the cells of your eye? Did they *know* they were destined to become part of an eye? If so, *when* did they know it, and how *set* were they in adopting this fate?

Cell differentiation

The generation of specialized cell types is called **differentiation**, a process during which a cell ceases to divide and develops specialized structural elements and distinct functional properties. Differentiation, though, is only the last, overt stage in a series of events that commit an undifferentiated cell of an embryo to become a particular cell type (**TABLE 2.1**). A red blood cell obviously differs radically in its protein composition and cell structure from a lens cell in the eye or a neuron in the brain. But these differences in cellular biochemistry and function are preceded by a process that commits the cell to a certain fate. During the course of **commitment**, the cell might not look different from its nearest or most distant neighbors in the embryo and show no visible signs of differentiation, but its developmental fate has become restricted.

Commitment

The process of commitment can be divided into two stages (Harrison 1933; Slack 1991). The first stage is **specification**. The fate of a cell or tissue is said to be specified when it is capable of differentiating autonomously (i.e., by itself) when placed in an environment that is neutral with respect to the developmental pathway, such as in a petri dish or test tube (**FIGURE 2.2A**). At the stage of specification, cell commitment is still labile (i.e., capable of being altered). If a specified cell is transplanted to a population of differently specified cells, the fate of the transplant will be altered by its interactions with its new neighbors (**FIGURE 2.2B**). It is not unlike many of you who may have entered

TABLE 2.1 Some differentiated cell types and their major products		
Type of cell	**Differentiated cell product**	**Specialized function**
Keratinocyte (epidermal cell)	Keratin	Protection against abrasion, desiccation
Erythrocyte (red blood cell)	Hemoglobin	Transport of oxygen
Lens cell	Crystallins	Transmission of light
B lymphocyte	Immunoglobulins	Synthesis of antibodies
T lymphocyte	Cytokines	Destruction of foreign cells; regulation of immune response
Melanocyte	Melanin	Pigment production
Pancreatic islet (β) cell	Insulin	Regulation of carbohydrate metabolism
Leydig cell (♂)	Testosterone	Male sexual characteristics
Chondrocyte (cartilage cell)	Chondroitin sulfate; type II collagen	Tendons and ligaments
Osteoblast (bone-forming cell)	Bone matrix	Skeletal support
Myocyte (muscle cell)	Actin and myosin	Muscle contraction
Hepatocyte (liver cell)	Serum albumin; numerous enzymes	Production of serum proteins and numerous enzymatic functions
Neurons	Neurotransmitters (acetylcholine, serotonin, etc.)	Transmission of communication signals in the nervous system
Tubule cell (♀) of hen oviduct	Ovalbumin	Egg white proteins for nutrition and protection of the embryo
Follicle cell (♀) of insect ovary	Chorion proteins	Eggshell proteins for protection of embryo

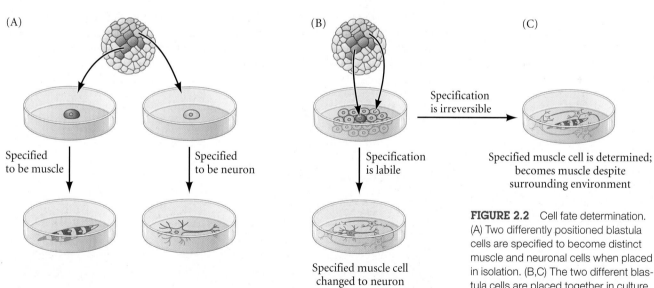

FIGURE 2.2 Cell fate determination. (A) Two differently positioned blastula cells are specified to become distinct muscle and neuronal cells when placed in isolation. (B,C) The two different blastula cells are placed together in culture. (B) In one scenario, the dark red cell was specified—but not determined—to form muscle. It adopts a neuronal fate due to its interactions with its neighbors. (C) If the red cell was committed and determined to become muscle at the time of culturing, it will continue to differentiate into a muscle cell type despite any interactions with its neighbors.

your developmental biology classroom interested in chemistry but, after being exposed to the awesomeness that is developmental biology, will be influenced to change your mind and become a developmental biologist.

The second stage of commitment is **determination**. A cell or tissue is said to be determined when it is capable of differentiating autonomously even when placed into another region of the embryo or a cluster of differently specified cells in a petri dish (**FIGURE 2.2C**). If a cell or tissue type is able to differentiate according to its specified fate even under these circumstances, it is assumed that commitment is irreversible. To continue our example from above, it would be similar to being unwaveringly determined to become a chemist no matter how awe-inspiring your developmental biology course might be.

In summary, then, during embryogenesis an undifferentiated cell matures through specific stages that cumulatively commit it to a specific fate: first specification, then determination, and finally differentiation. During specification, there are three major strategies that embryos can exhibit: autonomous, conditional, and syncytial. Embryos of different species use different combinations of these strategies.

Autonomous Specification

One major strategy of cell commitment is **autonomous specification**. Here, the blastomeres of the early embryo are apportioned a set of critical *determination* factors within the egg cytoplasm. In other words, the egg cytoplasm is not homogeneous; rather, different regions of the egg contain different **morphogenetic determinants** that will influence the cell's development. These determinants, as you will learn in Chapter 3, are molecules—often transcription factors—that regulate gene expression in a manner that directs the cell into a particular path of development. In autonomous specification, the cell "knows" very early what it is to become without interacting with other cells. For instance, even in the very early cleavage stages of the snail *Patella*, blastomeres that are presumptive trochoblast cells can be isolated in a petri dish. There, they will develop into the same ciliated cell types that they would give rise to in the embryo and with the same temporal precision (**FIGURE 2.3**). This continued commitment to the trochoblast fate suggests that these particular early blastomeres are already specified and determined to their fate.

Cytoplasmic determinants and autonomous specification in the tunicate

Tunicate (sea squirt) embryos exhibit some of the best examples of autonomous specification. In 1905, Edwin Grant Conklin, an embryologist working at the Woods Hole Marine Biological Laboratory, published a remarkable fate map of the tunicate *Styela partita*.[1] Upon careful examination of the developing embryo, Conklin noticed a visible

---

[1] Today, the most commonly researched tunicate is *Ciona intestinalis*, which has provided great insight into cell lineage maturation, vertebrate evolution and development, and, more recently, the physical properties governing neural tube closure, which is remarkably similar to that of humans.

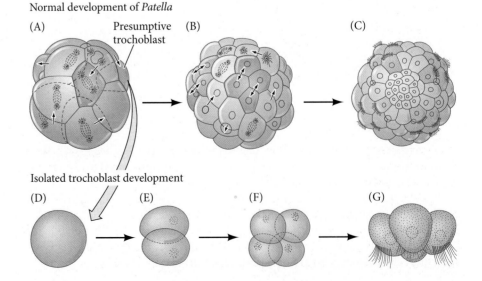

Normal development of *Patella*

(A)   Presumptive trochoblast   (B)   (C)

Isolated trochoblast development

(D)   (E)   (F)   (G)

FIGURE 2.3 Autonomous specification. (A–C) Differentiation of trochoblast (ciliated) cells of the snail *Patella*. (A) 16-Cell stage seen from the side; the presumptive trochoblast cells are shown in pink. (B) 48-Cell stage. (C) Ciliated larval stage, seen from the animal pole. (D–G) Differentiation of a *Patella* trochoblast cell isolated from the 16-cell stage and cultured in vitro. Even in isolated culture, the cells divide and become ciliated at the correct time. (After Wilson 1904.)

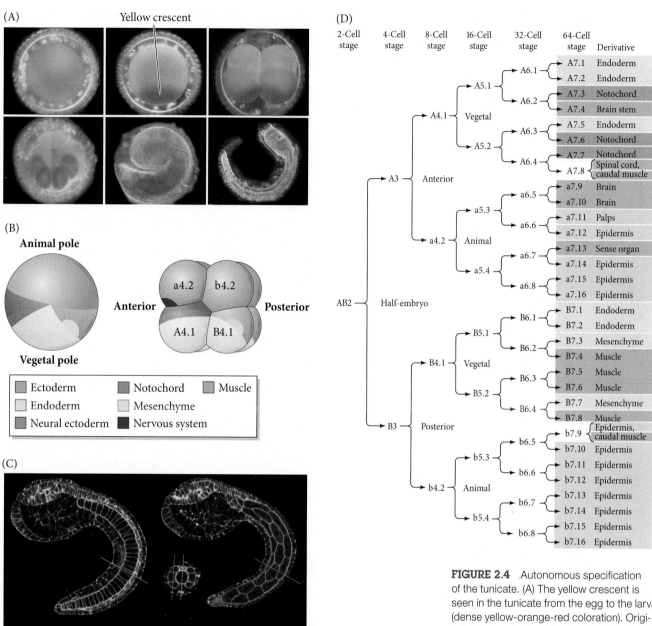

FIGURE 2.4 Autonomous specification of the tunicate. (A) The yellow crescent is seen in the tunicate from the egg to the larva (dense yellow-orange-red coloration). Original drawings by Conklin demonstrate his observations of the yellow crescent in egg and larva (golden color). (B) Zygote of *Styela partita* (left), shown shortly before the first cell division, with the fate of the cytoplasmic regions indicated. The 8-cell embryo on the right shows these regions after three cell divisions. (C) Confocal section through a larva of the tunicate *Ciona savignyi*. Different tissue types were pseudo colored. (D) A linear version of the *S. partita* fate map, showing the fates of each cell of the embryo. (A from Swalla 2004, courtesy of B. Swalla, K. Zigler, and M. Baltzley; B after Nishida 1987 and Reverberi and Minganti 1946; C from Veeman and Reeves 2015.)

yellow coloration that was partitioned within the egg cytoplasm and ultimately segregated to muscle lineages (**FIGURE 2.4**). Conklin meticulously followed the fates of each early cell and showed that "all the principle organs of the larva in their definitive positions and proportions are here marked out in the 2-cell stage by distinct kinds of protoplasm." The yellow pigment conveniently provided Conklin a means to trace the lineages of each blastomere. But is each blastomere determined as to its lineage? That is, *are they autonomously specified*?

The association of Conklin's fate map with autonomous specification was confirmed by cell-removal experiments. The muscle-forming cells of the *Styela* embryo always retain the yellow color and are easily seen to derive from a region

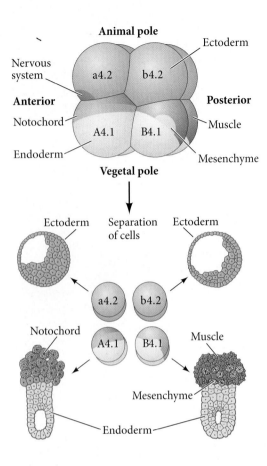

Animal pole

Nervous system — a4.2 — b4.2 — Ectoderm

Anterior **Posterior**

Notochord — A4.1 — B4.1 — Muscle

Endoderm — — Mesenchyme

Vegetal pole

↓

Separation of cells

Ectoderm — a4.2 — b4.2 — Ectoderm

Notochord — A4.1 — B4.1 — Muscle

Mesenchyme

Endoderm

FIGURE 2.5 Autonomous specification in the early tunicate embryo. When the four blastomere pairs of the 8-cell embryo are dissociated, each forms the structures it would have formed had it remained in the embryo. The tunicate nervous system, however, is conditionally specified. The fate map shows that the left and right sides of the tunicate embryo produce identical cell lineages. Here the muscle-forming yellow cytoplasm is colored red to conform to its association with mesoderm. (After Reverberi and Minganti 1946.)

of cytoplasm found in the B4.1 blastomeres. In fact, removal of the B4.1 cells (which according to Conklin's map should produce all the tail musculature) resulted in a larva with no tail muscles (Reverberi and Minganti 1946). This result supports the conclusion that *only those cells derived from the early B4.1 blastomeres possess the capacity to develop into tail muscle.* Further supporting a mode of autonomous specification, each blastomere will form most of its respective cell types even when separated from the remainder of the embryo (**FIGURE 2.5**). Moreover, if the yellow cytoplasm of the B4.1 cells is placed into other cells, those cells will form tail muscles (Whittaker 1973; Nishida and Sawada 2001). These results taken together suggest that critical factors that control cell fate are present and differentially segregated in the cytoplasm of early blastomeres.

In 1973, J. R. Whittaker provided dramatic biochemical confirmation of the cytoplasmic segregation of tissue determinants in early tunicate embryos. When Whittaker removed the pair of B4.1 blastomeres and placed them in isolation, they produced muscle tissue; however, no other blastomere was able to form muscles when separated. Interestingly, contained in the yellow-pigmented cytoplasm is mRNA for a muscle-specific transcription factor appropriately called Macho, and only those blastomeres that acquire this region of yellow cytoplasm (and thus the Macho factor) give rise to muscle cells (**FIGURE 2.6A**; Nishida and Sawada 2001; reviewed by Pourquié 2001). Functionally, Macho is required for tail muscle development in *Styela*; loss of *Macho* mRNA leads to a loss of muscle differentiation of the B4.1 blastomeres, whereas microinjection of *Macho* mRNA into other blastomeres promotes

(A)

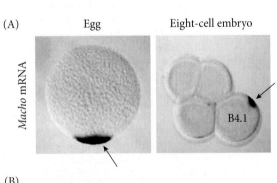

Egg   Eight-cell embryo

Macho mRNA

B4.1

Developing Questions

Look closely at the localization of *Macho* mRNA in the tunicate embryo (see Figure 2.6A). Is it evenly spread throughout the cell, or is it localized to only a small region? Once you have decided on its spatial distribution, contemplate whether this distribution is consistent with a mode of autonomous specification for the muscle lineage. From a cell biological perspective, how do you think this distribution of a specific mRNA is established?

(B)

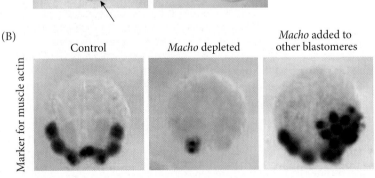

Control   *Macho* depleted   *Macho* added to other blastomeres

Marker for muscle actin

FIGURE 2.6 The *Macho* gene regulates muscle development in the tunicate. (A) Like the yellow crescent, *Macho* transcript is localized to the vegetal-most extent of the egg and differentially expressed only in the B4.1 blastomere. (B) Knockdown of *Macho* function by incorporation of targeting antisense oligonucleotides causes reductions in muscle differentiation, whereas ectopic misexpression of *Macho* in other blastomeres results in expanded muscle differentiation. (From Nishida and Sawada 2001.)

ectopic muscle differentiation (**FIGURE 2.6B**). Thus, the tail muscles of these tunicates are formed autonomously by acquiring and retaining the *Macho* mRNA from the egg cytoplasm with each round of mitosis.

> **WATCH DEVELOPMENT 2.1**   The Four-Dimensional Ascidian Body Atlas use real 3D data sets collected over time to offer an interactive way to view the Ascidian embryo.

Conditional Specification

We have just learned how most of the cells of an early tunicate embryo are determined by autonomous specification; however, even the tunicate embryo is not fully specified this way—its nervous system arises conditionally. Conditional specification is the process by which cells achieve their respective fates by interacting with other cells. Here, what a cell becomes is specified by the array of interactions it has with its neighbors, which may include cell-to-cell contacts (juxtacrine factors), secreted signals (paracrine factors), or the physical properties of its local environment (mechanical stress), mechanisms we will explore in detail in Chapter 4. For example, if cells from one region of a vertebrate blastula (e.g., frog, zebrafish, chick, or mouse) whose fates have been mapped to give rise to the dorsal region of the embryo are transplanted into the presumptive ventral region of another embryo, the transplanted "donor" cells will change their fates and differentiate into ventral cell types (**FIGURE 2.7** and **WATCH DEVELOPMENT 2.2**). Moreover, the dorsal region of the donor embryo where cells were extracted also ends up developing normally.

> **WATCH DEVELOPMENT 2.2**   Watch Dr. Barresi perform a similar gastrula-staged cell transplantation in zebrafish. The donor cells adopt their new location (see Figure 2.7A).

In one of the ironies of research, conditional specification was demonstrated by attempts to disprove it. In 1888, August Weismann proposed the first testable model of cell specification, the germ plasm theory, in which each cell of the embryo would develop autonomously. He boldly proposed that the sperm and egg provided equal chromosomal contributions, both quantitatively and qualitatively, to the new organism. Moreover, he postulated that the chromosomes carried the inherited potentials of this new organism.[2] However, not all the determinants on the chromosomes were

---

[2] Embryologists were thinking in terms of chromosomal mechanisms of inheritance some 15 years before the rediscovery of Mendel's work. Weismann (1882, 1893) also speculated that these nuclear determinants of inheritance functioned by elaborating substances that became active in the cytoplasm!

FIGURE 2.7 Conditional specification. (A) What a cell becomes depends on its position in the embryo. Its fate is determined by interactions with neighboring cells. (B) If cells are removed from the embryo, the remaining cells can regulate and compensate for the missing part.

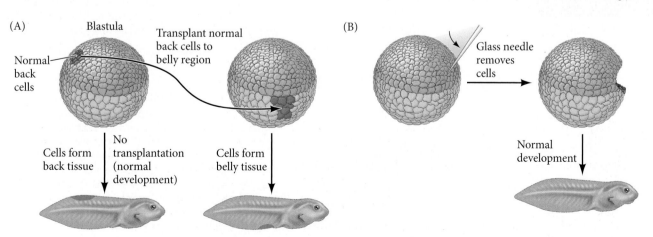

(A)

Blastula

Transplant normal back cells to belly region

Normal back cells

Cells form back tissue

No transplantation (normal development)

Cells form belly tissue

(B)

Glass needle removes cells

Normal development

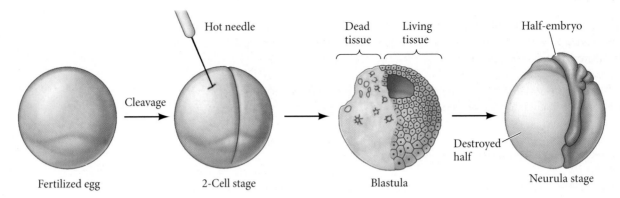

Fertilized egg · Cleavage · 2-Cell stage · Hot needle · Dead tissue · Living tissue · Blastula · Destroyed half · Half-embryo · Neurula stage

FIGURE 2.8 Roux's attempt to demonstrate autonomous specification. Destroying (but not removing) one cell of a 2-cell frog embryo resulted in the development of only one half of the embryo.

thought to enter every cell of the embryo. Instead of dividing equally, the chromosomes were hypothesized to divide in such a way that different determinants entered different cells. Whereas the fertilized egg was hypothesized to carry the full complement of determinants, certain somatic cells were considered to retain the "blood-forming" determinants, others retained the "muscle-forming" determinants, and so forth. (It sounds surprisingly similar to autonomous specification, doesn't it?) Only the nuclei of those cells destined to become germ cells (gametes) were postulated to contain all the different types of determinants.

Cell position matters: Conditional specification in the sea urchin embryo

In postulating his germ plasm model, Weismann proposed a hypothesis of development that could be tested immediately. Based on the fate map of the frog embryo, Weismann claimed that when the first cleavage division separated the future right half of the embryo from the future left half, there would be a separation of "right" determinants from "left" determinants in the resulting blastomeres. Wilhelm Roux tested Weismann's hypothesis by using a hot needle to kill one of the cells in a 2-cell frog embryo, and only the right or left half of a larva developed (**FIGURE 2.8**). Based on this result, Roux claimed that specification was autonomous and that all the instructions for normal development were present inside each cell.

Roux's colleague Hans Dreisch, however, obtained opposite results. Whereas Roux's studies were defect experiments that answered the question of how the embryo would develop when a subset of blastomeres was destroyed, Driesch (1892) sought to extend this research by performing isolation experiments (**FIGURE 2.9**). He separated sea urchin blastomeres from one another by vigorous shaking (or later, by placing

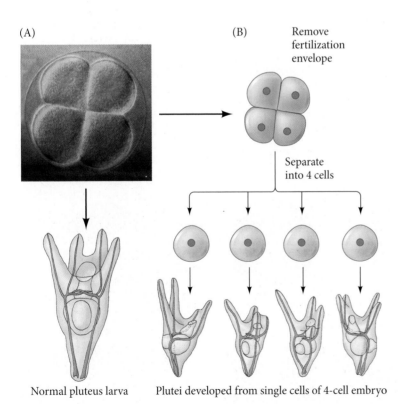

(A) Normal pluteus larva

(B) Remove fertilization envelope — Separate into 4 cells — Plutei developed from single cells of 4-cell embryo

FIGURE 2.9 Driesch's demonstration of conditional (regulative) specification. (A) An intact 4-cell sea urchin embryo generates a normal pluteus larva. (B) When one removes the 4-cell embryo from its fertilization envelope and isolates each of the four cells, each cell can form a smaller, but normal, pluteus larva. (All larvae are drawn to the same scale.) Note that the four larvae derived in this way are not identical, despite their ability to generate all the necessary cell types. Such variation is also seen in adult sea urchins formed in this way (see Marcus 1979). (A, photograph courtesy of G. Watchmaker.)

them in calcium-free seawater). To Driesch's surprise, each of the blastomeres from a 2-cell embryo developed into a complete larva. Similarly, when Driesch separated the blastomeres of 4- and 8-cell embryos, some of the isolated cells produced entire pluteus larvae. Here was a result drastically different from the predictions of Weismann and Roux. Rather than self-differentiating into its future embryonic part, each isolated blastomere regulated its development to produce a complete organism. These experiments provided the first experimentally observable evidence that a cell's fate depends on that of its neighbors. Driesch experimentally removed cells, which in turn changed the context for those cells still remaining in the embryo (they are now abutting new neighboring cells). As a result, all cell fates were altered and could support complete embryonic development. In other words, the cell fates were altered to suit the *conditions.* In conditional specification, interactions between cells determine their fates rather than cell fate being specified by some cytoplasmic factor particular to that type of cell.

Driesch confirmed conditional development in sea urchin embryos with an intricate recombination experiment. If in fact some nuclear determinant dictates a cell's fate (as proposed by Weismann and Roux), then changing how nuclei are partitioned during cleavages should result in deformed development. In sea urchin eggs, the first two cleavage planes are normally meridional, passing through both the animal and vegetal poles, whereas the third division is equatorial, dividing the embryo into four upper and four lower cells (**FIGURE 2.10A**). Driesch (1893) changed the direction of the third cleavage by gently compressing early embryos between two glass plates, thus causing the third division to be meridional like the preceding two. After he released the pressure, the fourth division was equatorial. This procedure reshuffled the nuclei, placing nuclei that normally would have been in the region destined to form endoderm into the presumptive ectoderm region. In other words, some nuclei that would normally have produced ventral structures were now found in the dorsal cells (**FIGURE 2.10B**). Driesch obtained normal larvae from these embryos. If segregation of nuclear determinants had occurred, then this recombination experiment should have resulted in a strangely disordered embryo. Thus, Driesch concluded that "the relative position of a blastomere within the whole will probably in a general way determine what shall come from it."

The consequences of these experiments were momentous, both for embryology and for Driesch, personally.[3] First, Driesch had demonstrated that the prospective potency of an isolated blastomere (i.e., those cell types that it was possible for it to form) is greater than

---

[3]The idea of nuclear equivalence and the ability of cells to interact eventually caused Driesch to abandon science. Driesch, who thought the embryo was like a machine, could not explain how the embryo could make its missing parts or how a cell could change its fate to become another cell type.

FIGURE 2.10 Driesch's pressure-plate experiment for altering the distribution of nuclei. (A) Normal cleavage in 8- to 16-cell sea urchin embryos, seen from the animal pole (upper sequence) and from the side (lower sequence). The nuclei are numbered. (B) Abnormal cleavage planes formed under pressure, seen from the animal pole and from the side. (After Huxley and de Beer 1934.)

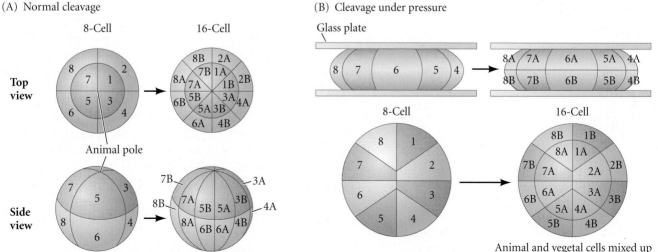

(A) Normal cleavage

8-Cell   16-Cell

Top view

Animal pole

Side view

(B) Cleavage under pressure

Glass plate

8-Cell   16-Cell

Animal and vegetal cells mixed up

the blastomere's actual prospective fate (those cell types that it would normally give rise to over the unaltered course of its development). According to Weismann and Roux, the prospective potency and the prospective fate of a blastomere should have been identical. Second, Driesch concluded that the sea urchin embryo is a "harmonious equipotential system" because all of its potentially independent parts interacted together to form a single organism. Driesch's experiment implies that *cell interaction is critical for normal development.* Moreover, if each early blastomere can form all the embryonic cells when isolated, it follows that in normal development the community of cells must prevent it from doing so (Hamburger 1997). Third, Driesch concluded that the fate of a nucleus depended solely on its location in the embryo. Interactions between cells determined their fates.

We now know (and will see in Chapters 10 and 11) that sea urchins and frogs alike use both autonomous and conditional specification of their early embryonic cells. Moreover, both animal groups use a similar strategy and even similar molecules during early development. In the 16-cell sea urchin embryo, a group of cells called the micromeres inherits a set of transcription factors from the egg cytoplasm. These transcription factors cause the micromeres to develop autonomously into the larval skeleton, but these same factors also activate genes for paracrine and juxtacrine signals that are then secreted by the micromeres and conditionally specify the cells around them.

Embryos (especially vertebrate embryos) in which most of the early blastomeres are conditionally specified have traditionally been called regulative embryos. But as we become more cognizant of the manner in which both autonomous and conditional specification are used in each embryo, the notions of "mosaic" and "regulative" embryos appear less and less tenable. Indeed, attempts to get rid of these distinctions were begun by the embryologist Edmund B. Wilson (1894, 1904) more than a century ago.

Syncytial Specification

In addition to autonomous and conditional specification, there is a third strategy that uses elements of both. A cytoplasm that contains many nuclei is called a syncytium,[4] and the specification of presumptive cells within such a syncytium is called **syncytial specification**. A notable example of an embryo that goes through a syncytial stage is found in insects, as illustrated by the fruit fly *Drosophila melanogaster.* During its early cleavage stages, nuclei divide through 13 cycles in the absence of any cytoplasmic cleavage. This division creates an embryo of many nuclei contained within one shared cytoplasm surrounded by one common plasma membrane. This embryo is called the **syncytial blastoderm** (**FIGURE 2.11** and **WATCH DEVELOPMENT 2.3**).

> **WATCH DEVELOPMENT 2.3** Observe the waves of nuclear divisions that occur during development of the syncytial blastoderm in the *Drosophila* early embryo.

It is within the syncytial blastoderm that the *identity* of future cells is established simultaneously across the entire embryo along the anterior-to-posterior axis of the blastoderm. Therefore, identity is established without any membranes separating nuclei into individual cells. Membranes do eventually form around each nucleus through a process called cellularization, which occurs after mitotic cycle 13 just prior to gastrulation (see Figure 2.11). A fascinating issue is how the cell fates—those cells **determined** to become the head, thorax, abdomen, and tail—are specified before cellularization. Are there determination factors segregated to discrete locations in the blastoderm to determine identity, as seen in autonomous specification? Or do nuclei in this syncytium obtain their identity from their position relative to neighboring nuclei, akin to conditional specification? The answer to both these questions is *yes.*

---

[4] Syncytia can be found in many organisms, from fungi to humans. Examples are the syncytium of the nematode germ cells (connected by cytoplasmic bridges), the multinucleated skeletal muscle fiber, and the giant cancer cells derived from fused immune cells.

(A) Nuclear cycles during *Drosophila* early development

(B) Global wave of nuclei division during cycle 13

Pre-blastoderm cycles 1–9

Cycles 4–6

Cycles 8–9
Nuclear migration toward cortex

Pole cells

Cycles 10–13
Syncytial blastoderm cycles

Cellularization interphase 14

Cellular blastoderm

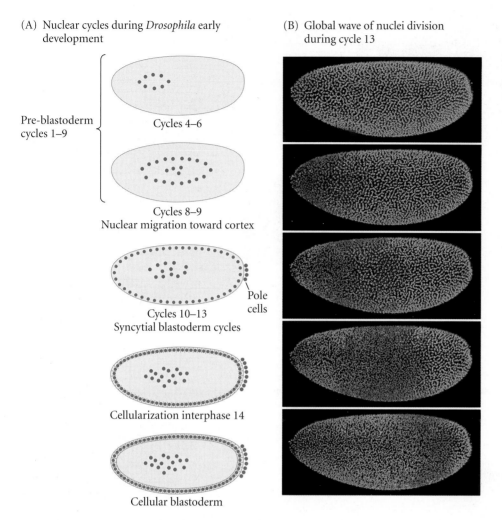

FIGURE 2.11 The syncytial blastoderm in *Drosophila melanogaster*. (A) Schematic of the progression of blastoderm cellularization in *Drosophila* (nuclei are red). (B) Still frames from a time lapse movie of a developing *Drosophila* embryo with nuclei that are premitotic (blue) and actively dividing nuclei in mitosis (purple). (A after Mazumdar and Mazumdar 2002; B from Tomer et al. 2012.)

Opposing axial gradients define position

What has emerged from numerous studies is that, just as we've seen in other eggs, the cytoplasm of the *Drosophila* egg is not uniform. Instead, it contains gradients of positional information that dictate cell fate along the egg's anterior-posterior axis (reviewed in Kimelman and Martin 2012). In the syncytial blastoderm, nuclei in the anterior part of the cell are exposed to **cytoplasmic determination factors** that are not present in the posterior part of the cell and vice versa. It is the interaction between nuclei and the differing amounts of determination factors that specify cell fate. It is important that these gradients of determination factors are established during maturation of the egg prior to fertilization. After fertilization, as nuclei undergo synchronous waves of division (see Figure 2.11B), each nucleus becomes positioned at specific coordinates along the anterior-posterior axis and experiences unique concentrations of determination factors.

How do the nuclei maintain a position within the syncytial blastoderm? They do so through the action of their own cytoskeletal machinery: their centrosome, affiliated microtubules, actin filaments, and interacting proteins (Kanesaki et al. 2011; Koke et al. 2014). Specifically, when the nuclei are in between divisions (in interphase), each nucleus radiates *dynamic* microtubule extensions organized by their centrosome that establish an "orbit" and exert force on the orbits of other nuclei (**FIGURE 2.12A** and **WATCH DEVELOPMENT 2.4**). Each time the nuclei divide, this radial microtubule array is reestablished to exert force on neighboring nuclear orbits, ensuring regular spacing of nuclei across the syncytial blastoderm. Maintaining the positional relationships between nuclei across the early embryo is essential for successful syncytial specification.

FIGURE 2.12 Nuclei positioning and morphogens during syncytial specification in *Drosophila melanogaster*. Nuclei are dynamically ordered within the syncytium of the early embryo, holding their positions using the cytoskeletal elements associated with them. (A) Interphase stage of nuclear cycle 13 of the *Drosophila* syncytium. (Left) EB1-GFP illuminates microtubules associated with each nuclei, which shows the aster arrays defining nuclear orbits that have some overlap with neighboring asters. (See also Watch Development 2.4.) (Right) An illustration of how the nuclei maintain their positions during interphase to establish orbits. This pattern of nuclei and cytoplasmic arrays was generated through computational modeling. (B) Expression of Bicoid protein in the early embryo is shown in green. (C) Quantification of Bicoid distribution along the anterior-posterior axis demonstrates the highest concentrations are found anteriorly and diminish posteriorly. (D) Anterior-posterior specification originates from morphogen gradients in the egg cytoplasm, specifically of the transcription factors Bicoid and Caudal. The concentrations and ratios of these two proteins distinguish each position along the axis from any other position. When nuclear division occurs, the amounts of each morphogen differentially activate transcription of the various nuclear genes that specify the segment identities of the larval and the adult fly. (As we will see in Chapter 9, the Caudal gradient is itself constructed by interactions between constituents of the egg cytoplasm.) (A from Kanesaki et al. 2011; B from Koke et al. 2014; C from Sample and Shvartsman 2010.)

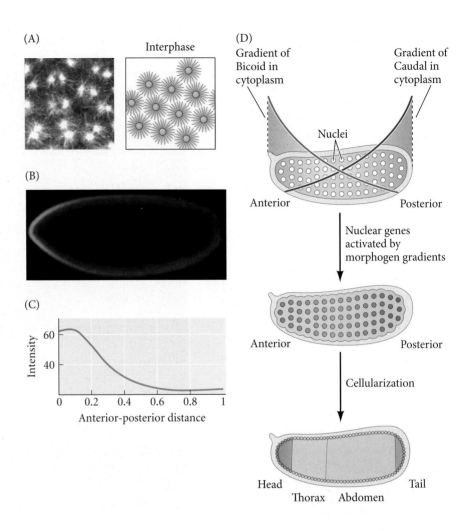

WATCH DEVELOPMENT 2.4 This movie demonstrates the microtubule dynamics associated with nuclear divisions in the syncytial blastoderm of *Drosophila*.

Keeping nuclear position stable during early development allows each nucleus to be exposed to different amounts of the determination factors distributed in gradients throughout the shared cytoplasmic environment. A nucleus can interpret its position (whether to become part of the anterior, midsection, or posterior part of the body) based on the concentration of cytoplasmic determinants it experiences. Each nucleus thereby becomes genetically programmed toward a particular identity. The determinants are **transcription factors**, proteins that bind DNA and regulate gene transcription. In Chapter 3, we will go into great detail about the role that transcription factors play in development.

As we will detail in Chapter 9, the anteriormost portion of the *Drosophila* embryo produces a transcription factor called Bicoid, with a concentration of both mRNA and protein that is highest in the anterior region of the egg and declines toward the posterior (**FIGURE 2.12B,C**; Gregor et al. 2007; Sample and Shvartsman 2010; Little et al. 2011). The Bicoid concentration gradient across the syncytium is the combined result of diffusion with a mechanism of protein and mRNA degradation. In addition, the posteriormost portion of the egg forms a posterior-to-anterior gradient of the transcription factor Caudal. Thus, the long axis of the *Drosophila* egg is spanned by opposing gradients: Bicoid from the anterior and Caudal from the posterior (**FIGURE 2.12D**). Bicoid and Caudal are considered **morphogens** because they occur in a concentration gradient and are capable of regulating different genes at different threshold concentrations. We will discuss morphogens in great detail in Chapter 4, but their repeated use in embryonic development will merit their inclusion throughout the textbook.

As morphogens, Bicoid and Caudal proteins activate different sets of genes in the syncytial nuclei. Those nuclei in regions containing high amounts of Bicoid and little Caudal are instructed to activate the genes that produce the head. Nuclei in regions with slightly less Bicoid and a small amount of Caudal are instructed to activate genes that generate the thorax. In regions with little or no Bicoid but plenty of Caudal, the activated genes form abdominal structures (Nüsslein-Volhard et al. 1987). Thus, when the syncytial nuclei are eventually incorporated into cells, these cells will have their *general* fate specified. Afterward, the *specific* fate of each cell will become determined both autonomously (from transcription factors acquired after cellularization) and conditionally (from interactions between the cell and its neighbors).

A Rainbow of Cell Identities

Each of the three major strategies of cell specification summarized in **TABLE 2.2** offers a different way of providing each embryonic cell with a set of determinants (often transcription factors) that will activate specific genes and cause the cell to differentiate into a particular cell type. Is the designation of a "cell type" the most precise way of identifying a cell? To answer this question, we would have to be able to watch and analyze individual cells in an embryo over time. In Chapter 1, we discussed using fate mapping techniques, which enable the marking of a single cell with something such

Developing Questions

If a mechanism of opposing concentration gradients of Bicoid and Caudal determine specification of the anterior-to-posterior axis in *Drosophila melanogaster*, could this same mechanism work in a fly embryo that is larger or has different proportions per body segment, or would it need some modification? How precise is the actual gradient, and how precise does it actually have to be to set a nucleus/cell upon a lineage-specific path of maturation?

TABLE 2.2   Modes of cell type specification

AUTONOMOUS SPECIFICATION

Predominates in most invertebrates.

Specification by differential acquisition of certain cytoplasmic molecules present in the egg.

Invariant cleavages produce the same lineages in each embryo of the species; blastomere fates are generally invariant.

Cell type specification precedes any large-scale embryonic cell migration.

Results in "mosaic" development: cells cannot change fate if a blastomere is lost.

CONDITIONAL SPECIFICATION

Predominates in vertebrates and a few invertebrates.

Specification by interactions among cells. Positions of cells relative to each other are key.

Variable cleavages, no invariant fate assignment to cells.

Massive cell rearrangements and migrations precede or accompany specification.

Capacity for "regulative" development allows cells to acquire different functions as a result of interactions with neighboring cells.

SYNCYTIAL SPECIFICATION

Predominates in most insect classes.

Specification of body regions by interactions between cytoplasmic regions prior to cellularization of the blastoderm.

Variable cleavage produces no rigid cell fates for particular nuclei.

After cellularization, both autonomous and conditional specification are seen.

Source: After Davidson 1991.

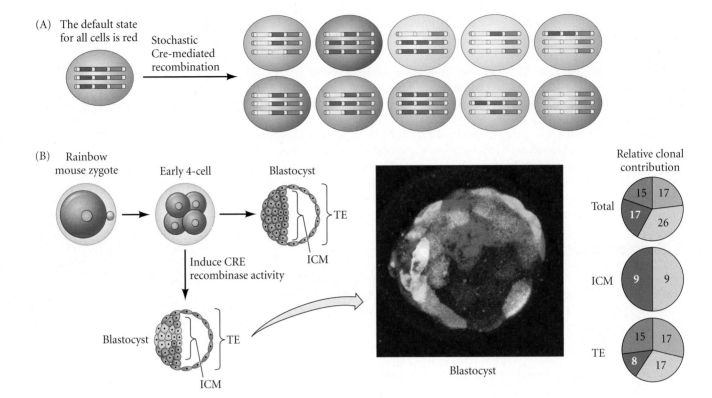

FIGURE 2.13 The Brainbow lineage tracing system. (A) The Brainbow genetic system is used to randomly fix cells with a distinct fluorescent color or hue and is accomplished by inserting multiple copies of different fluorescent genes into the organism's genome. Through Cre-recombinase activity, different combinations of these fluorescent genes can be activated to produce an array of different colors. In the example here, each cell will, by default, express red fluorescent protein; upon Cre-mediated recombination, however, cyan, yellow, or green fluorescent proteins begin to be expressed in a stochastic manner (in this example, 10 differently colored cells are labeled). (B) The Rainbow mouse system is a version of Brainbow and works similarly. In this experiment, recombination was initiated during early mouse blastocyst development to permanently mark different cells within the trophectoderm (TE) and inner cell mass (ICM) with unique colors. Those colors were then followed over time and the populations quantified (pie charts), which revealed a statistically significant distribution demonstrating clonal origins from the earlier labeled cells. (A after Weissman and Pan 2015; B after Tabansky et al. 2013.)

as a dye that can be traced through development to determine the cell's fate (Klein and Moody 2016). A genetic approach to fate mapping has been developed to label cells with a seeming *rainbow* of possible colors, which can be used to identify each individual cell in a tissue or even a whole embryo (Livet et al. 2007). This method was named **Brainbow** because the initial study focused on characterizing cells of the developing mouse brain. It can be applied to any organism, however, and has been called different names, such as "Flybow" and "dBrainbow" for its use in *Drosophila*, "Rainbow" and "Confetti" for its use in mouse, and "Zebrabow" for its use in zebrafish (Weissman and Pan 2015).

The Brainbow system triggers the expression of different combinations and amounts of distinct fluorescent proteins (green, red, blue, etc.; see Weissman and Pan 2015). The resulting stochastic distribution of fluorescent protein combinations gives each cell a distinct color that is stably inherited by all its progeny. How is that achieved? The answer is that genes for each fluorescent protein are engineered into the genome of the organism being studied in such a way that they are initially inactive; upon exposure to Cre-recombinase (an enzyme that catalyzes recombination events at specific sites in the DNA), however, a random combination of fluorescent genes can become active (**FIGURE 2.13A**). Different cells are then distinguishable based on the hue of fluorescence created by the different combinations of fluorescent proteins active in each cell.

Brainbow enables researchers to study the morphology of cells and their interactions in any tissue at any age and allows us to chart the developmental lineage of an individual cell from the early embryo through its progeny to their final destinations. For instance, Kevin Eggan's research team has used the "Rainbow" system to label cells of the early cleavage stages of the mouse embryo to address the following question (Tabansky et al. 2013): Is the first lineage choice of becoming an embryonic cell or an extraembryonic cell a random or a regulated process? They discovered that it is nonrandom (**FIGURE 2.13B**). This example illustrates how powerful this innovative technology is at providing new insights into the life history of individual cells within a community of cells within whole embryos.

Next Step Investigation

You have now learned that strategically positioned cytoplasmic determinants and cell-cell interactions directly regulate the progression of cell maturation and differentiation toward a specific cell type. What if every cell type were like a species within the animal kingdom? This analogy suggests that there is still a lot of diversity on the level of the individual cell. Therefore, the next step is to determine whether the individual cells within a population of cells of a specific type are truly "individuals," possessing a unique identity. Pick your favorite cell type and imagine using the Brainbow system to label the individual cells within this tissue. What would you now do to determine if each cell is distinct from its neighbors, even though they look identical morphologically? What would you be looking for? What data could you collect that would distinguish one cell from another? Answers to these questions will certainly vary, but Chapter 3 ("Differential Gene Expression") should provide you with many clues.

Closing Thoughts on the Opening Photo

A crowd of individuals or a gang of clones? That was the question asked of the multicolored Brainbow-labeled neurons illuminated in this section of a mouse hippocampus, an image made by Tamily Weissman and Jeff Lichtman (Weissman and Pan 2015). The philosopher Søren Kierkegaard once wrote of the truth that is inherent in the individual that can become obscured by the noise and direction of the crowd. Right now, the field of developmental biology has largely defined differentiation on the order of broad cell-type categories, and researchers are curious as to how much "truth" we may be missing on the individual cell level. In this image, each cell was experimentally marked in a random fashion with different fluorescent proteins, which gives the illusion that these neurons are different. Are they different? If so, how much? How might one even define a cell as different if cells look the same morphologically? Most exciting is how, with new techniques like Brainbow, the study of cell specification is moving closer and closer to refining the differences underlying distinct individual cell identities. So, the next time you are in a class of students or perhaps cheering or rallying with a crowd, reflect on the commonalities and differences that may exist between the individuals that make up this group. Someday soon, we may have the information necessary to similarly reflect on the identity of cells at the level of an individual. (Photograph courtesy of T. Weissman and Y. Pan.)

Snapshot Summary
Specifying Identity

1. Cell differentiation is the process by which a cell acquires the structural and functional properties unique to a given cell type.

2. From an undifferentiated cell to a postmitotic differentiated cell type, a cell goes through a process of maturation that experiences different levels of commitment toward its end fate.

3. A cell is first specified toward a given fate, suggesting that it would develop into this cell type even in isolation.

4. A cell is committed or determined to a given fate if it maintains its developmental maturation toward this cell type even when placed in a new environment.

5. There are three different modes of cell specification: autonomous, conditional, and syncytial.

6. Autonomous specification refers to cells in an embryo that possess the necessary cytoplasmic determinants that function to commit that cell toward a specific fate.

Such cells will mature into their determined cell types even when isolated, as best exemplified by cells of the tunicate embryo.

7. Conklin first observed the yellow crescent in the tunicate embryo and showed that cells with this yellow crescent gave rise to muscle. The muscle cell fate in tunicates is dependent on the transcription factor Macho.

8. Conditional specification is the acquisition of a given cell identity based on its position or, more specifically, on the interactions that cell has with the other cells and molecules it comes in contact with. An extreme example of conditional specification was demonstrated by the complete, normal development of sea urchin larvae from single isolated blastomeres.

9. Most species have cells that develop via autonomous specification as well as cells that develop via conditional specification.

10. Patterns of cell fate can also be laid out in a syncytium of nuclei—called syncytial specification—as in the *Drosophila* blastoderm.

11. Cytoskeletal arrangements maintain positioning of nuclei in the syncytium, which enables specification of

these nuclei by opposing morphogen gradients, namely Bicoid and Caudal.

12. Genetic techniques like Brainbow enable scientists to follow the developmental history of individual cells and help further define what cell identity means.

Further Reading

Klein, S. L. and S. A. Moody. 2016. When family history matters: the importance of lineage analyses and fate maps for explaining animal development. *Curr. Top. Dev. Biol.* 117: 93–112.

Little, S., G. Tkačik, T. B. Kneeland, E. F. Wieschaus, and T. Gregor. 2011. The formation of the Bicoid morphogen gradient requires protein movement from anteriorly localized mRNA. *PLoS Biol.* 3: e1000596.

Livet, J. and 7 others. 2007. Transgenic strategies for combinatorial expression of fluorescent proteins in the nervous system. *Nature* 450: 56–62.

Nishida, H. and K. Sawada. 2001. macho-1 encodes a localized mRNA in ascidian eggs that specifies muscle fate during embryogenesis. *Nature* 409: 724–729.

Tabansky, I. and 11 others. 2013. Developmental bias in cleavage-stage mouse blastomeres. *Curr. Biol.* 23: 21–31.

Weissman, T. A. and Pan Y. A. 2015. Brainbow: new resources and emerging biological applications for multicolor genetic labeling and analysis. *Genetics* 199: 293–306.

Wieschaus, E. 2016. Positional Information and cell fate determination in the early drosophila embryo. *Curr. Top. Dev. Biol.* 117: 567–579.

GO TO WWW.DEVBIO.COM...

...for Web Topics, Scientists Speak interviews, Watch Development videos, Dev Tutorials, and complete bibliographic information for all literature cited in this chapter.

Differential Gene Expression
Mechanisms of Cell Differentiation

What underlies cell differentiation?

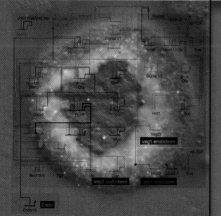

FROM ONE CELL COME MANY, and of many different types. That is the seemingly miraculous phenomenon of embryonic development. How is it possible that such a diversity of cell types within a multicellular organism can be derived from a single cell, the fertilized egg? Cytological studies done at the start of the twentieth century established that the chromosomes in each cell of an organism's body are the mitotic descendants of the chromosomes established at fertilization (Wilson 1896; Boveri 1904). In other words, each somatic cell nucleus has the same chromosomes—and therefore the same set of genes—as all other somatic cell nuclei. This fundamental concept, known as **genomic equivalence**, presented a significant conceptual dilemma. If every cell in the body contains the genes for hemoglobin and insulin, for example, why are hemoglobin proteins made only in red blood cells and insulin proteins only in certain pancreatic cells? Based on the embryological evidence for genomic equivalence (as well as on bacterial models of gene regulation), a consensus emerged in the 1960s that the answer lies in **differential gene expression**.

Defining Differential Gene Expression

Differential gene expression is the process by which cells become different from one another based upon the unique combination of genes that are active or "expressed." By expressing

The Punchline

The selective production of different proteins within cells creates cellular diversity. As the single-celled zygote divides to start the generation of all the cells making up an organism, differences in the expression of genes in these cells govern maturation toward distinct cell types. Many regulatory mechanisms targeting DNA access, RNA production and processing, and protein synthesis and modification lead to this differential gene expression. They include using a specific repertoire of transcription factors that bind gene promoters to enhance or repress transcription, modifying histones to modulate the accessibility of chromatin, and degrading and alternative splicing of RNA to change the coded message for different protein construction. In addition, translational controls and posttranslational modifications of proteins as well as changes in protein transport affect what proteins are created and where they function. Use of these numerous mechanisms at different times and in different cells fuels the creation of different cell types as the embryo develops.

different genes, cells can create different proteins that lead to the differentiation of different cell types. There are three postulates of differential gene expression:

1. Every somatic cell nucleus of an organism contains the complete genome established in the fertilized egg. In molecular terms, the DNAs of all differentiated cells are identical.

2. The unused genes in differentiated cells are neither destroyed nor mutated; they retain the potential for being expressed.

3. Only a small percentage of the genome is expressed in each cell, and a portion of the RNA synthesized in each cell is specific for that cell type.

By the late 1980s, it was established that gene expression can be regulated at four levels such that different cell types synthesize different sets of proteins:

1. *Differential gene transcription* regulates which of the nuclear genes are transcribed into nuclear RNA.

2. *Selective nuclear RNA processing* regulates which of the transcribed RNAs (or which parts of such a nuclear RNA) are able to enter into the cytoplasm and become messenger RNAs.

3. *Selective messenger RNA translation* regulates which of the mRNAs in the cytoplasm are translated into proteins.

4. *Differential protein modification* regulates which proteins are allowed to remain and/or function in the cell.

Some genes (such as those coding for the globin protein subunits of hemoglobin) are regulated at all these levels.

 DEV TUTORIAL *Differential Gene Expression*   In this tutorial, Dr. Michael Barresi discusses the basics of gene regulation and how differences in this regulation can lead to unique developmental patterns.

WEB TOPIC 3.1   **DOES THE GENOME OR THE CYTOPLASM DIRECT DEVELOPMENT?** The geneticists versus the embryologists. Geneticists were certain that genes controlled development, whereas embryologists generally favored the cytoplasm. Both sides had excellent evidence for their positions.

WEB TOPIC 3.2   **THE ORIGINS OF DEVELOPMENTAL GENETICS**   The first hypotheses for differential gene expression came from C. H. Waddington, Salome Gluecksohn-Waelsch, and other scientists who understood both embryology and genetics.

Quick Primer on the Central Dogma

To properly comprehend all the mechanisms regulating the differential expression of a gene, you must first understand the principles of the central dogma of biology. The **central dogma** pertains to the sequence of events that enables the use and transfer of information to make the proteins of a cell (**FIGURE 3.1**). *Central* to this theory is the sequenced order of deoxyribonucleotides in double-stranded DNA that provides the informative code or *blueprints* for the precise combination of amino acids needed to build specific proteins. Proteins are not made directly from DNA, however; rather, the information laid out in the sequence of DNA bases is first copied or *transcribed* into a single-stranded polymer of similar molecules called a nuclear ribonucleic acid (nRNA). The process of copying DNA into RNA is called **transcription**, and the RNA produced from a given gene is often referred to as a transcript. Although the transcribed nRNA includes the information to code for a protein, it can also hold non-protein-coding (simply called "noncoding") information. The nRNA strand will undergo processing to excise the noncoding domains and protect the ends of the strand to yield a **messenger RNA (mRNA)** molecule. mRNA is transported out of the nucleus into the cytoplasm

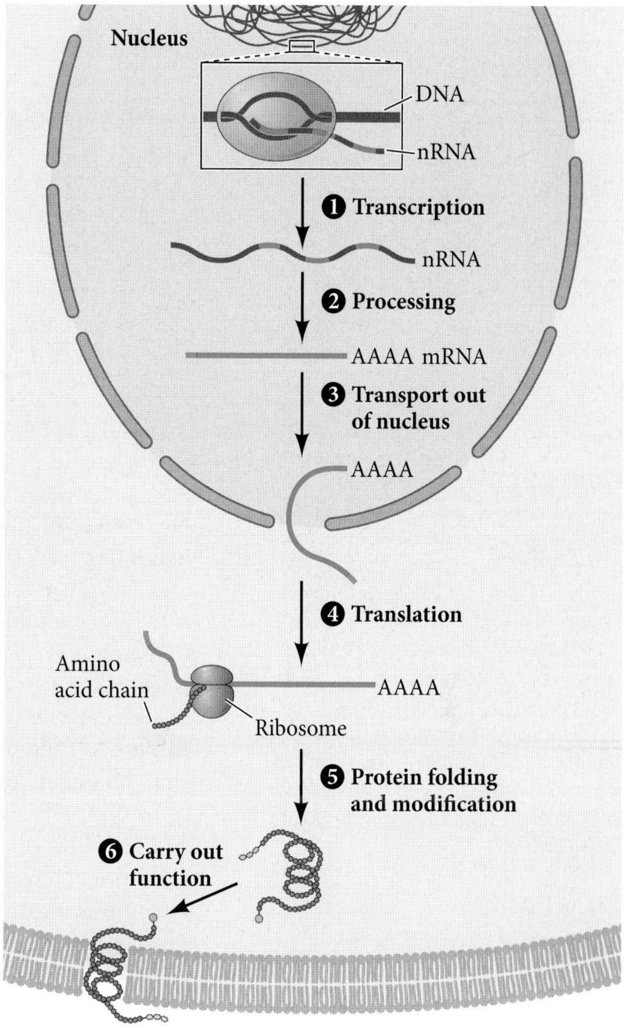

FIGURE 3.1 The central dogma of biology. A simplified schematic of the key steps in the process of gene and protein expression. (1) Transcription. In the nucleus, a region of the genomic DNA is seen accessible to a RNA polymerase, which transcribes an exact complementary copy of the gene in the form of a single-stranded nuclear RNA molecule. The gene is now said to be "expressed." (2) Processing. The nRNA transcript undergoes processing to make a finalized messenger RNA strand, which is transported out of the nucleus (3). (4) Translation. mRNA complexes with a ribosome, and its information is translated into an ordered polymer of amino acids. (5) Protein folding and modification. This polypeptide adopts secondary and tertiary structures through proper folding and potential modifications (such as the addition of a carbohydrate group as seen here). (6) Carry out function. The protein is now said to be "expressed" and can carry out its specific function (such as functioning as a transmembrane receptor).

where it can interact with a ribosome and convey its *message* for the synthesis of a specific protein. mRNA unveils the complementary sequence of DNA three bases at a time, each triplet being called a codon. Each codon calls for a specific amino acid that will be covalently attached to its neighboring amino acid denoted by the codon next in line. In this manner, **translation** leads to the synthesis of a polypeptide chain that will undergo protein folding and potential modification by the addition of various functional moieties such as carbohydrates, phosphates, or cholesterol groups. The completed protein is now ready to carry out its specific function serving to support the structural or functional properties of the cell. Cells that express different proteins will therefore possess different structural and functional properties, making it a distinct type of cell.

Evidence for Genomic Equivalence

Until the mid-twentieth century, genomic equivalence was not so much proved as it was assumed (because every cell is the mitotic descendant of the fertilized egg). One of the first tasks of developmental genetics was to determine whether every cell of an organism indeed does have the same **genome**—that is, the same set of genes—as every other cell.

Evidence that every cell in the body has the same genome originally came from the analysis of *Drosophila* chromosomes, in which the DNA of certain larval tissues undergoes numerous rounds of DNA replication without separation such that the structure of the chromosomes can be seen. In these **polytene** (Greek, "many strands") **chromosomes**, no structural differences were seen between cells; however, different regions of the chromosomes were "puffed up" at different times and in different cell types, which suggested that these areas were actively making RNA (**FIGURE 3.2A**; Beermann 1952). When Giemsa dyes allowed such observations to be made in mammalian chromosomes, it was also found that no chromosomal regions were lost in most cells. These observations, in turn, were confirmed by nucleic acid in situ hybridization studies, a technique that enables the visualization of the spatial and temporal pattern of specific gene (mRNA) expression in the embryo (see Figure 3.35). For instance, the mRNA of the *odd-skipped* gene is present in cells that display a segmented pattern in the *Drosophila* embryo, a pattern that changes over time (**FIGURE 3.2B**). Similarly, the mouse homolog of *odd-skipped*, called *odd-skipped related 1*, is differentially expressed in cells of specific structures such as the segmented branchial arches, the limb buds, and the

(A)

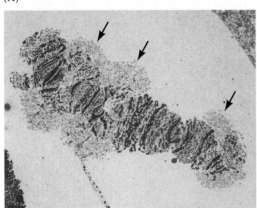

(B) *odd-skipped* (stage 5)

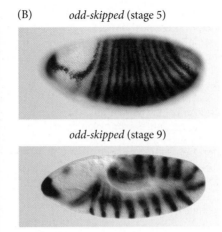

odd-skipped (stage 9)

(C) *odd-skipped related 1*

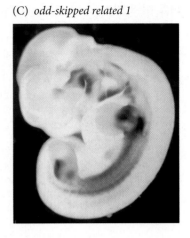

FIGURE 3.2 Gene expression.
(A) Transmission electron micrograph of a polytene chromosome from a salivary gland cell of *Chironomus tentans* showing three giant puffs indicating active transcription in these regions (arrows). (B) mRNA expression of the *odd-skipped* gene in a stage 5 and a stage 9 *Drosophila* embryo (blue). (C) mRNA expression of the *odd-skipped related 1* gene in a 11.5 days post-conception mouse embryo (blue). (A from Daneholt, 1975; B from Weiszmann et al. 2009; C from So and Danielian, 1999.)

heart (**FIGURE 3.2C**). Is the DNA in an organism's cells that is now expressing different genes truly still the same, however? Does it still possess the same potential to make any cell? The ultimate test of whether the nucleus of a differentiated cell has undergone irreversible functional restriction is to have that nucleus generate every other type of differentiated cell in the body. If each cell's nucleus is identical to the zygote nucleus, each cell's nucleus should also be capable of directing the entire development of the organism when transplanted into an activated enucleated egg. Although such experiments had been proposed in the 1930s, the first demonstration that a nucleus from an adult mammalian somatic cell could direct the development of an entire animal didn't come until 1997, when Dolly the sheep was cloned.

Ian Wilmut and colleagues took cells from the mammary gland of a 6-year-old pregnant ewe and placed them in culture (**FIGURE 3.3A**; Wilmut et al. 1997). The culture medium was formulated to keep the cell nuclei at the intact diploid stage (G1) of the cell cycle; this cell-cycle stage turned out to be critical. The researchers then obtained oocytes from a different strain of sheep and removed their nuclei. These oocytes had to be in the second meiotic metaphase, the stage at which they are usually fertilized. The donor cell and the enucleated oocyte were brought together, and electric pulses were sent through them, thereby destabilizing the cell membranes and allowing the cells to fuse. The same electric pulses that fused the cells activated the egg to begin development. The resulting embryos were eventually transferred into the uteri of pregnant sheep.

WEB TOPIC 3.3 **THE 2012 NOBEL PRIZE FOR PHYSIOLOGY OR MEDICINE: CLONING AND NUCLEAR EQUIVALENCE** The final "proof" of genomic equivalence was the demonstration that the nuclei of differentiated somatic cells could generate any cell type in the body.

Of the 434 sheep oocytes originally used in this experiment, only one survived: Dolly[1] (**FIGURE 3.3B**). DNA analysis confirmed that the nuclei of Dolly's cells were derived from the strain of sheep from which the donor nucleus was taken (Ashworth et al. 1998; Signer et al. 1998). Cloning of adult mammals has been confirmed in guinea pigs, rabbits, rats, mice, dogs, cats, horses, and cows. In 2003, a cloned mule became the first sterile animal to be so reproduced (Woods et al. 2003). Thus, it appears that

---

[1] The creation of Dolly was the result of a combination of scientific and social circumstances. These circumstances involved job security, people with different areas of expertise meeting one another, children's school holidays, international politics, and who sits near whom in a pub. The complex interconnections giving rise to Dolly are told in *The Second Creation* (Wilmut et al. 2000), a book that should be read by anyone who wants to know how contemporary science actually works. As Wilmut acknowledged (p. 36), "The story may seem a bit messy, but that's because life is messy, and science is a slice of life."

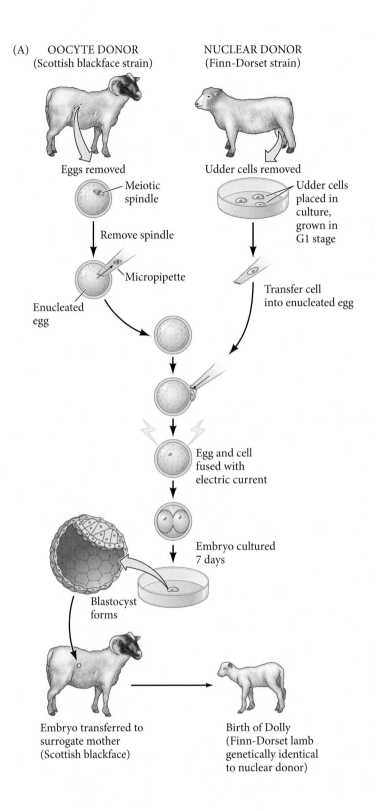

(A)   OOCYTE DONOR
(Scottish blackface strain)

NUCLEAR DONOR
(Finn-Dorset strain)

Eggs removed

Meiotic
spindle

Remove spindle

Micropipette

Enucleated
egg

Udder cells removed

Udder cells
placed in
culture,
grown in
G1 stage

Transfer cell
into enucleated egg

Egg and cell
fused with
electric current

Embryo cultured
7 days

Blastocyst
forms

Embryo transferred to
surrogate mother
(Scottish blackface)

Birth of Dolly
(Finn-Dorset lamb
genetically identical
to nuclear donor)

(B)

FIGURE 3.3   Cloning a mammal using nuclei from adult somatic cells. (A) Procedure used for cloning sheep. (B) Dolly, the adult sheep on the left, was derived by fusing a mammary gland cell nucleus with an enucleated oocyte, which was then implanted in a surrogate mother (of a different breed of sheep) that gave birth to Dolly. Dolly later gave birth to a lamb (Bonnie, at right) by normal reproduction. (A after Wilmut et al. 2000; B photograph by Roddy Field © Roslin Institute.)

the nuclei of vertebrate adult somatic cells contain all the genes needed to generate an adult organism. No genes necessary for development have been lost or mutated in the somatic cells; their *nuclei are equivalent*.[2]

---

[2]Although all the organs were properly formed in the cloned animals, many of the clones developed debilitating diseases as they matured (Humphreys et al. 2001; Jaenisch and Wilmut 2001; Kolata 2001). As we will see shortly, this problem is due in large part to the differences in methylation between the chromatin of the zygote and the differentiated cell.

(((● SCIENTISTS SPEAK 3.1   Listen to Sir Ian Wilmut discuss cloning and cellular
reprogramming.

Modulating Access to Genes

So how does the same genome give rise to different cell types? To address this question, we need to first understand the anatomy of genes. A fundamental difference distinguishing most eukaryotic genes from prokaryotic genes is that eukaryotic genes are contained within a complex of DNA and protein called **chromatin**. The protein component constitutes about half the weight of chromatin and is composed largely of **histones**. The **nucleosome** is the basic unit of chromatin structure (**FIGURE 3.4A,B**). It is composed of an octamer of histone proteins (two molecules each of histones H2A, H2B, H3, and H4) wrapped with two loops containing approximately 147 base pairs of DNA (Kornberg and Thomas 1974). Histone H1 is bound to the 60 to 80 or so base pairs of "linker" DNA between the nucleosomes (Weintraub 1984, 1985). There are more than a dozen contacts between the DNA and the histones (Luger et al. 1997; Bartke et al. 2010), which function to enable the remarkable packaging of more than *6 feet* of DNA into the approximately 6 micrometer (in diameter) nucleus of each human cell (Schones and Zhao 2008).

Whereas classical geneticists have likened genes to "beads on a string," molecular geneticists liken genes to "string on the beads," an image in which the beads are nucleosomes. Much of the time, the nucleosomes appear to be wound into tight structures called **solenoids** that are stabilized by histone H1 (**FIGURE 3.4C**). This H1-dependent conformation of nucleosomes inhibits the transcription of genes in somatic cells by packing adjacent nucleosomes together into tight arrays that prevent transcription factors and RNA polymerases from gaining access to the genes (Thoma et al. 1979; Schlissel and Brown 1984). Chromatin regions that are tightly packed are called **heterochromatin**, and regions loosely packed are called **euchromatin**. One way to achieve differential gene expression is by regulating how tightly packed a given region of chromatin may be, thereby regulating whether genes are even accessible for transcription.

Loosening and tightening chromatin: Histones as gatekeepers

Histones are critical because they appear to be responsible for either facilitating or forbidding gene expression (**FIGURE 3.4D**). Repression and activation are controlled to a large extent by modifying the "tails" of histones H3 and H4 with two small organic groups: methyl (CH_3) and acetyl ($COCH_3$) residues. In general, **histone acetylation**—the addition of negatively charged acetyl groups to histones—neutralizes the basic charge of lysine and loosens the histones, which activates transcription. Enzymes known as **histone acetyltransferases** place acetyl groups on histones (especially on lysines in H3 and H4), destabilizing the nucleosomes so that they come apart easily (become more *euchromatic*). As might be expected, then, enzymes that *remove* acetyl groups—**histone deacetylases**—stabilize the nucleosomes (which become more *heterochromatic*) and prevent transcription.

Histone methylation is the addition of methyl groups to histones by enzymes called **histone methyltransferases**. Although histone methylation more often results in heterochromatic states and transcriptional repression, it can also activate transcription depending on the amino acid being methylated and the presence of other methyl or acetyl groups in the vicinity (see Strahl and Allis 2000; Cosgrove et al. 2004). For instance, acetylation of the tails of H3 and H4 along with the addition of three methyl groups on the lysine at position four of H3 (i.e., H3K4me3; remember that K is the abbreviation for lysine) is usually associated with actively transcribed chromatin. In contrast, a combined lack of acetylation of the H3 and H4 tails and methylation of the lysine in the ninth position of H3 (H3K9) is usually associated with highly repressed chromatin (Norma et al. 2001). Indeed, lysine methylations at H3K9, H3K27, and H4K20 are often

(A)

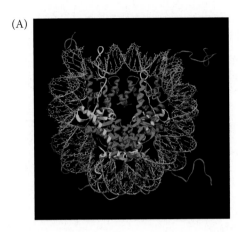

FIGURE 3.4 Nucleosome and chromatin structure. (A) Model of nucleosome structure as seen by X-ray crystallography at a resolution of 1.9 Å. Histones H2A and H2B are yellow and red, respectively; H3 is purple; and H4 is green. The DNA helix (gray) winds around the protein core. The histone "tails" that extend from the core are the sites of acetylation and methylation, which may disrupt or stabilize, respectively, the formation of nucleosome assemblages. (B) Histone H1 can draw nucleosomes together into compact forms. About 147 base pairs of DNA encircle each histone octamer, and about 60 to 80 base pairs of DNA link the nucleosomes together. (C) Model for the arrangement of nucleosomes in the highly compacted solenoidal chromatin structure. Histone tails protruding from the nucleosome subunits allow for the attachment of chemical groups. (D) Methyl groups condense nucleosomes more tightly, preventing access to promoter sites and thus preventing gene transcription. Acetylation loosens nucleosome packing, exposing the DNA to RNA polymerase II and transcription factors that will activate the genes. (A after Davey et al. 2002.)

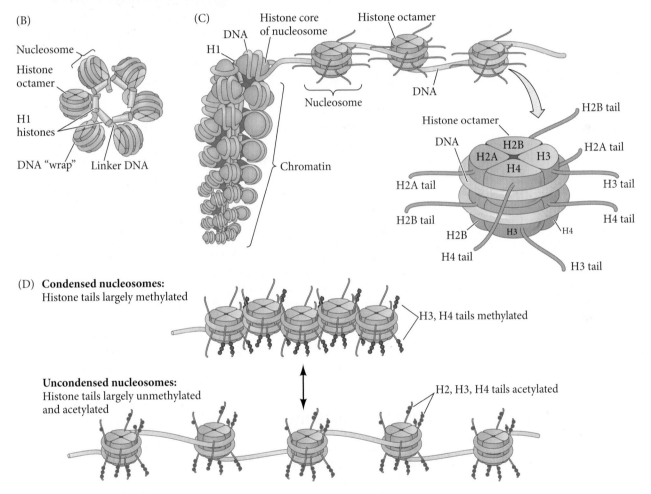

associated with highly repressed chromatin. **FIGURE 3.5** depicts a nucleosome with lysine residues on its H3 tail. Modifications of such residues regulate transcription.

If methyl groups at specific places on histones repress transcription, getting rid of these methyl moieties should be expected to permit transcription. That has been shown to be the case in the activation of the Hox genes, a family of genes that are critical in giving cells their identities along the anterior-posterior body axis. In early development, Hox genes are repressed by H3K27 trimethylation (the lysine at position 27 on histone 3 has three methyl groups: H3K27me3). In differentiated cells, however, a demethylase specific for H3K27me3 is recruited to these regions, eliminating the methyl groups and

FIGURE 3.5 Histone methylations on histone H3. The tail of histone H3 (its amino-terminal sequence, at the beginning of the protein) sticks out from the nucleosome and is capable of being methylated or acetylated. Here, lysines can be methylated and recognized by particular proteins. Methylated lysine residues at positions 4, 38, and 79 are associated with gene activation, whereas methylated lysines at positions 9 and 27 are associated with repression. The proteins binding these sites (not shown to scale) are represented above the methyl group. (After Kouzarides and Berger 2007.)

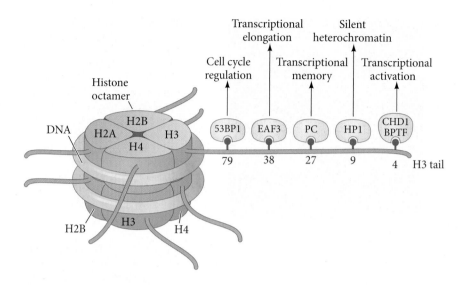

enabling access to the gene for transcription (Agger et al. 2007; Lan et al. 2007). The effects of methylation in controlling gene transcription are extensive.

Maintaining a memory of methylation

The modifications of histones can also signal the recruitment of proteins that retain the *memory* of the transcriptional state from generation to generation as cells go through mitosis. They are the proteins of the **Trithorax** and **Polycomb** families. When bound to the nucleosomes of active genes, Trithorax proteins keep these genes active, whereas Polycomb proteins, which bind to condensed nucleosomes, keep the genes in a repressed state.

The Polycomb proteins fall into two categories that act sequentially in repression. The first set has histone methyltransferase activities that methylate lysines H3K27 and H3K9 to repress gene activity. In many organisms, this repressed state is stabilized by the activity of a second set of Polycomb factors, which bind to the methylated tails of histone 3 and keep the methylation active and also methylate adjacent nucleosomes, thereby forming tightly packed repressive complexes (Grossniklaus and Paro 2007; Margueron et al. 2009).

The Trithorax proteins help retain the memory of activation; they act to counter the effect of the Polycomb proteins. Trithorax proteins can modify the nucleosomes or alter their positions on the chromatin, allowing transcription factors to bind to the DNA previously covered by them. Other Trithorax proteins keep the H3K4 lysine trimethylated (preventing its demethylation into a dimethylated, repressed state; Tan et al. 2008).

Anatomy of the Gene

So far, we have documented that modulating the access to a gene, largely by *histone* methylation, affects gene expression. Later in this chapter, we will discuss the exciting research on the direct control of transcription by *DNA* methylation. Now that we understand that modifying histones can grant access to regions of the genome, we can ask, what mechanisms exist to influence gene transcription more directly? More simply, once a gene is accessible, how can it be turned on and off? Before we answer, we need a basic understanding of the parts that make up a gene and how those parts can influence gene expression.

Exons and introns

A fundamental feature that distinguishes eukaryotic from prokaryotic genes (along with eukaryotic genes being contained within chromatin) is that eukaryotic genes are not co-linear with their peptide products. Rather, the single nucleic acid strand of eukaryotic mRNA comes from noncontiguous regions on the chromosome. **Exons** are

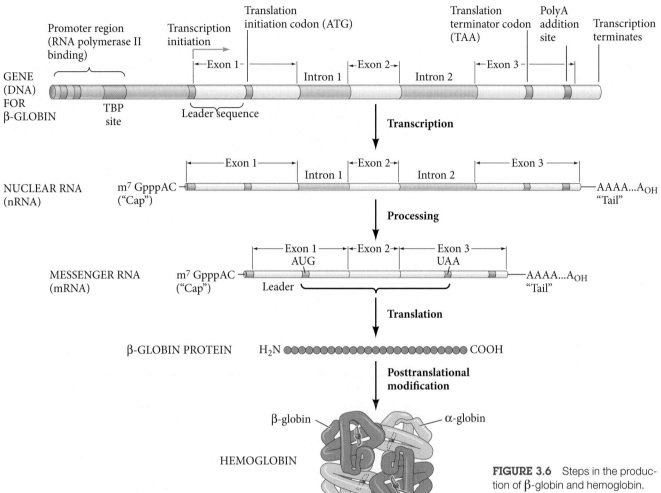

FIGURE 3.6 Steps in the production of β-globin and hemoglobin. Transcription of the β-globin gene creates a nuclear RNA containing exons and introns as well as the cap, tail, and 3′ and 5′ untranslated regions. Processing the nuclear RNA into messenger RNA removes the introns. Translation on ribosomes uses the mRNA to encode a protein. The β-globin protein is inactive until it is modified and complexed with α-globin and heme to become active hemoglobin (bottom).

the regions of DNA that code for parts of a protein[3]; between exons, however, are intervening sequences called **introns** that have nothing whatsoever to do with the amino acid sequence of the protein. To help illustrate the structural components of a typical eukaryotic gene, we highlight the anatomy of the human β-globin gene (**FIGURE 3.6**). This gene, which encodes part of the hemoglobin protein of the red blood cells, consists of the following elements:

- A **promoter region**, where RNA polymerase II binds to initiate transcription. The promoter region of the human β-globin gene has three distinct units and extends from 95 to 26 base pairs before ("upstream from")[4] the transcription initiation site (i.e., from −95 to −26). Some promoters have the DNA sequence TATA (called the "TATA-box"), which binds the basal or general transcription factor (TATA-binding protein, TBP) that helps anchor RNA polymerase II to the promoter.

- The **transcription initiation site**, which for human β-globin is ACATTTG. This site is often called the **cap sequence** because it is the DNA sequence

---

[3]The term *exon* refers to a nucleotide sequence whose RNA "exits" the nucleus. It has taken on the functional definition of a protein-encoding nucleotide sequence. Leader sequences and 3′ UTR sequences are also derived from exons, even though they are not translated into protein.

[4]By convention, upstream, downstream, 5′, and 3′ directions are specified in relation to the RNA. Thus, the promoter is upstream of the gene, near to and "before" its 5′ end.

that will code for the addition of a modified nucleotide "cap" at the 5' end of the RNA soon after it is transcribed. The specific cap sequence varies among genes. This cap sequence begins the first exon.

- The **5' untranslated region** (**5' UTR**), also called the **leader sequence**. In the human β-globin gene, it is the sequence of 50 base pairs intervening between the initiation points of transcription and translation. The 5' UTR can determine the rate at which translation is initiated.

- The **translation initiation site**, **ATG**. This codon (which becomes AUG in mRNA) is located 50 base pairs after the transcription initiation site in the human β-globin gene (this distance differs greatly among different genes). The ATG translation start sequence is the same in every gene.

- The protein-encoding portion of the first exon, which contains 90 base pairs coding for amino acids 1–30 of human β-globin protein.

- An intron containing 130 base pairs with no coding sequences for β-globin. The structure of this intron, however, is important in enabling the RNA to be processed into mRNA and exit from the nucleus.

- An exon containing 222 base pairs coding for amino acids 31–104.

- A large intron—850 base pairs—having nothing to do with β-globin protein structure.

- An exon containing 126 base pairs coding for amino acids 105–146 of the protein.

- A **translation termination codon**, **TAA**. This codon becomes UAA in the mRNA. When a ribosome encounters this codon, the ribosome dissociates, and the protein is released. Translation termination can also be represented by the TAG or TGA codon sequences in other genes.

- A **3' untranslated region** (**3' UTR**) that, although transcribed, is not translated into protein. This region includes the sequence AATAAA, which is needed for **polyadenylation**, the insertion of a "tail" of some 200–300 adenylate residues on the RNA transcript, about 20 bases downstream of the AAUAAA sequence. This polyA tail (1) confers stability on the mRNA, (2) allows the mRNA to exit the nucleus, and (3) permits the mRNA to be translated into protein.

- A **transcription termination sequence**. Transcription continues beyond the AATAAA site for about 1000 nucleotides before being terminated.

The original transcription product is called **nuclear RNA** (**nRNA**) or, sometimes, *heterogeneous nuclear RNA* (hnRNA) or *pre-messenger RNA* (pre-mRNA). Nuclear RNA contains the cap sequence, the 5' UTR, exons, introns, and the 3' UTR. Both ends of these transcripts are modified before these RNAs leave the nucleus. A cap consisting of methylated guanosine is placed on the 5' end of the RNA in opposite polarity to the RNA itself, which means that there is no free 5' phosphate group on the nRNA. The 5' cap is necessary for the binding of mRNA to the ribosome and for subsequent translation (Shatkin 1976). The 3' terminus is usually modified in the nucleus by the addition of a polyA tail. The adenylate residues in this tail are added to the transcript enzymatically; they are not part of the gene sequence. Both the 5' and 3' modifications may protect the mRNA from exonucleases that would otherwise digest it (Sheiness and Darnell 1973; Gedamu and Dixon 1978). The modifications thus stabilize the message and its precursor.

Before the nRNA leaves the nucleus, its introns are removed and the remaining exons spliced together. In this way, the coding regions of the mRNA—that is, the exons—are brought together to form a single uninterrupted transcript, and this transcript is translated into a protein. The protein can be further modified to make it functional (see Figure 3.6).

Cis regulatory elements: The on, off, and dimmer switches of a gene

In addition to the protein-encoding region of the gene, regulatory sequences can be located on either end of the gene (or even within it). These regulatory sequences—the *promoter, enhancers*, and *silencers*—are necessary for controlling where, when, and how actively a particular gene is transcribed. When located on the same chromosome as the gene (and they usually are), they can be referred to as **cis-regulatory elements**.[5]

Promoters are sites where RNA polymerase II binds to the DNA sequence to initiate transcription. Promoters of genes that synthesize messenger RNAs (i.e., those genes that encode proteins[6]) are typically located immediately upstream from the site where RNA polymerase II initiates transcription. Most of these promoters contain a stretch of about 1000 base pairs that is rich in the sequence CG, often referred to as CpG (a **C** and a **G** connected through the normal **p**hosphate bond). These regions are called **CpG islands** (Down and Hubbard 2002; Deaton and Bird 2011). The reason transcription is initiated near CpG islands is thought to involve proteins called **basal transcription factors**, which are present in every cell and specifically bind to the CpG-rich sites. These basal transcription factor proteins form a "saddle" that can recruit RNA polymerase II and position it appropriately for the polymerase to begin transcription (Kostrewa et al. 2009).

RNA polymerase II does not bind to every promoter in the genome at the same time, however. Rather, it is recruited to and stabilized on the promoters by DNA sequences called **enhancers** that signal where and when a promoter can be used and how much gene product to make. In other words, enhancers control the efficiency and rate of transcription from a specific promoter (see Ong and Corces 2011). In contrast, DNA sequences called **silencers** can prevent promoter use and inhibit gene transcription. **Transcription factors** are proteins that bind DNA with precise sequence recognition for specific promoters, enhancers, or silencers. Transcription factors that bind enhancers can activate a gene by (1) recruiting enzymes (such as histone acetyltransferases) that break up the nucleosomes in the area or (2) stabilizing the transcription initiation complex as described above. Thus, transcription factors usually work in two nonexclusive ways:

1. Once bound, transcription factors can bind cofactors that recruit nucleosome-modifying proteins (such as histone methyltransferases and acetyltransferases) that make that area of the genome accessible for RNA polymerase II to bind and enable the chromatin in that vicinity to be unwound and transcribed.

2. Transcription factors can form bridges, looping the chromatin such that the transcription factors (and their histone-modifying enzymes) on enhancers can be brought into the vicinity of the promoter. In the activation of mammalian β-globin

---

[5] *Cis-* and *trans*-regulatory elements are so named by analogy with *E. coli* genetics and organic chemistry. Therefore, *cis*-elements are regulatory elements that reside on the same chromosome (*cis-*, "on the same side as"), whereas *trans*-elements are those that could be supplied from another chromosome (*trans-*, "on the other side of"). The term *cis-regulatory elements* now refers to those DNA sequences that regulate a gene on the same stretch of DNA (i.e., the promoters and enhancers). *Trans*-regulatory factors are soluble molecules whose genes are located elsewhere in the genome and that bind to the *cis*-regulatory elements. They are usually transcription factors or microRNAs. Some evidence points to the ability of an enhancer to activate a *trans*-promoter (i.e., a promoter on another chromosome), but such cases appear to be exceptional and rare events (Noordermeer et al. 2011).

[6] In the case of protein-encoding genes, RNA polymerase II is used for transcription. There are several types of RNA that do not encode proteins, including the ribosomal RNAs and transfer RNAs (which are used in protein synthesis) and the small nuclear RNAs (which are used in RNA processing). In addition, there are regulatory RNAs (such as the microRNAs and long noncoding RNAs that we will discuss later in this chapter) that are involved in regulating gene expression and are not translated into peptides. These regulatory RNAs often are transcribed by other RNA polymerases.

(A)

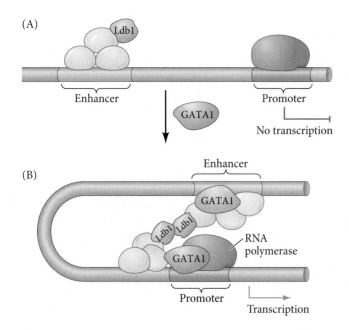

GATA1

No transcription

(B)

Enhancer

GATA1

Ldb1 Ldb1

GATA1

RNA polymerase

Promoter

Transcription

FIGURE 3.7 The bridge between enhancer and promoter can be made by transcription factors. Certain transcription factors bind to DNA on the promoter (where RNA polymerase II will initiate transcription), whereas other transcription factors bind to the enhancer (which regulates when and where transcription can occur). Other transcription factors do not bind to the DNA; rather, they link the transcription factors that have bound to the enhancer and promoter sequences. In this way, the chromatin loops to bring the enhancer to the promoter. The example shown here is the mouse β-globin gene. (A) Transcription factors assemble on the enhancer, but the promoter is not used until the GATA1 transcription factor binds to the promoter. (B) GATA1 can recruit several other factors, including Ldb1, which forms a link uniting the enhancer-bound factors to the promoter-bound factors. (After Deng et al. 2012.)

genes, such a bridge uniting the promoter and enhancer is formed by proteins that bind to transcription factors on both the enhancer and promoter sequences. These proteins recruit the nucleosome-modifying enzymes and transcription-associated factors (TAFs) that stabilize RNA polymerase II (**FIGURE 3.7**; Gurdon 2016; Deng et al. 2012; Noordermeer and Duboule 2013).

THE MEDIATOR COMPLEX: LINKING ENHANCER AND PROMOTER In many genes, a bridge between enhancer and promoter is made by a large, multimeric complex called the **Mediator**, whose nearly 30 protein subunits connect RNA polymerase II to enhancer regions that relay developmental signals (Malik and Roeder 2010). This bridge forms the **pre-initiation complex** at the promoter. Therefore, the Mediator helps create a chromatin loop, bringing the enhancers to the promoter. This chromatin loop is stabilized by the protein **cohesin**, which wraps around portions of this loop like a ring upon association with the Mediator after the Mediator is bound by transcription factors (**FIGURE 3.8**).

Although the Mediator may help bring the RNA polymerase II to the promoter, for transcription to take place the connection between the Mediator and the RNA polymerase II has to be broken, and RNA polymerase II must be released from the promoter. The release of RNA polymerase II is accomplished by a **transcription elongation complex (TEC)**, which is made up of several transcription factors and enzymes (e.g., Ikaros, NuRD, and P-TEFb[7]; Bottardi et al. 2015). This release coincides with the capping of the transcript, phosphorylation of the polymerase, and elongation of the transcript. In some instances (discussed later in the chapter), however, the RNA polymerase II either does not dissociate from the Mediator, or it dissociates but only transcribes a short stretch of nucleotides before it pauses. In the latter case, a **transcription elongation suppressor** (such as NELF) functions to prevent the TEC from associating with the polymerase, and the RNA polymerase II is paused, held in readiness for a new developmental signal.

ENHANCER FUNCTIONING One of the principal methods of identifying enhancer sequences is to clone DNA sequences flanking the gene of interest and fuse them to reporter genes whose products are both readily identifiable and not usually made in the organism being studied. Researchers can insert constructs of possible enhancers

---

[7]Ikaros is a type of zinc-finger transcription factor that binds the histone deacetylase NuRD, which recruits P-TEFb (Positive transcription elongation factor b) to form a complex that breaks transcriptional pausing and promotes nRNA elongation (Bottardi et al. 2015). Interestingly, the repertoire of bound factors can be gene specific. For example, progenitor blood cells expressing high levels of Ikaros differentiate into various types of white blood cells, and those expressing low levels differentiate mostly into red blood cells (Frances et al. 2011).

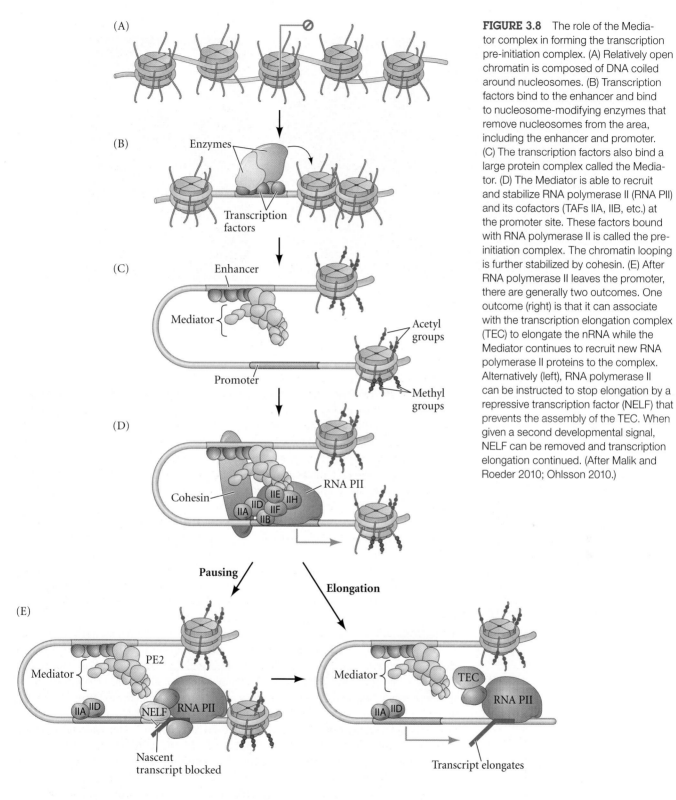

FIGURE 3.8 The role of the Mediator complex in forming the transcription pre-initiation complex. (A) Relatively open chromatin is composed of DNA coiled around nucleosomes. (B) Transcription factors bind to the enhancer and bind to nucleosome-modifying enzymes that remove nucleosomes from the area, including the enhancer and promoter. (C) The transcription factors also bind a large protein complex called the Mediator. (D) The Mediator is able to recruit and stabilize RNA polymerase II (RNA PII) and its cofactors (TAFs IIA, IIB, etc.) at the promoter site. These factors bound with RNA polymerase II is called the pre-initiation complex. The chromatin looping is further stabilized by cohesin. (E) After RNA polymerase II leaves the promoter, there are generally two outcomes. One outcome (right) is that it can associate with the transcription elongation complex (TEC) to elongate the nRNA while the Mediator continues to recruit new RNA polymerase II proteins to the complex. Alternatively (left), RNA polymerase II can be instructed to stop elongation by a repressive transcription factor (NELF) that prevents the assembly of the TEC. When given a second developmental signal, NELF can be removed and transcription elongation continued. (After Malik and Roeder 2010; Ohlsson 2010.)

with reporter genes into embryos and then monitor the spatial and temporal pattern of expression displayed by the visible protein product of the reporter gene (such as *green fluorescent protein, GFP*; **FIGURE 3.9A**). If the sequence contains an enhancer, the reporter gene should become active at particular times and places. For instance, the *E. coli* gene for β-galactosidase (the *lacZ* gene) can be used as a reporter gene and fused to

(A)

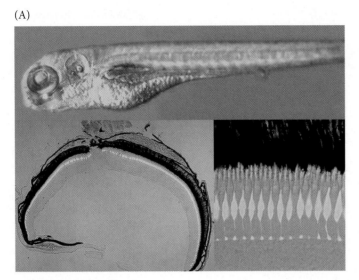

(B)

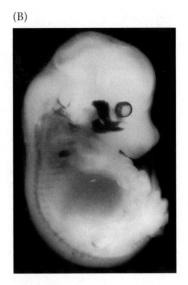

FIGURE 3.9 The genetic elements regulating tissue-specific transcription can be identified by fusing reporter genes to suspected enhancer regions of the genes expressed in particular cell types. (A) The *GFP* gene is fused to a zebrafish gene that is active only in certain cells of the retina. The result is expression of green fluorescent protein in the larval retina (below left), specifically in the cone cells (below right). (B) The enhancer region of the gene for the muscle-specific protein Myf5 is fused to a β-galactosidase reporter gene and incorporated into a mouse embryo. When stained for β-galactosidase activity (darkly staining region), the 13.5-day mouse embryo shows that the reporter gene is expressed in the muscles of the eye, face, neck, and forelimb and in the segmented myotomes (which give rise to the back musculature). (A from Takechi et al. 2003, courtesy of S. Kawamura, T. Hamaoka, and M. Takechi; B courtesy of A. Patapoutian and B. Wold.)

(1) a promoter that can be activated in any cell and (2) an enhancer that directs expression of a particular gene (*Myf5*) only in mouse muscles. When the resulting transgene is injected into a newly fertilized mouse egg and becomes incorporated into its DNA, β-galactosidase protein reveals the expression pattern of that muscle-specific gene (**FIGURE 3.9B**). More recently, genomic techniques such as ChIP-Seq (discussed later in the chapter) have enabled researchers to identify enhancer elements by sequencing the DNA regions bound by specific transcription factors.

Enhancers generally activate only *cis*-linked promoters (i.e., promoters on the same chromosome); therefore, they are sometimes called *cis*-regulatory elements. Because of DNA folding, however, enhancers can regulate genes at great distances (some as great as a million bases away) from the promoter (Visel et al. 2009). Moreover, enhancers do not need to be on the 5′ (upstream) side of the gene; they can be at the 3′ end and can be located in the introns (Maniatis et al. 1987). As we will see in Chapter 19, an important enhancer for a gene involved in specifying the "pinky" of each of our limbs is found in an intron of *another* gene, some one million base pairs away from its promoter (Lettice et al. 2008). In each cell, the enhancer becomes associated with particular transcription factors, binds nucleosome regulators and the Mediator complex, and engages with the promoter to transcribe the gene in that particular type of cell (**FIGURE 3.10A**).

ENHANCER MODULARITY The enhancer sequences on the DNA are the same in every cell type; what differs is the combination of transcription factor proteins that the enhancers experience. Once bound to enhancers, transcription factors are able to enhance or suppress the ability of RNA polymerase II to initiate transcription. Several transcription factors can bind to an enhancer, and it is the specific *combination* of transcription factors present that allows a gene to be active in a particular cell type. That is, the same transcription factor, in conjunction with different combinations of factors, will activate different promoters in different cells. Moreover, the same gene can have several enhancers, with each enhancer binding transcription factors that enable that same gene to be expressed in different cell types.

The mouse *Pax6* gene (which is expressed in the lens, cornea, and retina of the eye, in the neural tube, and in the pancreas) has several enhancers (**FIGURE 3.10B,C**). The 5′ regulatory regions of the mouse *Pax6* gene were discovered by taking regions from its 5′ flanking sequence and introns and fusing them to a *lacZ* reporter gene. Each of these transgenes was then microinjected into newly fertilized mouse pronuclei, and the resulting embryos were stained for β-galactosidase (**FIGURE 3.10D**; Kammandel et al. 1998; Williams et al. 1998). Analysis of the results revealed that the enhancer farthest

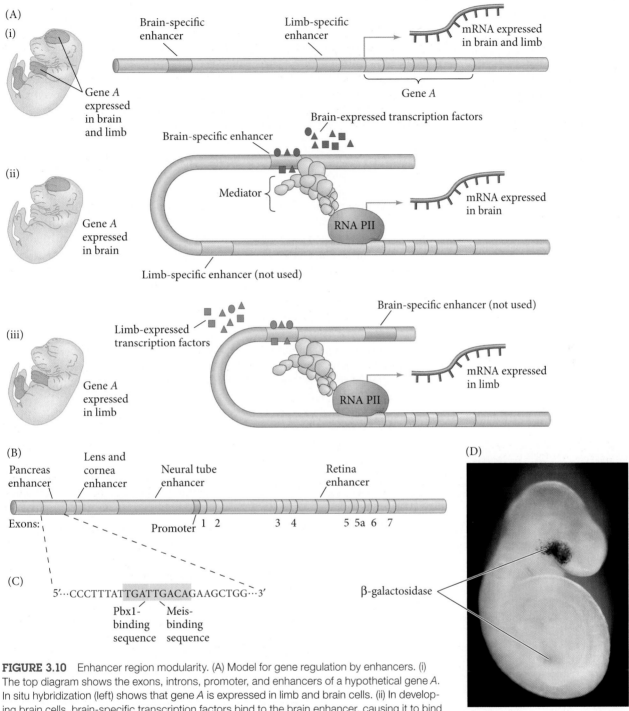

FIGURE 3.10 Enhancer region modularity. (A) Model for gene regulation by enhancers. (i) The top diagram shows the exons, introns, promoter, and enhancers of a hypothetical gene *A*. In situ hybridization (left) shows that gene *A* is expressed in limb and brain cells. (ii) In developing brain cells, brain-specific transcription factors bind to the brain enhancer, causing it to bind to the Mediator, stabilize RNA polymerase II at the promoter, and modify the nucleosomes in the region of the promoter. The gene is transcribed in the brain cells only; the limb enhancer does not function. (iii) An analogous process allows for transcription of the same gene in the cells of the limbs. The gene is not transcribed in any cell type whose transcription factors the enhancers cannot bind. (B) The Pax6 protein is critical in the development of several widely different tissues. Enhancers direct *Pax6* gene expression (yellow exons 1–7) differentially in the pancreas, the lens and cornea of the eye, the retina, and the neural tube. (C) A portion of the DNA sequence of the pancreas-specific enhancer element. This sequence has binding sites for the Pbx1 and Meis transcription factors; both must be present to activate *Pax6* in the pancreas. (D) When the β-galactosidase reporter gene is fused to the *Pax6* enhancers for expression in the pancreas and lens/cornea, the enzyme is seen in those tissues. (A after Visel et al. 2009; D from Williams et al. 1998, courtesy of R. A. Lang.)

Developing Questions

What are the consequences of enhancer modularity to a developing individual? To a species? How might a mutation in an enhancer affect development? For instance, what might occur in an embryo if there were a mutation in the enhancer region of the *Pax6* gene? Could such a mutation have evolutionary importance? *Hint*: It does, and it's profound!

upstream from the promoter contains the regions necessary for *Pax6* expression in the pancreas, whereas a second enhancer activates *Pax6* expression in surface ectoderm (lens, cornea, and conjunctiva). A third enhancer resides in the leader sequence; it contains the sequences that direct *Pax6* expression in the neural tube. A fourth enhancer, located in an intron shortly downstream of the translation initiation site, determines the expression of *Pax6* in the retina. The *Pax6* gene illustrates the principle of enhancer modularity, wherein genes having multiple, separate enhancers allow a protein to be expressed in several different tissues but not expressed at all in others.

COMBINATORIAL ASSOCIATION   Although there is modularity *among* enhancers, there are codependent units *within* each enhancer. Enhancers contain regions of DNA that bind transcription factors, and it is this *combination* of transcription factors that activates the gene. For instance, the pancreas-specific enhancer of the *Pax6* gene has binding sites for the Pbx1 and Meis transcription factors (see Figure 3.10C). Both need to be present for the enhancer to activate *Pax6* in the pancreas cells (Zhang et al. 2006).

Moreover, the product of the *Pax6* gene encodes a transcription factor that works in combinatorial partnerships with other transcription factors. Figure 3.11 shows two gene enhancer regions that bind *Pax6*. The first is that of the chick δ1 lens *crystallin* gene (**FIGURE 3.11A**; Cvekl and Piatigorsky 1996; Muta et al. 2002). This gene encodes crystallin, a lens protein that is transparent and allows light to reach the retina. A promoter in the *crystallin* gene contains binding sites for TBP and Sp1 (basal transcriptional factors that recruit RNA polymerase II to the DNA). The gene also has an enhancer in its third intron that controls the time and place of crystallin expression. This enhancer has two Pax6-binding sites. The Pax6 protein works with the Sox2 and L-Maf transcription factors to activate the *crystallin* gene only in those head cells that are going to become lens. As we will see in Chapter 16, this means that the cell (1) must be head ectoderm (which expresses Pax6), (2) must be in the region of the ectoderm capable of forming eyes (expressing L-Maf), and (3) must be in contact with the future retinal cells (which induce Sox2 expression; see Kamachi et al. 1998).

Meanwhile, Pax6 also regulates the transcription of the genes encoding insulin, glucagon, and somatostatin in the pancreas (**FIGURE 3.11B**). Here, Pax6 works in cooperation with other transcription factors such as Pdx1 (specific for the pancreatic region of the endoderm) and Pbx1 (Andersen et al. 1999; Hussain and Habener 1999). So, in the absence of Pax6, the eye fails to form, and the endocrine cells of the pancreas do not develop properly; these improperly developed endocrine cells produce deficient amounts of their hormones (Sander et al. 1997; Zhang et al. 2002).

Other genes are activated by Pax6 binding, and one of them is the *Pax6* gene itself. Pax6 protein can bind to a *cis*-regulatory element of the *Pax6* gene (Plaza et al. 1993). So, once the *Pax6* gene is turned on, it will continue to be expressed, even if the signal that originally activated it is no longer present.

FIGURE 3.11   Modular transcriptional regulatory regions using Pax6 as an activator. (A) Promoter and enhancer of the chick δ1 lens *crystallin* gene. Pax6 interacts with two other transcription factors, Sox2 and L-Maf, to activate this gene. The protein δEF3 binds factors that permit this interaction; δEF1 binds factors that inhibit it. (B) Promoter and enhancer of the rat *somatostatin* gene. Pax6 activates this gene by cooperating with the Pbx1 and Pdx1 transcription factors. (A after Cvekl and Piatigorsky 1996; B after Andersen et al. 1999.)

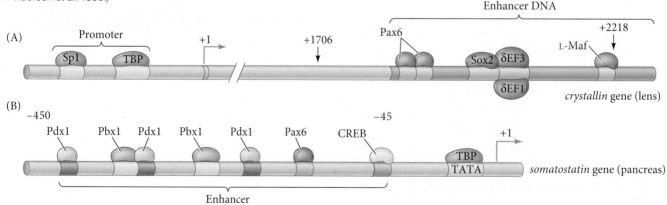

SILENCERS   Silencers are DNA regulatory elements that actively repress the transcription of a particular gene. They can be viewed as "negative enhancers," and they can silence gene expression spatially (in particular cell types) or temporally (at particular times). In the mouse, for instance, there is a DNA sequence that prevents a promoter's activation in any tissue *except* neurons. This sequence, given the name **neural restrictive silencer element** (**NRSE**), has been found in several mouse genes whose expression is limited to the nervous system: those encoding synapsin I, sodium channel type II, brain-derived neurotrophic factor, Ng-CAM, and L1. The protein that binds to the NRSE is a transcription factor called **neural restrictive silencer factor** (**NRSF**, sometimes called **REST**). NRSF appears to be expressed in every cell that is *not* a mature neuron (Chong et al. 1995; Schoenherr and Anderson 1995). When NRSE is deleted from particular neural genes, these genes are expressed in non-neural cells (**FIGURE 3.12**; Kallunki et al. 1995, 1997). Thus, neural-specific genes are actively repressed in non-neural cells.

A recently discovered "temporal silencer" may play a role in regulating the human globin genes. In most people, a fetal globin gene is active from about week 12 until birth. Then, around the time of birth, the fetal globin gene is turned off, and the adult globin gene is activated. Some families, however, show a hereditary persistence of fetal hemoglobin, with the fetal globin genes remaining active in the adults. Some of these families have a mutation in a region of DNA that usually silences the fetal globin gene at birth. In most people, this silencer contains binding sites for the transcription factors GATA1 and BCL11A, whose combination on the DNA recruits histone modification enzymes. This action causes the formation of deacetylated and repressive (H3K27me3-containing) nucleosomes (Sankaran et al. 2011).

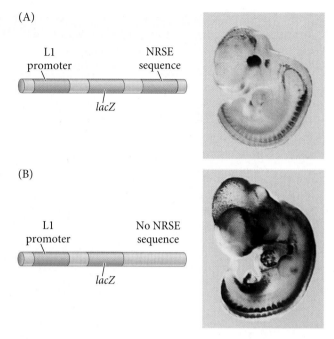

FIGURE 3.12   A silencer represses gene transcription. (A) Mouse embryo containing a transgene composed of the L1 promoter, a portion of the neuron-specific *L1* gene, and a *lacZ* gene fused to the *L1* second exon, which contains the NRSE sequence. (B) Same-stage embryo with a similar transgene but lacking the NRSE sequence. Dark areas reveal the presence of β-galactosidase (the *lacZ* product). (Photographs from Kallunki et al. 1997.)

GENE REGULATORY ELEMENTS: SUMMARY   Enhancers and silencers enable genes for specific proteins to use numerous transcription factors in various combinations to control their expression. Thus, *enhancers and silencers are modular* such that, for example, the *Pax6* gene is regulated by enhancers that enable it to be expressed in the eye, pancreas, and nervous system, as seen in Figure 3.10B; this is the Boolean "OR" function. But *within each cis-regulatory module, transcription factors work in a combinatorial fashion* such that Pax6, L-Maf, and Sox2 proteins are all needed for the transcription of crystallin in the lens (see Figure 3.11A); that is the Boolean "AND" function. The combinatorial association of transcription factors on enhancers leads to the spatiotemporal output of any particular gene (see Peter and Davidson 2015; Zinzen et al. 2009). This "AND" function may be extremely important in activating entire groups of genes simultaneously.

Transcription factor function

FAMILIES AND OTHER ASSOCIATIONS   The science journalist Natalie Angier (1992) wrote that "a series of new discoveries suggests that DNA is more like a certain type of politician, surrounded by a flock of protein handlers and advisers that must vigorously massage it, twist it, and on occasion, reinvent it before the grand blueprint of the body can make any sense at all." These "handlers and advisers" are the transcription factors. During development, transcription factors play essential roles in every aspect of embryogenesis, controlling differential gene expression leading to differentiation. When in doubt, it is usually a transcription factor's fault, a sentiment that is often used by politicians, too.

TABLE 3.1 Some major transcription factor families and subfamilies

Family	Representative transcription factors	Some functions
Homeodomain:		
Hox	Hoxa1, Hoxb2, etc.	Axis formation
POU	Pit1, Unc-86, Oct-2	Pituitary development; neural fate
Lim	Lim1, Forkhead	Head development
Pax	Pax1, 2, 3, 6, etc.	Neural specification; eye development
Basic helix-loop-helix (bHLH)	MyoD, MITF, daughterless	Muscle and nerve specification; *Drosophila* sex determination; pigmentation
Basic leucine zipper (bZip)	C/EBP, AP1, MITF	Liver differentiation; fat cell specification
Zinc-finger:		
Standard	WT1, Krüppel, Engrailed	Kidney, gonad, and macrophage development; *Drosophila* segmentation
Nuclear hormone receptors	Glucocorticoid receptor, estrogen receptor, testosterone receptor, retinoic acid receptors	Secondary sex determination; craniofacial development; limb development
Sry-Sox	Sry, SoxD, Sox2	Bend DNA; mammalian primary sex determination; ectoderm differentiation

Transcription factors can be grouped together in families based on similarities in DNA-binding domains (**TABLE 3.1**). The transcription factors in each family share a common framework in their DNA-binding sites, and slight differences in the amino acids at the binding site can cause the binding site to recognize different DNA sequences.

As we have already seen, DNA regulatory elements such as enhancers and silencers function by binding transcription factors, and each element can have binding sites for several transcription factors. Transcription factors bind to the DNA of the regulatory element using one site on the protein and other sites to interact with other transcription factors and proteins, leading to the recruitment of histone-modifying enzymes. For example, the association of the Pax6, Sox2, and L-Maf transcription factors in lens cells recruits a histone acetyltransferase that can transfer acetyl groups to the histones and dissociate the nucleosomes in that area (Yang et al. 2006). Similarly, when MITF,[8] a transcription factor essential for ear development and pigment production, binds to its specific DNA sequence, it also binds to (different) histone acetyltransferase that facilitates the dissociation of nucleosomes (Ogryzko et al. 1996; Price et al. 1998). In addition, the Pax7 transcription factor that activates muscle-specific genes binds to the enhancer region of these genes within the muscle precursor cells. Pax7 then recruits a histone methyltransferase that methylates the lysine in the fourth position of histone H3 (H3K4), resulting in the trimethylation of this lysine and the activation of transcription (McKinnell et al. 2008). The displacement of nucleosomes along the DNA makes it possible for other transcription factors to find their binding sites and regulate expression (Adkins et al. 2004; Li et al. 2007).

In addition to recruiting histone-modifying enzymes, transcription factors can also work by stabilizing the transcription pre-initiation complex that enables RNA polymerase II to bind to the promoter (see Figures 3.7 and 3.8). For instance, MyoD, a transcription factor that is critical for muscle cell development, stabilizes TAF IIB, which supports RNA polymerase II at the promoter site (Heller and Bengal 1998). Indeed, MyoD plays several roles in activating gene expression because it also can bind histone acetyltransferases that initiate nucleosome remodeling and dissociation (Cao et al. 2006).

[8]MITF stands for **mi**crophthalmia-associated **t**ranscription **f**actor.

One important consequence of the combinatorial association of transcription factors is **coordinated gene expression**. The simultaneous expression of many cell-specific genes can be explained by the binding of transcription factors by the enhancer elements. For example, many genes that are specifically activated in the lens contain an enhancer that binds Pax6. So, all the other transcription factors might be assembled at the enhancer, but until Pax6 binds, they cannot activate the gene. Similarly, many of the coexpressed muscle-specific genes contain enhancers that bind the Mef2 transcription factor, and the enhancers on genes encoding pigment-producing enzymes bind MITF (see Davidson 2006). In some instances, entire ensembles of transcription factors appear to direct simultaneous gene transcription. Junion and colleagues have shown, for example, that a particular ensemble of five transcription factors is bound on hundreds of enhancers that are active in the developing *Drosophila* heart muscle cells (Junion et al. 2012).

TRANSCRIPTION FACTOR DOMAINS   Transcription factors have three major domains. The first is a **DNA-binding domain** that recognizes a particular DNA sequence in the enhancer. There are several different types of DNA-binding domains, and they often designate the major family classifications for transcription factors. Some of the most common protein domains that convey DNA binding are the Homeodomain, Zinc Finger, Leucine Zipper, Helix-Loop-Helix, and Helix-Turn-Helix (see Table 3.1). For instance, the homeodomain transcription factor Pax6[9] uses its paired DNA-binding sites to recognize the enhancer sequence, CAATTAGTCACGCTTGA (Askan and Goding 1998; Wolf et al. 2009). In contrast, the MITF transcription factor involved in ear and pigment cell development contains both leucine zipper and helix-loop-helix domains, and it recognizes shorter DNA sequences called the E-box (CACGTG) and the M-box (CATGTG; Pogenberg et al. 2012).[10] These sequences for MITF binding have been found in the regulatory regions of genes encoding several pigment-cell-specific enzymes of the tyrosinase family (Bentley et al. 1994; Yasumoto et al. 1994, 1997). Without MITF, these proteins are not synthesized properly, and melanin pigment is not made.

The second domain is a ***trans-activating domain*** that activates or suppresses the transcription of the gene whose promoter or enhancer it has bound. Usually, this *trans*-activating domain enables the transcription factor to interact with the proteins involved in binding RNA polymerase II (such as TAF IIB or TAF IIE; see Sauer et al. 1995) or with enzymes that modify histones. MITF contains such a domain of amino acids in the center of the protein. When the MITF dimer is bound to its target sequence in the enhancer, the *trans*-activating region is able to bind a transcription-associated factor (TAF), p300/CBP. The p300/CBP protein is a histone acetyltransferase enzyme that can transfer acetyl groups to each histone in the nucleosomes (Ogryzko et al. 1996; Price et al. 1998). Acetylation of the nucleosomes destabilizes them and allows the genes for pigment-forming enzymes to be expressed.

Finally, there is usually a **protein-protein interaction domain** that allows the transcription factor's activity to be modulated by TAFs or other transcription factors. MITF has a protein-protein interaction domain that enables it to dimerize with another MITF protein (Ferré-D'Amaré et al. 1993). The resulting homodimer (i.e., two identical protein molecules bound together) is the functional protein that binds to enhancer DNA of certain genes and activates transcription (**FIGURE 3.13**).

INSULATORS   The boundaries of gene expression appear to be set by DNA sequences called **insulators**. Insulator sequences limit the range in which an enhancer can activate

---

[9] Pax stands for "paired box," and "box" refers to its DNA-binding domain. Pax proteins are homeodomain transcription factors that contain a paired domain for binding to DNA. Studies on *Drosophila* have shown that the loss of a *homeodomain* transcription factor causes dramatic *homeotic* transformations in structures, such as the transformation of an antenna into a leg.

[10] E-box and M-box refer to "Enhancer" and "Myc" respectively, with "box" meaning DNA-binding domain.

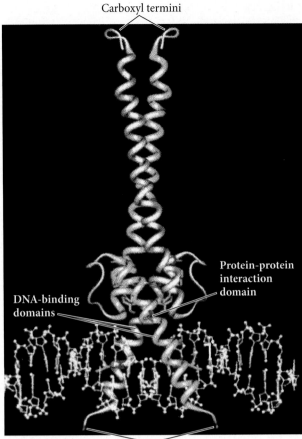

Carboxyl termini

Protein-protein interaction domain

DNA-binding domains

Amino termini

FIGURE 3.13 Three-dimensional model of the homodimeric transcription factor MITF (one protein shown in red, the other in blue) binding to a promoter element in DNA (white). The amino termini are located at the bottom of the figure and form the DNA-binding domains that recognize an 11-base-pair sequence of DNA having the core sequence CATGTG. The protein-protein interaction domain is located immediately above. MITF has the basic helix-loop-helix structure found in many transcription factors. The carboxyl end of the molecule is thought to be the *trans*-activating domains that bind the p300/CBP transcription-associated factor (TAF). (From Steingrímsson et al. 1994, courtesy of N. Jenkins.)

gene expression. They thereby "insulate" a promoter from being activated by another gene's enhancers. Some insulator DNA regions have been found to bind a zinc-finger transcription factor called CTCF,[11] which functions to alter the three-dimensional conformation of chromatin and thereby separate (or insulate) enhancer elements from the promoter (Yusufzai et al. 2004; Kim and Kaang 2015). CTCF is ubiquitously expressed in eukaryotes and has been charted to bind tens of thousands of sites on the genome (Chen et al. 2012). Mechanistically, CTCF physically interacts with cohesin, a ring-shaped complex of multiple subunits that function to stabilize chromatin loop structures (see the discussion of the Mediator complex on p. 56). It is hypothesized that CTCF uses its 11 zinc-finger domains to selectively bind DNA, often insulator elements, to create loop structures that distance enhancers from promoters. For instance, the chick β-globin gene has been shown to form a complex with cohesin (Wendt et al. 2008; Wood et al. 2010). This CTCF-cohesin complex may bind to the enhancer-bound Mediator, thereby preventing the enhancer from activating the adjacent promoter.

PIONEER TRANSCRIPTION FACTORS: BREAKING THE SILENCE   Finding an enhancer is not easy because the DNA is usually so wound up that the enhancer sites are not accessible. Given that the enhancer might be covered by nucleosomes, how can a transcription factor find its binding site? That is the job of certain transcription factors that penetrate repressed chromatin and bind to their enhancer DNA sequences (Cirillo et al. 2002; Berkes et al. 2004). They have been called "pioneer" transcription factors, and they appear to be critical in establishing certain cell lineages. One of these transcription factors is FoxA1, which binds to certain enhancers and opens up the chromatin to allow other transcription factors access to the promoter (Lupien et al. 2008; Smale 2010). FoxA1 is extremely important in specifying liver cells, remaining bound to the DNA during mitosis, and providing a mechanism to reestablish normal transcription in presumptive liver cells (Zaret et al. 2008). Another pioneer transcription factor is the Pax7 protein mentioned above. It activates muscle-specific gene transcription in a population of muscle stem cells by binding to its DNA recognition sequence and being stabilized there by dimethylated H3K4 on the nucleosomes. It then recruits the histone methyltransferase that converts the dimethylated H3K4 into the trimethylated H3K4 associated with active transcription (McKinnell et al. 2008).

MASTER REGULATORY TRANSCRIPTION FACTORS   The phrase "master regulator" has been used to describe certain transcription factors that seem to have the power to control

Developing Questions

The precise binding of transcription factors to *cis*-regulatory elements drives differential gene expression both spatially and temporally in the developing embryo. Is a cell's identity determined by one transcription factor complex binding to one regulatory element, leading to the expression of one gene? How many genes are required to establish a specific cell's fate?

---

[11] CTCF stands for CC**CTC**-binding **F**actor. Although we highlight its role as an insulating factor, CTCF can also contribute to chromatin architecture and in some cases activate transcription by bringing enhancers in contact with promoters. (See Kim and Kaang 2015.)

cell differentiation, but can one transcription factor really direct a progenitor cell down a specific path of maturation or even more dramatically change the fate of a differentiated cell? To be called a **master regulator**, a transcription factor must (1) be expressed when the specification of a cell type begins, (2) regulate the expression of genes specific to that cell type, and (3) be able to redirect a cell's fate to this cell type (Chan and Kyba 2013).

Early evidence of master regulatory power came from some of the original cloning experiments, in which Briggs and King (1952) and John Gurdon (1962) were able to *reprogram* the nuclei of larval frog fibroblasts or gut cells to support embryonic development. They replaced the nucleus of a frog egg with the nucleus of a *terminally* differentiated cell (fibroblast or gut cell), and the egg went on to develop into a normal frog. These experiments provided the first significant support for nuclear equivalence (further shown by the cloning of Dolly) but they did not show what proteins in the egg cytoplasm were responsible for this reprogramming. Clues came in 2006 when Shinya Yamanaka compiled a list of genes implicated in maintaining cells of the early mouse embryo in an immature state. These immature cells were from the inner cell mass of the blastula (discussed in later chapters). Yamanaka's lab experimentally expressed only four of these genes (*Oct3/4*, *Sox2*, *c-Myc*, and *Klf4*) in differentiated mouse fibroblasts and found that the fibroblasts *dedifferentiated* into inner cell mass-like cells (**FIGURE 3.14**; Takahashi and Yamanaka 2006). All four of these genes code for transcription factors, making them good candidates for being master regulators. The dedifferentiated cells have since been shown to be able to generate any cell type of the embryo. This means, they can function as pluripotent stem cells, and because they were induced to this state, they are called **induced pluripotent stem cells** (**iPSC**). Yamanaka shared the 2012 Nobel Prize in Physiology or Medicine with Gurdon for their discoveries, and iPSCs are now being used to study human development and disease in ways never before possible (further discussed in Chapter 5).

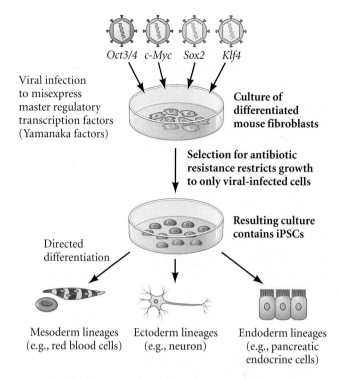

Cells from many lineages can stem from iPSC cells

FIGURE 3.14 From differentiated fibroblast to induced pluripotent stem cell with four transcription factors. If the "Yamanaka factors" (the *Oct3/4*, *cMyc*, *Sox2*, and *Klf4* transcription factors) are virally inserted into differentiated fibroblasts, these cells will dedifferentiate into *induced* pluripotent stem cells (iPSCs). Like embryonic stem cells, iPSCs can give rise to progeny of all three germ layers (mesoderm, ectoderm, and endoderm).

 SCIENTISTS SPEAK 3.2 Watch a Developmental Documentary on cellular reprogramming.

 SCIENTISTS SPEAK 3.3 Enjoy a question and answer session with Dr. Derrick Rossi on the generation of iPSC with mRNA.

Another example of possible master regulators came from Doug Melton's lab, which tested whether select transcription factors could convert pancreatic cells of a diabetic mouse into insulin-producing β cells. The researchers infected the pancreatic cells with harmless viruses containing the genes for three transcription factors: Pdx1, Ngn3, and MafA (**FIGURE 3.15**; Zhou et al. 2008; Cavelti-Weder et al. 2014; Melton 2016). In early development, Pdx1 protein stimulates the outgrowth of the digestive tube that results in the pancreatic buds. This transcription factor is found throughout the pancreas and is critical for specifying that organ's endocrine (hormone-secreting) cells and activating genes that encode endocrine proteins. Ngn3 is a transcription factor found in endocrine, but not exocrine (digestive-enzyme secreting), pancreatic cells. MafA, a transcription factor regulated by glucose levels, is found only in insulin-secreting β cells and activates transcription of the insulin gene. During normal development, Pdx1, Ngn3, and MafA activate other transcription factors that together work to turn a pancreatic endodermal cell into an insulin-secreting β cell. After experimentally inducing production of

FIGURE 3.15 Pancreatic lineage, transcription factors, and direct conversion of β cells to treat diabetes. (A) New pancreatic β cells arise in adult mouse pancreas in vivo after viral delivery of three transcription factors (*Pdx1*, *Ngn3*, and *MafA*) into a diabetic mouse model. Virally infected exocrine cells (shown in the photo) are detected by their expression of nuclear green fluorescent protein. Newly induced β cells are detected by insulin staining (red). Their overlap (co-expressed cells) produces yellow. The nuclei of all pancreatic cells are stained blue. (B) Simplified depiction of the role of transcription factors in pancreatic islet β cell development. Pdx1 protein is critical for specifying a certain group of endoderm cells as pancreas precursors (dark purple lineage). Those descendants of Pdx1-expressing cells that express Ngn3 become the endocrine (hormone-secreting) lineages (shades of purple), whereas those that do not express Ngn3 become the exocrine (digestive enzyme-secreting) lineage of the pancreas (gold). Types of hormone-secreting cells in the pancreatic islets include the insulin-secreting β cells, the somatostatin-secreting δ cells, and the glucagon-secreting α cells. Those cells destined to become β cells express the Nkx6.1 transcription factor, which in turn will activate the gene for the MafA transcription factor found in the insulin-producing β cells. (Photograph courtesy of D. Melton.)

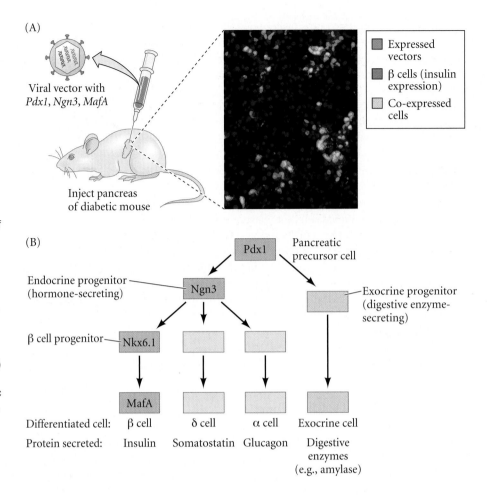

VADE MECUM

Movies in the Fruit Fly segment describe Ernst Hadorn's discovery of transdetermination and Walter Gehring's pioneering study of homeotic mutants, changing body parts into eyes through transcription factors.

these three transcription factors in the pancreas cells of the diabetic mice, Zhou and colleagues saw that the non-insulin-secreting cells had been converted into insulin-secreting β cells. The converted cells looked identical to normal β islet cells and cured the mice of their diabetes.

These studies have opened the door to a new field of regenerative medicine, illustrating the possibilities of changing one adult cell type into another by using the transcription factors that had made the new cell type in the embryo. In some instances, the developmental histories of the cells can be very distant. For instance, adult mouse skin fibroblasts (the mesodermally derived connective tissue of the skin) can be transformed into endodermal hepatocyte-like cells by the addition of only two liver transcription factors (Hnf4α and FoxA1). These induced hepatocytes make several liver-specific proteins and are able to substitute for liver cells in adult mice (Sekiya and Suzuki 2011). Indeed, several laboratories (Caiazzo et al. 2011; Pfisterer et al. 2011; Qiang et al. 2011) have been able to "re-program" adult human and mouse fibroblasts into functional dopaminergic neurons (i.e., the type of nerve cell that degenerates in Parkinson disease) by the addition of three particular transcription factor genes to adult skin cells. Other laboratories (Son et al. 2011) have used a different mix of transcription factors to convert adult human fibroblasts into functional spinal motor neurons (of the type that degenerate in Lou Gehrig disease). These "induced neurons" had the electrophysiological signatures of spinal nerves and formed synapses with muscle cells. The cell type conversions in these studies have helped to reveal the role master regulatory transcription factors play in differential gene expression. *How is it that only a few transcription factors can initiate cell type specific gene expression? Who controls gene expression? When in doubt, who do you blame?*

The Gene Regulatory Network: Defining an Individual Cell

At this point in the chapter, we hope it is clear that different cell types are the result of differentially expressed genes. Although master regulatory genes are necessary for the process, they are not sufficient for implementing an entire genomic program on their own.

Studies on sea urchin development have begun to demonstrate ways in which DNA can be regulated to specify cell type and direct morphogenesis of the developing organism. Eric Davidson's group has pioneered a network model approach in which they envision *cis*-regulatory elements (such as promoters and enhancers) in a logic circuit connected by transcription factors (**FIGURE 3.16**; see http://sugp.caltech.edu/endomes; Davidson and Levine 2008; Oliveri et al. 2008). The network receives its first inputs from maternal transcription factors in the egg cytoplasm; from then on, the network

(A)

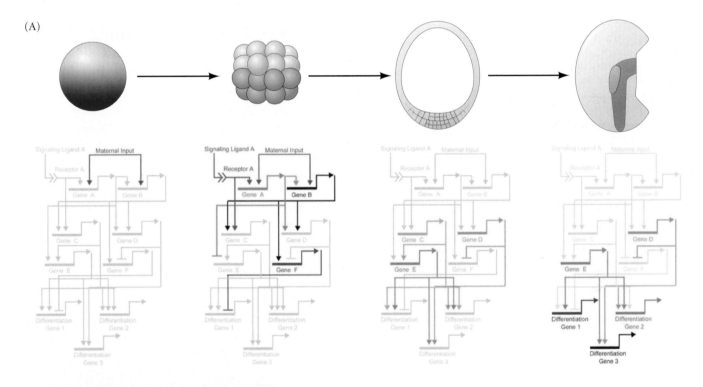

(B)

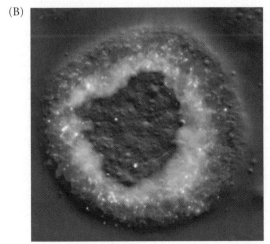

FIGURE 3.16 Gene regulatory networks of endodermal lineages in the sea urchin embryo. (A) Schematics of the sea urchin embryo across four developmental stages showing the progressive specification of endodermal cell fates (top) and the corresponding gene regulatory model of this specification from maternal contributions and signals to master regulatory transcription factors leading to the final differentiation genes (bottom). (B) Double fluorescent in situ hybridization at 24 hours post fertilization showing the restricted expression of *hox11/13b* only in veg1-derived cells (red), whereas *foxa* expression is in the veg2-derived cells (green). (A after Hinman and Cheatle Jarvela 2014; B from Peter and Davidson 2011.)

self-assembles through (1) the ability of the maternal transcription factors to recognize *cis*-regulatory elements of particular genes that encode other transcription factors (*when in doubt . . .*) and (2) the ability of this new set of transcription factors to activate paracrine signaling pathways that activate or inhibit specific transcription factors in neighboring cells (see Figure 3.16A). The studies show the regulatory logic by which the genes of the sea urchin interact to specify and generate characteristic cell types. This set of interconnections among genes specifying cell types is referred to as a **gene regulatory network** (**GRN**), a term first coined by Davidson's group. *Therefore, each cell lineage, cell type, and likely each individual cell can be defined by the GRN that it possesses at that moment in time.*

> *Embryonic development is an enormous informational transaction, in which DNA sequence data generate and guide the system-wide deployment of specific cellular functions.*
>
> E. H. Davidson (2010)

 SCIENTISTS SPEAK 3.4   Listen to a question and answer session with Dr. Marianne Bronner-Fraser on neural crest GRNs in lamprey.

Mechanisms of Differential Gene Transcription

During the twentieth century, we found the actors in the drama of gene transcription, but not until the twenty-first century were their scripts discovered. How does one locate the places on the gene where a particular transcription factor binds, or where nucleosomes with specific modifications are localized? How does one determine the "regulatory architecture" of individual genes and of the entire genome? The recent ability to identify protein-specific DNA-binding sequences using ChIP-Seq technology showed that there are different types of promoters and that they use different scripts to transcribe their genes. **ChIP-Seq**, for *Chromatin Immunoprecipitation-Sequencing*, is a technique that enables a researcher to use known transcription factors as bait to isolate the DNA sequences they specifically recognize (Johnson et al. 2007; Jothi et al. 2008). We elaborate on the ChIP-Seq methodology in the "Tools" section of this chapter (see Figure 3.37) and describe below the insight that ChIP-Seq has provided on differential gene expression.

Differentiated proteins from high and low CpG-content promoters

ChIP-Seq has overturned many of our hypotheses concerning the mechanisms by which promoters and enhancers regulate differential gene expression. It turns out that not all promoters are the same. Rather, there are two general classes of promoters that use different methods for controlling transcription. These promoter types are catalogued as having either a relatively high or a relatively low number of CpG sequences at which DNA methylation can occur.

- **High CpG-content promoters** (**HCPs**) are usually found in "developmental control genes," where they regulate synthesis of the transcription factors and other developmental regulatory proteins used in the *construction* of the organism (Zeitlinger and Stark 2010; Zhou et al. 2011). The default state of these promoters is "on," and they have to be actively repressed by *histone* methylation (**FIGURE 3.17A**).

- **Low CpG-content promoters** (**LCPs**) are usually found in those genes whose products characterize mature cells (e.g., the globins of red blood cells, the hormones of pancreatic cells, and the enzymes that carry out the normal maintenance functions of the cell). The default state of these promoters is "off," but they can be activated by transcription factors (**FIGURE 3.17B**). The nucleosomes on these promoters have relatively few modified histones in the repressed state. Rather, their CpG sites on the DNA are usually methylated, and this methylation is critical for preventing transcription. When the DNA becomes unmethylated, the histones become modified with H3K4me3 and disperse so that RNA polymerase II can bind.

(A) High CpG-content promoters (HCPs)

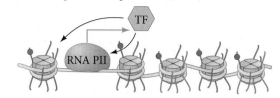

(B) Low CpG-content promoters (LCPs)

Active

"Open" chromatin RNAPII initiation (default)

Selective use

Poised
(intermediate
state)

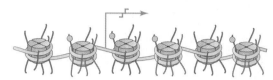

Bivalent chromatin modifications

H3K4me2 chromatin modifications

Repressed
(inactive)

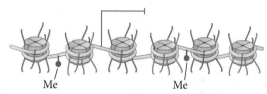

Repressed by histone modification

Me   Me
DNA methylation, no transcription (default)

● H3K4me3   ○ H3K4me2   ● H3K27me3

FIGURE 3.17 Chromatin regulation in HCPs and LCPs. Promoters with high and low CpG content have different modes of regulation. (A) HCPs are typically in an *active* state, with unmethylated DNA and nucleosomes rich in H3K4me3. The open chromatin allows RNA polymerase II (RNA PII) to bind. The *poised* state of HCPs is bivalent, having both activating (H3K4me3) and repressive (H3K27me3) modifications of the nucleosomes. RNA polymerase II can bind but not transcribe. The *repressed* state is characterized by repressive histone modification, but not by extensive DNA methylation. (B) *Active* LCPs, like HCPs, have nucleosomes rich in H3K4me3 and low methylation but require stimulation by transcription factors (TF). *Poised* LCPs are capable of being activated by transcription factors and have relatively unmethylated DNA and nucleosomes enriched in H3K4me2. In their usual state, LCPs are *repressed* by methylated DNA nucleosomes rich in H3K27me3. (After Zhou et al. 2011.)

DNA methylation, another key on/off switch of transcription

Earlier in this chapter, we discussed *histone* methylation and its importance for transcription. Now we look at how the *DNA itself* can be methylated to regulate transcription. Generally speaking, the promoters of inactive genes are methylated at certain cytosine residues, and the resulting methylcytosine stabilizes nucleosomes and prevents transcription factors from binding. This characteristic is especially important in the LCP promoters.

It is often assumed that a gene contains exactly the same nucleotides whether it is active or inactive; that is, a β-globin gene that is activated in a red blood cell precursor has the same nucleotides as the inactive β-globin gene in a fibroblast or retinal cell of the same animal. There is a subtle difference, however. In 1948, R. D. Hotchkiss discovered a "fifth base" in DNA, **5-methylcytosine**. In vertebrates, this base is made enzymatically after DNA is replicated. At this time, about 5% of the cytosines in mammalian DNA are converted to 5-methylcytosine (**FIGURE 3.18A**). This conversion can occur only when the cytosine residue is followed by a guanosine; in other words, it can only occur *at a CpG sequence* (as we will soon see, this restriction is important). Numerous studies have shown that the degree to which the cytosines of a gene are methylated can control the level of the gene's transcription. Cytosine methylation appears to be a major mechanism of transcriptional regulation in many phyla, but the amount of DNA methylation greatly varies among species. For instance, the plant *Arabidopsis*

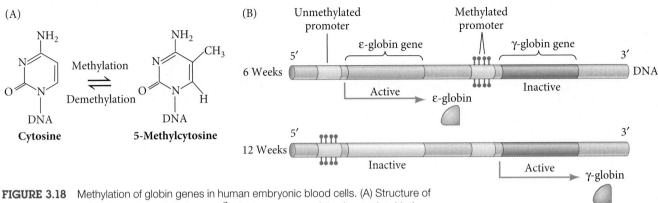

FIGURE 3.18 Methylation of globin genes in human embryonic blood cells. (A) Structure of 5-methylcytosine. (B) The activity of the human β-globin genes correlates inversely with the methylation of their promoters. (After Mavilio et al. 1983.)

thaliana has among the highest percentages of methylated cytosines at 14%, the mouse at 7.6%, and the bacterium *E. coli* at 2.3% (Capuano et al. 2014). Interestingly, for years researchers thought that the model organisms *Drosophila* and *C. elegans* did not have methylated cytosines, yet recent studies using more sensitive methods[12] have detected low levels of DNA methylation at cytosines (0.034% in *Drosophila* and 0.0019–0.0033% in *C. elegans*; Capuano et al. 2014; Hu et al. 2015). Currently, using these same high-resolution methods, no cytosine methylation has been found in yeast. Why such varied amounts of DNA methylation exist among species remains an open question.

In vertebrates, the presence of methylated cytosines in a gene's promoter correlates with the repression of transcription from that gene. In developing human and chick red blood cells, for example, the DNA of the globin gene promoters is almost completely unmethylated, whereas the same promoters are highly methylated in cells that do not produce globins. Moreover, the methylation pattern changes during development (**FIGURE 3.18B**). The cells that produce hemoglobin in the human embryo have unmethylated promoters in the genes encoding the ε-globins ("embryonic globin chains") of embryonic hemoglobin. These promoters become methylated in the fetal tissue as the genes for fetal-specific γ-globin (rather than the embryonic chains) become activated (van der Ploeg and Flavell 1980; Groudine and Weintraub 1981; Mavilio et al. 1983). Similarly, when fetal globin gives way to adult (β) globin, promoters of the fetal (γ) globin genes become methylated.

MECHANISMS BY WHICH DNA METHYLATION BLOCKS TRANSCRIPTION DNA methylation appears to act in two ways to repress gene expression. First, it can block the binding of transcription factors to enhancers. Several transcription factors can bind to a particular sequence of unmethylated DNA, but they cannot bind to that DNA if one of its cytosines is methylated (**FIGURE 3.19**). Second, a methylated cytosine can recruit the binding of proteins that facilitate the methylation or deacetylation of histones, thereby stabilizing the nucleosomes. For instance, methylated cytosines in DNA

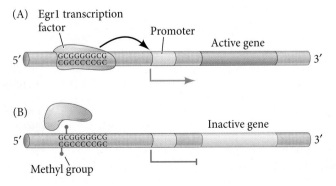

Methyl group

[12]The method used in Capuano et al. (2014) was liquid chromatography-tandem mass spectrometry (LC–MS/MS), which enabled specific detection of DNA-derived 5-methylcytosine as opposed to potential detection of methylated RNA.

FIGURE 3.19 DNA methylation can block transcription by preventing transcription factors from binding to the enhancer region. (A) The Egr1 transcription factor can bind to specific DNA sequences such as 5'…GCGGGGGCG…3', helping activate transcription of those genes. (B) If the first cytosine residue is methylated, however, Egr1 will not bind, and the gene will remain repressed. (After Weaver et al. 2005.)

can bind particular proteins such as MeCP2.[13] Once connected to a methylated cytosine, MeCP2 binds to histone deacetylases and histone methyltransferases, which, respectively, remove acetyl groups (**FIGURE 3.20A**) and add methyl groups (**FIGURE 3.20B**) on the histones. As a result, the nucleosomes form tight complexes with the DNA and do not allow other transcription factors and RNA polymerases to find the genes. Other proteins, such as HP1 and histone H1, will bind and aggregate methylated histones (Fuks 2005; Rupp and Becker 2005). In this way, repressed chromatin becomes associated with regions where there are methylated cytosines.

INHERITANCE OF DNA METHYLATION PATTERNS   Another enzyme recruited to the chromatin by MeCP2 is DNA methyltransferase-3 (Dnmt3). This enzyme methylates previously unmethylated cytosines on the DNA. In this way, a relatively large region can be repressed. The newly established methylation pattern is then transmitted to the next generation by DNA methyltransferase-1 (Dnmt1). This enzyme recognizes methyl cytosines on one strand of DNA and places methyl groups on the newly synthesized strand opposite it (**FIGURE 3.21**; see Bird 2002; Burdge et al. 2007). That is why it is necessary for the C to be next to a G in the sequence. Thus, in each cell division, the pattern of DNA methylation can be maintained. The newly synthesized (unmethylated) strand will become properly methylated when Dnmt1 binds to a methyl C on the old CpG sequence and methylates the cytosine of the CpG sequence on the complementary strand. In this way, once the DNA methylation pattern is established in a cell, it can be stably inherited by all the progeny of that cell.

GENOMIC IMPRINTING AND DNA METYHLATION   DNA methylation has explained at least one very puzzling phenomenon, that of genomic imprinting (Ferguson-Smith 2011). It is usually assumed that the genes one inherits from one's father and the genes one inherits from one's mother are equivalent. In fact, the basis for Mendelian ratios (and the Punnett square analyses used to teach them) is that it does not matter whether the genes came from the sperm or from the egg. In mammals, however, there are about 100 genes for which it *does* matter (International Human Epigenome Consortium).[14] In these cases, the chromosomes from the male and the female are not equivalent; only the sperm-derived or only the egg-derived allele of the gene is expressed. Thus, a severe or lethal condition may arise if a mutant allele is derived from one parent, but that same mutant allele will have no deleterious effects if inherited from the other parent. In some of these cases, the nonfunctioning gene has been rendered inactive by DNA methylation. (This means that a mammal must have both a male parent and a female parent. Unlike sea urchins, flies, and even some turkeys, mammals cannot experience parthenogenesis, or "virgin birth.") The methyl groups are placed on the DNA during spermatogenesis and oogenesis by a series of enzymes that first take the existing methyl groups off the chromatin and then place new sex-specific ones on the DNA (Ciccone et al. 2009; Gu et al. 2011).

---

[13] Loss of MeCP2 in humans is the leading cause of an X-linked syndrome resulting in encephalopathy (brain disorder) and early death in males, but Rett syndrome (a neurological disorder that displays symptoms within the autism spectrum disorder) in females. The mechanism by which MeCP2 is linked to these pathological conditions is not yet known, but some studies suggest that it acts through a signaling pathway (mTOR) to affect synaptic plasticity (Pohodich and Zoghbi 2015; Tsujimura et al. 2015).

[14] A list of imprinted mouse genes is maintained at www.mousebook.org/all-chromosomes-imprinting-chromosome-map

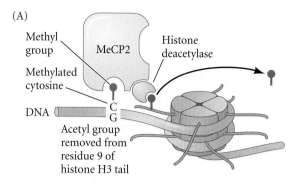

(A)

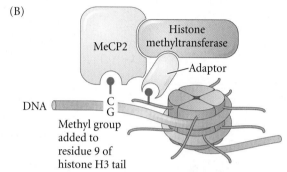

(B)

FIGURE 3.20   Modifying nucleosomes through methylated DNA. MeCP2 recognizes the methylated cytosines of DNA. It binds to the DNA and is thereby able to recruit (A) histone deacetylases (which take acetyl groups off the histones) or (B) histone methyltransferases (which add methyl groups to the histones). Both modifications promote the stability of the nucleosome and the tight packing of DNA, thereby repressing gene expression in these regions of DNA methylation. (After Fuks 2005.)

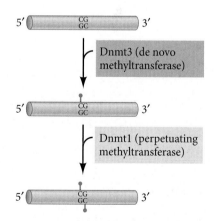

FIGURE 3.21   Two DNA methyltransferases are critically important in modifying DNA. The "de novo" methyltransferase Dnmt3 can place a methyl group on unmethylated cytosines. The "perpetuating" methyltransferase, Dnmt1, recognizes methylated Cs on one strand and methylates the C on the CG pair on the opposite strand.

As described in this chapter, methylated DNA is associated with stable DNA silencing either (1) by interfering with the binding of gene-activating transcription factors or (2) by recruiting repressor proteins that stabilize nucleosomes in a restrictive manner along the gene. The presence of a methyl group in the minor groove of DNA can prevent certain transcription factors from binding to the DNA, thereby preventing the gene from being activated (Watt and Molloy 1988).

For example, during early embryonic development in mice, the *Igf2* gene (for insulin-like growth factor) is transcribed only from the sperm-derived (paternal) chromosome 7. The egg-derived (maternal) *Igf2* gene does not function during embryonic development because the CTCF protein is an inhibitor that can block the promoter from getting activation signals from enhancers. The CTCF protein binds to a region near the *Igf2* gene in females because this region is not methylated. Once bound, it prevents the maternally derived *Igf2* gene from functioning. In the paternally derived chromosome 7, the region where CTCF would bind is methylated. CTCF cannot bind, and the gene is not inhibited from functioning (**FIGURE 3.22**; Bartolomei et al. 1993; Ferguson-Smith et al. 1993; Bell and Felsenfeld 2000).

In humans, misregulation of *IGF2* methylation causes Beckwith-Wiedemann growth syndrome. Although DNA methylation is the mechanism for imprinting this gene in both mice and humans, the mechanisms responsible for the differential *Igf2* methylation between sperm and egg appear to be very different in the two species (Ferguson-Smith et al. 2003; Walter and Paulsen 2003). Differential methylation is one of the most important mechanisms of epigenetic changes and is a reminder that an organism cannot be explained solely by its genes. One needs knowledge of developmental parameters (such as whether the gene was modified by the gamete transmitting it) as well as genetic ones.

WEB TOPIC 3.4 **POISED CHROMATIN** Learn more about the poised state of chromatin, which uses high CpG-content promoters for rapid transcriptional responses to developmental signals.

WEB TOPIC 3.5 **CHROMATIN DIMINUTION** The inactivation or elimination of entire chromosomes is not uncommon among invertebrates and is sometimes used as a mechanism of sex determination. In some organisms, portions of the chromosomes condense and break off such that only the germ cells have the full chromatin complement.

WEB TOPIC 3.6 **THE NUCLEAR ENVELOPE'S ROLE IN GENE REGULATION** There is evidence that many genes are regulated by enzymes that are localized to the nuclear envelope. The inner portion of the nuclear envelope (the nuclear lamina) may be critical in activating and silencing transcription.

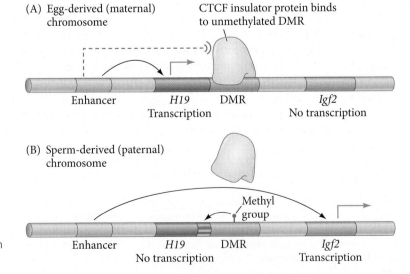

FIGURE 3.22 Regulation of the imprinted *Igf2* gene in the mouse. This gene is activated by an enhancer element it shares with the *H19* gene. The differentially methylated region (DMR) is a sequence located between the enhancer and the *Igf2* gene and is found on both sperm- and egg-derived chromosomes. (A) In the egg-derived chromosome, the DMR is unmethylated. The CTCF insulator protein binds to the DMR and blocks the enhancer signal. (B) In the sperm-derived chromosome, the DMR is methylated. The CTCF insulator protein cannot bind to the methylated sequence, and the signal from the enhancer is able to activate *Igf2* transcription.

Differential RNA Processing

The regulation of gene expression is not confined to the differential transcription of DNA. Even if a particular RNA transcript is synthesized, there is no guarantee that it will create a functional protein in the cell. To become an active protein, the nuclear RNA must be (1) processed into messenger RNA by the removal of introns, (2) translocated from the nucleus to the cytoplasm, and (3) translated by the protein-synthesizing apparatus. In some cases, even the newly synthesized protein is not in its mature form and must be (4) posttranslationally modified to become active. Regulation during development can occur at any of these steps.

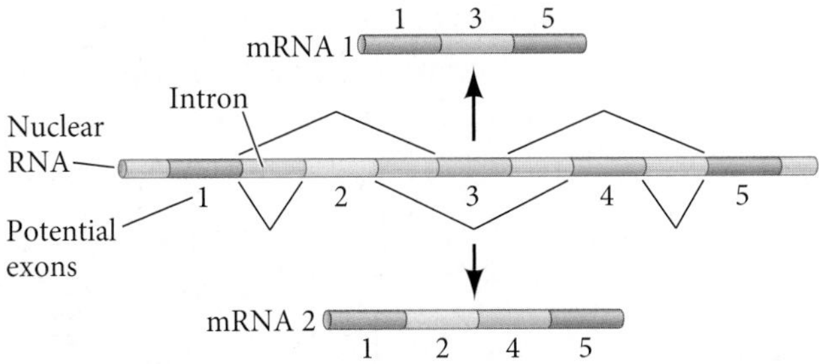

FIGURE 3.23 Differential RNA processing. By convention, splicing paths are shown by fine V-shaped lines. Differential splicing can process the same nuclear RNA into different mRNAs by selectively using different exons.

In bacteria, differential gene expression can be effected at the levels of transcription, translation, and protein modification. In eukaryotes, however, another possible level of regulation exists: control at the level of RNA processing and transport. **Differential RNA processing** is the *splicing* of mRNA precursors into messages that specify different proteins by using different combinations of potential exons. If an mRNA precursor had five potential exons, one cell type might use exons 1, 2, 4, and 5; a different cell type might use exons 1, 2, and 3; and yet another cell type might use all five (**FIGURE 3.23**). Thus, a single gene can produce an entire family of proteins. The different proteins encoded by the same gene are called **splicing isoforms** of the protein.

Creating families of proteins through differential nRNA splicing

Alternative nRNA splicing is a means of producing a wide variety of proteins from the same gene, and most vertebrate genes make nRNAs that are alternatively spliced[15] (Wang et al. 2008; Nilsen and Graveley 2010). The average vertebrate nRNA consists of several relatively short exons (averaging about 140 bases) separated by introns that are usually much longer. Most mammalian nRNAs contain numerous exons. By splicing together different sets of exons, different cells can make different types of mRNAs, and hence, different proteins. *Recognizing* a sequence of nRNA as either an exon or an intron is a crucial step in gene regulation.

Alternative nRNA splicing is based on the determination of which sequences will be spliced out as introns, which can occur in several ways. Most genes contain **consensus sequences** at the 5′ and 3′ ends of the introns. These sequences are the "splice sites" of the intron. The splicing of nRNA is mediated through complexes known as **spliceosomes** that bind to the splice sites. Spliceosomes are made up of small nuclear RNAs (snRNAs) and proteins called **splicing factors** that bind to splice sites or to the areas adjacent to them. By their production of specific splicing factors, cells can differ in their ability to recognize a sequence as an intron. That is to say, a sequence that is an *exon* in one cell type may be an *intron* in another (**FIGURE 3.24A,B**). In other instances, the factors in one cell might recognize different 5′ sites (at the beginning of the intron) or different 3′ sites (at the end of the intron; **FIGURE 3.24C,D**).

The 5′ splice site is normally recognized by *small nuclear RNA* U1 (U1 snRNA) and splicing factor 2 (SF2; also known as alternative splicing factor). The choice of alternative 3′ splice sites is often controlled by which splice site can best bind a protein called U2AF. The spliceosome forms when the proteins that accumulate at the 5′ splice site contact those proteins bound to the 3′ splice site. Once the 5′ and 3′ ends are brought together, the intervening intron is excised, and the two exons are ligated together.

In some instances, alternatively spliced RNAs yield proteins that play similar yet distinguishable roles in the same cell. Different isoforms of the WT1 protein perform

---

[15] Mutations can generate species-specific splicing events, and tissue-specific differences in nRNA splicing among vertebrate species occur 10 to 100 times more frequently than changes in gene transcription (Barbosa-Morais et al. 2012; Merkin et al. 2012).

(A) Cassette exon: Type II procollagen

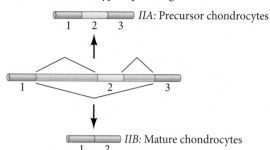

(B) Mutually exclusive exons: FgfR2

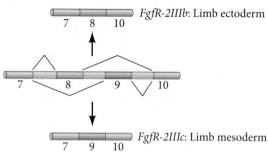

(C) Alternative 5′ splice site: *Bcl-x*

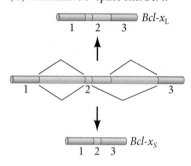

(D) Alternative 3′ splice site: Chordin

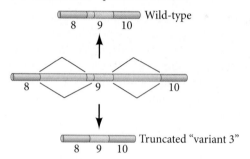

FIGURE 3.24 Some examples of alternative RNA splicing. Blue and colored portions of the bars represent exons; gray represents introns. Alternative splicing patterns are shown with V-shaped lines. (A) A "cassette" (yellow) that can be used as an exon or removed as an intron distinguishes the type II collagen types of chondrocyte precursors and mature chondrocytes (cartilage cells). (B) Mutually exclusive exons distinguish fibroblast growth factor receptors found in the limb ectoderm from those found in the limb mesoderm. (C) Alternative 5′ splice site selection, such as that used to create the large and small isoforms of the protein Bcl-X. (D) Alternative 3′ splice sites are used to form the normal and truncated forms of Chordin. (After McAlinden et al. 2004.)

different functions in the development of the gonads and kidneys. The isoform without the extra exon functions as a transcription factor during kidney development, whereas the isoform containing the extra exon appears to be critical in testis development (Hammes et al. 2001; Hastie 2001).

The *Bcl-x* gene provides a good example of how alternative nRNA splicing can make a huge difference in a protein's function. If a particular DNA sequence is used as an exon, the "large Bcl-X protein," or $Bcl-X_L$, is made (see Figure 3.24C). This protein inhibits programmed cell death. If this sequence is seen as an intron, however, the "small Bcl-X protein" ($Bcl-X_S$) is made, and this protein *induces* cell death. Many tumors have a higher than normal amount of $Bcl-X_L$.

If you get the impression from this discussion that a gene with dozens of introns could create literally thousands of different, related proteins through differential splicing, you are probably correct. The current champion at making multiple proteins from the same gene is the *Drosophila Dscam*[16] gene. This gene encodes a membrane adhesion protein that prevents dendrites from the same neuron from interacting (Wu et al. 2012). *Dscam* contains 115 exons. Moreover, a dozen different adjacent DNA sequences can be selected to be exon 4, and more than 30 mutually exclusive adjacent DNA sequences can become exons 6 and 9, respectively (**FIGURE 3.25A**; Schmucker et al. 2000). If all possible combinations of exons are used, this one gene can produce 38,016 different proteins, and random searches for these combinations indicate that a large fraction of them are, in fact, made. The nRNA of *Dscam* has been found to be alternatively spliced in different neurons, and when two dendrites from the same *Dscam* expressing neuron touch each other, they are repelled (Wu et al. 2012; **FIGURE 3.25B**). This repulsion promotes the extensive branching of the dendrites and ensures that axon-dendrite synapses occur appropriately between neurons. It appears that the thousands of splicing

---

[16]*DSCAM* (*Down syndrome cell adhesion molecule*) is a gene found within the "Down syndrome" region of chromosome 21. It encodes a cell adhesion molecule that functions through homophilic binding important for axon guidance.

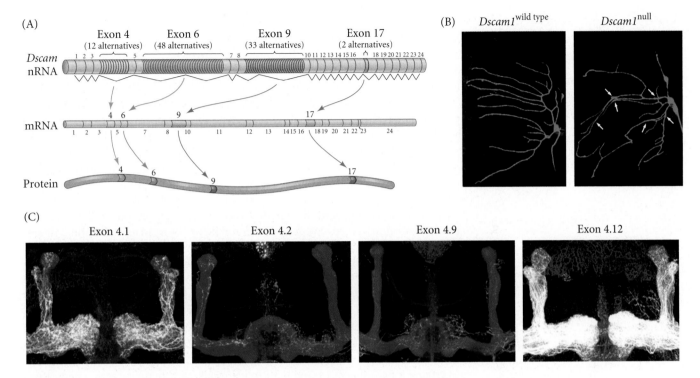

FIGURE 3.25 The *Dscam* gene of *Drosophila* can produce 38,016 different types of proteins by alternative nRNA splicing. (A) The gene contains 24 exons. Exons 4, 6, 9, and 17 are encoded by sets of mutually exclusive possible sequences. Each messenger RNA will contain one of the 12 possible exon 4 sequences, one of the 48 possible exon 6 alternatives, one of the 33 possible exon 9 alternatives, and one of the 2 possible exon 17 sequences. The *Drosophila Dscam* gene is homologous to a DNA sequence on human chromosome 21 that is expressed in the nervous system. Disturbances of this gene in humans may contribute to the neurological defects of Down syndrome. (B) *Dscam* is required for self-avoidance between dendrites that fosters a dispersed pattern of dendrites (left). Loss of *Dscam* in *Drosophila*, however, causes crossing and fasciculated growth of dendrites from the same neuron (right; arrows). (C) Expression of alternatively spliced forms of *Dscam* (4.1, 4.2, 4.9, 4.12) in isolated populations of mushroom body neurons (white) in midpupal brains of the fly. The full mushroom body lobes and associated Kenyon cells are seen with antibodies to anti-Fasciclin II and anti-Dachshund, respectively (blue). (A after Yamakawa et al. 1998, Saito et al. 2000; B from Wu et al. 2012; C from Miura et al. 2013.)

isoforms are needed to ensure that each neuron acquires a unique identity (**FIGURE 3.25C**; Schmucker 2007; Millard and Zipursky 2008; Miura et al. 2013). Moreover, the combination of expressed Dscam1 isoforms can change in a given neuron with each new round of RNA synthesis! Such timely changes in alternative splicing may be in response to neuron-neuron interactions during the process of dendritic arborization. The *Drosophila* genome is thought to contain only 14,000 genes, but here is a single gene that encodes three times that number of proteins!

WEB TOPIC 3.7   **CONTROL OF EARLY DEVELOPMENT BY NUCLEAR RNA SELECTION**
In addition to alternative nRNA splicing, the nuclear RNA to mRNA stage can also be regulated by RNA "censorship"—selecting which nuclear transcripts are processed into cytoplasmic messages. Different cells select different nuclear transcripts to be processed and sent to the cytoplasm as messenger RNA.

WEB TOPIC 3.8   **SO YOU THINK YOU KNOW WHAT A GENE IS?**   Different scientists have different definitions, and nature has given us some problematic examples of DNA sequences that may or may not be considered genes.

Developing Questions

About 92% of human genes are thought to produce multiple types of mRNA. Therefore, even though the human genome may contain 20,000 genes, its proteome— the number and type of proteins encoded by the genome—is far larger and more complex. "Human genes are multitaskers," notes Christopher Burge, one of the scientists who calculated this figure (Ledford 2008). This fact explains an important paradox. Homo sapiens has around 20,000 genes in each nucleus; so does the nematode *C. elegans*, a tubular creature with only 959 cells. We have more cells and cell types in the shaft of a hair than *C. elegans* has in its entire body. What's this worm doing with approximately the same number of genes as we have?

Splicing enhancers and recognition factors

The mechanisms of differential RNA processing involve both *cis*-acting sequences on the nRNA and *trans*-acting protein factors that bind to these regions (Black 2003). The *cis*-acting sequences on nRNA are usually close to their potential 5′ or 3′ splice sites. These sequences are called **splicing enhancers** because they promote the assembly of spliceosomes at RNA cleavage sites. (Conversely, these same sequences can be "splicing silencers" if they act to exclude exons from an mRNA sequence.) These sequences are recognized by *trans*-acting proteins, most of which can recruit spliceosomes to that area. Some *trans*-acting proteins, however, like the polypyrimidine tract-binding proteins (PTPs), repress spliceosome formation where they bind. Indeed, different PTPs can control the splicing of batteries of nRNAs. For example, PTPb prevents the adult neuron-specific splicing of the neural nRNAs controlling cell fate, cell proliferation, and actin cytoskeleton, thereby keeping the neuronal precursors in a proliferating, immature state (Licatalosi et al. 2012).

The selection of particular exons is determined not only by the spliceosome-binding consensus sequences but also by numerous sequence elements that are recognized by regulatory factors that can regulate spliceosome binding (Ke and Chasin 2011). The splicing enhancers on the RNA sequence regulate whether a spliceosome can form on a particular splicing consensus sequence. As might be expected, some splicing enhancers appear to be specific for certain tissues. Muscle-specific splicing enhancers have been found around those exons characterizing muscle cell messages. They are recognized by certain proteins that are found in the muscle cells early in their development (Ryan and Cooper 1996; Charlet-B et al. 2002). Their presence is able to compete with the PTP that would otherwise prevent the inclusion of the muscle-specific exon into the mature message. In this way, an entire battery of muscle-specific isoforms can be generated. The context dependency of splicing is too complex to delineate by merely comparing sequences, however. Computational studies—in which the computer is asked to identify (1) the combination of sequence elements, (2) the proximity of these sequences to the splice junctions, and (3) the differences of splicing outcomes in different cell types—are providing our first look at a "splicing code" that may allow us to predict which exons will persist in one cell and not in others (Barash et al. 2010).

Mutations in the splicing sites can lead to alternative developmental phenotypes. Most splice site mutations lead to nonfunctional proteins and serious diseases. For instance, a single base change at the 5′ end of intron 2 in the human β-globin gene prevents splicing from occurring and generates a nonfunctional mRNA (Baird et al. 1981). That causes the absence of any β-globin from this gene and thus a severe (and often life-threatening) type of anemia. Similarly, a mutation in the *Dystrophin* gene at a particular splice site causes the skipping of that exon and a severe form of muscular dystrophy (Sironi et al. 2001). In at least one such case of aberrant splicing, the splice site mutation was not dangerous and actually gave the patient greater strength. In this case, Schuelke and colleagues (2004) described a family in which individuals in four generations had a splice site mutation in the *myostatin* gene. Among the family members were professional athletes and a 4-year-old toddler who was able to hold two 3-kg dumbbells with his arms fully extended. The product of the normal *myostatin* gene is a factor that tells muscle precursor cells to stop dividing; that is, it is a negative regulator. In mammals (including humans and mice) with the mutation, the factor is nonfunctional, and the muscle precursors are not told to differentiate until they have undergone many more rounds of cell division; the result is larger muscles (**FIGURE 3.26**).

Control of Gene Expression at the Level of Translation

The splicing of nuclear RNA is intimately connected with its export through the nuclear pores and into the cytoplasm. As the introns are removed, specific proteins bind to the spliceosome and attach the spliceosome-RNA complex to nuclear pores (Luo et al.

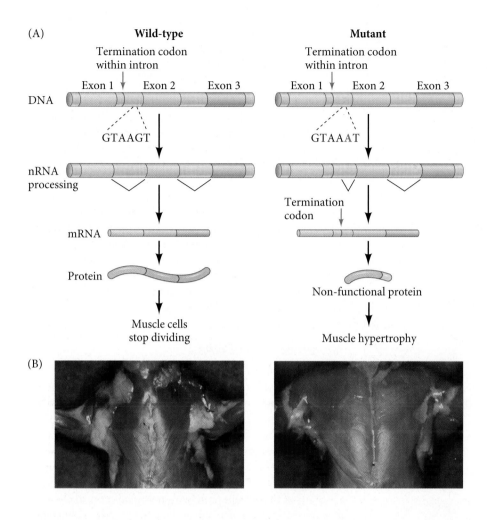

(A)

Wild-type

Termination codon
within intron

Exon 1   Exon 2   Exon 3

DNA

GTAAGT

nRNA
processing

mRNA

Protein

Muscle cells
stop dividing

Mutant

Termination codon
within intron

Exon 1   Exon 2   Exon 3

GTAAAT

Termination
codon

Non-functional protein

Muscle hypertrophy

(B)

FIGURE 3.26 Muscle hypertrophy through mispliced RNA. This mutation results in a deficiency of the negative growth regulator myostatin in the muscle cells. (A) Molecular analysis of the mutation. There is no mutation in the coding sequence of the gene, but in the first intron, a mutation from a G to an A creates a new (and widely used) splicing site, which causes aberrant nRNA splicing and the inclusion of an early protein synthesis termination codon into the mRNA. Thus, proteins made from that message are short and nonfunctional. (B) Pectoral musculature of a "mighty mouse" with the mutation (right) compared with the muscles of a wild-type mouse (left). (A after Schuelke et al. 2004; B from McPherron et al. 1997, courtesy of A. C. McPherron.)

2001; Strässer and Hurt 2001). The proteins coating the 5′ and 3′ ends of the RNA also change. The nuclear cap binding protein at the 5′ end is replaced by *eukaryotic translation initiation factor eIF4E,* and the polyA tail becomes bound by the cytoplasmic polyA binding protein. Although both of these changes facilitate the initiation of translation, there is no guarantee that the RNA will be translated once it reaches the cytoplasm. The control of gene expression at the level of translation can occur by many means; some of the most important of them are described below.

Differential mRNA longevity

The longer an mRNA persists, the more protein can be translated from it. If a message with a relatively short half-life were selectively stabilized in certain cells at certain times, it would make large amounts of its particular protein only at those times and places.

    The stability of a message often depends on the length of its polyA tail. The length, in turn, depends largely on sequences in the 3′ untranslated region, certain ones of which allow longer polyA tails than others. If these 3′ UTRs are experimentally traded, the half-lives of the resulting mRNAs are altered: long-lived messages will decay rapidly, whereas normally short-lived mRNAs will remain around longer (Shaw and Kamen 1986; Wilson and Treisman 1988; Decker and Parker 1995).

    In some instances, mRNAs are selectively stabilized at specific times in specific cells. The mRNA for casein, the major protein of milk, has a half-life of 1.1 hours in rat mammary gland tissue. During periods of lactation, however, the presence of the hormone prolactin increases this half-life to 28.5 hours (**FIGURE 3.27**; Guyette et al. 1979). In the development of the nervous system, a set of RNA binding proteins called **Hu proteins** (HuA, HuB, HuC, and HuD) stabilizes two groups of mRNAs that would otherwise

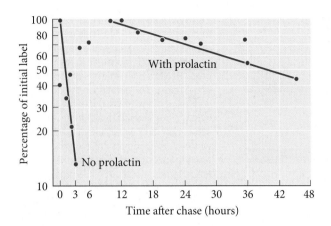

FIGURE 3.27 Degradation of casein mRNA in the presence and absence of prolactin. Cultured rat mammary cells were given radioactive RNA precursors (pulse) and, after a given time, were washed and given nonradioactive precursors (chase). This procedure labeled the casein mRNA synthesized during the pulse time. Casein mRNA was then isolated at different times following the chase and its radioactive label measured. In the absence of prolactin, the labeled (i.e., newly synthesized) casein mRNA decayed rapidly, with a half-life of 1.1 hours. When the same experiment was done in a medium containing prolactin, the half-life was extended to 28.5 hours. (After Guyette et al. 1979.)

perish quickly (Perrone-Bizzozero and Bird 2013). One group of target mRNAs encodes proteins that stop neuronal precursor cells from dividing, and the second group of mRNAs encodes proteins that initiate neuronal differentiation (Okano and Darnell 1997; Deschênes-Furry et al. 2006, 2007). Thus, once the Hu proteins are made, the neuronal precursor cells can become neurons.[17]

Stored oocyte mRNAs: Selective inhibition of mRNA translation

Some of the most remarkable cases of translational regulation of gene expression occur in the oocyte. Prior to meiosis, the oocyte often makes and stores mRNAs that will be used only after fertilization occurs. These messages stay in a dormant state until they are activated by ion signals (discussed in Chapters 6 and 7) that spread through the egg during ovulation or fertilization.

Some of these stored mRNAs encode proteins that will be needed during cleavage, when the embryo makes enormous amounts of chromatin, cell membranes, and cytoskeletal components. These maternal mRNAs include the messages for histone proteins, the transcripts for the actin and tubulin proteins of the cytoskeleton, and the mRNAs for the cyclin proteins that regulate the timing of early cell division (Raff et al. 1972; Rosenthal et al. 1980; Standart et al. 1986). The stored mRNAs and proteins are referred to as **maternal contributions** (produced from the maternal genome), and in many species (including sea urchins, *Drosophila*, and zebrafish), maintenance of the normal rate and pattern of early cell divisions does not require DNA or even a nucleus! Rather, it requires continued protein synthesis from the stored maternally contributed mRNAs (**FIGURE 3.28**; Wagenaar and Mazia 1978; Edgar et al. 1994; Dekens et al. 2003). Stored mRNA also encodes proteins that determine the fates of cells. They include the *bicoid*, *caudal*, and *nanos* messages that provide information in the *Drosophila* embryo for the production of its head, thorax, and abdomen. So, at some point, each of us should give a shout-out to our moms for giving us those transcripts early on.

(A) Wild-type   (B) *Futile cycle* mutant

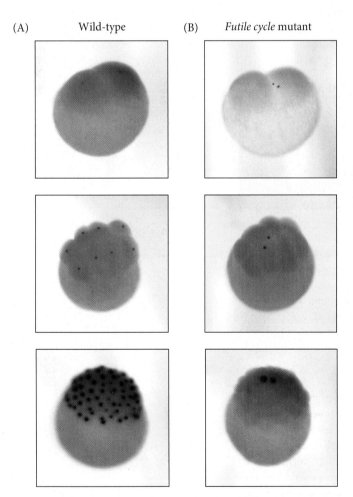

FIGURE 3.28 Maternal contributions to DNA replication in the zebrafish blastula. (A) Wild-type blastulae show BrdU-labeled nuclei (blue) in all cells. (B) Although the correct number of cells is present in *futile cycle* mutants, they consistently show only two labeled nuclei, indicating that these mutants fail to undergo pronuclear fusion. Even in the absence of any zygotic DNA, early cleavages progress perfectly well due to the presence of maternal contributions. However, *futile cycle* mutants arrest at the onset of gastrulation. (From Dekens et al. 2003.)

---

[17]Interestingly, several alternatively spliced isoforms have been discovered for mouse HuD that show differential expression, different subcellular positions (posttranslational regulatory mechanism), and different functional consequences for neuronal survival and differentiation (Hayashi et al. 2015).

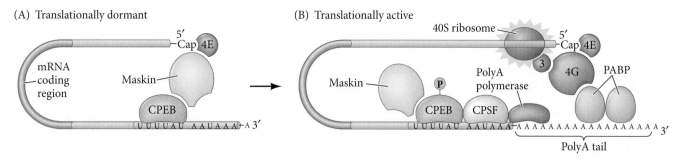

FIGURE 3.29 Translational regulation in oocytes. (A) In *Xenopus* oocytes, the 3′ and 5′ ends of the mRNA are brought together by maskin, a protein that binds CPEB on the 3′ end and eukaryotic initiation factor 4E (eIF4E) on the 5′ end. Maskin blocks the initiation of translation by preventing eIF4E from binding eIF4G. (B) When stimulated by progesterone during ovulation, a kinase phosphorylates CPEB, which can then bind CPSF. CPSF can bind polyA polymerase and initiate growth of the polyA tail. PolyA binding protein (PABP) can bind to this tail and then bind eIF4G in a stable manner. This initiation factor can then bind eIF4E and, through its association with eIF3, position a 40S ribosomal subunit on the mRNA. (After Mendez and Richter 2001.)

Most translational regulation in oocytes is negative because the "default state" of the maternal mRNA is to be available for translation. Therefore, there must be inhibitors preventing the translation of these mRNAs in the oocyte, and these inhibitors must somehow be removed at the appropriate times around fertilization. The 5′ cap and the 3′ UTR seem especially important in regulating the accessibility of mRNA to ribosomes. If the 5′ cap is not made or if the 3′ UTR lacks a polyA tail, the message probably will not be translated. The oocytes of many species have "used these ends as means" to regulate the translation of their mRNAs. For instance, the oocyte of the tobacco hornworm moth makes some of its mRNAs without their methylated 5′ caps. In this state, they cannot be efficiently translated. At fertilization, however, a methyltransferase completes the formation of the caps, and these mRNAs can then be translated (Kastern et al. 1982).

In amphibian oocytes, the 5′ and 3′ ends of many mRNAs are brought together to form repressive loop structures by a protein called **maskin** (Stebbins-Boaz et al. 1999; Mendez and Richter 2001). Maskin links the 5′ and 3′ ends into a circle by binding to two other proteins, each at opposite ends of the message. First, it binds to the **cytoplasmic polyadenylation-element-binding protein (CPEB)** attached to the UUUUAU sequence in the 3′ UTR; second, maskin also binds to the eIF4E factor that is attached to the cap sequence. In this configuration, the mRNA cannot be translated (**FIGURE 3.29A**). The binding of eIF4E to maskin is thought to prevent the binding of eIF4E to eIF4G, a critically important translation initiation factor that brings the small ribosomal subunit to the mRNA.

Mendez and Richter (2001) proposed an intricate scenario to explain how mRNAs bound together by maskin become translated at about the time of fertilization. At ovulation (when the hormone progesterone stimulates the last meiotic divisions of the oocyte and the oocyte is released for fertilization), a kinase activated by progesterone phosphorylates the CPEB protein. The phosphorylated CPEB can now bind to the cleavage and polyadenylation specificity factor, CPSF (Mendez et al. 2000; Hodgman et al. 2001). The bound CPSF protein sits on the 3′ UTR and complexes with a polymerase that elongates the polyA tail of the mRNA. The important aspect of this model is that the length of the polyA tail is what is being manipulated to control translation. In oocytes, a message having a short polyA tail is not degraded, yet such messages are also not translated. Once the tail is extended, however, molecules of the polyA binding protein (PABP) can attach to the growing tail. PABP stabilizes the eIF4G to eIF4E interaction (outcompeting maskin) to facilitate ribosomal assembly around the mRNA and initiate translation (**FIGURE 3.29B**).

In the *Drosophila* oocyte, Bicoid protein initiates head and thorax formation. Bicoid can act both as a transcription factor (activating genes such as *hunchback* that are necessary for forming the fly anterior) and as a translational inhibitor of those genes such as *caudal* that are critical for making the fly posterior (see Chapters 2 and 9). Bicoid inhibits *caudal* mRNA translation by binding to a "bicoid recognition element," a series of nucleotides in the 3′ UTR of the *caudal* message. Once there, Bicoid can bind with and recruit another protein, d4EHP, which can compete with eIF4E protein for the cap. Without eIF4E, there is no association with eIF4G, and *caudal* mRNA becomes untranslatable. As a result, the *caudal* message is not translated in the anterior of the embryo (where Bicoid is abundant) but is active in the posterior portion of the embryo.

FIGURE 3.30   Model of ribosomal heterogeneity in mice. (A) Ribosomes have slightly different proteins depending on the tissue in which they reside. Ribosomal protein Rpl38 (i.e., protein 38 of the large ribosomal subunit) is concentrated in those ribosomes found in the somites that give rise to the vertebrae. (B) A wild-type embryo (left) has normal vertebrae and normal Hox gene translation. Mice deficient in Rpl38 have an extra pair of vertebrae, tail deformities, and reduced Hox gene translation. (After Kondrashov et al. 2011.)

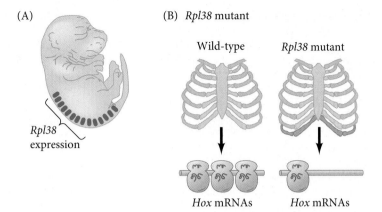

Ribosomal selectivity: Selective activation of mRNA translation

It has long been assumed that ribosomes do not show favoritism toward translating certain mRNAs. After all, eukaryotic messages can be translated even by *E. coli* ribosomes, and ribosomes from immature red blood cells have long been used to translate mRNAs for any source. However, evidence has shown that ribosomal proteins are not the same in all cells and that some ribosomal proteins are necessary for translating certain messages. When Kondrashov and colleagues (2011) mapped the gene that caused numerous axial skeleton deformities in mice, they found that the mutation was not in one of the well-known genes that control skeletal polarity. Rather, it was in ribosomal protein Rpl38. When this protein is mutated, the ribosomes can still translate most messages, but the ribosomes in the skeletal precursors cannot translate the mRNA from a specific subset of Hox genes. The Hox transcription factors, as we will see in Chapters 12 and 17, specify the type of vertebrae at each particular axial level (ribbed thoracic vertebrae, unribbed abdominal vertebrae, etc.). Without functioning Rpl38, vertebral cells are unable to form the initiation complex with mRNA from the appropriate Hox genes, and the skeleton is deformed (**FIGURE 3.30**). Mutations in other ribosomal proteins have also been found to produce deficient phenotypes (Terzian and Box 2013; Watkins-Chow et al. 2013).

microRNAs: Specific regulation of mRNA translation and transcription

If proteins can bind to specific nucleic acid sequences to block transcription or translation, you would think that RNA would do the job even better. After all, RNA can be made specifically to complement and bind a particular sequence. Indeed, one of the most efficient means of regulating the translation of a specific message is to make a small *antisense* RNA complementary to a portion of a particular transcript. Such a naturally occurring antisense RNA was first seen in *C. elegans* (Lee et al. 1993; Wightman et al. 1993). Here, the *lin-4* gene was found to encode a 21-nucleotide RNA that bound to multiple sites in the 3′ UTR of the *lin-14* mRNA (**FIGURE 3.31**). The *lin-14* gene encodes the LIN-14 transcription factor that is important during the first larval phase of *C. elegans* development. It is not needed afterward, and *C. elegans* is able to inhibit synthesis of LIN-14 from these messages by the small *lin-4* antisense RNA. The binding of these *lin-4* transcripts to the *lin-14* mRNA 3′ UTR causes degradation of the *lin-14* message (Bagga et al. 2005).

The *lin-4* RNA is now thought to be the "founding member" of a very large group of **microRNAs** (**miRNAs**). Computer analysis of the human genome predicts that we have more than 1000 miRNA loci and that these miRNAs probably modulate *50%* of the protein-encoding genes in our bodies (Berezikov and Plasterk 2005; Friedman et al. 2009). These miRNAs usually contain only 22 nucleotides and are made from longer precursors. These precursors can be in independent transcription units (the *lin-4* gene is far apart from the *lin-14* gene), or they can reside in the introns of other genes (Aravin

FIGURE 3.31 Hypothetical model of the regulation of *lin-14* mRNA translation by *lin-4* RNAs. The *lin-4* gene does not produce an mRNA. Rather, it produces small RNAs that are complementary to a repeated sequence in the 3' UTR of the *lin-14* mRNA, which bind to it and prevent its translation. (After Wickens and Takayama 1995.)

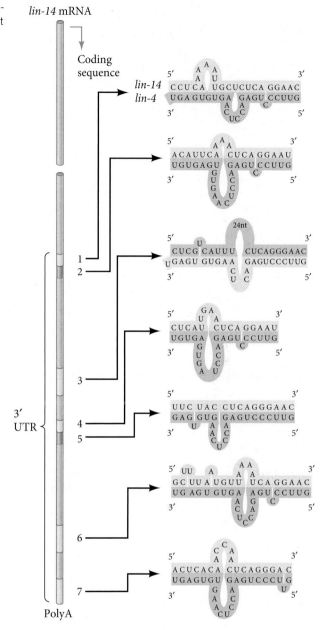

et al. 2003; Lagos-Quintana et al. 2003). The initial RNA transcript (which may contain several repeats of the miRNA sequence) forms hairpin loops wherein the RNA finds complementary structures within its strand. Because short double-stranded RNA molecules can resemble pathogenic viral genomes, the cell has a mechanism to both recognize these structures and use them as guides for their eradication (Wilson and Doudna 2013). Interestingly, this protective mechanism has been co-opted to be used as yet another way that the cell can differentially regulate the expression of endogenous genes. The process by which miRNAs inhibit expression of specific genes by degrading their mRNAs is called **RNA interference** (Guo and Kemphues 1995; Sen and Blau 2006; Wilson and Doudna 2013), the characterization of which garnered Andrew Fire and Craig Mello the Nobel Prize in Physiology or Medicine in 2006 (Fire et al. 1998).

The miRNA double-stranded stem-loop structures are processed by a set of RNases (Drosha and Dicer) to make single-stranded microRNA (**FIGURE 3.32**). The microRNA is then packaged with a series of proteins to make an **RNA-induced silencing complex** (**RISC**). Proteins of the Argonaute family are particularly important members of this complex. Such small regulatory RNAs can bind to the 3' UTR of messages and inhibit their translation. In some cases (especially when the binding of the miRNA to the 3' UTR is perfect), the RNA is cleaved. More often, however, several RISCs attach to sites on the 3' UTR and physically block the message from being translated (see Bartel 2004; He and Hannon 2004). The binding of microRNAs and their associated RISCs to the 3' UTR can regulate translation in two ways (Filipowicz et al. 2008). First, this binding can block initiation of translation, preventing the binding of initiation factors or ribosomes. The Argonaute proteins, for instance, have been found to bind directly to the methylated guanosine cap at the 5' end of the mRNA message (Djuranovic et al. 2010, 2011). Second, this binding can recruit endonucleases that digest the mRNA, usually starting with the polyA tail (Guo et al. 2010). The latter seems to be commonly used in mammalian cells.

 SCIENTISTS SPEAK 3.5 Listen to a question and answer with Dr. Ken Kempues. See the follow-up question associated with Question 4 to hear about the first demonstration of double-stranded RNA in *C. elegans*.

 SCIENTISTS SPEAK 3.6 Hear a question and answer with Dr. Craig Mello on his shared Nobel Prize–winning discovery of RNA interference.

MicroRNAs can be used to "clean up" and fine-tune the level of gene products. We mentioned those maternal RNAs in the oocyte that allow early development to occur. How does the embryo get rid of maternal RNAs once they have been used and the embryonic cells are making their own mRNAs? In zebrafish, this cleanup operation is assigned to microRNAs such as *miR430*. That is one of the first genes transcribed by the fish embryonic cells, and there are about 90 copies of this gene in the zebrafish genome.

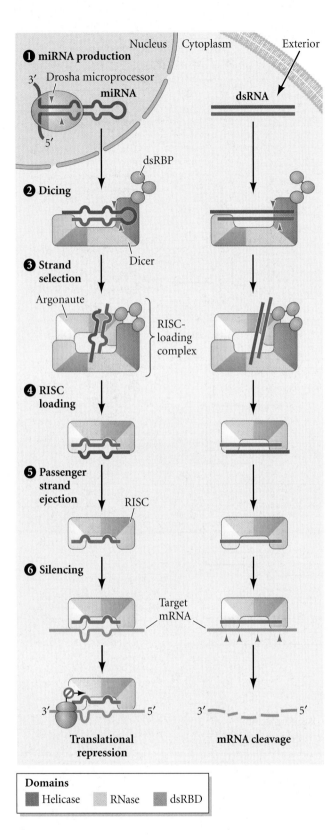

Domains

| Helicase | RNase | dsRBD |

FIGURE 3.32 Model for RNA interference from siRNA and miRNA. Double-stranded siRNA or miRNA that is added to a cell or produced through transcription and processed by the Drosha RNAase (1) will interact with the RNA-induced silencing complex (RISC) made up primarily of Dicer and Argonaute that prepares the RNA to be used as a guide for targeted mechanisms of interference. Specifically, (1) transcription of siRNA or miRNA form several hairpin regions where the RNA finds nearby complementary bases with which to pair. The pri-miRNA is processed into individual pre-miRNA "hairpins" by the Drosha RNAase (as are the siRNAs), and they are exported from the nucleus. (2–4) Once in the cytoplasm, these double-stranded RNAs are recognized by and form the RISC complex with Argonaute and the RNAase, Dicer. (5) Dicer also acts as a helicase to separate the strands of the double-stranded RNA. (6) One strand (probably recognized by placement of Dicer) will be used to bind to the 3′ UTRs of target mRNAs to block translation or to induce cleavage of the target transcript, depending (at least in part) on the strength of the complementarity between the miRNA and its target. siRNA is best known for the targeting of transcript degradation. (After He and Hannon 2004; Wilson and Doudna 2013.)

So, the level of *miR430* goes up very rapidly. This microRNA has hundreds of targets (about 40% of the maternal RNA types), and when it binds to the 3′ UTR of these target mRNAs, these mRNAs lose their polyA tails and are degraded (**FIGURE 3.33**; Giraldez et al. 2006; Giraldez 2010). In addition, *miR430* represses initiation of translation prior to promoting mRNA decay (Bazzini et al. 2012).

 SCIENTISTS SPEAK 3.7 Listen to a question and answer discussion with Dr. Antonio Giraldez on the role of *miR430* in the clearance of maternal contributions.

Although the microRNA is usually 22 bases long, it recognizes its target primarily through a "seed" region of about 5 bases in the 5′ end of the microRNA (usually at positions 2–7). This seed region recognizes targets in the 3′ UTR of the message. What happens, then, if an mRNA has a mutated 3′ UTR? Such a mutation appears to have given rise to the Texel sheep, a breed with a large and well-defined musculature that is the dominant meat-producing sheep in Europe. Genetic techniques mapped the basis of the sheep's meaty phenotype to the *myostatin* gene. We have already seen that a mutation in the *myostatin* gene that prevents the proper splicing of the nRNA can produce a large-muscled phenotype (see Figure 3.26). Another way of reducing the levels of myostatin involves a mutation in its 3′ UTR sequence. In the Texel breed, there has been a G-to-A transition in the 3′ UTR of the gene for myostatin, creating a target for the *miR1* and *miR206* microRNAs that are abundant in skeletal muscle (Clop et al. 2006). This mutation causes the depletion of *myostatin* messages and the increase in muscle mass characteristic of these sheep.

(A)

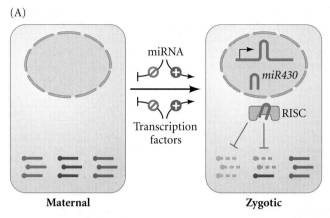

(B)

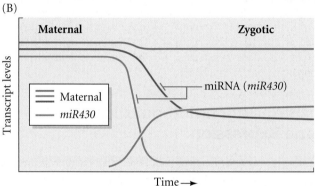

FIGURE 3.33 The role of *miR430* during the maternal-to-zygotic transition in zebrafish. (A) Numerous mRNAs derived from maternal contributions fuel development during the cleavage stages, but transitioning into the gastrula requires active transcription of the zygotic genome. miRNAs play a major role in clearing these maternally derived transcripts during this transition. (B) *miR430* has been discovered to play a major role in the interference of a majority of maternal transcripts in the zebrafish blastula as it transitions to zygotic control during gastrulation. In this graph, the different curves denote the reduction in three specific transcripts, two genes of which (purple and red) are differentially degraded by *miR430* (green). (After Giraldez 2010.)

Control of RNA expression by cytoplasmic localization

Not only is the timing of mRNA translation regulated, but so is the place of RNA expression. A majority of mRNAs (about 70% in *Drosophila* embryos) are localized to specific places in the cell (Lécuyer et al. 2007). Just like the selective repression of mRNA translation, the selective localization of messages is often accomplished through their 3′ UTRs. There are three major mechanisms for the localization of an mRNA (see Palacios 2007):

1. *Diffusion and local anchoring.* Messenger RNAs such as *nanos* diffuse freely in the cytoplasm. When they diffuse to the posterior pole of the *Drosophila* oocyte, however, they are trapped there by proteins that reside particularly in these regions. These proteins also activate the mRNA, allowing it to be translated (**FIGURE 3.34A**).

2. *Localized protection.* Messenger RNAs such as those encoding the *Drosophila* heat shock protein hsp83 (which helps protect the embryos from thermal extremes) also float freely in the cytoplasm. Like *nanos* mRNA, *hsp83* accumulates at the posterior pole, but its mechanism for getting there is different. Throughout the embryo, the mRNA is degraded. Proteins at the posterior pole, however, protect the *hsp83* mRNA from being destroyed (**FIGURE 3.34B**).

3. *Active transport along the cytoskeleton.* Active transport is probably the most widely used mechanism for mRNA localization. Here, the 3′ UTR of the mRNA is recognized by proteins that can bind these messages to "motor proteins" that travel along the cytoskeleton to their final destination (**FIGURE 3.34C**). These motor proteins are usually ATPases such as dynein or kinesin that split ATP for their motive force. We will see in Chapter 9 that this mechanism is very important for localizing transcription factor mRNAs into different regions of the *Drosophila* oocyte.

(A) Diffusion and local anchoring

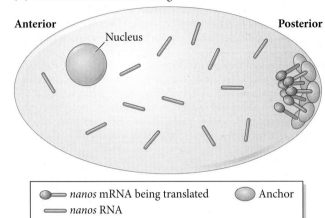

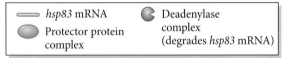

(B) Localized protection

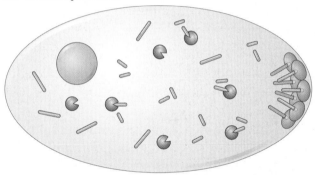

(C) Active transport along cytoskeleton

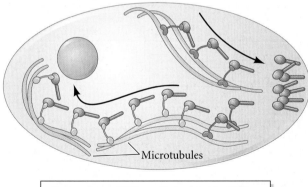

FIGURE 3.34 Localization of mRNAs. (A) Diffusion and local anchoring. *nanos* mRNA diffuses through the *Drosophila* egg and is bound (in part by the Oskar protein) at the posterior end of the oocyte. This anchoring allows the *nanos* mRNA to be translated. (B) Localized protection. The mRNA for *Drosophila* heat shock protein (hsp83) will be degraded unless it binds to a protector protein (in this case, also at the posterior end of the oocyte). (C) Active transport on the cytoskeleton, causing the accumulation of mRNA at a particular site. Here, *bicoid* mRNA is transported to the anterior of the oocyte by dynein and kinesin motor proteins. Meanwhile, *oskar* mRNA is brought to the posterior pole by transport along microtubules by kinesin ATPases. (After Palacios 2007.)

WEB TOPIC 3.9 **STORED mRNA IN BRAIN CELLS** One of the most important areas of local translational regulation may be in the brain. The storage of long-term memory requires new protein synthesis, and the local translation of mRNAs in the dendrites of brain neurons has been proposed as a control point for increasing the strength of synaptic connections.

Posttranslational Regulation of Gene Expression

The story is not over when a protein is synthesized. Once a protein is made, it becomes part of a larger level of organization. It may become part of the structural framework of the cell, for instance, or it may become involved in one of the many enzymatic pathways for the synthesis or breakdown of cellular metabolites. In any case, the individual protein is now part of a complex "ecosystem" that integrates it into a relationship with numerous other proteins. Several changes can still take place that determine whether or not the protein will be active.

Some newly synthesized proteins remain inactive until certain inhibitory sections are cleaved away. That is what happens when insulin is made from its larger protein precursor. Some proteins must be "addressed" to their specific intracellular destinations to function. Proteins are often sequestered in certain regions of the cell, such as membranes, lysosomes, nuclei, or mitochondria. Some proteins need to assemble with other proteins to form a functional unit. The hemoglobin protein, the microtubule, and the ribosome are all examples of multiple proteins joining together to form a functional unit. In addition, some proteins are not active unless they bind an ion (such as Ca^{2+}) or are modified by the covalent addition of a phosphate or acetate group. The importance of this type of protein modification will become obvious in Chapter 4 because many of the critical proteins in embryonic cells just sit there until some signal activates them. Finally, even when a protein may be actively translated and ready to function, the cell can immediately transport this protein to the proteasome for degradation. Why would a cell expend energy synthesizing a protein only to degrade it? If a cell needed a protein to function with rapid response at a precise moment in time, it might consider the energy expenditure worth it. For instance, a neuron searching for its synaptic target extends a long axonal process in search of this target in a process called axon guidance (described in Chapter 15). Pathfinding neurons synthesize certain receptor proteins only to immediately degrade them until the cell has reached an environment where a directional guidance decision is required. Signals in this location cause the cell to suspend the receptor degradation, enabling the receptors to be transported to the membrane, and immediately function to guide the axon onward toward its target.

All the processes we have discussed in this chapter—histone modification, interacting transcription factors, binding of RNA polymerase II to the promoter, elongation of the mRNAs, kinetics of RNA splicing, and half-lives of mRNAs—are stochastic events. They depend on the concentrations of the interacting proteins (Cacace et al. 2012; Murugan and Kreiman 2012; Costa et al. 2013; Neuert et al. 2013). Thus, each organism is a unique "performance" coordinated by interactions that tell the individual cells which genes are to be expressed and which are to remain silent. Chapter 4 will detail the mechanisms by which cells communicate to orchestrate this differential expression of genes.

The Basic Tools of Developmental Genetics

Characterizing gene expression

Differential gene transcription is critical in development. To know the specific time and place of gene expression, one needs to use procedures that locate a particular type of messenger RNA or protein within a cell. These techniques include northern blots, RT-PCR, in situ hybridization, microarray technology for transcripts, and western blots and immunocytochemistry for proteins. To ascertain the function of genes once they are located, scientists are using new techniques, such as CRISPR/Cas9-mediated knockouts, antisense, RNA interference, **morpholinos** (knockdowns), Cre-lox analysis (which allows the message to be made or destroyed in particular cell types), and ChIP-Seq techniques (which allow the identification of proteins bound to specific DNA sequences and active chromatin). In addition, "high-throughput" RNA analysis by microarrays, macroarrays, and RNAseq enables researchers to compare thousands of mRNAs, and computer-aided synthetic techniques can predict interactions between proteins and mRNAs. Descriptions for a majority of these procedures can be found on devbio.com. In addition, some of the techniques most relevant to today's experimental methods are described below.

IN SITU HYBRIDIZATION   In **whole mount in situ hybridization,** the entire embryo (or a part thereof) can be stained for certain mRNAs. The main principle is to take advantage of the single-stranded nature of mRNA and introduce a complementary sequence to the target mRNA that enables visualization. This technique uses dyes to allow researchers to look at entire embryos (or their organs) without sectioning them, thereby observing large regions of gene expression next to regions devoid of expression. **FIGURE 3.35A** shows an **in situ** hybridization targeting mRNA from the *odd-skipped* gene performed on a fixed, intact *Drosophila* embryo. First an mRNA detection probe—the in situ probe—had to be created. The probe is an antisense RNA molecule that can typically vary in length from 200bp to 2000bp. More important is that the uridine triphosphate (UTP) nucleosides in this RNA strand are conjugated with digoxigenin (**FIGURE 3.35B**). Digoxigenin—a compound made by particular groups of plants and not found in animal cells—does not interfere with the coding properties of the resulting mRNA, but it does make it recognizably different from any other RNA in the cell. During the procedure, the embryo is permeabilized by lipid solvents and proteinases so that the probe can get in and out of its cells. Once in the cells, *hybridization* occurs between the probe anti-sense RNA and the targeted mRNA. To visualize the cells in which hybridization has occurred, researchers apply an antibody that specifically recognizes digoxigenin. This antibody, however, has been artificially conjugated to an enzyme, such as alkaline phosphatase. After incubation in the antibody and repeated washes to remove all unbound antibodies, the embryo is bathed in a solution containing a substrate for the enzyme (traditionally NTB/BCIP for alkaline phosphatase) that

FIGURE 3.35 In situ hybridization. (A) Whole mount in situ hybridization for *odd-skipped* mRNA (blue) in a stage 9 *Drosophila* embryo. (B) Antisense RNA probe with uracil conjugated to digoxigenin (DIG). (C) Illustration of two cells at the border of the *odd-skipped* expression pattern seen in (A; box). The antisense DIG-labeled probe with complementarity to the *odd-skipped* gene becomes hybridized to any cell expressing *odd-skipped* transcripts. The cell on the left is not expressing *odd-skipped*, whereas the expression of *odd-skipped* in the cell on the right is revealed by a blue precipitate. NBT and BCIP are typically used as the substrate compounds that create the blue precipitate. Following probe hybridization, anti-DIG antibodies conjugated to the enzyme alkaline phosphatase are used to localize the NBT/BCIP reactions to produce the blue precipitate found only in those cells expressing *odd-skipped*. (A from So and Danielian 1999.)

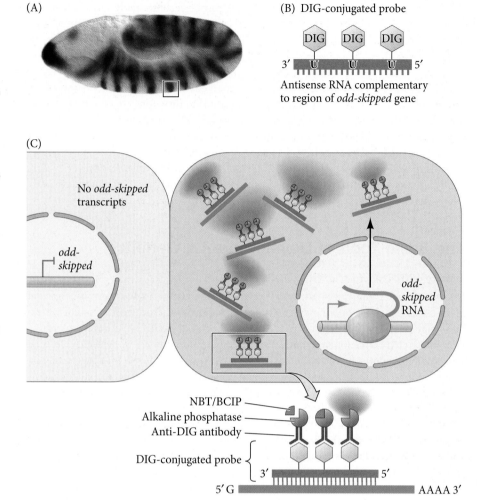

(A)

(B) DIG-conjugated probe

3′   U   U   U   5′

Antisense RNA complementary to region of *odd-skipped* gene

(C)

No *odd-skipped* transcripts

odd-skipped

odd-skipped RNA

NBT/BCIP
Alkaline phosphatase
Anti-DIG antibody

DIG-conjugated probe

3′   5′
5′ G   AAAA 3′

can be converted into a colored product by the enzyme. The enzyme should be present only where the digoxigenin is present, and the digoxigenin should be present only where the specific complementary mRNA is found. Thus, in **FIGURE 3.35C**, the dark blue precipitate formed by the enzyme indicates the presence of the target mRNA.

CHROMATIN IMMUNOPRECIPITATION-SEQUENCING ChIP-Seq is based on two highly specific interactions. One is the binding of a transcription factor or a modified nucleosome to very particular sequences of DNA (such as enhancer elements), and the other is the binding of antibody molecules specifically to the transcription factor or modified histone being studied (**FIGURE 3.36**; Liu et al. 2010).

In the first step of ChIP-Seq, chromatin is isolated, and the proteins are crosslinked (usually by glutaraldehyde or formaldehyde) to the DNA to which they are bound. This process prevents the nucleosome or transcription factors from dissociating from the DNA. After crosslinking, the DNA is fragmented (usually by sonication, but sometimes by enzymes) into pieces about 500 nucleotides long. The next step is to bind these proteins with an antibody that recognizes only that particular protein. Indeed, these antibodies are so specific that an antibody that recognizes histone 3 when it is dimethylated at position 4 will not recognize histone 3 that is trimethylated at that same position. The antibodies can be precipitated out of solution (often with magnetic beads that bind to antibodies), and they will bring down to the bottom of the test tube any DNA fragments bound by the protein of interest. These DNA fragments, once separated from the proteins, are amplified and can be sequenced and mapped to the entire

FIGURE 3.36 Chromatin immunoprecipitation-sequencing (ChIP-Seq). Chromatin is isolated from the cell nuclei. The chromatin proteins are crosslinked to their DNA-binding sites, and the DNA is fragmented into small pieces. Antibodies bind to specific chromatin proteins, and the antibodies—with whatever is bound to them—are precipitated out of solution. The DNA fragments associated with the precipitated complexes are purified from the proteins and sequenced. These sequences can be compared with the genome maps to give a precise localization of what genes these proteins may be regulating. (After Szalkowski and Schmid 2011.)

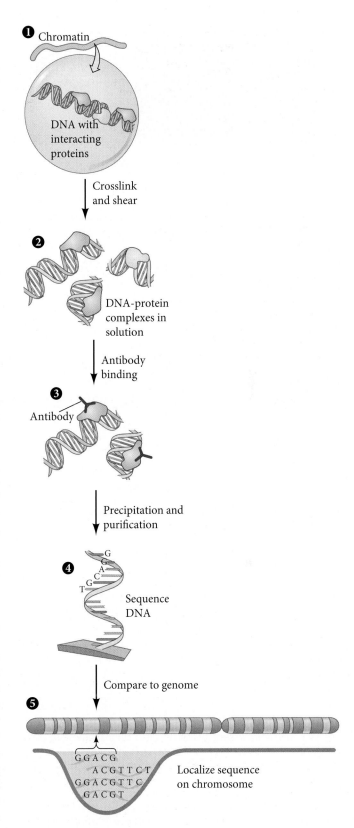

genome. In this way, the DNA sequences bound specifically by particular transcription factors or nucleosomes containing modified histones can be identified very precisely. As you will see throughout this text, researchers use these identified enhancer regions to generate transgenic reporter constructs and organisms that enable visualization of gene expression in live cells and organisms.

DEEP SEQUENCING: RNA-SEQ   As emphasized in this chapter, it is the full repertoire of genes expressed by a cell that establishes the gene regulatory network controlling cell identity. Major improvements in sequencing technology have enabled whole genomes to be sequenced, but a genome does not equal the cell's transcriptome. To move closer to the identification of all the transcripts present in a given embryo, tissue, or even single cell, **RNA-Seq** was developed. RNA-Seq takes advantage of the high throughput capabilities of *next-generation sequencing* technology to sequence and quantify the RNA present in a cell (**FIGURE 3.37**). Specifically, RNA is isolated from samples and converted to complementary DNA (cDNA) with standard procedures using *reverse transcriptase.* This cDNA is broken up into smaller fragments, and known adaptor sequences are added to the ends. These adaptors allow immobilization and PCR-based amplification of these transcripts. Next-generation sequencing can analyze these transcripts for both nucleotide sequence and quantity (Goldman and Domschke 2014). RNA-seq has been particularly powerful for comparing transcriptomes between identical samples differing only in select experimental parameters. One can ask, how does the array of transcripts differ between tissues located in different regions of the embryo, or the same tissue at different times of development, or the same tissue treated or untreated with a specific compound? These comparisons only scratch the surface of what is possible and what we can learn from differences in transcriptomes. The advent of fluorescence activated cell sorting (typically spoken as FACS for short) and microdissection has allowed for the precise isolation of tissues and individual cells, and recent advances in RNA-seq sensitivity has permitted transcriptomics of single cells.

A common experimental approach has been to design a targeted deep-sequencing experiment to arrive at a list of genes associated with a given condition. Researchers then use bioinformatics and an understanding of developmental

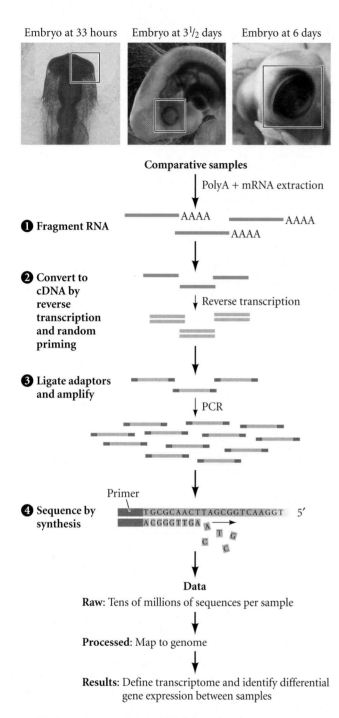

Embryo at 33 hours   Embryo at 3½ days   Embryo at 6 days

Comparative samples

PolyA + mRNA extraction

❶ **Fragment RNA**

AAAA   AAAA
AAAA

❷ **Convert to cDNA by reverse transcription and random priming**

Reverse transcription

❸ **Ligate adaptors and amplify**

PCR

❹ **Sequence by synthesis**

Primer

TGCGCAACTTAGCGGTCAAGGT   5′
ACGGGTTGA   A T G
C C

Data

Raw: Tens of millions of sequences per sample

Processed: Map to genome

Results: Define transcriptome and identify differential gene expression between samples

FIGURE 3.37 Deep sequencing: RNA-Seq. (Top) Researchers begin with specific sorts of tissues, often comparing different conditions, such as embryos of different ages (chick embryos, as shown here), isolated tissues (such as the eye; boxed regions) or even single cells, samples from different genotypes, or experimental paradigms. (1) RNA is isolated to obtain only those genes that are actively expressed; (2) these transcripts are then fragmented into smaller stretches and used to create cDNA with reverse transcriptase. (3) Specialized adaptors are ligated to the cDNA ends to enable PCR amplification and immobilization for (4) subsequent sequencing. (After Goldman and Domschke 2014; Malone and Oliver 2011; left photo © Ed Reschke/Getty Images; center, right photos © Oxford Scientific/Getty Images.)

biology to select gene candidates from the list to test the function of these genes in their system.

Testing Gene Function

Developmental biologists have used an array of methods to eliminate genes to determine their functions. These methods fall into two categories: forward genetics and reverse genetics. In **forward genetics**, an organism is exposed to an agent that causes unbiased, random mutations, and the resulting phenotypes are screened for ones that affect development. Individual mutations can be maintained either as homozygotes or as heterozygotes if the mutation seriously affects survival. The identities of the mutated loci are typically determined only after the initial phenotypic analysis. Two important forward genetics mutagenesis screens were done on *Drosophila* and zebrafish by Christiane Nüsslein-Volhard and colleagues (Nüsslein-Volhard and Wieschaus 1980, 1996; an entire issue of *Development* was devoted to the zebrafish screen). These screens have contributed immensely to the identification and functional characterization of many of the genes and pathways we know today to be important in development and disease.

In contrast to forward genetics, in **reverse genetics** you start with a gene in mind that you want to manipulate and then either knock down or knock out the expression of that gene. Using an RNAi or morpholino specific for a given gene, you can target its mRNA for degradation or block its splicing or translation, respectively (see Figure 3.32). These tools inhibit gene function but not always completely and only for a limited period of time because the RNAi or morpholino becomes diluted and degraded over the course of development (hence only a "knockdown" and not a "knockout"). Researchers can take advantage of that and use different amounts of RNAi or morpholinos to achieve a dose response effect.

Targeted gene knockouts, on the other hand, have been notable for completely eliminating the function of targeted genes. Such elimination has been done effectively in the mouse, where researchers have used embryonic stem cells for inserting a DNA construct called a *neomycin cassette* into a specific gene through a process of homologous recombination. This insertion both mutates the gene and provides an antibiotic selection mechanism for identifying mutated cells. These cells are injected into blastocysts, which develop into chimeric mice in which only some of the cells carry the mutation. These mice are bred to obtain homozygous mutant mice in which there is complete loss of the targeted gene's function.[18]

CRISPR/CAS9 GENOME EDITING The technique of CRISPR/Cas9 genome editing has had an enormous effect on genetic research, making gene editing faster and less expensive than ever and making it relatively simple in organisms from *E. coli* to primates (Jansen et al. 2002). This technique uses a system that

[18] Additional details about these and other loss-of-function methods can be found on devbio.com.

occurs naturally in prokaryotes for defending against invading viruses (Barrangou et al. 2007). In prokaryotes, CRISPR (**c**lustered **r**egularly **i**nterspaced **s**hort **p**alindromic **r**epeats) is a stretch of DNA containing short regions that when transcribed into RNA serve as guides (short-guide RNAs or sgRNAs) for recognizing segments of viral DNA. The RNA also binds to an endonuclease called Cas9 (**C**RISPR **as**sociated enzyme **9**). When the sgRNA binds to viral DNA, the RNA brings Cas9 with it, which catalyzes a double-strand break in the foreign DNA, disabling the virus.

Researchers in Jennifer Doudna's lab at University of California, Berkeley and Emmanuelle Charpentier's lab at the Friedrich Miescher Institute in Switzerland wondered whether, if the sgRNA can recognize specific viral sequences, it could be engineered to recognize any gene. Could we create a CRISPR/Cas9 unit that could target any gene, in any species, and disable it? In 2012, these researchers demonstrated that the answer to this question is, unequivocally, yes (Jinek et al. 2012). When CRISPR sgRNAs specific for a gene are introduced into cells along with Cas9, the Cas9 protein is guided by the CRISPR to the gene of interest and causes a double-strand break in the DNA. This technique is highly successful at creating gene mutations (**FIGURE 3.38**). Cells will naturally try to repair double-strand breaks through a process called non-homologous end joining (NHEJ). In an effort to reconnect the DNA rapidly and avoid catastrophic DNA damage, however, NHEJ is often imperfect in its repairs, resulting in **indels** (an insertion or deletion of DNA bases). Whether the indel is an insertion or a deletion, there is a significant chance that it will cause a frameshift in the gene and consequently create a premature stop codon somewhere downstream of the mutation; hence, there will be a loss of gene function.[19]

The CRISPR/Cas9 system has been used successfully in a variety of species, such as *Drosophila*, zebrafish, and mouse, with some mutation rates exceeding 80% (Bassett et al. 2013). Researchers have been able to push CRISPR even further by using multiple sgRNAs to target several genes simultaneously, yielding double and triple knockouts. In addition, the system can be used to precisely edit a genome by including short DNA fragments with the CRISPR/Cas9. These DNA pieces are engineered to have sequence homology on their 5′ and 3′ ends to encourage homologous recombination flanking the double-stranded breaks (see Figure 3.38). This *directed homology repair* is now being tested to repair locations of known human mutations and has potential for treating numerous genetic diseases, such as muscular dystrophy (Nelson 2015). Finally, recent developments using deactivated Cas9 protein lacking nuclease activity are being explored to deliver a cargo fused to this *dead* Cas9

[19]Zinc finger nucleases (ZFNs) and TALENs (transcription activator-like effector nucleases) are also methods of creating double-strand breaks in DNA at precise locations. CRISPR differs from ZFNs and TALENs in how it recognizes gene targets. ZFNs and TALENs both use protein- and DNA-binding domains that can be identified within genes and can be highly specific, but generating the correct array of artificial protein domains paired with nuclease activity can be laborious and expensive.

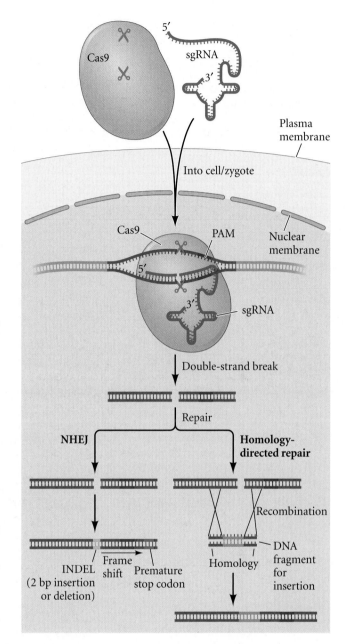

FIGURE 3.38 CRISPR/Cas9-mediated gene editing. The CRISPR/Cas9 system is used to cause targeted indel formation or insertional mutagenesis within a gene of interest. A gene-specific "short guide RNA" (sgRNA) is designed and co-injected with the nuclease Cas9, often into the single cell of a newly fertilized zygote. The sgRNA will bind to the genome with complementarity and will recruit Cas9 to this same location to induce a double-strand break. Non-homologous end joining (NHEJ) is the cell's DNA repair mechanism that often results in small insertions or deletions (approximately 2–30 base pairs), which can cause the establishment of a premature stop codon and potential loss of the protein's function. In addition, plasmid insertions with homology to regions surrounding the sgRNA target sites are used to foster the insertion of known sequences. Such methods are being explored as a way to repair mutations.

(A)

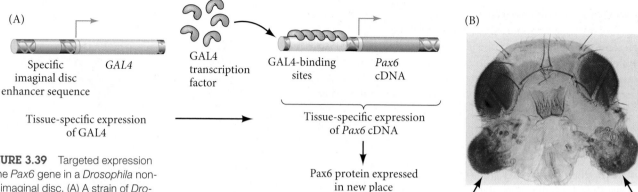

(B)

Specific imaginal disc enhancer sequence   *GAL4*

Tissue-specific expression of GAL4

GAL4 transcription factor

GAL4-binding sites   *Pax6* cDNA

Tissue-specific expression of *Pax6* cDNA

Pax6 protein expressed in new place

FIGURE 3.39 Targeted expression of the *Pax6* gene in a *Drosophila* non-eye imaginal disc. (A) A strain of *Drosophila* was constructed wherein the gene for the yeast GAL4 transcription factor was placed downstream from an enhancer sequence that normally stimulates gene expression in the imaginal discs for mouthparts. If the embryo also contains a transgene that places GAL4-binding sites upstream of the *Pax6* gene, the *Pax6* gene will be expressed in whichever imaginal disc the GAL4 protein is made. (B) *Drosophila* ommatidia (compound eyes) emerging from the mouthparts of a fruit fly in which the *Pax6* gene was expressed in the labial (jaw) discs. (Photograph courtesy of W. Gehring and G. Halder.)

without breaking the DNA. For instance, GFP fused to dead Cas9 is opening the door to better visualizing chromatin architecture in live cells. CRISPR/Cas9 is rapidly proving to be a remarkably versatile method for genome editing to further both research and therapeutic objectives across species. One of the immediate benefits is that CRISPR/Cas9 appears to be successful in all organisms. This universal utility has the potential to start a new frontier for functional gene analysis in species in which genetic approaches have previously been an insurmountable obstacle.

GAL4-UAS SYSTEM One of the most powerful uses of this genetic technology has been to activate regulatory genes such as *Pax6* in new places. Using *Drosophila* embryos, Halder and colleagues (1995) placed a gene encoding the yeast GAL4 transcriptional activator protein downstream from an enhancer that was known to function in the labial imaginal discs (those parts of the *Drosophila* larva that become the adult mouth parts). In other words, the gene for the GAL4 transcription factor was placed next to an enhancer for genes normally expressed in the developing jaw. Therefore, *GAL4* should be expressed in jaw tissue. Halder and colleagues then constructed a second transgenic fly, placing the cDNA for the *Drosophila Pax6* regulatory gene downstream from a sequence composed of five GAL4-binding sites. The GAL4 protein should be made only in a particular group of cells destined to become the jaw, and when that protein is made, it should cause the transcription of *Pax6* in those particular cells (**FIGURE 3.39A**). In flies in which *Pax6* was expressed in the incipient jaw cells, part of the jaw gave rise to eyes (**FIGURE 3.39B**). In *Drosophila* and frogs (but not in mice), *Pax6* is able to turn several developing tissue types into eyes (Chow et al. 1999). It appears that in *Drosophila*, *Pax6* not only activates those genes that are necessary for the construction of eyes, but also represses those genes that are used to construct other organs.

CRE-LOX SYSTEM An important experimental use of enhancers has been the conditional elimination of gene expression in certain cell types. For example, the transcription factor Hnf4α is expressed in liver cells, but it is also expressed prior to liver formation in the visceral endoderm of the yolk sac. If this gene is deleted from mouse embryos, the embryos die before the liver can even form. So, if you wanted to study the consequence of eliminating this gene's function in the liver, you would need to create a mutation that would be *conditional*; that is, you would need a mutation that would appear only in the liver and nowhere else. How can that be done? Parviz and colleagues (2002) accomplished it using a site-specific recombinase technology called Cre-lox.

The **Cre-lox** technique uses homologous recombination to place two Cre-recombinase recognition sites (*loxP* sequences) within the gene of interest, usually flanking important exons (see Kwan 2002). Such a gene is said to be "floxed" ("*loxP*-flanked"). For example, using cultured mouse embryonic stem (ES) cells, Parviz and colleagues

In most cells: No recombination

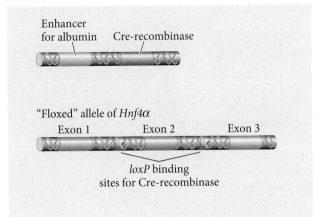

In liver cells only (expressing albumin)

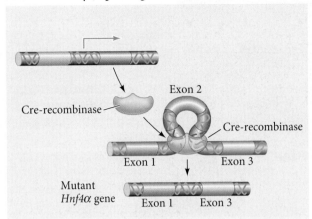

FIGURE 3.40 The Cre-lox technique for conditional mutagenesis, by which gene mutations can be generated in specific cells only. Mice are made wherein wild-type alleles (in this case, the genes encoding the Hnf4α transcription factor) have been replaced by alleles in which the second exon is flanked by *loxP* sequences. These mice are mated with mice having the gene for Cre-recombinase transferred onto a promoter that is active only in particular cells. In this case, the promoter is that of an albumin gene that functions early in liver development. In mice with both these altered alleles, Cre-recombinase is made only in the cells where that promoter was activated (i.e., in these cells synthesizing albumin). The Cre-recombinase binds to the *loxP* sequences flanking exon 2 and removes that exon. Thus, in the case depicted here, only the developing liver cells lack a functional *Hnf4α* gene.

(2002) placed two *loxP* sequences around the second exon of the mouse *Hnf4α* gene (**FIGURE 3.40**). These ES cells were then used to generate mice that had this floxed allele. A second strain of mice was generated that had a gene encoding bacteriophage Cre-recombinase (the enzyme that recognizes the *loxP* sequence) attached to the promoter of an albumin gene that is expressed very early in liver development. Thus, during mouse development, Cre-recombinase would be made only in the liver cells. When the two strains of mice were crossed, some of their offspring carried both additions. In these double-marked mice, Cre-recombinase (made only in the liver cells) bound to its recognition sites—the *loxP* sequences—flanking the second exon of the *Hnf4α* genes. It then acted as a recombinase and deleted this second exon. The resulting DNA would encode a nonfunctional protein because the second exon has a critical function in *Hnf4α*. Thus, the *Hnf4α* gene was "knocked out" only in liver cells.

The Cre-lox system allows for control over the spatial and temporal pattern of a gene knockout and gene misexpression. Researchers have inserted stop codons flanked with *loxP* sites to prevent transcription of a given gene until the stop codon is removed by Cre-recombinase. Moreover, Cre-recombinase expression can be controlled with greater temporal control through the use of an estrogen-responsive element sensitive to tamoxifen exposure. This control allows researchers to introduce genes for specific proteins, such as reporter proteins like GFP, that are kept inactive until a timed treatment with tamoxifen.

WEB TOPIC 3.10 **TECHNIQUES OF RNA AND DNA ANALYSIS** Familiarize yourself with more specifics on a variety of commonly used methodology in developmental genetics.

Next Step Investigation

In this chapter, you have learned that the compilation of active proteins of a cell confer upon it its phenotype and identity. We also discussed a variety of mechanisms that control the gene expression necessary to arrive at this identity. What can be done with this knowledge? If every cell is defined by the gene regulatory network it expresses, can any cell type be created in the laboratory simply by matching its network? How important are a cell's neighbors to maintaining its GRN and consequently its fate? From a cell to a tissue to an organism to a species, how do the mechanisms of differential gene expression lead to different morphologies? These questions can be applied to your favorite cell type and species. For instance, what approaches might be taken to generate in culture or regenerate in a brain the dopamine-secreting neurons needed to repair the deficits seen in Parkinson's disease? What evolutionary insights might you gain if you compare the transcriptomes of cells from the limb buds of human and non-human primates?

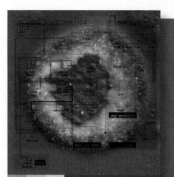

Closing Thoughts on the Opening Photo

What underlies cell differentiation? Here you see an image of a 24 hours post-fertilization sea urchin embryo differentially expressing *hox11/13b* and *foxa* in different cells. This image is overlaid on the gene regulatory network determined to "underlie" the development of endoderm. The gene regulatory network represents the combinatorial interactions that occur between genes to establish the specific array of differentially expressed genes. Networks like this one use the myriad molecular mechanisms discussed in this chapter to control gene expression and ultimately provide the most comprehensive definition of a given cell's identity. This chapter is dedicated to the memory of Dr. Eric H. Davidson, and the seemingly infinite network of contributions he made to the field of Developmental Biology. (Photograph from I. Peter and E. Davidson 2011.)

Snapshot Summary
Differential Gene Expression

1. Evidence from molecular biology, cell biology, and somatic cell nuclear cloning has shown that each cell of the body (with very few exceptions) carries the same nuclear genome.

2. Differential gene expression from genetically identical nuclei creates different cell types. Differential gene expression can occur at the levels of gene transcription, nuclear RNA processing, mRNA translation, and protein modification. Notice that RNA processing can occur while the RNA is still being transcribed from the gene.

3. Chromatin is made of DNA and proteins. The histone proteins form nucleosomes, and the methylation and acetylation of specific histone residues can activate or repress gene transcription.

4. Histone methylation is often used to silence gene expression. Histones can be methylated by histone methyltransferases and can be demethylated by histone demethylases.

5. Acetylated histones are often associated with active gene expression. Histone acetyltransferases add acetyl groups to histones, whereas histone deacetylases remove them.

6. Eukaryotic genes contain promoter sequences to which RNA polymerase II can bind to initiate transcription. To do so, the eukaryotic RNA polymerases are bound by a series of proteins called transcription-associated factors, including TFIID and TFIIB.

7. Eukaryotic genes expressed in specific cell types contain enhancer sequences that regulate their transcription in time and space. Enhancers activate genes on the same chromosome. Enhancer sequences can be within introns or the 3′ UTR; they can even be millions of base pairs away from the gene they activate. Enhancers can also act as silencers to suppress the transcription of a gene in inappropriate cell types.

8. Specific transcription factors can recognize specific sequences of DNA in the promoter and enhancer regions. These proteins activate or repress transcription from the genes to which they have bound.

9. Enhancers work in a combinatorial fashion. The binding of several transcription factors can act to promote or inhibit transcription from a certain promoter. In some cases, transcription is activated only if both factor A and factor B are present; in other cases, transcription is activated if either factor A or factor B is present.

10. Enhancers work in a modular fashion. A gene can contain several enhancers, each directing the gene's expression in a particular cell type.

11. A gene encoding a transcription factor can maintain itself in the activated state if the transcription factor it encodes also activates its own promoter. Thus, a transcription factor gene can have one set of enhancer sequences to initiate its activation and a second set of enhancer sequences (which bind the encoded transcription factor) to maintain its activation.

12. Transcription factors act in different ways to regulate RNA synthesis. Some transcription factors stabilize RNA polymerase II binding to the DNA, and some disrupt nucleosomes, increasing the efficiency of transcription.

13. The Mediator complex often serves as the bridge between the enhancer and promoter.

14. Transcription elongation complexes enable the RNA polymerase II to be released from the pre-initiation complex and continue transcribing the DNA.

15. A transcription factor usually has three domains: a sequence-specific DNA-binding domain, a *trans*-activating domain that enables the transcription factor to recruit histone remodeling enzymes, and a protein-protein interaction domain that enables it to interact with other proteins on the enhancer or promoter.

16. Even differentiated cells can be converted into another cell type by the activation of a different set of transcription factors.

17. In low CpG-content promoters, transcription correlates with a lack of DNA methylation on the promoter and enhancer regions of genes.

18. In high CpG-content promoters, the nucleosomes often allow transcription to start but do not permit the elongation of the nRNA.

19. Differences in DNA methylation can account for genomic imprinting, wherein a gene transmitted through the sperm is expressed differently than the same gene transmitted through the egg. Some genes are active only if inherited from the sperm or the egg. The imprinting marks appear to be CpG sites that are methylated on either the maternally inherited or paternally inherited locus.

20. Maintaining active gene expression is often accomplished by Trithorax proteins, whereas active repression is maintained by Polycomb protein complexes that contain histone methyltransferases.

21. Insulators are DNA sequences that bind CTCF protein. Insulators limit the range over which an enhancer can activate a promoter.

22. DNA methylation can block transcription by preventing the binding of certain transcription factors or by recruiting histone methyltransferases or histone deacetylases to the chromatin.

23. Some chromatin is "poised" to respond quickly to developmental signals. The mRNA of poised chromatin has begun to be transcribed, and its histones have both activating and repressive marks.

24. Differential RNA splicing can create a family of related proteins by causing different regions of the nRNA to be read as exons or introns. What is an exon in one set of circumstances may be an intron in another.

25. Alternative RNA splicing can create several different proteins from the same pre-mRNA transcript. These proteins (splicing isoforms) can play different roles.

26. Alternative pre-mRNA slicing is accomplished by splicing site recognition factors that can be different in different cell types. Mutations in splice sites can lead to alternative phenotypes and disease.

27. Some messages are translated only at certain times. The oocyte, in particular, uses translational regulation to set aside certain messages that are transcribed during egg development but used only after the egg is fertilized. This activation is often accomplished either by the removal of inhibitory proteins or by the polyadenylation of the message.

28. MicroRNAs can act as translational inhibitors, binding to the 3′ UTR of the RNA. The microRNA recruits an RNA-induced silencing complex that either prevents translation or leads to the degradation of the mRNA.

29. Many mRNAs are localized to particular regions of the oocyte or other cells. This localization appears to be regulated by the 3′ UTR of the mRNA.

30. Ribosomes can differ in different cell types, and ribosomes in one cell may be more efficient at translating certain mRNAs than ribosomes in other cells.

31. Differential gene expression is more like interpreting a musical score than decoding a code script. It is a stochastic phenomenon in which there are numerous events that have to take place, each having numerous interactions between component parts.

32. A variety of molecular tools have enabled the study of differentially expressed genes, such as in situ hybridization for gene expression, ChIP/Seq to identify regulatory regions of the DNA that proteins bind to, and gene knockdown (RNA interference) and knockout (CRISPR/Cas9) to test gene function.

Further Reading

Core, L. J. and J. T. Lis. 2008. Transcriptional regulation through promoter-proximal pausing of RNA polymerase II. *Science* 319: 1791–1792.

Fire, A., S. Q. Xu, M. K. Montgomery, S. A. Kostas, S. E. Driver and C. C. Mello. 1998. Potent and specific genetic interference by double-stranded RNA in *Caenorhabditis elegans*. *Nature* 39: 806–811.

Giraldez, A. J. and 7 others. 2006. Zebrafish MiR-430 promotes deadenylation and clearance of maternal mRNAs. *Science* 312: 75–79.

Gurdon, J. B. 2016. Cell fate determination by transcription factors. *Curr. Top. Dev. Biol.* 116: 445–454.

Jinek, M., K. Chylinski, I. Fonfara, M. Hauer, J. A. Doudna and E. Charpentier. 2012. A programmable dual-RNA-guided DNA endonuclease in adaptive bacterial immunity. *Science* 337: 816–821.

Jothi, R., S. Cuddapah, A. Barski, K. Cui and K. Zhao. 2008. Genome-wide identification of the in vivo protein-DNA binding sites from ChIP-Seq data. *Nucl. Acids. Res.* 36: 5221–5231.

Melton, D. A. 2016. Applied developmental biology: making human pancreatic beta cells for diabetics. *Curr. Top. Dev. Biol.* 117: 65–73.

Miura, S. K., A. Martins, K. X. Zhang, B. R. Graveley and S. L. Zipursky. 2013. Probabilistic splicing of *Dscam1* establishes identity at the level of single neurons. *Cell* 155: 1166–1177.

Muse, G. W. and 7 others. 2007. RNA polymerase is poised for activation across the genome. *Nature Genet.* 39: 1507–1511.

Nelson, C. E. and 14 others. 2016. In vivo genome editing improves muscle function in a mouse model of Duchenne muscular dystrophy. *Science* 351: 403–407.

Nüsslein-Volhard, C. and E. Wieschaus. 1980. Mutations affecting segment number and polarity in *Drosophila*. *Nature* 287: 795–801.

Ong, T.-C. and V. G. Corces. 2011. Enhancer function: New insights into the regulation of tissue-specific gene expression. *Nature Rev. Genet.* 12: 283–293.

Palacios, I. M. 2007. How does an mRNA find its way? Intracellular localization of transcripts. *Sem. Cell Dev. Biol.* 163–170.

Peter, I. and E. H. Davidson. 2015. *Genomic Control Process: Development and Evolution*. Academic Press, Cambridge.

Takahashi, K. and S. Yamanaka. 2006. Induction of pluripotent stem cells from mouse embryonic and adult fibroblast cultures by defined factors. *Cell* 126: 663–676.

Wilmut, I., K. Campbell and C. Tudge. 2001. *The Second Creation: Dolly and the Age of Biological Control.* Harvard University Press, Cambridge, MA.

Wilson, R. C. and J. A. Doudna. 2013. Molecular mechanisms of RNA interference. *Annu. Rev. Biophys.* 42: 217–239.

Yasumoto, K., K. Yokoyama, K. Shibata, Y. Tomita and S. Shibahara. 1994 Microphthalmia-associated transcription factor as a regulator for melanocyte-specific transcription of the human tyrosinase gene. *Mol. Cell Biol.* 12: 8058–8070.

Zhou, Q., J. Brown, A. Kanarek, J. Rajagopal and D. A. Melton. 2008. In vivo reprogramming of adult pancreatic exocrine cells to β cells. *Nature* 455: 627–632.

Zhou, V. W., A. Goren and B. E. Bernstein. 2011. Charting histone modifications and the functional organization of mammalian genomes. *Nature Rev. Genet.* 12: 7–18.

Zinzen, R. P., C. Girardot, J. Gagneur, M. Braun and E. E. Furlong. 2009. Combinatorial binding predicts spatio-temporal *cis*-regulatory activity. *Nature* 462: 65–70.

GO TO WWW.DEVBIO.COM…

…for Web Topics, Scientists Speak interviews, Watch Development videos, Dev Tutorials, and complete bibliographic information for all literature cited in this chapter.

4

Cell-to-Cell Communication
Mechanisms of Morphogenesis

DEVELOPMENT IS MORE THAN JUST DIFFERENTIATION. The different cell types of an organism do not exist as random arrangements. Rather, they form organized structures such as limbs and hearts. Moreover, the types of cells that constitute our fingers—bone, cartilage, neurons, blood cells, and others—are the same cell types that make up our pelvis and legs. Somehow, the cells must be ordered to create different shapes and make different connections. This construction of organized form is called **morphogenesis**, and it has been one of the great sources of wonder for humankind.

The twelfth-century rabbi and physician Maimonides framed the question of morphogenesis beautifully when he noted that the pious men of his day (around 1190 CE) believed that an angel of God had to enter the womb to form the organs of the embryo. That act, the man said, was a miracle. How much more powerful a miracle would life be, Maimonides asked, if the Deity had made matter such that it could generate this remarkable order without a matter-molding angel having to intervene in every pregnancy? The problem addressed today is the secular version of Maimonides' question: *How can matter alone construct itself into the organized tissues of the embryo?*

In the mid-twentieth century, E. E. Just (1939) and Johannes Holtfreter (Townes and Holtfreter 1955) predicted that embryonic cells could have differences in their cell

Could this be a cell's antenna? For what?

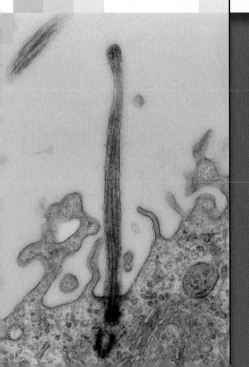

The Punchline

Communication between cells is achieved by informational molecules that are either secreted or positioned in the cell's membrane. When these molecules bind to receptors on neighboring cells, they set off a cascade of intracellular reactions that result in changes in gene expression, enzymatic activity, and cytoskeletal arrangements, affecting cell fate, cell behavior, and cell shape. Differential adhesion of cells to one another can influence the spatial organization of cells within the embryo and organs; it is often mediated by the homophilic binding of cadherin receptors. Epithelial cells sometimes transition into migrating mesenchymal cells, an important cell behavior both for development and for the spread of cancer. Specialized protrusions from cells, such as nonmotile cilia and long filopodia-like extensions, also play major roles in cell communication. Secreted signaling proteins like FGFs, Hedgehog, Wnts, and BMPs function as morphogens that induce changes in gene expression depending on their concentrations. Morphogen gradients are used to pattern cell fates across whole axes of an embryo or tissue. Lastly, cell-adjacent juxtacrine signaling can influence polarized cell patterning across tissues. All these mechanisms together direct cell-fate patterning and morphogenesis in the embryo.

membrane components that would enable the formation of organs. In the late twentieth century, these membrane components—the molecules by which embryonic cells are able to adhere to, migrate over, and induce gene expression in neighboring cells—began to be discovered and described. Today these pathways and networks are being modeled, and we are beginning to understand how the cell integrates the information from its nucleus and from its surroundings to take its place in the community of cells in a way that fosters unique morphogenetic events.

As we discussed in Chapter 1, the cells of an embryo are either epithelial or mesenchymal (see Table 1.1). Epithelial cells adhere to one another and can form sheets and tubes, whereas mesenchymal cells often migrate individually and form extensive extracellular matrices that can keep individual cells separate. An organ is formed from an epithelium and an underlying mesenchyme. There appear to be only a few processes through which cells create structured organs (Newman and Bhat 2008), and all these processes involve the cell surface. This chapter will concentrate on three behaviors requiring cell-to-cell communication via the cell surface: cell adhesion, cell shape, and cell signaling.

A Primer on Cell-to-Cell Communication

An embryo at any stage is held together, organized, and formed by the interactions that occur between cells. The interactions exhibited by cells define their methods of *communication*. For communication to occur successfully between humans, there needs to be some initial "voice" or signal from one person that is "heard" or received by the other person, which results in a specific response (a change in mood, a hug, or perhaps a sarcastic remark back), much like friends conversing. Molecular communication between cells is largely carried out through highly diverse and specific protein-protein interactions, which have evolved to elicit an array of cellular responses, from changes in gene transcription and glucose metabolism to cell migration and cell death. Interactions (or *communication*) between cells and between cells and their environment begin at the plasma membrane, with proteins that are housed within, anchored to, or secreted through the membrane.

In an embryo, communication between cells can occur across short distances, such as between two neighboring cells in direct contact, called **juxtacrine signaling**, or across long distances through the secretion of proteins into the extracellular matrix, called **paracrine signaling** (**FIGURE 4.1**). Proteins that are secreted from a cell and designed to communicate a response in another cell are generally referred to as signaling proteins (generally called **ligands**), while the proteins within a membrane that function to bind either other membrane-associated proteins or signaling proteins are called **receptors**. A receptor in the membrane of one cell that binds the same type of receptor in another cell represents a **homophilic binding**. In contrast, **heterophilic binding** occurs between different receptor types (see Figure 4.1A).

Binding to a receptor of any kind generally alters the shape, or *conformation*, of the receptor. This conformational change on

FIGURE 4.1 Local and long-range modes of cell-to-cell communication. (A) Local cell signaling is carried out via membrane receptors that bind to proteins in the extracellular matrix (ECM) or directly to receptors from a neighboring cell in a process called juxtacrine signaling. (B) One mechanism for long-range signaling is through paracrine signaling, such that one cell secretes a signaling protein (ligand) into the environment and across the distance of many cells. Only those cells expressing this ligand's corresponding receptor can respond, either rapidly through chemical reactions in the cytosol, or more slowly through the process of gene and protein expression.

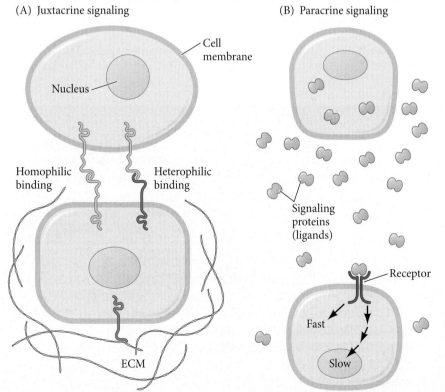

(A) Juxtacrine signaling

Cell membrane

Nucleus

Homophilic binding

Heterophilic binding

ECM

(B) Paracrine signaling

Signaling proteins (ligands)

Receptor

Fast

Slow

the outside of the cell affects the shape of the receptor inside the cell, and this latter change can give the intracellular portion of the receptor a new property. It now has the ability to activate the enzymatic reactions that constitute a signal transduction pathway. Often the "signal" is relayed or "transduced" through successive conformational changes in the molecules of the pathway, changes orchestrated through the binding of phosphate groups or other small molecules (cAMP, Ca^{2+}) that eventually lead to cellular responses. Signal transduction pathways that culminate in activating gene expression in the nucleus are typically slower than those that enzymatically activate biochemical pathways or regulate cytoskeletal proteins, thereby affecting physiological functions or movement, respectively. These signal transduction pathways are fundamental to animal development.

Adhesion and Sorting: Juxtacrine Signaling and the Physics of Morphogenesis

How are separate tissues formed from populations of cells and organs constructed from tissues? How do organs form in particular locations and migrating cells reach their destinations? For example, how do bone cells stick to other bone cells to create a bone rather than merging with adjacent capillary cells or muscle cells? What keeps the mesoderm separate from the ectoderm such that the skin has both a dermis and an epidermis? Why do eyes only form in the head? How do some cells—such as the precursors of our pigment cells and germ cells—travel long distances to reach their final destinations?

Could there be a simple common answer to all these questions? After all, an embryo, from its molecular strands of RNA to its systemic vasculature, develops within the same physical constraints that define our universe. Consider a snowman made out of sand (**FIGURE 4.2**). The thermodynamic properties governing the surface tension between water molecules and the grains of sand serve to hold the parts of Olaf together. Moreover, the sunlight on this sand sculpture establishes differential temperatures and associated water evaporation on the surface compared to the inner composition; consequently, the adhesion between sand grains on the surface rapidly becomes reduced, whereas more centrally located grains hold tight (that is, until the tide changes). Could these same thermodynamic principles govern the connections between cells that support morphogenesis of the embryo?

Differential cell affinity

The experimental analysis of morphogenesis began with the experiments of Townes and Holtfreter in 1955. Taking advantage of the discovery that amphibian tissues become dissociated into single cells when placed in alkaline solutions, they prepared single-cell suspensions from each of the three germ layers of amphibian embryos soon after the neural tube had formed. Two or more of these single-cell suspensions could be combined in various ways. When the pH of the solution was normalized, the cells adhered to one another, forming aggregates on agar-coated petri dishes. By using embryos from species having cells of different sizes and colors, Townes and Holtfreter were able to follow the behavior of the recombined cells.

The results of their experiments were striking. Townes and Holtfreter found that reaggregated cells become spatially segregated. That is, instead of two cell types remaining mixed, each type sorts

FIGURE 4.2 Adhesion between sand grains holds this sand sculpture of the Disney character Olaf together.

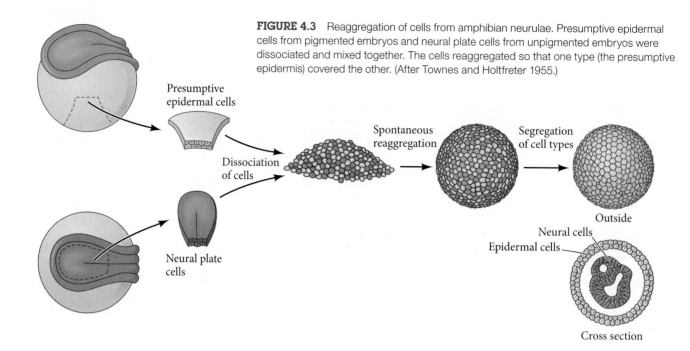

FIGURE 4.3 Reaggregation of cells from amphibian neurulae. Presumptive epidermal cells from pigmented embryos and neural plate cells from unpigmented embryos were dissociated and mixed together. The cells reaggregated so that one type (the presumptive epidermis) covered the other. (After Townes and Holtfreter 1955.)

out into its own region. Thus, when epidermal (ectodermal) and mesodermal cells are brought together in a mixed aggregate, the epidermal cells move to the periphery of the aggregate, and the mesodermal cells move to the inside (**FIGURE 4.3**). Importantly, the researchers found that the final positions of the reaggregated cells reflect their respective positions in the embryo. The reaggregated mesoderm migrates centrally with respect to the epidermis, adhering to the inner epidermal surface (**FIGURE 4.4A**). The mesoderm also migrates centrally with respect to the gut or endoderm (**FIGURE 4.4B**). When the three germ layers are mixed together, however, the endoderm separates from the ectoderm and mesoderm and is then enveloped by them (**FIGURE 4.4C**). In the final configuration, the ectoderm is on the periphery, the endoderm is internal, and the mesoderm lies in the region between them.

Holtfreter interpreted this finding in terms of **selective affinity**. The inner surface of the ectoderm has a positive affinity for mesodermal cells and a negative affinity for the endoderm, whereas the mesoderm has positive affinities for both ectodermal and endodermal cells. Mimicry of normal embryonic structure by cell aggregates is also seen in the recombination of epidermis and neural plate cells (**FIGURE 4.4D**). The presumptive epidermal cells migrate to the periphery as before; the neural plate cells migrate inward, forming a structure reminiscent of the neural tube. When axial mesoderm (notochord) cells are added to a suspension of presumptive epidermal and presumptive neural cells, cell segregation results in an external epidermal layer, a centrally located neural tissue, and a layer of mesodermal tissue between them (**FIGURE 4.4E**). *Somehow, the cells are able to sort out into their proper embryonic positions.* Holtfreter and colleagues concluded that selective affinities change during development. For development to occur, cells must interact differently with other cell populations at specific times. Such changes in cell affinity are extremely important in the processes of morphogenesis.

The thermodynamic model of cell interactions

Cells, then, do not sort randomly, but they can actively move to create tissue organization. What forces direct cell movement during morphogenesis? In 1964, Malcolm Steinberg proposed the **differential adhesion hypothesis**, a model that sought to explain patterns of cell sorting based on thermodynamic principles. Using cells derived from trypsinized embryonic tissues, Steinberg showed that certain cell types

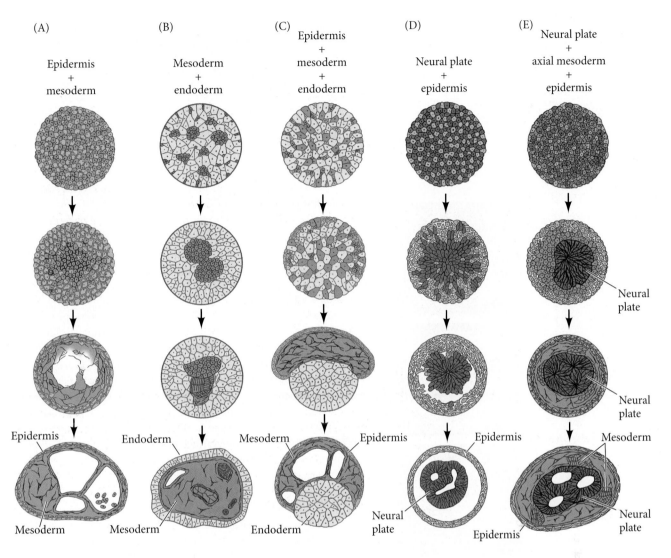

(A)

Epidermis
+
mesoderm

(B)

Mesoderm
+
endoderm

(C)

Epidermis
+
mesoderm
+
endoderm

(D)

Neural plate
+
epidermis

(E)

Neural plate
+
axial mesoderm
+
epidermis

Neural
plate

Neural
plate

Epidermis

Mesoderm

Endoderm

Mesoderm

Mesoderm

Mesoderm

Epidermis

Endoderm

Epidermis

Neural
plate

Epidermis

Mesoderm

Neural
plate

FIGURE 4.4 Sorting out and reconstruction of spatial relationships in aggregates of embryonic amphibian cells. (After Townes and Holtfreter 1955.)

migrate centrally when combined with some cell types, but migrate peripherally when combined with others. Such interactions form a hierarchy (Steinberg 1970). If the final position of cell type A is internal to a second cell type B and if the final position of B is internal to a third cell type C, the final position of A will always be internal to C (**FIGURE 4.5A**; Foty and Steinberg 2013). For example, pigmented retina cells migrate internally to neural retina cells, and heart cells migrate internally to pigmented retina cells. Therefore, heart cells migrate internally to neural retina cells. This observation led Steinberg to propose that cells interact so as to form an aggregate with the smallest interfacial free energy. In other words, the cells rearrange themselves into the most thermodynamically stable pattern. If cell types A and B have different strengths of adhesion and if the strength of A-A connections is greater than the strength of A-B or B-B connections, sorting will occur, with the A cells becoming central. However, if the strength of A-A connections is less than or equal to the strength of A-B connections, the aggregate will remain as a random mix of cells. Finally, if the strength of A-A connections is far greater than the strength of A-B connections or, in other words, if A and B cells show essentially no adhesivity toward one another, A cells and B cells will form separate aggregates. According to this hypothesis, the early embryo can be viewed as existing in an equilibrium state until some change in the adhesive properties of the cell's plasma membrane changes. The movements that result seek to restore the cells

(A)

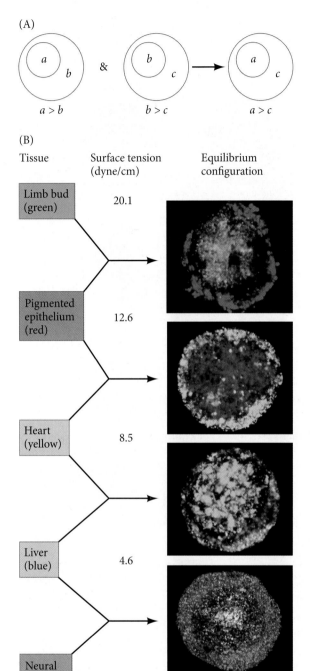

$a > b$ & $b > c$ → $a > c$

(B)

Tissue	Surface tension (dyne/cm)	Equilibrium configuration
Limb bud (green)	20.1	
Pigmented epithelium (red)	12.6	
Heart (yellow)	8.5	
Liver (blue)	4.6	
Neural retina (orange)	1.6	

FIGURE 4.5 Hierarchy of cell sorting of decreasing surface tensions. (A) Simple schematic demonstrating a logic statement for the properties of differential cell adhesion. (B) The equilibrium configuration reflects the strength of cell cohesion, with the cell types having the greater cell cohesion segregating inside the cells with less cohesion. These images were obtained by sectioning the aggregates and assigning colors to the cell types by computer. Black areas represent cells whose signal was edited out in the program of image optimization. (From Foty et al. 1996, courtesy of M. S. Steinberg and R. A. Foty.)

to a new equilibrium configuration. All that is required for sorting to occur is that cell types differ in the strengths of their adhesion; differential adhesion is caused by changes in the amount or repertoire of cell surface molecules.

In several meticulous experiments using numerous tissue types, researchers showed that those cell types that had greater surface cohesion migrated centrally compared to cells that had less surface tension (**FIGURE 4.5B**; Foty et al. 1996; Krens and Heisenberg 2011). In the simplest form of this model, all cells could have the same type of "glue" on the cell surface. The amount of this "glue," or the cellular architecture that allows such a substance to be differentially distributed across the surface, could create a difference in the number of stable contacts made between cell types. In a more specific version of this model, the thermodynamic differences could be caused by different types of adhesion molecules (see Moscona 1974). When Holtfreter's studies were revisited using modern techniques, Davis and colleagues (1997) found that the tissue surface tensions of the individual germ layers were precisely those required for the sorting patterns observed both in vitro and in vivo.

Cadherins and cell adhesion

Evidence shows that boundaries between tissues can indeed be created by different cell types having both different types and different amounts of cell adhesion molecules. Several classes of molecules can mediate cell adhesion, but the major cell adhesion molecules appear to be the cadherins.

As their name suggests, **cadherins** are *ca*lcium-*d*ependent ad*he*sion molecules. They are critical for establishing and maintaining intercellular connections, and they appear to be crucial to the spatial segregation of cell types and to the organization of animal form (Takeichi 1987). Cadherins are transmembrane proteins that interact with other cadherins on adjacent cells. The cadherins are anchored inside the cell by a complex of proteins called **catenins** (**FIGURE 4.6**), and the cadherin-catenin complex forms the classic adherens junctions that help hold epithelial cells together. Moreover, because the cadherins and the catenins bind to the actin (microfilament) cytoskeleton of the cell, they integrate the epithelial cells into a mechanical unit. Blocking cadherin *function* (by antibodies that bind and inactivate cadherin) or blocking cadherin *synthesis* (with antisense RNA that binds cadherin messages and prevents their translation) can prevent the formation of epithelial tissues and cause the cells to disaggregate (Takeichi et al. 1979).

Cadherins perform several related functions. First, their external domains serve to adhere cells together. Second, cadherins link to and help assemble the actin cytoskeleton, thereby providing the mechanical forces for forming sheets and tubes. Third, cadherins can serve to initiate and transduce signals that can lead to changes in a cell's gene expression.

VADE MECUM

Movies depict how pioneering experiments by Townes and Holtfreter and by Malcolm Steinberg demonstrated how cell surface adhesion molecules can direct cell sorting behaviors.

In vertebrate embryos, several major cadherin types have been identified. For example, **E-cadherin** is expressed on all early mammalian embryonic cells, even at the zygote stage. In the zebrafish embryo, E-cadherin is needed for the formation and migration of the epiblast as a sheet of cells during gastrulation. Loss of E-cadherin in the "*half-baked*" zebrafish mutant results in a failure of deep epiblast cells to move radially into the more superficial epiblast layer, an in vivo cell sorting process known as **radial intercalation** that helps power epiboly during gastrulation (**FIGURE 4.7**; see also Chapter 11 and Kane et al. 2005). Later in development, this E-cadherin is restricted to epithelial tissues of embryos and adults.

In mammals, **P-cadherin** is found predominantly on the placenta, where it helps the placenta stick to the uterus (Nose and Takeichi 1986; Kadokawa et al. 1989). **N-cadherin** becomes highly expressed on the cells of the developing central nervous system (Hatta and Takeichi 1986), and it may play a role in mediating neural signals. **R-cadherin** is critical in retina formation (Babb et al. 2005). A class of cadherins called **protocadherins** (Sano et al. 1993) lacks the attachment to the actin cytoskeleton through catenins. Expressing similar protocadherins is an important means of keeping migrating epithelial cells together, and expressing dissimilar protocadherins is an important way of separating tissues (as when

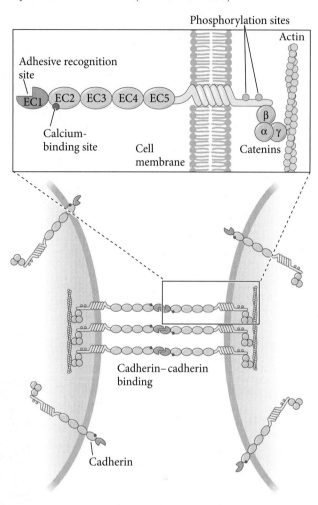

FIGURE 4.6 Simplified scheme of cadherin linkage to the cytoskeleton via catenins. (After Takeichi 1991.)

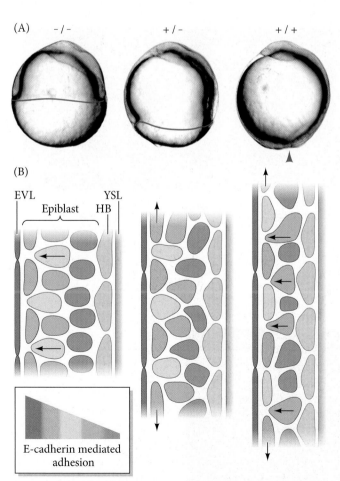

Epiboly (thinning of epiblast) over time by radial intercalation

FIGURE 4.7 E-cadherin is required for epiboly in zebrafish. (A) Wild-type embryos (right), and embryos heterozygous (center) and homozygous (left) for the E-cadherin mutation called *half-baked*. During normal gastrulation, cells merge into a thinner but more expansive epiblast layer that envelops the entire yolk (the red arrowhead points to the location of final yolk enclosure in the wild-type). E-cadherin mutants fail to complete epiboly, which is most severely impaired in the homozygous mutant (red lines denote the leading edge of epiblast). (B) Schematic of radial intercalating cell movements in the zebrafish epiblast over time during gastrulation. Cells move toward the superficial enveloping layer in relationship to increasing expression of E-cadherin. E-cadherin is expressed at higher levels in the more superficial layers of the epiblast, including the enveloping layer, and it is this differential expression (and consequently differential adhesion) that powers the radial movement of deep cells to the periphery. EVL, enveloping layer; HB, hypoblast; YSL, yolk syncytial layer. (Data and images based on Kane et al. 2005, courtesy of R. Warga.)

the mesoderm forming the notochord separates from the surrounding mesoderm that will form somites).

Differences in cell surface tension and the tendency of cells to bind together depend on the strength of cadherin interactions (Duguay et al. 2003). This strength can be achieved quantitatively (the more cadherins on the apposing cell surfaces, the tighter the adhesion) or qualitatively (some cadherins will bind to different cadherin types, whereas other cadherins will not bind to different types).

QUANTITY AND COHESION The ability of cells to sort themselves based on the *amount* of cadherin expression was first shown when Steinberg and Takeichi (1994) collaborated on an experiment using two cell lines that were identical except that they synthesized different amounts of P-cadherin. When these two groups of cells, each expressing a different amount of cadherin, were mixed, the cells that expressed more P-cadherin had a higher surface cohesion and migrated internally to the lower-expressing group of cells. Foty and Steinberg (2005) demonstrated that this quantitative cadherin-dependent sorting directly correlated with surface tension (**FIGURE 4.8A,B**). The surface tensions of these homotypic aggregates (all cells have same type of cadherin) are linearly related to the amount of cadherin they express on the cell surface. The cell sorting hierarchy is strictly dependent on the amount of cadherin interactions between the cells. This thermodynamic principle also applies to heterotypic aggregates, in which the relative amounts of different cadherin types still predict cell-sorting behavior in vitro (Foty and Steinberg 2013) (**FIGURE 4.8C**).

FIGURE 4.8 Importance of the amount of cadherin for correct morphogenesis. (A) Aggregate surface tension correlates with the number of cadherin molecules on the cell membranes. (B) Sorting out of two subclones having different amounts of cadherin on their cell surfaces. The green-stained cells had 2.4 times as many N-cadherin molecules in their membrane as did the other cells. (These cells had no normal cadherin genes being expressed.) At 4 hours of incubation (left), the cells are randomly distributed, but after 24 hours of incubation (right), the red cells (with a surface tension of about 2.4 erg/cm²) have formed an envelope around the more tightly cohering (5.6 erg/cm²) green cells. (C) Sorting can occur based on cadherin number even if the two cells express different cadherin proteins (i.e., are heterotypic). Red indicates P-cadherin, green E-cadherin. (A,B from Foty and Steinberg 2005; C from Foty and Steinberg 2013.)

(A)

(B)

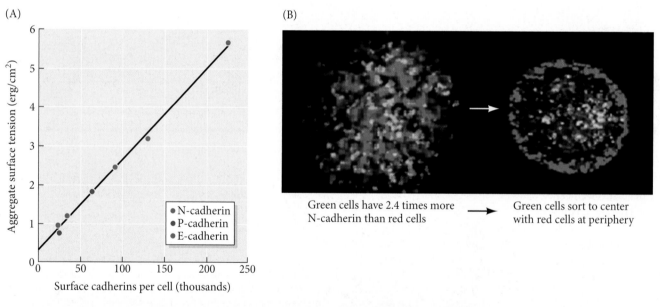

Green cells have 2.4 times more N-cadherin than red cells ⟶ Green cells sort to center with red cells at periphery

(C) P-cadherin > E-cadherin   P-cadherin = E-cadherin   P-cadherin < E-cadherin

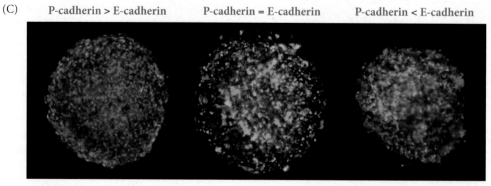

(A)

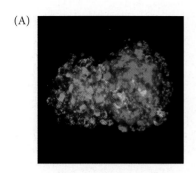

FIGURE 4.9   Importance of the types of cadherin for correct morphogenesis. (A) The type of cadherin expressed can result in different sorting behaviors, as seen when cells expressing R-cadherin (red stain) are mixed together with an equal number of cells expressing B-cadherin (green stain). The cells form two distinct mounds with one common boundary of contact. (B) Cross section of a mouse embryo showing the domains of E-cadherin expression (left) and N-cadherin expression (right). N-cadherin is critical for separation of presumptive epidermal and neural tissues during organogenesis. (C) The neural tube separates cleanly from surface epidermis in wild-type zebrafish embryos but not in mutant embryos where N-cadherin fails to be made. In these images, the cell outlines are stained green with antibodies to β-catenin, while the cell interiors are stained blue. (A from Duguay et al. 2003, photographs courtesy of R. Foty; B photographs by K. Shimamura and H. Matsunami, courtesy of M. Takeichi; C from Hong and Brewster 2006, courtesy of R. Brewster.)

(B)   E-cadherin expression   N-cadherin expression

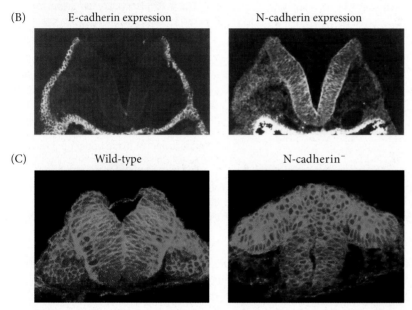

(C)   Wild-type   N-cadherin⁻

TYPE, TIMING, AND BORDER FORMATION   The quantitative effects of cadherins are crucial, but *qualitative* interactions—that is, the *type* and *timing* of cadherin expression—also can be important. The timing of particular developmental events can depend on cadherin expression. For instance, N-cadherin appears in the mesenchymal cells of the developing chick leg just before these cells condense and form nodules of cartilage (which are the precursors of the limb skeleton). N-cadherin is not seen prior to condensation, nor is it seen afterward. If the limbs are injected just prior to condensation with antibodies that block N-cadherin, the mesenchyme cells fail to condense and cartilage fails to form (Oberlender and Tuan 1994). It therefore appears that the signal to begin cartilage formation in the chick limb is the appearance of N-cadherin.

The type of cadherin can matter as well. Duguay and colleagues (2003) showed, for instance, that R-cadherin and B-cadherin do *not* bind well to each other. When two populations of cells expressing either R-cadherin or B-cadherin at equal levels are mixed together, they sort out into two opposing mounds of cells with a distinct border between them (**FIGURE 4.9A**). The formation of boundaries is a critical physical achievement necessary for many morphogenetic events. For instance in the developing ectoderm, the expression of N-cadherin is important in separating the precursors of the neural cells from the precursors of the epidermal cells (**FIGURE 4.9B**). Initially, all early embryonic cells contain E-cadherin, but those cells destined to become the neural tube lose E-cadherin and gain N-cadherin. If epidermal cells are experimentally made to express N-cadherin or if N-cadherin synthesis is blocked in prospective neural cells, the border between the skin and the nervous system fails to form properly (**FIGURE 4.9C**; Kintner et al. 1992). Thus, through the differential expression of two different cadherin types,

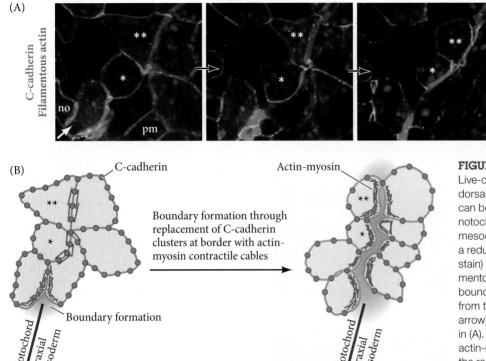

(A)

C-cadherin
Filamentous actin

** * no pm

(B)

C-cadherin

**

*

Boundary formation

Notochord
Paraxial mesoderm

Boundary formation through replacement of C-cadherin clusters at border with actin-myosin contractile cables

Actin-myosin

**

*

Notochord
Paraxial mesoderm

FIGURE 4.10 Boundary formation. (A) Live-cell imaging of explants of *Xenopus* dorsal mesoderm cells. Boundary formation can be seen to occur over time between notochord cells (no; asterisks) and paraxial mesodermal cells (pm) commensurate with a reduction of C-cadherin expression (green stain) and an increasing accumulation of filamentous actin (red stain) at the presumptive boundary. Boundary formation progresses from the lower left to the upper right (white arrow). (B) Schematized drawing of the cells in (A). Relative levels of C-cadherin and actin-myosin contractile units are indicated; the resulting boundary is shown in blue. (After Fagotto et al. 2013.)

Developing Questions

The underlying actin cytoskeleton appears to be crucial in organizing cadherins for forming stable linkages between cells. Although the energetic value of cadherin-cadherin binding is remarkably strong—about 3400 kcal/mole, or some 200 times stronger than most metabolic protein-protein interactions—actin-myosin contractile forces are also important for establishing the tensile forces of a cell. Recently, a "differential interfacial tension hypothesis" proposed that cell cortex contractility governs cell sorting more than cell-to-cell adhesion. As better in vivo tools are developed to quantitatively measure forces on the cellular and molecular levels, it will be exciting to learn how differential adhesion and differential interfacial tension cooperatively regulate morphogenesis. In the coming years, keep an eye out for a building understanding of the role biophysical properties play in mechanisms of morphogenesis.

different tissues can become separated by the formation of a border at the cell membrane occupying the weaker heterophilic interaction (Fagotto 2014).

Another example of boundary formation in the embryo occurs within the mesoderm to separate the axial (notochordal) mesoderm from the paraxial (somitic) mesoderm. The primary mechanism for forming this boundary rests in the reduction of C-cadherins in the apposing membranes of the border cells (Fagotto et al. 2013). Fagotto and colleagues examined this mechanism in live *Xenopus laevis* embryos and found that actin-myosin contractile cables line up parallel to the border interface and are required for both C-cadherin reduction and boundary formation (**FIGURE 4.10**).

WEB TOPIC 4.1 **SHAPE CHANGE AND EPITHELIAL MORPHOGENESIS: "THE FORCE IS STRONG IN YOU"** The ability of epithelial cells to form sheets and tubes depends on cell shape changes that usually involve cadherins and the actin cytoskeleton.

The Extracellular Matrix as a Source of Developmental Signals

Cell-to-cell interactions do not happen in the absence of an environment; rather, they occur in coordination with and often due to the environmental conditions surrounding the cells. This environment is called the **extracellular matrix**, which is an insoluble network consisting of macromolecules secreted by cells. These macromolecules form a region of noncellular material in the interstices between the cells. Cell adhesion, cell migration, and the formation of epithelial sheets and tubes all depend on the ability of cells to form attachments to extracellular matrices. In some cases, as in the formation of epithelia, these attachments have to be extremely strong. In other instances, as when cells migrate, attachments have to be made, broken, and made again. In some cases, the extracellular matrix merely serves as a permissive substrate to which cells can adhere or on which they can migrate. In other cases, it provides the directions for cell movement or the signal for a developmental event. Extracellular matrices are made up of the matrix protein collagen, proteoglycans, and a variety of specialized glycoprotein molecules such as fibronectin and laminin.

Proteoglycans play critically important roles in the delivery of the paracrine factors. These large molecules consist of core proteins (such as syndecan) with covalently attached glycosaminoglycan polysaccharide side chains. Two of the most widespread proteoglycans are heparan sulfate and chondroitin sulfate. Heparan sulfate can bind many members of different paracrine families, and it appears to be essential for presenting the paracrine factor in high concentrations to its receptors. In *Drosophila, C. elegans,* and mice, mutations that prevent proteoglycan protein or carbohydrate synthesis block normal cell migration, morphogenesis, and differentiation (García-García and Anderson 2003; Hwang et al. 2003; Kirn-Safran et al. 2004).

The large glycoproteins are responsible for organizing the matrix and the cells into an ordered structure. **Fibronectin** is a very large (460-kDa) glycoprotein dimer synthesized by numerous cell types. One function of fibronectin is to serve as a general adhesive molecule, linking cells to one another and to other substrates such as collagen and proteoglycans. Fibronectin has several distinct binding sites, and their interaction with the appropriate molecules results in the proper alignment of cells with their extracellular matrix (**FIGURE 4.11A**). Fibronectin also has an important role in cell migration because the "roads" over which certain migrating cells travel are paved with this protein. Fibronectin paths lead germ cells to the gonads and heart cells to the midline of the embryo. If chick embryos are injected with antibodies to fibronectin, the heart-forming cells fail to reach the midline, and two separate hearts develop (Heasman et al. 1981; Linask and Lash 1988).

((• **SCIENTISTS SPEAK 4.1** A question-and-answer session with Dr. Doug DeSimone and Dr. Tania Rozario about the role of fibronectin during *Xenopus* gastrulation.

Laminin (another large glycoprotein) and **type IV collagen** are major components of a type of extracellular matrix called the **basal lamina**. The basal lamina is characterized by closely knit sheets that underlie epithelial tissue (**FIGURE 4.11B**). The adhesion of epithelial cells to laminin (on which they sit) is much greater than the affinity of mesenchymal cells for fibronectin (to which they must bind and release if they are to migrate). Like fibronectin, laminin plays a role in assembling the extracellular matrix, promoting cell adhesion and growth, changing cell shape, and permitting cell migration (Hakamori et al. 1984; Morris et al. 2003).

FIGURE 4.11 Extracellular matrices in the developing embryo. (A) Fluorescent antibodies to fibronectin show fibronectin deposition as a green band in the *Xenopus* embryo during gastrulation. The fibronectin will orient the movements of the mesoderm cells. (B) Fibronectin links together migrating cells, collagen, heparan sulfate, and other extracellular matrix proteins. This scanning electron micrograph shows the extracellular matrix at the junction of the epithelial cells (above) and mesenchymal cells (below). The epithelial cells synthesize a tight, laminin-based basal lamina, whereas the mesenchymal cells secrete a loose reticular lamina made primarily of collagen. (A courtesy of M. Marsden and D. W. DeSimone; B courtesy of R. L. Trelsted.)

(A)

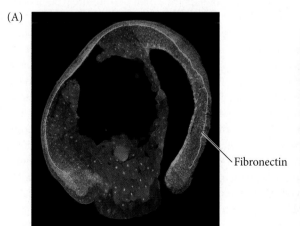

Fibronectin

(B)

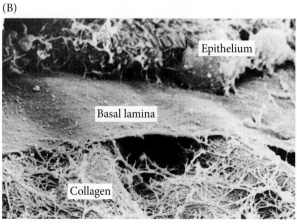

Epithelium

Basal lamina

Collagen

Integrins: Receptors for extracellular matrix molecules

The ability of a cell to bind to adhesive glycoproteins such as laminin or fibronectin depends on its expressing membrane receptors for the cell-binding sites of these large molecules. The fibronectin receptors were identified by using antibodies that block the attachment of cells to fibronectin (Chen et al. 1985; Knudsen et al. 1985). The main fibronectin receptor was found to be an extremely large protein that could bind fibronectin on the outside of the cell, span the membrane, and bind cytoskeletal proteins on the inside of the cell (**FIGURE 4.12**).

This family of receptor proteins are called integrins because they *integrate* the extracellular and intracellular scaffolds, allowing them to work together (Horwitz et al. 1986; Tamkun et al. 1986). On the extracellular side, integrins bind to the amino acid sequence arginine-glycine-aspartate (RGD), found in several extracellular matrix adhesive proteins, including fibronectin, vitronectin (found in the basal lamina of the eye), and laminin (Ruoslahti and Pierschbacher 1987). On the cytoplasmic side, integrins bind to talin and α-actinin, two proteins that connect to actin microfilaments. This dual binding enables the cell to move by contracting the actin microfilaments against the fixed extracellular matrix.

Integrins can also signal from the outside of the cell to the inside of the cell, altering gene expression (Walker et al. 2002). Bissell and colleagues (Bissell et al. 1982; Martins-Green and Bissell 1995) have shown that integrin is critical for inducing specific gene expression in developing tissues, especially those of the liver, testis, and mammary gland. In the mammary gland, extracellular laminin is able to signal the expression of estrogen receptor and casein protein genes through the integrin proteins (Streuli et al. 1991; Notenboom et al. 1996; Muschler et al. 1999; Novaro et al. 2003).

The presence of bound integrin prevents the activation of genes that promote apoptosis, or programmed cell death (Montgomery et al. 1994; Frisch and Ruoslahti 1997). For instance, the chondrocytes that produce the cartilage of our vertebrae and limbs can survive and differentiate only if they are surrounded by an extracellular matrix and are joined to that matrix through their integrins (Hirsch et al. 1997). If chondrocytes from the developing chick sternum are incubated with antibodies that block the binding of integrins to the extracellular matrix, they shrivel up and die. Indeed, when focal adhesions linking an epithelial cell to its extracellular matrix are broken, the caspase-dependent apoptosis pathway is activated, and the cell dies. Such "death-on-detachment" is a special type of apoptosis called **anoikis**, and it appears to be a major weapon against cancer (Frisch and Francis 1994; Chiarugi and Giannoni 2008).

Although the mechanisms by which bound integrins inhibit apoptosis remain controversial, the extracellular matrix is obviously an important source of signals that can be transduced into the nucleus to produce specific gene expression. Some of the genes induced by matrix attachment are being identified. When plated onto tissue culture plastic, mouse mammary gland cells will divide (**FIGURE 4.13**). Indeed, genes for cell division (*c-myc, cyclinD1*) are expressed, whereas genes for differentiated products of the mammary gland (casein, lactoferrin, whey acidic protein) are not expressed. If the same cells are plated onto plastic coated with a basal lamina, the cells stop dividing, and the genes of differentiated mammary gland cells are expressed. That happens only after the integrins of the mammary gland cells bind to the laminin of the basal lamina. Then the gene for lactoferrin is expressed, as is the gene for p21, a cell division inhibitor. The *c-myc* and *cyclinD1*

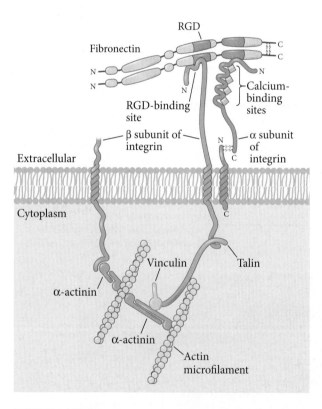

FIGURE 4.12 Simplified diagram of the fibronectin receptor complex. The integrins of the complex are membrane-spanning receptor proteins that bind fibronectin on the outside of the cell while binding cytoskeletal proteins on the inside of the cell. (After Luna and Hitt 1992.)

FIGURE 4.13 Basal lamina-directed gene expression in mammary gland tissue. (A) Mouse mammary gland tissue divides when placed on tissue culture plastic (no basal lamina). The genes encoding cell division proteins are on, and the genes capable of synthesizing the differentiated products of the mammary gland—lactoferrin, casein, and whey acidic protein (WAP)—are off. (B) When these cells are placed on a basal lamina, the genes for cell division proteins are turned off, while the genes encoding inhibitors of cell division (such as p21) and the gene for lactoferrin are turned on. (C,D) The mammary gland cells wrap the basal lamina around them, forming a secretory epithelium. The genes for casein and WAP are sequentially activated. (After Bissell et al. 2003.)

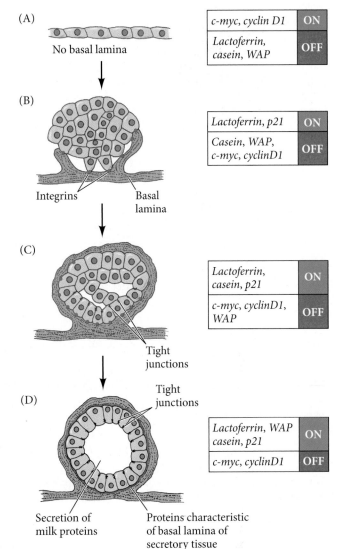

genes become silent. Eventually, all the genes for the developmental products of the mammary gland are expressed, and the cell division genes remain turned off. By this time, the mammary gland cells have enveloped themselves in a basal lamina, forming a secretory epithelium reminiscent of the mammary gland tissue. The binding of integrins to laminin is essential for transcription of the casein gene, and the integrins act in concert with prolactin (see Figure 4.27) to activate that gene's expression (Roskelley et al. 1994; Muschler et al. 1999).

The Epithelial-Mesenchymal Transition

One important developmental phenomenon, the **epithelial-mesenchymal transition**, or **EMT**, integrates all the processes we have discussed so far in this chapter. EMT is an orderly series of events whereby epithelial cells are transformed into mesenchymal cells. In this transition, a polarized stationary epithelial cell, which normally interacts with basal lamina through its basal surface, becomes a migratory mesenchymal cell that can invade tissues and help form organs in new places (**FIGURE 4.14A**; see Sleepman and Thiery 2011). EMT is usually initiated when paracrine factors from neighboring cells activate gene expression in the target cells, thereby instructing the target cells to downregulate their cadherins, release their attachment to laminin and other basal lamina components, rearrange their actin cytoskeleton, and secrete new extracellular matrix molecules characteristic of mesenchymal cells.

The epithelial-mesenchymal transition is critical during development (**FIGURE 4.14B,C**). Examples of developmental processes in which this transition is active include (1) the formation of neural crest cells from the dorsalmost region of the neural tube; (2) the formation of mesoderm in chick embryos, wherein cells that had been part of an epithelial layer become mesodermal and migrate into the embryo; and (3) the formation of vertebrae precursor cells from the somites, wherein these cells detach from the somite and migrate around the developing spinal cord. EMT is also important in adults, in whom it is needed for wound healing. The most critical adult form of EMT, however, is seen in cancer metastasis, wherein cells that have been part of a solid tumor mass leave that tumor to invade other tissues and form secondary tumors elsewhere in the body. It appears that in metastasis, the processes that generated the cellular transition in the embryo are reactivated, allowing cancer cells to migrate and become invasive. Cadherins are downregulated, the actin cytoskeleton is reorganized, and the cells secrete enzymes such as metalloproteinases to degrade the basal lamina and mesenchymal extracellular matrix while also undergoing cell division (Acloque et al. 2009; Kalluri and Weinberg 2009).

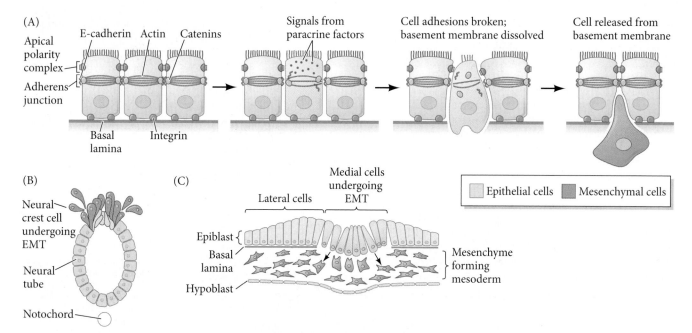

FIGURE 4.14 Epithelial-mesenchymal transition, or EMT. (A) Normal epithelial cells are attached to one another through adherens junctions containing cadherin, catenins, and actin rings. They are attached to the basal lamina through integrins. Paracrine factors can repress the expression of genes that encode these cellular components, causing the cell to lose polarity, lose attachment to the basal lamina, and lose cohesion with other epithelial cells. Cytoskeletal remodeling occurs, as well as the secretion of proteases that degrade the basal lamina and extracellular matrix molecules, enabling the migration of the newly formed mesenchymal cell. (B,C) EMT is seen in vertebrate embryos during the normal formation of neural crest from the dorsal region of the neural tube (B) and during the formation of the mesoderm by mesenchymal cells delaminating from the epiblast (C).

Cell Signaling

We have just learned how cell-to-cell adhesion (a juxtacrine interaction) can influence how cells position themselves within an embryo, and in previous chapters, we discussed the importance that a cell's position in the embryo can have on regulating its fate. What is so special about a given position in the embryo that it can determine a cell's fate? As you know, the experiences one has in early life greatly influence the type of person one becomes as an adult in terms of personality, career choice, or food preferences. Similarly, the experiences a cell has in its embryonic position influence the gene regulatory network under which it develops. Therefore, the real question is, in a given location, what defines the cell's experience?

Induction and competence

From the earliest stages of development through the adult, cell behaviors such as adhesion, migration, differentiation, and division are regulated by signals from one cell being received by another cell. Indeed, these interactions (which are often reciprocal, as we will describe later) are what allow organs to be constructed. The development of the vertebrate eye is a classic example used to describe the modus operandi of tissue organization via intercellular interactions.

In the vertebrate eye, light is transmitted through the transparent corneal tissue and focused by the lens tissue (the diameter of which is controlled by muscle tissue), eventually impinging on the tissue of the neural retina. The precise arrangement of tissues in the eye cannot be disturbed without impairing its function. Such coordination in the construction of organs is accomplished by one group of cells changing the behavior of an adjacent set of cells, thereby causing them to change their shape, mitotic rate, or cell fate. This kind of interaction at close range between two or more cells or tissues of different histories and properties is called **induction**.

DEFINING INDUCTION AND COMPETENCE There are at least two components to every inductive interaction. The first component is the **inducer**, the tissue that produces a signal (or signals) that changes the cellular behavior of the other tissue. Often this signal is a secreted protein called a paracrine factor. **Paracrine factors** are proteins made by a cell or a group of cells that alter the behavior or differentiation of adjacent cells. In contrast to endocrine factors (hormones), which travel through

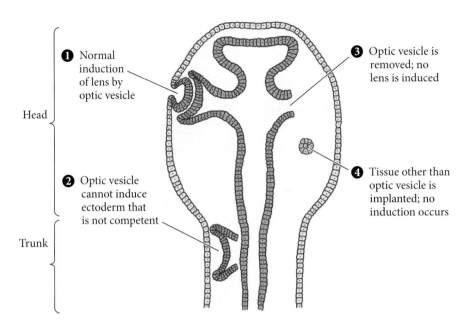

❶ Normal induction of lens by optic vesicle

❸ Optic vesicle is removed; no lens is induced

❷ Optic vesicle cannot induce ectoderm that is not competent

❹ Tissue other than optic vesicle is implanted; no induction occurs

Head

Trunk

FIGURE 4.15 Ectodermal competence and the ability to respond to the optic vesicle inducer in *Xenopus*. The optic vesicle is able to induce lens formation in the anterior portion of the ectoderm (1) but not in the presumptive trunk and abdomen (2). If the optic vesicle is removed (3), the surface ectoderm forms either an abnormal lens or no lens at all. Most other tissues are not able to substitute for the optic vesicle (4).

the blood and exert their effects on cells and tissues far away, paracrine factors are secreted into the extracellular space and influence their close neighbors. The second component, the **responder**, is the cell or tissue being induced. Cells of the responding tissue must have both a receptor protein for the inducing factor and the *ability* to respond to the signal. The ability to respond to a specific inductive signal is called **competence** (Waddington 1940).

BUILDING THE VERTEBRATE EYE In the initiation of the vertebrate eyes, paired regions of the brain bulge out and approach the surface ectoderm of the head. The head ectoderm is competent to respond to the paracrine factors made by these brain bulges (the **optic vesicles**), and the head ectoderm receiving these paracrine factors is induced to form the lens of the eye. The genes for lens proteins become induced in the head ectoderm cells and are expressed in these cells. The Rho-family GTPases are activated to control the elongation and curvature of the lens fibers (see Chapter 16; Maddala et al. 2008). Moreover, the prospective lens cells secrete paracrine factors that instruct the optic vesicle to form the retina. Thus, the two major parts of the eye co-construct each other, and the eye forms from reciprocal paracrine interactions. The head ectoderm is the only region capable of responding to the optic vesicle. If an optic vesicle from a *Xenopus laevis* embryo is placed underneath head ectoderm in a different part of the head from where the frog's optic vesicle normally occurs, the vesicle will induce that ectoderm to form lens tissue; trunk ectoderm, however, will not respond to the optic vesicle (**FIGURE 4.15**; Saha et al. 1989; Grainger 1992). Only head ectoderm is *competent* to respond to the signals from the optic vesicle by producing a lens.

Often, one induction will give a tissue the competence to respond to another inducer. Studies on amphibians suggest that the first inducers of the lens may be the foregut endoderm and heart-forming mesoderm that underlie the lens-forming ectoderm during the early and mid gastrula stages (Jacobson 1963, 1966). The anterior neural plate may produce the next signals, including a signal that promotes the synthesis of the Paired box 6 (Pax6) transcription factor in the anterior ectoderm which is required for the competence to respond to the optic vesicle's signals (**FIGURE 4.16**; Zygar et al. 1998). Thus, although the optic vesicle appears to be *the* lens inducer, the anterior ectoderm has already been induced by at least two other tissues. The optic vesicle's situation is like that of the player who kicks the "winning" goal in a soccer match, yet many others helped to position that ball for the final kick!

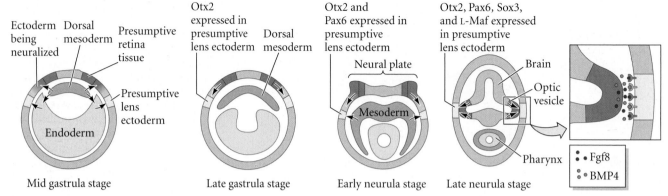

Ectoderm being neuralized | Dorsal mesoderm | Presumptive retina tissue

Presumptive lens ectoderm

Endoderm

Mid gastrula stage

Otx2 expressed in presumptive lens ectoderm | Dorsal mesoderm

Late gastrula stage

Otx2 and Pax6 expressed in presumptive lens ectoderm

Neural plate

Mesoderm

Early neurula stage

Otx2, Pax6, Sox3, and L-Maf expressed in presumptive lens ectoderm

Brain

Optic vesicle

Pharynx

Late neurula stage

Fgf8

BMP4

FIGURE 4.16 Sequence of amphibian lens induction postulated by experiments on embryos of the frog *Xenopus laevis*. Unidentified inducers (possibly from the foregut endoderm and cardiac mesoderm) cause the synthesis of the Otx2 transcription factor in the head ectoderm during the late gastrula stage. As the neural folds rise, inducers from the anterior neural plate (including the region that will form the retina) induce Pax6 expression in the anterior ectoderm that can form lens tissue. Expression of Pax6 protein may constitute the competence of the surface ectoderm to respond to the optic vesicle during the late neurula stage. The optic vesicle secretes BMP and FGF family paracrine factors (see signals in higher magnification of boxed area) that induce the synthesis of the Sox transcription factors and initiate observable lens formation. (After Grainger 1992.)

The optic vesicle appears to secrete two paracrine factors, one of which is BMP4 (Furuta and Hogan 1998), a protein that is received by the lens cells and induces the production of the Sox transcription factors (see Figure 4.16, right-most panels). The other is Fgf8, a secreted signal that induces the appearance of the L-Maf transcription factor (Ogino and Yasuda 1998; Vogel-Höpker et al. 2000). As we saw in Chapter 3, the combination of Pax6, Sox2, and L-Maf in the ectoderm is needed for the production of the lens and the activation of lens-specific genes such as δ-*crystallin*. Pax6 is important in providing the competence for the ectoderm to respond to the inducers from the optic cup (Fujiwara et al. 1994). If *Pax6* is lost, whether it is in fruit flies, frogs, rats, or humans, it results in a complete loss or reduction of the eyes (Quiring et al. 1994). Experiments recombining surface ectoderm with the optic vesicle from wild-type and *Pax6* mutant rat embryos demonstrated that Pax6 must be functional in the surface ectoderm for it to form a lens (**FIGURE 4.17A,B**). In humans, a spectrum of eye malformations have been associated with a variety of *Pax6* mutations. These malformations include aniridia, in which the iris is reduced or lacking (**FIGURE 4.17C**); *Pax6* mutations in *Xenopus* have revealed remarkably similar aniridia-like symptoms, enabling researchers to model and further investigate the developmental role of Pax6 in this human disease (Nakayama et al. 2015).

Reciprocal induction

Another feature of induction is the reciprocal nature of many inductive interactions. To continue the above example, once the lens has formed, it induces other tissues. One of these responding tissues is the optic vesicle itself; thus, the inducer becomes the induced. Under the influence of factors secreted by the lens, the optic vesicle becomes the optic cup, and the wall of the optic cup differentiates into two layers: the pigmented retina and the neural retina (see Figure 16.8; Cvekl and Piatigorsky 1996; Strickler et al. 2007). Such interactions are called **reciprocal inductions**.

Another principle can be seen in such reciprocal inductions: a structure does not need to be fully differentiated to have a function. As we will detail in Chapter 16, the optic vesicle induces the surface ectoderm to become a lens before the optic vesicle has become the retina. Similarly, the developing lens reciprocates by inducing the optic vesicle before the lens forms its characteristic fibers. Thus, before a tissue has its "adult" functions, it has critically important transient functions in building the organs of the embryo.

INSTRUCTIVE AND PERMISSIVE INTERACTIONS Howard Holtzer (1968) distinguished two major modes of inductive interaction. In **instructive interaction**, a signal from the inducing cell is *necessary* for initiating new gene expression in the responding cell. Without the inducing cell, the responding cell is not capable of differentiating in that particular way. For example, one instructive interaction is when a *Xenopus* optic vesicle

(A)

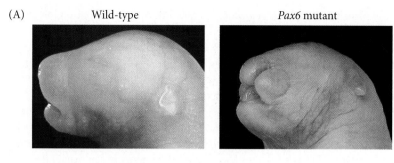

Wild-type   *Pax6* mutant

(B)

Optic vesicles	Surface ectoderm	Lens induction
Wild-type	Wild-type	Yes
Pax6⁻/Pax6⁻	Wild-type	Yes
Wild-type	*Pax6⁻/Pax6⁻*	No
Pax6⁻/Pax6⁻	*Pax6⁻/Pax6⁻*	No

Lens

(C)

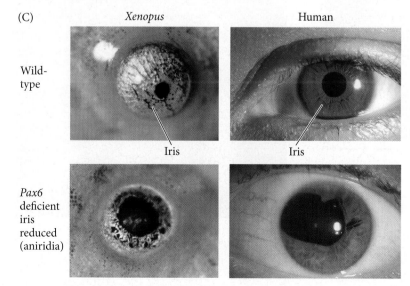

Wild-type   *Xenopus*   Human

Iris   Iris

Pax6 deficient iris reduced (aniridia)

FIGURE 4.17   The *Pax6* gene is similarly required for eye development in frogs, rats, and humans. (A) Loss of *Pax6* in rats results in the failure to form eyes as well as significant reductions in nasal structures. (B) An analysis of lens induction following recombination experiments of the optic vesicle and surface ectoderm between wild-type and *Pax6* null rat embryos. Pax6 is required only in the surface ectoderm for proper lens induction. (C) Mutations in the *Pax6* gene in *Xenopus* and humans result in similar reductions in the iris of the eye as compared to wild-type individuals. This phenotype is characteristic of aniridia. (A from Fujiwara et al. 1994; B photographs courtesy of M. Fujiwara; C from Nakayama et al. 2015, courtesy of R. M. Grainger.)

Developing Questions

Although rebuilding a decellularized heart is clearly an example of permissive interactions, could there be instructive interactions too? Recently, iPSC-derived cardiovascular progenitor cells successfully seeded a decellularized mouse heart and differentiated into cardiomyoctyes, smooth muscle and endothelial cells (Lu et al. 2013). What could the ECM be providing to directly influence the differentiation of progenitor cells into these varied cell types?

experimentally placed under a new region of head ectoderm causes that region of the ectoderm to form a lens.

The second type of inductive interaction is **permissive interaction**. Here, the responding tissue has already been specified and needs only an environment that allows the expression of these traits. For instance, many tissues need an extracellular matrix to develop. The extracellular matrix does not alter the type of cell that is produced, but it enables what has already been determined to be expressed.[1] A dramatic example of permissive interactions at work comes from the regenerative medicine field, in which an extracellular matrix scaffold can promote the differentiation and rebuilding of a beating heart. Doris Taylor's research group used detergents to remove all the cells from a cadaveric rat heart, which leaves behind the natural extracellular matrix (**FIGURE 4.18A**; Ott et al. 2008). Proteins like fibronectin, collagen, and laminin held together the rest of the ECM and maintained the intricate shape of the heart. The researchers then infused this ECM scaffold with cardiomyocytes. Surprisingly, these cells differentiated and organized into a functionally contracting "recellularized" heart (**FIGURE 4.18B**). Therefore, the environmental conditions of the decellularized ECM were permissive in allowing the cardiomyocytes to recreate contracting heart muscle. You will be reading more about regenerative medicine in Chapter 5.

SCIENTISTS SPEAK 4.2 Dr. Doris Taylor discusses the use of decellularized organs for regeneration.

Epithelial-mesenchymal interactions

Some of the best-studied cases of induction involve the interactions of sheets of epithelial cells with adjacent mesenchymal cells. All organs consist of an epithelium and an associated mesenchyme, so these interactions are among the most important phenomena in nature. Some examples are listed in **TABLE 4.1**.

---

[1]It is easy to distinguish permissive and instructive interactions using an analogy. This textbook is made possible by both permissive and instructive interactions. A reviewer can convince us to change the material in the chapters, which is an instructive interaction because the information expressed in the book is changed from what it would have been. However, the information in the book could not be expressed at all without permissive interactions with the publisher and printer.

(A) Decellularization

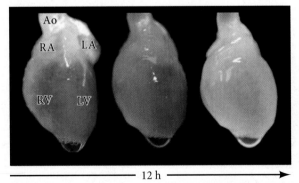

← 12 h →

(B) Recellularized beating heart

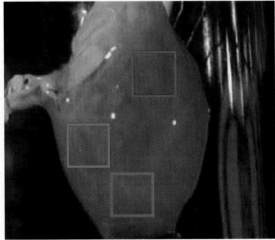

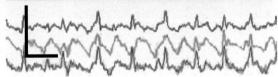

FIGURE 4.18 Reconstructing a decellularized rat heart. (A) Whole hearts from rat cadavers were decellularized (all cells removed) over the course of 12 hours using the detergent SDS. Progression of decellularization is seen here from left to right. Ao, aorta; LA, left atrium; LV, left ventricle; RA, right atrium; RV, right ventricle. (B) A decellularized heart was mounted into a bioreactor and recellularized with neonatal cardiac cells, which developed into self-contracting cardiomyocytes and powered the beating of the heart construct. Regional ECG tracings indicate synchronous contractions of the indicated heart regions (blue, green, and red plots). (From Ott et al. 2008.)

TABLE 4.1 Some epithelial-mesenchymal interactions		
Organ	**Epithelial component**	**Mesenchymal component**
Cutaneous structures (hair, feathers, sweat glands, mammary glands)	Epidermis (ectoderm)	Dermis (mesoderm)
Limb	Epidermis (ectoderm)	Mesenchyme (mesoderm)
Gut organs (liver, pancreas, salivary glands)	Epithelium (endoderm)	Mesenchyme (mesoderm)
Foregut and respiratory-associated organs (lungs, thymus, thyroid)	Epithelium (endoderm)	Mesenchyme (mesoderm)
Kidney	Ureteric bud (mesoderm)	Mesenchyme epithelium (mesoderm)
Tooth	Jaw epithelium (ectoderm)	Mesenchyme (neural crest)

REGIONAL SPECIFICITY OF INDUCTION   Using the induction of cutaneous (skin) structures as our examples, we will look at the properties of epithelial-mesenchymal interactions. The first of these properties is the regional specificity of induction. Skin is composed of two main tissues: an outer epidermis (an epithelial tissue derived from ectoderm) and a dermis (a mesenchymal tissue derived from mesoderm). The chick epidermis secretes proteins that signal the underlying dermal cells to form condensations, and the condensed dermal mesenchyme responds by secreting factors that cause the epidermis to form regionally specific cutaneous structures (Nohno et al. 1995; Ting-Berreth and Chuong 1996). These structures can be the broad feathers of the wing, the narrow feathers of the thigh, or the scales and claws of the feet (**FIGURE 4.19**). The dermal mesenchyme is responsible for the regional specificity of induction in the competent epidermal epithelium. Researchers can separate the embryonic epithelium and mesenchyme from each other and recombine them in different ways (Saunders et al. 1957). The same epithelium develops cutaneous structures according to the region from which the mesenchyme was taken. Here, the mesenchyme plays an instructive role, calling into play different sets of genes in the responding epithelial cells.

GENETIC SPECIFICITY OF INDUCTION   The second property of epithelial-mesenchymal interactions is the genetic specificity of induction. Whereas the mesenchyme may instruct the epithelium as to what sets of genes to activate, the responding epithelium can comply with these instructions only so far as its genome permits. This property was

FIGURE 4.19   Feather induction in the chick. (A) In situ hybridization of a 10-day chick embryo shows Sonic hedgehog expression (dark spots) in the ectoderm of the developing feathers and scales. (B) When cells from different regions of the chick dermis (mesenchyme) are recombined with wing epidermis (epithelium), the type of cutaneous structure made by the epidermal epithelium is determined by the source of the mesenchyme. (A courtesy of W.-S. Kim and J. F. Fallon; after Saunders 1980.)

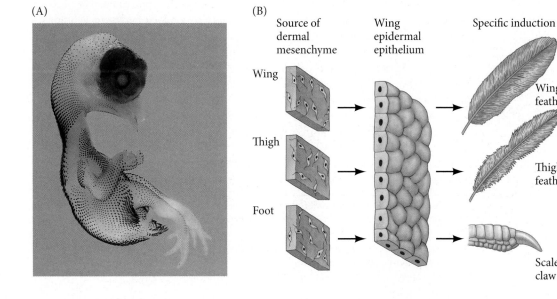

(A)

(B)

Source of dermal mesenchyme

Wing epidermal epithelium

Specific induction

Wing

Thigh

Foot

Wing feather

Thigh feather

Scales, claw

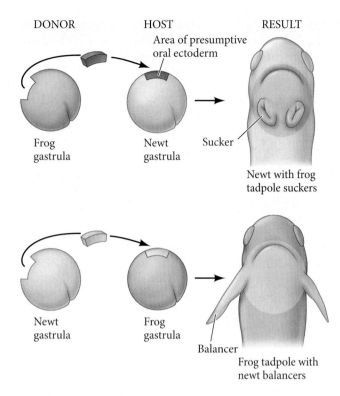

DONOR | HOST | RESULT

Area of presumptive oral ectoderm

Frog gastrula

Newt gastrula

Sucker

Newt with frog tadpole suckers

Newt gastrula

Frog gastrula

Balancer

Frog tadpole with newt balancers

FIGURE 4.20 Genetic specificity of induction in amphibians. Reciprocal transplantation between the presumptive oral ectoderm regions of salamander and frog gastrulae leads to newts with tadpole suckers and tadpoles with newt balancers. (After Hamburgh 1970.)

discovered through experiments involving the transplantation of tissues from one species to another.

In one of the most dramatic examples of interspecific induction, Hans Spemann and Oscar Schotté (1932) transplanted flank ectoderm from an early frog gastrula to the region of a newt gastrula destined to become parts of the mouth. Similarly, they placed presumptive flank ectodermal tissue from a newt gastrula into the presumptive oral regions of frog embryos. The structures of the mouth region differ greatly between salamander and frog larvae. The salamander larva has club-shaped balancers beneath its mouth, whereas the frog tadpole produces mucus-secreting glands and suckers. The frog tadpole also has a horny jaw without teeth, whereas the salamander has a set of calcareous teeth in its jaw. The larvae resulting from the transplants were chimeras. The salamander larvae had frog-like mouths, and the frog tadpoles had salamander teeth and balancers (**FIGURE 4.20**). In other words, the mesenchymal cells instructed the ectoderm to make a mouth, but the ectoderm responded by making the only kind of mouth it "knew" how to make, no matter how inappropriate.[2]

Thus, the instructions sent by the mesenchymal tissue can cross species barriers. Salamanders respond to frog inducers, and chick tissue responds to mammalian inducers. The response of the epithelium, however, is species-specific. So, whereas organ-type specificity (e.g., feather or claw) is usually controlled by the mesenchyme, species specificity is usually controlled by the responding epithelium. As we will see in Chapter 26, major evolutionary changes in the phenotype can be brought about by changing the response to a particular inducer.

The insect trachea: Combining inductive signals with cadherin regulation

Earlier in this chapter we talked about the shared role of cadherins and actinomyosin cortical contraction in mediating cell-to-cell adhesions involved in tissue morphogenesis. Instructions from outside the cell can influence cell shape change through modulation of the cadherin-actinomyosin mechanism. For instance, the tracheal (respiratory) system in *Drosophila* embryos develops from epithelial sacs. The approximately 80 cells in each of these sacs become reorganized into primary, secondary, and tertiary branches without any cell division or cell death (Ghabrial and Krasnow 2006). This reorganization is initiated when nearby cells secrete a protein called Branchless, which acts as a **chemoattractant** (usually a diffusible molecule that attracts a cell to migrate along an increasing concentration gradient toward the source secreting the factor).[3] Branchless binds to a receptor on the cell membranes of the epithelial cells. The cells receiving the most Branchless protein lead the rest, whereas the followers (connected to one another by cadherins) receive a signal from the leading cells to form the tracheal tube (**FIGURE 4.21**). It is the lead cell that will change its shape (by rearranging its actin-myosin cytoskeleton via a Rho GTPase-mediated process) to migrate and form the secondary branches. During this migration, cadherin proteins are regulated such that the epithelial

[2]Spemann is reported to have put it this way: "The ectoderm says to the inducer, 'you tell me to make a mouth; all right, I'll do so, but I can't make your kind of mouth; I can make my own and I'll do that.' " (Quoted in Harrison 1933.)

[3]There are also *chemorepulsive* factors that send the migrating cells in an opposite direction. Generally speaking, chemotactic factors—soluble factors that cause cells to move in a particular direction—are assumed to be chemoattractive unless otherwise described.

cells can migrate over one another to form a tube while keeping their integrity as an epithelium (Cela and Llimagas 2006).

Another external force is also at work, however. The dorsalmost secondary branches of the sacs move along a groove that forms between the developing muscles. These tertiary cell migrations cause the trachea to become segmented around the musculature (Franch-Marro and Casanova 2000). In this way, the respiratory tubes are placed close to the larval musculature.

Paracrine Factors: Inducer Molecules

How are the signals between inducer and responder transmitted? While studying the mechanisms of induction that produce the kidney tubules and teeth, Grobstein (1956) and others (Saxén et al. 1976; Slavkin and Bringas 1976) found that some inductive events could occur despite a filter separating the epithelial and mesenchymal cells. Other inductions, however, were blocked by the filter. The researchers therefore concluded that some of the inducers were soluble molecules that could pass through the small pores of the filter and that other inductive events required physical contact between the epithelial and mesenchymal cells.

When membrane proteins on one cell surface interact with receptor proteins on adjacent cell surfaces (as seen with cadherins), the event is called a **juxtacrine interaction** (since the cell membranes are *juxtaposed*). When proteins synthesized by one cell can diffuse over small distances to induce changes in neighboring cells, the event is called a **paracrine interaction**. Paracrine factors are diffusible molecules that work in a range of about 15 cell diameters, or about 40–200 μm (Bollenbach et al. 2008; Harvey and Smith 2009).

A specific type of paracrine interaction is the **autocrine interaction**. Autocrine interactions occur when the same cells that secrete the paracrine factors also respond to them. In other words, the cell synthesizes a molecule for which it has its own receptor. Although autocrine regulation is not common, it is seen in placental cytotrophoblast cells; these cells synthesize and secrete platelet-derived growth factor, whose receptor is on the cytotrophoblast cell membrane (Goustin et al. 1985). The result is the explosive proliferation of that tissue.

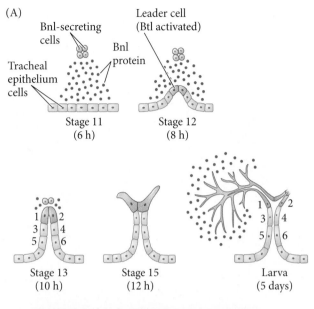

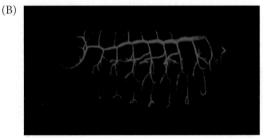

FIGURE 4.21 Tracheal development in *Drosophila.* (A) Diagram of dorsal tracheal branch budding from tracheal epithelium. Nearby cells secrete Branchless protein (Bnl; blue dots), which activates Breathless protein (Btl) on tracheal cells. The activated Btl induces migration of the leader cells and tube formation; the dorsal branch cells are numbered 1 through 6. Branchless also induces unicellular secondary branches (stage 15). (B) Larval *Drosophila* tracheal system visualized with a fluorescent red antibody. Note the intercalated branching pattern. (A after Ghabrial and Krasnow 2006; B from Casanova 2007.)

Morphogen gradients

One of the most important mechanisms governing cell fate specification involves gradients of paracrine factors that regulate gene expression; such signaling molecules are called morphogens. A **morphogen** (Greek, "form-giver") is a diffusable biochemical molecule that can determine the fate of a cell by its concentration.[4] That is, cells exposed to high levels of a morphogen activate different genes than those cells exposed to lower levels. Morphogens can be transcription factors produced within a syncytium of nuclei as in the *Drosophila* blastoderm (see Chapter 2). They can also be paracrine factors that are produced in one group of cells and then travel to another population of cells, specifying the target cells to have similar or different fates according to the concentration of the morphogen. Uncommitted cells exposed to high concentrations of the morphogen

---

[4]Although there is overlap in the terminology, a *morphogen* specifies cells in a quantitative ("more or less") manner, whereas a *morphogenetic determinant* specifies cells in a qualitative ("present or absent") way. Morphogens are analog; morphogenetic determinants are digital.

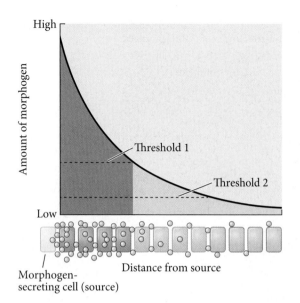

High

Amount of morphogen

Threshold 1

Threshold 2

Low

Distance from source

Morphogen-
secreting cell (source)

FIGURE 4.22 Specification of uniform cells into three cell types by a morphogen gradient. A morphogenetic paracrine factor (yellow dots) is secreted from source cells (yellow) and forms a concentration gradient within the responsive tissue. Cells exposed to morphogen concentrations above threshold 1 activate certain genes (red). Cells exposed to intermediate concentrations (between thresholds 1 and 2) activate a different set of genes (pink) and also inhibit the genes induced at the higher concentrations. Those cells encountering low concentrations of morphogen (below threshold 2) activate a third set of genes (blue). (After Rogers and Schier 2011.)

(nearest its source of production) are specified as one cell type. When the morphogen's concentration drops below a certain threshold, a different cell fate is specified. When the concentration falls even lower, a cell that initially was of the same uncommitted type is specified in yet a third distinct manner (**FIGURE 4.22**).

DEV TUTORIAL *Morphogen Signaling* A lecture and demonstration by Dr. Michael Barresi of some ways in which morphogen signaling operates.

Regulation by gradients of paracrine factor concentration was elegantly demonstrated by the specification of different mesodermal cell types in the frog *Xenopus laevis* by activin, a paracrine factor of the TGF-β family (**FIGURE 4.23**; Green and Smith 1990; Gurdon et al. 1994). Activin-secreting beads were placed on unspecified cells from an early *Xenopus* embryo. The activin then diffused from the beads. At high concentrations (about 300 molecules/cell), activin induced expression of the *goosecoid* gene, whose product is a transcription factor that specifies the frog's dorsal-most structures. At slightly lower concentrations of activin (about 100 molecules per cell), the same tissue activated the *Xbra* gene and was specified to become muscle. At still lower concentrations, these genes were not activated, and the "default" gene expression instructed the cells to become blood vessels and heart (Dyson and Gurdon 1998).

The range of a paracrine factor (and thus the shape of its morphogen gradient) depends on several aspects of that factor's synthesis, transport, and degradation. In some cases, cell surface molecules stabilize the paracrine factor and aid in its diffusion, while in other cases, cell surface moieties retard diffusion and enhance degradation. Such diffusion-regulating interactions between morphogens and extracellular matrix factors are very important in coordinating organ growth and shape (Ben Zvi et al. 2010, 2011).

The induction of numerous organs is effected by a relatively small set of paracrine factors that often function as morphogens. The embryo inherits a rather compact genetic "tool kit" and uses many of the same proteins to construct the heart, kidneys, teeth, eyes, and other organs. Moreover, the same proteins are used throughout the animal kingdom; for instance, the factors active in creating the *Drosophila* eye or heart are very similar to those used in generating mammalian organs. Many paracrine factors can be grouped into one of four major families on the basis of their structure:

1. The fibroblast growth factor (FGF) family
2. The Hedgehog family
3. The Wnt family
4. The TGF-β superfamily, encompassing the TGF-β family, the activin family, the bone morphogenetic proteins (BMPs), the Nodal proteins, the Vg1 family, and several other related proteins

Signal transduction cascades: The response to inducers

For a ligand to induce a cellular response in a cell, it must bind to a receptor, which starts a cascade of events within the cell that ultimately regulate a response. Paracrine

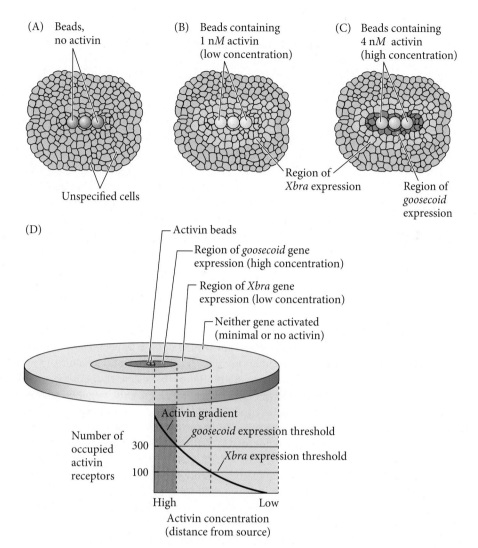

(A) Beads, no activin

Unspecified cells

(B) Beads containing 1 n*M* activin (low concentration)

(C) Beads containing 4 n*M* activin (high concentration)

Region of *Xbra* expression

Region of *goosecoid* expression

(D)

Activin beads

Region of *goosecoid* gene expression (high concentration)

Region of *Xbra* gene expression (low concentration)

Neither gene activated (minimal or no activin)

Activin gradient

goosecoid expression threshold

Xbra expression threshold

Number of occupied activin receptors

300

100

High   Low

Activin concentration (distance from source)

FIGURE 4.23 A gradient of the paracrine factor activin, a morphogen, causes concentration-dependent expression differences of two genes in unspecified amphibian cells. (A) Beads containing no activin did not elicit expression (i.e., mRNA transcription) of either the *Xbra* or *goosecoid* gene. (B) Beads containing 1 n*M* activin elicited *Xbra* expression in nearby cells. (C) Beads containing 4 n*M* activin elicited *Xbra* expression, but only at a distance of several cell diameters from the beads. A region of *goosecoid* expression is seen near the source bead, however. Thus, it appears that *Xbra* is induced at particular concentrations of activin and that *goosecoid* is induced at higher concentrations. (D) Interpretation of the *Xenopus* activin gradient. High concentrations of activin activate *goosecoid*, whereas lower concentrations activate *Xbra*. A threshold value appears to exist that determines whether a cell will express *goosecoid*, *Xbra*, or neither gene. In addition, Brachyury (the *Xbra* protein product in *Xenopus*) inhibits the expression of *goosecoid*, thereby creating a distinct boundary. This pattern correlates with the number of activin receptors occupied on individual cells. (After Gurdon et al. 1994; Dyson and Gurdon 1998.)

factors function by binding to a receptor that initiates a series of enzymatic reactions within the cell. These enzymatic reactions have as their end point either the regulation of transcription factors (such that different genes are expressed in the cells reacting to these paracrine factors) and/or the regulation of the cytoskeleton (such that the cells responding to the paracrine factors alter their shape or are permitted to migrate). These pathways of responses to the paracrine factor often have several end points and are called **signal transduction cascades**.

The major signal transduction pathways all appear to be variations on a common and rather elegant theme, exemplified in **FIGURE 4.24**. Each receptor spans the cell membrane and has an extracellular region, a transmembrane region, and a cytoplasmic region. When a paracrine factor binds to its receptor's extracellular domain, the paracrine factor induces a conformational change in the receptor's structure. This shape change is transmitted through the membrane and alters the shape of the receptor's cytoplasmic domain, giving that domain the ability to activate cytoplasmic proteins. Often such a conformational change confers enzymatic activity on the domain, usually a kinase activity that can use ATP to phosphorylate specific tyrosine residues of particular proteins. Thus, this type of receptor is often called a **receptor tyrosine kinase (RTK)**. The active receptor can now catalyze reactions that phosphorylate other proteins, and this phosphorylation in turn activates their latent activities. Eventually, the *cascade* of phosphorylation activates a dormant transcription factor or a set of cytoskeletal proteins.

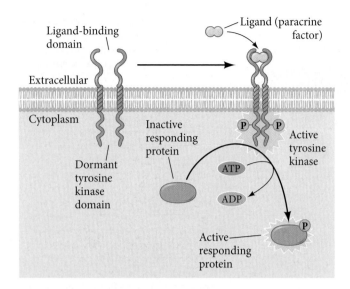

FIGURE 4.24 Structure and function of a receptor tyrosine kinase. The binding of a paracrine factor (such as Fgf8) by the extracellular portion of the receptor protein activates the dormant tyrosine kinase, whose enzyme activity phosphorylates its reciprocal receptor partner followed by specific tyrosine residues of certain intracellular proteins.

FIGURE 4.25 Fgf8 in the developing chick. (A) *Fgf8* gene expression pattern in the 3-day chick embryo, shown by in situ hybridization. Fgf8 protein (dark areas) is seen in the distalmost limb bud ectoderm (1), in the somitic mesoderm (the segmented blocks of cells along the anterior-posterior axis (2), in the branchial arches of the neck (3), at the boundary between the midbrain and hindbrain (4), in the optic vesicle of the developing eye (5), and in the tail (6). (B) In situ hybridization of *Fgf8* in the optic vesicle. The *Fgf8* mRNA (purple) is localized to the presumptive neural retina of the optic cup and is in direct contact with the outer ectoderm cells that will become the lens. (C) Ectopic expression of L-Maf in competent ectoderm can be induced by the optic vesicle (above) and by an Fgf8-containing bead (below). (A courtesy of E. Laufer, C.-Y. Yeo, and C. Tabin; B,C courtesy of A. Vogel-Höpker.)

Below we will describe some of the major characteristics of the four families of paracrine factors, their modes of secretion, gradient manipulation, and the mechanisms underlying transduction in the responding cells. The distinctive roles of each paracrine factor in a variety of developmental processes will be discussed throughout the book, however.

Fibroblast growth factors and the RTK pathway

The **fibroblast growth factor** (**FGF**) family of paracrine factors comprises nearly two dozen structurally related members, and the FGF genes can generate hundreds of protein isoforms by varying their RNA splicing or initiation codons in different tissues (Lappi 1995). Fgf1 protein is also known as acidic FGF and appears to be important during regeneration (Yang et al. 2005), Fgf2 is sometimes called basic FGF and is very important in blood vessel formation, and Fgf7 sometimes goes by the name of keratinocyte growth factor and is critical in skin development. Although FGFs can often substitute for one another, the expression patterns of the FGFs and their receptors give them separate functions.

One member of this family, Fgf8, is especially important during segmentation, limb development, and lens induction. Fgf8 is usually made by the optic vesicle that contacts the outer ectoderm of the head (**FIGURE 4.25A**; Vogel-Höpker et al. 2000). After contact with the outer ectoderm occurs, *Fgf8* gene expression becomes concentrated in the region of the presumptive neural retina (the tissue directly apposed to the presumptive lens) (**FIGURE 4.25B**). Moreover, if Fgf8-containing beads[5] are placed adjacent to head ectoderm, this ectopic Fgf8 will induce this ectoderm to produce ectopic lenses and express the lens-associated transcription factor L-Maf (**FIGURE 4.25C**). FGFs often work by activating a set of receptor tyrosine kinases called the **fibroblast growth factor receptors** (**FGFRs**). For instance, the Branchless protein is an FGFR in *Drosophila*.

When an FGFR binds an FGF ligand (and *only* when it binds an FGF ligand), the dormant kinase is activated and phosphorylates certain proteins (including other FGFRs) within the

---

[5]Synthetic beads can be coated with proteins and placed into the tissue of an embryo. These proteins are released from the bead slowly and then diffuse radially, creating concentration gradients.

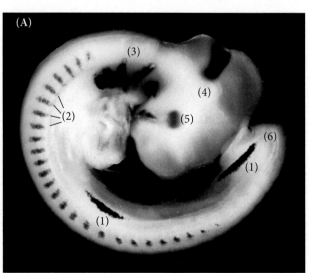

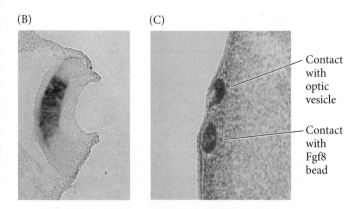

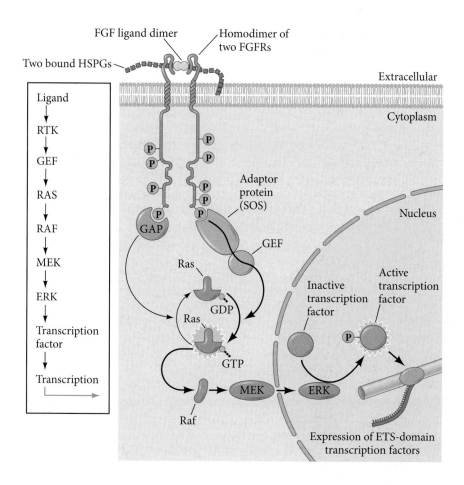

Two bound HSPGs

FGF ligand dimer

Homodimer of two FGFRs

Extracellular

Cytoplasm

Ligand
↓
RTK
↓
GEF
↓
RAS
↓
RAF
↓
MEK
↓
ERK
↓
Transcription factor
↓
Transcription

Adaptor protein (SOS)

Nucleus

GAP

GEF

Ras

GDP

Ras

GTP

Raf

MEK

ERK

Inactive transcription factor

Active transcription factor

Expression of ETS-domain transcription factors

FIGURE 4.26 The widely used RTK signal transduction pathway can be activated by fibroblast growth factor. The receptor tyrosine kinase is dimerized by the ligand (a paracrine factor, such as FGF) along with heparan sulfate proteoglycans (HSPG), which together cause the dimerization and autophosphorylation of the RTKs. The adaptor protein recognizes the phosphorylated tyrosines on the RTK and activates an intermediate protein, GEF, which activates the Ras G-protein by allowing phosphorylation of the GDP-bound Ras. At the same time, the GAP protein stimulates hydrolysis of this phosphate bond, returning Ras to its inactive state. The active Ras activates the Raf protein kinase C, which in turn phosphorylates a series of kinases (such as MEK). Eventually, the activated kinase ERK alters gene expression in the nucleus of the responding cell by phosphorylating certain transcription factors (which can then enter the nucleus to change the types of genes transcribed) and certain translation factors (which alter the level of protein synthesis). In many cases, this pathway is reinforced by the release of calcium ions. A simplified version of the pathway is shown on the left.

responding cell. These proteins, once activated, can perform new functions. The **RTK pathway** was one of the first signal transduction pathways to unite various areas of developmental biology (**FIGURE 4.26**). Researchers studying *Drosophila* eyes, nematode vulvae, and human cancers found that they were all studying the same genes!

Fibroblast growth factors, epidermal growth factors, platelet-derived growth factors, and stem cell factor are all paracrine factors that bind to receptor tyrosine kinase (RTK). Each RTK can bind only one (or one small set) of these ligands, and stable binding requires an additional element, heparan sulfate proteoglycans, or HSPG (Mohammadi et al. 2005; Bökel and Brand 2013). When RTK-ligand binding occurs, RTK undergoes a conformational change that enables it to dimerize with another RTK. This conformational change stimulates the latent kinase activity of each RTK, and these receptors phosphorylate each other on particular tyrosine residues (see Figure 4.26). Thus, the binding of the paracrine factor to its RTK causes a cascade of autophosphorylation of the cytoplasmic domain of the receptor partners. The phosphorylated tyrosine on the receptor is then recognized by an adaptor protein that serves as a bridge linking the phosphorylated RTK to a powerful intracellular signaling system.

While binding to the phosphorylated RTK through one of the RTK's cytoplasmic domains, the adaptor protein also activates a G protein, such as Ras. Normally, the G protein is in an inactive, GDP-bound state. The activated receptor stimulates the adaptor protein to activate the **GTP exchange factor** (**GEF**; also called **guanine nucleotide releasing factor**, or **GNRP**). GEF catalyzes the exchange of GDP with GTP. The GTP-bound G protein is an active form that transmits the signal to the next molecule. After the signal is delivered, the GTP on the G protein is hydrolyzed back into GDP. This catalysis is greatly stimulated by the complexing of the Ras protein with the **GTPase-activating protein** (**GAP**). In this way, the G protein is returned to its inactive state, where it can await further signaling. Without the GAP protein, Ras protein cannot efficiently

FIGURE 4.27 A JAK-STAT pathway: casein gene activation. The gene for casein is activated during the final (lactogenic) phase of mammary gland development, and its activating signal is the secretion of the hormone prolactin from the anterior pituitary gland. Prolactin causes the dimerization of prolactin receptors in the mammary duct epithelial cells. A particular JAK protein (Jak2) is "hitched" to the cytoplasmic domain of these receptors. When the receptors bind prolactin and dimerize, the JAK proteins phosphorylate each other and the dimerized receptors, activating the dormant kinase activity of the receptors. The activated receptors add a phosphate group to a tyrosine residue (Y) of a particular STAT protein, which in this case is Stat5. This addition allows Stat5 to dimerize, be translocated into the nucleus, and bind to particular regions of DNA. In combination with other transcription factors (which presumably have been waiting for its arrival), the Stat5 protein activates transcription of the casein gene. GR is the glucocorticoid receptor, OCT1 is a general transcription factor, and TBP is the major promoter-binding protein that anchors RNA polymerase II (see Chapter 2) and is responsible for binding RNA polymerase II. A simplified diagram is shown on the left. (For details, see Groner and Gouilleux 1995.)

catalyze GTP and so remains in its active configuration for a longer time (Cales et al. 1988; McCormick 1989). Mutations in the *RAS* gene account for a large proportion of cancerous human tumors (Shih and Weinberg 1982), and the mutations of *RAS* that make it oncogenic all inhibit the binding of the GAP protein.

The active Ras G protein associates with a kinase called Raf. The G protein recruits the inactive Raf kinase to the cell membrane, where it becomes active (Leevers et al. 1994; Stokoe et al. 1994). Raf kinase activates the MEK protein by phosphorylating it. MEK is itself a kinase, which activates the ERK protein by phosphorylation. In turn, ERK is a kinase that enters the nucleus and phosphorylates certain transcription factors, many of which belong to the Pea3/Etv4-subfamily (Raible and Brand 2001; Firnberg and Neubüser 2002; Brent and Tabin 2004; Willardsen et al. 2014). The end point of the RTK-signaling pathway is the regulation of expression of a variety of different genes, including but not limited to ones involved in the cell cycle.

FGFs and the JAK-STAT pathway

Fibroblast growth factors can also activate the JAK-STAT cascade. This pathway is extremely important in the differentiation of blood cells, the growth of limbs, and the activation of the casein gene during milk production (**FIGURE 4.27**; Briscoe et al. 1994; Groner and Gouilleux 1995). The cascade starts when a paracrine factor is bound by the extracellular domain of a receptor that spans the cell membrane, with the cytoplasmic domain of the receptor being linked to **JAK** (*Janus kinase*) proteins. The binding of paracrine factor to the receptor activates the JAK kinases and causes them to phosphorylate the **STAT** (*signal transducers and activators of transcription*) family of transcription factors (Ihle 1996, 2001). The phosphorylated STAT is a transcription factor that can now enter into the nucleus and bind to its enhancers.

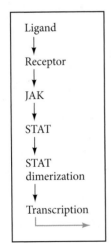

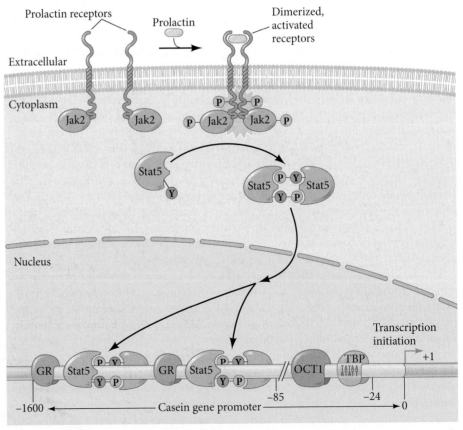

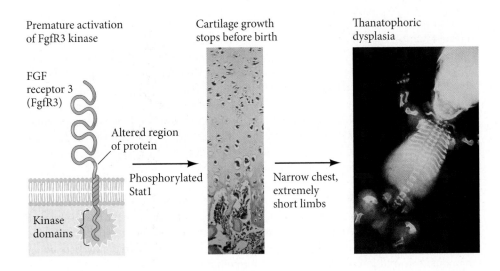

Premature activation of FgfR3 kinase

FGF receptor 3 (FgfR3)

Altered region of protein

Phosphorylated Stat1

Kinase domains

Cartilage growth stops before birth

Narrow chest, extremely short limbs

Thanatophoric dysplasia

FIGURE 4.28 A mutation in the gene for FgfR3 causes the premature constitutive activation of the STAT pathway and the production of phosphorylated Stat1 protein. This transcription factor activates genes that cause the premature termination of chondrocyte cell division. The result is thanatophoric dysplasia, a condition of failed bone growth that results in the death of the newborn infant because the thoracic cage cannot expand to allow breathing. (After Gilbert-Barness and Opitz 1996.)

The JAK-STAT pathway is critically important in regulating human fetal bone growth. Mutations that prematurely activate the STAT pathway have been implicated in some severe forms of dwarfism, such as the lethal condition thanatophoric dysplasia, in which the growth plates of the rib and limb bones fail to proliferate. The short-limbed newborn dies because its ribs cannot support breathing. The genetic lesion responsible is in *FGFR3*, the gene encoding fibroblast growth factor receptor 3 (**FIGURE 4.28**; Rousseau et al. 1994; Shiang et al. 1994). *FGFR3* is expressed in the cartilage precursor cells (chondrocytes) in the growth plates of the long bones. Normally, the FgfR3 protein (a receptor tyrosine kinase) is activated by a fibroblast growth factor and signals the chondrocytes to stop dividing and begin differentiating into cartilage. This signal is mediated by the Stat1 protein, which is phosphorylated by activated FgfR3 and then translocated into the nucleus. Inside the nucleus, Stat1 activates the genes encoding a cell cycle inhibitor, the p21 protein (Su et al. 1997). Thus, the mutations causing thanatophoric dwarfism result from a gain-of-function mutation in the *FGFR3* gene. The mutant receptor gene is active constitutively; that is, it is without the need to be activated by an FGF signal (Deng et al. 1996; Webster and Donoghue 1996). Chondrocytes stop proliferating shortly after they are formed and the bones fail to grow. Other mutations that activate *FGFR3* prematurely but to a lesser degree produce achondroplasic (short-limbed) dwarfism (Legeai-Mallet et al. 2004).

SCIENTISTS SPEAK 4.3 Dr. Francesca Mariani talks about the role of FGF signaling during limb bud outgrowth.

WEB TOPIC 4.2 **FGF RECEPTOR MUTATIONS** Mutations of human FGF receptors have been associated with several skeletal malformation syndromes, including syndromes in which skull, rib, or limb cartilages fail to grow or differentiate.

The Hedgehog family

The proteins of the **Hedgehog family** of paracrine factors are multifunctional signaling proteins that act in the embryo through signal transduction pathways to induce particular cell types and through other means to influence cell guidance. The original *hedgehog* gene was found in *Drosophila*, in which genes are named after their mutant phenotypes: the loss-of-function *hedgehog* mutation causes the fly larva to be covered with pointy denticles on its cuticle (hair-like structures), thus resembling a hedgehog. Vertebrates have at least three homologues of the *Drosophila hedgehog* gene: *sonic hedgehog* (*shh*), *desert hedgehog* (*dhh*), and *indian hedgehog* (*ihh*). The Desert hedgehog protein is found in the Sertoli cells of the testes, and mice homozygous for a null allele of *dhh* exhibit defective spermatogenesis. Indian hedgehog is expressed in the gut and cartilage and is important in postnatal

FIGURE 4.29 Hedgehog processing and secretion. Translation of the *hedgehog* gene in the endoplasmic reticulum produces a Hedgehog protein with autoproteolytic activity that cleaves off the carboxyl terminus (C) to reveal a signal sequence that marks the protein for secretion. The freed C-terminal segment is not involved in signaling and is often degraded, whereas the amino-terminal portion (N) of the molecule becomes the active Hedgehog protein intended for secretion. Secretion requires the addition of cholesterol and palmitic acid to the Hedgehog protein (Briscoe and Thérond 2013). Interactions between the cholesterol moiety and a transmembrane protein called Dispatched enables Hedgehog to be secreted and diffuse as monomers; both cholesterol and palmitic acid are required for multimeric assembly. In addition, Hedgehog interactions with a class of membrane-associated heparan sulfate proteoglycans (HSPGs) foster the congregation and secretion of Hedgehog molecules as lipoprotein assemblies (Breitling 2007; Guerrero and Chiang 2007). Similar clustering of Hedgehog can be used to transport Hedgehog out of the cell within exovesicles.

bone growth (Bitgood and McMahon 1995; Bitgood et al. 1996). Sonic hedgehog[6] has the greatest number of functions of the three vertebrate Hedgehog homologues. Among other important functions, Sonic hedgehog is responsible for assuring that motor neurons come only from the ventral portion of the neural tube (see Chapter 13), that a portion of each somite forms the vertebrae (see Chapter 17), that the feathers of the chick form in their proper places (see Figure 4.19), and that our pinkies are always our most posterior digits (see Chapter 19). Hedgehog signaling is capable of regulating these many developmental events because they function as morphogens; Hedgehog proteins are secreted from a cellular source, displayed in a spatial gradient, and induce differential gene expression at different threshold concentrations that result in distinct cell identities.

HEDGEHOG SECRETION   Different modes of processing and assembly of Hedgehog proteins can significantly alter the amount secreted and the gradient that is formed (**FIGURE 4.29**). By cleaving off its carboxyl terminus and associating with both cholesterol and palmitic acid moieties, Hedgehog protein can be processed and secreted as monomers or multimers, packaged as lipoprotein assemblies, or even transported out of the cell within exovesicles.

In the mouse limb bud, it was shown that if Shh lacks the cholesterol modification, it diffuses too quickly and dissipates into the surrounding space (Li et al. 2006). These lipid modifications are also required for stable concentration gradients of Hedgehog and pathway activation. Through these varied protein processing and transport mechanisms, stable gradients of Hedgehog can be established over distances of several hundred microns (about 30 cell diameters in the mouse limb).

---

[6]Yes, it is named after the Sega Genesis character. The vertebrate *Hedgehog* genes were discovered by searching vertebrate gene libraries (chick, rat, zebrafish) with probes that found sequences similar to that of the fruit fly *hedgehog* gene. Riddle and colleagues (1993) discovered three genes homologous to *Drosophila hedgehog*. Two were named after existing species of hedgehog, and the third was named after the animated character. Two other *hedgehog* genes, found only in fish, were originally named *echidna hedgehog* (possibly after Sonic's cartoon friend) and *tiggywinkle hedgehog* (after Beatrix Potter's fictional hedgehog), but they are now referred to as *ihh-b* and *shh-b*, respectively.

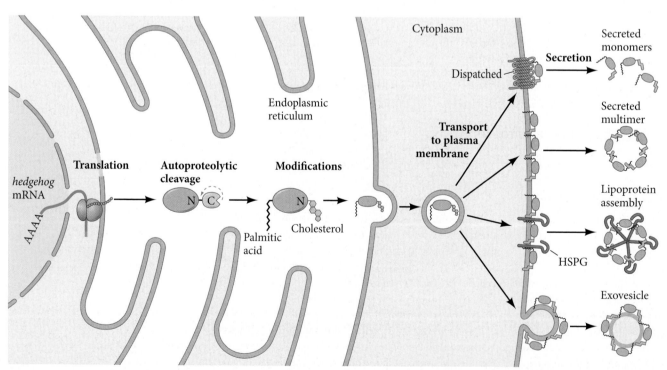

THE HEDGEHOG PATHWAY   The cholesterol moiety on Hedgehog is not only important for modulating its extracellular transport; it is also critical for Hedgehog to anchor to its receptor on the receiving cell's plasma membrane (Grover et al. 2011). The Hedgehog binding receptor is called Patched, which is a large, 12-pass transmembrane protein (**FIGURE 4.30**). Patched, however, is not a signal transducer. Rather, the Patched protein represses the function of another transmembrane receptor called Smoothened. *In the absence of Hedgehog* binding to Patched, Smoothened is inactive and degraded, and a

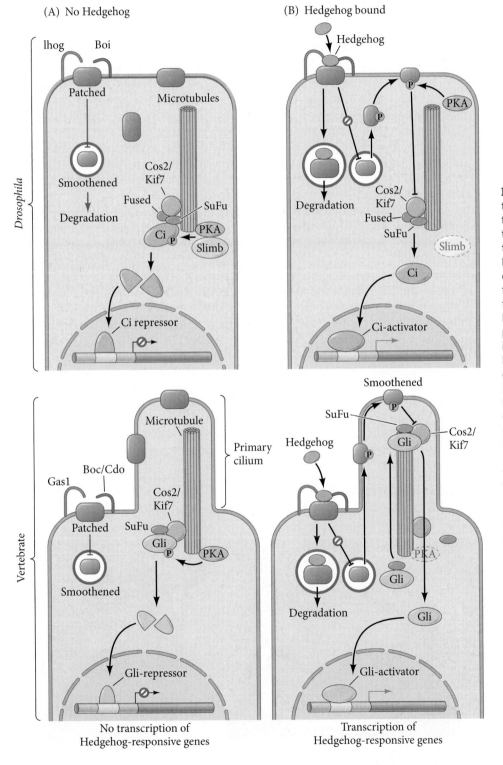

FIGURE 4.30 Hedgehog signal transduction pathway. Patched protein in the cell membrane is an inhibitor of the Smoothened protein. (A) In the absence of Hedgehog binding to Patched, Patched inhibits Smoothened and in *Drosophila melanogaster* the Ci protein remains tethered to the microtubules by the Cos2 and Fused proteins. This tether allows the proteins PKA and Slimb to cleave Ci into a transcriptional repressor that blocks the transcription of particular genes. (B) When Hedgehog binds to Patched, its conformational changes release the inhibition of the Smoothened protein. Smoothened then releases Ci from the microtubules, inactivating the cleavage proteins PKA and Slimb. The Ci protein enters the nucleus, and acts as a transcriptional activator of particular genes. In vertebrates (lower panels), the homologues of Ci are the *Gli* genes, which function similarly as transcriptional activators or repressors when a hedgehog ligand is bound to Patched or absent, respectively. Additionally in vertebrates, for Smoothened to positively regulate Gli processing into an activator form, it needs to gain access into the primary cilium— hedgehog ligand binding to patched enables the transport of Smoothened into the primary cilium. Lastly, several co-receptors such as Gas1 and Boc function to enhance hedgehog signaling. (After Johnson and Scott 1998; Briscoe and Thérond 2013; Yao and Chuang 2015.)

transcription factor—Cubitus interruptus (Ci) in *Drosophila* or one of its vertebrate homologues Gli1, Gli2, and Gli3—is tethered to the microtubules of the responding cell. Although tethered to the microtubules, Ci/Gli is cleaved in such a way that a portion of it enters the nucleus and acts as a transcriptional repressor. This cleavage reaction is catalyzed by several proteins that include Fused, Suppressor of fused (SuFu), and Protein kinase A (PKA). *When Hedgehog is present,* the responding cells express several additional co-receptors (Ihog/Cdo, Boi/Boc, and Gas1) that together foster strong Hedgehog-Patched interactions. Upon binding, the Patched protein's shape is altered such that it no longer inhibits Smoothened, and Patched enters an endocytic pathway for degradation. Smoothened releases Ci/Gli from the microtubules (probably by phosphorylation), and the full-length Ci/Gli protein can now enter the nucleus to act as a transcriptional *activator* of the same genes the cleaved Ci/Gli used to repress (see Figure 4.30; Yao and Chuang 2015; Briscoe and Thérond 2013; Lum and Beachy 2004).

There are other targets for Hedgehog signaling independent of Gli transcription factors, and they involve the fast remodeling of the actin cytoskeleton, resulting in directed migration of the responding cells. For instance, the Charron lab has shown that pathfinding axons in the neural tube can sense the presence of a gradient of Sonic hedgehog emanating from the floorplate, which will serve to attract commissural neurons to turn toward the midline and cross to the other hemisphere of the nervous system (Yam et al. 2009; Sloan et al. 2015). We will discuss the mechanisms of axon guidance in greater detail in Chapter 15.

The Hedgehog pathway is extremely important in vertebrate limb patterning, neural differentiation and pathfinding, retinal and pancreas development, and craniofacial morphogenesis, among many other processes (**FIGURE 4.31A**; McMahon et al. 2003). When mice were made homozygous for a mutant allele of *Sonic hedgehog,* they had major limb and facial abnormalities. The midline of the face was severely reduced, and a single eye formed in the center of the forehead, a condition known as cyclopia after the one-eyed Cyclops of Homer's *Odyssey* (**FIGURE 4.31B**; Chiang et al. 1996). Some human cyclopia syndromes are caused by mutations in genes that encode either Sonic hedgehog or the enzymes that synthesize cholesterol (Kelley et al. 1996; Roessler et al. 1996; Opitz and Furtado 2013). Moreover, certain chemicals that induce cyclopia do so by interfering with the Hedgehog pathway (Beachy et al. 1997; Cooper et al. 1998). Two teratogens[7] known to cause cyclopia in vertebrates are jervine and cyclopamine. Both are alkaloids found in the plant *Veratrum californicum* (corn lily), and both directly bind to and inhibit Smoothened function (see Figure 4.31B; Keeler and Binns 1968).

In later development, Sonic hedgehog is critical for feather formation in the chick embryo, for hair formation in mammals, and, when misregulated, for the formation of

---

[7]A *teratogen* is an exogenous compound capable of causing malformations in embryonic development; see Chapters 1 and 24.

FIGURE 4.31 (A) Sonic hedgehog is shown by in situ hybridization to be expressed in the nervous system (red arrow), gut (blue arrow), and limb bud (black arrow) of a 3-day chick embryo. (B) Head of a cyclopic lamb born of a ewe that ate *Veratrum californicum* early in pregnancy. The cerebral hemispheres fused, resulting in the formation of a single, central eye and no pituitary gland. The jervine alkaloid made by this plant inhibits cholesterol synthesis, which is needed for Hedgehog production and reception. (A courtesy of C. Tabin; B courtesy of L. James and USDA Poisonous Plant Laboratory.)

(A)

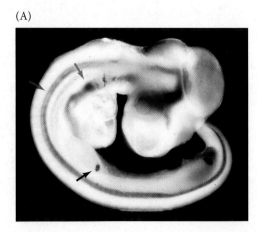

(B)

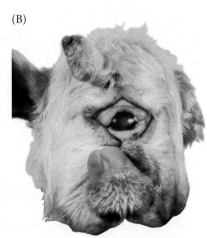

skin cancer in humans (Harris et al. 2002; Michino et al. 2003). Although mutations that inactivate the Hedgehog pathway can cause malformations, mutations that activate the pathway ectopically can have mitogenic effects and cause cancers. If the Patched protein is mutated in somatic tissues such that it can no longer inhibit Smoothened, it can cause tumors of the basal cell layer of the epidermis (basal cell carcinomas). Heritable mutations of the *patched* gene cause basal cell nevus syndrome, a rare autosomal dominant condition characterized by both developmental anomalies (fused fingers; rib and facial abnormalities) and multiple malignant tumors (Hahn et al. 1996; Johnson et al. 1996). Interestingly, vismodegib, a compound that inhibits Smoothened function similar to cyclopamine, is currently in clinical trials as a therapy to combat basal cell carcinomas (Dreno et al. 2014; Erdem et al. 2015). (What do you think the warnings for pregnancy should be on this drug?)

 SCIENTISTS SPEAK 4.4 Dr. James Briscoe answers questions on the role of Hedgehog signaling during neural tube development.

 SCIENTISTS SPEAK 4.5 Dr. Marc Tessier-Lavigne speaks on the role of Hedgehog as a noncanonical axon guidance cue.

> **VADE MECUM**
>
> The segment on zebrafish development demonstrates how alcohol can induce cyclopia in these embryos.

The Wnt family

The Wnts are paracrine factors that make up a large family of cysteine-rich glycoproteins with at least 11 conserved Wnt members among vertebrates (Nusse and Varmus 2012); 19 separate *Wnt* genes are found in humans![8] The Wnt family was originally discovered and named *wingless* during a forward genetic screen in *Drosophila melanogaster* in 1980 by Christiane Nüsslein-Volhard and Eric Wieschaus, when mutations in this locus prevented the formation of the wing. The Wnt name is a fusion of the *Drosophila* segment polarity gene *wingless* with the name of one of its vertebrate homologues, *integrated*. The enormous array of different *Wnt* genes across species speaks to their importance in an equally large number of developmental events. For example, Wnt proteins are critical in establishing the polarity of insect and vertebrate limbs, in promoting the proliferation of stem cells, in regulating cell fates along axes of various tissues, in development of the mammalian urogenital system (**FIGURE 4.32**), and in guiding the migration of mesenchymal cells and pathfinding axons. How is it that Wnt signaling is capable of mediating such diverse processes as cell division, cell fate, and cell guidance?

WNT SECRETION Similar to the building of the functional Hedgehog proteins, Wnt proteins are synthesized in the endoplasmic reticulum and modified by the addition of lipids (palmitic and palmitoleic acid). These lipid modifications are catalyzed by the *O*-acetyltransferase Porcupine. (*How do you think this enzyme received this name?*[9]) It is interesting that loss of the *Porcupine* gene results in reduced Wnt secretion paired with its build up in the endoplasmic reticulum (van den Heuvel et al. 1993; Kadowaki et al. 1996), indicating that adding lipids to Wnt is important for transporting it to the plasma membrane. Once at the plasma membrane, Wnt can be secreted by the same mechanisms we saw for Hedgehog

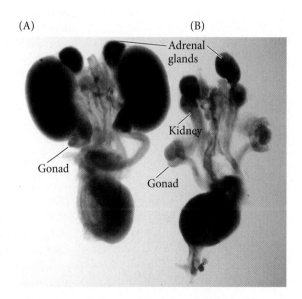

FIGURE 4.32 Wnt4 is necessary for kidney development and for female sex determination. (A) Urogenital rudiment of a wild-type newborn female mouse. (B) Urogenital rudiment of a newborn female mouse with targeted knockout of *Wnt4* shows that the kidney fails to develop. In addition, the ovary starts synthesizing testosterone and becomes surrounded by a modified male duct system. (Courtesy of J. Perasaari and S. Vainio.)

---

[8]A comprehensive summary of all the Wnt proteins and Wnt signaling components can be found at http://web.stanford.edu/group/nusselab/cgi-bin/wnt/.

[9]In flies, the mutated *Porcupine* gene results in segmentation defects creating denticles resembling porcupine spines in the larva (Perrimon et al. 1989). Do you recall the naming of Hedgehog? Porcupine is specific to Wnt palmitoylation, whereas Hedgehog is palmitoylated by a similar enzyme called Hhat.

FIGURE 4.33 Notum antagonism of Wnt. (A) Structures of Notum (gray) and Wnt3A (green) bound together. The active site of Notum is visualized in this cutaway view demonstrating the precise binding with the palmitoleic acid moiety of Wnt3A (orange). (B) Once bound, Notum possesses the enzymatic hydrolase activity to cleave this lipid off of Wnt3A, rendering it unable to interact with the Frizzled receptor. The data shown here demonstrate the requirement of this hydrolase function for appropriate delipidation of Wnt3A. Notum lacking its enzymatic activity is unable to remove the lipid group from Wnt3A (Delipidated, purple bars) as compared to wild-type Notum. (C) Model of extracellular regulation of Wnt. Lipidated Wnt can bind both its Frizzled receptor and glypicans (heparan sulfate proteoglycans). Active Wnt signaling leads to the upregulation of Notum, which is secreted and interacts with glypicans, where it will also bind to and cleave off the palmitoleic acid portions of Wnt proteins. In this way, Wnt signaling leads to a Notum-mediated negative feedback mechanism. (A created by Matthias Zebisch; data from Kakugawa et al. 2015; courtesy of Yvonne Jones and Jean-Paul Vincent.)

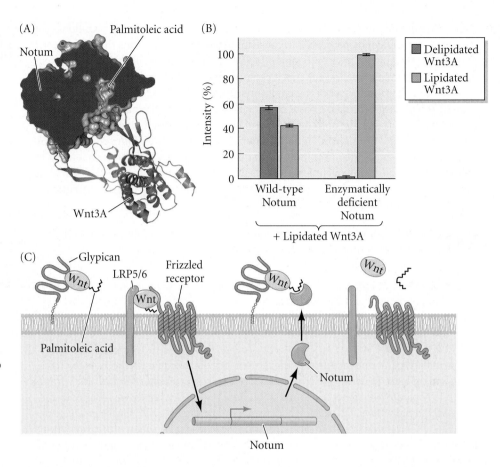

protein: by free diffusion, by being transported in exosomes, or by being packaged in lipoprotein particles (Tang et al. 2012; Saito-Diaz et al. 2013; Solis et al. 2013).

Upon secretion, Wnt proteins associate with glypicans (a type of heparan sulfate proteoglycan) in the extracellular matrix, which restricts diffusion and leads to a greater accumulation of Wnt closer to the source of production. When Wnt attaches to the Frizzled receptor on a responding cell, the cell secretes Notum, a hydrolase that associates with glypican and then cleaves off Wnt's attached lipids in a process of *deacylation* or *delipidation* (Kakugawa et al. 2015). This process reduces Wnt signaling because the lipids are essential for Wnt to bind to Frizzled, which creates a negative feedback mechanism for preventing excessive Wnt signaling. The Frizzled receptor possesses a unique hydrophobic cleft adapted to interact with lipidated Wnts, a binding conformation mimicked in Notum's structure as well (**FIGURE 4.33A,B**). Overexpression of Notum in the *Drosophila* imaginal wing disc causes a reduction in Wnt/Wg target gene expression; in contrast, clonal loss of *Notum* yields to expanded Wnt target gene expression. Interestingly, *Notum* gene expression is upregulated in Wnt-responsive cells, creating a mechanism of negative feedback (**FIGURE 4.33C**; Kakugawa et al. 2015; Nusse 2015). Notum is not alone in functioning to inhibit binding of Wnt to its receptor; numerous antagonists exist, including the Secreted frizzled-related protein (Sfrp), Wnt inhibitory factor (Wif), and members of the Dickkopf (Dkk) family (Niehrs 2006). Together, the multiple modes of Wnt secretion, glypican-mediated restriction, secreted ligand inhibitors, and negative feedback establish stable gradients of Wnt ligands and pathway response.

THE CANONICAL WNT PATHWAY (β-CATENIN DEPENDENT) The first Wnt pathway to be characterized was the canonical "Wnt/β-catenin pathway," which represents the signaling events that culminate in the activation of the β-catenin transcription factor and modulation of specific gene expression (**FIGURE 4.34A**; Chien et al. 2009; Clevers

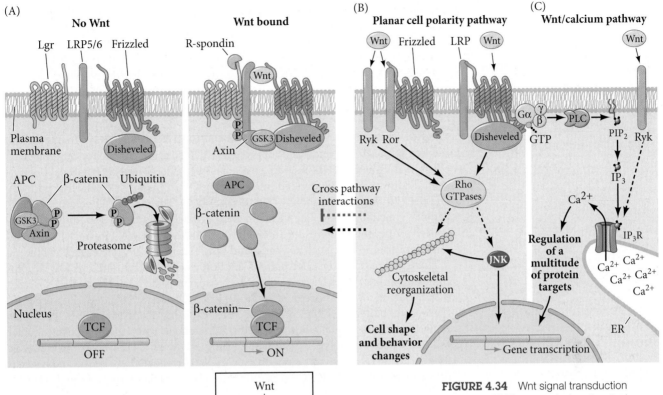

FIGURE 4.34 Wnt signal transduction pathways. (A) The canonical, or β-catenin-dependent, Wnt pathway. The Wnt protein binds to its receptor, a member of the Frizzled family, but it often does so in combination with interactions with LRP5/6 and Lgr receptors. During periods of Wnt absence, β-catenin interacts with a complex of proteins, including GSK3, APC, and Axin, that target Wnt for protein degradation in the proteasome. The downstream transcriptional effector of Wnt signaling is the β-catenin transcription factor. In the presence of certain Wnt proteins, Frizzled then activates Disheveled, allowing Disheveled to become an inhibitor of glycogen synthase kinase 3 (GSK3). GSK3, if it were active, would prevent the dissociation of β-catenin from the APC protein. So, by inhibiting GSK3, the Wnt signal frees β-catenin to associate with it's co-factors (LEF or TCF) and become an active transcription factor. (B,C) Alternatively, noncanonical (β-catenin-independent) Wnt signaling pathways can regulate cell morphology, division, and movement. (B) Certain Wnt proteins can similarly signal through Frizzled to activate Disheveled but in a way that leads to the activation of Rho GTPases, like Rac and RhoA. These GTPases coordinate changes in cytoskeleton organization and also through janus kinase (JNK) regulate gene expression. (C) In a third pathway, certain Wnt proteins activate Frizzled and Ryk receptors in a way that releases calcium ions and can result in Ca²⁺-dependent gene expression. (After MacDonald et al. 2009.)

and Nusse 2012; Nusse 2012; Saito-Diaz et al. 2013). In Wnt/β-catenin signaling, lipidated Wnt family members interact with a pair of transmembrane receptor proteins: one from the Frizzled family and one large transmembrane protein called LRP5/6 (Logan and Nusse 2004; MacDonald et al. 2009). *In the absence of Wnts*, the transcriptional cofactor β-catenin is constantly being degraded by a protein degradation complex containing several proteins (such as axin and APC) as well as **glycogen synthase kinase 3 (GSK3)**. GSK3 phosphorylates β-catenin so that it will be recognized and degraded by proteosomes. The result is Wnt-responsive genes being repressed by the LEF/TCF transcription factor, which functionally complexes with at least two other proteins, including a histone deacetylase.

When Wnts come into contact with a cell, they bring together the Frizzled and LRP5/6 receptors to form a multimeric complex. This linkage enables LRP5/6 to bind both Axin and GSK3, and enables the Frizzled protein to bind Disheveled—all of which occurs on the intracellular side of the plasma membrane. Disheveled keeps Axin and GSK3 bound to the cell membrane and thereby prevents β-catenin from being phosphorylated by GSK3. This process stabilizes β-catenin, which accumulates and enters the nucleus (see Figure 4.34A). Here it binds to the LEF/TCF transcription factor and converts this former repressor into a transcriptional activator, thereby activating Wnt-responsive genes (Cadigan and Nusse 1997; Niehrs 2012).

This model is undoubtedly an oversimplification because different cells use the pathway in different ways (see McEwen and Peifer 2001; Clevers and Nusse 2012; Nusse 2012; Saito-Diaz et al. 2013). One overriding principle already evident in both the Wnt and Hedgehog pathways, however, is that *activation is often accomplished by inhibiting an inhibitor.*

THE NONCANONICAL WNT PATHWAYS (β-CATENIN INDEPENDENT) In addition to sending signals to the nucleus, Wnt proteins can also cause changes within the cytoplasm that influence cell function, shape, and behavior. These alternative or *noncanonical* pathways can be divided into two types: the "planar cell polarity" pathway and the "Wnt/calcium" pathway (**FIGURE 4.34B,C**). The planar cell polarity, or PCP, pathway functions to regulate the actin and microtubule cytoskeleton, thus influencing cell shape, and often results in bipolar protrusive behaviors necessary for a cell to migrate. Certain Wnts (such as Wnt5a and Wnt11) can activate Disheveled by binding to a different receptor (Frizzled paired with Ror instead of Lrp5), and this Ror receptor complex phosphorylates Disheveled in a way that allows it to interact with Rho GTPases (Grumolato et al. 2010; Green et al. 2014). Rho GTPases are colloquially viewed as the "master builders" of the cell because they can activate an array of other proteins (kinases and cytoskeletal binding proteins) that remodel cytoskeletal elements to alter cell shape and movement. Wnt signaling through the PCP pathway is most notable for instructing cell behaviors along the same spatial plane within a tissue and hence is called *planar polarity*. Wnt/PCP signaling through cytoskeleton control can direct cells to divide in the same plane (rather than forming upper and lower tissue compartments) and to move within that same plane (Shulman et al. 1998; Winter et al. 2001; Ciruna et al. 2006; Witte et al. 2010; Sepich et al. 2011; Ho et al. 2012; Habib et al. 2013). In vertebrates, this regulation of cell division and migration is important for establishing germ layers and for anterior-posterior axis extension during gastrulation and neurulation.

As its name implies, the Wnt/calcium pathway leads to the release of calcium stored within cells, and this released calcium acts as an important *secondary messenger* to modulate the function of many downstream targets. In this pathway, Wnt binding to the receptor protein (possibly Ryk, alone or in concert with Frizzled) activates a phospholipase (PLC) whose enzyme activities release a compound that in turn releases calcium ions from the smooth endoplasmic reticulum (see Figure 4.34C). The released calcium can activate enzymes, transcription factors, and translation factors. In zebrafish, Ryk deficiency impairs Wnt-directed calcium release from internal stores and as a result impairs directional cell movement (Lin et al. 2010; Green et al. 2014). Ryk has been demonstrated to be cleaved and transported into the nucleus, where it plays roles in mammalian neural development and *C. elegans* vulval development (Lyu et al. 2008; Poh et al. 2014).

Although each of the three Wnt pathways—β-catenin, PCP, and calcium—possess primary functions that are different from one another, mounting evidence suggests that there are significant cross interactions between these pathways (van Amerongen and Nusse 2009; Thrasivoulou et al. 2013). For instance, Wnt5-mediated calcium signaling has been shown to *antagonize* the Wnt/β-catenin pathway during vertebrate gastrulation and limb development (Ishitani et al. 2003; Topol et al. 2003; Westfall et al. 2003).

The TGF-β superfamily

There are more than 30 structurally related members of the **TGF-β superfamily**,[10] and they regulate some of the most important interactions in development (**FIGURE 4.35**). The TGF-β superfamily includes the TGF-β family, the Nodal and activin families, the

---

[10]TGF stands for *Transforming Growth Factor*. The designation "superfamily" is often applied when each of the different classes of molecules constitutes a family. The members of a superfamily all have similar structures but are not as similar as the molecules within each family are to one another.

> ### Developing Questions
>
> How different are the pathways of Wnt/β-catenin, calcium, and PCP? Arguably the most significant challenge to understanding Wnt signaling is figuring out how the different pathways interact. Perhaps we need a more integrated comprehension of signal transduction, one that can predict interactions not only between canonical and noncanonical Wnt signaling pathways, but among those for all the paracrine factors (Wnt, Hedgehog, FGF, BMP, etc.). What do you think? How would you go about trying to examine meaningful pathway interactions?

bone morphogenetic proteins (BMPs), the Vg1 family, and other proteins, including glial-derived neurotrophic factor (GDNF; necessary for kidney and enteric neuron differentiation) and anti-Müllerian hormone (AMH), a paracrine factor involved in mammalian sex determination. Below we summarize three of these families most widely used throughout development: TGF-βs, BMPs and Nodal/Activin.

- Among members of the **TGF-β family**, TGF-β1, 2, 3, and 5 are important in regulating the formation of the extracellular matrix between cells and for regulating cell division (both positively and negatively). TGF-β1 increases the amount of extracellular matrix that epithelial cells make (both by stimulating collagen and fibronectin synthesis and by inhibiting matrix degradation). TGF-β proteins may be critical in controlling where and when epithelia branch to form the ducts of kidneys, lungs, and salivary glands (Daniel 1989; Hardman et al. 1994; Ritvos et al. 1995). The effects of the individual TGF-β family members are difficult to sort out because members of the TGF-β family appear to function similarly and can compensate for losses of the others when expressed together.

- The members of the **BMP family** can be distinguished from other members of the TGF-β superfamily by having seven (rather than nine) conserved cysteines in the mature polypeptide. Because they were originally discovered by their ability to induce bone formation, they were given the name **bone morphogenetic proteins**. It turns out, though, that bone formation is only one of their many functions; the BMPs are extremely multifunctional.[11] They have been found to regulate cell division, apoptosis (programmed cell death), cell migration, and differentiation (Hogan 1996). They include proteins such as BMP4 (which in some tissues causes bone formation, in other tissues specifies epidermis, and in other instances causes cell proliferation or cell death) and BMP7 (which is important in neural tube polarity, kidney development, and sperm formation). The BMP4 homologue in *Drosophila* is critically involved in forming appendages, including the limbs, wings, genitalia, and antennae. Indeed, the malformations of 15 such structures have given this homologue the name Decapentaplegic (DPP). As it (rather oddly) turns out, BMP1 is not a member of the BMP family at all; rather, it is a protease. BMPs are thought to work by diffusion from the cells producing them (Ohkawara et al. 2002). Inhibitors such as Noggin and Chordin that bind directly to BMP reduce BMP-receptor interactions. We will cover this morphogenetic mechanism more directly when we discuss dorsoventral axis specification in the gastrula.

- The **Nodal** and **activin** proteins are extremely important in specifying the different regions of the mesoderm and for distinguishing the left and right sides of the vertebrate body axis. The left-right asymmetry of bilateral organisms is strongly influenced by a gradient of Nodal from right to left across the embryo. In vertebrates, this Nodal gradient appears to be created by the beating of motile cilia that promotes the graded flow of Nodal across the midline (Babu and Roy 2013; Molina et al. 2013; Blum et al. 2014; Su 2014).

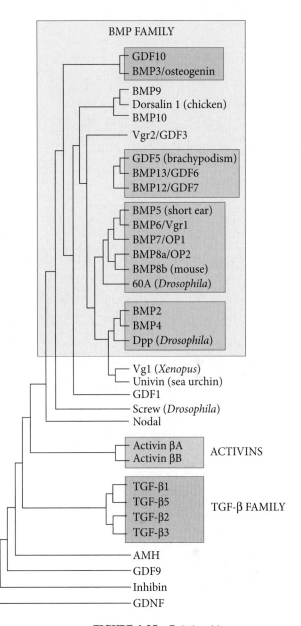

FIGURE 4.35 Relationships among members of the TGF-β superfamily. (After Hogan 1996.)

---

[11] One of the many reasons humans do not seem to need an enormous genome is that the gene products—proteins—involved in our construction and development often have many functions. Many of the proteins that we are familiar with in adults (such as hemoglobin, keratins, and insulin) do have only one function, which led to the erroneous conclusion that this situation is the norm.

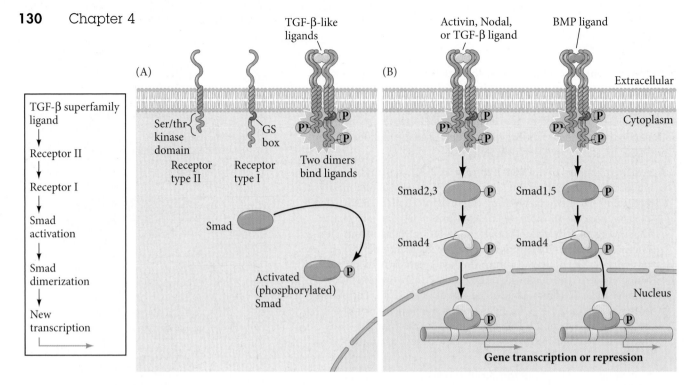

FIGURE 4.36 The Smad pathway is activated by TGF-β superfamily ligands. (A) An activation complex is formed by the binding of the ligand by the type I and type II receptors, which allows the type II receptor to phosphorylate the type I receptor on particular serine or threonine residues. The phosphorylated type I receptor protein can now phosphorylate the Smad proteins. (B) Those receptors that bind TGF-β family proteins or members of the activin family phosphorylate Smads 2 and 3. Those receptors that bind to BMP family proteins phosphorylate Smads 1 and 5. These Smads can complex with Smad4 to form active transcription factors. A simplified version of the pathway is shown on the left.

THE SMAD PATHWAY   Members of the TGF-β superfamily activate members of the **Smad family** of transcription factors (Heldin et al. 1997; Shi and Massagué 2003). The TGF-β ligand binds to a type II TGF-β receptor, which allows that receptor to bind to a type I TGF-β receptor. Once the two receptors are in close contact, the type II receptor phosphorylates a serine or threonine on the type I receptor, thereby activating it. The activated type I receptor can now phosphorylate the Smad[12] proteins (**FIGURE 4.36A**). Smads 1 and 5 are activated by the BMP family of TGF-β factors, whereas the receptors binding activin, Nodal, and the TGF-β family phosphorylate Smads 2 and 3. These phosphorylated Smads bind to Smad4 and form the transcription factor complexes that will enter the nucleus (**FIGURE 4.36B**).

Other paracrine factors

Although most paracrine factors are members of one of the four families described above (FGF, Hedgehog, Wnt, or the TGF-β superfamily), some paracrine factors have few or no close relatives. Epidermal growth factor, hepatocyte growth factor, neurotrophins, and stem cell factor are not included among these four groups, but each plays important roles during development. In addition, there are numerous paracrine factors involved almost exclusively with developing blood cells: erythropoietin, the cytokines, and the interleukins. Another class of paracrine factors was first characterized for their role in cell/axon guidance and include members of the Netrin, Semaphorin, and Slit families. These classic guidance molecules (such as netrins) are now being shown to regulate gene expression as well. We will discuss all these paracrine factors in the context of their developmental relevance later in the book.

The Cell Biology of Paracrine Signaling

We have been discussing cell membrane dynamics and cell signaling as if they were two separate entities, but their functioning is closely related. Paracrine factors can rearrange the cell surface, and the cell surface is critical in regulating paracrine factor synthesis, flow, and function. The actions of paracrine signals often change the composition of the cell membrane.

---

[12]Researchers named the Smad proteins by merging the names of the first identified members of this family: the *C. elegans* SMA protein and the *Drosophila* Mad protein.

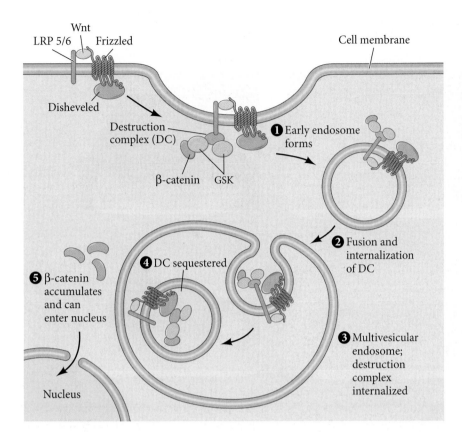

Wnt
LRP 5/6   Frizzled

Disheveled

Destruction
complex (DC)

β-catenin   GSK

Cell membrane

❶ Early endosome
forms

❷ Fusion and
internalization
of DC

❹ DC sequestered

❺ β-catenin
accumulates
and can
enter nucleus

❸ Multivesicular
endosome;
destruction
complex
internalized

Nucleus

FIGURE 4.37   A Wnt pathway: packaging the β-catenin destruction apparatus into endosomes. A major mechanism for separating β-catenin from enzymes that would otherwise destroy it is to package the complex in membrane-bound vesicles called endosomes. When Wnt binds to Frizzled, Frizzled can bind the destruction complex; the entire complex (including the bound Wnt and its receptor) is internalized, allowing β-catenin to accumulate rather than being degraded. (After Taelman et al. 2010.)

ENDOSOME INTERNALIZATION   The type and number of receptors that a cell displays at its cell surface present its potential for response. Endocytosis is one mechanism used to eliminate a receptor at the membrane. Recent studies are revealing that internalization of ligand-receptor complexes into membrane-bound vesicles called **endosomes** is a common mechanism in paracrine signaling. When Wnt binds to its receptors, the β-catenin destruction complex binds to the receptor, and the entire complex (including the receptor and its bound Wnt) is internalized in endosomes (**FIGURE 4.37**; Taelman et al. 2010; Niehrs 2012). This process removes the complex, targets it for degradation, and enables the survival of β-catenin. The internalization of the signaling complex appears to be critical for the accumulation of β-catenin, and proteins that aid in this endocytosis (such as R-spondins; see Figure 4.34) make the Wnt pathway more efficient (Ohkawara et al. 2011). Similarly, Hedgehog-Patched complexes and FGF-FGFR complexes are also internalized in endosomes and targeted for degradation, a process that is required for proper limb development (Briscoe and Thérond 2013; Handschuh et al. 2014; Hsia et al. 2015).

DIFFUSION OF PARACRINE FACTORS   Paracrine factors do not flow freely through the extracellular space. Rather, factors can be bound by cell membranes and extracellular matrices of the tissues. In some cases, such binding can impede the spread of a paracrine morphogen and even target the paracrine factor for degradation (Capurro et al. 2008; Schwank et al. 2011). Wnt proteins, for instance, do not diffuse far from the cells secreting them unless helped by other proteins. Thus, the range of Wnt factors is significantly extended when the nearby cells secrete proteins that bind to the paracrine factor and prevent it from binding prematurely to the target tissue (**FIGURE 4.38**; Mulligan et al. 2012). Similarly, as we have mentioned above, **heparan sulfate proteoglycans (HSPGs)** in the extracellular matrix often modulate the stability, reception, diffusion rate, and concentration gradient of FGF, BMP, and Wnt proteins (Akiyama et al. 2008; Yan and Lin 2009; Berendsen et al. 2011; Christian 2011; Müller and Schier 2011; Nahmad and Lander 2011).

FIGURE 4.38 Wnt diffusion is affected by other proteins. (A) Diffusion of Wingless (Wg, a Wnt paracrine factor) throughout the developing wing of wild-type *Drosophila* (above) is enhanced by Swim, a protein that stabilizes Wg and that is made by some of the wing cells. When Swim is not present, as in the mutant below, Wg does not disperse but is confined to the narrow band of *Wg*-expressing cells. (B) Similarly, Wingless usually activates the *Distal-less* gene (green) in much of the wild-type wing (seen above). However, in *swim*-mutant flies, the range of *Distal-less* expression is confined to those areas near the band of *Wg*-expressing cells. (From Mulligan et al. 2012.)

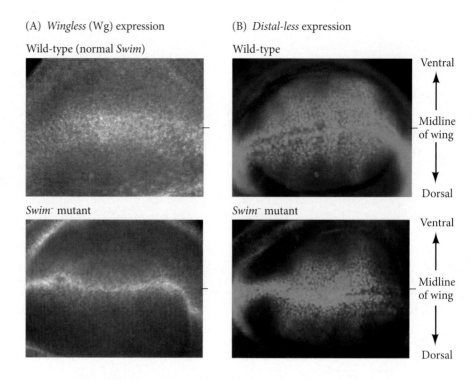

(A) *Wingless* (Wg) expression

Wild-type (normal *Swim*)

Swim⁻ mutant

(B) *Distal-less* expression

Wild-type

Ventral

Midline of wing

Dorsal

Swim⁻ mutant

Ventral

Midline of wing

Dorsal

FGF secretion represents a comprehensive example of the ways that HSPGs can influence paracrine factor diffusion. Cells secrete FGFs into the extracellular matrix, where the FGFs can interact with a diversity of HSPGs that function to both modulate the diffusion of FGF and influence FGF-FGFR binding (Balasubramanian and Zhang 2015). Like all proteoglycans, HSPGs possess chains of sugar molecules that vary in length and type, and different forms of HSPG-FGF interactions can differently shape the FGF gradient. Specifically, the morphogen gradient of Fgf8 is thought to be established through a source-sink model (also known as a "secretion-diffusion-clearance" mechanism; Yu et al. 2009). In this model, cells secreting Fgf8 are the source of the morphogen, and the receiving cells provide the sink through mechanisms of binding, internalization, or protein degradation for clearance of Fgf8 (Balasubramanian and Zhang 2015). Michael Brand's lab tested this model in the zebrafish gastrula by microinjecting a cluster of cells with Fgf8 fused with GFP, quantifying the amount of Fgf8 in the extracellular space at varying distances from the microinjected cells using fluorescence correlation spectroscopy (**FIGURE 4.39A,B**). Remarkably, the researchers were able to visualize an Fgf8-GFP gradient that differed under different circumstances (**FIGURE 4.39C**): free diffusion of the ligand achieved the greatest distance traveled; "directed diffusion" along HSPG fibers fostered rapid movement over several cell distances; "confined clustering" of Fgf8 on dense HSPG matrices significantly restricted diffusion; and endocytosis internalized the Fgf8-FGFR complex for lysosomal degradation in receiving cells (Yu et al. 2009; Bökel and Brand 2013). Thus, the target tissue is not passive. It can promote diffusion, retard diffusion, or degrade the paracrine factor.

CILIA AS SIGNAL RECEPTION CENTERS In many cases, the reception of paracrine factors is not uniform throughout the cell membrane; rather, receptors are often congregated asymmetrically. For instance, the reception of Hedgehog proteins in vertebrates occurs on the primary cilium, a focal extension of the cell membrane made by microtubules (Huangfu et al. 2003; Goetz and Anderson 2010). The primary cilium should not be confused with motile cilia such as those found lining the trachea or in the node of a gastrulating embryo. The primary cilium is much shorter than motile cilia and largely went unnoticed until we realized its direct role in numerous human diseases. In fact,

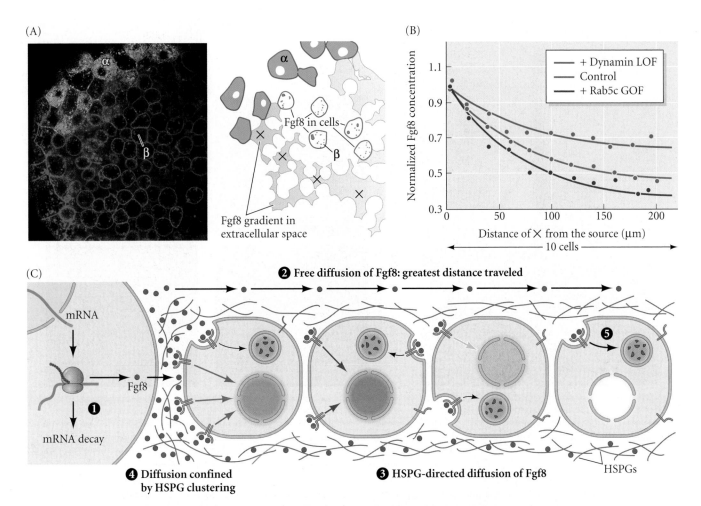

FIGURE 4.39 The Fgf8 gradient. (A) Zebrafish blastulae were injected with DNA encoding Fgf8-GFP (green stain) and mRFP-glycosyl phosphatidylinositol (GPI; red stain) to visualize, respectively, Fgf8 expression and the cell membrane. The confocal image is of a resulting zebrafish gastrula, showing Fgf8 protein being produced by and secreted away from isolated GFP-labeled cells (green). On the right is a schematic representation of select cells and the Fgf8 expression seen in the confocal image (compare α and β identifiers). Fgf8 is seen in a gradient in the extracellular matrix as well as being internalized in receiving cells. (B) Quantification of Fgf8 protein at different locations in (A), indicated by "X" marks in schematic. Manipulation of endocytosis causes predictable changes in the range of Fgf8 secretion. Inhibition of endocytosis with the dominant negative GTPase dynamin causes a shallower Fgf8 gradient over a longer distance (green plot) (LOF, loss of function), whereas increased endocytosis with the overexpression of the endosomal protein Rab5c (GOF, gain of function) yields a steeper and shorter Fgf8 gradient (blue plot). (C) Five primary mechanisms for shaping the Fgf8 gradient. (1) The difference in the rate of *fgf8* transcription and *fgf8* mRNA decay can influence the amount of Fgf8 protein ultimately secreted from a producing cell. Once secreted, Fgf8 can (2) freely diffuse or (3) travel rapidly along HSPG fibers for directed diffusion. (4) In contrast, however, dense areas of HSPGs can also confine and restrict Fgf8 diffusion. (5) The Fgf8-FGFR complex can also be internalized by endocytosis and targeted for lysosomal degradation. Together these different mechanisms result in the displayed gradient of Fgf8, and differential responses in cells that experience different concentrations of Fgf8 signaling (different colored nuclei). (A courtesy of Michael Brand; B after Yu et al. 2009; C after Bökel and Brand 2013; Balasubramanian and Zhang 2015.)

some of these "ciliopathies," such as Bardet-Biedl syndrome, are suspected to be due to an indirect effect on Hedgehog signaling (Nachury 2014). In unstimulated cells, the Patched protein (the Hedgehog receptor; see Figure 4.30) is located in the primary cilium membrane , whereas the Smoothened protein is in the cell membrane close to the cilium or part of an endosome being targeted for degradation. Patched inhibits Smoothened function by preventing it from entering the primary cilium (Milenkovic et al. 2009; Wang et al. 2009). When Hedgehog binds to Patched, however, Smoothened is allowed to join it on the ciliary cell membrane, where it inhibits the PKA and SuFu proteins that make the repressive form of the Gli transcription factor (**FIGURE 4.40**). The microtubules of these

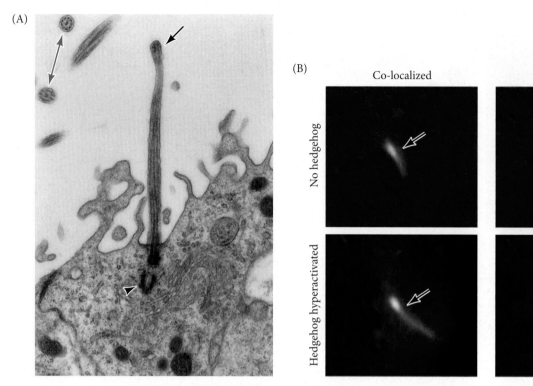

FIGURE 4.40 The primary cilium for Hedgehog reception. (A) Transmission electron micrograph showing a longitudinal section of the primary cilium (black arrow) of a "B-type cell," a neural stem cell in the adult mammalian brain (see Chapter 5). The centriole at the base of this cilium is visible (arrowhead); the microtubules in this primary cilium form an 8+0 structure (other types of cilia, such as motile cilia, typically form a 9+2 arrangement; seen in upper left corner in cross-section [red arrows]). (B) Activation of the Hedgehog pathway requires the transport of Smoothened into the primary cilium. Seen here is the primary cilium (arrow; immunofluorescence stained for acetylated tubulin, blue) on a fibroblast in culture. The ciliary protein Evc (stained green) co-localizes with Smoothened (red) upon hyperactivation of Hedgehog signaling by the drug SAG. Compare the co-localized labeling on the left with the overlays on the right, which have been shifted to show each individual marker. Activation of the Evc-Smo complex in the primary cilium leads to full-length Gli signaling. (A from Alvarez-Bulla et al. 1998; B from Caparrós-Martín et al. 2013.)

cilia provide a scaffold for motor proteins to transport Patched and Smoothened as well as activated Gli proteins, and mutations that knock out cilia formation or their transport mechanism also prevent Hedgehog signaling (Mukhopadhyay and Rohatgi 2014).

Focal membrane protrusions as signaling sources

We have discussed the roles of secreted growth factors for both short- and long-range cell-to-cell communication. But is there a mechanism to present a signal without secreting it? In such a scenario, the producing cell itself *physically reaches out* and presents the signal. Here we highlight emerging ideas of how two types of dynamic membrane extensions can facilitate intercellular communication, and even produce long-range gradients.

LAMELLIPODIA In tunicates, an asymmetric division of a single precardiac founder cell gives rise to the heart progenitors. Although both daughter cells are exposed to the inductive signal Fgf9, only the smaller of the two responds to generate the heart progenitor lineage. During asymmetric division, localized protrusions (**lamellipodia**) form on the ventral-anterior side of the founder cell (Cooley et al. 2011). These protrusions are actin-rich (unlike the microtubule-rich cilia) and result from the polarized localization of a Rho GTPase (Cdc42) in this region. It is possible that the underlying extracellular matrix of the ventral epidermis stimulates this localization. At the same time, FGF receptor activity becomes concentrated in the lamellipodia. When the cell

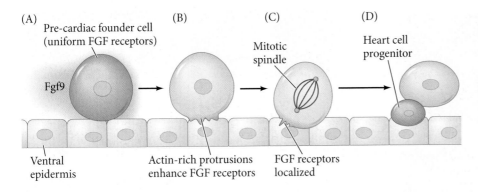

(A) Pre-cardiac founder cell (uniform FGF receptors)

Fgf9

Ventral epidermis

(B) Actin-rich protrusions enhance FGF receptors

(C) Mitotic spindle

FGF receptors localized

(D) Heart cell progenitor

FIGURE 4.41 Model for differential specification of the tunicate heart progenitor lineage. (A) Uniform exposure to Fgf9 leads to uniform FGF receptor occupancy on all parts of the founder cell membrane. (B) Actin-rich protrusions on the ventral-anterior membrane of the cell are associated with high FGF receptor activation. (C) As the progenitor cell enters mitosis, invasive protrusions of the ventral-anterior cell membrane restrict FGF receptors to this region. (D) Following asymmetric cell division, the FGF-activated MAPK pathway is restricted to the ventral daughter cell, leading to differential expression of heart progenitor genes. (After Cooley et al. 2011.)

divides, the smaller daughter inherits these localized, activated FGF receptors, leading to differential activation of the genes that will form heart muscle (**FIGURE 4.41**).

THE FILOPODIAL CYTONEME What if the molecules we thought were diffusible paracrine factors moving through the extracellular matrix were actually transferred from one cell to another at synapse-like connections? There is now significant evidence to support the existence of specialized filopodial projections called **cytonemes**, which stretch out remarkable distances (more than 100 μm) from either the target cells or the signal-producing cells, like long membrane conduits connecting the two types of cells (Roy and Kornberg 2015). Under this model, ligand-receptor binding would initially occur at the tips of cytonemes projecting from the target cells when the tips are positioned in direct apposition to the producing cell's plasma membrane. The ligand-receptor complex would then be transported down the cytoneme to the target cell body.

Cytoneme-mediated morphogen signaling was first described by Thomas Kornberg's laboratory studying development of the air sac and wing disc in *Drosophila* (Roy et al. 2011). A cluster of cells called the air sac primordium (ASP) develops along the basal surface of the wing disc in response to DPP (a BMP homologue) and FGF morphogen gradients in the wing disc (**FIGURE 4.42A,B**). The Kornberg lab discovered that the ASP cells extend cytonemes toward the DPP- and FGF-expressing cells, and that these cytonemes contain receptors for these morphogens—separate receptors in separate cytonemes. Moreover, DPP bound to its receptor on ASP cells has been documented traveling along a cytoneme to the cell body. Anterior-posterior patterning of the wing disc by a gradient of Hedgehog (Hh) signaling also appears to be accomplished through cytonemes (**FIGURE 4.42C**). Hedgehog coming from posterior cells is delivered through cytonemes that extend from the basolateral surface of anterior cells to the Hh-producing posterior cells (**FIGURE 4.42D,E**; Bischoff et al. 2013).

Recent investigations have shown that vertebrates use cytonemes as well. Work in Michael Brand's lab and recent work by Steffen Scholpp's lab have shown that the same gastrulating cells also transport the morphogen Wnt8a along cytoneme-like extensions. In this case, the signal-producing cells are extending the cytonemes, transporting the Wnt8a morphogen to target cells (**FIGURE 4.42F**; Luz et al. 2014; Stanganello et al. 2015). Cytoneme-like interactions are also suspected in one of the classic examples of morphogen signaling, that of anterior-posterior specification in the tetrapod limb bud. Here, a posterior-to-anterior gradient of Sonic hedgehog (Shh) in the limb bud leads to the correct patterning of digits (see Chapter 19). In the chick limb bud, both the Sonic hedgehog-expressing cells and the anterior target cells extend filopodial projections toward each other and make contact at points where Sonic hedgehog (Patch) receptors are localized (**FIGURE 4.42G**; Sanders et al. 2013).

))) **SCIENTISTS SPEAK 4.6** An iBiology Seminar by Dr. Thomas Kornberg of the University of California, San Francisco, discusses cytoneme-directed transport and direct transfer models.

Developing Questions

Are all the molecules that we have considered to be paracrine factors distributed solely by contact through filopodial cytoneme processes, as opposed to diffusion through the extracellular matrix? This question is increasingly coming up in debates among developmental biologists. Where do you stand? Are you a "diffusionist" or a "cytonemist"? Is there room for both mechanisms, or perhaps even a developmental need for both?

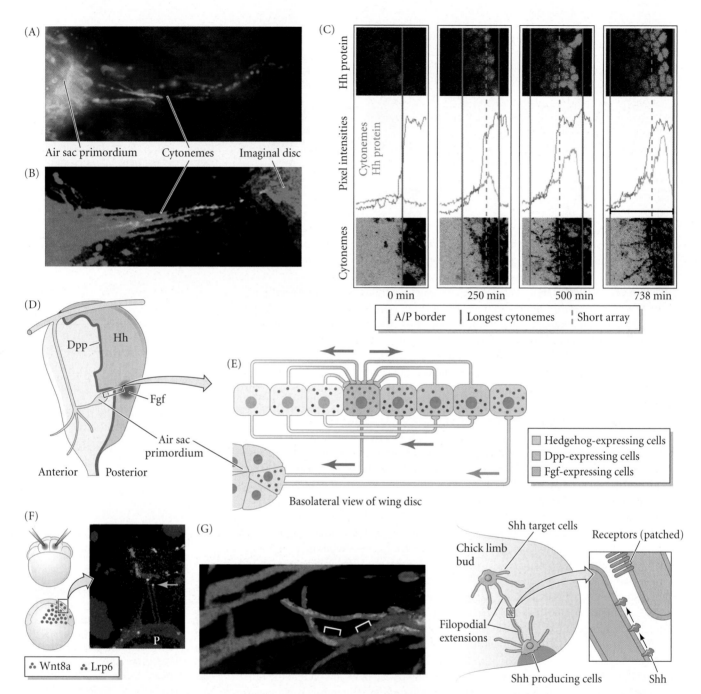

FIGURE 4.42 Filopodia-transported morphogens. (A) Cytonemes from the air sac primordium (ASP) extend toward the epithelium of the wing imaginal disc in *Drosophila* to shuttle the FGF (green) and DPP (red) morphogens. (B) Transported DPP receptor binds DPP produced by the wing disc cells, which gets transported back down the cytoneme to the ASP. (C) This system of cytonemes in the *Drosophila* wing disc is capable of establishing a gradient of Hedgehog (Hh) protein (green in top panels and in plot) over the course of filopodial extension (black processes in lower panels and red plot line). (D) Illustration of the *Drosophila* wing imaginal disc during its interactions with tracheal cells, namely the air sac primordium. Hh, Dpp, and Fgf expressing cells are represented as blue, red, and green domains. (E) Magnified cross section of the boxed region in (D). Cytoneme extensions from the air sac primordium as well as between cells of the wing disc are illustrated along with the morphogens produced and transported along these cytonemes (arrows). (F) Wnt8a (red) and its receptor Lrp6 (green) were micro-injected into two different cells of an early-stage zebrafish blastula. Live cell imaging of these cells at the gastrula stage revealed Wnt8a interactions with the Lrp6 receptor at the tips of filopodial extensions from the producer cells (P, yellow arrow). (G) In the chick limb bud, long, thin filopodial protrusions have been documented extending both from Sonic hedgehog-producing cells in the posterior region (purple cell with green Shh protein in left image) and from the target cells in the anterior limb bud (red cells). These opposing filopodia directly interact (brackets, left image), and at this point of interaction, it is proposed that Shh and its receptor Patch can bind (right illustration). (A from Roy and Kornberg 2011; B from Roy et al. 2014; C from Bischoff et al. 2013; F from Stanganello et al. 2015; G from Sanders et al. 2013.)

Juxtacrine Signaling for Cell Identity

In juxtacrine interactions, proteins from the inducing cell interact with receptor proteins of adjacent responding cells without diffusing from the cell producing it. Three of the most widely used families of juxtacrine factors are the **Notch proteins** (which bind to a family of ligands exemplified by the Delta protein); **cell adhesion molecules** such as cadherins; and the **eph receptors** and their **ephrin ligands**. When an ephrin on one cell binds with the eph receptor on an adjacent cell, signals are sent to each of the two cells (Davy et al. 2004; Davy and Soriano 2005). These signals are often those of either attraction or repulsion, and ephrins are often seen where cells are being told where to migrate or where boundaries are forming. We will see the ephrins and the eph receptors functioning in the formation of blood vessels, neurons, and somites. We will now look more closely at the Notch proteins and their ligands, as well as discussing cell adhesion molecules as part of an important developmental pathway called Hippo signaling.

The Notch pathway: Juxtaposed ligands and receptors for pattern formation

Although most known regulators of induction are diffusible proteins, some inducing proteins remain bound to the inducing cell surface. In one such pathway, cells expressing the Delta, Jagged, or Serrate proteins in their cell membranes activate neighboring cells that contain Notch protein in their cell membranes (see Artavanis-Tsakakonas and Muskavitch 2010). Notch extends through the cell membrane, and its external surface contacts Delta, Jagged, or Serrate proteins extending out from an adjacent cell. When complexed to one of these ligands, Notch undergoes a conformational change that enables a part of its cytoplasmic domain to be cut off by the presenilin-1 protease. The cleaved portion enters the nucleus and binds to a dormant transcription factor of the CSL family. When bound to the Notch protein, the CSL transcription factors activate their target genes (**FIGURE 4.43**; Lecourtois and Schweisguth 1998; Schroeder et al. 1998; Struhl and Adachi 1998). This activation is thought to involve the recruitment of histone acetyltransferases (Wallberg et al. 2002). Thus, Notch can be considered as a transcription

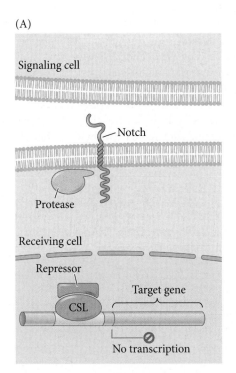
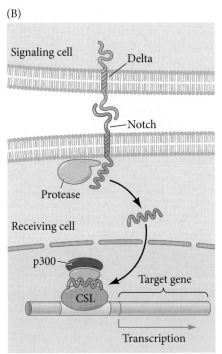

FIGURE 4.43 Mechanism of Notch activity. (A) Prior to Notch signaling, a CSL transcription factor (such as Suppressor of hairless or CBF1) is on the enhancer of Notch-regulated genes. The CSL binds repressors of transcription. (B) Model for the activation of Notch. A ligand (Delta, Jagged, or Serrate protein) on one cell binds to the extracellular domain of the Notch protein on an adjacent cell. This binding causes a shape change in the intracellular domain of Notch, which activates a protease. The protease cleaves Notch and allows the intracellular region of the Notch protein to enter the nucleus and bind the CSL transcription factor. This intracellular region of Notch displaces the repressor proteins and binds activators of transcription, including the histone acetyltransferase p300. The activated CSL can then transcribe its target genes. (After K. Koziol-Dube, personal communication.)

factor tethered to the cell membrane. When the attachment is broken, Notch (or a piece of it) can detach from the cell membrane and enter the nucleus (Kopan 2002).

Notch proteins are involved in the formation of numerous vertebrate organs—kidney, pancreas, and heart—and they are extremely important receptors in the nervous system. In both the vertebrate and *Drosophila* nervous systems, the binding of Delta to Notch tells the receiving cell not to become neural (Chitnis et al. 1995; Wang et al. 1998). In the vertebrate eye, the interactions between Notch and its ligands regulate which cells become optic neurons and which become glial cells (Dorsky et al. 1997; Wang et al. 1998).

WEB TOPIC 4.3 **NOTCH MUTATIONS** Humans have genes for more than one Notch protein and more than one ligand. Their interactions are critical in neural development, and mutations in Notch genes can cause nervous system abnormalities.

Induction does indeed occur on the cell-to-cell level, and one of the best examples is the formation of the vulva in the nematode worm *C. elegans*. Remarkably, the signal transduction pathways involved turn out to be the same as those used in the formation of retinal receptors in *Drosophila*; only the targeted transcription factors are different. In both cases, an epidermal growth-factor-like inducer activates the RTK pathway, leading to the differential regulation of Notch-Delta signaling.

Paracrine and juxtacrine signaling in coordination: Vulval induction in C. elegans

Most *C. elegans* individuals are hermaphrodites. In their early development, they are male, and the gonad produces sperm, which are stored for later use. As they grow older, they develop ovaries. The eggs "roll" through the region of sperm storage, are fertilized inside the nematode, and then pass out of the body through the vulva (see Chapter 8; Barkoulas et al. 2013). The formation of the vulva occurs during the larval stage from six cells called the **vulval precursor cells** (**VPCs**). The cell connecting the overlying gonad to the vulval precursor cells is called the **anchor cell** (**FIGURE 4.44**). The anchor cell secretes LIN-3 protein, a paracrine factor (similar to mammalian epidermal growth factor, or EGF) that activates the RTK pathway (Hill and Sternberg 1992). If the anchor cell is destroyed (or if the *lin-3* gene is mutated), the VPCs will not form a vulva and instead become part of the hypodermis or skin (Kimble 1981).

The six VPCs influenced by the anchor cell form an **equivalence group**. Each member of this group is competent to become induced by the anchor cell and can assume any of three fates, depending on its proximity to the anchor cell. The cell directly beneath the anchor cell divides to form the central vulval cells. The two cells flanking that central cell divide to become the lateral vulval cells, whereas the three cells farther away from the anchor cell generate hypodermal cells. If the anchor cell is destroyed, all six cells of the equivalence group divide once and contribute to the hypodermal tissue. If the three central VPCs are destroyed, the three outer cells, which normally form hypodermis, generate vulval cells instead.

LIN-3 secreted from the anchor cell forms a concentration gradient, in which the VPC closest to the anchor cell (i.e., the P6.p cell) receives the highest concentration of LIN-3 and generates the central vulval cells. The two adjacent VPCs (P5.p and P7.p) receive lower amounts of LIN-3 and become the lateral vulval cells. VPCs farther away from the anchor cell do not receive enough LIN-3 to have an effect, so they become hypodermis (Katz et al. 1995).

NOTCH-DELTA AND LATERAL INHIBITION We have discussed the reception of the EGF-like LIN-3 signal by the cells of the equivalence group that forms the vulva. Before this induction occurs, however, an earlier interaction has formed the anchor cell. The formation of the anchor cell is mediated by *lin-12*, the *C. elegans* homologue of the *Notch* gene. In

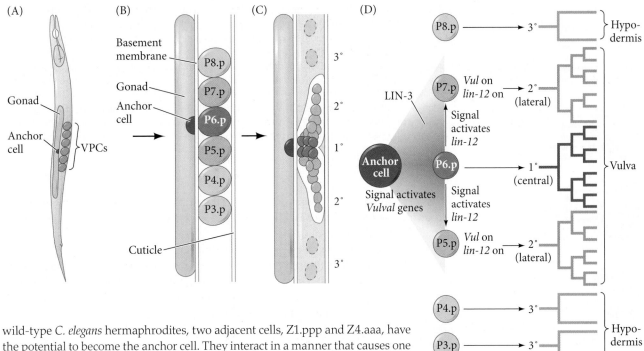

FIGURE 4.44 *C. elegans* vulval precursor cells (VPCs) and their descendants. (A) Location of the gonad, anchor cell, and VPCs in the second instar larva. (B,C) Relationship of the anchor cell to the six VPCs and their subsequent lineages. Primary (1°) lineages result in the central vulval cells, secondary (2°) lineages constitute the lateral vulval cells, and tertiary (3°) lineages generate hypodermal cells. (C) Outline of the vulva in the fourth instar larva. The circles represent the positions of the nuclei. (D) Model for the determination of vulval cell lineages in *C. elegans*. The LIN-3 signal from the anchor cell causes the determination of the P6.p cell to generate the central vulval lineage (dark purple). Lower concentrations of LIN-3 cause the P5.p and P7.p cells to form the lateral vulval lineages. The P6.p (central lineage) cell also secretes a short-range juxtacrine signal that induces the neighboring cells to activate the LIN-12 (Notch) protein. This signal prevents the P5.p and P7.p cells from generating the primary central vulval cell lineage. (After Katz and Sternberg 1996.)

wild-type *C. elegans* hermaphrodites, two adjacent cells, Z1.ppp and Z4.aaa, have the potential to become the anchor cell. They interact in a manner that causes one of them to become the anchor cell while the other one becomes the precursor of the uterine tissue. In loss-of-function *lin-12* mutants, both cells become anchor cells, whereas in gain-of-function mutations, both cells become uterine precursors (Greenwald et al. 1983). Studies using genetic mosaics and cell ablations have shown that this decision is made in the second larval stage, and that the *lin-12* gene needs to function only in that cell destined to become the uterine precursor cell. The presumptive anchor cell does not need it. Seydoux and Greenwald (1989) speculate that these two cells originally synthesize both the signal for uterine differentiation (the LAG-2 protein, homologous to Delta) and the receptor for this molecule (the LIN-12 protein, homologous to Notch; Wilkinson et al. 1994).

During a particular time in larval development, the cell that, by chance, is secreting more LAG-2 causes its neighbor to cease its production of this differentiation signal and to increase its production of LIN-12. The cell secreting LAG-2 becomes the gonadal anchor cell, while the cell receiving the signal through its LIN-12 protein becomes the ventral uterine precursor cell (**FIGURE 4.45**). Thus, the two cells are thought to determine each other prior to their respective differentiation events. When LIN-12 is used again during vulva formation, it is activated by the primary vulval lineage to stop the lateral vulval cells from forming the central vulval phenotype (see Figure 4.44). Thus, the anchor cell/ventral uterine precursor decision illustrates two important aspects of determination in two originally equivalent cells. First, the initial difference between the two cells is created by chance. Second, this initial difference is reinforced by feedback. This Notch-Delta mediated mechanism of restricting adjacent cell fates is called **lateral inhibition**.

Hippo: An integrator of pathways

Most of the signal transduction pathways that we have discussed are named for the players involved in the initial signaling event at the cell membrane. The Hippo signal transduction pathway does not have a dedicated ligand or receptor, however. Hippo stands for one of several important kinases that are critical for organ size control. It was first identified in *Drosophila*, where its loss resulted in a "hippopotamus"-shaped phenotype due to excessive growth (Hansen et al. 2015).

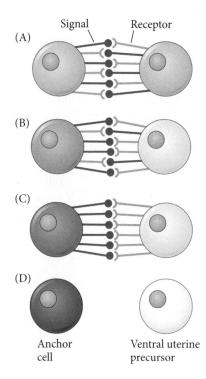

Signal   Receptor

(A)

(B)

(C)

(D)

Anchor cell   Ventral uterine precursor

FIGURE 4.45 Model for the generation of two cell types (anchor cell and ventral uterine precursor cell) from two equivalent cells (Z1.ppp and Z4.aaa) in *C. elegans*. (A) The cells start off as equivalent, producing fluctuating amounts of signal and receptor. The *lag-2* gene is thought to encode the signal, and the *lin-12* gene is thought to encode the receptor. Reception of the signal turns down LAG-2 (Delta) production and upregulates LIN-12 (Notch). (B) A stochastic (chance) event causes one cell to produce more LAG-2 than the other cell at some particular critical time, which stimulates more LIN-12 production in the neighboring cell. (C) This difference is amplified because the cell producing more LIN-12 produces less LAG-2. Eventually, just one cell is delivering the LAG-2 signal, and the other cell is receiving it. (D) The signaling cell becomes the anchor cell, and the receiving cell becomes the ventral uterine precursor cell. (After Greenwald and Rubin 1992.)

Loss of Hippo (or overexpression of its main transcriptional effector, Yorkie) causes cells to divide significantly faster while slowing apoptosis (**FIGURE 4.46**; Justice et al. 1995; Xu et al. 1995; Huang et al. 2005).

The essential players in the Hippo signaling cascade begin at the cell membrane with cell-to-cell interactions involving cell adhesion molecules such as E-cadherin or Crumbs (see Figure 5.7B). These cell adhesion molecules interact with the F-actin binding protein angiomotin, which initiates activation of the Hippo kinase cascade (Hansen et al. 2015). The main kinase in this cascade is the Large tumor suppressor 1/2 (Lats1/2; Warts is the *Drosophila* homologue), which functions to phosphorylate Yorkie or its mammalian homologue Yap/Taz. When phosphorylated, Yap/Taz will either be retained in the cytoplasm or degraded, whereas lack of Hippo signaling frees Yap/Taz to enter the nucleus and function as a transcription co-activator of Tead (Scalloped homologue). There are a number of ways in which Hippo signaling components can regulate the pathways of other paracrine factors such as Wnts, EGF, TGF-β, and BMP. Likewise, these pathways can modulate Hippo signaling, typically operating through Yap/Taz. Thus, the Hippo pathway is emerging as a major crossroad for the biochemical pathways of the cell, heightening our attention to the long-unsolved problem of understanding how all these conceptually linear pathways are truly integrated.

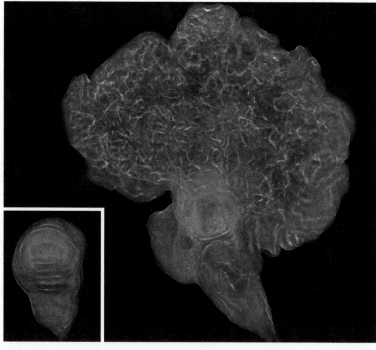

Wild-type   Overexpression of *yorkie*

FIGURE 4.46 Hippo signaling is critical for controlling organ size. Overexpression of *yorkie* (the main transcriptional effector of the Hippo kinase) in *Drosophila* resulted in an extremely overgrown ("hippopotamus") wing imaginal disc compared to the same stage wild-type wing disc. (Photograph from Huang et al. 2005.)

Next Step Investigation

How do cells communicate, interact, and understand their place in the embryo? This chapter covered many of the mechanisms at play that facilitate cell-to-cell attachments, relay chemical signals, and respond to environmental cues. There are many exciting next steps to investigate, from the biophysics of morphogenesis to the role of cytonemes in morphogen gradients. These types of mechanisms are easy to comprehend on the scale of the cell and tissue, and we are sure that you can propose some logical and exciting experimental designs to further test such mechanisms. This field is lacking a significant understanding of how

morphogenesis is coordinated on the scale of the entire embryo, however. How might you begin to apply your understanding of cell-to-cell communication toward a more comprehensive understanding of coordinated development across the embryo? Do you think there could be a kind of global oversight of timing, size, pattern, movement, and differentiation? Please know there are no correct answers to these questions at the back of the book, hiding in your professor's notes, or buried in Google search results. The answers reside in the completion of your own ideas and experiments.

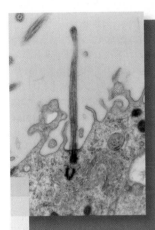

Closing Thoughts on the Opening Photo

Is this a cell's antenna? If so, what is its purpose? It's for cells to communicate! This image shows a primary cilium on a neural stem cell in the brain, a structure that is in fact used like an antenna, enabling the cell to receive signals from its environment. We discussed the critical role of select signaling proteins that convey a myriad of information about position, adhesion, cell specification, and migration. New mechanisms of cell communication—such as the essential role of the primary cilium emphasized in this image; the potential reach of cytonemes, which may change our understanding of morphogen delivery; the modifying and potentially instructive roles of the extracellular matrix; and how the physical properties of cell adhesion can both sort different cells and regulate organ size—are rapidly emerging. (Photograph courtesy of Alvarez-Bullya et al. 1998.)

4 Snapshot Summary
Cell-to-Cell Communication

1. The sorting out of one cell type from another results from differences in the cell membrane.

2. The membrane structures responsible for cell sorting out are often cadherin proteins that change the surface tension properties of the cells. Cadherins can cause cells to sort by both quantitative (different amounts of cadherin) and qualitative (different types of cadherin) differences. Cadherins appear to be critical during certain morphological changes.

3. Cell migration occurs through changes in the actin cytoskeleton. These changes can be directed by internal instructions (from the nucleus) or by external instructions (from the extracellular matrix or chemoattractant molecules).

4. Inductive interactions involve inducing and responding tissues. The ability to respond to inductive signals depends on the competence of the responding cells. The specific response to an inducer is determined by the genome of the responding tissue.

   - Reciprocal induction occurs when the two interacting tissues are both inducers and are competent to respond to each other's signals.

   - Cascades of inductive events are responsible for organ formation.

   - Regionally specific inductions can generate different structures from the same responding tissue.

5. Paracrine interactions occur when a cell or tissue secretes proteins that induce changes in neighboring cells. Juxtacrine interactions are inductive interactions that take place between the cell membranes of adjacent cells or between a cell membrane and an extracellular matrix secreted by another cell.

6. Paracrine factors are secreted by inducing cells. These factors bind to cell membrane receptors in competent responding cells. Competence is the ability to bind and to respond to inducers, and it is often the result of a prior induction. Competent cells respond to paracrine factors through signal transduction pathways.

7. Morphogens are secreted signaling molecules that affect gene expression differently at different concentrations.

8. Signal transduction pathways begin with a paracrine or juxtacrine factor causing a conformational change in

its cell membrane receptor. The new shape can result in enzymatic activity in the cytoplasmic domain of the receptor protein. This activity allows the receptor to phosphorylate other cytoplasmic proteins. Eventually, a cascade of such reactions activates a transcription factor (or set of factors) that activates or represses specific gene activity.

9. The differentiated state can be maintained by positive feedback loops involving transcription factors, autocrine factors, or paracrine factors.

10. The extracellular matrix is both a source of signals and serves to modify how such signals may be secreted across cells to influence differentiation and cell migration.

11. Cells can convert from being epithelial to being mesenchymal and vice versa. The epithelial-mesenchymal transition (EMT) is a series of transformations involved in the dispersion of neural crest cells and the creation of vertebrae from somitic cells. In adults, EMT is involved in wound healing and cancer metastasis.

12. The cell surface is intimately involved with cell signaling. Proteoglycans and other membrane components can expand or restrict the diffusion of paracrine factors.

13. Specializations of the cell surface, including cilia and lamellipodia, may concentrate receptors for paracrine and extracellular matrix proteins. Newly discovered filopodia-like extensions called cytonemes can be involved in transferring morphogens between signaling and responding cells and may be a major component of cell signaling.

14. Juxtacrine signaling involves local protein interactions between receptors. Examples include Notch-Delta signaling that patterns cell fates through lateral inhibition and Hippo signaling that influences organ size.

Further Reading

Ananthakrishnan, R. and A. Ehrlicher. 2007. The forces behind cell movement. *Int. J. Biol. Sci.* 3: 303–317.

Balasubramanian, R. and X. Zhang. 2015. Mechanisms of FGF gradient formation during embryogenesis. *Semin. Cell Dev. Biol.* doi:10.1016/j.semcdb.2015.10.004.

Bischoff, M. and 6 others. 2013. Cytonemes are required for the establishment of a normal Hedgehog morphogen gradient in *Drosophila* epithelia. *Nature Cell Biol.* 11: 1269–1281.

Briscoe, J. and P. P. Thérond. 2013. The mechanisms of Hedgehog signalling and its roles in development and disease. *Nat. Rev. Mol. Cell Biol.* 7: 416–429.

Fagotto, F., N. Rohani, A. S. Touret and R. Li. 2013. A molecular base for cell sorting at embryonic boundaries: Contact inhibition of cadherin adhesion by ephrin/ Eph-dependent contractility. *Dev. Cell* 27: 72–87.

Foty, R. A. and M. S. Steinberg. 2013. Differential adhesion in model systems. *Wiley Interdiscip. Rev. Dev. Biol.* 2: 631–645.

Hansen, C. G., T. Moroishi and K. L. Guan. 2015. YAP and TAZ: A nexus for Hippo signaling and beyond. *Trends Cell Biol.* 25: 499–513.

Heldin, C.-H., K. Miyazono and P. ten Dijke. 1997. TGF-β signaling from cell membrane to nucleus through SMAD proteins. *Nature* 390: 465–471.

Huangfu, D., A. Liu, A. S. Rakeman, N. S. Murcia, L. Niswander and K. V. Anderson. 2003. Hedgehog signalling in the mouse requires intraflagellar transport proteins. *Nature* 426: 83–87.

Kakugawa, S. and 11 others. 2015. Notum deacylates Wnt proteins to suppress signalling activity. *Nature* 519: 187–192.

Molina, M. D., N. de Crozé, E. Haillot and T. Lepage. 2013. Nodal: Master and commander of the dorsal-ventral and left-right axes in the sea urchin embryo. *Curr. Opin. Genet. Dev.* 23: 445–453.

Müller P. and A. F. Schier. 2011. Extracellular movement of signaling molecules. *Dev. Cell* 21: 145–158.

Nahmad, M. and A. D. Lander. 2011. Spatiotemporal mechanisms of morphogen gradient interpretation. *Curr. Opin. Genet. Dev.* 21: 726–731.

Roy, S. and T. B. Kornberg. 2015. Paracrine signaling mediated at cell-cell contacts. *Bioessays* 37: 25–33.

Saito-Diaz, K. and 6 others. 2013. The way Wnt works: Components and mechanism. *Growth Factors* 31: 1–31.

Stanganello, E. and 8 others. 2015. Filopodia-based Wnt transport during vertebrate tissue patterning. *Nature Commun.* 6: 5846.

van Amerongen, R. and R. Nusse. 2009. Towards an integrated view of Wnt signaling in development. *Development* 136: 3205–3214.

van den Heuvel, M., C. Harryman-Samos, J. Klingensmith, N. Perrimon and R. Nusse. 1993. Mutations in the segment polarity genes *wingless* and *porcupine* impair secretion of the wingless protein. *EMBO J.* 12: 5293–5302.

Yu, S. R. and 7 others. 2009. Fgf8 morphogen gradient forms by a source-sink mechanism with freely diffusing molecules. *Nature* 461: 533–536.

GO TO WWW.DEVBIO.COM...

...for Web Topics, Scientists Speak interviews, Watch Development videos, Dev Tutorials, and complete bibliographic information for all literature cited in this chapter.

Stem Cells
Their Potential and Their Niches

WE HAVE COMPLETED AN ANALYSIS of cell maturation through the levels of cell specification, commitment, and ultimately differentiation, all of which are driven by cell-to-cell communication and the regulation of gene expression. There is no better example that encapsulates this entire process than the stem cell.

A **stem cell** retains the ability to divide and re-create itself while also having the ability to generate progeny capable of specializing into a more differentiated cell type. Stem cells are sometimes referred to as "undifferentiated" due to this maintenance of proliferative properties. There are many different types of stem cells, however, and their status as "undifferentiated" really only pertains to the retained ability to divide. Because they maintain the ability to proliferate and differentiate, stem cells hold great potential to transform modern medicine.

Currently, there are few topics in developmental biology that can rival stem cells in the pace at which new knowledge is being generated. In this chapter, we will address some of the fundamental questions regarding stem cells. What are the mechanisms governing stem cell division, self-renewal, and differentiation? Where are stem cells found, and how do they differ when in an embryo, an adult, or a culture dish? How are scientists and clinicians using stem cells to study and treat disease?

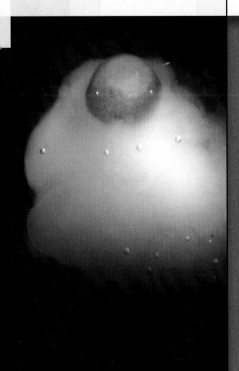

Is that really an eye and a brain in a dish?

The Punchline

Stem cells retain the ability to divide while also generating differentiating progeny. The differences among stem cell types is based on their potential for cell fate derivation. Because embryonic stem cells are pluripotent, they can make every cell of the body, whereas an adult stem cell is multipotent and usually can give rise only to the different cell types of its residing tissue. Stem cells reside within a "stem cell niche," which provides a microenvironment of local and long-range signals that regulate whether the stem cell is in a state of quiescence, division, or differentiation. A common mechanism of regulation in the niche involves modulating changes in cell adhesion molecules that link the stem cell to its niche. Loss of adhesion leads to the movement of the stem cell or its progeny away from quiescence-promoting signals (often paracrine factors), thus fostering division and differentiation. Isolation or derivation of human pluripotent and multipotent stem cells offers opportunities to study the mechanisms of human development and disease as never before. Precise regulation of stem cells helps build the embryo, maintain and regenerate tissues, and potentially could provide cell-based therapies to treat disease.

The Stem Cell Concept

A cell is a stem cell if it can divide and in doing so produce a replica of itself (a process called **self-renewal**), as well as a daughter cell that can undergo further development and differentiation. It thus has the power, or **potency**, to produce many different types of differentiated cells.

 DEV TUTORIAL *Stem Cells* Dr. Michael Barresi's lecture covers the basics of stem cell biology.

Division and self-renewal

Upon division, a stem cell may produce a daughter cell that can mature into a terminally differentiated cell type. Cell division can occur either symmetrically or asymmetrically. If a stem cell divides symmetrically, it could produce two self-renewing stem cells or two daughter cells that are committed to differentiate, resulting in, respectively, the expansion or reduction of the resident stem cell population. In contrast, if the stem cell divides asymmetrically, it could stabilize the stem cell pool as well as generate a daughter cell that goes on to differentiate. This strategy, in which two types of cells (a stem cell and a developmentally committed cell) are produced at each division, is called the **single stem cell asymmetry** mode and is seen in many types of stem cells (**FIGURE 5.1A**).

An alternative (but not mutually exclusive) mode of retaining cell homeostasis is the **population asymmetry** mode of stem cell division. Here, some stem cells are more prone to produce differentiated progeny, which is compensated for by another set of stem cells that divide symmetrically to maintain the stem cell pool within this population (**FIGURE 5.1B**; Watt and Hogan 2000; Simons and Clevers 2011).

(A) Single-cell asymmetry

Stem cell   Committed cell

(B) Population asymmetry (symmetrical differentiation)

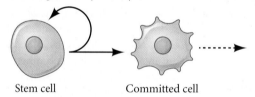

Stem cells   Stem cell   Committed cells

FIGURE 5.1 The stem cell concept. (A) The fundamental notion of a stem cell is that it can make more stem cells while also producing cells committed to undergoing differentiation. This process is called asymmetric cell division. (B) A population of stem cells can also be maintained through population asymmetry. Here a stem cell is shown to have the ability to divide symmetrically to produce *either* two stem cells (thus increasing the stem cell pool by one) *or* two committed cells (thus decreasing the pool by one). This is called symmetrical renewing or symmetrical differentiating. (C) In many organs, stem cell lineages pass from a multipotent stem cell (capable of forming numerous types of cells) to a committed stem cell that makes one or very few types of cells to a progenitor cell (also known as a transit-amplifying cell) that can proliferate for multiple rounds of divisions but is transient in its life and is committed to becoming a particular type of differentiated cell.

(C) Adult stem cell lineage

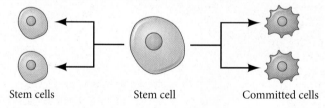

Multipotent   Committed   Progenitor (transit-   Differentiated
stem cell   stem cell   amplifying) cell   cells

Potency defines a stem cell

The diversity of cell types that a stem cell can generate in vivo defines its natural potency. A stem cell capable of producing all the cell types of a lineage is said to be **totipotent**. In organisms such as hydra, each individual cell is totipotent (see Chapter 22). In mammals, only the fertilized egg and first 4 to 8 cells are totipotent, which means that they can generate both the embryonic lineages (that form the body and germ cells) and the extraembryonic lineages (that form the placenta, amnion, and yolk sac) (**FIGURE 5.2**). Shortly after the 8-cell stage, the mammalian embryo develops an outer layer (which becomes the fetal portion of the placenta), and an inner cell mass that generates the embryo. The cells of the inner cell mass are thus said to be **pluripotent**, or capable of producing all the cells of the embryo. When these inner cells are removed from the embryo and cultured in vitro, they establish a population of pluripotent **embryonic stem cells**.

As cell populations within each germ layer expand and differentiate, resident stem cells are maintained within these developing tissues. These stem cells are **multipotent** and function to generate cell types with restricted specificity for the tissue in which they reside (**FIGURE 5.1C** and see Figure 5.2). From the embryonic gut to the adult small intestine or from the neural tube to the adult brain, multipotent stem cells play critical roles in fueling organogenesis in the embryo and regeneration in adult tissues.

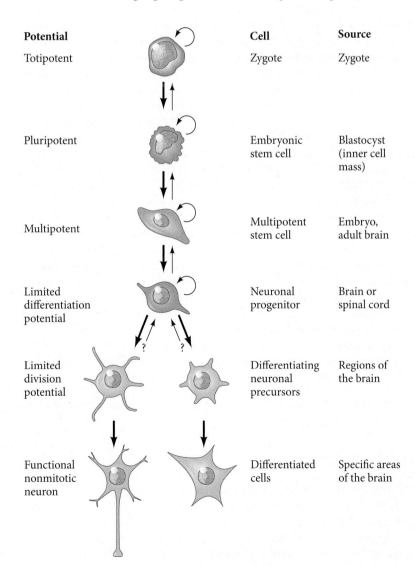

Potential		Cell	Source
Totipotent		Zygote	Zygote
Pluripotent		Embryonic stem cell	Blastocyst (inner cell mass)
Multipotent		Multipotent stem cell	Embryo, adult brain
Limited differentiation potential		Neuronal progenitor	Brain or spinal cord
Limited division potential		Differentiating neuronal precursors	Regions of the brain
Functional nonmitotic neuron		Differentiated cells	Specific areas of the brain

FIGURE 5.2 An example of the maturational series of stem cells. The differentiation of neurons is illustrated here. (After http://thebrain.mcgill.ca/)

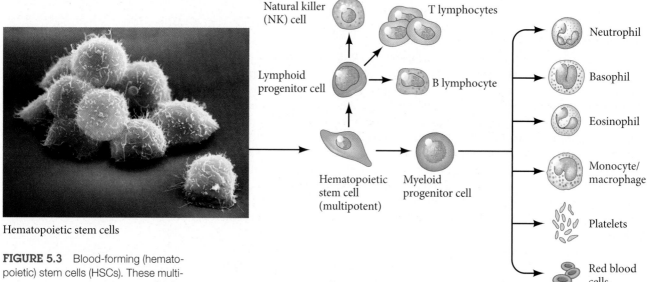

Hematopoietic stem cells

FIGURE 5.3 Blood-forming (hematopoietic) stem cells (HSCs). These multipotent stem cells generate blood cells throughout an individual's life. HSCs from human bone marrow (photo) can divide to produce more HSCs. Alternatively, HSC daughter cells are capable of becoming either lymphoid progenitor cells (which divide to form the cells of the adaptive immune system) or myeloid progenitor cells (which become the other blood cell precursors). The lineage path each cell takes is regulated by the HSC's microenvironment, or niche (see Figure 5.15). (After http://stemcells.nih.gov/; photograph © SPL/Photo Researchers, Inc.)

Numerous adult organs possess **adult stem cells**, which in most cases are multipotent. In addition to the known hematopoietic stem cells that function to generate all the cells of the blood, biologists have also discovered adult stem cells in the epidermis, brain, muscle, teeth, gut, and lung, among other locations. Unlike pluripotent stem cells, adult or multipotent stem cells in culture not only have a restricted array of cell types that they can create, but they also have a finite number of generations for self-renewal. This limited renewal of adult stem cells may contribute to aging (Asumda 2013).

When a multipotent stem cell divides asymmetrically, its maturing daughter cell often goes through a transition stage as a **progenitor** or **transit-amplifying cell**, as is seen in the formation of blood cells, sperm, and neurons (see Figures 5.1C and 5.2). Progenitor cells are not capable of unlimited self-renewal; rather, they have the capacity to divide only a few times before differentiating (Seaberg and van der Kooy 2003). Although limited, this proliferation serves to *amplify* the pool of progenitors before they terminally differentiate. Cells within this progenitor pool can mature along different but related paths of specification. As an example, the hematopoietic stem cell generates blood and lymphoid progenitor cells that further develop into the differentiated cell types of the blood, such as red blood cells, neutrophils, and lymphocytes (cells of the immune response), as shown in **FIGURE 5.3**. Yet another term, **precursor cell** (or simply **precursors**), is widely used to denote any ancestral cell type (either stem cell or progenitor cell) of a particular lineage; it is often used when such distinctions do not matter or are not known (see Tajbakhsh 2009). Some adult stem cells, such as spermatogonia, are referred to as **unipotent stem cells** because they function in the organism to generate only one cell type, the sperm cell in this example. Precise control of the division and differentiation of these varied stem cell types is necessary for building the embryo as well as maintaining and regenerating tissues in the adult.

((• SCIENTISTS SPEAK 5.1 Developmental documentaries from 2009 cover both embryonic and adult stem cells.

Stem Cell Regulation

As discussed above, the basic functions of stem cells revolve around self-renewal and differentiation. But how are stem cells regulated between these different states in a coordinated way to meet the patterning and morphogenetic needs of the embryo and mature tissue? Regulation is highly influenced by the microenvironment that surrounds a stem cell and is known as the **stem cell niche** (Schofield 1978). There is growing

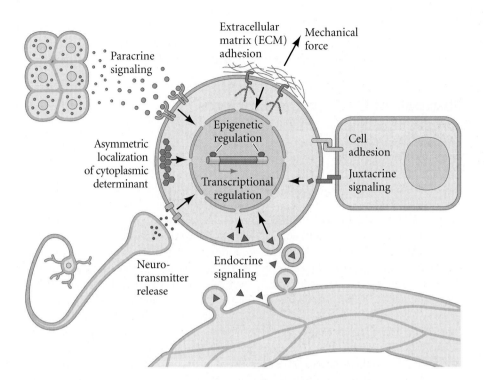

FIGURE 5.4 To divide or not to divide: an overview of stem cell regulatory mechanisms. Shown here are some of the more general external and internal molecular mechanisms that can influence the quiescent, proliferative, or differentiative behaviors of a stem cell.

evidence that all tissue types possess a unique stem cell niche, and despite many differences among the niche architecture of different tissues, several common principles of stem cell regulation can be applied to all environments. These principles involve *extracellular* mechanisms leading to *intracellular* changes that regulate stem cell behavior (**FIGURE 5.4**). Extracellular mechanisms include:

- *Physical mechanisms* of influence, including structural and adhesion factors within the extracellular matrix that support the cellular architecture of the niche. Differences in cell-to-cell and cell-to-matrix adhesions as well as the cell density within the niche can alter the mechanical forces that influence stem cell behavior.

- *Chemical regulation* of stem cells takes the form of secreted proteins from surrounding cells that influence stem cell states and progenitor differentiation through endocrine, paracrine, or juxtacrine mechanisms (Moore and Lemischka 2006; Jones and Wagers 2008). In many cases, these signaling factors maintain the stem cell in an uncommitted state. Once stem cells become positioned farther from the niche, however, these factors cannot reach them, and differentiation commences.

Intracellular regulatory mechanisms include:

- *Regulation by cytoplasmic determinants*, the partitioning of which occurs at cytokinesis. As a stem cell divides, factors determining cell fate are either selectively partitioned to one daughter cell (asymmetric differentiating division) or shared evenly between daughter cells (symmetrical division).

- *Transcriptional regulation* occurs through a network of transcription factors that keep a stem cell in its quiescent or proliferative state, as well as promoting maturation of daughter cells toward a particular fate.

- *Epigenetic regulation* occurs at the level of chromatin. Different patterns of chromatin accessibility influence gene expression related to stem cell behavior.

The types of the intracellular mechanisms used by a given stem cell are in part the downstream net result of the extracellular stimuli in its niche. Just as important, however, is the stem cell's developmental history within its niche. Below are descriptions of some of the better-known stem cell niches, highlighting their developmental origins and the specific extracellular and intracellular mechanisms important for regulating stem cell behavior.

Pluripotent Cells in the Embryo

Cells of the inner cell mass

The pluripotent stem cells of the mammalian **inner cell mass** (**ICM**) are one of the most studied types of stem cells. Following cleavages of the mammalian zygote and formation of the morula, the process of cavitation creates the blastocyst,[1] which consists of a spherical layer of **trophectoderm cells** surrounding the inner cell mass and a fluid-filled cavity called the **blastocoel** (**FIGURE 5.5**). In the early mouse blastocyst, the ICM is a cluster of approximately 12 cells adhering to one side of the trophectoderm (Handyside 1981; Fleming 1987). The ICM will subsequently develop into a cluster of cells called the epiblast and a layer of primitive endoderm (yolk sac) cells that establish a barrier between the epiblast and the blastocoel. The epiblast develops into the embryo proper, generating all the cell types (more than 200) of the adult mammalian body including the primordial germ cells (see Shevde 2012), whereas the trophectoderm and primitive endoderm give rise to extraembryonic structures, namely the embryonic side of the placenta, chorion, and yolk sac (Stephenson et al. 2012; Artus et al. 2014). Importantly, cultured cells[2] of the ICM or epiblast produce **embryonic stem cells** (**ESCs**), which retain pluripotency and similarly can generate all cell types of the body (Martin 1980; Evans and Kaufman 1981). In contrast to the in vivo behavior of ICM cells, however, ESCs can self-renew seemingly indefinitely in proper culture conditions. We discuss the properties and use of ESCs later in this chapter. Here we will focus on the mammalian blastocyst as its own stem cell niche for the development of the only cells in the embryo that are at least transiently pluripotent.

Mechanisms promoting pluripotency of ICM cells

Essential to the transient pluripotency of the ICM is expression of the transcription factors Oct4,[3] Nanog, and Sox2 (Shi and Jin 2010). These three regulatory transcription factors are necessary to maintain the uncommitted stem cell-like state and functional pluripotency of the ICM, enabling ICM cells to give rise to the epiblast and all associated derived cell types (Pardo et al. 2010; Artus and Chazaud 2014; Huang and Wang

---

[1] This description is a generalization; not all mammals are treated equally during early blastocyst development. For instance, marsupials do not form an inner cell mass; rather, they create a flattened layer of cells called the pluriblast that gives rise to an equivalent epiblast and hypoblast. See Kuijk et al. 2015 for further reading on the surprising divergence during early development across species.

[2] Most ESC lines begin as co-cultures of multiple cells from the ICM, after which isolated cells can be propagated as clonal lines.

[3] Oct4 is also known as Oct3, Oct3/4, and Pou5f1. Mice deficient in *Oct4* fail to develop past the blastocyst stage. They lack a pluripotent ICM, and all cells differentiate into trophectoderm (Nichols et al. 1998; Le Bin et al. 2014). Oct4 expression is also necessary for the sustained pluripotency of derived primordial germ cells.

☐ Trophectoderm
☐ ICM
☐ Primitive endoderm
☐ Epiblast → Embryo

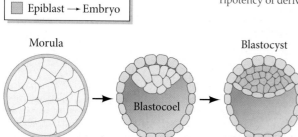

Morula

Blastocyst

Blastocoel

FIGURE 5.5 Establishment of the inner cell mass (the ICM, which will become the embryo) in the mouse blastocyst. From morula to blastocyst, the three principal cell types—trophectoderm, ICM, and primitive endoderm—are illustrated.

2014). It is interesting that expression of these three transcription factors is normally lost from the ICM as the epiblast differentiates (Yeom et al. 1996; Kehler et al. 2004). In contrast, the transcription factor Cdx2 is upregulated in the *outer* cells of the morula to promote trophectoderm differentiation and repress epiblast development (Strumpf et al. 2005; Ralston et al. 2008; Ralston et al. 2010).

What mechanisms are at work to control the temporal and spatial expression patterns of genes within the presumptive ICM and trophectoderm? Cell-to-cell interactions set the foundation for initial specification and architecture of these layers. First, cellular polarity along the **apicobasal axis** (apical-to-basal, or outer side of embryo to inside embryo) creates a mechanism by which symmetrical or asymmetrical divisions can produce two different cells. Perpendicularly positioned, asymmetrical divisions along the apicobasal axis would yield daughter cells segregated to the outside and inside of the embryo, corresponding to the development of the trophectoderm and ICM, respectively. In contrast, symmetrical divisions parallel to the apicobasal axis would distribute cytoplasmic determinants evenly to both daughter cells, further propagating cells only within either the outer trophectoderm layer or the ICM (**FIGURE 5.6**).

Asymmetric localization of factors along the apicobasal axis occurs at the morula stage in the outer cells of the presumptive trophectoderm. Well-known proteins in the partitioning defective (PAR) and atypical protein kinase C (aPKC) families become asymmetrically localized along the apicobasal axis. One outcome of these *partitioning proteins* is the recruitment of the cell adhesion molecule E-cadherin to the basolateral membrane where outer cells contact underlying ICM cells (**FIGURE 5.7A**; see Chapter 4; Stephenson et al. 2012; Artus and Chazaud 2014). Experimentally eliminating E-cadherin disrupts both apicobasal polarity and the specification of the ICM and trophectoderm lineages (Stephenson et al. 2010). How does E-cadherin influence these cell lineages?

Research has shown that the presence of E-cadherin activates the Hippo pathway, but only in the ICM. As discussed in Chapter 4, activated Hippo signaling represses the Yap-Taz-Tead transcriptional complex, and in the ICM, the result is the maintenance of pluripotent ICM development through Oct4. In the outer cells, the apically positioned partitioning proteins inhibit Hippo signaling, leading to an active Yap-Taz-Tead transcriptional complex, an upregulation of *cdx2*, and the trophectoderm fate (**FIGURE 5.7B**; Hirate et al. 2013). Thus, differential localization of specific proteins within the cell can lead to the activation of different gene regulatory networks within neighboring cells and the acquisition of different cell fates.

Adult Stem Cell Niches

Many adult tissues and organs contain stem cells that undergo continual renewal. These include but are not limited to germ cells across species; and brain, epidermis, hair follicles, intestinal villi, and blood in mammals. Also, multipotent adult stem cells play major roles in organisms with high regenerative capabilities, such as hydra, axolotl, and zebrafish. Adult stem cells must maintain the long-term ability to divide, be able to produce some differentiated daughter cells, and still repopulate the stem cell pool. The adult stem cell is housed in and controlled by its own **adult stem cell niche**, which regulates stem cell self-renewal, survival, and differentiation of those progeny that leave the niche (**TABLE 5.1**). Below we describe some of the better-characterized niches, which include those for the *Drosophila* germ stem cells and mammalian neural, gut epithelial, and hematopoietic stem cells. This list is obviously not exhaustive, but it highlights some universal mechanisms that control stem cell development.

Developing Questions

Is there such a thing as "stem-cellness"? Is being a stem cell an intrinsic property of the cell, or is it a property acquired through interactions with the stem cell niche? Is the niche making the stem cell? What approaches might you use to determine which of these conditions exist in a particular organ?

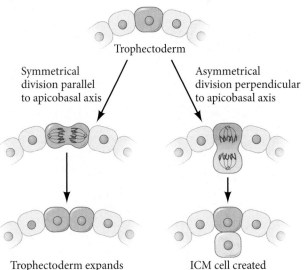

FIGURE 5.6 Divisions about the apicobasal axis. Depending on the axis of cell division in the trophectoderm, the trophectoderm layer can be expanded (left), or the ICM can be seeded (right).

(A)

Morula   Blastocyst

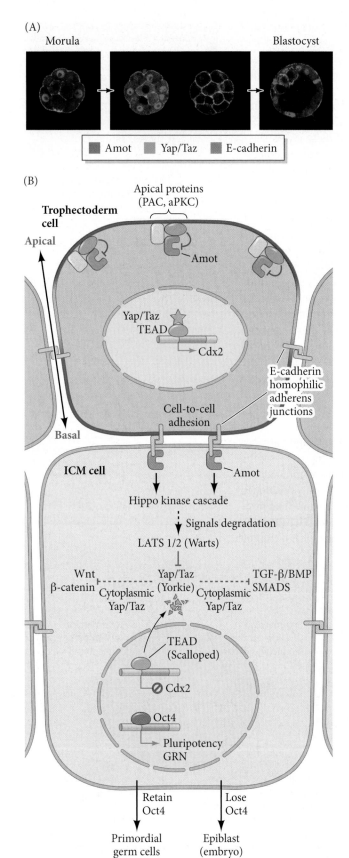

■ Amot ■ Yap/Taz ■ E-cadherin

FIGURE 5.7 Hippo signaling and ICM development. (A) Immunolocalization of the Hippo pathway components Amot (angiomotin; red stain) and Yap (green stain)—as well as E-cadherin—from morula to blastocyst. Activated Yap is localized to the trophectoderm nuclei, while E-cadherin (purple) is restricted to the trophectoderm-ICM membrane contacts. (B) Hippo signaling in trophectoderm (top) and ICM (bottom) cells. Hippo signaling is activated through E-cadherin binding with Amot and, as a result, Yap is degraded in the ICM cell. Names in parentheses are the *Drosophila* homologues. (A from Hirate et al. 2013.)

Stem cells fueling germ cell development in *Drosophila*

***DROSOPHILA* TESTES STEM CELL NICHE** Stem cell niches in the testes of male *Drosophila* illustrate the importance of local signals, cell-to-cell adhesion, and asymmetric cell division. The stem cells for sperm reside in a regulatory microenvironment called the **hub** (**FIGURE 5.8**). The hub consists of about 12 somatic testes cells and is surrounded by 5–9 **germ stem cells** (**GSCs**). The division of a sperm stem cell is asymmetric, always producing one cell that remains attached to the hub and one unattached cell. The daughter cell attached to the hub is maintained as a stem cell, whereas the cell that is not touching the hub becomes a **gonialblast**, a committed progenitor cell that will divide to become the precursors of the sperm cells. The somatic cells of the hub create this asymmetric proliferation by secreting the paracrine factor Unpaired onto the cells attached to them. Unpaired protein activates the JAK-STAT pathway in the adjacent germ stem cells to specify their self-renewal. Cells that are distant from the paracrine factor do not receive this signal and begin their differentiation into the sperm cell lineage (Kiger et al. 2001; Tulina and Matunis 2001).

Physically, this asymmetric division involves the interactions between the sperm stem cells and the somatic cells. In the division of the stem cell, one centrosome remains attached to the cortex at the contact site between the stem cell and the somatic cells. The other centrosome moves to the opposite side, thus establishing a mitotic spindle that will produce one daughter cell attached to the hub and one daughter cell away from it (Yamashita et al. 2003). (We will see a similar positioning of centrosomes in the division of mammalian neural stem cells.) The cell adhesion molecules linking the hub and stem cells together are probably involved in retaining one of the centrosomes in the region where the two cells touch. Here we see stem cell production using asymmetric cell division.

***DROSOPHILA* OVARIAN STEM CELL NICHE** Similar to sperm, the *Drosophila* oocyte is derived from a germ stem cell. These GSCs are held within the ovarian stem cell niche, and positional secretion of paracrine factors influences stem cell self-renewal and oocyte differentiation in a concentration-dependent manner. Egg production in the adult fly ovary occurs in more than 12 egg tubes or **ovarioles**, each one housing identical GSCs (usually two per ovariole) and several somatic cell types that construct

TABLE 5.1   Some stem cell niches of adult humans

Stem cell type	Niche location	Cellular components of niche
LOW TURNOVER[a]		
Brain (neurons and glia)	Ventricular-subventricular zone (V-SVZ; see Figure 5.10), subgranular zone	Ependymal cells, blood vessel epithelium
Skeletal muscle	Between basal lamina and muscle fibers	Muscle fiber cells
HIGH TURNOVER[a]		
Mesenchymal stem cells (MSCs)	Bone marrow, adipose tissue, heart, placenta, umbilical cord	Probably blood vessel epithelium
Intestine	Base of small intestinal crypts (see Figure 5.13)	Paneth cells, MSCs
Hematopoietic (blood-forming) stem cells (HSCs)	Bone marrow (see Figure 5.15)	Macrophages, T_{reg} cells, osteoblasts, pericytes, glia, neurons, MSCs
Epidermis (skin)	Basal layer of epidermis	Dermal fibroblasts
Hair follicle	Bulge (see Figure 16.17)	Dermal papillae, adipocyte precursors, subcutaneous fat, keratin
Sperm	Testes	Sertoli cells (see Figure 6.21)

[a] Niches with low rates of cell turnover generate stem cells for repair, slow growth, and (in the case of neurons) learning. Niches with high turnover are constantly producing new cells for bodily maintenance.

the niche known as the **germarium** (Lin and Spradling 1993). As a GSC divides, it self-renews and produces a **cystoblast** that (like the sperm gonialblast progenitor cell) will mature as it moves farther out of the stem cell niche—beyond the reach of the niche's regulatory signals—and becomes an oocyte surrounded by follicle cells (**FIGURE 5.9A**; Eliazer and Buszczak 2011; Slaidina and Lehmann 2014).

Although the GSCs are within the stem cell niche, they are in contact with Cap cells. Upon division of the GSC perpendicular to the Cap cells, one daughter cell remains tethered to the Cap cell by E-cadherin and maintains its self-renewal identity, whereas

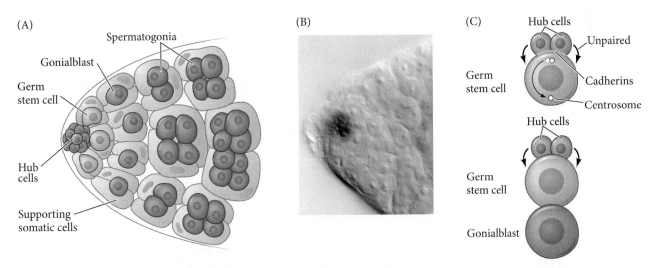

FIGURE 5.8 Stem cell niche in *Drosophila* testes. (A) The apical hub consists of about 12 somatic cells, to which are attached 5–9 germ stem cells. The germ stem cells divide asymmetrically to form another germ stem cell (which remains attached to the somatic hub cells), and a gonialblast that will divide to form the sperm precursors (the spermatogonia and the spermatocyte cysts where meiosis is initiated). (B) Reporter β-galactosidase inserted into the gene for Unpaired reveals that this protein is transcribed in the somatic hub cells. (C) Cell division pattern of the germline stem cells, wherein one of the two centrosomes remains in the cortical cytoplasm near the site of hub cell adhesion while the other migrates to the opposite pole of the germ stem cell. The result is one cell remaining attached to the hub and the other cell detaching from the hub and differentiating. (After Tulina and Matunis 2001; photograph courtesy of E. Matunis.)

FIGURE 5.9 *Drosophila* ovarian stem cell niche. (A) Immunolabeling of different cell types within the *Drosophila* germarium. Germ stem cells (GSCs) are identified by the presence of spectrosomes. Differentiating germ cells (cystoblasts) are stained blue. Bam-expressing (cyst) cells are green. (B) The interactions between Cap cells and GSCs in the germarium. See text for a description of the interactions between the regulatory components. (A from Slaidina and Lehmann 2014.)

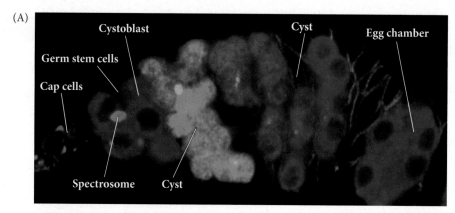

(A)

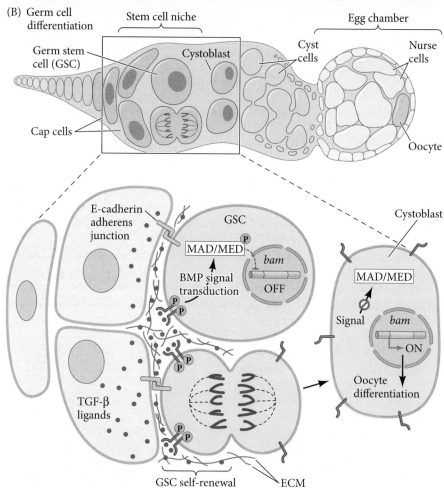

(B)

the displaced daughter cell begins oocyte differentiation (Song and Xie 2002). Cap cells affect GSCs by secreting TGF-β family proteins, which activate the BMP signal transduction pathway in the GSCs, and, as a result, prevent GSC differentiation (**FIGURE 5.9B**). Extracellular matrix components like collagen and heparan sulfate proteoglycan restrict the diffusion of the TGF-β family proteins such that only the tethered GSCs receive significant amounts of these TGF-β signals (Akiyama et al. 2008; Wang et al. 2008; Guo et al. 2009; Hayashi et al. 2009).[4] Activation of BMP signal transduction in

---

[4] Gain or loss of function of the TGF-β proteins results in tumor-like expansion of the GSC population or loss of the GSCs, respectively (Xie and Spradling 1998).

the GSC prevents differentiation by repressing transcription of genes that promote differentiation, primarily that of *bag of marbles* (*bam*). When *bam* is expressed, the cell goes on to differentiate into an oocyte (see Figure 5.9).

In conclusion, in both the testis and ovary of *Drosophila*, coordinated cell division paired with adhesion and paracrine-mediated repression of differentiation controls GSC renewal and progeny differentiation. New insights into the epigenetic regulation of GSC development are beginning to emerge, for example the histone methyltransferase Set1 has been discovered to play an essential role in GSC self-renewal (Yan et al. 2014; see Scientists Speak 5.2). Moreover, many of the structural and mechanistic factors at play in the *Drosophila* germ stem cell niche are similar in other species. For example, the distal tip cell of the *C. elegans* gonad is much like the *Drosophila* Cap cells in that it provides the niche signals that regulate the nematode's germ stem cells (see Chapter 8). *C. elegans* does not use Unpaired or BMP as in the fly GSC niche; rather, Notch signaling from the distal tip cell is used to similarly suppress differentiation of the GSCs. Those cells leaving the range of Notch signaling start differentiating into germ cells (see the comparative review of GSCs by Spradling et al. 2011).

 SCIENTISTS SPEAK 5.2   Dr. Norbert Perrimon answers questions on defining the gene regulatory network for germ stem cell self-renewal in *Drosophila*.

Adult Neural Stem Cell Niche of the V-SVZ

Despite the first reports of adult neurogenesis in the postnatal rat in 1969 and in songbirds in 1983, the doctrine that "no new neurons are made in the adult brain" held for decades. At the turn of the twenty-first century, however, a flurry of investigations, primarily in the adult mammalian brain, began to mount strong support for continued neurogenesis throughout life (Gage 2002). This acceptance of **neural stem cells** (**NSCs**) in the adult central nervous system (CNS) marks an exciting time in the field of developmental neuroscience and has significant implications for both our understanding of brain development and the treatment of neurological disorders.

Whether in fish or humans, adult NSCs[5] retain much of the cellular morphology and molecular characteristics of their embryonic progenitor cell, the radial glial cell. Radial glia and adult NSCs are polarized epithelial cells spanning the full apicobasal axis of the CNS (Grandel and Brand 2013). The development of radial glia and the embryonic origins of the adult mammalian neural stem cell niche are covered in Chapter 13. In anamniotes such as teleosts (the bony fishes), radial glia function as NCSs throughout life, occurring in numerous neurogenic zones (at least 12) in the adult brain (Than-Trong and Bally-Cuif 2015). In the adult mammalian brain, however, NSCs have been characterized only in two principal regions of the cerebrum: the **subgranular zone** (**SGZ**) of the hippocampus and the **ventricular-subventricular zone** (**V-SVZ**) of the lateral ventricles (Faigle and Song 2013; Urbán and Guillemot 2014). There are similarities and differences between these mammalian neurogenic niches such that each NSC has characteristics reminiscent of its radial glial origin, yet only the NSC of the V-SVZ maintains contact with the cerebral spinal fluid. During development of the adult V-SVZ, radial glia-like NSCs transition into **type B cells** that fuel the generation of specific types of neurons in both the olfactory bulb and striatum, as has been shown in both the mouse and human brain (**FIGURE 5.10**; Curtis et al. 2012; Lim and Alvarez-Buylla 2014).

WEB TOPIC 5.1   **THE SUBGRANULAR ZONE NICHE**   Delve deeper into the other neural stem cell niche of the mammalian brain.

---

[5] Most NSCs exhibit astroglial characteristics, although there are exceptions. Self-renewing neuroepithelial-like cells persist in the zebrafish telencephalon and function as neural progenitors that lack typical astroglial gene expression. Consider the work of Michael Brand's lab for further study. (Kaslin et al. 2009; Ganz et al. 2010; Ganz et al. 2012.)

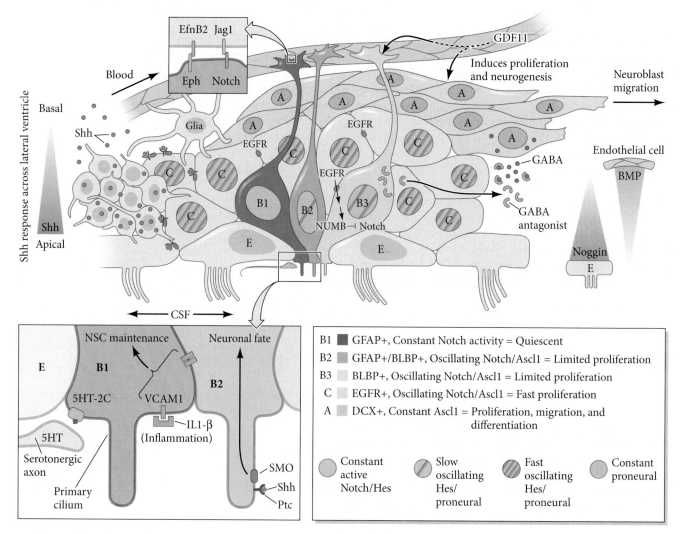

FIGURE 5.10 Schematic of the ventricular-subventricular zone (V-SVZ) stem cell niche and its regulation. Multiciliated ependymal cells (E; light gray) line the ventricle and contact the apical surface of V-SVZ NSCs (blue). Typically quiescent B1-type NSCs (dark blue) give rise to activated B2 and B3 cells (lighter shades of blue) that possess limited proliferation. The B3 cells generate the C cells (green), which, after three rounds of division, give rise to migrating neuroblasts (A cells; orange). The niche is penetrated by endothelial cell-built blood vessels that are in part enwrapped by the basal endfeet of B cells. Maintenance of the stem cell pool is regulated by VCAM1 adhesion and Notch signaling (changes in Notch pathway oscillations are depicted as color changes in the nuclei). Clusters of neurons in the ventral region of the lateral ventricle express Sonic hedgehog (Shh) that influence different neuronal cell differentiation from the niche. Antagonistic signaling between BMP and Noggin from endothelial cells and ependymal cells, respectively, balance neurogenesis along this gradient. Serotonergic (5HT) axons lace the ventricular surface, and—along with IL1-β and GDF11 from the cerebral spinal fluid (CSF) and blood, respectively—play roles as external stimuli to regulate the niche. Non-niche neurons, astrocytes, and glia can be found within the niche and influence its regulations. Glial fibrillary acidic protein (GFAP); brain lipid binding protein (BLBP); double cortin (DCX). (Based on various sources, including Basak et al. 2012; Giachino et al. 2014; Lim and Alvarez-Buylla 2014; and Ottone et al. 2014.)

The neural stem cell niche of the V-SVZ

In the V-SVZ, B cells project a primary cilium (see Chapter 4) from their apical surface into the cerebrospinal fluid of the ventricular space, and a long basal process terminates with an endfoot tightly contacting blood vessels (akin to the astrocytic endfeet that contribute to the blood-brain barrier). The fundamental cell constituents of the V-SVZ niche include four cell types: (1) a layer of ependymal cells, E-cells, along the ventricular wall; (2) the neural stem cell called the B cell; (3) progenitor (transit-amplifying) C cells; and (4) migrating neuroblast A cells (see Figure 5.10). Small clusters of B cells are surrounded by the multiciliated E-cells, forming a pinwheel-like rosette structure (**FIGURE 5.11A**; Mirzadeh et al. 2008). Cell generation within the V-SVZ begins in its central core with a dividing B cell, which directly gives rise to a C cell. These type C progenitor cells proliferate and develop into type A neural precursors that stream into the olfactory bulb for final neuronal differentiation (see Figure 5.10). The B cell has been further categorized into three subtypes (B1, B2, and B3) based on differences in proliferative states that correlate with distinct radial glial gene expression patterns (Codega et al. 2014; Giachino et al. 2014). It is important to note that in the NSC niche, *type 1 B cells are quiescent or inactive,* whereas *types 2 and 3 B cells represent actively proliferating neural stem cells* (Basak et al. 2012).[6]

 SCIENTISTS SPEAK 5.3   Dr. Arturo Alvarez-Buylla describes the adult V-SVZ neural stem cell niche.

Maintaining the NSC pool with cell-to-cell interactions

Maintaining the stem cell pool is a critical responsibility of any stem cell niche because too many symmetrical differentiating and progenitor-generating divisions can deplete the stem cell pool. The V-SVZ niche is designed structurally and is equipped with signaling systems to ensure that its B cells are not lost during calls for neurogenic growth or repair in response to injury.

VCAM1 AND ADHERENCE TO THE ROSETTE NICHE   The rosette or pinwheel architecture is a distinctive physical characteristic of the V-SVZ niche. It is maintained at least in part by a specific cell adhesion molecule, VCAM1 (Kokovay et al. 2012). The rosette pattern is not unique to the NSC niche; it is a repeated structural element throughout development (Harding et al. 2014). However, whereas the early developmental uses of the rosette structure are transient, V-SVZ pinwheels are maintained throughout adult life. As the mammalian brain ages, both the number of observed pinwheel structures and the number of neural stem cells in those pinwheels decreases, which correlates with a reduction in neurogenic potency in later life (Mirzadeh et al. 2008; Mirzadeh et al. 2010; Sanai et al. 2011; Shoo et al. 2012; Shook et al. 2014). Much like football players huddled around the quarterback, ependymal cells surround the type B cells; yet, unlike the directing quarterback, the B cell is listening to the ependymal cells (and other niche signals) for instructions either to remain quiescent or to become active. The B cells most tightly associated with ependymal cells are the more quiescent B1 cells. The more loosely packed B cells are actively proliferating B2 and B3 cells. (Doetsch et al. 1997). Experimental inhibition of VCAM1, an adhesion protein specifically localized to the apical process of B cells, disrupts the pinwheel pattern and causes a loss of NSC quiescence while promoting differentiation of progenitors (**FIGURE 5.11B**; Kokovay et al. 2012). *The tighter the hold, the more quiescent the stem cell.*

NOTCH, THE TIMEPIECE TO DIFFERENTIATION   Notch signaling has been found to play an important role in the maintenance of the pool of B type stem cells (Pierfelice et al. 2011; Giachino and Taylor 2014). Notch family members function as transmembrane receptors, and through cell-to-cell interactions, the Notch intracellular domain (NICD) is cleaved and released to function as part of a transcription

[6] In the mouse V-SVZ, one B cell can yield 16 to 32 A cells: each C cell that a B cell produces will divide three times, and their A cell progeny typically divide once, yielding 16 cells, but can also divide twice, yielding 32 cells (Ponti et al. 2013).

(A)

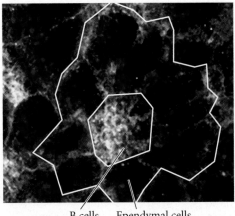

B cells   Ependymal cells

(B)
Control

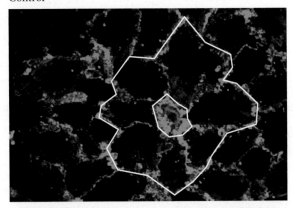

VCAM1 blocked

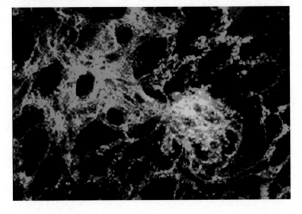

FIGURE 5.11   VCAM1 and pinwheel architecture. (A) The pinwheel arrangement of cells in the V-SVZ of the NSC niche is revealed with membrane labeling. Immunolabeling for VCAM1 (red) shows its co-localization with GFAP (green) in the B cells at the pinwheel core. The blue stain shows the presence of β-catenin; pinwheel organization is outlined in white. (B) Blocking adhesion using antibodies to VCAM1 disrupts the pinwheel organization of B cells and ependymal cells. In these photos, red visualizes GFAP; green indicates the presence of β-catenin. (After Kokovay et al. 2012.)

Developing Questions

We know quite a bit about *Notch/Delta* and *Hes* gene oscillations during embryonic development. What do oscillations of these genes in the adult neural stem cell niche actually look like? How do these oscillations result in the progression from stem cell to neuron?

factor complex typically repressing proneural gene expression (see Figure 5.10 and Chapter 4). Higher levels of NICD activity support stem cell quiescence, whereas decreasing levels of Notch pathway activity promote progenitor proliferation and maturation toward neural fates.[7] The NICD is more active in the type B1 cell than in the other cells of the V-SVZ niche, and it functions with other transcription factors to repress gene expression associated with both proliferation and differentiation, thereby promoting quiescence and maintaining the number of NSCs (Ables et al. 2011; Pierfelice et al. 2011; Giachino and Taylor 2014; Urbán and Guillemot 2014).

Notch1 is actually expressed in all major cell types in the V-SVZ niche (B cells, progenitor [transit-amplifying] cells, and type A migrating neuroblasts; Basak et al. 2012), which raises the question, how can differentiation begin in the presence of Notch? An important part of the Notch regulatory mechanism of neurogenesis lies in the downstream transcriptional targets, namely the *Hairy and Enhancer of Split* (*Hes*)-related genes. *Hes* genes primarily function to repress proneural gene expression. As seen in Chapter 17, Notch-Delta signaling and its Hes targets can show temporally oscillating patterns of gene expression established through negative feedback loops in which upregulation of *Hes* by Notch leads to Hes-mediated repression of Notch. A growing hypothesis is that constant activity of Notch signaling promotes quiescence, whereas the oscillating expression of *Hes* genes—and, consequently, the anti-oscillation periods of proneural genes (such as *Ascl1/Mash1*)—supports proliferative states until proneural gene expression is sustained and the cell differentiates (see Figure 5.10; Imayoshi et al. 2013).

Promoting differentiation in the V-SVZ niche

The main purpose of a stem cell niche is to produce new progenitor cells capable of differentiating toward specific cell types. In the V-SVZ niche, a number of factors are involved.

EGF REPRESSES NOTCH   As discussed above, active (and constant) Notch signaling encourages quiescence and represses differentiation; therefore, one mechanism to promote neurogenesis is to attenuate (and oscillate) Notch activity. The type C progenitor cells do that by using epidermal growth factor receptor (EGFR) signaling, which upregulates NUMB, which in turn inhibits NICD (see Figure 5.10; Aguirre et al. 2010). Therefore, EGF signaling promotes the use of the stem cell pool for neurogenesis by counterbalancing Notch signaling (McGill and McGlade 2003; Kuo et al. 2006; Aguirre et al. 2010).

BONE MORPHOGENIC PROTEIN SIGNALING AND THE NSC NICHE   Further movement toward differentiation is driven by additional factors, such as BMP signaling, which promotes gliogenesis in the V-SVZ as well as other regions of the mammalian brain (Lim et al. 2000; Colak et al. 2008; Gajera et al. 2010; Morell et al. 2015). BMP signaling from endothelial cells is kept high at the basal side of niche, whereas ependymal cells at the apical border secrete the BMP inhibitor Noggin, keeping BMP levels in this region low. Therefore, as B3 cells transition into type C-progenitor cells and then move closer to the basal border of the niche, they leave the reach of BMP inhibitors and experience increasing levels of BMP signaling, which promotes neurogenesis with a preference toward glial cells (see Figure 5.10).

Environmental influences on the NSC niche

The adult NSC niche has to react to changes in the organism, such as injury and inflammation, exercise, and changes in circadian rhythms. How might the NSC niche

---

[7] Many of the roles that Notch signaling plays in neurogenesis in the adult brain are similar to its regulation of radial glia in the embryonic brain, but some important differences are beginning to emerge. For a direct comparison of Notch signaling in embryonic versus adult neurogenesis and across species, see Pierfelice et al. 2011 and Grandel and Brand 2013.

respond to such changes? The cerebral spinal fluid (CSF), neural networks, and vasculature are in direct contact with the niche, and they can influence NSC behavior through paracrine release into the CSF, electrophysical activity from the brain, and endocrine signaling delivered through the circulatory system.

NEURAL ACTIVITY   Intrinsic to the niche, migrating neural precursors secrete the neurotransmitter GABA to negatively feedback upon progenitor cells and attenuate their rates of proliferation. In opposition to this action, B cells secrete a competitive inhibitor to GABA (diazepam-binding inhibitor protein) to increase proliferation in the niche (Alfonso et al. 2012). Extrinsic inputs have also been discovered from serotonergic axons densely contacting both the ependymal and type B cells (Tong et al. 2014). Type B cells express serotonin receptors, and activation or repression of the serotonin pathway in B1 cells increases or decreases, respectively, proliferation in the V-SVZ (see Figure 5.10). Additional neural activity from dopaminergic axons and a population of choline acetyltransferase neurons residing in the niche have also been found to promote proliferation and neurogenesis (see the references cited in Lim and Alvarez-Buylla 2014).

SONIC HEDGEHOG SIGNALING AND THE NSC NICHE   Similar to neural tube patterning in the embryo (which we will describe in Chapter 13), the creation of different neuronal cell types from the V-SVZ is in part patterned by a gradient of Sonic hedgehog (Shh) signaling along the apicobasal axis of the niche, with highest levels of Shh in the apical region[8] (Goodrich et al. 1997; Bai et al. 2002; Ihrie et al. 2011). When the *Shh* gene is knocked out, the loss of Shh signaling results in specific reductions in apically derived olfactory neurons (Ihrie et al. 2011). This result implies that cells derived from NSC clusters in the more apical positions of the niche will adopt different neuronal fates compared to cells derived from NSCs in more basal positions, based on differences in Shh signaling (see Figure 5.10).

COMMUNICATION WITH THE VASCULATURE   Another external source of influence on NSC activity in the brain comes from the vasculature that infiltrates this stem cell niche: from blood vessel cells (endothelial, smooth muscle, pericytes), from the associated extracellular matrix, and from substances in the blood (Licht and Keshet 2015; Ottone and Parrinello 2015). Despite the far distance that the apical surface and bodies of B cells can be from blood vessels, the basal endfoot is quite intimately associated with the vasculature (see Figure 5.10). This physical characteristic puts cells of the blood vessel in direct contact with the NSC. As discussed earlier, Notch signaling is fundamental in controlling B1 cell quiescence. Notch receptors in the B cell's endfoot bind to the Jagged1 (Jag1) transmembrane receptor in endothelial cells, which causes Notch to be processed into its NICD transcription factor, and B1 cell quiescence is maintained as a result (Ottone et al. 2014). As the B2 and B3 cells transition into type C progenitor cells, their basal connections with endothelial cells are lost; consequently, NICD is reduced, enabling the progenitor cells to mature.

For a blood-borne substance to influence neurogenesis, it must cross the tight blood-brain barrier. The use of fluorescent tracer compounds in the blood has demonstrated that the blood-brain barrier in the NSC niche is "leakier" than in other brain regions (see Figure 5.10; Tavazoie et al. 2008). A variety of blood-borne substances that influence the adult NSC niche are rapidly being identified, however, and one of the most intriguing molecules identified to date is growth differentiation factor 11 (GDF11, also known as BMP11), which appears to ward off some of the symptoms of aging in the brain. Like humans, elderly mice show a significantly reduced neurogenic potential. Researchers realized that something in the circulation of young mice could prevent this decline when they surgically connected the circulation of a young mouse to that of an

---

[8] The gradient of Shh in the brain is more accurately described as being oriented along the dorsal-to-ventral axis; for simplification, however, we have restricted our discussion to its presence only along the apical-to-basal axis.

FIGURE 5.12 Young blood can rejuvenate an old mouse. (A) Parabiosis—fusion of the circulatory systems of two individuals—was done using mice of similar (isochronic) or different (heterochronic) ages. When an old mouse was parabiosed to a young mouse, the result was an increase in the amount of vasculature (stained green in the photographs) as well as the amount of proliferative neural progeny in the old mouse. (B) Administering GDF11 into the circulatory system of an old mouse was sufficient to similarly increase both vasculature (green in photographs) and the population of neural progenitors in the V-SVZ (outlined red population in photographs and quantified SOX2+ cells in graph). (After Katsimpardi et al. 2014.)

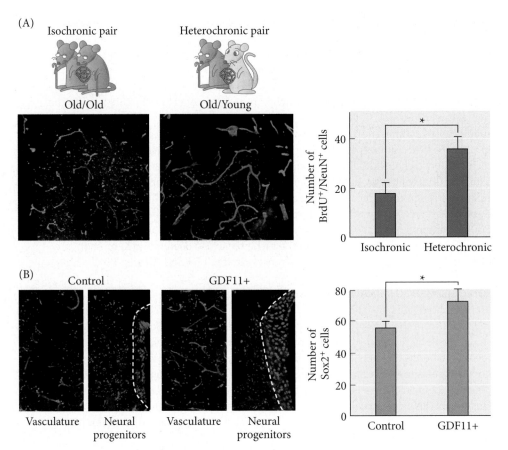

Developing Questions

What are the cellular and molecular mechanisms receiving and interpreting the GDF11 signal in the V-SVZ niche to stimulate neurogenesis? Most intriguing is that the original parabiosis experiment demonstrated that substances in the blood of a young mouse can, by themselves, rejuvenate the old mouse. What besides GDF11 may be playing a role in this healing process? Could it be the increased presence of hematopoietic stem cells in the blood (see below)?

old mouse (heterochronic parabiosis). Doing so caused increased vasculature to develop in the brain of the heterochronic old mouse (**FIGURE 5.12A**), followed by increased NSC proliferation that restored neurogenesis and cognitive functions (Katsimpardi et al. 2014). The researchers then showed that they could similarly restore neurogenic potential in the old mouse brain using a single circulating factor, GDF11; moreover, GDF11 is known to decrease with age[9] (**FIGURE 5.12B**; Loffredo et al. 2013; Poggioli et al. 2015). These results strongly suggest that communication between the NSC and its surrounding vasculature is a major regulatory mechanism of neurogenesis in the adult brain, and that changes in this communication over time may underlie some of the cognitive deficits associated with aging.

The Adult Intestinal Stem Cell Niche

As discussed above, the neural stem cell is part of a specialized epithelium. Not all epithelial stem cell niches are the same, however. The epithelium of the mammalian intestine is organized into a very different stem cell niche. The epithelial lining of the intestine projects millions of finger-like villi into the lumen for nutrient absorption, and the base of each villus sinks into a steep valley called a **crypt** that connects with adjacent villi (**FIGURE 5.13A**). Critical to understanding the evolved function of the intestinal stem cell (ISC) niche is appreciating the rapid rate of cell turnover in the intestine.

[9] One recent study (Egerman et al. 2015) reported that GDF11 levels do not decline with age. In addition, despite research claiming the muscle rejuvenation capacity of GDF11 (Sinha et al. 2014), this study also states that GDF11 (like its protein cousin myostatin) inhibits muscle growth. The age-related drop in GDF11 levels has recently been confirmed (Poggioli et al. 2015), however, and GDF11's effect on neurogenesis was never originally contested by Egerman and colleagues.

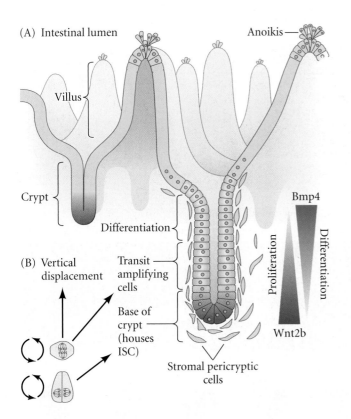

(A) Intestinal lumen

Anoikis

Villus

Crypt

Differentiation

(B) Vertical displacement

Transit amplifying cells

Base of crypt (houses ISC)

Proliferation

Differentiation

Bmp4

Wnt2b

Stromal pericryptic cells

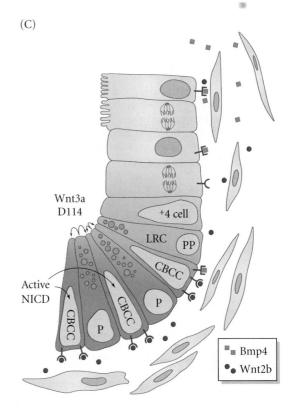

(C)

Wnt3a
D114

Active NICD

+4 cell

LRC

PP

CBCC

CBCC

P

CBCC

P

■ Bmp4
● Wnt2b

Clonal renewal in the crypt

Cell generation occurs in the crypts, whereas cell removal largely happens at the tips of villi. Through this upward movement from cell source to cell sink, a turnover of intestinal absorptive cells occurs approximately every 2 to 3 days[10] (Darwich et al. 2014).

Several stem cells reside at the base of each crypt in the mouse small intestine; some daughter cells remain in the crypts as stem cells, whereas others become progenitor cells and divide rapidly (**FIGURE 5.13B**; Lander et al. 2012; Barker 2014; Krausova and Korinek 2014; Koo and Clevers 2014). Division of stem cells within the crypt and of the progenitor cells drives cell displacement vertically up the crypt toward the villus, and as cells become positioned farther from the crypt base, they progressively differentiate into the cells characteristic of the small intestine epithelium: enterocytes, goblet cells, and enteroendocrine cells. Upon reaching the tip of the intestinal villus, they are shed and undergo **anoikis**, a process of programmed cell death (apoptosis) caused by a loss of attachment, in this case, loss of contact with the other villus epithelial cells and extracellular matrix (see Figure 5.13A).[11]

Lineage-tracing studies (Barker et al. 2007; Snippert et al. 2010; Sato et al. 2011) have shown that intestinal stem cells (expressing the Lgr5 protein) can generate all the differentiated cells of the intestinal epithelium. Due to their specific location at the very base of the crypt, these Lgr5+ stem cells are referred to as crypt base columnar cells (CBCC) and are found in a checkered pattern with the differentiated Paneth cells, which are also restricted to the base of the crypt (**FIGURE 5.13C**; Sato et al. 2011). One of the most convincing demonstrations that CBCC cells represent "active stem cells" is that a single CBCC cell can completely repopulate the crypt over time (**FIGURE 5.14**; Snippert

---

[10] This figure was determined through a meta-analysis of six species, including mouse and humans.

[11] This process is highly reminiscent of growth in the hydra, where each cell is formed at the animal's base, migrates to become part of the differentiated body, and is eventually shed from the tips of the arms (see Chapter 22).

FIGURE 5.13 The ISC niche and its regulators. (A) The intestinal epithelium is composed of long, finger-like villi that project into the lumen, and at the base of the villi, the epithelium extends into deep pits called crypts. The ISC and progenitors reside at the very bottom of the crypts (red), and cell death through anoikis occurs at the apex of the villi. (B) Along the proximodistal axis (crypt to villus), the crypt epithelium can be functionally divided into three regions: the base of the crypt houses ISC, the proliferative zone is made of transit amplifying cells, and the differentiation zone characterizes the maturation of epithelial cell types. Pericryptal stromal cells surround the basal surface of the crypt and secrete opposing morphogenic gradients of Wnt2b and Bmp4 that regulate stemness and differentiation, respectively. (C) Higher magnification of the cells residing in the base of the crypt. Paneth cells (P) secrete Wnt3a and D114, which stimulates proliferation of the Lrg5+, crypt base columnar cells (CBCC) in part through activation of the notch intracellular domain (NICD). (LRC, label-retaining cell; PP, Paneth progenitor cell.)

FIGURE 5.14 Clonogenic nature of the intestinal stem cell niche. (A) Cre responsive transgenic mice using the Lgr5 promoter and the Rosa26-LacZ reporter mark discrete clones of ISCs at the base of the crypt (blue). Retention of LacZ in cell descendants over time shows the progressive movement up the villus. (B) Mosaic labeling of ISCs in the intestinal crypt with transgenic "confetti" mice demonstrates a stochastic (predictable randomness) progression toward monoclonal (visualized as one color) crypts over time. This same progression can be mathematically modeled and simulated to produce a similar coarsening of color patterns, as seen below the photographs. (A from Barker et al. 2007; B after Snippert et al. 2010 and Klein and Simons 2011.)

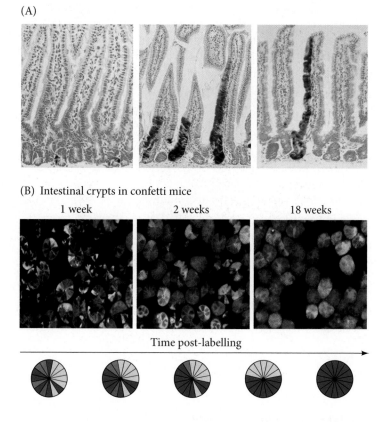

(A)

(B) Intestinal crypts in confetti mice

1 week   2 weeks   18 weeks

Time post-labelling

Developing Questions

Is there a true quiescent stem cell in the crypt? This question is hotly debated. The employment of transgenic tools to study discrete cell populations in the base of the crypt has opened the door to defining a quiescent stem cell based on the length of time that an induced cell label remains in a cell. Buczacki and colleagues) found a subset of Lrg5-expressing crypt cells retaining the induced label for weeks (Buczacki et al. 2013). So, are these label-retaining cells a quiescent, reserve population of stem cells? Take note of Hans Clevers' comments on this work (Clevers 2013); consider also the recent intravital imaging of the crypt (Ritsma et al. 2014), and debate it yourself.

et al. 2010). After CBCC symmetrical division, one daughter cell will (by chance) be adjacent to a Paneth cell, while the other daughter cell is pushed away from the base to progress through the transit-amplifying (progenitor) fate. In this manner, the neutral competition for the Paneth cells' surfaces dictates which will remain as a stem cell and which will mature (Klein and Simons 2011).

Why might the longevity of an ISC be left to chance? It seems like a very non-Darwinian approach of the intestinal crypt to not favor the survival of the fittest stem cell, wouldn't you agree? Consider what would make an ISC particularly "fit." It might include high proliferative capability, particularly in a tissue that is designed for rapid turnover like the intestinal epithelium; the paradox, however, is that enhanced proliferation is also a precursor for tumorogenesis and cancer. A current idea is that ISC evolution through "neutral drift" promotes the retention of random CBCCs based more on location than on genes. Such retention would decrease the probability of fixing "favorable" mutations that may also lead to cancer (Kang and Shibata 2013; Walther and Graham 2014).

Regulatory mechanisms in the crypt

The Paneth cell plays an important role in the intestine's immune response because it is a secretory cell housing many granules for the release of anti-microbial substances. In addition, with nearly 80% of the intestinal stem cell's surface in direct contact with the Paneth cell, it is a vital contributor to stem cell regulation. Each niche contains about 15 Paneth cells and an equal number of CBCCs. Deleting the Paneth cells destroys the ability of the stem cells to generate other cells. Paneth cells express several paracrine and juxtacrine factors, including but not limited to Wnt3a and Delta-like4 (Dll4), an activator of Notch (Sato et al. 2009; Barker 2014; Krausova and Korinek 2014). When Dll4 binds to Notch receptors on the intestinal stem cells, it is interpreted as a signal for sustained proliferation and lineage specification toward more secretory over absorptive cell fates (see Figure 5.13C; Fre et al. 2011; Pellegrinet et al. 2011).

The stromal cells below the crypt epithelium also help regulate the intestinal stem cell niche. They secrete Wnt2b with highest levels at the base of the crypt, whereas an opposing gradient of Bmp4 is most abundant at the top of the crypt (see Figure 5.13C). CBCCs, expressing both the Frizzled7 and BMPR1a receptors for Wnt2b and Bmp4, respectively, can be affected by both factors (He et al. 2004; Farin et al. 2012; Flanagan et al. 2015). The currently accepted model is that Wnt signaling promotes survival and proliferation of the CBCCs and progenitor cells, whereas the opposing BMP signals promote differentiation in the crypt with maturation progressing in the direction of the villus (Carulli et al. 2014; Krausova and Korinek 2014).

There exists another small population of intestinal stem cells called the "+4 cells" due to their location next to the fourth Paneth cell from the base of the crypt (see Figure 5.13C; Potten et al. 1978; Potten et al. 2002; Clevers 2013). Like CBCCs, +4 cells can generate all the cell types of the intestine. Some reports indicate that +4 cells divide at a slower rate than CBCCs, which suggests that they may be the quiescent stem cell of the crypt. Minimally, it is undisputable that the +4 cells make important contributions to intestinal homeostasis; significant debate still surrounds the notion that they represent the niche's quiescent stem cell, however (Carulli 2014).

 SCIENTISTS SPEAK 5.4   There are similarities between the ISC niche with that of the lung. Dr. Brigid Hogan talks about the role of stem cells in lung development and disease.

Stem Cells Fueling the Diverse Cell Lineages in Adult Blood

The hematopoietic stem cell niche

Every day in your blood, more than 100 billion cells are replaced with new cells. Whether the needed cell type is for gas exchange or for immunity, the **hematopoietic stem cell (HSC)** is at the top of the hierarchical lineage powering the amazing cell generating machine that is the HSC niche (see Figure 18.24). The importance of HSCs cannot be overstated, for both its importance to the organism and its history of discovery. Since the late 1950s, stem cell therapy with HSCs has been used to treat blood-based diseases through the use of bone marrow transplantation.[12] In addition, the "niche hypothesis" of a stem cell residing in and being controlled by a specialized microenvironment was first inspired by the HSC (Schofield 1978).

The success of the bone marrow transplant is evidence of the location of the HSC niche, the cavities in bones where the bone marrow resides (**FIGURE 5.15**). In the highly vascularized tissue of the bone marrow, HSCs are in close proximity to the bone cells (osteocytes), the endothelial cells that line the blood vessels, and the connective stromal cells. Were HSCs somehow born from bone to then reside in the marrow? The answer to this question is *no*. Primitive hematopoiesis first occurs in the embryonic yolk sac; "definitive hematopoietic stem cells" (dHSCs), however, are born in the developing aortic portion of the aorta-gonad-mesonephros (AGM). Through developed vasculature, HSCs migrate to the fetal liver, where they rapidly proliferate and begin to generate progeny of the hematopoietic lineages (Mikkola and Orkin 2006; Al-Drees et al. 2015; Boulais and Frenette 2015). During this period, bones are taking shape and becoming vascularized, which establishes a pathway for HSCs to find their way to the bone marrow. The remarkable ability of HSCs to migrate through the circulatory system and find their tissue-specific destination is called **homing**. HSCs recognize the bone marrow as the environment to seed through the HSCs' CXCL4 receptor sensing the

---

[12]The first successful bone marrow transplantation was between two identical twins, one of whom had leukemia. It was conducted by Dr. E. Donnall Thomas, whose continued research in stem cell transplantations won him the Nobel Prize in Physiology or Medicine in 1990.

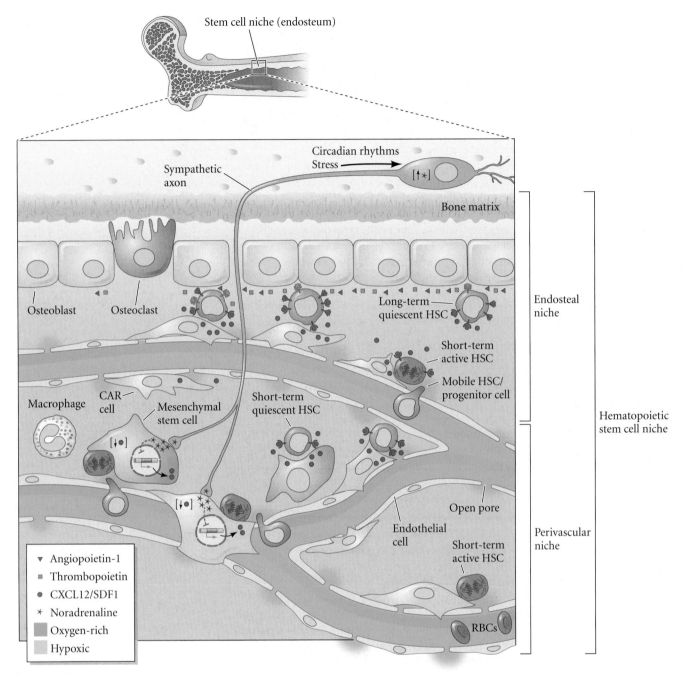

FIGURE 5.15 Model of adult HSC niche. Housed within the bone marrow, the HSC niche can be divided into two subniches: the endosteal and the perivascular. HSCs in the endosteal niche that are adhered to osteoblasts are long-term HSCs (purple) typically in the quiescent state, whereas short-term active HSCs (red) can be seen associated with blood vessels (green) at oxygen-rich pores. Stromal cells—that is, the CAR cells (yellow) and mesenchymal cells—interact directly with mobile HSCs and progenitor cells, which can be stimulated by sympathetic connections.

chemokine CXCL12 (also known as Stromal-Derived Factor 1, or SDF1) expressed by osteoblasts and stromal cells of the marrow (Moll and Ransohoff 2010). A variety of adhesion proteins, such as E-selectins and VCAM1, also support HSC homing to the niche (Al-Drees et al. 2015).

The hematopoietic niche can be subdivided into two regions, the endosteal niche and the perivascular[13] niche (see Figure 5.15). HSCs in the endosteal niche are often in direct contact with the osteoblasts lining the inner surface of the bone, and HSCs in the perivascular niche are in close contact with cells lining or surrounding blood

---

[13] *Peri* is Latin for "around." *Perivascular* refers to cells that are located on the periphery of blood vessels. The perivascular niche is also called the vascular niche, and the endosteal niche is also called the osteoblastic niche.

vessels (endothelial cells and stromal cells). With the different physical and cellular properties of these two niches come differential regulation of the HSCs (Wilson et al. 2007). In addition, there are two subpopulations of HSCs within these niches: one population can divide rapidly in response to immediate needs, while a quiescent population is held in reserve and possesses the greatest potential for self-renewal (Wilson et al. 2008, 2009). Depending on physiological conditions, stem cells from one subpopulation can enter the other subpopulation.

Regulatory mechanisms in the endosteal niche

HSCs found within the endosteal niche tend to be the most quiescent population, with long-term self-renewal serving to sustain the stem cell population for the life of the organism (Wilson et al. 2007). In contrast, more active HSCs tend to reside in the perivascular niche, exhibiting faster cycles of renewal and sustaining progenitor development for a shorter period of time (see Figure 5.15). A complex cocktail of cell adhesion molecules, paracrine factors, extracellular matrix components, hormonal signals, pressure changes from blood vessels, and sympathetic neural inputs all combine to influence the proliferative states of the HSCs (Spiegel et al. 2008; Malhotra and Kincade 2009; Cullen et al. 2014).

In the endosteal niche, HSCs interact intimately with osteoblasts, and manipulation of osteoblast number causes similar increases or decreases in the presence of HSCs (Zhang et al. 2003; Visnjic et al. 2004; Lo Celso et al. 2009; Al-Drees et al. 2015; Boulais and Frenette 2015). Moreover, osteoblasts promote HSC quiescence by binding to the HSCs and secreting angiopoietin-1 and thrombopoietin, which keep these stem cells on reserve for long-term hematopoiesis (Arai et al. 2004; Qian et al. 2007; Yoshihara et al. 2007). Improved imaging techniques have revealed that the endosteal niche is permeated with sinusoidal microvessels[14] (Nombela-Arrieta et al. 2013), and some of the HSCs (*cKit+*) and progenitor cells are intimately associated with this highly permeable microvasculature (**FIGURE 5.16**). It has always been assumed that the endosteal niche was more hypoxic than the perivascular niche, but these microvessels undoubtedly aid in bringing oxygen to the endosteal regions, making the microlocales immediately surrounding sinusoids less hypoxic. It even has been proposed that the HSCs may use differences in oxygen content in the niche as a cue for assessing where blood vessels are (Nombela-Arrieta et al. 2013).

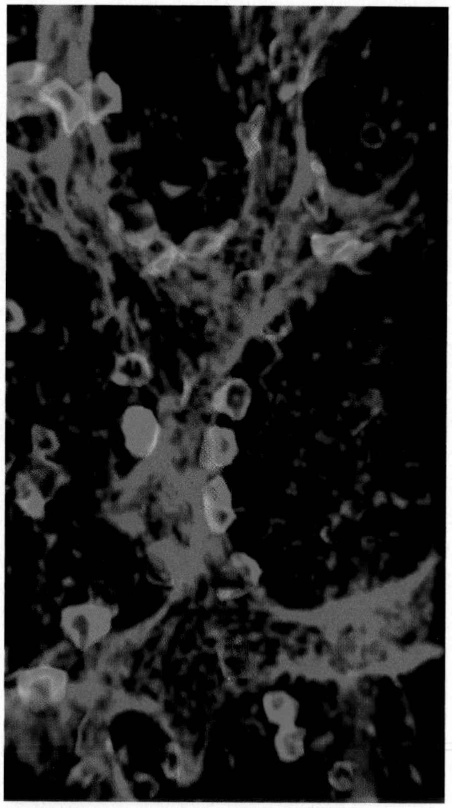

FIGURE 5.16 HSCs sit adjacent to microvasculature in the bone marrow. The c-Kit receptor (green) is a marker for HSCs and progenitors, which are seen in direct contact with the sinusoidal microvasculature in the niche (stained with anti-laminin, red). HSCs are associated with all types of vasculature in the niche. Watch Development 5.1 shows this image being projected in 3-D. (From Nombela-Arrieta et al. 2013.)

WATCH DEVELOPMENT 5.1   Watch a rotating projection of HSCs associated with the perivasculature.

Regulatory mechanisms in the perivascular niche

HSCs are also associated with vasculature in the perivascular niche. CXCL12 is secreted by several cell types, such as endothelial and CXCL12-abundant reticular (CAR) cells (see Figure 5.15; Sugiyama et al. 2006). Although loss of CXCL12 in CAR cells does not seem to affect HSCs, it does cause a significant movement of hematopoietic progenitor cells into the bloodstream and concurrent losses in progenitors of B lymphocytes (cells that secrete antibodies). Other cells in this niche that express CXCL12 are the **mesenchymal stem cells (MSCs)**, which play a major regulatory role in the HSC niche (see Figure 5.15; Méndez-Ferrer et al. 2010). Selective knockout of CXCL12 in these MSCs does lead to a loss of HSCs (Greenbaum et al. 2013).

Cell-specific modulation of CXCL12 seems to be an important mechanism governing quiescence and retention of HSCs and progenitor cells in the perivascular niche. It is a complex story that is related to daily fluctuations in the rate that progenitor

---

[14] Sinusoidal microvessels are small capillaries that are rich in open pores, enabling significant permeability between the capillary and the tissue it resides in.

Developing Questions

We discussed two distinct regions in the hematopoietic stem cell niche, but could there be more? It has been proposed that the MSCs in the bone marrow exert unique control over the HSCs and represent their own niche within a niche. What do you think? How is cell communication and HSC movement among the endosteal, perivascular, and (potentially) MSC niches orchestrated?

cells are mobilized into the bloodstream; there is greater cell division of HSCs at night and increased migration of progenitor cells into the bloodstream during the day. This circadian pattern of mobilization is controlled by the release of noradrenaline from sympathetic axons infiltrating the bone marrow (see Figure 5.15; Méndez-Ferrer et al. 2008; Kollet et al. 2012). Receptors on stromal cells respond to this neurotransmitter by downregulating the expression of *CXCL12*, which temporarily reduces the hold that these stromal cells have on HSCs and progenitor cells, freeing them to circulate. Although circadian rhythms stimulate a normal round of HSC proliferation, chronic stress leads to increased release of noradrenaline (Heidt et al. 2014). This release lowers CXCL12 levels, which reduces HSC proliferation and increases their mobilization into the circulation. So, the next time you wake up, know that your sympathetic nervous system is telling your hematopoietic stem cells to wake up, too.

Additional signaling factors (Wnt, TGF-β, Notch/Jagged1, stem cell factor, and integrins; reviewed in Al-Drees et al. 2015 and Boulais and Frenette 2015) influence the production rates of different types of blood cells under different conditions; examples are an increased production of white blood cells during infections and increased red blood cells when you climb to high altitudes. When the system is misregulated, it can cause diseases such as the different types of blood cancers. Myeloproliferative disease is one such cancer that results from a failure of proper signals for blood cell differentiation (Walkley et al. 2007a,b). It stems from a failure of the osteoblasts to function properly; as a result, HSCs proliferate rapidly without differentiation (Raaijmakers et al. 2010, 2012).

The Mesenchymal Stem Cell: Supporting a Variety of Adult Tissues

Most adult stem cells are restricted to forming only a few cell types (Wagers et al. 2002). For example, when HSCs marked with green fluorescent protein were transplanted into a mouse, their labeled descendants were found throughout the animal's blood but not in any other tissue[15] (Alvarez-Dolado et al. 2003). Some adult stem cells, however, appear to have a surprisingly large degree of plasticity. These multipotent MSCs are sometimes called **bone marrow-derived stem cells** (**BMDCs**), and their potency remains a controversial subject (Bianco 2014).

Originally found in bone marrow (Friedenstein et al. 1968; Caplan 1991), multipotent MSCs have also been found in numerous adult tissues (such as dermis of the skin, bone, fat, cartilage, tendon, muscle, thymus, cornea, and dental pulp) as well as in the umbilical cord and placenta (see Gronthos et al. 2000; Hirata et al. 2004; Traggiai et al. 2004; Perry et al. 2008; Kuhn and Tuan 2010; Nazarov et al. 2012; Via et al. 2012). Indeed, the finding that human umbilical cords and deciduous ("baby") teeth contain MSCs has led some physicians to propose that parents freeze cells from their child's umbilical cord or shed teeth so that these cells will be available for transplantation later in life.[16] Whether MSCs can pass the test of pluripotency—the ability to generate cells of all germ layers when inserted into a blastocyst—has not yet been shown.

Much of the controversy surrounding MSCs rests in their "split personality" as supportive stromal cells on the one hand and stem cells on the other. Morphologically, MSCs resemble fibroblasts, a cell type secreting the extracellular matrix of connective tissues (stroma). In culture, however, MSCs behave differently from fibroblasts. A single

---

[15] Initial attempts at such transplants did show incorporation of HSCs in a variety of tissues, even the brain. It turns out, however, that this finding was due to fusion events rather than actual lineage derivation from HSCs. See Alvarez-Dolado et al. 2003 and an affiliated web conference with Arturo Alvarez-Buylla in 2005 for further investigation.

[16] Another argument for saving umbilical cord cells is that they contain hematopoietic stem cells that might be transplanted into the child should he or she later develop leukemia (see Goessling et al. 2011).

MSC in culture can self-renew to produce a clonal population of cells that can go on to form organs in vitro that contain a diversity of cell types (**FIGURE 5.17**; Sacchetti et al. 2007; Méndez-Ferrer et al. 2010; reviewed in Bianco 2014). As seen in bone marrow, MSCs in other tissues may play roles as both progenitor cells and regulators of the resident niche stem cell, possibly through paracrine signaling (Gnecchi et al. 2009; Kfoury and Scadden 2015).

Regulation of MSC development

Certain paracrine factors appear to direct development of the MSC into specific lineages. Platelet-derived growth factor (PDGF) is critical for fat formation and chondrogenesis, TGF-β signaling is also crucial for chondrogenesis, and fibroblast growth factor (FGF) signaling is necessary for the differentiation into bone cells (Pittenger et al. 1999; Dezawa et al. 2005; Ng et al. 2008; Jackson et al. 2010). Such paracrine signaling factors may underlie not only MSC differentiation but also their modulation of the resident niche stem cell. For instance, MSCs have been shown to play important dual roles as multipotent progenitor cells and stem cell niche regulators during hair follicle and skeletal muscle development and regeneration (Kfoury and Scadden 2015). The rapid turnover of epidermis and associated hair follicles in skin requires robust activation of resident stem cells (see Chapter 16). Immature adipose progenitor cells that surround the base of the growing follicle are both necessary and sufficient to trigger hair stem cell activation during growth and regeneration of the skin through a PDGF paracrine mechanism (Festa et al. 2011).

Similarly, a mesenchymal cell type called fibroadipogenic progenitor (FAP) in skeletal muscle tissue functions to generate white fat cells (as the *adipogenic* part of the name implies). In response to muscle injury, however, FAP cells increase the rate of promyogenic differentiation of myosatellite stem cells (Joe et al. 2010; Pannérec et al. 2013; Formicola et al. 2014). In fact, the increased presence of FAP cells in the muscle stem cell niche has been suggested to serve anti-aging functions and reduce the effects of Duchenne muscular dystrophies (Formicola et al. 2014). This hypothesis is further supported by the link between MSCs and the premature aging syndrome Hutchinson-Gilford progeria (see Figure 23.1B), which appears to be caused by the inability of MSCs to differentiate into certain cell types, such as fat cells (Scaffidi and Misteli 2008). These findings lead to speculation that the loss either of MSCs themselves or of their ability to differentiate may be a component of the normal aging syndrome.

The differentiation of MSCs is dependent on not only paracrine factors but also cell matrix molecules in the stem cell niche. Certain cell matrix components, especially laminin, appear to keep MSCs in a state of undifferentiated "stemness" (Kuhn and Tuan 2010). Researchers have taken advantage of the influence that the physical matrix has on MSC regulation to achieve a repertoire of derived cell types in vitro by growing stem cells on different surfaces. For example, if human MSCs are grown on soft matrices of collagen, they differentiate into neurons, a cell type that these cells do not appear to form in vivo. If instead MSCs are grown on a moderately elastic matrix of collagen, they become muscle cells, and if grown on harder matrices, they differentiate into bone cells (**FIGURE 5.18**; Engler et al. 2006). It is not yet known whether this range of potency is found normally in the body. As technology improves, answers may come from gaining a better understanding of the properties of different MSC niches.

Other stem cells supporting adult tissue maintenance and regeneration

This chapter has focused on several well-defined adult stem cell niches. It is important to understand, however, that many more adult stem cell niches have been discovered and are providing new insights into the molecular regulation of the adult stem cell. Adult stem cells can be found in tissues of teeth, eye, fat, muscle, kidney, liver, and lung. There are interesting instances of some animals having evolutionarily lost a

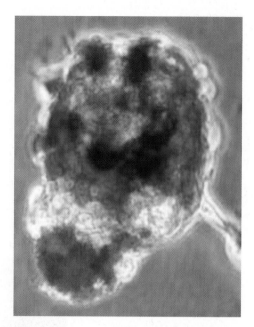

FIGURE 5.17 A mesensphere containing two derived cell types. Mesenchymal stem cells placed in culture form mesenspheres that can produce different cell types. Here a mesensphere contains osteoblasts (bone-forming cells; teal) and adipocytes (fat-forming cells; red). (From Méndez-Ferrer et al. 2010.)

Developing Questions

What molecular mechanisms may govern the change of MSCs from being a progenitor at one moment to regulating other stem cells at another?

FIGURE 5.18 Mesenchymal stem cell differentiation is influenced by the elasticity of the matrices upon which the cells sit. On collagen-coated gels having elasticity similar to that of the brain (about 0.1–1 kPa), human MSCs differentiated into cells containing neural markers (such as β3-tubulin) but not into cells containing muscle cell markers (MyoD) or bone cell markers (CBFα1). As the gels became stiffer, the MSCs generated cells exhibiting muscle-specific proteins, and even stiffer matrices elicited the differentiation of cells with bone markers. Differentiation of the MSC on any matrix could be abolished with blebbistatin, which inhibits microfilament assembly at the cell membrane. (After Engler et al. 2006; photographs courtesy of J. Shields.)

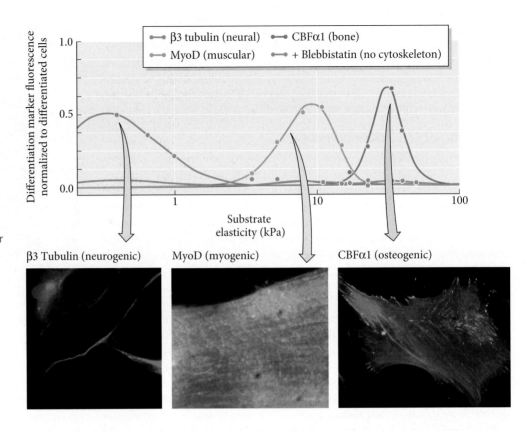

stem cell niche while related animals retained the niche. Rodent incisors, for instance, differ from mammalian incisors, including your own, in that they continue to grow throughout the lifetime of the animal. In the mouse, each incisor has two stem cell niches, one on the "inside," facing into the mouth (lingual), and one on the "outside," facing the lips (labial) (**FIGURE 5.19**). Because most other mammals lack these incisor stem cell niches, their teeth do not regenerate. We will describe various other stem cell lineages throughout the rest of the book.

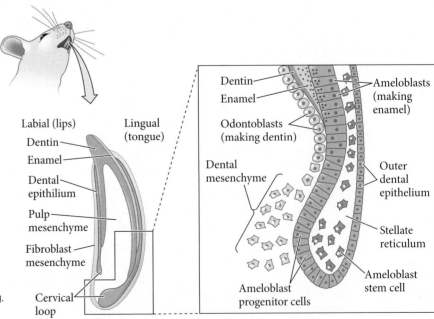

FIGURE 5.19 The cervical loop of the mouse incisor is a stem cell niche for the enamel-secreting ameloblast cells. These cells migrate from the base of the stellate reticulum into the enamel layer, allowing the teeth to keep growing. (After Wang et al. 2007.)

The Human Model System to Study Development and Disease

Up to this point, we have focused on the in vivo life of stem cells. The properties of self-renewal and differentiation that define a stem cell, however, also enable their manipulation in vitro. Before we were able to culture human embryonic stem cells (Thomson 1998), researchers studying human cell development used immortalized tumor cells or cells from teratocarcinomas, which are cancers that arise from germ cells (Martin 1980). The most investigated human cell has been the HeLa cell, a line of cultured cells that were derived from the cervical cancer of *Henrietta Lacks* (a cancer that took her life in 1951 and a cell line that was isolated without her or her family's knowledge or consent[17]). None of these cells represents a model of normal human cells. With our present ability to grow embryonic and adult human stem cells in the lab and induce them to differentiate into different cell types, however, we finally have a tractable model system for studying human development and disease in vitro.

 SCIENTISTS SPEAK 5.5   A Developmental Documentary on modeling diseases using stem cells.

Pluripotent stem cells in the lab

EMBRYONIC STEM CELLS   Pluripotent embryonic cells are a special case because these stem cells can generate all the cell types needed to produce the adult mammalian body (see Shevde 2012). In the laboratory, pluripotent embryonic cells are derived from two major sources (**FIGURE 5.20**). As reviewed earlier in this chapter, one source is ICM of the early blastocyst, whose cells can be maintained in culture as a clonal line of ESCs (Thomson et al. 1998). The second source is primordial germ cells that have not yet differentiated into sperm or eggs. When isolated from the embryo and grown in culture, they are called **embryonic germ cells**, or **EGCs** (Shamblott et al. 1998).

 SCIENTISTS SPEAK 5.6   Dr. Janet Rossant answers questions about the differences between mouse and human ESCs.

As in the ICM of the embryo, the pluripotency of ECSs in culture is maintained by the same core of three transcription factors: Oct4, Sox2, and Nanog. Acting in concert, these factors activate the gene regulatory network required to maintain pluripotency and repress those genes whose products would lead to differentiation (Marson et al. 2008; Young 2011). Are all pluripotent stem cells created equal, however? Although the

---

[17] The story of Henrietta Lacks, HeLa cells in science, and social policy are beautifully articulated in Rebecca Skloot's 2010 book, *The Immortal Life of Henrietta Lacks*.

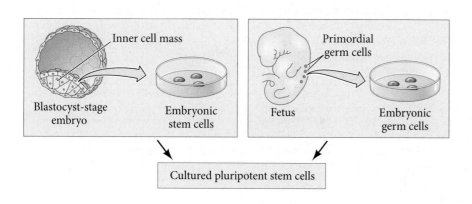

FIGURE 5.20   Major sources of pluripotent stem cells from the early embryo. Embryonic stem cells (ES cells) arise from culturing the inner cell mass of the early embryo. Embryonic germ cells (EG cells) are derived from primordial germ cells that have not yet reached the gonads.

Developing Questions

What is possible now that naïve human ESCs can be isolated and maintained? Proof of these cells' pluripotency was displayed when naïve human ESCs were transplanted into the mouse morula and differentiated into many cell types of an interspecies chimeric humanized mouse embryo (Gafni et al. 2013). Although federal funds cannot be used to create human-mouse chimeras in the United States, such regulations do not exist in other countries. It is theoretically plausible to create a human ICM from naïve human ESCs that is supported by a mouse trophectoderm. Minimally, doing so could enable the first direct study of human gastrulation. Should the human gastrula be studied in this way? What, if any, ethical concerns could such studies raise?

years of experimentation with both mouse and human ESCs have demonstrated clear pluripotency (Martin 1981; Evans and Kaufman 1981; Thomson et al. 1998), they have also revealed differences in their degrees of self-renewal, the types of cells they can form, and their cellular characteristics (Martello and Smith 2014; Fonseca et al. 2015; Van der Jeught et al. 2015). It appears that these differences may be based on slight differences in the developmental stage of the original ICM cells from which the cultures were derived, which has led to recognizing two different pluripotent states of an ESC: **naïve** and **primed**.[18] The naïve ESC represents the most immature, undifferentiated ESC with the greatest potential for pluripotency. In contrast, the primed ESC represents an ICM cell with some maturation toward the epiblast lineage; hence, it is "primed," or ready for differentiation.

Therefore, the growing consensus is that most of the existing mouse ESC lines represent the naïve state, whereas much of the research conducted with human ESC lines captured more of the primed state of pluripotency. Different methods of derivation for the maintenance of naïve human ESCs from ICM cells or even from primed ESCs are emerging (Van der Jeught et al. 2015). As an example, leukemia inhibitory factor (LIF) has been used in combination with with at least two kinase inhibitors (called 2i) that are associated with the MAPK/ErK pathway inhibitor (MEKi) and glycogen synthase kinase 3 inhibitor (GSK3i); as an example, see Theunissen et al. 2014. These factors, along with additional conditions, serve to prevent differentiation and maintain the ESCs in the naïve, or ground, state.

Researchers are now studying the gene networks, epigenetic modulators, paracrine factors, and adhesion molecules required for the differentiation of ESCs. These cells can respond to specific combinations and sequential application of growth factors to coax their differentiation toward specific cell fates associated with the three germ layers (**FIGURE 5.21**; Murry and Keller 2008). For instance, applying a chemically defined growth medium to a monolayer of ESCs can push their specification toward a mesodermal fate; when followed by a period of Wnt activation and then Wnt inhibition, the cells differentiate into contracting heart muscle cells (Burridge et al. 2012, 2014). In contrast, ESCs pushed toward an ectodermal fate by inhibiting Bmp4, Wnt, and activin can be subsequently induced by fibroblast growth factors (FGFs) to become neurons (see Figure 5.21; Kriks et al. 2011).

WATCH DEVELOPMENT 5.2   Watch ESC-derived cardiomyocytes beat in a petri dish.

 SCIENTISTS SPEAK 5.7   Watch the 2011 developmental documentary "Stem Cells and Regenerative Medicine."

The physical constraints of the environment in which ESCs are cultured can also profoundly influence their differentiation. Constraining the area of cell growth to small disc shapes[19] can alone initiate a pattern of differential gene expression in the colony of cells that correlates to that of the early embryo (**FIGURE 5.22**; Warmflash et al. 2014). These results demonstrate that an incredible amount of patterning can be initiated solely by the geometry and size of the growth landscape. These discoveries are enabling further research into the structure and function of specific human cell types and their use in medical applications.

---

[18]As you examine past ESCs literature, it will be important to critically consider the pluripotent state of the ESCs being depicted in each study. Are the ESCs naïve or primed, and what implications may that have on the authors' interpretations of their results? Also be aware that naïve ESCs have also been referred to as being in the "ground state."

[19]Researchers applied a micropattern of adhesive substrate to a glass plate, which restricted cell growth to a defined size and shape for systematic analysis (Warmflash et al. 2014). In a different study, lined grid substrates promoted ESC differentiation into dopamine neurons (Tan et al. 2015).

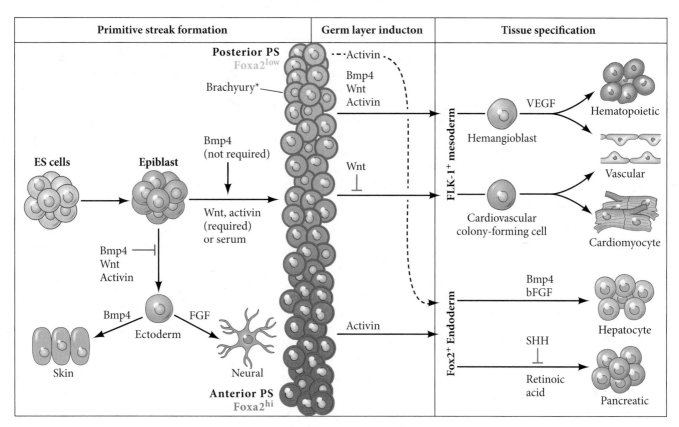

| Primitive streak formation | Germ layer inducton | Tissue specification |

FIGURE 5.21 Inducing stem cell differentiation from ESCs. Similar to the steps of differentiation epiblast cells take during their maturation in the mammalian embryo, ESCs in culture can be coaxed with the same developmental factors (paracrine and transcription factors, among others) to differentiate into the cell types of each germ layer. With the inhibition of several growth factors, ESCs can make ectoderm lineages; for mesoderm or endoderm, however, ESCs are first induced to become primitive streak-like cells (PS) with paracrine factors such as Wnt, Bmp4, or activin, depending on the desired differentiated cell type. (After Murry and Keller 2008.)

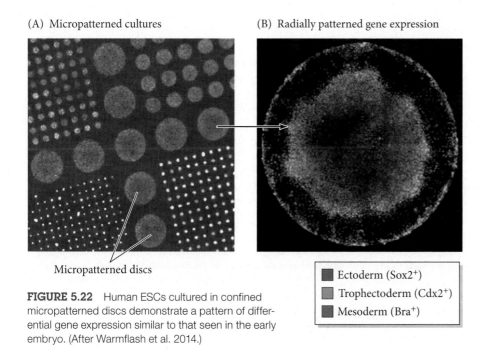

(A) Micropatterned cultures

Micropatterned discs

(B) Radially patterned gene expression

■ Ectoderm (Sox2+)
■ Trophectoderm (Cdx2+)
■ Mesoderm (Bra+)

FIGURE 5.22 Human ESCs cultured in confined micropatterned discs demonstrate a pattern of differential gene expression similar to that seen in the early embryo. (After Warmflash et al. 2014.)

ESCs AND REGENERATIVE MEDICINE   A major hope for human stem cell research is that it will yield therapies for treating diseases and repairing injuries. In fact, pluripotent stem cells have opened an entirely new field of therapy called **regenerative medicine** (Wu and Hochelinger 2011; Robinton and Daley 2012). The therapeutic possibilities for ESCs lie in their ability to differentiate into any cell type, especially for treatment of human conditions in which adult cells degenerate (such as Alzheimer disease, Parkinson disease, diabetes, and cirrhosis of the liver). For instance, Kerr and colleagues (2003) found that human EGCs were able to cure motor neuron injuries in adult rats both by differentiating into new neurons and by producing paracrine factors (BDNF and TGF-α) that prevent the death of existing neurons. Similarly, precursor cells for dopamine-secreting neurons derived from ESCs (Kriks et al. 2011) were able to complete their differentiation into dopaminergic neurons and cure a Parkinson-like condition when engrafted into the brains of mice, rats, and even monkeys.

Although great excitement surrounds the potential of therapies using stem cells, another line of research is aimed at understanding the development of disease and assessing the effectiveness of pharmaceuticals. Such studies have already advanced our understanding of rare blood-based diseases such as Fanconi anemia, which causes bone marrow failure and consequent loss of both red and white blood cells (Zhu et al. 2011). Often, diseases like Fanconi anemia are caused by **hypomorphic mutations**—mutations that merely reduce gene function, as opposed to a "null" mutation that results in the total loss of a protein's function. Researchers used human ESCs to create a model of Fanconi anemia by using RNAi to knock *down* (not knock *out*) specific isoforms of the Fanconi anemia genes (Tulpule et al. 2010). The results gave new insights into the role of the Fanconi anemia genes during the initial steps of embryonic hematopoiesis.

 SCIENTISTS SPEAK 5.8   Dr. George Daley talks about modeling Fanconi anemia and other blood diseases. A Developmental Documentary also covers the modeling of rare blood disorders.

Unfortunately, there are real challenges to expanding the use of ESCs to model human diseases. One reason is that ESCs are only found at such an early stage of development; another is that human diseases involve cells that have a long history of differentiation events and are often multigenic (caused by the interplay of many genes). Further complicating matters is the risk of immune rejection by the patients receiving ESCs as part of a treatment. Transplanted cells derived from ESCs are from another individual and are therefore not the same genotype as the patient, so, just like any other tissue transplant, they can be rejected by the patient's immune system.[20] Also, various social and ethical issues are raised by the use of ESCs in therapies because they are derived from human blastocysts, also known as embryos (Gilbert et al. 2005; Siegel 2008; NSF 2012).[21] If we could obtain similarly pluripotent stem cells from individuals diagnosed with known diseases, perhaps those cells could be used to study these diseases and identify new therapies. When hunting for pluripotent cells, finding a way to induce them could be an answer.

 SCIENTISTS SPEAK 5.9   Watch a Web conference with Dr. Bernard Siegel on stem cell and cloning ethics and public policy. A 2011 documentary also covers stem cell ethics and government policy.

---

[20] One reason diseases of the brain are being targeted is that the brain and the eyes are among the few places where immune rejection is not a big problem. The blood-brain barrier of the brain's endothelial cells keeps the brain and the eyes shielded from the immune system.

[21] In 2010, two stem cell scientists filed a lawsuit against the U.S. government to ban federal funding for human ESC research. This lawsuit halted all human ESC research in the United States for months. Consider reading Wadman 2011 as well as listening to a Web conference with one of the plantiffs, Theresa Deisher, recorded in 2011 while the court case *Sherley v. Sebelius* was ongoing.

Induced pluripotent stem cells

Although we know that the nuclei of differentiated somatic cells retain copies of an individual's entire genome, biologists have long thought that potency was like going down a steep hill with no return. Once differentiated, we believed, a cell could not be restored to an immature and more plastic state. Our newfound knowledge of the transcription factors needed to maintain pluripotency, however, has illuminated a startlingly easy way to reprogram somatic cells into embryonic stem cell-like cells.

In 2006, Kazutoshi Takahashi and Shinya Yamanaka of Kyoto University demonstrated that by inserting activated copies of four genes that encoded some of these critical transcription factors, nearly any cell in the adult mouse body could be made into an **induced pluripotent stem cell** (**iPSC**) with the pluripotency of an embryonic stem cell. These genes were *Sox2* and *Oct4* (which activated Nanog and other transcription factors that established pluripotency and blocked differentiation), c-*Myc* (which opened up chromatin and made the genes accessible to Sox2, Oct4, and Nanog), and *Klf4* (which prevents cell death; see Figure 3.14).

 SCIENTISTS SPEAK 5.10   Developmental documentaries from 2009 and 2011 on cellular reprogramming.

Within 6 months of the publication of this work (Takahashi and Yamanaka 2006), three groups of scientists reported that the same or similar transcription factors could induce pluripotency in a variety of differentiated human cells (Takahashi et al. 2007; Yu et al. 2007; Park et al. 2008). Like embryonic stem cells, iPS cell lines can be propagated indefinitely and, whether in culture or in a teratoma, can form cell types representative of all three germ layers. By 2012, modifications of the culture techniques made it possible for the gene expression of mouse iPSCs to become nearly identical to that of mouse embryonic stem cells (Stadtfeld et al. 2012). Most important was that entire mouse embryos could be generated from single iPSCs, showing complete pluripotency. Although iPSCs are functionally pluripotent, they are best at generating the cell types of the organ from which the parent somatic cell originated (Moad et al. 2013). These data suggest that, like naïve versus primed ESC, not all iPSCs are the same and that they may retain an epigenetic memory of their past home.

 SCIENTISTS SPEAK 5.11   Question and answer sessions with Dr. Rudolf Jaenisch on iPSCs and Dr. Derrick Rossi on generating iPSCs with mRNA.

APPLYING iPSCs TO HUMAN DEVELOPMENT AND DISEASE   Using iPSCs provides medical researchers with the ability to experiment on diseased human tissue while avoiding the complications introduced by using human embryonic stem cells. Currently, there are four major medical uses for iPSCs: (1) making patient-specific iPSCs for studying disease pathology, (2) combining gene therapy with patient-specific iPSCs to treat disease, (3) using patient-specific iPSC-derived progenitor cells in cell transplants without the complications of immune rejection, and (4) using differentiated cells derived from patient-derived iPSCs for screening drugs.

Transplanting cells derived from mouse iPSCs back into the same donor mouse does not elicit immune rejection (Guha et al. 2013), suggesting that iPSC-based cell replacement may, in fact, be a promising therapy in the future.[22] So far, the most significant advances with iPSCs have been in modeling human diseases. Following a major study (Park et al. 2008) that created iPSCs from patients associated with 10 different diseases, numerous studies have leveraged iPSC technology to model a diverse array of diseases, including Down syndrome, diabetes, and more (Singh et al. 2015).

---

[22]At this time, the cost and scalability of iPSC-derived cell types to achieve the cell numbers required for effective cell replacement therapy are significant obstacles to the progress of this approach as a medical intervention.

FIGURE 5.23 Lung epithelium derived from mouse iPS cells. The lung transcription factor Nxk2.1 is stained red, indicating that the iPS cells cells have become lung epithelia. The tubulin of the epithelial cilia, whose functions are disturbed in patients with cystic fibrosis, is stained green. Nuclei are blue. (Photograph courtesy of J. Rajagopal.)

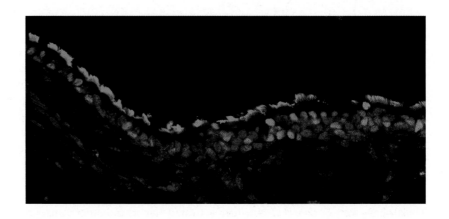

Disease modeling is of particular importance for diseases that are not easily modeled in non-human organisms. Mice, for instance, do not get the same type of cystic fibrosis—a disease that severely compromises lung function—that humans get. After discovering what factors caused mouse iPSCs to differentiate into lung tissue (**FIGURE 5.23**; Mou at al. 2012), researchers made iPSCs from a person with cystic fibrosis and turned them into lung epithelium that showed the characteristics of human cystic fibrosis. Knowing that cystic fibrosis is often caused by mutations within a single gene (the gene for CF transmembrane conductance regulator, which encodes a chloride channel; Riordan et al. 1989; Kerem et al. 1989), researchers sought to repair the human mutation in these iPSCs. Crane and colleagues (2015) accomplished this task in iPSCs derived from a cystic fibrosis patient that made functional chloride channels in differentiated epithelium. The next step will be to test this approach in a non-human animal model to see if it might be used to treat cystic fibrosis in humans.

The benefits of combining the use of iPSCs and gene correction was eloquently demonstrated by Rudolf Jaenisch's lab in 2007 to cure a mouse model of sickle-cell anemia. This disease is caused by a mutation in the gene for hemoglobin. The Jaenisch lab generated iPSCs from this mouse, corrected the hemoglobin mutation, and then differentiated the iPSCs into hematopoietic stem cells that, when implanted in the mouse, cured its sickle-cell phenotype (**FIGURE 5.24**; Hanna et al. 2007). Ongoing studies are attempting to determine if similar therapies could cure human conditions such as diabetes, macular degeneration, spinal cord injury, Parkinson disease, and Alzheimer disease as well as liver disease and heart disease. Other studies have shown that iPSCs can be induced to form numerous cell types that are functional when transplanted back into the organism from which they were derived. Even sperm and oocytes have been generated from mouse iPSCs. First, skin fibroblasts were induced to form iPSCs, and these iPSCs were then induced to form primordial germ cells (PGCs). When these induced PGCs were aggregated with gonadal tissues, the cells proceeded through meiosis and became functional gametes (Hayashi et al. 2011; Hayashi et al. 2012). This work could become significant in circumventing many types of infertility as well as in allowing scientists to study the details of meiosis.

MODELING MULTIGENIC HUMAN DISEASES WITH IPSCs One challenge in studying a human disease is that individuals differ in the repertoire of genes associated with a disease as well as the timing of onset and progression of the disease. Fortunately, iPSCs have provided a new tool to help unravel this complexity. Here we highlight the use of iPSCs to study two particularly complex and multigenic diseases of the nervous system that fall at opposite ends of the developmental calendar: autism spectrum disorders and amyotrophic lateral sclerosis (ALS).

 SCIENTISTS SPEAK 5.12 A Developmental Documentary from 2012 on modeling diseases of the nervous system.

Autism spectrum disorders present a range of neural dysfunctions typically affecting social and cognitive abilities that are not clearly apparent until around 3 years of age.[23] Disorders that fall within this spectrum include classic autism, Asperger syndrome, fragile-X syndrome, and Rett syndrome. Rett syndrome appears to be associated with a single gene (*methyl CpG binding protein-2, or MeCP2*). In contrast, autism is truly multi-allelic, with some children being non-syndromic (autism with no known cause) and likely possessing sporadic mutations (Iossifov et al. 2014; Ronemus et al. 2014; De Rubeis and Buxbaum 2015). In fact, the causative agents (genetics and environmental factors) may be unique to each autistic child, which presents significant challenges to researching autism.

One approach has been to generate iPSCs from as many children on the autism spectrum as possible to establish a more comprehensive understanding of the associated genes. This approach has been facilitated through a program called the Tooth Fairy Project, through which donations of children's deciduous (baby) teeth provide sufficient dental pulp for deriving iPSCs.[24] In using the iPSCs from a child with nonsyndromic autism, researchers created a culture of neurons and found a mutation in the TRPC6 calcium channel gene that impaired the structure and function of these neurons (Griesi-Oliveira et al. 2014). They further demonstrated improved neuronal function after exposing these cells to hyperforin, a compound found in St. John's wort and known to stimulate calcium influx. It turns out that *TRPC6* expression can be regulated by MeCP2, which confirms a direct genetic association between autism and Rett syndrome. Remarkably, the medical intervention for this child was changed to now include St. John's wort, which highlights the potential for patient-specific precision medicine in the future. This finding shows that iPSCs can play an important role in modeling a complex disease to research mechanisms that can lead to direct patient intervention.

Amyotrophic lateral sclerosis (ALS), or Lou Gehrig's disease, is an adult-onset degenerative motor neuron disease that is multi-allelic through familial inheritance as well as sporadic mutation; unfortunately, it has no cure or treatment. Some of the first disease-specific iPSCs were derived from ALS patients in 2008 by Kevin Eggan's lab (Dimos et al. 2008). ALS-derived iPSCs can be coaxed to differentiate into motor neurons and non-neuronal cell types such as astrocytes, which are cells implicated in the ALS phenotype. More recently, motor neurons differentiated from patient-derived

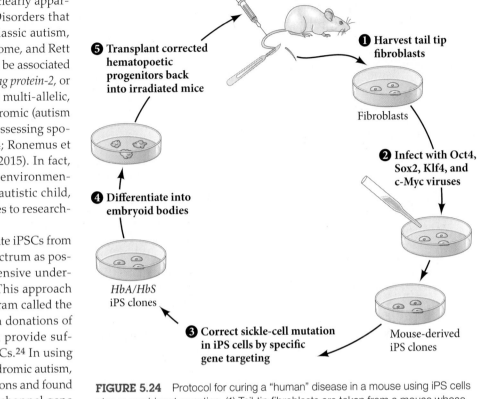

FIGURE 5.24 Protocol for curing a "human" disease in a mouse using iPS cells plus recombinant genetics. (1) Tail-tip fibroblasts are taken from a mouse whose genome contains the human alleles for sickle-cell anemia (*HbS*) and no mouse genes for this protein. (2) The cells are cultured and infected with viruses containing the four transcription factors known to induce pluripotency. (3) The iPS cells are identified by their distinctive shapes and are given DNA containing the wild-type allele of human globin (*HbA*). (4) The embryos are allowed to differentiate in culture. They form "embryoid bodies" that contain blood-forming stem cells. (5) Hematopoietic progenitor and stem cells from these embryoid bodies are injected into the original mouse and cure its sickle-cell anemia. (After Hanna et al. 2007.)

---

[23] Although signs of some autism spectrum disorders are not overtly apparent early on, subtle early indicators—such as gazing at geometric shapes in preference to people's faces—are being identified.

[24] See Dr. Alysson Muotri describe his research and the Tooth Fairy Project at https://www.cirm.ca.gov/our-progress/video/reversing-autism-lab-help-stem-cells-and-tooth-fairy. You can also access a BioWeb conference (see Scientists Speak 5.13) in which Dr. Muotri discusses iPSC modeling of ALS and autism.

iPSCs harboring a known ALS familial mutation exhibited typical hallmarks of ALS cellular pathology (Egawa et al. 2012). The researchers used these differentiated motor neurons to screen for drugs that might improve motor neuron health, and they identified a histone acetyltransferase inhibitor capable of reducing the ALS cellular phenotypes. Thus, experimentation with iPSCs has revealed new insights into how ALS could be epigenetically regulated and possibly treated.

 SCIENTISTS SPEAK 5.13 Web conferences with Dr. Carol Marchetto on modeling autism with iPSCs, and with Dr. Alysson Muotri on modeling ALS with iPSCs.

Organoids: Studying human organogenesis in a culture dish

We have discussed the many ways in which pluripotent stem cells (ESCs and iPSCs) can be used to better understand human development and disease at the level of the cell, but there is a vast difference between cells in culture and cells in the embryo. Human blastocysts are routinely used to research early human development and interventions for treating infertility; using human embryos for studying human organogenesis, however, has been both technically impossible and viewed as unethical by most. Through recent advances in pluripotent cell culturing techniques, though, researchers have been able to grow rudimentary organs from pluripotent stem cells. To date, the most complex structures that have been created are the optic cup of the eye, mini-guts, kidney tissues, liver buds, and even brain regions (**FIGURE 5.25A**; Lancaster and Knoblich 2014).

These **organoids**, as they are called, are generally the size of a pea and can be maintained in culture for more than a year. The striking feature of organoids is that they actually mimic embryonic organogenesis. Pluripotent cells often self-organize into aggregates based on differential adhesion between cells (much like during gastrulation; see also Chapter 4), leading to cell sorting and the differentiation of cells with different fates that interact to form the tissues of an organ (**FIGURE 5.25B**). Organoids have been made from both ESCs and iPSCs derived from healthy and diseased individuals. Therefore, the same therapeutic approaches that we discussed for ESCs and iPSCs can also be applied to the organoid system. Although speculative at this point, creating organoids may prove to be a viable procedure for growing autologous[25] structures not just for patient-specific cell replacement therapy but also for tissue replacement. As an example, we highlight below some of the remarkable features associated with the development of the cerebral organoid and its use in modeling a congenital brain disease.

THE CEREBRAL ORGANOID The human cerebral cortex is arguably the most sophisticated tissue in the animal kingdom, so trying to build even parts of this structure may seem daunting. Ironically, neural differentiation from pluripotent cells seems to be a sort of "default state," similar to the presumptive neural forming cells of the gastrula. Many previous studies characterizing the development of pluripotent stem cells into neural tissues have paved the way to growing multiregional brain organoids (Eiraku et al. 2008; Muguruma et al. 2010; Danjo et al. 2011; Eiraku and Sasai 2012; Mariani et al. 2012). In relatively simple growth conditions, pluripotent cells will self-organize into small spherical clusters of cells called embryoid bodies, and cells within these bodies will differentiate into a stratified neuroepithelium, similar to the neural epithelium of an embryo. The "self-organizing" ability of pluripotent cells to form three-dimensional neuroepithelial structures strongly suggests that robust intrinsic mechanisms exist that are primed for neural development (Harris et al. 2015). As seen in most adult neural stem cell niches, this neuroepithelium is polarized along the apical-to-basal axis and is capable of developing into brain tissue.

---

[25]*Autologous* means derived from the same individual. In this case, cells from a patient are reprogrammed into iPSCs that are developed into a specific organoid. Cells and whole tissues from the organoid can be transplanted back into the same patient without concern of immune rejection.

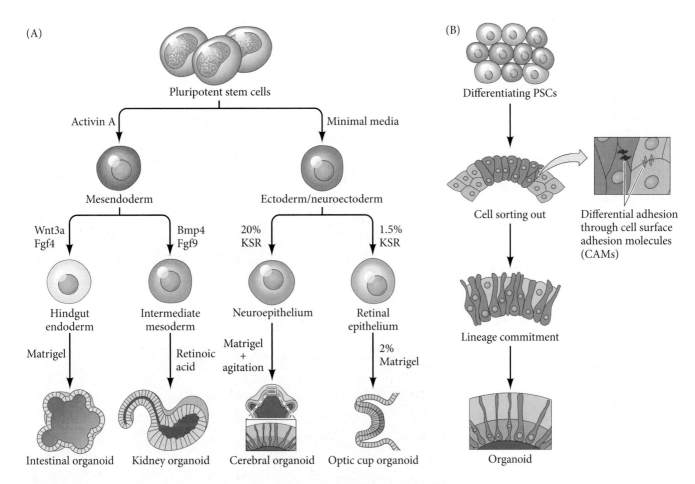

FIGURE 5.25 Organoid derivation. (A) Schematic represents the various strategies used to promote the morphogenesis of specific tissue-type organoids. In most cases, a three-dimensional matrix (Matrigel) is used. KSR is a knockout serum replacement. (B) Early progression of organoid formation begins with differential gene expression, leading to cells with different cell adhesion molecules that confer self-organizing properties (see Chapter 4). Once sorted, cells continue to mature toward distinct lineages that interact to build a functional tissue. (After Lancaster and Knoblich 2014.)

In a landmark study, researchers took brain tissue organoids to the next level of complexity (Lancaster et al. 2013). They placed embryoid bodies into droplets of Matrigel (a matrix made from solubilized basement membrane, the ECM normally at the basal side of an epithelium) to provide a three-dimensional architecture. They next moved these neuroepithelial buds into a media-filled spinning bioreactor (**FIGURE 5.26A**; see also Lancaster and Knoblich 2014). The movement of the organoid in this three-dimensional matrix served to increase nutrient uptake, which supported the substantial growth required for multiregional cerebral organoid development. The resulting cerebral organoid showed characteristically layered tissue for a variety of brain regions, including appropriate neuronal and glial cell markers (**FIGURE 5.26B**). These cerebral organoids possessed radial glial cells adjacent to ventricular-like structures, similar to the developing neural tube and even the adult neural stem niche discussed earlier (**FIGURE 5.26C**). These human radial glial cells within the cerebral organoid displayed all patterns of mitotic behaviors: symmetrical division for stem cell expansion and asymmetrical divisions for self-renewal and differentiation (Lancaster et al. 2013).

Knoblich's group also generated iPSCs from fibroblast samples of a patient with severe microcephaly in the hope that they could study the pathologies associated with this disease (Lancaster et al. 2013). Microcephaly is a congenital disease characterized

FIGURE 5.26 The cerebral organoid. (A) Schematic showing the process over time for the creation of a cerebral organoid from initial cell suspension to bioreactor spinning. Representative light microscopic images of the developing organoid are shown below each step. (B) Section of a cerebral organoid labeled for neural progenitors (red; Sox2), neurons (green; Tuji), and nuclei (blue), which reveals the multilayered organization characteristic of the developing cerebral cortex. (C) Radial glial cell labeled with p-Vimentin (green) undergoes division and shows its characteristic morphology with a long basal process and its apical membrane at the ventricular-like lumen (dashed white line). (From Lancaster and Knoblich 2014a,b.)

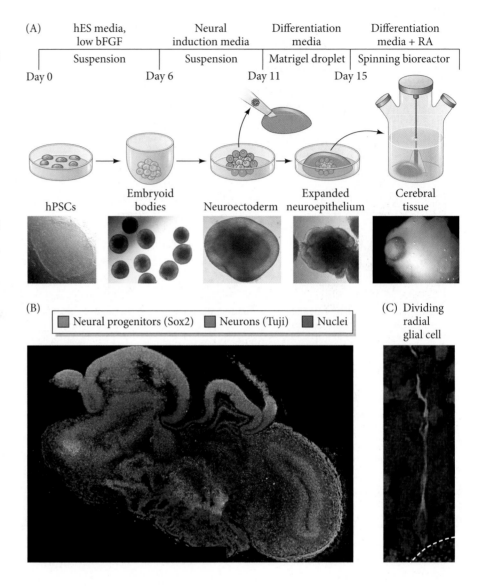

by a significant reduction in brain size (**FIGURE 5.27A**). Remarkably, cerebral organoids from this patient did show smaller developed tissues, but outer layers of the cortex-like tissues showed increased numbers of neurons compared to control organoids (**FIGURE 5.27B**). The researchers discovered that this patient had a mutation in the gene for CDK5RAP2,[26] a protein involved in mitotic spindle function. Moreover, the radial glial cells in this cerebral organoid exhibited abnormally low levels of symmetrical division (**FIGURE 5.27C**). Recall that one of the most basic functions of a stem cell is cell division. It appears that CDK5RAP2 is required for the cell division needed for expansion of the stem cell pool. Lack of symmetrical divisions leads to premature neuronal differentiation, which explains the increased number of neurons in this patient-derived organoid despite the smaller size of its tissues (Lancaster et al. 2013).

Stem Cells: Hope or Hype?

The ability to induce, isolate, and manipulate stem cells offers a vision of regenerative medicine wherein patients can have their diseased organs regrown and replaced

---

[26] Cdk5 regulatory subunit-associated protein 2 (CDK5RAP2) encodes a centrosomal protein that interacts with the mitotic spindle during division.

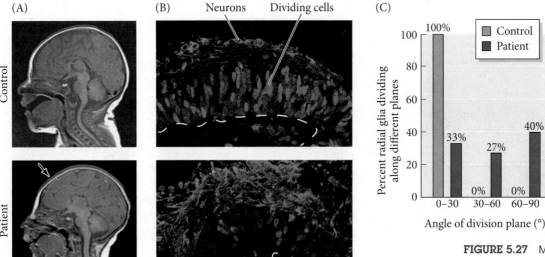

FIGURE 5.27 Modeling human microcephaly with a patient-specific cerebral organoid. (A) Sagittal views of magnetic resonance imaging scans from age-matched control (top) and patient brains at birth. The patient has a smaller brain and reduced brain folding (arrow). (B) Immunolabeling of control and patient-derived cerebral organoids. Neurons (green) and dividing cells (red) are labeled with DCX and BrdU, respectively. There is decreased proliferation and an increase in neuron numbers in the patient-derived organoid. (C) Quantification of the number of radial glial cells undergoing mitotic divisions along specific planes relative to the apical-to-basal axis of the organoid. Due to a loss of CDK5RAP2, patient radial glial cells divide randomly along all axes. (After Lancaster et al. 2013.)

using their own stem cells. Stem cells also offer fascinating avenues for the treatment of numerous diseases. Indeed, when one thinks about the mechanisms of aging, the replacement of diseased body tissues, and even the enhancement of abilities, the line between medicine and science fiction becomes thin. Developmental biologists have to consider not only the biology of stem cells, but also the ethics, economics, and justice behind their use (see Faden et al. 2003; Dresser 2010; Buchanan 2011).

Several years ago, stem cell therapy protocols were being tested in only a few human trials (Normile 2012; Cyranoski 2013). A simple search of stem cell therapies at clinicaltrials.gov will reveal a growing list of ongoing testing with stem cells. Although a majority of current clinical trials are associated with adult stem cells, progenitors derived from human ESCs and iPSCs are being conducted in the United States and elsewhere. Of significant concern is the increase in fraudulent stem cell therapies being offered. The International Society for Stem Cell Research (www.isscr.org) provides valuable resources to learn about stem cells and identify qualified stem cell therapies being used today.

Stem cell research may be the beginning of a revolution that will be as important for medicine (and as transformative for society) as the research on infectious microbes was a century ago. Beyond the potential for medical applications, however, stem cells can tell us a great deal about how the body is constructed and how it maintains its structure. Stem cells certainly give credence to the view that "development never stops."

Next Step Investigation

Can our behavior affect neurogenesis in our brains or the number of immune cells in our blood? It has been shown that exercise can increase neurogenesis in the brain, whereas stress has the opposite effect. This amazing response begs the question, what else can affect cell genesis throughout our bodies? Are certain stem cells responsive to particular types of environmental stimuli and could we harness this knowledge to improve health and tissue regeneration? For instance, could certain diets promote a healthier renewal of cells in the gut epithelium or increase neurogenesis in our brains? What about healthy sleep patterns, social interactions, reading, watching happy versus sad movies, or playing the piano? Could these activities stimulate healthy stem cell development? How would you test for that possibility?

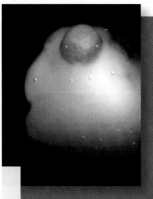

Closing Thoughts on the Opening Photo

This chapter started out with the question "Is that really an eye and a brain in a dish?" Three-dimensional tissue construction from stem cells in a plate is a remarkable example of the "potential" that stem cells hold for the study of development and disease. Yes, that image is of a pigmented epithelium of the retina growing over the neural epithelium of a brain-like cerebral organoid. Although these organoids are certainly providing a new platform to study human organogenesis and affiliated diseases, its generated excitement must be accompanied with objectivity to understand the limitations these systems also present. What is this cerebral organoid currently lacking? Ponder these structures: blood vessels, the flow of cerebral spinal fluid, and the pituitary. Whether brain, kidney, or intestinal organoid, they are not yet complete. Perhaps in the future it will be your experiment that generates the first fully functional organ from stem cells in a plate. (Photograph courtesy of Lancaster et al. 2013.)

5 Snapshot Summary
Stem Cells

1. A stem cell maintains the ability to divide to produce a copy of itself as well as generating progenitor cells capable of maturing into different cell types.

2. Stem cell potential refers to the range of cell types a stem cell can produce. A totipotent stem cell can generate all cell types of both the embryonic and extraembryonic lineages. Pluripotent and multipotent stem cells produce restricted lineages of just the embryo and of only select tissues or organs, respectively.

3. Adult stem cells reside in microenvironments called stem cell niches. Most organs and tissues possess stem cell niches, such as the germ cell, hematopoietic, gut epithelial, and ventricular-subventricular niches.

4. The niche employs a variety of mechanisms of cell-to-cell communication to regulate the quiescent, proliferative, and differentiative states of the resident stem cell.

5. Inner cell mass cells of the mouse blastocyst are maintained in a pluripotent state through E-cadherin interactions with trophectoderm cells that activate the Hippo kinase cascade and repress the function of Yap/Taz as transcriptional regulators of *Cdx2*.

6. Cadherin links the germ stem cells of the *Drosophila* oocyte and testes to the niche keeping them within fields of TGF-β and Unpaired signaling, respectively. Asymmetric divisions push daughter cells out of this niche to promote cell differentiation of germ cells.

7. The ventricular-subventricular zone (V-SVZ) of the mammalian brain represents a complex niche architecture of B type stem cells arranged in a "pinwheel" organization with a primary cilium at the apical surface and long radial processes that terminate with a basal endfoot.

8. Constant Notch activity in the V-SVZ niche keeps B cells in the quiescent state, whereas increasing oscillations of Notch activity versus proneural gene expression progressively promote maturation of B cells to transit-amplifying C cells and then into migrating neural progenitors (A cells).

9. Additional signals—from neural activity and substances like GDF11 from blood vessels to gradients of Shh, BMP4, and Noggin—all influence cell proliferation and differentiation of B cells in the V-SVZ niche.

10. The base columnar cells located at the base of the intestinal crypt serve as clonogenic stem cells for the gut epithelium, which generates transit-amplifying epithelial cells that slowly differentiate as they are pushed further up the villus.

11. Wnt signals at the base of the crypt maintain stem cell proliferation, whereas opposing gradients of BMP from the cells at the top of the crypt induce differentiation.

12. Adhesion to osteoblasts keeps the hematopoietic stem cell (HSC) quiescent in the endosteal niche. Increased exposure to CXC12 signals from CAR cells and mesenchymal stem cells can transition HSCs into proliferative behavior, yet downregulation of *CXCL12* in the perivascular niche encourages migration of short-term active HSCs into the oxygen-rich blood vessels.

13. Mesenchymal stem cells can be found in a variety of tissues, including connective tissues, muscle, eye, teeth, bone, and more. They play dual roles as supportive stromal cells as well as being multipotent stem cells.

14. Embryonic and induced pluripotent stem cells can be maintained in culture indefinitely and, when exposed to certain combinations of factors and/or constrained by the physical growth substrate, can be coaxed to differentiate into potentially any cell type of the body.

15. ESCs and iPSCs are being used to study human cell development and diseases. The use of stem cells to study patient-specific cell differentiation of the rare blood disorder Fanconia anemia or disorders of the nervous system like autism and ALS have already started to provide novel insight into disease mechanisms.

16. Pluripotent stem cells can also be used in regenerative medicine to rebuild tissues and to make structures called organoids, which seem to possess many of the multicellular hallmarks of human organs. Organoids are being used to study human organogenesis and patient-specific disease progression on the tissue level, all in vitro.

Further Reading

Ables, J. L., J. J. Breunig, A. J. Eisch and P. Rakic. 2011. Not(ch) just development: Notch signalling in the adult brain. *Nature Rev. Neurosci.* 12: 269–283.

Al-Drees, M. A., J. H. Yeo, B. B. Boumelhem, V. I. Antas, K. W. Brigden, C. K. Colonne and S. T. Fraser. 2015. Making blood: The haematopoietic niche throughout ontogeny. *Stem Cells Int.* doi: 10.1155/2015/571893.

Barker, N. 2014. Adult intestinal stem cells: Critical drivers of epithelial homeostasis and regeneration. *Nature Rev. Mol. Cell Biol.* 15: 19–33.

Bianco, P. 2014. "Mesenchymal" stem cells. *Annu. Rev. Cell Dev. Biol.* 677–704.

Boulais, P. E. and P. S. Frenette. 2015. Making sense of hematopoietic stem cell niches. *Blood* 125: 2621–2629.

Dimos, J. T. and 12 others. 2008. Induced pluripotent stem cells generated from patients with ALS can be differentiated into motor neurons. *Science* 32: 1218–1221.

Fonseca, S. A., R. M. Costas and L. V. Pereira. 2015. Searching for naïve human pluripotent stem cells. *World J. Stem Cells* 7: 649–656.

Freitas, B. C., C. A. Trujillo, C. Carromeu, M. Yusupova, R. H. Herai and A. R. Muotri. 2014. Stem cells and modeling of autism spectrum disorders. *Exp. Neurol.* 260: 33–43.

Gafni, O. and 26 others. 2013. Derivation of novel human ground state naive pluripotent stem cells. *Nature* 504: 282–286.

Greenbaum A. and 7 others. 2013. CXCL12 in early mesenchymal progenitors is required for haematopoietic stem-cell maintenance. *Nature* 495: 227–230.

Harris, J., G. S. Tomassy and P. Arlotta. 2015. Building blocks of the cerebral cortex: From development to the dish. *Wiley Interdiscip. Rev. Dev. Biol.* 4: 529–544.

Imayoshi, I. and 8 others. 2013. Oscillatory control of factors determining multipotency and fate in mouse neural progenitors. *Science* 342: 1203–1208.

Katsimpardi, L. and 9 others. 2014. Vascular and neurogenic rejuvenation of the aging mouse brain by young systemic factors. *Science* 344: 630–634.

Kim, N. G., E. Koh, X. Chen and B. M. Gumbiner. 2011. E-cadherin mediates contact inhibition of proliferation through Hippo signaling-pathway components. *Proc. Natl. Acad. Sci. USA* 108: 11930–11935.

Lancaster, M. A. and J. A. Knoblich. 2014. Organogenesis in a dish: Modeling development and disease using organoid technologies. *Science.* doi: 10.1126/science.1247125.

Lancaster, M. A. and 9 others. 2013. Cerebral organoids model human brain development and microcephaly. *Nature* 501: 373–379.

Le Bin, G. C. and 11 others. 2014. Oct4 is required for lineage priming in the developing inner cell mass of the mouse blastocyst. *Development* 141: 1001–1010.

Lim, D. A. and A. Alvarez-Buylla. 2014. Adult neural stem cells stake their ground. *Trends Neurosci.* 37: 563–571.

Méndez-Ferrer, S. and 10 others. 2010. Mesenchymal and haematopoietic stem cells form a unique bone marrow niche. *Nature* 466: 829–834.

Ottone, C. and S. Parrinello. 2015. Multifaceted control of adult SVZ neurogenesis by the vascular niche. *Cell Cycle* 14: 2222–2225.

Spradling, A., M. T. Fuller, R. E. Braun and S. Yoshida. 2011. Germline stem cells. *Cold Spring Harbor Perspect. Biol.* doi: 10.1101/cshperspect.a002642.

Snippert, H. J. and 10 others. 2010. Intestinal crypt homeostasis results from neutral competition between symmetrically dividing Lgr5 stem cells. *Cell* 143: 134–144.

Stephenson, R. O., J. Rossant and P. P. Tam. 2012. Intercellular interactions, position, and polarity in establishing blastocyst cell lineages and embryonic axes. *Cold Spring Harbor Perspect. Biol.* doi: 10.1101/cshperspect.a008235.

Xie, T. and A. C. Spradling. 1998. *decapentaplegic* is essential for the maintenance and division of germline stem cells in the *Drosophila* ovary. *Cell* 94: 251–260.

Yan, D. and 16 others. 2014. A regulatory network of *Drosophila* germline stem cell self-renewal. *Dev. Cell* 28: 459–473.

GO TO WWW.DEVBIO.COM...

...for Web Topics, Scientists Speak interviews, Watch Development videos, Dev Tutorials, and complete bibliographic information for all literature cited in this chapter.

Sex Determination and Gametogenesis

"SEXUAL REPRODUCTION IS … THE MASTERPIECE OF NATURE," wrote Erasmus Darwin in 1791. Male and female offspring are generated by equivalent, equally active, gene-directed processes, neither being "higher" or "lower" or "greater" or "lesser" than the other. In mammals and flies, the sex of the individual is determined when the gametes—sperm and egg—come together. As we will see, however, there are other schemes of sex determination where animals of certain species are both male and female (making both sperm and eggs), and schemes where the environment determines an individual's sex. The gametes are the product of a **germ line** that is separate from the somatic cell lineages that divide mitotically to generate the differentiated somatic cells of the developing individual. Cells in the germ line undergo meiosis, a remarkable process of cell division by which the chromosomal content of a cell is halved so that the union of two gametes in fertilization restores the full chromosomal complement of the new organism. Sexual reproduction means that each new organism receives genetic material from two distinct parents, and the mechanisms of meiosis provide an incredible amount of genomic variation upon which evolution can work.

Gametogenesis and fertilization are both the end and the beginning of the circle of life. This chapter describes how the sex of an individual organism is determined, which in turn will determine whether that individual's gametes will become sperm or eggs.

How can this chicken become half hen and half rooster?

The Punchline

In vertebrates and arthropods, sex is determined by chromosomes. In mammals, the *Sry* gene on the Y chromosome transforms the bipotential gonad into a testis (and prevents ovary development), while inheritance of two X chromosomes activates β-catenin, transforming the bipotential gonad into an ovary (and preventing testis formation). In flies, the number of X chromosomes regulates the *Sxl* gene, enabling differential splicing of particular nuclear RNAs into male- or female-specific mRNAs. In mammals, the testes secrete hormones such as testosterone and anti-Müllerian hormone. The first builds the male phenotype, the second blocks the female phenotype. The ovaries synthesize estrogen that builds the female phenotype; they also secrete progesterone to maintain pregnancy. In all species, the gonads instruct gametogenesis, the development of the germ cells. Mammalian germ cells entering the ovaries initiate meiosis while in the embryo and become oocytes. Germ cells entering the mammalian testes are prevented from entering meiosis and instead divide to produce a stem cell population that at puberty will generate the sperm. There are also animal species whose sex is determined by environmental factors such as temperature.

Chromosomal Sex Determination

There there are several ways chromosomes can determine the sex of an embryo. In *mammals*, the presence of either a second X chromosome or a Y chromosome determines whether the embryo will be female (XX) or male (XY). In *birds*, the situation is reversed (Smith and Sinclair 2001): the male has the two similar sex chromosomes (ZZ) and the female has the unmatched pair (ZW). In *flies*, the Y chromosome plays no role in sex determination, but the number of X chromosomes appears to determine the sexual phenotype. In other insects (especially hymenopterans such as bees, wasps, and ants), fertilized, diploid eggs develop into females, while unfertilized, haploid eggs become males (Beukeboom 1995; Gempe et al. 2009). This chapter will discuss only two of the many chromosomal modes of sex determination: sex determination in placental mammals and sex determination in the fruit fly *Drosophila*.

WEB TOPIC 6.1   **SEX DETERMINATION AND SOCIAL PERCEPTIONS**   In the not-so-distant past, femaleness was considered a "default state," while maleness was thought of as "something more," acquired by genes that propelled development farther.

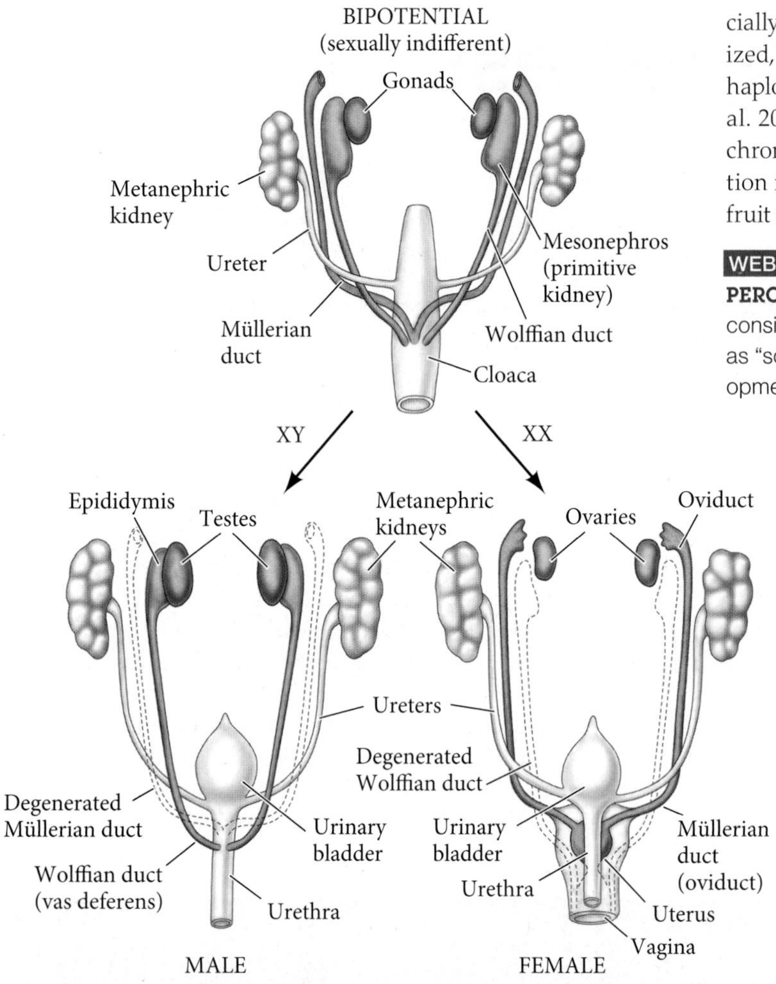

The Mammalian Pattern of Sex Determination

Mammalian sex determination is governed by the gonad-forming genes and by the hormones elaborated by the gonads. **Primary sex determination** is *the determination of the gonads*—the egg-forming ovaries or sperm-forming testes. **Secondary sex determination** is *the determination of the male or female phenotype by the hormones produced by the gonads*. The formation both of ovaries and of testes is an active, gene-directed process. Both the male and female gonads diverge from a common precursor, the **bipotential gonad** (sometimes called the **indifferent gonad**) (FIGURE 6.1).

GONADS		
Gonadal type	Testis	Ovary
Germ cell location	Inside testis cords (in medulla of testis)	Inside follicles of ovarian cortex
DUCTS		
Remaining duct	Wolffian	Müllerian
Duct differentiation	Vas deferens, epididymis, seminal vesicle	Oviduct, uterus, cervix, upper portion of vagina
UROGENITAL SINUS	Prostate	Skene's glands
LABIOSCROTAL FOLDS	Scrotum	Labia majora
GENITAL TUBERCLE	Penis	Clitoris

FIGURE 6.1   Development of gonads and their ducts in mammals. Originally, a bipotential (indifferent) gonad develops, with undifferentiated Müllerian ducts (female) and Wolffian ducts (male) ducts both present. If XY, the gonads becomes testes and the Wolffian duct persists. If XX, the gonads become ovaries and the Müllerian duct persists. Hormones from the gonads will cause the external genitalia to develop either in the male direction (penis, scrotum) or the female direction (clitoris, labia majora).

In mammals, primary sex determination is dictated by whether an organism has an XX or an XY karyotype. In most cases, the female's karyotype is XX and the male's is XY. Every individual must carry at least one X chromosome. Since the diploid female is XX, each of her haploid eggs has a single X chromosome. The male, being XY, generates two populations of haploid sperm: half will bear an X chromosome, half a Y. If at fertilization the egg receives a second X chromosome from the sperm, the resulting individual is XX, forms ovaries, and is female; if the egg receives a Y chromosome from the sperm, the individual is XY, forms testes, and is male (**FIGURE 6.2A**; Stevens 1905; Wilson 1905; see Gilbert 1978).

The Y chromosome carries a gene that encodes a **testis-determining factor** that organizes the bipotential gonad into a testis. This was demonstrated in 1959 when karyotyping showed that XXY individuals (a condition known as Klinefelter syndrome) are male (despite having two X chromosomes), and that individuals having only one X chromosome (XO, sometimes called Turner syndrome) are female (Ford et al. 1959;

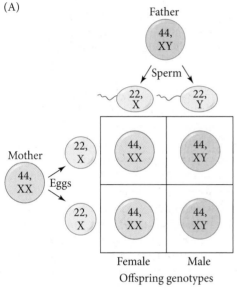

(A)

FIGURE 6.2 Sex determination in placental mammals. (A) Mammalian chromosomal sex determination results in approximately equal numbers of male and female offspring. (B) Postulated cascades leading to male and female phenotypes in mammals. The conversion of the genital ridge into the bipotential gonad requires, among others, the *Sf1*, *Wt1*, and *Lhx9* genes; mice lacking any of these genes lack gonads. The bipotential gonad appears to be moved into the female pathway (ovary development) by the *Foxl2*, *Wnt4*, and *Rspo1* genes and into the male pathway (testis development) by the *Sry* gene (on the Y chromosome), which triggers the activity of *Sox9*. (Lower levels of Wnt4 are also present in the male gonad.) The ovary makes thecal cells and granulosa cells, which together are capable of synthesizing estrogen. Under the influence of estrogen (first from the mother, then from the fetal gonads), the Müllerian duct differentiates into the female reproductive tract, the internal and external genitalia develop, and the offspring develops the secondary sex characteristics of a female. The testis makes two major hormones involved in sex determination. The first, anti-Müllerian hormone (AMH), causes the Müllerian duct to regress. The second, testosterone, causes differentiation of the Wolffian duct into the male internal genitalia. In the urogenital region, testosterone is converted into dihydrotestosterone (DHT), which causes the morphogenesis of the penis and prostate gland. (B after Marx 1995; Birk et al. 2000.)

(B)

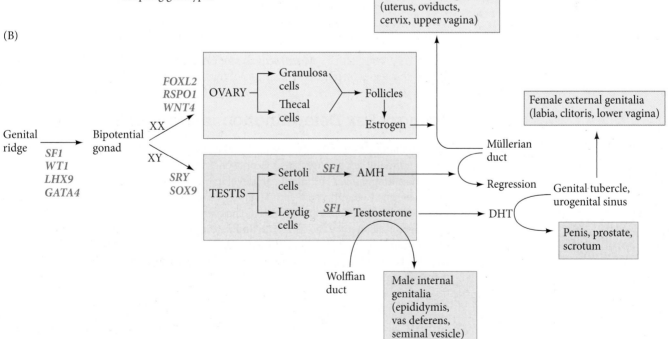

Jacobs and Strong 1959). XXY men have functioning testes. Women with a single X chromosome begin making ovaries, but the ovarian follicles cannot be maintained without the second X chromosome. Thus, a second X chromosome completes the ovaries, whereas the presence of a Y chromosome (even when multiple X chromosomes are present) initiates the development of testes.

The reason the Y chromosome is able to direct testis formation even when more than one X chromosome is present may be a matter of timing. It appears there is a crucial window of opportunity during gonad development during which the testis-determining factor (now known to be the product of the *Sry* gene) can function. If the *Sry* gene is present, it usually acts during this duration to promote testis formation and to inhibit ovary formation. If the *Sry* gene is not present (or if it fails to act at the appropriate time), the ovary-forming genes are the ones that will function (**FIGURE 6.2B**; Hiramatsu et al. 2009; Kashimada and Koopman 2010).

Once primary (chromosomal) determination has established the gonads, the gonads begin to produce the hormones and paracrine factors that govern secondary sex determination—development of the sexual phenotype outside the gonads. This includes the male or female duct systems and the external genitalia. A male mammal has a penis, scrotum (testicle sac), seminal vesicles, and prostate gland. A female mammal has a uterus, oviducts, cervix, vagina, clitoris, labia, and mammary glands.[1] In many species, each sex also has a sex-specific body size, vocal cartilage, and musculature. Secondary sex characteristics are usually determined by hormones and paracrine factors secreted from the gonads. In the absence of gonads, it appears the female phenotype is generated. When Jost (1947, 1953) removed fetal rabbit gonads before they had differentiated, the resulting rabbits had a female phenotype, regardless of whether their genotype was XX or XY.

The general scheme of primary sex determination is shown in Figure 6.2B. If the embryonic cells have two X chromosomes and no Y chromosome, the gonadal primordia develop into ovaries. The ovaries produce **estrogen**, a hormone that enables the development of the **Müllerian duct** into the uterus, oviducts, cervix, and upper portion of the vagina (Fisher et al. 1998; Couse et al. 1999; Couse and Korach 2001). If embryonic cells contain both an X and a Y chromosome, testes form and secrete two major factors. The first is a TGF-β family paracrine factor called **anti-Müllerian hormone** (**AMH**; sometimes called **Müllerian-inhibiting factor, MIF**). AMH destroys the Müllerian duct, thus preventing formation of the uterus and oviducts. The second factor is the steroid hormone **testosterone**. Testosterone masculinizes the fetus, stimulating formation of the penis, male duct system, scrotum, and other portions of the male anatomy, as well as inhibiting development of the breast primordia.

 DEV TUTORIAL *Mammalian sex determination* Scott Gilbert outlines the sex determination schemes of mammals.

Primary Sex Determination in Mammals

Mammalian gonads embody a unique embryological situation. All other organ rudiments normally can differentiate into only one type of organ—a lung rudiment can only become a lung, a liver rudiment can develop only into a liver. The gonadal rudiment, however, has two options: it can develop into either an ovary or a testis, two organs with very different tissue architectures. The path of differentiation taken by the gonadal rudiment is dictated by the genotype and determines the future sexual development of the organism (Lillie 1917). But before this decision is made, the mammalian gonad first develops through a bipotential, or indifferent, stage during which it has neither female nor male characteristics (see Figure 6.1).

---

[1] The naturalist Carolus Linnaeus named the mammals after this female secondary sexual trait in the seventeenth century. The politics of this decision is discussed in Schiebinger 1993.

The developing gonads

In humans, two gonadal rudiments appear during week 4 and remain sexually indifferent until week 7. These gonadal precursors are paired regions of the mesoderm adjacent to the developing kidneys (Tanaka and Nishinakamura 2014; **FIGURE 6.3A,B**). The **germ cells**—the precursors of either sperm or eggs—migrate into the gonads during week 6 and are surrounded by the mesodermal cells.

If the fetus is XY, the mesodermal cells continue to proliferate through week 8, when a subset of these cells initiate their differentiation into **Sertoli cells**. During embryonic development, the developing Sertoli cells secrete the anti-Müllerian hormone that blocks development of the female ducts. These same Sertoli epithelial cells will also form the seminiferous tubules that will support the development of sperm throughout the lifetime of the male mammal.

During week 8, the developing Sertoli cells surround the incoming germ cells and organize themselves into the **testis cords**. These cords form loops in the central region of the developing testis and are connected to a network of thin canals, called the **rete testis**, located near the developing kidney duct (**FIGURE 6.3C,D**). Thus, when germ cells enter the male gonads, they will develop within the testis cords, *inside* the organ. Later in development (at puberty in humans; shortly after birth in mice, which procreate much faster), the testis cords mature to form the **seminiferous tubules**. The germ cells migrate to the periphery of these tubules, where they establish the spermatogonial stem cell population that produces sperm throughout the lifetime of the male (see Figure 6.21). In the mature seminiferous tubule, sperm are transported from the inside of the testis through the rete testis, which joins the **efferent ducts**. The efferent ducts are the remodeled tubules of the developing kidney. During male development, the Wolffian duct differentiates to become the **epididymis** (adjacent to the testis) and the **vas deferens** (the tube through which sperm pass into the urethra and out of the body). Note that both sperm and urine will use the urethra to exit the body.

Meanwhile, the other group of mesoderm cells (those that did not form the Sertoli epithelium) differentiate into a mesenchymal cell type, the testosterone-secreting **Leydig cells**. Thus, the fully developed testis will have epithelial tubes made of Sertoli cells that enclose the germ cells, as well as a mesenchymal cell population, the Leydig cells, that secrete testosterone. Each incipient testis is surrounded by a thick extracellular matrix, the tunica albuginea, which helps protect it.

If the fetus is XX, the sex cords in the center of the developing gonad degenerate, leaving sex cords at the surface (cortex) of the gonad. Each germ cell gets enveloped by a separate cluster of sex cord epithelial cells (**FIGURE 6.3E,F**). The germ cells will become **ova** (eggs), and the surrounding cortical epithelial cells will differentiate into **granulosa cells**. The remaining mesenchyme cells of the developing ovary differentiate into **thecal cells**. Together, the thecal and granulosa cells form **follicles** that envelop the germ cells and secrete steroid hormones such as estrogens and (when pregnant) progesterone. Each follicle contains a single germ cell—an **oogonium** (egg precursor)—which will enter meiosis at this time.

There is a reciprocal relationship between the germ cells and the somatic cells of the gonads. The germ cells are originally bipotential and can become either sperm or eggs. Once in the male or female sex cords, however, they are instructed to either (1) begin meiosis and become eggs, or (2) remain meiotically dormant and become spermatogonia (McLaren 1995; Brennan and Capel 2004). In XX gonads, germ cells are essential for the maintenance of ovarian follicles. Without germ cells, the follicles degenerate into cordlike structures and express male-specific markers. In XY gonads, the germ cells help support the differentiation of Sertoli cells, although testis cords will form even without the germ cells, albeit a bit later (McLaren 1991). When an ovary is being formed, the Müllerian duct remains intact (there is no AMH to destroy it), and it differentiates into the oviducts, uterus, cervix, and upper vagina. In the absence of adequate testosterone, the Wolffian duct degenerates (see Figures 6.1 and 6.2).

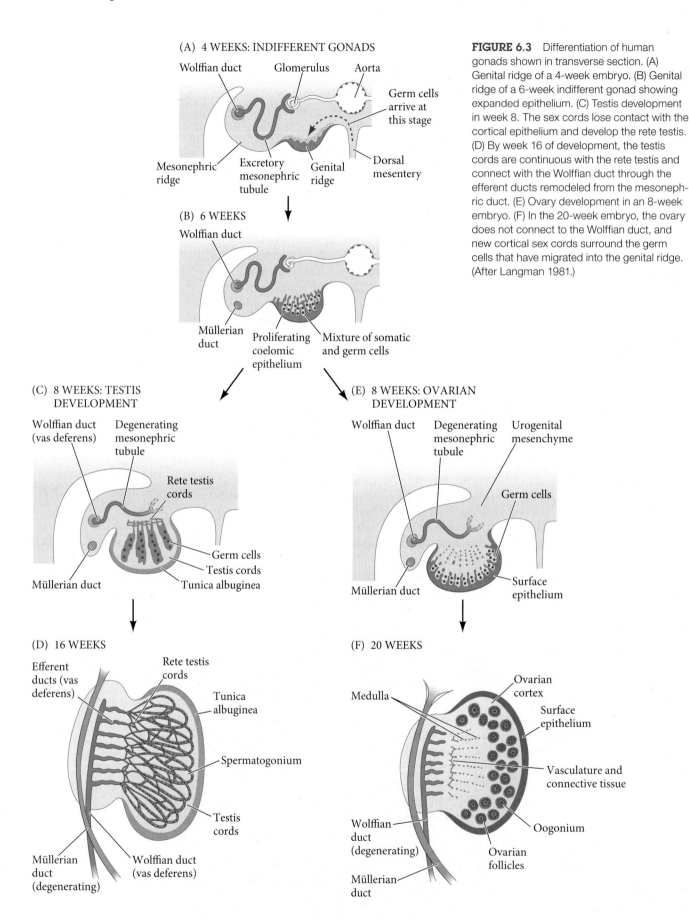

(A) 4 WEEKS: INDIFFERENT GONADS

Wolffian duct
Glomerulus
Aorta
Germ cells arrive at this stage
Dorsal mesentery
Mesonephric ridge
Excretory mesonephric tubule
Genital ridge

(B) 6 WEEKS

Wolffian duct
Müllerian duct
Proliferating coelomic epithelium
Mixture of somatic and germ cells

(C) 8 WEEKS: TESTIS DEVELOPMENT

Wolffian duct (vas deferens)
Degenerating mesonephric tubule
Rete testis cords
Germ cells
Testis cords
Tunica albuginea
Müllerian duct

(D) 16 WEEKS

Efferent ducts (vas deferens)
Rete testis cords
Tunica albuginea
Spermatogonium
Testis cords
Müllerian duct (degenerating)
Wolffian duct (vas deferens)

(E) 8 WEEKS: OVARIAN DEVELOPMENT

Wolffian duct
Degenerating mesonephric tubule
Urogenital mesenchyme
Germ cells
Surface epithelium
Müllerian duct

(F) 20 WEEKS

Ovarian cortex
Surface epithelium
Medulla
Vasculature and connective tissue
Oogonium
Ovarian follicles
Wolffian duct (degenerating)
Müllerian duct

FIGURE 6.3 Differentiation of human gonads shown in transverse section. (A) Genital ridge of a 4-week embryo. (B) Genital ridge of a 6-week indifferent gonad showing expanded epithelium. (C) Testis development in week 8. The sex cords lose contact with the cortical epithelium and develop the rete testis. (D) By week 16 of development, the testis cords are continuous with the rete testis and connect with the Wolffian duct through the efferent ducts remodeled from the mesonephric duct. (E) Ovary development in an 8-week embryo. (F) In the 20-week embryo, the ovary does not connect to the Wolffian duct, and new cortical sex cords surround the germ cells that have migrated into the genital ridge. (After Langman 1981.)

Genetic mechanisms of primary sex determination: Making decisions

Several human genes have been identified whose function is necessary for normal sexual differentiation. Because the phenotype of mutations in sex-determining genes is often sterility, clinical infertility studies have been useful in identifying those genes that are active in determining whether humans become male or female. Experimental manipulations to confirm the functions of these genes can then be done in mice.

The story starts in the bipotential gonad that has not yet been committed to the male or female direction. The genes for transcription factors Wt1, Lhx9, GATA4, and Sf1 are expressed, and the loss of function of any one of them will prevent the normal development of either male or female gonads. Then the decision is made:

- *If no Y chromosome is present,* these transcription and paracrine factors are thought to activate further expression of Wnt4 protein (already expressed at low levels in the genital epithelium) and of a small soluble protein called R-spondin1 (Rspo1). Rspo1 binds to its cell membrane receptor and further stimulates the Disheveled protein of the Wnt pathway, making the Wnt pathway more efficient at producing the transcriptional regulator β-catenin. One of the several functions of β-catenin in gonadal cells is to further activate the genes for Rspo1 and Wnt4, creating a positive feedback loop between these two proteins. A second role of β-catenin is to initiate the ovarian pathway of development by activating those genes involved in granulosa cell differentiation. Its third role is to prevent the production of Sox9, a protein crucial for testis determination (Maatouk et al. 2008; Bernard et al. 2012).

- *If a Y chromosome is present,* the same set of factors in the bipotential gonad activates the *Sry* gene on the Y chromosome. Sry protein binds to the enhancer of the *Sox9* gene and elevates expression of this key gene in the testis-determining pathway (Bradford et al. 2009b; Sekido and Lovell-Badge 2009). Sox9 and Sry also act to block the ovary-forming pathway, possibly by blocking β-catenin (Bernard et al. 2008; Lau and Li 2009).

FIGURE 6.4 shows one possible model of how primary sex determination can be initiated. Here we see an important rule of animal development: a pathway for cell specification often has two components, with one branch that says "Make A" and another branch that says "… and *don't* make B." In the case of the gonads, the male pathway says "Make testes and don't make ovaries," while the female pathway says "Make ovaries and don't make testes."

FIGURE 6.4 Possible mechanism for the initiation of primary sex determination in mammals. While we do not know the specific interactions involved, this model attempts to organize the data into a coherent sequence. If Sry is *not present* (pink region), the interactions between paracrine and transcription factors in the developing genital ridge activate *Wnt4* and *Rspo1*. Wnt4 activates the canonical Wnt pathway, which is made more efficient by Rspo1. The Wnt pathway causes the accumulation of β-catenin, and large accumulation of β-catenin stimulates further Wnt4 activity. This continual production of β-catenin both induces the transcription of ovary-producing genes and blocks the testis-determining pathway by interfering with *Sox9* activity. If Sry is *present* (blue region), it may block β-catenin signaling (thus halting ovary generation) and, along with Sf1, activate the *Sox9* gene. Sox9 activates Fgf9 synthesis, which stimulates testis development and promotes further Sox9 synthesis. Sox9 also prevents β-catenin's activation of ovary-producing genes. Sry may also activate other genes (such as *TCF21* and *NT3*) that help generate Sertoli cells. In summary, then, a Wnt4/β-catenin loop specifies the ovaries, whereas a Sox9/Fgf9 loop specifies the testes. One of the targets of the Wnt pathway is the *follistatin* gene, whose product organizes the granulosa cells of the ovary. Transcription factor Foxl2, which is activated (in a still unknown way) in the ovary, is also involved in inducing follistatin synthesis. The XY pathway appears to have an earlier initiation; if it does not function, the XX pathway takes over. (After Sekido and Lovell-Badge 2009; McClelland et al. 2012.)

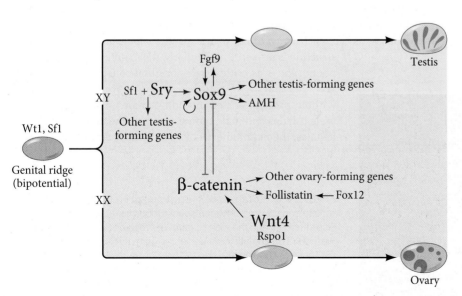

The ovary pathway: Wnt4 and R-spondin1

In mice, the paracrine factor **Wnt4** is expressed in the bipotential gonads, but its expression becomes undetectable in XY gonads as they become testes, whereas it is maintained in XX gonads as they begin to form ovaries. In XX mice that lack the *Wnt4* gene, the ovary fails to form properly, and the cells transiently express testis-specific markers, including Sox9, testosterone-producing enzymes, and AMH (Vainio et al. 1999; Heikkilä et al. 2005). Thus, Wnt4 appears to be an important factor in ovary formation, although it is not the only determining factor.

R-spondin1 (Rspo1) is also critical in ovary formation, since in human case studies, several XX individuals with *RSPO1* gene mutations became phenotypic males (Parma et al. 2006). Rspo1 acts in synergy with Wnt4 to produce β-catenin, which appears to be critical both in activating further ovarian development and in blocking the synthesis of a testis-determining factor, Sox9 (Maatouk et al. 2008; Jameson et al. 2012). In XY individuals with a duplication of the region on chromosome 1 that contains both the *WNT4* and *RSPO1* genes, the pathways that make β-catenin override the male pathway, resulting in a male-to-female sex reversal. Similarly, if XY mice are induced to overexpress β-catenin in their gonadal rudiments, they form ovaries rather than testes. Indeed, β-catenin appears to be a key "pro-ovarian/anti-testis" signaling molecule in all vertebrate groups, as it is seen in the female (but not the male) gonads of birds, mammals, and turtles. These three groups have very different modes of sex determination, yet Rspo1 and β-catenin are made in the ovaries of each of them (**FIGURE 6.5**; Maatouk et al. 2008; Cool and Capel 2009; Smith et al. 2009).

Certain transcription factors whose genes are activated by β-catenin are found exclusively in the ovaries. One possible target for β-catenin is the gene encoding TAFII-105 (Freiman et al. 2002). This transcription factor subunit (which helps bind RNA polymerase to promoters) is seen only in ovarian follicle cells. Female mice lacking this subunit have small ovaries with few, if any, mature follicles. The transcription factor Foxl2 is another protein that is strongly upregulated in ovaries, and XX mice homozygous for mutant *Foxl2* alleles develop male-like gonad structure and upregulate *Sox9* gene expression and testosterone production. Both Foxl2 and β-catenin are critical for activation of the *Follistatin* gene (Ottolenghi et al. 2005; Kashimada et al. 2011; Pisarska et al. 2011). Follistatin, an inhibitor of TGF-β family of paracrine factors, is thought to be the protein responsible for organizing the epithelium into the granulosa cells of the ovary (Yao et al. 2004). XX mice lacking follistatin in the developing gonad undergo a partial sex reversal, forming testicle-like structures. Numerous other transcription factors are upregulated by the Wnt4/R-spondin signal (Naillat et al. 2015), and we are just beginning to figure out how the components of the ovary-forming pathway are integrated.

As important as the *construction* of the ovaries is, the *maintenance* of the ovarian structure is also critical. Similarly, the maintenance of testicular phenotype is as critical as its original construction. Remarkably, gonadal organization is not stable throughout life, and without proper gene expression, female follicles can become male tubules and male tubules can become female follicles. In females, the maintainer of ovarian identity appears to be Foxl2

(A)   (B)

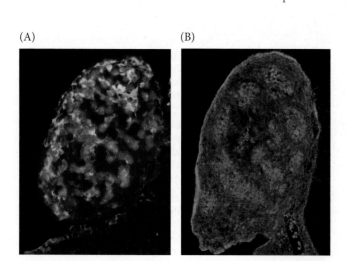

FIGURE 6.5 Localization of Rspo1 protein in embryonic day 14.5 mouse gonads. Immunofluorescent probes were used to identify Rspo1 (green) and the meiotic germ cell marker, Scp3 (red). (A) Rspo1 was found on somatic cells and at the germ cell surface of the ovaries. (B) These antibodies revealed neither Rspo1 nor Scp3 in the developing testes. (The germ cells in the male gonads have not entered meiosis at this point in development, whereas the ovarian germ cells have.) (From Smith et al. 2008; photograph courtesy of C. Smith.)

(Uhlenhaut et al. 2009). When Foxl2 is deleted in adult-stage ovaries, the Sox9 gene becomes active and the ovary is transformed into a testis.

The testis pathway: Sry and Sox9

SRY: THE Y CHROMOSOME SEX DETERMINANT   In humans, the major gene for testis determination resides on the short arm of the Y chromosome. By analyzing the DNA of rare XX men and XY women (i.e., individuals who are genotypically one sex but phenotypically the other), the position of the testis-determining gene was narrowed down to a 35,000-base-pair region of the Y chromosome found near the tip of the short arm. In this region, Sinclair and colleagues (1990) found a male-specific DNA sequence that encodes a peptide of 223 amino acids. This gene is called **Sry** (**sex-determining region of the Y chromosome**), and there is extensive evidence that it is indeed the gene that encodes the human testis-determining factor.

Sry is found in normal XY males and also in the rare XX males; it is absent from normal XX females and from many XY females. Approximately 15% of human XY females have the *SRY* gene, but their copies of the gene contain point or frameshift mutations that prevent Sry protein from binding to DNA (Pontiggia et al. 1994; Werner et al. 1995). If the *SRY* gene actually does encode the major testis-determining factor, one would expect it to act in the indifferent gonad immediately before or during testis differentiation. This prediction has been found to be the case in studies of the homologous gene in mice. The mouse *Sry* gene also correlates with the presence of testes; it is present in XX males and absent in XY females (Gubbay et al. 1990). *Sry* is expressed in the somatic cells of the bipotential gonads of XY mice immediately before the differentiation of these cells into Sertoli cells; its expression then disappears (Koopman et al. 1990; Hacker et al. 1995; Sekido et al. 2004).

The most impressive evidence for *Sry* being the gene for testis-determining factor comes from transgenic mice. If *Sry* induces testis formation, then inserting *Sry* DNA into the genome of a normal XX mouse zygote should cause that XX mouse to form testes. Koopman and colleagues (1991) took the 14-kilobase region of DNA that includes the *Sry* gene (and presumably its regulatory elements) and microinjected this sequence into the pronuclei of newly fertilized mouse zygotes. In several instances, XX embryos injected with this sequence developed testes, male accessory organs, and a penis[2] (**FIGURE 6.6**). Therefore, we conclude that *Sry/SRY* is the major gene on the Y chromosome for testis determination in mammals.

 SCIENTISTS SPEAK 6.1   Dr. Robin Lovell-Badge discusses his research showing how the *SRY* gene promotes testis formation in humans.

---

[2]These embryos did not form functional sperm—but they were not expected to. The presence of two X chromosomes prevents sperm formation in XXY mice and men, and the transgenic mice lacked the rest of the Y chromosome, which contains genes needed for spermatogenesis.

(A)

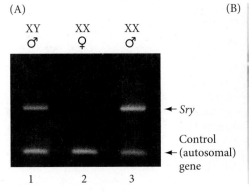

(B)

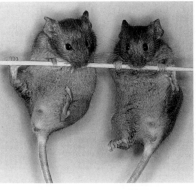

FIGURE 6.6   An XX mouse transgenic for *Sry* is male. (A) Polymerase chain reaction followed by electrophoresis shows the presence of the *Sry* gene in normal XY males and in a transgenic XX/*Sry* mouse. The gene is absent in a female XX littermate. (B) The external genitalia of the transgenic mouse are male (right) and are essentially the same as those in an XY male (left). (From Koopman et al. 1991; photographs courtesy of the authors.)

SOX9*: AN AUTOSOMAL TESTIS-DETERMINING GENE** For all its importance in sex determination, the *Sry* gene is probably active for only a few hours during gonadal development in mice. During this time, it synthesizes the Sry transcription factor, whose primary role appears to be to activate the ***Sox9 gene (Sekido and Lovell-Badge 2008; for other targets of Sry, see Web Topic 6.2). *Sox9* is an autosomal gene involved in several developmental processes, most notably bone formation. In the gonadal rudiments, however, *Sox9* induces testis formation. XX humans who have an extra activated copy of *SOX9* develop as males even if they have no *SRY* gene, and XX mice transgenic for *Sox9* develop testes (**FIGURE 6.7A–D**; Huang et al. 1999; Qin and Bishop 2005). Knocking out the *Sox9* gene in the gonads of XY mice causes complete sex reversal (Barrionuevo et al. 2006). Thus, even if *Sry* is present, mouse gonads cannot form testes if *Sox9* is absent, so it appears that *Sox9* can replace *Sry* in testis formation. This is not altogether surprising; although the *Sry* gene is found specifically in mammals, *Sox9* is found throughout the vertebrate phyla.

Indeed, *Sox9* appears to be the older and more central sex determination gene in vertebrates (Pask and Graves 1999). In mammals, it is activated by Sry protein; in birds, frogs, and fish, it appears to be activated by the dosage of the transcription factor Dmrt1; and in those vertebrates with temperature-dependent sex determination, it is often activated (directly or indirectly) by the male-producing temperature. Expression of the *Sox9* gene is specifically upregulated by the combined expression of Sry and Sf1 proteins in Sertoli cell precursors (**FIGURE 6.7E–H**; Sekido et al. 2004; Sekido and Lovell-Badge 2008). Thus, Sry may act merely

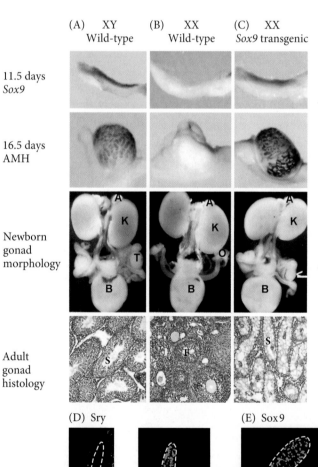

(A) XY Wild-type (B) XX Wild-type (C) XX *Sox9* transgenic

11.5 days *Sox9*

16.5 days AMH

Newborn gonad morphology

Adult gonad histology

(D) Sry

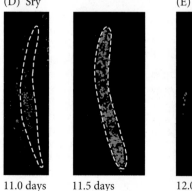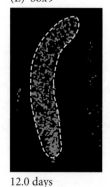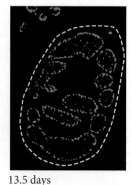

11.0 days   11.5 days

(E) Sox9

12.0 days   13.5 days

FIGURE 6.7 Ability of Sox9 protein to generate testes. (A) A wild-type XY mouse embryo expresses the *Sox9* gene in the genital ridge at 11.5 days postconception, anti-Müllerian hormone in the embryonic gonad Sertoli cells at 16.5 days, and eventually forms descended testes with seminiferous tubules. K, kidneys; A, adrenal glands; B, bladder; T, testis; O, ovary; S, seminiferous tubule; F, follicle cell. (B) The wild-type XX embryo shows neither *Sox9* expression nor AMH. It constructs ovaries with mature follicle cells. (C) An XX embryo with the *Sox9* transgene inserted expresses *Sox9* and has AMH in 16.5-day Sertoli cells. It has descended testes, but the seminiferous tubules lack sperm (due to the presence of two X chromosomes in the Sertoli cells). (D,E) Chronological sequence from the expression of *Sry* in the genital ridge to that of *Sox9* in the Sertoli cells. (D) *Sry* expression. At day 11.0, Sry protein (green) is seen in the center of the genital ridge. At day 11.5, the domain of *Sry* expression increases and *Sox9* expression is activated. (E) By day 12.0, Sox9 protein (green) is seen in the same cells that earlier expressed *Sry*. By day 13.5, Sox9 is seen in those cells of the testis tubule that will become Sertoli cells. (A–C from Vidal et al. 2001, photographs courtesy of A. Schedl; D,E from Kashimada and Koopman, 2010, courtesy of P. Koopman.)

as a "switch" operating during a very short time to activate *Sox9*, and the Sox9 protein may initiate the conserved evolutionary pathway to testis formation. So, borrowing Eric Idle's phrase, Sekido and Lovell-Badge (2009) propose that Sry initiates testis formation by "a wink and a nudge."

The Sox9 protein has several functions. First, it appears to be able to activate its own promoter, thereby allowing it to be transcribed for long periods of time. Second, it blocks the ability of β-catenin to induce ovary formation, either directly or indirectly (Wilhelm et al. 2009). Third, it binds to *cis*-regulatory regions of numerous genes necessary for testis production (Bradford et al. 2009a). Fourth, Sox9 binds to the promoter site on the gene for anti-Müllerian hormone, providing a critical link in the pathway toward a male phenotype (Arango et al. 1999; de Santa Barbara et al. 2000). Fifth, Sox9 promotes the expression of the gene encoding Fgf9, a paracrine factor critical for testis development. Fgf9 is also essential for maintaining *Sox9* gene transcription, thereby establishing a positive feedback loop driving the male pathway (Kim et al. 2007).

WEB TOPIC 6.2   **FINDING THE ELUSIVE TESTIS-DETERMINING FACTOR**   As one editor wrote, "The search for TDF has been a long and hard one."

FIBROBLAST GROWTH FACTOR 9   When the gene for **fibroblast growth factor 9 (Fgf9)** is knocked out in mice, the homozygous mutants are almost all female. Fgf9 protein, whose expression is dependent on Sox9 (Capel et al. 1999; Colvin et al. 2001), plays several roles in testis formation:

1. Fgf9 causes proliferation of the Sertoli cell precursors and stimulates their differentiation (Schmahl et al. 2004; Willerton et al. 2004).

2. It activates the migration of blood vessel cells from the adjacent kidney duct into the XY gonad. While this is normally a male-specific process, incubating XX gonads in Fgf9 leads to the migration of endothelial cells into XX gonads (**FIGURE 6.8**). These blood vessel cells form the major artery of the testis and play an instructive role in inducing the Sertoli cell precursors to form the testis cords; in their absence, testis cords do not form (Brennan et al. 2002; Combes et al. 2009).

3. It is required for maintaining *Sox9* expression in the presumptive Sertoli cells and directs their formation into tubules. Moreover, since it can act as both an autocrine and a paracrine factor, Fgf9 may coordinate Sertoli cell development by reinforcing *Sox9* expression in all the cells of the tissue (Hiramatsu et al. 2009). Such a "community effect" may be important in achieving the integrated assembly of testis tubules (Palmer and Burgoyne 1991; Cool and Capel 2009).

4. It represses Wnt4 signaling, which would otherwise direct ovarian development (Maatouk et al. 2008; Jameson et al. 2012).

5. Finally, Fgf9 appears to help coordinate the sex determination of the gonad with that of the germ cells. As we will see later in this chapter, those mammalian germ cells destined to become eggs enter meiosis quickly upon entering the gonad, whereas germ cells destined to become sperm delay their entry into meiosis until puberty. Fgf9 is one of the factors that blocks the immediate entry of germ cells into meiosis, thereby placing them onto the sperm-forming pathway (Barrios et al. 2010; Bowles et al. 2010).

SF1: A CRITICAL LINK BETWEEN SRY AND THE MALE DEVELOPMENTAL PATHWAYS
The transcription factor **steroidogenic factor 1 (Sf1)** is necessary to make the bipotential gonad. But whereas Sf1 levels decline in the genital ridge of XX mouse embryos, they remain high in the developing testis. It is thought that Sry either directly or indirectly maintains *Sf1* gene expression. Sf1 protein appears to be active in masculinizing both the Leydig and the Sertoli cells. In the Sertoli cells, Sf1 works in collaboration with Sry to activate *Sox9* (Sekido and Lovell-Badge 2008) and then, working with Sox9, elevates levels of anti-Müllerian hormone transcription (Shen et al. 1994; Arango et al. 1999). In the Leydig cells, Sf1 activates genes encoding the enzymes that make testosterone.

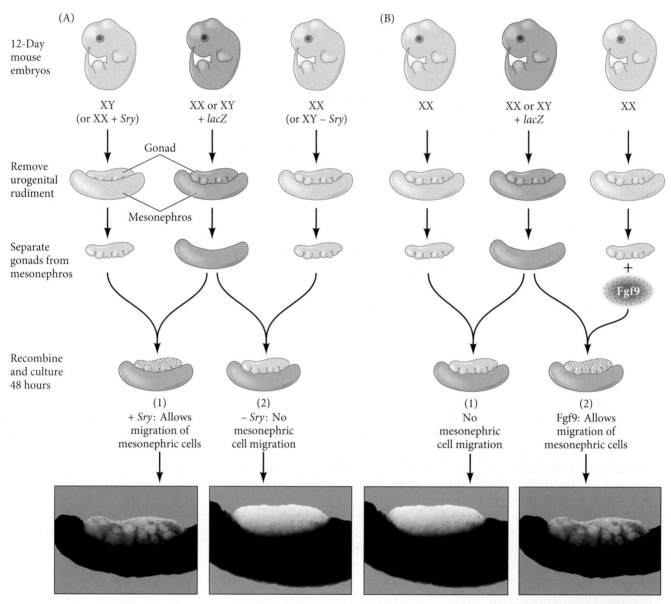

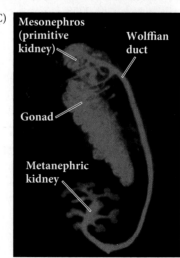

FIGURE 6.8 Migration of mesonephric endothelial cells into *Sry*⁺ gonadal rudiments. In the experiment diagrammed, urogenital ridges (containing both primitive mesonephric kidneys and bipotential gonadal rudiments) were collected from 12-day embryonic mice. Some of the mice were marked with a β-galactosidase transgene (*lacZ*) that is active in every cell. Thus, every cell of these mice turned blue when stained for β-galactosidase. The gonad and mesonephros were separated and recombined, using gonadal tissue from unlabeled mice and mesonephros from labeled mice. (A) Migration of mesonephric cells into the gonad was seen (1) when the gonadal cells were XY or when they were XX with a *Sry* transgene. No migration of mesonephric tissue into the gonad was seen (2) when the gonad contained either XX cells or XY cells in which the Y chromosome had a deletion in the *Sry* gene. The sex chromosomes of the mesonephros did not affect the migration. (B) Gonadal rudiments for XX mice could induce mesonephric cell migration if these rudiments had been incubated with Fgf9. (C) Intimate relation between the Wolffian duct and the developing gonad in a 16-day male mouse embryo. The mesonephric duct of the primitive kidney will form the efferent ducts of the testes and the Wolffian duct that leads to the ureter. The ducts and gonad have been stained for cytokeratin-8. (A,B after Capel et al. 1999, photographs courtesy of B. Capel; C from Sariola and Saarma 1999, courtesy of H. Sariola.)

SCIENTISTS SPEAK 6.2   Dr. Blanche Capel discusses her work on the sex determination pathways of mammals.

The right time and the right place

Having the right genes doesn't necessarily mean you'll get the organ you expect. Studies of mice have shown that the *Sry* gene of some strains of mice failed to produce testes when bred onto a different genetic background (Eicher and Washburn 1983; Washburn and Eicher 1989; Eicher et al. 1996). This failure can be attributed either to a delay in *Sry* expression, or to the failure of the protein to accumulate to the critical threshold level required to trigger *Sox9* expression and launch the male pathway. By the time *Sox9* gets turned on, it is too late—the gonad is already well along the path to become an ovary (Bullejos and Koopman 2005; Wilhelm et al. 2009).

The importance of timing was confirmed when Hiramatsu and collaborators (2009) were able to place the mouse *Sry* gene onto the regulatory sequences of a heat-sensitive gene, allowing them to activate *Sry* at any time in mouse development by merely raising the embryo's temperature. When they delayed *Sry* activation by as little as 6 hours, testis formation failed and ovaries started to develop (**FIGURE 6.9**). Thus, there appears to be a brief window during which the testis-forming genes can function. If this window of opportunity is missed, the ovary-forming pathway is activated.

Hermaphrodites are individuals in which both ovarian and testicular tissues exist; they have either ovotestes (gonads containing both ovarian and testicular tissue) or an ovary on one side and a testis on the other.[3] As seen in Figure 6.9, ovotestes can be generated when the *Sry* gene is activated later than normal. Hermaphrodites can also result in those very rare instances when a Y chromosome is translocated to an X chromosome. In those tissues where the translocated Y is on the active X chromosome, the Y chromosome will be active and the *SRY* gene will be transcribed; in cells where the Y chromosome is on the inactive X chromosome, the Y chromosome will also be inactive (Berkovitz et al. 1992; Margarit et al. 2000). Such gonadal mosaicism for expressing *SRY* can lead to the formation of a testis, an ovary, or an ovotestis, depending on the percentage of cells expressing *SRY* in the Sertoli cell precursors (see Brennan and Capel 2004; Kashimada and Koopman 2010).

---

[3]This anatomical phenotype is named for Hermaphroditos, a young man in Greek mythology whose beauty inflamed the ardor of the water nymph Salmacis. She wished to be united with him forever, and the gods, in their literal fashion, granted her wish. Hermaphroditism is often considered to be one of the "intersex" conditions discussed later in the chapter.

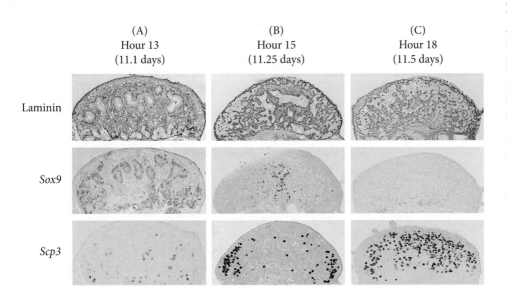

FIGURE 6.9 Experimental delay of *Sry* gene activation by 6 hours leads to failure of testis development and the initiation of ovary development. Genital ridges were removed from XX mice carrying a heat-inducible *Sry* gene. These tissues were then heat-shocked at different times to activate *Sry* and then allowed to mature. (A) Those genital tissues experiencing *Sry* induction at 11.1 days of development (when *Sry* is normally activated) produced testes. Their laminin distribution showed Sertoli cells, *Sox9* (a marker of testis development) was active, and *Scp3*, a marker of ovary development, was absent. (B) Three hours later, the activation of *Sry* caused a central testicular area to form, with ovary-like structures forming peripherally. *Sox9* was present in the central testicular region, while *Scp3* was found in the periphery. (C) If *Sry* was activated in the genital tissues 6 hours later, the structures formed ovarian tissue, *Sox9* was absent, and *Scp3* was seen throughout the tissue. (After Hiramatsu et al. 2009.)

And just as the *Foxl2* gene is critical for the maintenance of ovarian gonadal function throughout life, the **Dmrt1** gene is needed for maintaining testicular structure. The deletion of *Dmrt1* in adult mice leads to the transformation of Sertoli cells into ovarian granulosa cells. Moreover, overexpression of *Dmrt1* in female mouse ovaries can reprogram the ovarian tissue into Sertoli-like cells (Lindeman et al. 2015; Zhao et al. 2015). Dmrt1 protein is probably the major male sex inducer across the entire animal kingdom, having been found in flies, cnidarians, fish, reptiles, and birds (Murphy et al. 2015; Picard et al. 2015). In mammals, SRY has taken over this function. However, these recent results show that *Dmrt1* has retained an important role in male sex determination, even in mammals.

 SCIENTISTS SPEAK 6.3 Dr. David Zarkower discusses his studies showing Dmrt1 to be a major player in the male sex determination pathway.

Secondary Sex Determination in Mammals: Hormonal Regulation of the Sexual Phenotype

Primary sex determination—the formation of either an ovary or a testis from the bipotential gonad—does not result in the complete sexual phenotype. In mammals, secondary sex determination is the development of the female and male phenotypes in response to hormones secreted by the ovaries and testes. Both female and male secondary sex determination have two major temporal phases. The first phase occurs within the embryo during organogenesis; the second occurs at puberty.

During embryonic development, hormones and paracrine signals coordinate the development of the gonads with the development of secondary sexual organs. In females, the Müllerian ducts persist and, through the actions of estrogen, differentiate to become the uterus, cervix, oviducts, and upper vagina (see Figure 6.2). The **genital tubercle** becomes differentiated into the clitoris, and the **labioscrotal folds** become the labia majora. The Wolffian ducts require testosterone to persist, and thus they atrophy in females. In females, the portion of the **urogenital sinus** that does not become the bladder and urethra becomes Skene's glands, paired organs that make secretions similar to those of the prostate.

The coordination of the male phenotype involves the secretion of two testicular factors. The first of these is anti-Müllerian hormone, a BMP-like paracrine factor made by the Sertoli cells, which causes the degeneration of the Müllerian duct. The second is the steroid hormone testosterone, an **androgen** (masculinizing substance) secreted from the fetal Leydig cells. Testosterone causes the Wolffian ducts to differentiate into sperm-carrying tubes (the epididymis and vas deferens) as well as the seminal vesicle (which emerges as an outpocketing of the vas deferens), and it causes the **genital tubercle** (the precursor of the external genitalia) to develop into the penis and the labioscrotal folds to develop into the scrotum. In males, the urogenital sinus, in addition to forming the bladder and urethra, also forms the prostate gland.

The mechanism by which testosterone (and, as we shall see, its more powerful derivative dihydrotestosterone) masculinizes the genital tubercle is thought to involve its interaction with the Wnt pathway (**FIGURE 6.10**). The Wnt pathway, which in the bipotential gonad activates the female trajectory, acts in the genital tubercle to activate male development (Mazahery et al. 2013). The Wnt antagonist Dickkopf is made in the urogenital swellings and can be downregulated by testosterone and upregulated by anti-androgens. This finding led to a model wherein the urogenital swellings of XX individuals make Dickkopf, thus preventing the activity of Wnt in the mesenchyme, blocking further growth and leading to the feminization of the genital tubercle by estrogens (Holderegger and Keefer 1986; Miyagawa et al. 2009). In females, then, the genital tubercle becomes the clitoris and the labioscrotal folds become the labia majora. In males, however, testosterone and dihydrotestosterone bind to the androgen

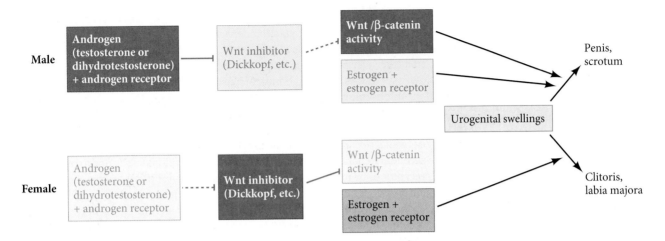

FIGURE 6.10 Model for the formation of external genitalia. In this schema, the mesen-chyme in the urogenital swellings secretes inhibitors of Wnt signaling. In the absence of Wnt signaling, estrogen modifies the genital tubercle into the clitoris and the labioscrotal folds into the labia majora surrounding the vagina. In males, however, androgens (such as testosterone and dihydrotestosterone) bind to the androgen receptor in the mesenchymal cells and prevent the synthesis of the Wnt inhibitors. Wnt signaling is permitted, and it causes the genital tubercle to become the penis and the labioscrotal folds to become the scrotum. (After Miyagawa et al. 2009.)

(testosterone) receptor in the mesenchyme and prevent the expression of Wnt inhibi-tors (thus permitting Wnt expression in the mesenchyme). With the influence of these Wnts, male urogenital swellings are converted into the penis and the scrotum.

WEB TOPIC 6.3 **THE ORIGINS OF GENITALIA** The cells that give rise to the penis and clitoris have only recently been identified. Their identity helps explain how male snakes get two penises and a female hyena develops a clitoris nearly as large as the male's penis.

The genetic analysis of secondary sex determination

The existence of separate and independent AMH and testosterone pathways of masculinization is demonstrated by people with **androgen insensitivity syndrome**. These XY individuals, being chromosomally males, have the *SRY* gene and thus have testes that make testosterone and AMH. However, they have a mutation in the gene encoding the androgen *receptor* protein that binds testosterone and brings it into the nucleus. Therefore, these individuals cannot respond to the testosterone made by their testes (Meyer et al. 1975; Jääskeluäinen 2012). They can, however, respond to the estrogen made by their adrenal glands (which is normal for both XX and XY indi-viduals), so they develop female external sex characteristics (**FIGURE 6.11**). Despite their distinctly female appearance, these XY individuals have testes, and even though they cannot respond to testosterone, they produce and respond to AMH. Thus, their Müllerian ducts degenerate. Persons with androgen insensitivity syndrome develop as normal-appearing but sterile women, lacking a uterus and oviducts and having internal testes in the abdomen.

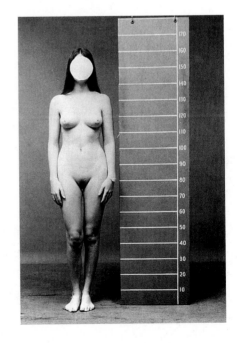

FIGURE 6.11 Androgen insensitivity syndrome. Despite having the XY karyotype, individu-als with this syndrome appear female. They cannot respond to testosterone but can respond to estrogen, so they develop female secondary sex characteristics (i.e., labia and a clitoris rather than a scrotum and a penis). Internally, they lack the Müllerian duct derivatives and have undescended testes. (Courtesy of C. B. Hammond.)

Although in most people correlation of the genetic and anatomical sexual phenotypes is high, about 0.4–1.7% of the population departs from the strictly dimorphic condition (Blackless et al. 2000; Hull 2003; Hughes et al. 2006). Phenotypes in which male and female traits are seen in the same individual are called **intersex** conditions.[4] Androgen insensitivity syndrome is one of several intersex conditions that have traditionally been labeled **pseudohermaphroditism**. In pseudohermaphrodites, there is only one type of gonad (as contrasted with true hermaphroditism, in which individuals have the gonads of both sexes), but the secondary sex characteristics differ from what would be expected from the gonadal sex. In humans, male pseudohermaphroditism (male gonadal sex with female secondary characteristics) can be caused by mutations in the androgen (testosterone) receptor or by mutations affecting testosterone synthesis (Geissler et al. 1994).

Female pseudohermaphroditism, in which the gonadal sex is female but the person is outwardly male, can be the result of overproduction of androgens in the ovary or adrenal gland. The most common cause of this latter condition is **congenital adrenal hyperplasia**, in which there is a genetic deficiency of an enzyme that metabolizes cortisol steroids in the adrenal gland. In the absence of this enzyme, testosterone-like steroids accumulate and can bind to the androgen receptor, thus masculinizing the fetus (Migeon and Wisniewski 2000; Merke et al. 2002).

TESTOSTERONE AND DIHYDROTESTOSTERONE   Although testosterone is one of the two primary masculinizing factors, there is evidence that it is *not* the active masculinizing hormone in certain tissues. Although testosterone is responsible for promoting the formation of the male structures that develop from the Wolffian duct primordium, testosterone does not directly masculinize the urethra, prostate, penis, or scrotum. These latter functions are controlled by **5α-dihydrotestosterone**, or **DHT** (**FIGURE 6.12**). Siiteri and Wilson (1974) showed that testosterone is converted to DHT in the urogenital sinus and swellings, but not in the Wolffian duct. DHT appears to be a more potent hormone than testosterone. It is most active prenatally and in early childhood.[5]

The importance of DHT in the early development of the male gonads was demonstrated by Imperato-McGinley and her colleagues (1974) when they studied a phenotypically remarkable syndrome in several inhabitants of a small community in the Dominican Republic. Individuals with this syndrome were found to lack a functional gene for the enzyme 5α-ketosteroid reductase 2—the enzyme that converts testosterone to DHT (Andersson et al. 1991; Thigpen et al. 1992). Chromosomally XY children with this syndrome have functional testes, but the testes remain inside the abdomen and do not descend before birth. These children appear to be girls and are raised as such. Their internal anatomy, however, is male: they have Wolffian duct development and Müllerian duct degeneration, along with their functional testes. At puberty, when the testes produce high levels of testosterone (which

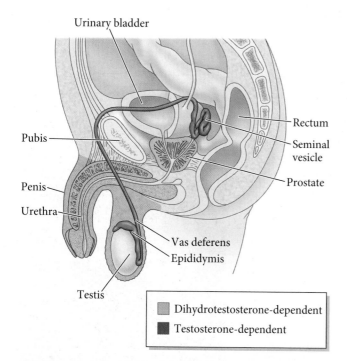

Urinary bladder
Pubis
Penis
Urethra
Testis
Rectum
Seminal vesicle
Prostate
Vas deferens
Epididymis

◼ Dihydrotestosterone-dependent
◼ Testosterone-dependent

FIGURE 6.12   Testosterone- and dihydrotestosterone-dependent regions of the human male genital system. (After Imperato-McGinley et al. 1974.)

---

[4]The "intersex" language used to group these conditions is being debated. Some activists, physicians, and parents wish to eliminate the term "intersex" to avoid confusion of these anatomical conditions with identity issues such as homosexuality. They prefer to call these conditions "disorders of sex development." In contrast, other activists do not want to medicalize this condition and find the "disorder" category offensive to individuals who do not feel there is anything wrong with their health. For a more detailed analysis of intersexuality, see Gilbert et al. 2005, Austin et al. 2011, and Dreger 2000.

[5]There's a reason the label on some hair-restoring drugs warns pregnant women not to handle them. Finasteride, an active ingredient in these products, blocks the metabolism of testosterone into dihydrotestosterone and thus could interfere with the gonadal development of a male fetus.

appears to compensate for the lack of DHT), their external genitalia are able to respond to the hormone and differentiate. The penis enlarges, the scrotum descends, and the person originally believed to be a girl is revealed to be a young man. Studies of this condition led to the current perception that the formation of the external genitalia is under the control of dihydrotestosterone, whereas Wolffian duct differentiation is controlled by testosterone itself.

WEB TOPIC 6.4 **DESCENT OF THE TESTES** The descent of the testes is initiated around week 10 of human pregnancy by dihydrotestosterone and another hormone from the Leydig cells, insulin-like hormone.

ANTI-MÜLLERIAN HORMONE Anti-Müllerian hormone, a member of the TGF-β family of growth and differentiation factors, is secreted from the fetal Sertoli cells and causes the degeneration of the Müllerian duct (Tran et al. 1977; Cate et al. 1986). AMH is thought to bind to the mesenchyme cells surrounding the Müllerian duct, causing these cells to secrete factors that induce apoptosis in the duct's epithelium and breaks down the basal lamina surrounding the duct (Trelstad et al. 1982; Roberts et al. 1999, 2002).

ESTROGEN The steroid hormone estrogen is needed for complete postnatal development of both the Müllerian and the Wolffian ducts, and is necessary for fertility in both males and females. In females, estrogen induces the differentiation of the Müllerian duct into the uterus, oviducts, cervix, and upper vagina. In female mice whose genes for estrogen receptors are knocked out, the germ cells die in the adult, and the granulosa cells that had enveloped them start developing into Sertoli-like cells (Couse et al. 1999). Male mice with knockouts of estrogen receptor genes produce few sperm. One of the functions of the male efferent duct cells (which bring the sperm from the seminiferous vesicles into the epididymis) is to absorb most of the water from the lumen of the rete testis. This absorption, which is regulated by estrogen, concentrates the sperm, giving them a longer life span and providing more sperm per ejaculate. If estrogen or its receptor is absent in male mice, water is not absorbed and the mouse is sterile (Hess et al. 1997). Although blood concentrations of estrogen are in general higher in females than in males, the concentration of estrogen in the rete testis is higher than in female blood.

In summary, primary sex determination in mammals is regulated by the chromosomes, which results in the production of testes in XY individuals and ovaries in XX individuals. This type of sex determination appears to be a "digital" (either/or) phenomenon. With chromosomal sex established, the gonads then produce the hormones that coordinate the different parts of the body to have a male or female phenotype. This secondary sex determination is more "analogue," where differing levels of hormones and responses to hormones can create different phenotypes. Secondary sex determination is thus usually, but not always, coordinated with the primary sex determination.

WEB TOPIC 6.5 **BRAIN SEX AND GENDER** In addition to the physical aspects of secondary sex determination, there are also behavioral attributes. The brain is an organ that differs between males and females; but does it generate a different pattern of human behaviors?

 SCIENTISTS SPEAK 6.4 Neuroscientist Dr. Daphna Joel discusses her research showing that male and female brains are remarkably similar.

Chromosomal Sex Determination in *Drosophila*

Although both mammals and fruit flies produce XX females and XY males, the ways in which their chromosomes achieve these ends are very different. In mammals, the Y chromosome plays a pivotal role in determining the male sex. In *Drosophila*, the Y chromosome is not involved in determining sex. Rather, in flies, the Y chromosome

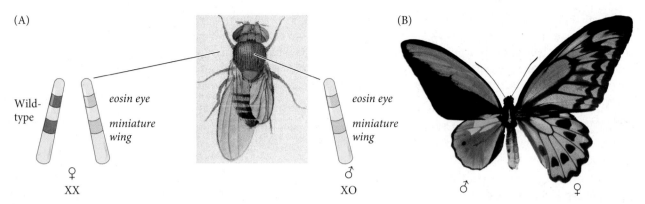

(A)

Wild-type

eosin eye

miniature wing

♀
XX

♂
XO

(B)

eosin eye

miniature wing

♂   ♀

FIGURE 6.13 Gynandromorph insects. (A) *D. melanogaster* in which the left side is female (XX) and the right side is male (XO). The male side has lost an X chromosome bearing the wild-type alleles of eye color and wing shape, thereby allowing expression of the recessive alleles *eosin eye* and *miniature wing* on the remaining X chromosome. (B) Birdwing butterfly *Ornithopera croesus*. The smaller male half is red, black, and yellow, while the female half is larger and brown. (A, drawing by Edith Wallace from Morgan and Bridges 1919; B, Montreal Insectarium, photograph by the author.)

seems to be a collection of genes that are active in forming sperm in adults, but not in sex determination.

A fruit fly's sex is determined predominantly by the number of X chromosomes in each cell. If there is only one X chromosome in a diploid cell, the fly is male. If there are two X chromosomes in a diploid cell, the fly is female. Should a fly have two X chromosomes and three sets of autosomes, it is a **mosaic**, where some of the cells are male and some of the cells are female. Thus, while XO mammals are sterile females (no Y chromosome, thus no *Sry* gene), XO *Drosophila* are sterile males (one X chromosome per diploid set).

In *Drosophila*, and in insects in general, one can observe gynandromorphs—animals in which certain regions of the body are male and other regions are female (**FIGURE 6.13**). Gynandromorph fruit flies result when an X chromosome is lost from one embryonic nucleus. The cells descended from that cell, instead of being XX (female), are XO (male). The XO cells display male characteristics, whereas the XX cells display female traits, suggesting that, in *Drosophila*, each cell makes its own sexual "decision." Indeed, in their classic discussion of gynandromorphs, Morgan and Bridges (1919) concluded, "Male and female parts and their sex-linked characters are strictly self-determining, each developing according to its own aspiration," and each sexual decision is "not interfered with by the aspirations of its neighbors, nor is it overruled by the action of the gonads." Although there are organs that are exceptions to this rule (notably the external genitalia), it remains a good general principle of *Drosophila* sexual development.

The Sex-lethal gene

Although it had long been thought that a fruit fly's sex was determined by the X-to-autosome (X:A) ratio (Bridges 1925), this assessment was based largely on flies with aberrant numbers of chromosomes. Recent molecular analyses suggest that X chromosome number alone is the primary sex determinant in normal diploid insects (Erickson and Quintero 2007). The X chromosome contains genes encoding transcription factors that activate the critical gene in *Drosophila* sex determination, the X-linked locus **Sex-lethal** (**Sxl**). The Sex-lethal protein is a splicing factor that initiates a cascade of RNA processing events that will eventually lead to male-specific and female-specific transcription factors (**FIGURE 6.14**). These transcription factors (the Doublesex proteins) then differentially activate the genes involved to produce either the male phenotype (testes, sex combs, pigmentation) or the female phenotype (ovaries, yolk proteins, pigmentation).

ACTIVATING *SEX-LETHAL*   The number of X chromosomes appears to have only a single function: activating (or not activating) the early expression of *Sex-lethal*.[6] *Sxl* encodes an RNA splicing factor that will regulate gonad development and will also

---

[6]This gene's gory name is derived from the fact that mutations of this gene can result in aberrant dosage compensation of X-linked genes (see Web Topic 6.6). As a result, there is inadequate transcription of those genes encoded on the X chromosome, and the embryo dies.

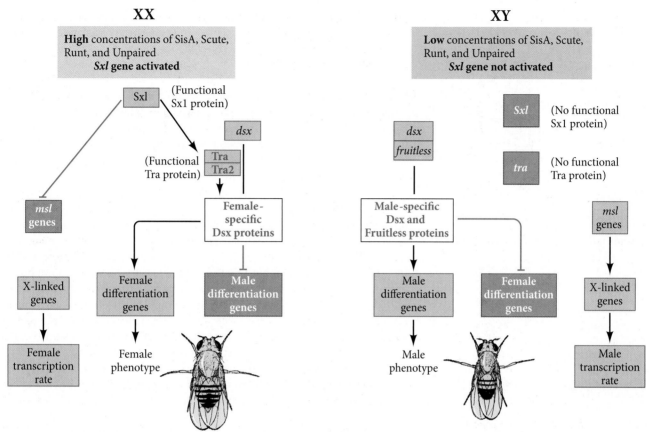

FIGURE 6.14 Proposed regulatory cascade for *Drosophila* somatic sex determination. Transcription factors from the X chromosomes activate the *Sxl* gene in females (XX) but not in males (XY). The Sex-lethal protein performs three main functions. First, it activates its own transcription, ensuring further Sxl production. Second, it represses the translation of *msl2* mRNA, a factor that facilitates transcription from the X chromosome. This equalizes the amount of transcription from the two X chromosomes in females with that of the single X chromosome in males. Third, Sxl enables the splicing of the *transformer-1* (*tra1*) pre-mRNA into functional proteins. The Tra proteins process *doublesex* (*dsx*) pre-mRNA in a female-specific manner that provides most of the female body with its sexual fate. They also process the *fruitless* pre-mRNA in a female-specific manner, giving the fly female-specific behavior. In the absence of Sxl (and thus the Tra proteins), *dsx* and *fruitless* pre-mRNAs are processed in the male-specific manner. (The *fruitless* gene is discussed in Web Topic 6.7.) (After Baker et al. 1987.)

regulate the amount of gene expression from the X chromosome. The gene has two promoters. The early promoter is active only in XX cells; the later promoter is active in both XX and XY cells. The X chromosome appears to encode four protein factors that activate the early promoter of *Sxl*. Three of these proteins are transcription factors— SisA, Scute, and Runt—that bind to the early promoter to activate transcription. The fourth protein, Unpaired, is a secreted factor that reinforces the other three proteins through the JAK-STAT pathway (Sefton et al. 2000; Avila and Erickson 2007). If these factors accumulate so they are present in amounts above a certain threshold, the *Sxl* gene is activated through its early promoter (Erickson and Quintero 2007; Gonzáles et al. 2008; Mulvey et al. 2014). The result is the transcription of *Sxl* early in XX embryos, during the syncytial blastoderm stage.

The *Sxl* pre-RNA transcribed from the *early* promoter of XX embryos lacks exon 3, which contains a stop codon. Thus, Sxl protein that is made early is spliced in a manner such that exon 3 is absent, so early XX embryos have complete and functional Sxl protein (**FIGURE 6.15**). In XY embryos, the early promoter of *Sxl* is not active and no functional Sxl protein is present. However, later in development, as cellularization is taking place, the *late* promoter becomes active and the *Sxl* gene is transcribed in both males and females. In XX cells, Sxl protein from the early promoter can bind to its own pre-mRNA and splice it in a "female" direction. In this case, Sxl binds to and blocks the splicing complex on exon 3 (Johnson et al. 2010; Salz 2011). As a result, exon 3 is skipped and is not included in the *Sxl* mRNA. Thus, early production ensures that functional full-length (354-amino acid) Sxl protein is made if the cells are XX (Bell et al. 1991; Keyes et al. 1992). In XY cells, however, the early promoter is not active (because the X-encoded transcription factors haven't reached the threshold to activate the promoter) and there is no early Sxl protein. Therefore, the *Sxl* pre-mRNA of XY cells is spliced in a manner that *includes* exon 3 and its termination codon. Protein synthesis ends at the third exon (after amino acid 48), and the Sxl is nonfunctional.

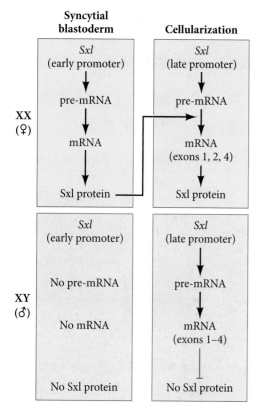

FIGURE 6.15 Differential RNA splicing and sex-specific expression of *Sex-lethal*. In the syncytial blastoderm of XX flies, transcription factors from the two X chromosomes are sufficient to activate the early promoter of the *Sxl* gene. This "early" transcript is spliced into an mRNA lacking exon 3 and makes a functional Sxl protein. The early promoter of XY flies is not activated, and males lack functional Sxl. By the cellularizing blastoderm stage, the late promoter of *Sxl* is active in both XX and XY flies. In XX flies, Sxl already present in the embryo prevents the splicing of exon 3 into mRNA and functional Sxl protein is made. Sxl then binds to its own promoter to keep it active; it also functions to splice downstream pre-mRNAs. In XY embryos, no Sxl is present and exon 3 is spliced into the mRNA. Because of the termination codon in exon 3, males do not make functional Sxl. (After Salz 2011.)

WEB TOPIC 6.6 **DOSAGE COMPENSATION** If the cells of female flies, nematodes, and mammals have twice the number of X chromosomes as male cells, how are the genes on the X chromosome regulated? The three groups offer three different solutions to the problem.

TARGETS OF SEX-LETHAL The protein made by the female-specific *Sxl* transcript contains regions that are important for binding to RNA. There appear to be three major RNA targets to which the female-specific *Sxl* transcript binds. One of these is the pre-mRNA of *Sxl* itself. Another target is the *msl2* gene that controls dosage compensation (see below). Indeed, if the *Sxl* gene is nonfunctional in a cell with two X chromosomes, the dosage compensation system will not work, and the result is cell death (hence the gene's name). The third target is the pre-mRNA of *transformer* (*tra*)—the next gene in the cascade (**FIGURE 6.16**; Nagoshi et al. 1988; Bell et al. 1991).

The pre-mRNA of *transformer* (so named because loss-of-function mutations turn females into males) is spliced into a functional mRNA by Sxl protein. The *tra* pre-mRNA is made in both male and female cells; however, in the presence of Sxl, the *tra* transcript is alternatively spliced to create a female-specific mRNA, as well as a nonspecific mRNA that is found in both females and males. Like the male *Sxl* message, the nonspecific *tra* mRNA message contains an early termination codon that renders the protein nonfunctional (Boggs et al. 1987). In *tra*, the second exon of the nonspecific mRNA contains the termination codon and is not utilized in the female-specific message (see Figures 6.14 and 6.16).

How is it that females and males make different mRNAs? The female-specific Sxl protein activates a 3′ splice site that causes *tra* pre-mRNA to be processed in a way that splices out the second exon. To do this, Sxl protein blocks the binding of splicing factor U2AF to the nonspecific splice site of the *tra* message by specifically binding to the polypyrimidine tract adjacent to it (Handa et al. 1999). This causes U2AF to bind to the lower-affinity (female-specific) 3′ splice site and generate a female-specific mRNA (Valcárcel et al. 1993). The female-specific Tra protein works in concert with the product of the *transformer-2* (*tra2*) gene to help generate the female phenotype by splicing the *doublesex* gene in a female-specific manner.

Doublesex: The switch gene for sex determination

The *Drosophila* **doublesex** (**dsx**) gene is active in both males and females, but its primary transcript is processed in a sex-specific manner (Baker et al. 1987). This alternative RNA processing is the result of the action of the *tra* and *tra2* gene products on the *dsx* gene (see Figures 6.14 and 6.16). If the Tra2 and female-specific Tra proteins are both present, the *dsx* transcript is processed in a female-specific manner (Ryner and Baker 1991). The female splicing pattern produces a female-specific protein that activates female-specific genes (such as those of the yolk proteins) and inhibits male development. If no functional Tra is produced, a male-specific *dsx* transcript is made; this transcript encodes a transcription factor that inhibits female traits and promotes male traits. In the embryonic gonad, Dsx regulates all known aspects of sexually dimorphic gonad cell fate.

In XX flies, the female Doublesex protein (Dsx^F) combines with the product of the *intersex* gene (*Ix*) to make a transcription factor complex that is responsible for promoting female-specific traits. This "Doublesex complex" activates the *Wingless* (*Wg*) gene, whose Wnt-family product promotes growth of the female portions of the genital disc. It also represses the *Fgf* genes responsible for making male accessory organs, activates the genes responsible for making yolk proteins, promotes the growth of the sperm storage duct, and modifies *bricabrac* (*bab*) gene expression to give the female-specific pigmentation profile. In contrast, the male Doublesex protein (Dsx^M) acts directly as a transcription

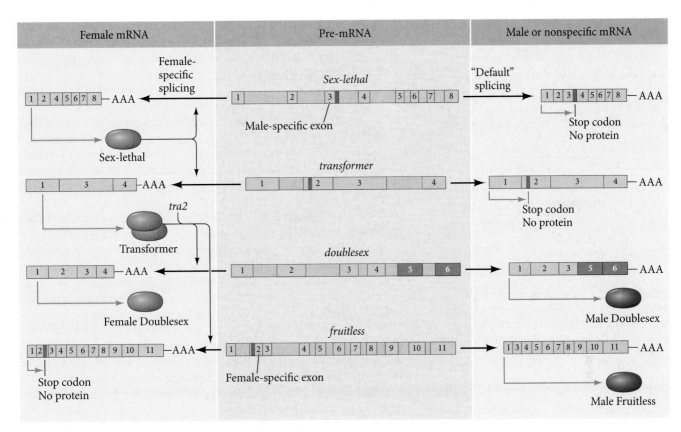

Female mRNA	Pre-mRNA	Male or nonspecific mRNA

FIGURE 6.16 Sex-specific RNA splicing in four major *Drosophila* sex-determining genes. The pre-mRNAs (shown in the center of diagram) are identical in both male and female nuclei. In each case, the female-specific transcript is shown at the left, while the default transcript (whether male or nonspecific) is shown to the right. Exons are numbered, and the positions of termination codons are marked. *Sex-lethal*, *transformer*, and *doublesex* are all part of the genetic cascade of primary sex determination. The transcription pattern of *fruitless* determines the secondary characteristic of courtship behavior. (After Baker 1989; Baker et al. 2001.)

factor and directs the expression of male-specific traits. It causes the male region of the genital disc to grow at the expense of the female disc regions. It activates the BMP homologue *Decapentaplegic* (*Dpp*), as well as stimulating *Fgf* genes to produce the male genital disc and accessory structures. Dsx^M also converts certain cuticular structures into claspers and modifies the *bricabrac* gene to produce the male pigmentation pattern (Ahmad and Baker 2002; Christiansen et al. 2002).

According to this model, the result of the sex determination cascade summarized in Figure 6.14 comes down to the type of mRNA processed from the *doublesex* transcript. If there are two X chromosomes, the transcription factors activating the early promoter of *Sxl* reach a critical concentration, and *Sxl* makes a splicing factor that causes the *transformer* gene transcript to be spliced in a female-specific manner. This female-specific protein interacts with the *tra2* splicing factor, causing *dsx* pre-mRNA to be spliced in a female-specific manner. If the *dsx* transcript is not acted on in this way, it is processed in a "default" manner to make the male-specific message. Interestingly, the *doublesex* gene of flies is very similar to the *Dmrt1* gene of vertebrates, and the two types of sex determination may have some common denominators.

WEB TOPIC 6.7 **BRAIN SEX IN *DROSOPHILA*** In addition to the "doublesex" mechanism for creating sexual phenotypes in *Drosophila*, a separate "brain sex" pathway characterized by the *fruitless* gene provides individuals with the appropriate set of courtship and aggression behaviors.

Environmental Sex Determination

In many organisms, sex is determined by environmental factors such as temperature, location and the presence of other members of the species. Chapter 25 will discuss the importance of environmental factors on normal development; here we will just discuss one of these systems, temperature-dependent sex determination in turtles.

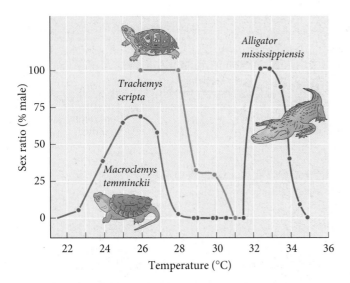

FIGURE 6.17 Temperature-dependent sex determination in three species of reptiles: the American alligator (*Alligator mississippiensis*), red-eared slider turtle (*Trachemys scripta elegans*), and alligator snapping turtle (*Macroclemys temminckii*). (After Crain and Guillette 1998.)

While the sex of most snakes and lizards is determined by sex chromosomes at the time of fertilization, the sex of most turtles and all species of crocodylians is determined *after* fertilization, by the embryonic environment. In these reptiles, the temperature of the eggs during a certain period of development is the deciding factor in determining sex, and small changes in temperature can cause dramatic changes in the sex ratio (Bull 1980; Crews 2003). Often, eggs incubated at low temperatures produce one sex, whereas eggs incubated at higher temperatures produce the other. There is only a small range of temperatures that permits both males and females to hatch from the same brood of eggs.[7]

FIGURE 6.17 shows the abrupt temperature-induced change in sex ratios for the red-eared slider turtle. If a brood of eggs is incubated at a temperature below 28°C, all the turtles hatching from the eggs will be male. Above 31°C, every egg gives rise to a female. At temperatures in between, the brood gives rise to individuals of both sexes. Variations on this theme also exist. The eggs of the snapping turtle *Macroclemys*, for instance, become female at either cool (22°C or lower) or hot (28°C or above) temperatures. Between these extremes, males predominate.

One of the best-studied reptiles is the European pond turtle, *Emys orbicularis*. In laboratory studies, incubating *Emys* eggs at temperatures above 30°C produces all females, whereas temperatures below 25°C produce all-male broods. The threshold temperature (at which the sex ratio is even) is 28.5°C (Pieau et al. 1994). The developmental "window" during which sex determination occurs can be discovered by incubating eggs at the male-producing temperature for a certain amount of time and then shifting them to an incubator at the female-producing temperature (and vice versa). In *Emys*, the middle third of development appears to be the most critical for sex determination, and it is believed that the turtles cannot reverse their sex after this period.

The expression of sex-determining genes (*Sox9* and *Sry* in males; *β-catenin* in females) are seen to correlate with male- or female-producing temperatures (see Mork and Chapel 2013; Bieser and Wibbels 2014). However, it is not known whether these genes are the temperature-sensitive components of sex determination. Recently, genetic studies on the sensitivity of temperature-induced sex determination have pointed to CIRBP (cold-induced RNA-binding protein) as the agent responding to temperature differences (Schroeder et al. 2016). The gene for CIRBP is expressed at the time of sex determination in snapping turtles, and different alleles give different sex ratio biases. This protein may act by repressing the splicing or translation of certain messages at certain temperatures. Another temperature-sensitive protein that may regulate sex determination is TRPV4, a Ca^{2+} channel whose activity correlates with activating testes-forming genes (Yatsui et al. 2015). The mechanisms of environmentally induced sex determination have yet to be elucidated.

Mammalian Gametogenesis

One of the most important events in sex determination is the determination of the germ cells to undergo **gametogenesis**, the formation of gametes (sperm and egg). As in the case of the genital ridges, the mammalian **primordial germ cells** (**PGCs**) are bipotential and can become either sperm or eggs; if they reside in the ovaries they become eggs, and if they reside in the testes they become sperm. All of these decisions are coordinated by factors produced by the developing gonads.

[7]The evolutionary advantages and disadvantages of temperature-dependent sex determination are discussed in Chapter 26.

First and importantly, the cells that generate the sperm or eggs do not originally form inside the gonads. Rather, they form in the posterior portion of the embryo and migrate into the gonads (Anderson et al. 2000; Molyneaux et al. 2001; Tanaka et al. 2005). This pattern is common throughout the animal kingdom: the germ cells are "set aside" from the rest of the embryo and the cells' transcription and translation are shut down while they migrate from peripheral sites into the embryo and to the gonad. It is as if the germ cells were a separate entity, reserved for the next generation, and repressing gene expression makes them insensitive to the intercellular commerce going on all around them (Richardson and Lehmann 2010; Tarbashevich and Raz 2010).

Although the mechanisms used to specify the germ cells vary enormously across the animal kingdom, the proteins expressed by germ cells to suppress gene expression are remarkably conserved. These proteins, which include the Vasa, Nanos, Tudor, and Piwi family proteins, can be seen in the germ cells of cnidarians, flies, and mammals (Ewen-Campen et al. 2010; Leclére et al. 2012). **Vasa** proteins are required for germ cells in nearly all animals studied. They are involved in binding RNA and most likely activate germ-cell-specific messages. In chickens, experimentally induced Vasa can direct embryonic stem cells toward a germ cell fate (Lavial et al. 2009). **Nanos** proteins bind to their partner, Pumilio, to form a very potent repressive dimer. Nanos can block RNA translation, and Pumilio binds to the 3' UTRs of specific mRNAs. In *Drosophila*, Nanos and Pumilio repress the translation of numerous mRNAs, and in so doing they (1) prevent the cell from becoming part of any germ layer; (2) prevent the cell cycle from continuing; and (3) prevent apoptosis (Kobayashi et al. 1996; Asaoka-Taguchi et al. 1999; Hayashi et al. 2004). **Tudor** proteins were discovered in *Drosophila*, in which females carrying these genes are sterile[8] and do not form pole cells (Boswell and Mahowald 1985). It appears that Tudor proteins interact with those **Piwi** proteins that are involved in transcriptionally silencing portions of the genome, especially active transposons.

WEB TOPIC 6.8 **THEODOR BOVERI AND THE FORMATION OF THE GERM LINE** In the early 1900s, Boveri's studies on the development of roundworms demonstrated that the cytoplasm of the cell destined to be the germ cell precursor was different from the cytoplasm of other cells.

The newly formed PGCs first enter into the hindgut (**FIGURE 6.18A**) and eventually migrate forward and into the bipotential gonads, multiplying as they migrate. From the time of their specification until they enter the genital ridges, the PGCs are surrounded by cells secreting stem cell factor (SCF). SCF is necessary for PGC motility and survival. Moreover, the cluster of SCF-secreting cells appears to migrate with the PGCs, forming a "traveling niche" of cells that support the persistence, the division, and movement of the PGCs (Gu et al. 2009).

The PGCs that migrate to the gonads do not make their own decision to become either sperm or eggs. That decision is made by the gonad in which they reside; it is signals from the gonad that create the profound differences between spermatogenesis and oogenesis (**TABLE 6.1**). One of the most fundamental differences involves the *timing* of meiosis. In females, meiosis begins in the *embryonic* gonads. In males, meiosis is not initiated until puberty. The "gatekeeper" for meiosis appears to be the Stra8 transcription factor, which promotes a new round of DNA synthesis and meiotic initiation in the germ cells. In the developing ovaries, Stra8 is *upregulated* by two factors—Wnt4 and retinoic acid—coming from the adjacent kidney (Baltus et al. 2006; Bowles et al. 2006; Naillat et al. 2010; Chassot et al. 2011). In the developing testes, however, Stra8 is *downregulated* by Fgf9, and the retinoic acid produced by the mesonephros is degraded by the testes' secretion of the RA-degrading enzyme Cyp26b1 (**FIGURE 6.19**; Bowles et al. 2006; Koubova et al. 2006). During male puberty, however, retinoic acid is synthesized

---

[8] Tudor and Vasa are both named after European royal houses that came to an end with female monarchs (Elizabeth of England and Christina of Sweden) who had no heirs.

FIGURE 6.18 Primordial germ cell migration in the mouse. (A) On embryonic day 8, PGCs established in the posterior epiblast migrate into the definitive endoderm of the embryo. The photo shows four large PGCs (stained for alkaline phosphatase) in the hindgut of a mouse embryo. (B) The PGCs migrate through the gut and, dorsally, into the genital ridges. (C) Alkaline phosphatase-staining cells are seen entering the genital ridges around embryonic day 11. (A from Heath 1978; C from Mintz 1957, courtesy of the authors.)

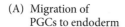
(A) Migration of PGCs to endoderm

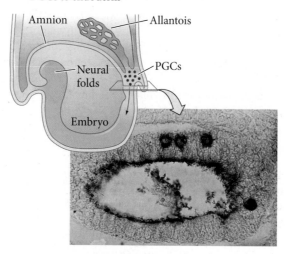

(B) Migration of PGCs into gonad

(C)

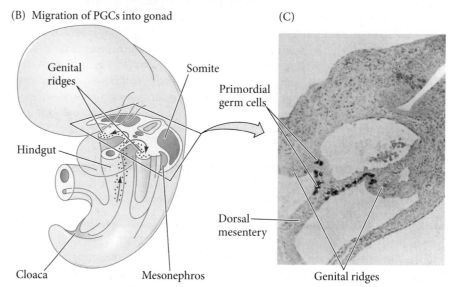

TABLE 6.1 Sexual dimorphism in mammalian meioses	
Female oogenesis	**Male spermatogenesis**
Meiosis initiated once in a finite population of cells	Meiosis initiated continuously in a mitotically dividing stem cell population
One gamete produced per meiosis	Four gametes produced per meiosis
Completion of meiosis delayed for months or years	Meiosis completed in days or weeks
Meiosis arrested at first meiotic prophase and reinitiated in a smaller population of cells	Meiosis and differentiation proceed continuously without cell cycle arrest
Differentiation of gamete occurs while diploid, in first meiotic prophase	Differentiation of gamete occurs while haploid, after meiosis ends
All chromosomes exhibit equivalent transcription and recombination during meiotic prophase	Sex chromosomes excluded from recombination and transcription during first meiotic prophase

Source: After Handel and Eppig 1998.

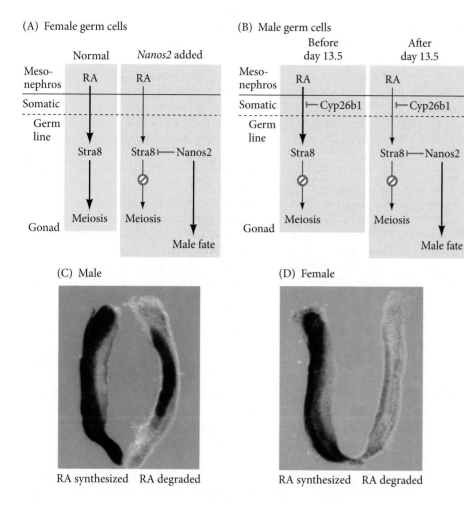

(A) Female germ cells

Normal   *Nanos2* added

Meso-
nephros

Somatic

Germ
line

Gonad

(B) Male germ cells

Before
day 13.5

After
day 13.5

Meso-
nephros

Somatic

Germ
line

Gonad

(C) Male

RA synthesized   RA degraded

(D) Female

RA synthesized   RA degraded

FIGURE 6.19 Retinoic acid (RA) determines the timing of meiosis and sexual differentiation of mammalian germ cells. (A) In female mouse embryos, RA secreted from the mesonephros reaches the gonad and triggers meiotic initiation via the induction of Stra8 transcription factor in female germ cells (beige). However, if activated *Nanos2* genes are added to female germ cells, they suppress Stra8 expression, leading the germ cells into a male pathway (gray). (B) In embryonic testes, Cyp26b1 blocks RA signaling, thereby preventing male germ cells from initiating meiosis until embryonic day 13.5 (left panel). After embryonic day 13.5, when Cyp26b1 expression is decreased, Nanos2 is expressed and prevents meiotic initiation by blocking Stra8 expression. This induces male-type differentiation in the germ cells (right panel). (C,D) Day 12 mouse embryos stained for mRNAs encoding the RA-synthesizing enzyme Aldh1a2 (left gonad) and the RA-degrading enzyme Cyp26b1 (right gonad). The RA-synthesizing enzyme is seen in the mesonephros of both the male (C) and female (D); the RA-degrading enzyme is seen only in the male gonad. (A,B from Saga 2008; C,D from Bowles et al. 2006, courtesy of P. Koopman.)

in the Sertoli cells and induces Stra8 in sperm stem cells. Once Stra8 is present, the sperm stem cells become committed to meiosis (Anderson et al. 2008; Mark et al. 2008). Thus, the timing of retinoic acid synthesis appears to control Stra8, and Stra8 commits germ cells to meiosis. Fgf9, which downregulates Stra8, also appears to be critical in keeping the male germ cells in a stem cell-like condition (Bowles et al. 2010).

The structure of the mammalian gonad plays a critical role as well. The Sertoli cells, Leydig cells, and blood vessels of the seminiferous tubules constitute a stem cell niche (Hara et al. 2014; Manku and Culty 2015.) The primordial germ cells that enter the developing testis will remain in a stem cell-like condition that enables them to mitotically produce sperm precursors. The follicle cells of the ovary, however, do not constitute a stem cell niche. Rather, each primordial germ cell will be surrounded by the follicle cells, and usually only one egg will mature from each follicle.

Meiosis: The intertwining of life cycles

Meiosis is perhaps the most revolutionary invention of eukaryotes, for it is the mechanism for transmission of genes from one generation to the next and for the recombination of sperm- and egg-derived genes into new combinations of alleles. Van Beneden's 1883 observations that the divisions of germ cells caused the resulting gametes to contain half the diploid number of chromosomes "demonstrated that the chromosomes of the offspring are derived in equal numbers from the nuclei of the two conjugating germ-cells and hence equally from the two parents" (Wilson 1924). Meiosis is a critical starting and ending point in the cycle of life. The body senesces and dies, but the gametes formed by meiosis survive the death of their parents and form the next generation.

Sexual reproduction, evolutionary variation, and the transmission of traits from one generation to the next all come down to meiosis. So to understand what germ cells do, we must first understand meiosis.

Meiosis is the means by which the gametes halve the number of their chromosomes. In the haploid condition, each chromosome is represented by only one copy, whereas diploid cells have two copies of each chromosome. Meiotic division differ from mitotic division in that (1) meiotic cells undergo two cell divisions without an intervening period of DNA replication, and (2) homologous chromosomes pair together and recombine genetic material.

After the germ cell's final mitotic division, a period of DNA synthesis occurs, so that the cell initiating meiosis doubles the amount of DNA in its nucleus. In this state, each chromosome consists of two sister **chromatids** attached at a common kinetochore.[9] (In other words, the diploid nucleus contains four copies of each chromosome.) In the first of the two meiotic divisions (meiosis I), homologous chromosomes (for example, the two copies of chromosome 3 in the diploid cell) come together and are then separated into different cells. Hence the first meiotic division *splits two homologous chromosomes between two daughter cells* such that each daughter cell has only one copy of each chromosome. But each of the chromosomes has already replicated (i.e., each has two chromatids), so the second division (meiosis II) *separates the two sister chromatids from each other*. The net result of meiosis is four cells, each of which has a single (haploid) copy of each chromosome.

The first meiotic division begins with a long prophase, which is subdivided into four stages (**FIGURE 6.20**). During the **leptotene** (Greek, "thin thread") stage, the chromatin of the chromatids is stretched out very thinly, and it is not possible to identify individual chromosomes. DNA replication has already occurred, however, and each chromosome consists of two parallel chromatids. At the **zygotene** (Greek, "yoked threads") stage, homologous chromosomes pair side by side. This pairing, called **synapsis**, is characteristic of meiosis; such pairing does not occur during mitotic divisions. Although the mechanism whereby each chromosome recognizes its homologue is not known (see Barzel and Kupiec 2008; Takeo et al. 2011), synapsis seems to require the presence of

---

[9] The terms *centromere* and *kinetochore* are often used interchangeably, but in fact the kinetochore is the complex protein structure that assembles on a sequence of DNA known as the centromere.

FIGURE 6.20 Meiosis, emphasizing the synaptonemal complex. Before meiosis, unpaired homologous chromosomes are distributed randomly within the nucleus. (A) At leptotene, telomeres have attached along the nuclear envelope. The chromosomes "search" for homologous chromosomes, and synapsis, the association of homologous chromosomes, begins at zygotene, where the first evidence of the synaptonemal complex (SC) can be seen. During pachytene, homologue alignment is seen along the entire length of the chromosomes and produces a bivalent structure. Paired homologs can recombine with each (cross over) other during zygotene and pachytene. The synaptonemal complex dissolves at diplotene, when recombination is completed. (B) In diakinesis, chromosomes condense further and then form a metaphase plate. Segregation of the homologous chromosomes occurs at anaphase I. Only one pair of sister chromatids is shown here in meiosis II, where sister chromatids align at metaphase II and then in anaphase II segregate to opposite poles. (After Tsai and McKee 2011.)

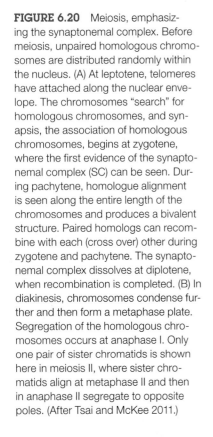
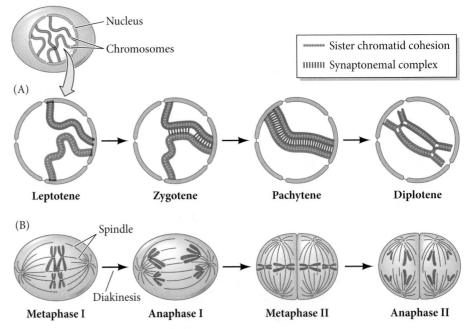

the nuclear envelope and the formation of a proteinaceous ribbon called the **synaptone-mal complex**. In many species, the nuclear envelope probably serves as an attachment site for the prophase chromosomes to bind and thereby reduces the complexity of the search for the other homologous chromosome (Comings 1968; Scherthan 2007; Tsai and McKee 2011). The synaptonemal complex is a ladderlike structure with a central element and two lateral bars (von Wettstein 1984; Yang and Wang 2009). The homologous chromosomes become associated with the two lateral bars, and the chromosomes are thus joined together. The configuration formed by the four chromatids and the synaptonemal complex is referred to as a **tetrad** or a **bivalent**.

During the next stage of meiotic prophase, **pachytene** (Greek, "thick thread"), the chromatids thicken and shorten. Individual chromatids can now be distinguished under the light microscope, and crossing-over may occur. **Crossing-over** represents an exchange of genetic material whereby genes from one chromatid are exchanged with homologous genes from another. Crossing-over may continue into the next stage, **diplotene** (Greek, "double threads"). During diplotene, the synaptonemal complex breaks down and the two homologous chromosomes start to separate. Usually, however, they remain attached at various points called **chiasmata**, which are thought to represent regions where crossing-over is occurring. The diplotene stage is characterized by a high level of gene transcription.

Meiotic metaphase begins with **diakinesis** (Greek, "moving apart") of the chromosomes (Figure 6.20B). The nuclear envelope breaks down and the chromosomes migrate to form a metaphase plate. Anaphase of meiosis I does not commence until the chromosomes are properly aligned on the mitotic spindle fibers. This alignment is accomplished by proteins that prevent cyclin B from being degraded until after all the chromosomes are securely fastened to microtubules.

During anaphase I, the homologous chromosomes separate from each other in an independent fashion. This stage leads to telophase I, during which two daughter cells are formed, each cell containing one partner of each homologous chromosome pair. After a brief **interkinesis**, the second meiotic division takes place. During meiosis II, the kinetochore of each chromosome divides during anaphase so that each of the new cells gets one of the two chromatids, the final result being the creation of four haploid cells. Note that meiosis has also reassorted the chromosomes into new groupings. First, each of the four haploid cells has a different assortment of chromosomes. Humans have 23 different chromosome pairs; thus 2^{23} (nearly 10 million) different haploid cells can be formed from the genome of a single person. In addition, the crossing-over that occurs during the pachytene and diplotene stages of first meiotic metaphase further increases genetic diversity and makes the number of potential different gametes incalculably large.

This organization and movement of meiotic chromosomes is choreographed by a ring of **cohesin proteins** that encircles the sister chromatids. Cohesin rings resist the pulling forces of the spindle microtubules, thereby keeping the sister chromatids attached during meiosis I (Haering et al. 2008; Brar et al. 2009). The cohesins also recruit other sets of proteins that help promote pairing between homologous chromosomes and allow recombination to occur (Pelttari et al. 2001; Villeneuve and Hillers 2001; Sakuno and Watanabe 2009). At the second meotic division, the cohesin ring is cleaved and the kinetochores can separate from each other (Schöckel et al. 2011).

WEB TOPIC 6.9 **MODIFICATIONS OF MEIOSIS** In many organisms, females can reproduce themselves without males by modifying meiosis. They can produce diploid eggs and activate them by some means other than sperm entry.

Gametogenesis in mammals: Spermatogenesis

Spermatogenesis—the developmental pathway from germ cell to mature sperm—begins at puberty and occurs in the recesses between the Sertoli cells (**FIGURE 6.21**). Spermatogenesis is divided into three major phases (Matson et al. 2010):

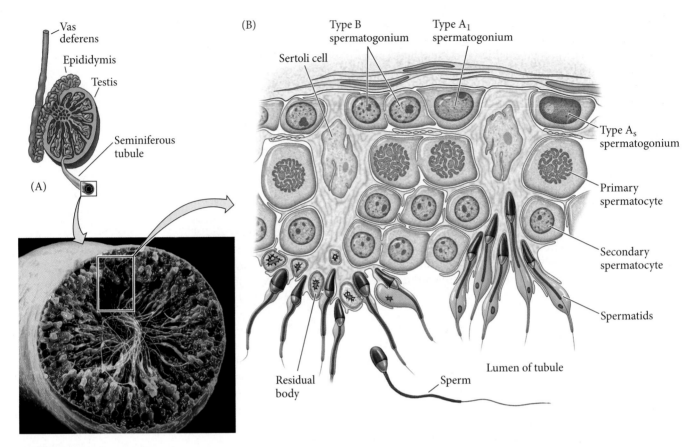

FIGURE 6.21 Sperm maturation. (A) Cross section of the seminiferous tubule. Spermatogonia are blue, spermatocytes are lavender, and the mature sperm appear yellow. (B) Simplified diagram of a portion of the seminiferous tubule, illustrating relationships between spermatogonia, spermatocytes, and sperm. As these germ cells mature, they progress toward the lumen of the seminiferous tubule. (See also Figure 7.1.) (A photograph courtesy of R. Wagner; B based on Dym 1977.)

1. A proliferative phase where sperm stem cells (**spermatogonia**) increase by mitosis.
2. A meiotic phase, involving the two divisions that create the haploid state.
3. A postmeiotic "shaping" phase called **spermiogenesis**, during which the round cells (spermatids) eject most of their cytoplasm and become the streamlined sperm.

The proliferative phase begins when the mammalian PGCs arrive at the genital ridge of a male embryo. Here they are called **gonocytes** and become incorporated into the sex cords that will become the seminiferous tubules (Culty 2009). The gonocytes become undifferentiated spermatogonia residing near the basal end of the tubular cells (Yoshida et al. 2007, 2016). These are true stem cells in that they can reestablish spermatogenesis when transferred into mice whose sperm production was eliminated by toxic chemicals. Spermatogonia appear to take up residence in stem cell niches at the junction of the Sertoli cells (the epithelium of the seminiferous tubules), the interstitial (testosterone-producing) Leydig cells, and the testicular blood vessels. Adhesion molecules join the spermatogonia directly to the Sertoli cells, which will nourish the developing sperm (Newton et al. 1993; Pratt et al. 1993; Kanatsu-Shinohara et al. 2008). The mitotic proliferation of these stem cells amplifies this small population into a population of differentiating spermatogonia (**type A spermatogonia**) that can generate more than 1000 sperm per second in adult human males (Matson et al. 2010).

As the spermatogonia divide, they remain attached to each other by cytoplasmic bridges. But these bridges are fragile, and when one cell splits from the others, it can become an undifferentiated spematogonia again (Hara et al. 2014.) The meiotic phase of spermatogenesis during puberty is regulated by several factors. Glial cell line-derived neurotrophic factor, (GDNF, a paracrine factor) is made by the Sertoli cells and by the myoid cells that surround the tubules and give them strength and elasticity. GDNF helps keep the spermatogonia dividing as stem cells (Chen et al. 2016a). As mentioned earlier, at puberty retinoic acid levels activate the Stra8 transcription factor, and levels of the BMP8b paracrine factor reach a critical concentration. BMP8b is thought to instruct the spermatogonia to produce receptors that enable them to respond to proteins such as stem cell factor (SCF). Indeed, mice lacking BMP8b do not initiate spermatogenesis at puberty (Zhao et al. 1996; Carlomagno et al. 2010). The transition between mitotically dividing spermatogonia and the spermatocytes that initiate meiosis appears to be mediated by the opposing influences of GDNF and SCF, both of which are secreted by the Sertoli cells. SCF promotes the transition to spermatogenesis, while GDNF promotes the division of spermatogonial stem cells (Rossi and Dolci 2013).

THE MEIOTIC PHASE: HAPLOID SPERMATIDS   Spermatogonia with high levels of Stra8 and responding to SCF divide to become **type B spermatogonia**. (**FIGURE 6.22**; de Rooij and Russell 2000; Nakagawa 2010; Griswold et al. 2012). Type B spermatogonia are the precursors of the spermatocytes and are the last cells of the line that undergo mitosis. They divide once to generate the **primary spermatocytes**—the cells that enter meiosis. Each primary spermatocyte undergoes the first meiotic division to yield a pair of **secondary spermatocytes**, which complete the second division of meiosis. The

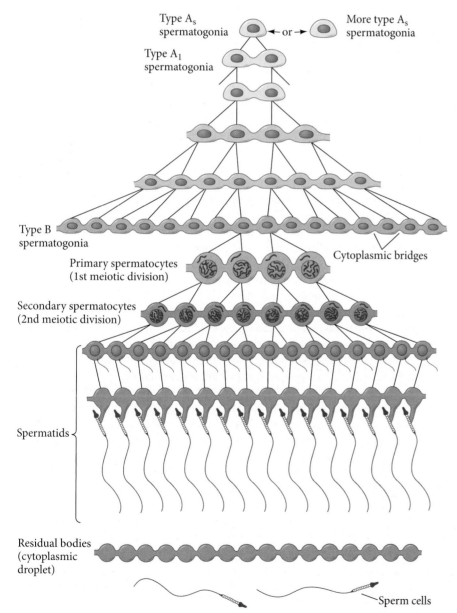

FIGURE 6.22   Formation of syncytial clones of human male germ cells. (After Bloom and Fawcett 1975.)

haploid cells thus formed are called **spermatids**, and they are still connected to one another through their cytoplasmic bridges. The spermatids that are connected in this manner have haploid nuclei but are functionally diploid, since a gene product made in one cell can readily diffuse into the cytoplasm of its neighbors (Braun et al. 1989).

During the divisions from undifferentiated spermatogonia to spermatids, the cells move farther and farther away from the basal lamina of the seminiferous tubule and closer to its lumen (see Figure 6.21; Siu and Cheng 2004). As the spermatids move toward border of the lumen, they lose their cytoplasmic connections and differentiate into spermatozoa. In humans, the progression from spermatogonial stem cell to mature spermatozoa takes 65 days (Dym 1994).

SPERMIOGENESIS: THE DIFFERENTIATION OF THE SPERM   The mammalian haploid spermatid is a round, unflagellated cell that looks nothing like the mature vertebrate sperm. The next step in sperm maturation, then, is **spermiogenesis** (sometimes called spermateliosis), the differentiation of the sperm cell. For fertilization to occur, the sperm has to meet and bind with an egg, and spermiogenesis prepares the sperm for these functions of motility and interaction. The process of mammalian sperm differentiation is shown in Figure 7.1. The first step is the construction of the acrosomal vesicle from the Golgi apparatus, a process about which we know very little (see Berruti and Paiardi 2011). The acrosome forms a cap that covers the sperm nucleus. As the acrosomal cap is formed, the nucleus rotates so that the cap faces the basal lamina of the seminiferous tubule. This rotation is necessary because the flagellum, which is beginning to form from the centriole on the other side of the nucleus, will extend into the lumen of the seminiferous tubule. During the last stage of spermiogenesis, the nucleus flattens and condenses, the remaining cytoplasm (the residual body, or cytoplasmic droplet; see Figure 6.22) is jettisoned, and the mitochondria form a ring around the base of the flagellum.

During spermiogenesis, the histones of the spermatogonia are often replaced by sperm-specific histone variants, and widespread nucleosome dissociation takes place. This remodeling of nucleosomes might also be the point at which the PGC pattern of methylation is removed and the male genome-specific pattern of methylation is established on the sperm DNA (see Wilkins 2005). As spermiogenesis ends, the histones of the haploid nucleus are eventually replaced by protamines.[10] This replacement results in the complete shutdown of transcription in the nucleus and facilitates the nucleus assuming an almost crystalline structure (Govin et al. 2004). The resulting sperm then enter the lumen of the seminiferous tubule.

Unexpectedly, the sperm continue to develop after they leave the testes. When being transported from the testes, sperm reside in the epididymis. During this residence, the epididymal cells release exosomes that fuse with the sperm. These exosomes have been shown to contain small ncRNAs and other factors that can activate and repress certain genes, and the sperm will bring these agents into the egg (Sharma et al. 2016; Chen et al. 2016). And the sperm still isn't fully mature, even when it exits the urethra. The final differentiation of the sperm, as we will see in Chapter 7, occurs in the reproductive tract of the female. Here, secretions from the oviducts will change the sperm cell membrane so that it can fuse with the membrane of the egg cell. Thus, the full differentiation of the sperm take place in two different organisms.

In the mouse, development from stem cell to spermatozoon takes 34.5 days: the spermatogonial stages last 8 days, meiosis lasts 13 days, and spermiogenesis takes another 13.5 days. Human sperm development takes nearly twice as long. Each day,

---

[10] *Protamines* are relatively small proteins that are over 60% arginine. Transcription of the genes for protamines is seen in the early haploid spermatids, although translation is delayed for several days (Peschon et al. 1987). The replacement, however, is not complete, and "activating" nucleosomes, having trimethylated H3K4, cluster around developmentally significant loci, including Hox gene promoters, certain microRNAs, and imprinted loci that are paternally expressed (Hammoud et al. 2009).

some 100 million sperm are made in each human testicle, and each ejaculation releases 200 million sperm. Unused sperm are either resorbed or passed out of the body in urine. During his lifetime, a human male can produce 10^{12} to 10^{13} sperm (Reijo et al. 1995).

Gametogenesis in mammals: Oogenesis

Mammalian oogenesis (egg production) differs greatly from spermatogenesis. The eggs mature through an intricate coordination of hormones, paracrine factors, and tissue anatomy. Mammalian egg maturation can be seen as having four stages. First, there is the stage of proliferation. In the human embryo, the thousand or so PGCs reaching the developing ovary divide rapidly from the second to the seventh month of gestation. They generate roughly 7 million **oogonia** (**FIGURE 6.23**). While most of these oogonia die soon afterward, the surviving population, under the influence of retinoic acid, enter then next step and initiate the first meiotic division. They become **primary oocytes.** This first meiotic division does not proceed very far, and the primary oocytes and remain in the diplotene stage of first meiotic prophase (Pinkerton et al. 1961). This prolonged diplotene stage is sometimes referred to as the **dictyate resting stage**. This may last from 12 to 40 years. With the onset of puberty, groups of oocytes periodically resume meiosis. At that time, **luteneizing hormone (LH)** from the pituitary gland releases this block and permits these oocytes to resume meiotic division (Lomniczi et al. 2013). They complete first meiotic division and proceed to second meiotic metaphase. This LH surge causes the oocyte to mature. The oocyte begins to synthesize the proteins that make it competent to fuse with the sperm cell and that enable the first cell divisions of the early embryo. This maturation involves the cross-talk of paracrine factors between the oocyte and its follicular cells, both of which are maturing during this phase. The follicle cells activate the translation of stored oocyte mRNA encoding proteins such as the sperm-binding proteins that will be used for fertilization and the cyclins that control embryonic cell division (Chen et al. 2013; Cakmak et al. 2016). After the secondary oocyte is released from the ovary, meiosis will resume only if fertilization occurs. At fertilization, calcium ions are released in the egg, and these calcium ions release the inhibitory block and allow the haploid nucleus to form.

WEB TOPIC 6.10 **THE BIOCHEMISTRY OF OOCYTE MATURATION** The maturation of the oocyte is intimately connected to several hormones produced by the brain. The effects of these hormones are mediated by the follicle cells of the ovary in fascinating ways.

OOGENIC MEIOSIS Oogenic meiosis in mammals differs from spermatogenic meiosis not only in its timing but in the placement of the metaphase plate. When the primary oocyte divides, its nuclear envelope, breaks down, and the metaphase spindle migrates to the periphery of the cell (see Severson et al. 2016). This asymmetric cytokinesis is directed through a cytoskeletal network composed chiefly of filamentous actin that cradles the mitotic spindle and brings it to the oocyte cortex by myosin-mediated contraction (Schuh and Ellenberg 2008). At the cortex, an oocyte-specific tubulin mediates the separation of chromosomes, and mutations in this tubulin have been found to cause infertility (Feng et al. 2016). At telophase, one of the two daughter cells contains hardly any cytoplasm, while the other daughter cell retains nearly the entire volume of cellular constituents (**FIGURE 6.24**). The smaller cell becomes the **first polar body**, and the larger cell is referred to as the **secondary oocyte.**

A similar unequal cytokinesis takes place during the second division of meiosis. Most of the cytoplasm is retained by the mature egg (the ovum), and a second polar body forms but receives little more than a haploid nucleus. (In humans, the first polar body usually does not divide. It undergoes apoptosis around 20 hours after the first

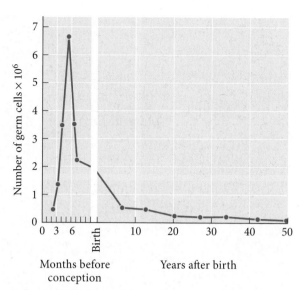

FIGURE 6.23 The number of germ cells in the human ovary changes over the life span. (After Baker 1970.)

VADE MECUM

A segment on gametogenesis in mammals has movies and photographs that illustrate the streamlining of the sperm and remarkable growth of the egg, taking you deeper into the mammalian gonad with each step.

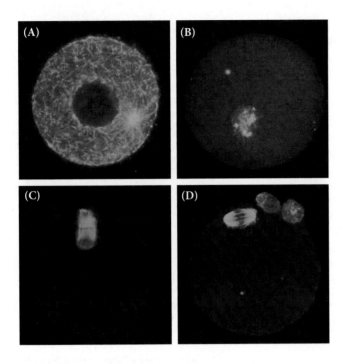

FIGURE 6.24 Meiosis in the mouse oocyte. The tubulin of the microtubules is stained green; the DNA is stained blue. (A) Mouse oocyte in meiotic prophase. The large haploid nucleus (the germinal vesicle) is still intact. (B) The nuclear envelope of the germinal vesicle breaks down as metaphase begins. (C) Meiotic anaphase I, wherein the spindle migrates to the periphery of the egg and releases a small polar body. (D) Meiotic metaphase II, wherein the second polar body is given off (the first polar body has also divided). (From De Vos 2002, courtesy of L. De Vos.)

meiotic division.) Thus, oogenic meiosis conserves the volume of oocyte cytoplasm in a single cell rather than splitting it equally among four progeny (Longo 1997; Schmerler and Wessel 2011).

OOCYTES AND AGE The retention of the oocyte in the ovary for decades has profound medical implications. A large proportion, perhaps even a majority, of fertilized human eggs have too many or too few chromosomes to survive. Genetic analysis has shown that usually such **aneuploidy** (incorrect number of chromosomes) is due primarily to errors in oocyte meiosis (Hassold et al. 1984; Munné et al. 2007). Indeed, the percentage of babies born with aneuploidies increases greatly with maternal age. Women in their 20s have only a 2–3% chance of bearing a fetus whose cells contain an extra chromosome. This risk goes to 35% in women who become pregnant in their 40s (**FIGURE 6.25A**; Hassold and Chiu 1985; Hunt and Hassold 2010). The reasons for this appear to be at least twofold. The first reason concerns the breakdown of cohesin proteins (Chiang et al. 2010; Lister et al. 2010; Revenkova et al. 2010). Once made and assembled, cohesins remain on the chromosomes for decades, but they are gradually lost as the cell ages (**FIGURE 6.25B,C**). This loss of protein and function is accelerated as the cells become physiologically senescent. The second reasons concerns the fact that human meiotic metaphase is remarkably long (16 hours to assemble a meiotic spindle in humans compared to 4 hours in mice) and the linkage between the kinetochore and the spindle does not seem very stable (Holubcová et al. 2015).

Coda

Thus the sex-determining mechanisms have assembled either ovaries or testes, and their respective gametes, the egg and sperm, have been made. When the sperm and egg are released from their gonads, they are cells on the verge of death. However, if they meet, an organism with a lifespan of decades can be generated. The stage is now set for one of the greatest dramas of the cycle of life—fertilization.

Nature has many variations on her masterpiece. In some species, including most mammals and insects, sex is determined by chromosomes; in other species, sex is a matter of environmental conditions. In yet other species, both environmental and genotypic sex determination can function, often in different geographical areas. Different environmental or genetic stimuli may trigger sex determination through a series of conserved pathways. As Crews and Bull (2009) have reflected, "it is possible that

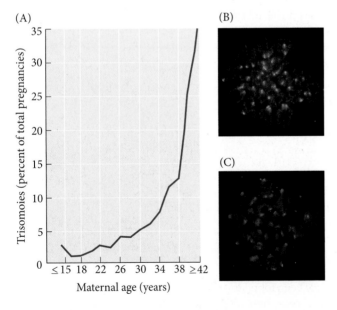

FIGURE 6.25 Chromosomal nondisjunction and meiosis. (A) Maternal age affects the incidence of trisomies in human pregnancy. (B,C) Reduction of chromosome-associated cohesin in aged mice. DNA (white) and cohesin (green) stained in oocyte nuclei of (B) 2-month-old (young) and (C) 14-month-old (aged, for a mouse) ovaries. A significant loss of cohesin can be seen (especially around the kinetochores) in aged mice. (A after Hunt and Hassold 2010; B,C after Lister 2010.)

the developmental decision of male versus female does not flow through a single gene but is instead determined by a 'parliamentary' system involving networks of genes that have simultaneous inputs to several components of the downstream cascade." We are finally beginning to understand the mechanisms by which this "masterpiece of nature" is created.

Next Step Investigation

Our knowledge of sex determination and gametogenesis is remarkably incomplete. First, we know very little about the fundamental processes of meiosis, namely, homologue pairing and how chromosomes are separated at first meiotic metaphase. These are processes fundamental to genetics, development, and evolution, yet we know little about them. We also need to know a great deal more about the cellular and tissue-level processes of gonad formation. We know many of the genes involved, but we are still relatively ignorant about how the testes form such that the germ cells are inside the organ and the ovaries form with their germ cells on the outside. And, of course, the relationship of developmental biology to sexual behaviors is in its infancy.

Closing Thoughts on the Opening Photo

This hermaphroditic chicken is split into a male (rooster) half with coxcomb, waddle, and light coloring, and a female (hen) half with darker coloring. Half the cells are ZW and half are ZZ (recall that birds have ZW/ZZ chromosomal sex determination), probably resulting from the egg's failure to extrude a polar body during meiosis and its subsequent fertilization by a separate sperm. In chickens, each cell makes its own sexual decision. In mammals, hormones play a much larger role in making a unified phenotype, and such man/woman chimeras don't arise (see Zhao et al. 2010). (Photograph courtesy of Michael Clinton.)

6 Snapshot Summary
Sex Determination and Gametogenesis

1. In mammals, primary sex determination (the determination of gonadal sex) is a function of the sex chromosomes. XX individuals are usually females, XY individuals are usually males.

2. The mammalian Y chromosome plays a key role in male sex determination. XY and XX mammals both have a bipotential gonad. In XY animals, Sertoli cells differentiate and enclose the germ cells within testis cords. The interstitial mesenchyme generates other testicular cell types, including the testosterone-secreting Leydig cells.

3. In XX mammals, the germ cells become surrounded by follicle cells in the cortex of the gonadal rudiment. The epithelium of the follicles becomes the granulosa cells; the mesenchyme generates the thecal cells.

4. In humans, the *SRY* gene encodes the testis-determining factor on the Y chromosome. *SRY* synthesizes a nucleic acid-binding protein that functions as a transcription factor to activate the evolutionarily conserved *SOX9* gene.

5. The *SOX9* gene product can also initiate testis formation. Functioning as a transcription factor, it binds to the gene encoding anti-Müllerian hormone and other genes. Fgf9 and Sox9 proteins have a positive feedback loop that activates testicular development and suppresses ovarian development.

6. Wnt4 and Rspo1 are involved in mammalian ovary formation. These proteins upregulate production of β-catenin; the functions of β-catenin include promoting the ovarian pathway of development while blocking the testicular pathway of development. The Foxl2 transcription factor is also required and appears to act in parallel with the Wnt4/Rspo1 pathway.

7. Secondary sex determination in mammals involves the factors produced by the developing gonads. In male mammals, the Müllerian duct is destroyed by the AMH produced by the Sertoli cells, while testosterone produced by the Leydig cells enables the Wolffian duct to

differentiate into the vas deferens and seminal vesicle. In female mammals, the Wolffian duct degenerates with the lack of testosterone, whereas the Müllerian duct persists and is differentiated by estrogen into the oviducts, uterus, cervix, and upper portion of the vagina. Individuals with mutations of these hormones or their receptors may have a discordance between their primary and secondary sex characteristics.

8. The conversion of testosterone to dihydrotestosterone in the genital rudiment and prostate gland precursor enables the differentiation of the penis, scrotum, and prostate gland.

9. In *Drosophila*, sex is determined by the number of X chromosomes in the cell; the Y chromosome does not play a role in sex determination. There are no sex hormones, so each cell makes a sex-determination "decision." However, paracrine factors play important roles in forming the genital structures.

10. The *Drosophila Sex-lethal* gene is activated in females (by the accumulation of proteins encoded on the X chromosomes), but the protein does not form in males because of translational termination. Sxl protein acts as an RNA splicing factor to splice an inhibitory exon from the *transformer* (*tra*) transcript. Therefore, female flies have an active Tra protein but males do not.

11. The Tra protein also acts as an RNA splicing factor to splice exons from the *doublesex* (*dsx*) transcript. The *dsx* gene is transcribed in both XX and XY cells, but its pre-mRNA is processed to form different mRNAs, depending on whether Tra protein is present. The proteins translated from both *dsx* messages are active, and they activate or inhibit transcription of a set of genes involved in producing the sexually dimorphic traits of the fly.

12. Sex determination of the brain may have different downstream agents than in other regions of the body. *Drosophila* Tra proteins also activate the *fruitless* gene in males (but not in females); in mammals, the *Sry* gene may activate brain sexual differentiation independently from the hormonal pathways.

13. In turtles and alligators, sex is often determined by the temperature experienced by the embryo during the time of gonad determination. Because estrogen is necessary for ovary development in these species, it is possible that differing levels of aromatase (an enzyme that can convert testosterone into estrogen) distinguish male from female patterns of gonadal differentiation.

14. The precursors of the gametes are the primordial germ cells (PGCs). In most species (*C. elegans* being an exception), the PGCs form outside the gonads and migrate into the gonads during development.

15. The cytoplasm of the PGCs in many species contains inhibitors of transcription and translation, such that they are both translationally and transcriptionally silent.

16. In most organisms studied, the coordination of germline sex (sperm/egg) is coordinated to somatic sex (male/female) by signals coming from the gonad (testis/ovary).

17. In humans and mice, germ cells entering ovaries initiate meiosis while in the embryo; germ cells entering testes do not initiate meiosis until puberty.

18. The first division of meiosis separates the homologous chromosomes. The second division of meiosis splits the kinetochore and separates the chromatids.

19. Spermatogenic meiosis in mammals is characterized by the production of four gametes per meiosis and by the absence of meiotic arrest. Oogenic meiosis is characterized by the production of one gamete per meiosis and by a prolonged first meiotic prophase to allow the egg to grow.

20. In male mammals, the PGCs generate stem cells that last for the life of the organism. PGCs do not become stem cells in female mammals (although in many other animal groups, PGCs do become germ stem cells in the ovaries).

21. In female mammals, germ cells initiate meiosis and are retained in the first meiotic prophase (dictyate stage) until ovulation. In this stage, they synthesize mRNAs and proteins that will be used for gamete recognition and early development of the fertilized egg.

22. In some species, meiosis is modified such that a diploid egg is formed. Such species can produce a new generation parthenogenetically, without fertilization.

Further Reading

Bell, L. R., J. I. Horabin, P. Schedl and T. W. Cline. 1991. Positive autoregulation of *Sex-lethal* by alternative splicing maintains the female determined state in *Drosophila*. *Cell* 65: 229–239.

Cunha G. R. and 17 others. 2014. Development of the external genitalia: Perspectives from the spotted hyena (*Crocuta crocuta*). *Differentiation* 87: 4–22.

Erickson, J. W. and J. J. Quintero. 2007. Indirect effects of ploidy suggest X chromosome dose, not the X:A ratio, signals sex in *Drosophila*. *PLoS Biol*. Dec. 5(12):e332.

Hiramatsu, R. and 9 others. 2009. A critical time window of *Sry* action in gonadal sex determination in mice. *Development* 136: 129–138.

Imperato-McGinley, J., L. Guerrero, T. Gautier and R. E. Peterson. 1974. Steroid 5α-reductase deficiency in man: An inherited form of male pseudohermaphroditism. *Science* 186: 1213–1215.

Jordan-Young, R. M. 2010. *Brainstorm: The Flaws in the Science of Sex Differences*. Harvard University Press, Cambridge, MA.

Ikami, K., M. Tokue, R. Sugimoto, C. Noda, S. Kobayashi, K. Hara and S. Yoshida. 2015. Hierarchical differentiation competence in response to retinoic acid ensures stem cell maintenance during mouse spermatogenesis. *Development* 142: 1582–1592.

Joel, D. and 13 others. 2015. Sex beyond the genitalia: The human brain mosaic. *Proc. Natl. Acad. Sci. USA* 112: 15468–15473.

Koopman, P., J. Gubbay, N. Vivian, P. Goodfellow and R. Lovell-Badge. 1991. Male development of chromosomally female mice transgenic for *Sry. Nature* 351: 117–121.

Maatouk, D. M., L. DiNapoli, A. Alvers, K. L. Parker, M. M. Taketo and B. Capel. 2008. Stabilization of β-catenin in XY gonads causes male-to-female sex-reversal. *Hum. Mol. Genet.* 17: 2949–2955.

Miyamoto, Y., H. Taniguchi, F. Hamel, D. W. Silversides and R. S. Viger. 2008. GATA4/WT1 cooperation regulates transcription of genes required for mammalian sex determination and differentiation. *BMC Mol. Biol.* 29: 9–44.

Sekido, R. and R. Lovell-Badge. 2008. Sex determination involves synergistic action of Sry and Sf1 on a specific *Sox9* enhancer. *Nature* 453: 930–934.

Sekido, R. and R. Lovell-Badge. 2009. Sex determination and *SRY*: Down to a wink and a nudge? *Trends Genet.* 25: 19–29.

Severson, A. F., G. von Dassow and B. Bowerman. 2016. Oocyte meiotic spindle assembly and function. *Curr. Top. Dev. Biol.* 116: 65–98.

GO TO WWW.DEVBIO.COM...

...for Web Topics, Scientists Speak interviews, Watch Development videos, Dev Tutorials, and complete bibliographic information for all literature cited in this chapter.

Fertilization
Beginning a New Organism

FERTILIZATION IS THE PROCESS WHEREBY THE GAMETES—sperm and egg—fuse together to begin the creation of a new organism. Fertilization accomplishes two separate ends: sex (the combining of genes derived from two parents) and reproduction (the generation of a new organism). Thus, the first function of fertilization is to transmit genes from parent to offspring, and the second is to initiate in the egg cytoplasm those reactions that permit development to proceed.

Although the details of fertilization vary from species to species, it generally consists of four major events:

1. *Contact and recognition* between sperm and egg. In most cases, this ensures that the sperm and egg are of the same species.
2. *Regulation* of sperm entry into the egg. Only one sperm nucleus can ultimately unite with the egg nucleus. This is usually accomplished by allowing only one sperm to enter the egg and actively inhibiting any others from entering.
3. *Fusion* of the genetic material of sperm and egg.
4. *Activation* of egg metabolism to start development.

This chapter will describe how these steps are accomplished in two groups of organisms: sea urchins (whose fertilization we know the best) and mammals.

How do the sperm and egg nuclei find each other?

The Punchline

During fertilization, the egg and sperm must meet, the genetic material of the sperm must enter the egg, and the fertilized egg must initiate cell division and the other processes of development. Sperm and egg must travel toward each other, and chemicals from the eggs can attract the sperm. Gamete recognition occurs when proteins on the sperm cell membrane meet proteins on the extracellular coating of the egg. In preparation for this meeting, the sperm cell membrane is altered significantly by exocytotic events. The sperm activates development by releasing calcium ions (Ca^{2+}) from within the egg. These ions stimulate the enzymes needed for DNA synthesis, RNA synthesis, protein synthesis, and cell division. The sperm and egg pronuclei travel toward one another and the genetic material of the gametes combines to form the diploid chromosome content carrying the genetic information for the development of a new organism.

Structure of the Gametes

A complex dialogue exists between egg and sperm. The egg activates the sperm metabolism that is essential for fertilization, and the sperm reciprocates by activating the egg metabolism needed for the onset of development. But before we investigate these aspects of fertilization, we need to consider the structures of the sperm and egg—the two cell types specialized for fertilization.

Sperm

Sperm were discovered in the 1670s, but their role in fertilization was not discovered until the mid-1800s. It was only in the 1840s, after Albert von Kölliker described the formation of sperm from cells in the adult testes that fertilization research could really begin. Even so, von Kölliker denied that there was any physical contact between sperm and egg. He believed that the sperm excited the egg to develop in much the same way a magnet communicates its presence to iron. The first description of fertilization was

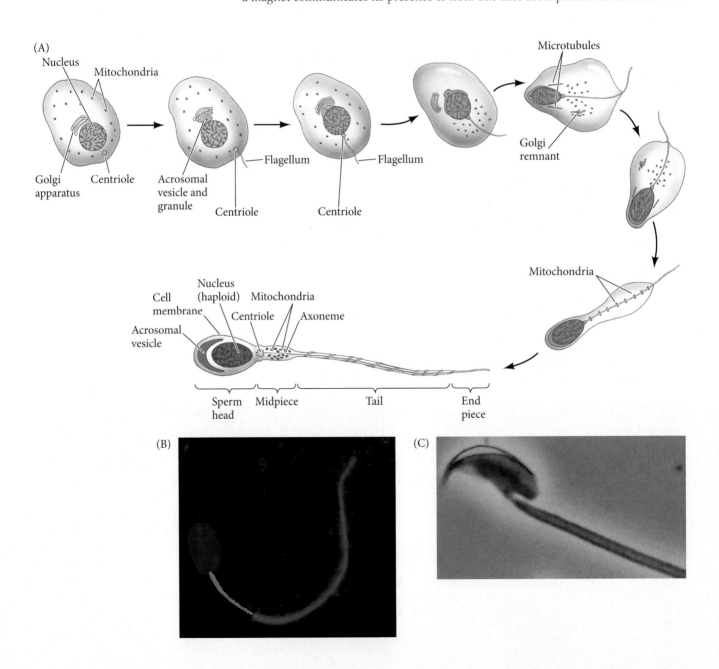

published in 1847 by Karl Ernst von Baer, who showed the union of sperm and egg in sea urchins and tunicates (Raineri and Tammiksaar 2013). He described the fertilization envelope, the migration of the sperm nucleus to the center of the egg, and the subsequent early cell divisions of development. In the 1870s, Oscar Hertwig and Herman Fol repeated this work and detailed the union of the two cells' nuclei.

WEB TOPIC 7.1   **THE ORIGINS OF FERTILIZATION RESEARCH**   Our knowledge of fertilization is relatively recent. Although sperm was discovered in the 1670s, it didn't have a known job until 200 years later.

SPERM ANATOMY   Each sperm cell consists of a haploid nucleus, a propulsion system to move the nucleus, and a sac of enzymes that enable the nucleus to enter the egg. In most species, almost all of the cell's cytoplasm is eliminated during sperm maturation, leaving only certain organelles that are modified for spermatic function (**FIGURE 7.1A,B**). During the course of maturation, the sperm's haploid nucleus becomes very streamlined and its DNA becomes tightly compressed. In front or to the side of this compressed haploid nucleus lies the **acrosomal vesicle**, or **acrosome** (**FIGURE 7.1C**). The acrosome is derived from the cell's Golgi apparatus and contains enzymes that digest proteins and complex sugars. Enzymes stored in the acrosome can digest a path through the outer coverings of the egg. In many species, a region of globular actin proteins lies between the sperm nucleus and the acrosomal vesicle. These proteins are used to extend a fingerlike **acrosomal process** from the sperm during the early stages of fertilization. In sea urchins and numerous other species, recognition between sperm and egg involves molecules on the acrosomal process. Together, the acrosome and nucleus constitute the **sperm head**.

The means by which sperm are propelled vary according to how the species has adapted to environmental conditions. In most species, an individual sperm is able to travel by whipping its **flagellum**. The major motor portion of the flagellum is the **axoneme**, a structure formed by microtubules emanating from the centriole at the base of the sperm nucleus. The core of the axoneme consists of two central microtubules surrounded by a row of nine doublet microtubules. These microtubules are made exclusively of the dimeric protein **tubulin**.

Although tubulin is the basis for the structure of the flagellum, other proteins are also critical for flagellar function. The force for sperm propulsion is provided by **dynein**, a protein attached to the microtubules. Dynein is an ATPase—an enzyme that hydrolyzes ATP, converting the released chemical energy into mechanical energy that propels the sperm.[1] This energy allows the active sliding of the outer doublet microtubules,

---

[1] The importance of dynein can be seen in individuals with a genetic syndrome known as the Kartagener triad. These individuals lack functional dynein in all their ciliated and flagellated cells, rendering these structures immotile (Afzelius 1976). Thus, males with Kartagener triad are sterile (immotile sperm). Both men and women affected by this syndrome are susceptible to bronchial infections (immotile respiratory cilia) and have a 50% chance of having the heart on the right side of the body (a condition known as *situs inversus*, the result of immotile cilia in the center of the embryo).

◀ **FIGURE 7.1**   Modification of a germ cell to form a mammalian sperm. (A) The centriole produces a long flagellum at what will be the posterior end of the sperm. The Golgi apparatus forms the acrosomal vesicle at the future anterior end. Mitochondria collect around the flagellum near the base of the haploid nucleus and become incorporated into the midpiece ("neck") of the sperm. The remaining cytoplasm is jettisoned, and the nucleus condenses. The size of the mature sperm has been enlarged relative to the other stages. (B) Mature bull sperm. The DNA is stained blue, mitochondria are stained green, and the tubulin of the flagellum is stained red. (C) The acrosomal vesicle of this mouse sperm is stained green by the fusion of proacrosin with green fluorescent protein (GFP). (A after Clermont and Leblond 1955; B from Sutovsky et al. 1996, courtesy of G. Schatten; C courtesy of K.-S. Kim and G. L. Gerton.)

causing the flagellum to bend (Ogawa et al. 1977; Shinyoji et al. 1998). The ATP needed to move the flagellum and propel the sperm comes from rings of mitochondria located in the **midpiece** of the sperm (see Figure 7.1B). In many species (notably mammals), a layer of dense fibers has interposed itself between the mitochondrial sheath and the cell membrane. This fiber layer stiffens the sperm tail. Because the thickness of this layer decreases toward the tip, the fibers probably prevent the sperm head from being whipped around too suddenly. Thus, the sperm cell has undergone extensive modification for the transport of its nucleus to the egg.

In mammals, the differentiation of sperm is not completed in the testes. Although they are able to move, the sperm released during ejaculation do not yet have the capacity to bind to and fertilize an egg. The final stages of sperm maturation, cumulatively referred to as **capacitation**, do not occur in mammals until the sperm has been inside the female reproductive tract for a certain period of time.

The egg

CYTOPLASM AND NUCLEUS   All the material necessary to begin growth and development must be stored in the egg, or **ovum**.[2] Whereas the sperm eliminates most of its cytoplasm as it matures, the developing egg (called the **oocyte** before it reaches the stage of meiosis at which it is fertilized) not only conserves the material it has, but actively accumulates more. The meiotic divisions that form the oocyte conserve its cytoplasm rather than giving half of it away; at the same time, the oocyte either synthesizes or absorbs proteins such as yolk that act as food reservoirs for the developing embryo. Birds' eggs are enormous single cells, swollen with accumulated yolk (see Figure 12.2). Even eggs with relatively sparse yolk are large compared to sperm. The volume of a sea urchin egg is about 200 picoliters (2×10^{-4} mm^3), more than 10,000 times the volume of sea urchin sperm (**FIGURE 7.2**). So even though sperm and egg have equal haploid

---

[2] Eggs over easy: the terminology used in describing the female gamete can be confusing. In general, an *egg*, or *ovum*, is a female gamete capable of binding sperm and being fertilized. An *oocyte* is a developing egg that cannot yet bind sperm or be fertilized (Wessel 2009). The problems in terminology come from the fact that the eggs of different species are in different stages of meiosis (see Figure 7.3). The human egg, for example, is in second meiotic metaphase when it binds sperm, whereas the sea urchin egg has completed all of its meiotic divisions when it binds sperm. The contents of the egg also vary greatly from species to species.

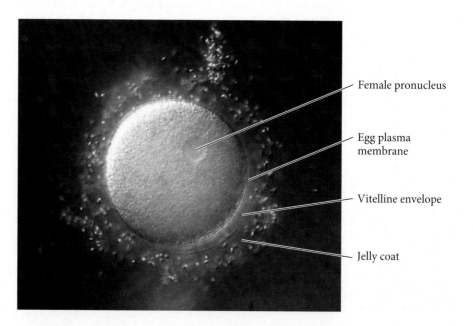

Female pronucleus

Egg plasma membrane

Vitelline envelope

Jelly coat

FIGURE 7.2 Structure of the sea urchin egg at fertilization. Sperm can be seen in the jelly coat and attached to the vitelline envelope. The female pronucleus is apparent within the egg cytoplasm. (Photograph by Kristina Yu © Exploratorium www.exploratorium.edu.)

nuclear components, the egg accumulates a remarkable cytoplasmic storehouse during its maturation. This cytoplasmic trove includes the following:

- **Nutritive proteins**. The early embryonic cells must have a supply of energy and amino acids. In many species, this is accomplished by accumulating yolk proteins in the egg. Many of these yolk proteins are made in other organs (e.g., liver, fat bodies) and travel through the maternal blood to the oocyte.

- **Ribosomes and tRNA**. The early embryo must make many of its own structural proteins and enzymes, and in some species there is a burst of protein synthesis soon after fertilization. Protein synthesis is accomplished by ribosomes and tRNA that exist in the egg. The developing egg has special mechanisms for synthesizing ribosomes; certain amphibian oocytes produce as many as 10^{12} ribosomes during their meiotic prophase.

- **Messenger RNAs**. The oocyte not only accumulates proteins, it also accumulates mRNAs that encode proteins for the early stages of development. It is estimated that sea urchin eggs contain thousands of different types of mRNA that remain repressed until after fertilization.

- **Morphogenetic factors**. Molecules that direct the differentiation of cells into certain cell types are present in the egg. These include transcription factors and paracrine factors. In many species, they are localized in different regions of the egg and become segregated into different cells during cleavage.

- **Protective chemicals**. The embryo cannot run away from predators or move to a safer environment, so it must be equipped to deal with threats. Many eggs contain ultraviolet filters and DNA repair enzymes that protect them from sunlight, and some eggs contain molecules that potential predators find distasteful. The yolk of bird eggs contains antibodies that protect the embryo against microbes.

Within the enormous volume of egg cytoplasm resides a large nucleus (see Figure 7.2). In a few species (such as sea urchins), this **female pronucleus** is already haploid at the time of fertilization. In other species (including many worms and most mammals), the egg nucleus is still diploid—the sperm enters before the egg's meiotic divisions are completed (**FIGURE 7.3**). In these species, the final stages of egg meiosis will take place after the sperm's nuclear material—the **male pronucleus**—is already inside the egg cytoplasm.

FIGURE 7.3 Stages of egg maturation at the time of sperm entry in different animal species. Note that in most species, sperm entry occurs before the egg nucleus has completed meiosis. The germinal vesicle is the name given to the large diploid nucleus of the primary oocyte. The polar bodies are nonfunctional cells produced by meiosis (see Chapter 6). (After Austin 1965.)

WEB TOPIC 7.2   **THE EGG AND ITS ENVIRONMENT**   Most eggs are fertilized in the wild, not in the laboratory. In addition to the developmental processes in the egg, there are also factors in the egg that help the developing embryo cope with environmental stresses.

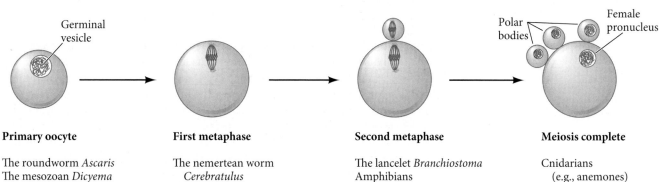

Primary oocyte	First metaphase	Second metaphase	Meiosis complete
The roundworm *Ascaris*	The nemertean worm *Cerebratulus*	The lancelet *Branchiostoma*	Cnidarians (e.g., anemones)
The mesozoan *Dicyema*	The polychaete worm *Chaetopterus*	Amphibians	Sea urchins
The sponge *Grantia*	The mollusk *Dentalium*	Most mammals	
The polychaete worm *Myzostoma*	The core worm *Pectinaria*	Fish	
The clam worm *Nereis*	Many insects		
The clam *Spisula*	Starfish		
The echiuroid worm *Urechis*			
Dogs and foxes			

(A)   (B)

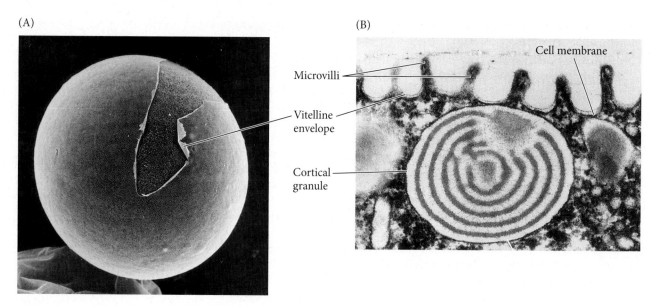

FIGURE 7.4 Sea urchin egg cell surfaces. (A) Scanning electron micrograph of an egg before fertilization. The cell membrane is exposed where the vitelline envelope has been torn. (B) Transmission electron micrograph of an unfertilized egg, showing microvilli and cell membrane, which are closely covered by the vitelline envelope. A cortical granule lies directly beneath the cell membrane. (From Schroeder 1979, courtesy of T. E. Schroeder.)

CELL MEMBRANE AND EXTRACELLULAR ENVELOPE The membrane enclosing the egg cytoplasm regulates the flow of specific ions during fertilization and must be capable of fusing with the sperm cell membrane. Outside this egg cell membrane is an extracellular matrix that forms a fibrous mat around the egg and is often involved in sperm-egg recognition (Wasserman and Litscher 2016). In invertebrates, this structure is usually called the **vitelline envelope** (**FIGURE 7.4A**). The vitelline envelope contains several different glycoproteins. It is supplemented by extensions of membrane glycoproteins from the cell membrane and by proteinaceous "posts" that adhere the vitelline envelope to the cell membrane (Mozingo and Chandler 1991). The vitelline envelope is essential for the species-specific binding of sperm. Many types of eggs also have a layer of **egg jelly** outside the vitelline envelope. This glycoprotein meshwork can have numerous functions, but most commonly it is used either to attract or to activate sperm. The egg, then, is a cell specialized for receiving sperm and initiating development.

Lying immediately beneath the cell membrane of most eggs is a thin layer (about 5 μm) of gel-like cytoplasm called the **cortex**. The cytoplasm in this region is stiffer than the internal cytoplasm and contains high concentrations of globular actin molecules. During fertilization, these actin molecules polymerize to form long cables of actin **microfilaments**. Microfilaments are necessary for cell division. They are also used to extend the egg surface into small projections called **microvilli**, which may aid sperm entry into the cell (**FIGURE 7.4B**). Also within the cortex are the **cortical granules** (see Figures 7.4B). These membrane-bound, Golgi-derived structures contain proteolytic enzymes and are thus homologous to the acrosomal vesicle of the sperm. However, whereas a sea urchin sperm contains just one acrosomal vesicle, each sea urchin egg contains approximately 15,000 cortical granules. In addition to digestive enzymes, the cortical granules contain mucopolysaccharides, adhesive glycoproteins, and hyalin protein. As we will soon describe, the enzymes and mucopolysaccharides help prevent polyspermy—that is, they prevent additional sperm from entering the egg after the first sperm has entered—while hyalin and the adhesive glycoproteins surround the early embryo, providing support for cleavage-stage blastomeres.

In mammalian eggs, the extracellular envelope is a separate, thick matrix called the **zona pellucida**. The mammalian egg is also surrounded by a layer of cells called the **cumulus** (**FIGURE 7.5**), which is made up of the ovarian follicular cells that were nurturing the egg at the time of its release from the ovary. Mammalian sperm have to get past these cells to fertilize the egg. The innermost layer of cumulus cells, immediately adjacent to the zona pellucida, is called the **corona radiata**.

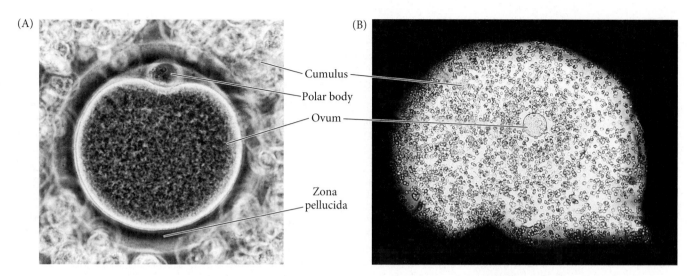

FIGURE 7.5 Mammalian eggs immediately before fertilization. (A) The hamster egg, or ovum, is encased in the zona pellucida, which in turn is surrounded by the cells of the cumulus. A polar body cell, produced during meiosis, is visible within the zona pellucida. (B) At lower magnification, a mouse oocyte is shown surrounded by the cumulus. Colloidal carbon particles (India ink, seen here as the black background) are excluded by the hyaluronidate matrix. (Courtesy of R. Yanagimachi.)

Recognition of egg and sperm

The interaction of sperm and egg generally proceeds according to five steps (**FIGURE 7.6**; Vacquier 1998):

1. *Chemoattraction* of the sperm to the egg by soluble molecules secreted by the egg

2. *Exocytosis* of the sperm acrosomal vesicle and release of its enzymes

3. *Binding of the sperm* to the extracellular matrix (vitelline envelope or zona pellucida) of the egg

4. *Passage of the sperm* through this extracellular matrix

5. *Fusion* of the egg and sperm cell membranes

After these steps are accomplished, the haploid sperm and egg nuclei can meet and the reactions that initiate development can begin. In this chapter, we will focus on these events in two well-studied organisms: sea urchins, which undergo external fertilization; and mice, which undergo internal fertilization. Some variations of fertilization events will be described in subsequent chapters as we study the development of particular organisms.

External Fertilization in Sea Urchins

Many marine organisms release their gametes into the environment. That environment may be as small as a tide pool or as large as an ocean and is shared with other species that may shed their gametes at the same time. Such organisms are faced with two problems: How can sperm and eggs meet in such a dilute concentration, and how can sperm be prevented from attempting to fertilize eggs of another species?

FIGURE 7.6 Summary of events leading to the fusion of egg and sperm cell membranes in sea urchin fertilization, which is external. (1) The sperm is chemotactically attracted to and activated by the egg. (2, 3) Contact with the egg jelly triggers the acrosome reaction, allowing the acrosomal process to form and release proteolytic enzymes. (4) The sperm adheres to the vitelline envelope and lyses a hole in it. (5) The sperm adheres to the egg cell membrane and fuses with it. The sperm pronucleus can now enter the egg cytoplasm.

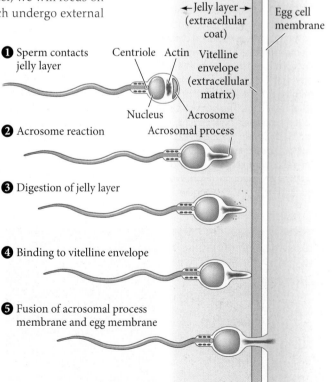

In addition to simply producing enormous numbers of gametes, two major mechanisms have evolved to solve these problems: species-specific **sperm attraction** and species-specific **sperm activation**. Here we describe these events as they occur in sea urchins.

Sperm attraction: Action at a distance

Species-specific sperm attraction has been documented in numerous species, including cnidarians, mollusks, echinoderms, amphibians, and urochordates (Miller 1985; Yoshida et al. 1993; Burnett et al. 2008). In many species, sperm are attracted toward eggs of their species by **chemotaxis**—that is, by following a gradient of a chemical secreted by the egg. These oocytes control not only the type of sperm they attract, but also the time at which they attract them, releasing the chemotactic factor only after they reach maturation (Miller 1978).

The mechanisms of chemotaxis differ among species (see Metz 1978; Eisenbach 2004), and chemotactic molecules are different even in closely related species. In sea urchins, sperm motility is acquired only after the sperm are spawned. As long as sperm cells are in the testes, they cannot move because their internal pH is kept low (about pH 7.2) by the high concentrations of CO_2 in the gonad. However, once sperm are spawned into seawater, their pH is elevated to about 7.6, resulting in the activation of the dynein ATPase. The splitting of ATP provides the energy for the flagella to wave, and the sperm begin swimming vigorously (Christen et al. 1982).

But the ability to move does not provide the sperm with a direction. In echinoderms, direction is provided by small chemotactic peptides called **sperm-activating peptides** (**SAPs**). One such SAP is **resact**, a 14-amino acid peptide that has been isolated from the egg jelly of the sea urchin *Arbacia punctulata* (Ward et al. 1985). Resact diffuses readily from the egg jelly into seawater and has a profound effect at very low concentrations when added to a suspension of *Arbacia* sperm. When a drop of seawater containing *Arbacia* sperm is placed on a microscope slide, the sperm generally swim in circles about 50 μm in diameter. Within seconds after a small amount of resact is injected, sperm migrate into the region of the injection and congregate there (**FIGURE 7.7**). As resact diffuses from the area of injection, more sperm are recruited into the growing cluster.

Resact is specific for *A. punctulata* and does not attract sperm of other urchin species. (An analogous compound, speract, has been isolated from the purple sea urchin, *Strongylocentrotus purpuratus*.) *A. punctulata* sperm have receptors in their cell membranes that bind resact (Ramarao and Garbers 1985; Bentley et al. 1986). When the extracellular side of the receptor binds resact, it activates latent guanylyl cyclase in the cytoplasmic side of the receptor (**FIGURE 7.8**). Active guanylyl cyclase causes the sperm cell to produce more cyclic GMP (cGMP), a compound that activates a calcium channel in the cell membrane of the sperm tail, allowing the influx of calcium ions (Ca^{2+}) from the seawater into the tail (Nishigaki et al. 2000; Wood et al. 2005). These sperm-specific calcium channels are encoded by CatSper genes—the same genes that control the direction of sperm migration in mice and humans (Seifert et al. 2014). The increases in cGMP and Ca^{2+} activate both the mitochondrial ATP-generating apparatus and the dynein ATPase that

FIGURE 7.7 Sperm chemotaxis in the sea urchin *Arbacia punctulata*. One nanoliter of a 10-n*M* solution of resact is injected into a 20-microliter drop of sperm suspension. (A) A 1-second photographic exposure showing sperm swimming in tight circles before the addition of resact. The position of the injection pipette is shown by the white lines. (B–D) Similar 1-second exposures showing migration of sperm to the center of the resact gradient 20, 40, and 90 seconds after injection. (From Ward et al. 1985, courtesy of V. D. Vacquier.)

(A)   (B)   (C)   (D)

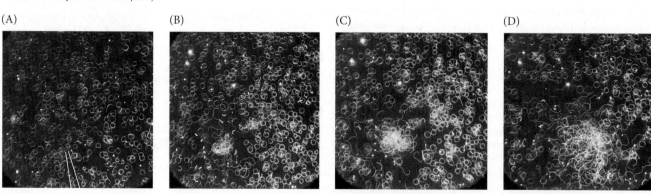

(A)

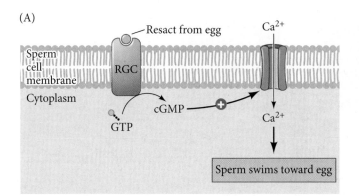

FIGURE 7.8 Model for chemotactic peptides in sea urchin sperm. (A) Resact from *Arbacia* egg jelly binds to its receptor on the sperm. This activates the receptor's guanylyl cyclase (RGC) activity, forming intracellular cGMP in the sperm. The cGMP opens calcium channels in the sperm cell membrane, allowing Ca^{2+} to enter the sperm. The influx of Ca^{2+} activates sperm motility, and the sperm swims up the resact gradient toward the egg. (B) Ca^{2+} levels in different regions of *Strongylocentrotus purpuratus* sperm after exposure to 125 n*M* speract (the *S. purpuratus* analog of resact). Red indicates the highest level of Ca^{2+}, blue the lowest. The sperm head reaches its peak Ca^{2+} levels within 1 second. (A after Kirkman-Brown et al. 2003; B from Wood et al. 2003, courtesy of M. Whitaker.)

(B)

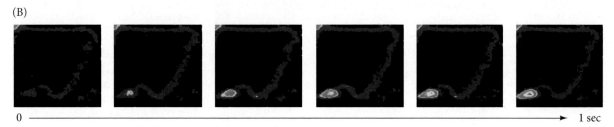

0 ——————————————————————————————————▶ 1 sec

stimulates flagellar movement in the sperm (Shimomura et al. 1986; Cook and Babcock 1993). In addition, the sperm sense the SAP gradient by curving their tails, interspersing straight swimming with a "turn" to sense the environment (Guerrero et al. 2010). The binding of a single resact molecule may be enough to provide direction for the sperm, which swim up a concentration gradient of this compound until they reach the egg (Kaupp et al. 2003; Kirkman-Brown et al. 2003). Thus, resact functions as a sperm-*attracting* peptide as well as a sperm-activating peptide. (In some organisms, the functions of sperm attraction and sperm activation are performed by different compounds.)

The acrosome reaction

A second interaction between sperm and egg jelly results in the **acrosome reaction**. In most marine invertebrates, the acrosome reaction has two components: the fusion of the acrosomal vesicle with the sperm cell membrane (an exocytosis that results in the release of the contents of the acrosomal vesicle), and the extension of the acrosomal process (Dan 1952; Colwin and Colwin 1963). The acrosome reaction in sea urchins is initiated by contact of the sperm with the egg jelly. Contact causes the exocytosis of the sperm's acrosomal vesicle. The proteolytic enzymes and proteasomes (protein-digesting complexes) thus released digest a path through the jelly coat to the egg cell surface. Once the sperm reaches the egg surface, the acrosomal process adheres to the vitelline envelope and tethers the sperm to the egg. It is possible that proteasomes from the acrosome coat the acrosomal process, allowing it to digest the vitelline envelope at the point of attachment and proceed toward the egg (Yokota and Sawada 2007).

In sea urchins, the acrosome reaction is initiated by sulfate-containing polysaccharides in the egg jelly that bind to specific receptors located directly above the acrosomal vesicle on the sperm cell membrane. These polysaccharides are often highly species-specific, and egg jelly factors from one species of sea urchin generally fail to activate the acrosome reaction even in closely related species (**FIGURE 7.9**; Hirohashi and Vacquier 2002; Hirohashi et al. 2002; Vilela-Silva et al. 2008). Thus, activation of the acrosome reaction serves as a barrier to interspecies (and thus unviable) fertilizations. This is important when numerous species inhabit the same habitat and when their spawning seasons overlap.

In *Strongylocentrotus purpuratus*, the acrosome reaction is initiated by a repeating polymer of fucose sulfate. When this sulfated polysaccharide binds to its receptor on the

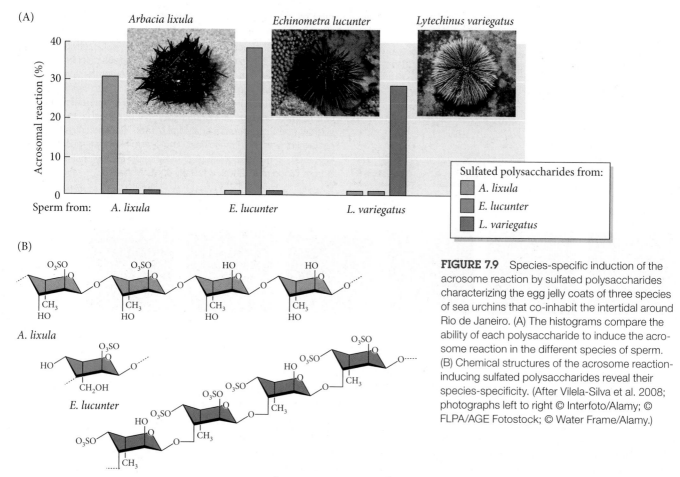

FIGURE 7.9 Species-specific induction of the acrosome reaction by sulfated polysaccharides characterizing the egg jelly coats of three species of sea urchins that co-inhabit the intertidal around Rio de Janeiro. (A) The histograms compare the ability of each polysaccharide to induce the acrosome reaction in the different species of sperm. (B) Chemical structures of the acrosome reaction-inducing sulfated polysaccharides reveal their species-specificity. (After Vilela-Silva et al. 2008; photographs left to right © Interfoto/Alamy; © FLPA/AGE Fotostock; © Water Frame/Alamy.)

sperm, the receptor activates three sperm membrane proteins: (1) a calcium transport channel that allows Ca^{2+} to enter the sperm head; (2) a sodium-hydrogen exchanger that pumps sodium ions (Na^+) into the sperm as it pumps hydrogen ions (H^+) out; and (3) a phospholipase enzyme that makes another second messenger, the phosopholipid **inositol 1,4,5-trisphosphate** (**IP$_3$**, of which we will hear much more later in the chapter). IP$_3$ is able to release Ca^{2+} from *inside* the sperm, probably from within the acrosome itself (Domino and Garbers 1988; Domino et al. 1989; Hirohashi and Vacquier 2003). The elevated Ca^{2+} level in a relatively basic cytoplasm triggers the fusion of the acrosomal membrane with the adjacent sperm cell membrane (**FIGURE 7.10A–C**), releasing enzymes that can lyse a path through the egg jelly to the vitelline envelope.

The second part of the acrosome reaction involves the extension of the acrosomal process by the polymerization of globular actin molecules into actin filaments (**FIGURE 7.10D**; Tilney et al. 1978). The influx of Ca^{2+} is thought to activate the protein RhoB in the acrosomal region and midpiece of the sperm (Castellano et al. 1997; de la Sancha et al. 2007). This GTP-binding protein helps organize the actin cytoskeleton in many types of cells and is thought to be active in polymerizing actin to make the acrosomal process.

Recognition of the egg's extracellular coat

The sea urchin sperm's contact with an egg's jelly coat provides the first set of species-specific recognition events (i.e., sperm attraction, activation, and acrosome reaction). Another critical species-specific binding event must occur once the sperm has penetrated the egg jelly and its acrosomal process contacts the surface of the egg (**FIGURE 7.11A**). The acrosomal protein mediating this recognition in sea urchins is an insoluble, 30,500-Da protein called **bindin**. In 1977, Vacquier and co-workers isolated bindin from the acrosome of *Strongylocentrotus purpuratus* and found it to be capable of binding to dejellied eggs of the same species. Further, sperm bindin, like egg jelly polysaccharides,

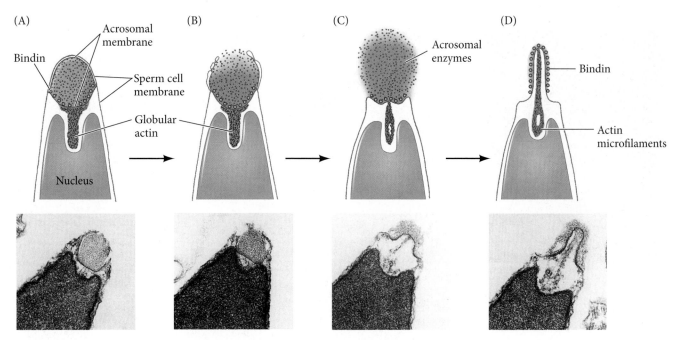

FIGURE 7.10 Acrosome reaction in sea urchin sperm. (A–C) The portion of the acrosomal membrane lying directly beneath the sperm cell membrane fuses with the cell membrane to release the contents of the acrosomal vesicle. (D) The actin molecules assemble to produce microfilaments, extending the acrosomal process outward. Actual photographs of the acrosome reaction in sea urchin sperm are shown below the diagrams. (After Summers and Hylander 1974; photographs courtesy of G. L. Decker and W. J. Lennarz.)

is usually species-specific: bindin isolated from the acrosomes of *S. purpuratus* binds to its own dejellied eggs but not to those of *S. franciscanus* (**FIGURE 7.11B**; Glabe and Vacquier 1977; Glabe and Lennarz 1979).

Biochemical studies have confirmed that the bindins of closely related sea urchin species have different protein sequences. This finding implies the existence of species-specific **bindin receptors** on the egg vitelline envelope (**FIGURE 7.12A**). Indeed, a 350-kDa glycoprotein that displays the properties expected of a bindin receptor, has been isolated

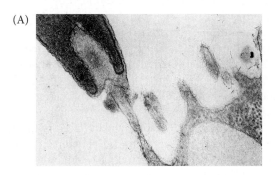

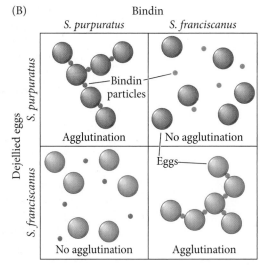

FIGURE 7.11 Species-specific binding of the acrosomal process to the egg surface in sea urchins. (A) Actual contact of a sperm acrosomal process with an egg microvillus. (B) In vitro model of species-specific binding. The agglutination of dejellied eggs by bindin was measured by adding bindin particles to a plastic well containing a suspension of eggs. After 2–5 minutes of gentle shaking, the wells were photographed. Each bindin bound to and agglutinated only eggs from its own species. (A from Epel 1977, courtesy of F. D. Collins and D. Epel; B based on photographs in Glabe and Vacquier 1978.)

(A)

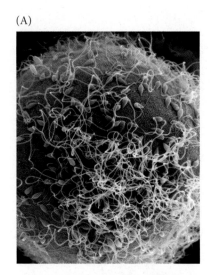

(B)

(C) DAB precipitate
(indicates bindin present)

(D)

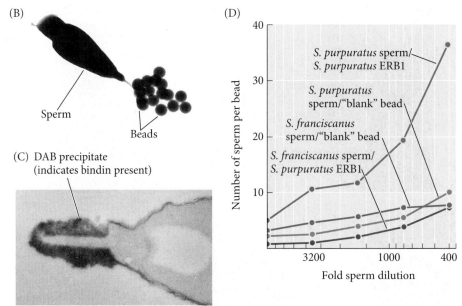

FIGURE 7.12 Bindin receptors on the sea urchin egg. (A) Scanning electron micrograph of sea urchin sperm bound to the vitelline envelope of an egg. Although this egg is saturated with sperm, there appears to be room on the surface for more sperm, implying the existence of a limited number of bindin receptors. (B) *Strongylocentrotus purpuratus* sperm bind to polystyrene beads that have been coated with purified bindin receptor protein. (C) Immunochemically labeled bindin (the label manifests as a dark precipitate of diaminobenzidine, DAB) is seen to be localized to the acrosomal process after the acrosome reaction. (D) Species-specific binding of sea urchin sperm to ERB1. *S. purpuratus* sperm bound to beads coated with ERB1 bindin receptor purified from *S. purpuratus* eggs, but *S. franciscanus* sperm did not. Neither sperm bound to uncoated "blank" beads. (A © Mia Tegner/SPL/Science Source; B from Foltz et al. 1993; C from Moy and Vacquier 1979, courtesy of V. Vacquier; D after Kamei and Glabe 2003.)

from sea urchin eggs (**FIGURE 7.12B**; Kamei and Glabe 2003). These bindin receptors are thought to be aggregated into complexes on the vitelline envelope, and hundreds of such complexes may be needed to tether the sperm to the egg. The receptor for sperm bindin on the egg vitelline envelope appears to recognize the protein portion of bindin on the acrosome (**FIGURE 7.12C**) in a species-specific manner. Closely related species of sea urchins (i.e., different species in the same genus) have divergent bindin receptors, and eggs will adhere only to the bindin of their own species (**FIGURE 7.12D**). Thus, species-specific recognition of sea urchin gametes can occur at the levels of sperm attraction, sperm activation, the acrosome reaction, and sperm adhesion to the egg surface.

Bindin and other gamete recognition proteins are among the fastest evolving proteins known (Metz and Palumbi 1996; Swanson and Vacquier 2002). Even when closely related urchin species have near-identity of every other protein, their bindins and bindin receptors may have diverged significantly.

Fusion of the egg and sperm cell membranes

Once the sperm has traveled to the egg and undergone the acrosome reaction, the fusion of the sperm cell membrane with the egg cell membrane can begin (**FIGURE 7.13**). Sperm-egg fusion appears to cause the polymerization of actin in the egg to form a **fertilization cone** (Summers et al. 1975). Homology between the egg and the sperm is again demonstrated, since the sperm's acrosomal process also appears to be formed by the polymerization of actin. Actin from the gametes forms a connection that widens the cytoplasmic bridge between the egg and sperm. The sperm nucleus and tail pass through this bridge.

Fusion is an active process, often mediated by specific "fusogenic" proteins. In sea urchins, bindin plays a second role as a fusogenic protein. In addition to recognizing the egg, bindin contains a long stretch of hydrophobic amino acids near its amino terminus, and this region is able to fuse phospholipid vesicles in vitro (Ulrich et al. 1999; Gage et al. 2004). Under the ionic conditions present in the mature unfertilized egg, bindin can cause the sperm and egg membranes to fuse.

One egg, one sperm

As soon as one sperm enters the egg, the fusibility of the egg membrane—which was necessary to get the sperm inside the egg—becomes a dangerous liability. In the normal case—**monospermy**—only one sperm enters the egg, and the haploid sperm nucleus

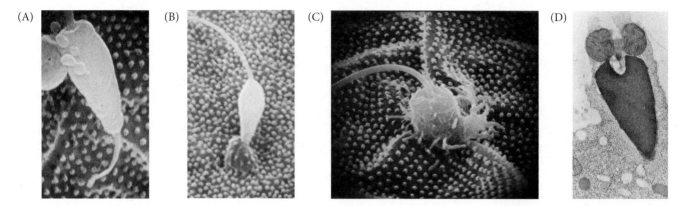

(A)   (B)   (C)   (D)

combines with the haploid egg nucleus to form the diploid nucleus of the fertilized egg (zygote), thus restoring the chromosome number appropriate for the species. During cleavage, the centriole provided by the sperm divides to form the two poles of the mitotic spindle while the egg-derived centriole is degraded.

In most animals, any sperm that enters the egg can provide a haploid nucleus and a centriole. The entrance of multiple sperm—**polyspermy**—leads to disastrous consequences in most organisms. In sea urchins, fertilization by two sperm results in a triploid nucleus, in which each chromosome is represented three times rather than twice. Worse, each sperm's centriole divides to form the two poles of a mitotic apparatus, so instead of a bipolar mitotic spindle separating the chromosomes into two cells, the triploid chromosomes may be divided into as many as four cells, with some cells receiving extra copies of certain chromosomes while other cells lack them (**FIGURE 7.14**). Theodor Boveri demonstrated in 1902 that such cells either die or develop abnormally.

FIGURE 7.13 Scanning electron micrographs of the entry of sperm into sea urchin eggs. (A) Contact of sperm head with egg microvillus through the acrosomal process. (B) Formation of fertilization cone. (C) Internalization of sperm within the egg. (D) Transmission electron micrograph of sperm internalization through the fertilization cone. (A–C from Schatten and Mazia 1976, courtesy of G. Schatten; D courtesy of F. J. Longo.)

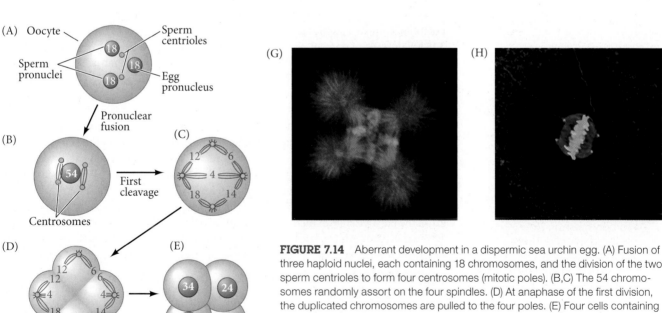

FIGURE 7.14 Aberrant development in a dispermic sea urchin egg. (A) Fusion of three haploid nuclei, each containing 18 chromosomes, and the division of the two sperm centrioles to form four centrosomes (mitotic poles). (B,C) The 54 chromosomes randomly assort on the four spindles. (D) At anaphase of the first division, the duplicated chromosomes are pulled to the four poles. (E) Four cells containing different numbers and types of chromosomes are formed, thereby causing (F) the early death of the embryo. (G) First metaphase of a dispermic sea urchin egg akin to (D). The microtubules are stained green; the DNA stain appears orange. The triploid DNA is being split into four chromosomally unbalanced cells instead of the normal two cells with equal chromosome complements. (H) Human dispermic egg at first mitosis. The four centrioles are stained yellow, while the microtubules of the spindle apparatus (and of the two sperm tails) are stained red. The three sets of chromosomes divided by these four poles are stained blue. (A–F after Boveri 1907; G courtesy of J. Holy; H from Simerly et al. 1999, courtesy of G. Schatten.)

FIGURE 7.15 Membrane potential of sea urchin eggs before and after fertilization. (A) Before the addition of sperm, the potential difference across the egg cell membrane is about –70 mV. Within 1–3 seconds after the fertilizing sperm contacts the egg, the potential shifts in a positive direction. (B,C) *Lytechinus* eggs photographed during first cleavage. (B) Control eggs developing in 490 m*M* Na⁺. (C) Polyspermy in eggs fertilized in similarly high concentrations of sperm in 120 m*M* Na⁺ (choline was substituted for sodium). (D) Table showing the rise of polyspermy with decreasing Na⁺ concentration. Salt water is about 600 m*M* Na⁺. (After Jaffe 1980; B,C courtesy of L. A. Jaffe.)

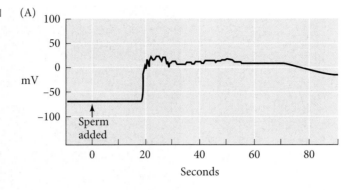

(B)

Na⁺	Polyspermic eggs (%)
490	22
360	26
120	97
50	100

The fast block to polyspermy

The most straightforward way to prevent the union of more than two haploid nuclei is to prevent more than one sperm from entering the egg. Different mechanisms to prevent polyspermy have evolved, two of which are seen in the sea urchin egg. An initial, fast reaction, accomplished by an electric change in the egg cell membrane, is followed by a slower reaction caused by the exocytosis of the cortical granules (Just 1919).

The **fast block to polyspermy** is achieved by a change in the electric potential of the egg cell membrane that occurs immediately upon the entry of a sperm. Once one sperm has fused with the egg, soluble material from that sperm (probably nicotinic acid adenine dinucleotide phosphate, NAADP) acts to change the egg cell membrane (McCulloh and Chambers 1992; Wong and Wessel 2013). Sodium channels are closed, thereby preventing the entry of sodium ions (Na⁺) into the egg, and the egg cell membrane maintains an electrical voltage gap between the interior of the egg and its environment. This **resting membrane potential** is generally about 70 mV, which is expressed as –70 mV because *the inside of the cell is negatively charged with respect to the exterior*. Within 1–3 seconds after the binding of the first sperm, the membrane potential shifts to a *positive* level—about +20 mV—with respect to the exterior (**FIGURE 7.15A**; Jaffe 1980; Longo et al. 1986). The shift from negative to positive is the result of a small influx of Na⁺ into the egg through newly opened sodium channels. Sperm cannot fuse with egg cell membranes that have a positive resting potential, so the shift means that no more sperm can fuse to the egg.

The importance of Na⁺ and the change in resting potential from negative to positive was demonstrated by Laurinda Jaffe and colleagues. They found that polyspermy can be induced if an electric current is applied to artificially keep the sea urchin egg membrane potential negative. Conversely, fertilization can be prevented entirely by artificially keeping the membrane potential of eggs positive (Jaffe 1976). The fast block to polyspermy can also be circumvented by lowering the concentration of Na⁺ in the surrounding water (**FIGURE 7.15B**). If the supply of sodium ions is not sufficient to cause the positive shift in membrane potential, polyspermy occurs (Gould-Somero et al. 1979; Jaffe 1980). An electric block to polyspermy also occurs in frogs (Cross and Elinson 1980; Iwao et al. 2014), but probably not in most mammals (Jaffe and Cross 1983).

WEB TOPIC 7.3   BLOCKS TO POLYSPERMY The work of Theodor Boveri and E. E. Just were critical in elucidating the blocks against multiple sperm entry.

The slow block to polyspermy

The fast block to polyspermy is transient, since the membrane potential of the sea urchin egg remains positive for only about a minute. This brief potential shift is not sufficient to prevent polyspermy permanently, and polyspermy can still occur if the sperm bound to the vitelline envelope are not somehow removed (Carroll and Epel 1975). This sperm removal is accomplished by the **cortical granule reaction**, also known as the **slow block to polyspermy**. This slower, mechanical block to polyspermy

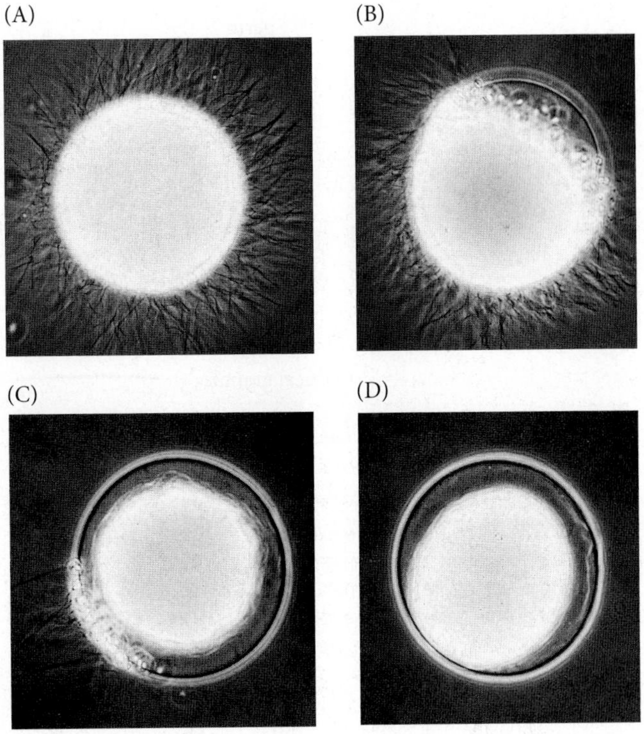

FIGURE 7.16 Formation of the fertilization envelope and removal of excess sperm. To create these photographs, sperm were added to sea urchin eggs, and the suspension was then fixed in formaldehyde to prevent further reactions. (A) At 10 seconds after sperm addition, sperm surround the egg. (B,C) At 25 and 35 seconds after insemination, respectively, a fertilization envelope is forming around the egg, starting at the point of sperm entry. (D) The fertilization envelope is complete, and excess sperm have been removed. (From Vacquier and Payne 1973, courtesy of V. D. Vacquier.)

becomes active about a minute after the first successful sperm-egg fusion (Just 1919). This reaction is found in many animal species, including sea urchins and most mammals.

Directly beneath the sea urchin egg cell membrane are about 15,000 cortical granules, each about 1 μm in diameter (see Figure 7.4B). Upon sperm entry, cortical granules fuse with the egg cell membrane and release their contents into the space between the cell membrane and the fibrous mat of vitelline envelope proteins. Several proteins are released by cortical granule exocytosis. One of these, the enzyme cortical granule serine protease, cleaves the protein posts that connect the vitelline envelope proteins to the egg cell membrane; it also clips off the bindin receptors and any sperm attached to them (Vacquier et al. 1973; Glabe and Vacquier 1978; Haley and Wessel 1999, 2004).

The components of the cortical granules bind to the vitelline envelope to form a **fertilization envelope**. The fertilization envelope starts to form at the site of sperm entry and continues its expansion around the egg. This process starts about 20 seconds after sperm attachment and is complete by the end of the first minute of fertilization (**FIGURE 7.16**; Wong and Wessel 2004, 2008).

> **WATCH DEVELOPMENT 7.1**   See the fertilization envelope rise from the egg surface.

The fertilization envelope is elevated from the cell membrane by mucopolysaccharides released by the cortical granules. These viscous compounds absorb water to expand the space between the cell membrane and the fertilization envelope, so that the envelope moves radially away from the egg. The fertilization envelope is then stabilized by crosslinking adjacent proteins through egg-specific peroxidase enzymes and a transglutaminase released from the cortical granules (**FIGURE 7.17**; Foerder and Shapiro 1977; Wong et al. 2004; Wong and Wessel 2009). This crosslinking allows the egg and early embryo to resist the shear forces of the ocean's intertidal waves. As this is happening, a fourth set of cortical granule proteins, including hyalin, forms a coating around the egg (Hylander and Summers 1982). The egg extends elongated microvilli whose tips attach to this **hyaline layer**, which provides support for the blastomeres during cleavage.

Calcium as the initiator of the cortical granule reaction

The mechanism of cortical granule exocytosis is similar to that of the exocytosis of the acrosome, and it may involve many of the same molecules. Upon fertilization, the concentration of free Ca^{2+} in the egg cytoplasm increases greatly. In this high-calcium environment, the cortical granule membranes fuse with the egg cell membrane, releasing their contents (see Figure 7.17A). Once the fusion of the cortical granules begins near the point of sperm entry, a wave of cortical granule exocytosis propagates around the cortex to the opposite side of the egg.

VADE MECUM

The two blocks to polyspermy were discovered in the early 1900s by the African-American embryologist Ernest Everett Just. The sea urchin segment contains videos of Just's work on these embryos.

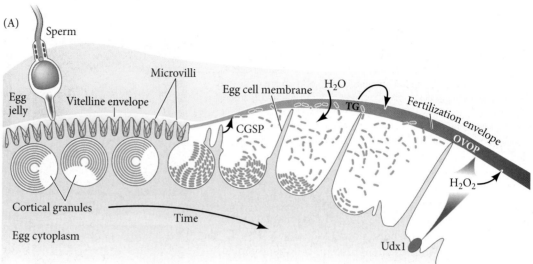

(A)

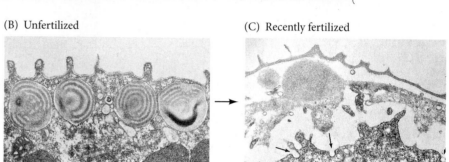

FIGURE 7.17 Cortical granule exocytosis and formation of the sea urchin fertilization envelope. (A) Schematic diagram of events leading to the formation of the fertilization envelope. As cortical granules undergo exocytosis, they release cortical granule serine protease (CGSP), an enzyme that cleaves the proteins linking the vitelline envelope to the cell membrane. Mucopolysaccharides released by the cortical granules form an osmotic gradient, causing water to enter and swell the space between the vitelline envelope and the cell membrane. The enzyme Udx1 in the former cortical granule membrane catalyzes the formation of hydrogen peroxide (H_2O_2), the substrate for soluble ovoperoxidase (OVOP). OVOP and transglutaminases (TG) harden the vitelline envelope, now called the fertilization envelope. (B,C) Transmission electron micrographs of the cortex of an unfertilized sea urchin egg and the same region of a recently fertilized egg. The raised fertilization envelope and the points at which the cortical granules have fused with the egg cell membrane of the egg (arrows) are visible in (C). (A after Wong et al. 2008; B,C from Chandler and Heuser 1979, courtesy of D. E. Chandler.)

In sea urchins and mammals, the rise in Ca^{2+} concentration responsible for the cortical granule reaction is not due to an influx of calcium into the egg, but comes from within the egg itself. The release of calcium from intracellular storage can be monitored visually using calcium-activated luminescent dyes such as aequorin (a protein that, like GFP, is isolated from luminescent jellyfish) or fluorescent dyes such as fura-2. These dyes emit light when they bind free Ca^{2+}. When a sea urchin egg is injected with dye and then fertilized, a striking wave of calcium release propagates across the egg and is visualized as a band of light that starts at the point of sperm entry and proceeds actively to the other end of the cell (**FIGURE 7.18**; Steinhardt et al. 1977; Hafner et al. 1988). The entire release of Ca^{2+} is complete within roughly 30 seconds, and free Ca^{2+} is re-sequestered shortly after being released.

WATCH DEVELOPMENT 7.2 This video of sea urchin fertilization shows waves of calcium ions starting at the point of sperm attachment and traversing the sea urchin egg.

Several experiments have demonstrated that Ca^{2+} is directly responsible for propagating the cortical granule reaction, and that these ions are stored within the egg itself. The drug A23187 is a calcium ionophore—a compound that allows the diffusion of ions such as Ca^{2+} across lipid membranes, permitting them to travel across otherwise impermeable barriers. Placing unfertilized sea urchin eggs into seawater containing A23187 initiates the cortical granule reaction and the elevation of the fertilization envelope. Moreover, this reaction occurs in the absence of any Ca^{2+} in the surrounding water; thus the A23187 must be stimulating the release of Ca^{2+} that is already sequestered in organelles within the egg (Chambers et al. 1974; Steinhardt and Epel 1974).

In sea urchins and vertebrates (but not snails and worms), the Ca^{2+} responsible for the cortical granule reaction is stored in the endoplasmic reticulum of the egg (Eisen and Reynolds 1985; Terasaki and Sardet 1991). In sea urchins and frogs, this reticulum

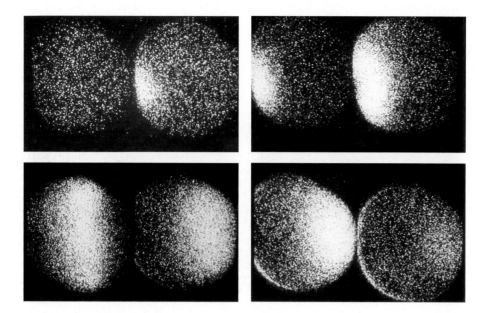

FIGURE 7.18 Calcium release across a sea urchin egg during fertilization. The egg is pre-loaded with a dye that fluoresces when it binds Ca²⁺. When a sperm fuses with the egg, a wave of calcium release is seen, beginning at the site of sperm entry and propagating across the egg. The wave does not simply diffuse but travels actively, taking about 30 seconds to traverse the egg. (Courtesy of G. Schatten.)

is pronounced in the cortex and surrounds the cortical granules (**FIGURE 7.19**; Gardiner and Grey 1983; Luttmer and Longo 1985). The cortical granules are themselves tethered to the cell membrane by a series of integral membrane proteins that facilitate calcium-mediated exocytosis (Conner et al. 1997; Conner and Wessel 1998). Thus, as soon as Ca²⁺ is released from the endoplasmic reticulum, the cortical granules fuse with the cell membrane above them. Once initiated, the release of calcium is self-propagating. Free calcium is able to release sequestered calcium from its storage sites, thus causing a wave of Ca²⁺ release and cortical granule exocytosis.

Activation of Egg Metabolism in Sea Urchins

Although fertilization is often depicted as nothing more than the means to merge two haploid nuclei, it has an equally important role in initiating the processes that begin development. These events happen in the cytoplasm and occur without the involvement of the parental nuclei.[3] In addition to initiating the slow block to polyspermy (through cortical granule exocytosis), the release of Ca²⁺ that occurs when the sperm enters the egg is critical for activating the egg's metabolism and initiating development. Calcium ions release the inhibitors from maternally stored messages, allowing these mRNAs to be translated; they also release the inhibition of nuclear division, thereby allowing

FIGURE 7.19 Endoplasmic reticulum surrounding cortical granules in sea urchin eggs. (A) The endoplasmic reticulum has been stained to allow visualization by transmission electron microscopy. The cortical granule is seen to be surrounded by dark-stained endoplasmic reticulum. (B) An entire egg stained with fluorescent antibodies to calcium-dependent calcium release channels. The antibodies show these channels in the cortical endoplasmic reticulum. (A from Luttmer and Longo 1985, courtesy of S. Luttmer; B from McPherson et al. 1992, courtesy of F. J. Longo.)

[3] In certain salamanders, this function of fertilization (i.e., initiating development of the embryo) has been totally divorced from the genetic function. The silver salamander *Ambystoma platineum* is a hybrid subspecies consisting solely of females. Each female produces an egg with an unreduced chromosome number. This egg, however, cannot develop on its own, so the silver salamander mates with a male Jefferson salamander (*A. jeffersonianum*). The sperm from the Jefferson salamander merely stimulates the egg's development; it does not contribute genetic material (Uzzell 1964). For details of this complex mechanism of procreation, see Bogart et al. 1989, 2009.

(A)

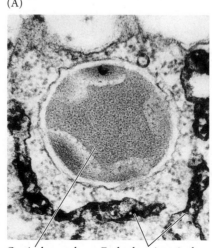

Cortical granule   Endoplasmic reticulum

(B)

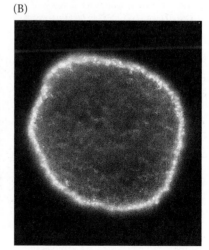

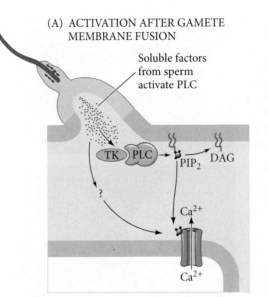

(A) ACTIVATION AFTER GAMETE MEMBRANE FUSION

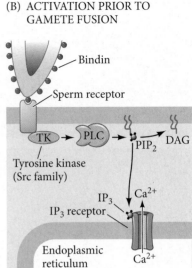

(B) ACTIVATION PRIOR TO GAMETE FUSION

FIGURE 7.20 Probable mechanisms of egg activation. In both cases, a phospholipase C (PLC) is activated and makes IP$_3$ and diacylglycerol (DAG). (A) Ca^{2+} release and egg activation by activated PLC directly from the sperm, or by a substance from the sperm that activates egg PLC. This may be the mechanism in mammals. (B) The bindin receptor (perhaps acting through a G protein) activates tyrosine kinase (TK, an Src kinase), which activates PLC. This is probably the mechanism used by sea urchins.

cleavage to occur. Indeed, throughout the animal kingdom, calcium ions are used to activate development during fertilization.

Release of intracellular calcium ions

The way Ca^{2+} is released varies from species to species (see Parrington et al. 2007). One way, first proposed by Jacques Loeb (1899, 1902), is that a soluble factor from the sperm is introduced into the egg at the time of cell fusion, and this substance activates the egg by changing the ionic composition of the cytoplasm (**FIGURE 7.20A**). This mechanism, as we will see later, probably works in mammals. The other mechanism, proposed by Loeb's rival Frank Lillie (1913), is that the sperm binds to receptors on the egg cell surface and changes their conformation, thus initiating reactions within the cytoplasm that activate the egg (**FIGURE 7.20B**). This is probably what happens in sea urchins.

IP$_3$: THE RELEASER OF CA^{2+} If Ca^{2+} from the egg's endoplasmic reticulum is responsible for the cortical granule reaction and the reactivation of development, what releases Ca^{2+}? Throughout the animal kingdom, it has been found that **inositol 1,4,5-trisphosphate (IP$_3$)** is the primary agent for releasing Ca^{2+} from intracellular storage.

The IP$_3$ pathway is shown in **FIGURE 7.21**. The membrane phospholipid **phosphatidylinositol 4,5-bisphosphate (PIP$_2$)** is split by the enzyme **phospholipase C (PLC)** to yield two active compounds: IP$_3$ and **diacylglycerol (DAG)**. IP$_3$ is able to release Ca^{2+} into the cytoplasm by opening the calcium channels of the endoplasmic reticulum. DAG activates protein kinase C, which in turn activates a protein that exchanges sodium ions for hydrogen ions, raising the pH of the egg (Nishizuka 1986; Swann and Whitaker 1986). This Na$^+$-H$^+$ exchange pump also requires Ca^{2+}. The result of PLC activation is therefore the liberation of Ca^{2+} and the alkalinization of the egg, and both of the compounds this activation creates—IP$_3$ and DAG—are involved in the initiation of development.

WATCH DEVELOPMENT 7.3 See a movie of fertilization with and without PLC activation.

In sea urchin eggs, IP$_3$ is formed initially at the site of sperm entry and can be detected within seconds of sperm-egg attachment. Inhibiting IP$_3$ synthesis prevents Ca^{2+} release (Lee and Shen 1998; Carroll et al. 2000), whereas injected IP$_3$ can release sequestered Ca^{2+} and lead to cortical granule exocytosis (Whitaker and Irvine 1984; Busa et al. 1985). Moreover, these IP$_3$-mediated effects can be thwarted by pre-injecting the egg with calcium-chelating agents (Turner et al. 1986).

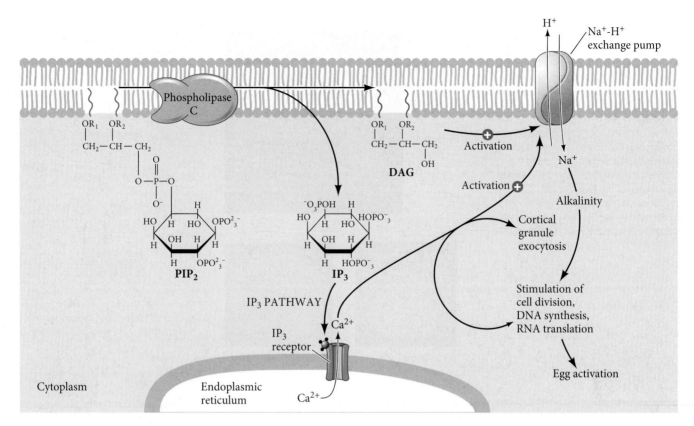

IP$_3$-responsive calcium channels have been found in the egg endoplasmic reticulum. The IP$_3$ formed at the site of sperm entry is thought to bind to IP$_3$ receptors in these calcium channels, effecting a local release of Ca^{2+} (Ferris et al. 1989; Furuichi et al. 1989). Once released, Ca^{2+} can diffuse directly, or it can facilitate the release of more Ca^{2+} by binding to *calcium-triggered calcium-release receptors*, also located in the cortical endoplasmic reticulum (McPherson et al. 1992). These receptors release stored Ca^{2+} when they bind Ca^{2+}, so binding Ca^{2+} releases more Ca^{2+}, which binds to more receptors, and so on. The resulting wave of calcium release is propagated throughout the cell, starting at the point of sperm entry (see Figure 7.18). The cortical granules, which fuse with the cell membrane in the presence of high Ca^{2+} concentrations, respond with a wave of exocytosis that follows the calcium wave. Mohri and colleagues (1995) have shown that IP$_3$-released Ca^{2+} is both necessary and sufficient for initiating the wave of calcium release.

PHOSPHOLIPASE C: THE GENERATOR OF IP$_3$ If IP$_3$ is necessary for Ca^{2+} release and phospholipase C is required in order to generate IP$_3$, the question then becomes, What activates PLC? This question has not been easy to address since (1) there are numerous types of PLC that (2) can be activated through different pathways, and (3) different species use different mechanisms to activate PLC. Results from studies of sea urchin eggs suggest that the active PLC in echinoderms is a member of the γ (gamma) family of PLCs (Carroll et al. 1997, 1999; Shearer et al. 1999). Inhibitors that specifically block PLCγ inhibit IP$_3$ production as well as Ca^{2+} release. Moreover, these inhibitors can be circumvented by microinjecting IP$_3$ into the egg. How PLCγ is activated by sperm is still a matter of controversy, although inhibitor studies have shown that membrane-bound kinases (Src kinases) and GTP-binding proteins play critical roles (**FIGURE 7.22**; Kinsey and Shen 2000; Giusti et al. 2003; Townley et al. 2009; Voronina and Wessel 2003, 2004). One possibility is that NAADP brought in by the sperm to initiate electrical depolarization also activates the enzyme cascade leading to IP$_3$ production and calcium release (Churchill et al. 2003; Morgan and Galione 2007).

FIGURE 7.21   Roles of inositol phosphates in releasing calcium from the endoplasmic reticulum and the initiation of development. Phospholipase C splits PIP$_2$ into IP$_3$ and DAG. IP$_3$ releases calcium from the endoplasmic reticulum, and DAG, with assistance from the released Ca^{2+}, activates the sodium-hydrogen exchange pump in the membrane.

FIGURE 7.22 G protein involvement in Ca²⁺ entry into sea urchin eggs. (A) Mature sea urchin egg immunologically labeled for the cortical granule protein hyaline (red) and the G protein Gαq (green). The overlap of signals produces the yellow color. Gαq is localized to the cortex. (B) A wave of Ca²⁺ appears in the control egg (computer-enhanced to show relative intensities, with red being the highest), but not in the egg injected with an inhibitor of the Gαq protein. (C) Possible model for egg activation by the influx of Ca²⁺. (After Voronina and Wessel 2003; photographs courtesy of G. M. Wessel.)

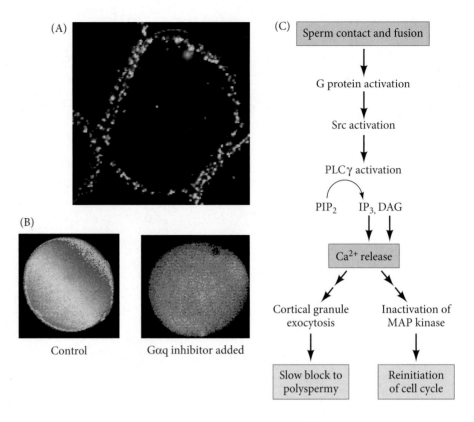

(A)

(B)

Control Gαq inhibitor added

(C)

Sperm contact and fusion

↓

G protein activation

↓

Src activation

↓

PLCγ activation

PIP₂ IP₃, DAG

↓ ↓

Ca²⁺ release

↙ ↘

Cortical granule exocytosis Inactivation of MAP kinase

↓ ↓

Slow block to polyspermy Reinitiation of cell cycle

Effects of calcium release

The flux of calcium across the egg activates a preprogrammed set of metabolic events. The responses of the sea urchin egg to the sperm can be divided into "early" responses, which occur within seconds of the cortical granule reaction, and "late" responses, which take place several minutes after fertilization begins (**TABLE 7.1**).

EARLY RESPONSES   As we have seen, contact or fusion of a sea urchin sperm and egg activates two major blocks to polyspermy: the fast block, mediated by sodium influx into the cell; and the cortical granule reaction, or slow block, mediated by the intracellular release of Ca²⁺. The same release of Ca²⁺ responsible for the cortical granule reaction is also responsible for the re-entry of the egg into the cell cycle and the reactivation of egg protein synthesis. Ca²⁺ levels in the egg increase from 0.05 to between 1 and 5 µM, and in almost all species this occurs as a wave or succession of waves that sweep across the egg beginning at the site of sperm-egg fusion (see Figure 7.18; Jaffe 1983; Terasaki and Sardet 1991; Stricker 1999).

Calcium release activates a series of metabolic reactions that initiate embryonic development (**FIGURE 7.23**). One of these is the activation of the enzyme NAD⁺ kinase, which converts NAD⁺ to NADP⁺ (Epel et al. 1981). Since NADP⁺ (but not NAD⁺) can be used as a coenzyme for lipid biosynthesis, such a conversion has important consequences for lipid metabolism and thus may be important in the construction of the many new cell membranes required during cleavage. Udx1, the enzyme responsible for the reduction of oxygen to crosslink the fertilization envelope, is also NADPH-dependent (Heinecke and Shapiro 1989; Wong et al. 2004). Lastly, NADPH helps regenerate glutathione and ovothiols, molecules that may be crucial scavengers of free radicals that could otherwise damage the DNA of the egg and early embryo (Mead and Epel 1995).

 DEV TUTORIAL   *Find it/lose it/move it*   The basic pattern of biological evidence—find it/lose it/move it—can be followed in the discoveries involving gamete adhesion and calcium activation of the egg.

TABLE 7.1 Events of sea urchin fertilization	
Event	**Approximate time postinsemination[a]**
EARLY RESPONSES	
Sperm-egg binding	0 sec
Fertilization potential rise (fast block to polyspermy)	within 1 sec
Sperm–egg membrane fusion	within 1 sec
Calcium increase first detected	10 sec
Cortical granule exocytosis (slow block to polyspermy)	15–60 sec
LATE RESPONSES	
Activation of NAD kinase	starts at 1 min
Increase in $NADP^+$ and NADPH	starts at 1 min
Increase in O_2 consumption	starts at 1 min
Sperm entry	1–2 min
Acid efflux	1–5 min
Increase in pH (remains high)	1–5 min
Sperm chromatin decondensation	2–12 min
Sperm nucleus migration to egg center	2–12 min
Egg nucleus migration to sperm nucleus	5–10 min
Activation of protein synthesis	starts at 5–10 min
Activation of amino acid transport	starts at 5–10 min
Initiation of DNA synthesis	20–40 min
Mitosis	60–80 min
First cleavage	85–95 min

Main sources: Whitaker and Steinhardt 1985; Mohri et al. 1995.

[a]Approximate times based on data from *S. purpuratus* (15–17°C), *L. pictus* (16–18°C), *A. punctulata* (18–20°C), and *L. variegatus* (22–24°C). The timing of events within the first minute is best known for *L. variegatus*, so times are listed for that species.

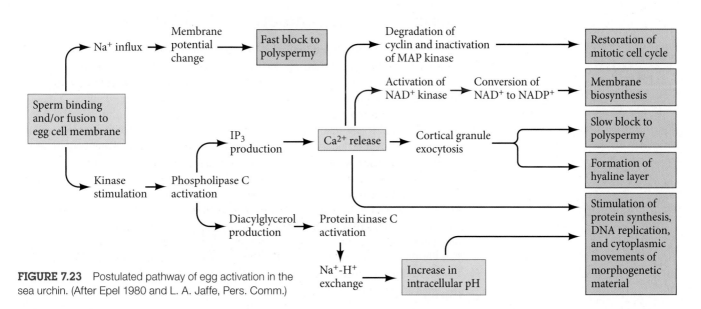

FIGURE 7.23 Postulated pathway of egg activation in the sea urchin. (After Epel 1980 and L. A. Jaffe, Pers. Comm.)

WEB TOPIC 7.4 **RULES OF EVIDENCE** The "find it/lose it/move it" pattern for experimentation fits into a larger system of scientific evidence, as shown by these examples from fertilization research.

LATE RESPONSES: RESUMPTION OF PROTEIN AND DNA SYNTHESIS The late responses of fertilization include the activation of a new burst of DNA and protein synthesis. In sea urchins, the fusion of egg and sperm causes the intracellular pH to increase. This rise in intracellular pH begins with a second influx of Na^+ from seawater, which results in a 1:1 exchange between these sodium ions and hydrogen ions (H^+) from inside the egg. This loss of H^+ causes the pH within the egg to rise (Shen and Steinhardt 1978; Michael and Walt 1999).

It is thought that pH increase and Ca^{2+} elevation act together to stimulate new DNA and protein synthesis (Winkler et al. 1980; Whitaker and Steinhardt 1982; Rees et al. 1995). If one experimentally elevates the pH of an unfertilized egg to a level similar to that of a fertilized egg, DNA synthesis and nuclear envelope breakdown ensue, just as if the egg were fertilized (Miller and Epel 1999). Calcium ions are also critical to new DNA synthesis. The wave of free Ca^{2+} inactivates the enzyme MAP kinase, converting it from a phosphorylated (active) to an unphosphorylated (inactive) form, thus removing an inhibition on DNA synthesis (Carroll et al. 2000). DNA synthesis can then resume.

In sea urchins, a burst of protein synthesis usually occurs within several minutes after sperm entry. This protein synthesis does not depend on the synthesis of new messenger RNA, but uses mRNAs already present in the oocyte cytoplasm. These mRNAs encode proteins such as histones, tubulins, actins, and morphogenetic factors that are used during early development. Such a burst of protein synthesis can be induced by artificially raising the pH of the cytoplasm using ammonium ions (Winkler et al. 1980).

One mechanism for this global rise in the translation of messages stored in the oocyte appears to be the release of inhibitors from the mRNA. In Chapter 2 we discussed maskin, an inhibitor of translation in the unfertilized amphibian oocyte. In sea urchins, a similar inhibitor binds translation initiation factor eIF4E at the 5′ end of several mRNAs and prevents these mRNAs from being translated. Upon fertilization, however, this inhibitor—the eIF4E-binding protein—becomes phosphorylated and is degraded, thus allowing eIF4E to complex with other translation factors and permit protein synthesis from the stored sea urchin mRNAs (Cormier et al. 2001; Oulhen et al. 2007). One of the mRNAs "freed" by the degradation of eIF4E-binding protein is the message encoding cyclin B protein (Salaun et al. 2003, 2004). Cyclin B combines with Cdk1 to create **mitosis-promoting factor (MPF)**, which is required to initiate cell division.

Fusion of Genetic Material in Sea Urchins

After the sperm and egg cell membranes fuse, the sperm nucleus and its centriole separate from the mitochondria and flagellum. The mitochondria and the flagellum disintegrate inside the egg, so very few, if any, sperm-derived mitochondria are found in developing or adult organisms. Thus, although each gamete contributes a haploid genome to the zygote, the *mitochondrial* genome is transmitted primarily by the maternal parent. Conversely, in almost all animals studied (the mouse being the major exception), the centrosome needed to produce the mitotic spindle of the subsequent divisions is derived from the sperm centriole (see Figure 7.14; Sluder et al. 1989, 1993).

Fertilization in sea urchin eggs occurs after the second meiotic division, so there is already a haploid female pronucleus present when the sperm enters the egg cytoplasm. Once inside the egg, the sperm nucleus undergoes a dramatic transformation as it decondenses to form the haploid male pronucleus. First, the nuclear envelope degenerates, exposing the compact sperm chromatin to the egg cytoplasm (Longo and Kunkle 1978; Poccia and Collas 1997). Kinases from the egg cytoplasm phosphorylate the sperm-specific histone proteins, allowing them to decondense. The decondensed histones are then replaced by egg-derived, cleavage-stage histones (Stephens et al.

FIGURE 7.24 Nuclear events in the fertilization of the sea urchin. (A) Sequential photographs showing the migration of the egg pronucleus and the sperm pronucleus toward each other in an egg of *Clypeaster japonicus*. The sperm pronucleus is surrounded by its aster of microtubules. (B) The two pronuclei migrate toward each other on these microtubular processes. (The pronuclear DNA is stained blue by Hoechst dye.) The microtubules (stained green with fluorescent antibodies to tubulin) radiate from the centrosome associated with the (smaller) male pronucleus and reach toward the female pronucleus. (C) Fusion of pronuclei in the sea urchin egg. (A from Hamaguchi and Hiramoto 1980, courtesy of the authors; B from Holy and Schatten 1991, courtesy of J. Holy; C courtesy of F. J. Longo.)

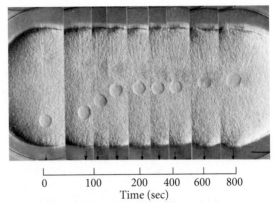

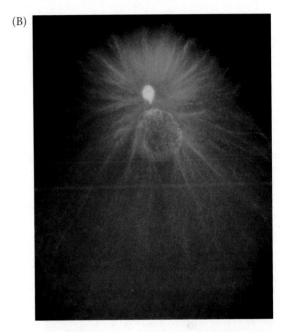

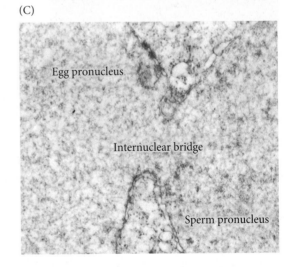

2002; Morin et al. 2012). This exchange permits the decondensation of the sperm chromatin. Once decondensed, the DNA adheres to the nuclear envelope, where DNA polymerase can initiate replication (Infante et al. 1973).

But how do the sperm and egg pronuclei find each other? After the sea urchin sperm enters the egg cytoplasm, the sperm nucleus separates from the tail and rotates 180° so that the sperm centriole is between the developing male pronucleus and the egg pronucleus. The sperm centriole then acts as a microtubule organizing center, extending its own microtubules and integrating them with egg microtubules to form an aster. Microtubules extend throughout the egg and contact the female pronucleus, at which point the two pronuclei migrate toward each other. Their fusion forms the diploid zygote nucleus (**FIGURE 7.24**). DNA synthesis can begin either in the pronuclear stage or after the formation of the zygote nucleus, and depends on the level of Ca^{2+} released earlier in fertilization (Jaffe et al. 2001).

At this point, the diploid nucleus has formed. DNA synthesis and protein synthesis have commenced, and the inhibitions to cell division have been removed. The sea urchin can now begin to form a multicellular organism. We will describe the means by which sea urchins achieve multicellularity in Chapter 10.

WATCH DEVELOPMENT 7.4 Two movies show the sperm pronucleus and egg pronucleus travelling toward each other and fusing.

Internal Fertilization in Mammals

It is very difficult to study any interactions between the mammalian sperm and egg that take place prior to these gametes making contact. One obvious reason for this is that mammalian fertilization occurs inside the oviducts of the female. Although it is relatively easy to mimic the conditions surrounding sea urchin fertilization using natural or artificial seawater, we do not yet know the components of the various natural environments that mammalian sperm encounter as they travel to the egg.

A second reason why it is difficult to study mammalian fertilization is that the sperm population ejaculated into the female is probably heterogeneous, containing spermatozoa at different stages of maturation. Out of the 280×10^6 human sperm normally ejaculated during coitus, only about 200 reach the vicinity of the egg (Ralt et al. 1991). Thus, since fewer than 1 in 10,000 sperm even gets close to the egg, it is difficult to assay those molecules that might enable the sperm to swim toward the egg and become activated.

A third reason why it has been difficult to elucidate the details of mammalian fertilization is the recent discovery that there may be multiple mechanisms (discussed later in the chapter) by which mammalian sperm can undergo the acrosome reaction and bind to the zona pellucida (see Clark 2011).

WATCH DEVELOPMENT 7.5 A video from the laboratory of Dr. Yasayuki Mio shows the events of human fertilization and early development in vitro.

Getting the gametes into the oviduct: Translocation and capacitation

The female reproductive tract is not a passive conduit through which sperm race, but a highly specialized set of tissues that actively regulate the transport and maturity of both gametes. Both the male and female gametes use a combination of small-scale biochemical interactions and large-scale physical propulsion to get to the **ampulla**, the region of the oviduct where fertilization takes place.

TRANSLOCATION The meeting of sperm and egg must be facilitated by the female reproductive tract. Different mechanisms are used to position the gametes at the right place at the right time. A mammalian oocyte just released from the ovary is surrounded by a matrix containing cumulus cells. (Cumulus cells are the cells of the ovarian follicle to which the developing oocyte was attached; see Figure 7.5.) If this matrix is experimentally removed or significantly altered, the fimbriae of the oviduct will not "pick up" the oocyte-cumulus complex (see Figure 12.11), nor will the complex be able to enter the oviduct (Talbot et al. 1999). Once it is picked up, a combination of ciliary beating and muscle contractions transport the oocyte-cumulus complex to the appropriate position for its fertilization in the oviduct.

The sperm must travel a longer path. In humans, about 300 million sperm are ejaculated into the vagina, but only one in a million enters the Fallopian tubes (Harper 1982; Cerezales et al. 2015). The translocation of sperm from the vagina to the oviduct involves several processes that work at different times and places.

• *Sperm motility.* Motility (flagellar action) is probably important in getting sperm through the cervical mucus and into the uterus. Interestingly, in those mammals where the female is promiscuous (mating with several males in rapid succession), sperm from the same male will often form "trains" or aggregates where the combined propulsion of the flagella make the sperm faster (**FIGURE 7.25**). This strategy probably evolved for competition between males. In those species without such female promiscuity, the sperm usually remain individual (Fisher and Hoeckstra 2010; Foster and Pizzari 2010; Fisher et al. 2014).

• *Uterine muscle contractions.* Sperm are found in the oviducts of mice, hamsters, guinea pigs, cows, and humans within 30 minutes of sperm deposition in the vagina—a time "too short to have been attained by even the most Olympian sperm relying on their own flagellar power" (Storey 1995). Rather, sperm appear to be transported to the oviduct by the muscular activity of the uterus.

• *Sperm rheotaxis.* Sperm also receive long-distance directional cues from the flow of liquid from the oviduct to the uterus. Sperm display rheotaxis— that is, they will migrate against the direction of the flow—using CatSper calcium channels (like sea urchin sperm) to sense calcium influx and monitor the direction of the current (Miki and Clapham 2013). Such sperm rheotaxis has been observed in mice and in humans.

(A)

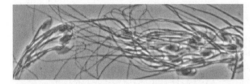

(B)

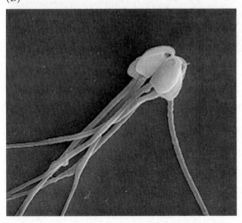

FIGURE 7.25 Sperm associations can occur in species where females mate with several males in a brief timespan. (A) The "sperm train" of the wood mouse *Apodemus sylvaticus*. Sperm are joined by their acrosomal caps. (B) Close-up of the sperm heads of the field mouse *Peromyscus maniculatus*, showing hook-to-hook attachment. (A from Foster and Pizzari 2010, courtesy of T. Pizzari and H. Moore; B from Fischer et al. 2014, courtesy of H. S. Fischer and H. Hoekstra.)

CAPACITATION During the trek from the vagina to the ampullary region of the oviduct, the sperm matures such that it has the capacity to fertilize the egg when the two finally meet. Unlike the sperm of frogs or sea urchins, newly ejaculated mammalian sperm are immature and cannot fertilize the egg; they are unable to undergo the acrosome reaction or to sense the cues that will eventually guide them to the egg. To achieve such competence, the sperm must undergo a suite of sequential physiological changes called **capacitation**. These changes are accomplished only after a sperm has resided for some time in the female reproductive tract (Chang 1951; Austin 1952). Sperm that are not capacitated are "held up" in the cumulus matrix and are unable to reach the egg (Austin 1960; Corselli and Talbot 1987).

 DEV TUTORIAL *Capacitation* **The knowledge that recently ejaculated mammalian sperm could not fertilize an egg was a critical breakthrough in the development of successful in vitro fertilization techniques.**

Contrary to popular belief, the race is not always to the swift. A study by Wilcox and colleagues (1995) found that nearly all human pregnancies result from sexual intercourse during a 6-day period ending on the day of ovulation. This means that the fertilizing sperm could have taken as long as 6 days to make the journey to the oviduct. Although some human sperm reach the ampulla of the oviduct within half an hour of intercourse, "speedy" sperm may have little chance of fertilizing the egg because they have not undergone capacitation. Eisenbach (1995) proposed a hypothesis wherein capacitation is a transient event, and sperm are given a relatively brief window of competence during which they can successfully fertilize the egg. As the sperm reach the ampulla, they acquire competence—but they lose it if they stay around too long.

The molecular processes of capacitation prepare the sperm for the acrosome reaction and enable the sperm to become hyperactive (**FIGURE 7.26**). Although the details of these processes still await description (they are notoriously difficult to study), two sets of molecular changes are considered to be important:

1. *Lipid changes.* The sperm cell membrane is altered by the removal of cholesterol by albumin proteins in the female reproductive tract (Cross 1998). The cholesterol efflux from the sperm cell membrane is thought to change the location of its "lipid rafts," isolated regions that often contain receptor proteins that can bind the zona pellucida and participate in the acrosome reaction (Bou Khalil et al. 2006; Gadella et al. 2008). Originally located throughout the sperm cell membrane, after cholesterol efflux lipid rafts are clustered over the anterior sperm head. The outer acrosomal membrane changes and comes into contact with the sperm cell membrane in a way that prepares it for the acrosome reaction (Tulsiani and Abou-Haila 2004).

2. *Protein changes.* Particular proteins or carbohydrates on the sperm surface

FIGURE 7.26 Hypothetical model for mammalian sperm capacitation. The pathway is modulated by the removal of cholesterol from the sperm cell membrane, which allows the influx of bicarbonate ions (HCO_3^-) and calcium ions (Ca^{2+}). These ions activate adenylate kinase (SACY), thereby elevating cAMP concentrations. The high cAMP levels then activate protein kinase A (PKA). Active PKA phosphorylates several tyrosine kinases, which in turn phosphorylate several sperm proteins, leading to capacitation. Increased intracellular Ca^{2+} also activates the phosphorylation of these proteins, as well as contributing to the hyperactivation of the sperm. (After Visconti et al. 2011.)

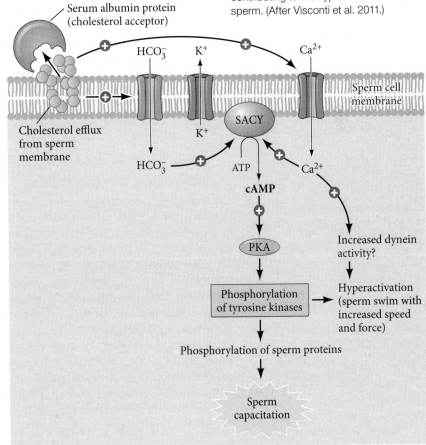

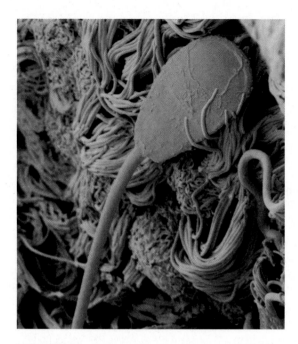

FIGURE 7.27 Scanning electron micrograph (artificially colored) showing bull sperm as it adheres to the membranes of epithelial cells in the oviduct of a cow prior to entering the ampulla. (From Lefebvre et al. 1995, courtesy of S. Suarez.)

are lost during capacitation (Lopez et al. 1985; Wilson and Oliphant 1987). It is possible that these compounds block the recognition sites for the sperm proteins that bind to the zona pellucida. It has been suggested that the unmasking of these sites might be one of the effects of cholesterol depletion (Benoff 1993). The membrane potential of the sperm cell becomes more negative as potassium ions leave the sperm. This change in membrane potential may allow calcium channels to be opened and permit calcium to enter the sperm. Calcium and bicarbonate ions are critical in activating cAMP production and in facilitating the membrane fusion events of the acrosome reaction (Visconti et al. 1995; Arnoult et al. 1999). The influx of bicarbonate ions (and possibly other ions) alkalinizes the sperm, raising its pH. This will be critical in the subsequent activation of calcium channels (Navarro et al. 2007). As a result of cAMP formation, protein phosphorylation occurs (Galantino-Homer et al. 1997; Arcelay et al. 2008). Once they are phosphorylated, some proteins migrate to the surface of the sperm head. One of these proteins is Izumo, which is critical in sperm-egg fusion (see Figure 7.30; Baker et al. 2010).

There may be an important connection between sperm translocation and capacitation. Smith (1998) and Suarez (1998) have documented that before entering the ampulla of the oviduct, the uncapacitated sperm bind actively to the membranes of the oviduct cells in the narrow passage (the isthmus) preceding it (**FIGURE 7.27**; see also Figure 12.11). This binding is temporary and appears to be broken when the sperm become capacitated. Moreover, the life span of the sperm is significantly lengthened by this binding. This restriction of sperm entry into the ampulla during capacitation, and the expansion of sperm life span may have important consequences (Töpfer-Petersen et al. 2002; Gwathmey et al. 2003). The binding action may function as a block to polyspermy by preventing many sperm from reaching the egg at the same time (if the oviduct isthmus is excised in cows, a much higher rate of polyspermy results). In addition, slowing the rate of sperm capacitation and extending the active life of sperm may maximize the probability that sperm will still be available to meet the egg in the ampulla.

In the vicinity of the oocyte: Hyperactivation, thermotaxis, and chemotaxis

Toward the end of capacitation, sperm become hyperactivated—they swim at higher velocities and generate greater force. Hyperactivation appears to be mediated by the opening of a sperm-specific calcium channel in the sperm tail (see Figure 7.26; Ren et al. 2001; Quill et al. 2003). The symmetric beating of the flagellum is changed into a rapid asynchronous beat with a higher degree of bending. The power of the beat and the direction of sperm head movement are thought to release the sperm from their binding with the oviduct epithelial cells. Indeed, only hyperactivated sperm are seen to detach and continue their journey to the egg (Suarez 2008a,b). Hyperactivation may enable sperm to respond differently to the fluid current. Uncapacitated sperm move in a planar direction, allowing more time for the sperm head to attach to the oviduct epithelial cells. Capacitated sperm rotate around their long axis, probably enhancing the detachment of the sperm from the epithelia (Miki and Clapham 2013). Hyperactivation, along with a hyaluronidase enzyme on the outside of the sperm cell membrane, enables the sperm to digest a path through the extracellular matrix of the cumulus cells (Lin et al. 1994; Kimura et al. 2009).

An old joke claims that the reason a man has to release so many sperm at each ejaculation is that no male gamete is willing to ask for directions. So what *does* provide the sperm with directions? Heat is one cue: there is a thermal gradient of 2°C between the isthmus of the oviduct and the warmer ampullary region (Bahat et al. 2003, 2006). Capacitated mammalian sperm can sense thermal differences as small as 0.014°C over a millimeter and tend to migrate toward the higher temperature (Bahat et al. 2012). This ability to sense temperature difference and preferentially swim from cooler to warmer sites (**thermotaxis**) is found only in capacitated sperm.

By the time the sperm are in the ampullary region, most of them have undergone the acrosome reaction (La Spina et al. 2016; Muro et al. 2016). Now, a second sensing mechanism, **chemotaxis**, may come into play. It appears that the oocyte and its accompanying cumulus cells secrete molecules that attract capacitated (and only capacitated) sperm toward the egg during the last stages of sperm migration (Ralt et al. 1991; Cohen-Dayag et al. 1995; Eisenbach and Tur-Kaspa 1999; Wang et al. 2001). The identity of these chemotactic compounds is being investigated, but one of them appears to be the hormone **progesterone**, which is made by the cumulus cells. Guidobaldi and colleagues (2008) have shown that rabbit sperm binds progesterone secreted from the cumulus cells surrounding the oocyte and uses the hormone as a directional cue. In humans, progesterone has been shown to bind to a receptor that activates Ca^{2+} channels in the cell membrane of the sperm tail, leading to sperm hyperactivity (Lishko et al. 2011; Strünker et al. 2011). Mouse cumulus cells also secrete a substance, CRISP1, that attracts sperm and hyperactivates them through CatSper channels (Ernesto et al. 2015). The human cumulus also appears to make a substance (or substances) that attracts sperm, and it appears to form a gradient permitting the sperm to move through the cumulus toward the egg (Sun et al. 2005; Williams et al. 2015). Whether these are the same chemoattractants or different ones has yet to be resolved. This activation takes place only after the sperm's intracellular pH has increased, which may help explain why capacitation is needed in order for sperm to reach and fertilize the egg (Navarro et al. 2007).

Thus it appears that, just as in the case of sperm-activating peptides in sea urchins, progesterone both provides direction and activates sperm motility. Moreover, as in certain invertebrate eggs, it appears that the human egg secretes a chemotactic factor only when it is capable of being fertilized, and that sperm are attracted to such a compound only when they are capable of fertilizing the egg.

In summary, three calcium-mediated sensing processes get the mammalian sperm to the egg: rheotaxis (long-range), thermotaxis (moderate range), and finally chemotaxis, which works within millimeters of the egg.

The acrosome reaction and recognition at the zona pellucida

Before the mammalian sperm can bind to the oocyte, it must first bind to and penetrate the egg's zona pellucida. The zona pellucida in mammals plays a role analogous to that of the vitelline envelope in invertebrates; the zona, however, is a far thicker and denser structure than the vitelline envelope. The mouse zona pellucida is made of three major glycoproteins—**ZP1**, **ZP2**, and **ZP3 (zona proteins 1, 2,** and **3)**—along with accessory proteins that bind to the zona's integral structure. The human zona pellucida has four major glycoproteins—ZP1, ZP2, ZP3, and ZP4.

The binding of sperm to the zona is relatively, but not absolutely, species-specific, and a species may use multiple mechanisms to achieve this binding. Early evidence from rabbits and hamsters (Huang et al. 1981; Yanagimachi and Phillips 1984) suggested that the sperm arriving at the egg had already undergone the acrosome reaction. More recently, Jin and colleagues (2011) showed that the mouse acrosome reaction occurs prior to the sperm binding to the zona (**FIGURE 7.28A**). They found that "successful" sperm—i.e., those that actually fertilized an egg—had already undergone the acrosome reaction by the time they were first seen in the cumulus. Sperm that underwent the acrosome reaction on the zona were almost always unsuccessful.

Thus it appears that most sperm undergo the acrosome reaction in or around the cumulus. Moreover they probably bind to the egg through ZP2 on the zona pellucida. In a gain-of-function experiment, ZP2 was shown to be critical for human sperm-egg

---

Developing Questions

Sometimes the egg and sperm fail to meet and conception does not take place. What are the leading causes of infertility in humans, and what procedures are being used to circumvent these blocks?

(A)

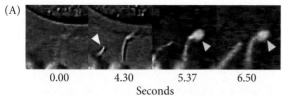

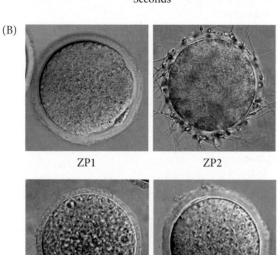

ZP1 ZP2

ZP3 ZP4

FIGURE 7.28 Acrosome-reacted mouse sperm bind to the zona and are successful at fertilizing the egg. (A) The acrosomes of mouse sperm were labeled with GFP such that intact acrosomes fluoresced green; sperm tails were labeled with red fluorescent markers. When the labeled sperm were allowed to interact with a mouse egg and cumulus, the resulting video revealed that the fertilizing sperm (arrowhead at 4.30 seconds) showed no green fluorescence when it reached the surface of the zona pellucida at 6.20 seconds—indicating that it had undergone the acrosome reaction before that time. An adjacent sperm did fluoresce green, meaning its acrosome remained intact. Such acrosome-intact sperm remain bound to the zona without undergoing the acrosome reaction or progressing to the egg cell membrane. (B) Gain-of-function experiment demonstrating that human sperm bind to ZP2. Of the four human zona pellucida proteins, only ZP4 is *not* found in mouse zona. Transgenic mouse oocytes were constructed that expressed the three normal mouse zona proteins and also one of the four human zona proteins. When human sperm were added to the mouse oocytes, they bound only to those transgenic oocytes that expressed human ZP2. Human sperm did not bind to cells expressing human ZP1, ZP3, or ZP4. (A from Jin et al. 2011, courtesy of N. Hirohashi; B from Baibakov et al. 2012.)

binding. Human sperm does not bind to the zona of mouse eggs, so Baibakov and colleagues (2012) added the different human zona proteins separately to the zona of mouse eggs. Only those mouse eggs with *human* ZP2 bound human sperm (**FIGURE 7.28B**). Using mutant forms of ZP2, Avella and colleagues (2014) demonstrated that there is a particular region of the mouse ZP2 protein (between amino acids 51 and 149) that bound the sperm. This region is seen in human ZP2 and may be responsible for sperm-zona binding in humans as well. ZP3 was the other candidate for binding sperm; however, Gahlay and colleagues (2010) provided evidence that mouse eggs with mutations in ZP3 were still fertilized.

In mice, there is also evidence that acrosome-intact sperm can bind to ZP3, and that ZP3 can cause the acrosome reaction directly on the zona (Bleil and Wassarman 1980, 1983). And in humans, there is evidence that the reaction can also be induced by the zona proteins, perhaps by all of them acting in concert (Gupta 2015). Indeed, there may be several means to initiate the acrosome reaction and to bind to and penetrate the zona pellucida. These mechanisms may act simultaneously, or perhaps one mechanism is used for acrosome-intact sperm and another for acrosome-reacted sperm. Given that the zona's biochemical composition differs in different species, the mechanisms that predominate in one species need not be the same in another. The sperm receptor that binds to the zona proteins has not yet been identified. It is probably a complex containing several proteins that bind to both the protein and carbohydrate portions of the zona glycoproteins (Chiu et al. 2014).

(A)

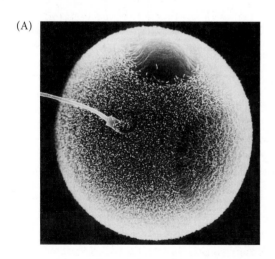

(B)

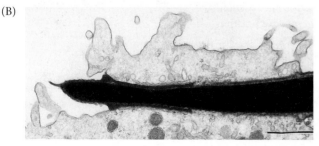

FIGURE 7.29 Entry of sperm into a golden hamster egg. (A) Scanning electron micrograph of sperm fusing with egg. The "bald" spot (without microvilli) is where the polar body has budded off. Sperm do not bind there. (B) Transmission electron micrograph of the sperm fusing parallel to the egg cell membrane. (From Yanagimachi and Noda 1970 and Yanagimachi 1994, courtesy of R. Yanagimachi.)

Gamete fusion and the prevention of polyspermy

In mammals, it is not the tip of the sperm head that makes contact with the egg (as happens in the perpendicular entry of sea urchin sperm) but the side of the sperm head (**FIGURE 7.29**). The acrosome reaction, in addition to expelling the enzymatic contents of the acrosome, also exposes the inner acrosomal membrane to the outside. The junction between this inner acrosomal membrane and the sperm cell membrane is called the **equatorial region**, and this is where membrane fusion between sperm and egg begins (**FIGURE 7.30A**). As in sea urchin gamete fusion, the sperm is bound to regions of the egg where actin polymerizes to extend microvilli to the sperm (Yanagimachi and Noda 1970).

The mechanism of mammalian gamete fusion is still controversial (see Lefèvre et al. 2010; Chalbi et al. 2014). On the sperm side of the mammalian gamete fusion process, Inoue and colleagues (2005) have implicated an immunoglobulin-like protein, named Izumo after a Japanese shrine dedicated to marriage. This protein is originally found in the membrane of the acrosomal granule (**FIGURE 7.30B**). However, after the acrosome reaction, Izumo redistributes along on the surface of acrosome-reacted sperm, where it is found primarily in the equatorial section, where mammalian sperm-egg binding takes place (see Figure 7.30A; Satouh et al. 2012). Sperm from mice carrying loss-of-function mutations in the *Izumo* gene are able to bind and penetrate the zona pellucida, but they are not able to fuse with the egg cell membrane. Human sperm also contain Izumo protein, and antibodies directed against Izumo prevent sperm-egg fusion in humans as well. There are other candidates for sperm fusion proteins, and there may be several sperm-egg binding systems operating, each of which may be necessary but not sufficient to ensure proper gamete binding and fusion.

Izumo binds to an oocyte protein called Juno (after the Roman goddess of marriage and fertility), and eggs deficient in Juno cannot bind or fuse with acrosome-reacted sperm (Bianchi et al. 2014). The interaction of Izumo and Juno recruits the egg membrane protein CD9 to the area of sperm-egg adhesion (Chalbi et al. 2014.) This protein appears to be involved with sperm-egg fusion, since female mice with the *CD9* gene knocked out are infertile due to fusion defects (Kaji et al. 2002; Runge et al. 2006). It is not known exactly how these proteins facilitate membrane fusion, but CD9 protein is also known to be critical for the fusion of myocytes (muscle cell precursors) to form striated muscle (Tachibana and Hemler 1999).

Polyspermy is a problem for mammals just as it is for sea urchins. In mammals, no electrical fast block to polyspermy has yet been detected; it may not be needed, given the limited number of sperm that reach the ovulated egg (Gardner and Evans 2006). However, a *slow* block to polyspermy occurs when enzymes released by the cortical granules modify the zona pellucida sperm receptor proteins such that they can no longer bind sperm (Bleil and Wassarman 1980). ZP2 is clipped by the protease ovastacin and loses its ability to bind sperm (**FIGURE 7.31**; Moller and Wassarman 1989). Ovastacin is found in the cortical granules of unfertilized eggs and is released during cortical granule fusion. Indeed, polyspermy occurs more frequently in mouse eggs bearing mutant ZP2 that cannot be cleaved by ovastacin (Gahlay et al. 2010; Burkart et al. 2012).

Another slow block to polyspermy involves the Juno protein (Bianchi and Wright 2014). As the sperm and egg membranes fuse, Juno protein appears to be released from the plasma membrane. Thus, the docking site for sperm would be removed. Moreover, this soluble Juno protein can bind sperm in the perivitelline space between the zona pellucida and oocyte,

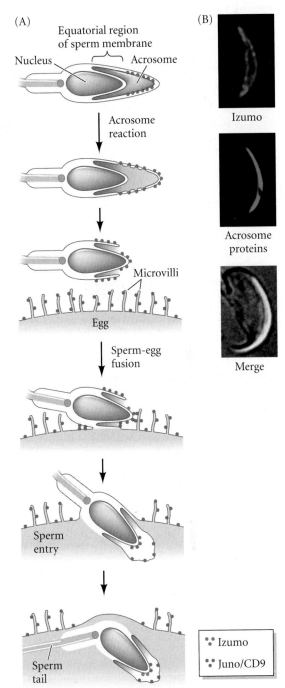

FIGURE 7.30 Izumo protein and membrane fusion in mouse fertilization. (A) Localization of Izumo to the inner and outer acrosomal membrane. Izumo is stained red, acrosomal proteins green. (B) Diagram of sperm-egg cell membrane fusion. During the acrosome reaction, Izumo localized on the acrosomal becomes translocated to the sperm cell membrane. There it meets the complex of Juno and CD9 proteins on the egg microvilli, initiating membrane fusion and the entry of the sperm into the egg. (After Satouh et al. 2012; photographs courtesy of M. Okabe.)

FIGURE 7.31 Cleaved ZP2 is necessary for the block to polyspermy in mammals. Eggs and embryos were visualized by fluorescence microscopy (to see sperm nuclei; top row) and brightfield microscopy (differential interference contrast, to see sperm tails; bottom row). Sperm bound normally to eggs containing a mutant ZP2 that could not be cleaved. However, the egg with normal (i.e., cleavable) ZP2 got rid of sperm by the 2-cell stage, whereas the egg with the mutant (uncleavable) ZP2 retained sperm. (From Gahlay et al. 2010, photograph courtesy of J. Dean.)

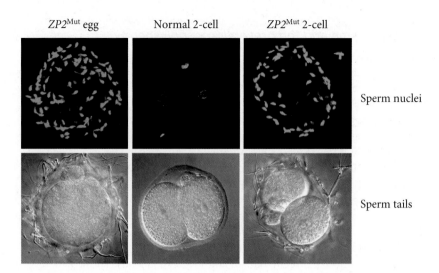

$ZP2^{Mut}$ egg   Normal 2-cell $ZP2^{Mut}$ 2-cell

Sperm nuclei

Sperm tails

Developing Questions

One of the goals of modern pharmacology is to create a male contraceptive. Reviewing the steps of fertilization, what steps do you think it might be possible to block pharmacologically in order to produce such a contraceptive for males?

thereby preventing the sperm from seeing any Juno protein that may still reside on the oocyte membrane.

Fusion of genetic material

As in sea urchins, the single mammalian sperm that finally enters the egg carries its genetic contribution in a haploid pronucleus. In mammals, however, the process of pronuclear migration takes about 12 hours, compared with less than 1 hour in the sea urchin. The DNA of the sperm pronucleus is bound by protamines—basic proteins that are tightly compacted through disulfide bonds. Glutathione in the egg cytoplasm reduces these disulfide bonds and allows the sperm chromatin to uncoil (Calvin and Bedford 1971; Kvist et al. 1980; Sutovsky and Schatten 1997).

The mammalian sperm enters the oocyte while the oocyte nucleus is "arrested" in metaphase of its second meiotic division (**FIGURE 7.32A,B**; see also Figure 7.3). As described for the sea urchin, the calcium oscillations brought about by sperm entry inactivate MAP kinase and allow DNA synthesis. But unlike the sea urchin egg, which is already in a haploid state, the chromosomes of the mammalian oocyte are still in the middle of meiotic metaphase. Oscillations in the level of Ca^{2+} activate another kinase that leads to the proteolysis of cyclin (thus allowing the cell cycle to continue) and securin (the protein that holds the metaphase chromosomes together), eventually resulting in a haploid female pronucleus (Watanabe et al. 1991; Johnson et al. 1998).

DNA synthesis occurs separately in the male and female pronuclei. The centrosome produced by the male pronucleus generates asters (largely from microtubule proteins stored in the oocyte). The microtubules join the two pronuclei and enable them to migrate toward one another. Upon meeting, the two nuclear envelopes break down. However, instead of producing a common zygote nucleus (as in sea urchins), the chromatin condenses into chromosomes that orient themselves on a common mitotic spindle (**FIGURE 7.32C,D**). Thus, in mammals a true diploid nucleus is seen for the first time not in the zygote, but at the 2-cell stage.

Each sperm brings into the egg not only its nucleus but also its mitochondria, its centriole, and a tiny amount of cytoplasm. The sperm mitochondria and their DNA are degraded in the egg cytoplasm, so all of the new individual's mitochondria are derived from its mother. The egg and embryo appear to get rid of the paternal mitochondria both by dilution and by actively targeting them for destruction (Cummins et al. 1998; Shitara et al. 1998; Schwartz and Vissing 2002). In most mammals, however, the sperm centriole not only survives but appears to serve as the organizing agent for making the new mitotic spindle. Moreover, the sperm cytoplasm has recently been found to contain enzymes that activate egg metabolism, as well as RNA fragments that may alter gene expression (Sharma et al. 2016).

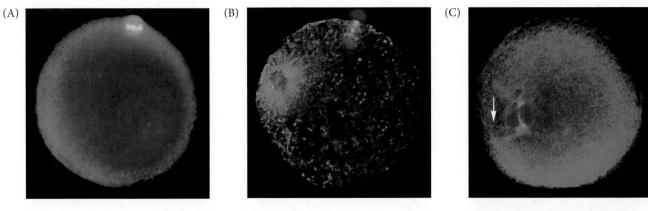

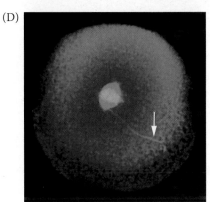

FIGURE 7.32 Pronuclear movements during human fertilization. Microtubules are stained green, DNA is dyed blue. Arrows point to the sperm tail. (A) The mature unfertilized oocyte completes the first meiotic division, budding off a polar body. (B) As the sperm enters the oocyte (left side), microtubules condense around it as the oocyte completes its second meiotic division at the periphery. (C) By 15 hours after fertilization, the two pronuclei have come together, and the centrosome splits to organize a bipolar microtubule array. The sperm tail is still seen (arrow). (D) At prometaphase, chromosomes from the sperm and egg intermix on the metaphase equator and a mitotic spindle initiates the first mitotic division. The sperm tail can still be seen. (From Simerly et al. 1995, courtesy of G. Schatten.)

WEB TOPIC 7.5 **THE NON-EQUIVALENCE OF MAMMALIAN PRONUCLEI**
Countering Mendelian expectations, some genes are only active when they come from the sperm while others are active only when they come from the egg. These are known as "imprinted" genes.

Activation of the mammalian egg

As in every other animal group studied, a transient rise in cytoplasmic Ca^{2+} is necessary for egg activation in mammals. The sperm induces a series of Ca^{2+} waves that can last for hours, terminating in egg activation (i.e., resumption of meiosis, cortical granule exocytosis, and release of the inhibition on maternal mRNAs) and the formation of the male and female pronuclei. And, again as in sea urchins, fertilization triggers intracellular Ca^{2+} release through the production of IP_3 by the enzyme phospholipase C (Swann et al. 2006; Igarashi et al. 2007).

However, the mammalian PLC responsible for egg activation and pronucleus formation may in fact come from the sperm rather than from the egg. Some of the first observations for a sperm-derived PLC came from studies of intracytoplasmic sperm injection (ICSI), an experimental treatment for curing infertility. Here, sperm are directly injected into oocyte cytoplasm, bypassing any interaction with the egg cell membrane. To the surprise of many biologists (who had assumed that sperm *binding* to an egg receptor protein was critical for egg activation), this treatment worked. The human egg was activated and pronuclei formed. Injecting mouse sperm into mouse eggs will also induce fertilization-like Ca^{2+} oscillations in the egg and lead to complete development (Kimura and Yanagimachi 1995).

It appeared that an activator of Ca^{2+} release was stored in the sperm head (see Figure 7.20A). This activator turned out to be a soluble sperm PLC enzyme, **PLCζ** (zeta), that is delivered to the egg by gamete fusion. In mice, expression of PLCζ mRNA in the egg produces Ca^{2+} oscillations, and removing PLCζ from mouse sperm (by antibodies or RNAi) abolishes the sperm's calcium-inducing activity (Saunders et al. 2002; Yoda et al. 2004; Knott et al. 2005). Human sperm that are unsuccessful in ICSI have been shown to have little or no functional PLCζ. In fact, normal human sperm can activate Ca^{2+} oscillations when injected into mouse eggs, but sperm lacking PLCζ do not (Yoon et al. 2008).

Whereas sea urchin eggs usually are activated as a single wave of Ca^{2+} crosses from the point of sperm entry, the mammalian egg is traversed by numerous waves of calcium ions (Miyazaki et al. 1992; Ajduk et al. 2008; Ducibella and Fissore 2008). The extent (amplitude, duration, and number) of these Ca^{2+} oscillations appears to regulate the timing of mammalian egg activation events (Ducibella et al. 2002; Ozil et al. 2005; Toth et al. 2006). In this way, cortical granule exocytosis occurs just before resumes meiosis and much before the translation of maternal mRNAs.

In mammals, the Ca^{2+} released by IP_3 binds to a series of proteins including calmodulin-activated protein kinase (which will be important in eliminating the inhibitors of mRNA translation), MAP kinase (which allows the resumption of meiosis), and synaptotagmin (which helps initiate cortical granule fusion). Unused Ca^{2+} is pumped back into the endoplasmic reticulum, and additional Ca^{2+} is acquired from outside the cell. This recruitment of extracellular Ca^{2+} appears to be necessary for the egg to complete meiosis. If Ca^{2+} influx is blocked, the second polar body does not form; instead, the result is two nonviable (triploid) egg pronuclei (Maio et al. 2012; Wakai et al. 2013).

 DEV TUTORIAL   *Legends of the sperm*   The stories people tell about fertilization are often at odds with the actual data of biology.

WEB TOPIC 7.6   **A SOCIAL CRITIQUE OF FERTILIZATION RESEARCH**   How we envision fertilization says a lot about us as well as about the science.

Coda

Fertilization is not a moment or an event, but a process of carefully orchestrated and coordinated events including the contact and fusion of gametes, the fusion of nuclei, and the activation of development. It is a process whereby two cells, each at the verge of death, unite to create a new organism that will have numerous cell types and organs. It is just the beginning of a series of cell-cell interactions that characterize animal development.

Next Step Investigation

Fertilization is a field ripe with important questions to be answered. Some of the most important involve the physiological changes that render the gametes "fertilization-competent." The mechanisms by which sperm become hyperactive and sense the egg are just beginning to become known, as are the mechanisms of sperm capacitation. Meiosis is resumed in mammalian oocytes, but the physiological mechanisms for this resumption remain largely unexplored. How is the polar body formed in a way that the oocyte retains most of the cytoplasm? And how do the gamete recognition proteins interact with cell fusion proteins to allow the sperm to enter the egg? Even the ways by which sperm activate the internal calcium ion channels is an open question. About 6% of American men and women between 15 and 44 years of age are infertile, and this makes the answering these questions extremely important.

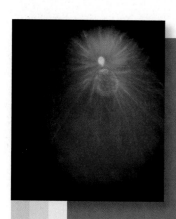

Closing Thoughts on the Opening Photo

When Oscar Hertwig (1877) discovered fertilization in sea urchins, he delighted in seeing what he called "the sun in the egg." This was evidence that the fertilization was going to be successful. This glorious projection turns out to be the microtubular array generated by the sperm centrosome. This set of microtubules reaches out and finds the female pronucleus, and the two pronuclei migrate toward one another on these microtubular tracks. In this micrograph, the DNA of the pronuculei is stained blue, and the female pronucleus is much larger than that derived from the sperm. The microtubules are stained green. (Photograph courtesy of J. Holy and G. Schatten.)

7 Snapshot Summary
Fertilization

1. Fertilization accomplishes two separate activities: sex (the combining of genes derived from two parents) and reproduction (the creation of a new organism).

2. The events of fertilization usually include (1) contact and recognition between sperm and egg; (2) regulation of sperm entry into the egg; (3) fusion of genetic material from the two gametes; and (4) activation of egg metabolism to start development.

3. The sperm head consists of a haploid nucleus and an acrosome. The acrosome is derived from the Golgi apparatus and contains enzymes needed to digest extracellular coats surrounding the egg. The midpiece of the sperm contain mitochondria and the centriole that generates the microtubules of the flagellum. Energy for flagellar motion comes from mitochondrial ATP and a dynein ATPase in the flagellum.

4. The female gamete can be an egg (with a haploid nucleus, as in sea urchins) or an oocyte (in an earlier stage of development, as in mammals). The egg (or oocyte) has a large mass of cytoplasm storing ribosomes and nutritive proteins. Some mRNAs and proteins that will be used as morphogenetic factors are also stored in the egg. Many eggs also contain protective agents needed for survival in their particular environment.

5. Surrounding the egg cell membrane is an extracellular layer often used in sperm recognition. In most animals, this extracellular layer is the vitelline envelope. In mammals, it is the much thicker zona pellucida. Cortical granules lie beneath the egg's cell membrane.

6. Neither the egg nor the sperm is the "active" or "passive" partner; the sperm is activated by the egg, and the egg is activated by the sperm. Both activations involve calcium ions and membrane fusions.

7. In many organisms, eggs secrete diffusible molecules that attract and activate the sperm.

8. Species-specific chemotactic molecules secreted by the egg can attract sperm that are capable of fertilizing it. In sea urchins, the chemotactic peptides resact and speract have been shown to increase sperm motility and provide direction toward an egg of the correct species.

9. The acrosome reaction releases enzymes exocytotically. These proteolytic enzymes digest the egg's protective coating, allowing the sperm to reach and fuse with the egg cell membrane. In sea urchins, this reaction in the sperm is initiated by compounds in the egg jelly. Globular actin polymerizes to extend the acrosomal process. Bindin on the acrosomal process is recognized by a protein complex on the sea urchin egg surface.

10. Fusion between sperm and egg is probably mediated by protein molecules whose hydrophobic groups can merge the sperm and egg cell membranes. In sea urchins, bindin may mediate gamete recognition and fusion.

11. Polyspermy results when two or more sperm fertilize an egg. It is usually lethal, since it results in blastomeres with different numbers and types of chromosomes.

12. Many species have two blocks to polyspermy. The fast block is immediate and causes the egg membrane resting potential to rise. Sperm can no longer fuse with the egg. In sea urchins this is mediated by the influx of sodium ions. The slow block, or cortical granule reaction, is physical and is mediated by calcium ions. A wave of Ca^{2+} propagates from the point of sperm entry, causing the cortical granules to fuse with the egg cell membrane. The released contents of these granules cause the vitelline envelope to rise and harden into the fertilization envelope.

13. The fusion of sperm and egg results in the activation of crucial metabolic reactions in the egg. These reactions include re-initiation of the egg's cell cycle and subsequent mitotic division, and the resumption of DNA and protein synthesis.

14. In all species studied, free Ca^{2+}, supported by the alkalinization of the egg, activates egg metabolism, protein synthesis, and DNA synthesis. Inositol trisphosphate (IP_3) is responsible for releasing Ca^{2+} from storage in the endoplasmic reticulum. DAG (diacylglycerol) is thought to initiate the rise in egg pH.

15. IP_3 is generated by phospholipases. Different species may use different mechanisms to activate the phospholipases.

16. Genetic material is carried in a male and a female pronucleus, which migrate toward each other. In sea urchins, the male and female pronuclei merge and a diploid zygote nucleus is formed. DNA replication occurs after pronuclear fusion.

17. Mammalian fertilization takes place internally, within the female reproductive tract. The cells and tissues of the female reproductive tract actively regulate the transport and maturity of both the male and female gametes.

18. The translocation of sperm from the vagina to the egg is regulated by the muscular activity of the uterus, by the binding of sperm in the isthmus of the oviduct, and by directional cues from the oocyte and/or the cumulus cells surrounding it.

19. Mammalian sperm must be capacitated in the female reproductive tract before they are capable of fertilizing the egg. Capacitation is the result of biochemical changes in the sperm cell membrane and the alkalinization of its cytoplasm. Capacitated mammalian sperm can penetrate the cumulus and bind the zona pellucida.

20. In one model of sperm-zona binding, the acrosome-intact sperm bind to ZP3 on the zona, and ZP3 induces the sperm to undergo the acrosome reaction on the zona pellucida. In a more recent model, the acrosome reaction is induced in the cumulus, and the acrosome-reacted sperm bind to ZP2.

21. In mammals, blocks to polyspermy include modification of the zona proteins by the contents of the cortical granules so that sperm can no longer bind to the zona.

22. The rise in intracellular free Ca^{2+} at fertilization in amphibians and mammals causes the degradation of cyclin and the inactivation of MAP kinase, allowing the second meiotic metaphase to be completed and the formation of the haploid female pronucleus.

23. In mammals, DNA replication takes place as the pronuclei are traveling toward each other. The pronuclear membranes disintegrate as the pronuclei approach each other, and their chromosomes gather around a common metaphase plate.

Further Reading

Bartolomei, M. S. and A. C. Ferguson-Smith. 2011. Mammalian genomic imprinting. *Cold Spring Harbor Persp. Biol.* doi: 10.1101/chsperspect.a002592.

Boveri, T. 1902. On multipolar mitosis as a means of analysis of the cell nucleus. [Translated by S. Gluecksohn-Waelsch.] *In* B. H. Willier and J. M. Oppenheimer (eds.), *Foundations of Experimental Embryology.* Hafner, New York, 1974.

Briggs, E. and G. M. Wessel. 2006. In the beginning: Animal fertilization and sea urchin development. *Dev. Biol.* 300: 15–26.

Gahlay, G., L. Gauthier, B. Baibakov, O. Epifano and J. Dean. 2010. Gamete recognition in mice depends on the cleavage status of an egg's zona pellucida protein. *Science* 329: 216–219.

Glabe, C. G. and V. D. Vacquier. 1978. Egg surface glycoprotein receptor for sea urchin sperm bindin. *Proc. Natl. Acad. Sci. USA* 75: 881–885.

Jaffe, L. A. 1976. Fast block to polyspermy in sea urchins is electrically mediated. *Nature* 261: 68–71.

Jin, M. and 7 others. 2011. Most fertilizing mouse spermatozoa begin their acrosome reaction before contact with the zona pellucida during in vitro fertilization. *Proc. Natl. Acad. Sci. USA* 108: 4892–4896.

Just, E. E. 1919. The fertilization reaction in *Echinarachinus parma. Biol. Bull.* 36: 1–10.

Knott, J. G., M. Kurokawa, R. A. Fissore, R. M. Schultz and C. J. Williams. 2005. Transgenic RNA interference reveals role for mouse sperm phospholipase Cζ in triggering Ca^{2+} oscillations during fertilization. *Biol Reprod.* 72: 992–996.

Parrington, J., L. C. Davis, A. Galione and G. Wessel. 2007. Flipping the switch: How a sperm activates the egg at fertilization. *Dev. Dyn.* 236: 2027–2038.

Vacquier, V. D. and G. W. Moy. 1977. Isolation of bindin: The protein responsible for adhesion of sperm to sea urchin eggs. *Proc. Natl. Acad. Sci. USA* 74: 2456–2460.

Wasserman, P. M. and E. S. Litscher. 2016. A bespoke coat for eggs: Getting ready for fertilization. *Curr. Top. Dev. Biol.* 117: 539–552.

GO TO WWW.DEVBIO.COM...

...for Web Topics, Scientists Speak interviews, Watch Development videos, Dev Tutorials, and complete bibliographic information for all literature cited in this chapter.

Rapid Specification in Snails and Nematodes

FERTILIZATION GIVES THE ORGANISM a new genome and rearranges its cytoplasm. Once this is accomplished, the resulting zygote begins the production of a multicellular organism. During **cleavage**, rapid cell divisions divide the zygote cytoplasm into numerous cells. These cells undergo dramatic displacements during **gastrulation**, a process whereby the cells move to different parts of the embryo and acquire new neighbors. The different patterns of cleavage and gastrulation were described in Chapter 1 (see pp. 11–14).

During cleavage and gastrulation, the major body axes of most animals are determined and the embryonic cells begin to acquire their respective fates. Three axes must be specified: the anterior-posterior (head-tail) axis; the dorsal-ventral (back-belly) axis; and the left-right axis (see Figure 1.6). Different species specify these axes at different times, using different mechanisms. Cleavage always precedes gastrulation, but in some species body axis formation begins as early as oocyte formation (as in *Drosophila*). In other species, the axes begin forming during cleavage (as in tunicates), while in still others axis formation extends all the way through gastrulation (as in *Xenopus*).

The chapters in this unit will look at how representative species in several groups undergo cleavage, gastrulation, axis specification, and cell fate determination. With the exception of the human examples in Chapter 12, virtually all of the species and groups described (including snails, nematodes, fruit flies, sea urchins, frogs, zebrafish, chicks, and mice) have been important **model organisms** for developmental biologists. In other words, these species are easily maintained in the laboratory and have special properties that allow their mechanisms of development to be readily observed. These properties

How are snail embryos determined to coil either to the right or to the left?

The Punchline

Mode of development plays a major role in classifying animal groups. One major taxonomical criterion is whether the anterior end (mouth) or posterior end (anus) of the body develops first. The snails (gastropod mollusks) and nematodes (roundworms) form their mouths first and have evolved rapid specification of their body axes and cell fates, often by placing transcription factors into specific blastomeres during early cleavage. These transcription factors can determine cells autonomously, or they can initiate paracrine factor pathways that induce the determination of neighboring cells. In particular, the D-quadrant blastomeres of snails can work as "organizers" that structure the morphogenesis of the entire embryo. Its transparent cuticle, small cell number, and minute genome allow the study of *C. elegans* development as a model of how genes can act to control axis formation and cell specification.

include quick generation time, large litters, amenability to genetic and surgical manipulation, and the ability to develop under laboratory conditions. (This very ability to develop in the laboratory, however, sometimes precludes our asking certain questions concerning the relationship of development to an organism's natural habitat, as we will discuss in Chapter 25.)

This chapter will detail the early development of two groups of protostome invertebrates, the gastropod mollusks (represented by snails) and the nematodes (represented by *Caenorhabditis elegans*). We begin, however, by taking a brief look at animal evolution and classification as seen through the lens of development.

Developmental Patterns among the Metazoa

To be a **eukaryotic organism** means that the cell contains a nucleus and several distinct chromosomes that undergo mitosis. To be a **multicellular eukaryotic organism** (i.e., plant, fungus, or animal) means that the cells formed by mitosis remain together as a functional whole and that subsequent generations form the same coherent individuals composed of many cells. To be a **metazoan** means to be an animal, and to be an animal means to undergo gastrulation. All animals gastrulate, and animals are the only organisms that gastrulate.

Different groups of organisms undergo different patterns of development. When we say that there are 35 metazoan phyla, we are stating that there are 35 surviving patterns of animal development (see Davidson and Erwin 2009; Levin et al. 2016). These patterns of organization have not evolved in a straight line but in branching pathways. **FIGURE 8.1** shows the four major branches of metazoans: the basal phyla, the lophotrochozoan and ecdysozoan protostomes, and the deuterostomes.

Basal phyla

Animals that have two germ layers—ectoderm and endoderm but little or no mesoderm—are referred to as **diploblasts**. The diploblasts have traditionally included the cnidarians (jellyfish and hydras) and the ctenophores (comb jellies). Recent genomic studies have shown that the ctenophore clade—not the sponges, as had long been thought—is the sister group to all other animals (Ryan et al. 2013; Moroz et al. 2014). The sponges apparently have the genes to produce a nervous system, although no modern sponge group has one. This finding indicates that the nervous system has been lost in the sponge lineage, rather than never having evolved among these animals. The hypothesis that cnidarians rather than sponges are the most ancient extant metazoan lineage remains controversial, despite increasing evidence for this position (Borowiec et al. 2015; Chang et al. 2015; Pisani et al. 2015).

Moreover, it has long been thought that the diploblast cnidarians and ctenophores have radial symmetry and no mesoderm, whereas the **triploblast** phyla (all other animals) have bilateral symmetry and a third germ layer, the mesoderm. However, this clear-cut demarcation is now being questioned in regard to the cnidarians. Although certain cnidarians (such as *Hydra*) have no true mesoderm, others seem to have some mesoderm, and some display bilateral symmetry only at certain stages of their life cycle (Martindale et al. 2004; Martindale 2005). However, the mesoderm of cnidarians may have evolved independently of the mesoderm found in the protostomes and deuterostomes. We now are aware that jellyfish possess striated muscle (necessary for their propulsion movement), but their muscles do not seem related either molecularly or developmentally to the mesodermally derived muscles of vertebrates or insects (Steinmetz et al. 2012). This independent generation of contractile cells appears to represent a remarkable case of evolutionary convergence.

The triploblastic animals: Protostomes and deuterostomes

The vast majority of metazoan species have three germ layers and are thus triploblasts. The evolution of the mesoderm enabled greater mobility and larger bodies because it

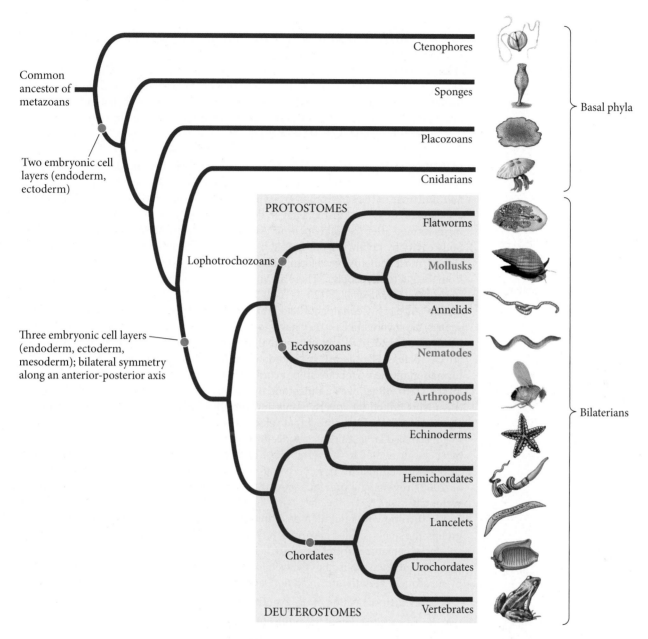

FIGURE 8.1 The tree of metazoan (animal) life. In this analysis, the ctenophores are the sister clade to (i.e., the group that branched off earliest from) the remainder of the animals. The four major groups of extant animals are the basal phyla, lophotrochozoan protostomes, ecdysozoan protostomes, and deuterostomes. Photographs of three protostomes—a gastropod mollusk (snail), the nematode *Caenorhabditis elegans*, and a fruit fly (*Drosophila*)—represent the organisms whose development is detailed here and in Chapter 9. Deuterostome organisms are covered in Chapters 10, 11, and 12. (Sources include Bourlat et al. 2006; Delsuc et al. 2006; Schierwater et al. 2009; Hejnol 2012; Ryan et al. 2013.)

became the animal's musculature and circulatory system. Triploblastic animals are also called **bilaterians** because they have bilateral symmetry—that is, they have right and left sides. Bilaterians are further classified as either **protostomes** or **deuterostomes**.

PROTOSTOMES Protostomes (Greek, "mouth first"), which include the mollusk, arthropod, and worm phyla, are so called because the mouth is formed first, at or near the opening to the gut that is produced during gastrulation. The anus forms later, at a different location. The protostome **coelom**, or body cavity, forms from the hollowing out of a previously solid cord of mesodermal cells in a process called **schizocoely**.

There are two major branches of protostomes. The **ecdysozoans** (Greek *ecdysis*, "to get out of" or "shed") are those animals that molt their exterior skeletons. The most prominent ecdysozoan group is Arthropoda, the arthropods, a well-studied phylum that includes the insects, arachnids, mites, crustaceans, and millipedes. Molecular analysis has also placed another molting group, the nematodes, in this clade. Members of the second major protostome group, the **lophotrochozoans**, are characterized by a common type of cleavage (spiral) and a common larval form (the trochophore). The trochophore (Greek *trochos*, "wheel") is a planktonic (free-swimming) larval form with characteristic bands of locomotive cilia. Adults of some lophotrochozoan species have a distinctive feeding apparatus, the lophophore. Lophotrochozoans include 14 of the 36 metazoan phyla, including the flatworms, annelids, and mollusks. The spiral cleavage program is so characteristic of this group that the term "spiralia" has become another way of describing this clade (Henry 2014).

DEUTEROSTOMES The major deuterostome lineages are the chordates (including the vertebrates) and the echinoderms. Although it may seem strange to classify humans, fish, and frogs in the same broad group as starfish and sea urchins, certain embryological features stress this kinship. First, in deuterostomes ("mouth second"), the oral opening is formed after the anal opening. Also, whereas protostomes generally form their body cavity by hollowing out a solid block of mesoderm (schizocoely, as mentioned earlier), most deuterostomes form their body cavity by extending mesodermal pouches from the gut (**enterocoely**). (There are many exceptions to this generalization, however; see Martín-Durán et al. 2012.)

The lancelets (Cephalochordata; amphioxus) and the tunicates (Urochordata; sea squirts) are invertebrates—they have no backbone. However, the *larvae* of these organisms have a notochord and pharyngeal arches (head structures), indicating that they are chordates. (The "chord" in "chordates" refers to the notochord, which induces the formation of the vertebrate spinal cord.) This discovery, made by Alexander Kowalevsky (1867, 1868), was a milestone in biology. The developmental stages of these organisms united the invertebrates and vertebrates into a single "animal kingdom." Darwin (1874) rejoiced, noting that vertebrates probably arose from a group of animals that resembled larval tunicates. Indeed, Urochordata is now considered to be the group most closely related to the vertebrates. This relationship has been demonstrated both by developmental affinities and by molecular analysis (Bourlat et al. 2006; Delsuc et al. 2006), reversing a previous view that cephalochordates were the sister group to vertebrates.

We turn now to a detailed description of early development in two protostome groups: the snails (shelled gastropod mollusks) and *C. elegans* (an extremely well-studied species of nematode worm). Despite their differences, the early development of both of these invertebrate groups has evolved for rapid development to a larval stage, followed by subsequent growth into an adult (Davidson 2001). Their common factors include:

- Immediate activation of the zygotic genes
- Rapid specification of the products of cleavage (the blastomeres) by the products of the zygotic genes and by maternally active genes
- A relatively small number of cells (several hundred or fewer) present at the start of gastrulation

Early Development in Snails

Snails have a long history as model organisms in developmental biology. They are abundant along the shores of all continents, they grow well in the laboratory, and they show variations in their development that can be correlated with their environmental needs. Some snails also have large eggs and develop rapidly, specifying cell types very early in development. Although each organism uses both autonomous and regulative modes of cell specification (see Chapter 2), snails provide some of the best examples of autonomous (mosaic) development, where the loss of an early blastomere causes the loss of an entire structure. Indeed, in snail embryos, the cells responsible for certain organs can be localized to a remarkable degree. The results of experimental embryology can now be extended (and explained) by molecular analyses, leading to fascinating syntheses of development and evolution (see Conklin 1897 and Henry et al. 2014).

Cleavage in Snail Embryos

"[T]he spiral is the fundamental theme of the molluscan organism. They are animals that twisted over themselves" (Flusser 2011). Indeed, the shells of snails are spirals, their larvae undergo a 180° torsion that brings the anus anteriorly above the head, and (most importantly) the cleavage of their early embryos is spiral. **Spiral holoblastic cleavage** (see Figure 1.5) is characteristic of several animal groups, including annelid worms, platyhelminth flatworms, and most mollusks (Lambert 2010; Hejnol 2010). The cleavage planes of spirally cleaving embryos are not parallel or perpendicular to the animal-vegetal axis of the egg; rather, cleavage is at oblique angles, forming a spiral arrangement of daughter blastomeres. The blastomeres are in intimate contact with each other, producing thermodynamically stable packing arrangements, much like clusters of soap bubbles. Moreover, spirally cleaving embryos usually undergo relatively fewer divisions before they begin gastrulation, making it possible to follow the fate of each cell of the blastula. When the fates of the individual blastomeres from annelid, flatworm, and mollusk embryos were compared, many of the same cells were seen in the same places, and their general fates were identical (Wilson 1898; Hejnol et al. 2010). Blastulae produced by spiral cleavage typically have very small or no blastocoel and are called **stereoblastulae**.

WEB TOPIC 8.1   **DEVELOPMENT PROVIDED EARLY INSIGHTS INTO ANIMAL TAXONOMY**   In 1898, long before molecular data confirmed that annelids, polyclad flatworms, and mollusks were linked as lophotrochozoans, E. B. Wilson concluded that these groups were related. Not only was spiral cleavage homologous among these groups, but so were the fates of many of their cells (including the 4d blastomere, of which we will speak at length). See what embryology was a century ago.

FIGURE 8.2 depicts the cleavage pattern typical of many molluscan embryos. The first two cleavages are nearly meridional, producing four large macromeres (labeled A, B, C, and D). In many species, these four blastomeres are different sizes (D being the largest), a characteristic that allows them to be individually identified. In each successive cleavage, each **macromere** buds off a small **micromere** at its animal pole. Each successive quartet of micromeres is displaced to the right or to the left of its sister macromere, creating the characteristic spiral pattern. Looking down on the embryo from

FIGURE 8.2 Spiral cleavage of the snail *Trochus* viewed from the animal pole (A) and from one side (B). Cells derived from the A blastomere are shown in color. The mitotic spindles, sketched in the early stages, divide the cells unequally and at an angle to the vertical and horizontal axes. Each successive quartet of micromeres (lowercase letters) is displaced clockwise or counterclockwise relative its sister macromere (uppercase letters), creating the characteristic spiral pattern.

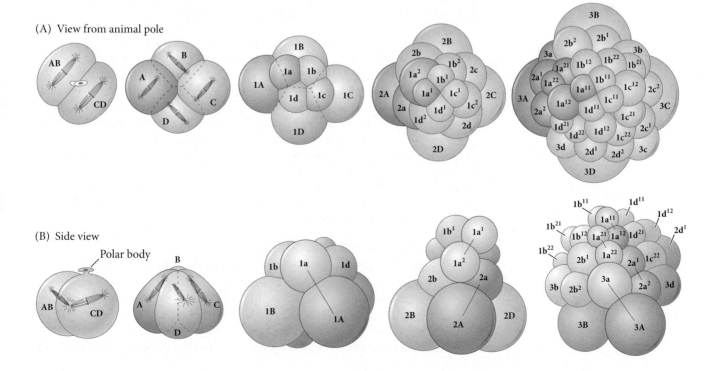

(A) View from animal pole

(B) Side view

(A)

FIGURE 8.3 Spiral cleavage in mollusks. (A) The spiral nature of third cleavage can be seen in the confocal fluorescence micrograph of the 4-cell embryo of the clam *Acila castrenis*. Microtubules stain red, RNA stains green, and DNA stains yellow. Two cells and a portion of a third cell are visible; a polar body can be seen at the top of the micrograph. (B–D) Cleavage in the mud snail *Ilyanassa obsoleta*. The D blastomere is larger than the others, allowing the identification of each cell. Cleavage is dextral. (B) 8-cell stage. PB, polar body (a remnant of meiosis). (C) Mid-fourth cleavage (12-cell embryo). The macromeres have already divided into large and small spirally oriented cells; 1a–d have not divided yet. (D) 32-cell embryo. (A courtesy of G. von Dassow and the Center for Cell Dynamics; B–D from Craig and Morrill 1986, courtesy of the authors.)

(B)   (C)   (D)

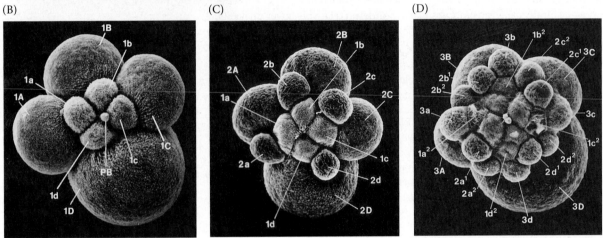

the animal pole, the upper ends of the mitotic spindles appear to alternate clockwise and counterclockwise (**FIGURE 8.3**). This arrangement causes alternate micromeres to form obliquely to the left and to the right of their macromeres.

> **WATCH DEVELOPMENT 8.1**   Video from the laboratory of Dr. Deirdre Lyons shows the first two micromere quartets forming in the snail *Crepidula fornicata*.

At the third cleavage, the A macromere gives rise to two daughter cells, macromere 1A and micromere 1a. The B, C, and D cells behave similarly, producing the first quartet of micromeres. In most species, these micromeres are clockwise (to the *right*) of their macromeres (looking down on the animal pole). At the fourth cleavage, macromere 1A divides to form macromere 2A and micromere 2a, and micromere 1a divides to form two more micromeres, $1a^1$ and $1a^2$ (see Figure 8.2). The micromeres of this second quartet are to the left of the macromeres. Further cleavage yields blastomeres 3A and 3a from macromere 2A, and micromere $1a^2$ divides to produce cells $1a^{21}$ and $1a^{22}$. In normal development, the first-quartet micromeres form the head structures, while the second-quartet micromeres form the statocyst (balance organ) and shell. These fates are specified both by cytoplasmic localization and by induction (Cather 1967; Clement 1967; Render 1991; Sweet 1998).

Maternal regulation of snail cleavage

The orientation of the cleavage plane to the left or to the right is controlled by cytoplasmic factors in the oocyte. This was discovered by analyzing mutations of snail coiling. Some snails have their coils opening to the right of their shells (**dextral coiling**), whereas the coils of other snails open to the left (**sinistral coiling**). Usually the direction of coiling is the same for all members of a given species, but occasional mutants are found (i.e., in a population of right-coiling snails, a few individuals will be found with coils that open on the left). Crampton (1894) analyzed the embryos of aberrant snails and found that their early cleavage differed from the norm (**FIGURE 8.4**). The orientation of the cells after the second cleavage was different in the sinistrally coiling snails as a result of a different orientation of the mitotic apparatus. You can see in Figure 8.4 that the position of the 4d blastomere is different in the right-coiling and left-coiling snail embryos. This 4d blastomere is rather special. It is often called the **mesentoblast**, since its progeny include most of the mesodermal organs (heart, muscles, primordial germ cells) and endodermal organs (gut tube).

In snails such as *Lymnaea*, the direction of snail shell coiling is controlled by a single pair of genes (Sturtevant 1923; Boycott et al. 1930; Shibazaki 2004). In *Lymnaea peregra*, mutants exhibiting sinistral coiling were found and mated with wild-type, dextrally coiling snails. These matings showed that the right-coiling allele, *D*, is dominant to the left-coiling allele, *d*. However, the direction of cleavage is determined not by the genotype of the developing snail but by the genotype of the snail's *mother*. This is called a **maternal effect**. (We'll see other important maternal effect genes when we discuss *Drosophila* development.) A *dd* female snail can produce only sinistrally coiling offspring, even if the offspring's genotype is *Dd*. A *Dd* individual will coil either left or right, depending on the genotype of its mother. Such matings produce a chart like this:

Genotype		Phenotype	
DD female × *dd* male	→	*Dd*	All right-coiling
DD male × *dd* female	→	*Dd*	All left-coiling
Dd × *Dd*	→	1*DD*:2*Dd*:1*dd*	All right-coiling

Thus it is the genotype of the *ovary* in which the oocyte develops that determines which orientation cleavage will take. The genetic factors involved in coiling are brought to the embryo in the oocyte cytoplasm. When Freeman and Lundelius (1982) injected a small amount of cytoplasm from dextrally coiling snails into the eggs of *dd* mothers, the resulting embryos coiled to the right. Cytoplasm from sinistrally coiling snails, however, did *not* affect right-coiling embryos. These findings confirmed that the wild-type mothers were placing a factor into their eggs that was absent or defective in the *dd* mothers. Working with similar populations, Davison and colleagues have identified and mapped a gene encoding a formin protein that is active in the eggs of mothers who carry the *D* allele, but not in the eggs of *dd* mothers (Liu et al. 2013; Davison et al. 2016). Thus, *DD* and *Dd* mothers produce active formin proteins. In *dd* females, however, the *formin* gene has a frameshift mutation in the coding region that renders its mRNA nonfunctional, so its message is rapidly degraded. When the egg contains functional *formin* mRNA from the mother's *D* allele, this message becomes asymmetrically positioned in the embryo as early as the two-cell stage. The formin protein encoded by the mRNA message binds to actin and helps align the cytoskeleton. These findings are upheld by studies showing that drugs that inhibit formins cause eggs from *DD* mothers to develop into left-coiling embryos.

The first indication that the cells will divide sinistrally rather than dextrally is a helical deformation of the cell membranes at the dorsal tip of the macromeres (**FIGURE 8.5A**). Once the third cleavage takes place, the Nodal protein (a TGF-β family paracrine factor) activates genes on the right side of dextrally coiling embryos and on the left side of sinistrally coiling embryos (**FIGURE 8.5B**). Using glass needles to change the direction

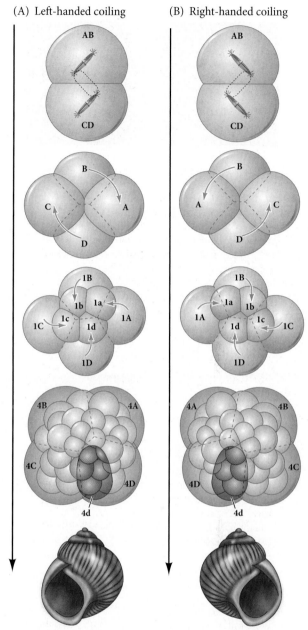

FIGURE 8.4 Dextral and sinistral snail coiling. Looking down on the animal pole of left-coiling (A) and right-coiling (B) snails. The origin of sinistral and dextral coiling can be traced to the orientation of the mitotic spindle at the third cleavage. Left- and right-coiling snails develop as mirror images of each other. (After Morgan 1927.)

(A)

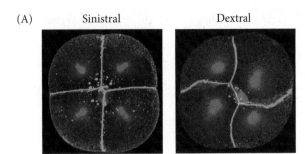

FIGURE 8.5 Mechanisms of right-and left-handed snail coiling. (A) Left- and right-handed coiling at third cleavage. Staining for actin (green) and microtubules (red) visualizes the helical deformation in right-handed cleavage. (B) In the embryo, Nodal is activated on the left side of sinestral embryos and on the right side of dextral embryos. (C) The Pitx1 transcription factor, seen expressed in the embryo (above) is responsible for organ formation, as seen in the ventral view of the adults (below). The positions of the pneumostome (breathing pore; po) and gonad (go) are indicated. (From Kuroda 2014, courtesy of the R. Kuroda.)

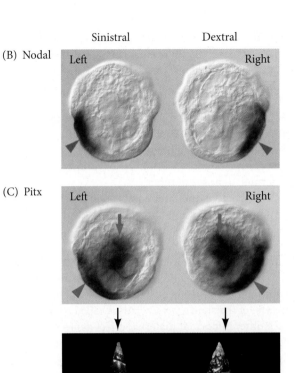

of cleavage at the 8-cell stage changes the location of *Nodal* gene expression (Grande and Patel 2009; Kuroda et al. 2009; Abe et al. 2014). Nodal appears to be expressed in the C-quadrant micromere lineages (which give rise to the ectoderm) and induces asymmetric expression of the gene for the Pitx1 transcription factor (a target of Nodal in vertebrate axis formation as well) in the neighboring D-quadrant blastomeres (**FIGURE 8.5C**).

WEB TOPIC 8.2 **A CLASSIC PAPER LINKS GENES AND DEVELOPMENT** In a masterful thought experiment from 1923, Alfred Sturtevant applied Mendelian genetics to the process of snail coiling. His were some of the first studies to link the power of genetics and the study of embryology.

WATCH DEVELOPMENT 8.2 See dextral and sinistral snail coiling in video from the laboratory of Dr. Reiko Kuroda.

The snail fate map

Detailed fate maps (see Chapters 1 and 2) have greatly advanced our knowledge of spiralian development. The fate maps of the gastropods *Ilyanassa obsoleta* and *Crepidula fornicata* were constructed by injecting large polymers conjugated to fluorescent dyes into specific micromeres (Render 1997; Hejnol et al. 2007). The fluorescence is maintained over the period of embryogenesis and can be seen in larval tissues derived from the injected cells. More specific results, showing divergences between snail species, have been obtained using live imaging (**FIGURE 8.6**; Chan and Lambert 2014; Lyons et al. 2015).

In general, the first quartet micromeres (1a–1d), supplemented by some cells from the second and third quartets, generate the head ectoderm, while the nervous system comes largely from the first- and second-quartet cells (see Figure 8.6A). Fate maps

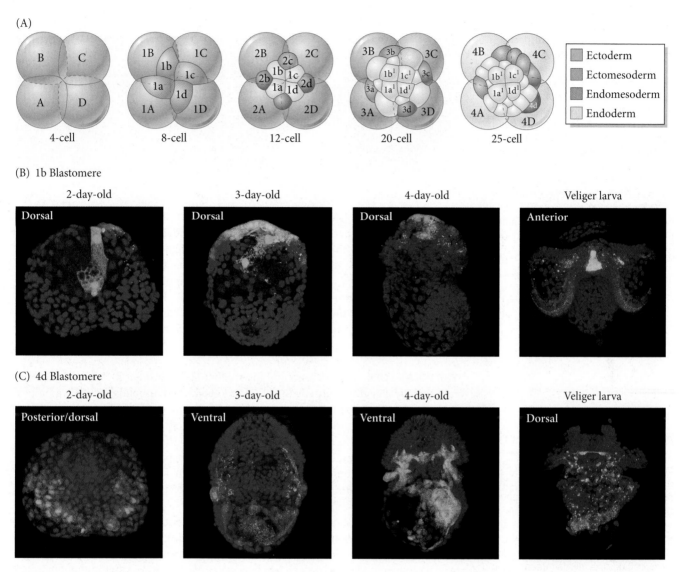

(A)

4-cell   8-cell   12-cell   20-cell   25-cell

Ectoderm
Ectomesoderm
Endomesoderm
Endoderm

(B) 1b Blastomere

2-day-old — Dorsal   3-day-old — Dorsal   4-day-old — Dorsal   Veliger larva — Anterior

(C) 4d Blastomere

2-day-old — Posterior/dorsal   3-day-old — Ventral   4-day-old — Ventral   Veliger larva — Dorsal

confirm that the mouth forms in the same location as the blastopore. The endoderm comes from macromeres A, B, C, and D. The mesoderm comes from two sources: cells from the second and third quartets contribute the larval and adult musculature (ectomesoderm), while the majority of the mesoderm—the larval kidney, heart, primordial germ cells, and retractor muscles—comes from a particularly remarkable cell, the 4d blastomere (see Figure 8.6C). This highly conserved spiralian blastomere is critical for the establishing the placement of the mesoderm and for inducing the formation of other cell types (Lyons et al. 2012). It is the cell most directly affected by the coiling mutation mentioned earlier.

Cell specification and the polar lobe

Mollusks provide some of the most impressive examples of autonomous (mosaic) development, in which the blastomeres are specified by morphogenetic determinants located in specific regions of the oocyte (see Chapter 2). Autonomous specification of early blastomeres is especially prominent in those groups of animals having spiral cleavage, all of which initiate gastrulation at the vegetal pole when only a few dozen cells have formed (Lyons et al. 2015). In mollusks, the mRNAs for some transcription factors and paracrine factors are placed in particular cells by associating with certain centrosomes (**FIGURE 8.7**; Lambert

FIGURE 8.6   (A) Generalized fate map of the snail embryo. The first two cleavages establish the A, B, C, and D quadrants. The micromere quartets (1a–1d and 3a–3d) are generated and divide to produce a "micromere cap" atop the yolky macromeres. The macromeres (3A–3D) produce most of the endoderm, while the apical micromeres generate the ectoderm. Specific tier 2 and tier 3 micromeres (this differs in different species) form the ectomesoderm, while the endomesoderm (heart and kidney) are generated by the 4d cell, which is often formed when the 3D cell divides ahead of the other macromeres. (B,C) Fate maps can be made by injecting tiny beads containing fluorescent dye into individual blastomeres. When the embryos develop into larvae, the descendants of each blastomere are identifiable by their fluorescence. (B) Results when the 1b blastomere of the snail *Ilyanassa* was injected with GFP. (C) Results of injecting *Ilyanassa* blastomere 4d. (A after Lyons and Henry 2014; B from Chan and Lambert 2014, photographs courtesy of the authors.)

FIGURE 8.7 Association of *decapentaplegic* (*dpp*) mRNA with specific centrosomes of *Ilyanassa*. (A) In situ hybridization of the mRNA for Dpp in the 4-cell snail embryo shows no Dpp accumulation. (B) At prophase of the 4- to 8-cell stage, *dpp* mRNA (black) accumulates at one centrosome of the pair forming the mitotic spindle. (The DNA is light blue.) (C) As mitosis continues, *dpp* mRNA is seen to attend the centrosome in the macromere rather than the centrosome in the micromere of each cell. The BMP-like paracrine factor encoded by *dpp* is critical to molluscan development. (From Lambert and Nagy 2002, courtesy of L. Nagy.)

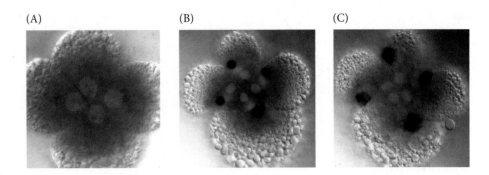

(A)   (B)   (C)

and Nagy 2002; Kingsley et al. 2007; Henry et al. 2010b,c). This association allows the mRNA to enter specifically into one of the two daughter cells. In many instances, the mRNAs that get transported together into a particular tier of blastomeres have 3′ tails that form very similar shapes, thus suggesting that the identity of the micromere tiers may be controlled largely by the 3′ untranslated regions (3′ UTRs) of the mRNAs that attach to the centrosomes at each division (**FIGURE 8.8**; Rabinowitz and Lambert 2010). In other cases, the patterning molecules [of still unknown identities] appear to be bound to a certain region of the egg that will form a unique structure called the polar lobe.

THE POLAR LOBE E. B. Wilson and his student H. E. Crampton observed that certain spirally cleaving embryos (mostly in mollusks and annelids) extrude a bulb of cytoplasm—the **polar lobe**—immediately before first cleavage. In some species of snails, the region uniting the polar lobe to the rest of the egg becomes a fine tube. The first cleavage splits the zygote asymmetrically, so that the polar lobe is connected only to the CD blastomere (**FIGURE 8.9A**). In several species, nearly one-third of the total cytoplasmic volume is contained in this anucleate lobe, giving it the appearance of another cell (**FIGURE 8.9B**). The resulting three-lobed structure is often referred to as the trefoil stage embryo (**FIGURE 8.9C**).

Crampton (1896) showed that if one removes the polar lobe at the trefoil stage, the remaining cells divide normally. However, the resulting larva is incomplete (**FIGURE 8.10**), wholly lacking its intestinal endoderm and mesodermal kidney and heart), as well as some ectodermal organs (such as eyes). Moreover, Crampton demonstrated that the same type of abnormal larva can be produced by removing the D blastomere from the 4-cell embryo. Crampton thus concluded that the polar lobe cytoplasm contains the heart and intestinal-forming determinants and that these determinants (as

FIGURE 8.8 Importance of the 3′ UTR for association of mRNAs with specific centrosomes. In *Ilyanassa*, the *R5LE* message is usually segregated into the first tier of micromeres. The message binds to one side of the centrosome complex (the side that will be in the small micromere.) (A) Normal *R5LE* mRNA distribution from the 2-cell through the 24-cell stage. The mRNA (green) associates with the centrosomic region (blue) that will generate the micromere tier and becomes localized to particular blastomeres by the 24-cell stage. (B) Hairpin loop of the 3′ UTR of the *R5LE* message. (After Rabinowitz and Lambert 2010.)

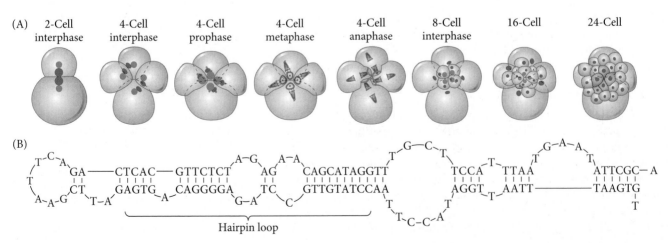

(A) 2-Cell interphase   4-Cell interphase   4-Cell prophase   4-Cell metaphase   4-Cell anaphase   8-Cell interphase   16-Cell   24-Cell

(A)

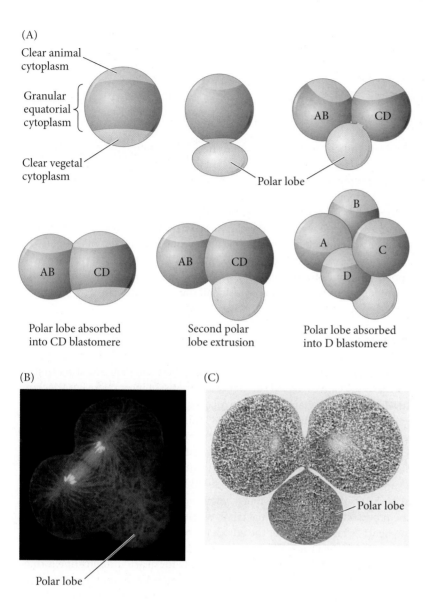

Clear animal cytoplasm

Granular equatorial cytoplasm

Clear vegetal cytoplasm

Polar lobe

AB   CD

B

A   C

D

AB   CD

AB   CD

Polar lobe absorbed into CD blastomere

Second polar lobe extrusion

Polar lobe absorbed into D blastomere

FIGURE 8.9   Polar lobe formation. (A) During cleavage, extrusion, and reincorporation of the polar lobe occur twice. The CD blastomere absorbs the polar lobe material but extrudes it again prior to second cleavage. After this division, the polar lobe is attached only to the D blastomere, which absorbs its material. From this point on, no polar lobe is formed. (B) Late in the first division of a scallop embryo, the anucleate polar lobe (lower right) contains nearly one-third of the cytoplasmic volume. Microtubules are stained red, RNA is green, and the chromosomal DNA appears yellow. (C) Section through first-cleavage, or trefoil-stage, embryo of *Dentalium*. (A after Wilson 1904; B courtesy of G. von Dassow and the Center for Cell Dynamics; C courtesy of M. R. Dohmen.)

(B)

Polar lobe

(C)

Polar lobe

(A)

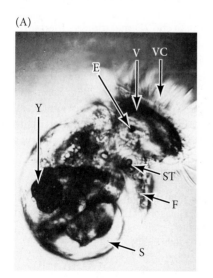

V   VC

E

Y

ST

F

S

(B)

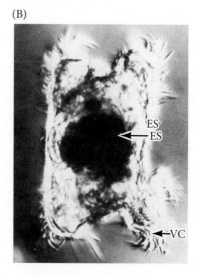

ES
ES

VC

FIGURE 8.10   Importance of the polar lobe in the development of *Ilyanassa*. (A) Normal trochophore larva. (B) Abnormal larva, typical of those produced when the polar lobe of the D blastomere is removed. (E, eye; F, foot; S, shell; ST, statocyst; V, velum; VC, velar cilia; Y, residual yolk; ES, everted stomodeum; DV, disorganized velum.) (From Newrock and Raff 1975, courtesy of K. Newrock.)

well as its inducing ability) is transferred to the D blastomere.[1] Crampton also showed that the localization of these endomesodermal determinants is established shortly after fertilization.

> **WATCH DEVELOPMENT 8.3** You can watch the polar lobe develop in this video of a non-molluscan lophotrochozoan, the annelid worm *Chaetopterus*.

Centrifugation studies have demonstrated that the morphogenetic determinants sequestered in the polar lobe are probably located in the lobe's cytoskeleton or cortex, not in its diffusible cytoplasm (Clement 1968). Van den Biggelaar (1977) obtained similar results when he removed the cytoplasm from the polar lobe with a micropipette. Cytoplasm from other regions of the cell flowed into the polar lobe, replacing the portion he removed, and subsequent development of these embryos was normal. In addition, when he added the diffusible polar lobe cytoplasm to the B blastomere, no duplicated structures were seen (Verdonk and Cather 1983). Therefore, the diffusible part of the polar lobe cytoplasm does not contain the morphogenetic determinants; these as-yet unidentified factors probably reside in the nonfluid cortical cytoplasm or on the cytoskeleton.

THE D BLASTOMERE The development of the D blastomere can be traced in Figure 8.3B–D. This macromere, having received the contents of the polar lobe, is larger than the other three (Clement 1962). When one removes the D blastomere or its first or second macromere derivatives (i.e., 1D or 2D), one obtains an incomplete larva lacking heart, intestine, velum, shell gland, eyes, and foot. This is essentially the same phenotype seen when one removes the polar lobe (see Figure 8.10B). Since the D blastomeres do not directly contribute cells to many of these structures, it appears that the D-quadrant macromeres are involved in inducing other cells to have these fates.

When one removes the 3D blastomere shortly after the division of the 2D cell to form the 3D and 3d blastomeres, the larva produced looks similar to those formed by the removal of the D, 1D, or 2D macromeres. However, ablation of the 3D blastomere at a later time produces an almost normal larva, with eyes, foot, velum, and some shell gland, but no heart or intestine. After the 4d cell is given off (by the division of the 3D blastomere), removal of the D derivative (the 4D cell) produces no qualitative difference in development. In fact, all the essential determinants for heart and intestine formation are now in the 4d blastomere (also called the mesentoblast, as mentioned earlier), and removal of that cell results in a heartless and gutless larva (Clement 1986). The 4d blastomere is responsible for forming (at its next division) the two bilaterally paired blastomeres that give rise to both the mesodermal (heart) and endodermal (intestine) organs (Lyons et al. 2012; Lambert and Chan, 2014).

The mesodermal and endodermal determinants of the 3D macromere, therefore, are transferred to the 4d blastomere. At least two morphogenetic determinants are involved in regulating the development of 4d. First, the cell appears to be specified by the presence of transcription factor **β-catenin**, which enters into the nucleus of the 4d mesentoblast and its immediate progeny (**FIGURE 8.11A**; Henry et al. 2008; Rabinowitz et al. 2008). When translational inhibitors suppressed β-catenin protein synthesis, the 4d cell underwent a normal pattern of early cell divisions but failed to differentiate into heart, muscles, or hindgut; and gastrulation also failed to occur in those embryos

---

[1]Although this looks like a great case for autonomous specification, it is possible that signaling from the micromeres is needed to activate the cytoplasmic determinants brought to the D blastomere by the polar lobe (Gharbiah et al. 2014; Henry 2014). In addition to these roles in cell differentiation, the material in the polar lobe is responsible for specifying the dorsal-ventral polarity of the embryo. When polar lobe material is forced to pass into the AB blastomere as well as into the CD blastomere, twin larvae form that are joined at their ventral surfaces (Guerrier et al. 1978; Henry and Martindale 1987).

(A)

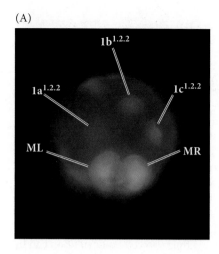

(B)

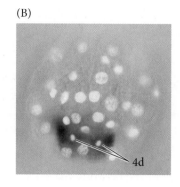

FIGURE 8.11 Morphogenetic determinants in the 4d snail blastomere. (A) β-Catenin expression in ML and MR, the progeny of the 4d blastomere of *Crepidula*. (B) *Nanos* mRNA localization (purple) in the dividing 4d blastomere and in its right and left progeny, 4dL and 4dR, of *Ilyanassa*. (A from Henry et al. 2010; B from Rabinowitz et al. 2008.)

(Henry et al. 2010b). Indeed, β-catenin may have an evolutionarily conserved role in mediating autonomous specification and specifying endomesodermal fates throughout the animal kingdom; in subsequent chapters we will see a similar role for this protein in both sea urchin and frog embryos.

The 4d blastomere also contains the protein and mRNA for the translation suppressor *Nanos* (**FIGURE 8.11B**). As with β-catenin, blocking translation of *Nanos* mRNA prevents formation of the larval muscles, heart, and intestine from the 4d blastomere (Rabinowitz et al. 2008). In addition, the germline cells (sperm and egg progenitors) do not form. As we will see throughout the book, the Nanos protein is often involved in specification of germ cell progenitors.

But the 4d blastomere not only develops autonomously, it also induces other cell lineages. The Notch signaling pathways may be critical for these inductive events of the 4d blastomere. Blocking Notch signaling after the 4d blastomere has formed causes the larva to resemble those formed when the 4d cell is removed; while the autonomous fates of the 4d cell (such as larval kidneys) are not disturbed (Gharbiah et al. 2014). The D set of blastomeres is thus the "organizer" of snail embryos. Experiments have demonstrated that the nondiffusible polar lobe [cortical] cytoplasm that is localized to the D blastomere is extremely important in normal molluscan development for several reasons:

- It contains the determinants for the proper cleavage rhythm and the cleavage orientation of the D blastomere.

- It contains certain determinants (those entering the 4d blastomere and hence leading to the mesentoblasts) for autonomous mesodermal and intestinal differentiation.

- It is responsible for permitting the inductive interactions (through the material entering the 3D blastomere) leading to the formation of the shell gland and eye.

Altering evolution by altering cleavage patterns: An example from a bivalve mollusk

Darwin's theory of evolution stated that biodiversity arose through descent with modification. This explanation united and explained both the commonalities of form (such as the same bones types in the arms of humans and the flippers of seals) as having evolved from a common ancestor. It also explained how natural selection results in changes that can enable an organism to better survive in its particular environment. We see both these principles in snail development. As shown above, E. B. Wilson demonstrated that

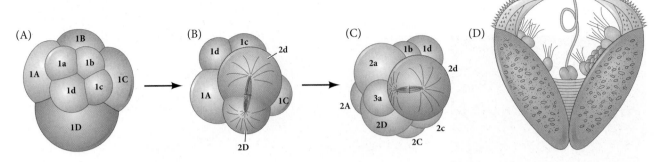

(A) 1B 1a 1b 1A 1d 1c 1C 1D

(B) 1d 1c 2d 1A 1C 2D

(C) 1b 1d 2a 2d 2A 3a 2c 2D 2C

(D)

FIGURE 8.12 Formation of a glochidium larva by modification of spiral cleavage. After the 8-cell embryo is formed (A), the placement of the mitotic spindle causes most of the D cytoplasm to enter the 2d blastomere (B). This large 2d blastomere divides (C), eventually giving rise to the large "bear trap" shell of the larva (D). (After Raff and Kaufman 1983.)

snails, annelids, and flatworms have spiral cleavage and that the similar roles played by the embryonic cells could best be explained by all these animal groups having evolved from a common ancestor.

The same year, the embryologist Frank R. Lillie showed that a new structure can evolve by changing the pattern of development. He thus showed that evolution can be the result of hereditary alterations in embryonic development. One such modification, discovered by Lillie in 1898, is brought about by an alteration of the typical pattern of molluscan spiral cleavage in a family of bivalve mollusks, the unionid clams. Unlike most clams, *Unio* and its relatives live in swift-flowing streams. Streams create a problem for the dispersal of larvae: because the adults are sedentary, free-swimming larvae would always be carried downstream by the current. *Unio* clams have adapted to this environment via two modifications of their development. The first is an alteration in embryonic cleavage. In typical molluscan cleavage, either all the macromeres are equal in size or the 2D macromere is the largest cell at that embryonic stage. However, cell division in *Unio* is such that the 2d "micromere" gets the largest amount of cytoplasm (**FIGURE 8.12**). This cell then divides to produce most of the larval structures, including a gland capable of producing a large shell. The resulting larva is called a **glochidium** and resembles a tiny bear trap. Glochidia have sensitive hairs that cause the valves of the shell to snap shut when they are touched by the gills or fins of a wandering fish. The larvae can thus attach themselves to the fish and "hitchhike" until they are ready to drop off and metamorphose into adult clams. In this manner, they can spread upstream as well as downstream.

In some unionid species, glochidia are released from the female's brood pouch (marsupium) and then wait passively for a fish to swim by. Some other species, such as *Lampsilis altilis*, have increased the chances of their larvae finding a fish by yet another developmental modification. Many clams develop a thin mantle that flaps around the shell and surrounds the brood pouch. In some unionids, the shape of the brood pouch and the undulations of the mantle mimic the shape and swimming behavior of a minnow (Welsh 1969). To make the deception even better, the clams develop a black "eyespot" on one end and a flaring "tail" on the other (**FIGURE 8.13**). When a predatory fish is lured within range of this "prey," the clam discharges the glochidia from the brood pouch and the larvae attach to the fish's gills. Thus, the modification of existing developmental patterns has permitted unionid clams to survive in challenging environments.

FIGURE 8.13 Phony fish atop the unionid clam *Lampsilis altilis.* The "fish" is actually the brood pouch and mantle of the clam. The "eyes" and flaring "tail" attract predatory fish, and the glochidium larvae attach to the fish's gills. (Courtesy of Wendell R. Haag/USDA Forest Service.)

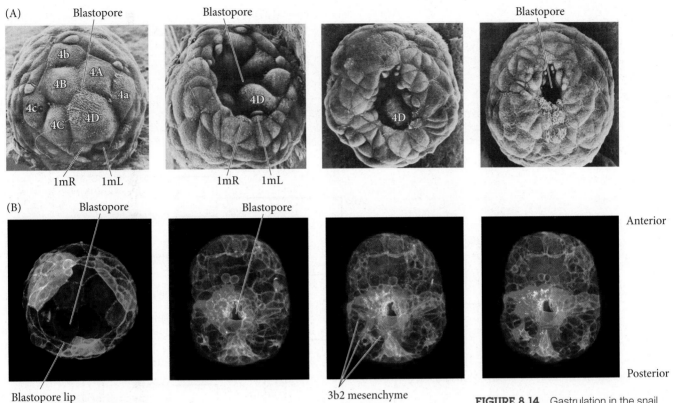

FIGURE 8.14 Gastrulation in the snail *Crepidula*. (A) Scanning electron micrographs focusing on the blastopore region show internalization of the endoderm, which is derived from the macromeres plus the fourth tier of micromeres. The 4d cell has divided into 1mR and 1mL (the right and left mesendoderm cells, respectively). The ectoderm undergoes epiboly from the animal pole and envelops the other cells of the embryo. (B) Live cell labeling of *Crepidula* embryos shows gastrulation occurring by epiboly. Cells derived from the 3b micromere are stained red. (A from Van den Biggelaar and Dictus 2004; B from Lyons et al. 2015, courtesy of D. Lyons.)

Gastrulation in Snails

The snail stereoblastula is relatively small, and its cell fates have already been determined by the D series of macromeres. Gastrulation is accomplished by a combination of processes, including the invagination of the endoderm to form the primitive gut, and the epiboly of the animal cap micromeres that multiply and "overgrow" the vegetal macromeres (Collier 1997; van den Biggelaar and Dictus 2004; Lyons and Henry 2014). Eventually, the micromeres cover the entire embryo, leaving a small blastopore slit at the vegetal pole (**FIGURE 8.14A**). The first through third quartet micromeres form an epithelial animal cap that expands to cover vegetal endomesodermal precursors. As the blastopore narrows, cells derived from $3a^2$ and $3b^2$ undergo epithelial-to-mesenchymal transition and move into the archenteron. Posteriorly, cells derived from $3c^2$ and $3d^2$ undergo convergence and extension that involves zippering of cells and their intercalation across the ventral midline (**FIGURE 8.14B**; Lyons et al. 2015).

During snail gastrulation, the mouth forms from cells all around the circumference of the blastopore, and the anus arises from the $2d^2$ cells, which are only very briefly part of the blastopore lip. The anus forms at 12 days postfertilization, as a separate hole, and not related to the blastopore. Thus, these animals are protostomes, forming their mouths in the area where the blastopore is first seen.

WATCH DEVELOPMENT 8.4 The epiboly of the snail micromeres and the internalization of macromeres is shown in two videos from the laboratory of Dr. Deirdre Lyons.

The Nematode *C. Elegans*

Unlike the snail, with its long embryological pedigree, the nematode *Caenorhabditis elegans* (usually referred to as *C. elegans*) is a thoroughly modern model system, uniting developmental biology with molecular genetics. In the 1970s, Sydney Brenner and

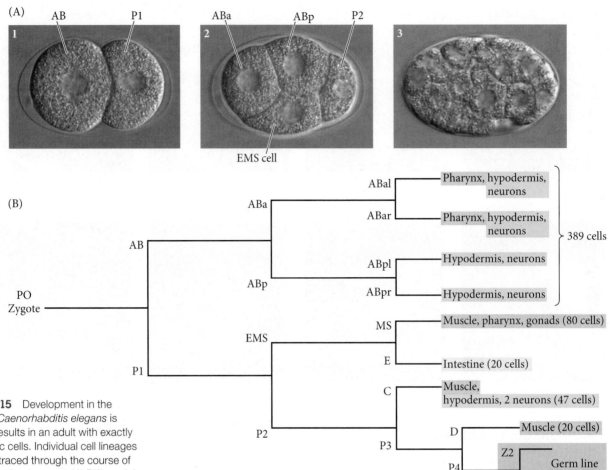

FIGURE 8.15 Development in the nematode *Caenorhabditis elegans* is rapid and results in an adult with exactly 959 somatic cells. Individual cell lineages have been traced through the course of the animal's development. (A) Differential interference micrographs of the cleaving embryo. (1) The AB cell (left) and the P1 cell (right) are the result of the first asymmetric division. Each will give rise to a different cell lineage. (2) The 4-cell embryo shows ABa, ABp, P2, and EMS cells. (3) Gastrulation is initiated by the movement of E-derived cells toward the center of the embryo. (B) Abbreviated cell lineage chart. The germ line segregates into the posterior portion of the most posterior (P) cell. The first three cell divisions produce the AB, C, MS, and E lineages. The number of derived cells (in parentheses) refers to the 558 cells present in the newly hatched larva. Some of these continue to divide to produce the 959 somatic cells of the adult. (A courtesy of D. G. Morton and K. Kemphues; B after Pines 1992, based on Sulston and Horvitz 1977 and Sulston et al. 1983.)

his students sought an organism wherein it might be possible to identify each gene involved in development as well as to trace the lineage of each and every cell (Brenner 1974). Nematode roundworms seemed like a good group to start with because embryologists such as Richard Goldschmidt and Theodor Boveri had already shown that several nematode species have a relatively small number of chromosomes and a small number of cells with invariant cell lineages.

Brenner and his colleagues eventually settled on *C. elegans,* a small (1 mm long), free-living (i.e., nonparasitic) soil nematode with relatively few cell types. *C. elegans* has a rapid period of embryogenesis—about 16 hours—that it can accomplish in a petri dish (**FIGURE 8.15A**). Moreover, its predominant adult form is hermaphroditic, with each individual producing both eggs and sperm. These roundworms can reproduce either by self-fertilization or by cross-fertilization with the infrequently occurring males.

The body of an adult *C. elegans* hermaphrodite contains exactly 959 somatic cells, and the entire cell lineage has been traced through its transparent cuticle (**FIGURE 8.15B**; Sulston and Horvitz 1977; Kimble and Hirsh 1979). It has what is called an **invariant cell lineage**, which means that each cell gives rise to the same number and type of cells in every embryo. This allows one to know which cells have the same precursor cells. Thus, for each cell in the embryo, we can say where it came from (i.e., which cells in earlier embryonic stages were its progenitors) and which tissues it will contribute to forming. Furthermore, unlike vertebrate cell lineages, the *C. elegans* lineage is almost entirely invariant from one individual to the next; there is little room for randomness (Sulston et al. 1983). It also has a very compact genome. The *C. elegans* genome was the first complete sequence ever obtained for a multicellular organism (*C. elegans* Sequencing Consortium 1999). Although it has about the same number of genes as humans (18,000–20,000 genes, whereas *Homo sapiens* has 20,000–25,000), the nematode has only about 3% the number of nucleotides in its genome (Hodgkin 1998, 2001).

C. elegans displays the rudiments of nearly all the major bodily systems (feeding, nervous, reproductive, etc., although it has no skeleton), and it exhibits an aging phenotype before it dies. Neurobiologists celebrate its minimal nervous system (302 neurons), and each one of its 7,600 synapses (neuronal connections) has been identified (White et al. 1986; Seifert et al. 2006). In addition, *C. elegans* is particularly friendly to molecular biologists. DNA injected into *C. elegans* cells is readily incorporated into their nuclei, and *C. elegans* can take up antisense RNA from its culture medium.

Cleavage and Axis Formation in *C. elegans*

Fertilization in *C. elegans* is a not your typical sperm-meets-egg story. Most *C. elegans* individuals are hermaphrodites, producing both sperm and eggs, and fertilization occurs within a single adult individual. The egg becomes fertilized by rolling through a region of the embryo (the spermatheca) containing mature sperm (**FIGURE 8.16A,B**). The sperm are not the typical long-tailed, streamlined cells, but are small, round, unflagellated cells that travel slowly by amoeboid motion. When a sperm fuses with the egg cell membrane, polyspermy is prevented by the rapid synthesis of chitin (the protein comprising the cuticle) by the newly fertilized egg (Johnston et al. 2010). The fertilized egg undergoes early divisions and is extruded through the vulva.

Developing Questions

Humans have trillions of cells, a regionalized brain, and intricate limbs. The nematode has 959 cells and can fit under a fingernail. However, humans and *C. elegans* have nearly the same number of genes, leaving Jonathan Hodgkin, curator of the *C. elegans* gene map, to ask "What does a worm want with 20,000 genes?" Any suggestions?

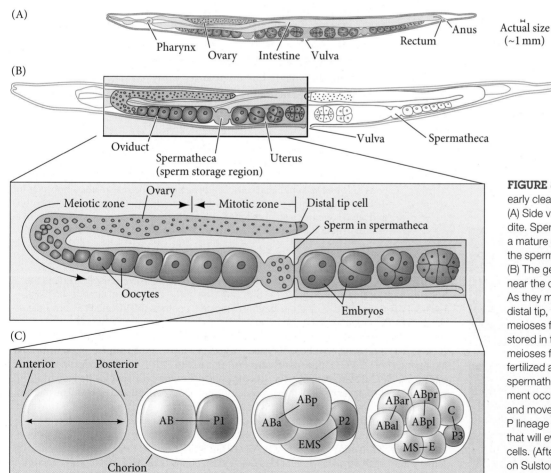

(A) Pharynx Ovary Intestine Vulva Rectum Anus Actual size (~1 mm)

(B) Oviduct Spermatheca (sperm storage region) Uterus Vulva Spermatheca

Ovary Meiotic zone Mitotic zone Distal tip cell Sperm in spermatheca Oocytes Embryos

(C) Anterior Posterior Chorion AB P1 ABp ABa EMS P2 ABar ABpr ABal ABpl MS E C P3

FIGURE 8.16 Fertilization and early cleavages in *C. elegans*. (A) Side view of adult hermaphrodite. Sperm are stored such that a mature egg must pass through the sperm on its way to the vulva. (B) The germ cells undergo mitosis near the distal tip of the gonad. As they move farther from the distal tip, they enter meiosis. Early meioses form sperm, which are stored in the spermatheca. Later meioses form eggs, which are fertilized as they roll through the spermatheca. (C) Early development occurs as the egg is fertilized and moves toward the vulva. The P lineage consists of stem cells that will eventually form the germ cells. (After Pines 1992, based on Sulston and Horvitz 1977 and Sulston et al. 1983.)

(A)

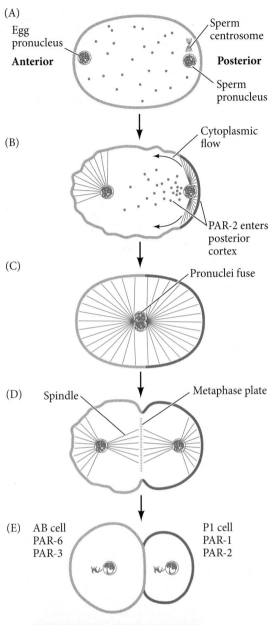

Egg
pronucleus

Anterior

Sperm
centrosome

Posterior

Sperm
pronucleus

(B)

Cytoplasmic
flow

PAR-2 enters
posterior
cortex

(C)

Pronuclei fuse

(D) Spindle

Metaphase plate

(E) AB cell
PAR-6
PAR-3

P1 cell
PAR-1
PAR-2

(F)

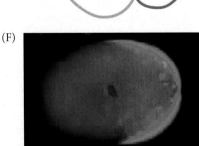

(G)

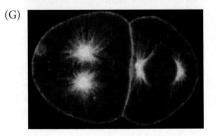

WATCH DEVELOPMENT 8.5 These are some excellent videos of developing *C. elegans* embryos, including those prepared in the laboratory of Dr. Bob Goldstein.

Rotational cleavage of the egg

The zygote of *C. elegans* exhibits rotational holoblastic cleavage (**FIGURE 8.16C**). During early cleavage, each asymmetrical division produces one founder cell (denoted AB, E, MS, C, and D) that produces differentiated descendants, and one stem cell (the P1–P4 lineage). The anterior-posterior axis is determined before the first cell division, and the cleavage furrow is located asymmetrically along this axis of the egg, closer to what will be the posterior pole. The first cleavage forms an anterior founder cell (AB) and a posterior stem cell (P1). The dorsal-ventral axis is determined during the second division. The founder cell (AB) divides equatorially (longitudinally, 90° to the anterior-posterior axis), while the P1 cell divides meridionally (transversely) to produce another founder cell (EMS) and a posterior stem cell (P2). The EMS cell marks the ventral region of the developing embryo. The stem cell lineage always undergoes meridional division to produce (1) an anterior founder cell and (2) a posterior cell that will continue the stem cell lineage. The right-left axis is seen at the transition between the 4- and 8-cell stage. Here, the locations of two "granddaughters" of the AB cell (ABal and ABpl) are on the left side, while two others (ABar and ABpr) are on the right (see Figure 8.16C).

Anterior-posterior axis formation

The decision as to which end of the egg will become the anterior and which the posterior seems to reside with the position of the sperm pronucleus (**FIGURE 8.17**). When the sperm pronucleus enters the oocyte cytoplasm, the oocyte has no polarity. However, the oocyte does have a distinct arrangement of "partitioning-defective," or **PAR proteins**,[2] in its cytoplasm (Motegi and Seydoux 2014). PAR-3 and PAR-6, interacting with the protein kinase PKC-3 (mutations of which cause defective

---

[2]Although originally discovered in *C. elegans*, many species use the PAR proteins in establishing cell polarity. They are critical for forming the anterior and posterior regions of *Drosophila* oocytes, and they distinguish the basal and apical ends of *Drosophila* epithelial cells. *Drosophila* PAR proteins are also important in distinguishing which product of a neural stem cell division becomes the neuron and which remains a stem cell. PAR-1 homologues in mammals also appear to be critical in neural polarity (Goldstein and Macara 2007; Nance and Zallen 2011).

FIGURE 8.17 PAR proteins and the establishment of polarity. (A) When sperm enters the egg, the egg nucleus is undergoing meiosis (left). The cortical cytoplasm (orange) contains PAR-3, PAR-6, and PKC-3, and the internal cytoplasm contains PAR-2 and PAR-1 (purple dots). (B,C) Microtubules of the sperm centrosome initiate contraction of the actin-based cytoskeleton toward the future anterior side of the embryo. These sperm microtubules also protect PAR-2 protein from phosphorylation, allowing it to enter the cortex along with its binding partner, PAR-1. PAR-1 phosphorylates PAR-3, causing PAR-3 and its binding partners PAR-6 and PKC-3 to leave the cortex. (D) The posterior of the cell becomes defined by PAR-2 and PAR-1, while the anterior of the cell becomes defined by PAR-6 and PAR-3. The metaphase plate is asymmetric, as the microtubules are closer to the posterior pole. (E) The metaphase plate separates the zygote into two cells, one having the anterior PARs and one the posterior PARs. (F) In this dividing *C. elegans* zygote, PAR-2 protein is stained green; DNA is stained blue. (G) In second division, the AB cell and the P1 cell divide perpendicularly (90° differently from each other). (A–E after Bastock and St. Johnston 2011; F, photograph courtesy of J. Ahrenger; G, photograph courtesy of J. White.)

partitioning), are uniformly distributed in the cortical cytoplasm. PKC-3 restricts PAR-1 and PAR-2 to the internal cytoplasm by phosphorylating them. The sperm centrosome (microtubule-organizing center) contacts the cortical cytoplasm through its microtubules and initiates cytoplasmic movements that push the male pronucleus to the nearest end of the oblong oocyte. That end becomes the posterior pole (Goldstein and Hird 1996). Moreover, these microtubules locally protect PAR-2 from phosphorylation, thereby allowing PAR-2 (and its binding partner, PAR-1) into the cortex nearest the centrosome. Once PAR-1 is in the cortical cytoplasm, it phosphorylates PAR-3, causing PAR-3 (and its binding partner, PKC-3) to leave the cortex. At the same time, the sperm microtubules induce the contraction of the actin-myosin cytoskeleton toward the anterior, thereby clearing PAR-3, PAR-6, and PKC-3 from the posterior of the 1-cell embryo. During first cleavage, the metaphase plate is closer to the posterior, and the fertilized egg is divided into two cells, one having the anterior PARs (PAR-6 and PAR-3) and one having the posterior PARs (PAR-2 and PAR-1) (Goehring et al. 2011; Motegi et al. 2011; Rose and Gönczy 2014).

 SCIENTISTS SPEAK 8.1   Michael Barresi interviews Dr. Kenneth Kemphues, who talks about his work on the *PAR* genes and RNAi.

Dorsal-ventral and right-left axis formation

The dorsal-ventral axis of *C. elegans* is established in the division of the AB cell. As the cell divides, it becomes longer than the eggshell is wide. This squeezing causes the daughter cells to slide, one becoming anterior and one posterior (hence their respective names, ABa and ABp; see Figure 8.16C). The squeezing also causes the ABp cell to take a position above the EMS cell that results from the division of the P1 blastomere. The ABp cell thus defines the future dorsal side of the embryo, while the EMS cell—the precursor of the muscle and gut cells—marks the future ventral surface of the embryo.

The left-right axis is not readily seen until the 12-cell stage, when the MS blastomere (from the division of the EMS cell) contacts half the "granddaughters" of the ABa cell, distinguishing the right side of the body from the left side (Evans et al. 1994). This asymmetric signaling sets the stage for several other inductive events that make the right side of the larva differ from the left (Hutter and Schnabel 1995). Indeed, even the different neuronal fates seen on the left and right sides of the *C. elegans* brain can be traced back to that single change at the 12-cell stage (Poole and Hobert 2006). Although readily seen at the 12-cell stage, the first indication of left-right asymmetry probably occurs at the zygote stage. Just prior to first cleavage, the embryo rotates 120° inside its vitelline envelope. This rotation is always in the same direction relative to the already established anterior-posterior axis, indicating that the embryo already has a left-right chirality. If cytoskeleton proteins or the PAR proteins are inhibited, the direction of the rotation and subsequent chirality become random (Wood and Schonegg 2005; Pohl 2011).

Control of blastomere identity

C. elegans demonstrates both the conditional and autonomous modes of cell specification. Both modes can be seen if the first two blastomeres are experimentally separated (Priess and Thomson 1987). The P1 cell develops autonomously without the presence of AB, generating all the cells it would normally make, and the result is the posterior half of an embryo. However, the AB cell in isolation makes only a small fraction of the cell types it would normally make. For instance, the resulting ABa blastomere fails to make the anterior pharyngeal muscles that it would have made in an intact embryo. Therefore, the specification of the AB blastomere is conditional, and it needs to interact with the descendants of the P1 cell in order to develop normally.

AUTONOMOUS SPECIFICATION   The determination of the P1 lineages appears to be autonomous, with cell fates determined by internal cytoplasmic factors rather than by interactions with neighboring cells (see Maduro 2006). The SKN-1, PAL-1, and PIE-1 proteins encode transcription factors that act intrinsically to determine the fates of cells derived from the four P1-derived somatic founder cells (MS, E, C, and D).

The **SKN-1** protein is a maternally expressed transcription factor that controls the fate of the EMS blastomere, the cell that generates the posterior pharynx. After first cleavage, only the posterior blastomere—P1—has the ability to produce pharyngeal cells when isolated. After P1 divides, only EMS is able to generate pharyngeal muscle cells in isolation (Priess and Thomson 1987). Similarly, when the EMS cell divides, only one of its progeny, MS, has the intrinsic ability to generate pharyngeal tissue. These findings suggest that pharyngeal cell fate may be determined autonomously, by maternal factors residing in the cytoplasm that are parceled out to these particular cells.

Bowerman and co-workers (1992a,b, 1993) found maternal effect mutants lacking pharyngeal cells and were able to isolate a mutation in the *skn-1* (*skin excess*) gene. Embryos from homozygous *skn-1*-deficient mothers lack both pharyngeal mesoderm and endoderm derivatives of EMS (**FIGURE 8.18**). Instead of making the normal intestinal and pharyngeal structures, these embryos seem to make extra hypodermal (skin) and body wall tissue where their intestine and pharynx should be. In other words, the EMS blastomere appears to be respecified as C. Only those cells destined to form pharynx or intestine are affected by this mutation. The SKN-1 protein is a transcription factor that initiates the activation of those genes responsible for forming the pharynx and intestine (Blackwell et al. 1994; Maduro et al. 2001).

Another transcription factor, **PAL-1**, is also required for the differentiation of the P1 lineage. PAL-1 activity is needed for the normal development of the *somatic* (but not the germline) descendants of the P2 blastomere, where it specifies muscle production. Embryos lacking PAL-1 have no somatic cell types derived from the C and D stem cells (Hunter and Kenyon 1996). PAL-1 is regulated by the MEX-3 protein, an RNA-binding protein that appears to inhibit the translation of *pal-1* mRNA. Wherever MEX-3 is expressed, PAL-1 is absent. Thus, in *mex-3*-deficient mutants, PAL-1 is seen in every blastomere. SKN-1 also inhibits PAL-1 (thereby preventing it from becoming active in the EMS cell). But what keeps *pal-1* from functioning in the prospective germ cells and turning them into muscles? In the germ line, PAL-1 synthesis is prevented by the

FIGURE 8.18 Deficiencies of intestine and pharynx in *skn-1* mutants of *C. elegans*. Embryos derived from wild-type animals (A,C) and from animals homozygous for mutant *skn-1* (B,D) were tested for the presence of pharyngeal muscles (A,B) and gut-specific granules (C,D). A pharyngeal muscle-specific antibody labels the pharynx musculature of those embryos derived from wild-type (A) but does not bind to any structure in the embryos from *skn-1* mutants (B). Similarly, the gut granules characteristic of embryonic intestines (C) are absent from embryos derived from the *skn-1* mutants (D). (From Bowerman et al. 1992a, courtesy of B. Bowerman.)

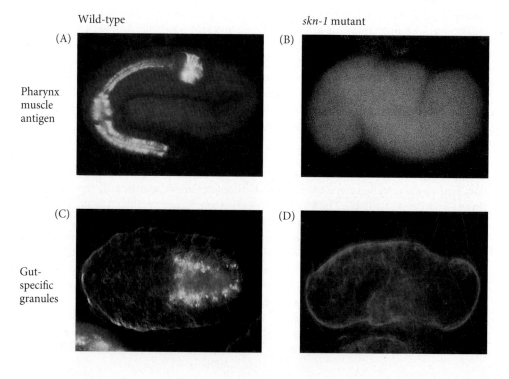

Wild-type   *skn-1* mutant

(A)   (B)

Pharynx muscle antigen

(C)   (D)

Gut-specific granules

PUF-8 protein, which binds to the 3′ UTR of *pal-1* mRNA and blocks its translation (Mainpal et al. 2011).

A third transcription factor, **PIE-1**, is necessary for germline cell fate. PIE-1 is placed into the P blastomeres through the action of the PAR-1 protein (**FIGURE 8.19**), and it appears to inhibit both SKN-1 and PAL-1 function in the P2 and subsequent germline cells (Hunter and Kenyon 1996). Mutations of the maternal *pie-1* gene result in germline blastomeres adopting somatic fates, with the P2 cell behaving similarly to a wild-type EMS blastomere. The localization and the genetic properties of PIE-1 suggest that it represses the establishment of somatic cell fate and preserves the totipotency of the germ cell lineage (Mello et al. 1996; Seydoux et al. 1996).

CONDITIONAL SPECIFICATION As mentioned earlier, the *C. elegans* embryo uses both autonomous and conditional modes of specification. Conditional specification can be seen in the development of the endoderm cell lineage. At the 4-cell stage, the EMS cell requires a signal from its neighbor (and sister cell), the P2 blastomere. Usually, the EMS cell divides into an MS cell (which produces mesodermal muscles) and an E cell (which produces the intestinal endoderm). If the P2 cell is removed at the early 4-cell stage, the EMS cell will divide into two MS cells, and no endoderm will be produced. If the EMS cell is recombined with the P2 blastomere, however, it will form endoderm; it will not do so, however, when combined with ABa, ABp, or both AB derivatives (Goldstein 1992). Specification of the MS cell begins with maternal SKN-1 activating the genes encoding transcription factors such as MED-1 and MED-2. The POP-1 signal (which encodes the TCF protein that binds β-catenin to the DNA) blocks the pathway to the E (endodermal) fate in the prospective MS cell to become MS by blocking the ability of MED-1 and MED-2 to activate the *tbx-35* gene (**FIGURE 8.20**; Broitman-Maduro et al. 2006; Maduro 2009). Throughout the animal kingdom, TBX proteins are known to be active in mesoderm formation; TBX-35 acts to activate the mesodermal genes in the pharynx (*pha-4*) and muscles (*hlh-1*) of *C. elegans*.

The P2 cell produces a signal that interacts with the EMS cell and instructs the EMS daughter next to it to become the E cell. This message is transmitted through the Wnt signaling cascade (**FIGURE 8.21**; Rocheleau et al. 1997; Thorpe et al. 1997; Walston et al. 2004). The P2 cell produces the MOM-2 protein, a *C. elegans* Wnt protein. MOM-2 is received in the EMS cell by the MOM-5 protein, a *C. elegans* version of the Wnt receptor protein Frizzled. The result of this signaling cascade is to downregulate the expression of the *pop-1* gene in the EMS daughter destined to become the E cell. In *pop-1*-deficient

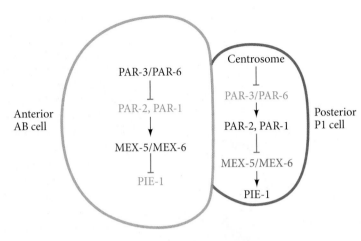

FIGURE 8.19 Segregation of PIE-1 determinant into the P1 blastomere at the 2-cell stage. The sperm centrosome inhibits the presence of the PAR-3/PAR-6 complex in the posterior of the egg. This allows the function of PAR-2 and PAR-1, which inhibit the MEX-5 and MEX-6 proteins that would degrade PIE-1. So while PIE-1 is degraded in the resulting anterior AB cell, it is preserved in the posterior P1 cell. (After Gönczy and Rose 2005.)

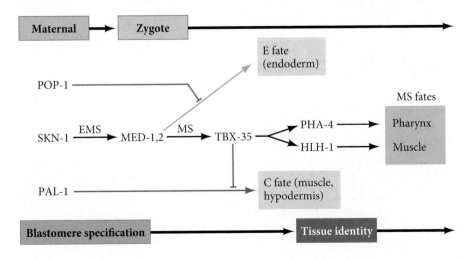

FIGURE 8.20 Model for specification of the MS blastomere. Maternal SKN-1 activates GATA transcription factors MED-1 and MED-2 in the EMS cell. The POP-1 signal prevents these proteins from activating the endodermal transcription factors (such as END-1) and instead activates the *tbx-35* gene. The TBX-35 transcription factor activates mesodermal genes in the MS cell, including *pha-4* in the pharynx lineage and *hlh-1* (which encodes a myogenic transcription factor) in muscles. TBX-35 also inhibits *pal-1* gene expression, thereby preventing the MS cell from acquiring the C-blastomere fates. (After Broitman-Maduro et al. 2006.)

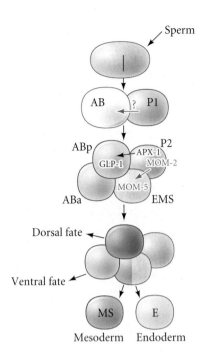

FIGURE 8.21 Cell-cell signaling in the 4-cell embryo of *C. elegans*. The P2 cell produces two signals: (1) the juxtacrine protein APX-1 (a Delta homologue), which is bound by GLP-1 (Notch) on the ABp cell, and (2) the paracrine protein MOM-2 (Wnt), which is bound by the MOM-5 (Frizzled) protein on the EMS cell. (After Han 1998.)

embryos, both EMS daughter cells become E cells (Lin et al. 1995; Park et al. 2004). Thus, the Wnt pathway is seen to be critical in establishing the anterior-posterior axis of *C. elegans*. Remarkably, as we will see, Wnt signaling appears to specify the A-P axis throughout the animal kingdom.

The P2 cell is also critical in giving the signal that distinguishes ABp from its sister, ABa (see Figure 8.21). ABa gives rise to neurons, hypodermis, and the anterior pharynx cells, while ABp makes only neurons and hypodermal cells. However, if one experimentally reverses the positions of these two cells, their fates are similarly reversed and a normal embryo forms. In other words, ABa and ABp are equivalent cells whose fates are determined by their positions in the embryo (Priess and Thomson 1987). Transplantation and genetic studies have shown that ABp becomes different from ABa through its interaction with the P2 cell. In an unperturbed embryo, both ABa and ABp contact the EMS blastomere, but only ABp contacts the P2 cell. If the P2 cell is killed at the early 4-cell stage, the ABp cell does not generate its normal complement of cells (Bowerman et al. 1992a,b). Contact between ABp and P2 is essential for the specification of ABp cell fates, and the ABa cell can be made into an ABp-type cell if it is forced into contact with P2 (Hutter and Schnabel 1994; Mello et al. 1994).

This interaction is mediated by the GLP-1 protein on the ABp cell and the APX-1 (anterior pharynx excess) protein on the P2 blastomere. In embryos whose mothers have mutant *glp-1*, ABp is transformed into an ABa cell (Hutter and Schnabel 1994; Mello et al. 1994). The GLP-1 protein is a member of a widely conserved family called the Notch proteins, which serve as cell membrane receptors in many cell-cell interactions; it is seen on both the ABa and ABp cells (Evans et al. 1994).[3] One of the most important ligands for Notch proteins such as GLP-1 is the cell surface protein Delta. In *C. elegans*, the Delta-like protein is APX-1, and it is found on the P2 cell (Mango et al. 1994a; Mello et al. 1994). This APX-1 signal breaks the symmetry between ABa and ABp, since it stimulates the GLP-1 protein solely on the AB descendant that it touches—namely, the ABp blastomere. In doing this, the P2 cell establishes the dorsal-ventral axis of *C. elegans* and confers on the ABp blastomere a fate different from that of its sister cell.

INTEGRATION OF AUTONOMOUS AND CONDITIONAL SPECIFICATION: DIFFERENTIATION OF THE *C. ELEGANS* PHARYNX It should become apparent from the above discussion that the pharynx is generated by two sets of cells. One group of pharyngeal precursors comes from the EMS cell and is dependent on the maternal *skn-1* gene. The second group of pharyngeal precursors comes from the ABa blastomere and is dependent on GLP-1 signaling from the EMS cell. In both cases, the pharyngeal precursor cells (and only these cells) are instructed to activate the *pha-4* gene (Mango et al. 1994b). The *pha-4* gene encodes a transcription factor that resembles the mammalian HNF3β protein. Microarray studies by Gaudet and Mango (2002) revealed that the PHA-4 transcription factor activates almost all of the pharynx-specific genes. It appears that the PHA-4 transcription factor may be the node that takes the maternal inputs and transforms them into a signal that transcribes the zygotic genes necessary for pharynx development.

Gastrulation in *C. elegans*

Gastrulation in *C. elegans* starts extremely early, just after the generation of the P4 cell in the 26-cell embryo (**FIGURE 8.22**; Skiba and Schierenberg 1992). At this time, the two daughters of the E cell (Ea and Ep) migrate from the ventral side into the center of the embryo. There they divide to form a gut consisting of 20 cells. There is a very small

[3] GLP-1 protein is localized in the ABa and ABp blastomeres, but the maternally encoded *glp-1* mRNA is found throughout the embryo. Evans and colleagues (1994) have postulated that there might be some translational determinant in the AB blastomere that enables the *glp-1* message to be translated in its descendants. The *glp-1* gene is also active in regulating post-embryonic cell-cell interactions. It is used later by the distal tip cell of the gonad to control the number of germ cells entering meiosis; hence the name GLP, for "germ line proliferation."

and transient blastocoel prior to the movement of the Ea and Ep cells, and their inward migration creates a tiny blastopore. The next cell to migrate through this blastopore is the P4 cell, the precursor of the germ cells. It migrates to a position beneath the gut primordium. The mesodermal cells move in next: the descendants of the MS cell migrate inward from the anterior side of the blastopore, and the C- and D-derived muscle precursors enter from the posterior side. These cells flank the gut tube on the left and right sides (Schierenberg 1997). Finally, about 6 hours after fertilization, the AB-derived cells that contribute to the pharynx are brought inside, while the hypoblast cells (precursors of the hypodermal skin cells) move ventrally by epiboly, eventually closing the blastopore. The two sides of the hypodermis are sealed by E-cadherin on the tips of the leading cells that meet at the ventral midline (Raich et al. 1999).

During the next 6 hours, the cells move and develop into organs, while the ball-shaped embryo stretches out to become a worm with 556 somatic cells and 2 germline stem cells (see Priess and Hirsh 1986; Schierenberg 1997). There is evidence (Schnabel et al. 2006) that, although these gastrulation movements provide a good first approximation of the final form, an additional "cell focusing" is used to move cells into functional arrangements. Here cells of the same fate sort out along the anterior-posterior axis. Other modeling takes place as well; an additional 115 cells undergo apoptosis (programmed cell death). After four molts, the worm is a sexually mature, hermaphroditic adult, containing exactly 959 somatic cells as well as numerous sperm and eggs.

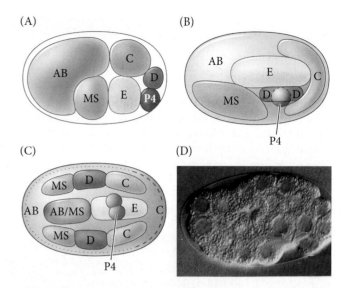

FIGURE 8.22 Gastrulation in *C. elegans*. (A) Positions of founder cells and their descendants at the 26-cell stage, at the start of gastrulation. (B) 102-cell stage, after the migration of the E, P4, and D descendants. (C) Positions of the cells near the end of gastrulation. The dotted and dashed lines represent regions of the hypodermis contributed by AB and C, respectively. (D) Early gastrulation, as the two E cells start moving inward. (After Schierenberg 1997; photograph courtesy of E. Schierenberg.)

WATCH DEVELOPMENT 8.6   Video from the Goldstein laboratory at the University of North Carolina beautifully depicts *C. elegans* gastrulation.

One characteristic that distinguishes *C. elegans* development from that of most other well-studied organisms is the prevalence of cell fusion. During *C. elegans* gastrulation, about one-third of all the cells fuse together to form syncytial cells containing many nuclei. The 186 cells that comprise the hypodermis (skin) of the nematode fuse into 8 syncytial cells, and cell fusion is also seen in the vulva, uterus, and pharynx. The functions of these fusion events can be determined by observing mutations that prevent syncytia from forming (Shemer and Podbilewicz 2000, 2003). It seems that fusion prevents individual cells from migrating beyond their normal borders. In the vulva (see Chapter 2), fusion prevents hypodermis cells from adopting a vulval fate and making an ectopic (and nonfunctional) vulva.

The *C. elegans* research program integrates genetics, cell biology, embryology, and even ecology to provide an understanding of the networks that govern cell differentiation and morphogenesis. In addition to providing some remarkable insights into how gene expression can change during development, studies of *C. elegans* have also humbled us by demonstrating how complex these networks are. Even in an organism as "simple" as *C. elegans*, with only a few genes and cell types, the right side of the body is made in a different manner from the left. The identification of the genes mentioned above is just the beginning of our effort to understand the complex interacting systems of development.

WEB TOPIC 8.3   **HETEROCHRONIC GENES AND THE CONTROL OF LARVAL STAGES**
C. elegans undergoes four larval stages before becoming an adult. These stages are regulated by microRNAs that control the translation of particular messages.

It is remarkable, especially in light of how much we know about vertebrate development, how little we know about even the most basic phenomena of invertebrate development. For instance, we do not know the identity of the morphogenetic determinants in the polar lobe or how they get there. We do not know what how the 4d cell gets its ability to produce both mesoderm and endoderm. We do not know how the non-gastropod mollusks—including squids, octopodes, clams, and chitons—develop, and how

their mode of cell specification, cleavage, and gastrulation relate to that of the gastropod mollusks such as snails. Moreover, we have scant knowledge of the mechanisms of molluscan metamorphosis, the mechanisms by which their larvae become juveniles. Even though the genetics of *C. elegans* is so remarkably complete, we are still looking for what localizes the PARs and what causes the cytoplasmic flow in the one-cell *C. elegans* embryo. These are examples of basic problems of development waiting to be solved.

Closing Thoughts on the Opening Photo

In 1923, Alfred Sturtevant identified left-coiling of snail shells as one of the first developmental mutations known. He was able to link the genetics of *Limnaea* snails with their coiling patterns, establishing that the left-coiling (sinistral) phenotype was a maternal effect (see p. 257). His work demonstrated in a highly visible manner the profound effect of genes on development. In 2016, the genetic basis of snail coiling may have been identified and the pathway leading to right-left asymmetry outlined (see Davison et al. 2016). Today the genetics of snail shell coiling informs our understanding of early development, illuminating such principles as the establishment of blastomere identity and how morphogenetic determinants affect cleavage and gastrulation patterns on the road to creating the organism's final phenotype. (Photograph courtesy of R. Kuroda.)

Snapshot Summary
Early Development in Snails and Nematodes

1. Body axes are established in different ways in different species. In some species the axes are established at fertilization through determinants in the egg cytoplasm. In others, such as nematodes and snails, the axes are established by cell interactions later in development.

2. Both snails and nematodes have holoblastic cleavage. In snails, cleavage is spiral; in nematodes, it is rotational.

3. In snails and *C. elegans*, gastrulation begins when there are relatively few cells. The blastopore becomes the mouth (the protostome mode of gastrulation).

4. Spiral cleavage in snails results in stereoblastulae (i.e., blastulae with no blastocoels). The direction of the cleavage spirals is regulated by a factor encoded by the mother and placed in the oocyte. Spiral cleavage can be modified by evolution, and adaptations of spiral cleavage have allowed some mollusks to survive in otherwise harsh environments.

5. The polar lobe of certain mollusks contains the morphogenetic determinants for mesoderm and endoderm. These determinants enter the D blastomere.

6. The soil nematode *Caenorhabditis elegans* was chosen as a model organism because it has a small number of cells, has a small genome, is easily bred and maintained, has a short life span, can be genetically manipulated, and has a cuticle through which one can see cell movements.

7. In the early divisions of the *C. elegans* zygote, one daughter cell becomes a founder cell (producing differentiated descendants) and the other becomes a stem cell (producing other founder cells and the germ line).

8. Blastomere identity in *C. elegans* is regulated by both autonomous and conditional specification.

Further Reading

Crampton, H. E. and E. B. Wilson. 1896. Experimental studies on gastropod development. *Arch. Entw. Mech.* 3: 1–18.

Davison, A. and 15 others. 2016. Formin is associated with left-right asymmetry in the pond snail and frog. *Curr. Biol.* 26: 654–660.

Henry, J. Q. 2014. Spiralian model systems. *Int. J. Dev. Biol.* 58: 389-401.

Hoso, M., Y. Kameda, S.-P. Wu, T. Asami, M. Kato and M. Hori. 2010. A speciation gene for left–right reversal in snails results in anti-predator adaptation. *Nature Communications* 1: 133. doi:10.1038/ncomms1133.

Lambert, J. D. 2010. Developmental patterns in spiralian embryos. *Curr. Biol.* 20: 272–277.

Resnick T. D., K. A. McCulloch and A. E. Rougvie. 2010. miRNAs give worms the time of their lives: Small RNAs and temporal control in *Caenorhabditis elegans. Dev. Dyn.* 239:1477–1489.

Shibazaki, Y., M. Shimizu and R. Kuroda. 2004. Body handedness is directed by genetically determined cytoskeletal dynamics in the early embryo. *Curr. Biol.* 14: 1462–1467.

Sturtevant, A. H. 1923. Inheritance of direction of coiling in *Limnaea. Science*, New Series, 58: 269-270.

Sulston, J. E., J. Schierenberg, J. White and N. Thomson. 1983. The embryonic cell lineage of the nematode *Caenorhabditis elegans. Dev. Biol.* 100: 64–119.

Wilson, E. B. 1898. Cell lineage and ancestral reminiscence. In *Biological Lectures from the Marine Biological Laboratories, Woods Hole, Massachusetts*, pp. 21–42. Ginn, Boston.

GO TO WWW.DEVBIO.COM...

...for Web Topics, Scientists Speak interviews, Watch Development videos, Dev Tutorials, and complete bibliographic information for all literature cited in this chapter.

The Genetics of Axis Specification in *Drosophila*

THANKS LARGELY TO STUDIES spearheaded by Thomas Hunt Morgan's laboratory during the first two decades of the twentieth century, we know more about the genetics of *Drosophila melanogaster* than that of any other multicellular organism. The reasons have to do with both the flies themselves and with the people who first studied them. *Drosophila* is easy to breed, hardy, prolific, and tolerant of diverse conditions. Moreover, in some larval cells, the DNA replicates several times without separating. This leaves hundreds of strands of DNA adjacent to each other, forming polytene (Greek, "many strands") chromosomes (**FIGURE 9.1**). The unused DNA is more condensed and stains darker than the regions of active DNA. The banding patterns were used to indicate the physical location of the genes on the chromosomes. Morgan's laboratory established a database of mutant strains, as well as an exchange network whereby any laboratory could obtain them.

Historian Robert Kohler noted in 1994 that "The chief advantage of *Drosophila* initially was one that historians have overlooked: it was an excellent organism for student projects." Indeed, undergraduates (starting with Calvin Bridges and Alfred Sturtevant) played important roles in *Drosophila* research. The *Drosophila* genetics program, says Kohler, was "designed by young persons to be a young person's game," and the students set the rules for *Drosophila* research: "No trade secrets, no monopolies, no poaching, no ambushes."

Jack Schultz (originally in Morgan's laboratory) and others attempted to relate the burgeoning supply of data on the genetics of *Drosophila* to its development. But *Drosophila* was a difficult organism on which to study embryology. Fly embryos proved complex and intractable, being neither large enough to manipulate experimentally nor transparent

What changes in development caused this fly to have four wings instead of two?

The Punchline

The development of the fruit fly is extremely rapid, and its body axes are specified by factors in the maternal cytoplasm even before the sperm enters the egg. The anterior-posterior axis is specified by proteins and mRNAs made in maternal nurse cells and transported into the oocyte, such that each region of the egg contains different ratios of anterior- and posterior-promoting proteins. Eventually, gradients of these proteins control a set of transcription factors—the homeotic proteins—that specify the structures to be formed by each segment of the adult fly. The dorsal-ventral axis is also initiated in the egg, which sends a signal to its surrounding follicle cells. The follicle cells respond by initiating a molecular cascade that leads both to cell-type specification and to gastrulation. Specific organs form at the intersection of the anterior-posterior axis and the dorsal-ventral axis.

(A)

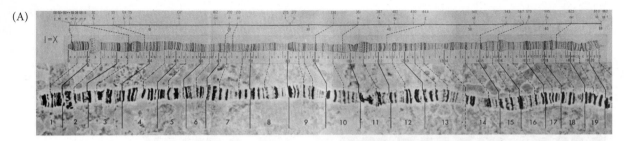

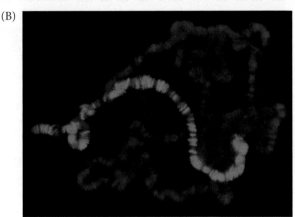

(B)

FIGURE 9.1   Polytene chromosomes of *Drosophila*. DNA in the larval salivary glands and other larval tissue replicates without separating. (A) Photograph of the *D. melanogaster* X chromosome. The chart above it was made by Morgan's student Calvin Bridges in 1935. (B) Chromosomes from salivary gland cells of a third instar *D. melanogaster* male. Each polytene chromosome has 1024 strands of DNA (blue stain). Here, an antibody (red) directed against the MSL transcription factor binds only to genes on the X chromosome. MSL accelerates gene expression in the single male X chromosome so it can match the amount of gene expression by females with their two X chromosomes. (A from Brody 1996; B photograph by A. A. Alekseyenko and M. I. Kuroda.)

enough to observe microscopically. It was not until the techniques of molecular biology allowed researchers to identify and manipulate the insect's genes and RNA that its genetics could be related to its development. And when that happened, a revolution occurred in the field of biology. This revolution is continuing, in large part because of the availability of the complete *Drosophila* genome sequence and our ability to generate transgenic flies at high frequency (Pfeiffer et al. 2009; del Valle Rodríguez et al. 2011). Researchers are now able to identify developmental interactions taking place in very small regions of the embryo, to identify enhancers and their transcription factors, and to mathematically model the interactions to a remarkable degree of precision (Hengenius et al. 2014).

Early *Drosophila* Development

We have already discussed the specification of early embryonic cells by cytoplasmic determinants stored in the oocyte. The cell membranes that form during cleavage establish the region of cytoplasm incorporated into each new blastomere, and the morphogenetic determinants in the incorporated cytoplasm then direct differential gene expression in each cell. But in *Drosophila* development, cell membranes do not form until after the thirteenth nuclear division. Prior to this time, the dividing nuclei all share a common cytoplasm and material can diffuse throughout the whole embryo. The specification of cell types along the anterior-posterior and dorsal-ventral axes is accomplished by the interactions of components *within* the single multinucleated cell. Moreover, these axial differences are initiated at an earlier developmental stage by the position of the egg within the mother's egg chamber. Whereas the sperm entry site may fix the axes in nematodes and tunicates, the fly's anterior-posterior and dorsal-ventral axes are specified by interactions between the egg and its surrounding follicle cells prior to fertilization.

VADE MECUM

The fruit fly segment has remarkable time-lapse sequences, including footage of cleavage and gastrulation. This segment also provides access to the fly life cycle.

WATCH DEVELOPMENT 9.1   The website "The Interactive Fly" features movies illustrating all aspects of *Drosophila* development.

Fertilization

Drosophila fertilization is a remarkable series of events and is quite different from fertilizations we've described previously.

- *The sperm enters an egg that is already activated.* Egg activation in *Drosophila* is accomplished at ovulation, a few minutes *before* fertilization begins. As the *Drosophila* oocyte squeezes through a narrow orifice, calcium channels open and Ca^{2+} flows in. The oocyte nucleus then resumes its meiotic divisions and the cytoplasmic mRNAs become translated without fertilization (Mahowald et al. 1983; Fitch and Wakimoto 1998; Heifetz et al. 2001; Horner and Wolfner 2008).

- *There is only one site where the sperm can enter the egg.* This is the **micropyle**, a tunnel in the chorion (eggshell) located at the future dorsal anterior region of the embryo. The micropyle allows sperm to pass through it one at a time and probably prevents polyspermy in *Drosophila*. There are no cortical granules to block polyspermy, although cortical changes are seen.

- By the time the sperm enters the egg, the egg already has begun to specify the body axes; thus the sperm enters an egg that is already organizing itself as an embryo.

- *The sperm and egg cell membranes do not fuse. Rather, the sperm enters into the egg intact.* The DNA of the male and female pronuclei replicate before the pronuclei have fused, and after the pronuclei fuse, the maternal and paternal chromosomes remain separate until the end of the first mitosis (Loppin et al. 2015).

WEB TOPIC 9.1 ***DROSOPHILA* FERTILIZATION** Fertilization of a *Drosophila* egg can only occur in the region of the oocyte that will become the anterior of the embryo. Moreover, the sperm tail appears to stay in this region.

Cleavage

Most insect eggs undergo **superficial cleavage**, wherein a large mass of centrally located yolk confines cleavage to the cytoplasmic rim of the egg (see Figure 1.5). One of the fascinating features of this cleavage pattern is that cells do not form until after the nuclei have divided several times. In the *Drosophila* egg, karyokinesis (nuclear division) occurs without cytokinesis (cell division) so as to create a **syncytium**, a single cell with many nuclei residing in a common cytoplasm (**FIGURE 9.2**). The zygote nucleus

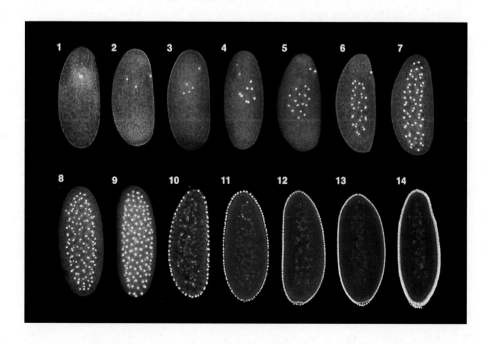

FIGURE 9.2 Laser confocal micrographs of stained chromatin showing syncytial nuclear divisions and superficial cleavage in a series of *Drosophila* embryos. The future anterior end is positioned upward; numbers refer to the nuclear division cycle. The early nuclear divisions occur centrally within a syncytium. Later, the nuclei and their cytoplasmic islands (energids) migrate to the periphery of the cell. This creates the syncytial blastoderm. After cycle 13, the cellular blastoderm forms by ingression of cell membranes between nuclei. The pole cells (germ cell precursors) form in the posterior. (Courtesy of D. Daily and W. Sullivan.)

FIGURE 9.3 Nuclear and cell division in *Drosophila* embryos. (A) Nuclear division (but not cell division) can be seen in a syncytial *Drosophila* embryo using a dye that stains DNA. The first region to cellularize, the pole region, can be seen forming the cells in the posterior region of the embryo that will eventually become the germ cells (sperm or eggs) of the fly. (B) Chromosomes dividing at the cortex of a syncytial blastoderm. Although there are no cell boundaries, actin (green) can be seen forming regions within which each nucleus divides. The microtubules of the mitotic apparatus are stained red with antibodies to tubulin. (C,D) Cross section of a part of a cycle 10 *Drosophila* embryo showing nuclei (green) at the cortex of the syncytial cell, adjacent to a layer of actin microfilaments (red). (C) Interphase nuclei. (D) Nuclei in anaphase, dividing parallel to the cortex and enabling the nuclei to stay in the cell periphery. (A from Bonnefoy et al. 2007; B from Sullivan et al. 1993, courtesy of W. Theurkauf and W. Sullivan; C,D from Foe 2000, courtesy of V. Foe.)

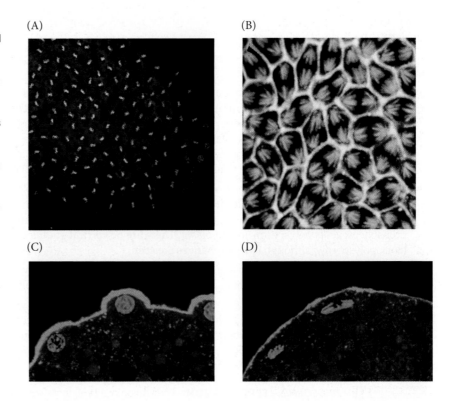

(A)   (B)

(C)   (D)

undergoes several nuclear divisions within the central portion of the egg; 256 nuclei are produced by a series of eight nuclear divisions averaging 8 minutes each (**FIGURE 9.3A,B**). This rapid rate of division is accomplished by repeated rounds of alternating S (DNA replication) and M (mitosis) phases in the absence of the gap (G) phases of the cell cycle. During the ninth division cycle, approximately five nuclei reach the surface of the posterior pole of the embryo. These nuclei become enclosed by cell membranes and generate the **pole cells** that give rise to the gametes of the adult. At cycle 10, the other nuclei migrate to the cortex (periphery) of the egg and the mitoses continue, albeit at a progressively slower rate (**FIGURE 9.3C,D**; Foe et al. 2000). During these stages of nuclear division, the embryo is called a **syncytial blastoderm**, since no cell membranes exist other than that of the egg itself.

Although the nuclei divide within a common cytoplasm, the cytoplasm itself is far from uniform. Karr and Alberts (1986) have shown that each nucleus within the syncytial blastoderm is contained within its own little territory of cytoskeletal proteins. When the nuclei reach the periphery of the egg during the tenth cleavage cycle, each nucleus becomes surrounded by microtubules and microfilaments. The nuclei and their associated cytoplasmic islands are called **energids**. Following division cycle 13, the cell membrane (which had covered the egg) folds inward between the nuclei, eventually partitioning off each energid into a single cell. This process creates the **cellular blastoderm**, in which all the cells are arranged in a single-layered jacket around the yolky core of the egg (Turner and Mahowald 1977; Foe and Alberts 1983; Mavrakis et al. 2009).

As with all cell formation, the formation of the cellular blastoderm involves a delicate interplay between microtubules and microfilaments (**FIGURE 9.4**). The membrane movements, nuclear elongation, and actin polymerization all appear to be coordinated by the microtubules (Riparbelli et al. 2007). The first phase of blastoderm cellularization is characterized by the invagination of cell membranes between the nuclei to form furrow canals. This process can be inhibited by drugs that block microtubules. After the furrow canals have passed the level of the nuclei, the second phase of cellularization occurs. The rate of invagination increases and the actin-membrane complex

(A)

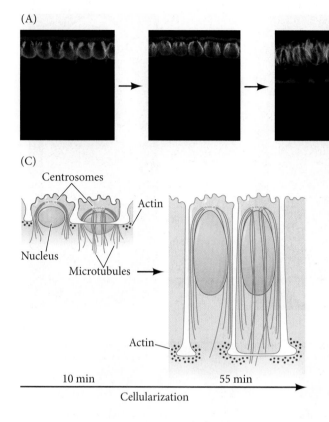

(B)

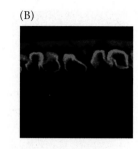

(C)

Centrosomes

Actin

Nucleus

Microtubules →

Actin

10 min   55 min

Cellularization

FIGURE 9.4 Formation of the cellular blastoderm in *Drosophila*. Nuclear shape change and cellularization are coordinated through the cytoskeleton. (A) Cellularization and nuclear shape change shown by staining the embryo for microtubules (green), microfilaments (blue), and nuclei (red). The red stain in the nuclei is due to the presence of the Kugelkern protein, one of the earliest proteins made from the zygotic nuclei. It is essential for nuclear elongation. (B) This embryo was treated with nocadozole to disrupt microtubules. The nuclei fail to elongate, and cellularization is prevented. (C) Diagrammatic representation of cell formation and nuclear elongation. (After Brandt et al. 2006; photographs courtesy of J. Grosshans and A. Brandt.)

begins to constrict at what will be the basal end of the cell (Foe et al. 1993; Schejter and Wieschaus 1993; Mazumdar and Mazumdar 2002). In *Drosophila*, the cellular blastoderm consists of approximately 6000 cells and is formed within 4 hours of fertilization.

WATCH DEVELOPMENT 9.2   The superficial cleavage of *Drosophila* in the syncytial embryo is shown in time-lapse video.

The mid-blastula transition

After the nuclei reach the periphery, the time required to complete each of the next four divisions becomes progressively longer. Whereas cycles 1–10 average 8 minutes each, cycle 13—the last cycle in the syncytial blastoderm—takes 25 minutes to complete. Cycle 14, in which the *Drosophila* embryo forms cells (i.e., after 13 divisions), is asynchronous. Some groups of cells complete this cycle in 75 minutes, other groups take 175 minutes (Foe 1989).

It is at this time that the genes of the nuclei become active. Before this point, the early development of *Drosophila* is directed by proteins and mRNAs placed into the egg during oogenesis. These are the products of the *mother's* genes, not the genes of the embryo's own nuclei. Such genes that are active in the mother to make products for the early development of the offspring are called **maternal effect genes**, and the mRNAs in the oocyte are often referred to as **maternal messages**. Zygotic gene transcription (i.e., the activation of the embryo's own genes) begins around cycle 11 and is is greatly enhanced at cycle 14. This slowdown of nuclear division, cellularization, and concomitant increase in new RNA transcription is often referred to as the **mid-blastula transition**. It is at this stage that the maternally provided mRNAs are degraded and control of development is handed over to the zygote's own genome (Brandt et al. 2006; De Renzis et al. 2007; Benoit et al. 2009). Such a **maternal-to-zygotic transition** is seen in the embryos of numerous vertebrate and invertebrate phyla.

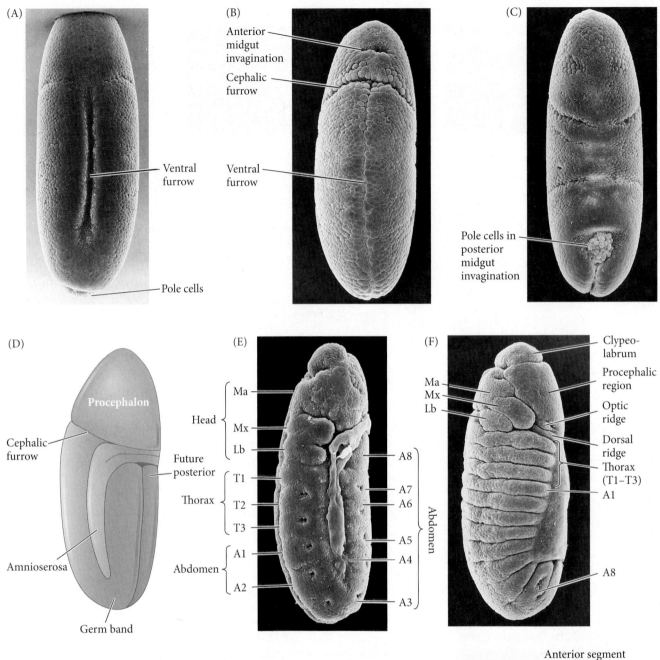

(A)

Ventral furrow

Pole cells

(B)

Anterior midgut invagination

Cephalic furrow

Ventral furrow

(C)

Pole cells in posterior midgut invagination

(D)

Procephalon

Cephalic furrow

Amnioserosa

Germ band

Future posterior

(E)

Head
{ Ma
Mx
Lb }

Thorax
{ T1
T2
T3 }

Abdomen
{ A1
A2 }

A8
A7
A6
A5
A4
A3
} Abdomen

(F)

Ma
Mx
Lb

Clypeo-labrum

Procephalic region

Optic ridge

Dorsal ridge

Thorax (T1–T3)

A1

A8

FIGURE 9.5 Gastrulation in *Drosophila*. The anterior of each gastrulating embryo points upward in this series of scanning electron micrographs. (A) Ventral furrow beginning to form as cells flanking the ventral midline invaginate. (B) Closing of ventral furrow, with mesodermal cells placed internally and surface ectoderm flanking the ventral midline. (C) Dorsal view of a slightly older embryo, showing the pole cells and posterior endoderm sinking into the embryo. (D) Schematic representation showing dorsolateral view of an embryo at fullest germ band extension, just prior to segmentation. The cephalic furrow separates the future head region (procephalon) from the germ band, which will form the thorax and abdomen. (E) Lateral view, showing fullest extension of the germ band and the beginnings of segmentation. Subtle indentations mark the incipient segments along the germ band. Ma, Mx, and Lb correspond to the mandibular, maxillary, and labial head segments; T1–T3 are the thoracic segments; and A1–A8 are the abdominal segments. (F) Germ band reversing direction. The true segments are now visible, as well as the other territories of the dorsal head, such as the clypeolabrum, procephalic region, optic ridge, and dorsal ridge. (G) Newly hatched first instar larva. (Photographs courtesy of F. R. Turner; D after Campos-Ortega and Hartenstein 1985.)

(G)

Anterior segment

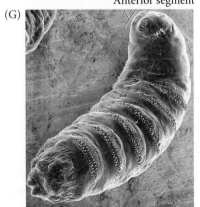

WEB TOPIC 9.2 **MECHANISMS OF THE *DROSOPHILA* MID-BLASTULA TRANSITION**
Coordination of the mid-blastula transition and the maternal-to-zygotic transition is controlled by several factors, including (1) the ratio of chromatin to cytoplasm; (2) Smaug protein; and (3) the Zelda transcription factor.

Gastrulation

The general body plan of *Drosophila* is the same in the embryo, the larva, and the adult, each of which has a distinct head end and a distinct tail end, between which are repeating segmental units. Three of these segments form the thorax, while another eight segments form the abdomen. Each segment of the adult fly has its own identity. The first thoracic segment, for example, has only legs; the second thoracic segment has legs and wings; and the third thoracic segment has legs and halteres (balancing organs).

Gastrulation begins shortly after the mid-blastula transition. The first movements of *Drosophila* gastrulation segregate the presumptive mesoderm, endoderm, and ectoderm. The prospective mesoderm—about 1000 cells constituting the ventral midline of the embryo—folds inward to produce the **ventral furrow** (**FIGURE 9.5A**). This furrow eventually pinches off from the surface to become a ventral tube within the embryo. The prospective endoderm invaginates to form two pockets at the anterior and posterior ends of the ventral furrow. The pole cells are internalized along with the endoderm (**FIGURE 9.5B,C**). At this time, the embryo bends to form the **cephalic furrow**.

The ectodermal cells on the surface and the mesoderm undergo convergence and extension, migrating toward the ventral midline to form the **germ band**, a collection of cells along the ventral midline that includes all the cells that will form the trunk of the embryo. The germ band extends posteriorly and, perhaps because of the egg case, wraps around the top (dorsal) surface of the embryo (**FIGURE 9.5D**). Thus, at the end of germ band formation, the cells destined to form the most posterior larval structures are located immediately behind the future head region (**FIGURE 9.5E**). At this time, the body segments begin to appear, dividing the ectoderm and mesoderm. The germ band then retracts, placing the presumptive posterior segments at the posterior tip of the embryo (**FIGURE 9.5F**). At the dorsal surface, the two sides of the epidermis are brought together in a process called **dorsal closure**. The amnioserosa (the extraembryonic layer that surrounds the embryo), which had been the most dorsal structure, interacts with the epidermal cells to stimulate their migration (reviewed in Panfilio 2008; Heisenberg 2009).

While the germ band is in its extended position, several key morphogenetic processes occur: organogenesis, segmentation (**FIGURE 9.6A**), and segregation of the imaginal discs.[1] The nervous system forms from two regions of ventral ectoderm. Neuroblasts (i.e., the neural

---

[1] Imaginal discs are cells set aside to produce the adult structures. Imaginal disc differentiation will be discussed as a part of metamorphosis in Chapter 21.

FIGURE 9.6 Axis formation in *Drosophila*. (A) Comparison of larval (left) and adult (right) segmentation. In the adult, the three thoracic segments can be distinguished by their appendages: T1 (prothorax) has legs only; T2 (mesothorax) has wings and legs; T3 (metathorax) has halteres (not visible) and legs. (B) During gastrulation, the mesodermal cells in the most ventral region enter the embryo, and the neurogenic cells expressing *Short gastrulation* (*Sog*) become the ventralmost cells of the embryo. *Sog*, blue; *ventral nervous system defective*, green; *intermediate neuroblast defective*, red. (B courtesy of E. Bier.)

(A)   Anterior

Head
Prothorax
Mesothorax
Metathorax

T1
T2
T3
A1
A2
A3
A4
A5
A6
A7 A8

T1
T2
T3
A1
A2
A3
A4
A5 A6
A7 A8

Dorsal

Ventral

Abdominal segments

Posterior

(B)

progenitor cells) differentiate from this neurogenic ectoderm and migrate inward within each segment (and also from the nonsegmented region of the head ectoderm). Therefore, in insects such as *Drosophila*, the nervous system is located ventrally rather than being derived from a dorsal neural tube as it is in vertebrates (**FIGURE 9.6B**; see also Figure 9.29).

WATCH DEVELOPMENT 9.3   Watch a video showing external and internal development in *Drosophila*.

The Genetic Mechanisms Patterning the *Drosophila* Body

Most of the genes involved in shaping the larval and adult forms of *Drosophila* were identified in the early 1980s using a powerful forward genetics approach (i.e., identifying the genes responsible for a particular phenotype). The basic strategy was to randomly mutagenize flies and then screen for mutations that disrupted the normal formation of the body plan. Some of these mutations were quite fantastic, including embryos and adult flies in which specific body structures were either missing or in the wrong place. These mutant collections were distributed to many different laboratories. The genes involved in the mutant phenotypes were sequenced and then characterized with respect to their expression patterns and their functions. This combined effort has led to a molecular understanding of body plan development in *Drosophila* that is unparalleled in all of biology, and in 1995 the work resulted in Nobel Prizes for Edward Lewis, Christiane Nüsslein-Volhard, and Eric Wieschaus.

The rest of this chapter details the genetics of *Drosophila* development as we have come to understand it over the past three decades. First we will examine how the anterior-posterior axis of the embryo is established by interactions between the developing oocyte and its surrounding follicle cells. Next we will see how dorsal-ventral patterning gradients are formed within the embryo, and how these gradients specify different tissue types. Finally, we will briefly show how the positioning of embryonic tissues along the two primary axes specifies these tissues to become particular organs.

Segmentation and the Anterior-Posterior Body Plan

The processes of embryogenesis may officially begin at fertilization, but many of the molecular events critical for *Drosophila* embryogenesis actually occur during oogenesis. Each oocyte is descended from a single female germ cell—the **oogonium**. Before oogenesis begins, the oogonium divides four times with incomplete cytokinesis, giving rise to 16 interconnected cells. These 16 germline cells, along with a surrounding epithelial layer of somatic follicle cells, constitute the **egg chamber** in which the oocyte will develop. These germline cells include 15 metabolically active **nurse cells** that make mRNAs and proteins that are transported into the single cell that will become the oocyte. As the oocyte precursor develops at the posterior end of the egg chamber, numerous mRNAs made in the nurse cells are transported along microtubules through the cellular interconnections into the enlarging oocyte.

The genetic screens pioneered by Nüsslein-Volhard and Wieschaus identified a hierarchy of genes that (1) establish anterior-posterior polarity and (2) divide the embryo into a specific number of segments, each with a different identity (**FIGURE 9.7**). This hierarchy is initiated by **maternal effect genes** that produce messenger RNAs localized to different regions of the egg. These mRNAs encode transcriptional and translational regulatory proteins that diffuse through the syncytial blastoderm and activate or repress the expression of certain zygotic genes.

The first such zygotic genes to be expressed are called **gap genes** because mutations in them cause gaps in the segmentation pattern. These genes are expressed in certain

(A)
Cytoplasmic polarity
(maternal effect)

Gap genes

Pair-rule genes

Segment polarity
genes

Homeotic
genes

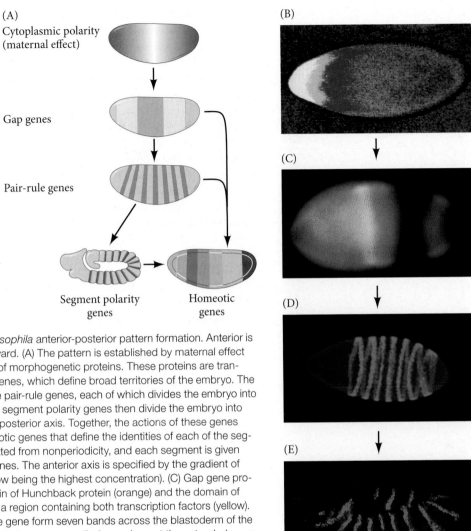

(B)

(C)

(D)

(E)

FIGURE 9.7 Generalized model of *Drosophila* anterior-posterior pattern formation. Anterior is to the left; the dorsal surface faces upward. (A) The pattern is established by maternal effect genes that form gradients and regions of morphogenetic proteins. These proteins are transcription factors that activate the gap genes, which define broad territories of the embryo. The gap genes enable the expression of the pair-rule genes, each of which divides the embryo into regions about two segments wide. The segment polarity genes then divide the embryo into segment-sized units along the anterior-posterior axis. Together, the actions of these genes define the spatial domains of the homeotic genes that define the identities of each of the segments. In this way, periodicity is generated from nonperiodicity, and each segment is given a unique identity. (B) Maternal effect genes. The anterior axis is specified by the gradient of Bicoid protein (yellow through red; yellow being the highest concentration). (C) Gap gene protein expression and overlap. The domain of Hunchback protein (orange) and the domain of Krüppel protein (green) overlap to form a region containing both transcription factors (yellow). (D) Products of the *fushi tarazu* pair-rule gene form seven bands across the blastoderm of the embryo. (E) Products of the segment polarity gene *engrailed*, seen here at the extended germ band stage. (B courtesy of C. Nüsslein-Volhard; C courtesy of C. Rushlow and M. Levine; D courtesy of D. W. Knowles; E courtesy of S. Carroll and S. Paddock.)

broad (about three segments wide), partially overlapping domains. Gap genes encode transcription factors, and differing combinations and concentrations of gap gene proteins regulate the transcription of **pair-rule genes**, which divide the embryo into periodic units. The transcription of the different pair-rule genes results in a striped pattern of seven transverse bands perpendicular to the anterior-posterior axis. The transcription factors encoded by the pair-rule genes activate the **segment polarity genes**, whose mRNA and protein products divide the embryo into 14-segment-wide units, establishing the periodicity of the embryo. At the same time, the protein products of the gap, pair-rule, and segment polarity genes interact to regulate another class of genes, the **homeotic selector genes**, whose transcription determines the developmental fate of each segment.

Anterior-posterior polarity in the oocyte

The anterior-posterior polarity of the embryo is established while the oocyte is still in the egg chamber, and it involves interactions between the developing egg cell and the follicular cells that enclose it. The follicular epithelium surrounding the developing oocyte is initially uniform with respect to cell fate, but this uniformity is broken by two signals organized by the oocyte nucleus. Interestingly, both of these signals involve the same gene, **gurken**. The *gurken* message appears to be synthesized in the nurse cells, but it is transported into the oocyte. Here it becomes localized between the oocyte nucleus and the cell membrane, and it is translated into Gurken protein (Cáceres and Nilson 2005). At this time the oocyte nucleus is very near what will become the posterior tip of the egg chamber, and the Gurken signal is received by the follicle cells at that position through a receptor protein encoded by the **torpedo** gene[2] (**FIGURE 9.8A**). This signal results in the "posteriorization" of these follicle cells (**FIGURE 9.8B**). The posterior follicle cells send a signal back into the oocyte. This signal, a lipid kinase, recruits the Par-1 protein to the posterior edge of the oocyte cytoplasm (**FIGURE 9.8C**; Doerflinger et al. 2006; Gervais et al. 2008). Par-1 protein organizes microtubules specifically with their minus (cap) and plus (growing) ends at the anterior and posterior ends of the oocyte, respectively (Gonzalez-Reyes et al. 1995; Roth et al. 1995; Januschke et al. 2006).

The orientation of the microtubules is critical, because different microtubule motor proteins will transport their mRNA or protein cargoes in different directions. The motor protein kinesin, for instance, is an ATPase that will use the energy of ATP to transport material to the plus end of the microtubule. Dynein, however, is a "minus-directed" motor protein that transports its cargo in the opposite direction. One of the messages transported by kinesin along the microtubules to the posterior end of the oocyte is **oskar** mRNA (Zimyanin et al. 2008). The *oskar* mRNA is not able to be translated until it reaches the posterior cortex, at which time it generates the Oskar protein. Oskar recruits more Par-1 protein, thereby stabilizing the microtubule orientation and allowing more material to be recruited to the posterior pole of the oocyte (Doerflinger et al. 2006; Zimyanin et al. 2007). The posterior pole will thereby have its own distinctive cytoplasm, called **pole plasm**, which contains the determinants for producing the abdomen and the germ cells.

This cytoskeletal rearrangement in the oocyte is accompanied by an increase in oocyte volume, owing to transfer of cytoplasmic components from the nurse cells. These components include maternal messengers such as the **bicoid** and **nanos** mRNAs. These mRNAs are carried by motor proteins along the microtubules to the anterior and posterior ends of the oocyte, respectively (**FIGURE 9.8D–F**). As we will soon see, the protein products encoded by *bicoid* and *nanos* are critical for establishing the anterior-posterior polarity of the embryo.

Maternal gradients: Polarity regulation by oocyte cytoplasm

PROTEIN GRADIENTS IN THE EARLY EMBRYO A series of ligation experiments (see Web Topic 9.3) showed that two organizing centers control insect development: a head-forming center anteriorly and a posterior-forming center in the rear of the embryo. These centers appeared to secrete substances that generated a head-forming gradient and a tail-forming gradient. In the late 1980s, this gradient hypothesis was united with a genetic approach to the study of *Drosophila* embryogenesis. If there were gradients, what were the morphogens whose concentrations changed over space? What were the genes that shaped these gradients? And did these morphogens act by activating or inhibiting certain genes in the areas where they were concentrated? Christiane Nüsslein-Volhard led a research program that addressed these questions. The researchers found that one set of genes encoded morphogens for the anterior part of the embryo, another set of genes encoded morphogens responsible for organizing the posterior region of the embryo, and a third set of genes encoded proteins that produced the terminal regions at both ends of the embryo, the acron and the tail (**TABLE 9.1**).

WEB TOPIC 9.3 **INSECT SIGNALING CENTERS** The two-signaling-center approach to insect patterning was largely a product of the German school of embryology. This web topic describes how the observations of experimental embryology were transformed into questions of molecular biology.

---

[2] Gurken protein is a member of the EGF (*epidermal growth factor*) family, and *torpedo* encodes a homologue of the vertebrate EGF receptor (Price et al. 1989; Neuman-Silberberg and Schüpbach 1993).

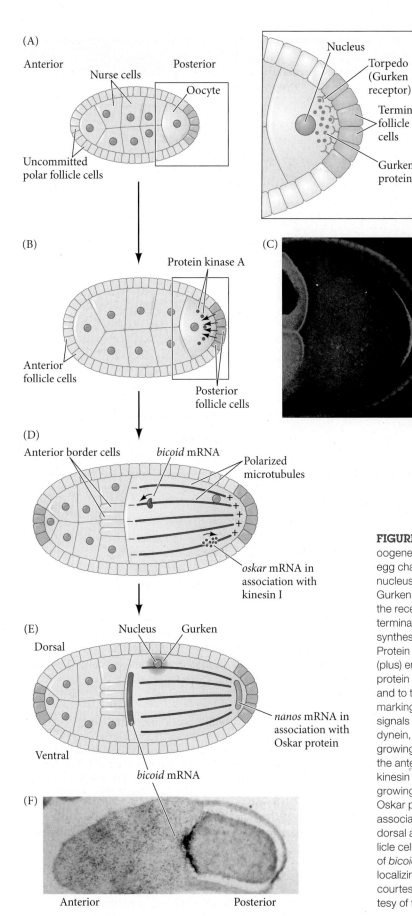

(A)

Anterior

Nurse cells

Posterior

Oocyte

Uncommitted polar follicle cells

Nucleus

Torpedo (Gurken receptor)

Terminal follicle cells

Gurken protein

(B)

Protein kinase A

Anterior follicle cells

Posterior follicle cells

(C)

(D)

Anterior border cells

bicoid mRNA

Polarized microtubules

oskar mRNA in association with kinesin I

(E)

Dorsal

Nucleus

Gurken

Ventral

nanos mRNA in association with Oskar protein

bicoid mRNA

(F)

Anterior

Posterior

FIGURE 9.8 The anterior-posterior axis is specified during oogenesis. (A) The oocyte moves into the posterior region of the egg chamber, while nurse cells fill the anterior portion. The oocyte nucleus moves toward the terminal follicle cells and synthesizes Gurken protein (green). The terminal follicle cells express Torpedo, the receptor for Gurken. (B) When Gurken binds to Torpedo, the terminal follicle cells differentiate into posterior follicle cells and synthesize a molecule that activates protein kinase A in the egg. Protein kinase A orients the microtubules such that the growing (plus) ends are at the posterior (depicted in panel D). (C) Par-1 protein (green) localizes to the cortical cytoplasm of nurse cells and to the posterior pole of the oocyte. (The Staufen protein marking the posterior pole is labeled red; the red and green signals combine to fluoresce yellow.) (D) *bicoid* mRNA binds to dynein, a "minus-directed" motor protein associated with the non-growing end of microtubules; dynein moves the *bicoid* mRNA to the anterior end of the egg. *oskar* mRNA becomes complexed to kinesin I, a "plus-directed" motor protein that moves it toward the growing end of the microtubules at the posterior region, where Oskar protein can bind *nanos* mRNA. (E) The nucleus (with its associated Gurken protein) migrates along the microtubules to the dorsal anterior region of the oocyte and induces the adjacent follicle cells to become the dorsal follicle cells. (F) Photomicrograph of *bicoid* mRNA (stained black) passing from the nurse cells and localizing to the anterior end of the oocyte during oogenesis. (C courtesy of H. Doerflinger; F from Stephanson et al. 1988, courtesy of the authors.)

TABLE 9.1 Maternal effect genes that establish the anterior-posterior polarity of the *Drosophila* embryo

Gene	Mutant phenotype	Proposed function
ANTERIOR GROUP		
bicoid (bcd)	Head and thorax deleted, replaced by inverted telson	Graded anterior morphogen; contains homeodomain; represses *caudal* mRNA
exuperantia (exu)	Anterior head structures deleted	Anchors *bicoid* mRNA
swallow (swa)	Anterior head structures deleted	Anchors *bicoid* mRNA
POSTERIOR GROUP		
nanos (nos)	No abdomen	Posterior morphogen; represses *hunchback* mRNA
tudor (tud)	No abdomen, no pole cells	Localization of *nanos* mRNA
oskar (osk)	No abdomen, no pole cells	Localization of *nanos* mRNA
vasa (vas)	No abdomen, no pole cells; oogenesis defective	Localization of *nanos* mRNA
valois (val)	No abdomen, no pole cells; cellularization defective	Stabilizes Nanos localization complex
pumilio (pum)	No abdomen	Helps Nanos protein bind *hunchback* message
caudal (cad)	No abdomen	Activates posterior terminal genes
TERMINAL GROUP		
torsolike	No termini	Possible morphogen for termini
trunk (trk)	No termini	Transmits Torsolike signal to Torso
fs(1)Nasrat[fs(1)N]	No termini; collapsed eggs	Transmits Torsolike signal to Torso
fs(1)polehole[fs(1)ph]	No termini; collapsed eggs	Transmits Torsolike signal to Torso

Source: After Anderson 1989.

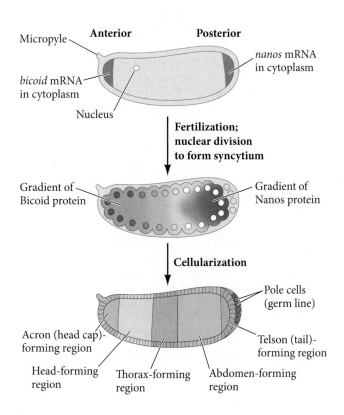

Two maternal messenger RNAs, *bicoid* and *nanos*, were found to correspond to the anterior and posterior signaling centers, and initiate the formation of the anterior-posterior axis. The *bicoid* mRNAs are located near the anterior tip of the unfertilized egg, and *nanos* messages are located at the posterior tip. These distributions occur as a result of the dramatic polarization of the microtubule networks in the developing oocyte (see Figure 9.8). After ovulation and fertilization, the *bicoid* and *nanos* mRNAs are translated into proteins that can diffuse in the syncytial blastoderm, forming gradients that are critical for anterior-posterior patterning (**FIGURE 9.9**; see also Figure 9.7B).

FIGURE 9.9 Syncytial specification in *Drosophila*. Anterior-posterior specification originates from morphogen gradients in the egg cytoplasm. *bicoid* mRNA is stabilized in the most anterior portion of the egg, while *nanos* mRNA is tethered to the posterior end. (The anterior can be recognized by the micropyle on the shell; this structure permits sperm to enter.) When the egg is laid and fertilized, these two mRNAs are translated into proteins. The Bicoid protein forms a gradient that is highest at the anterior end, and the Nanos protein forms a gradient that is highest at the posterior end. These two proteins form a coordinate system based on their ratios. Each position along the axis is thus distinguished from any other position. When the nuclei divide, each nucleus is given its positional information by the ratio of these proteins. The proteins forming these gradients activate the transcription of the genes specifying the segmental identities of the larva and the adult fly.

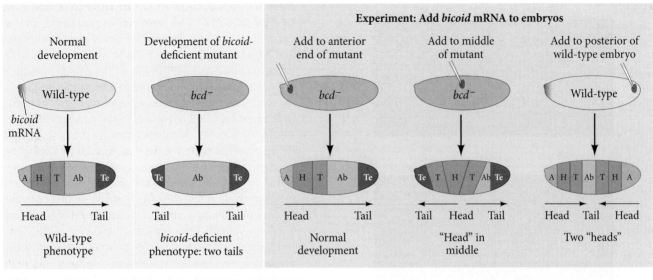

Experiment: Add *bicoid* mRNA to embryos

| Normal development | Development of *bicoid*-deficient mutant | Add to anterior end of mutant | Add to middle of mutant | Add to posterior of wild-type embryo |

FIGURE 9.10 Schematic representation of experiments demonstrating that the *bicoid* gene encodes the morphogen responsible for head structures in *Drosophila*. The phenotypes of *bicoid*-deficient and wild-type embryos are shown at left. When *bicoid*-deficient embryos are injected with *bicoid* mRNA, the point of injection forms the head structures. When the posterior pole of an early-cleavage wild-type embryo is injected with *bicoid* mRNA, head structures form at both poles. (After Driever et al. 1990.)

BICOID AS THE ANTERIOR MORPHOGEN That Bicoid was the head morphogen of *Drosophila* was demonstrated by a "find it, lose it, move it" experimentation scheme (see the Dev Tutorial in Chapter 7). Christiane Nüsslein-Volhard, Wolfgang Driever, and their colleagues (Driever and Nüsslein-Volhard 1988a,b; Driever et al. 1990) showed that (1) Bicoid protein was found in a gradient, highest in the anterior (head-forming) region; (2) embryos lacking Bicoid could not form a head; and (3) when *bicoid* mRNA was added to Bicoid-deficient embryos in different places, the place where *bicoid* mRNA was injected became the head (**FIGURE 9.10**). Moreover, the areas around the site of Bicoid injection became the thorax, as expected from a concentration-dependent signal. When injected into the anterior of *bicoid*-deficient embryos (whose mothers lacked *bicoid* genes), the *bicoid* mRNA "rescued" the embryos and they developed normal anterior-posterior polarity. If *bicoid* mRNA was injected into the center of an embryo, then that middle region became the head, with the regions on either side of it becoming thorax structures. If a large amount of *bicoid* mRNA was injected into the posterior end of a wild-type embryo (with its own endogenous *bicoid* message in its anterior pole), two heads emerged, one at either end (Driever et al. 1990).

BICOID mRNA LOCALIZATION IN THE ANTERIOR POLE OF THE OOCYTE The 3' untranslated region (3'UTR) of *bicoid* mRNA contains sequences that are critical for its localization at the anterior pole (**FIGURE 9.11**; Ferrandon et al. 1997; Macdonald and Kerr 1998; Spirov et al. 2009). These sequences interact with the Exuperantia and Swallow proteins while the messages are still in the nurse cells of the egg chamber (Schnorrer et al. 2000). Experiments in which fluorescently labeled *bicoid* mRNA was microinjected into living egg chambers of wild-type or mutant flies indicate that Exuperantia must be present in the nurse cells for anterior localization. But Exuperantia alone is not sufficient to bring the *bicoid* message into the oocyte (Cha et al. 2001; Reichmann and Ephrussi 2005). The *bicoid*-Exuperantia complex is transported out of the nurse cells and into the oocyte via microtubules, seeming to ride on a kinesin ATPase (Arn et al. 2003). Once inside the oocyte, *bicoid* mRNA attaches to dynein proteins that are maintained at the microtubule organizing center (the slower growing "minus end") at the anterior of the oocyte (see Figure 9.8; Cha et al. 2001). About 90% of the *bicoid* mRNA is localized to the anterior 20% of the embryo, with its concentration peaking at 7% egg length (Little et al. 2011).

(A) mRNA

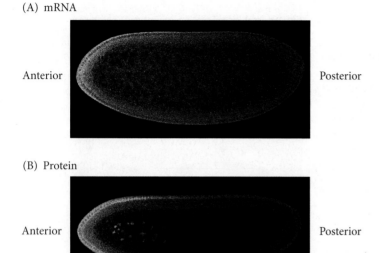

Anterior

Posterior

(B) Protein

Anterior

Posterior

FIGURE 9.11 The *bicoid* mRNA and protein gradients shown by in situ hybridization and confocal microscopy. (A) *bicoid* mRNA shows a steep gradient (here seen as red to blue) across the anterior portion of the oocyte. (B) When the mRNA is translated, the Bicoid protein gradient can be seen in the anterior nuclei. Anterior is to the left; the dorsal surface is upward. (After Spirov et al. 2009, courtesy of S. Baumgartner.)

NANOS mRNA LOCALIZATION IN THE POSTERIOR POLE OF THE OOCYTE The posterior organizing center is defined by the activities of the *nanos* gene (Lehmann and Nüsslein-Volhard 1991; Wang and Lehmann 1991; Wharton and Struhl 1991). While the *bicoid* message is actively transported and bound to the anterior end of the microtubules, the *nanos* message appears to get "trapped" in the posterior end of the oocyte by passive diffusion. The *nanos* message becomes bound to the cytoskeleton in the posterior region of the egg through its 3'UTR and its association with the products of several other genes (*oskar, valois, vasa, staufen,* and *tudor*).[3] If *nanos* (or any other of these maternal effect genes) is absent in the mother, no abdomen forms in the embryo (Lehmann and Nüsslein-Volhard 1986; Schüpbach and Wieschaus 1986). But before the *nanos* message can be localized in the posterior cortex, a *nanos* mRNA-specific "trap" has to be made; this trap is the Oskar protein (Ephrussi et al. 1991). The *oskar* message and the Staufen protein are transported to the posterior end of the oocyte by the kinesin motor protein (see Figure 9.8). There they become bound to the actin microfilaments of the cortex. Staufen allows the translation of the *oskar* message, and the resulting Oskar protein is capable of binding the *nanos* message (Brendza et al. 2000; Hatchet and Ephrussi 2004).

Most *nanos* mRNA, however, is not trapped. Rather, it is bound in the cytoplasm by the translation inhibitors Smaug and CUP. Smaug (yes, it's named after the dragon in *The Hobbit*) binds to the 3'UTR of *nanos* mRNA and recruits the CUP protein that prevents the association of the message with the ribosome as well as recruiting other proteins that deadenylate the message and target it for degradation (Rouget et al. 2010). If the *nanos*-Smaug-CUP complex reaches the posterior pole, however, Oskar can dissociate CUP from Smaug, allowing the mRNA to be bound at the posterior and ready for translation (Forrest et al. 2004; Nelson et al. 2004).

Thus, at the completion of oogenesis, the *bicoid* message is anchored at the anterior end of the oocyte and the *nanos* message is tethered to the posterior end (Frigerio et al. 1986; Berleth et al. 1988; Gavis and Lehmann 1992; Little et al. 2011). These two mRNAs are dormant until ovulation and fertilization, at which time they are translated. Since the Bicoid and Nanos *protein products* are not bound to the cytoskeleton, they diffuse toward the middle regions of the early embryo, creating the two opposing gradients that establish the anterior-posterior polarity of the embryo. Mathematical models indicate that these gradients are established by protein diffusion as well as by the active degradation of the proteins (Little et al. 2011; Liu and Ma 2011).

GRADIENTS OF SPECIFIC TRANSLATIONAL INHIBITORS Two other maternally provided mRNAs—*hunchback, hb*; and *caudal, cad*—are critical for patterning the anterior and posterior regions of the body plan, respectively (Lehmann et al. 1987; Wu and Lengyel 1998). These two mRNAs are synthesized by the nurse cells of the ovary and

---

[3] Like the placement of the *bicoid* message, localization of the *nanos* message is determined by its 3'UTR. If the *bicoid* 3'UTR is experimentally transferred to the protein-encoding region of *nanos* mRNA, the *nanos* message gets localized in the anterior of the egg. When this chimeric mRNA is translated, Nanos protein inhibits translation of *hunchback* and *bicoid* mRNAs and the embryo forms two abdomens—one in the anterior of the embryo and one in the posterior (Gavis and Lehmann 1992).

transported to the oocyte, where they are distributed ubiquitously throughout the syncytial blastoderm. But if they are not localized, how do they mediate their localized patterning activities? It turns out that translation of the *hb* and *cad* mRNAs is repressed by the diffusion gradients of Nanos and Bicoid proteins, respectively.

In the anterior region, Bicoid protein prevents translation of the *caudal* message. Bicoid binds to a specific region of *caudal*'s 3′UTR. Here, it binds Bin3, a protein that stabilizes an inhibitory complex that prevents the binding of the mRNA 5′ cap to the ribosome. By recruiting this translational inhibitor, Bicoid prevents translation of *caudal* in the anterior of the embryo (**FIGURE 9.12**; Rivera-Pomar et al. 1996; Cho et al. 2006; Signh et al. 2011). This suppression is necessary; if Caudal protein is made in the embryo's anterior, the head and thorax do not form properly. Caudal activates the genes responsible for the invagination of the hindgut and thus is critical in specifying the posterior domains of the embryo.

In the posterior region, Nanos protein prevents translation of the *hunchback* message. Nanos in the posterior of the embryo forms a complex with several other ubiquitous proteins, including Pumilio and Brat. This complex binds to the 3′ UTR of the *hunchback* message, where it recruits d4EHP and prevents the *hunchback* message from attaching to ribosomes (Tautz 1988; Cho et al. 2006).

The result of these interactions is the creation of four maternal protein gradients in the early embryo (**FIGURE 9.13**):

- An anterior-to-posterior gradient of Bicoid protein
- An anterior-to-posterior gradient of Hunchback protein
- A posterior-to-anterior gradient of Nanos protein
- A posterior-to-anterior gradient of Caudal protein

The stage is now set for the activation of zygotic genes in the insect's nuclei, which were busy dividing while these four protein gradients were being established.

SCIENTISTS SPEAK 9.1 Dr. Eric Wieschaus discusses the patterning of anterior-posterior development in *Drosophila*.

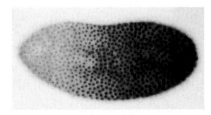

Anterior   Posterior

FIGURE 9.12 Caudal protein gradient of a wild-type *Drosophila* embryo at the syncytial blastoderm stage. Anterior is to the left. The protein (stained darkly) enters the nuclei and helps specify posterior fates. Compare with the complementary gradient of Bicoid protein in Figure 9.22. (From Macdonald and Struhl 1986, courtesy of G. Struhl.)

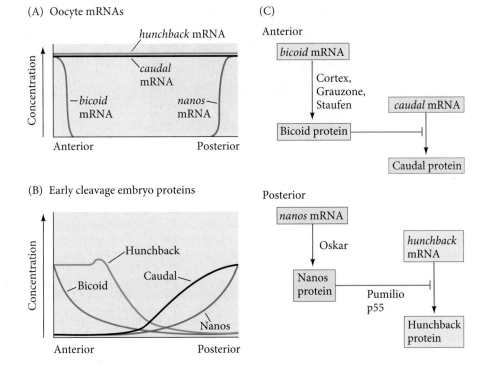

FIGURE 9.13 Model of anterior-posterior pattern generation by *Drosophila* maternal effect genes. (A) The *bicoid*, *nanos*, *hunchback*, and *caudal* mRNAs are deposited in the oocyte by the ovarian nurse cells. The *bicoid* message is sequestered anteriorly; the *nanos* message is localized to the posterior pole. (B) Upon translation, the Bicoid protein gradient extends from anterior to posterior, while the Nanos protein gradient extends from posterior to anterior. Nanos inhibits the translation of the *hunchback* message (in the posterior), while Bicoid prevents the translation of the *caudal* message (in the anterior). This inhibition results in opposing Caudal and Hunchback gradients. The Hunchback gradient is secondarily strengthened by transcription of the *hunchback* gene in the anterior nuclei (since Bicoid acts as a transcription factor to activate *hunchback* transcription). (C) Parallel interactions whereby translational gene regulation establishes the anterior-posterior patterning of the *Drosophila* embryo. (C after Macdonald and Smibert 1996.)

The anterior organizing center: The Bicoid and Hunchback gradients

In *Drosophila*, the phenotype of *bicoid* mutants provides valuable information about the function of morphogenetic gradients (**FIGURE 9.14A–C**). Instead of having anterior structures (acron, head, and thorax) followed by abdominal structures and a telson, the structure of a *bicoid* mutant is telson-abdomen-abdomen-telson (**FIGURE 9.14D**). It would appear that these embryos lack whatever substances are needed for the formation of the anterior structures. Moreover, one could hypothesize that the substance these mutants lack is the one postulated by Sander and Kalthoff to turn on genes for the anterior structures and turn off genes for the telson structures.

Bicoid protein appears to act as a morphogen (i.e., a substance that differentially specifies the fates of cells by different concentrations; see Chapter 4). High concentrations of Bicoid produce anterior head structures. Slightly less Bicoid tells the cells to become jaws. A moderate concentration of Bicoid is responsible for instructing cells to become the thorax, whereas the abdomen is characterized as lacking Bicoid. How might a gradient of Bicoid protein control the determination of the anterior-posterior axis? Bicoid's primary function is to act as a transcription factor that activates the expression of target genes in the anterior part of the embryo.[4] The first target of Bicoid to be discovered was the *hunchback* (*hb*) gene. In the late 1980s, two laboratories independently demonstrated that Bicoid binds to and activates *hb* (Driever and Nüsslein-Volhard 1989; Struhl et al.

---

[4] *bicoid* appears to be a relatively "new" gene that evolved in the Dipteran lineage (two-winged insects such as flies); it has not been found in other insect lineages. The anterior determinant in other insect groups includes the Orthodenticle and Hunchback proteins, both of which can be induced in the anterior of the *Drosophila* embryo by Bicoid (Wilson and Dearden 2011).

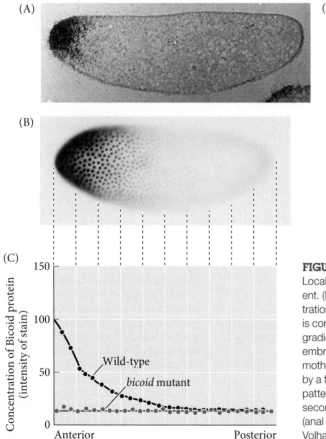

(A)

(B)

(C)

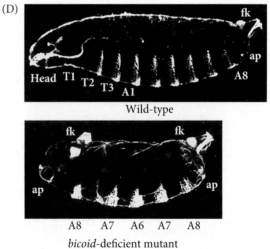

(D)

FIGURE 9.14 Bicoid protein gradient in the early *Drosophila* embryo. (A) Localization of *bicoid* mRNA to the anterior tip of the embryo in a steep gradient. (B) Bicoid protein gradient shortly after fertilization. Note that the concentration is greatest anteriorly and trails off posteriorly. Notice also that Bicoid is concentrated in the nuclei. (C) Densitometric scan of the Bicoid protein gradient. The upper curve (black) represents the Bicoid gradient in wild-type embryos. The lower curve (red) represents Bicoid in embryos of *bicoid* mutant mothers. (D) Phenotype of cuticle from a strongly affected embryo produced by a female fly deficient in the *bicoid* gene compared with the wild-type cuticle pattern. The head and thorax of the *bicoid* mutant have been replaced by a second set of posterior telson structures, abbreviated fk (filzkörper) and ap (anal plates). (A from Kaufman et al. 1990; B,C from Driever and Nüsslein-Volhard 1988b; D from Driever et al. 1990, courtesy of the authors.)

1989; Wieschaus 2016). Bicoid-dependent transcription of *hb* is seen only in the anterior half of the embryo—the region where Bicoid is found. Driever and co-workers (1989) also predicted that Bicoid must activate other anterior genes besides *hb*. First, deletions of *hb* produced only some of the defects seen in the *bicoid* mutant phenotype. Second, head formation required higher Bicoid concentrations than thorax formation. Bicoid is now known to activate head-forming target genes such as *buttonhead*, *empty spiracles*, and *orthodenticle*, which are expressed in specific subregions of the anterior part of the embryo (Cohen and Jürgens 1990; Finkelstein and Perrimon 1990; Grossniklaus et al. 1994). Driever and co-workers (1989) also predicted that the promoters of such head-specific genes would have low-affinity binding sites for Bicoid protein, causing them to be activated only at extremely high concentrations of Bicoid—that is, near the anterior tip of the embryo. In addition to needing high Bicoid levels for activation, transcription of these genes also requires the presence of Hunchback protein (Simpson-Brose et al. 1994; Reinitz et al. 1995). Bicoid and Hunchback act synergistically at the enhancers of these "head genes" to promote their transcription in a feedforward manner.

 SCIENTISTS SPEAK 9.2   In two separate videos, Dr. Eric Wieschaus discusses the stability of the Bicoid gradient and its role throughout fly evolution.

In the posterior half of the embryo, the Caudal protein gradient also activates a number of zygotic genes, including the gap genes *knirps* (*kni*) and *giant* (*gt*), which are critical for abdominal development (Rivera-Pomar et al. 1995; Schulz and Tautz 1995). Since a second function of Bicoid protein is to inhibit the translation of *caudal* mRNA, Caudal protein is absent from the anterior portion of the embryo. Thus, the posterior-forming genes are not activated in this region.

The terminal gene group

In addition to the anterior and posterior morphogens, there is a third set of maternal genes whose proteins generate the unsegmented extremities of the anterior-posterior axis: the **acron** (the terminal portion of the head that includes the brain) and the **telson** (tail). Mutations in these terminal genes result in the loss of both the acron and most anterior head segments *and* the telson and most posterior abdominal segments (Degelmann et al. 1986; Klingler et al. 1988).

WEB TOPIC 9.4   **THE TERMINAL GENE GROUP**   Further details about these genes are provided, including how the two ends are specified differently from the central trunk segments, and how Bicoid helps determine which terminal becomes anterior.

Summarizing early anterior-posterior axis specification in Drosophila

The anterior-posterior axis of the *Drosophila* embryo is specified by three sets of genes:

1. **Genes that define the anterior organizing center.** Located at the anterior end of the embryo, the anterior organizing center acts through a gradient of Bicoid protein. Bicoid functions both as a *transcription factor* to activate anterior-specific gap genes and as a *translational repressor* to suppress posterior-specific gap genes.

2. **Genes that define the posterior organizing center.** The posterior organizing center is located at the posterior pole. This center acts *translationally* through the Nanos protein to inhibit anterior formation, and *transcriptionally* through the Caudal protein to activate those genes that form the abdomen.

3. **Genes that define the terminal boundary regions.** The boundaries of the acron and telson are defined by the product of the *torso* gene, which is activated at the tips of the embryo.

The next step in development will be to use these gradients of transcription factors to activate specific genes along the anterior-posterior axis.

Segmentation Genes

Cell fate commitment in *Drosophila* appears to have two steps: specification and determination (Slack 1983). Early in fly development, the fate of a cell depends on cues provided by protein gradients. This specification of cell fate is flexible and can still be altered in response to signals from other cells. Eventually, however, the cells undergo a transition from this loose type of commitment to an irreversible determination. At this point, the fate of a cell becomes cell-intrinsic.[5]

The transition from specification to determination in *Drosophila* is mediated by **segmentation genes** that divide the early embryo into a repeating series of segmental primordia along the anterior-posterior axis. Segmentation genes were originally defined by zygotic mutations that disrupted the body plan, and these genes were divided into three groups based on their mutant phenotypes (**TABLE 9.2**; Nüsslein-Volhard and Wieschaus 1980):

- *Gap mutants* lack large regions of the body (several contiguous segments; **FIGURE 9.15A**).
- *Pair-rule mutants* lack portions of every other segment (**FIGURE 9.15B**).
- *Segment polarity mutants* show defects (deletions, duplications, polarity reversals) in every segment (**FIGURE 9.15C**).

Segments and parasegments

Mutations in segmentation genes result in *Drosophila* embryos that lack certain segments or parts of segments. However, early researchers found a surprising aspect of these mutations: many of them did not affect actual adult segments. Rather, they affected the posterior compartment of one segment and the anterior compartment of the immediately posterior segment (**FIGURE 9.16**). These "transegmental" units were named **parasegments** (Martinez-Arias and Lawrence 1985).

Once the means to detect gene expression patterns were available, it was discovered that the expression patterns in the early embryo are delineated by parasegmental boundaries, not by the boundaries of the segments. Thus, the parasegment appears to be the fundamental unit of *embryonic* gene expression. Although parasegmental organization is also seen in the nerve cord of adult *Drosophila*, it is not seen in the adult epidermis (the most obvious manifestation of segmentation), nor is it

[5]Aficionados of information theory will recognize that the process by which the anterior-posterior information in morphogenetic gradients is transferred to discrete and different parasegments represents a transition from analog to digital specification. Specification is analog, determination digital. This process enables the transient information of the gradients in the syncytial blastoderm to be stabilized so that it can be used much later in development (Baumgartner and Noll 1990).

(A) Gap: *Krüppel* (as an example)

Early embryo (normal)   Later embryo (normal)   Larva (normal)   Larva (lethal mutant)

Area of gene action   Area of gene action

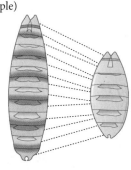

(B) Pair-rule: *fushi tarazu* (as an example)

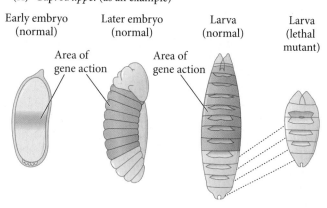

(C) Segment polarity: *engrailed* (as an example)

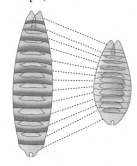

FIGURE 9.15   Three types of segmentation gene mutations. The left side shows the early-cleavage embryo (yellow), with the region where the particular gene is normally transcribed in wild-type embryos shown in blue. These areas are deleted as the mutants develop into late-stage embryos.

found in the adult musculature. These adult structures are organized along the segmental pattern. In *Drosophila*, segmental grooves appear in the epidermis when the germ band is retracted; the muscle-forming mesoderm becomes segmental later in development.

One can think about the segmental and parasegmental organization schemes as representing different ways of organizing the compartments along the anterior-posterior axis of the embryo. The cells of one compartment do not mix with cells of neighboring compartments, and parasegments and segments are out of phase by one compartment.[6]

The gap genes

The gap genes are activated or repressed by the maternal effect genes, and are expressed in one or two broad domains along the anterior-posterior axis. These expression patterns correlate quite well with the regions of the embryo that are missing in gap mutations. For example, *Krüppel* is expressed primarily in parasegments 4–6, in the center of the embryo (see Figures 9.7C and Figure 9.15A); in the absence of the Krüppel protein, the embryo lacks parasegments from these regions.

Deletions caused by mutations in three gap genes—*hunchback*, *Krüppel*, and *knirps*—span the entire segmented region of the *Drosophila* embryo. The gap gene *giant* overlaps with these three, and the gap genes *tailless* and *huckebein* are expressed in domains near the anterior and posterior ends of the embryo. Taken together, the four gap genes of the trunk have enough specificity to define a cell's location with an error of only around 1% along the embryo's anterior/posterior axis. With the interactions between these gap gene products, each cell appears to be given a unique spatial identity (Dubuis et al. 2013).

The expression patterns of the gap genes are highly dynamic. These genes usually show low levels of transcriptional activity across

TABLE 9.2	Major genes affecting segmentation pattern in *Drosophila*
Category	**Gene name**
Gap genes	*Krüppel (Kr)* *knirps (kni)* *hunchback (hb)* *giant (gt)* *tailless (tll)* *huckebein (hkb)* *buttonhead (btd)* *empty spiracles (ems)* *orthodenticle (otd)*
Pair-rule genes (primary)	*hairy (h)* *even-skipped (eve)* *runt (run)*
Pair-rule genes (secondary)	*fushi tarazu (ftz)* *odd-paired (opa)* *odd-skipped (odd)* *sloppy-paired (slp)* *paired (prd)*
Segment polarity genes	*engrailed (en)* *wingless (wg)* *cubitus interruptus (ci)* *hedgehog (hh)* *fused (fu)* *armadillo (arm)* *patched (ptc)* *gooseberry (gsb)* *pangolin (pan)*

---

[6] The two modes of segmentation may be required for the coordination of movement in the adult fly. In arthropods, the ganglia of the ventral nerve cord are organized by parasegments, but the cuticle grooves and musculature are segmental. This shift in frame by one compartment allows the muscles on both sides of any particular epidermal segment to be coordinated by the same ganglion. This, in turn, allows rapid and coordinated muscle contractions for locomotion (Deutsch 2004). A similar situation occurs in vertebrates, where the posterior portion of the anterior somite combines with the anterior portion of the next somite.

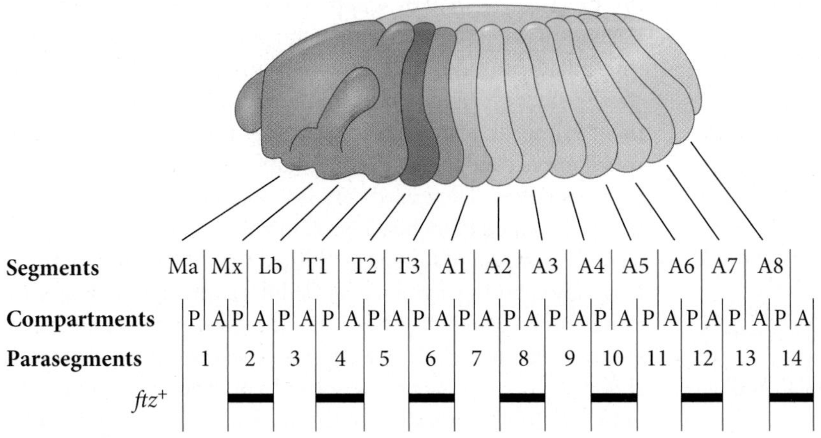

FIGURE 9.16 Parasegments in the *Drosophila* embryo are shifted one compartment forward in relation to the segments. Ma, Mx, and Lb are the mandibular, maxillary, and labial head segments; T1–T3 are the thoracic segments; and A1–A8 are abdominal segments. Each segment has an anterior (A) and a posterior (P) compartment. Each parasegment (numbered 1–14) consists of the posterior compartment of one segment and the anterior compartment of the segment in the next posterior position. Black bars indicate the boundaries of *ftz* gene expression; these regions are missing in the *fushi tarazu* (*ftz*) mutant (see Figure 9.15B). (After Martinez-Arias and Lawrence 1985.)

the entire embryo that become consolidated into discrete regions of high activity as nuclear divisions continue (Jäckle et al. 1986). The Hunchback gradient is particularly important in establishing the initial gap gene expression patterns. By the end of nuclear division cycle 12, Hunchback is found at high levels across the anterior part of the embryo. Hunchback then forms a steep gradient through about 15 nuclei near the middle of the embryo (see Figures 9.7C and 9.13B). The posterior third of the embryo has undetectable Hunchback levels at this time.

The transcription patterns of the anterior gap genes are initiated by the different concentrations of the Hunchback and Bicoid proteins. High levels of Bicoid and Hunchback induce the expression of *giant,* while the *Krüppel* transcript appears over the region where Hunchback begins to decline. High levels of Hunchback (in the absence of Bicoid) also prevent the transcription of the posterior gap genes (such as *knirps* and *giant*) in the embryo's anterior (Struhl et al. 1992). It is thought that a gradient of the Caudal protein, highest at the posterior pole, is responsible for activating the abdominal gap genes *knirps* and *giant* in the posterior part of the embryo. The *giant* gene thus has two methods of activation (Rivera-Pomar 1995; Schulz and Tautz 1995): one for its anterior expression band (through Bicoid and Hunchback), and one for its posterior expression band (by Caudal).

After the initial gap gene expression patterns have been established by the maternal effect gradients and Hunchback, they are stabilized and maintained by repressive interactions between the different gap gene products themselves. (These interactions are facilitated by the fact that they occur within a syncytium, in which the cell membranes have not yet formed.) These boundary-forming inhibitions are thought to be directly mediated by the gap gene products, because all four major gap genes (*hunchback, giant, Krüppel,* and *knirps*) encode DNA-binding proteins (Knipple et al. 1985; Gaul and Jäckle 1990; Capovilla et al. 1992). One such model, established by genetic experiments, biochemical analyses, and mathematical modeling, is presented in **FIGURE 9.17A** (Papatsenko and Levine 2011). The model depicts a network with three major toggle switches (**FIGURE 9.17B–D**). Two of these switches are the strong mutual inhibition between Hunchback and Knirps, and the strong mutual inhibition between Giant and Krüppel (Jaeger et al. 2004). The third is the concentration-dependent interaction between Hunchback and Krüppel. At high doses, Hunchback inhibits the production of Krüppel protein, but at moderate doses (at about 50% of the embryo length) Hunchback promotes Krüppel formation (see Figure 9.17C).

The end result of these repressive interactions is the creation of a precise system of overlapping mRNA expression patterns. Each domain serves as a source for diffusion of gap proteins into adjacent embryonic regions. This creates a significant overlap (at least eight nuclei, which accounts for about two segment primordia) between adjacent gap protein domains. This was demonstrated in a striking manner by Stanojević and coworkers (1989). They fixed cellularizing blastoderms (see Figure 9.2), stained Hunchback protein with an antibody carrying a red dye, and simultaneously stained Krüppel protein with an antibody carrying a green dye. Cellularizing regions that contained both proteins bound both antibodies and stained bright yellow (see Figure 9.7C). Krüppel overlaps with Knirps in a similar manner in the posterior region of the embryo (Pankratz et al. 1990). The precision of these patterns is maintained by

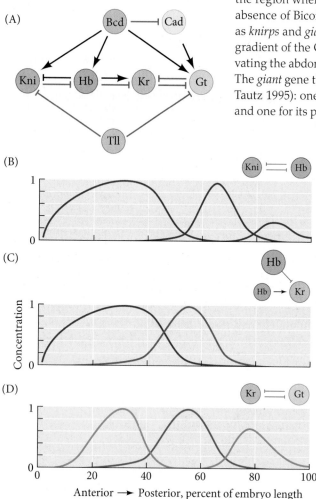

FIGURE 9.17 Architecture of the gap gene network. These interactions are supported by mathematical modeling, genetic data, and biochemical analyses. (A) The anterior-posterior gradient of Bicoid (Bcd) and Caudal (Cad) regulates expression of Knirps (Kni), Hunchback (Hb), Krüppel (Kr; weakly activated by both Bicoid and Caudal proteins), and Giant (Gt). Tailless (Tll) prevents these patterning pathways at the terminal ends of the embryo. (B–D) The three "toggle switches" activated along the anterior-posterior axis to establish gap gene domains. (B) The mutual inhibition of Knirps and Hunchback positions the Knirps protein domain at around 60–80% along the anterior-posterior axis. (C) Hunchback inhibits Krüppel expression at high concentrations but promotes it at intermediate concentrations. (D) Krüppel and Giant mutually inhibit each other's synthesis. (After Papatsenko and Levine 2011.)

having redundant enhancers; if one of these enhancers fails to work, there is a high probability that the other will still function (Perry et al. 2011).

The pair-rule genes

The first indication of segmentation in the fly embryo comes when the pair-rule genes are expressed during nuclear division cycle 13, as the cells begin to form at the periphery of the embryo. The transcription patterns of these genes divide the embryo into regions that are precursors of the segmental body plan. As can be seen in **FIGURE 9.18** (and in Figure 9.7D), one vertical band of nuclei (the cells are just beginning to form) expresses a pair-rule gene, the next band of nuclei does not express it, and then the next band expresses it again. The result is a "zebra stripe" pattern along the anterior-posterior axis, dividing the embryo into 15 subunits (Hafen et al. 1984). Eight genes are currently known to be capable of dividing the early embryo in this fashion, and they overlap one another so as to give each cell in the parasegment a specific set of transcription factors (see Table 6.2).

The primary pair-rule genes include *hairy, even-skipped,* and *runt,* each of which is expressed in seven stripes. All three build their striped patterns from scratch, using distinct enhancers and regulatory mechanisms for each stripe. These enhancers are often modular: control over expression in each stripe is located in a discrete region of the DNA, and these DNA regions often contain binding sites recognized by gap proteins. Thus, it is thought that the different concentrations of gap proteins determine whether or not a pair-rule gene is transcribed.

One of the best-studied primary pair-rule genes is *even-skipped* (**FIGURE 9.19**). Its enhancer region is composed of modular units arranged such that each enhancer

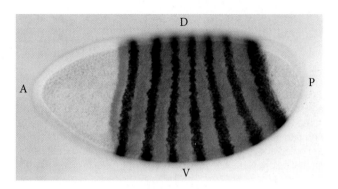

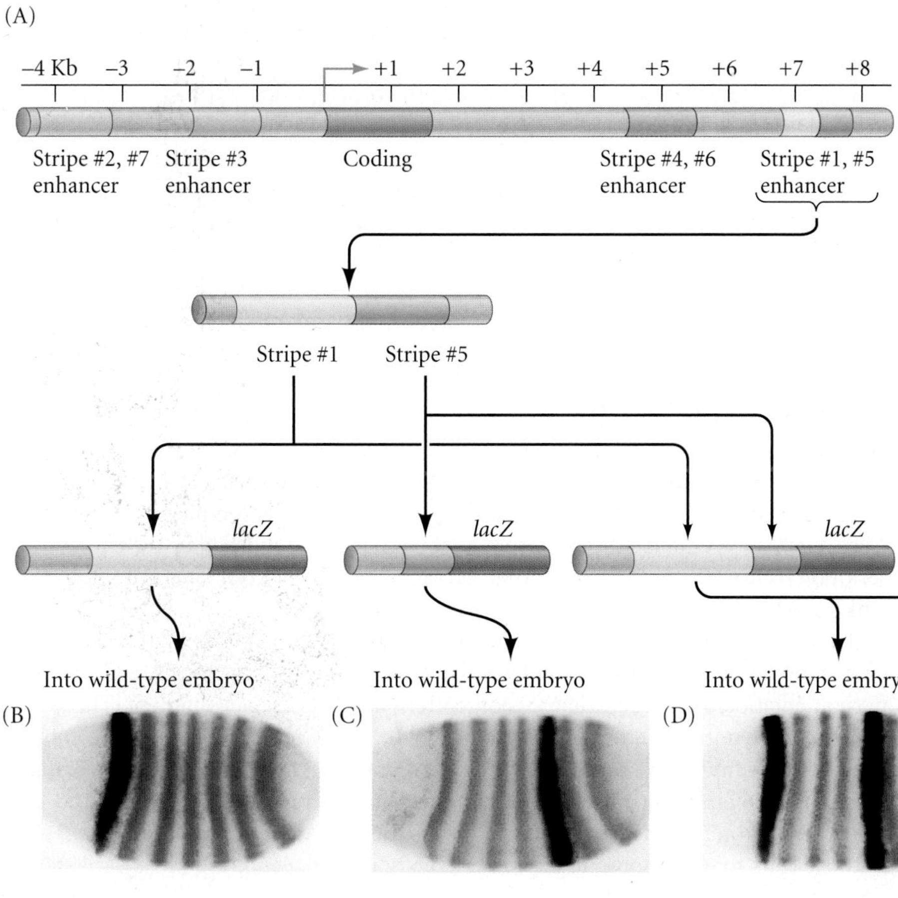

FIGURE 9.18 Messenger RNA expression patterns of two pair-rule genes, *even-skipped* (red) and *fushi tarazu* (black) in the *Drosophila* blastoderm. Each gene is expressed as a series of seven stripes. Anterior is to the left, dorsal is up. (Courtesy of S. Small.)

(A)

−4 Kb	−3	−2	−1	+1	+2	+3	+4	+5	+6	+7	+8

Stripe #2, #7 enhancer   Stripe #3 enhancer   Coding   Stripe #4, #6 enhancer   Stripe #1, #5 enhancer

Stripe #1   Stripe #5

lacZ   *lacZ*   *lacZ*

Into wild-type embryo   Into wild-type embryo   Into wild-type embryo   Into *giant*-deficient embryo

(B)   (C)   (D)   (E)

FIGURE 9.19 Specific promoter regions of the *even-skipped* (*eve*) gene control specific transcription bands in the embryo. (A) Partial map of the *eve* promoter, showing the regions responsible for the various stripes. (B–E) A reporter β-galactosidase gene (*lacZ*) was fused to different regions of the *eve* promoter and injected into fly embryos. The resulting embryos were stained (orange bands) for the presence of Even-skipped protein. (B–D) Wild-type embryos that were injected with *lacZ* transgenes containing the enhancer region specific for stripe 1 (B), stripe 5 (C), or both regions (D). (E) The enhancer region for stripes 1 and 5 was injected into an embryo deficient in *giant*. Here the posterior border of stripe 5 is missing. (After Fujioka et al. 1999 and Sackerson et al. 1999; photographs courtesy of M. Fujioka and J. B. Jaynes.)

FIGURE 9.20 Model for formation of the second stripe of transcription from the *even-skipped* gene. The enhancer element for stripe 2 regulation contains binding sequences for several maternal and gap gene proteins. Activators (e.g., Bicoid and Hunchback) are noted above the line; repressors (e.g., Krüppel and Giant) are shown below. Note that nearly every activator site is closely linked to a repressor site, suggesting competitive interactions at these positions. (Moreover, a protein that is a repressor for stripe 2 may be an activator for stripe 5; it depends on which proteins bind next to them.) B, Bicoid; C, Caudal; G, Giant; H, Hunchback; K, Krüppel; N, Knirps; T, Tailless. (After Janssens et al. 2006.)

regulates a separate stripe or a pair of stripes. For instance, *even-skipped* stripe 2 is controlled by a 500-bp region that is activated by Bicoid and Hunchback and repressed by both Giant and Krüppel proteins (**FIGURE 9.20**; Small et al. 1991, 1992; Stanojević et al. 1991; Janssens et al. 2006). The anterior border is maintained by repressive influences from Giant, while the posterior border is maintained by Krüppel. DNase I footprinting showed that the minimal enhancer region for this stripe contains five binding sites for Bicoid, one for Hunchback, three for Krüppel, and three for Giant. Thus, this region is thought to act as a switch that can directly sense the concentrations of these proteins and make on/off transcriptional decisions.

The importance of these enhancer elements can be shown by both genetic and biochemical means. First, a mutation in a particular enhancer can delete its particular stripe and no other. Second, if a reporter gene (such as *lacZ*, which encodes β-galactosidase) is fused to one of the enhancers, the reporter gene is expressed only in that particular stripe (see Figure 9.19; Fujioka et al. 1999). Third, placement of the stripes can be altered by deleting the gap genes that regulate them. Thus, stripe placement is a result of (1) the modular *cis*-regulatory enhancer elements of the pair-rule genes and (2) the *trans*-regulatory gap gene and maternal gene proteins that bind to these enhancer sites.

Once initiated by the gap gene proteins, the transcription pattern of the primary pair-rule genes becomes stabilized by interactions among their products (Levine and Harding 1989). The primary pair-rule genes also form the context that allows or inhibits expression of the later-acting secondary pair-rule genes, such as *fushi tarazu* (*ftz*; **FIGURE 9.21**). The eight known pair-rule genes are all expressed in striped patterns, but the patterns are not coincident with each other. Rather, each row of nuclei within a parasegment has its own array of pair-rule products that distinguishes it from any other row. These products activate the next level of segmentation genes, the segment polarity genes.

The segment polarity genes

So far our discussion has described interactions between molecules within the syncytial embryo. But once cells form, interactions take place between the cells. These interactions are mediated by the segment polarity genes, and they accomplish two important tasks. First, they reinforce the parasegmental periodicity established by the earlier transcription factors. Second, through this cell-to-cell signaling, cell fates are established within each parasegment.

The segment polarity genes encode proteins that are constituents of the Wnt and Hedgehog signaling pathways (Ingham 2016). Mutations in these genes lead to defects in segmentation and in gene expression pattern across each parasegment. The development of the normal pattern relies on the fact that only one row of cells in each parasegment is permitted to express the Hedgehog protein, and only one row of cells in each parasegment is permitted to express the Wingless protein. (Wingless is the *Drosophila* Wnt protein.) The key to this pattern is the activation of the *engrailed* (*en*) gene in those cells that are going to express Hedgehog. The *engrailed* gene is activated in

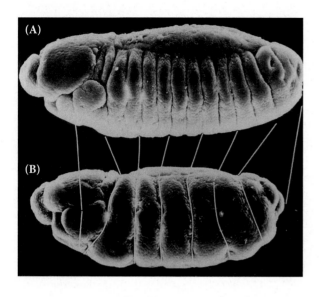

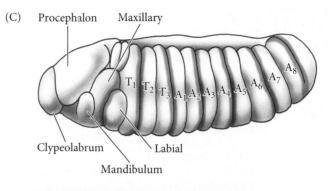

cells that have high levels of the Even-skipped, Fushi tarazu, or Paired transcription factors; *engrailed* is repressed in those cells with high levels of Odd-skipped, Runt, or Sloppy-paired proteins. As a result, the Engrailed protein is found in 14 stripes across the anterior-posterior axis of the embryo (see Figure 9.7E). (Indeed, in *ftz*-deficient embryos, only seven bands of *engrailed* are expressed.)

These stripes of *engrailed* transcription mark the anterior compartment of each parasegment (and the posterior compartment of each segment). The *wingless* (*wg*) gene is activated in those bands of cells that receive little or no Even-skipped or Fushi tarazu protein, but which do contain Sloppy-paired. This pattern causes *wingless* to be transcribed solely in the column of cells directly anterior to the cells where *engrailed* is transcribed (**FIGURE 9.22A**).

Once *wingless* and *engrailed* expression patterns are established in adjacent cells, this pattern must be maintained to retain the parasegmental periodicity of the body plan. It should be remembered that the mRNAs and proteins involved in initiating these patterns are short-lived, and that the patterns must be maintained after their initiators are no longer being synthesized. The maintenance of these patterns is regulated by reciprocal interaction between neighboring cells: cells secreting Hedgehog protein activate *wingless* expression in their neighbors, and the Wingless protein signal, which is received by the cells that secreted Hedgehog, serves to maintain *hedgehog* (*hh*) expression (**FIGURE 9.22B**). Wingless protein also acts in an autocrine fashion, maintaining its own expression (Sánchez et al. 2008).

In the cells transcribing the *wingless* gene, *wingless* mRNA is translocated by its 3'UTR to the apex of the cell (Simmonds et al. 2001; Wilkie and Davis 2001). At the apex, the *wingless* message is translated and secreted from the cell. The cells expressing *engrailed* can bind this protein because they contain Frizzled, which is the *Drosophila* membrane receptor protein for Wingless (Bhanot et al. 1996). Binding of Wingless to Frizzled activates the Wnt signal transduction pathway, resulting in the continued expression of *engrailed* (Siegfried et al. 1994). In this way, the transcription pattern of these two types of cells is stabilized. This interaction creates a stable boundary, as well as a signaling center from which Hedgehog and Wingless proteins diffuse across the parasegment.

The diffusion of these proteins is thought to provide the gradients by which the cells of the parasegment acquire their identities. This process can be seen in the dorsal

FIGURE 9.21 Defects seen in the *fushi tarazu* mutant. Anterior is to the left; dorsal surface faces upward. (A) Scanning electron micrograph of a wild-type embryo, seen in lateral view. (B) A *fushi tarazu*–mutant embryo at the same stage. The white lines connect the homologous portions of the segmented germ band. (C) Diagram of wild-type embryonic segmentation. The areas shaded in purple show the parasegments of the germ band that are missing in the mutant embryo. (D) Transcription pattern of the *fushi tarazu* gene. (After Kaufman et al. 1990; A,B courtesy of T. Kaufman; D courtesy of T. Karr.)

FIGURE 9.22 Model for transcription of the segment polarity genes *engrailed* (*en*) and *wingless* (*wg*). (A) Expression of *wg* and *en* is initiated by pair-rule genes. The *en* gene is expressed in cells that contain high concentrations of either Even-skipped or Fushi tarazu proteins. The *wg* gene is transcribed when neither *eve* nor *ftz* genes are active, but when a third gene (probably *sloppy-paired*) is expressed. (B) The continued expression of *wg* and *en* is maintained by interactions between the Engrailed- and Wingless-expressing cells. Wingless protein is secreted and diffuses to the surrounding cells. In those cells competent to express Engrailed (i.e., those having Eve or Ftz proteins), Wingless protein is bound by the Frizzled and Lrp6 receptor proteins, which enables the activation of the *en* gene via the Wnt signal transduction pathway. (Armadillo is the *Drosophila* name for β-catenin.) Engrailed protein activates the transcription of the *hedgehog* gene and also activates its own (*en*) gene transcription. Hedgehog protein diffuses from these cells and binds to the Patched receptor protein on neighboring cells. The Hedgehog signal enables the transcription of the *wg* gene and the subsequent secretion of the Wingless protein. For a more complex view, see Sánchez et al. 2008.

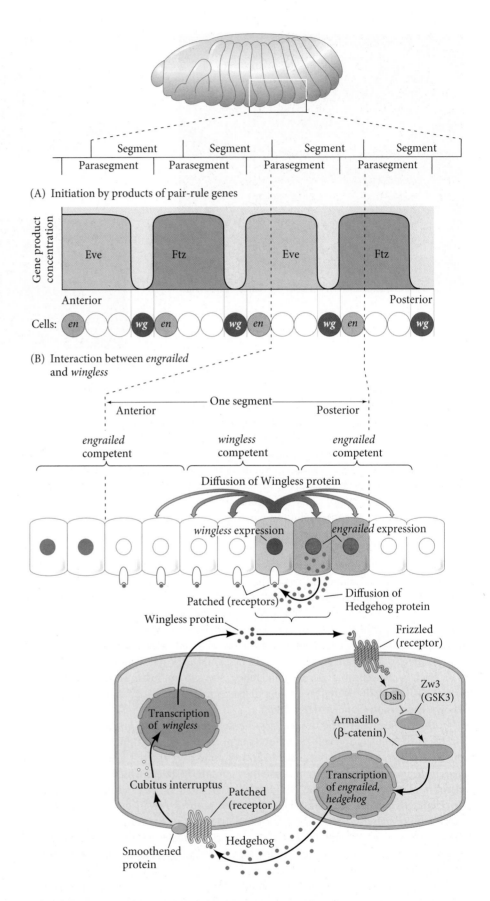

FIGURE 9.23   Cell specification by the Wingless/Hedgehog signaling center. (A) Dark-field photograph of wild-type *Drosophila* embryo, showing the position of the third abdominal segment. Anterior is to the left; the dorsal surface faces upward. (B) Close-up of the dorsal area of the A3 segment, showing the different cuticular structures made by the 1°, 2°, 3°, and 4° rows of cells. (C) A model for the roles of Wingless and Hedgehog. Each signal is responsible for roughly half the pattern. Either each signal acts in a graded manner (shown here as gradients decreasing with distance from their respective sources) to specify the fates of cells at a distance from these sources, or each signal acts locally on the neighboring cells to initiate a cascade of inductions (shown here as sequential arrows). (After Heemskerk and DiNardo 1994; photographs courtesy of the authors.)

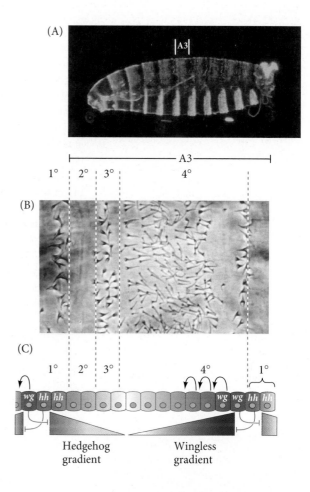

epidermis, where the rows of larval cells produce different cuticular structures depending on their position in the segment. The 1° row of cells consists of large, pigmented spikes called denticles. Posterior to these cells, the 2° row produces a smooth epidermal cuticle. The next two cell rows have a 3° fate, making small, thick hairs; they are followed by several rows of cells that adopt the 4° fate, producing fine hairs (**FIGURE 9.23**).

The Homeotic Selector Genes

After the segmental boundaries are set, the pair-rule and gap genes interact to regulate the homeotic selector genes, which specify the characteristic structures of each segment (Lewis 1978). By the end of the cellular blastoderm stage, each segment primordium has been given an individual identity by its unique constellation of gap, pair-rule, and homeotic gene products (Levine and Harding 1989). Two regions of *Drosophila* chromosome III contain most of these homeotic genes (**FIGURE 9.24**). The first region, known as the

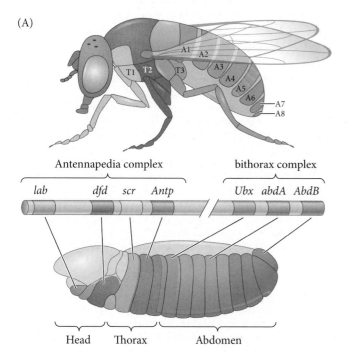

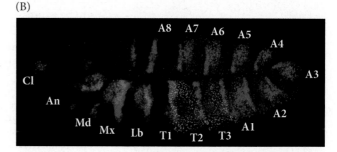

FIGURE 9.24   Homeotic gene expression in *Drosophila*. (A) Expression map of the homeotic genes. In the center are the genes of the Antennapedia and bithorax complexes and their functional domains. Below and above the gene map, the regions of homeotic gene expression (both mRNA and protein) in the blastoderm of the *Drosophila* embryo and the regions that form from them in the adult fly are shown. (B) In situ hybridization for four genes at a slightly later stage (the extended germ band). The *engrailed* (blue) expression pattern separates the body into segments; *Antennapedia* (green) and *Ultrabithorax* (purple) separate the thoracic and abdominal regions; *Distal-less* (red) shows the placement of jaws and the beginnings of limbs. (A after Kaufman et al. 1990 and Dessain et al. 1992; B courtesy of D. Kosman.)

Antennapedia complex, contains the homeotic genes *labial* (*lab*), *Antennapedia* (*Antp*), *sex combs reduced* (*scr*), *deformed* (*dfd*), and *proboscipedia* (*pb*). The *labial* and *deformed* genes specify the head segments, while *sex combs reduced* and *Antennapedia* contribute to giving the thoracic segments their identities. The *proboscipedia* gene appears to act only in adults, but in its absence, the labial palps of the mouth are transformed into legs (Wakimoto et al. 1984; Kaufman et al. 1990; Maeda and Karch 2009).

The second region of homeotic genes is the **bithorax complex** (Lewis 1978; Maeda and Karch 2009). Three protein-coding genes are found in this complex: *Ultrabithorax* (*Ubx*), which is required for the identity of the third thoracic segment; and the *abdominal A* (*abdA*) and *Abdominal B* (*AbdB*) genes, which are responsible for the segmental identities of the abdominal segments (Sánchez-Herrero et al. 1985). The chromosome region containing both the Antennapedia complex and the bithorax complex is often referred to as the **homeotic complex**, or **Hom-C**.

Because the homeotic selector genes are responsible for the specification of fly body parts, mutations in them lead to bizarre phenotypes. In 1894, William Bateson called these organisms **homeotic mutants**, and they have fascinated developmental biologists for decades.[7] For example, the body of the normal adult fly contains three thoracic segments, each of which produces a pair of legs. The first thoracic segment does not produce any other appendages, but the second thoracic segment produces a pair of wings in addition to its legs. The third thoracic segment produces a pair of legs and a pair of balancers known as **halteres**. In homeotic mutants, these specific segmental identities can be changed. When the *Ultrabithorax* gene is deleted, the third thoracic segment (characterized by halteres) is transformed into another second thoracic segment. The result is a fly with four wings (**FIGURE 9.25**)—an embarrassing situation for a classic dipteran.[8]

Similarly, Antennapedia protein usually specifies the second thoracic segment of the fly. But when flies have a mutation wherein the *Antennapedia* gene is expressed in the head (as well as in the thorax), legs rather than antennae grow out of the head sockets (**FIGURE 9.26**). This is partly because, in addition to promoting the formation of thoracic structures, the Antennapedia protein binds to and represses the enhancers of at least two genes, *homothorax* and *eyeless*, which encode transcription factors that are critical for antenna and eye formation, respectively (Casares and Mann 1998; Plaza et al. 2001). Therefore, one of Antennapedia's functions is to repress the genes that would trigger antenna and eye development. In the recessive mutant of *Antennapedia*, the gene fails to be expressed in the second thoracic segment, and antennae sprout in the leg positions (Struhl 1981; Frischer et al. 1986; Schneuwly et al. 1987).

The major homeotic selector genes have been cloned and their expression analyzed by in situ hybridization (Harding et al. 1985; Akam 1987). Transcripts from each gene

(A)

Second thoracic segment

(B)

Second thoracic   Third thoracic
segment   segment

[7] *Homeo*, from the Greek, means "similar." *Homeotic mutants* are mutants in which one structure is replaced by another (as where an antenna is replaced by a leg). *Homeotic genes* are those genes whose mutation can cause such transformations; thus, homeotic genes are genes that specify the identity of a particular body segment. The *homeobox* is a conserved DNA sequence of about 180 base pairs that is shared by many homeotic genes. This sequence encodes the 60-amino acid *homeodomain*, which recognizes specific DNA sequences. The homeodomain is an important region of the transcription factors encoded by homeotic genes. However, not all genes containing homeoboxes are homeotic genes.

[8] Dipterans—two-winged insects such as flies—are thought to have evolved from four-winged insects, and it is possible that this change arose via alterations in the bithorax complex. Chapter 26 includes further speculation on the relationship between the homeotic complex and evolution.

FIGURE 9.25   (A) Wings of the wild-type fruit fly emerge from the second thoracic segment. (B) A four-winged fruit fly constructed by putting together three mutations in *cis*-regulators of the *Ultrabithorax* gene. These mutations effectively transform the third thoracic segment into another second thoracic segment (i.e., transform halteres into wings). (Courtesy of Nipam Patel.)

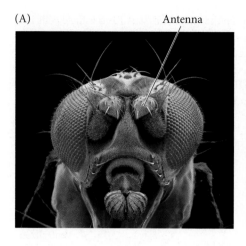

(A)   Antenna

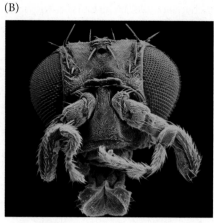

(B)

can be detected in specific regions of the embryo (see Figure 9.24B) and are especially prominent in the central nervous system.

SCIENTISTS SPEAK 9.3 Listen to this interview with Dr. Walter Gehring, who spearheaded investigations that unified genetics, development, and evolution, leading to the discovery of the homeobox and its ubiquity throughout the animal kingdom.

WEB TOPIC 9.5 **INITIATION AND MAINTENANCE OF HOMEOTIC GENE EXPRESSION** Homeotic genes make specific boundaries in the *Drosophila* embryo. Moreover, the protein products of the homeotic genes activate batteries of other genes, specifying the segment.

> **Developing Questions**
>
> Homeobox genes specify the anterior-posterior body axis in both *Drosophila* and humans. How come we do not see homeotic mutations that result in extra sets of limbs in humans, as can happen in flies?

Generating the Dorsal-Ventral Axis

Dorsal-ventral patterning in the oocyte

As oocyte volume increases, the oocyte nucleus is pushed by the growing microtubules to an anterior dorsal position (Zhao et al. 2012). Here the *gurken* message, which had been critical in establishing the anterior-posterior axis, initiates the formation of the dorsal-ventral axis. The *gurken* mRNA becomes localized in a crescent between the oocyte nucleus and the oocyte cell membrane, and its protein product forms an anterior-posterior gradient along the dorsal surface of the oocyte (**FIGURE 9.27**; Neuman-Silberberg and Schüpbach 1993). Since it can diffuse only a short distance, Gurken protein reaches only those follicle cells closest to the oocyte nucleus, and it signals those cells to become the more columnar dorsal follicle cells (Montell et al. 1991; Schüpbach et al. 1991). This establishes the dorsal-ventral polarity in the follicle cell layer that surrounds the growing oocyte.

Maternal deficiencies of either the *gurken* or the *torpedo* gene cause ventralization of the embryo. However, *gurken* is active only in the oocyte, whereas *torpedo* is active only in the somatic follicle cells

(A)

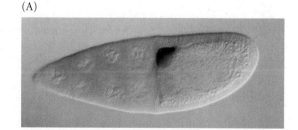

(B)

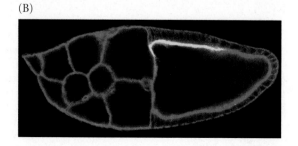

FIGURE 9.27 Expression of Gurken between the oocyte nucleus and the dorsal anterior cell membrane. (A) The *gurken* mRNA is localized between the oocyte nucleus and the dorsal follicle cells of the ovary. Anterior is to the left; dorsal faces upward. (B) A more mature oocyte shows Gurken protein (yellow) across the dorsal region. Actin is stained red, showing cell boundaries. As the oocyte grows, follicle cells migrate across the top of the oocyte, where they become exposed to Gurken. (A from Ray and Schüpbach 1996, courtesy of T. Schüpbach; B courtesy of C. van Buskirk and T. Schüpbach.)

(Schüpbach 1987). The Gurken-Torpedo signal that specifies dorsalized follicle cells initiates a cascade of gene activity that creates the dorsal-ventral axis of the embryo. The activated Torpedo receptor protein activates Mirror, a transcription factor that represses expression of the *pipe* gene (Andreu et al. 2012; Fuchs et al. 2012). As a result, Pipe is made only in the ventral follicle cells (**FIGURE 9.28A**; Sen et al. 1998; Amiri and Stein 2002). Pipe protein modifies the ventral vitelline envelope by sulfating its proteins. This

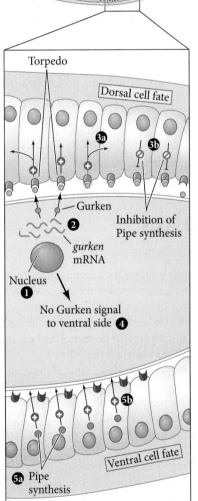

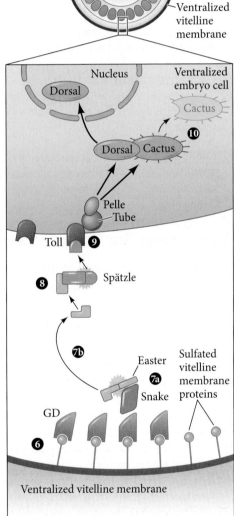

❶ Oocyte nucleus travels to anterior dorsal side of oocyte where it localizes *gurken* mRNA.

❷ Translated Gurken is received by Torpedo proteins.

❸ₐ Torpedo signal causes follicle cells to differentiate to a dorsal morphology.

❸ᵦ Pipe synthesis is inhibited in dorsal follicle cells.

❹ Gurken does not diffuse to ventral follicle cells.

❺ₐ Ventral follicle cells synthesize Pipe.

❺ᵦ Pipe signal sulfates ventral vitelline proteins.

❻ Sulfated vitelline membrane proteins bind Gastrulation-defective (GD).

❼ₐ GD cleaves Snake to its active form and forms a complex with Snake and uncleaved Easter proteins.

❼ᵦ Easter protein is cleaved into its active form.

❽ Cleaved Easter binds to and cleaves Spätzle; activated Spätzle binds to Toll receptor protein.

❾ Toll activation activates Tube and Pelle, which phosphorylate the Cactus protein. Cactus is degraded, releasing it from Dorsal.

❿ Dorsal protein enters the nucleus and ventralizes the cell.

FIGURE 9.28 Generating dorsal-ventral polarity in *Drosophila*. (A) The nucleus of the oocyte travels to what will become the dorsal side of the embryo. The *gurken* genes of the oocyte synthesize mRNA that becomes localized between the oocyte nucleus and the cell membrane, where it is translated into Gurken protein. The Gurken signal is received by the Torpedo receptor protein made by the follicle cells (see Figure 9.8). Given the short diffusibility of the signal, only the follicle cells closest to the oocyte nucleus (i.e., the dorsal follicle cells) receive the Gurken signal, which causes the follicle cells to take on a characteristic dorsal follicle morphology and inhibits the synthesis of Pipe protein. Therefore, Pipe protein is made only by the *ventral* follicle cells. (B) The ventral region, at a slightly later stage of development. Sulfated proteins on the ventral region of the vitelline envelope recruit Gastrulation-defective (GD), which in turn complexes with other proteins, initiating a cascade that results in the cleaved Spätzle protein binding to the Toll receptor. The resulting cascade ventralizes the cell. (After van Eeden and St. Johnston 1999; Cho et al. 2010.)

allows the Gastrulation-defective protein to bind to the vitelline envelope (only in the ventral region) and to recruit other proteins to make a complex that will cleave the Easter protein into its active protease form (**FIGURE 9.28B**; Cho et al. 2010, 2012). Easter then cleaves the Spätzle protein (Chasan et al. 1992; Hong and Hashimoto 1995; LeMosy et al. 2001), and the cleaved Spätzle protein is the ligand that binds to and activates the Toll receptor. It is important that the cleavage of Spätzle be limited to the most ventral portion of the embryo. This is accomplished by the secretion of a protease inhibitor from the follicle cells of the ovary. This can inhibit any small amounts of proteases that might be expected on the margins (Hashimoto et al. 2003; Ligoxygakis et al. 2003).

Toll protein is a maternal product that is evenly distributed throughout the cell membrane of the egg (Hashimoto et al. 1988, 1991), but it becomes activated only by binding Spätzle, which is produced only on the ventral side of the egg. The ventral Toll receptors bind the mature Spätzle protein, and the membrane containing activated Toll protein undergoes endocytosis. Signaling from the Toll receptor is believed to occur in these cytoplasmic endosomes rather than on the cell surface (Lund et al. 2010). Therefore, the Toll receptors on the ventral side of the egg are transducing a signal into the egg, whereas the Toll receptors on the dorsal side are not. This localized activation establishes the dorsal-ventral polarity of the oocyte.

 SCIENTISTS SPEAK 9.4 Two videos featuring Dr. Trudi Schüpbach show how the anchoring and regulation of the Gurken protein are accomplished in the *Drosophila* embryo.

Generating the dorsal-ventral axis within the embryo

The protein that distinguishes dorsum (back) from ventrum (belly) in the fly embryo is the product of the *dorsal* gene. The Dorsal protein is a transcription factor that activates the genes that generate the ventrum. (Note that this is another *Drosophila* gene named after its mutant phenotype: the *dorsal* gene product is a morphogen that ventralizes the region in which it is present.) The mRNA transcript of the mother's *dorsal* gene is deposited in the oocyte by the nurse cells. However, Dorsal protein is not synthesized from this maternal message until about 90 minutes after fertilization. When Dorsal is translated, it is found throughout the embryo, not just on the ventral or dorsal side. How can this protein act as a morphogen if it is located everywhere in the embryo?

The answer to this question was unexpected (Roth et al. 1989; Rushlow et al. 1989; Steward 1989). Although Dorsal protein is found throughout the syncytial blastoderm of the early *Drosophila* embryo, it is translocated into nuclei only in the ventral part of the embryo. In the nucleus, Dorsal acts as a transcription factor, binding to certain genes to activate or repress their transcription. If Dorsal does not enter the nucleus, the genes responsible for specifying ventral cell types are not transcribed, the genes responsible for specifying dorsal cell types are not repressed, and all the cells of the embryo become specified as dorsal cells.

This model of dorsal-ventral axis formation in *Drosophila* is supported by analyses of maternal effect mutations that give rise to an entirely dorsalized or an entirely ventralized phenotype, where there is no "back" to the larval embryo, which soon dies (Anderson and Nüsslein-Volhard 1984). In mutants in which all the cells are dorsalized (evident from their dorsal-specific exoskeleton), Dorsal does not enter the nucleus of any cell. Conversely, in mutants in which all cells have a ventral phenotype, Dorsal protein is found in every cell nucleus (**FIGURE 9.29A**).

Establishing a nuclear Dorsal gradient

So how does Dorsal protein enter into the nuclei only of the ventral cells? When Dorsal is first produced, it is complexed with a protein called Cactus in the cytoplasm of the syncytial blastoderm. As long as Cactus is bound to it, Dorsal remains in the cytoplasm. Dorsal enters ventral nuclei in response to a signaling pathway that frees it from Cactus (see Figure 9.28B). This separation of Dorsal from Cactus is initiated by

(A)

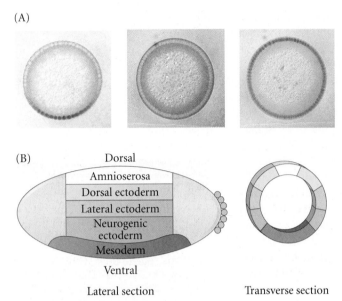

(B)

Dorsal

| Amnioserosa |
| Dorsal ectoderm |
| Lateral ectoderm |
| Neurogenic ectoderm |
| Mesoderm |

Ventral

Lateral section   Transverse section

FIGURE 9.29 Specification of cell fate by the Dorsal protein. (A) Transverse sections of embryos stained with antibody to show the presence of Dorsal protein (dark area). The wild-type embryo (left) has Dorsal protein only in the ventralmost nuclei. A dorsalized mutant (center) has no localization of Dorsal protein in any nucleus. In the ventralized mutant (right), Dorsal protein has entered the nucleus of every cell (B) Fate maps of cross sections through the *Drosophila* embryo at division cycle 14. The most ventral part becomes the mesoderm; the next higher portion becomes the neurogenic (ventral) ectoderm. The lateral and dorsal ectoderm can be distinguished in the cuticle, and the dorsalmost region becomes the amnioserosa, the extraembryonic layer that surrounds the embryo. The translocation of Dorsal protein into ventral, but not lateral or dorsal, nuclei produces a gradient whereby the ventral cells with the most Dorsal protein become mesoderm precursors. (C) Dorsal-ventral patterning in *Drosophila*. The readout of the Dorsal gradient can be seen in the trunk region of a whole-mount stained embryo. The expression of the most ventral gene, *ventral nervous system defective* (blue), is from the neurogenic ectoderm. The *intermediate neuroblast defective* gene (green) is expressed in lateral ectoderm. Red represents the *muscle-specific homeobox* gene, expressed in the mesoderm above the intermediate neuroblasts. The dorsalmost tissue expresses *decapentaplegic* (yellow). (A from Roth et al. 1989, courtesy of the authors; B after Rushlow et al. 1989; C from Kosman et al. 2004, courtesy of D. Kosman and E. Bier.)

(C)

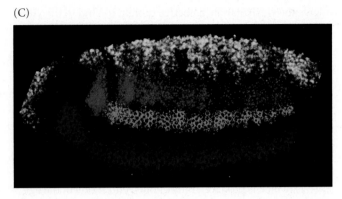

the ventral activation of the Toll receptor. When Spätzle binds to and activates the Toll protein, Toll activates a protein kinase called Pelle. Another protein, Tube, is probably necessary for bringing Pelle to the cell membrane, where it can be activated (Galindo et al. 1995). The activated Pelle protein kinase (probably through an intermediate) can phosphorylate Cactus. Once phosphorylated, Cactus is degraded and Dorsal can enter the nucleus (Kidd 1992; Shelton and Wasserman 1993; Whalen and Steward 1993; Reach et al. 1996). Since Toll is activated by a gradient of Spätzle protein that is highest in the most ventral region, there is a corresponding gradient of Dorsal translocation in the ventral cells of the embryo, with the highest concentrations of Dorsal in the most ventral cell nuclei, which become the mesoderm (**FIGURE 9.29B**).

The Dorsal protein signals the first morphogenetic event of *Drosophila* gastrulation. The 16 ventralmost cells of the embryo—those cells containing the highest amount of Dorsal in their nuclei—invaginate into the body and form the mesoderm (**FIGURE 9.30**). All of the body muscles, fat bodies, and gonads derive from these mesodermal cells (Foe 1989). The cells that will take their place at the ventral midline will become the nerves and glia.

WEB TOPIC 9.6   EFFECTS OF THE DORSAL PROTEIN GRADIENT   Dorsal works as a transcription factor initiating a cascade that sets up the conditions for the specification of the mesoderm, endoderm, and ectoderm.

Axes and Organ Primordia: The Cartesian Coordinate Model

The anterior-posterior and dorsal-ventral axes of *Drosophila* embryos form a coordinate system that can be used to specify positions within the embryo (**FIGURE 9.31A**). Theoretically, cells that are initially equivalent in developmental potential can respond to their position by expressing different sets of genes. This type of specification has been demonstrated in the formation of the salivary gland rudiments (Panzer et al. 1992; Bradley et al. 2001; Zhou et al. 2001).

Drosophila salivary glands form only in the strip of cells defined by the activity of the *sex combs reduced* (*scr*) gene along the anterior-posterior axis (parasegment 2). No salivary glands form in *scr*-deficient mutants. Moreover, if *scr* is experimentally expressed throughout the embryo, salivary gland primordia form in a ventrolateral stripe along most of the length of the embryo. The formation of salivary glands along the dorsal-ventral axis is repressed by both Decapentaplegic and Dorsal proteins, which inhibit salivary gland formation both dorsally and ventrally. Thus, the salivary glands form at the intersection of the vertical *scr* expression band (parasegment 2) and the horizontal region in the middle of the embryo's circumference that has neither Decapentaplegic nor Dorsal (**FIGURE 9.31B**). The cells that form the salivary glands are directed to do so by the intersecting gene activities along the anterior-posterior and dorsal-ventral axes.

A similar situation is seen in neural precursor cells found in every segment of the fly. Neuroblasts arise from 10 clusters of 4 to 6 cells each that form on each side in every segment in the strip of neural ectoderm at the midline of the embryo (Skeath and Carroll 1992). The cells in each cluster interact (via the Notch pathway discussed in Chapter 4) to generate a single neural cell from each cluster. Skeath and colleagues (1993) have shown that the pattern of neural gene transcription is imposed by a coordinate system. Their expression is repressed along the dorsal-ventral axis by the Decapentaplegic and Snail proteins, while positive enhancement by pair-rule genes along the anterior-posterior axis causes neural gene repetition in each half-segment. It is very likely, then, that the positions of organ primordia in the fly are specified via a two-dimensional coordinate system based on the intersection of the anterior-posterior and dorsal-ventral axes.

WEB TOPIC 9.7 **THE RIGHT-LEFT AXES** *Drosophila* has a right-left axis extending from the center of the embryo to its sides. Each set of limbs develops on the right and left sides, but the embryo is not entirely symmetrical.

Coda

Genetic studies of the *Drosophila* embryo have uncovered numerous genes that are responsible for specification of the anterior-posterior and dorsal-ventral axes. Mutations of *Drosophila* genes have given us our first glimpses of the multiple levels of pattern regulation in a complex organism and have enabled us to isolate these genes and their products. Most importantly, however, as we will see in forthcoming chapters, the insights arising from work on *Drosophila* genes have been pivotal in helping us understand the general mechanism of pattern formation used not only by insects but throughout the animal kingdom.

WEB TOPIC 9.8 **EARLY DEVELOPMENT OF OTHER INSECTS** *Drosophila melanogaster* and its relatives are highly derived species. Other insect species develop very differently from the "standard" fruit fly.

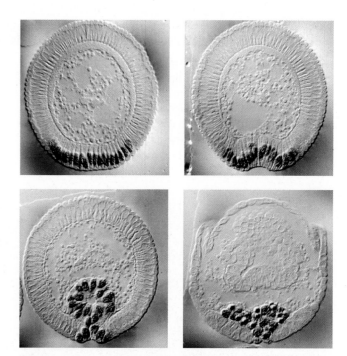

FIGURE 9.30 Gastrulation in *Drosophila*. In this cross section, the mesodermal cells at the ventral portion of the embryo buckle inward, forming the ventral furrow (see Figure 9.5A,B). This furrow becomes a tube that invaginates into the embryo and then flattens and generates the mesodermal organs. The nuclei are stained with antibody to the Twist protein, a marker for the mesoderm. (From Leptin 1991a, courtesy of M. Leptin.)

(A)

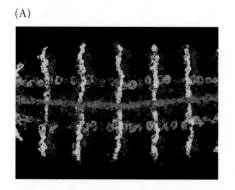

(B)

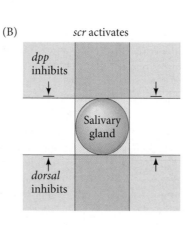

FIGURE 9.31 Cartesian coordinate system mapped out by gene expression patterns. (A) A grid (ventral view, looking "up" at the embryo) formed by the expression of *short-gastrulation* (red), *intermediate neuroblast defective* (green), and *muscle segment homeobox* (magenta) along the dorsal-ventral axis, and by the expression of *wingless* (yellow) and *engrailed* (purple) transcripts along the anterior-posterior axis. (B) Coordinates for the expression of genes giving rise to *Drosophila* salivary glands. These genes are activated by the protein product of the *sex combs reduced* (*scr*) homeotic gene in a narrow band along the anterior-posterior axis, and they are inhibited in the regions marked by *decapentaplegic* (*dpp*) and *dorsal* gene products along the dorsal-ventral axis. This pattern allows salivary glands to form in the midline of the embryo in the second parasegment. (A courtesy of D. Kosman; B after Panzer et al. 1992.)

The precision of *Drosophila* transcription patterning is remarkable, and a transcription factor may specify whole regions or small parts. Some of the most important regulatory genes in *Drosophila*, such as the gap genes, have been found to have "shadow enhancers," secondary enhancers that may be quite distant from the gene. These shadow enhancers seem to be critical for the fine-tuning of gene expression, and they may cooperate or compete with the main enhancer. Some of these shadow enhancers may work under particular physiological stresses. New studies are showing that the robust phenotypes of flies may result from an entire series of secondary enhancers that are able to improvise for different conditions (Bothma et al. 2015).

Closing Thoughts on the Opening Photo

In the fruit fly, inherited genes produce proteins that interact to specify the normal orientation of the body, with the head at one end and the tail at the other. As you studied this chapter, you should have observed how these interactions result in the specification of entire blocks of the fly's body as modular units. A patterned array of homeotic proteins specifies the structures to be formed in each segment of the adult fly. Mutations in the genes for these proteins, called homeotic mutations, can change the structure specified, resulting in wings where there should have been halteres, or legs where there should have been antennae (see pp. 302–303). Remarkably, the proximal-distal orientation of the mutant appendages corresponds to the original appendage's proximal-distal axis, indicating that the appendages follow similar rules for their extension. We now know that many mutations affecting segmentation of the adult fly in fact work on the embryonic modular unit, the parasegment (see pp. 293–294). You should keep in mind that, in both invertebrates and vertebrates, the units of embryonic construction often are not the same units we see in the adult organism. (Photograph courtesy of Nipam Patel.)

Snapshot Summary
Drosophila *Development and Axis Specification*

1. *Drosophila* cleavage is superficial. The nuclei divide 13 times before forming cells. Before cell formation, the nuclei reside in a syncytial blastoderm. Each nucleus is surrounded by actin-filled cytoplasm.

2. When the cells form, the *Drosophila* embryo undergoes a mid-blastula transition, wherein the cleavages become asynchronous and new mRNA is made. At this time, there is a transfer from maternal to zygotic control of development.

3. Gastrulation begins with the invagination of the most ventral region (the presumptive mesoderm), which causes the formation of a ventral furrow. The germ band expands such that the future posterior segments curl just behind the presumptive head.

4. The genes regulating pattern formation in *Drosophila* operate according to certain principles:

   • There are *morphogens*—such as Bicoid and Dorsal—whose gradients determine the specification of different cell types. In syncytial embryos, these morphogens can be transcription factors.

   • *Boundaries* of gene expression can be created by the interaction between transcription factors and their gene targets. Here, the transcription factors transcribed earlier regulate the expression of the next set of genes.

   • *Translational control* is extremely important in the early embryo, and localized mRNAs are critical in patterning the embryo.

   • *Individual cell fates* are not defined immediately. Rather, there is a stepwise specification wherein a given field is divided and subdivided, eventually regulating individual cell fates.

5. There is a temporal order wherein different classes of genes are transcribed, and the products of one gene often regulate the expression of another gene.

6. Maternal effect genes are responsible for the initiation of anterior-posterior polarity. *bicoid* mRNA is bound by its 3′ UTR to the cytoskeleton in the future anterior pole; *nanos* mRNA is sequestered by its 3′ UTR in the future posterior pole. *hunchback* and *caudal* messages are seen throughout the embryo.

7. Dorsal-ventral polarity is regulated by the entry of Dorsal protein into the nucleus. Dorsal-ventral polarity is initiated when the nucleus moves to the dorsal-anterior of the oocyte and sequesters the *gurken* message, enabling it to synthesize proteins in the dorsal side of the egg.

8. Dorsal protein forms a gradient as it enters the various nuclei. Those nuclei at the most ventral surface incorporate the most Dorsal protein and become mesoderm; those more lateral become neurogenic ectoderm.

9. Bicoid and Hunchback proteins activate the genes responsible for the anterior portion of the fly; Caudal activates genes responsible for posterior development.

10. The unsegmented anterior and posterior extremities are regulated by the activation of Torso protein at the anterior and posterior poles of the egg.

11. The gap genes respond to concentrations of the maternal effect gene proteins. Their protein products interact with each other such that each gap gene protein defines specific regions of the embryo.

12. The gap gene proteins activate and repress the pair-rule genes. The pair-rule genes have modular enhancers such that they become activated in seven "stripes." Their boundaries of transcription are defined by the gap genes. The pair-rule genes form seven bands of transcription along the anterior-posterior axis, each one comprising two parasegments.

13. The pair-rule gene products activate *engrailed* and *wingless* expression in adjacent cells. The *engrailed*-expressing cells form the anterior boundary of each parasegment. These cells form a signaling center that organizes the cuticle formation and segmental structure of the embryo.

14. Homeotic selector genes are found in two complexes on chromosome III of *Drosophila*. Together, these regions are called Hom-C, the homeotic gene complex. The genes are arranged in the same order as their transcriptional expression. Genes of the Hom-C specify the individual segments, and mutations in these genes are capable of transforming one segment into another.

15. Organs form at the intersection of dorsal-ventral and anterior-posterior regions of gene expression.

Further Reading

Driever, W., V. Siegel, and C. Nüsslein-Volhard. 1990. Autonomous determination of anterior structures in the early *Drosophila* embryo by the Bicoid morphogen. *Development* 109: 811–820.

Dubuis, J. O., G. Tkacik, E. F. Wieschaus, T. Gregor and W. Bialek. 2013. Positional information, in bits. *Proc. Natl. Acad. Sci. USA* 110: 16301-–16308.

Ingham, P. W. 2016. *Drosophila* segment polarity mutants and the rediscovery of the Hedgehog pathway genes. *Curr. Top. Dev. Biol.* 116: 477–488.

Lehmann, R. and C. Nüsslein-Volhard. 1991. The maternal gene *nanos* has a central role in posterior pattern formation of the *Drosophila* embryo. *Development* 112: 679–691.

Lewis, E. B. 1978. A gene complex controlling segmentation in *Drosophila*. *Nature* 276: 565–570.

Maeda, R. K. and F. Karch. 2009. The Bithorax complex of *Drosophila*. *Curr. Top. Dev. Biol.* 88: 1–33.

Martinez-Arias, A. and P. A. Lawrence. 1985. Parasegments and compartments in the *Drosophila* embryo. *Nature* 313: 639–642.

McGinnis, W., R. L. Garber, J. Wirz, A. Kuroiwa and W. J. Gehring. 1984. A homologous protein-coding sequence in *Drosophila* homeotic genes and its conservation in other metazoans. *Cell* 37: 403–408.

Nüsslein-Volhard, C. and E. Wieschaus. 1980. Mutations affecting segment number and polarity in *Drosophila*. *Nature* 287: 795–801.

Pankratz, M. J., E. Seifert, N. Gerwin, B. Billi, U. Nauber and H. Jäckle. 1990. Gradients of *Krüppel* and *knirps* gene products direct pair-rule gene stripe patterning in the posterior region of the *Drosophila* embryo. *Cell* 61: 309–317.

Roth, S., D. Stein and C. Nüsslein-Volhard. 1989. A gradient of nuclear localization of the dorsal protein determines dorsoventral pattern in the *Drosophila* embryo. *Cell* 59: 1189–1202.

Schultz, J. 1935. Aspects of the relation between genes and development in *Drosophila*. *American Naturalist* 69: 30–54.

Schüpbach, T. 1987. Germline and soma cooperate during oogenesis to establish the dorsoventral pattern of egg shell and embryo in *Drosophila melanogaster*. *Cell* 49: 699–707.

Struhl, G. 1981. A homeotic mutation transforming leg to antenna in *Drosophila*. *Nature* 292: 635–638.

Wang, C. and R. Lehman. 1991. Nanos is the localized posterior determinate in *Drosophila*. *Cell* 66: 637–647.

Wieschaus, E. 2016. Positional information and cell fate determination in the early *Drosophila* embryo. *Curr. Top. Dev. Biol.* 117: 567–580.

GO TO WWW.DEVBIO.COM...

...for Web Topics, Scientists Speak interviews, Watch Development videos, Dev Tutorials, and complete bibliographic information for all literature cited in this chapter.

Sea Urchins and Tunicates
Deuterostome Invertebrates

HAVING DESCRIBED THE PROCESSES of early development in representative species from three protostome groups—mollusks, nematodes, and insects—we turn next to the deuterostomes. Although there are many fewer species of deuterostomes than there are of protostomes, these include the members of all the vertebrate groups—fish, amphibians, reptiles, birds, and mammals. Several invertebrate groups also follow the deuterostome pattern of development (in which the blastopore becomes the anus during gastrulation). These include the hemichordates (acorn worms), cephalochordates (amphioxus), echinoderms (sea urchins, starfish, sea cucumbers, and others), and urochordates (tunicates, also called sea squirts) (**FIGURE 10.1**). This chapter covers the early development of echinoderms (notably sea urchins) and tunicates, both of which have been the subjects of critically important studies in developmental biology.

Indeed, conditional specification ("regulative development") was first discovered in sea urchins, while tunicates provided the first evidence for autonomous specification ("mosaic development"). As we will see, it turns out that both groups use both modes of specification.

Early Development in Sea Urchins

Sea urchins have been exceptionally important organisms in studying how genes regulate the formation of the body. Hans Driesch discovered regulative development when he was studying sea urchins. He found that the early stages of sea urchin development had a

How do the fluorescing cells of this tunicate embryo proclaim its kinship with humans?

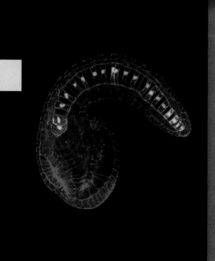

The Punchline

Sea urchins and tunicates are deuterostome invertebrates. They have no backbone, although tunicates have a notochord. Both groups use both conditional and autonomous specification schemes. Sea urchins are known for integrating these types of specification. The micromeres are specified autonomously through a gene regulatory network with a double-negative circuit that inhibits the inhibitor of skeleton development. Part of the resulting "micromere phenotype" is the ability to induce neighboring cells to become endoderm and secondary mesenchyme. Tunicates, on the other hand, are better known for their autonomous mode of development, wherein determinants such as the muscle cell transcription factor Macho are placed into specific blastomeres during oogenesis and early cleavage. Tunicates also display conditional specification; this mode is used to create organs such as the notochord, which links this group of invertebrates to the vertebrates.

FIGURE 10.1 The echinoderms and tunicates represent deuterostome invertebrates. The tunicates, however, are classified as chordates because their larvae possess a notochord, dorsal neural tube, and pharyngeal arches. The tunicates are referred to as urochordates, a name that emphasizes their affinity with the other chordate groups. The green sea urchin *Lytechinus variegatus* and the tunicate *Ciona intestinalis* are two widely studied model organisms. (*L. variegatus* photograph courtesy of David McIntyre; *C. intestinalis* photograph © Nature Picture Library/Alamy.)

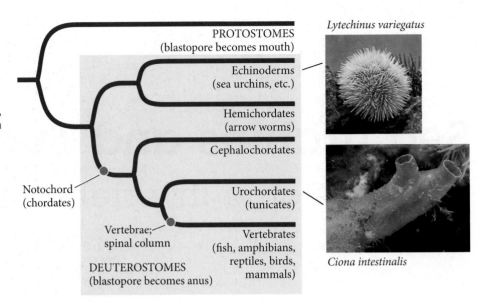

Lytechinus variegatus

Ciona intestinalis

regulatory mode, since a single blastomere isolated from the four-cell stage could form an entire sea urchin pluteus larva (Driesch 1891; also see Chapter 2). However, cells isolated from later stages could not become all the cells of the larval body.

Sea urchin embryos also provided the first evidence that chromosomes were needed for development, that DNA and RNA were present in every animal cell, that messenger RNAs directed protein synthesis, that stored messenger RNA provided the proteins for early embryonic development, that cyclins controlled cell division, and that enhancers were modular (Ernst 2011; McClay 2011). The first cloned eukaryotic gene encoded a sea urchin histone protein (Kedes et al. 1975), and the first evidence for chromatin remodeling concerned histone alterations during sea urchin development (Newrock et al. 1978). With the advent of new genetic techniques, sea urchin embryos continue to be critically important organisms for delineating the mechanisms by which genetic interactions specify different cell fates.

Early cleavage

Sea urchins exhibit **radial holoblastic cleavage** (**FIGURES 10.2** and **10.3**). Recall from Chapter 1 that this type of cleavage occurs in eggs with sparse yolk, and that holoblastic cleavage furrows extend through the entire egg (see Figure 1.5). In sea urchins, the first seven cleavage divisions are stereotypic in that the same pattern is followed in every individual of the same species. The first and second cleavages are both meridional and are perpendicular to each other (that is to say, the cleavage furrows pass through the animal and vegetal poles). The third cleavage is equatorial, perpendicular to the first two cleavage planes, and separates the animal and vegetal hemispheres from each other (see Figure 10.2A, top row, and Figure 10.3A–C). The fourth cleavage, however, is very different. The four cells of the animal tier divide meridionally into eight blastomeres, each with the same volume. These eight cells are called **mesomeres**. The vegetal tier, however, undergoes an unequal equatorial cleavage (see Figure 10.2B) to produce four large cells—the **macromeres**—and four smaller **micromeres** at the vegetal pole. In *Lytechinus variegatus*, a species often used for experimentation, the ratio of cytoplasm retained in the macromeres and micromeres is 95:5. As the 16-cell embryo cleaves, the eight "animal" mesomeres divide equatorially to produce two tiers—an_1 and an_2—one staggered above the other. The macromeres divide meridionally, forming a tier of eight cells below an_2 (see Figure 10.2A, bottom row). Somewhat later, the micromeres divide unequally, producing a cluster of four small micromeres at

(A)

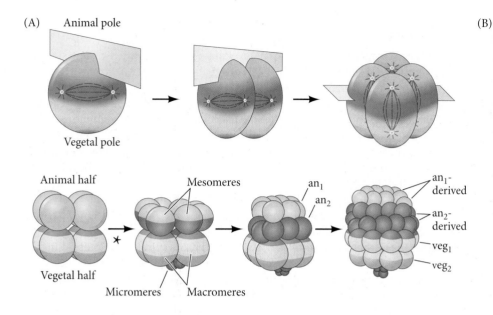

(B)

FIGURE 10.2 Cleavage in the sea urchin. (A) Planes of cleavage in the first three divisions, and the formation of tiers of cells in divisions 3–6. (B) Confocal fluorescence micrograph of the unequal cell division that initiates the 16-cell stage (asterisk in A), highlighting the unequal equatorial cleavage of the vegetal blastomeres to produce the micromeres and macromeres. (B courtesy of G. van Dassow and the Center for Cell Dynamics.)

the vegetal pole, beneath a tier of four large micromeres. The small micromeres divide once more, then stop dividing until the larval stage. At the sixth division, the animal hemisphere cells divide meridionally while the vegetal cells divide equatorially; this pattern is reversed in the seventh division (see Figure 10.2A, bottom row). At that time, the embryo is a 120-cell blastula[1] in which the cells form hollow sphere surrounding a central cavity called the **blastocoel** (see Figure 10.3F). From here on, the pattern of divisions becomes less regular.

---

[1] You might have been expecting a 128-cell blastula, but remember that the small micromeres stopped dividing.

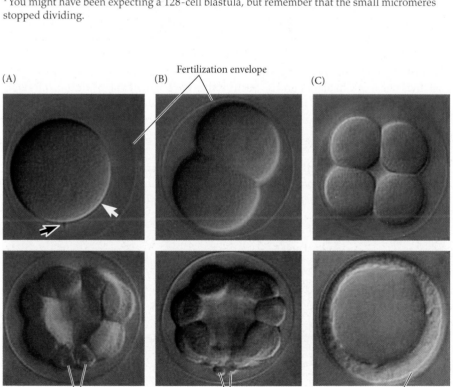

FIGURE 10.3 Micrographs of cleavage in live embryos of the sea urchin *Lytechinus variegatus*, seen from the side. (A) The 1-cell embryo (zygote). The site of sperm entry is marked with a black arrow; a white arrow marks the vegetal pole. The fertilization envelope surrounding the embryo is clearly visible. (B) 2-cell stage. (C) 8-cell stage. (D) 16-cell stage. Micromeres have formed at the vegetal pole. (E) 32-cell stage. (F) The blastula has hatched from the fertilization envelope. The vegetal plate is beginning to thicken. (Courtesy of J. Hardin.)

Blastula formation

By the blastula stage, all the cells of the developing sea urchin are the same size, the micromeres having slowed down their cell divisions. Every cell is in contact with the proteinaceous fluid of the blastocoel on the inside and with the hyaline layer on the outside. Tight junctions unite the once loosely connected blastomeres into a seamless epithelial sheet that completely encircles the blastocoel. As the cells continue to divide, the blastula remains one cell layer thick, thinning out as it expands. This is accomplished by the adhesion of the blastomeres to the hyaline layer and by an influx of water that expands the blastocoel (Dan 1960; Wolpert and Gustafson 1961; Ettensohn and Ingersoll 1992).

These rapid and invariant cell cleavages last through the ninth or tenth division, depending on the species. By this time, the fates of the cells have become specified (discussed in the next section), and each cell becomes ciliated on the region of the cell membrane farthest from the blastocoel. Thus, there is apical-basal (outside-inside) polarity in each embryonic cell, and there is evidence that PAR proteins (like those of the nematode) are involved in distinguishing the basal cell membranes (Alford et al. 2009). The ciliated blastula begins to rotate within the fertilization envelope. Soon afterward, differences are seen in the cells. The cells at the vegetal pole of the blastula begin to thicken, forming a **vegetal plate** (see Figure 10.3F). The cells of the animal hemisphere synthesize and secrete a hatching enzyme that digests the fertilization envelope (Lepage et al. 1992). The embryo is now a free-swimming **hatched blastula**.

WEB TOPIC 10.1 **URCHINS IN THE LAB**   Stanford University hosts a valuable and freely accessible site called VirtualUrchin that describes and outlines numerous ways of studying sea urchin development in the laboratory. The Swarthmore College Developmental Biology site also provides useful laboratory protocols.

Fate maps and the determination of sea urchin blastomeres

Early fate maps of the sea urchin embryo followed the descendants of each of the 16-cell-stage blastomeres. More recent investigations have refined these maps by following the fates of individual cells that have been injected with fluorescent dyes. These dye markers glow not only in the injected cell, but also in that cell's progeny for many cell divisions (see Chapters 1 and 2). Such studies have shown that by the 60-cell stage, most of the embryonic cell fates are specified but the cells are not irreversibly committed. In other words, particular blastomeres consistently produce the same cell types in each embryo, but these cells remain pluripotent and can give rise to other cell types if experimentally placed in a different part of the embryo.

A fate map of the 60-cell sea urchin embryo is shown in **FIGURE 10.4**. The animal half of the embryo consistently gives rise to the ectoderm—the larval skin and its neurons. The veg₁ layer produces cells that can enter into either the larval ectodermal or

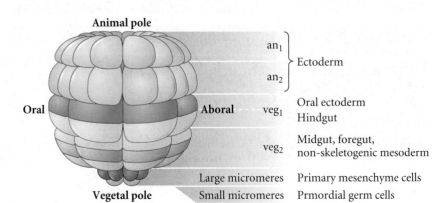

FIGURE 10.4 Fate map and cell lineage of the sea urchin *Strongylocentrotus purpuratus*. The 60-cell embryo is shown, with the left side facing the viewer. Blastomere fates are segregated along the animal-vegetal axis of the egg.

the endodermal organs. The veg$_2$ layer gives rise to cells that can populate three different structures—the endoderm, the coelom (internal mesodermal body wall), and the **non-skel-etogenic mesenchyme** (sometimes called **secondary mesenchyme**), which generates pigment cells, immunocytes, and muscle cells. The upper tier of micromeres (the large micromeres) produces the **skeletogenic mesenchyme** (also called **primary mesenchyme**), which forms the larval skeleton. The lower-tier micromeres (i.e., the small micromeres) play no role in embryonic development. Rather, they contribute cells to the larval coelom, from which the tissues of the adult are derived during metamorphosis (Logan and McClay 1997, 1999; Wray 1999). These small micromeres also contribute to producing the germline cells (Yajima and Wessel 2011.)

The fates of the different cell layers are determined in a two-step process:

1. The large micromeres are *autonomously* specified. They inherit maternal determinants that were deposited at the vegetal pole of the egg; these become incorporated into the large micromeres at fourth cleavage. These four micromeres are thus determined to become skeletogenic mesenchyme cells that will leave the blastula epithelium, enter the blastocoel, migrate to particular positions along the blastocoel wall, and then differentiate into the larval skeleton.

2. The autonomously specified large micromeres are now able to produce paracrine and juxtacrine factors that *conditionally* specify the fates of their neighbors. The micromeres produce signals that tell the cells above them to become endomesoderm (the endoderm and the secondary mesenchyme cells) and induces them to invaginate into the embryo.

The ability of the micromeres to produce signals that change the fates of neighboring cells is so pronounced that if micromeres are removed from the embryo and placed below an isolated **animal cap**—that is, below the top two animal tiers that usually become ectoderm—the animal cap cells will generate endoderm and a more or less normal larva will develop (**FIGURE 10.5**; Hörstadius 1939).

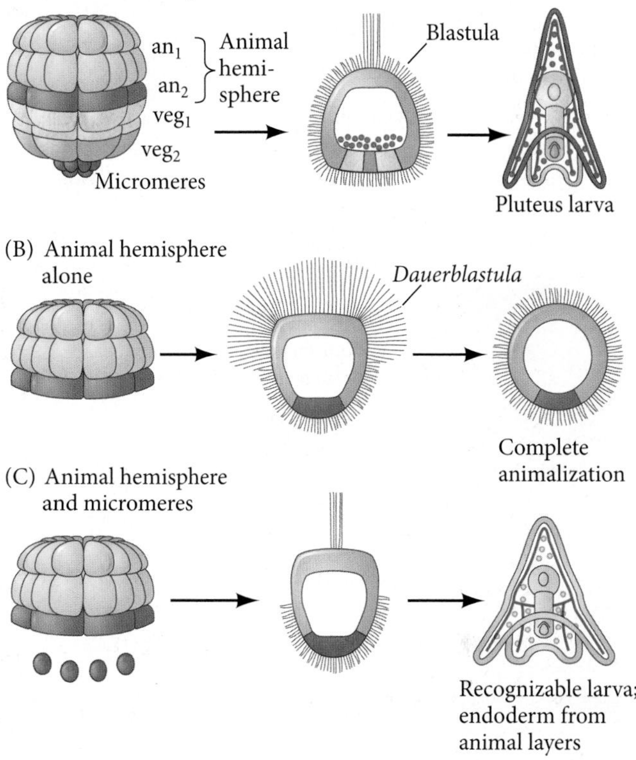

(A) Normal development

(B) Animal hemisphere alone

(C) Animal hemisphere and micromeres

FIGURE 10.5 Ability of micromeres to induce presumptive ecto-dermal cells to acquire other fates. (A) Normal development of the 60-cell sea urchin embryo, showing the fates of the different layers. (B) An isolated animal hemisphere becomes a ciliated ball of undifferentiated ectodermal cells called a *Dauerblastula* (permanent blastula). (C) When an isolated animal hemisphere is combined with isolated micromeres, a recognizable pluteus larva is formed, with all the endoderm derived from the animal hemisphere. (After Hörstadius 1939.)

These skeletogenic micromeres are the first cells whose fates are determined autonomously. If micromeres are isolated from the 16-cell embryo and placed in petri dishes, they will divide the appropriate number of times and produce the skeletal spicules (Okazaki 1975). Thus, the isolated micromeres do not need any other signals to generate their skeletal fates. Moreover, if skeletogenic micromeres are transplanted into the animal region of the blastula, not only will their descendants form skeletal spicules, but the transplanted micromeres will alter the fates of nearby cells by inducing a secondary site for gastrulation. Cells that would normally have produced ectodermal skin cells will be respecified as endoderm and will produce a secondary gut (**FIGURE 10.6**; Hörstadius 1973; Ransick and Davidson 1993). Therefore, the inducing ability of the micromeres is also established autonomously.

WATCH DEVELOPMENT 10.1 "A Sea-Biscuit Life" is a beautifully photographed and subtitled video that chronicles the development of the sand dollar (another echinoderm, basically a flattened sea urchin).

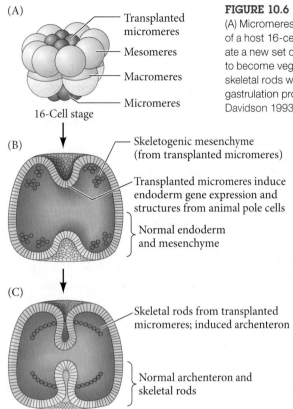

(A)

Transplanted micromeres

Mesomeres

Macromeres

Micromeres

16-Cell stage

(B)

Skeletogenic mesenchyme (from transplanted micromeres)

Transplanted micromeres induce endoderm gene expression and structures from animal pole cells

Normal endoderm and mesenchyme

(C)

Skeletal rods from transplanted micromeres; induced archenteron

Normal archenteron and skeletal rods

FIGURE 10.6 Ability of micromeres to induce a secondary axis in sea urchin embryos. (A) Micromeres are transplanted from the vegetal pole of a 16-cell embryo into the animal pole of a host 16-cell embryo. (B) The transplanted micromeres invaginate into the blastocoel to create a new set of skeletogenic mesenchyme cells, and they induce the animal cells next to them to become vegetal plate endoderm cells. (C) The transplanted micromeres differentiate into skeletal rods while the induced animal cap cells form a secondary archenteron. Meanwhile, gastrulation proceeds normally from the original vegetal plate of the host. (After Ransick and Davidson 1993.)

Gene regulatory networks and skeletogenic mesenchyme specification

According to the embryologist E. B. Wilson, heredity is the transmission from generation to generation of a particular pattern of development, and evolution is the hereditary alteration of such a plan. Wilson was probably the first scientist to write (in his 1895 analysis of sea urchin development) that the instructions for development were somehow stored in chromosomal DNA and were transmitted by the chromosomes at fertilization. However, he had no way of knowing how the chromosomal information organized matter into forming an embryo.

Studies from the sea urchin developmental biology community have begun to demonstrate how DNA can be regulated to specify the cells and direct the morphogenesis of the developing organism (McClay 2016). Eric Davidson's group has pioneered a network approach to development in which they envision *cis*-regulatory elements (such as promoters and enhancers) in a logic circuit connected to each other by transcription factors (Davidson and Levine 2008; Oliveri et al. 2008; Peter and Davidson 2015). The network receives its first inputs from transcription factors in the egg cytoplasm. From then on, the network self-assembles from (1) the ability of the maternal transcription factors to recognize *cis*-regulatory elements of particular genes that encode other transcription factors, and (2) the ability of this new set of transcription factors to activate paracrine signaling pathways that activate specific transcription factors in neighboring cells. The studies show the regulatory logic by which the genes of the sea urchin interact to specify and generate characteristic cell types. The researchers refer to such a set of interconnections among cell-type specifying genes as a **gene regulatory network**, or **GRN**.

WEB TOPIC 10.2 **THE DAVIDSON LABORATORY SEA URCHIN DEVELOPMENT PROJECT** This link to the Davidson Sea Urchin Development Project expands on the concept of the GRN and provides updated systems diagrams showing an hour-by-hour account of the specification of the sea urchin cell types during the early cleavage stages.

Here we will focus on the earliest part of one such GRN: the reactions by which the skeletogenic mesenchyme cells of the sea urchin embryo receive their developmental fate and inductive properties. Descended from the micromeres, skeletogenic mesenchyme cells are those cells that are autonomously specified to ingress into the blastocoel and become the larval skeleton. And, as we have seen, they are also the cells that conditionally induce their neighbors to become endoderm (gut) and non-skeletogenic mesenchyme (pigment; coelom) cells (see Figure 10.6).

DISHEVELED AND β-CATENIN: SPECIFYING THE MICROMERES The specification of the micromere lineage (and hence the rest of the embryo) begins inside the undivided egg. The initial regulatory inputs are two transcription regulators, Disheveled and β-catenin, both of which are found in the cytoplasm and are inherited by the micromeres as soon as they are formed (i.e., at the fourth cleavage). During oogenesis, Disheveled becomes located in the vegetal cortex of the egg (**FIGURE 10.7A**; Weitzel et al. 2004; Leonard and

Ettensohn 2007), where it prevents the degradation of β-catenin in the micromere and veg$_2$-tier macromere cells. The β-catenin then enters the nucleus, where it combines with the TCF transcription factor to activate gene expression from specific promoters.

Several pieces of evidence suggest that β-catenin specifies the micromeres. First, during normal sea urchin development, β-catenin accumulates in the nuclei of those cells fated to become endoderm and mesoderm (**FIGURE 10.7B**). This accumulation is autonomous and can occur even if the micromere precursors are separated from the rest of the embryo. Second, this nuclear accumulation appears to be responsible for specifying the vegetal half of the embryo. It is possible that levels of nuclear β-catenin accumulation help determine the mesodermal and endodermal fates of the vegetal cells (Kenny et al. 2003). Treating sea urchin embryos with lithium chloride allows β-catenin to accumulate in every cell and transforms presumptive ectoderm into endoderm (**FIGURE 10.7C**). Conversely, experimental procedures that inhibit β-catenin accumulation

FIGURE 10.7 Role of the Disheveled and β-catenin proteins in specifying the vegetal cells of the sea urchin embryo. (A) Localization of Disheveled (arrows) in the vegetal cortex of the sea urchin oocyte before fertilization (left) and in the region of a 16-cell embryo about to become the micromeres (right). (B) During normal development, β-catenin accumulates predominantly in the micromeres and somewhat less in the veg$_2$ tier cells. (C) In embryos treated with lithium chloride, β-catenin accumulates in the nuclei of all blastula cells (probably by LiCl's blocking the GSK3 enzyme of the Wnt pathway), and the animal cells become specified as endoderm and mesoderm. (D) When β-catenin is prevented from entering the nuclei (i.e., it remains in the cytoplasm), the vegetal cell fates are not specified, and the entire embryo develops as a ciliated ectodermal ball. (From Weitzel et al. 2004, courtesy of C. Ettensohn, and from Logan et al. 1998, courtesy of D. McClay.)

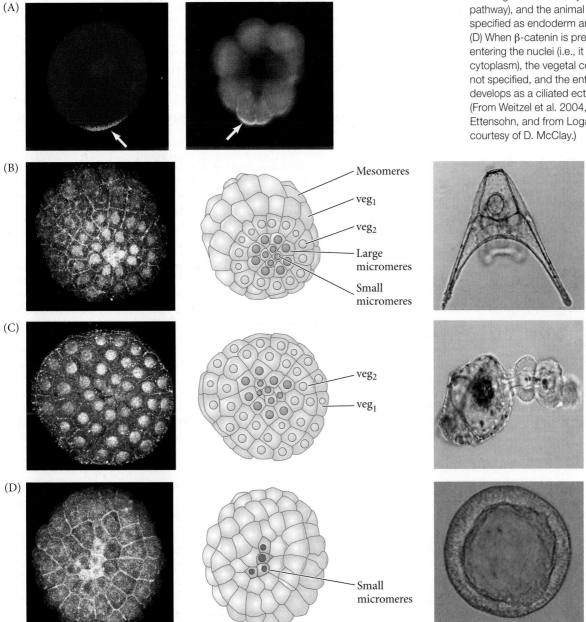

in the vegetal cell nuclei prevent the formation of endoderm and mesoderm (**FIGURE 10.7D**; Logan et al. 1998; Wikramanayake et al. 1998).

PMAR1 AND HESC: A DOUBLE-NEGATIVE GATE The next micromere regulatory input comes from the Otx transcription factor, which is also enriched in the micromere cytoplasm. Otx interacts with the β-catenin/TCF complex at the enhancer of the *Pmar1* gene to activate *Pmar1* transcription in the micromeres shortly after their formation (**FIGURE 10.8A**; Oliveri et al. 2008). Pmar1 protein is a repressor of *HesC*, a gene that encodes another repressive transcription factor. *HesC* is expressed in every nucleus of the sea urchin embryo *except* those of the micromeres.[2]

In the micromeres, where *Pmar1* is activated, the *HesC* gene is repressed. This mechanism, whereby a repressor locks the genes of specification and these genes can be unlocked by the repressor of that repressor (in other words, when activation occurs by the repression of a repressor), is called a **double-negative gate** (**FIGURES 10.8B** and **10.9A**). Such a gate allows for tight regulation of fate specification: it *promotes* the expression of these genes where the input occurs, and it *represses* the same genes in every other cell type (Oliveri et al. 2008).

The genes repressed by HesC are those involved in micromere specification and differentiation: *Alx1, Ets1, Tbr, Tel,* and *SoxC*. Each of these genes can be activated by ubiquitous transcription factors, but these positive transcription factors cannot work while HesC repressor protein binds to their respective enhancers. When the Pmar1 protein is present, it represses *HesC*, and all these genes become active (Revilla-i-Domingo et al. 2007). The newly activated genes synthesize transcription factors that activate another

---

[2]This is an oversimplification of the skeletogenic gene regulatory network. In the pathway to micromere specification, other transcription factors must be expressed, and the maternal transcription factor SoxB1 has to be eliminated from the micromeres or it will inhibit the activation of *Pmar1*. In addition, the cytoskeletal processes partitioning the cells and anchoring certain factors are not considered here. For complete details of the model, see the continually updated website cited in Web Topic 10.2.

(A)

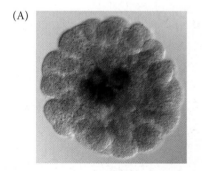

(B)

FIGURE 10.8 Simplified illustration of the double-negative gated "circuit" for micromere specification. (A) In situ hybridization reveals the accumulation of *Pmar1* mRNA (dark purple) in the micromeres. (B) OTX is a general transcription factor, and β-catenin from the maternal cytoplasm is concentrated at the vegetal pole of the egg. These transcriptional regulators are inherited by the micromeres and activate the *Pmar1* gene. *Pmar1* encodes a repressor of *HesC*, which in turn encodes a repressor (hence the "double-negative") of several genes involved in micromere specification (e.g., *Alx1, Tbr,* and *Ets*). Genes encoding signaling proteins (e.g., *Delta*) are also under the control of HesC. In the micromeres, where activated Pmar1 protein represses the *HesC* repressor, the micromere specification and signaling genes are active. In the veg₂ cells, *Pmar1* is not activated and the HesC gene product shuts down the skeletogenic genes; however, those cells containing Notch can respond to the Delta signal from the skeletogenic mesenchyme. The gene expression patterns are seen below. U represents ubiquitous activating transcription factors. (After Oliveri et al. 2008; photograph courtesy of P. Oliveri.)

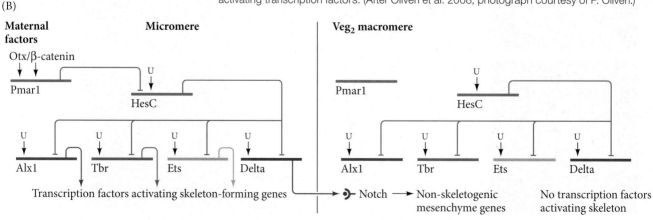

(A)

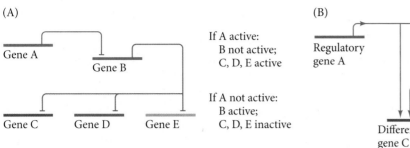

If A active:
  B not active;
  C, D, E active

If A not active:
  B active;
  C, D, E inactive

(B)

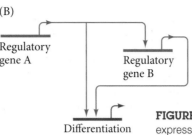

FIGURE 10.9 "Logic circuits" for gene expression. (A) In a double-negative gate, a single gene encodes a repressor of an entire battery of genes. When this repressor gene is repressed, the battery of genes is expressed. (B) In a feedforward circuit, gene product A activates both gene B and gene C, and gene B also activates gene C. Feedforward circuits provide an efficient way to amplify a signal in one direction.

set of genes, most of which are genes that activate skeletal determinants. Some of these transcription factors also activate each other's genes, so that once one factor is activated, it maintains the activity of the other skeletogenic genes. This stabilizes the regulatory state of the skeletogenic mesenchyme cells (see Peter and Davidson 2016).

Another way micromeres retain their specification is to secrete an autocrine factor, Wnt8 (Angerer and Angerer 2000; Wikramanayake et al. 2004). As soon as the micromeres form, maternal β-catenin and Otx activate the *Blimp1* gene, whose product (in conjunction with more β-catenin) activates the *Wnt8* gene. Wnt8 protein is then received by the same micromeres that made it (i.e., autocrine regulation), activating the micromeres' own genes for β-catenin. Because β-catenin activates *Blimp1*, this autocrine regulation sets up a positive feedback loop between Blimp1 and Wnt8 that establishes a source of β-catenin for the micromere nuclei. Equally important, this cross-regulatory loop serves to lock both genes "on" and can amplify their expression levels.

In contrast to the double-negative gate that *specifies* the micromeres, control of the genes that *differentiate* the cells of the sea urchin skeleton operates on a feedforward process (**FIGURE 10.9B**). Here, regulatory gene A produces a transcription factor that is needed for the differentiation of gene C and also activates regulatory gene B, which produces a transcription factor also needed for differentiation of gene C. This feedforward process stabilizes gene expression and makes the resulting cell type irreversible.

EVOLUTION BY SUBROUTINE COOPTION   The skeletogenic portion of the micromere gene regulatory network shown in Figure 10.8B appears to have arisen from the recruitment of a network subroutine that in most echinoderms (including sea urchins) is used in making the adult skeleton (Gao and Davidson 2008; Erkenbrack and Davidson 2015). The cooption of subroutines by a new lineage is one of the ways evolution occurs. It happens that the GRN of sea urchin micromeres is very different from that in other echinoderm embryos. Only in sea urchins has the skeletogenic subroutine come under the control of the genes that specify cells to the micromere lineage; in all other echinoderms skeletogenesis is activated late in development. The most important evolutionary events were those placing the skeletogenic genes *Alx1* and *Ets1* (necessary for adult skeletal development) and *Tbr* (used in later larval skeleton formation) under the regulation of the *Pmar1-HesC* double-negative gate. This occurred through mutations in the *cis*-regulatory regions of these genes. Thus, the skeletogenic properties that distinguish the sea urchin micromeres appear to have arisen through the recruitment of a preexisting skeletogenic regulatory system by the micromere lineage gene regulatory system.

Specification of the vegetal cells

The skeletogenic micromeres also produce signals that can induce changes in other tissues. One of these signals is the TGF-β family paracrine factor activin. Expression of the gene for activin is also under the control of the *Pmar1-HesC* double-negative gate, and activin secretion appears to be critical for endoderm formation (Sethi et al. 2009). Indeed, if *Pmar1* mRNA is injected into an animal cell, that animal cell will develop into a skeletogenic mesenchyme cell, and the cells adjacent to it will start developing like a macromere (Oliveri et al. 2002). If the activin signal is blocked, the adjacent cells

Developing Questions

Evolution is accomplished by changes in development. Such developmental changes, in turn, can be accomplished by changes in GRNs. In considering the evolution of two closely related species, how might the micromere GRN of a starfish differ from that of sea urchins?

do not become endoderm[3] (Ransick and Davidson 1995; Sherwood and McClay 1999; Sweet et al. 1999).

Another cell-specifying signal from the micromeres is the juxtacrine protein Delta, also a factor that is controlled by the double-negative gate. Delta functions by activating Notch proteins on the adjacent veg_2 cells and later will act on the adjacent small micromeres. Delta causes the veg_2 cells to become the non-skeletogenic mesenchyme cells by activating the Gcm transcription factor and repressing the FoxA transcription factor (which activates the endoderm-specific genes). The upper veg_2 cells, since they do not receive the Delta signal, retain FoxA expression, and this pushes them in the direction of becoming endodermal cells (Croce and McClay 2010).

In sum, the genes of the sea urchin micromeres specify their cell fates autonomously and also specify the fates of their neighbors conditionally. The original inputs come from the maternal cytoplasm and activate genes that unlock repressors of a specific cell fate. Once the maternal cytoplasmic factors accomplish their functions, the nuclear genome takes over.

Sea Urchin Gastrulation

Architect Frank Lloyd Wright wrote in 1905 that "Form and function should be one, joined in a spiritual union." While Wright never used sea urchin skeletons as inspiration, other architects (such as Antoni Gaudi) may have; the characteristic sea urchin **pluteus larvae** is a feeding structure in which form and function are remarkably well integrated.

The late blastula of the sea urchin is a single layer of about 750 epithelial cells that form a hollow ball, somewhat flattened at the vegetal end. These blastomeres are derived from different regions of the zygote and have different sizes and properties. The cells that are destined to become the endoderm (gut) and mesoderm (skeleton) are still on the outside and need to be brought inside the embryo through gastrulation.

 SCIENTISTS SPEAK 10.1   Dr. Jeff Hardin offers a brief tutorial on sea urchin gastrulation.

Ingression of the skeletogenic mesenchyme

FIGURE 10.10 illustrates development of the blastula through gastrulation to the pluteus larva stage (hour 24). Shortly after the blastula hatches from its fertilization envelope, the descendants of the large micromeres undergo an epithelial-to-mesenchymal transition. The epithelial cells change their shape, lose their adhesions to their neighboring cells, and break away from the epithelium to enter the blastocoel as skeletogenic mesenchyme cells (Figure 10.10, 9–10 hours). The skeletogenic mesenchyme cells then begin extending and contracting long, thin (250 nm in diameter and 25 μm long) processes called **filopodia**. At first the cells appear to move randomly along the inner blastocoel surface, actively making and breaking filopodial connections to the wall of the blastocoel. Eventually, however, they become localized within the prospective ventrolateral region of the blastocoel. Here they fuse into syncytial cables that will form the axis of the calcium carbonate spicules of the larval skeletal rods. This is coordinated through the same GRN that specified the skeletogenic mesenchyme cells.

---

[3] Recall the experiments in Figure 10.6, which demonstrated that the micromeres are able to induce a second embryonic axis when transplanted to the animal hemisphere. However, micromeres in which β-catenin is prevented from entering the nucleus are unable to induce the animal cells to form endoderm, and no second axis forms (Logan et al. 1998). β-Catenin also accumulates in macromeres, but by a different means, and the *Pmar1* gene is not activated in macromeres (possibly due to the presence of SoxB1; see Kenny et al. 2003 and Lhomond et al. 2012).

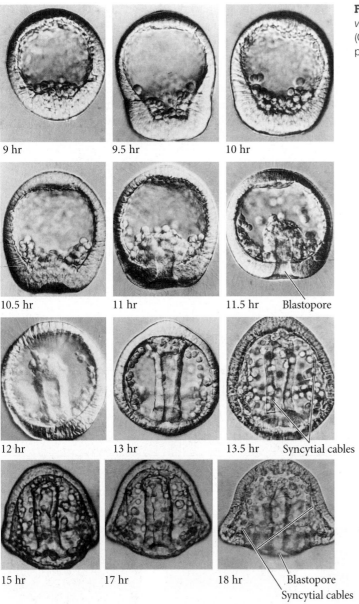

9 hr   9.5 hr   10 hr

10.5 hr   11 hr   11.5 hr   Blastopore

12 hr   13 hr   13.5 hr   Syncytial cables

15 hr   17 hr   18 hr   Blastopore
Syncytial cables   24 hr

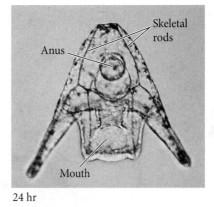

FIGURE 10.10 Entire sequence of gastrulation in *Lytechinus variegatus*. Times show the length of development at 25°C. (Courtesy of J. Morrill; pluteus larva seen from the original vegetal pole courtesy of G. Watchmaker.)

Skeletal rods

Anus

Mouth

EPITHELIAL-TO-MESENCHYMAL TRANSITION The ingression of the large micromere descendants into the blastocoel is a result of their losing their affinity for their neighbors and for the hyaline membrane; instead these cells acquire a strong affinity for a group of proteins that line the blastocoel. Initially, all the cells of the blastula are connected on their outer surface to the hyaline layer, and on their inner surface to a basal lamina secreted by the cells. On their lateral surfaces, each cell has another cell for a neighbor. Fink and McClay found that the prospective ectoderm and endoderm cells (descendants of the mesomeres and macromeres, respectively) bind tightly to one another and to the hyaline layer, but adhere only loosely to the basal lamina. The micromeres initially display a similar pattern of binding. However, the micromere pattern changes at gastrulation. Whereas the other cells retain their tight binding to the hyaline layer and to their neighbors, the skeletogenic mesenchyme precursors lose their affinities for these structures (which drop to about 2% of their original value), while their affinity for components of the basal lamina and extracellular matrix increases a hundredfold. This accomplishes an **epithelial-to-mesenchymal transition** (**EMT**), where cells that had formerly been part of an epithelium lose their attachments and become individual, migrating cells (**FIGURE 10.11A**; see also Chapter 4).

EMTs are important events throughout animal development, and the pathways to EMT are revisited in cancer cells, where the EMT is often necessary for the formation of secondary tumor sites. There appear to be five distinct processes in this EMT, and all of these events are regulated by the same micromere GRN that specifies and forms the skeletogenic mesenchyme. However, each of these processes is controlled by a different subset of transcription factors. Even more surprisingly, none of the transcription factors was the "master regulator" of the EMT (Saunders and McClay 2014). These five events are:

1. *Apical-basal polarity.* The vegetal cells of the blastula elongate to form a thickened "vegetal plate" epithelium (Figure 10.10, 9 hours).

2. *Apical constriction of the micromeres.* The cells alter their shape, wherein the apical end (away from the blastocoel) becomes constricted. Apical constriction is seen during gastrulation and neurulation in both vertebrates and invertebrates, and it is one of the most important cell shape changes associated with morphogenesis (Sawyer et al. 2010).

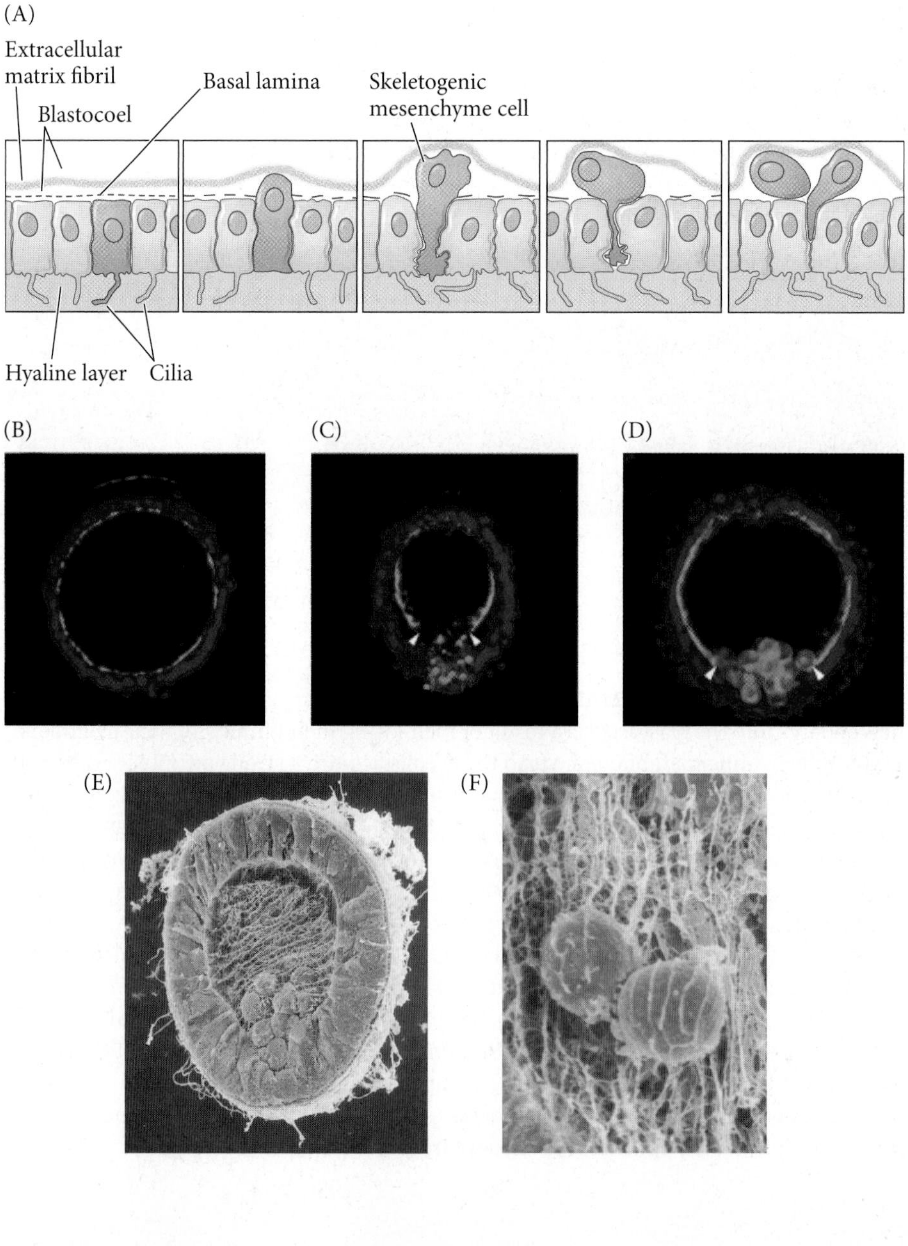

FIGURE 10.11 Ingression of skeletogenic mesenchyme cells. (A) Interpretative depiction of changes in the adhesive affinities of the presumptive skeletogenic mesenchyme cells (pink). These cells lose their affinities for hyalin and for their neighboring blastomeres while gaining an affinity for the proteins of the basal lamina. Nonmesenchymal blastomeres retain their original high affinities for the hyaline layer and neighboring cells. (B-D) Skeletogenic mesenchyme cells breaking through extracellular matrix. The matrix laminin is stained pink, the mesenchyme cells are green and cell nuclei are blue. (B) Laminin matrix is uniformly spread throughout the lining of the blastocoel. (C) Hole is made in blastocoel laminin above the vegetal cells, and the mesenchyme begins to pass through it into the blastocoel. (D) Within an hour, cells are in the blastocoel. (E) Scanning electron micrograph of skeletogenic mesenchyme cells enmeshed in the extracellular matrix of an early *Strongylocentrotus* gastrula. (F) Gastrula-stage mesenchyme cell migration. The extracellular matrix fibrils of the blastocoel lie parallel to the animal-vegetal axis and are intimately associated with the skeletogenic mesenchyme cells. (B–D courtesy of David McClay; E,F from Cherr et al. 1992, courtesy of the authors.)

3. *Basement membrane remodeling.* The cells must pass through the laminin-containing basement membrane. Originally, this membrane is uniform around the blastocoel. However, the micromere cells secrete proteases (protein-digesting enzymes) that digest a hole in this membrane, shortly before the first mesenchymal cells are seen inside the blastocoel (**FIGURE 10.11B–D**).

4. *De-adhesion.* The cadherins that couple epithelial cells together are degraded, thereby allowing the cells to become free from their neighbors. Downregulation of cadherins is controlled by the transcription factor Snail. The gene for *snail* is activated by the Alx1 transcription factor, which in turn is regulated by the double-negative gate of the gene regulatory network (Wu et al. 2007). The Snail transcription factor is involved in de-adhesion throughout the animal kingdom (including cancers).

5. *Cell motility.* The transcription factors of the GRN activate those proteins causing the active migration of the cells out of the epithelium and into the blastocoel. One of the most critical ones is *Foxn2/3*. This transcription factor is also seen in regulation of the motility of neural crest cells after their EMT (to form the face in vertebrates). The cells bind to and travel on extracellular matrix proteins within the blastocoel (**FIGURE 10.11E,F**).

The proteins necessary for selective migration, cell fusion, and skeleton formation (e.g., biomineralization proteins) are also regulated by the transcription factors (such as Alx1, Ets1, and Tbr) activated by the double-negative gate (Rafiq et al. 2014). At two sites near the future ventral side of the larva, many skeletogenic mesenchyme cells cluster together, fuse with one another, and initiate spicule formation (Hodor and Ettensohn 1998; Lyons et al. 2015). If a labeled micromere from another embryo is injected into the blastocoel of a gastrulating sea urchin embryo, it migrates to the correct location and contributes to the formation of the embryonic spicules (Ettensohn 1990; Peterson and McClay 2003). It is thought that the necessary positional information is provided by the prospective ectodermal cells and their basal laminae (**FIGURE 10.12A**; Harkey and Whiteley 1980; Armstrong et al. 1993; Malinda and Ettensohn 1994). Only the skeletogenic mesenchyme cells (and not other cell types or latex beads) are capable of responding to these patterning cues (Ettensohn and McClay 1986). The extremely fine filopodia on the skeletogenic mesenchyme cells explore and sense the blastocoel wall and appear to be sensing dorsal-ventral and animal-vegetal patterning cues from the ectoderm (**FIGURE 10.12B**; Malinda et al. 1995; Miller et al. 1995).

Two signals secreted by the blastula wall appear to be critical for this migration. VEGF paracrine factors are emitted from two small regions of the ectoderm where the skeletogenic mesenchyme cells will congregate (Duloquin et al. 2007), and a fibroblast growth factor (FGF) paracrine factor is made in the equatorial belt between endoderm

FIGURE 10.12 Positioning of skeletogenic mesenchyme cells in the sea urchin. (A) Positioning of the micromeres to form the calcium carbonate skeleton is determined by the ectodermal cells. Skeletogenic mesenchyme cells are stained green; β-catenin is red; skeletogenic mesenchyme cells appear to accumulate in those regions characterized by high β-catenin concentrations. (B) Nomarski videomicrograph showing a long, thin filopodium extending from a skeletogenic mesenchyme cell to the ectodermal wall of the gastrula, as well as a shorter filopodium extending inward from the ectoderm. Mesenchymal filopodia extend through the extracellular matrix and directly contact the cell membrane of the ectodermal cells. (C) Seen in cross section through the archenteron (top), the surface ectoderm expresses FGF in the particular locations where skeletogenic micromeres congregate. Moreover, the ingressing skeletal micromeres (bottom; longitudinal section) express the FGF receptor. When FGF signaling is suppressed, the skeleton does not form properly. (B from Miller et al. 1995, photographs courtesy of J. R. Miller and D. McClay; C from Röttinger et al. 2008, photographs courtesy of T. Lepage.)

(A)
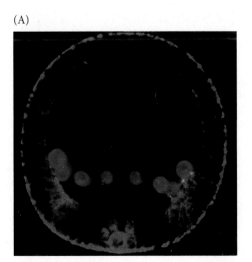

(B)
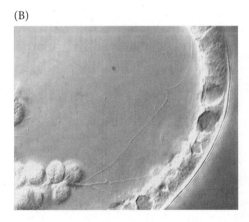

(C)
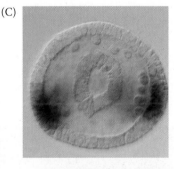
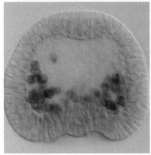

FIGURE 10.13 Formation of syncytial cables by skeletogenic mesenchyme cells of the sea urchin. (A) Skeletogenic mesenchyme cells in the early gastrula align and fuse to lay down the matrix of the calcium carbonate spicule (arrows). (B) Scanning electron micrograph of the syncytial cables formed by the fusing of skeletogenic mesenchyme cells. (A from Ettensohn 1990; B from Morrill and Santos 1985.)

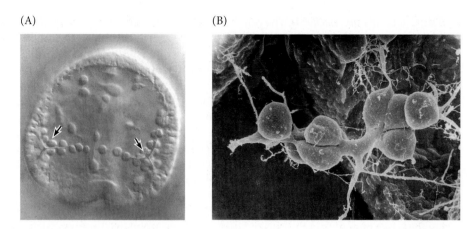

(A)

(B)

and ectoderm, becoming defined into the lateral domains where the skeletogenic mesenchyme cells collect (**FIGURE 10.12C**; Röttinger et al. 2008; McIntyre et al. 2014). The skeletogenic mesenchyme cells migrate to these points of VEGF and FGF synthesis and arrange themselves in a ring along the animal-vegetal axis (**FIGURE 10.13**). The receptors for these paracrine factors appear to be specified by the double-negative gate (Peterson and McClay 2003).

> **WEB TOPIC 10.3** **AXIS SPECIFICATION IN SEA URCHIN EMBRYOS** The Nodal protein determines the oral-aboral axis of the sea urchin pluteus larva during early gastrulation. During late gastrulation and the early larval period, Nodal specifies the right half of the larva.

Invagination of the archenteron

FIRST STAGE OF ARCHENTERON INVAGINATION As the skeletogenic mesenchyme cells leave the vegetal region of the spherical embryo, important changes are occurring in the cells that remain there. These cells thicken and flatten to form a vegetal plate, changing the shape of the blastula (see Figure 10.10, 9 hours). The vegetal plate cells remain bound to one another and to the hyaline layer of the egg, and they move to fill the gaps caused by the ingression of the skeletogenic mesenchyme. The vegetal plate involutes inward by altering its cell shape, then invaginates about one-fourth to one-half of the way into the blastocoel before invagination suddenly ceases. The invaginated region is called the **archenteron** (primitive gut), and the opening of the archenteron at the vegetal pole is the **blastopore** (**FIGURE 10.14A**; see also Figure 10.10, 10.5–11.5 hours).

FIGURE 10.14 Invagination of the vegetal plate. (A) Vegetal plate invagination in *Lytechinus variegatus*, seen by scanning electron microscopy of the external surface of the early gastrula. The blastopore is clearly visible. (B) Fate map of the vegetal plate of the sea urchin embryo, looking "upward" at the vegetal surface. The central portion becomes the non-skeletogenic mesenchyme cells. The concentric layers around it become the foregut, midgut, and hindgut, respectively. The boundary where the endoderm meets the ectoderm marks the anus. The non-skeletogenic mesenchyme and foregut come from the veg_2 layer; the midgut comes from veg_1 and veg_2 cells; the hindgut and the ectoderm surrounding it come from the veg_1 layer. (A from Morrill and Santos 1985, courtesy of J. B. Morrill; B after Logan and McClay 1999.)

(A)

(B)

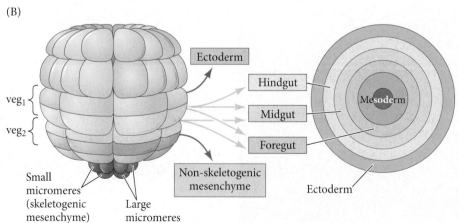

The movement of the vegetal plate into the blastocoel appears to be initiated by shape changes in the vegetal plate cells and in the extracellular matrix underlying them (see Kominami and Takata 2004). Actin microfilaments collect in the apical ends of the vegetal cells, causing these ends to constrict, forming bottle-shaped vegetal cells that pucker inward (Kimberly and Hardin 1998; Beane et al. 2006). Destroying these cells with lasers retards gastrulation. In addition, the hyaline layer at the vegetal plate buckles inward due to changes in its composition, directed by the vegetal plate cells (Lane et al. 1993).

At the stage when the skeletogenic mesenchyme cells begin ingressing into the blastocoel, the fates of the vegetal plate cells have already been specified (Ruffins and Ettensohn 1996). The non-skeletogenic mesenchyme is the first group of cells to invaginate, forming the tip of the archenteron and leading the way into the blastocoel. The non-skeletogenic mesenchyme will form the pigment cells, the musculature around the gut, and contribute to the coelomic pouches. The endodermal cells adjacent to the macromere-derived secondary mesenchyme become foregut, migrating the farthest into the blastocoel. The next layer of endodermal cells becomes midgut, and the last circumferential row to invaginate forms the hindgut and anus (**FIGURE 10.14B**).

SECOND AND THIRD STAGES OF ARCHENTERON INVAGINATION   After a brief pause following initial invagination, the second phase of archenteron formation begins. The archenteron elongates dramatically, sometimes tripling in length. In this process of extension, the wide, short gut rudiment is transformed into a long, thin tube (**FIGURE 10.15A**; see also Figure 10.10, 12 hours). To accomplish this extension, numerous cellular phenomena work together. First, the endoderm cells proliferate as they enter into the embryo. Second, the clones derived from these cells slide past one another, like

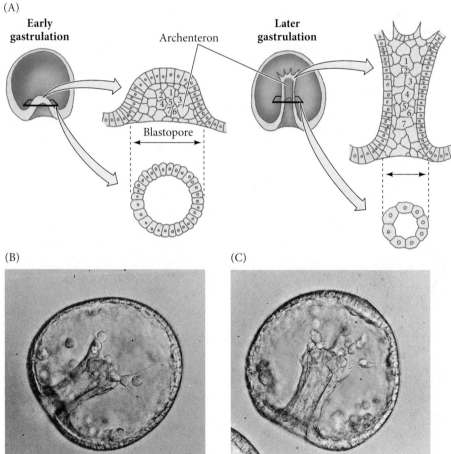

FIGURE 10.15 Extension of the archenteron in sea urchin embryos. (A) Cell rearrangement during extension of the archenteron in sea urchin embryos. In this species, the early archenteron has 20–30 cells around its circumference. Later in gastrulation, the archenteron has a circumference made by only 6–8 cells. (B) Later, at the mid-gastrula stage of *Lytechinus pictus*, filopodial processes extend from non-skeletogenic mesenchyme. (B) Non-skeletogenic mesenchyme cells extend filopodia (arrows) from the tip of the archenteron. (C) Filopodial cables connect the blastocoel wall to the archenteron tip. The tension of the cables can be seen as they pull on the blastocoel wall at the point of attachment (arrows) (A After Hardin 1990; B, C Courtesy of C. Ettensohn.)

the extension of a telescope. And lastly, the cells rearrange themselves by intercalating between one another, like lanes of traffic merging (Ettensohn 1985; Hardin and Cheng 1986; Martins et al. 1998; Martik et al. 2012). This phenomenon, where cells intercalate to narrow the tissue and at the same time lengthen it, is called **convergent extension**.

The final phase of archenteron elongation is initiated by the tension provided by non-skeletogenic mesenchyme cells, which form at the tip of the archenteron and remain there. These cells extend filopodia through the blastocoel fluid to contact the inner surface of the blastocoel wall (Dan and Okazaki 1956; Schroeder 1981). The filopodia attach to the wall at the junctions between the blastomeres and then shorten, pulling up the archenteron (**FIGURE 10.15 B,C**; see also Figure 10.12, 12 and 13 hours). Hardin (1988) ablated non-skeletogenic mesenchyme cells of *Lytechinus pictus* gastrulae with a laser, with the result that the archenteron could elongate only to about two-thirds of the normal length. If a few non-skeletogenic mesenchyme cells were left, elongation continued, although at a slower rate. Thus, in this species the non-skeletogenic mesenchyme cells play an essential role in pulling the archenteron upward to the blastocoel wall during the last phase of invagination.

But can the filopodia of non-skeletogenic mesenchyme cells attach to any part of the blastocoel wall, or is there a specific target in the animal hemisphere that must be present for attachment to take place? Is there a region of the blastocoel wall that is already committed to becoming the ventral side of the larva? Studies by Hardin and McClay (1990) show that there is indeed a specific target site for filopodia that differs from other regions of the animal hemisphere. The filopodia extend, touch the blastocoel wall at random sites, and then retract. However, when filopodia contact a particular region of the wall, they remain attached and flatten out against this region, pulling the archenteron toward it. When Hardin and McClay poked in the other side of the blastocoel wall so that contacts were made most readily with that region, the filopodia continued to extend and retract after touching it. Only when the filopodia found their target tissue did they cease these movements. If the gastrula was constricted so that the filopodia never reached the target area, the non-skeletogenic mesenchyme cells continued to explore until they eventually moved off the archenteron and found the target as freely migrating cells. There appears, then, to be a target region on what is to become the ventral side of the larva that is recognized by the non-skeletogenic mesenchyme cells, and which positions the archenteron near the region where the mouth will form. Thus, as is characteristic of deuterostomes, the blastopore marks the position of the anus.

As the top of the archenteron meets the blastocoel wall in the target region, many of the non-skeletogenic mesenchyme cells disperse into the blastocoel, where they proliferate to form the mesodermal organs (see Figure 10.10, 13.5 hours). Where the archenteron contacts the wall, a mouth eventually forms. The mouth fuses with the archenteron to create the continuous digestive tube of the pluteus larva. The remarkable metamorphosis from pluteus larva to adult sea urchin will be described in Chapter 21.

WATCH DEVELOPMENT 10.2   A video chronicles archenteron formation in a sea urchin.

Early Development in Tunicates

Tunicates (also known as ascidians or "sea squirts") are fascinating animals for several reasons, but the foremost is that they are closest evolutionary relatives of the vertebrates. As Lemaire (2009) has written, "looking at an adult ascidian, it is difficult, and slightly degrading, to imagine that we are close cousins to these creatures." Even though tunicates lack vertebrae at all stages of their life cycles, the free-swimming tunicate larva, or "tadpole," has a notochord and a dorsal nerve cord, making these animals invertebrate chordates (see Figure 10.1). When the tadpole undergoes metamorphosis,

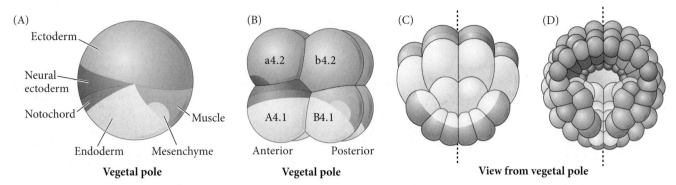

(A) Ectoderm
Neural ectoderm
Notochord
Endoderm
Mesenchyme
Muscle
Vegetal pole

(B) a4.2 b4.2
A4.1 B4.1
Anterior Posterior
Vegetal pole

(C)

(D)
View from vegetal pole

FIGURE 10.16 Bilateral symmetry in the egg of the ascidian tunicate *Styela partita*. (The cell lineages of *Styela* are shown in Figure 1.12C.) (A) Uncleaved egg. The regions of cytoplasm destined to form particular organs are labeled here and coded by color throughout the diagrams. (B) 8-Cell embryo, showing the blastomeres and the fates of various cells. The embryo can be viewed as left and right 4-cell halves; from here on, each division on the right side of the embryo has a mirror-image division on the left. (C,D) Views of later embryos from the vegetal pole. The dashed line shows the plane of bilateral symmetry. (A after Balinsky 1981.)

its nerve cord and notochord degenerate, and it secretes a cellulose tunic that is the source of the tunicates' name.

Cleavage

Tunicates have **bilateral holoblastic cleavage** (**FIGURE 10.16**). The most striking feature of this type of cleavage is that the first cleavage plane establishes the earliest axis of symmetry in the embryo, separating the embryo into its future right and left sides. Each successive division is oriented to this plane of symmetry, and the half-embryo formed on one side of the first cleavage plane is the mirror image of the half-embryo on the other side.[4] The second cleavage is meridional like the first but, unlike the first division, it does not pass through the center of the egg. Rather, it creates two large anterior cells (the A and a blastomeres) and two smaller posterior cells (the B and b blastomeres). Each side now has a large and a small blastomere.

Indeed, from the 8- through the 64-cell stages, every cell division is asymmetrical, such that the posterior blastomeres are always smaller than the anterior blastomeres (Nishida 2005; Sardet et al. 2007). Prior to each of these unequal cleavages, the posterior centrosome in the blastomere migrates toward the **centrosome-attracting body** (**CAB**), a macroscopic subcellular structure composed of endoplasmic reticulum. The CAB connects to the cell membrane through a network of PAR proteins that position the centrosomes asymmetrically in the cell (as in *C. elegans*; see Figure 8.17), resulting in one large and one small cell at each of these three divisions. The CAB also attracts particular mRNAs in such a way that these messengers are placed in the posteriormost (i.e., smaller) cell of each division (Hibino et al. 1998; Nishikata et al. 1999; Patalano et al. 2006). In this way, the CAB integrates cell patterning with cell determination.[5] At the 64-cell stage, a small blastocoel is formed and gastrulation begins from the vegetal pole.

The tunicate fate map

Most early tunicate blastomeres are specified autonomously, each cell acquiring a specific type of cytoplasm that will determine its fate. In many tunicate species, the different regions of cytoplasm have distinct pigmentation, so the cell fates can easily be seen to correspond to the type of cytoplasm taken up by each cell. These cytoplasmic regions are apportioned to the egg during fertilization.

Figure 2.4 shows the fate map and cell lineages of the tunicate *Styela partita*. In the unfertilized egg, a central gray cytoplasm is enveloped by a cortical layer containing yellow lipid inclusions (**FIGURE 10.17A**). During meiosis, the breakdown of the nucleus releases a clear substance that accumulates in the animal hemisphere of the egg. Within

---

[4] This conclusion turns out to be a good first approximation—indeed, it has lasted over a century. However, new labeling techniques have shown that there is some left-right asymmetry in later tunicate embryos (B. Davidson, personal communication).

[5] This description should remind you of the discussion of the posterior cytoplasm of *Drosophila* eggs in Chapter 9. Indeed, mRNAs are localized in the CAB by their 3' UTRs, the CAB is enriched with vesicles, and some of the mRNAs of the CAB become partitioned into the germ cells while others help construct the anterior-posterior axis (Makabe and Nishida 2012).

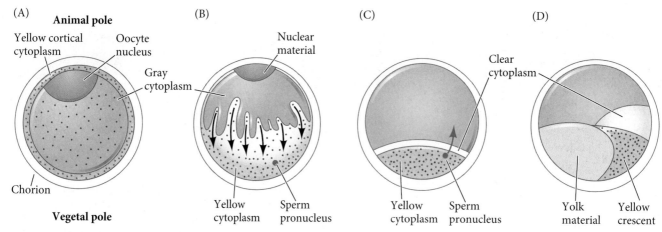

FIGURE 10.17 Cytoplasmic rearrangement in the fertilized egg of *Styela partita*. (A) Before fertilization, yellow cortical cytoplasm surrounds the gray (yolky) inner cytoplasm. (B) After sperm entry into the vegetal hemisphere of the oocyte, the yellow cortical cytoplasm and the clear cytoplasm derived from the breakdown of the oocyte nucleus contract vegetally toward the sperm. (C) As the sperm pronucleus migrates animally toward the newly formed egg pronucleus, the yellow and clear cytoplasms move with it. (D) The final position of the yellow cytoplasm marks the location where cells give rise to tail muscles. (After Conklin 1905.)

5 minutes of sperm entry, the inner clear and cortical yellow cytoplasms contract into the vegetal (lower) hemisphere of the egg (Prodon et al. 2005, 2008; Sardet et al. 2005). As the male pronucleus migrates from the vegetal pole to the equator of the cell along the future posterior side of the embryo, the yellow lipid inclusions migrate with it. This migration forms the **yellow crescent**, extending from the vegetal pole to the equator (**FIGURE 10.17B–D**); this region will produce most of the tail muscles of the tunicate larva. The movement of these cytoplasmic regions depends on microtubules that are generated by the sperm centriole and on a wave of calcium ions that contracts the animal pole cytoplasm (Sawada and Schatten 1989; Speksnijder et al. 1990; Roegiers et al. 1995).

Edwin Conklin (1905) took advantage of the differing coloration of these regions of cytoplasm to follow each of the cells of the tunicate embryo to its fate in the larva (see Web Topic 1.3). Conklin found that cells receiving clear cytoplasm become ectoderm; those containing yellow cytoplasm give rise to mesodermal cells; those incorporating slate gray inclusions become endoderm; and light gray cells become the neural tube and notochord. The cytoplasmic regions are arrayed bilaterally on both sides of the plane of symmetry, so they are bisected by the first cleavage furrow into the right and left halves of the embryo. The second cleavage causes the prospective mesoderm to lie in the two posterior cells, while the prospective neural ectoderm and chordamesoderm (notochord) will be formed from the two anterior cells (**FIGURE 10.18**). The third division further partitions these cytoplasmic regions such that the mesoderm-forming cells are confined to the two vegetal posterior blastomeres, while the chordamesoderm cells are restricted to the two vegetal anterior cells.

Autonomous and conditional specification of tunicate blastomeres

The autonomous specification of tunicate blastomeres was one of the first observations in the field of experimental embryology (Chabry 1888). Cohen and Berrill (1936) confirmed Chabry's and Conklin's results, and by counting the number of notochord and muscle cells, they demonstrated that larvae derived from only one of the first two blastomeres (either the right or the left) had half the expected number of cells.[6] When the 8-cell embryo is separated into its four doublets (the right and left sides being equivalent), both autonomous and conditional specification are seen (Reverberi and Minganti 1946). Autonomous specification is seen in the gut endoderm, muscle mesoderm, and

---

[6] Chabry and Driesch each seem to have gotten the results the other desired (see Fisher 1991). Driesch, who saw the embryo as a machine, expected autonomous specification but showed conditional specification (regulative development). Chabry, a Socialist who believed everyone starts off equally endowed, expected to find conditional specification but instead discovered autonomous specification (mosaic development). Recent research on gene regulatory networks has begun to provide a molecular basis for this regulation (Peter et al. 2012).

(A)　　　　　　　　　　(B)　　　　　　　　　　(C)

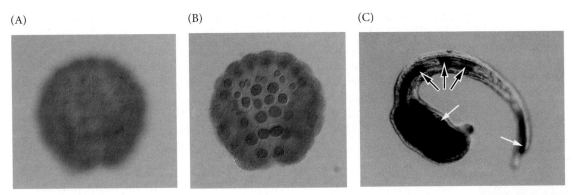

skin ectoderm (see Lemaire 2009). Conditional specification (by induction) is seen in the formation of the brain, notochord, heart, and mesenchyme cells. Indeed, a majority of tunicate cell lineages involve some inductions.

AUTONOMOUS SPECIFICATION OF THE MYOPLASM: THE YELLOW CRESCENT AND MACHO-1 From the cell lineage studies of Conklin and others, it was known that only one pair of blastomeres (posterior vegetal; B4.1) in the 8-cell embryo is capable of producing tail muscle tissue (Whittaker 1982). These cells contain the yellow crescent cytoplasm. As mentioned in Chapter 2, when yellow crescent cytoplasm is transferred from the B4.1 (muscle-forming) blastomere to the b4.2 or a4.2 (ectoderm-forming) blastomeres of an 8-cell tunicate embryo, the ectoderm-forming blastomeres generate muscle cells as well as their normal ectodermal progeny. Nishida and Sawada (2001) showed that this muscle-forming determinant was an mRNA that encoded a transcription factor they named Macho-1. They correlated the presence of *macho-1* mRNA with the ability of the cell to form muscles: when they deleted *macho-1*, the cells did not make muscles, and when they added this mRNA to nonmuscle progenitor cells, those cells were able to make muscles.

The Macho-1 protein is a transcription factor that is required for the activation of several mesodermal genes, including *muscle actin, myosin, tbx6,* and *snail* (Sawada et al. 2005; Yagi et al. 2004). Of these gene products, only the Tbx6 protein produced muscle differentiation (as Macho-1 did) when expressed ectopically. Macho-1 thus appears to directly activate a set of *tbx6* genes, and Tbx6 proteins activate the rest of muscle development (Yagi et al. 2005; Kugler et al. 2010). Thus, the *macho-1* message is found at the right place and at the right time to be the long-sought muscle cell determinant of the yellow crescent, and these experiments suggest that Macho-1 protein is both necessary and sufficient to promote muscle differentiation in certain tunicate cells.

Macho-1 and Tbx6 also appear to activate (possibly in a feedforward manner) the muscle-specific gene *snail*. Snail protein is important in preventing *Brachyury* expression in presumptive muscle cells, and thereby prevents the muscle precursors from becoming notochord cells.[7] It appears, then, that Macho-1 is a critical transcription factor of the tunicate yellow crescent, muscle-forming cytoplasm. Macho-1 activates a transcription factor cascade that promotes muscle differentiation while at the same time inhibiting notochord specification.

WEB TOPIC 10.4 **THE SEARCH FOR THE MYOGENIC FACTOR** The yellow crescent cytoplasm was one of the most tantalizing regions of all known eggs. Before molecular biology was able to identify individual RNA species, some elegant experiments detailed the properties of this cytoplasm's myogenic determinant.

FIGURE 10.18 Antibody staining of β-catenin protein shows its involvement with endoderm formation. (A) No β-catenin is seen in the animal pole nuclei of a 110-cell *Ciona* embryo. (B) In contrast, β-catenin is readily seen in the nuclei of the vegetal endoderm precursors at the 110-cell stage. (C) When β-catenin is expressed in notochordal precursor cells, those cells will become endoderm and express endodermal markers such as alkaline phosphatase. The white arrows show normal endoderm; the black arrows show notochordal cells that are expressing endodermal enzymes. (From Imai et al. 2000, courtesy of H. Nishida and N. Satoh.)

---

[7] We will see the importance of *Brachyury* in vertebrate notochord formation as well. Indeed, the notochord is the "cord" that links the tunicates with the vertebrates, and *Brachyury* appears to be the gene that specifies the notochord (Satoh et al. 2012). As we will also see, *Tbx6* (which is closely related to *Brachyury*) is important in forming vertebrate musculature.

AUTONOMOUS SPECIFICATION OF ENDODERM: β-CATENIN Presumptive endoderm originates from the vegetal A4.1 and B4.1 blastomeres. The specification of these cells coincides with the localization of β-catenin, discussed earlier in regard to sea urchin endoderm specification. Inhibition of β-catenin results in the loss of endoderm and its replacement by ectoderm in the tunicate embryo (**FIGURE 10.18**; Imai et al. 2000). Conversely, increasing β-catenin synthesis causes an increase in the endoderm at the expense of the ectoderm (just as in sea urchins). The β-catenin transcription factor appears to function by activating the synthesis of the homeobox transcription factor Lhx3. Inhibition of the *lhx3* message prohibits differentiation of endoderm (Satou et al. 2001).

WEB TOPIC 10.5 **SPECIFICATION OF THE LARVAL AXES IN TUNICATE EMBRYOS**
Unlike many other embryos, all of the embryonic axes of tunicates are determined by the cytoplasm of the zygote *prior* to first cleavage.

CONDITIONAL SPECIFICATION OF MESENCHYME AND NOTOCHORD BY THE ENDO-DERM Although most tunicate muscles are specified autonomously from the yellow crescent cytoplasm, the most posterior muscle cells form through conditional specification by interactions with the descendants of the A4.1 and b4.2 blastomeres (Nishida 1987, 1992a,b). Moreover, the notochord, brain, heart, and mesenchyme also form through inductive interactions. In fact, the notochord and mesenchyme appear to be induced by FGFs secreted by the endoderm cells (Nakatani et al. 1996; Kim et al. 2000; Imai et al. 2002). These FGF proteins induce the *Brachyury* gene, which binds to the *cis*-regulatory elements that specify notochord development (**FIGURE 10.19**; Davidson and Christiaen 2006).

Interestingly, those genes that are turned on early by the Brachyury transcription factor have multiple binding sites for this protein and need all these sites occupied for maximal effect. Those genes that are turned on a bit later (also by Brachyury) have only one site, and this site may not bind as well as those on genes that are activated early. Notochord genes that are activated even later are activated indirectly. In the last instance, Brachyury activates a second transcription factor, which then will bind to activate these later genes (Katikala et al. 2013; José-Edwards et al. 2015). In this way, the timing of gene expression in the notochord can be carefully regulated.

The presence of Macho-1 in the posterior vegetal cytoplasm causes those posterior cells that will become mesenchyme to respond differently to the FGF signal than do the cells that will form neural structures (**FIGURE 10.20**; Kobayashi et al. 2003). Macho-1 prevents notochord induction in the mesenchymal cell precursors by activating the *snail* gene (which will in turn suppress the gene for Brachyury). Thus, Macho-1 is not only a muscle-activating determinant, it is also a factor that distinguishes cell responses to the FGF signal. These FGF-responding cells do not become muscle because FGFs

Developing Questions

The tunicate nervous system is present during the larval stage but degenerates during metamorphosis. Consider the neural tube of a vertebrate such as fish and how it becomes divided into forebrain, midbrain, hindbrain, and spinal cord portions (see Chapter 13). How would one determine whether the neural tubes of tunicates and vertebrates are similar? Are they homologous or analogous (see Chapter 26)?

FIGURE 10.19 Simplified version of the gene network leading to notochord development in the early tunicate embryo. (A,B) Vegetal views of 32- and 64-cell *Ciona* embryos. (A) β-Catenin accumulation leads to expression of the *FoxD* gene. FoxD protein helps specify the cells to become endoderm and to secrete FGFs. (B) FGFs induce *Brachyury* expression in neighboring cells; these are the cells that will become the notochord. (C) Dorsal views. Brachyury activates regulators of cellular activity such as Prickle, which regulates cell polarity, leading to convergent extension of the notochord in the gastrula and neurula stages. (After Davidson and Christiaen 2006.)

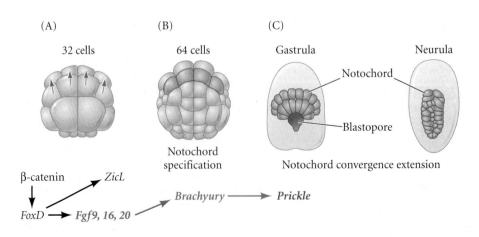

(A) 32 cells (B) 64 cells (C) Gastrula Neurula

Notochord specification

Notochord

Blastopore

Notochord convergence extension

β-catenin → ZicL
FoxD → Fgf9, 16, 20
Brachyury → Prickle

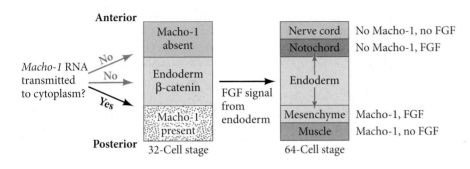

Anterior

Macho-1 RNA transmitted to cytoplasm?

No → Macho-1 absent / Endoderm β-catenin

No →

Yes → Macho-1 present

Posterior

32-Cell stage

FGF signal from endoderm →

Nerve cord	No Macho-1, no FGF
Notochord	No Macho-1, FGF
Endoderm	
Mesenchyme	Macho-1, FGF
Muscle	Macho-1, no FGF

64-Cell stage

FIGURE 10.20 The two-step process for specifying the marginal cells of the tunicate embryo. The first step involves the cells' acquisition (or nonacquisition) of the Macho-1 transcription factor. The second step involves the reception (or nonreception) of the FGF signal from the endoderm. (After Kobayashi et al. 2003.)

activate cascades that block muscle formation (a role for these factors that is conserved in vertebrates). As seen in Figure 10.21, the presence of Macho-1 changes the responses to endodermal FGFs, causing the anterior cells to form notochord while the posterior cells become mesenchyme.

WEB TOPIC 10.6 **GASTRULATION IN TUNICATES** Tunicate embryos follow the deuterostome pattern of gastrulation, but do so very differently than sea urchins.

Next Step Investigation

Sea urchin embryology can be said to have given birth to the field of immunology when in 1882 Elie Metchnikoff discovered the innate immune system of the sea urchin larva. Today the mechanisms by which the primitive larval immune system forms and mediates the growth of the larva in its environment are being given increasing study. And, in the field of evolution, the tunicate notochord represents a new cell type. How do new cell types come into being? This is a question much under study, and we will come back to it in Chapter 26.

Closing Thoughts on the Opening Photo

The cells of this tunicate larva are fluorescing because they express a labeled target gene for the Brachyury protein, thus identifying them as the cells that will form the notochord (see p. 330). In tunicate and vertebrate embryos, the notochord is a primitive backbone that instructs the ectodermal cells above it to become the neural tube. Thus the tunicate is an invertebrate (it has no spinal cord) chordate (it does have a notochord.) When Alexander Kowalevsky discovered this in 1866, Charles Darwin was thrilled, realizing that tunicates are an evolutionary link between the invertebrate and vertebrate phyla. Today the Brachyury transcription factor is known to be important in both tunicate and vertebrate notochord formation, thus providing molecular support for Kowalevsky's finding. (Image courtesy of J. H. Imai and A. Di Gregorio.)

10 Snapshot Summary
Early Development in Sea Urchins and Tunicates

1. In both sea urchins and tunicates, the blastopore becomes the anus and the mouth is formed elsewhere; this deuterostome mode of gastrulation is also characteristic of chordates (including vertebrates).

2. Sea urchin cleavage is radial and holoblastic. At fourth cleavage, however, the vegetal tier divides into large macromeres and small micromeres. The animal pole divides to form the mesomeres.

3. Sea urchin cell fates are determined both by autonomous and conditional modes of specification. The micromeres are specified autonomously and become a major signaling center for conditional specification of other lineages. Maternal β-catenin is important for the autonomous specification of the micromeres.

4. Differential cell adhesion is important in regulating sea urchin gastrulation. The micromeres detach first from

the vegetal plate and move into the blastocoel. They form the skeletogenic mesenchyme, which becomes the skeletal rods of the pluteus larva. The vegetal plate invaginates to form the endodermal archenteron, with a tip of non-skeletogenic mesenchyme cells. The archenteron elongates by convergent extension and is guided to the future mouth region by the non-skeletogenic mesenchyme.

5. The large micromeres become the skeleton of the larva; the small micromeres contribute to the coelomic pouches and the germ cells of the adult.

6. The micromeres regulate the fates of their neighboring cells through juxtacrine and paracrine pathways. They can convert animal cells into endoderm.

7. Gene regulatory networks are modules that work by logic circuits to integrate inputs into coherent cellular outputs. The micromeres integrate maternal components such that the placement of Disheveled at the vegetal pole enables the formation of β-catenin to help activate the *Pmar1* gene, whose products inhibit the *HesC* gene. The product of *HesC* inhibits skeletogenic genes. Thus, by locally inhibiting the inhibitor, the most vegetal cells become committed to skeleton production. This is called a double-negative gate.

8. The Nodal protein determines the oral-aboral axis of the sea urchin pluteus larva during early gastrulation. During late gastrulation and the early larval period, it specifies the right half of the larva.

9. The ingression of the skeletogenic mesenchyme is accomplished through an epithelial-mesenchymal transition in which these cells lose cadherins and gain affinity to adhere to the matrix within the blastocoel.

10. Archenteron invagination and growth are coordinated by cell shape changes, cell proliferation, and convergent extension. In the final stage in invagination, the tip of the archenteron is actively pulled to the blastocoel roof by the non-skeletogenic mesenchyme cells.

11. The adult sea urchin rudiment, called the imaginal rudiment, forms from the left coelomic pouch under the influence of BMPs.

12. The tunicate embryo divides holoblastically and bilaterally.

13. The yellow cytoplasm contains muscle-forming determinants that act autonomously. The heart and nervous system are formed conditionally by signaling interactions between blastomeres.

14. Macho-1 is the tunicate muscle determinant (a transcription factor that is sufficient to activate muscle-specifying genes). The notochord and mesenchyme are generated conditionally by paracrine factors such as FGFs.

15. FGFs induce expression of *Brachyury* in neighboring cells, inducing these cells to become the notochord.

16. The transcription factor Nodal appears to specify the left-right axis in tunicates, in which it is expressed solely on the left side of the body (which is also the case in snails and vertebrates).

Further Reading

Lemaire, P. 2009. Unfolding a chordate developmental program, one cell at a time. *Dev. Biol.* 332: 48–60.

Lyons, D. C., M. L. Martik, L. R. Saunders and D. R. McClay. 2014. Specification to biomineralization: Following a single cell type as it constructs a skeleton. *Integr. Comp. Biol.* 54: 723–733.

McClay, D. R. 2016. Sea urchin morphogenesis. *Curr. Top. Dev. Biol.* 117: 15–30.

Nishida, H. and K. Sawada. 2001. *Macho-1* encodes a localized mRNA in ascidian eggs that specifies muscle fate during embryogenesis. *Nature* 409: 724–729.

Peter, I. S. and E. H. Davidson. 2016. Implications of developmental gene regulatory networks inside and outside developmental biology. *Curr. Top. Dev. Biol.* 117: 237–252.

Revilla-i-Domingo, R., P. Oliveri and E. H. Davidson. 2010. A missing link in the sea urchin embryo gene regulatory network: *hesC* and the double-negative specification of micromeres. *Proc. Natl. Acad. Sci. USA* 104: 12383–12388.

Saunders, L. R. and D. R. McClay. 2014. Sub-circuits of a gene regulatory network control a developmental epithelial-mesenchymal transition. *Development* 141: 1503–1513.

Wu, S. Y., M. Ferkowicz and D. R. McClay. 2010. Ingression of primary mesenchyme cells of the sea urchin embryo: A precisely timed epithelial-mesenchymal transition. *Birth Def. Res. C Embryol. Today* 81: 241–252.

GO TO WWW.DEVBIO.COM...

...for Web Topics, Scientists Speak interviews, Watch Development videos, Dev Tutorials, and complete bibliographic information for all literature cited in this chapter.

Amphibians and Fish

DESPITE VAST DIFFERENCES IN ADULT MORPHOLOGY, early development in each of the vertebrate groups is very similar. Fish and amphibians are among the most easily studied vertebrates. In both cases, hundreds of eggs are laid externally and fertilized simultaneously. Fish and amphibians are **anamniotic** vertebrates (**FIGURE 11.1**), meaning that they do not form the amnion that permits embryonic development to take place on dry land. However, developing amphibians and fish employ many of the same processes and genes used by other vertebrates (including humans) to generate body axes and organs.

Early Amphibian Development

Amphibian embryos once dominated the field of experimental embryology. With their large cells and rapid development, salamander and frog embryos were excellently suited for transplantation experiments. However, amphibian embryos fell out of favor during the early days of developmental genetics, in part because these animals undergo a long period of growth before they become fertile and because their chromosomes are often found in several copies, precluding easy mutagenesis. But with the advent of molecular techniques such as in situ hybridization, antisense oligonucleotides, chromatin immunoprecipitation, and dominant-negative proteins, researchers have returned to the study of amphibian embryos and have been able to integrate their molecular analyses with earlier experimental findings. The results have been spectacular, revealing new vistas of how vertebrate

This zebrafish embryo has two body axes. How can this happen, and what are some implications for vertebrate development?

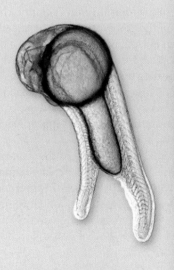

The Punchline

Amphibians have played major roles in experimental embryology since the earliest days of the field. In the frog, fertilization induces protein movements in the egg cytoplasm, leading to the accumulation of β-catenin in the egg's future dorsal region. In this future dorsal portion of the embryo, β-catenin activates genes that establish a crucial embryonic structure dubbed "the organizer." The organizer secretes proteins that block the activities of paracrine factors (mainly BMPs) that would ventralize the mesoderm and turn ectoderm into epidermis. As a result, the ectoderm adjacent to the organizer become specified as neural tissue. Inhibition of Wnt signaling is critical to successful formation of the head, and Wnt protein inhibitors are produced in the anterior portion of the organizer tissues. Although the cleavage and gastrulation patterns of fish are very different from those of frogs, the same pattern of axis-specifying gene activity appears to be working in both groups of vertebrates.

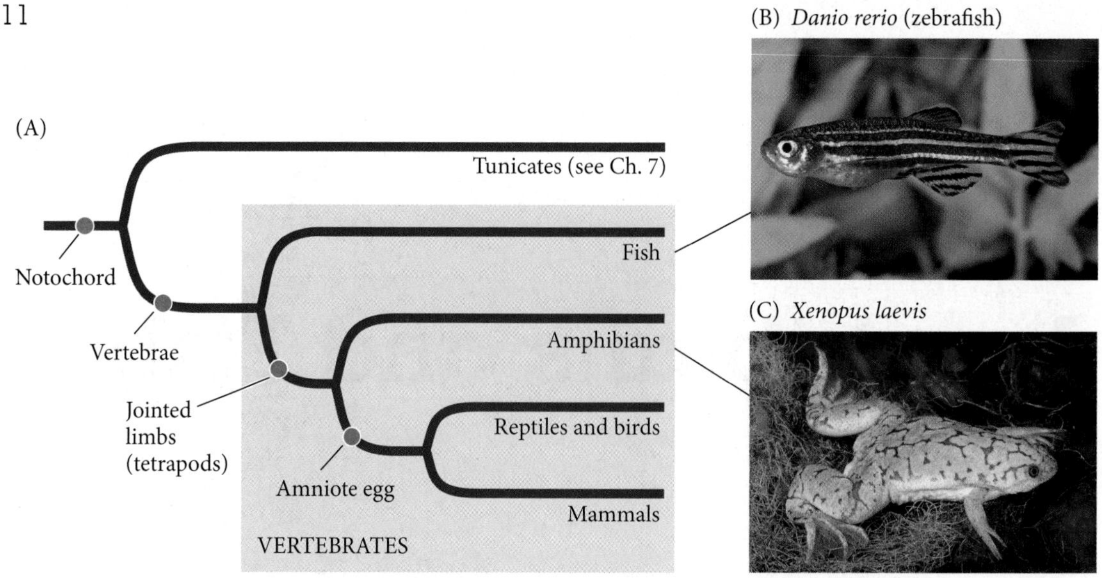

(A)

(B) *Danio rerio* (zebrafish)

Tunicates (see Ch. 7)

Notochord

Fish

Vertebrae

Amphibians

Jointed
limbs
(tetrapods)

Reptiles and birds

Amniote egg

Mammals

VERTEBRATES

(C) *Xenopus laevis*

FIGURE 11.1 (A) Phylogenetic tree of the chordates showing the relationship of the vertebrate groups. The embryonic development of fish and amphibians must be carried out in moist environments. The evolution of the shelled amniote egg permitted development to proceed on dry land for the reptiles and their descendants, as we will see in Chapter 12. (B) The zebrafish (*Danio rerio*) has become a popular model organism for the study of development. It is the first vertebrate species to be subjected to mutagenesis studies similar to those that have been carried out on *Drosophila*. (C) *Xenopus laevis*, the African clawed frog, is one of the most studied amphibians because it has the rare property of not having a breeding season and thus can generate embryos year-round. (B courtesy of D. McIntyre; C © Michael Redmer/Visuals Unlimited.)

VADE MECUM

The Vade Mecum segment "Amphibians" has video of the events of frog development from fertilization through the metamorphosed adult, using three-dimensional models as well as living *Xenopus*.

---

bodies are patterned and structured. As Jean Rostand wrote in 1960, "Theories come and theories go. The frog remains."

WATCH DEVELOPMENT 11.1 A two-minute video provides a speedy overview of frog development from fertilization to larvae.

Fertilization, Cortical Rotation, and Cleavage

Most frogs have external fertilization, with the male fertilizing the eggs as the female is laying them. Even before fertilization, the frog egg has polarity, in that the dense yolk is at the vegetal (bottom) end, whereas the animal part of the egg (the upper half) has very little yolk. As we will also see, certain proteins and mRNAs are already localized in specific regions of the unfertilized egg.

Fertilization can occur anywhere in the animal hemisphere of the amphibian egg. The point of sperm entry is important because it determines dorsal-ventral polarity. The point of sperm entry marks the ventral (belly) side of the embryo, while the site 180° opposite the point of sperm entry will mark the dorsal (spinal) side. The sperm centriole, which enters the egg with the sperm nucleus, organizes the microtubules of the egg into parallel tracks in the vegetal cytoplasm, separating the outer cortical cytoplasm from the yolky internal cytoplasm (**FIGURE 11.2A,B**). These microtubular tracks allow the cortical cytoplasm to rotate with respect to the inner cytoplasm. Indeed, these parallel arrays are first seen immediately before rotation starts, and they disappear when rotation ceases (Elinson and Rowning 1988; Houliston and Elinson 1991).

In the zygote, the cortical cytoplasm rotates about 30° with respect to the internal cytoplasm (**FIGURE 11.2C**). In some cases, this exposes a region of gray-colored inner cytoplasm directly opposite the sperm entry point (**FIGURE 11.2D**; Roux 1887; Ancel and Vintenberger 1948). This region, the **gray crescent**, is where gastrulation will begin. Even in *Xenopus* eggs, which do not expose a gray crescent, cortical rotation occurs and

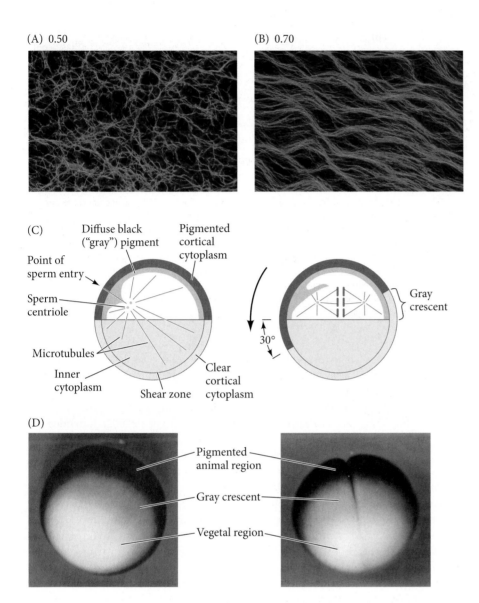

(A) 0.50

(B) 0.70

(C)

Diffuse black ("gray") pigment

Pigmented cortical cytoplasm

Point of sperm entry

Sperm centriole

Microtubules

Inner cytoplasm

Shear zone

Clear cortical cytoplasm

30°

Gray crescent

(D)

Pigmented animal region

Gray crescent

Vegetal region

FIGURE 11.2 Reorganization of the cytoplasm and cortical rotation produce the gray crescent in frog eggs. (A,B) Parallel arrays of microtubules (visualized here using fluorescent antibodies to tubulin) form in the vegetal hemisphere of the egg, along the future dorsal-ventral axis. (A) With the first cell cycle 50% complete, microtubules are present, but they lack polarity. (B) By 70% completion, the vegetal shear zone is characterized by a parallel array of microtubules; cortical rotation begins at this time. At the end of rotation, the microtubules will depolymerize. (C) Schematic cross section of cortical rotation. At left, the egg is shown midway through the first cell cycle. It has radial symmetry around the animal-vegetal axis. The sperm nucleus has entered at one side and is migrating inward. At right, 80% into first cleavage, the cortical cytoplasm has rotated 30° relative to the internal cytoplasm. Gastrulation will begin in the gray crescent—the region opposite the point of sperm entry, where the greatest displacement of cytoplasm occurs. (D) Gray crescent of *Rana pipiens*. Immediately after cortical rotation (left), lighter gray pigmentation is exposed beneath the heavily pigmented cortical cytoplasm. The first cleavage furrow (right) bisects this gray crescent. (A,B from Cha and Gard 1999, courtesy of the authors; C after Gerhart et al. 1989; D courtesy of R. P. Elinson.)

cytoplasmic movements can be seen (Manes and Elinson 1980; Vincent et al. 1986). Gastrulation begins at the part of the egg opposite the point of sperm entry, and this region will become the dorsal portion of the embryo. The microtubular arrays organized by the sperm centriole at fertilization plays a large role in initiating these movements. Therefore, the dorsal-ventral axis of the larva can be traced back to the point of sperm entry.

WATCH DEVELOPMENT 11.2   See the microtubular arrays cortical rotation during the first cleavage of *Xenopus* eggs.

Unequal radial holoblastic cleavage

Cleavage in most frog and salamander embryos is radially symmetrical and holoblastic, like echinoderm cleavage. The amphibian egg, however, is much larger than echinoderm eggs and contains much more yolk. This yolk is concentrated in the vegetal hemisphere and is an impediment to cleavage. Thus, the first division begins at the animal pole and slowly extends down into the vegetal region (**FIGURE 11.3A**). In those species having a gray crescent (especially salamanders and frogs of the genus *Rana*), the first cleavage usually bisects the gray crescent (see Figure 11.2D).

(A)

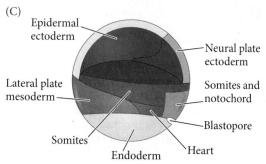

Animal

1 | 2 1 | 2 3

1

Vegetal

(B)

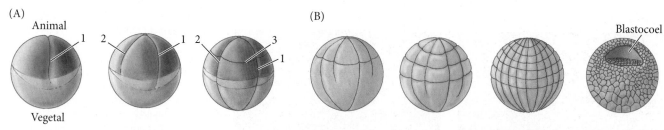

Blastocoel

(C)

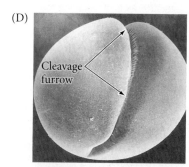

Epidermal
ectoderm

Neural plate
ectoderm

Lateral plate
mesoderm

Somites and
notochord

Blastopore

Somites

Heart

Endoderm

FIGURE 11.3 Cleavage of a *Xenopus* egg. (A) The first three cleavage furrows, numbered in order of appearance. Because the vegetal yolk impedes cleavage, the second division begins in the animal region of the egg before the first division has divided the vegetal cytoplasm completely. The third division is displaced toward the animal pole. (B) As cleavage progresses, the vegetal hemisphere ultimately contains larger and fewer blastomeres than the animal hemisphere. The final drawing shows a cross section through a mid-blastula stage embryo. (C) Fate map of the *Xenopus* embryo superimposed on the mid-blastula stage. (D) SEMs of the first, second, and fourth cleavages. Note the size discrepancies of the animal and vegetal cells after third cleavage. (A,B after Carlson 1981; C after Lane and Smith 1999 and Newman and Kreig 1999; D from Beams and Kessel 1976, courtesy of the authors and L. Biedler.)

(D)

Cleavage
furrow

While the first cleavage furrow is still cleaving the yolky cytoplasm of the vegetal hemisphere, the second cleavage has already started near the animal pole. This cleavage is at right angles to the first one and is also meridional. The third cleavage is equatorial. However, because of the vegetally placed yolk, the third cleavage furrow is not at the equator but is displaced toward the animal pole (Valles et al. 2002). It divides the amphibian embryo into four small animal blastomeres (micromeres) and four large blastomeres (macromeres) in the vegetal region. Despite their unequal sizes, the blastomeres continue to divide at the same rate until the twelfth cell cycle (with only a small delay of the vegetal cleavages). As cleavage progresses, the animal region becomes packed with numerous small cells, while the vegetal region contains a relatively small number of large, yolk-laden macromeres. An amphibian embryo containing 16 to 64 cells is commonly called a **morula** (plural *morulae*; Latin, "mulberry," whose shape it vaguely resembles). At the 128-cell stage, the blastocoel becomes apparent, and the embryo is considered a blastula (**FIGURE 11.3B**).

Numerous cell adhesion molecules keep the cleaving blastomeres together. One of the most important of these is EP-cadherin. The mRNA for this protein is supplied in the oocyte cytoplasm. If this message is destroyed by antisense oligonucleotides so that no EP-cadherin protein is made, the adhesion between blastomeres is dramatically reduced, resulting in the obliteration of the blastocoel (Heasman et al. 1994a,b). Membrane adhesions may have a further role; it is probable that the coordination of cell division is mediated through waves of membrane contractions (Chang and Ferrell 2013).

Although amphibian development differs from species to species (see Hurtado and De Robertis 2007), in general the animal hemisphere cells will give rise to the ectoderm, the vegetal cells will give rise to the endoderm, and the cells beneath the blastocoel cavity will become mesoderm (**FIGURE 11.3C**). The cells opposite the point of sperm entry will become the neural ectoderm, the notochord mesoderm, and the pharyngeal (head) endoderm (Keller 1975, 1976; Landstrom and Løvtrup 1979).

The amphibian blastocoel serves two major functions. First, it can change its shape such that cell migration can occur during gastrulation; and second, it prevents the cells beneath it from interacting prematurely with the cells above it. When Nieuwkoop (1973) took embryonic newt cells from the roof of the blastocoel in the animal hemisphere (a region often called the **animal cap**) and placed them next to the yolky vegetal cells from the base of the blastocoel, the animal cap cells differentiated into mesodermal tissue instead of ectoderm. Thus, the blastocoel prevents the premature contact of the vegetal cells with the animal cap cells, and keeps the animal cap cells undifferentiated.

The mid-blastula transition: Preparing for gastrulation

An important precondition for gastrulation is the activation of the zygotic genome (that is, the genes within each nucleus of the embryo). In *Xenopus laevis*, only a few genes appear to be transcribed during early cleavage. For the most part, nuclear genes are not activated until late in the twelfth cell cycle (Newport and Kirschner 1982a,b; Yang et al. 2002). At that time, the embryo experiences a **mid-blastula transition**, or **MBT**, as different genes begin to be transcribed in different cells, the cell cycle acquires gap phases, and the blastomeres acquire the capacity to become motile. It is thought that some factor in the egg is being absorbed by the newly made chromatin because (as in *Drosophila*) the time of this transition can be changed experimentally by altering the ratio of chromatin to cytoplasm in the cell (Newport and Kirschner 1982a,b).

Some of the events that trigger the mid-blastula transition involve chromatin modification. First, certain promoters are demethylated, allowing transcription of these genes. During the late blastula stages, there is a loss of methylation on the promoters of genes that are activated at MBT. This demethylation is not seen on promoters that are not activated at MBT, nor is it observed in the coding regions of MBT-activated genes. The methylation of lysine-4 on histone H3 (forming a trimethylated lysine associated with active transcription) is also seen on the 5′ ends of many genes during MBT. It appears, then, that modification of certain promoters and their associated nucleosomes may play a pivotal role in regulating the timing of gene expression at the mid-blastula transition (Stancheva et al. 2002; Akkers et al. 2009; Hontelez et al. 2015).

It is thought that once the chromatin at the promoters has been remodeled, various transcription factors (such as the VegT protein, formed in the vegetal cytoplasm from localized maternal mRNA) bind to the promoters and initiate new transcription. For instance, the vegetal cells (under the direction of the VegT protein) become the endoderm and begin secreting factors that induce the cells above them to become the mesoderm (see Figure 11.10).

Amphibian Gastrulation

The study of amphibian gastrulation is both one of the oldest and one of the newest areas of experimental embryology (see Beetschen 2001; Braukmann and Gilbert 2005). Even though amphibian gastrulation has been studied extensively since the 1870s, new live-imaging techniques have emerged that are giving us a new appreciation of the intricacies of these cell movements (Papan et al. 2007; Moosmann et al. 2013). Moreover, most of our theories concerning the mechanisms of gastrulation and axis specification have been revised over the past two decades. The study of these developmental movements has also been complicated by the fact that there is no single way amphibians gastrulate; different species employ different means to achieve the same goal. In recent years, the most intensive investigations have focused on *Xenopus laevis*, so we will concentrate on the mechanisms of gastrulation in that species.

WATCH DEVELOPMENT 11.3   See *Xenopus* gastrulation from the outside, looking at the blastopore lip.

Vegetal rotation and the invagination of the bottle cells

Amphibian blastulae are faced with the same tasks as the invertebrate blastulae we followed in Chapters 8 through 10—namely, to bring inside the embryo those areas destined to form the endodermal organs; to surround the embryo with cells capable of forming the ectoderm; and to place the mesodermal cells in the proper positions between the ectoderm and the endoderm. The cell movements of gastrulation that will accomplish this are initiated on the future dorsal side of the embryo, just below the equator, in the region of the gray crescent (i.e., the region opposite the point of sperm entry; see Figure 11.2C). Here the cells invaginate to form the slitlike blastopore. These **bottle cells** change their shape dramatically. The main body of each cell is displaced toward the inside of the embryo while maintaining contact with the outside surface by way of a slender neck. As in sea urchins, the bottle cells will initiate the formation

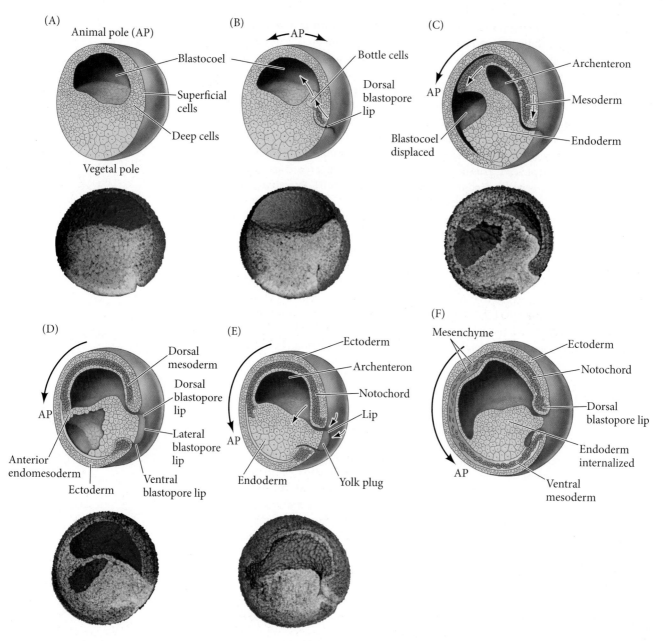

FIGURE 11.4 Cell movements during frog gastrulation. The drawings show meridional sections cut through the middle of the embryo and positioned so that the vegetal pole is tilted toward the observer and slightly to the left. The major cell movements are indicated by arrows, and the superficial animal hemisphere cells are colored so that their movements can be followed. Below the drawings are corresponding micrographs imaged with a surface imaging microscope (see Ewald et al. 2002). (A,B) Early gastrulation. The bottle cells of the margin move inward to form the dorsal lip of the blastopore, and the mesodermal precursors involute under the roof of the blastocoel. AP marks the position of the animal pole, which

will change as gastrulation continues. (C,D) Mid-gastrulation. The archenteron forms and displaces the blastocoel, and cells migrate from the lateral and ventral lips of the blastopore into the embryo. The cells of the animal hemisphere migrate down toward the vegetal region, moving the blastopore to the region near the vegetal pole. (E,F) Toward the end of gastrulation, the blastocoel is obliterated, the embryo becomes surrounded by ectoderm, the endoderm has been internalized, and the mesodermal cells have been positioned between the ectoderm and endoderm. (Drawings after Keller 1986; micrographs courtesy of Andrew Ewald and Scott Fraser.)

of the archenteron (primitive gut).[1] However, unlike sea urchins, gastrulation in the frog begins not in the most vegetal region but in the **marginal zone**—the region surrounding the equator of the blastula, where the animal and vegetal hemispheres meet (**FIGURE 11.4A,B**). Here the endodermal cells are not as large or as yolky as the most vegetal blastomeres.

But cell involution is not a passive event. At least 2 hours before the bottle cells form, internal cell rearrangements propel the cells of the dorsal floor of the blastocoel toward the animal cap. This **vegetal rotation** places the prospective pharyngeal endoderm cells adjacent to the blastocoel and immediately above the involuting mesoderm (see Figure 11.5D). These cells then migrate along the basal surface of the blastocoel roof, traveling toward the future anterior of the embryo (**FIGURE 11.4C–E**; Nieuwkoop and Florschütz 1950; Winklbauer and Schürfeld 1999; Ibrahim and Winklbauer 2001). The superficial layer of marginal cells is pulled inward to form the endodermal lining of the archenteron, merely because it is attached to the actively migrating deep cells. Although experimentally removing the bottle cells does not affect the involution of the deep or superficial marginal zone cells into the embryo, removal of the deep **involuting marginal zone (IMZ)** cells stops archenteron formation.

INVOLUTION AT THE BLASTOPORE LIP   After the bottle cells have brought the involuting marginal zone into contact with the blastocoel wall, the IMZ cells involute into the embryo. As the migrating marginal cells reach the lip of the blastopore, they turn inward and travel along the inner surface of the outer animal hemisphere cells (i.e., the blastocoel roof; **FIGURE 11.4D–F**). The order of the march into the embryo is determined by the vegetal rotation that abuts the prospective pharyngeal endoderm against the inside of the animal cap tissue (Winklebauer and Damm 2011). Meanwhile, the animals cells undergo epiboly, producing a stream of cells that converge at and become the **dorsal blastopore lip**.

The first cells to compose the dorsal blastopore lip and enter into the embryo are the cells of the prospective pharyngeal endoderm of the foregut (**FIGURE 11.5**). These cells migrate anteriorly beneath the surface ectoderm of the blastocoel.[2] These anterior endoderm cells transcribe the *hhex* gene, which encodes a transcription factor that is critical for forming the head and heart (Rankin et al. 2011). As these first cells pass into the interior of the embryo, the dorsal blastopore lip becomes composed of cells that involute into the embryo to become the **prechordal plate**, the precursor of the head mesoderm. Prechordal plate cells transcribe the *goosecoid* gene, whose product is a transcription factor that activates numerous genes controlling head formation. It achieves this activation indirectly, *by repressing those genes (e.g., Wnt8) that repress head development.* This phenomenon—the activation of genes by repressing their repressors—is a major feature of animal development, as we saw in the double-negative gate that specifies sea urchin micromeres.

---

[1] Ray Keller and his students showed that the peculiar shape change of the bottle cells is needed to initiate gastrulation in *Xenopus*. It is the constriction of these cells that forms the blastopore, and it brings subsurface marginal cells into contact with the basal region of the surface blastomeres. Once this contact is made, the marginal cells begin to migrate along the extracellular matrix on the basal region of these surface cells. When such involution movements are underway, the bottle cells are no longer essential. At this point, they have done their job and can be removed without stopping gastrulation (Keller 1981; Hardin and Keller 1988). Thus, in *Xenopus*, the major factor in the movement of cells into the embryo appears to be the involution of the subsurface cells rather than the invagination of superficial marginal bottle cells.

[2] The pharyngeal endoderm and head mesoderm cannot be separated experimentally at this stage, so they are sometimes referred to collectively as the head endomesoderm. The notochord is the basic unit of the dorsal mesoderm, but it is thought that the dorsal portion of the somites may have similar properties to the notochord. X-ray phase-contrast and in vitro microtomography have shown new details of gastrulation that have yet to be integrated into molecular models (Moosmann et al. 2013).

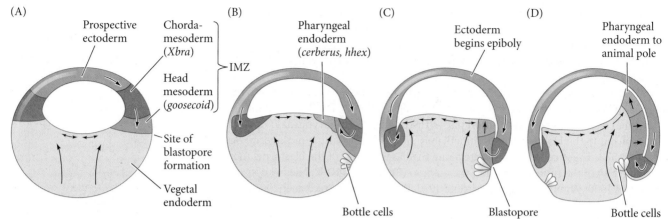

FIGURE 11.5 Early movements of *Xenopus* gastrulation. (A) At the beginning of gastrulation, the involuting marginal zone (IMZ) forms. Pink represents the prospective head mesoderm (*goosecoid* expression). Chordamesoderm (*Xbra* expression) is red. (B) Vegetal rotation (arrows) pushes the prospective pharyngeal endoderm (orange; specified by *hhex* and *cerberus* expression) to the side of the blastocoel. (C,D) The vegetal endoderm (yellow) movements push the pharyngeal endoderm forward, driving the mesoderm passively into the embryo and toward the animal pole. The ectoderm (blue) begins epiboly. (After Winklbauer and Schürfeld 1999.)

Next to involute through the dorsal blastopore lip are the cells of the **chordamesoderm**. These cells will form the **notochord**, the transient mesodermal rod that plays an important role in inducing and patterning the nervous system. Chordamesoderm cells express the *Xbra* (*Brachyury*) gene, whose product (as we saw in the previous chapter) is a transcription factor critical for notochord formation. Thus, the cells constituting the dorsal blastopore lip are constantly changing as the original cells migrate into the embryo and are replaced by cells migrating downward, inward, and upward.

As the new cells enter the embryo, the blastocoel is displaced to the side opposite the dorsal lip. Meanwhile, the lip expands laterally and ventrally as bottle cell formation and involution continue around the blastopore. The widening blastopore "crescent" develops lateral lips and, finally, a ventral lip over which additional mesodermal and endodermal precursor cells pass (**FIGURE 11.6**). These cells include the precursors of the heart and kidney. With the formation of the ventral lip, the blastopore has formed a ring around the large endodermal cells that remain exposed on the vegetal surface.

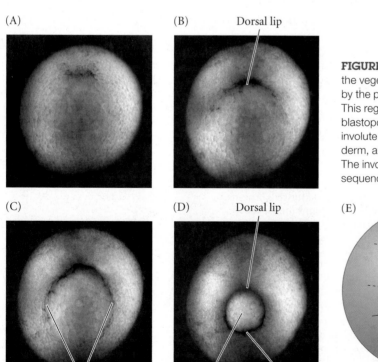

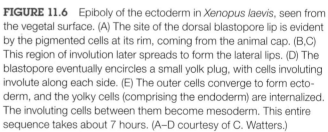

FIGURE 11.6 Epiboly of the ectoderm in *Xenopus laevis*, seen from the vegetal surface. (A) The site of the dorsal blastopore lip is evident by the pigmented cells at its rim, coming from the animal cap. (B,C) This region of involution later spreads to form the lateral lips. (D) The blastopore eventually encircles a small yolk plug, with cells involuting involute along each side. (E) The outer cells converge to form ectoderm, and the yolky cells (comprising the endoderm) are internalized. The involuting cells between them become mesoderm. This entire sequence takes about 7 hours. (A–D courtesy of C. Watters.)

This remaining patch of endoderm is called the **yolk plug**; it, too, is eventually internalized (at the site of the anus). At that point, all the endodermal precursors have been brought into the interior of the embryo, the ectoderm has encircled the surface, and the mesoderm has been brought between them. The first cells into the blastopore become the most anterior.

WATCH DEVELOPMENT 11.4 Watch the video by Dr. Christopher Watters from which the photographs in Figure 11.6 were taken.

CONVERGENT EXTENSION OF THE DORSAL MESODERM Figure 11.7 depicts the behavior of the involuting marginal zone cells at successive stages of *Xenopus* gastrulation (Keller and Schoenwolf 1977; Hardin and Keller 1988). The IMZ is originally several layers thick. Shortly before their involution through the blastopore lip, the several layers of deep IMZ cells intercalate radially to form one thin, broad layer. This intercalation further extends the IMZ vegetally (**FIGURE 11.7A**). At the same time, the superficial cells spread out by dividing and flattening. When the deep cells reach the blastopore lip, they involute into the embryo and initiate a second type of intercalation. This intercalation causes a convergent extension along the mediolateral axis that integrates several mesodermal streams to form a long, narrow band (**FIGURE 11.7B**). The anterior part of this band migrates toward the animal cap. Thus, the mesodermal stream continues to migrate toward the animal pole, and the overlying layer of superficial cells (including the bottle cells) is passively pulled toward the animal pole, thereby forming the endodermal roof of the archenteron (see Figures 11.4 and 11.7). The radial and mediolateral intercalations of the deep layer of cells appear to be responsible for the continued movement of mesoderm into the embryo.

Several forces appear to drive convergent extension. The first force is a polarized cell cohesion, wherein the involuted mesodermal cells send out protrusions to contact one another. These "reachings out" are not random, but occur toward the midline of the embryo and require an extracellular fibronectin matrix (Goto et al. 2005; Davidson et al. 2008). These intercalations, both mediolaterally and radially, are stabilized by the planar cell polarity (PCP) pathway that is initiated by Wnts (Jessen et al. 2000; Shindo and Wallingford 2014; Ossipova et al. 2015). The second force is differential cell cohesion. During gastrulation, the genes encoding the adhesion proteins **paraxial protocadherin** and **axial protocadherin** become expressed specifically in the paraxial (somiteforming; see Chapter 17) mesoderm and the notochord, respectively. An experimental dominant-negative form of axial protocadherin prevents the presumptive notochord cells from sorting out from the paraxial mesoderm and blocks normal axis formation.

FIGURE 11.7 *Xenopus* gastrulation continues. (A) The deep marginal cells flatten, and the formerly superficial cells form the wall of the archenteron. (B) Radial intercalation, looking down at the dorsal blastopore lip from the dorsal surface. In the noninvoluting marginal zone (NIMZ) and the upper portion of the IMZ, deep (mesodermal) cells are intercalating radially to make a thin band of flattened cells. This thinning of several layers into a few causes convergent extension (white arrows) toward the blastopore lip. Just above the lip, mediolateral intercalation of the cells produces stresses that pull the IMZ over the lip. After involuting over the lip, mediolateral intercalation continues, elongating and narrowing the axial mesoderm. (After Wilson and Keller 1991 and Winklbauer and Schürfeld 1999.)

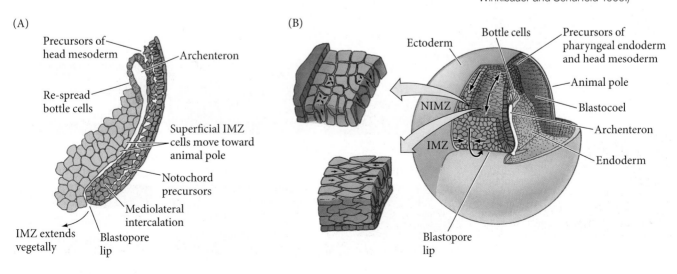

A dominant-negative paraxial protocadherin (which is secreted instead of being bound to the cell membrane) prevents convergent extension[3] (Kim et al. 1998; Kuroda et al. 2002). Moreover, the expression domain of paraxial protocadherin characterizes the trunk mesodermal cells, which undergo convergent extension, distinguishing them from the head mesodermal cells, which do not undergo convergent extension. A third factor regulating convergent extension is calcium flux. Wallingford and colleagues (2001) found that dramatic waves of calcium ions (Ca^{2+}) surge across the dorsal tissues undergoing convergent extension, causing waves of contraction within the tissue. Ca^{2+} is released from intracellular stores and is required for convergent extension. If Ca^{2+} release is blocked, normal cell specification still occurs, but the dorsal mesoderm neither converges nor extends. These findings support a model for convergent extension wherein regulatory proteins cause changes in the outer surface of the tissue and generate mechanical traction forces that either prevent or encourage cell migration (Beloussov et al. 2006; Davidson et al. 2008; Kornikova et al. 2009).

Those mesodermal cells entering through the dorsal lip of the blastopore give rise to the central dorsal mesoderm (notochord and somites), while the remainder of the body mesoderm (which forms the heart, kidneys, bones, and parts of several other organs) enters through the ventral and lateral blastopore lips to create the **mesodermal mantle**. The endoderm is derived from the superficial cells of the involuting marginal zone that form the lining of the archenteron roof and from the subblastoporal vegetal cells that become the archenteron floor (Keller 1986). The remnant of the blastopore—where the endoderm meets the ectoderm—now becomes the anus. As gastrulation expert Ray Keller famously remarked, "Gastrulation is the time when a vertebrate takes its head out of its anus."

WEB TOPIC 11.1 **MIGRATION OF THE MESODERMAL MANTLE** Different growth rates coupled with the intercalation of cell layers allows the mesoderm to expand in a tightly coordinated fashion.

Epiboly of the prospective ectoderm

During gastrulation, the animal cap and noninvoluting marginal zone (NIMZ) cells expand by epiboly to cover the entire embryo (see Figure 11.7B). These cells will form the surface ectoderm. One important mechanism of epiboly in *Xenopus* gastrulation appears to be an increase in cell number (through division) coupled with a concurrent integration of several deep layers into one (**FIGURE 11.8**; Keller and Schoenwolf 1977; Keller and Danilchik 1988; Saka and Smith 2001). A second mechanism of *Xenopus*

---

[3] Dominant-negative proteins are mutated forms of the wild-type protein that interfere with the normal functioning of the wild-type protein. Thus, a dominant-negative protein will have an effect similar to a loss-of-function mutation in the gene that encodes the protein.

FIGURE 11.8 Epiboly of the ectoderm is accomplished by cell division and intercalation. Scanning electron micrographs of the *Xenopus* blastocoel roof, showing the changes in cell shape and arrangement. Stages 8 and 9 are blastulae; stages 10–11.5 represent progressively later gastrulae. (From Keller 1980, courtesy of R. E. Keller.)

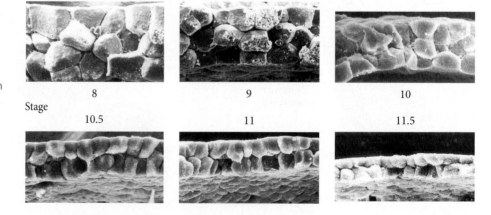

Stage

8   9   10

10.5   11   11.5

epiboly involves the assembly of fibronectin into fibrils by the blastocoel roof. This fibrillar fibronectin is critical in allowing the vegetal migration of the animal cap cells and enclosure of the embryo (Rozario et al. 2009). In *Xenopus* and many other amphibians, it appears that the involuting mesodermal precursors migrate toward the animal pole by traveling on an extracellular lattice of fibronectin secreted by the presumptive ectoderm cells of the blastocoel roof (**FIGURE 11.9A,B**).

Confirmation of fibronectin's importance for the involuting mesoderm came from experiments with a chemically synthesized peptide fragment that was able to compete with fibronectin for the binding sites of embryonic cells (Boucaut et al. 1984). If fibronectin were essential for cell migration, then cells binding this synthesized peptide fragment instead of extracellular fibronectin should stop migrating. Unable to find their "road," these prospective mesodermal cells should cease involution. That is precisely what happened, and the mesodermal precursors remained outside the embryo, forming a convoluted cell mass (**FIGURE 11.9C,D**). Thus, the fibronectin-containing extracellular matrix appears to provide both a substrate for adhesion as well as cues for the direction of cell migration.

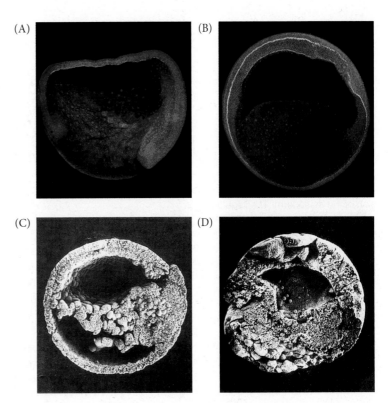

FIGURE 11.9 Fibronectin and amphibian gastrulation. (A,B) Sagittal section of *Xenopus* embryos at early (A) and late (B) gastrulation. The fibronectin lattice on the blastocoel roof is identified by fluorescent antibody labeling (yellow), while the embryonic cells are counterstained red. (C) Scanning electron micrograph of a normal salamander embryo injected with a control solution at the blastula stage. (D) Salamander embryo at the same stage but injected with a synthesized cell-binding fragment that competes with fibronectin. The archenteron has failed to form, and the mesodermal precursors have not undergone involution and remain on the embryo's surface. (A,B from Marsden and DeSimone 2001, photographs courtesy of the authors; C,D from Boucaut et al. 1984, courtesy of J.-C. Boucaut and J.-P. Thiery.)

Progressive Determination of the Amphibian Axes

Specification of the germ layers

As we have seen, the unfertilized amphibian egg has polarity along the animal-vegetal axis, and the germ layers can be mapped onto the oocyte even before fertilization. The animal hemisphere blastomeres will become the cells of the ectoderm (skin and nerves); the vegetal hemisphere cells become the cells of the gut and associated organs (endoderm); and the equatorial cells form the mesoderm (bone, muscle, heart). This general fate map is thought to be imposed on the embryo by the vegetal cells, which have two major functions: (1) to differentiate into endoderm and (2) to induce the cells immediately above them to become mesoderm.

The mechanism for this "bottom-up" specification of the frog embryo resides in a set of mRNAs that are tethered to the vegetal cortex. This includes the mRNA for the transcription factor **VegT**, which becomes apportioned to the vegetal cells during cleavage. VegT is critical in generating both the endodermal and mesodermal lineages. When VegT transcripts are destroyed by antisense oligonucleotides, the entire embryo becomes epidermis, with no mesodermal or endodermal components (Zhang et al. 1998; Taverner et al. 2005). *VegT* mRNA is translated shortly after fertilization. Its product activates a set of genes prior to the mid-blastula transition. One of the genes activated by this VegT protein encodes the Sox17 transcription factor. Sox17, in turn, is critical for activating the genes that specify cells to be endoderm. Thus, the fate of the vegetal cells is to become endodermal.

Another set of early genes activated by VegT encodes Nodal paracrine factors that instruct the cell layers *above* them to become mesoderm (Skirkanich et al. 2011). Nodal secreted from the vegetal cells in the nascent endoderm and signal the cells above them to accumulate phosphorylated Smad2. Phosphorylated Smad2 helps activate the *eomesodermin* and *Brachyury* (*Xbra*) genes in those cells, causing the cells to become specified as mesoderm. The Eomesodermin and Smad2 proteins working together can activate the zygotic genes for the VegT proteins, thus creating a positive feedforward loop that

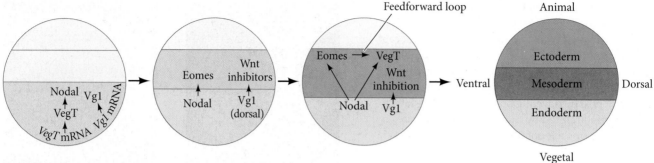

FIGURE 11.10 Model for the specification of the mesoderm. The vegetal region of the oocyte has accumulated mRNA for the transcription factor VegT and (in the future dorsal region) mRNA for the Nodal paracrine factor Vg1. At the late blastula stage, the *Vg1* mRNA is translated and Vg1 induces the future dorsal mesoderm to transcribe the genes for several Wnt antagonists (such as Dickkopf). The *VegT* message is also translated, and VegT activates nuclear genes encoding Nodal proteins. These TGF-β family members activate the expression of the transcription factor Eomesodermin in the presumptive mesoderm. Eomesodermin, with the help of activated Smad2 from the Nodal proteins, activates nuclear genes encoding VegT. In this way, VegT expression has gone from maternal mRNAs in the presumptive endoderm to nuclear expression in the presumptive mesoderm. (After Fukuda et al. 2010.)

is critical in sustaining the mesoderm (**FIGURE 11.10**). In the absence of such induction, cells become ectoderm (Fukuda et al. 2010).

In addition, the *Vg1* mRNA that has been stored in the vegetal cytoplasm is also translated. The production of Vg1 (another Nodal-like protein) is needed to activate other genes in the dorsal mesoderm. If either Nodal or Vg1 signaling is blocked, there is little or no mesoderm induction (Kofron et al. 1999; Agius et al. 2000; Birsoy et al. 2006). Thus, by the late blastula stage, the fundamental germ layers are becoming specified. The vegetal cells are specified as endoderm through transcription factors such as Sox17. The equatorial cells are specified as mesoderm by transcription factors such as Eomesodermin. And the animal cap—which has not begun receiving signals yet—becomes specified as ectoderm (see Figure 11.10).

The dorsal-ventral and anterior-posterior axes

Although animal-vegetal polarity initiates specification of the germ layers, the anterior-posterior, dorsal-ventral, and left-right axes are specified by events triggered at fertilization but not realized until gastrulation. In *Xenopus* (and in other amphibians), the formation of the anterior-posterior axis is inextricably linked to the formation of the dorsal-ventral axis. This, as we will see, is predicated on fertilization events that will place the transcription factor β-catenin in the region of the egg opposite the point of sperm entry and will specify that region of the egg to be the dorsal region of the embryo.

Once β-catenin is localized in this region of the egg, the cells containing β-catenin will induce expression of certain genes and thus initiate the movement of the involuting mesoderm. This movement will establish the anterior-posterior axis of the embryo. The first mesodermal cells to migrate over the dorsal blastopore lip will induce the ectoderm above them to produce anterior structures such as the forebrain; mesoderm that involutes later will signal the ectoderm to form more posterior structures, such as the hindbrain and spinal cord. This process, whereby the central nervous system forms through interactions with the underlying mesoderm, has been called *primary embryonic induction* and is one of the principal ways in which the vertebrate embryo become organized. Indeed, its discoverers called the dorsal blastopore lip and its descendants "the organizer," and found that this region is different from all the other parts of the embryo. In the early twentieth century, experiments by Hans Spemann and his students at the University of Freiburg, Germany, framed the questions that experimental embryologists would continue to ask for most of the rest of the century and resulted in a Nobel Prize for Spemann in 1935 (see Hamburger 1988; De Robertis and Aréchaga 2001; Sander and Fässler 2001).

The Work of Hans Spemann and Hilde Mangold

Autonomous specification versus inductive interactions

The experiment that began the Spemann laboratory's research program was performed in 1903, when Spemann demonstrated that early newt blastomeres have identical nuclei, each capable of producing an entire larva. His procedure was ingenious: Shortly

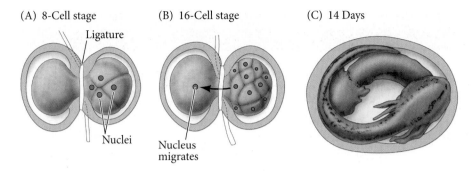

(A) 8-Cell stage (B) 16-Cell stage (C) 14 Days

Ligature

Nuclei

Nucleus migrates

FIGURE 11.11 Spemann's demonstration of nuclear equivalence in newt cleavage. (A) When the fertilized egg of the newt *Triturus taeniatus* was constricted by a ligature, the nucleus was restricted to one half of the embryo. The cleavage on that side of the embryo reached the 8-cell stage, while the other side remained undivided. (B) At the 16-cell stage, a single nucleus entered the as-yet undivided half, and the ligature was further constricted to complete the separation of the two halves. (C) After 14 days, each side had developed into a normal embryo. (After Spemann 1931.)

after fertilizing a newt egg, Spemann used a baby's hair (taken from his infant daughter) to "lasso" the zygote in the plane of the first cleavage. He then partially constricted the egg, causing all the nuclear divisions to remain on one side of the constriction. Eventually—often as late as the 16-cell stage—a nucleus would escape across the constriction into the non-nucleated side. Cleavage then began on this side too, whereupon Spemann tightened the lasso until the two halves were completely separated. Twin larvae developed, one slightly more advanced than the other (**FIGURE 11.11**). Spemann concluded from this experiment that early amphibian nuclei were genetically identical and that each cell was capable of giving rise to an entire organism.

However, when Spemann performed a similar experiment with the constriction still longitudinal but perpendicular to the plane of the first cleavage (i.e., separating the future dorsal and ventral regions rather than the right and left sides), he obtained a different result altogether. The nuclei continued to divide on both sides of the constriction, but only one side—the future dorsal side of the embryo—gave rise to a normal larva. The other side produced an unorganized tissue mass of ventral cells, which Spemann called the *Bauchstück* ("belly piece"). This tissue mass was a ball of epidermal cells (ectoderm) containing blood cells and mesenchyme (mesoderm) and gut cells (endoderm), but it contained no dorsal structures such as nervous system, notochord, or somites.

Why did these two experiments have such different results? One possibility was that when the egg was divided perpendicular to the first cleavage plane, some *cytoplasmic* substance was not equally distributed into the two halves. Fortunately, the salamander egg was a good organism to test that hypothesis. As we saw earlier in this chapter (see Figure 11.2), there are dramatic movements in the cytoplasm following the fertilization of amphibian eggs, and in some amphibians these movements expose a gray, crescent-shaped area of cytoplasm in the region directly opposite the point of sperm entry. The first cleavage plane normally splits this gray crescent equally between the two blastomeres (see Figure 11.2D). If these cells are then separated, two complete larvae develop (**FIGURE 11.12A**). However, should this cleavage plane be aberrant (either in the rare natural event or in an experiment), the gray crescent material passes into only one of the two blastomeres. Spemann's work revealed that when two blastomeres are separated such that only one of the two cells contains the crescent, only the blastomere containing the gray crescent develops normally (**FIGURE 11.12B**).

It appeared, then, that *something in the region of the gray crescent was essential for proper embryonic development.* But how did it function? What role did it play in normal development? The most

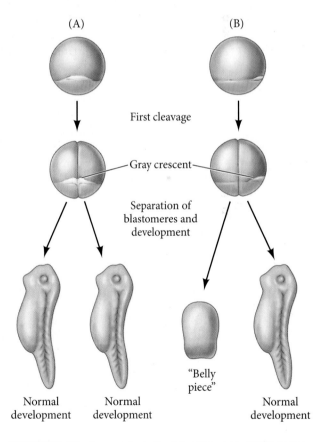

(A) (B)

First cleavage

Gray crescent

Separation of blastomeres and development

Normal development   Normal development

"Belly piece"

Normal development

FIGURE 11.12 Asymmetry in the amphibian egg. (A) When the egg is divided along the plane of first cleavage into two blastomeres, each of which gets half of the gray crescent, each experimentally separated cell develops into a normal embryo. (B) When only one of the two blastomeres receives the entire gray crescent, it alone forms a normal embryo. The other blastomere produces a mass of unorganized tissue lacking dorsal structures. (After Spemann 1931.)

TABLE 11.1 Results of tissue transplantation during early- and late-stage newt gastrulae

Donor region	Host region	Differentiation of donor tissue	Conclusion
EARLY GASTRULA			
Prospective neurons	Prospective epidermis	Epidermis	Conditional development
Prospective epidermis	Prospective neurons	Neurons	Conditional development
LATE GASTRULA			
Prospective neurons	Prospective epidermis	Neurons	Autonomous development
Prospective epidermis	Prospective neurons	Epidermis	Autonomous development

important clue came from fate maps, which showed that the gray crescent region gives rise to those cells that form the dorsal lip of the blastopore. These dorsal lip cells are committed to invaginate into the blastula, initiating gastrulation and the formation of the head endomesoderm and notochord. Because all future amphibian development depends on the interaction of cells that are rearranged during gastrulation, Spemann speculated that the importance of the gray crescent material lies in its ability to initiate gastrulation, and that crucial changes in cell potency occur during gastrulation. In 1918, he performed experiments that showed both statements to be true. He found that the cells of the *early* gastrula were uncommitted, but that the fates of *late* gastrula cells were determined.

Spemann's demonstration involved exchanging tissues between the gastrulae of two species of newts whose embryos were differently pigmented—the darkly pigmented *Triturus taeniatus* and the nonpigmented *T. cristatus*. When a region of prospective epidermal cells from an early gastrula of one species was transplanted into an area in an early gastrula of the other species and placed in a region where neural tissue normally formed, the transplanted cells gave rise to neural tissue. When prospective neural tissue from early gastrulae was transplanted to the region fated to become belly skin, the neural tissue became epidermal (**FIGURE 11.13A; TABLE 11.1**). Thus, cells of the early newt gastrula exhibit conditional (induction-dependent) specification: their ultimate fate depends on their location in the embryo.

However, when the same interspecies transplantation experiments were performed on *late* gastrulae, Spemann obtained completely different results. Rather than differentiating in accordance with their new location, the transplanted cells exhibited *autonomous* (mosaic, independent) development. Their prospective fate was *determined*, and the cells developed independently of their new embryonic location. Specifically, prospective neural cells now developed into brain tissue even when placed in the region of prospective epidermis (**FIGURE 11.13B**), and prospective epidermis formed skin even in the region of the prospective neural tube. Within the time separating early

(A) Transplantation in early gastrula

Presumptive neural ectoderm

Presumptive epidermis

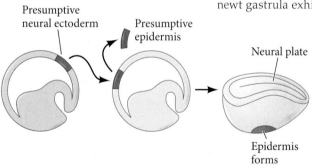

Neural plate

Epidermis forms

(B) Transplantation in late gastrula

Presumptive neural ectoderm

Presumptive epidermis

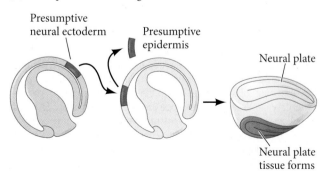

Neural plate

Neural plate tissue forms

FIGURE 11.13 Determination of ectoderm during newt gastrulation. Presumptive neural ectoderm from one newt embryo is transplanted into a region in another embryo that normally becomes epidermis. (A) When the tissues are transferred between early gastrulae, the presumptive neural tissue develops into epidermis, and only one neural plate is seen. (B) When the same experiment is performed using late-gastrula tissues, the presumptive neural cells form neural tissue, thereby causing two neural plates to form on the host. (After Saxén and Toivonen 1962.)

and late gastrulation, the potencies of these groups of cells had become restricted to their eventual paths of differentiation. Something caused them to become committed to epidermal and neural fates. What was happening?

Primary embryonic induction

The most spectacular transplantation experiments were published by Spemann and his doctoral student Hilde Mangold in 1924.[4] They showed that, of all the tissues in the early gastrula, only one has its fate autonomously determined. This self-determining tissue is the dorsal lip of the blastopore—the tissue derived from the gray crescent cytoplasm opposite the point of sperm entry. When this tissue was transplanted into the presumptive belly skin region of another gastrula, it not only continued to be dorsal blastopore lip but also initiated gastrulation and embryogenesis in the surrounding tissue!

WATCH DEVELOPMENT 11.5   See the Spemann-Mangold experiment performed by Dr. Eddy De Robertis.

In these experiments, Spemann and Mangold once again used the differently pigmented embryos of *Triturus taeniatus* and *T. cristatus* so they could identify host and donor tissues on the basis of color. When the dorsal lip of an early *T. taeniatus* gastrula was removed and implanted into the region of an early *T. cristatus* gastrula fated to become ventral epidermis (belly skin), the dorsal lip tissue invaginated just as it would normally have done (showing self-determination) and disappeared beneath the vegetal cells (**FIGURE 11.14A**). The pigmented donor tissue then continued to self-differentiate into the chordamesoderm (notochord) and other mesodermal structures that normally form from the dorsal lip (**FIGURE 11.14B**). As the donor-derived mesodermal cells moved forward, host cells began to participate in the production of a new embryo, becoming organs that normally they never would have formed. In this secondary embryo, a somite could be seen containing both pigmented (donor) and nonpigmented (host) tissue. Even more spectacularly, the dorsal lip cells were able to interact with the host tissues to form a complete neural plate from host ectoderm. Eventually, a secondary embryo formed, conjoined face to face with its host (**FIGURE 11.14C**). The results of these technically difficult experiments have been confirmed many times and in many amphibian species, including *Xenopus* (**FIGURE 11.14D,E**; Capuron 1968; Smith and Slack 1983; Recanzone and Harris 1985).

Spemann referred to the dorsal lip cells and their derivatives (notochord and head endomesoderm) as the **organizer** because (1) they induced the host's ventral tissues to change their fates to form a neural tube and dorsal mesodermal tissue (such as somites), and (2) they organized host and donor tissues into a secondary embryo with clear anterior-posterior and dorsal-ventral axes. He proposed that during normal development, these cells "organize" the dorsal ectoderm into a neural tube and transform the flanking mesoderm into the anterior-posterior body axis (Spemann 1938). It is now known (thanks largely to Spemann and his students) that the interaction of the chordamesoderm and ectoderm is not sufficient to organize the entire embryo. Rather, it initiates a series of sequential inductive events. Because there are numerous inductions during embryonic development, this key induction—in which the progeny of dorsal lip cells induce the dorsal axis and the neural tube—is traditionally called the **primary embryonic induction**. This classical term has been a source of confusion, however, because the induction of the neural tube by the notochord is no longer considered the

---

[4] Hilde Proescholdt Mangold died in a tragic accident in 1924, when her kitchen's gasoline heater exploded. She was 26 years old, and her paper was just about to be published. Hers is one of the very few doctoral theses in biology that have directly resulted in the awarding of a Nobel Prize. For more information about Hilde Mangold, her times, and the experiments that identified the organizer, see Hamburger 1984, 1988, and Fässler and Sander 1996.

FIGURE 11.14 Organization of a secondary axis by dorsal blastopore lip tissue. (A–C) Spemann and Mangold's 1924 experiments visualized the process by using differently pigmented newt embryos. (A) Dorsal lip tissue from an early *T. taeniatus* gastrula is transplanted into a *T. cristatus* gastrula in the region that normally becomes ventral epidermis. (B) The donor tissue invaginates and forms a second archenteron, and then a second embryonic axis. Both donor and host tissues are seen in the new neural tube, notochord, and somites.(C) Eventually, a second embryo forms, joined to the host. (D) Live twinned *Xenopus* larvae generated by transplanting a dorsal blastopore lip into the ventral region of an early-gastrula host embryo. (E) Similar twinned larvae are seen from below and stained for notochord; the original and secondary notochords can be seen. (A–C after Hamburger 1988; D,E photographs by A. Wills, courtesy of R. Harland.)

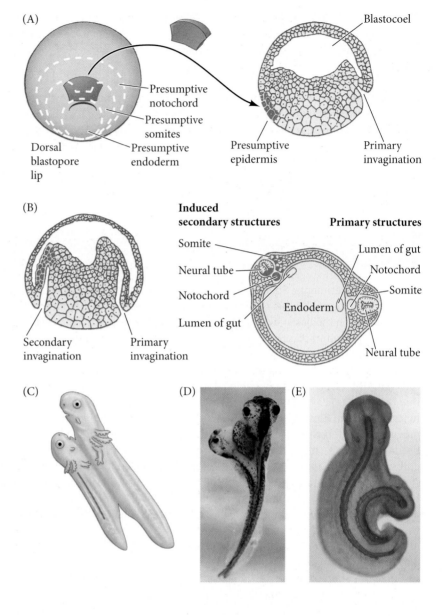

first inductive process in the embryo. We will soon discuss inductive events that precede this "primary" induction.

WEB TOPIC 11.2 **SPEMANN, MANGOLD, AND THE ORGANIZER** Spemann did not see the importance of this work the first time he and Mangold did it. This web topic provides a more detailed account of why Spemann and Mangold performed this particular experiment.

Molecular Mechanisms of Amphibian Axis Formation

The experiments of Spemann and Mangold showed that the dorsal lip of the blastopore, along with the dorsal mesoderm and pharyngeal endoderm that form from it, constituted an "organizer" able to instruct the formation of embryonic axes. But the mechanisms by which the organizer itself was constructed and through which it operated remained a mystery. Indeed, it is said that Spemann and Mangold's landmark paper posed more questions than it answered. Among those questions were:

- How did the organizer get its properties? What caused the dorsal blastopore lip to differ from any other region of the embryo?

- What factors were being secreted from the organizer to cause the formation of the neural tube and to create the anterior-posterior, dorsal-ventral, and left-right axes?

- How did the different parts of the neural tube become established, with the most anterior becoming the sensory organs and forebrain and the most posterior becoming spinal cord?

Spemann and Mangold's description of the organizer was the starting point for one of the first truly international scientific research programs (see Gilbert and Saxén 1993; Armon 2012). Researchers from Britain, Germany, France, the United States, Belgium, Finland, Japan, and the Soviet Union all joined in the search for the remarkable substances responsible for the organizer's ability. R. G. Harrison referred to the amphibian gastrula as the "new Yukon to which eager miners were now rushing to dig for gold around the blastopore" (see Twitty 1966, p. 39). Unfortunately, their early picks and shovels proved too blunt to uncover the molecules involved. The proteins responsible for induction were present in concentrations too small for biochemical analyses, and the large quantity of yolk and lipids in the amphibian egg further interfered with protein purification (Grunz 1997). The analysis of organizer molecules had to wait until recombinant DNA technologies enabled investigators to make cDNA clones from blastopore lip mRNA, thus allowing them to see which of these clones encoded factors that could dorsalize the embryo (Carron and Shi 2016). We are now able to take up each of the above questions in turn.

How does the organizer form?

Why are the dozen or so initial cells of the organizer positioned opposite the point of sperm entry, and what determines their fate so early? Recent evidence provides an unexpected answer: these cells are in the right place at the right time, at a point where two signals converge. The first signal tells the cells that they are dorsal. The second signal says that these cells are mesoderm. These signals interact to create a polarity within the mesoderm that is the basis for specifying the organizer and for creating dorsal-ventral polarity.

THE DORSAL SIGNAL: β-CATENIN   Experiments by Pieter Nieuwkoop and Osamu Nakamura showed that the organizer receives its special properties from signals coming from the prospective endoderm beneath it. Nakamura and Takasaki (1970) showed that the mesoderm arises from the marginal (equatorial) cells at the border between the animal and vegetal poles. The Nakamura and Nieuwkoop laboratories then demonstrated that the properties of this newly formed mesoderm can be induced by the vegetal (presumptive endoderm) cells underlying them. Nieuwkoop (1969, 1973, 1977) removed the equatorial cells (i.e., presumptive mesoderm) from a blastula and showed that neither the animal cap (presumptive ectoderm) nor the vegetal cap (presumptive endoderm) produced any mesodermal tissue. However, when the two caps were recombined, the animal cap cells were induced to form mesodermal structures such as notochord, muscles, kidney cells, and blood cells. The polarity of this induction (i.e., whether the animal cells formed dorsal mesoderm or ventral mesoderm) depended on whether the endodermal (vegetal) fragment was taken from the dorsal or the ventral side: ventral and lateral vegetal cells (those closer to the site of sperm entry) induced ventral (mesenchyme, blood) and intermediate (kidney) mesoderm, while the dorsalmost vegetal cells specified dorsal mesoderm components (somites, notochord)—including those having the properties of the organizer. These dorsalmost vegetal cells of the blastula, which are capable of inducing the organizer, have been called the **Nieuwkoop center** (Gerhart et al. 1989).

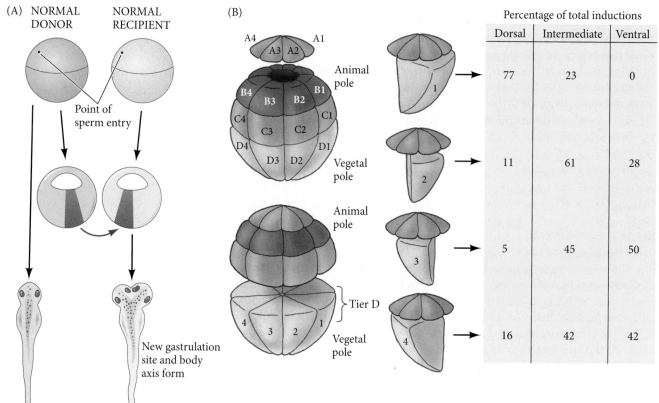

FIGURE 11.15 Transplantation and recombination experiments on *Xenopus* embryos demonstrate that the vegetal cells underlying the prospective dorsal blastopore lip region are responsible for initiating gastrulation. (A) Formation of a new gastrulation site and body axis by the transplantation of the dorsalmost vegetal cells of a 64-cell embryo into the ventralmost vegetal region of another embryo. (B) The regional specificity of mesoderm induction demonstrated by recombining blastomeres of 32-cell *Xenopus* embryos. Animal pole cells were labeled with fluorescent polymers so their descendants could be identified, then combined with individual vegetal blastomeres. The inductions resulting from these recombinations are summarized at the right. D1, the dorsalmost vegetal blastomere, was the most likely to induce the animal pole cells to form dorsal mesoderm. These dorsalmost vegetal cells constitute the Nieuwkoop center. (A after Gimlich and Gerhart 1984; B after Dale and Slack 1987.)

The Nieuwkoop center was demonstrated in the *Xenopus* embryo by transplantation and recombination experiments. First, Gimlich and Gerhart (Gimlich and Gerhart 1984; Gimlich 1985, 1986) performed an experiment analogous to the Spemann and Mangold studies, except that they used early *Xenopus* blastulae rather than newt gastrulae. When they transplanted the dorsalmost vegetal blastomere from one blastula into the ventral vegetal side of another blastula, two embryonic axes formed (**FIGURE 11.15A**). Second, Dale and Slack (1987) recombined single vegetal blastomeres from a 32-cell *Xenopus* embryo with the uppermost animal tier of a fluorescently labeled embryo of the same stage. The dorsalmost vegetal cell, as expected, induced the animal pole cells to become dorsal mesoderm. The remaining vegetal cells usually induced the animal cells to produce either intermediate or ventral mesodermal tissues (**FIGURE 11.15B**). Holowacz and Elinson (1993) found that cortical cytoplasm from the dorsal vegetal cells of the 16-cell *Xenopus* embryo was able to induce the formation of secondary axes when injected into ventral vegetal cells. Thus, *dorsal vegetal cells can induce animal cells to become dorsal mesodermal tissue.*

So one important question became, What gives the dorsalmost vegetal cells their special properties? The major candidate for the factor that forms the Nieuwkoop center in these vegetal cells was β-catenin. We saw in Chapter 10 that β-catenin is responsible for specifying the micromeres of the sea urchin embryo. This multifunctional protein also proved to be a key player in the formation of the dorsal amphibian tissues. Experimental depletion of this molecule results in the lack of dorsal structures (Heasman et al. 1994a), while injection of exogenous β-catenin into the *ventral* side of an embryo produces a secondary axis (Funayama et al. 1995; Guger and Gumbiner 1995).

In *Xenopus* embryos, β-catenin is initially synthesized throughout the embryo from maternal mRNA (Yost et al. 1996; Larabell et al. 1997). It begins to accumulate in the dorsal region of the egg during the cytoplasmic movements of fertilization and continues to accumulate preferentially at the dorsal side throughout early cleavage. This accumulation is seen in the nuclei of the dorsal cells and appears to cover both the Nieuwkoop center and organizer regions (**FIGURE 11.16**; Schneider et al. 1996; Larabell et al. 1997).

(A)

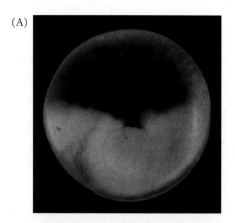

(B)

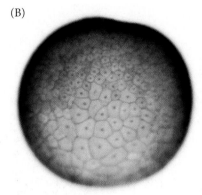

(C)

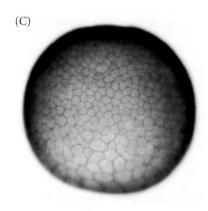

(D)

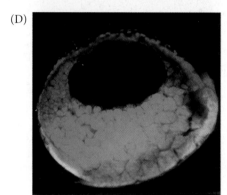

FIGURE 11.16   Role of Wnt pathway proteins in dorsal-ventral axis specification. (A–D) Differential translocation of β-catenin into *Xenopus* blastomere nuclei. (A) Early 2-cell stage, showing β-catenin (orange) predominantly at the dorsal surface. (B) Presumptive dorsal side of a blastula stained for β-catenin shows nuclear localization. (C) Such nuclear localization is not seen on the ventral side of the same embryo. (D) Dorsal localization of β-catenin persists through the gastrula stage. (A,D courtesy of R. T. Moon; B,C from Schneider et al. 1996, courtesy of P. Hausen.)

If β-catenin is originally found throughout the embryo, how does it become localized specifically to the side opposite sperm entry? The answer appears to reside in the localizations of three proteins in the egg cortical cytoplasm. The proteins Wnt11, GSK3-binding protein (GBP), and Disheveled (Dsh) all are translocated from the vegetal pole of the egg to the future dorsal side of the embryo during fertilization. From research on the Wnt pathway, we have learned that β-catenin is targeted for destruction by glycogen synthase kinase 3 (GSK3; see Chapter 4). Indeed, activated GSK3 stimulates degradation of β-catenin and blocks axis formation when added to the egg, and if endogenous GSK3 is knocked out by a dominant-negative form of GSK3 in the ventral cells of the early embryo, a second axis forms (see Figure 11.17F; He et al. 1995; Pierce and Kimelman 1995; Yost et al. 1996).

GSK3 can be inactivated by GBP and Disheveled. These two proteins release GSK3 from the degradation complex and prevent it from binding β-catenin and targeting it for destruction. During the first cell cycle, when the microtubules form parallel tracts in the vegetal portion of the egg, GBP travels along the microtubules by binding to kinesin, an ATPase motor protein that travels on microtubules. Kinesin always migrates toward the growing end of the microtubules, and in this case, that means moving to the point opposite sperm entry, i.e., the future dorsal side (**FIGURE 11.17A–C**). Disheveled, which is originally found in the vegetal pole cortex, grabs onto the GPB, and it too becomes translocated along the microtubular monorail (Miller et al. 1999; Weaver et al. 2003). The cortical rotation is probably important in orienting and straightening the microtubular array and in maintaining the direction of transport when the kinesin complexes occasionally jump the track (Weaver and Kimelman 2004). Once at the site opposite the point of sperm entry, GBP and Dsh are released from the microtubules. Here, on the future dorsal side of the embryo, they inactivate GSK3, allowing β-catenin to accumulate on the dorsal side while ventral β-catenin is degraded (**FIGURE 11.17D,E**).

But the mere translocation of these proteins to the dorsal side of the embryo does not seem to be sufficient for protecting β-catenin. It appears that a Wnt paracrine factor

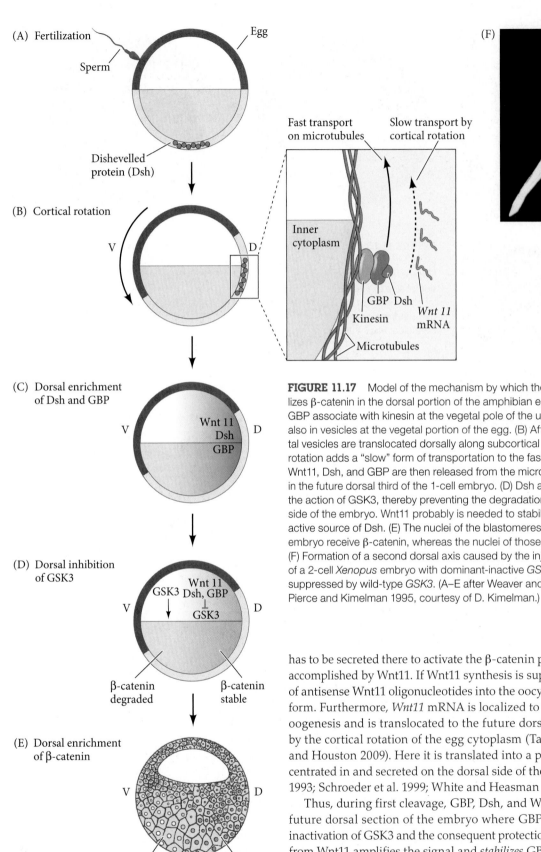

(A) Fertilization

Sperm

Egg

Dishevelled protein (Dsh)

(B) Cortical rotation

V — D

Fast transport on microtubules

Slow transport by cortical rotation

Inner cytoplasm

GBP Dsh

Kinesin

Wnt 11 mRNA

Microtubules

(C) Dorsal enrichment of Dsh and GBP

V — Wnt 11 Dsh GBP — D

(D) Dorsal inhibition of GSK3

GSK3 — Wnt 11 Dsh, GBP — GSK3

V — D

β-catenin degraded

β-catenin stable

(E) Dorsal enrichment of β-catenin

V — D

No β-catenin in ventral nuclei

β-catenin in dorsal nuclei

(F)

FIGURE 11.17 Model of the mechanism by which the Disheveled protein stabilizes β-catenin in the dorsal portion of the amphibian egg. (A) Disheveled (Dsh) and GBP associate with kinesin at the vegetal pole of the unfertilized egg. Wnt11 is also in vesicles at the vegetal portion of the egg. (B) After fertilization, these vegetal vesicles are translocated dorsally along subcortical microtubule tracks. Cortical rotation adds a "slow" form of transportation to the fast-track microtubule ride. (C) Wnt11, Dsh, and GBP are then released from the microtubules and are distributed in the future dorsal third of the 1-cell embryo. (D) Dsh and GBP bind to and block the action of GSK3, thereby preventing the degradation of β-catenin on the dorsal side of the embryo. Wnt11 probably is needed to stabilize this reaction, keeping an active source of Dsh. (E) The nuclei of the blastomeres in the dorsal region of the embryo receive β-catenin, whereas the nuclei of those in the ventral region do not. (F) Formation of a second dorsal axis caused by the injection of both blastomeres of a 2-cell *Xenopus* embryo with dominant-inactive *GSK3*. Dorsal fate is actively suppressed by wild-type *GSK3*. (A–E after Weaver and Kimelman 2004; F from Pierce and Kimelman 1995, courtesy of D. Kimelman.)

has to be secreted there to activate the β-catenin protection pathway; this is accomplished by Wnt11. If Wnt11 synthesis is suppressed (by the injection of antisense Wnt11 oligonucleotides into the oocytes), the organizer fails to form. Furthermore, *Wnt11* mRNA is localized to the vegetal cortex during oogenesis and is translocated to the future dorsal portion of the embryo by the cortical rotation of the egg cytoplasm (Tao et al. 2005; Cuykendall and Houston 2009). Here it is translated into a protein that becomes concentrated in and secreted on the dorsal side of the embryo (Ku and Melton 1993; Schroeder et al. 1999; White and Heasman 2008).

Thus, during first cleavage, GBP, Dsh, and Wnt11 are brought into the future dorsal section of the embryo where GBP and Dsh can *initiate* the inactivation of GSK3 and the consequent protection of β-catenin. The signal from Wnt11 amplifies the signal and *stabilizes* GBP and Dsh and organizes them to protect β-catenin; β-catenin can associate with other transcription factors, giving these factors new properties. It is known, for example, that *Xenopus* β-catenin can combine with a ubiquitous transcription factor

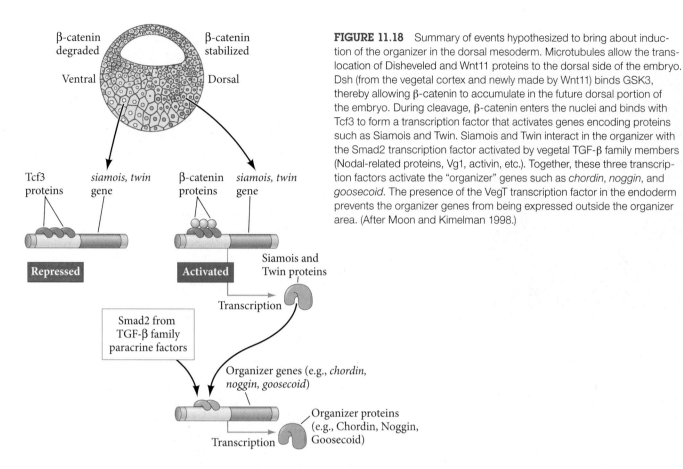

FIGURE 11.18 Summary of events hypothesized to bring about induction of the organizer in the dorsal mesoderm. Microtubules allow the translocation of Disheveled and Wnt11 proteins to the dorsal side of the embryo. Dsh (from the vegetal cortex and newly made by Wnt11) binds GSK3, thereby allowing β-catenin to accumulate in the future dorsal portion of the embryo. During cleavage, β-catenin enters the nuclei and binds with Tcf3 to form a transcription factor that activates genes encoding proteins such as Siamois and Twin. Siamois and Twin interact in the organizer with the Smad2 transcription factor activated by vegetal TGF-β family members (Nodal-related proteins, Vg1, activin, etc.). Together, these three transcription factors activate the "organizer" genes such as *chordin*, *noggin*, and *goosecoid*. The presence of the VegT transcription factor in the endoderm prevents the organizer genes from being expressed outside the organizer area. (After Moon and Kimelman 1998.)

known as Tcf3, converting the Tcf3 repressor into an activator of transcription. Expression of a mutant form of Tcf3 that lacks the β-catenin binding domain results in embryos without dorsal structures (Molenaar et al. 1996).

The β-catenin /Tcf3 complex binds to the promoters of several genes whose activity is critical for axis formation. Two of these genes, *twin* and *siamois*, encode homeodomain transcription factors and are expressed in the organizer region immediately following the mid-blastula transition. If these genes are ectopically expressed in the ventral cells, a secondary axis emerges on the former ventral side of the embryo; and if cortical microtubular polymerization is prevented, *siamois* expression is eliminated (Lemaire et al. 1995; Brannon and Kimelman 1996). The Tcf3 protein is thought to inhibit *siamois* and *twin* transcription when it binds to those genes' promoters in the absence of β-catenin. However, when β-catenin binds to Tcf3, the repressor is converted into an activator, and *twin* and *siamois* are transcibed (**FIGURE 11.18**).

The proteins Siamois and Twin bind to the enhancers of several genes involved in organizer function (Fan and Sokol 1997; Bae et al. 2011). These include genes encoding the transcription factors Goosecoid and Xlim1 (which are critical in specifying the dorsal mesoderm) and the paracrine factor antagonists Noggin, Chordin, Frzb, and Cerberus (which specify the ectoderm to become neural; Laurent et al. 1997; Engleka and Kessler 2001). In the vegetal cells, Siamois and Twin appear to combine with vegetal transcription factors to help activate endodermal genes (Lemaire et al. 1998). Thus, one could expect that if the dorsal side of the embryo contained β-catenin, this β-catenin would allow the region to express Twin and Siamois, which in turn would initiate formation of the organizer.

THE VEGETAL NODAL-RELATED SIGNAL   Yet another factor appears to be critical in activating the genes that characterize the organizer cells. This other factor is the phosphorylated Smad2 transcription factor (discussed earlier), which is essential in forming the

(A)

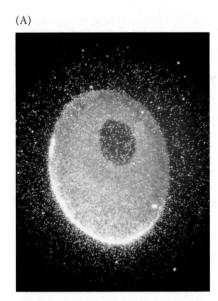

(B) Stage 8   (C) Stage 9   (D) Stage 10

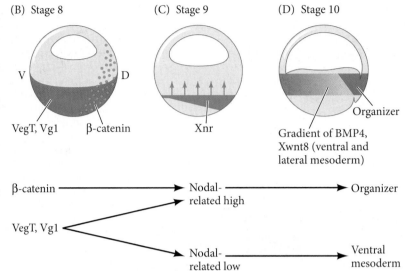

FIGURE 11.19 Vegetal induction of mesoderm. (A) The maternal RNA encoding Vg1 (bright white crescent) is tethered to the vegetal cortex of a *Xenopus* oocyte. The message (along with the maternal VegT message) will be translated at fertilization. Both proteins appear to be crucial for the ability of vegetal cells to induce cells above them to become mesoderm. (B–D) Model for mesoderm induction and organizer formation by the interaction of β-catenin and TGF-β proteins. (B) At late blastula stages, Vg1 and VegT are found in the vegetal hemisphere; β-catenin is located in the dorsal region. (C) β-Catenin acts synergistically with Vg1 and VegT to activate the *Xenopus nodal-related* (*Xnr*) genes. This creates a gradient of Xnr proteins across the endoderm, highest in the dorsal region. (D) The mesoderm is specified by the Xnr gradient. Mesodermal regions with little or no Xnr have high levels of BMP4 and Xwnt8; they become ventral mesoderm. Those having intermediate concentrations of Xnr become lateral mesoderm. Where there is a high concentration of Xnr, *goosecoid* and other dorsal mesodermal genes are activated and the mesodermal tissue becomes the organizer. (A courtesy of D. Melton; B–D after Agius et al. 2000.)

mesoderm. Smad2 is activated in the mesodermal cells when it becomes phosphorylated in response to Nodal-related paracrine factors secreted by the vegetal cells beneath the mesoderm (Brannon and Kimelman 1996; Engleka and Kessler 2001). Activated Smad2 usually binds with a partner to form a complex that acts as a transcription factor.

At the late blastula stages, there is a gradient of Nodal-related proteins across the endoderm, with low concentrations ventrally and high concentrations dorsally (Onuma et al. 2002; Rex et al. 2002; Chea et al. 2005). Because Vg1 and the Nodal-related proteins act through the same pathway (i.e., by activating the Smad2 transcription factor), we would expect them to produce an additive signal (Agius et al. 2000). Indeed, this appears to be the case.

The Nodal-related gradient is produced in large part by β-catenin. High β-catenin levels activate Nodal-related gene expression (**FIGURE 11.19**). In the most dorsal (Nieuwkoop center) blastomeres, β-catenin cooperates with the VegT transcription factor to activate the *Xenopus nodal-related 1, 5,* and *6* (*Xnr1, 5,* and *6*) genes even before the mid-blastula transition. The more ventral blastomeres in the endoderm lack the expression of these Nodal-related genes. In the region that will become the most anterior portion of the organizer—the pharyngeal endoderm—higher levels of Nodal-related proteins produce higher concentrations of activated Smad2. Smad2 can bind to the promoter of the *hhex* gene, and in concert with Twin and Siamois (induced by β-catenin), Hhex activates genes that specify pharyngeal endoderm cells to become foregut endoderm and to induce anterior brain development (Smithers and Jones 2002; Rankin et al. 2011). Slightly lower levels of Smad2 are believed to activate *goosecoid* expression in the cells that will become the prechordal mesoderm and notochord. Even lower amounts of Smad2 result in the formation of lateral and ventral mesoderm.

In summary, then, the formation of the dorsal mesoderm and the organizer originates through the activation of critical transcription factors by intersecting pathways. The first pathway is the Wnt/β-catenin pathway that activates genes encoding the Siamois and Twin transcription factors. The second pathway is the vegetal pathway that activates the expression of Nodal-related paracrine factors, which in turn activate the Smad2 transcription factor in the mesodermal cells above them. The high levels of Smad2 and Siamois/Twin transcription factor proteins work within the dorsal mesoderm cells and activate the genes that give these cells their "organizer" properties (Germain et al. 2000; Cho 2012; review Figures 11.17–11.19).

SCIENTISTS SPEAK 11.1 Hear Daniel Kessler discuss the molecular mechanisms of primary embryonic induction in amphibians.

Functions of the organizer

While the Nieuwkoop center cells remain endodermal, the cells of the organizer become the dorsal mesoderm and migrate underneath the dorsal ectoderm. The cells of the organizer ultimately contribute to four cell types: pharyngeal endoderm, head mesoderm (prechordal plate), dorsal mesoderm (primarily the notochord), and the dorsal blastopore lip (Keller 1976; Gont et al. 1993). The pharyngeal endoderm and prechordal plate lead the migration of the organizer tissue and induce the forebrain and midbrain. The dorsal mesoderm induces the hindbrain and trunk. The dorsal blastopore lip remaining at the end of gastrulation eventually becomes the chordaneural hinge that induces the tip of the tail. The properties of the organizer tissue can be divided into four major functions:

1. The ability to self-differentiate into dorsal mesoderm (prechordal plate, chordamesoderm, etc.)

2. The ability to dorsalize the surrounding mesoderm into paraxial (somite-forming) mesoderm when it otherwise would form ventral mesoderm

3. The ability to dorsalize the ectoderm and induce formation of the neural tube

4. The ability to initiate the movements of gastrulation

WEB TOPIC 11.3 EARLY ATTEMPTS TO LOCATE THE ORGANIZER MOLECULES
Although Spemann did not believe that molecules alone could organize the embryo, his students began a long quest to identify these factors in the region of the organizer.

Induction of neural ectoderm and dorsal mesoderm: BMP inhibitors

Evidence from experimental embryology showed that one of the most critical properties of the organizer was its production of soluble factors. The evidence for such diffusible signals from the organizer came from several sources. First, Hans Holtfreter (1933) showed that if the notochord fails to migrate beneath the ectoderm, the ectoderm will not become neural tissue (and will become epidermis). More definitive evidence for the importance of soluble factors came later from the transfilter studies of Finnish investigators (Saxén 1961; Toivonen et al. 1975; Toivonen and Wartiovaara 1976). Here, newt dorsal lip tissue was placed on one side of a filter fine enough so that no processes could fit through the pores, and competent gastrula ectoderm was placed on the other side. After several hours, neural structures were observed in the ectodermal tissue (**FIGURE 11.20**). The identities of the factors diffusing from the organizer, however, took another quarter of a century to find.

It turned out that scientists were looking for the wrong mechanism. They were searching for a molecule secreted by the organizer and received by the ectoderm that would then convert the ectoderm into neural tissue. However, molecular studies led to a remarkable and non-obvious conclusion: *it is the epidermis that is induced to form, not the neural tissue.* The ectoderm is induced to become epidermal tissue by binding **bone morphogenetic proteins (BMPs)**, whereas the nervous system forms from that region of the ectoderm that is *protected* from epidermal induction by BMP-inhibiting molecules (Hemmati-Brivanlou and Melton 1994, 1997). In other words, (1) the "default fate" of the ectoderm is to become neural tissue; (2) certain parts of the embryo induce the ectoderm to become epidermal tissue by secreting BMPs; and (3) the organizer tissue acts by secreting molecules that block BMPs, thereby allowing the ectoderm "protected" by these BMP inhibitors to become neural tissue.

Thus, BMPs induce naïve ectodermal cells to become epidermal, while the Organizer produces substances that block this induction (Wilson and Hemmati-Brivanlou 1995; Piccolo et al. 1996; Zimmerman et al. 1996; Iemura et al. 1998). In *Xenopus*, the major epidermal inducers are BMP4 and its close relatives BMP2, BMP7, and ADMP. Initially, BMPs such as BMP4 are initially

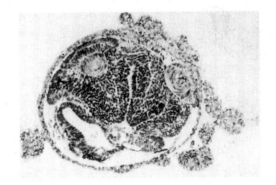

FIGURE 11.20 Neural structures induced in presumptive ectoderm by newt dorsal lip tissue, separated from the ectoderm by a nucleopore filter with an average pore diameter of 0.05 mm. Anterior neural tissues are evident, including some induced eyes. (From Toivonen 1979, courtesy of L. Saxén.)

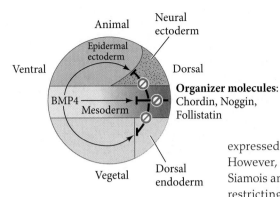

Animal

Neural
ectoderm

Epidermal
ectoderm

Ventral

Dorsal

BMP4

Organizer molecules:
Chordin, Noggin,
Follistatin

Mesoderm

Vegetal

Dorsal
endoderm

FIGURE 11.21   Model for the action of the organizer. (A) BMP4 (along with certain other molecules) is a powerful ventralizing factor. Organizer proteins such as Chordin, Noggin, and Follistatin block the action of BMP4; their inhibitory effects can be seen in all three germ layers. (After Dosch et al. 1997.)

expressed throughout the ectodermal and mesodermal regions of the late blastula. However, during gastrulation, transcription factors (such as Goosecoid) induced by Siamois and Twin prevent the transcription of *bmp4* in the dorsal region of the embryo, restricting their expression to the ventrolateral marginal zone (Blitz and Cho 1995; Yao and Kessler 2001; Hemmati-Brivanlou and Thomsen 1995; Northrop et al. 1995; Steinbeisser et al. 1995). In the ectoderm, BMPs repress the genes (such as *Foxd4* and *neurogenin*) involved in forming neural tissue, while activating other genes involved in epidermal specification (Lee et al. 1995). In the mesoderm, it appears that graded levels of BMP4 activate different sets of mesodermal genes: an absence of BMP4 specifies the dorsal mesoderm; a low amount specifies the intermediate mesoderm; and a high amount specifies the ventral mesoderm (**FIGURE 11.21**; Gawantka et al. 1995; Hemmati-Brivanlou and Thomsen 1995; Dosch et al. 1997).

The organizer acts by blocking the BMPs. Three of the major BMP inhibitors secreted by the organizer are Noggin, Chordin, and Follistatin. The genes encoding these proteins are some of the most critical genes activated by Smad2 and Siamois/Twin (Carnac et al. 1996; Fan and Sokol 1997; Kessler 1997). A fourth BMP inhibitor, Norrin, appears to be stored in the animal pole of the oocyte and functions to block BMPs in the dorsal ectoderm (Xu et al. 2015).

NOGGIN   In 1992, Smith and Harland constructed a cDNA plasmid library from dorsalized (lithium chloride-treated) gastrulae. Messenger RNAs synthesized from sets of these plasmids were injected into ventralized embryos (having no neural tube) produced by irradiating early embryos with ultraviolet light. Those plasmid sets whose mRNAs rescued dorsal structures in these embryos were split into smaller sets, and so on, until single-plasmid clones were isolated whose mRNAs were able to restore the dorsal tissue in such embryos. One of these clones contained the gene for the protein Noggin (**FIGURE 11.22A**). Injection of *noggin* mRNA into 1-cell, UV-irradiated embryos completely rescued dorsal development and allowed the formation of a complete embryo.

Noggin is a secreted protein that is able to accomplish two of the major functions of the organizer: it induces dorsal ectoderm to form neural tissue, and it dorsalizes mesoderm cells that would otherwise contribute to the ventral mesoderm (Smith et al. 1993). Smith and Harland showed that newly transcribed *noggin* mRNA is first localized in the dorsal blastopore lip region and then becomes expressed in the notochord (**FIGURE 11.22B**). Noggin binds to BMP4 and BMP2 and inhibits their binding to receptors (Zimmerman et al. 1996).

CHORDIN   Chordin protein was isolated from clones of cDNA whose mRNAs were present in dorsalized, but not in ventralized, embryos (Sasai et al. 1994). These cDNA clones were tested by injecting them into ventral blastomeres and seeing whether they induced secondary axes. One of the clones capable of inducing a secondary neural tube contained the *chordin* gene; *chordin* mRNA was found to be localized in the dorsal blastopore lip and later in the notochord (**FIGURE 11.23**). Morpholino antisense oligomers directed against the *chordin* message blocked the ability of an organizer graft to induce a secondary central nervous system (Oelgeschläger et al. 2003). Of all organizer genes observed, *chordin* is the one most acutely activated by β-catenin (Wessely et al. 2004). Like Noggin,

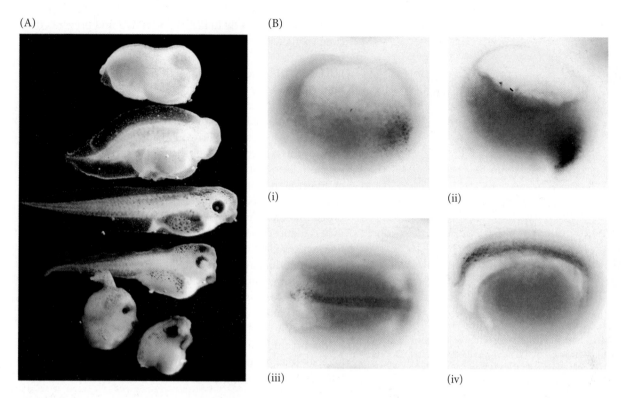

FIGURE 11.22 The soluble protein Noggin dorsalizes the amphibian embryo. (A) Rescue of dorsal structures by Noggin protein. When *Xenopus* eggs are exposed to ultraviolet radiation, cortical rotation fails to occur, and the embryos lack dorsal structures (top). If such an embryo is injected with *noggin* mRNA, it develops dorsal structures in a dosage-related fashion (top to bottom). If too much *noggin* message is injected, the embryo produces dorsal anterior tissue at the expense of ventral and posterior tissue, becoming little more than a head (bottom). (B) Localization of *noggin* mRNA in the organizer tissue, shown by in situ hybridization. At gastrulation (i), *noggin* mRNA (dark areas) accumulates in the dorsal marginal zone. When cells involute (ii), *noggin* mRNA is seen in the dorsal blastopore lip. During convergent extension (iii), *noggin* is expressed in the precursors of the notochord, prechordal plate, and pharyngeal endoderm, which (iv) extend beneath the ectoderm in the center of the embryo. (Courtesy of R. M. Harland.)

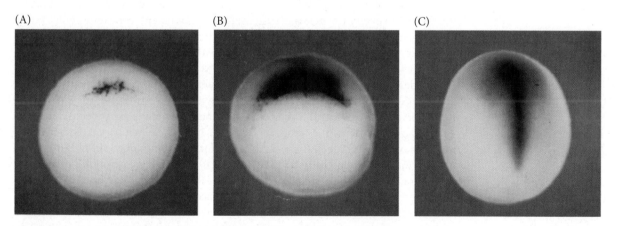

FIGURE 11.23 Localization of *chordin* mRNA. (A) Whole-mount in situ hybridization shows that just prior to gastrulation, *chordin* mRNA (dark area) is expressed in the region that will become the dorsal blastopore lip. (B) As gastrulation begins, *chordin* is expressed at the dorsal blastopore lip. (C) In later stages of gastrulation, the *chordin* message is seen in the organizer tissues. (From Sasai et al. 1994, courtesy of E. De Robertis.)

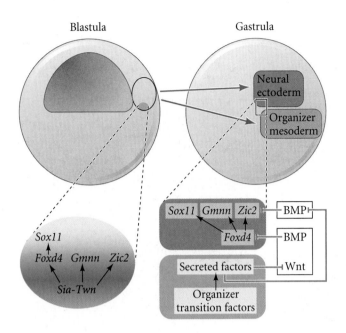

Blastula   Gastrula

FIGURE 11.24 Schematic diagram of Siamois (Sia) and Twin (Twn) inducing the activation of the neuroepithelium genes. During the blastula stage, cells expected to give rise to both Organizer mesoderm and the neural ectoderm express both Sia and Twn, These genes activate the neuroectoderm genes *Foxd4*, *Gmnn*, and *Zic2*. These genes encode transcription factors that will activate other neural genes such as *Sox11*. During the gastrula stage, the descendants of these cells have become the Organizer mesoderm and the neural ectoderm. Here, the neuroepithelial genes are up-regulated by the factors secreted by the organizer that inhibit the BMP and Wnt pathways. If BMPs and Wnts are not blocked, *Sox11*, *Gmnn*, *Foxd4*, and *Zic2* transcription declines. (After Klein and Moody 2015.)

Chordin binds directly to BMP4 and BMP2 and prevents their complexing with their receptors (Piccolo et al. 1996).

FOLLISTATIN   The mRNA for a third organizer-secreted protein, Follistatin, is also transcribed in the dorsal blastopore lip and notochord. Follistatin was found in the organizer as an unexpected result of an experiment that was looking for something else. Ali Hemmati-Brivanlou and Douglas Melton (1992, 1994) wanted to see whether the protein activin was required for mesoderm induction. In searching for the mesoderm inducer, they found that Follistatin, an inhibitor of both activin and BMPs, caused ectoderm to become neural tissue. They then proposed that under normal conditions, ectoderm becomes neural unless induced to become epidermal by the BMPs. This model was supported by, and explained, certain cell dissociation experiments that had also produced odd results. Three 1989 studies—by Grunz and Tacke, Sato and Sargent, and Godsave and Slack—showed that when whole embryos or their animal caps were dissociated, they formed neural tissue. This result would be explainable if the "default state" of the ectoderm was not epidermal but neural, so that tissue had to be induced to have an epidermal phenotype. Thus, we conclude that the organizer blocks epidermalizing induction by inactivating BMPs.

ECTODERMAL BIAS   The ectoderm above the notochord also appears to have been biased to become neural ectoderm by β-catenin that extended into the periphery of the egg. This causes the expression of Siamois and Twin proteins in the cells that will become the neural ectoderm. Here, these transcription factors perform two critical functions. First, they activate those genes (such as *Foxd4* and *Sox11*) that will permit these cells to become neural ectoderm (**FIGURE 11.24**). However, these genes can be suppressed by BMPs, so as a second step during gastrulation, the organizer mesoderm produces proteins that block the BMP signal from reaching the ectoderm (Klein and Moody 2015).

In 2005, two important sets of experiments confirmed the importance of blocking BMPs to specify the nervous system. First, Khokha and colleagues (2005) used antisense morpholinos to eliminate three BMP antagonists (Noggin, Chordin, and Follistatin) in *Xenopus*. The resulting embryos had catastrophic failure of dorsal development and lacked neural plates and dorsal mesoderm (**FIGURE 11.25A,B**). Second, Reversade and colleagues blocked BMP activity with antisense morpholinos (Reversade et al. 2005; Reversade and De Robertis 2005). When they simultaneously blocked the formation of BMPs 2, 4, and 7, the neural tube became greatly expanded, taking over a much larger region of the ectoderm (**FIGURE 11.25C**). When they did a quadruple inactivation of the three BMPs *and* ADMP (another protein of the BMP family), the entire ectoderm became neural (**FIGURE 11.25D**). Thus, the epidermis is instructed by BMP signaling, and the organizer specifies the ectoderm above it to become neural by blocking that BMP signal from reaching the adjacent ectoderm.

In the absence of BMP signaling, the Foxd4 transcription factor becomes expressed in the presumptive neural ectoderm. It initiates a pathway that leads to the stabilization of neural identity in most of the induced ectodermal cells, while allowing the formation of an immature, stem cell-like state in other induced cells (see Rogers et al. 2009; Klein and Moody 2015).

Remarkably, the ability of BMPs to induce the skin ectoderm and BMP antagonists to specify the neural ectoderm has been seen across the animal kingdom. In *Drosophila*,

FIGURE 11.25 Control of neural specification by levels of BMPs. (A,B) Lack of dorsal structures in *Xenopus* embryos whose BMP-inhibitor genes *chordin*, *noggin*, and *follistatin* were eliminated by antisense morpholino oligonucleotides. (A) Control embryo with neural folds stained for the expression of the neural gene *Sox2*. (B) Lack of neural tube and *Sox2* expression in an embryo treated with the morpholinos against three BMP inhibitors. (C,D) Expanded neural development. (C) The neural tube, visualized by *Sox2* staining, is greatly enlarged in embryos treated with antisense morpholinos that destroy BMPs 2, 4, and 7. (D) Complete transformation of the entire ectoderm into neural ectoderm (and loss of the dorsal-ventral axis) by inactivation of ADMP as well as BMPs 2, 4, and 7. (A,B from Khokha et al. 2005, courtesy of R. Harland; C,D from Reversade and De Robertis 2005.)

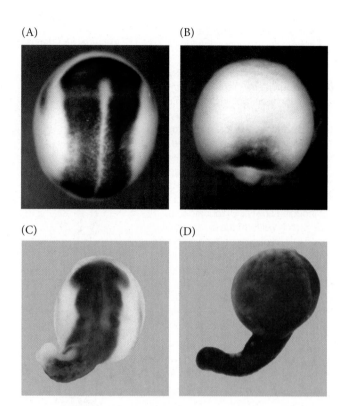

(A) (B)

(C) (D)

the BMP homologue Decapentaplegic (Dpp) specifies the hypodermis (skin), while the BMP antagonist Short gastrulation (Sog) blocks the actions of Dpp and specifies the neural system. Sog protein is a homologue of Chordin. These insect homologues not only appear to be similar to their vertebrate counterparts, they can actually substitute for each other. When *sog* mRNA is injected into ventral regions of *Xenopus* embryos, it induces the amphibian notochord and neural tube. Injecting *chordin* mRNA into *Drosophila* embryos produces ventral nervous tissue. Although Chordin dorsalizes the *Xenopus* embryo, it ventralizes *Drosophila*. In *Drosophila*, Dpp is made dorsally; in *Xenopus*, BMP4 is made ventrally. In both cases, Sog/Chordin helps specify neural tissue by blocking the effects of Dpp/BMP4 (Hawley et al. 1995; Holley et al. 1995; De Robertis et al. 2000; Bier and De Robertis 2015). Thus, arthropods appear to be upside-down vertebrates—a fact the French anatomist Geoffroy Saint-Hilaire pointed out in his attempts to convince other anatomists of the unity of the animal kingdom in the 1840s (see Appel 1987; Genikhovich et al. 2015; De Robertis and Moriyama 2016).

SCIENTISTS SPEAK 11.2   Dr. Richard Harland discusses gastrulation and neural induction in *Xenopus*.

Regional Specificity of Neural Induction along the Anterior-Posterior Axis

Just as the dorsal-ventral axis across the animal kingdom is predicated on BMP and its inhibitors (the neural region being the area of lowest BMPs), the specification of the anterior-posterior axis is predicated on a gradient of Wnt proteins, with the head being characterized by the lowest concentrations of Wnts (Petersen and Reddien 2009). There are a few exceptions (such as *Drosophila*) where the Wnt gradient does not provide major patterning cues; but even in these cases, vestigial patterns of Wnt are still seen (Vorwald-Denholtz and De Robertis 2011).

In vertebrates, one of the most important phenomena along the anterior-posterior axis is the regional specificity of the neural structures that are produced. Forebrain, hindbrain, and spinocaudal regions of the neural tube must be properly organized in an anterior-to-posterior direction. The organizer tissue not only induces the neural tube, it also specifies the regions of the neural tube. This region-specific induction was demonstrated by Hilde Mangold's husband, Otto Mangold, in 1933. He transplanted four successive regions of the archenteron roof of late-gastrula newt embryos into the blastocoels of early-gastrula embryos. The most anterior portion of the archenteron roof (containing head mesoderm) induced balancers and portions of the oral apparatus; the

Developing Questions

Chordin and BMPs seem to be homologous between flies and vertebrates—but are the *processing pathways* of Chordin and BMP homologous as well? How might the Chordin-BMP axis allow the regulation of the embryonic axes such the same pattern always occurs, even if the embryo is much smaller or larger?

FIGURE 11.26 Regional and temporal specificity of induction. (A–D) Regional specificity of structural induction can be demonstrated by implanting different regions (color) of the archenteron roof into early *Triturus* gastrulae. The resulting embryos develop secondary dorsal structures. (A) Head with balancers. (B) Head with balancers, eyes, and forebrain. (C) Posterior part of head, diencephalon, and otic vesicles. (D) Trunk-tail segment. (E,F) Temporal specificity of inducing ability. (E) Young dorsal lips (which will form the anterior portion of the organizer) induce anterior dorsal structures when transplanted into early newt gastrulae. (F) Older dorsal lips transplanted into early newt gastrulae produce more posterior dorsal structures. (A–D after Mangold 1933; E,F after Saxén and Toivonen 1962.)

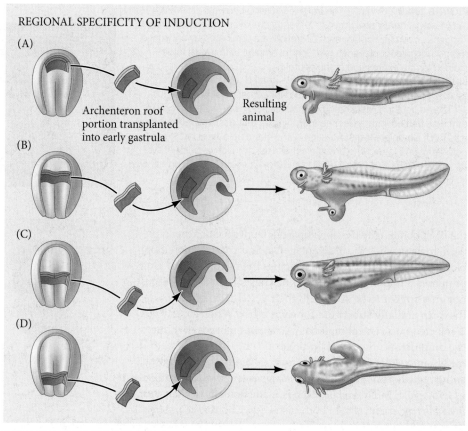

REGIONAL SPECIFICITY OF INDUCTION

(A)

Archenteron roof portion transplanted into early gastrula

Resulting animal

(B)

(C)

(D)

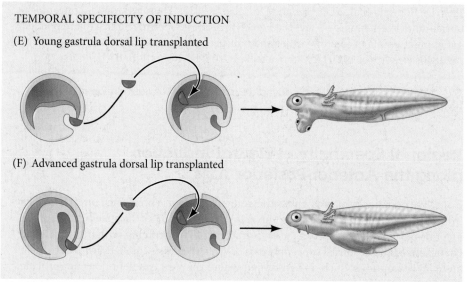

TEMPORAL SPECIFICITY OF INDUCTION

(E) Young gastrula dorsal lip transplanted

(F) Advanced gastrula dorsal lip transplanted

next most anterior section induced the formation of various head structures, including nose, eyes, balancers, and otic vesicles; the third section (including the notochord) induced the hindbrain structures; and the most posterior section induced the formation of dorsal trunk and tail mesoderm[5] (**FIGURE 11.26A–D**).

---

[5]The induction of dorsal mesoderm—rather than the dorsal ectoderm of the nervous system—by the posterior end of the notochord was confirmed by Bijtel (1931) and Spofford (1945), who showed that the posterior fifth of the neural plate gives rise to tail somites and the posterior portions of the pronephric kidney duct.

In further experiments, Mangold demonstrated that when dorsal blastopore lips from early salamander gastrulae were transplanted into other *early* salamander gastrulae, they formed secondary heads. When dorsal lips from *later* gastrulas were transplanted into early salamander gastrulae, however, they induced the formation of secondary tails (**FIGURE 11.26E,F**; Mangold 1933). These results show that the first cells of the organizer to enter the embryo induce the formation of brains and heads, while those cells that form the dorsal lip of later-stage embryos induce the cells above them to become spinal cords and tails.

The question then became, What are the molecules being secreted by the organizer in a regional fashion such that the first cells involuting through the blastopore lip (the endomesoderm) induce head structures, whereas the next portion of involuting mesoderm (notochord) produces trunk and tail structures? **FIGURE 11.27** shows a possible model for these inductions, the elements of which we will now describe in detail.

The head inducer: Wnt antagonists

The most anterior regions of the head and brain are underlain not by notochord but by pharyngeal endoderm and head (prechordal) mesoderm (see Figures 11.4C,D and 11.27A). This endomesodermal tissue constitutes the leading edge of the dorsal blastopore lip. Recent studies have shown that these cells not only induce the most anterior head structures, but that they do it by blocking the Wnt pathway as well as by blocking BMPs. The Wnt antagonists appear to be induced by the high levels of phosphorylated Smad2 in response to Nodal and Vg1 secreted by the vegetal cells (Agius et al. 2000; Bisroy et al. 2006).

CERBERUS: AN ALL-PURPOSE PARACRINE INHIBITOR FOR HEAD PRODUCTION   The induction of trunk structures may be caused by the blockade of BMP signaling from the notochord, while Wnt signals are allowed to proceed. However, to produce a head, both

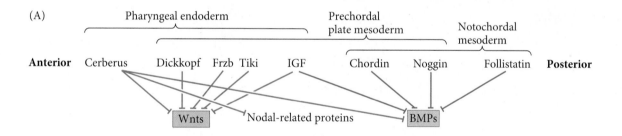

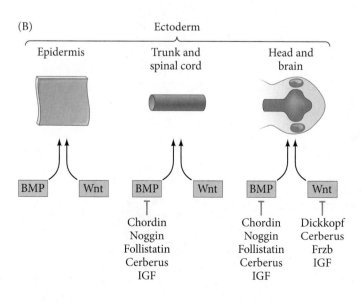

FIGURE 11.27 Paracrine factor antagonists from the organizer are able to block specific paracrine factors to distinguish head from tail. (A) The pharyngeal endoderm that underlies the head secretes Dickkopf, Frzb, and Cerberus. Dickkopf and Frzb block Wnt proteins; Cerberus blocks Wnts, Nodal-related proteins, and BMPs. The prechordal plate secretes the Wnt blockers Dickkopf and Frzb, as well as BMP blockers Chordin and Noggin. The notochord contains the BMP blockers Chordin, Noggin, and Follistatin but does not secrete Wnt blockers. Insulin-like growth factor (IGF) from the head endomesoderm probably acts at the junction of the notochord and prechordal mesoderm. (B) Summary of paracrine antagonist function in the ectoderm. Brain formation requires inhibiting both the Wnt and BMP pathways. Spinal cord neurons are produced when Wnt functions without the presence of BMPs. Epidermis is formed when both the Wnt and BMP pathways are operating.

the BMP signal and the Wnt signal must be blocked. This Wnt blockade comes from the endomesoderm, the most anterior portion of the organizer (Glinka et al. 1997). In 1996, Bouwmeester and colleagues showed that the induction of the most anterior head structures could be accomplished by a secreted protein called Cerberus (named after the three-headed dog that guarded the entrance to Hades in Greek mythology). When *cerberus* mRNA was injected into a vegetal ventral *Xenopus* blastomere at the 32-cell stage, ectopic head structures were formed (**FIGURE 11.28A**). These head structures arose from the injected cell as well as from neighboring cells.

The *cerberus* gene is expressed in the pharyngeal endomesoderm cells that arise from the deep cells of the early dorsal lip. Cerberus protein can bind BMPs, Nodal-related proteins, and Xwnt8 (see Figure 11.27A and 11.30; Piccolo et al. 1999). When Cerberus synthesis is blocked, the levels of BMP, Nodal-related proteins, and Wnts all rise in the anterior of the embryo, and the ability of the anterior endomesoderm to induce a head is severely diminished (Silva et al. 2003).

FRZB, DICKKOPF, NOTUM, AND TIKI: MORE WAYS TO BLOCK WNTS Shortly after the attributes of Cerberus were demonstrated, two other proteins, Frzb and Dickkopf, were

(A)
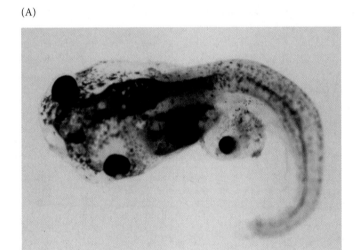

(B)
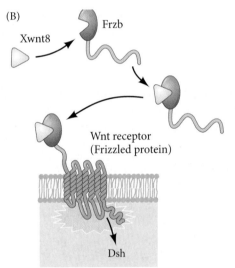

(C)
(D)
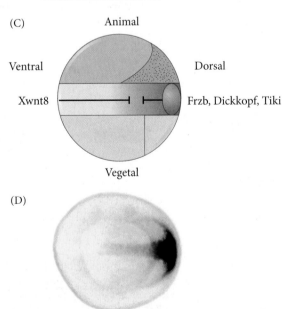

FIGURE 11.28 Inhibiting Wnt signaling enables head formation. (A) Injecting *cerberus* mRNA into a single D4 (ventral vegetal) blastomere of a 32-cell *Xenopus* embryo induces head structures as well as a duplicated heart and liver. The secondary eye (a single cyclopic eye) and olfactory placode can be readily seen. Xwnt8 is capable of ventralizing the mesoderm and preventing anterior head formation in the ectoderm. (B) Frzb protein is secreted by the anterior region of the organizer. It must bind to Xwnt8 before that inducer can bind to its receptor. Frzb resembles the Wnt-binding domain of the Wnt receptor (the Frizzled protein), but Frzb is a soluble molecule. (C) Xwnt8 is made throughout the marginal zone. (D) Double in situ hybridization localizing Frzb (dark stain) and Chordin (reddish stain) messages. The *frzb* mRNA is transcribed in the head endomesoderm of the organizer, but not in the notochord (where *chordin* is expressed). (A from Bouwmeester et al. 1996; D from Leyns et al. 1997; photographs courtesy of E. M. De Robertis.)

found to be synthesized in the involuting endomesoderm. Frzb (pronounced "frisbee") is a small, soluble form of Frizzled (the Wnt receptor) that is capable of binding Wnt proteins in solution (**FIGURE 11.28B,C**; Leyns et al. 1997; Wang et al. 1997). Frzb is synthesized predominantly in the endomesoderm cells beneath the prospective brain (**FIGURE 11.28D**). If embryos are made to synthesize excess Frzb, Wnt signaling fails to occur throughout the embryo; such embryos lack ventral posterior structures and become "all head." The Dickkopf (German, "thick head," "stubborn") protein also appears to interact directly with the Wnt receptors, preventing Wnt signaling (Mao et al. 2001, 2002). Injection of antibodies against Dickkopf causes the resulting embryos to have small, deformed heads with no forebrain (Glinka et al. 1998).

Two other organizer proteins, Tiki and Notum, have recently been found to bind to Wnt proteins during gastrulation. Tiki not only prevents Wnts binding to their receptors, it cleaves the Wnt to render it nonfunctional. Tiki is synthesized primarily in the anterior regions of the organizer and is crucial for head formation in *Xenopus* (Zhang et al. 2012.) And, to make certain that Wnt doesn't stop brain development, the ectoderm itself makes a membrane-tethered Wnt inhibitor, Notum, that acts by removing the lipid moiety that keeps the Wnt proteins from forming inactive dimers with one another (Zhang et al. 2015).

INSULIN-LIKE GROWTH FACTORS AND FIBROBLAST GROWTH FACTORS   All the above-mentioned Wnt inhibitors are extracellular. In addition to these, the head region contains yet another set of proteins that prevent BMP and Wnt signals from reaching the nucleus. **Fibroblast growth factors** (**FGFs**) and **insulin-like growth factors** (**IGFs**) are also required for inducing the brain and sensory placodes (Pera et al. 2001, 2013). IGFs and FGFs are especially prominent in the anterior region of the embryo, and they both initiate the receptor tyrosine kinase (RTK) signal transduction cascade (see Chapter 4). These tyrosine kinases interfere with the signal transduction pathways of both BMPs and Wnts (Richard-Parpaillon et al. 2002; Pera et al. 2013). When injected into ventral mesodermal blastomeres, mRNA for IGFs causes the formation of ectopic heads, while blocking IGF receptors in the anterior results in the lack of head formation.

Trunk patterning: Wnt signals and retinoic acid

Toivonen and Saxén provided evidence for a gradient of a posteriorizing factor that would act to specify the trunk and tail tissues of the amphibian embryo[6] (Toivonen and Saxén 1955, 1968; reviewed in Saxén 2001). This factor's activity would be highest in the posterior of the embryo and weakened anteriorly. Recent studies have extended this model and have proposed that Wnt proteins, especially Wnt8, as the posteriorizing molecules (Domingos et al. 2001; Kiecker and Niehrs 2001). In *Xenopus*, an endogenous gradient of Wnt signaling and β-catenin is highest in the posterior and absent in the anterior (**FIGURE 11.29A**). Moreover, if Xwnt8 is added to developing embryos, spinal cord-like neurons are seen more anteriorly in the embryo, and the most anterior markers of the forebrain are absent. Conversely, suppressing Wnt signaling (by adding Frzb or Dickkopf to the developing embryo) leads to the expression of the anteriormost markers in more posterior neural cells. Therefore, there appear to be two major gradients in the amphibian gastrula—a BMP gradient that specifies the dorsal-ventral axis and a Wnt gradient specifying the anterior-posterior axis (**FIGURE 11.29B**). It must be remembered, too, that both of these axes are established by the initial axes of

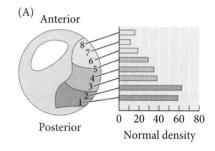

(A)

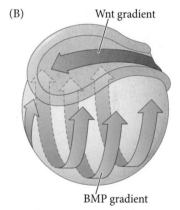

(B)

FIGURE 11.29   Signaling gradients and axis specification. (A) A Wnt signaling pathway posteriorizes the neural tube. Gastrulating embryos were stained for β-catenin and the density of the stain compared between regions of the ectodermal cells, revealing a gradient of β-catenin in the presumptive neural plate. (B) The Wnt gradient specifies posterior-anterior polarity and the BMP gradient specifies dorsal-ventral polarity. This double-gradient interaction, first discovered in amphibians, has now been shown to be characteristic of animal development. (After Saxén and Toivonen 1962; Kiecker and Niehrs 2001 and Niehrs 2004.)

---

[6] The tail inducer was initially thought to be part of the trunk inducer, since transplantation of the late dorsal blastopore lip into the blastocoel often produced larvae with extra tails. However, it appears that tails are normally formed by interactions between the neural plate and the posterior mesoderm during the neurula stage (and thus are generated outside the organizer). Here, Wnt, BMPs, and Nodal signaling all seem to be required (Tucker and Slack 1995; Niehrs 2004). Interestingly, all three of these signaling pathways have to be inactivated if the head is to form.

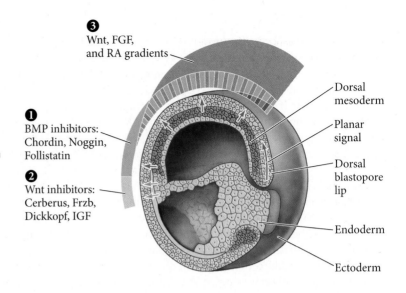

FIGURE 11.30 Model of organizer function and axis specification in the *Xenopus* gastrula. (1) BMP inhibitors from organizer tissue (dorsal mesoderm and pharyngeal mesendoderm) block the formation of epidermis, ventro-lateral mesoderm, and ventrolateral endoderm. (2) Wnt inhibitors in the anterior of the organizer (pharyngeal endo-mesoderm) allow the induction of head structures. (3) A gradient of caudalizing factors (Wnts, FGFs, and retinoic acid) results in the regional expression of Hox genes, which specify the regions of the neural tube.

❸
Wnt, FGF,
and RA gradients

Dorsal
mesoderm

Planar
signal

Dorsal
blastopore
lip

Endoderm

Ectoderm

❶
BMP inhibitors:
Chordin, Noggin,
Follistatin

❷
Wnt inhibitors:
Cerberus, Frzb,
Dickkopf, IGF

Nodal-like TGF-β paracrine factors and β-catenin across the vegetal cells. The basic model of neural induction, then, looks like the diagram in **FIGURE 11.30**.

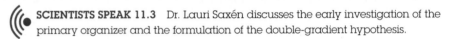 **SCIENTISTS SPEAK 11.3** Dr. Lauri Saxén discusses the early investigation of the primary organizer and the formulation of the double-gradient hypothesis.

While the Wnt proteins play a major role in specifying the anterior-posterior axis, they are probably not the only agents involved. Fibroblast growth factors appear to be critical in allowing the cells to respond to the Wnt signal (Holowacz and Sokol 1999; Domingos et al. 2001). Retinoic acid (RA) also is seen to have a gradient highest at the posterior end of the neural plate, and RA can also posteriorize the neural tube in a concentration-dependent manner (Cho and De Robertis 1990; Sive and Cheng 1991; Chen et al. 1994; Pera et al. 2013). RA signaling appears to be especially important in patterning the hindbrain, as it appears to interact with FGF signals to activate the posterior Hox genes (Kolm et al. 1997; Dupé and Lumsden 2001; Shiotsugu et al. 2004). RA is also critical in allowing the growth of the most posterior part of the tadpole, the tail. The receptors for RA are transcription factors, and when they are not bound to retinoic acid, they bind co-repressors that inhibit the activity of certain genes. However, when RA binds to the receptors, the co-repressors are exchanged for co-activators, and these genes become actively transcribed (Chakravarti et al. 1996). The unbound RA receptors in the tailbud of the developing tadpole maintain a pool of undifferentiated caudal mesoderm cells. As the wavefront of RA synthesis extends posteriorly, the retinoic acid binds to these receptors, after which these cells express posterior Hox genes and differentiate into somatic tissue and allow the outgrowth of the tail (Janesick et al. 2014). Together, the posterior-to-anterior Wnt, FGF, and RA gradients function to determine the boundaries of the Hox genes along the anterior-posterior axis (Wacker et al. 2004; Durston et al. 2010a,b).

WEB TOPIC 11.4 **GRADIENTS AND HOX GENE EXPRESSION** The mechanisms by which the gradients of Wnt, RA, and FGFs specify the Hox genes in neural ectoderm are still not agreed upon.

Specifying the Left-Right Axis

Although the developing tadpole outwardly appears symmetrical, several internal organs, such as the heart and the gut tube, are not evenly balanced on the right and left sides. In other words, in addition to its dorsal-ventral and anterior-posterior axes, the embryo has a left-right axis. In all vertebrates studied so far, the crucial event in left-right axis formation is the expression of a Nodal gene in the lateral plate mesoderm on the left side of the embryo. In *Xenopus*, this gene is *Xnr1* (*Xenopus* nodal-related 1). If *Xnr1* expression is reversed (so as to be solely on the right-hand side), the position of the heart (normally found on the left) is reversed, as is the coiling of the gut. If *Xnr1* is expressed on both sides, coiling and heart placement is random.

But what limits *Xnr1* expression to the left-hand side? In Chapter 8, we saw that left-right patterning in snails was controlled by Pitx2 and Nodal and was regulated by cytoskeletal proteins that were active during the first cleavage cycles. The case is similar in frogs, where Pitx2 and Nodal appear to be regulated by the cytoskeleton. In frogs, the cytoskeleton (especially the tubulin of microtubules) has been implicated in right-left patterning, During the first cleavage of *Xenopus*, embryos, tubulin-associated proteins are critical for distributing maternal products differentially into the future left- and right-side cells (Lobikin et al. 2012). Mutations of tubulin genes cause laterality defects in frogs (and in *C. elegans*), and the injection of dominant-negative tubulin mutant proteins randomizes heart and guts.

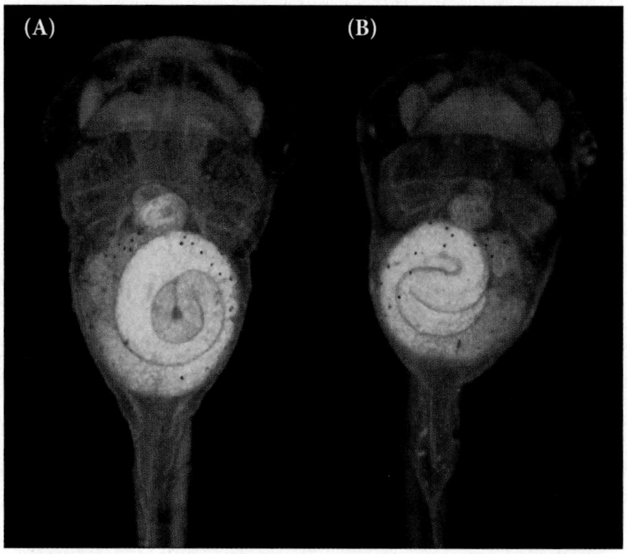

This original lateralization may be strengthened by the clockwise rotation of cilia found in the organizer region. In *Xenopus*, these specific cilia are formed at the dorsal blastopore lip during the later stages of gastrulation (i.e., after the original specification of the mesoderm) (Schweickert et al. 2007; Blum et al. 2009). That is, they are located in the posterior region of the embryo, at the site where the archenteron is still forming. If rotation of these cilia is blocked, *Xnr1* expression fails to occur in the mesoderm and laterality defects result (Walentek et al. 2013).

One of the key genes activated by Xnr1 protein appears to encode the transcription factor Pitx2, which normally is expressed only on the left side of the embryo. Pitx2 persists on the left side as the heart and gut develop, controlling their respective positions. If Pitx2 is injected into the right side of an embryo, it is expressed there as well and heart placement and gut coiling are randomized (**FIGURE 11.31**; Ryan et al. 1998). As we will see, the pathway through which Nodal protein establishes left-right polarity by activating Pitx2 on the left side is conserved throughout all vertebrate lineages.

FIGURE 11.31 Pitx2 determines the direction of heart looping and gut coiling. (A) A wild-type *Xenopus* tadpole viewed from the ventral side, showing rightward heart looping and counterclockwise gut coiling. (B) If an embryo is injected with Pitx2 so that this protein is present in the mesoderm of both the right and left sides (instead of just the left side), heart looping and gut coiling are random with respect to each other. Sometimes this treatment results in complete reversals, as in this embryo, in which the heart loops to the left and the gut coils in a clockwise manner. (From Ryan et al. 1998, courtesy of J. C. Izpisúa-Belmonte.)

Early Zebrafish Development

In recent years, the teleost (bony) fish *Danio rerio*, commonly known as the zebrafish, has joined *Xenopus* as a widely studied model of vertebrate development (see Figure 11.1B). Despite differences in their cleavage patterns (*Xenopus* eggs are holoblastic, dividing the entire egg, whereas the yolky zebrafish egg is meroblastic, wherein only a small portion of the yolky cytoplasm forms cells), *Xenopus* and *Danio* form their body axes and specify their cells in very similar ways.

Zebrafish have large broods, breed all year, are easily maintained, have transparent embryos that develop outside the mother (an important feature for microscopy), and can be raised so that mutants can be readily discovered and propagated in the laboratory. In addition, these fish develop rapidly. By 24 hours after fertilization, the embryo has already formed most of its organ primordia and displays a characteristic tadpole-like form (**FIGURE 11.32**; see Granato and Nüsslein-Volhard 1996; Langeland and Kimmel 1997). Furthermore, the ability to microinject fluorescent dyes into single blastomeres and to generate transgenes driving cell-type-specific fluorescent protein expression has allowed scientists to follow individual living cells as an organ develops.

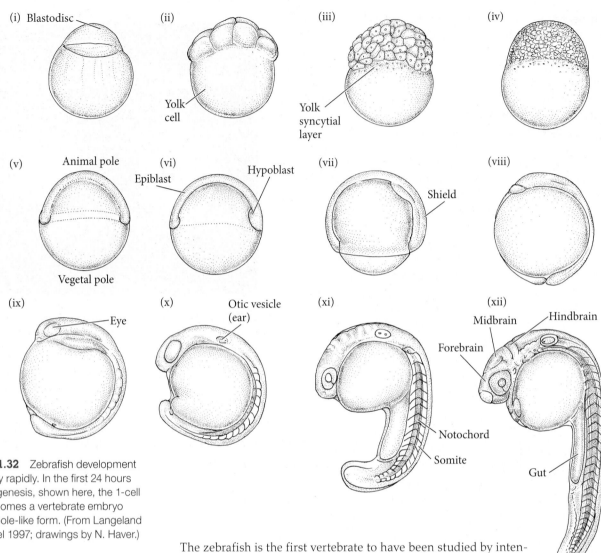

FIGURE 11.32 Zebrafish development occurs very rapidly. In the first 24 hours of embryogenesis, shown here, the 1-cell zygote becomes a vertebrate embryo with a tadpole-like form. (From Langeland and Kimmel 1997; drawings by N. Haver.)

The zebrafish is the first vertebrate to have been studied by intensive mutagenesis screens. By treating parents with mutagens and selectively breeding the progeny, scientists have found thousands of mutations whose normally functioning genes are critical for development. The traditional method of genetic screening (modeled after large-scale screens in *Drosophila*) begins when the male parental fish are treated with a chemical mutagen that will cause random mutations in their germ cells. Each mutagenized male is then mated with a wild-type female fish to generate F_1 fish. Individuals in the F_1 generation carry the mutations inherited from their father. If the mutation is dominant, it will be expressed in the F_1 generation. If the mutation is recessive, the F_1 fish will not show a mutant phenotype, since the wild-type dominant allele will mask the mutation. The F_1 fish are then mated with wild-type fish to produce an F_2 generation that includes both males and females that carry the mutant allele. When two F_2 parents carry the same recessive mutation, there is a 25% chance that their offspring will show the mutant phenotype (**FIGURE 11.33**). Since zebrafish development occurs in the open (as opposed to within an opaque shell or inside the mother's body), abnormal developmental stages can be readily observed, and the defects in development can often be traced to changes in a particular group of cells (Driever et al. 1996; Haffter et al. 1996). Recently, high-throughput methods of gene analysis and the CRISPR genome editing system have propelled the analysis of zebrafish development, enabling mutations in particular genes to be rapidly generated, identified, and bred (see Gonzales and Yeh 2014; Vashney et al. 2015).

Parents

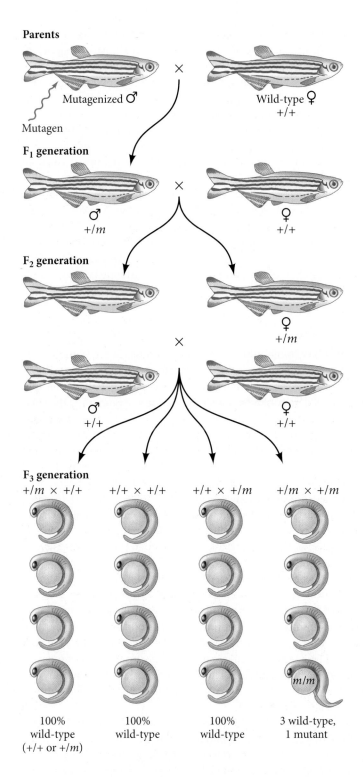

FIGURE 11.33 Screening protocol for identifying mutations of zebrafish development. The male parent is mutagenized and mated with a wild-type (+/+) female. If some of the male's sperm carry a recessive mutant allele (*m*), then some of the F₁ progeny of the mating will inherit that allele. F₁ individuals (here shown as a male carrying the mutant allele (*m*) are then mated with wild-type partners. This creates an F₂ generation wherein some males and some females carry the recessive mutant allele. When the F₂ fish are mated, approximately 25% of their progeny will show the mutant phenotype. (After Haffter et al. 1996.)

Like *Xenopus* embryos, zebrafish embryos are susceptible to morpholino antisense molecules (Zhong et al. 2001), and researchers can use this method to test whether a particular gene is required for a particular function. Furthermore, the green fluorescent protein reporter gene can be fused with specific zebrafish promoters and enhancers and inserted into the fish embryos. The resulting transgenic fish express GFP at the same times and places as the endogenous proteins controlled by these regulatory sequences. The amazing thing is that one can observe the reporter protein in living transparent embryos (**FIGURE 11.34**).

(A)

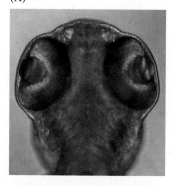

(B)

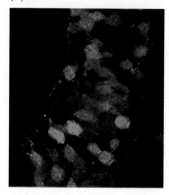

FIGURE 11.34 The gene for green fluorescent protein (GFP) was fused to the regulatory region of a zebrafish *sonic hedgehog* gene. As a result, GFP was synthesized wherever Hedgehog protein is normally expressed in the fish embryo. (A) In the head, GFP is seen in the developing retina and nasal placodes. (B) Because GFP is expressed by individual cells, scientists can see precisely which cells make GFP, and thus which cells normally transcribe the gene of interest (in this case, *sonic hedgehog* in the retina). (Photographs courtesy of U. Strahle and C. Neumann.)

The similarity of developmental mechanisms among all vertebrates and the ability of *Danio rerio* to be genetically manipulated has given this small fish an important role in investigating the genes that operate during human development (Mudbhary and Sadler 2011). When developmental biologists screened zebrafish mutants for cystic kidney disease, they found 12 different genes. Two of these genes were already known to cause human cystic kidney disease, but the other 10 were as-yet unknown genes that were found to interact with the first two in a common pathway. Moreover, that pathway, which involves the synthesis of cilia, was not what had been expected. Thus, the zebrafish studies disclosed an important and previously unknown pathway to explore human birth defects (Sun et al. 2004).

Zebrafish embryos are also permeable to small molecules placed in the water—a property that allows us to test drugs that may be deleterious to vertebrate development. For instance, zebrafish development can be altered by the addition of ethanol or retinoic acid, both of which produce malformations in the fish that resemble human developmental syndromes known to be caused by these molecules (Blader and Strähle 1998). As one zebrafish researcher joked, "Fish really are just little people with fins" (Bradbury 2004).

Cleavage

The eggs of most bony fish are **telolecithal**, meaning that most of the cytoplasm is occupied by yolk. Cleavage can take place only in the **blastodisc**, a thin region of yolk-free cytoplasm at the animal pole. The cell divisions do not completely divide the egg, so this type of cleavage is called **meroblastic** (Greek *meros*, "part"). Since only the blastodisc becomes the embryo, this type of meroblastic cleavage is referred to as **discoidal**.

Scanning electron micrographs show beautifully the incomplete nature of discoidal meroblastic cleavage in fish eggs (**FIGURE 11.35**). The calcium waves initiated at fertilization stimulate the contraction of the actin cytoskeleton to squeeze non-yolky cytoplasm into the animal pole of the egg. This process converts the spherical egg into a pear-shaped structure with an apical blastodisc (Leung et al. 1998, 2000). In fish, there are many waves of calcium release, and they orchestrate the processes of cell division. The calcium ions are critical for coordinating mitosis. They integrate the movements of the mitotic spindle with those of the actin cytoskeleton, deepen the cleavage furrow, and heal the membrane after the separation of the blastomeres (Lee et al. 2003).

The first cell divisions follow a highly reproducible pattern of meridional and equatorial cleavages. These divisions are rapid, taking only about 15 minutes each. The first 10 divisions occur synchronously, forming a mound of cells that sits at the animal pole of a large **yolk cell**. This mound of cells constitutes the **blastoderm**. Initially, all the cells maintain some open connection with one another and with the underlying yolk cell, so that moderately sized (17 kDa) molecules can pass freely from one blastomere to the next (Kane and Kimmel 1993; Kimmel and Law 1985). Remarkably, as the daughter cells migrate away from one another, they often retain these bridges through long tunnels connecting the cells (Caneparo et al. 2011).

Maternal effect mutations have shown the importance of oocyte proteins and mRNAs in embryonic polarity, cell division, and axis formation (Dosch et al. 2004; Langdon and Mullins 2011). As in frogs, the microtubules are important roads along which morphogenetic determinants travel, and maternal mutants affecting the formation of the microtubule cytoskeleton prevent the normal positioning of the cleavage furrow and of mRNAs in the early embryo (Kishimoto et al. 2004).

Fish embryos, like many other embryos, undergo a mid-blastula transition (seen around the tenth cell division in zebrafish) when zygotic gene transcription begins, cell divisions slow, and cell movements become evident (Kane and Kimmel 1993). At this time, three distinct cell populations can be distinguished. The first of these is the **yolk syncytial layer**, or **YSL** (Agassiz and Whitman 1884; Carvalho and Heisenberg

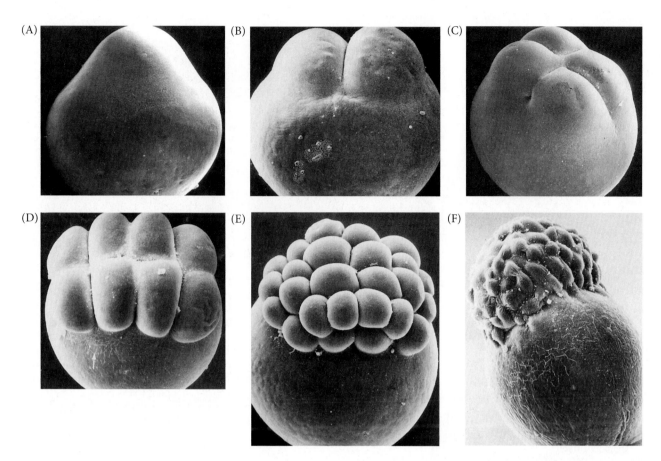

FIGURE 11.35 Discoidal meroblastic cleavage in a zebrafish egg. (A) 1-Cell embryo. The mound atop the cytoplasm is the blastodisc. (B) 2-Cell embryo. (C) 4-Cell embryo. (D) 8-Cell embryo, wherein two rows of four cells are formed. (E) 32-Cell embryo. (F) 64-Cell embryo, wherein the blastodisc can be seen atop the yolk cell. (From Beams and Kessel 1976, courtesy of the authors.)

2010). The YSL will not contribute cells or nuclei to the embryo, but it is critical for generating the fish organizer, patterning the mesoderm, and leading the epiboly of the ectoderm over the embryo (Chu et al. 2012). The YSL is formed at the tenth cell cycle, when the cells at the vegetal edge of the blastoderm fuse with the underlying yolk cell. This fusion produces a ring of nuclei in the part of the yolk cell cytoplasm that sits just beneath the blastoderm. Later, as the blastoderm expands vegetally to surround the yolk cell, some of the yolk syncytial nuclei will move under the blastoderm to form the **internal YSL** (**iYSL**), and others will move vegetally, staying ahead of the blastoderm margin, to form the **external YSL** (**eYSL**; **FIGURE 11.36A,B**). The YSL will be important for directing some of the cell movements of gastrulation.

The second cell population distinguished at the mid-blastula transition is the **enveloping layer** (**EVL**). It is made up of the most superficial cells from the blastoderm, which form an epithelial sheet a single cell layer thick. The EVL is a protective covering that is sloughed off after 2 weeks. It allows the embryo to develop in a hypotonic solution (such as fresh water) that would otherwise burst the cells (Fukazawa et al. 2010). Between the EVL and the YSL is the third set of blastomeres, the **deep cells**, that give rise to the embryo proper.

The fates of the early blastoderm cells are not determined, and cell lineage studies (in which a nondiffusible fluorescent dye is injected into a cell so that its descendants can be followed) show that there is much cell mixing during cleavage. Moreover, any one of these early blastomeres can give rise to an unpredictable variety of tissue descendants (Kimmel and Warga 1987; Helde et al. 1994). A fate map of the blastoderm cells can be made shortly before gastrulation begins. At this time, cells in specific regions of the embryo give rise to certain tissues in a highly predictable manner (**FIGURE 11.36C**; see also Figure 1.11), although they remain plastic, and cell fates can change if tissue is grafted to a new site.

VADE MECUM

The segment on zebrafish development includes time-lapse movies of the beautiful and rapid development of this organism.

(A)

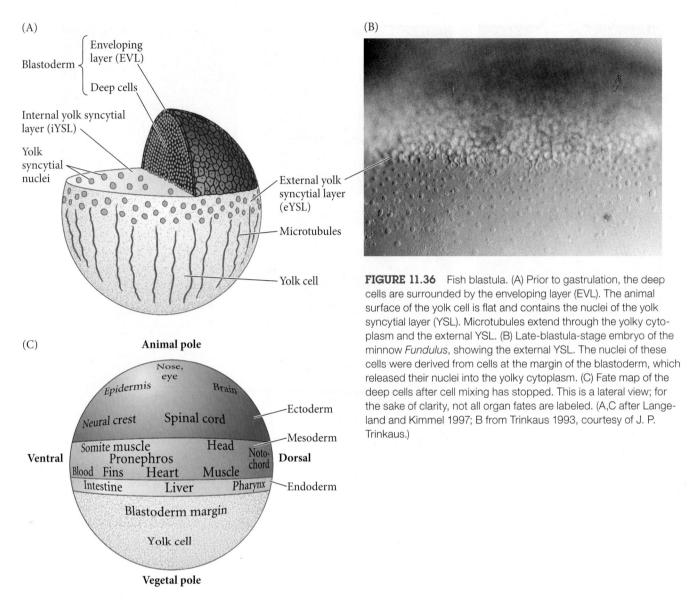

Blastoderm { Enveloping layer (EVL)

Deep cells

Internal yolk syncytial layer (iYSL)

Yolk syncytial nuclei

External yolk syncytial layer (eYSL)

Microtubules

Yolk cell

(B)

(C)

Animal pole

Nose, eye

Epidermis

Brain

Neural crest

Spinal cord

Ectoderm

Somite muscle

Head

Mesoderm

Ventral

Pronephros

Noto-chord

Dorsal

Blood Fins

Heart

Muscle

Intestine

Liver

Pharynx

Endoderm

Blastoderm margin

Yolk cell

Vegetal pole

FIGURE 11.36 Fish blastula. (A) Prior to gastrulation, the deep cells are surrounded by the enveloping layer (EVL). The animal surface of the yolk cell is flat and contains the nuclei of the yolk syncytial layer (YSL). Microtubules extend through the yolky cytoplasm and the external YSL. (B) Late-blastula-stage embryo of the minnow *Fundulus*, showing the external YSL. The nuclei of these cells were derived from cells at the margin of the blastoderm, which released their nuclei into the yolky cytoplasm. (C) Fate map of the deep cells after cell mixing has stopped. This is a lateral view; for the sake of clarity, not all organ fates are labeled. (A,C after Langeland and Kimmel 1997; B from Trinkaus 1993, courtesy of J. P. Trinkaus.)

Gastrulation and Formation of the Germ Layers

All three layers of the zebrafish blastoderm undergo epiboly. The first cell movement of fish gastrulation is the epiboly of the blastoderm cells over the yolk, and this is thought to be controlled both by maternal proteins (such as Eomesodermin) and by new proteins transcribed from the YSL nuclei (Du et al. 2012). In the initial phase of this movement, the deep cells of the blastoderm move outward to intercalate with the cells closer to the surface of the embryo, and the yolk cell (with its syncytial nuclei) pushes upward (Warga and Kimmel 1990). This intercalation of cells causes a flattening of the "dome" of the blastoderm cells (**FIGURE 11.37A**).

PROGRESSION OF EPIBOLY When about half the yolk is covered, a new set of movements is initiated. The YSL nuclei divide such that some nuclei (constituting the external YSL, or eYSL) remain in the upper cortex of the yolk cell, while the iYSL (internal YSL) lies beneath the blastoderm. The enveloping layer is tightly joined to the iYSL by E-cadherin and tight junctions (Shimizu et al. 2005a; Siddiqui et al. 2010) and is dragged ventrally as the iYSL nuclei migrate "downward." That the vegetal migration of the blastoderm margin is dependent on the epiboly of the YSL can be demonstrated by severing the attachments between the YSL and the EVL. When this is done, the

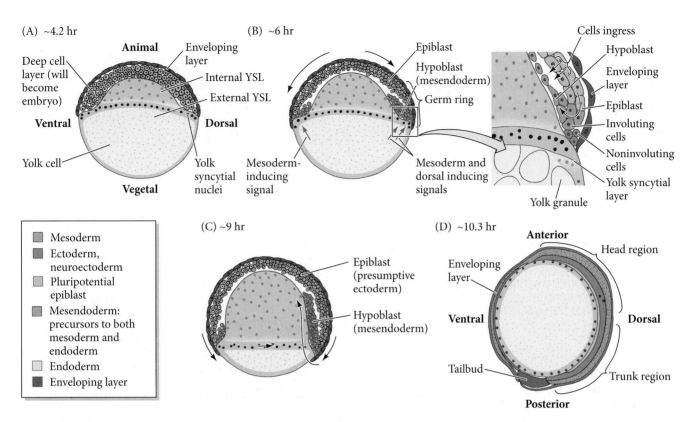

EVL and the deep cells spring back to the top of the yolk, while the YSL continues its expansion around the yolk cell (Trinkaus 1984, 1992).

The migration of the YSL ventrally depends partially on the expansion of this layer by cell division and intercalation, and partly on the cytoskeletal network within the yolk cell (see Lepage and Bruce 2010). An actomyosin band forms in the eYSL, at the boundary between the YSL and the EVL. This pulls down the YSL/EVL at its vegetal connection by means of contraction and friction (Behrndt et al. 2012). Meanwhile, the eYSL nuclei appear to migrate along the microtubles aligned along the animal-vegetal axis of the yolk cell, presumably pulling the iYSL and its accompanying EVL over the yolk cell. (Radiation or drugs that block the polymerization of tubulin slow epiboly; Strahle and Jesuthasan 1993; Solnica-Krezel and Driever 1994.) At the end of gastrulation, the entire yolk cell is covered by the blastoderm.

INTERNALIZATION OF THE HYPOBLAST After the blastoderm cells have covered about half the zebrafish yolk cell, a thickening occurs throughout the margin of the deep cells. This thickening, called the **germ ring**, is composed of a superficial layer, the epiblast (which will become the ectoderm); and an inner layer, the **hypoblast** (which will become endoderm and mesoderm). The hypoblast forms in a synchronous "wave" of internalization (Keller et al. 2008) that has some characteristics of ingression (especially in the dorsal region; see Carmany-Rampey and Schier 2001) and some elements of involution (especially in the future ventral regions). Thus, as the cells of the blastoderm undergo epiboly around the yolk, they are also internalizing cells at the blastoderm margin to form the hypoblast. The epiblast cells (presumptive ectoderm) do not involute, whereas the deep cells—the future mesoderm and endoderm—do (Figure 11.37 B,C). As the hypoblast cells internalize, the future mesoderm cells (the majority of the hypoblast cells) initially migrate vegetally while proliferating to make new mesoderm cells. Later, they alter direction and proceed toward the animal pole. The endodermal precursors, however, appear to move randomly over the yolk (Pézeron et al. 2008). The coordination of migration and cell specification

FIGURE 11.37 Cell movements during zebrafish gastrulation. (A) The blastoderm at 30% completion of epiboly (about 4.7 hours). (B) Formation of the hypoblast, either by involution of cells at the margin of the epibolizing blastoderm or by delamination and ingression of cells from the epiblast (6 hr). A close-up of the marginal region is at the right. (C) As ectodermal epiboly nears completion, the hypoblast, carrying the mesoderm and endoderm precursors, begins to cover the yolk. (D) Completion of gastrulation (10.3 hr). The germ layers (yellow endoderm, blue ectoderm, red mesoderm) are present. (After Driever 1995; Langeland and Kimmel 1997; Carvalho and Heisenberg 2010; Lepage and Bruce 2010.)

FIGURE 11.38 Stretching the zebrafish epiblast cells generates mesoderm. (A) During epiboly, cells at the border undergo structural changes and involute. As they do, mesodermal genes (red) are activated. (B) When the cortical cytoskeleton is prevented from contracting, the animal cap cells remain ectodermal and do not involute. (C) However, if these cells are pulled by a magnetic field, the mesodermal genes become expressed. (D,E) Circumpolar views of B and C, respectively, visualize expression of the mesodermal gene *Notail* (the zebrafish homologue of the *Brachyury* gene). (D) *Notail* expression is blocked by the lack of involution. (E) *Notail* expression induced by stretching and subsequent involution. (A–C after Piccolo 2013; D,E from Brunet et al. 2013.)

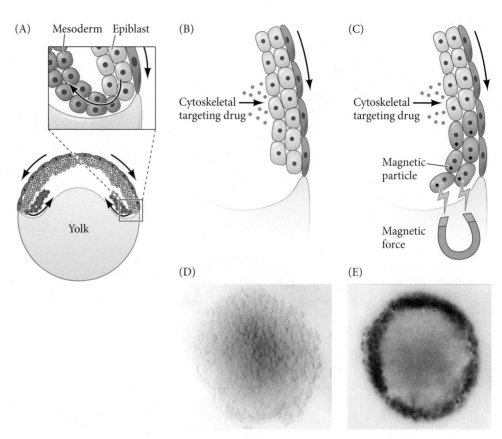

is accomplished by physical forces rather than by chemicals. When the cortical cytoskeleton is disrupted by drugs, the cells fail to turn and the mesodermal genes are not activated. However, if the cells are injected with magnetic particles before being hit with the drugs, they can be mechanically towed around the embryo. The cells don't involute, but the mesodermal genes do turn on. Thus, during normal development, epiboly and cell specification may be coordinated by the mechanical stress of involution (**FIGURE 11.38**; Brunet et al. 2013).

THE EMBRYONIC SHIELD AND THE NEURAL KEEL Once the hypoblast has formed, cells of the epiblast and hypoblast intercalate on the future dorsal side of the embryo to form a localized thickening, the **embryonic shield** (Schmitz and Campos-Ortega 1994). Here, the cells converge and extend anteriorly, eventually narrowing along the dorsal midline (**FIGURE 11.39A**). This convergent extension in the hypoblast forms the chordamesoderm, the precursor of the notochord (Trinkaus 1992; **FIGURE 11.39B,C**). This convergent extension is similar to that discussed in *Xenopus*, and is similarly accomplished by the Wnt-mediated planar cell polarity pathway (see Vervenne et al. 2008).

WATCH DEVELOPMENT 11.6 Watch convergent extension happen as the "ball" of cells is converted into a structure with a definite elongated anterior-posterior axis.

As we will see, the embryonic shield is functionally equivalent to the dorsal blastopore lip of amphibians, since it can organize a secondary embryonic axis when transplanted to a host embryo (Oppenheimer 1936; Ho 1992). The cells adjacent to the chordamesoderm—the paraxial mesoderm cells—are the precursors of the mesodermal somites (see Chapter 17). Concomitant convergence and extension in the epiblast bring presumptive neural cells from the epiblast into the dorsal midline, where they form the **neural keel**. The neural keel, a band of neural precursors that extends over the axial

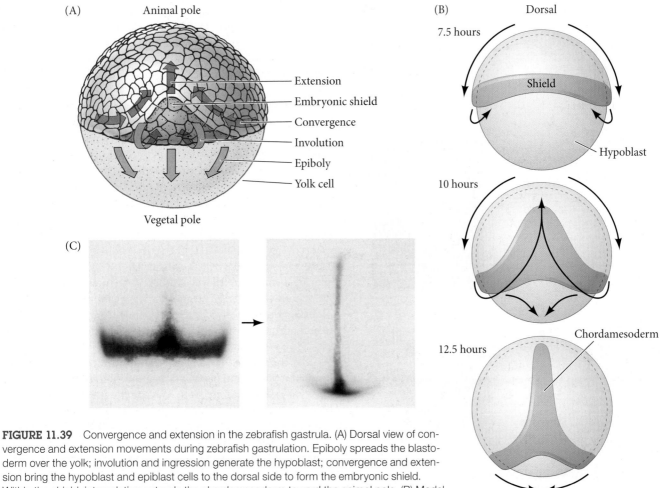

FIGURE 11.39 Convergence and extension in the zebrafish gastrula. (A) Dorsal view of convergence and extension movements during zebrafish gastrulation. Epiboly spreads the blastoderm over the yolk; involution and ingression generate the hypoblast; convergence and extension bring the hypoblast and epiblast cells to the dorsal side to form the embryonic shield. Within the shield, intercalation extends the chordamesoderm toward the animal pole. (B) Model of mesendoderm (hypoblast) formation. Numbers indicate hours after fertilization. On the future dorsal side, the internalized cells undergo convergent extension to form the chordamesoderm (notochord) and the paraxial (somitic) mesoderm adjacent to it. On the ventral side, the hypoblast cells migrate with the epibolizing epiblast toward the vegetal pole, eventually converging there. (C) Convergent extension of the chordamesoderm of the hypoblast cells. These cells are marked by their expression of the *no-tail* gene (dark areas) encoding a T-box transcription factor. (A,C from Langeland and Kimmel 1997, courtesy of the authors; B after Keller et al. 2001.)

and paraxial mesoderm, eventually develops a slitlike lumen to become the neural tube and to enter into the embryo.[7] Those cells remaining in the epiblast become the epidermis. On the ventral side (see Figure 11.39B), the hypoblast ring moves toward the vegetal pole, migrating directly beneath the epiblast that is epibolizing itself over the yolk cell. Eventually, the ring closes at the vegetal pole, completing the internalization of those cells that will become mesoderm and endoderm (Keller et al. 2008).

WATCH DEVELOPMENT 11.7 Watch two separate views of zebrafish neurulation, as well as Dr. Rolf Karlson's stunning video showing zebrafish development.

By different mechanisms, the *Xenopus* egg and zebrafish egg have reached the same state: they have become multicellular; they have undergone gastrulation; and they have

---

[7] This is different from the formation of the neural tube in frog embryos and is probably equivalent to "secondary" neural tube formation in the posterior of mammalian embryos (see Chapter 13).

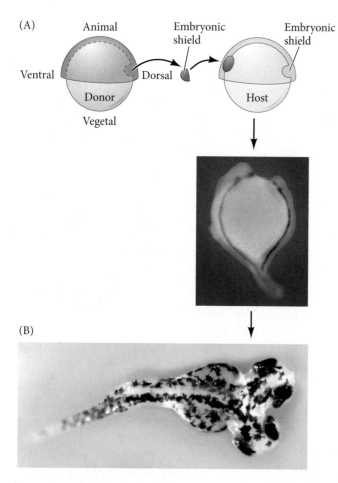

(A)

Animal

Ventral

Donor

Dorsal

Vegetal

Embryonic shield

Embryonic shield

Host

(B)

FIGURE 11.40 The embryonic shield as organizer in the fish embryo. (A) A donor embryonic shield (about 100 cells from a stained embryo) is transplanted into a host embryo at the same early-gastrula stage. The result is two embryonic axes joined to the host's yolk cell. In the photograph, both axes have been stained for *sonic hedgehog* mRNA, which is expressed in the ventral midline. (The embryo to the right is the secondary axis.) (B) The same effect can be achieved by activating nuclear β-catenin in embryos at sites opposite where the embryonic shield will form. (A after Shinya et al. 1999, photograph courtesy of the authors; B courtesy of J. C. Izpisúa-Belmonte.)

positioned their germ layers such that the ectoderm is on the outside, the endoderm is on the inside, and the mesoderm lies between them. We will now see that zebrafish form their body axes in ways very similar to those of *Xenopus*, and using very similar molecules.

Dorsal-ventral axis formation

As mentioned above, the embryonic shield of fish is homologous to the dorsal blastopore lip of amphibians, and it is critical in establishing the dorsal-ventral axis. Shield tissue can convert lateral and ventral mesoderm (blood and connective tissue precursors) into dorsal mesoderm (notochord and somites), and it can cause the ectoderm to become neural rather than epidermal. This transformative capacity was shown by transplantation experiments in which the embryonic shield of an early-gastrula embryo was transplanted to the ventral side of another (**FIGURE 11.40**; Oppenheimer 1936; Koshida et al. 1998). Two axes formed, sharing a common yolk cell. Although the prechordal plate and notochord were derived from the donor embryonic shield, the other organs of the secondary axis came from host tissues that would normally form ventral structures. The new axis had been induced by the donor cells.

Like the amphibian blastopore lip, the embryonic shield forms the prechordal plate and the notochord of the developing embryo. The prechordal plate cells are the first to involute, and they migrate toward the animal pole (Dumortier et al. 2012). The presumptive prechordal plate and notochord are responsible for inducing ectoderm to become neural ectoderm, and they appear to do this in a manner very much like the homologous structures in amphibians.[8] Like amphibians, fish induce the epidermis by BMPs (especially BMP2B) and Wnt proteins (especially Wnt8) made in the ventral and lateral regions of the embryo (see Schier 2001; Tucker et al. 2008). The notochords of both zebrafish and *Xenopus* secrete factors (the homologues of *chordin*, *noggin*, and *follistatin*) that block this induction, thereby allowing the ectoderm to become neural (Dal-Pra 2006). Like in amphibians, FGFs made in the dorsal side of the embryo also inhibit BMP gene expression (Fürthauer et al. 2004; Tsang et al. 2004; Little and Mullins 2006). In the caudal region of the embryo, FGF signaling is probably the predominant neural specifier (Kudoh et al. 2004). And as in *Xenopus*, insulin-like growth factors (IGFs) also play a role in the production of the anterior neural plate. Zebrafish IGFs appear to upregulate *chordin* and *goosecoid* while restricting the expression of *bmp2b*. Although IGFs appear to be made throughout the embryo, during gastrulation the IGF *receptors* are found predominantly in the anterior portion of the embryo (Eivers et al. 2004). Also, Wnt inhibitors appear to play roles in head formation. When antisense morpholinos are used to downregulate Wnt3a and Wnt8 throughout gastrulating zebrafish embryos, the trunk structures become anteriorized (Shimizu et al. 2005b).

But in fish, there may be another important source of organization: the entire blastopore lip. Recall that the blastopore of a fish extends around the entire yolk cell. The dorsal lip (the shield) will induce head structures when placed into the ventral region of the blastopore margin. However, it will not induce any structures from the neighboring

---

[8] Another similarity between the amphibian and fish organizers is that they can be duplicated by rotating the egg and changing the orientation of the microtubules (Fluck et al. 1998). One difference in the axial development of these groups is that in amphibians, the prechordal plate is necessary for inducing the anterior brain to form. In zebrafish, although the prechordal plate appears to be necessary for forming ventral neural structures, the anterior regions of the brain can form in its absence (Schier et al. 1997; Schier and Talbot 1998).

tissue when placed onto the animal cap of a blastula, which contains thoroughly undifferentiated cells. When a graft from the *ventral* blastopore lip is placed on animal cap cells, a well-organized tail structure is formed, having epidermis, somites, neural tube, but no dorsal mesoderm (Agathon et al. 2003). Much of this structure is induced from the host tissue. So the ventral blastopore lip in zebrafish is a "tail organizer." Cells from the *lateral* blastopore lips will induce trunk and posterior head structures, having notochord tissue. Moreover, these transplanted tissues do not express BMPs, Wnts, or their antagonists.

So it appears that, in addition to the classical shield organizer, the entire blastopore lip appears involved in forming posterior head, trunk and tail, through another means. This second set of axis-determining factors appears to be a dual gradient of Nodal and BMP proteins (Fauny et al. 2009; Thisse and Thisse 2015). Along the blastopore lip, from the ventral to the dorsal margin, there forms a continuous gradation in the ratio of BMP to Nodal activity. BMP is highest at the ventral margin, low dorsolaterally, and approaches 0 in the dorsalmost domain, where only Nodal is active. Thus, each region of the blastopore lip is characterized by a specific BMP/Nodal ratio of activity. Remarkably, an entire ectopic axis can be made by injecting one of the animal cap blastomeres with Nodal mRNA and another animal cap cell with BMP mRNA (Xu et al., 2014). A gradient is formed between them, and the neighboring cells respond by constructing a new axis (**FIGURE 11.41**).

Moreover, by injecting different amounts of BMP and Nodal mRNAs into a single blastula animal cap cells, one can mimic the effect of the blastopore lip. Injections of mRNAs with a high BMP to Nodal ratio induce the formation of new tails growing from the animal pole of the embryo. Wnt8, a posterior morphogen, is produced in these cells. Injection of mRNAs with decreasing BMP to Nodal ratios induces the formation of secondary trunks from these animal pole cells. When Nodal and BMP are injected in the same amounts, a posterior head is induced (Thisse et al. 2000). As mentioned in *Xenopus*, Nodal proteins are critical for the formation of the Organizer; and in zebrafish, ectopic expression of Nodal in the ventral blastopore margin will convert the ventral blastopore lip into a shield, inducing an entire secondary axis. The shield in zebrafish may be the "head organizer," while the cells 180° away become the "tail organizer."

The engine for integrating the BMP-Chordin and BMP-Nodal axes appears to be β-catenin. As in *Xenopus*, β-catenin activates the *Nodal* genes. In addition, β-catenin activates the genes encoding FGFs and other factors that repress BMP and Wnt expression on the dorsal side of the embryo while activating the genes for *goosecoid*, *noggin*, and *dickkopf* there (Solnica-Krezel and Driever 2001; Sampath et al. 1998; Gritsman et al. 2000; Schier and Talbot 2001; Fürthauer et al. 2004; Tsang et al. 2004). As in *Xenopus*, β-catenin accumulates specifically in the nuclei destined to become the dorsal cells (Langdon and Mullins 2011). And, as

FIGURE 11.41 Correlation between the relative position of BMP- and Nodal-secreting clones and the orientation of the secondary embryonic axis induced at the animal cap. (A,B) When the Nodal-BMP vector (yellow arrow in A) is parallel to the D-V axis (white arrow) of the embryonic margin (where Nodal is strong dorsally and BMP strong ventrally), the original axis (blue arrow in B) and the secondary axis (red arrow) axes are parallel. (C,D) When the Nodal-BMP vector is perpendicular to the original dorsal-ventral axis, the secondary embryonic axis grows perpendicular to the primary axis. (E,F) When the Nodal-BMP vector is against that of the original dorsal-ventral axis, the primary and secondary axes grow in opposite directions. sh, embryonic shield. A, C, and E are animal pole views at the shield stage; B, D, and F are lateral views at 30 hours after fertilization. (From Xu et al. 2014, courtesy of C. Thisse.)

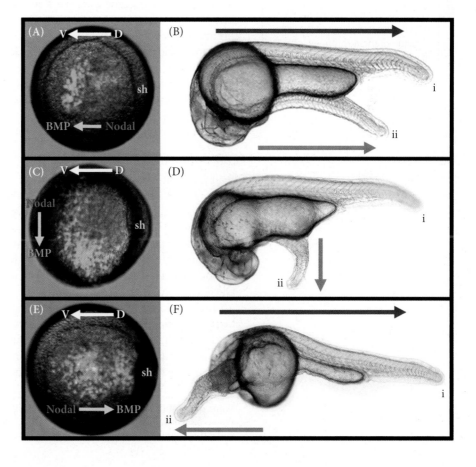

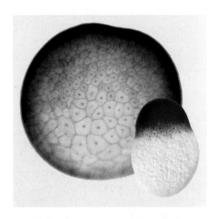

FIGURE 11.42 β-Catenin activates organizer genes in the zebrafish. (A) Nuclear localization of β-catenin marks the dorsal side of the *Xenopus* blastula (larger image) and helps form its Nieuwkoop center beneath the organizer. In the zebrafish late blastula (smaller image), nuclear localization of β-catenin is seen in the yolk syncytial layer nuclei beneath the future embryonic shield. (A courtesy of S. Schneider.)

VADE MECUM

Within the segment on zebrafish development you can see a visualization of the teratogenic effects of retinoic acid.

in *Xenopus*, this appears to be regulated by a maternal Wnt protein, in this case Wnt8a (Lu et al. 2011). The presence of β-catenin distinguishes dorsal YSL from the lateral and ventral YSL regions[9] (**FIGURE 11.42**; Schneider et al. 1996), and inducing β-catenin accumulation on the *ventral* side of the egg results in its dorsalization and a second embryonic axis (Kelly et al. 1995).

 SCIENTISTS SPEAK 11.4 Dr. Bernard Thisse discusses experiments leading to the notion that the dorsal-ventral axis of the zebrafish is specified by a gradient of Nodal and BMP. Dr. Christine Thisse discusses the evidence that an entire embryo can be generated from pluripotent cells having two opposite gradient activities.

Anterior-posterior axis formation

The patterning of the neural ectoderm along the anterior-posterior axis in the zebrafish appears to be the result of the interplay of FGFs, Wnts, and retinoic acid, similar to that seen in *Xenopus*. In fish embryos, there seem to be two separate processes. First a Wnt signal represses the expression of anterior genes; then Wnts, retinoic acid, and FGFs are required to activate the posterior genes.

This regulation of anterior-posterior identity appears to be coordinated by **retinoic acid-4-hydroxylase**, an enzyme that degrades RA (Kudoh et al. 2002; Dobbs-McAuliffe et al. 2004). The gene encoding this enzyme, *cyp26*, is expressed specifically in the region of the embryo destined to become the anterior end. Indeed, this gene's expression is first seen during the late blastula stage, and by the time of gastrulation it defines the presumptive anterior neural plate. Retinoic acid-4-hydroxylase prevents the accumulation of RA at the embryo's anterior end, blocking the expression of the posterior genes there. This inhibition is reciprocated, since the posteriorly expressed FGFs and Wnts inhibit the expression of the *cyp26* gene, as well as inhibiting the expression of the head-specifying gene *Otx2*. This mutual inhibition creates a border between the zone of posterior gene expression and the zone of anterior gene expression. As epiboly continues, more and more of the body axis is specified to become posterior.

Retinoic acid acts as a morphogen, regulating cell properties depending on its concentration. Cells receiving very little RA express anterior genes; cells receiving high levels of RA express posterior genes; and those cells receiving intermediate levels of RA express genes characteristic of cells between the anterior and posterior regions. This morphogen is extremely important in the hindbrain, where different levels of RA specify different types of cells along the anterior-posterior axis (White et al. 2007).

Left-right axis formation

In all vertebrates studied, the right and left sides differ both anatomically and developmentally. In fish, the heart is on the left side and there are different structures in the left and right regions of the brain. Moreover, as in other vertebrates, the cells on the left side of the body are given that information by Nodal signaling and by the Pitx2 transcription factor. The ways the different vertebrate classes accomplish this asymmetry differ, but recent evidence suggests that the currents produced by motile cilia in the node may be responsible for left-right axis formation in all the vertebrate classes (Okada et al. 2005).

In zebrafish, the Nodal structure housing the cilia that control left-right asymmetry is a transient fluid-filled organ called **Kupffer's vesicle**. As mentioned earlier, Kupffer's vesicle arises from a group of dorsal cells near the embryonic shield shortly after gastrulation. Essner and colleagues (2002, 2005) were able to inject small beads into Kupffer's vesicle and see their translocation from one side of the vesicle to the other. Blocking

---

[9] Some of the endodermal cells that accumulate β-catenin will become the precursors of the ciliated cells of Kupffer's vesicle (Cooper and D'Amico 1996). As we will discuss in the final section of this chapter, these cells are critical in determining the left-right axis of the embryo.

ciliary function by preventing the synthesis of dynein or by ablating the precursors of the ciliated cells resulted in abnormal left-right axis formation. Cilia are responsible for the left-side specific activation of the Nodal signaling cascade. Nodal target genes are critically important in instructing asymmetric organ migration and morphogenesis in the body (Rebagliati et al. 1998; Long et al. 2003).

WATCH DEVELOPMENT 11.8   See the rotary motion of cilia in the Kupffer's vesicle of zebrafish.

Next Step Investigation

The BMP-Nodal gradient so vital to amphibian and fish development may be critically important in other vertebrates (including humans) as well. Moreover, can *any* field of pluripotent cells (such as human embryonic stem cells) respond to gradients of BMP and Nodal signals? If this is the case, it may be possible to induce morphogenesis in vitro and to organize the pluripotent cells into fully functional structures. Knowing the events that generate patterned organs from the cues in gradients could be an important breakthrough for regenerative medicine.

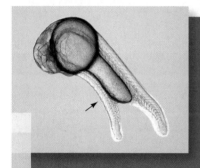

Closing Thoughts on the Opening Photo

The zebrafish embryo in the photo has two body axes, a normal one and a second axis (arrow) that was induced by adding a region of an embryo containing high amounts of Nodal (see p. 375). New theories concerning conjoined twinning hypothesize that during gastrulation, ectopic expression of signaling molecules such as Nodal might lead to new axis formation. Human conjoined twins will be discussed in more detail in Chapter 12. (Photograph courtesy of Christine Thisse.)

Snapshot Summary
Early Development in Amphibians and Fish

1. Amphibian cleavage is holoblastic, but it is unequal because of the presence of yolk in the vegetal hemisphere.

2. Amphibian gastrulation begins with the invagination of the bottle cells, followed by the coordinated involution of the mesoderm and the epiboly of the ectoderm. Vegetal rotation plays a significant role in directing the involution.

3. The driving forces for ectodermal epiboly and the convergent extension of the mesoderm are the intercalation events in which several tissue layers merge. Fibronectin plays a critical role in enabling the mesodermal cells to migrate into the embryo.

4. The dorsal lip of the blastopore forms the organizer tissue of the amphibian gastrula. This tissue dorsalizes the ectoderm, transforming it into neural tissue, and it transforms ventral mesoderm into lateral and dorsal mesoderm.

5. The organizer consists of pharyngeal endoderm, head mesoderm, notochord, and dorsal blastopore lip tissues. The organizer functions by secreting proteins (Noggin, Chordin, and Follistatin) that block the BMP signal that would otherwise ventralize the mesoderm and activate the epidermal genes in the ectoderm.

6. Dorsal-ventral specification begins with maternal messages and proteins stored in the vegetal cytoplasm. These include Nodal-like paracrine factors, transcription factors (such as VegT), and agents that protect β-catenin from degradation.

7. The organizer is itself induced by the Nieuwkoop center, located in the dorsalmost vegetal cells. This center is formed by the translocation of the Disheveled protein and Wnt11 to the dorsal side of the egg to stabilize β-catenin in the dorsal cells of the embryo.

8. The Nieuwkoop center is formed by the accumulation of β-catenin, which can complex with Tcf3 to form a transcription factor complex that can activate the transcription of the *siamois* and *twin* genes on the dorsal side of the embryo.

9. The Siamois and Twin proteins collaborate with activated Smad2 transcription factors generated by the TGF-β pathway (Nodal, Vg1) to activate genes encoding BMP inhibitors. These inhibitors include the secreted factors Noggin, Chordin, and Follistatin, as well as the transcription factor Goosecoid.

10. In the presence of BMP inhibitors, ectodermal cells form neural tissue. The action of BMP on ectodermal cells causes them to become epidermis.

11. In the head region, an additional set of proteins (Cerberus, Frzb, Dickkopf, Tiki) blocks the Wnt signal from the ventral and lateral mesoderm.

12. Wnt signaling causes a gradient of β-catenin along the anterior-posterior axis of the neural plate that appears to specify the regionalization of the neural tube.

13. Insulin-like growth factors (IGFs) help transform the neural tube into anterior (forebrain) tissue.

14. The left-right axis appears to be initiated by the activation of a Nodal protein solely on the left side of the embryo. In *Xenopus*, as in other vertebrates, Nodal protein activates expression of *pitx2*, which is critical in distinguishing left-sidedness from right-sidedness.

15. Cleavage in fish is meroblastic. The deep cells of the blastoderm form between the yolk syncytial layer and the enveloping layer. These deep cells migrate over the top of the yolk, forming the hypoblast and epiblast.

16. On the future dorsal side, the hypoblast and epiblast intercalate to form the embryonic shield, a structure homologous to the amphibian organizer. Transplantation of the embryonic shield into the ventral side of another embryo will cause a second embryonic axis to form.

17. In both amphibians and fish, neural ectoderm is permitted to form where the BMP-mediated induction of epidermal tissue is prevented. The fish embryonic shield, like the amphibian dorsal blastopore lip, secretes the BMP antagonists. Like the amphibian organizer, the shield receives its abilities by being induced by β-catenin and by underlying endodermal cells expressing Nodal-related paracrine factors.

Further Reading

Carron, C. and D. L. Shi. 2016. Specification of anterioposterior axis by combinatorial signaling during *Xenopus* development. *Wiley Interdiscip. Rev. Dev. Biol.* 5: 150–168.

Cho, K. W. Y., B. Blumberg, H. Steinbeisser and E. De Robertis. 1991. Molecular nature of Spemann's organizer: The role of the *Xenopus* homeobox gene *goosecoid*. *Cell* 67: 1111–1120.

De Robertis, E. M. 2006. Spemann's organizer and self-regulation in amphibian embryos. *Nature Rev. Mol. Cell Biol.* 7: 296–302.

Essner, J. J., J. D. Amack, M. K. Nyholm, E. B. Harris and H. J. Yost. 2005. Kupffer's vesicle is a ciliated organ of asymmetry in the zebrafish embryo that initiates left-right development of the brain, heart and gut. *Development* 132: 1247–1260.

Hontelez, S. and 6 others. 2015. Embryonic transcription is controlled by maternally defined chromatin state *Nature Commun.* 6: 10148.

Khokha, M. K., J. Yeh, T. C. Grammer and R. M. Harland. 2005. Depletion of three BMP antagonists from Spemann's organizer leads to catastrophic loss of dorsal structures. *Dev. Cell* 8: 401–411.

Langdon, Y. G. and M. C. Mullins. 2011. Maternal and zygotic control of zebrafish dorsoventral axial patterning. *Annu. Rev. Genet.* 45: 357–377.

Larabell, C. A. and 7 others. 1997. Establishment of the dorsal-ventral axis in *Xenopus* embryos is presaged by early asymmetries in β-catenin which are modulated by the Wnt signaling pathway. *J. Cell Biol.* 136: 1123–1136.

Lepage, E. S. and A. E. Bruce. 2010. Zebrafish epiboly: Mechanics and mechanisms. *Int. J. Dev. Biol.* 54: 1213–12211.

Niehrs, C. 2004. Regionally specific induction by the Spemann-Mangold organizer. *Nature Rev. Genet.* 5: 425–434.

Piccolo, S. and 6 others. 1999. The head inducer Cerberus is a multifunctional antagonist of Nodal, BMP, and Wnt signals. *Nature* 397: 707–710.

Reversade, B., H. Kuroda, H. Lee, A. Mays, and E. M. De Robertis. 2005. Deletion of BMP2, BMP4, and BMP7 and Spemann organizer signals induces massive brain formation in *Xenopus* embryos. *Development* 132: 3381–3392.

Spemann, H. and H. Mangold. 1924. Induction of embryonic primordia by implantation of organizers from a different species. (Trans. V. Hamburger.) In B. H. Willier and J. M. Oppenheimer (eds.), *Foundations of Experimental Embryology.* Hafner, New York, pp. 144–184. Reprinted in *Int. J. Dev. Biol.* 45: 13–311.

Winklbauer, R. and E. W. Damm. 2011. Internalizing the vegetal cell mass before and during amphibian gastrulation: Vegetal rotation and related movements. *WIREs Dev Biol.* Doi:10.1002/wdev.26.

Xu, P. F., N. Houssin, K. F. Ferri-Lagneau, B. Thisse and C. Thisse. 2014. Construction of a vertebrate embryo from two opposing morphogen gradients. *Science* 344: 87–89.

GO TO WWW.DEVBIO.COM...

...for Web Topics, Scientists Speak interviews, Watch Development videos, Dev Tutorials, and complete bibliographic information for all literature cited in this chapter.

Birds and Mammals

THIS FINAL CHAPTER ON THE PROCESSES OF EARLY DEVELOPMENT extends our survey of vertebrate development to include the **amniotes**—those vertebrates whose embryos form an amnion, or water sac (i.e., the reptiles, birds, and mammals). Birds and reptiles follow a very similar pattern of development (Gilland and Burke 2004; Coolen et al. 2008), and birds are considered by modern taxonomists to be a reptilian clade (**FIGURE 12.1A**).

The **amniote egg** is characterized by a set of membranes that together enable the embryo to survive on land (**FIGURE 12.1B**). First, the **amnion**, for which the amniote egg is named, is formed early in embryonic development and enables the embryo to float in a fluid environment that protects it from desiccation. Another cell layer derived from the embryo, the **yolk sac**, enables nutrient uptake and the development of the circulatory system. The **allantois**, developing at the posterior end of the embryo, stores waste products. The **chorion** contains blood vessels that exchange gases with the outside environment. In birds and most reptiles, the embryo and its membranes are enclosed in a hard or leathery shell within which the embryo develops outside the mother's body. Cleavage in bird and reptile eggs, like that of the bony fishes described in the last chapter, is meroblastic, with only a small portion of the egg cytoplasm being used to make the cells of the embryo. The vast majority of the large egg is composed of yolk that will nourish the growing embryo.

In most mammals, holoblastic cleavage is modified to accommodate the formation of a **placenta**, an organ containing tissues and blood vessels from both the embryo and the

How did this mammalian embryo determine which end is its head and which its tail?

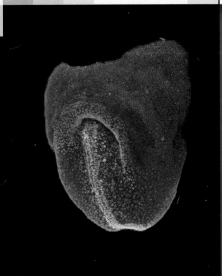

The Punchline

Birds and mammals begin development differently. Birds have a meroblastic cleavage, where mammalian cleavage is holoblastic. The chick forms a layer of cells over a large body of yolk, whereas the relatively yolk-free eggs of mammals form a blastocyst containing an outer layer (that becomes part of the placenta) and an inner cell mass composed of embryonic stem cells (that will form all the cells of the embryo). Gastrulation is initiated at the node, a site that is most likely determined by physical as well as chemical cues. Nodal and Wnt proteins are especially important in determining where this takes place as well as specifying the anterior-posterior polarity of the embryo. The node extends into the primitive streak, and the cells of the upper layer travel to and through this structure. The cells migrating into and through the streak become the mesoderm and the endoderm. Those remaining on the surface become the ectoderm. The node is very similar to the dorsal blastopore lip of amphibians, and similar molecules are involved in its formation.

(A)

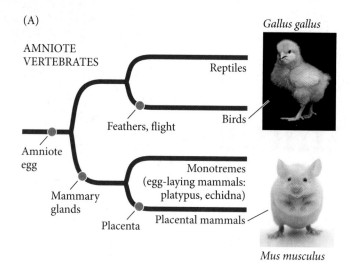

AMNIOTE
VERTEBRATES

Gallus gallus

Reptiles

Feathers, flight

Birds

Amniote
egg

Mammary
glands

Placenta

Monotremes
(egg-laying mammals:
platypus, echidna)

Placental mammals

Mus musculus

FIGURE 12.1 The membranes of the amniote egg characterize reptiles, birds, and mammals. (A) Phylogenetic relationships of the amniotes. Note that birds are considered reptiles by most modern taxonomists, but for physiological studies they are often treated as separate taxa. (Other flying and feathered reptile groups have not survived to the present day.) The domestic chicken (*Gallus gallus*) is the most widely studied bird species. Among mammals, the development of the laboratory mouse *Mus musculus* is the most widely studied. Both avian and mouse studies contribute to our understanding of human development. (B) The shelled amniote egg (as exemplified by the chicken egg on the left) permitted animals to develop away from bodies of water. The amnion provides a "water sac" in which the embryo develops; the allantois stores wastes; and the blood vessels of the chorion exchange gases and nutrients from the yolk sac. In mammals (right), this arrangement is modified such that the blood vessels acquire nutrients and exchange gases via a placenta joined to the mother's uterus rather than from the yolk sac. (Chick photograph courtesy of D. McIntyre; mouse photograph © Antagain/iStock.)

(B)

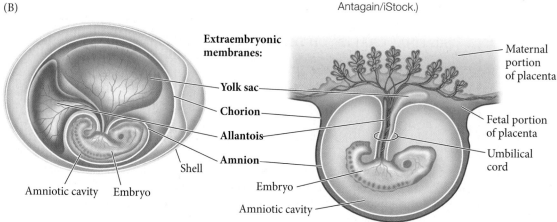

Extraembryonic
membranes:

Yolk sac

Chorion

Allantois

Amnion

Shell

Amniotic cavity   Embryo

Maternal
portion
of placenta

Fetal portion
of placenta

Umbilical
cord

Embryo

Amniotic cavity

mother. Gas exchange, nutrient uptake, and waste elimination take place through the placenta, enabling the embryo to develop inside another organism.

WEB TOPIC 12.1 **THE EXTRAEMBRYONIC MEMBRANES** The amniote embryo is supported by a variety of membranes that provide it with nourishment, protection, and waste disposal services.

Ever since Aristotle first observed and recorded the details of its 3-week-long development, the domestic chicken (*Gallus gallus*) has been a favorite organism for embryological studies. It is accessible year-round and is easily maintained. Moreover, at any particular temperature, its developmental stage can be accurately predicted, so large numbers of same-stage embryos can be obtained and manipulated. Chick organ formation is accomplished by genes and cell movements similar to those of mammalian organ formation, and the chick is one of the few organisms whose embryos are amenable to both surgical and genetic manipulations (Stern 2005a). Thus, the chick embryo has often served as a model for human embryos, as has the ubiquitous laboratory mouse.

The mouse is the mammalian model organism of choice and is the subject of many studies involving genetic and surgical manipulation. In addition, the mouse was the first mammalian genome to be sequenced, and when it was first published many scientists felt it was more valuable than knowing the human genome sequence. Their reasoning was that "working on mouse models allows the manipulation of each and every gene to determine their functions" (Gunter and Dhand 2002). We cannot do that with humans. Human development is a subject of medical as well as general scientific interest, however, and the latter sections of this chapter will cover early human development, illustrating the application of many of the principles we have described in model organisms.

Early Development in Birds

Avian Cleavage

Fertilization of the chick egg occurs in the hen's oviduct, before the albumin ("egg white") and shell are secreted to cover it. Cleavage occurs during the first day of development, while the egg is still inside the hen, during which time the embryo progresses from a zygote through late blastula stages (Sheng 2014). Like the egg of the zebrafish, the chick egg is telolecithal, with a small disc of cytoplasm—the **blastodisc**—sitting on top of a large yolk (**FIGURE 12.2A**). Like fish eggs, the yolky eggs of birds undergo **discoidal meroblastic cleavage**. Cleavage occurs only in the blastodisc, which is about 2–3 mm in diameter and is located at the animal pole of the egg. The first cleavage furrow appears centrally in the blastodisc; other cleavages follow to create a **blastoderm** (**FIGURE 12.2B,C**). As in the fish embryo, the cleavages do not extend into the yolky cytoplasm, so the early-cleavage cells are continuous with one another and with the yolk at their

FIGURE 12.2 Discoidal meroblastic cleavage in a chick egg. (A) Avian eggs include some of the largest cells known (inches across), but cleavage takes place in only a small region. The yolk fills up the entire cytoplasm of the egg cell, with the exception of a small blastodisc in which cleavage and development will take place. The chalaza are protein strings that keep the yolky egg cell centered in the shell. The albumin (egg white) is secreted onto the egg in its passage out of the oviduct. (B) Early cleavage stages viewed from the animal pole (the future dorsal side of the embryo). In the micrographs, the tightly apposed cell membranes have been stained with phalloidin (green). (C) Schematic view of cellularization in the chick egg during the day it is fertilized and still inside the hen. The numbers refer to the layers of cells. (A,B after Bellairs et al. 1978, photographs from Lee et al. 2013, courtesy of J. Y. Han; C after Nagai et al. 2015.)

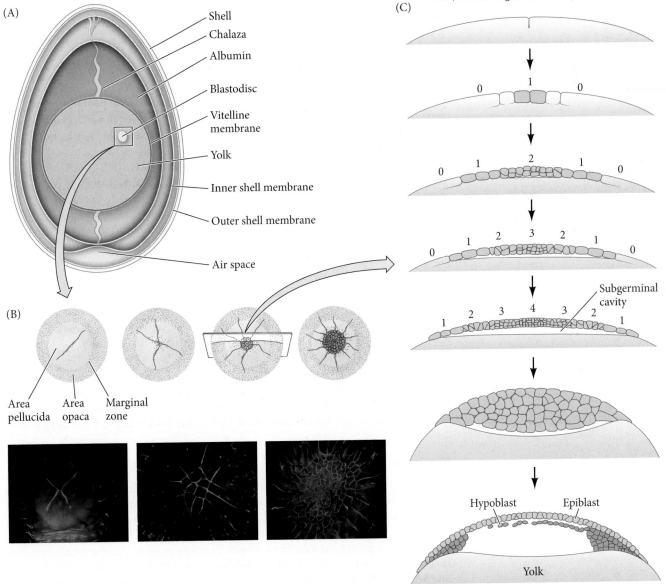

(A)
Shell
Chalaza
Albumin
Blastodisc
Vitelline membrane
Yolk
Inner shell membrane
Outer shell membrane
Air space

(B)
Area pellucida   Area opaca   Marginal zone

(C)
Subgerminal cavity
Hypoblast   Epiblast
Yolk

bases. Thereafter, equatorial and vertical cleavages divide the blastoderm into a tissue about 4 cell layers thick, with the cells linked together by tight junctions (see Figure 12.2C; Bellairs et al. 1978; Eyal-Giladi 1991; Nagai et al. 2015). The switch from maternal to zygotic gene expression occurs at about the seventh or eighth division, when there are around 128 cells (Nagai et al. 2015).

Between the blastoderm and the yolk of avian eggs is a space called the **subgerminal cavity**, which is created when the blastoderm cells absorb water from the albumin ("egg white") and secrete the fluid between themselves and the yolk (New 1956). At this stage, the deep cells in the center of the blastoderm appear to be shed and die, leaving behind a 1-cell-thick **area pellucida**; this part of the blastoderm forms most of the actual embryo. The peripheral ring of blastoderm cells that have not shed their deep cells constitutes the **area opaca**. Between the area pellucida and the area opaca is a thin layer of cells called the **marginal zone** (Eyal-Giladi 1997; Arendt and Nübler-Jung 1999). Some marginal zone cells become very important in determining cell fate during early chick development.

Gastrulation of the Avian Embryo

The hypoblast

By the time a hen has laid an egg, its blastoderm contains some 50,000 cells. At this time, most of the cells of the area pellucida remain at the surface, forming an "upper layer" called the **epiblast**. Shortly after the egg is laid, a local thickening of the epiblast, called **Koller's sickle**, is formed at the posterior edge of the area pellucida. In between the area opaca and Koller's sickle is a beltlike region called the **posterior marginal zone** (**PMZ**). A sheet of cells at the posterior boundary between the area pellucida and marginal zone migrates anteriorly beneath the surface. Meanwhile, cells in more anterior regions of the epiblast have delaminated and stay attached to the epiblast, to form hypoblast "islands," an archipelago of disconnected clusters of 5–20 cells each that migrate and become the **primary hypoblast** (**FIGURE 12.3A,B**). The sheet of cells that grows anteriorly from Koller's sickle combines with the primary hypoblast to form the complete hypoblast layer, also called the **secondary hypoblast** or **endoblast** (**FIGURE 12.3C–E**; Eyal-Giladi et al. 1992; Bertocchini and Stern 2002; Khaner 2007a,b). The resulting two-layered blastoderm (epiblast and hypoblast) is joined together at the marginal zone of the area opaca, and the space between the layers forms a blastocoel-like cavity Thus, although the shape and formation of the avian blastodisc differs from those of the amphibian, fish, or echinoderm blastula, the overall spatial relationships are retained.

The avian embryo comes entirely from the epiblast; the hypoblast does not contribute any cells to the developing embryo (Rosenquist 1966, 1972). Rather, the hypoblast cells form portions of the extraembryonic membranes (see Figure 12.1B), especially the yolk sac and the stalk linking the yolk mass to the endodermal digestive tube. Hypoblast cells also provide chemical signals that specify the migration of epiblast cells. However, the three germ layers of the embryo proper (plus the amnion, chorion, and allantois extraembryonic membranes) are formed solely from the epiblast (Schoenwolf 1991).

The primitive streak

Although many reptile groups initiate gastrulation by migration through an amphibian-like blastopore, avian and mammalian gastrulation takes place through the **primitive streak**. This can be considered the equivalent of an elongated blastopore lip of amphibian embryos (Alev et al. 2013; Bertocchini et al. 2013; Stower et al. 2015). Dye-marking experiments and time-lapse cinemicrography indicate that the primitive streak first arises from Koller's sickle and the epiblast above it (Bachvarova et al. 1998; Lawson and Schoenwolf 2001a,b; Voiculescu et al. 2007). As cells converge to form the primitive

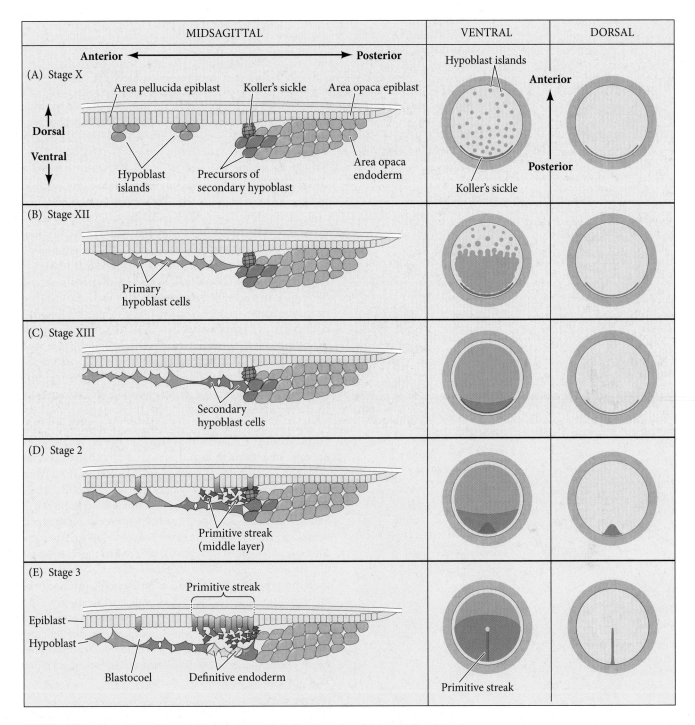

FIGURE 12.3 Formation of the chick blastoderm. The left column is a diagrammatic midsagittal section through part of the blastoderm. The middle column depicts the entire embryo viewed from the ventral side, showing the migration of the primary hypoblast and the secondary hypoblast (endoblast) cells. The right column shows the entire embryo seen from the dorsal side. (A–C) Events prior to laying of the shelled egg. (A) Stage X embryo, where islands of hypoblast cells can be seen, as well as a congregation of hypoblast cells around Koller's sickle. (B) By stage XII, a sheet of cells that grows anteriorly from Koller's sickle combines with the hypoblast islands to form the complete hypoblast layer. (C) By stage XIII, just prior to primitive streak formation, the formation of the hypoblast just been completed. (D) By stage 2 (12–14 hours after the egg is laid), the primitive streak cells form a third layer that lies between the hypoblast and epiblast cells. (E) By stage 3 (15–17 hours post laying), the primitive streak has become a definitive region of the epiblast, with cells migrating through it to become the mesoderm and endoderm. (After Stern 2004.)

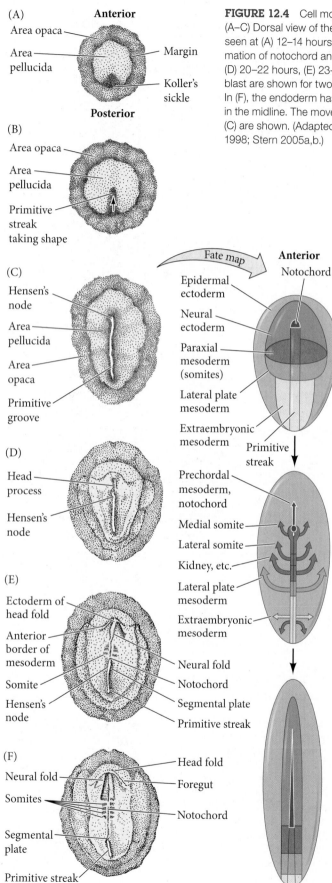

(A)
Anterior
Area opaca
Area pellucida
Margin
Koller's sickle
Posterior

(B)
Area opaca
Area pellucida
Primitive streak taking shape

(C)
Hensen's node
Area pellucida
Area opaca
Primitive groove

Fate map
Anterior
Notochord
Epidermal ectoderm
Neural ectoderm
Paraxial mesoderm (somites)
Lateral plate mesoderm
Extraembryonic mesoderm
Primitive streak

(D)
Head process
Hensen's node

Prechordal mesoderm, notochord
Medial somite
Lateral somite
Kidney, etc.
Lateral plate mesoderm
Extraembryonic mesoderm

(E)
Ectoderm of head fold
Anterior border of mesoderm
Somite
Hensen's node
Neural fold
Notochord
Segmental plate
Primitive streak

(F)
Neural fold
Somites
Segmental plate
Primitive streak
Head fold
Foregut
Notochord

FIGURE 12.4 Cell movements of the primitive streak and fate map of the chick embryo. (A–C) Dorsal view of the formation and elongation of the primitive streak. The blastoderm is seen at (A) 12–14 hours, (B) 15–17 hours, and (C) 18–20 hours after the egg is laid. (D–F) Formation of notochord and mesodermal somites as the primitive streak regresses, shown at (D) 20–22 hours, (E) 23–25 hours, and (F) the four-somite stage. Fate maps of the chick epiblast are shown for two stages, the definitive primitive streak stage (C) and neurulation (F). In (F), the endoderm has ingressed beneath the epiblast, and convergent extension is seen in the midline. The movements of the mesodermal precursors through the primitive streak at (C) are shown. (Adapted from several sources, especially Spratt 1946; Smith and Schoenwolf 1998; Stern 2005a,b.)

streak, a depression called the **primitive groove** forms within the streak. Most migrating cells pass through the primitive groove, which serves as a gateway into the deep layers of the embryo (**FIGURE 12.4**; Voiculescu et al. 2014). Thus, the primitive groove is homologous to the amphibian blastopore, and the primitive streak is homologous to the blastopore lip.

At the anterior end of the primitive streak is a regional thickening of cells called **Hensen's node** (also known as the **primitive knot**; see Figure 12.4C). The center of Hensen's node contains a funnel-shaped depression (sometimes called the **primitive pit**) through which cells can enter the embryo to form the notochord and prechordal plate. Hensen's node is the functional equivalent of the dorsal lip of the amphibian blastopore (i.e., the organizer)[1] and the fish embryonic shield (Boettger et al. 2001).

The primitive streak defines the major body axes of the avian embryo. It extends from posterior to anterior; migrating cells enter through its dorsal side and move to its ventral side; and it separates the left portion of the embryo from the right. The axis of the streak is equivalent to the dorsal-ventral axis of amphibians. The anterior end of the streak—Hensen's node—gives rise to the prechordal mesoderm, notochord, and medial part of the somites. Cells that ingress through the middle of the streak give rise to the lateral part of the somites and to the heart and kidneys. Cells in the posterior portion of the streak make the lateral plate and extra-embryonic mesoderm (Psychoyos and Stern 1996). After the ingression of the mesoderm cells, epiblast cells remaining outside of but close to the streak will form medial (dorsal) structures such as the neural plate, while those epiblast cells farther from the streak will become epidermis (see Figure 12.4, right-hand panels).

WEB TOPIC 12.2 **ORGANIZING THE CHICK NODE** FGFs and BMPs have major roles in determining the place where gastrulation is initiated.

ELONGATION OF THE PRIMITIVE STREAK As cells enter the primitive streak, they undergo an epithelial-to-mesenchymal transformation and the basal lamina beneath them breaks down. The streak elongates toward the future head region as more anterior cells migrate toward the center of

---

[1] Frank M. Balfour proposed the homology of the amphibian blastopore and the chick primitive streak in 1873, while he was still an undergraduate (Hall 2003). August Rauber (1876) provided further evidence for their homology.

the embryo. Convergent extension is responsible for the progression of the streak—a doubling in streak length is accompanied by a concomitant halving of its width (see Figure 12.4B). Cell division adds to the length produced by convergent extension, and some of the cells from the anterior portion of the epiblast contribute to the formation of Hensen's node (Streit et al. 2000; Lawson and Schoenwolf 2001b).

At the same time, the secondary hypoblast (endoblast) cells continue to migrate anteriorly from the posterior marginal zone of the blastoderm (see Figure 12.3E). The elongation of the primitive streak appears to be coextensive with the anterior migration of these secondary hypoblast cells, and the hypoblast directs the movement of the primitive streak (Waddington 1933; Foley et al. 2000). The streak eventually extends to 60–75% of the length of the area pellucida.

FORMATION OF ENDODERM AND MESODERM

The basic rule of amniote cell specification is that germ layer identity (ectoderm, mesoderm, or endoderm) is established before gastrulation starts (see Chapman et al. 2007), but the specification of cell type is controlled by inductive influences during and after migration through the primitive streak. As soon as the primitive streak has formed, epiblast cells begin to migrate through it and into the blastocoel. The streak thus has a continually changing cell population. Cells migrating through the anterior end pass down into the blastocoel and migrate anteriorly, forming the endoderm, head mesoderm, and notochord; cells passing through the more posterior portions of the primitive streak give rise to the majority of mesodermal tissues (**FIGURE 12.5**; Rosenquist et al. 1966; Schoenwolf et al. 1992).

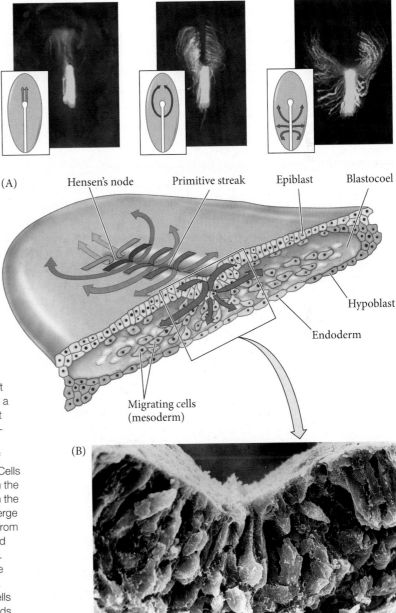

FIGURE 12.5 Migration of endodermal and mesodermal cells through the primitive streak. (A) Stereogram of a gastrulating chick embryo, showing the relationship of the primitive streak, the migrating cells, and the hypoblast and epiblast of the blastoderm. The lower layer becomes a mosaic of hypoblast and endodermal cells; the hypoblast cells eventually sort out to form a layer beneath the endoderm and contribute to the yolk sac. Above each region of the stereogram are micrographs showing the tracks of GFP-labeled cells at that position in the primitive streak. Cells migrating through Hensen's node travel anteriorly to form the prechordal plate and notochord; those migrating through the next anterior region of the streak travel laterally but converge near the midline to make notochord and somites; those from the middle of the streak form intermediate mesoderm and lateral plate mesoderm (see the fate maps in Figure 12.4). Farther posterior, the cells migrating through the primitive streak make the extraembryonic mesoderm (not shown). (B) This scanning electron micrograph shows epiblast cells passing into the blastocoel and extending their apical ends to become bottle cells. (A after Balinsky 1975, photographs from Yang et al. 2002; B from Solursh and Revel 1978, courtesy of M. Solursh and C. J. Weijer.)

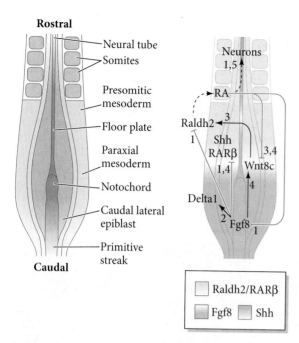

FIGURE 12.6 Signals that regulate axis extension in chick embryos. In the stage 10 chick embryo, Fgf8 inhibits expression of the retinoic acid (RA) synthesizing enzyme Raldh2 in the presomitic mesoderm (1) and the expression of the RA receptor RARβ in the neural ectoderm (4), thus preventing RA from triggering differentiation in the caudal-lateral epiblast cells (those cells adjacent to the node/streak border and which give rise to lateral and dorsal neural tube) and the caudalmost paraxial mesoderm (1,5). In addition, Fgf8 inhibits Sonic hedgehog (Shh) expression in the neural tube floorplate, controlling the onset of ventral patterning genes (1). FGF signaling is also required for expression of Delta1 in the medial portion of the caudal-lateral epiblast cells (2) and promotes expression of Wnt8c (4). As Fgf8 decays in the caudal paraxial mesoderm, Wnt signaling, most likely provided by Wnt8c, now acts to promote Raldh2 in the adjacent paraxial mesoderm (4). RA produced by Raldh2 activity represses Fgf8 (1) and Wnt8c (3,4). (After Wilson et al. 2009.)

The first cells to migrate through Hensen's node are those destined to become the pharyngeal endoderm of the foregut. Once deep within the embryo, these endodermal cells migrate anteriorly and eventually displace the hypoblast cells, causing the hypoblast cells to be confined to a region in the anterior portion of the area pellucida. This anterior region, the **germinal crescent**, does not form any embryonic structures, but it does contain the precursors of the germ cells, which later migrate through the blood vessels to the gonads.

The next cells entering through Hensen's node also move anteriorly, but they do not travel as far ventrally as the presumptive foregut endodermal cells. Rather, they remain between the endoderm and the epiblast to form the **prechordal plate mesoderm** (Psychoyos and Stern 1996). Thus, the head of the avian embryo forms anterior (rostral) to Hensen's node.

The next cells passing through Hensen's node become the **chordamesoderm**. The chordamesoderm has two components: the head process and the notochord. The most anterior part, the **head process**, is formed by central mesoderm cells migrating anteriorly, behind the prechordal plate mesoderm and toward the rostral tip of the embryo (see Figures 12.4 and 12.5). The head process underlies those cells that will form the forebrain and midbrain. As the primitive streak regresses, the cells deposited by the regressing Hensen's node will become the notochord. In the ectoderm, most of the initial neural plate corresponds to the future head region (from forebrain to the level of the future ear vesicle, which lies adjacent to Hensen's node at full primitive streak stage). A small region of neural ectoderm just lateral and posterior to the node (sometimes called the caudal lateral epiblast) will give rise to the rest of the nervous system, including the posterior hindbrain and all of the spinal cord. As the primitive streak regresses, this latter region regresses with Hensen's node and adds cells to the caudal end of the elongating neural plate. It appears that FGF signaling in the streak and paraxial (future somite) mesoderm keeps this region "young" and undifferentiated as it regresses, and that this is antagonized by retinoic acid (RA) activity as cells leave this zone (**FIGURE 12.6**; Diez del Corral et al. 2003).

SCIENTISTS SPEAK 12.1 Dr. Steven Oppenheimer lectures on the development of the chick embryo.

Molecular mechanisms of migration through the primitive streak

FORMATION OF THE PRIMITIVE STREAK The migration of chick epiblast cells to form the primitive streak was first analyzed by Ludwig Gräper, who in 1926 made time-lapse movies of labeled cells under the microscope. He wrote that these movements reminded him of the Polonaise, a courtly dance in which men and women move in parallel rows along the sides of the room, and the man and woman at the "posterior end" leave their respective lines to dance forward through the center. The mechanism for the cellular "dance" was revealed by Voiculescu and colleagues (2007), who used a modern version of cinemicrography (specifically, multiphoton time-lapse microscopy) that identified individual moving cells. They found that cells came down the sides of the epiblast to undergo a medially directed intercalation of cells in the posterior margin where the primitive streak was forming (**FIGURE 12.7**). And although the movement may look like a dance from far away, "at high power, it looks like a rush hour" (Stern 2007).

This rush to the center is mediated by the activation of the Wnt planar cell polarity pathway (see Chapter 4) in the epiblast next to Koller's sickle, at the posterior edge of the embryo. If this pathway is blocked, the mesoderm and endoderm form peripherally instead of centrally. The Wnt pathway in turn appears to be activated by **fibroblast growth factors** (**FGFs**) produced by the hypoblast. If the hypoblast is rotated, the orientation of the primitive streak follows it. Moreover, if FGF signaling is activated in the margin of the epiblast, Wnt signaling will occur there and the orientation of the primitive streak will change, as if the hypoblast had been placed there. The cell migrations that form the primitive streak thus appear to be regulated by FGFs coming from the hypoblast, which activates the Wnt planar cell polarity pathway in the epiblast.

MIGRATION THROUGH THE PRIMITIVE STREAK   Cells migrate to the primitive streak, and as they enter the embryo, the cells separate into two layers. The deep layer joins the hypoblast along its midline, displacing the hypoblast cells to the sides. These deep-moving cells give rise to the endodermal organs of the embryo, as well as to most of the extraembryonic membranes (the hypoblast and peripheral cells of the area opaca form the rest). The second migrating layer spreads to form a loose layer of cells between the endoderm and the epiblast. This middle layer of cells generates the mesodermal portions of the embryo and the mesoderm lining the extraembryonic membranes.

The migration of mesodermal cells through the anterior primitive streak and their condensation to form the chordamesoderm also appear to be controlled by FGF and Wnt signaling. Fgf8 is expressed in the primitive streak and repels migrating cells away from the streak. Yang and colleagues (2002) were able to follow the trajectories of cells as they migrated through the primitive streak (see Figure 12.5) and were able to deflect these normal trajectories by using beads that released Fgf8.

Once cells migrate away from the streak, further movement of the mesodermal precursors appears to be regulated by Wnt proteins. In the more posterior regions, Wnt5a is unopposed and directs the cells to migrate broadly and become lateral plate mesoderm (see Chapter 18). In the more anterior regions of the streak, however, Wnt5a is opposed by Wnt3a, which inhibits migration and causes the cells to form paraxial mesoderm (see Chapter 17). Indeed, the addition of Wnt3a-secreting pellets to the posterior primitive streak suppresses lateral migration and prevents the formation of lateral plate mesoderm (Sweetman et al. 2008). By 22 hours of incubation, most of the presumptive endodermal cells are in the interior of the embryo, although presumptive mesodermal cells continue to migrate inward for a longer time.

Regression of the primitive streak and epiboly of the ectoderm

Now a new phase of development begins. As mesodermal ingression continues, the primitive streak starts to regress, moving Hensen's node from near the center of the area pellucida to a more posterior position (**FIGURE 12.8**). The regressing streak leaves in its wake the dorsal axis of the embryo, including the notochord. The notochord is laid down in a head-to-tail direction, starting at the level where the ears and hindbrain form and extending caudally to the tailbud. As in the frog, the pharyngeal endoderm and head mesoendoderm will induce the anterior parts of the brain, while the notochord will induce the hindbrain and spinal cord. By this time, all the presumptive endodermal and mesodermal cells have entered the embryo and the epiblast is composed entirely of presumptive ectodermal cells.

While the presumptive mesodermal and endodermal cells are moving inward, the ectodermal precursors proliferate and migrate to surround the yolk by epiboly. The

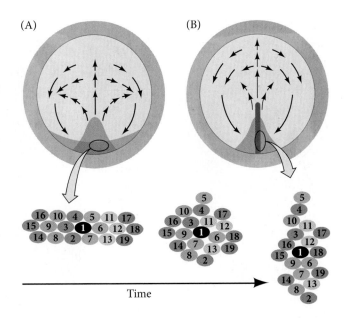

FIGURE 12.7   Mediolateral intercalation in the formation of the primitive streak. Chick embryos at (A) stage 13 (immediately prior to primitive streak formation) and (B) stage 2 (shortly after primitive streak formation). Arrows show cell displacement toward the streak and in front of it. The red area represents the streak-forming region; in (A), the original location of this region is shown in green. The circled areas are represented in the lower row. Each colored disc represents an individual cell, and the cells become mediolaterally intercalated as the primitive streak forms. (After Voiculescu et al. 2007.)

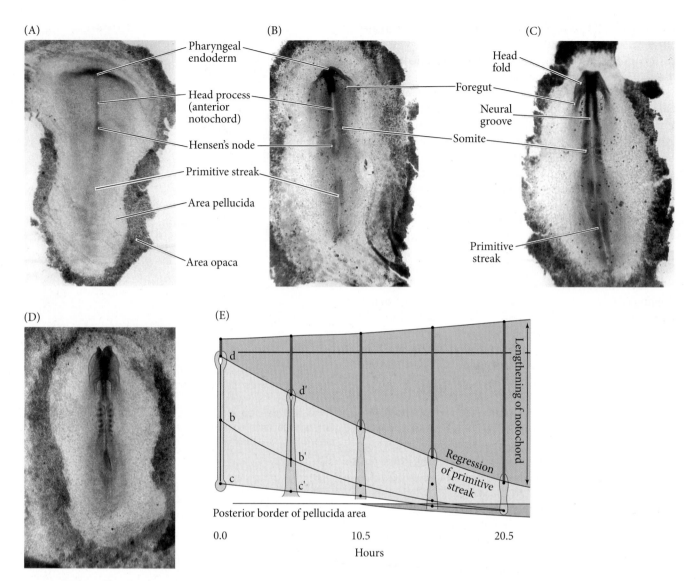

(A)

Pharyngeal endoderm

Head process (anterior notochord)

Hensen's node

Primitive streak

Area pellucida

Area opaca

(B)

Pharyngeal endoderm

Head process (anterior notochord)

Hensen's node

Primitive streak

(C)

Head fold

Foregut

Neural groove

Somite

Primitive streak

(D)

(E)

Lengthening of notochord

Regression of primitive streak

Posterior border of pellucida area

0.0   10.5   20.5

Hours

FIGURE 12.8 Chick gastrulation 24–28 hours after fertilization. (A) The primitive streak at full extension (24 hours). The head process (anterior notochord) can be seen extending from Hensen's node. (B) Two-somite stage (25 hours). Pharyngeal endoderm is seen anteriorly, while the anterior notochord pushes up the head process beneath it. The primitive streak is regressing. (C) Four-somite stage (27 hours). (D) At 28 hours, the primitive streak has regressed to the caudal portion of the embryo. (E) Regression of the primitive streak, leaving the notochord in its wake. Various points of the streak (represented by letters) were followed after it achieved its maximum length. The x axis (time) represents hours after achieving maximum length (the reference line is about 18 hours of incubation). (A–D courtesy of K. Linask; E after Spratt 1947.)

enclosure of the yolk by the ectoderm (again, reminiscent of the epiboly of the amphibian ectoderm) is a Herculean task that takes the greater part of 4 days to complete. It involves the continuous production of new cellular material and the migration of the presumptive ectodermal cells along the underside of the vitelline envelope (New 1959; Spratt 1963). Interestingly, only the cells of the outer margin of the area opaca attach firmly to the vitelline envelope. These cells are inherently different from the other blastoderm cells, as they can extend enormous (500 μm) cytoplasmic processes onto the vitelline envelope. These elongated filopodia are believed to be the locomotor apparatus of the marginal cells, by which the marginal cells pull other ectodermal cells around the yolk (Schlesinger 1958). The filopodia bind to fibronectin, a laminar protein that is a component of the chick vitelline envelope. If the contact between the marginal cells and the fibronectin is experimentally broken by adding a soluble polypeptide similar to fibronectin, the filopodia retract and ectodermal migration ceases (Lash et al. 1990).

Thus, as avian gastrulation draws to a close, the ectoderm has surrounded the embryo, the endoderm has replaced the hypoblast, and the mesoderm has positioned itself between these two regions. Although we have identified many of the processes involved in avian gastrulation, we are only beginning to understand the molecular mechanisms by which some of these processes are carried out.

WEB TOPIC 12.3 **EPIBLAST CELL HETEROGENEITY** Although the early epiblast appears uniform, different cells have different molecules on their cell surfaces. This variability allows some of them to remain in the epiblast while others migrate into the embryo.

Axis Specification and the Avian "Organizer"

As a consequence of the sequence in which the head endomesoderm and notochord are established, gastrulating avian (and mammalian) embryos exhibit a distinct anterior-to-posterior gradient. While cells of the posterior portions of the embryo are still part of a primitive streak and entering inside the embryo, cells at the anterior end are already starting to form organs (see Darnell et al. 1999). For the next several days, the anterior end of the embryo is more advanced in its development (having had a "head start," if you will) than the posterior end. Although the formation of the chick body axes is accomplished during gastrulation, axis specification begins earlier, during the cleavage stage.

The role of gravity and the PMZ

The conversion of the radially symmetrical blastoderm into a bilaterally symmetrical structure appears to be determined by gravity. As the ovum passes through the hen's reproductive tract, it is rotated for about 20 hours in the shell gland. This spinning, at a rate of 15 revolutions per hour, shifts the yolk such that its lighter components (probably containing stored maternal determinants for development) lie beneath one side of the blastoderm. This imbalance tips up one end of the blastoderm, and that end becomes the posterior marginal zone, where primitive streak formation begins (**FIGURE 12.9**; Kochav and Eyal-Giladi 1971; Callebaut et al. 2004).

It is not known what interactions cause this specific portion of the blastoderm to become the PMZ. Early on, the ability to initiate a primitive streak is found throughout the marginal zone; if the blastoderm is separated into parts, each with its own marginal zone, each part will form its own primitive streak (Spratt and Haas 1960; Bertocchini and Stern 2012). However, once the PMZ has formed, it controls the other regions of the margin. Not only do the cells of the PMZ initiate gastrulation, they also prevent other regions of the margin from forming their own primitive streaks (Khaner and Eyal-Giladi 1989; Eyal-Giladi et al. 1992; Bertocchini et al. 2004).

It now seems apparent that the PMZ contains cells that act as the equivalent of the amphibian Nieuwkoop center. When placed in the anterior region of the marginal zone, a graft of PMZ tissue (posterior to and including Koller's sickle) is able to induce a primitive streak and Hensen's node without contributing cells to either structure (Bachvarova et al. 1998; Khaner 1998). Current evidence suggests that the entire marginal zone produces Wnt8c (capable of inducing the accumulation of β-catenin) and that, like the amphibian Nieuwkoop center, the PMZ cells secrete Vg1, a member of the TGF-β family (Mitrani et al. 1990; Hume and Dodd 1993; Seleiro et al. 1996).

Wnt8c and Vg1 act together to induce expression of Nodal (another secreted TGF-β protein) in the future embryonic epiblast next to Koller's sickle and the PMZ (Skromne

FIGURE 12.9 Specification of the chick anterior-posterior axis by gravity. (A) Rotation in the shell gland results in (B) the lighter components of the yolk pushing up one side of the blastoderm. (C) This more elevated region becomes the posterior of the embryo. (After Wolpert et al. 1998.)

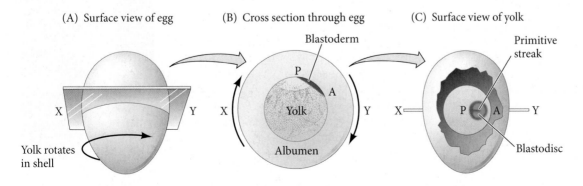

(A) Surface view of egg   (B) Cross section through egg   (C) Surface view of yolk

FIGURE 12.10 Model for generating left-right asymmetry in the chick embryo. (A) On the left side of Hensen's node, Sonic hedgehog (Shh) activates Cerberus, which stimulates BMPs to induce the expression of Nodal. In the presence of Nodal, the *Pitx2* gene is activated. Pitx2 protein is active in the various organ primordia and specifies which side will be the left. On the right side of the embryo, activin is expressed, along with activin receptor IIa. This activates Fgf8, a protein that blocks expression of the gene for Cerberus. In the absence of Cerberus, Nodal is not activated and thus Pitx2 is not expressed. (B) Whole-mount in situ hybridization of *Cerberus* mRNA. This view is from the ventral surface ("from below," so the expression seems to be on the right). Dorsally, the expression pattern would be on the left. (C) Whole-mount in situ hybridization using probes for the chick *Nodal* message (stained purple) shows its expression in the lateral plate mesoderm only on the left side of the embryo. This view is from the dorsal side. (D) Similar in situ hybridization, using the probe for *Pitx2* at a later stage of development. The embryo is seen from its ventral surface. At this stage, the heart is forming, and *Pitx2* expression can be seen on the left side of the heart tube (as well as symmetrically in more anterior tissues). (A after Raya and Izpisua-Belmonte 2004; B from Rodriguez-Esteban et al. 1999, courtesy of J. Izpisúa-Belmonte; C courtesy of C. Stern; D from Logan et al. 1998, courtesy of C. Tabin.)

and Stern 2002). Thus, the pattern appears similar to that of amphibian embryos. Recent studies suggest that Nodal activity is needed to initiate the primitive streak, and that it is the secretion of Cerberus—an antagonist of Nodal—by the primary hypoblast cells that prevents primitive streak formation (Bertocchini et al. 2004; Voiculescu et al. 2014). As the primary hypoblast cells move away from the PMZ, Cerberus protein is no longer present, allowing Nodal activity (and therefore formation of the primitive streak) in the posterior epiblast. Once formed, however, the streak secretes its own Nodal antagonist—the Lefty protein—thereby preventing any further primitive streaks from forming. Eventually, the Cerberus-secreting hypoblast cells are pushed to the future anterior of the embryo, where they contribute to ensuring that neural cells in this region become forebrain rather than more posterior structures of the nervous system.[2]

Left-right axis formation

The vertebrate body has distinct right and left sides. The heart and spleen, for instance, are generally on the left side of the body, whereas the liver is usually on the right. The distinction between the sides is primarily regulated by the left-sided expression of two proteins: the paracrine factor Nodal and the transcription factor Pitx2. However, the mechanism by which *Nodal* gene expression is activated in the left side of the body differs among the vertebrate classes. The ease with which chick embryos can be manipulated has allowed scientists to elucidate the pathways of left-right axis determination in birds more readily than in other vertebrates.

As the primitive streak reaches its maximum length, transcription of the *Sonic hedgehog* gene (*Shh*) becomes restricted to the left side of the embryo, controlled by activin and its receptor (**FIGURE 12.10A**). Activin signaling, along with BMP4, appears to block the expression of Sonic hedgehog protein and to activate expression of Fgf8 protein on the right side of the embryo. Fgf8 blocks expression of the paracrine factor Cerberus on the right-hand side; it may also activate a signaling cascade that instructs the mesoderm to have right-sided capacities (Schlueter and Brand 2009).

---

[2] Conjoined twins may be formed by having by having two sources of *Nodal* expression within the same blastodisc. Experimentation with chick embryos can produce two axes in the same blastodisc by circumventing the usual inhibition of *Nodal* by the Vg1-secreting posterior cells (Bertocchini et al. 2004). In mammals, multiple axes can also form if Nodal antagonists are blocked (Perea Gomez et al. 2002).

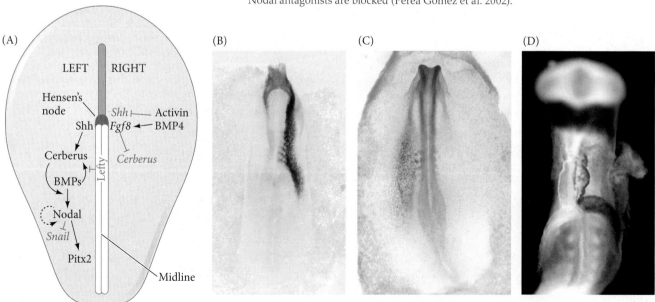

(A) LEFT RIGHT
Hensen's node
Shh ⊣— Activin
Shh *Fgf8* ← BMP4
Cerberus *Cerberus*
Lefty
BMPs
Nodal
Snail
Pitx2
Midline

(B) (C) (D)

Meanwhile, on the left side of the body, Shh protein activates Cerberus (**FIGURE 12.10B**), which in this case acts with BMP to stimulate the synthesis of Nodal protein (Yu et al. 2008). Nodal activates the *Pitx2* gene while repressing *Snail*. In addition, Lefty1 in the ventral midline prevents the Cerberus signal from passing to the right side of the embryo (**FIGURE 12.10C,D**). As in *Xenopus*, Pitx2 is crucial in directing the asymmetry of the embryonic structures. Experimentally induced expression of either Nodal or Pitx2 on the right side of the chick embryo reverses the asymmetry or causes randomization of asymmetry on the right or left sides[3] (Levin et al. 1995; Logan et al. 1998; Ryan et al. 1998).

The real mystery is, What processes create the original asymmetry of Shh and Fgf8? One important observation is that the first asymmetry seen during the formation of Hensen's node in chicks involves Fgf8- and Shh-expressing cells rearranging themselves to converge on the right-hand side of the node (Cui et al. 2009; Gros et al. 2009). Therefore, the differences in gene expression can be traced back to differences in cell migration to the right and left sides of the embryo. What establishes this initial asymmetry is still unknown, but it may be a physical displacement of cells around the node (Tsikolia et al. 2012; Otto et al. 2014).

Early Development in Mammals

Cleavage

Mammalian eggs are among the smallest in the animal kingdom, making them difficult to manipulate experimentally. The human zygote, for instance, is only 100 μm in diameter—barely visible to the eye and less than one-thousandth the volume of a *Xenopus laevis* egg. Also, mammalian zygotes are not produced in numbers comparable to sea urchin or frog zygotes; a female mammal usually ovulates fewer than 10 eggs at a given time, so it is difficult to obtain enough material for biochemical studies. As a final hurdle, the development of mammalian embryos is accomplished inside another organism rather than in the external environment (although early embryos prior to implantation can be cultured and observed in vitro). Most research on mammalian development has focused on the mouse, since mice are relatively easy to breed, have large litters, and are easily housed in laboratories.

The unique nature of mammalian cleavage

Prior to fertilization, the mammalian oocyte, wrapped in cumulus cells, is released from the ovary and swept by the fimbriae into the oviduct (**FIGURE 12.11**). Fertilization occurs in the **ampulla** of the oviduct, a region close to the ovary. Meiosis is completed after sperm entry, and the first cleavage begins about a day later (see Figure 7.32). The positioning of the first cleavage plane may depend on the point of sperm entry (Piotrowska and Zernicka-Goetz 2001), and in mice, a sperm-borne microRNA (miRNA-34c) is required to initiate this first cell division. This miRNA appears to bind and inhibit Bcl-2, a protein that prevents the cell from entering the S phase of the cell cycle (Liu et al. 2012). The two nuclei produced by this cleavage are the first nuclei to contain the entire genome, since the haploid pronuclei enter cell division upon meeting (see Chapter 7).

Cleavages in mammalian eggs are among the slowest in the animal kingdom, taking place some 12–24 hours apart. The cilia in the oviduct push the embryo toward the uterus, and the first cleavages occur along this journey. In addition to the slowness of cell division, several other features distinguish mammalian cleavage, including the unique orientation of mammalian blastomeres relative to one another. In many but not all mammalian embryos, the first cleavage is a normal meridional division; however,

---

[3] In humans, homozygous loss of *PITX2* causes Rieger's syndrome, a condition characterized by asymmetry anomalies. A similar condition is caused by knocking out the *Pitx2* gene in mice (Fu et al. 1998; Lin et al. 1999).

FIGURE 12.11 Development of a human embryo from fertilization to implantation. Compaction of the human embryo occurs on day 4, at the 10-cell stage. The embryo "hatches" from the zona pellucida upon reaching the uterus. During its migration to the uterus, the zona prevents the embryo from prematurely adhering to the oviduct rather than traveling to the uterus.

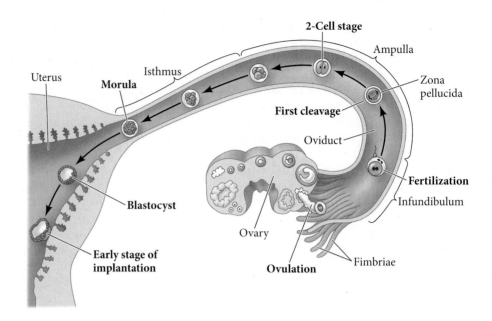

in the second cleavage, one of the two blastomeres divides meridionally and the other divides equatorially (**FIGURE 12.12**). This is called **rotational cleavage** (Gulyas 1975).

Another major difference between mammalian cleavage and that of most other embryos is the marked asynchrony of early cell division. Mammalian blastomeres do not all divide at the same time. Thus, mammalian embryos do not increase exponentially from 2 to 4 to 8 cells, but frequently contain odd numbers of cells. Furthermore, the mammalian genome, unlike the genomes of rapidly developing animals, is activated during early cleavage and zygotically transcribed proteins are necessary for cleavage and development. Maternally encoded proteins can persist through most of the cleavage stages and play important roles in early development. In the mouse and goat, the activation of zygotic (i.e., nuclear) genes begins in the late zygote and continues through the 2-cell stage (Zeng and Schultz 2005; Rother et al. 2011). In humans, the zygotic genes are activated slightly later, around the 8-cell stage (Piko and Clegg 1982; Braude et al. 1988; Dobson et al. 2004).

In order for the zygotic genes to be activated, the parental chromatin undergoes many changes. New histones are placed on the DNA during the early cell divisions, and the gamete-specific DNA methyl groups are removed (except for those on imprinted genes; see Chapter 3). In both mice and human embryos, the DNA methylation of sperm and egg chromatin is almost entirely removed. While some "imprinted gene" methylation remains, that concerned with cell differentiation appears to be removed. This allows an almost "clean slate" for the newly forming blastocyst cells. New DNA methylation patterns characteristic of totipotent and pluripotent cells are established (Abdalla et al. 2009; Guo et al. 2014; Smith et al. 2014). Thus, by the 16-cell stage, the genome of each cell is hypomethylated and each of these 16 cells appears to be pluripotent (Tarkowski et al. 2010). The stage is now set for cell differentiation to take place.

(A) Echinoderm and amphibian

(B) Mammal

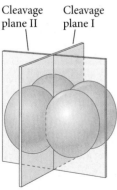

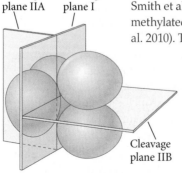

FIGURE 12.12 Comparison of early cleavage in (A) echinoderms and amphibians (radial cleavage) and (B) mammals (rotational cleavage). Nematodes also have a rotational form of cleavage, but they do not form the blastocyst structure characteristic of mammals. (After Gulyas 1975.)

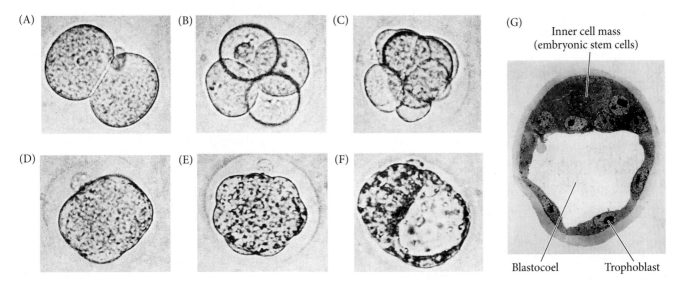

FIGURE 12.13   Cleavage of a single mouse embryo in vitro. (A) 2-Cell stage. (B) 4-Cell stage. (C) Early 8-cell stage. (D) Compacted 8-cell stage. (E) Morula. (F) Blastocyst. (G) Electron micrograph through the center of a mouse blastocyst. (A–F from Mulnard 1967, courtesy of J. G. Mulnard; G from Ducibella et al. 1975, courtesy of T. Ducibella.)

Compaction

One of the most crucial events of mammalian cleavage is **compaction**. Mouse blastomeres through the 8-cell stage form a loose arrangement (**FIGURE 12.13A–C**). Following the third cleavage, however, the blastomeres undergo a spectacular change in their behavior. Cell adhesion proteins such as E-cadherin become expressed, and the blastomeres gradually huddle together and form a compact ball of cells (**FIGURE 12.13D**; Peyrieras et al. 1983; Fleming et al. 2001). This tightly packed arrangement is stabilized by tight junctions that form between the outside cells of the ball, sealing off the inside of the sphere. The cells within the sphere form gap junctions, thereby enabling small molecules and ions to pass between them.

The cells of the compacted 8-cell embryo divide to produce a 16-cell **morula** (**FIGURE 12.13E**). The morula consists of a small group of internal cells surrounded by a larger group of external cells (Barlow et al. 1972). Most of the descendants of the external cells become **trophoblast** (trophectoderm) cells, whereas the internal cells give rise to the **inner cell mass** (**ICM**). The inner cell mass, which will give rise to the embryo, becomes positioned on one side of the ring of trophoblast cells; the resulting **blastocyst** is another hallmark of mammalian cleavage (**FIGURE 12.13F,G**; see also Figure 5.5).

The trophoblast cells produce no embryonic structures, but rather form the tissues of the chorion, the extraembryonic membrane and portion of the placenta that enables the fetus to get oxygen and nourishment from the mother. The chorion also secretes hormones that cause the mother's uterus to retain the fetus, and it produces regulators of the immune response so that the mother will not reject the embryo.

It is important to remember that a crucial outcome of these first divisions is the generation of cells that attach the embryo to the uterus. Thus, formation of the trophectoderm is the first differentiation event in mammalian development. The earliest blastomeres (such as each blastomere of a 2-cell embryo) can form both trophoblast cells and the embryo precursor cells of the ICM. These very early cells are said to be **totipotent** (Latin, "capable of everything"). The inner cell mass is said to be **pluripotent** (Latin, "capable of many things"). That is, each cell of the ICM can generate any cell type in the body but is no longer able to form the trophoblast. These pluripotent cells of the inner cell mass are the embryonic stem cells (see Chapter 5).

WATCH DEVELOPMENT 12.1   Two videos of mammalian development, from in vitro fertilization clinics, show the dynamics of early development.

FIGURE 12.14 Core transcriptional circuitry for the pluripotency of ES cells (A) Feedforward circuit in which Oct4/Sox2 dimers activate *Nanog* genes. Nanog protein then activates its own gene as well as genes promoting pluripotency. (B) The interconnected regulatory circuit whereby Oct4, Sox2, and Nanog each activate themselves and each other's synthesis. (After Boyer et al. 2005.)

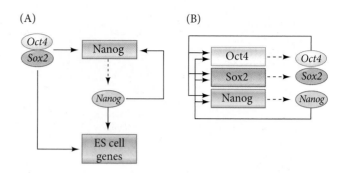

Developing Questions

We have been discussing eutherian mammals, those organisms such as mice and humans that retain the fetus during its development. But what about monotreme mammals (such as the platypus) that lay eggs, or marsupial mammals (such as kangaroos) that have extremely short pregnancies? Do their embryos have blastocysts?

Trophoblast or ICM? The first decision of the rest of your life

The philosopher and theologian Søren Kierkegaard wrote that we define ourselves by the choices we make. It seems that the embryo already knows this. The decision to become either trophoblast or inner cell mass is the first binary decision in mammalian life. Later in development, embryonic cells must lose their pluripotency and decide on what they are going to grow up to be. In the first decision, Oct4 mutually represses Cdx2 expression, enabling some cells to be trophoblast and other cells to become the pluripotent cells of the ICM. In the second decision, each of the cells of the ICM expresses either Nanog or Gata6, thereby retaining its pluripotency (Nanog) or becoming primitive endoderm (Gata6) (Ralston and Rossant 2005; Rossant 2016).

Prior to blastocyst formation, each embryonic blastomere expresses both the Cdx2 and Oct4 transcription factors (Niwa et al. 2005; Dietrch and Hiiragi 2007; Ralston and Rossant 2008) and appears to be capable of becoming either ICM or trophoblast (Hiiragi and Solter 2004; Motosugi et al. 2005; Kurotaki et al. 2007). However, once the decision to become either trophoblast or ICM is made, the cell expresses a set of genes specific to each region. The pluripotency of the ICM is maintained by a core of three transcription factors, Oct4, Sox2, and Nanog. These proteins bind to the enhancers of their own genes to maintain their expression while at the same time activating one another's enhancers (**FIGURE 12.14**). Thus, when one of these genes is activated, the other ones are too. Acting in concert, Sox2 and Oct4 form a dimer and often reside on enhancers adjacent to Nanog, activating those genes required to maintain pluripotency in embryonic stem (ES) cells and repressing those genes whose products would lead to differentiation (Marson et al. 2008; Young 2011). These transcription factors appear to work by recruiting RNA polymerase II to the promoters of those genes being activated while recruiting histone methyltransferases to those genes being repressed (Kagey et al. 2010; Adamo et al. 2011).

Only trophoblast cells synthesize the transcription factor Cdx2, which downregulates Oct4 and Nanog (Strumpf et al. 2005). The activation of the *Cdx2* gene in the trophoblast cells appears to be regulated by the Yap protein, which in turn is a cofactor for the transcription factor Tead4 (**FIGURE 12.15A**). Tead4 is found in the nuclei of both the inner and outer cells of the blastocyst, but it is activated by Yap only in the outer compartment. That is because Yap can enter the nucleus in the outer cells and thereby allow Tead4 to transcribe trophoblast-specifying genes such as *Cdx2* and *eomesodermin* (*Eomes*). In contrast, the inner cells, with each of their surfaces surrounded by other cells, activate the gene for Lats, a protein kinase that phosphorylates Yap (**FIGURE 12.15B**). Phosphorylated Yap cannot enter the nucleus and is degraded (Nishioka et al. 2009). Therefore, in the inner cells, Tead4 cannot function and *Cdx2* remains untranscribed (see Wu and Scholer 2016). Cdx2 blocks the expression of Oct4, and Oct4 blocks the expression of Cdx2. In this way, the two lineages become separated.

WEB TOPIC 12.4 **MECHANISMS OF COMPACTION AND FORMATION OF THE INNER CELL MASS** What determines whether a cell is to become a trophoblast cell or a member of the inner cell mass? It may just be a matter of chance. However, once the decision is made, different genes are switched on.

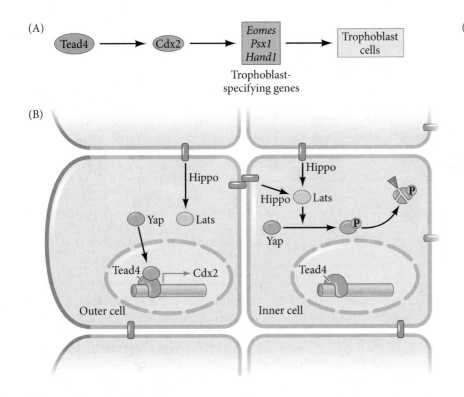

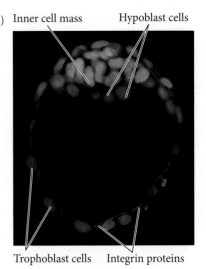

FIGURE 12.15 Possible pathway initiating the distinction between inner cell mass and trophoblast. (A) The Tead4 transcription factor, when active, promotes transcription of the *Cdx2* gene. Together, the Tead4 and Cdx2 transcription factors activate the genes that specify the outer cells to become the trophoblast. (B) Model for Tead4 activation. In the outer cells, the lack of cells surrounding the embryo sends a signal (as yet unknown) that blocks the Hippo pathway from activating the Lats protein. In the absence of functional Lats, the Yap transcriptional co-factor can bind with Tead4 to activate the *Cdx2* gene. In the inner cells, the Hippo pathway is active and the Lats kinase phosphorylates the Yap transcriptional co-activator. The phosphorylated form of Yap does not enter the nucleus and is targeted for degradation. (C) Mouse blastocyst in which the Oct4 protein in the ICM is stained orange. The extracellular lineages (trophoblast and hypoblast) are stained green. (A,B after Nishioka et al. 2009; C courtesy of J. Rossant.)

In mice, the embryo proper is derived from the inner cell mass of the 16-cell stage, supplemented by cells dividing from the outer cells of the morula during the transition to the 32-cell stage (Pedersen et al. 1986; Fleming 1987; McDole et al. 2011). The cells of the ICM give rise to the embryo and its associated yolk sac, allantois, and amnion. By the 64-cell stage, the ICM (comprising approximately 13 cells at that stage) and the trophoblast cells have become separate cell layers, neither of which contributes cells to the other group (Dyce et al. 1987; Fleming 1987). The ICM actively supports the trophoblast, secreting proteins that stimulate the trophoblast cells to divide (Tanaka et al. 1998).

Initially, the morula does not have an internal cavity. However, during a process called **cavitation**, the trophoblast cells secrete fluid into the morula to create a blastocoel. The membranes of trophoblast cells contain sodium pumps (an Na^+-K^+ ATPase and an Na^+-H^+ exchanger) that pump Na^+ into the central cavity. The subsequent accumulation of Na^+ draws in water osmotically, creating and enlarging the blastocoel (Borland 1977; Ekkert et al. 2004; Kawagishi et al. 2004). Interestingly, this sodium pumping activity appears to be stimulated by the oviduct cells on which the embryo is traveling toward the uterus (Xu et al. 2004). As the blastocoel expands, the inner cell mass becomes positioned on one side of the ring of trophoblast cells, resulting in the distinctive mammalian blastocyst.[4]

Escape from the zona pellucida and implantation

While the embryo is moving through the oviduct en route to the uterus, the blastocyst expands within the zona pellucida (the extracellular matrix of the egg that was essential for sperm binding during fertilization; see Chapter 7). During this time, the zona pellucida prevents the blastocyst from adhering to the oviduct walls. (If such adhering occurs—as it sometimes does in humans—it forms an ectopic, or "tubal," pregnancy, a dangerous condition because an embryo implanted in the oviduct can cause a life-threatening hemorrhage when it begins to grow.) When the embryo reaches the uterus, it must "hatch" from the zona so that it can adhere to the uterine wall.

---

[4]Although the mammalian blastocyst was discovered by Rauber in 1881, its first public display was probably in Gustav Klimt's 1907 painting *Danae*, in which blastocyst-like patterns are featured on the heroine's robe as she becomes impregnated by Zeus (Gilbert and Braukmann 2011).

(A)

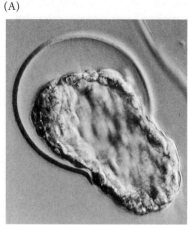

(B)

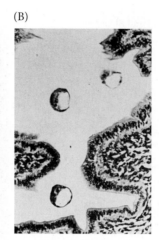

(C)

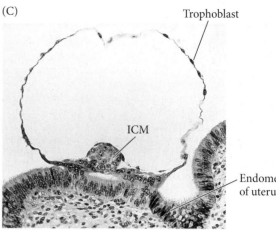

Trophoblast

ICM

Endometrium
of uterus

FIGURE 12.16 Hatching from the zona and implantation of the mammalian blastocyst in the uterus. (A) Mouse blastocyst hatching from the zona pellucida. (B) Mouse blastocysts entering the uterus. (C) Initial implantation of a rhesus monkey blastocyst. (A from Mark et al. 1985, courtesy of E. Lacy; B from Rugh 1967; C, Carnegie Institution of Washington, Chester Reather, photographer.)

The mouse blastocyst hatches from the zona pellucida by digesting a small hole in it and squeezing through the hole as the blastocyst expands (**FIGURE 12.16A**). A trypsin-like protease secreted by the trophoblast seems responsible for hatching the blastocyst from the zona (Perona and Wassarman 1986; O'Sullivan et al. 2001). Once outside the zona, the blastocyst can make direct contact with the uterus (**FIGURE 12.16B,C**). The **endometrium**—the epithelial lining of the uterus—has been altered by estrogen and progesterone hormones and has made an extensive extracellular matrix that "catches" the blastocyst. This extracellular matrix is composed of complex sugars, collagen, laminin, fibronectin, cadherins, hyaluronic acid, and heparan sulfate receptors (see Ramathal et al. 2011; Tu et al. 2014).

After the initial binding, several other adhesion systems appear to coordinate their efforts to keep the blastocyst tightly bound to the uterine lining. The trophoblast cells synthesize integrins that bind to the uterine collagen, fibronectin, and laminin, and they synthesize heparan sulfate proteoglycan precisely prior to implantation (see Carson et al. 1993). P-cadherins (see Chapter 4) on the trophoblast and uterine endometrium also help dock the embryo to the uterus. Once in contact with the endometrium, Wnt proteins (from the trophoblast, the endometrium, or from both) instruct the trophoblast to secrete a set of proteases, including collagenase, stromelysin, and plasminogen activator. These protein-digesting enzymes digest the extracellular matrix of the uterine tissue, enabling the blastocyst to bury itself within the uterine wall (Strickland et al. 1976; Brenner et al. 1989; Pollheimer et al. 2006).

Mammalian Gastrulation

Birds and mammals are both descendants of reptilian species (albeit different reptilian species). It is not surprising, therefore, that mammalian development parallels that of reptiles and birds. What *is* surprising is that the gastrulation movements of reptilian and avian embryos, which evolved as an adaptation to yolky eggs, are retained in the mammalian embryo even in the absence of large amounts of yolk. The mammalian inner cell mass can be envisioned as sitting atop an imaginary ball of yolk, following instructions that seem more appropriate to its reptilian ancestors.

Modifications for development inside another organism

The mammalian embryo obtains nutrients directly from its mother and does not rely on stored yolk. This adaptation has entailed a dramatic restructuring of the maternal anatomy (such as expansion of the oviduct to form the uterus) as well as the development of a fetal organ capable of absorbing maternal nutrients. The origins of early mammalian tissues are summarized in **FIGURE 12.17**. As we saw above, the first distinction is that between inner cell mass and trophoblast. The trophoblast develops through several

FIGURE 12.17 Tissue and germ layer formation in the early human embryo. Days 5–9: Implantation of the blastocyst. The inner cell mass delaminates hypoblast cells that line the blastocoel, forming the extraembryonic endoderm of the primitive yolk sac and a bilayered (epiblast and hypoblast) blastodisc. Days 10–12: The trophoblast divides into the cytotrophoblast, which will form the villi, and the syncytiotrophoblast, which will ingress into the uterine tissue to form the chorion. Days 12–15: Gastrulation and formation of primitive streak. Meanwhile, the epiblast splits into the amniotic ectoderm (which encircles the amniotic cavity) and the embryonic epiblast. The adult mammal (ectoderm, endoderm, mesoderm, and germ cells) forms from the cells of the embryonic epiblast. The extraembryonic endoderm forms the yolk sac. The actual size of the embryo at this stage is about that of the period at the end of this sentence.

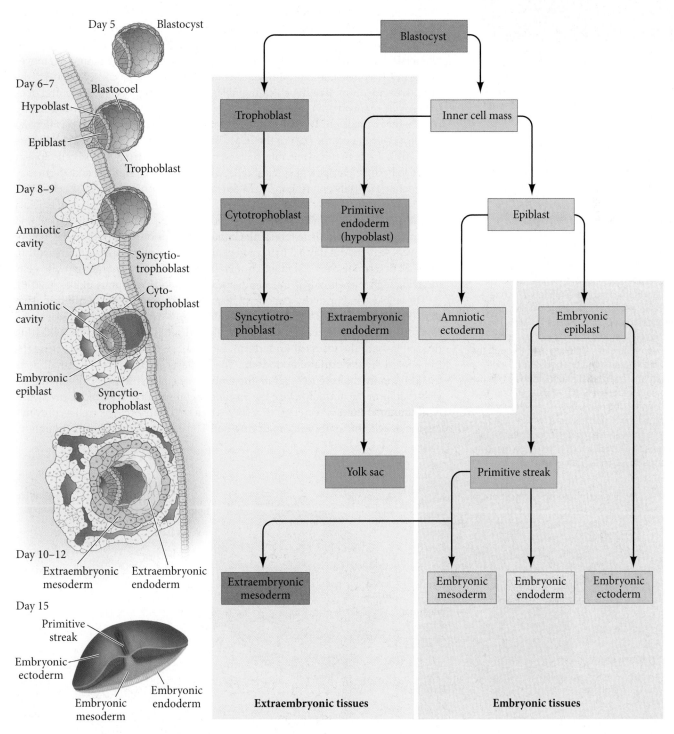

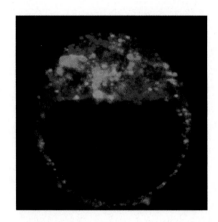

FIGURE 12.18 Mouse embryo at day 3.5 (early blastocyst), showing the random expression of Nanog (blue, for the epiblast) and Gata6 (red, for the visceral endoderm) in the inner cell mass. In another 24 hours, the cells will sort out: the hypoblast cells will abut the blastocoel, and the epiblast cells will be between the hypoblast cells and the trophoblast (as in Figure 12.15C). (Courtesy of J. Rossant.)

FIGURE 12.19 Amnion structure and cell movements during human gastrulation. (A,B) Human embryo and uterine connections at day 15 of gestation. (A) Sagittal section through the midline. (B) View looking down on the dorsal surface of the embryo. Movements of the epiblast cells through the primitive streak and the node and underneath the epiblast are superimposed on the dorsal surface view. (C) At days 14 and 15 the ingressing epiblast cells are thought to replace the hypoblast cells (which contribute to the yolk sac lining), and at day 16 the ingressing cells fan out to form the mesodermal layer. (After Larsen 1993.)

stages, eventually becoming the chorion, the embryonically derived portion of the placenta. Trophoblast cells also induce the mother's uterine cells to form the maternal portion of the placenta, the **decidua**. The decidua becomes rich in the blood vessels that will provide oxygen and nutrients to the embryo. The inner cell mass gives rise to the epiblast and the hypoblast (primitive endoderm). The hypoblast will generate yolk sac cells, while the epiblast will generate the embryo, the amnion, and the allantois.

THE PRIMITIVE ENDODERM: THE MAMMALIAN HYPOBLAST   The first segregation of cells within the inner cell mass forms two layers. The lower layer, in contact with the blastocoel, is called the **primitive endoderm**, and it is homologous to the hypoblast of the chick embryo. The remaining inner cell mass tissue above it is the epiblast. The primitive endoderm will form the yolk sac of the embryo, and like the chick hypoblast, will be used for positioning the site of gastrulation, regulating cell movements in the epiblast, and promoting the maturation of blood cells. Moreover, the primitive endoderm, like the chick hypoblast, is an extraembryonic layer and does not provide many (if any) cells to the actual embryo (see Stern and Downs 2012).

Whether a mouse ICM cell becomes epiblast or primitive endoderm may depend on *when* the cell became part of the ICM (Bruce and Zernicka-Goetz 2010; Morris et al. 2010). Cells that become internalized in the division from 8 to 16 cells appear biased to become pluripotent epiblast cells, while the future primitive endoderm may be generated by cells entering the ICM during the division from 16 to 32 cells (**FIGURE 12.18**). At that stage, the blastomeres of the ICM are a mosaic of future epiblast cells (expressing the Nanog transcription factor, which promotes pluripotency) and primitive endoderm cells (expressing Gata6 transcription factor) a full day before the layers segregate at day 4.5 (Chazaud et al. 2006). Levels of FGF signaling within the ICM determine the final identity of epiblast or primitive endoderm, with cells receiving higher levels of FGF becoming primitive endoderm (Yamanaka et al. 2010).

The epiblast and primitive endoderm form a structure called the **bilaminar germ disc** (**FIGURE 12.19A**). The primitive endoderm cells expand to line the blastocoel cavity, where they give rise to the yolk sac. The primitive endoderm cells contacting the epiblast are the **visceral endoderm**, while those yolk sac cells contacting the trophoblast are the **parietal endoderm**. The epiblast cell layer is split by small clefts that eventually coalesce to separate the embryonic epiblast from the other epiblast cells that form the amnion. Once the amnion is completed, the amniotic cavity fills with **amniotic fluid**, a secretion that serves as a shock absorber as well as preventing the developing embryo from drying out. The embryonic epiblast is thought to

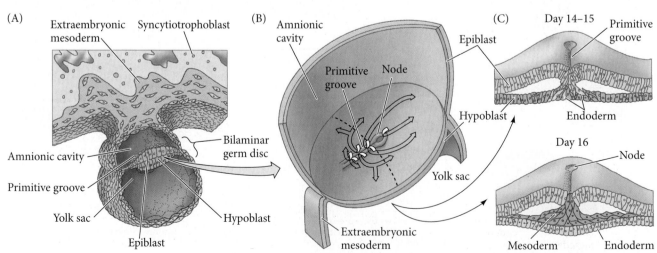

contain all the cells that will generate the actual embryo and is similar in many ways to the avian epiblast.

By labeling individual cells of the epiblast with horseradish peroxidase, Kirstie Lawson and her colleagues (1991) were able to construct a detailed fate map of the mouse epiblast (see Figure 1.11). Gastrulation begins at the posterior end of the embryo, and this is where the cells of the **node**[5] arise (**FIGURE 12,19B,C**). Like the chick epiblast cells, mammalian mesoderm and endoderm cells originate in the epiblast, undergo epithelial-mesenchymal transition, lose E-cadherin, and migrate through a primitive streak as individual mesenchymal cells (Burdsal et al. 1993). Those cells arising from the node give rise to the notochord. However, in contrast to notochord formation in the chick, the cells that form the mouse notochord are thought to become integrated into the endoderm of the primitive gut (Jurand 1974; Sulik et al. 1994). These cells can be seen as a band of small, ciliated cells extending rostrally from the node. They form the notochord by converging medially and "budding" off in a dorsal direction from the roof of the gut. The timing of these developmental events varies enormously in mammals. In humans, the migration of cells forming the mesoderm doesn't start until day 16—around the time that a mouse embryo is almost ready to be born (see Figure 12.19C; Larsen 1993).

Cell migration and specification are coordinated by fibroblast growth factors. The cells of the primitive streak appear to be capable of both synthesizing and responding to FGFs (Sun et al. 1999; Ciruna and Rossant 2001). In embryos that are homozygous for the loss of the *Fgf8* gene or its receptor, cells fail to emigrate from the primitive streak, and neither mesoderm nor endoderm are formed. Fgf8 (and perhaps other FGFs) probably control cell movement into the primitive streak by downregulating the E-cadherin that holds the epiblast cells together. Fgf8 may also control cell specification by regulating *snail*, *Brachyury*, and *Tbx6*, three genes that are essential (as they are in the chick embryo) for mesodermal migration, specification, and patterning.

The ectodermal precursors are located anterior and lateral to the fully extended primitive streak, as in the chick epiblast and (as in the chick embryo), a single cell can give rise to descendants in more than one germ layer. Thus, at the epiblast stage these lineages have not yet become fully separate from one another. Indeed, in mice, some of the visceral endoderm, which had been extraembryonic, is able to intercalate with the definitive endoderm and become part of the gut (Kwon et al. 2008).

 SCIENTISTS SPEAK 12.2   In two videos, Dr. Janet Rossant discusses her research on embryonic cell lineages in the mouse embryo.

WEB TOPIC 12.5 **PLACENTAL FORMATION AND FUNCTIONS**   In addition to providing nutrition, the placenta is an endocrine and immunological organ, producing hormones that enable the uterus to maintain the pregnancy and promote the development of the mother's mammary glands. Recent studies suggest that the placenta uses several mechanisms to block the mother's immune response against the developing fetus.

Mammalian Axis Formation

Biologist and poet Miroslav Holub (1990) remarked:

> *Between the fifth and tenth days the lump of stem cells differentiates into the overall building plan of the [mouse] embryo and its organs. It is a bit like a lump of iron turning into the space shuttle. In fact it is the profoundest wonder we can still imagine and accept, and at the same time so usual that we have to force ourselves to wonder about the wondrousness of this wonder.*

It is, indeed, wonderful, and we are just beginning to find out how really amazing it is.

---

[5] In mouse development, Hensen's node is usually just called "the node," despite the fact that Hensen discovered this structure in rabbit and guinea pig embryos.

(A)

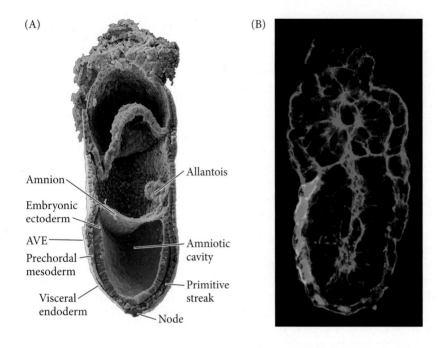

Amnion

Embryonic
ectoderm

AVE

Prechordal
mesoderm

Visceral
endoderm

Allantois

Amniotic
cavity

Primitive
streak

Node

(B)

FIGURE 12.20 Axis and notochord formation in the mouse. (A) In the 7-day mouse embryo, the dorsal surface of the epiblast (embryonic ectoderm) is in contact with the amniotic cavity. The ventral surface of the epiblast contacts the newly formed mesoderm. In this cuplike arrangement, the endoderm covers the surface of the embryo. The node is at the bottom of the cup, and it has generated chordamesoderm. The two signaling centers, the node and the anterior visceral endoderm (AVE), are located on opposite sides of the cup. Eventually, the notochord will link them. The caudal side of the embryo is marked by the presence of the allantois. (B) Confocal fluorescence image of *Cerberus* gene expression, with the *Cerberus* gene fused to a gene for GFP. At this stage, the Cerberus-synthesizing cells are migrating to the most anterior region of the visceral endoderm. (A courtesy of K. Sulik; B courtesy of J. Belo.)

The anterior-posterior axis: Two signaling centers

The formation of the mammalian anterior-posterior axis has been studied most extensively in mice. The structure of the mouse epiblast, however, differs from that of humans in that it is cup-shaped rather than disc-shaped. Whereas the human embryo looks very much like the chick embryo, the mouse embryo "drops" such that it looks like a droplet enclosed by the primitive endoderm (**FIGURE 12.20A**).

The mammalian embryo appears to have two signaling centers: one in the node (equivalent to Hensen's node and the trunk portion of the amphibian organizer), and one in the **anterior visceral endoderm** (**AVE**; Beddington and Robertson 1999; Foley et al. 2000). The node appears to be responsible for neural induction and for the patterning of most of the anterior-posterior axis, while the AVE is critical for positioning the primitive streak (see Bachiller et al. 2000).

The signals that initiate the primitive streak appear to come from interactions between the trophoblast-derived extraembryonic ectoderm and the epiblast. BMP4 originating from the extraembryonic ectoderm instructs the adjacent epiblast cells to make Wnt3a and Nodal. However, the AVE prevents Wnt3a and Nodal from having an effect on the anterior side of the embryo, by secreting antagonists of these paracrine factors, Lefty-1, Dickkopf, and Cerberus (**FIGURE 12.20B**; Brennan et al. 2001; Perea-Gomez et al. 2001; Yamamoto et al. 2004). As in amphibian embryos, the anterior region is protected from Wnt signals. Thus, Wnt3a activates the *Brachyury* gene in cells of the *posterior* but not the *anterior* epiblast, generating mesoderm cells (Bertocchini et al. 2002; Perea-Gomez et al. 2002). Once formed, the node secretes Chordin; the head process and notochord will later add Noggin. Mice missing *both* genes lack a forebrain, nose, and other facial structures.

But how does the mouse AVE form? The answer was unexpected. The AVE, and thus the anterior-posterior axis of the mammal, appears to be generated by an environmental force—the shape of the uterus. The uterus constrains the embryo such that growth occurs only in one direction. This stretching "downward" breaks the extracellular matrix and induces new gene expression in the distalmost epiblast cells (**FIGURE 12.21**). The products of these newly expressed genes cause the cell to migrate anteriorly and become the AVE. Hiramatsu and colleagues (2013) found that if an embryo grows in nonconstricted chambers, the A-P axis doesn't form. Thus, the mechanical

FIGURE 12.21 Formation of the AVE precursor cells by mechanical stress. (A) At day 5 of embryogenesis, growth is not restricted by the shape of the uterus and the embryo grows in several directions. (B) About 12 hours later, embryonic growth becomes restricted and the embryo grows only in the proximal-distal direction. The basement membrane at the distal region breaks, and epiblast cells enter the visceral endoderm layer (blue arrows), forming the precursors of the AVE. (After Hiramatsu et al. 2013.)

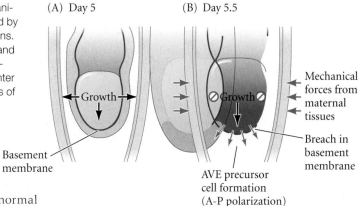

force generated by the uterus is critical in instructing normal development.

Anterior-posterior patterning by FGF and RA gradients

The head region of the mammalian embryo is devoid of Nodal signaling, and BMPs, FGFs, and Wnts are also inhibited. The posterior region is characterized by Nodal, BMPs, Wnts, FGFs, and retinoic acid. There appears to be a gradient of Wnt, BMP, and FGF proteins that is highest in the posterior and drops off strongly near the anterior region. Moreover, in the anterior half of the embryo, starting at the node, there is a high concentration of antagonists that prevent BMPs and Wnts from acting (**FIGURE 12.22A**). The Fgf8 gradient is created by the decay of mRNA: *Fgf8* is expressed at the growing posterior tip of the embryo, but its message is slowly degraded in the newly formed tissues. Thus there is a gradient of *Fgf8* mRNA across the posterior of the embryo, which is converted into an Fgf8 protein gradient (**FIGURE 12.22B**; Dubrulle and Pourquié 2004). As we will see in later chapters, the Fgf8 gradient mainly affects the development of the somatic mesoderm (forming muscles and vertebrae), while the gradient of Wnts affect the polarity of neural development.

In addition to FGFs, the late-stage gastrula has a gradient of retinoic acid, with RA levels high in the posterior regions and low in the anterior of the embryo. This gradient (like that of chick, frog, and fish embryos) appears to be controlled by the expression of RA-synthesizing enzymes in the embryo's posterior and RA-degrading enzymes in the anterior parts of the embryo (Sakai et al. 2001; Oosterveen et al. 2004). These will be important in distinguishing different regions of the brain.

The FGF gradient patterns the posterior portion of the embryo by working through the Cdx family of caudal-related genes (**FIGURE 12.22C**; Lohnes 2003).

FIGURE 12.22 Anterior-posterior patterning in the mouse embryo. (A) Concentration gradients of BMPs, Wnts, and FGFs in the late-gastrula mouse embryo (depicted as a flattened disc). The primitive streak and other posterior tissues are the sources of Wnt and BMP proteins, whereas the organizer and its derivatives (such as the notochord) produce antagonists. Fgf8 is expressed in the posterior tip of the gastrula and continues to be made in the tailbud. Its mRNA decays, creating a gradient across the posterior portion of the embryo. (B) Fgf8 gradient in the tailbud region of a 9-day mouse embryo. The highest amount of Fgf8 (red) is found near the tip. The gradient was determined by in situ hybridization of an Fgf8 probe and staining for increasing amounts of time. (C) Retinoic acid, Wnt3a, and Fgf8 each contribute to posterior patterning, but they are integrated by the Cdx family of proteins that regulates the activity of the Hox genes. (A after Robb and Tam 2004; B from Dubrulle and Pourquié 2004, courtesy of O. Pourquié; C after Lohnes 2003.)

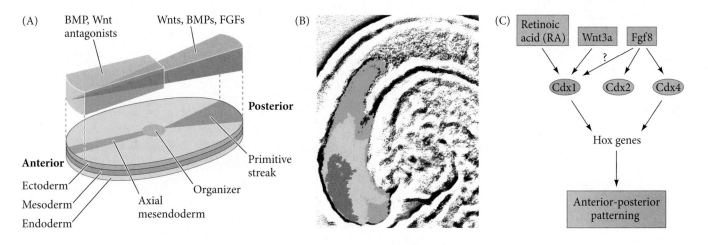

The Cdx genes, in turn, integrate the various posteriorization signals and activate particular Hox genes.

Anterior-posterior patterning: The Hox code hypothesis

In all vertebrates, anterior-posterior polarity becomes specified by the expression of Hox genes. Vertebrate Hox genes are homologous to the homeotic selector genes (Hom-C genes) of the fruit fly (see Chapter 9). The *Drosophila* homeotic gene complex on chromosome 3 contains the *Antennapedia* and *bithorax* clusters (see Figure 9.24) and can be seen as a single functional unit. (Indeed, in some other insects, such as the flour beetle *Tribolium*, it *is* a single physical unit.) All of the known mammalian genomes contain four copies of the Hox complex per haploid set, located on four different chromosomes (*Hoxa* through *Hoxd* in the mouse, *HOXA* through *HOXD* in humans; see Boncinelli et al. 1988; McGinnis and Krumlauf 1992; Scott 1992).

The order of these genes on their respective chromosomes is remarkably similar in insects and humans, as is the expression pattern of these genes. Those mammalian genes homologous to the *Drosophila labial*, *proboscipedia*, and *deformed* genes are expressed anteriorly and early, whereas those genes homologous to the *Drosophila AbdB* gene are expressed posteriorly and later. As in *Drosophila*, a separate set of genes in mice encodes the transcription factors that regulate head formation. In *Drosophila*, these are the *orthodenticle* and *empty spiracles* genes. In mice, the midbrain and forebrain are made through the expression of genes homologous to these—*Otx2* and *Emx* (see Kurokawa et al. 2004; Simeone 2004).

Mammalian Hox/HOX genes are numbered from 1 to 13, starting from the end of each complex that is expressed most anteriorly. **FIGURE 12.23** shows the relationships between the *Drosophila* and mouse homeotic gene sets. The equivalent genes in each

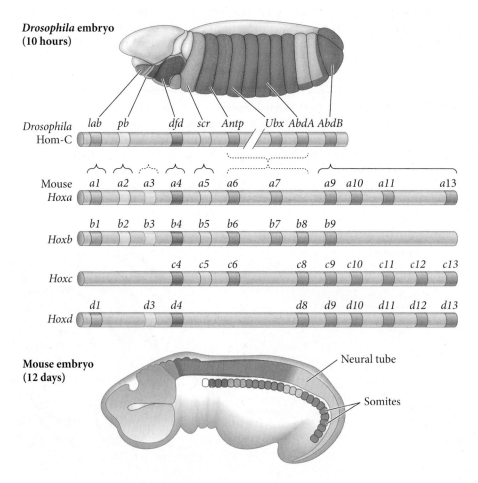

FIGURE 12.23 Evolutionary conservation of homeotic gene organization and transcriptional expression in fruit flies and mice is seen in the similarity between the Hom-C cluster on *Drosophila* chromosome 3 and the four Hox gene clusters in the mouse genome. Genes with similar structures occupy the same relative positions on each of the four chromosomes, and paralogous gene groups display similar expression patterns. The mouse genes in the higher-numbered groups are expressed later in development and more posteriorly. The comparison of the transcription patterns of the Hom-C and *Hoxb* genes of *Drosophila* and mice are shown above and below the chromosomes, respectively. (After Carroll 1995.)

mouse complex (such as *Hoxa4*, *b4*, *c4*, and *d4*) are **paralogues**—that is, it is thought that the four mammalian Hox complexes were formed by chromosome duplications. Because the correspondence between the *Drosophila* Hom-C genes and mouse Hox genes is not one-to-one, it is likely that independent gene duplications and deletions have occurred since these two animal groups diverged (Hunt and Krumlauf 1992). Indeed, the most posterior mouse Hox gene (equivalent to *Drosophila AbdB*) underwent its own set of duplications in some mammalian chromosomes.

Hox gene expression can be seen along the mammalian body axis (in the neural tube, neural crest, paraxial mesoderm, and surface ectoderm) from the anterior boundary of the hindbrain through the tail. The regions of expression are not in register, but the 3' Hox genes (homologous to *labial*, *proboscopedia*, and *deformed* of the fly) are expressed more anteriorly than the 5' Hox genes (homologous to *Ubx*, *abdA*, and *AbdB*). Thus, one generally finds the genes of paralogous group 4 expressed anteriorly to those of paralogous group 5, and so forth (see Figure 12.23; Wilkinson et al. 1989; Keynes and Lumsden 1990). Mutations in the Hox genes suggest that the regional identity along the anterior-posterior axis is determined primarily by the most posterior Hox gene expressed in that region.

EXPERIMENTAL ANALYSIS OF THE HOX CODE   The expression patterns of mouse Hox genes suggest a code whereby certain combinations of Hox genes specify a particular region of the anterior-posterior axis (Hunt and Krumlauf 1991). Particular sets of paralogous genes provide segmental identity along the anterior-posterior axis of the body. Evidence for such a code comes from two main sources: (1) comparative anatomy, in which the types of vertebrae in different vertebrate species are correlated with the constellation of Hox gene expression; and (2) gene targeting ("knockout") experiments, in which mice are constructed that lack both copies of one or more Hox genes.

COMPARATIVE ANATOMY AND HOX GENE EXPRESSION   A new type of comparative embryology is emerging based on the comparison of gene expression patterns that produce the phenotypes of different species. Gaunt (1994) and Burke and her collaborators (1995) have compared the vertebrae of the mouse and the chick (**FIGURE 12.24A**). Although the mouse and chick have a similar number of vertebrae, they apportion them differently. Mice (like all mammals, be they giraffes or whales) have 7 cervical (neck) vertebrae. These are followed by 13 thoracic (rib) vertebrae, 6 lumbar (abdominal) vertebrae, 4 sacral (hip) vertebrae, and a variable (20+) number of caudal (tail) vertebrae. The chick, by contrast, has 14 cervical vertebrae, 7 thoracic vertebrae, 12 or 13 (depending on the strain) lumbosacral vertebrae, and 5 coccygeal (fused tail) vertebrae. The researchers asked, Does the constellation of Hox gene expression correlate with the type of vertebra formed (e.g., cervical or thoracic) or with the relative position of the vertebrae (e.g., number 8 or 9)?

FIGURE 12.24 Schematic representation of the chick and mouse vertebral pattern along the anterior-posterior axis. (A) Axial skeletons stained with alcian blue at comparable stages of development. The chick has twice as many cervical vertebrae as the mouse. (B) The boundaries of expression of certain Hox gene paralogous groups (*Hox5/6* and *Hox9/10*) have been mapped onto the vertebral type domains. (A from Kmita and Duboule 2003, courtesy of M. Kmita and D. Duboule; B after Burke et al. 1995.)

(A)

Chick   Mouse

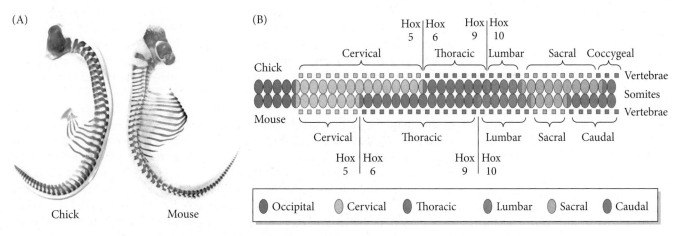

(B)

(A)

Wild type

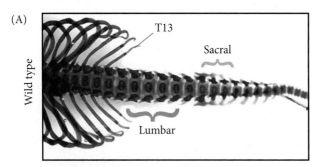

T13

Sacral

Lumbar

(B)

Hox10aaccdd⁻

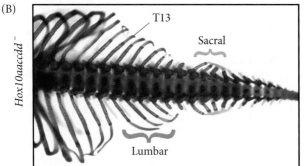

T13

Sacral

Lumbar

(C)

Hox11aaccdd⁻

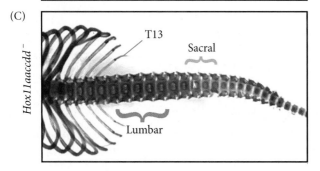

T13

Sacral

Lumbar

FIGURE 12.25 Axial skeletons of mice in gene knockout experiments. Each photograph is of an 18.5-day embryo, looking upward at the ventral region from the middle of the thorax toward the tail. (A) Wild-type mouse. (B) Complete knockout of *Hox10* paralogues (*Hox10aaccdd*) converts lumbar vertebrae (after the thirteenth thoracic vertebrae) into ribbed thoracic vertebrae. (C) Complete knockout of *Hox11* paralogues (*Hox11aaccdd*) transforms the sacral vertebrae into copies of lumbar vertebrae. (After Wellik and Capecchi 2003, courtesy of M. Capecchi.)

The answer is that the constellation of Hox gene expression predicts the type of vertebra formed. In the mouse, the transition between cervical and thoracic vertebrae is between vertebrae 7 and 8; in the chick, it is between vertebrae 14 and 15 (**FIGURE 12.24B**). In both cases, the *Hox5* paralogues are expressed in the last cervical vertebra, while the anterior boundary of the *Hox6* paralogues extends to the first thoracic vertebra. Similarly, in both animals, the thoracic-lumbar transition is seen at the boundary between the *Hox9* and *Hox10* paralogous groups. It appears there is a code of differing Hox gene expression along the anterior-posterior axis, and that code determines the type of vertebra formed.

GENE TARGETING As noted above, there is a specific pattern to the number and type of vertebrae in mice, and the Hox gene expression pattern dictates which vertebral type will form (**FIGURE 12.25A**). This was demonstrated when all six copies of the *Hox10* paralogous group (i.e., *Hoxa10, c10,* and *d10* in Figure 12.23) were knocked out and no lumbar vertebrae developed. Instead, the presumptive lumbar vertebrae formed ribs and other characteristics similar to those of thoracic vertebrae (**FIGURE 12.25B**). This was a homeotic transformation comparable to those seen in insects; however, the redundancy of genes in the mouse made it much more difficult to produce, because the existence of even one copy of the *Hox10* group genes prevented the transformation (Wellik and Capecchi 2003; Wellik 2009). Similarly, when all six copies of the *Hox11* group were knocked out, the thoracic and lumbar vertebrae were normal, but the sacral vertebrae failed to form and were replaced by lumbar vertebrae (**FIGURE 12.25C**). More recently, a *Hoxb6* gene was placed on a Delta enhancer, causing it to be expressed in every somite. The result was a "snakelike" mouse in which each somite had formed a rib-bearing thoracic vertebra (see Figure 17.7C; Guerreiro et al. 2013).

The left-right axis

The internal organs of the mammalian body are not symmetric, as the placement of the spleen, heart, and liver is determined along a right-left axis (**FIGURE 12.26A–C**). As in the chick embryo, the left-right axis appears to be due to the activation of Nodal proteins and the Pitx2 transcription factor on the left side of the lateral plate mesoderm, while Cerberus, an inhibitor of Nodal protein, is expressed on the right (see Figure 12.10; Collignon et al. 1996; Lowe et al. 1996; Meno et al. 1996). However, each amniote group may have different ways of initiating this pathway (Vanderberg and Levin 2013). In mammals, the distinction between left and right sides can be seen in the ciliary cells of the node (**FIGURE 12.26D**). The cilia cause fluid in the node to flow from right to left (clockwise when viewed from the ventral side). When Nonaka and colleagues (1998) knocked out a mouse gene encoding the ciliary motor protein dynein (see Chapter 7), the nodal cilia did not move and the lateral position of each asymmetrical organ was randomized. (This helped explain the clinical observation that humans with a dynein deficiency have immotile cilia and a random chance of having their heart on the left

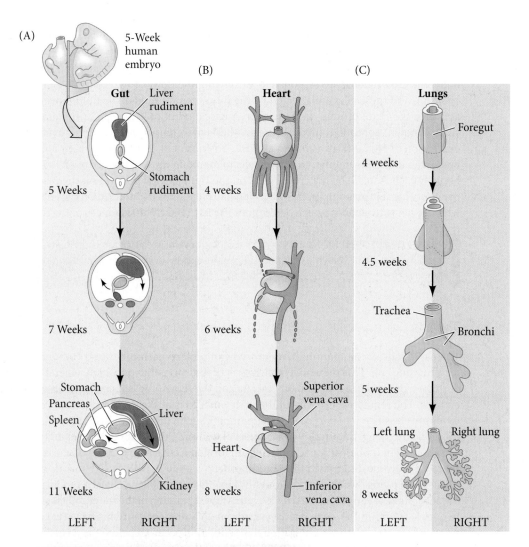

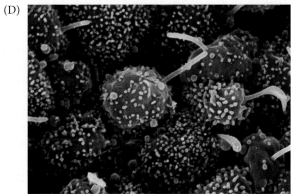

FIGURE 12.26 Left-right asymmetry in the developing human. (A) Abdominal cross sections show that the originally symmetrical organ rudiments acquire asymmetric positions by week 11. The liver moves to the right and the spleen moves to the left. (B) Not only does the heart move to the left side of the body, but the originally symmetrical veins of the heart regress differentially to form the superior and inferior venae cavae, which connect only to the right side of the heart. (C) The right lung branches into three lobes, while the left lung (near the heart) forms only two lobes. In human males, the scrotum also forms asymmetrically. (D) Ciliated cells of the mouse node, each with a cilium extending from the posterior ventral region of the cell. (A–C After Kosaki and Casey 1998; D courtesy of K. Sulik and G. C. Schoenwolf.)

or right side of the body; see Afzelius 1976). Moreover, when Nonaka and colleagues (2002) cultured early mouse embryos under an artificial flow of medium from left to right, they obtained a reversal of the left-right axis.

The mechanism for this rotation appears to be the placement of the basal body of the cilium on each of the 200 or so monociliated node cells. The basal body giving rise to each cilium is at the posterior side of each cell and extends out the ventral surface. Thus, the placement of cilia integrates information concerning the anterior-posterior and dorsal-ventral axes to construct the right-left axis (Guirao et al. 2010; Hashimoto et al. 2010). The placement of the cilia is governed by the planar cell polarity (PCP) pathway, possibly directed by a Wnt. Mutations in PCP pathway signaling molecules can random-ize localization of cilia in these cells, also causing randomization of the left-right axis.

But how does rotation generate a body axis? It appears that the cells neighboring the node, the **crown cells,** are responsible for sensing the flow. Crown cells have immo-bile cilia, and these cilia are affected by the movement of fluids. The fluid movement activates the Pkd2 protein on their cilia. A cascade initiated by Pdk2 (in a manner still unknown) appears to suppress the synthesis of Cerberus and thereby activate the expression of Nodal (Kawasumi et al. 2011; Yoshiba et al. 2012). Nodal is thought to bind in an autocrine manner to crown cells to maintain its own transcription; so Cer-berus (which is made by the crown cells on the right side) would inhibit the mainte-nance of Nodal expression. In this way, Nodal expression is maintained on the left side, where it can activate *Pitx*, which determines the left- and right-sidedness of the tissue.

WATCH DEVELOPMENT 12.2   View video on the several sites that focus on the week-by-week changes of human development, as well as sites detailing the molecular biology of mouse development, comparing it with that of humans.

Twins

The early cells of the mammalian embryo can replace each other and compensate for a missing cell. This regulative ability was first demonstrated in 1952, when Seidel destroyed one cell of a 2-cell rabbit embryo and the remaining cell produced an entire embryo. The regulative capacity of the early embryo is also seen in humans. Human twins are classified into two major groups: monozygotic (one-egg; identical) twins and dizygotic (two-egg; fraternal) twins. Fraternal twins are the result of two separate fer-tilization events, whereas identical twins are formed from a single embryo whose cells somehow become dissociated from one another.

Identical twins, which occur in roughly 1 in 400 human births, may be produced by the separation of early blastomeres, or even by the separation of the inner cell mass into two regions within the same blastocyst. About 33% of identical twins have two com-plete and separate chorions, indicating that separation occurred before the formation of the trophoblast tissue at day 5 (**FIGURE 12.27A**). Other identical twins share a common chorion, suggesting that the split occurred within the inner cell mass after the tropho-blast formed. By day 9, the human embryo has completed the construction of another extraembryonic layer, the lining of the amnion. If separation of the embryo comes after the formation of the chorion on day 5 but before the formation of the amnion on day 9, then the resulting embryos should have one chorion and two amnions (**FIGURE 12.27B**). This happens in about two-thirds of human identical twins. A small percent-age of identical twins are born within a single chorion and amnion (**FIGURE 12.27C**), meaning the division of the embryo came after day 12.

According to these studies of twins, each cell of the inner cell mass should be able to produce any cell of the body. This hypothesis has been confirmed, and it has important consequences for the study of mammalian development. When ICM cells are isolated and grown under certain conditions, they remain undifferentiated and continue to divide in culture (Evans and Kaufman 1981; Martin 1981). Such cells are embryonic stem cells (ES cells). When ES cells are injected into a mouse blastocyst, they can inte-grate into the host inner cell mass. The resulting embryo has cells from both the host and the donor tissue. This technique has become extremely important in determining the function of genes during mammalian development.

WEB TOPIC 12.6   **CHIMERISM**   The opposite of twins, chimeras occur when two distinct embryos are fused together to create a single individual.

Whereas fraternal twins are created during fertilization (two separate eggs meet-ing two separate sperm) and identical twins are formed during cleavage, **conjoined**

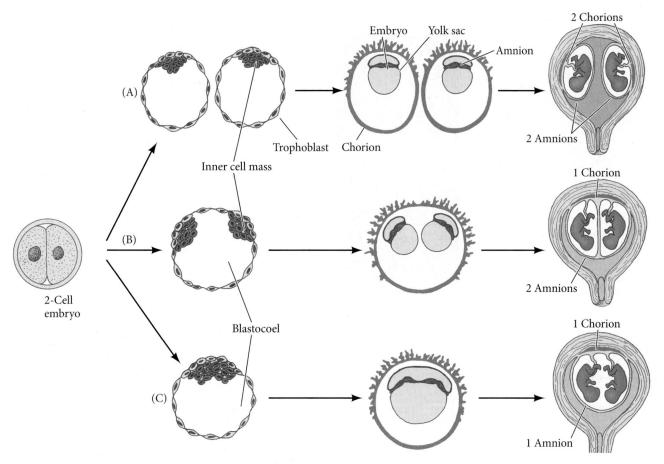

FIGURE 12.27 The timing of human monozygotic twinning with relation to extraembryonic membranes. (A) Splitting occurs before formation of the trophoblast, so each twin has its own chorion and amnion. (B) Splitting occurs after trophoblast formation but before amnion formation, resulting in twins having individual amniotic sacs but sharing one chorion. (C) Splitting after amnion formation leads to twins in one amniotic sac and a single chorion. (After Langman 1981.)

twins are probably created during gastrulation. Conjoined twins are identical (there has never been a documented case of conjoined twins being different sexes) and occur approximately once in every 200,000 live births. Spratt and Haas (1960) showed that if the chick epiblast is divided into four parts, each forms a primitive streak (**FIGURE 12.28A**). Moreover, if a second Hensen's node from a chick is placed into an epiblast, the two primitive streaks can fuse together. It appears that if there is a tear in the marginal zone (allowing a new center of Nodal expression to form), or if a second region of the marginal zone should express *Nodal*, then a second axis can form (**FIGURE 12.28B**; Bertocchini and Stern 2002; Perea Gomez et al. 2002; Torlopp et al. 2014). Since humans are thought to have the same molecular pathway to making the primitive streak as chicks do, it seems possible that conjoined twins could be the result of two areas of the margin producing Nodal. Levin (1999) has shown that this would explain the different type of conjoined twins (**FIGURE 12.28C**). We do not know how conjoined twins form, but the generation of multiple axes during gastrulation might begin to explain this phenomenon.

> **Developing Questions**
>
> In an episode of the U.S. television series *CSI: Crime Scene Investigation*, the DNA of the obvious perpetrator did not match the DNA of the cells at the crime scene. The episode was based on actual rare instances where a mammal has been found to have two different sets of DNA. How do you think such a situation might come about?

WEB TOPIC 12.7   CONJOINED TWINS   Although conjoined twins are a rare occurrence, the medical and social issues raised by conjoined twins provide a fascinating look at what people throughout history have considered "individuality."

(A)

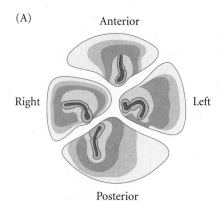

FIGURE 12.28 (A) Spratt and Haas (1960) showed that if the chick epiblast were divided into four parts, each would form a primitive streak. (B) If a second Hensen's node from a chick were placed into an epiblast, the two primitive streaks could fuse together. It appears that if there is a tear or in a second region of the marginal zone expresses *Nodal*, then a second axis can form. (C) Since humans are thought to have the same molecular pathway to making the primitive streak, it would seem possible that conjoined twins could be made by having two areas of the margin produce Nodal. This would explain the different type of conjoined twins (Levin 1999).

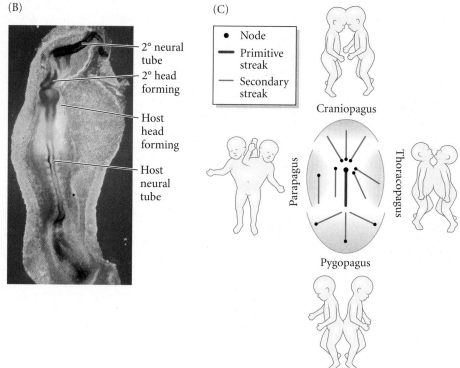

DEV TUTORIAL *Twinning* In this brief lecture, Scott Gilbert describes some of the recent theories of twin formation.

Coda

Variations on the important themes of development have evolved in the different vertebrate groups (**FIGURE 12.29**). The major themes of vertebrate gastrulation include:

- Internalization of the endoderm and mesoderm
- Epiboly of the ectoderm around the entire embryo
- Convergence of the internal cells to the midline
- Extension of the body along the anterior-posterior axis

Although fish, amphibian, avian, and mammalian embryos have different patterns of cleavage and gastrulation, they use many of the same molecules to accomplish the same goals. Each group uses gradients of Nodal and Wnt proteins to establish polarity along the anterior-posterior axis. In *Xenopus* and zebrafish, maternal factors induce Nodal proteins in the vegetal hemisphere or marginal zone. In the chick, Nodal expression is induced by Wnt and Vg1 emanating from the posterior marginal zone, while elsewhere Nodal activity is suppressed by the hypoblast. In the mouse, the hypoblast similarly restricts Nodal activity, using Cerberus in the chick embryo and Cerberus and Lefty1 in mammals.

Each of these vertebrate groups uses BMP inhibitors to specify the dorsal axis. Similarly, Wnt inhibition and Otx2 expression are important in specifying the anterior regions of the embryo, but different groups of cells may express these proteins. In all cases, the region of the body from the hindbrain to the tail is specified by Hox genes. Finally, the left-right axis is established through the expression of Nodal on the left-hand side of the embryo. Nodal activates *Pitx2*, leading to the differences between the left and right sides of the embryo. How Nodal becomes expressed on the left side appears to differ among the vertebrate groups. But overall, despite their initial differences in cleavage and gastrulation, the different vertebrate groups have maintained very similar ways of establishing the three body axes.

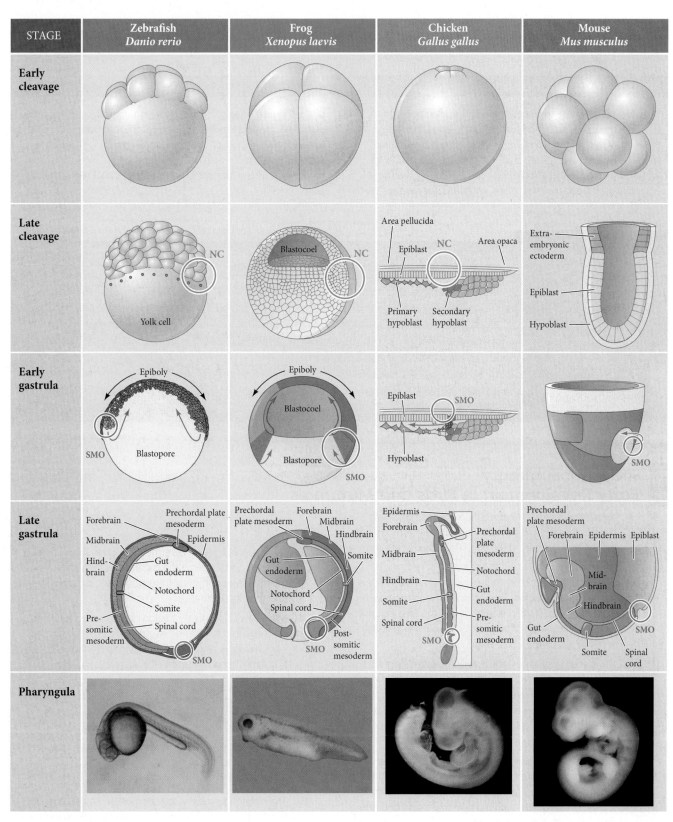

FIGURE 12.29 Early development in four vertebrates. Cleavage differs greatly among the four groups. Zebrafish and chicks have meroblastic discoidal cleavage; frogs have unequal holoblastic cleavage; and mammals have equal holoblastic cleavage. These cleavage patterns form different structures, but there are many conserved features, such as the Nieuwkoop center (NC; green circles). As gastrulation begins, each of the groups has cells equivalent to the Spemann-Mangold organizer (SMO; red circles). The SMO marks the beginning of the blastopore region, and the remainder of the blastopore is indicated by the red arrows extending from the organizer. By the late gastrula stage, the endoderm (yellow) is inside the embryo, the ectoderm (blue, purple) surrounds the embryo, and the mesoderm (red) is between the endoderm and ectoderm. The regionalization of the mesoderm has also begun. The bottom row shows the pharyngula stage that immediately follows gastrulation. This stage, with a pharynx, a central neural tube and notochord flanked by somites, and a sensory cephalic (head) region, characterizes the vertebrates. (After Solnica-Krezel 2005.)

Next Step Investigation

Induced pluripotent stem cells (see Chapter 5) provide us with an opportunity to study mammalian development as never before. Working with stem cells enables us to mutate particular genes and elucidate the pathways by which developmental anomalies occur. Some of the most important of these investigations enter territory where we will no doubt have to address the ethical implications and economic impacts of such research. Should people be able to regenerate old organs if they can afford to? Should we add genes to embryos to allow them to survive in different environments or to have greater neural plasticity or hardier bodies?

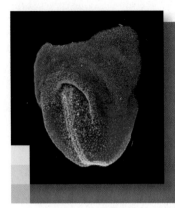

Closing Thoughts on the Opening Photo

This photograph of the primitive streak and head endoderm of the 7.5 day mouse embryo shows the beginnings of anterior specification. The nuclei of all cells are stained blue. The transcription factors Lhx1, Foxa2 (green), and Otx2 interact to regulate differentiation of the anterior mesendoderm. Brachyury (red) stains the central mesoderm. Foxa2 and Brachyury are co-expressed (yellow) in the anterior midline structure and node. This pattern is regulated by Nodal and Wnt proteins and establishes the regions of the embryo that will form the anterior (head) and posterior (tail) structures of the mammalian body. (Photograph courtesy of I. Costello and E. Robertson.)

Snapshot Summary
Early Development in Birds and Mammals

1. Reptiles and birds, like fish, undergo discoidal meroblastic cleavage, wherein the early cell divisions do not cut through the yolk of the egg. These early cells form a blastoderm.

2. In chick embryos, early cleavage forms an area opaca and an area pellucida. The region between them is the marginal zone. Gastrulation begins in the area pellucida next to the posterior marginal zone, as the hypoblast and primitive streak both start there.

3. The primitive streak is derived from epiblast cells and the central cells of Koller's sickle. As the primitive streak extends rostrally, Hensen's node is formed. Cells migrating out of Hensen's node become prechordal mesendoderm and are followed by the head process and notochord cells.

4. The prechordal plate helps induce formation of the forebrain; the chordamesoderm induces formation of the midbrain, hindbrain, and spinal cord. The first cells migrating laterally through the primitive streak become endoderm, displacing the hypoblast. The mesoderm cells then migrate through the primitive streak. Meanwhile, the surface ectoderm undergoes epiboly around the yolk.

5. In birds, gravity helps determine the position of the primitive streak, which points in a posterior-to-anterior direction and whose differentiation establishes the dorsal-ventral axis. The left-right axis is formed by the expression of Nodal protein on the left side of the embryo, which signals Pitx2 expression on the left side of developing organs.

6. The hypoblast helps determine the body axes of the embryo, and its migration determines the cell movements that accompany formation of the primitive streak and thus its orientation.

7. Mammals undergo a variation of holoblastic rotational cleavage that is characterized by a slow rate of cell division, a unique cleavage orientation, lack of divisional synchrony, and formation of a blastocyst.

8. The blastocyst forms after the blastomeres undergo compaction. It contains outer cells—the trophoblast cells—that become the chorion, and an inner cell mass that becomes the amnion and the embryo.

9. The inner cell mass cells are pluripotent and can be cultured as embryonic stem cells. They give rise to the epiblast and to the visceral endoderm (hypoblast).

10. The chorion forms the fetal portion of the placenta, which functions to provide oxygen and nutrition to the embryo, to provide hormones for the maintenance of pregnancy, and to block the potential immune response of the mother to the developing fetus.

11. Mammalian gastrulation is not unlike that of birds. There appear to be two signaling centers, one in the node and one in the anterior visceral endoderm. The latter center is critical for establishing the body axes, while the former is

critical in inducing the nervous system and in patterning axial structures caudally from the midbrain.

12. Hox genes pattern the anterior-posterior axis and help specify positions along that axis. If Hox genes are knocked out, segment-specific malformations can arise. Similarly, causing the ectopic expression of Hox genes can alter the body axis.

13. The homology of gene structure and the similarity of expression patterns between *Drosophila* and mammalian Hox genes suggest that this patterning mechanism is extremely ancient.

14. The mammalian left-right axis is specified similarly to that of the chick, but with some significant differences in the roles of certain genes.

15. In amniote gastrulation, the pluripotent epithelium, or epiblast, produces the mesoderm and endoderm (which migrate through the primitive streak), and the precursors of the ectoderm, which remain on the surface. By the end of gastrulation, the head and anterior trunk structures are formed. Elongation of the embryo continues through precursor cells in the caudal epiblast surrounding the posteriorized Hensen's node.

16. In each class of vertebrates, neural ectoderm is permitted to form where the BMP-mediated induction of epidermal tissue is prevented.

17. Fraternal twins arise from two separate fertilization events. Identical twins result from the splitting of the embryo into two cellular groups during stages where there are still pluripotent cells in the embryo. Experimental evidence suggests that conjoined twins may occur through the formation of two organizers within a common blastodisc.

Further Reading

Beddington, R. S. P. and E. J. Robertson. 1999. Axis development and early asymmetry in mammals. *Cell* 96: 195–2012.

Bertocchini, F. and C. D. Stern. 2002. The hypoblast of the chick embryo positions the primitive streak by antagonizing Nodal signaling. *Dev. Cell* 3: 735–744.

Boyer, L. A. and 13 others. 2005. Core transcriptional regulatory circuitry in human embryonic stem cells. *Cell* 122: 947–956.

Burke, A. C., A. C. Nelson, B. A. Morgan and C. Tabin. 1995. Hox genes and the evolution of vertebrate axial morphology. *Development* 121: 333–346.

Hiramatsu, R., T. Matsuoka, C. Kimura-Yoshida, S. W. Han, K. Mochida, T. Adachi, S. Takayama and I. Matsuo. 2013. External mechanical cues trigger the establishment of the anterior-posterior axis in early mouse embryos. *Dev. Cell* 27: 131–144.

Rossant, J. 2016. Making the mouse blastocyst: Past, present, and future. *Curr. Top. Dev. Biol.* 116: 275–288.

Stern, C. D. and K. M. Downs. 2012. The hypoblast (visceral endoderm): An evo-devo perspective. *Development* 139: 1059–10612.

Strumpf, D., C.-A. Mao, Y. Yamanaka, A. Ralston, K. Chawengsaksophak, F. Beck and J. Rossant. 2005. Cdx2 is required for correct cell fate specification and differentiation of trophectoderm in the mouse blastocyst. *Development* 132: 2093–2102.

Vanderberg, L. N. and M. Levin. 2013. A unified model for left-right asymmetry? Comparison and synthesis of molecular models of embryonic laterality. *Dev. Biol.* 379: 1–15.

GO TO WWW.DEVBIO.COM...

...for Web Topics, Scientists Speak interviews, Watch Development videos, Dev Tutorials, and complete bibliographic information for all literature cited in this chapter.

Neural Tube Formation and Patterning

"LIKE THE ENTOMOLOGIST IN SEARCH of brightly colored butterflies, my attention hunted, in the garden of the gray matter, cells with delicate and elegant forms, the mysterious butterflies of the soul." Thus reflected Santiago Ramón y Cajal, often referred to as the father of neuroscience, on his study of the brain. His 1937 quotation masterfully captures the fascination and mystery of the brain as part of a larger system that controls communication, consciousness, memory, emotion, motor control, digestion, sensory perceptions, sex, and so much more. How the development of this central organ is coordinated with the development of the rest of the organism for integrated connectivity will remain one of the most fundamental questions in developmental biology for the next century. The first pivotal event is the transformation of an epithelial sheet into a tube. This initial structure will provide the foundation for the regionalization and diversification of brain structures along the anterior-to-posterior axis, and, through strategic mechanisms of cell growth and differentiation, the elaborate and highly connected structure of the vertebrate central nervous system can be realized. Over the next three chapters, we will study the development of the nervous system, beginning in this chapter with the formation of the neural tube and the specification of cell fates within it (**FIGURE 13.1**). In Chapter 14, we will delve into the mechanisms governing cell fate patterning and neurogenesis along the dorsal-ventral axis of the CNS. Then, in Chapter 15, we will navigate the molecular

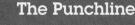

What is the time value of Sonic hedgehog?

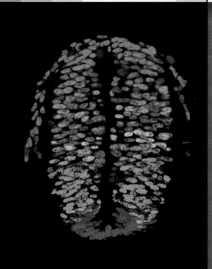

The Punchline

The vertebrate brain and spinal cord start their development as a flat plate of neuroepithelial cells that folds up along most of its length to form a tube. The process of forming this neural tube is called neurulation. Folding occurs at specific locations in the plate through asymmetric changes in the shapes of cells such that their apical sides contract, establishing hinge points of tissue bending. The folding brings the sides of the plate upward and toward each other until they fuse along the midline, as if the tube were being zipped up. The tube separates from the surface ectoderm via differential adhesion, and the central nervous system is born. The cells of the new neural tube become specialized as precursors of neurons and glial cells, and the different regions of the tube become specified along its dorsoventral axis. Morphogen gradients emanating from the dorsal surface ectoderm and ventral notochord establish positional cues for the induction of key regulatory transcription factors that initiate cell type-specific gene regulatory networks. TGF-β and Sonic hedgehog signals play important roles in both neurulation and cell fate patterning of the neural tube.

FIGURE 13.1 The major questions to be addressed in Chapters 13, 14, and 15. Questions of neurulation and cell fate specification (A,B) will be answered in this chapter. How the neural tube (NT) is expanded into the elaborate structures of the brain (C) will be covered in Chapter 14. The peripheral nervous system (D) is largely derived from neural crest cells (NCC) migrating out of the dorsal neural tube. In addition, newly born neurons must extend long processes to find their synaptic partners (E) and thus connect up the nervous system. Topics (D) and (E) will be covered in Chapter 15.

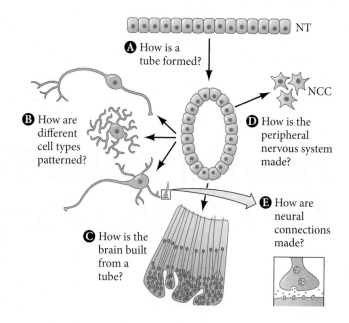

guidance mechanisms underlying the wiring of the nervous system and the development of neural crest lineages.

The vertebrate ectoderm, the outer germ layer covering the late-stage gastrula, has three major responsibilities (**FIGURE 13.2**):

1. One part of the ectoderm will become the **neural plate**, the presumptive neural tissue induced by the prechordal plate and notochord during gastrulation. The neural plate involutes into the body to form the **neural tube**, the precursor of the **central nervous system** (**CNS**)—the brain and spinal cord.

2. Another part of this germ layer will become the **epidermis**, the outer layer of the skin (which is the largest organ of the vertebrate body). The epidermis forms an elastic, waterproof, and constantly regenerating barrier between the organism and the outside world.

3. Between the compartments forming the epidermis and the central nervous system lies the presumptive **neural crest**. The cells of the neural crest delaminate from these epithelia at the dorsal midline and migrate away (between the neural tube and epidermis) to generate, among other things, the **peripheral nervous system** (all the nerves and neurons lying outside the CNS) and pigment cells (melanocytes).

The processes by which the three ectodermal regions are made physically and functionally distinct from one another is called **neurulation**, and an embryo undergoing these processes is called a **neurula** (**FIGURE 13.3**; Gallera 1971). As we saw in the preceding chapters, the specification of the ectoderm is accomplished during gastrulation, primarily by regulating the levels of BMP experienced by the ectodermal cells. High levels of BMP specify the cells to become epidermis. Very low levels specify the cells to become neural plate. Intermediate levels effect the formation of the neural crest cells. Neurulation directly follows gastrulation.

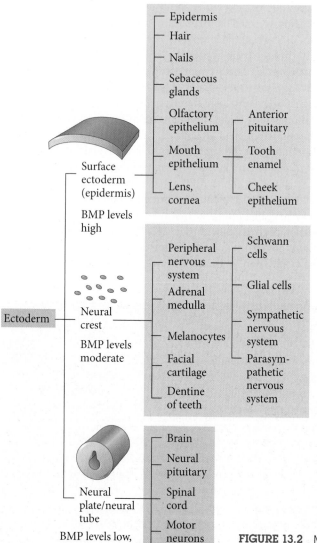

FIGURE 13.2 Major derivatives of the ectoderm germ layer. The ectoderm is divided into three major domains: the surface ectoderm (primarily epidermis), the neural crest (peripheral neurons, pigment, facial cartilage), and the neural tube (brain and spinal cord).

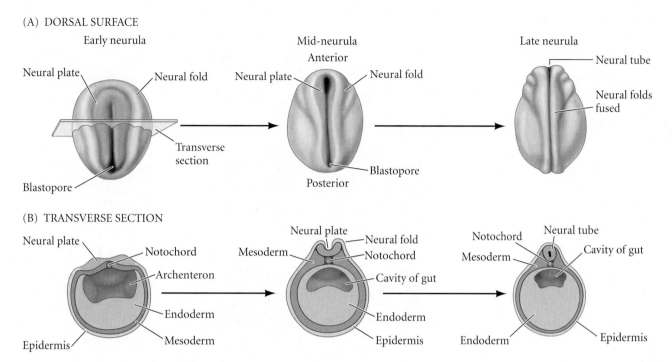

(A) DORSAL SURFACE

Early neurula

Neural plate

Neural fold

Transverse section

Blastopore

Mid-neurula

Anterior

Neural plate

Neural fold

Blastopore

Posterior

Late neurula

Neural tube

Neural folds fused

(B) TRANSVERSE SECTION

Neural plate

Notochord

Archenteron

Endoderm

Epidermis

Mesoderm

Neural plate

Mesoderm

Neural fold

Notochord

Cavity of gut

Endoderm

Epidermis

Notochord

Neural tube

Mesoderm

Cavity of gut

Endoderm

Epidermis

FIGURE 13.3 Two views of primary neurulation in an amphibian embryo, showing early (left), middle (center), and late (right) neurulae in each case. (A) Looking down on the dorsal surface of the whole embryo. (B) Transverse section through the center of the embryo. (After Balinsky 1975.)

Transforming the Neural Plate into a Tube: The Birth of the Central Nervous System

The cells of the neural plate are characterized by expression of the Sox family of transcription factors (Sox1, 2, and 3). These factors (1) activate the genes that specify cells to be neural plate and (2) inhibit the formation of epidermis and neural crest by blocking the transcription and signaling of BMPs (Archer et al. 2011). In this process, we see once again an important principle of development: *often the signals promoting the specification of one cell type also block the specification of an alternative cell type*. The expression of Sox transcription factors establishes the neural plate cells as neural precursors that can form all the cell types of the central nervous system (Wilson and Edlund 2001).

Although the neural plate lies on the surface of the embryo, the nervous system will not lie on the outside of the mature body. Somehow, the neural plate has to move inside the embryo and form a neural tube. This process is accomplished through neurulation, which occurs with some diversity across vertebrates (Harrington et al. 2009). There are two principal modes of neurulation. In **primary neurulation**, the cells surrounding the neural plate direct the neural plate cells to proliferate, invaginate into the body, and separate from the surface ectoderm to form an underlying hollow tube. In **secondary neurulation**, the neural tube arises from the aggregation of mesenchyme cells into a solid cord that subsequently forms cavities that coalesce to create a hollow tube. In many vertebrates, primary and secondary neurulation are divided spatially in the embryo such that primary neurulation forms the *anterior* portion of the neural tube and the *posterior* portion of the neural tube is the product of secondary neurulation (**FIGURE 13.4**).

In birds, primary neurulation generates the neural tube anterior to the hindlimbs (Pasteels 1937; Catala et al. 1996). In mammals, secondary neurulation begins at the level of the sacral vertebrae of the tail (Schoenwolf 1984; Nievelstein et al. 1993). In fish and amphibians (e.g., zebrafish and *Xenopus*), only the tail neural tube is derived from secondary neurulation (Gont et al. 1993; Lowery and Sive 2004). More basal chordates, such as *Amphioxus* and *Ciona*, only exhibit mechanisms of primary neurulation, suggesting that primary neurulation was the ancestral condition and that secondary

FIGURE 13.4 Primary and secondary neurulation and the transition zone between them. The bottom image is a lateral view of the neural tube surface. The illustrations above the neural tube correspond to transverse sections through the axial level indicated as the neural tube forms in a rostral-to-caudal direction. Different cell types are represented in different colors, as indicated in the key. (After Dady et al. 2014.)

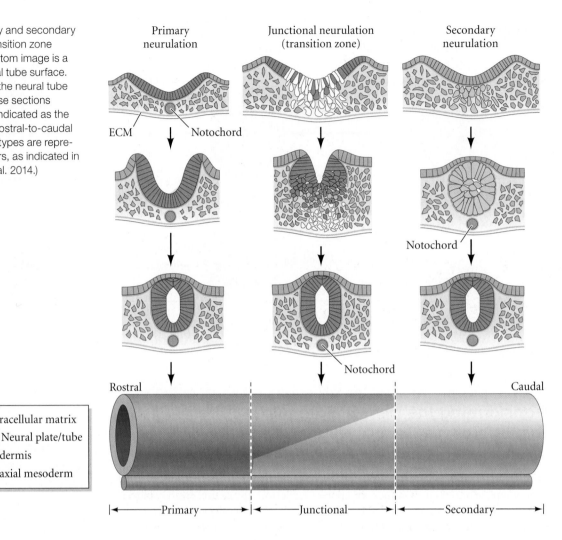

Extracellular matrix
Neural plate/tube
Epidermis
Paraxial mesoderm

neurulation evolved much like limbs—that is, as a vertebrate novelty associated with tail elongation (Handrigan 2003).

The neural tube is finally complete when these two separately formed tubes join together (Harrington et al. 2009). The size of the **transition zone** between the primary and secondary neural tubes varies among species, from relatively abrupt in the mouse, to a region spanning the thoracic vertebrae in the chick, to the thoracolumbar region in humans (Dady et al. 2014). Formation of the neural tube in this transition zone has been named **junctional neurulation** (Dady et al. 2014) because it involves a combination of mechanisms involved in both primary and secondary neurulation (see Figure 13.4).

▶ **DEV TUTORIAL** *Neurulation* Dr. Michael Barresi describes the cellular events and molecular mechanisms behind neural tube formation.

Primary neurulation

Although some species differences exist, the process of primary neurulation is relatively similar in all vertebrates.[1] To explore the mechanisms of neural plate folding, we will largely focus on the process of primary neurulation in amniotes. Shortly after the neural plate has formed in the chick, its edges thicken and move upward to form the **neural**

[1] In teleost (bony) fish such as zebrafish, the neural plate does not fold; rather, convergence at the midline generates the neural keel, and the lumen of the neural tube is formed through a process of cavitation (Lowery and Sive 2004; see also Harrington et al. 2009).

folds, and a U-shaped **neural groove** appears in the center of the plate, dividing the future right and left sides of the embryo (**FIGURE 13.5**). The neural folds on the lateral sides of the neural plate migrate toward the midline of the embryo, eventually fusing to form the neural tube beneath the overlying ectoderm.

Primary neurulation can be divided into four distinct but spatially and temporally overlapping stages:

1. *Elongation and folding of the neural plate.* Cell divisions within the neural plate are preferentially in the anterior-posterior direction (often referred to as the **rostral-caudal**, or beak-to-tail, direction), which fuels continued axial elongation associated with gastrulation. These events occur even if the neural plate tissue is isolated from the rest of the embryo. To roll into a neural tube, however, the presumptive

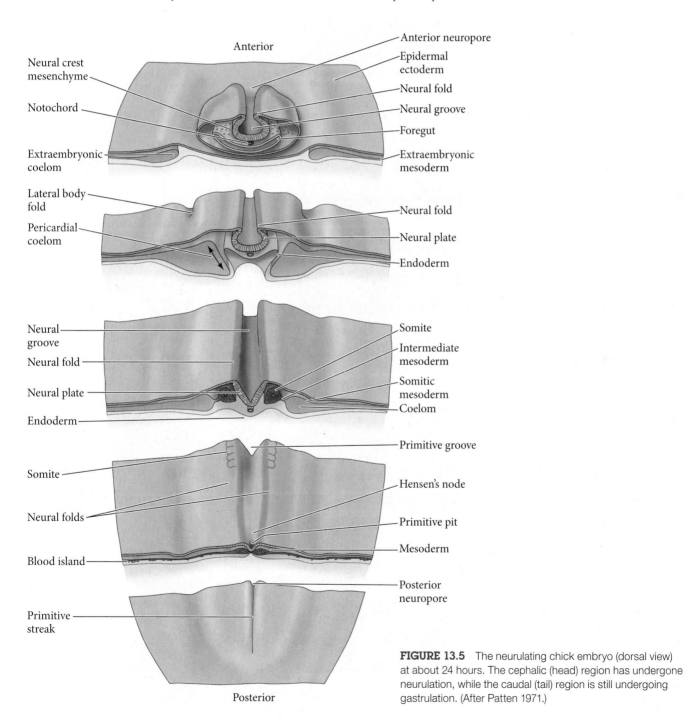

FIGURE 13.5 The neurulating chick embryo (dorsal view) at about 24 hours. The cephalic (head) region has undergone neurulation, while the caudal (tail) region is still undergoing gastrulation. (After Patten 1971.)

epidermis is also needed (**FIGURE 13.6A,B**; Jacobson and Moury 1995; Moury and Schoenwolf 1995; Sausedo et al. 1997).

2. *Bending of the neural plate.* The bending of the neural plate involves the formation of hinge regions where the neural plate contacts surrounding tissues. In birds and mammals, the cells at the midline of the neural plate form the **medial hinge point**, or **MHP** (Schoenwolf 1991a,b; Catala et al. 1996). MHP cells are reported to be firmly anchored to the notochord beneath them and form a hinge, which

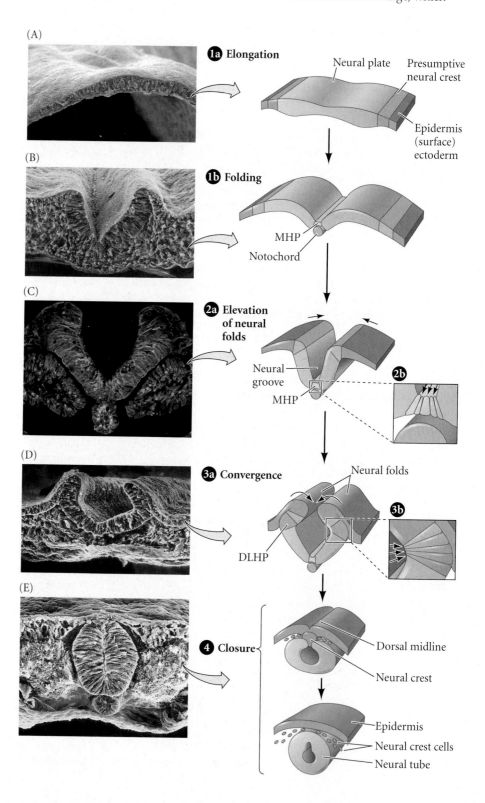

FIGURE 13.6 Primary neurulation: neural tube formation in the chick embryo. (A, 1a) Cells of the neural plate can be distinguished as elongated cells in the dorsal region of the ectoderm. (B, 1b) Folding begins as the medial hinge point (MHP) cells anchor to the notochord and change their shape while the presumptive epidermal cells move toward the dorsal midline. (C, 2a) The neural folds are elevated as the presumptive epidermis continues to move toward the dorsal midline. Asymmetric constriction of actin on the apical side changes cell shapes to promote MHP bending (B, C, 2b). (C) Elevated neural folds stained to show the extracellular matrix (green) and the actin cytoskeleton (red) concentrated in the apical portions of the neural plate cells. (D, 3a) Convergence of the neural folds occurs as the cells at the dorsolateral hinge point (DLHP) become wedge-shaped and the epidermal cells push toward the center. (D, 3b) Similar apical constriction occurs at the DLHP. (E, 4) The neural folds are brought into contact with one another. The neural crest cells disperse, leaving the neural tube separate from the epidermis. (Scanning electron micrographs courtesy of K. Tosney and G. Schoenwolf, drawings after Smith and Schoenwolf 1997; C courtesy of E. Marti Gorostiza and M. Angeles Rabadán.)

enables the creation of a furrow, or **neural groove**, at the dorsal midline (**FIGURE 13.6C**).

3. *Convergence of the neural folds.* Shortly thereafter, two **dorsolateral hinge points (DLHPs)** are induced by and anchored to the surface (epidermal) ectoderm. After the initial furrowing of the neural plate, the plate bends around the hinge regions. Each hinge acts as a pivot that directs the rotation of the cells around it (Smith and Schoenwolf 1991). Continued convergence of the surface ectoderm pushes toward the midline of the embryo, providing another motive force for bending the neural plate, causing the neural folds to converge (**FIGURE 13.6D**; Alvarez and Schoenwolf 1992; Lawson et al. 2001). This movement of the presumptive epidermis and the anchoring of the neural plate to the underlying mesoderm may also be important for ensuring that the neural tube invaginates inward, or into the embryo, and not outward (Schoenwolf 1991a).

4. *Closure of the neural tube.* The neural tube closes as the paired neural folds are brought in contact together at the dorsal midline. The folds adhere to each other, and the neural and surface ectoderm cells from one side fuse with their respective counterparts from the other side. During this fusion event, cells at the apex of the neural folds delaminate and become neural crest cells (**FIGURE 13.6E**).

REGULATION OF HINGE POINTS   To fold the neural plate means to bend a sheet of epithelial cells. How can a row of attached boxes be bent? While in the shape of a rectangular box (i.e., epithelial), they cannot; however, if the surface area of one side of each box in a region were reduced relative to its apposing side (creating the shape of a truncated pyramid), each of these cells should introduce a displacing angle with its neighboring cells and cause the row of boxes to bend. The MHP and two DLHPs are three regions of the neural plate where such cell shape changes occur (see Figure 13.6B–D). The epithelial cells in these locations adopt a "wedge-shaped" morphology along the apicobasal axis, one that is wider basally than apically (Schoenwolf and Franks 1984; Schoenwolf and Smith 1990). Similar to the bottle cells that initiate invagination during gastrulation (see Figure 11.4), localized contraction of actinomyosin complexes at the apical border reduces the size of the apical half of the cell relative to the basal compartment, a process known as **apical constriction**. This apical constriction pairs with the basal retention of nuclei to yield the wedge-shaped hinge point cells (see Figure 13.6C,D; Smith and Schoenwolf 1987, 1988). In addition, recent findings suggest that the division rates in the dorsolateral domains of the neural plate are significantly faster than in ventral regions; this increases the cell density in the neural folds and adds a force that is hypothesized to promote buckling at the DLHP (McShane et al. 2015). The physical forces exerted by different regions of the neural plate have yet to be quantified, but at the cellular level hinge points are formed by (1) apical constriction, (2) basal thickening with retention of the nucleus within the basal portion of cells, and (3) cell packing in the neural folds. What regulates these cellular changes in the correct locations of the neural plate?

It is known that the notochord induces the MHP cells to become wedge-shaped (see Figure 13.6B–D; van Straaten et al. 1988; Smith and Schoenwolf 1989). The morphogen Sonic hedgehog (Shh) is expressed in the notochord and is required for the induction of floor plate cells in the neural plate (Chiang et al. 1996), which in turn form the MHP. The persistence of the MHP in Shh knockout mice suggests that other notochord-derived signals may be required for its morphogenesis (Ybot-Gonzalez et al. 2002).

In the DLHP, *Noggin* appears to be critical for proper hinge formation. In mice, loss of *Noggin* results in a severe failure of neural tube closure (**FIGURE 13.7**; Stottman et al. 2006). *Noggin* is expressed in the neural folds, and this expression is sufficient to induce the DLHP to form; *Noggin* is also expressed transiently in the notochord (Ybot-Gonzalez et al. 2002, 2007). It is important to note that Noggin binds to and inhibits bone morphogenic proteins (BMPs). Could it be the inhibition of BMPs that results in cell shape changes in the DLHP? Several experiments indicate it is more complicated than that.

FIGURE 13.7 Activated BMP signaling leads to neural tube defects. (A) In the wild-type mouse, *Noggin*—a direct antagonist of BMP ligands—is expressed in the notochord and neural folds. The darkly stained region marks the dorsal neural folds and tube. (B) Loss of *Noggin* results in failure of neural tube closure (arrows), perhaps due in part to a lack of BMP inhibition. (From Stottman et al. 2006.)

(A) Wild-type

(B) *Noggin*$^{-/-}$

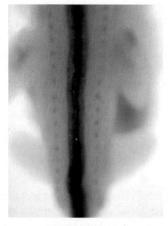
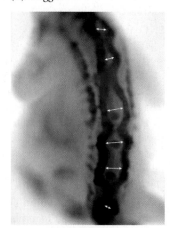

Noggin expressed;
neural tube closure

BMPs hyperactive,
neural tube fails to close

(A)

Constitutively active
BMP receptor 1

Control (normal
BMP signaling)

Dominant negative
BMP receptor 1

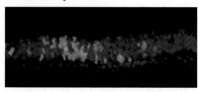
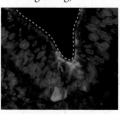
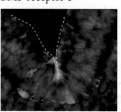

Relative level of BMP signaling (SMAD activation)

(B)

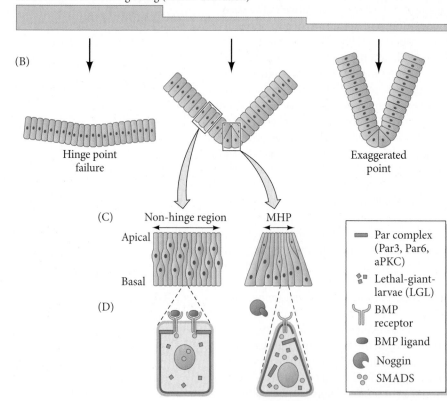

Hinge point
failure

Exaggerated
point

(C) Non-hinge region   MHP

Apical

Basal

(D)

Par complex
(Par3, Par6,
aPKC)

Lethal-giant-
larvae (LGL)

BMP
receptor

BMP ligand

Noggin

SMADS

FIGURE 13.8 BMP prevents MHP formation by regulating apical-basal polarity. This experiment demonstrates that high levels of BMP signaling inhibit MHP formation while low levels promote excessive folding at the midline. (A) Electroporation of a constitutively active BMP receptor (left) or a dominant negative form of BMP receptor (right) into the chick neural plate prior to folding. The normal (control) condition is seen in the center. Electroporated cells are visible through GFP expression (green); nuclei are visible (blue), as are apical cell membranes marked by Par3 in the middle and right sections (red). (B) Schematic representation of the results in A. (C) Illustrations of the cell shape and nuclear positions in the non-hinge region of the neural plate (left) and at the MHP (right). Preferential positioning of nuclei (blue) to the basal cell compartment paired with apical actin-myosin contraction promotes apical constriction. (D) Illustrations of a single cell in non-hinge (C, left) and hinge (D, right) regions of this epithelium, demonstrating the effect of BMP signaling on apical to basal polarity. BMP signaling on the apical surface leads to an apically stabilized Par complex that segregates basal defining components such as LGL, all of which promotes an equal epithelial morphology. Attenuation of BMP signaling by Noggin can disrupt the division of these the compartments, leading to an expansion of typical basal components and a loosening of junctional complexes, all of which enables apical constriction. (After Eom et al. 2011, 2012, 2013.)

When chick embryos are genetically manipulated to contain constitutively active BMP receptors in the neural tube, these receptors bind the BMPs secreted by the surface ectoderm, which is still intimately connected to the neural plate and folds, with the result that all hinge points are repressed from forming. In contrast, when manipulations cause the total loss of BMP signaling in the neural plate, the result is ectopic and exaggerated MHP and DLHPs (**FIGURE 13.8A**; Eom et al. 2011, 2012, 2013). It therefore seems that intermediate amounts of of BMP signaling in the neural plate are required for normal hinge point size and location (**FIGURE 13.8B**). The neural plate cells at this level of BMP signaling will undergo apical constriction and basal thickening to form the hinge points. This process occurs through a modification of the junctional complexes holding the cells together (**FIGURE 13.8C,D**). Specifically, when BMP signals are present in higher amounts, they promote the recruitment of proteins that serve to stabilize junctional proteins and maintain size equality between the apical and basal membranes, which prevents folding. In contrast, attenuation of BMP signaling (by Noggin) leads to a relaxation in these cell-to-cell junctions, which then permits apically restricted actinomyosin contractions and a shortening of the apical membrane. In summary, hinge point formation appears to center around the precise control of BMP signaling. BMP inhibits MHP and DLHP formation, whereas repression of BMP by Noggin enables DLHPs to form, and Shh from the notochord and floor plate prevent precocious and ectopic hinges from forming in the neural plate (**FIGURE 13.9**).

EVENTS OF NEURAL TUBE CLOSURE   Closure of the neural tube does not occur simultaneously throughout the neural ectoderm. This phenomenon is best seen in amniote vertebrates (reptiles, birds, and mammals), whose body axis is elongated prior to neurulation. In amniotes, induction occurs in an anterior-to-posterior fashion. So, in the 24-hour chick embryo, neurulation in the **cephalic** (head) region is well advanced, but the **caudal** (tail) region of the embryo is still undergoing gastrulation (see Figure 13.5). The two open ends of the neural tube are called the **anterior neuropore** and the **posterior neuropore**.

In chicks, neural tube closure is initiated at the level of the future midbrain and "zips up" in both directions. By contrast, in mammals, neural tube closure is initiated at several places along the anterior-posterior axis (**FIGURE 13.10**). In humans, there are probably five sites of neural tube closure (see Figure 13.5B; Nakatsu et al. 2000; O'Rahilly and Muller 2002; Bassuk and Kibar 2009), and the closure mechanism may differ at each site (Rifat et al. 2010). The rostral closure site (closure site 1) is located at the junction of the spinal cord and hindbrain and appears to close, as does the chick neural tube, by "zipping up" the neural folds. Similarly, at closure site 2, located at

Developing Questions

What induces medial hinge point formation? Two findings—that (1) an extra notochord can induce ectopic hinge point formation and (2) Sonic hedgehog represses the DLHP—suggest that some other factor(s) may be responsible. Could it be the early expression of Noggin in the notochord (and thus still be all about repressing BMPs)? Here is an additional fact to bear in mind: in the anteriormost neural plate only MHP forms, whereas in the posteriormost neural plate only DLHPs form. Only in the central axial domains are both types of hinge points present. Why are these hinge points located in different positions along the anterior-to-posterior axis, and how is this difference regulated?

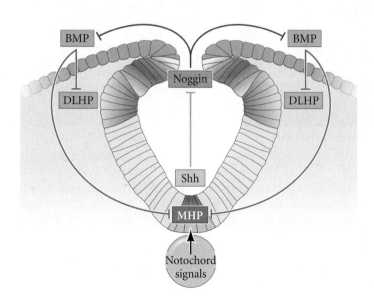

FIGURE 13.9 Morphogen regulation of hinge point formation. BMPs are expressed by the surface ectoderm (green), *Noggin* is expressed in the dorsal neural folds (blue), and *Shh* is expressed ventrally in the notochord and floor plate. The regulation of hinge points revolves around BMP as an antagonist to both DLHP and MHP formation. Shh is required for the specification of floor plate, while additional signals from the notochord induce MHP morphology. Noggin directly inhibits BMP ligands, thus alleviating BMP repression of the hinge points. The DLHPs, however, only form at the correct size and dorsal-ventral position, which is based on Noggin's distance from inhibitory Shh gradients ascending from the floor plate. Therefore, apical constriction occurs only in those cells experiencing low enough concentrations of both BMP (MHP and DLHP) and Shh (DLHP) morphogens.

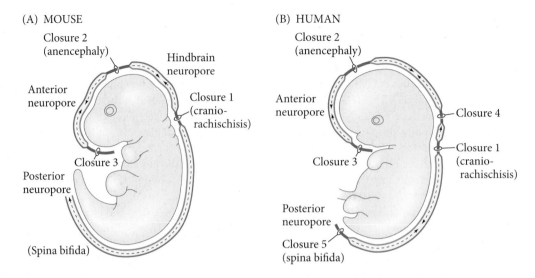

(A) MOUSE

Closure 2
(anencephaly)

Hindbrain
neuropore

Anterior
neuropore

Closure 1
(cranio-
rachischisis)

Closure 3

Posterior
neuropore

(Spina bifida)

(B) HUMAN

Closure 2
(anencephaly)

Anterior
neuropore

Closure 4

Closure 3

Closure 1
(cranio-
rachischisis)

Posterior
neuropore

Closure 5
(spina bifida)

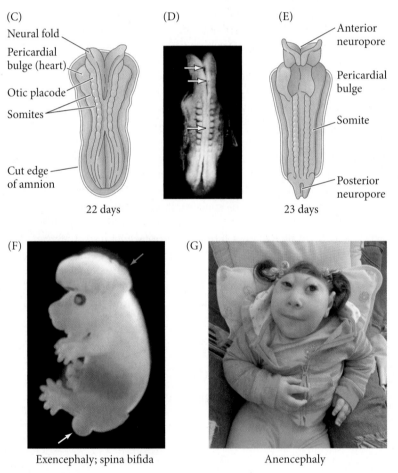

(C)

Neural fold

Pericardial
bulge (heart)

Otic placode

Somites

Cut edge
of amnion

22 days

(D)

(E)

Anterior
neuropore

Pericardial
bulge

Somite

Posterior
neuropore

23 days

(F)

Exencephaly; spina bifida

(G)

Anencephaly

FIGURE 13.10 Neural tube closure in the mammalian embryo. (A,B) Initiation sites for neural tube closure of mouse (A) and human (B) embryos. In addition to the three initiation sites found in mice, neural tube closure in humans also initiates at the posterior end of the hindbrain and in the lumbar region. (C) Dorsal view of a 22-day (8-somite) human embryo initiating neurulation. Both anterior and posterior neuropores are open to the amniotic fluid. (D) A 10-somite human embryo showing some of the major sites of neural tube closure (arrows). (E) Dorsal view of a 23-day neurulating human embryo with only its neuropores open. (F) Midbrain exencephaly and open spina bifida are seen in the mouse *curly tail* mutation, a hypomorphic mutation in the *grainhead-like3* gene. (G) Vitoria de Cristo, who lived with anencephaly for two and a half years. Anencephaly results when a failure to close the neural tube at sites 2 and 3 allows the forebrain to remain in contact with amniotic fluid and subsequently degenerate. (A,B after Bassuk and Kibar 2009; C from Nakatsu et al. 2000; F from Copp et al. 2003; G courtesy of Joana Schmitz Croxato; see also http://belovedvitoria.blogspot.com.)

the midbrain/forebrain boundary, a directional zipper-like mechanism paired with dynamic cell extension appears to be at work. At closure site 3 (the rostral forebrain), the dorsolateral hinge points appear to be fully responsible for the neural tube closure.

How do the apices of the neural folds zip up? Are there interlocking cell membranes and some mysterious force that sequentially puts them together one at a time along the anterior-to-posterior axis? One way to better understand a process as complex as neural tube closure is to simply watch it. Rather remarkable in toto live-cell imaging has been

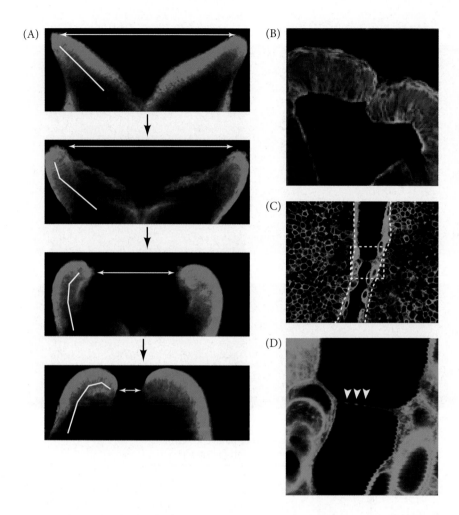

FIGURE 13.11 Neural tube closure at mouse site 2 (midbrain region). (A) Live imaging of a 15-somite stage embryo using a transgenic CAG:Venus^myr mouse to visualize all cell membranes. Optical dorsoventral (cross) sections seen from the top image to the bottom image show DLHP formation (curving of white line on left fold) to the point of near closure (decreasing size of double arrow). (B) Optical section through a mouse embryo as the neural folds are touching but not yet closed. The single layer of non-neural surface ectoderm (large, flattened cells; stained green) has wrapped itself around the neural ectoderm (stained blue) at the edge of the closing neural folds. (C) Dotted lines show the border between neural and non-neural ectoderm. Cellular bridges from the non-neural ectoderm connect the two juxtaposed neural folds. (D) A close-up of one of these bridges is seen at the right (arrowheads). (A from Massarwa and Niswander 2013; B–D from Pyrgaki et al. 2011; photographs courtesy of H. Ray and L. Niswander.)

conducted on mouse embryos in culture (Pyrgaki et al. 2010; Massarwa and Niswander 2013). During DLHP bending, dynamic cell processes extend from the juxtaposed tips of the neural folds (**FIGURE 13.11**; also see Watch Development 13.1). This cellular behavior is being displayed by non-neural surface ectoderm cells, which ultimately extend long filopodial processes toward the apposing fold. These filopodial extensions establish temporary "cellular bridges" whose functions are currently unknown.

WATCH DEVELOPMENT 13.1 Clarify your understanding of neural tube formation by watching three different stages of the live mouse embryo complete closure of the neural tube.

 SCIENTISTS SPEAK 13.1 Listen to a web conference in which Dr. Lee Niswander discusses neural tube closure.

Visualizing neural tube zipping in the mouse revealed potentially important cell behaviors at the time of fusion, but what are the forces driving this attachment of the apposing neural folds? To better quantify the mechanisms of neural tube zipping, we will examine a simpler vertebrate system, that of *Ciona intestinalis*. Tunicates (also called ascidians) such as *Ciona* form neural tubes through primary neurulation, which includes a very similar zipper-like fusion event that proceeds in a posterior-to-anterior direction (**FIGURE 13.12A**; Nicol and Meinertzhagen 1988a,b; Hashimoto et al. 2015). Live-cell imaging of the membrane junction points between epidermal and neural cells during neural tube closure revealed a mechanism of sequential exchange of junctions at apical membrane junctions (**FIGURE 13.12B**). The driving force for zipper advancement

FIGURE 13.12 Neural tube zipper advance in *Ciona*. (A) *Ciona* embryos stained with phalloidin to label cell membranes at the early neurula and early tailbud stages. Corresponding transverse sections showing the neural (blue) and non-neural (gray) ectoderm are shown to the sides of these embryos (teal arrows). Closure of the neural tube progresses in a posterior to anterior direction, which is represented by the yellow progression points and yellow arrow. (B) Time-lapse imaging of an embryo expressing GFP in the membranes of cells of the left hemisphere. The schematic on the left illustrates the region being imaged. Cell membranes are outlined in color to indicate important cell junctions. Epidermal-to-epidermal cell junctions are white; epithelial-to-neural cell junctions are in color. Arrow positions denote the location of the advancing rostral closure point. The critical observation is the correlation between zipper point advancement and the exchange of an epithelial-to-neural junction (solid colored lines) with a newly formed epithelial-to-epithelial junction (dashed colored lines). (C) Model for zipper advancement. Myosin contraction (red) pulls the zipper point anteriorly to the next cell junction (green), an event that occurs when the posterior attachments are finally released. (After Hashimoto et al. 2015.)

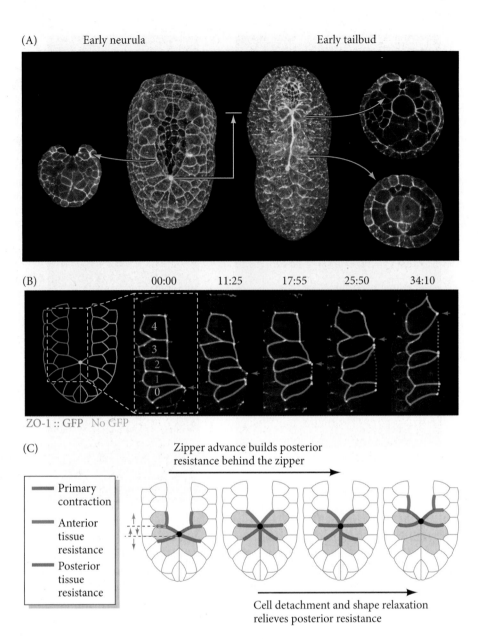

in *Ciona* may be the localized activation of actomyosin contraction (i.e., myosin moving on filamentous actin) in the apical membranes of epidermal cells lying immediately ahead of the zipper point (**FIGURE 13.12C**). Junctional tension is highest between the apical membrane of these epidermal cells and their adjacent neural ectodermal neighbors. Moreover, inhibition of myosin prevents zipper advancement. These data suggest that a stepwise exchange of cell junctions is initiated by the apical activation of actomyosin contraction, which is then followed by a release in the attachments of posterior junctions to the zipper point and consequently a reduction in posterior resistance. As a result of these junctional exchanges, epidermal-to-neural attachments are replaced with epidermal-to-epidermal adhesion, and neural tube closure advances (Hashimoto et al. 2015).

FUSION AND SEPARATION The neural tube eventually forms a closed cylinder that separates from the surface ectoderm. This separation appears to be mediated by the expression of different cell adhesion molecules. Although the cells that will become the neural tube originally express E-cadherin, they stop producing this protein as the

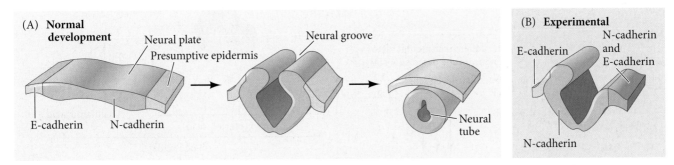

FIGURE 13.13 Expression of N- and E-cadherin adhesion proteins during neurulation in *Xenopus*. (A) Normal development. In the neural plate stage, N-cadherin is seen in the neural plate, whereas E-cadherin is seen on the presumptive epidermis. Eventually, the N-cadherin-bearing neural cells separate from the E-cadherin-containing epidermal cells. (Neural crest cells express neither N- nor E-cadherin, and they disperse.) (B) No separation of the neural tube occurs when one side of the frog embryo is injected with N-cadherin mRNA so that N-cadherin is expressed in the epidermal cells as well as in the presumptive neural tube.

neural tube forms and instead synthesize N-cadherin (**FIGURE 13.13A**). As a result, the surface ectoderm and neural tube tissues no longer adhere to each other. If the surface ectoderm is experimentally made to express N-cadherin (by injecting N-cadherin mRNA into one cell of a two-cell *Xenopus* embryo), the separation of the neural tube from the presumptive epidermis is dramatically impeded (**FIGURE 13.13B**; Detrick et al. 1990; Fujimori et al. 1990). Loss of the gene for N-cadherin in zebrafish also results in failure to form a neural tube (Lele et al. 2002). The Grainyhead transcription factors are especially important in this process (Rifat et al. 2010; Werth et al. 2010; Pyrgaki et al. 2011). Grainyhead-like2, for instance, controls a battery of cell adhesion molecules and downregulates E-cadherin synthesis in the neural folds. Mice with mutations in *Grainyhead-like2* or *Grainyhead-like3* genes have severe neural tube defects that include a split face, exencephaly, and spina bifida (see Figure 13.10F and Scientists Speak 13.1; Copp et al. 2003; Pyrgaki et al. 2011).

NEURAL TUBE CLOSURE DEFECTS   In humans, neural tube closure defects occur in about 1 in every 1000 live births. Failure to close the posterior neuropore (closure site 5) around day 27 of development results in a condition called **spina bifida**, the severity of which depends on how much of the spinal cord remains exposed. Failure to close site 2 or site 3 in the rostral neural tube keeps the anterior neuropore open, resulting in a usually lethal condition called **anencephaly**, in which the forebrain remains in contact with the amniotic fluid and subsequently degenerates. The fetal forebrain ceases development and the vault of the skull fails to form (see Figure 13.11G). The failure of the entire neural tube to close over the body axis is called **craniorachischisis**.

Failure to close the neural tube can result from both genetic and environmental causes (Fournier-Thibault et al. 2009; Harris and Juriloff 2010; Wilde et al. 2014). Mutations (first found in mice) in genes such as *Pax3*, *Sonic hedgehog*, *Grainyhead*, *Tfap2*, and *Openbrain* show that these genes are essential for the formation of the mammalian neural tube; in fact, more than 300 genes appear to be involved. Environmental factors including drugs, maternal dietary factors (such as cholesterol and folate, also known as folic acid or vitamin B_9), diabetes, obesity, and toxins can all influence human neural tube closure. How these factors lead to neural tube disorders is largely unknown. An emerging idea posits that a major outcome of environmental perturbations is the modification of the embryo's epigenome, which in turn causes transcription variability leading to neural tube defects (**FIGURE 13.14A**; Feil et al. 2012; Shyamasundar et al. 2013; Wilde et al. 2014). This idea is most associated with the potential downstream consequences of folic acid metabolism.

Developing Questions

What initiates the directionality of neural tube closure? Zipping proceeds in a posterior to anterior direction in *Ciona*, as well as in certain closure points in mammals, yet it proceeds in opposite directions to close other regions of the mammalian brain. Moreover, are the cell forces that seem to advance the zipper in the primitive chordate *Ciona* conserved throughout vertebrates?

VADE MECUM

Whole mounts and a complete set of serial cross sections of chick neurulation are included in the "Chick-Mid" segment and can be viewed in several formats, including as movies.

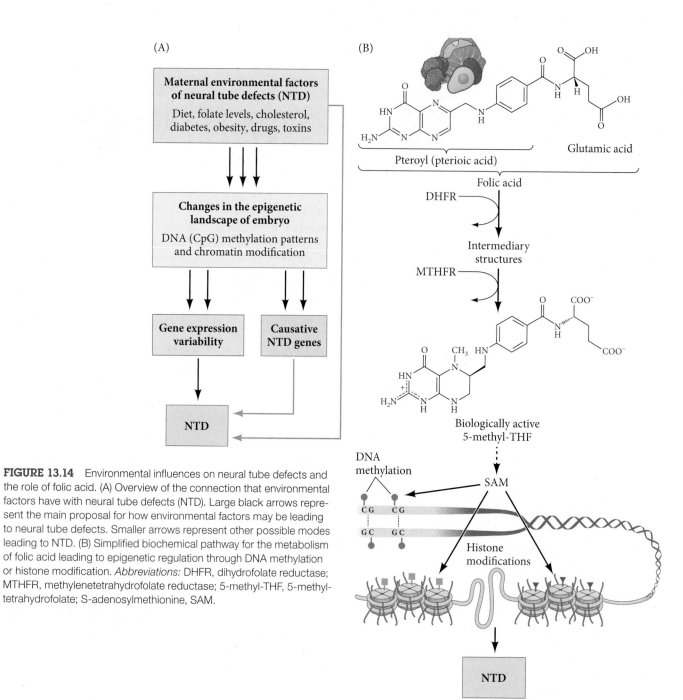

FIGURE 13.14 Environmental influences on neural tube defects and the role of folic acid. (A) Overview of the connection that environmental factors have with neural tube defects (NTD). Large black arrows represent the main proposal for how environmental factors may be leading to neural tube defects. Smaller arrows represent other possible modes leading to NTD. (B) Simplified biochemical pathway for the metabolism of folic acid leading to epigenetic regulation through DNA methylation or histone modification. *Abbreviations:* DHFR, dihydrofolate reductase; MTHFR, methylenetetrahydrofolate reductase; 5-methyl-THF, 5-methyl-tetrahydrofolate; S-adenosylmethionine, SAM.

Although the exact role of folate remains unknown, the early use of folic acid antagonists led to fetuses with neural tube defects. Since then, many large-scale human trials have demonstrated clear correlations of neural tube disorders with folic acid deficiency, which is the reason folic acid is not only recommended for pregnant women but also systematically fortified in foods (reviewed in Wilde et al. 2014). How folic acid deficiency leads to neural tube disorders is currently an active area of research. Folic acid is an important nutrient used for regulating DNA synthesis during cell division in the brain (Anderson et al. 2012) and is also critical in regulating DNA methylation (**FIGURE 13.14B**). Further evidence that epigenetic mechanisms are essential for proper neural tube development are the findings that functional manipulation of histone modifying enzymes (acetyltransferases, deacetylases, demethyltransferases) cause neural tube defects (Artama et al. 2005; Bu et al. 2007; Shpargel et al. 2012; Welstead et al. 2012;

Murko et al. 2013). Whatever the mechanisms, it has been estimated that 25–30% of human neural tube birth defects can be prevented if pregnant women take supplemental folate. Therefore, the U.S. Public Health Service recommends that women of childbearing age take 0.4 milligram of folate daily (Milunsky et al. 1989; Centers for Disease Control 1992; Czeizel and Dudas 1992).

Secondary neurulation

Secondary neurulation, which takes place in the most posterior region of the embryo during tailbud elongation, produces a neural tube through a very different process than primary neurulation (see Figure 13.4). Secondary neurulation involves the production of mesenchyme cells from the prospective ectoderm and mesoderm, followed by the condensation of these cells into a **medullary cord** beneath the surface ectoderm (**FIGURE 13.15A,B**). After this mesenchymal-epithelial transition, the central portion of this cord undergoes cavitation to form several hollow spaces, or **lumens** (**FIGURE 13.15C**); the lumens then coalesce into a single central cavity (**FIGURE 13.15D**; Schoenwolf and Delongo 1980).

We have seen that after Hensen's node has migrated to the posterior end of the embryo, the caudal region of the epiblast contains a precursor cell population that gives rise to both neural ectoderm and paraxial (somite) mesoderm as the embryo's trunk elongates (Tzouanacou et al. 2009). The ectodermal cells that will form the posterior (secondary) neural tube express the *Sox2* gene, whereas the ingressing mesodermal cells (which no longer encounter high levels of BMPs as they migrate beneath the epiblast) do not express *Sox2*. Rather, the ingressing mesodermal cells express *Tbx6* and

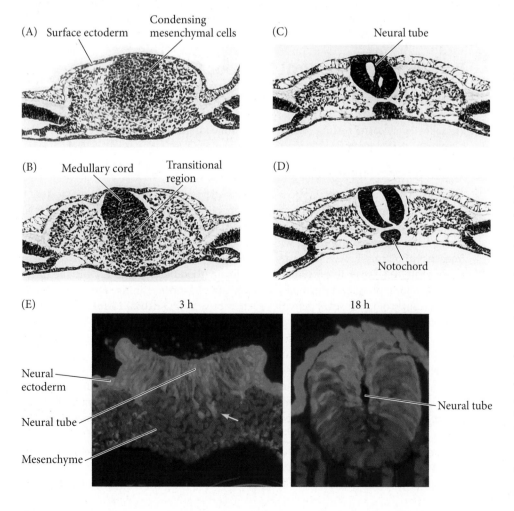

FIGURE 13.15 Secondary neurulation in the caudal region of a chick embryo. (A–D) A 25-somite chick embryo. (A) Mesenchymal cells condense to form the medullary cord at the most caudal end of the chick tailbud. (B) The medullary cord at a slightly more anterior position in the tailbud. (C) The neural tube is cavitating and the notochord forming; note the presence of separate lumens. (D) The lumens coalesce to form the central canal of the neural tube. (E) Tracing of cells from the superficial blastoderm in the junctional neural tube region. Superficial cells of the primitive streak are labeled with a cell permanent fluorescent green dye at the position where junctional neurulation was to take place. Nuclei are stained blue with DAPI. At 3 hours, the superficial cells can be seen ingressing from the overlying neural ectoderm into the mesenchyme (arrow), which yields a neural tube made of cells from both locations. (A–D from Catala et al. 1995, photographs courtesy of N. M. Le Douarin; E from Dady et al. 2014.)

form somites (see Chapter 17; Shimokita and Takahashi 2010; Takemoto et al. 2011). The ability of the Tbx6 transcription factor to repress neural-inducing *Sox2* expression explains the bizarre phenotype of homozygous *Tbx6* mouse mutants, which have three neural tubes, posteriorly (see Figure 17.4; Chapman and Papaioannou 1998; Takemoto et al. 2011). In these mutants, the two rods of paraxial mesoderm have become neural tubes that even express regionally appropriate genes (such as *Pax6*). Thus, the epiblast surrounding the rostral primitive streak (the caudal lateral epiblast; see Chapter 12) contains a common precursor pool for paraxial mesoderm and for the neural plate that forms the caudal hindbrain and spinal cord (Cambray and Wilson 2007; Wilson et al. 2009). This distinction also emphasizes another fundamental difference between primary and secondary neurulation. During primary neurulation, the surface ectoderm and neural ectoderm are intimately connected through the process of neural tube closure and fusion, whereas during secondary neurulation, these two tissues are essentially uncoupled and develop independently of one another.[2]

In human and chick embryos, there appears to be a transitional region at the junction of the anterior (primary) and posterior (secondary) neural tubes. As mentioned earlier, neural tube formation in this transition zone is referred to as junctional neurulation (see Figure 13.4). In human embryos, coalescing cavities are seen in the transitional region, but the neural tube also forms by the bending of neural plate cells. The junctional neural tube in the chick is a mosaic of both ventral mesenchyme cells and dorsal neural ectodermal cells. In addition to directly providing dorsal epithelial cells to the junctional neural tube, neural plate cells also undergo an epithelial-to-mesenchymal transition and ingress into the underlying mesenchyme pool (**FIGURE 13.15E**; Dady et al. 2014). Some posterior neural tube anomalies result when the two regions of the neural tube fail to coalesce (Saitsu et al. 2007). Given the prevalence of human posterior spinal cord malformations, further understanding of the mechanisms of secondary neurulation may have important clinical implications.

Patterning the Central Nervous System

The early development of most vertebrate brains is similar (**FIGURE 13.16A–D**). Because the human brain may be the most organized piece of matter in the solar system and is arguably the most interesting organ in the animal kingdom, we will concentrate on the development that is supposed to make *Homo* sapient.[3]

The anterior-posterior axis

The early mammalian neural tube is a straight structure, but even before the posterior portion of the tube has formed, the most anterior portion of the tube is undergoing drastic changes. In the anterior region, the neural tube balloons into the three primary vesicles: the forebrain (**prosencephalon**), which forms the cerebral hemispheres; the midbrain (**mesencephalon**), whose neurons are involved in motivation, movement, and depression (Niwa et al. 2013; Tye et al. 2013); and the hindbrain (**rhombencephalon**), which becomes the cerebellum, pons, and the medulla oblongata (the most primitive area of the brain and the center of involuntary activities such as breathing; **FIGURE 13.16E**). By the time the posterior end of the neural tube closes, secondary vesicles have formed. The forebrain becomes the telencephalon (which forms the cerebral hemispheres) and the diencephalon (which will form the optic vesicle that initiates eye development).

The rhombencephalon develops a segmental pattern that specifies the places where certain nerves originate. Periodic swellings called **rhombomeres** divide the rhombencephalon into smaller compartments. The rhombomeres represent separate "territories," in that the cells within each rhombomere mix freely within it but do not mix with cells

---

[2] Failure to form the nerve (medullary) cord in zebrafish does not prevent surface ectoderm from spanning the dorsal midline (Harrington et al. 2009).

[3] Our species name comes from the Latin *sapio*, meaning "to be capable of discerning."

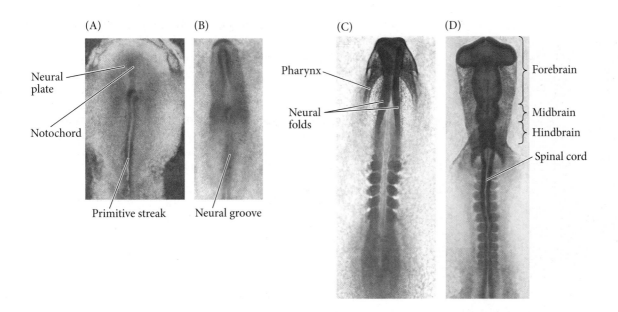

Adult derivatives

Olfactory lobes	–	Smell
Hippocampus	–	Memory storage
Cerebrum	–	Association ("intelligence")
Optic vesicle	–	Vision (retina)
Epithalamus	–	Pineal gland
Thalamus	–	Relay center for optic and auditory neurons
Hypothalamus	–	Temperature, sleep, and breathing regulation
Midbrain	–	Temperature regulation, motor control, motivation, and emotional control
Cerebellum	–	Coordination of complex muscular movements
Pons	–	Fiber tracts between cerebrum and cerebellum
Medulla	–	Reflex center of involuntary activities

(E)
Primary vesicles

Wall
Cavity

Forebrain (Prosencephalon) → Telencephalon
→ Diencephalon

Midbrain (Mesencephalon) → Mesencephalon

Hindbrain (Rhombencephalon) → Metencephalon
→ Myelencephalon

Secondary vesicles

Spinal cord

from adjacent rhombomeres (Guthrie and Lumsden 1991; Lumsden 2004). Each rhombomere expresses a unique combination of transcription factors, thereby generating rhombomere-specific patterns of neuronal differentiation. Thus, each rhombomere produces neurons with different fates. As we will see in Chapter 15, the neural crest cells derived from the rhombomeres will form **ganglia**, clusters of neuronal cell bodies whose axons form a nerve. Each rhombomeric ganglion produces a different type of nerve. The generation of the cranial nerves from the rhombomeres has been studied most extensively in the chick, in which the first neurons appear in the even-numbered rhombomeres r2, r4, and r6 (**FIGURE 13.17**; Lumsden and Keynes 1989). Neurons originating from r2 ganglia form the fifth (trigeminal) cranial nerve, those from r4 form the seventh (facial) and eighth (vestibuloacoustic) cranial nerves, and those from r6 form the ninth (glossopharyngeal) cranial nerve.

FIGURE 13.16 Early brain development and formation of the first brain chambers. (A–D) Chick brain development. (A) Flat neural plate with underlying notochord (head process). (B) Neural groove. (C) Neural folds begin closing at the dorsalmost region, forming the incipient neural tube. (D) Neural tube, showing the three brain regions and the spinal cord. The neural tube remains open at the anterior end, and the optic bulges (which become the retinas) have extended to the lateral margins of the head. (E) In humans, the three primary brain vesicles become further subdivided as development continues. At the right is a list of the adult derivatives formed by the walls and cavities of the brain along with some of their functions. (A–D courtesy of G. C. Schoenwolf; E after Moore and Persaud 1993.)

(A) (B)

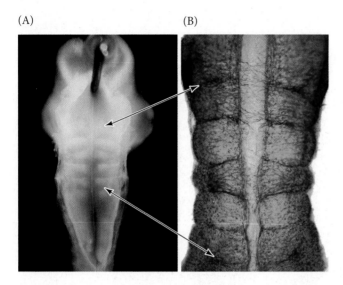

FIGURE 13.17 Rhombomeres of the chick hindbrain. (A) Hindbrain of a 3-day chick embryo. The roof plate has been removed so that the segmented morphology of the neural epithelium can be seen. The r1/r2 boundary is at the upper arrow, and the r6/r7 boundary is at the lower arrow. (B) A chick hindbrain at a similar stage stained with antibody to a neurofilament subunit. The rhombomere boundaries are emphasized because they serve as channels for neurons crossing from one side of the brain to the other. (From Lumsden 2004, courtesy of A. Lumsden.)

The anterior-to-posterior patterning of the hindbrain and spinal cord is controlled by a series of genes that include the Hox gene complexes. For more details on the mechanisms that pattern cell fates along the anterior-to-posterior axis, see Web Topics 13.1 and 13.2.

WEB TOPIC 13.1 DIVIDING THE CENTRAL NERVOUS SYSTEM
The physical division of the prospective brain from the prospective spinal cord is achieved by occluding the lumen of the neural tube at the boundary between these regions.

WEB TOPIC 13.2 SPECIFYING THE BRAIN BOUNDARIES
Pax transcription factors and the paracrine factor Fgf8 are critical in establishing the boundaries of the forebrain, midbrain, and hindbrain.

The dorsal-ventral axis

The neural tube is polarized along its dorsal-ventral axis. In the spinal cord, for instance, the dorsal region is the place where the spinal neurons receive input from sensory neurons, whereas the ventral region is where the motor neurons reside. In the middle are numerous interneurons that relay information between the sensory and motor neurons (**FIGURE 13.18**). These differentiated cell types organized along the dorsoventral axis arose from progenitor cell populations located adjacent to the cavities (ventricles) of the brain that run along the anterior-to-posterior axis (i.e., in the ventricular zone). Each progenitor domain can be defined by its expression of specific transcription factors (such as the products of Hox genes), which specify progeny to differentiate into the specific classes of neuronal and glial cells that make up the CNS (Catela et al. 2015). This presents a logical question:

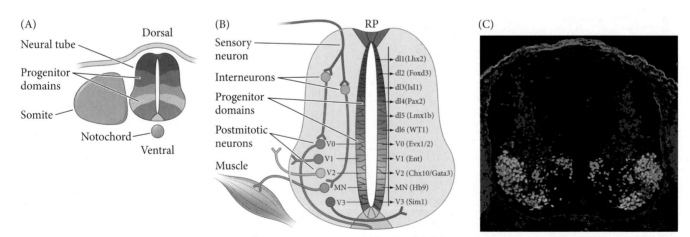

FIGURE 13.18 Differential expression of transcription factors define progenitor domains and derived cell types along the dorsoventral axis. (A) The early neural tube is made up of neuroepithelial progenitor cells that can be divided into discrete domains based on their unique repertoire of transcription factor expression. Pax3 and Pax7 define the most dorsal domain (dark blue), Nkx6.1 is expressed ventrally (red), and Pax6 is located in the central region of the neural tube (green). Overlap in expression of these different transcription factors creates further subdomains (yellow and light blue). (B) As the neural tube develops, these progenitor zones expand and continue to diversify with their maturing gene regulatory networks until the full differentiative program is adopted and derived cell types emerge (such as the different neuronal cell types illustrated). (C) Immunolabeling for the Isl1 (blue), Foxp1 (red), and Lhx3 (green) transcription factors in an embryonic (day 12.5) mouse spinal cord at the cervical level. (A,B after Catela et al. 2015; C courtesy of Jeremy Dasen.)

How does a cell sense its position within the neural tube such that it develops into the progenitor cell population that generates the precise type and properly positioned neurons and glia? Said another way, how is pattern created in the neural tube?

Opposing morphogens

The dorsal-ventral polarity of the neural tube is induced by morphogenetic signals coming from its immediate environment. The ventral pattern is imposed by the notochord, whereas the dorsal pattern is induced by the overlying epidermis (**FIGURE 13.19A–D**). Specification of the axis is initiated by two major paracrine factors: Sonic hedgehog

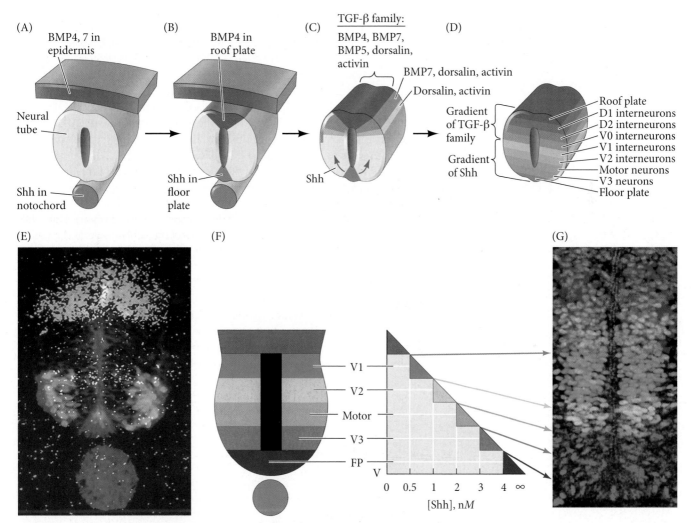

FIGURE 13.19 Dorsal-ventral specification of the neural tube. (A) The newly formed neural tube is influenced by two signaling centers. The roof of the neural tube is exposed to BMP4 and BMP7 from the epidermis, while the floor of the neural tube is exposed to Sonic hedgehog (Shh) from the notochord. (B) Secondary signaling centers are established in the neural tube. BMP4 is expressed and secreted from the roof plate cells; Shh is expressed and secreted from the floor plate cells. (C) BMP4 establishes a nested cascade of TGF-β factors spreading ventrally into the neural tube from the roof plate. Sonic hedgehog diffuses dorsally as a gradient from the floor plate cells. (D) The neurons of the spinal cord are given their identities by their exposure to these gradients of paracrine factors. The amounts and types of paracrine factors present cause different transcription factors to be activated in the nuclei of these cells,

depending on their position in the neural tube. (E) Chick neural tube showing areas of Shh (green) and the expression domain of the protein dorsalin (blue; dorsalin is a member of the TGF-β superfamily). Motor neurons induced by a particular concentration of Shh are stained orange/yellow. (F) Relationship between Sonic hedgehog concentrations, the generation of particular neuronal types in vitro, and distance from the notochord. Cells closest to the notochord become the floor plate neurons; motor neurons and V3 interneurons emerge on the ventrolateral sides. (G) In situ hybridization for three other transcription factors: Pax7 (blue; characteristic of the dorsal neural tube cells), Pax6 (green), and Nkx6.1 (red). Where Nkx6.1 and Pax6 overlap (yellow), motor neurons become specified. (E from Jessell 2000, courtesy of T. M. Jessell; F,G after Briscoe et al. 1999, photograph courtesy of J. Briscoe.)

protein (Shh) originating from the notochord, and TGF-β proteins originating from the dorsal ectoderm (**FIGURE 13.19E**). In both cases, these factors induce a second signaling center within the neural tube itself.

Sonic hedgehog secreted from the notochord induces the medial hinge point cells to become the **floor plate** of the neural tube. The floor plate cells also secrete Sonic hedgehog, which forms a gradient that is highest at the most ventral portion of the neural tube (**FIGURE 13.19B,C,E**). Cells experiencing the highest concentrations of Shh develop into the progenitor cells for motor neurons and a class of interneurons called V3 neurons, whereas moderate and lower levels of Shh induce increasingly more dorsal progenitor populations (**FIGURE 13.19D,F,G**; Roelink et al. 1995; Briscoe et al. 1999).

The importance of Sonic hedgehog in patterning the ventral portion of the neural tube has been confirmed by experiments that again demonstrate the principles of "find it, lose it, move it" (see Web Topic 7.4 and the associated Dev Tutorial). If a piece of notochord is removed from an embryo, the neural tube adjacent to the deleted region will have no floor plate cells (Placzek et al. 1990). Moreover, if notochord fragments are taken from one embryo and transplanted to the lateral side of a host neural tube, the result will be to induce another set of floor plate cells in the adjacent host neural tube and bilaterally positioned sets of ectopic motor neurons around the induced floor plate (**FIGURE 13.20**). The same results can be obtained if the notochord fragments are replaced by pellets of cultured cells secreting Sonic hedgehog, demonstrating that Shh alone is sufficient for the induction of floor plate and affiliated motor neurons (Echelard et al. 1993).

The dorsal fates of the neural tube are established by proteins of the TGF-β superfamily, especially BMPs 4 and 7, dorsalin, and activin (Liem et al. 1995, 1997, 2000). Initially, BMP4 and BMP7 are found in the epidermis. Just as the notochord establishes a secondary signaling center—the floor plate cells—on the ventral side of the neural tube, the epidermis establishes a secondary signaling center by inducing BMP4 expression in the **roof plate** cells of the neural tube. The BMP4 protein from the roof plate induces a cascade of TGF-β proteins in adjacent cells (see Figure 13.19). Dorsal sets of cells are thus exposed to higher concentrations of TGF-β proteins, and at earlier times, when compared with the more ventral neural cells. The importance of the TGF-β superfamily factors in patterning the dorsal portion of the neural tube was demonstrated by the phenotypes of zebrafish mutants. Those mutants deficient in certain BMPs lacked dorsal and intermediate types of neurons (Nguyen et al. 2000).

HOW MUCH AND HOW LONG? How do these morphogens ultimately confer positional information to cells in the neural tube? Recall that progenitor identity is determined

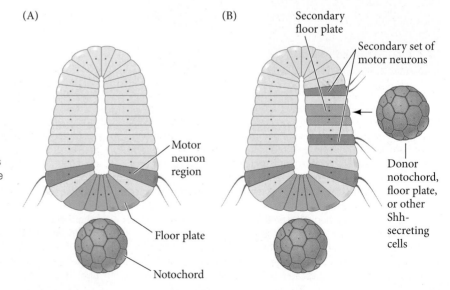

(A)

(B)

Secondary floor plate

Secondary set of motor neurons

Motor neuron region

Floor plate

Notochord

Donor notochord, floor plate, or other Shh-secreting cells

FIGURE 13.20 Notochord-derived Shh induces ventral neural tube structures. (A) Cells closest to the notochord become the floor plate neurons; motor neurons emerge on the ventrolateral sides. (B) If a second notochord, floor plate, or any other Sonic hedgehog-secreting cell is placed adjacent to the neural tube, it induces a second set of floor plate neurons as well as two other sets of motor neurons. (After Placzek et al. 1990.)

by the unique gene regulatory network it expresses. The system of differential gene expression a cell exhibits is dependent on the combination of its distance from, and duration of, exposure to the morphogenetic signaling centers. Cells adjacent to the floor plate that receive high concentrations of Sonic hedgehog synthesize the transcription factors Nkx6.1 and Nkx2.2 and become the ventral (V3) interneurons. The cells dorsal to them, exposed to slightly less Sonic hedgehog (and slightly more TGF-β), produce Pax6 and Olig2 and become motor neurons. The next two groups of cells, receiving progressively less Sonic hedgehog, express Pax6 alone and become the V2 and V1 interneurons. Finally, the cells at the dorsalmost segment of the neural tube express Pax7 and become dorsal progenitors (see Figure 13.18; Lee and Pfaff 2001; Muhr et al. 2001).

It had been thought that the intersecting gradients of Shh and TGF-β signals would be sufficient to instruct the synthesis of the various transcription factors, but the regulatory network is far more complex, and appears to integrate both the spatial and temporal distributions of morphogen signaling. If Pax7-expressing intermediate neural tube explants are exposed to increasing concentrations of Shh, they will stop expressing Pax7 and will express Olig2 and Nkx2.2 in a dose-dependent manner (**FIGURE 13.21A**). If these same explants are exposed to a constant concentration of Shh over an extended period of time, they first express Olig2, followed by increasing levels of Nkx2.2 expression (**FIGURE 13.21B,C**). These results support a model in which the *concentration* of Shh as well as *duration* of Shh signaling together induce differential gene expression and cell fate patterning in the neural tube (Dessaud et al. 2007). This experiment makes certain assumptions about the concentrations and duration of Shh signaling in the embryo, which often proves to be more complex.

In vertebrates, the main downstream effectors of Shh signaling are the Gli family of transcription factors, which function as repressors or activators based, respectively, on the absence or presence of Shh (see Figure 4.30 to review the Hedgehog pathway). Therefore, Shh from the notochord and floor plate is transduced into a ventral-to-dorsal gradient of Gli activators to Gli repressors. It is interesting that the pattern of cells with Gli activator function in the mouse neural tube was shown not only to be a gradient, but also to change over time (see Web Topic 13.3; Balaskas et al. 2012; Cohen et al. 2015). An early expansion of Gli activator function coincided with the initial induction of the broad and overlapping expression of progenitor cell transcription factors, yet Gli activity was not maintained over the course of cell differentiation in the neural tube. Despite this reduction in Shh signaling over time, the domains of progenitor-specific transcription factors still became highly refined, with tight borders between each domain (see part G of Web Topic 13.3). This result suggests that this level and duration of Shh are sufficient for cell specification, which is the case only in the context of a robust gene regulator network that maintains the Shh-induced pattern of gene expression (reviewed in Briscoe and Small 2015).

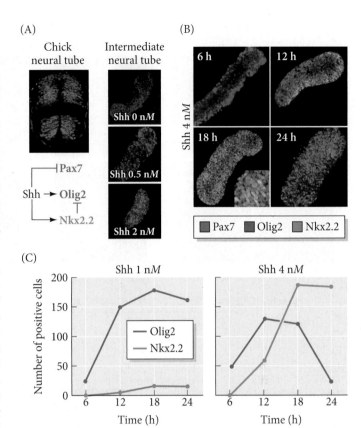

FIGURE 13.21 Neural tube gene expression responds to both concentration and duration of Shh. (A) The expression of three defining transcription factors—Pax7 (dorsal-most, blue), Olig2 (ventromedial, red), and Nkx2.2 (ventral-most, green)—are shown in a transverse section of the chick neural tube. Intermediate neural tube explants express Pax7 in the absence of Shh; when Shh is applied in increasing doses, however, Pax7 is lost, and Olig2 and Nkx2.2 are induced in a dose-dependent manner. This result suggests that Shh represses Pax7 while inducing Olig2 and Nkx2.2. It is known that Nkx2.2 also represses *Olig2* transcription. (B) Initial exposure of intermediate neural tube explants to 4nM Shh results in only Olig2 expression, yet over longer exposure times, Nkx2.2 expression becomes progressively induced. These data are quantified in (C). (From Dessaud et al. 2007.)

WEB TOPIC 13.3 **GLI ACTIVATION** Explore how Sonic hedgehog signaling establishes gradients of the Gli transcription factor along the dorsal-ventral axis.

((• SCIENTISTS SPEAK 13.2 Listen to a web conference with Dr. Andy McMahon on Gli activator targets.

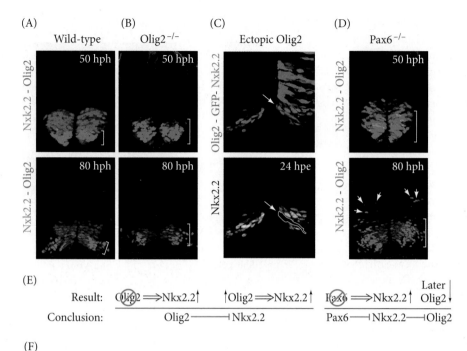

(A) Wild-type

(B) Olig2⁻/⁻

(C) Ectopic Olig2

(D) Pax6⁻/⁻

(E)

Result: ⊘Olig2 ⟹ Nkx2.2↑   ↑Olig2 ⟹ Nkx2.2↑   ⊘Pax6 ⟹ Nkx2.2↑   Later | Olig2↓

Conclusion: Olig2 ——⊣ Nkx2.2   Pax6 ——⊣ Nkx2.2 ——⊣ Olig2

(F)

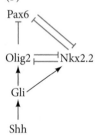

FIGURE 13.22 Transcriptional cross-repression in neural progenitor cells. (A–D) Transverse sections of the mouse neural tube labeled for Nkx2.2, Olig2, and GFP-expressing cells. (B) Loss of *Olig2* leads to expansion of the Nkx2.2 domain (compare the brackets in A and B). (C) Gain of *Olig2* function by ectopically expressing the gene through electroporation represses Nkx2.2 expression (arrow, yellow outline). (D) Loss of the *Pax6* gene leads to a marked expansion of Nkx2.2 at early and later time points (compare the brackets in A with the brackets and arrows in C and D) yet Olig2 is lost only at the later time point (80 hph). Abbreviations: hph, hours post headfold stage; hpe, hours post electroporation. (E) Shorthand explanation of the experimental manipulation and its results (above line) and the conclusion that can be drawn from these results (below line). (F) The gene regulatory network combining the Gli (activated by the Hedgehog pathway), Olig2, Nkx2.2, and Pax6 transcription factors. (From Balaskas et al. 2012.)

Transcriptional cross-repression

Gene regulatory networks of progenitor cells play a direct role in reinforcing, refining, and maintaining progenitor cell fates through the mechanism of transcriptional cross-repression, whereby transcription factors repress each other. Gain- and loss-of-function manipulations of transcription factors in progenitor cells have demonstrated that different transcription factors, such as Olig2 and Nkx2.2, which are expressed in adjacent domains, can mutually repress each other's expression, thereby helping define the borders between adjacent domains (**FIGURE 13.22**; Balaskas et al. 2012). Transcriptional cross-repression integrated into a model that includes Shh provides a mechanism for a cell to "remember" the Shh signal and, consequently, its position in the neural tube (**FIGURE 13.23**; see also Figure 13.22F).

((• SCIENTISTS SPEAK 13.3 Listen to a web conference with Dr. James Briscoe on neural tube patterning and Sonic hedgehog.

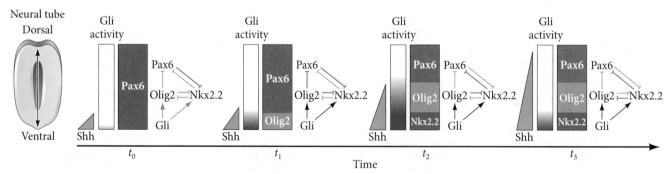

FIGURE 13.23 Model for interpreting the Shh morphogen gradient. Signal-mediated patterning of the ventral portion of the neural tube. At the earliest time of induction (t_0–t_1), Shh from the notochord (green triangles) induces Gli (purple) in the floor plate cells. This action is not sufficient to activate Olig2 or Nkx2.2 or to repress Pax6. As development ensues, Gli is able to induce Olig2, which inhibits Nkx2.2 and Pax6. As the most ventral cells experience higher concentrations of Shh for longer periods, Nkx2.2 is activated and suppresses Olig2. This pattern can be retained even when Gli levels decrease. (After Balaskas et al. 2012.)

All Axes Come Together

The model of dorsal-ventral patterning by TGF-β and Shh morphogens pertains to cell fates throughout the CNS along the rostral-caudal axis. Remember, though, that there are differences in how the anterior regions of the neural tube form and how the posteriormost region forms—differences that are defined by primary and secondary neurulation. Progenitor cells in the *anterior* regions of the neural tube (which become the brain and most of the spinal cord) adopt a proneural fate directly from the epiblast (Harland 2000; Stern 2005). Cells in the *posterior* begin as bipotential **neuromesodermal progenitors** (**NMPs**) that undergo a transition to become neural or somitic cell types, with the neural cells forming the caudal end of the neural tube (**FIGURE 13.24**). NMPs are born in the caudal lateral epiblast during tailbud elongation and are positively maintained by Fgf8 and Wnt signals (see Chapter 17 for details on axis elongation). In opposition to the caudal Fgf/Wnt signals, retinoic acid is expressed by somitic mesoderm and inhibits Fgf8 signaling. It is these antagonistic gradients of retinoic acid and Fgf/Wnt along the rostral-caudal axis that establish a "road" to NMP maturation. An NMP cell is born in the tailbud and enters the mesenchyme (undergoing secondary neurulation). NMP cells that enter the neural mesenchyme become preneural progenitor cells and are initially competent to respond to either Shh or BMP signals by differentiating into either floor plate or roof plate, respectively. As the tailbud elongates, these preneural NMP cells become positioned farther from Fgf/Wnt and closer to retinoic acid; this repositioning triggers a switch in their competency to respond to Shh/TGF-β signals, thus allowing patterning of the proneural progenitors along the dorsoventral axis of the maturing neural tube (Sasai et al. 2014; Gouti et al. 2015).

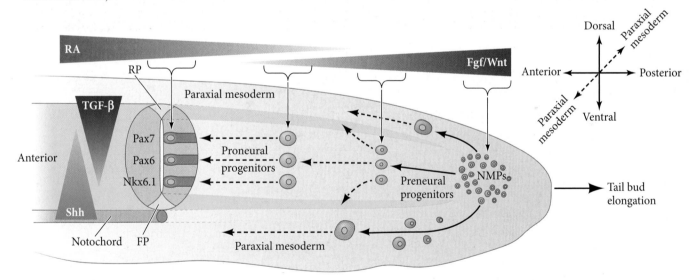

FIGURE 13.24 Model for converging signals for maturation and specification of neural progenitors in the developing caudal region of the spinal cord. During spinal cord development, the tail is undergoing elongation, in part fueled by the caudal lateral epiblast, which houses proliferative and the motile neuromesodermal progenitors (NMPs). NMPs leave the tailbud and either progress into the neural mesenchyme or the mesenchyme of the paraxial mesoderm, where they will give rise to the neural tube or somites respectively. (Dashed arrows indicate that those cells will contribute to those regions of the neural tube, but the cells are not actively migrating into those regions.) Opposing antagonistic morphogens of retinoic acid (RA) and Fgf/Wnt establish inverse gradients along the rostral-caudal axis, and setting up graded positional instructions along this axis. High Fgf/Wnt maintains the NMP pool. Moderate Fgf/Wnt and low RA promote early preneural progenitors to be competent to respond to TGF-β and Shh dorsoventral signals and develop into the roof plate (RP) and floor plate (FP), respectively. As the tailbud continues to elongate, preneural progenitors will experience low Fgf/Wnt and moderate RA, which broadens their competency to initiate gene regulatory programs specific for proneural progenitor populations. In this way, morphogens along all axes pattern the cell fates in the neural tube.

Next Step Investigation

It has been estimated that a significant proportion of human neural tube birth defects can be prevented by pregnant women taking supplemental folate. Neural tube closure occurs early during human gestation, often before a woman is even aware she is pregnant. Therefore, the U.S. Public Health Service recommends that women of child-bearing age take 0.4 milligram of folate daily (Milunsky et al. 1989; Centers for Disease Control 1992; Czeizel and

Dudas 1992). We do not understand the direct mechanisms by which folate deficiencies lead to neural tube defects, however. A better understanding of the mechanisms linking folic acid function with neural development could present new therapeutic opportunities. Until then, although apples are good, asparagus has 20 times as much folate. So we could say that a bowl of asparagus a day will keep the doctor away.

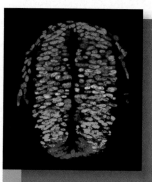

Closing Thoughts on the Opening Photo

This beautiful rainbow of a neural tube was generated by Elisa Martí Gorostiza's laboratory at the Institute of Molecular Biology in Barcelona. It represents a photomontage of transcription factor labeling from serial sections. The expression of each differently colored transcription factor occupies a discrete region along the dorsoventral axis. Based on their position along this axis, cells will interpret graded morphogens differently, and their interpretations will nudge the specification of these cells by activating a specific repertoire of regulatory transcription factors. *What is the time value of Sonic hedgehog?* We uncovered in this chapter that not only does the concentration of the morphogen Sonic hedgehog matter, but the duration of time that a cell experiences activation of the Hedgehog pathway is just as important. If there is value in being more ventral, it would be that the longer a cell "banks" Hedgehog activation, the more ventral the neural tube cell type it will become. (Photograph courtesy of E. M. Gorostiza.)

Snapshot Summary
Neural Tube Formation and Patterning

1. The neural tube forms from the shaping and folding of the neural plate. In primary neurulation, the surface ectoderm folds into a tube that separates from the surface ectoderm. In secondary neurulation, ectoderm and mesoderm cells coalesce as a mesenchyme to form first a cord and then a cavity (lumen) within the cord.

2. Primary neurulation is regulated both by intrinsic and extrinsic forces. Intrinsic wedging occurs within cells of the hinge regions, bending the neural plate. Extrinsic forces include the movement of the surface ectoderm toward the center of the embryo.

3. Neural tube closure is also the result of extrinsic and intrinsic forces. In humans, congenital anomalies can result if the neural tube fails to close. Folate is important in mediating neural tube closure.

4. After the node has reached the posterior of the epiblast, certain cells contribute to both the paraxial mesoderm and the neural tube.

5. The neural crest cells arise at the borders of the neural tube and surface ectoderm. They become located between the neural tube and surface ectoderm, and they migrate away from this region to become peripheral neural, glial, and pigment cells.

6. There is a gradient of maturity in many embryos (especially those of amniotes) such that the anterior develops earlier than the posterior.

7. The brain forms three primary vesicles, the prosencephalon (forebrain), mesencephalon (midbrain), and rhombencephalon (hindbrain). The prosencephalon and rhombencephalon become further subdivided.

8. Dorsal-ventral patterning of the neural tube is accomplished by proteins of the TGF-β superfamily secreted from the surface ectoderm and roof plate of the neural tube, and by Sonic hedgehog protein secreted from the notochord and floor plate cells. Temporal and spatial gradients of Shh trigger the synthesis of particular

transcription factors that specify the neuroepithelium. Some of these transcription factors show cross-repression, allowing discrete borders to form between regions along the dorsal-ventral axis.

9. Secondary neurulation in the caudal end of the neural plate gives rise to bipotential neuromesodermal progenitor cells (NMPs) that can become either neural or somitic cells. Exposure of the preneural progenitors to opposing gradients of Fgf/Wnt and retinoic acid leads to the patterning of the caudal neural tube along its dorsal-ventral axis.

Further Reading

Balaskas, N. and 7 others. 2012. Gene regulatory logic for reading the Sonic hedgehog signaling gradient in the vertebrate neural tube. *Cell* 14: 273–284.

Chiang, C., Y. Litingtung, E. Lee, K. E. Yong, J. L. Corden, H. Westphal and P. A. Beachy. 1996. Cyclopia and defective axial patterning in mice lacking Sonic hedgehog gene function. *Nature* 383: 407–413.

Cohen, M., A. Kicheva, A. Ribeiro, R. Blassberg, K. M. Page, C. P. Barnes, and J. Briscoe. 2015. Ptch1 and Gli regulate Shh signalling dynamics via multiple mechanisms. *Nature Commun.* 6: 6709.

Dady, A., E. Havis, V. Escriou, M. Catala, and J. L. Duband. 2014. Junctional neurulation: A unique developmental program shaping a discrete region of the spinal cord highly susceptible to neural tube defects. *J. Neurosci.* 34: 13208–13221.

Gouti, M., V. Metzis and J. Briscoe. 2015. The route to spinal cord cell types: A tale of signals and switches. *Trends Genet.* 31: 282–289.

Hashimoto, H., F. B. Robin, K. M. Sherrard and E. M. Munro. 2015. Sequential contraction and exchange of apical junctions drives zippering and neural tube closure in a simple chordate. *Dev. Cell* 32: 241–255.

Jessell, T. M. 2000. Neuronal specification in the spinal cord: Inductive signals and transcriptional codes. *Nature Rev. Genet.* 1: 20–29.

Lawson, A., H. Anderson and G. C. Schoenwolf. 2001. Cellular mechanisms of neural fold formation and morphogenesis in the chick embryo. *Anat. Rec.* 262: 153–168.

Massarwa, R. and L. Niswander. 2013. In toto live imaging of mouse morphogenesis and new insights into neural tube closure. *Development* 140: 226–236.

McShane, S. G., M. A. Molè, D. Savery, N. D. Greene, P. P. Tam and A. J. Copp. 2015. Cellular basis of neuroepithelial bending during mouse spinal neural tube closure. *Dev. Biol.* 404: 113–124.

Milunsky, A., H. Jick, S. S. Jick, C. L. Bruell, D. S. Maclaughlen, K. J. Rothman and W. Willett. 1989. Multivitamin folic acid supplementation in early pregnancy reduces the prevalence of neural tube defects. *J. Am. Med. Assoc.* 262: 2847–2852.

Sasai, N., E. Kutejova and J. Briscoe. 2014. Integration of signals along orthogonal axes of the vertebrate neural tube controls progenitor competence and increases cell diversity. *PLoS Biol.* 12(7):e1001907.

Wilde, J. J., J. R. Petersen and L. Niswander. 2014. Genetic, epigenetic, and environmental contributions to neural tube closure. *Annu. Rev. Genet.* 48: 583–611.

GO TO WWW.DEVBIO.COM...

...for Web Topics, Scientists Speak interviews, Watch Development videos, Dev Tutorials, and complete bibliographic information for all literature cited in this chapter.

Brain Growth

"WHAT IS PERHAPS THE MOST INTRIGUING QUESTION OF ALL is whether the brain is powerful enough to solve the problem of its own creation," declared Gregor Eichele in 1992. Determining how the brain—an organ that perceives, thinks, loves, hates, remembers, changes, deceives itself, and coordinates all our conscious and unconscious bodily processes—is constructed is undoubtedly the most challenging of all developmental enigmas. A combination of genetic, cellular, and systems level approaches is now giving us a very preliminary understanding of how the basic anatomy of the brain becomes ordered.

Differentiation of the neural tube into the various regions of the brain and spinal cord occurs simultaneously in three different ways. On the gross anatomical level, the neural tube and its lumen bulge and constrict to form the vesicles of the brain and spinal cord. At the tissue level, the cell populations in the wall of the neural tube arrange themselves into the different functional regions of the brain and spinal cord. Finally, on the cellular level, the neuroepithelial cells differentiate into the numerous types of nerve cells (**neurons**) and associated cells (**glia**) present in the body. In this chapter, we will concentrate on the development of the mammalian brain in general, as well as the human brain in particular, as we consider what makes us human.

The complexities of being human: How deep do they fold?

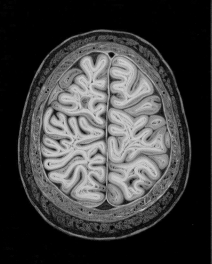

The Punchline

Brain growth begins with expansion of the newly formed neural tube along the apicobasal axis within three regions: the ventricular zone, the mantle or intermediate zone, and the marginal zone. Stem cells called radial glia span this neuroepithelium and proliferate, giving rise to progenitor cells and neurons. Newly born neurons use the radially oriented fibers of radial glia to migrate toward the marginal zone. In the cerebral cortex, a basally concentrated gradient of Reelin regulates the ordered "inside-to-outside" layering of migrating neurons. Bergmann glia act similarly to radial glia, but function in the cerebellum to generate Purkinje neurons. The self-renewal and neurogenic potential of these stem cells is influenced by numerous factors, including the orientation of the mitotic spindle, inheritance of the parental centriole and cilium, partitioning of Notch signaling, and mitogenic factors from the cerebral spinal fluid. The large and complex human brain has evolved through modification of the mechanisms controlling cerebellar neurogenesis, namely, the expansion of radial glial progenitor populations and the differential expression of unique neurogenesis genes. Neurogenesis does not end at birth but remains active in different ways throughout life.

Neuroanatomy of the Developing Central Nervous System

Your brain contains approximately 170 billion cells, an equal number of neurons and associated glial cells (Azevedo et al. 2009). There is a wide variety of neuronal and glial cell *types*, however, from the relatively small (e.g., granule cells) to the comparatively enormous (e.g., Purkinje neurons). All this diversity begins with the multipotent neuroepithelial cells of the neural tube.

The cells of the developing central nervous system

NEURAL STEM CELLS OF THE EMBRYO   Neuroepithelial cells are the first multipotent neural stem cells of the embryo. They make up the neural plate and early neural tube, and as epithelial cells, they are polarized along their apical to basal axis (**FIGURE 14.1A**). Once the plate closes into a neural tube, the apical surface of the neuroepithelium borders the internal cavity of the tube, which will become filled with cerebrospinal fluid. The basal surface of each cell terminates with an **endfoot**, or swelling of its basal membrane at the outer surface of the neural tube. The surface of the CNS is also referred to as the **pial surface**, after the pia mater that represents the fibrous membranes that surround nervous tissues. As stem cells, neuroepithelial cells are highly proliferative, generating progenitor cells for the first neuronal and glial cell types of the neural tube (Turner and Cepko 1987).

FIGURE 14.1   Cell types of the CNS. (A) Scanning electron micrograph of a newly formed chick neural tube, showing neuroepithelial cells at different stages of their cell cycles spanning the full width of the epithelium. (B) A Purkinje neuron with its elaborate dendritic processes. If you look carefully, those dendrites are not blurry; rather, postsynaptic membrane protrusions called spines are faintly visible. (C) Derived from a mouse hippocampus, a single oligodendrocyte (green) wrapping around multiple axons (purple) in co-culture. (D) Rat cerebral cortex with astroglial (yellow) endfeet wrapping around blood vessels (red). Cell nuclei are cyan. (A courtesy of K. Tosney; B courtesy of Boris Barbour; C from Fields 2013, courtesy of Doug Fields; D micrograph by Madelyn May, Honorable Mention, 2011 Olympus BioScapes Digital Imaging Competition.)

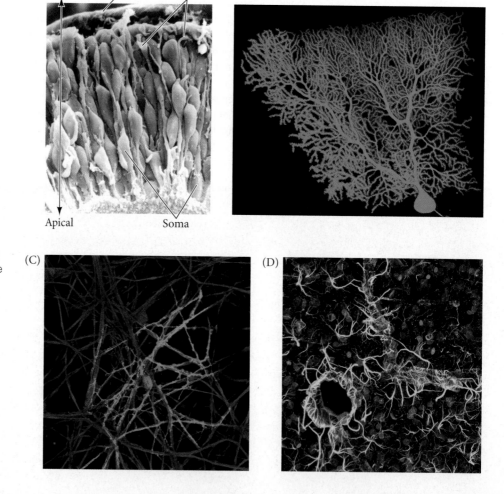

(A) Basal   Pial surface   Endfeet

Apical   Soma

(B)

(C)

(D)

Neuroepithelial cells are only present in the early embryo and eventually transform into **ventricular (ependymal) cells** and **radial glial cells**, or **radial glia**. Ependymal cells remain an integral component of the neural tube lining and secrete the cerebrospinal fluid. Radial glia[1] maintain a polarized morphology spanning the apicobasal axis of the central nervous system (CNS) and carry out two primary functions. First, they serve as the major neural stem cell throughout embryonic and fetal development, demonstrating self-renewal and the multipotent generation of both neurons and glial cells (Doetsch et al. 1999; Kriegstein and Alvarez-Buylla 2009); and second, they serve as a scaffold for the migration of other progenitor cells and newborn neurons (Bentivoglio and Mazzarello 1999). These two functions provide the foundational mechanism driving brain growth.

NEURONS AND NERVES   **Neurons** are cells that conduct electric potentials and transform these electric impulses into signals that coordinate our bodily functions, thoughts, sensations, and perceptions of the world. The fine, branching extensions of the neuron used to pick up electric impulses from other cells are called **dendrites** (**FIGURE 14.2A**). Some neurons develop only a few dendrites, whereas others (such as the Purkinje neurons; **FIGURE 14.1B**) develop extensive, branching **dendritic arbors**. Very few dendrites are found on cortical neurons at birth, and one of the amazing events of the first year of human life is the increase in the number of these receptive cellular processes. During this year, each cortical neuron develops enough dendritic surface to accommodate as many as 100,000 connections, or **synapses**, with other neurons. The average neuron in the highly developed cortex of the human cerebrum connects with 10,000 other neural cells, enabling the human cortex to function as the center for learning and reasoning.

Another important feature of a developing neuron is its **axon**. Whereas dendrites are often numerous and do not extend far from the neuronal cell body, or **soma**, axons may extend 2–3 feet (see Figure 14.2A). The pain receptors on your big toe, for example, must transmit messages all the way to your spinal cord. One of the fundamental concepts of neurobiology is that the axon is a continuous extension of the nerve cell body. The process by which neuronal connections between cell bodies are established from soma to soma through axons has been one of the most investigated events in neural development. As we will describe in Chapter 15, to "wire up" the embryonic brain, axons extend from the cell body, led by a motile growth cone at their tip that uses the cue-laden environment to navigate to its target for synaptic connection.

NEURONAL SIGNALING   A variety of different molecules known as **neurotransmitters** are critical in generating many action potentials. Axons are specialized for secreting specific neurotransmitters across a small gap—the **synaptic cleft**—that separates the axon of a signaling neuron from the dendrite or soma of its target cell. Some neurons develop the ability to synthesize and secrete acetylcholine (the first known neurotransmitter), whereas others develop the enzymatic pathways for making and secreting epinephrine, norepinephrine, octopamine, glutamate, serotonin, γ-aminobutyric acid (GABA), or dopamine, among other neurotransmitters. Each neuron must activate those genes responsible for making the enzymes that can synthesize its neurotransmitter. Thus, neuronal development involves both structural and molecular differentiation.

GLIAL CELLS   There are three categories of glial cells: oligodendrocytes, astroglia, and microglia. Neurons transmit information via electric impulses that travel from one region of the body to another along the axons. To prevent dispersal of the electric signal and to facilitate conduction to its target cell, axons in the CNS are insulated by **oligodendrocytes** (**FIGURE 14.1C**). The oligodendrocyte wraps itself around the developing

---

[1]Increasing evidence suggests that radial glia represent a heterogeneous population of neural stem cells and progenitor cells.

axon and then produces a specialized cell membrane called the **myelin sheath** (**FIGURE 14.2B**). In the peripheral nervous system (i.e., all the nerves and neurons outside of the central nervous system), myelination is accomplished by a similar type of glial cell, the **Schwann cell** (**FIGURE 14.2C**). Transplantation experiments have shown that the axon, and not the glial cell, controls the thickness of the myelin sheath by the amount of neuregulin-1 the axon secretes (Michailov et al. 2004).

The myelin sheath is essential for proper nerve function and also helps keep axons alive for decades. Loss of this sheath (demyelination) is associated with convulsions, paralysis, and certain debilitating afflictions such as multiple sclerosis (Emery 2010; Nave 2010). There are mouse mutants in which subsets of neurons are poorly myelinated. In the *trembler* mutant, the Schwann cells are unable to produce a particular protein component such that myelination is deficient in the peripheral nervous system but normal in the CNS. Conversely, in the mouse mutant *jimpy*, the CNS is deficient in myelin but the peripheral nerves are unaffected (Sidman et al. 1964; Henry and Sidman 1988).

Astroglial cells represent a diverse class of glial cells that include radial glia and a variety of differentiated subtypes of astrocytes (e.g., type I, type II, and reactive astrocytes) (**FIGURE 14.1D**). Astrocytes were originally named after their star (astral) shape appearance in a culture dish, and, historically astrocytes were presumed to function as the connective tissue of the nervous system, that is, its "glue." However, modern studies have revealed that astrocytes carry out an array of functions critical to the adult nervous system. These functions include establishing the blood-brain barrier, responding to inflammation in the CNS, and (most importantly) supporting synapse homeostasis and neural transmission.

A major marker for astroglia is an intermediate filament protein called glial fibrillary acidic protein (Gfap). Protein-misfolding mutations in

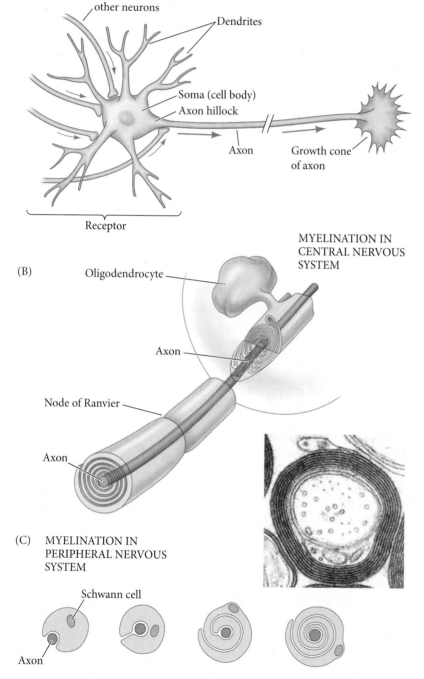

(A) Input via axons from other neurons

Dendrites

Soma (cell body)
Axon hillock

Axon Growth cone of axon

Receptor

MYELINATION IN CENTRAL NERVOUS SYSTEM

(B) Oligodendrocyte

Axon

Node of Ranvier

Axon

(C) MYELINATION IN PERIPHERAL NERVOUS SYSTEM

Schwann cell

Axon

FIGURE 14.2 Neural transmission and myelination. (A) A motor neuron. Electric impulses (red arrows) are received by the dendrites, and the stimulated neuron transmits impulses through its axon to its target tissue. The axon (which may be 2–3 feet long) is a cellular extension, or process, through which the neuron sends its signals. The axon's growth cone is both a locomotor and a sensory apparatus that actively explores the environment, picking up directional cues that tell it where to go. Eventually, the growth cone will form a connection, or synapse, with the axon's target tissue. (B,C) In the peripheral nervous system, Schwann cells wrap themselves around the axon; in the central nervous system, myelination is accomplished by the processes of oligodendrocytes. The micrograph shows an axon enveloped by the myelin membrane of a Schwann cell. (Micrograph courtesy of C. S. Raine.)

the human *gfap* gene can lead to Alexander disease, a neurodegenerative disease caused by fibrous protein aggregates that impair multiple functions of the nervous system (Brenner et al. 2001; Hagemann et al. 2006).

Microglia are often considered the "immune cells" of the central nervous system, since they function to engulf dying and dysfunctional neurons and glia. As their name implies, microglia are small relative to the other cell types of the nervous system. They are also very motile, with behaviors reminiscent of macrophages. In fact, microglia are not born in the nervous system, but are first generated by macrophage progenitor cells derived from the yolk sac (Wieghofer et al. 2015). These circulating microglial progenitors take root in the CNS prior to formation of the blood-brain barrier.

WATCH DEVELOPMENT 14.1   To learn more about glia, watch this introductory video, which describes the different glial cell types.

Tissues of the developing central nervous system

The neurons of the brain are organized into layers (**laminae**) and clusters (**nuclei**[2]), each having different functions and connections. The original neural tube is composed of a **germinal neuroepithelium**, a layer of rapidly dividing neural stem cells one cell layer thick. Sauer and colleagues (1935) showed that the cells of the germinal neuroepithelium span the full width of the epithelium, from the luminal surface of the neural tube to the outside surface. Over the course of evolution, adaptations have led to the germinal neuroepithelium producing a diversity of highly complex regions within the CNS. All of these regions, however, are elaborations of the same basic three-zone pattern of layers: ventricular (next to the lumen), mantle (intermediate), and marginal (outer) (**FIGURE 14.3**).

As the stem cells in the **ventricular zone** continue to divide, the migrating cells form a second layer around the original neural tube. This layer becomes progressively thicker as more cells are added to it from the germinal neuroepithelium. This new layer is the **mantle**, or **intermediate**, **zone**. The mantle zone cells differentiate into both neurons and glia. The neurons make connections among themselves and send forth axons away from the lumen, thereby creating a **marginal zone** poor in neuronal cell bodies. Eventually, oligodendrocytes cover many of the axons in the marginal zone in myelin sheaths, giving them a whitish appearance. Hence, the axonal marginal layer is often called **white matter**, while the mantle zone, containing the neuronal cell bodies, is referred to as **gray matter** (see Figure 14.3). The germinal epithelium of the ventricular zone will later shrink to become the **ependyma** that lines the brain cavity.

Here we will focus our investigation of CNS structure on the architecture associated with the spinal cord and medulla, cerebellum, and cerebrum.

SPINAL CORD AND MEDULLA ORGANIZATION   The basic three-zone pattern of ventricular (ependymal), mantle, and marginal layers is retained throughout development of the spinal cord and the medulla (the posterior region of the hindbrain). When viewed in cross section, the mantle gradually becomes a butterfly-shaped structure surrounded by the marginal zone or white matter, and both become encased in connective tissue. As the neural tube matures, a longitudinal groove—the **sulcus limitans**—divides it into dorsal and ventral halves. The dorsal portion receives input from sensory neurons, whereas the ventral portion is involved in effecting various motor functions (**FIGURE 14.4**). This developmental anatomy generates the basis of medullary and spinal cord physiology (such as the reflex arch).

---

[2]In neuroanatomy, the term *nucleus* refers to an anatomically discrete collection of neurons within the brain that typically serves a specific function. Note that it is a distinct structure from the cell nucleus.

Developing Questions

Glia—are they more than glue? The recent spotlight on glial cell function has revealed numerous ways in which these cells influence nervous system physiology, but how these roles manifest during CNS development remains unknown. Consider some of these questions: Is oligodendrocyte wrapping required for neuronal survival? Do astroglial cells regulate the targeting of synaptic partners? Do microglia help "sculpt" the brain during development?

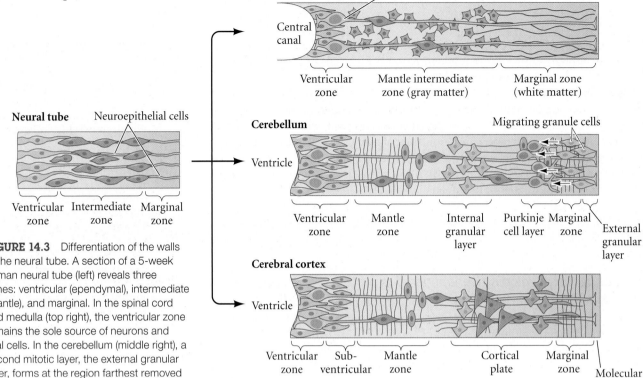

FIGURE 14.3 Differentiation of the walls of the neural tube. A section of a 5-week human neural tube (left) reveals three zones: ventricular (ependymal), intermediate (mantle), and marginal. In the spinal cord and medulla (top right), the ventricular zone remains the sole source of neurons and glial cells. In the cerebellum (middle right), a second mitotic layer, the external granular layer, forms at the region farthest removed from the ventricular zone. A type of neuron called granule cells migrate from this layer back into the intermediate zone to form the internal granular layer. In the cerebral cortex (bottom right), the migrating neurons and glioblasts form a cortical plate containing six layers. (After Jacobson 1991.)

CEREBELLAR ORGANIZATION In the cerebellum, cell migration and selective proliferation and cell death produce modifications of the three-zone pattern shown in Figure 14.3. Cerebellar development results in a highly folded cortex (outer region) composed of Purkinje neurons and granule neurons integrated into "nuclei" that control balance functions and relay information from the cerebellar cortex to other brain regions. In the development of the cerebellum, the critical event appears to be the migration of neural progenitor cells to the outer surface of the developing

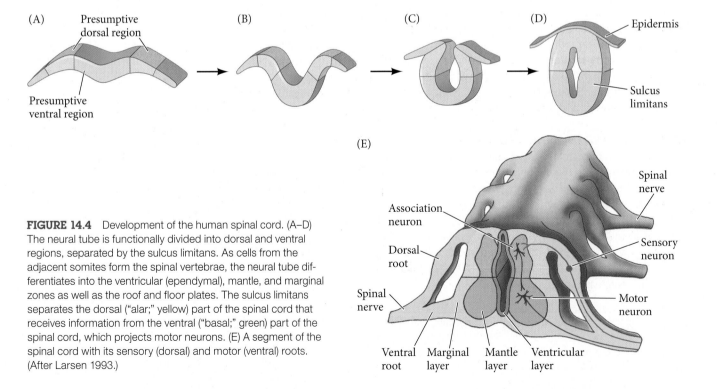

FIGURE 14.4 Development of the human spinal cord. (A–D) The neural tube is functionally divided into dorsal and ventral regions, separated by the sulcus limitans. As cells from the adjacent somites form the spinal vertebrae, the neural tube differentiates into the ventricular (ependymal), mantle, and marginal zones as well as the roof and floor plates. The sulcus limitans separates the dorsal ("alar;" yellow) part of the spinal cord that receives information from the ventral ("basal;" green) part of the spinal cord, which projects motor neurons. (E) A segment of the spinal cord with its sensory (dorsal) and motor (ventral) roots. (After Larsen 1993.)

cerebellum. Here they form a new germinal zone—the **external granular layer**—near the outer boundary of the neural tube.

At the outer boundary of the external granular layer, which is 1–2 cell bodies thick, neural progenitor cells proliferate and come into contact with cells that secrete bone morphogenetic proteins (BMPs). The BMPs specify the postmitotic cells derived from neural progenitor divisions to become a type of neuron called **granule cells** (Alder et al. 1999). Granule cells migrate back toward the ventricular (ependymal) zone, where they form a region called the **internal granular layer** (see Figure 14.3). Meanwhile, the original ventricular zone of the cerebellum generates a wide variety of neurons and glial cells, including the distinctive and large **Purkinje neurons**, the major cell type of the cerebellum (**FIGURE 14.5**). Purkinje neurons secrete Sonic hedgehog, which sustains the division of granule cell precursors in the external granular layer (Wallace 1999). Each Purkinje neuron has an enormous **dendritic arbor** that spreads like a tree above a bulb-like cell body (see Figure 14.1B). A typical Purkinje neuron may form as many as 100,000 synapses with other neurons—more connections than any other type of neuron studied. Each Purkinje neuron also sends out a slender axon, which connects to neurons in the deep cerebellar nuclei.

Purkinje neurons are critical in the electrical pathway of the cerebellum. All electric impulses eventually regulate their activity because Purkinje neurons are the only output neurons of the cerebellar cortex. Such regulation requires the proper cells to differentiate at the appropriate places and times. *How is this complicated series of events accomplished?* Contemplate a few ideas about what mechanisms might control the pattern of neuronal differentiation, as we will return to this overarching question later in the chapter.

CEREBRAL ORGANIZATION The three-zone arrangement of the neural tube is also seen, although modified, in the cerebrum. The cerebrum is organized in two distinct ways. First, like the cerebellum, it is organized radially into layers that interact with one another. Certain neural progenitor cells from the mantle zone migrate on radial glial processes toward the outer surface of the brain and accumulate in a new layer, the cortical plate (see Figure 14.3). This new layer of gray matter will become the **neocortex,** a distinguishing feature of the mammalian brain. Specification of the neocortex involves the Lhx2 transcription factor, which activates numerous other cerebral genes. In *Lhx2*-deficient mice, the cerebral cortex fails to form (**FIGURE 14.6**; Mangale et al. 2008; Chou et al. 2009).

The neocortex eventually stratifies into six layers of neuronal cell bodies; the adult forms of these layers are not fully mature until the middle of childhood. Each layer of the neocortex differs from the others in its functional properties, the types of neurons found there, and the sets of connections they make (**FIGURE 14.7**). For instance, neurons in cortical layer 4 *receive* their major input from the thalamus (a region that forms from the diencephalon), whereas neurons in layer 6 *send* their major output to the thalamus.

In addition to the six vertical layers, the cerebral cortex is organized horizontally into more than 40 regions that regulate anatomically and functionally distinct processes. For instance, neurons of the visual cortex in layer 6 project axons

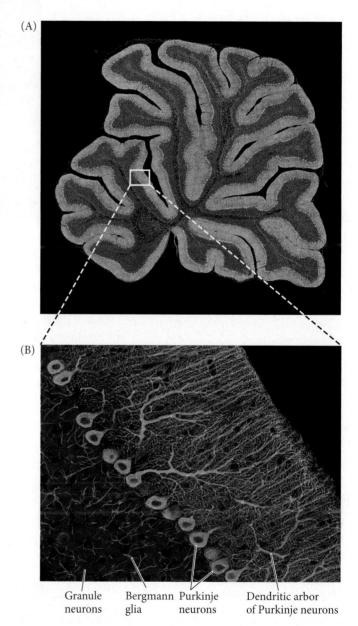

(A)

(B)

Granule neurons   Bergmann glia   Purkinje neurons   Dendritic arbor of Purkinje neurons

FIGURE 14.5 Cerebellar organization. (A) Sagittal section of a fluorescently labeled rat cerebellum photographed using dual-photon confocal microscopy. (B) Enlargement of the boxed area in (A) illustrates the highly structured organization of neurons and glial cells. Purkinje neurons are light blue with bright green processes, Bergmann glia are red, and granule cells are dark blue. (Courtesy of T. Deerinck and M. Ellisman, University of California, San Diego.)

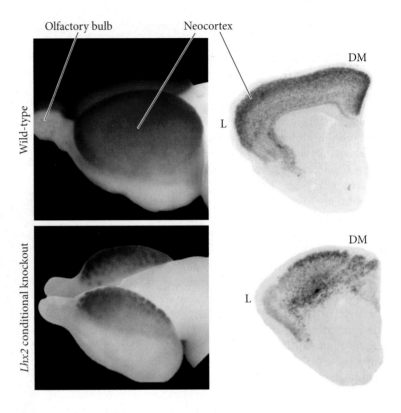

FIGURE 14.6 *Lhx2* is required for neocortex development. Whole mount and coronal sections of brains of wild-type and *Lhx2* conditional knockout mice, in which early stem cells experience the loss of *Lhx2*. The neocortex marker *Satb2* (brown) shows high expression in both dorsomedial (DM) and lateral (L) regions of the neocortex in the wild-type mouse, whereas in *Lhx2*-knockout mice, significant levels of expression of the *Satb2* marker are found only in the dorsomedial neocortex. (From Chou et al. 2009.)

to the lateral geniculate nucleus of the thalamus, which is involved in vision, whereas neurons of the auditory cortex of layer 6 (located more anteriorly than the visual cortex) project axons to the medial geniculate nucleus of the thalamus, which functions in hearing.

One of the major questions in developmental neurobiology is whether the different functional regions of the cerebral cortex are already specified in the ventricular region, or if specification is accomplished much later by synaptic connections between the regions. Evidence that specification is early (and that there might be some "protomap" of the cerebral cortex) is suggested by certain human mutations that destroy the

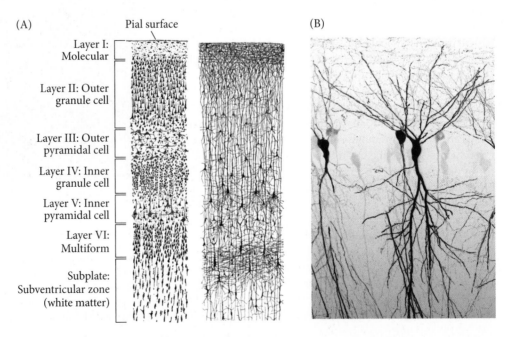

FIGURE 14.7 Different neuronal cell types are organized into the six layers of the neocortex. (A) Different cellular stains reveal neocortical layering in these exquisite drawings by Santiago Ramón y Cajal from his 1899 work "Comparative study of the sensory areas of the human cortex." (B) Pyramidal neurons of mouse hippocampus (postnatal day 7). (B micrograph by Joanna Szczurkowska, Honorable Mention, 2014 Olympus BioScapes Digital Imaging Competition.)

layering and functional abilities in only one part of the cortex, leaving the other regions intact (Piao et al. 2004). More direct evidence for the existence of a protomap in the embryonic cortex recently emerged when Fuentealba and colleagues (2015) followed ventricular radial glial cells from different regions of the embryonic mouse brain using retroviral barcoding until the cells' direct clonal descendants could be identified in the adult cortex (**FIGURE 14.8**). They discovered that the differentiated neurons of the cortex were descended from stem cells that resided in comparable areas in the embryo (which were themselves derived from radial glia from comparable ventricular zone areas). These results support a model in which ventricular zone radial glia are regionally specified in the embryo and give rise to similarly specified adult stem cells that propagate regionally restricted progeny.

 DEV TUTORIAL *Neurogenesis in the cerebral cortex* Dr. Michael J. F. Barresi describes the cellular and molecular processes governing the inside-out development of the cerebral cortex.

Developmental Mechanisms Regulating Brain Growth

Growing the vertebrate brain is much like constructing a multilevel, multicolored brick building. First, those bricks need to be made and an appropriate number of the correctly colored bricks supplied to the right locations. Second, scaffolding is used throughout the structure to transport the necessary bricks and supplies to their destined locations. The building is constructed from bottom to top, building outward in the various dimensions to create increasingly complex architecture. In the developing brain, precisely controlled cell division of stem and progenitor cells generates the necessary numbers and types of cells ("the bricks"). Radial glial cells not only serve as stem cells, they also provide the scaffolding required for movement of progenitor cells and newborn neurons to increasingly more superficial layers in a manner that effectively builds the brain from the inside outward.

Neural stem cell behaviors during division

INTERKINETIC NUCLEAR MIGRATION DURING DIVISION Sauer and colleagues' 1935 study of the germinal neuroepithelium not only indicated that the cells spanned the width of the epithelium, but also showed that the *cell* nuclei are at different heights in this tissue (see Figure 14.1A), and that the nuclei move as the cell goes through the cell cycle. During DNA synthesis (S phase of the cell cycle), the nucleus is near the basal end of the cell near the outside edge of the neural tube, and translocates toward the apical end of the cell as the cycle proceeds. By mitosis (M phase), the nucleus is at the cell's apical end, near the ventricular surface. Following mitosis (G1 phase), the nucleus slowly migrates basally again (**FIGURE 14.9**). This process, called **interkinetic nuclear migration**, is also seen in the radial glial cells and occurs in a broad range of vertebrates (Alexandre et al. 2010; Meyer et al. 2011; Spear and Erickson 2012). The mechanisms involved are not fully understood, but microtubules and motor proteins appear to be involved. When a gene for a motor protein that is important for mitotic spindle separation is mutated in zebrafish, radial glial cells can successfully initiate interkinetic

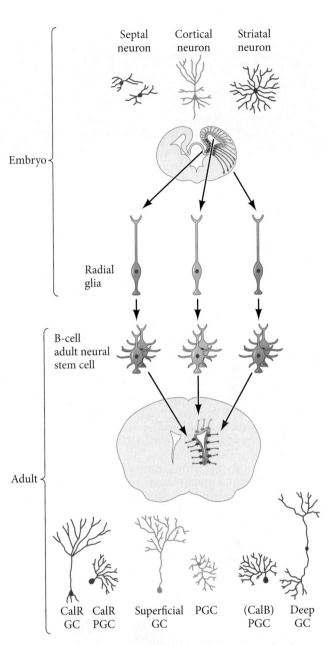

FIGURE 14.8 Regional specification of embryonic radial glia translates into restricted progenitor derivation. This schematic shows the positions of ventricular radial glia in the embryonic brain (above), and the clonal derivation of B type stem cells and their related differentiated neurons in the adult brain (below). (GC, granule cells; PGC, periglomerular cell; CalB, calbindin; CalR, calretinin.) (After Fuentealba et al. 2015.)

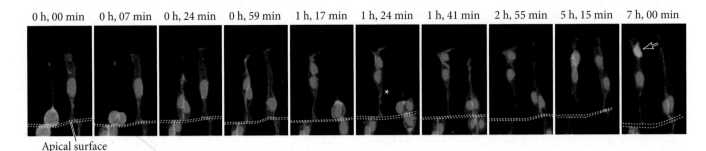

0 h, 00 min   0 h, 07 min   0 h, 24 min   0 h, 59 min   1 h, 17 min   1 h, 24 min   1 h, 41 min   2 h, 55 min   5 h, 15 min   7 h, 00 min

Apical surface

FIGURE 14.9 Live imaging of neuroepithelial cell interkinetic nuclear migration and division of neural stem cells in the zebrafish embryonic hindbrain. Two nearly adjacent progenitor cells in the germinal epithelium were recorded over 7 hours. Cells are labeled to show cell membranes (green) and nuclei (red). A reporter gene specifically marks neurons (yellow). The progenitor cell on the left underwent an asymmetrical division, generating a neuron (arrow at 7 h) and another progenitor (below the neuron). The cell on the right underwent symmetrical division, giving rise to two progenitor cells. The asterisk at 1 h 24 min indicates the point at which the neuronal daughter cell detached from the apical surface (white dotted double lines). Notice the translocation of the nucleus in a progenitor cell as it proceeds through the cell cycle. The cell is undergoing DNA synthesis (S phase) when its nucleus is toward the basal end of the cell (away from the dotted white lines) and is in mitosis (M phase) when its nucleus is near the apical end of the cell. (From Alexandre et al. 2010.)

Developing Questions

It has been said that a picture is worth a thousand words, in which case, a movie must be worth a million. Watching movies of interkinetic nuclear migration (see Watch Development 14.2), we challenge you to avoid seeing something new each time. Hidden in those movies are answers to many questions, including: Why does cytokinesis need to happen at the apical surface in the neuroepithelium? Do these cells maintain their basal process? What role do centrosomes—key organizing structures for microtubules and mitosis—play in this nuclear migration?

nuclear migration but fail to progress through mitosis, and the somas of these radial glia accumulate at the luminal (apical) surface over time (Johnson et al. 2016).

WATCH DEVELOPMENT 14.2 Watch the cell behaviors associated with the interkinetic nuclear migration of radial glial cells during division in the developing brains of zebrafish and chick.

SYMMETRY OF DIVISION When neuroepithelial cells or radial glial cells divide, what options do they have? Recall from Chapter 5 our descriptions of division in other stem cells (see Figure 5.1). A stem cell can divide symmetrically to produce two copies of itself, thus increasing the pool of stem cells. Alternatively, symmetrical division can produce two differentiating daughter cells, which depletes the stem cell pool. A stem cell can also divide asymmetrically to self-renew and yield a differentiating daughter cell. How might you investigate which of these divisions occurs in the neuroepithelium? Labeling the cells with a tracer such as radioactive thymidine that is incorporated only into dividing cells would allow you to trace cell lineages. When mammalian neuroepithelial cells are labeled in this way during early development, 100% of them incorporate the radioactive thymidine into their DNA, indicating they are all undergoing some form of division (Fujita 1964). Shortly thereafter, however, certain cells stop incorporating this thymidine analog, indicating that they are no longer dividing. These cells can then be seen to migrate away from the lumen of the neural tube and differentiate into neuronal and glial cells (Fujita 1966; Jacobson 1968). When a cell of the germinal neuroepithelium is ready to generate neurons (instead of more neural stem cells), the division plane often shifts to create an asymmetrical division (the arrow in Figure 14.9). Instead of both daughter cells remaining attached to the luminal surface, one of them becomes detached (the asterisk in Figure 14.9). The cell remaining connected to the luminal surface usually remains a stem cell, while the other cell migrates and differentiates into a neuron or other type of progenitor (Chenn and McConnell 1995; Hollyday 2001).

Neurogenesis: Building from the bottom up (or from the inside out)

In a 2008 paper, Nicholas Gaiano summarized neurogenesis:

> The construction of the mammalian neocortex is perhaps the most complex biological process that occurs in nature. A pool of seemingly homogeneous stem cells first undergoes proliferative expansion and diversification and then initiates the

production of successive waves of neurons. As these neurons are generated, they take up residence in the nascent cortical plate where they integrate into the developing neocortical circuitry. The spatial and temporal coordination of neuronal generation, migration, and differentiation is tightly regulated and of paramount importance to the creation of a mature brain capable of processing and reacting to sensory input from the environment and of conscious thought.

As the neural tube matures, the progeny of the neuroepithelial stem cells become radial glial cells. Only recently have cell lineage studies demonstrated that radial glia are neural stem cells that undergo symmetrical and asymmetrical divisions (Malatesta et al. 2000, 2003; Miyata et al. 2001; Noctor et al. 2001; Anthony et al. 2004; Casper and McCarthy 2006; Johnson et al. 2016). The divisions of the radial glia take place in the **ventricular zone** (the zone lining the ventricle and therefore in contact with the cerebrospinal fluid). In the cerebrum, as the progenitor cells delaminate from the ventricular zone, they form a **subventricular zone** basal to it. Together, these zones form the germinal strata that generate the neurons that migrate into the cortical plate and form the layers of neurons of the neocortex (**FIGURE 14.10A,B**; Frantz et al. 1994; for reviews, see Kriegstein and Alvarez-Buylla 2009; Lui et al. 2011; Kwan et al. 2012; Paridaen and Huttner 2014).

A single stem cell in the ventricular layer can give rise to both neurons and glial cells in any of the cortical layers (Walsh and Cepko 1988). There are three major progenitor cell types in the germinal strata: **ventricular radial glia (vRG)**, **outer radial glia (oRG)**, and **intermediate progenitor (IP)** cells. During the early stages of CNS development, neuroepithelial cells transform into ventricular radial glia that, as their name suggests, maintain contact with the luminal surface. The vRG serve as the parental stem cell type and, in addition to directly generating neurons, will give rise to both the oRG and IP cells (**FIGURE 14.10C,D**). Self-renewing, symmetrical divisions dominate early in

FIGURE 14.10 Summary model of neurogenesis in the cerebral cortex. (SVZ, subventricular zone; VZ, ventricular zone.) (Model based on Kriegstein and Alvarez-Buylla 2009; Kwan et al. 2012; Paridaen and Huttner 2014.)

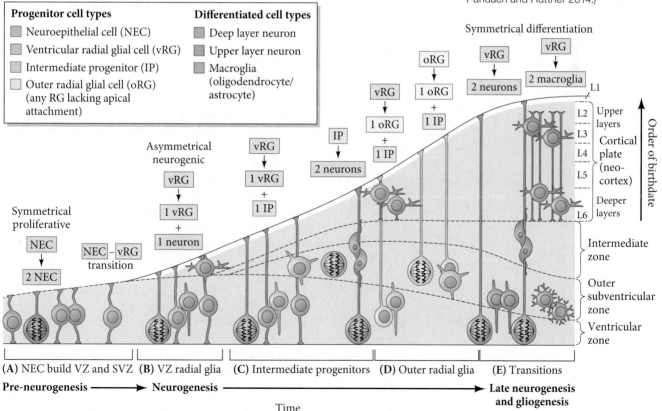

(A) NEC build VZ and SVZ   **(B)** VZ radial glia   **(C)** Intermediate progenitors   **(D)** Outer radial glia   **(E)** Transitions

Pre-neurogenesis ⟶ **Neurogenesis** ⟶ **Late neurogenesis and gliogenesis**

Time

neurogenesis to expand the progenitor pool, and then more asymmetrical divisions govern progenitor differentiation.

The oRG cells always maintain contact between their basal process and the pial surface; however, they are no longer tethered to the apical surface, and their soma reside in the subventricular zone and are therefore "outer" relative to vRG (Lui et al. 2011; Wang et al. 2011). Both vRG and oRG can divide to produce IP cells (see Figure 14.10C,D). IP cells have limited proliferative capacity, normally being able to undergo only a single round of division, yet during neurogenesis they play a pivotal role as a progenitor cell population for the specific expansion of particular lineages. It is generally thought that cell type potency (i.e., the cell types a progenitor can give rise to) becomes more restricted from vRG to oRG, with IP cells showing the most lineage restriction (Noctor et al. 2004; Lui et al. 2011).

 SCIENTISTS SPEAK 14.1 See a web conference with Dr. Arnold Kriegstein on oRG cells and the development of the neocortex.

Glia as scaffold for the layering of the cerebellum and neocortex

Different types of neurons and glial cells are born at different times. Labeling cells at different times during development of the cerebrum shows that the cells with the earliest birthdays migrate the shortest distances; those with later birthdays migrate farther to form the more superficial regions of the brain cortex. Subsequent differentiation depends on the positions the neurons occupy once outside the germinal neuroepithelium (Letourneau 1977; Jacobson 1991). What are the developmental mechanisms governing this pairing of neuronal birth with differentiation along the apicobasal axis of the brain?

It has been known for decades that radial glial cells guide neural progenitor cell migration from the inner (luminal) region to the outer zones throughout the CNS (Rakic 1971). Thus, the progenitor cells formed as the progeny of radial glia also use their "sister" stem cell's connection between luminal and outer surfaces to migrate to their appropriate positions. We will explore the mechanisms of radial glia-enabled migration in the cerebellum and cerebrum.

BERGMANN GLIA IN THE CEREBELLUM One mechanism thought to be important for positioning young neurons in the developing mammalian brain is **glial guidance** (Rakic 1972; Hatten 1990). Throughout the cortex, neurons are seen to ride a "glial monorail" to their respective destinations. In the cerebellum, the granule cell precursors travel on the long processes of the **Bergmann glia**, a type of radial glial cell that extends one to two thin processes throughout the germinative neuroepithelium (see Figure 14.5B; Rakic and Sidman 1973; Rakic 1975). As **FIGURE 14.11** illustrates, this neuron-glia interaction is a complex and fascinating series of events involving reciprocal recognition between glia and newborn neurons (Hatten 1990; Komuro and Rakic 1992).

It appears that the migration of newborn neurons involves the *loss* of those adhesion molecules that linked the neuron to the germinal layer cells and the *acquisition* of a set of adhesion molecules that attach it to the glia (Famulski et al. 2010). The molecules involved in this adhesion were discovered through a number of mouse mutants that could not keep their balance and were given names such as *reeler, staggerer,* and *weaver* that reflected their movement problems (Falconer 1951). In *reeler* brains, glial cells lack the extracellular matrix protein Reelin that permits the neurons to bind them. Another adhesion protein, astrotactin, is needed by granule cell neurons to maintain their adhesion to the glial process. If the astrotactin on a neuron is masked by antibodies to that protein, the neuron will fail to adhere to the glial processes (Edmondson et al. 1988; Fishell and Hatten 1991). The direction of this migration appears to be regulated by a complex series of events orchestrated by **brain-derived neurotrophic factor** (**BDNF**), a paracrine factor made by the internal granular layer (Zhou et al. 2007).

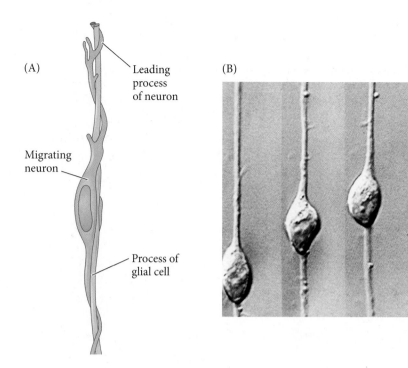

(A)

Leading
process
of neuron

Migrating
neuron

Process of
glial cell

(B)

FIGURE 14.11   Neuron-glia interaction in the mouse. (A) Diagram of a cortical neuron migrating on a glial cell process. (B) Sequential photographs of a neuron migrating on a cerebellar glial process. The leading process has several filopodial extensions. The neuron can reach speeds of about 40 mm per hour as it travels. (A after Rakic 1975; B from Hatten 1990, photograph courtesy of M. Hatten.)

RADIAL GLIA IN THE NEOCORTEX   In the developing cerebrum, most of the neurons generated in the ventricular zone migrate outward along radial glial processes to form the **cortical plate** near the outer surface of the brain, where they set up the six layers of the neocortex. As in the rest of the brain, those neurons with the earliest birthdays form the layer closest to the ventricle (**FIGURE 14.12A,B**). Subsequent neurons travel greater distances to form the more superficial layers of the cortex. This process forms an "inside-out" gradient of development (Rakic 1974). McConnell and Kaznowski (1991) have shown that the determination of laminar identity (i.e., which layer a cell migrates to) is made during the final cell division. Newly generated neuronal precursors transplanted after this last division from young brains (where they would form layer 6) into older brains, whose migratory neurons are forming layer 2, are committed to their fate and migrate only to layer 6. However, if these cells are transplanted prior to their final division (i.e., during mid-S phase), they are uncommitted and can migrate to layer 2 (**FIGURE 14.12C,D**). The fates of neuronal precursors from older brains are more fixed. The neuronal precursor cells formed early in development have the potential to become any neuron (at layer 2 or 6, for instance); later precursor cells give rise only to upper level (layer 2) neurons (Frantz and McConnell 1996). Once the cells arrive at their final destination, it is thought that they express specific adhesion molecules that organize them into brain nuclei (Matsunami and Takeichi 1995).

Signaling mechanisms regulating development of the neocortex

CAJAL-RETZIUS CELLS: "MOVING TARGETS" IN THE NEOCORTEX   How do migrating neural progenitors become segregated to the correct layer? As mentioned above, earlier-born neurons establish the deeper layers and later-born neurons form the more superficial layers. *Think about this.* It means that the cerebrum is growing from inside to outside. One outcome of such growth is that with each new expanding layer, the outer pial surface moves farther away from the ventricular surface. Therefore, the pial surface is an ever-expanding outer limit, and the neurons embarking on their outward trek have farther to travel than their predecessors. This important dynamic ultimately influences the layering of the brain (see Frotscher 2010).

When the luminal and pial surfaces are relatively close during early development of the neocortex, a newly born neuron extends basal filopodia toward the pial surface, establishes adhesive contact, and then simply displaces its nucleus and associated cytoplasm toward the pial surface, translocating the cell body from the apical to the basal

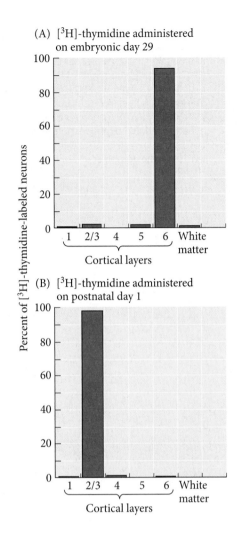

(A) [³H]-thymidine administered on embryonic day 29

(B) [³H]-thymidine administered on postnatal day 1

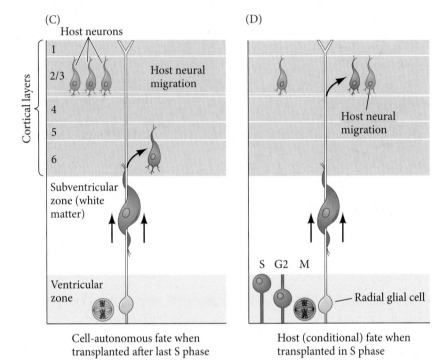

FIGURE 14.12 Determination of cortical laminar identity in the ferret cerebrum. (A) "Early" neuronal precursors (birthdays on embryonic day 29) migrate to layer 6. (B) "Late" neuronal precursors (birthdays on postnatal day 1) migrate farther, into layers 2 and 3. (C) When early neuronal precursors (dark blue) are transplanted into older ventricular zones after their last mitotic S phase, the neurons they form migrate to layer 6. (D) If these precursors are transplanted before or during their last S phase, they migrate with the host neurons to layer 2. (After McConnell and Kaznowski 1991.)

Cell-autonomous fate when transplanted after last S phase

Host (conditional) fate when transplanted in S phase

region of the cell. The basal attachment provides the required physical resistance and tension that enables this translocation (Miyata and Ogawa 2007). Thus, no actual cell migration is necessary. During later development, however, each progenitor cell needs to actively migrate along the radial glial cell's basal process until its own basal membrane makes contact with the outermost region of the cortical plate, at which point a similar translocation can complete the journey (**FIGURE 14.13A**).

The cells influencing this outward migration of progenitor cells are the **Cajal-Retzius cells**, which lie under the pial surface and secrete the extracellular protein Reelin—the same protein mentioned earlier as regulating layering in the cerebellum (D'Arcangelo et al. 1995, 1997). Translocating progenitor cells express transmembrane receptors for Reelin (Trommsdorff et al. 1999), and when these receptors bind to Reelin, they set off a series of signal transduction pathways mediated by the enzyme Disabled-1 (see Figure 14.13A, cell 1). As a result, the cells increase their expression of N-cadherin, allowing them to attach to other cells that also express N-cadherins. Cadherins are expressed with increasing intensity from the ventricular zone to highest levels in the marginal zone, overlapping Cajal-Retzius cells; thus, newly born neurons expressing N-cadherin become oriented toward regions of increasing adhesion (Franco et al. 2011; Jossin and Cooper 2011). The neurons also extend filopodia toward the fibronectin-rich extracellular matrix at the pial surface (Chai et al. 2009) and use transmembrane proteins called integrins to attach the filopodia to this extracellular matrix (Sekine et al. 2012). Once the filopodia are attached, Disabled-1-mediated regulation of the actin cytoskeleton powers contraction of the filopodia in a spring-like motion, pulling the cell body forward as the cell's apical end detaches (see Figure 14.13A, cell 2; Miyata and Ogawa 2007).

(A)

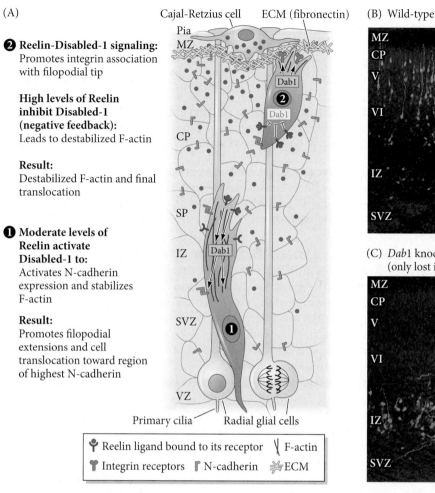

Cajal-Retzius cell   ECM (fibronectin)

❷ **Reelin-Disabled-1 signaling:**
Promotes integrin association
with filopodial tip

**High levels of Reelin
inhibit Disabled-1
(negative feedback):**
Leads to destabilized F-actin

Result:
Destabilized F-actin and final
translocation

❶ **Moderate levels of
Reelin activate
Disabled-1 to:**
Activates N-cadherin
expression and stabilizes
F-actin

Result:
Promotes filopodial
extensions and cell
translocation toward region
of highest N-cadherin

Primary cilia   Radial glial cells

🌱 Reelin ligand bound to its receptor   ╲ F-actin
🌱 Integrin receptors   ┌ N-cadherin   ❋ ECM

(B) Wild-type

(C) *Dab1* knockout
(only lost in green cells)

FIGURE 14.13 Model of Reelin regulation of directed neuronal migration. (A) Secreted from Cajal-Retzius cells, Reelin (red circles) is distributed in a gradient in the extracellular matrix. Reelin instructs the newly born migrating neurons (labeled 1 and 2) to extend filopodia from their basal membrane toward the pial surface. *Disabled-1* (*Dab1*) is activated by Reelin. The product of the *Dab1* gene stabilizes filamentous actin (F-actin) as well as upregulating N-cadherin expression. N-cadherin is also localized to the membranes of radial glial fibers and other cells throughout the epithelium, increasing to highest concentrations closest to the marginal zone. Initial Reelin-Dab1 signaling results in marginal zone-directed extension of the filopodium and translocation of neuron 1. In a migrating neuron approaching the marginal zone (cell 2), Dab1 upregulates integrin expression in the tip of the filopodium to anchor this cell to the fibronectin-rich extracellular matrix. However, at highest Reelin concentrations, a negative feedback mechanism is triggered that inhibits Dab1 by protein degradation (cell 2), ending migration and enabling cell differentiation within the specified cortical layer. (B,C) Conditional inactivation of *Dab1*

in newly born neurons and migrating progenitor cells. Two types of mice were used, wild type and a strain carrying a conditional *Dab1* mutation that is activated only when combined with a second gene (*CRE*). *CRE* has no affect on wild-type mice. A plasmid carrying *CRE* and *GFP* was introduced into progenitor cells of the two mouse strains. Cells that received the plasmid could be identified by their GFP expression (green). (B) The wild-type control shows that the treated progenitor cells successfully reached the layer of the cortical plate. (C) In the *Dab1* conditional mutant, *Dab1* is knocked out in the green (*CRE*-containing, GFP-expressing) cells. These cells were retained in the intermediate zone. Time-lapse imaging of a single cell demonstrates that a typical progenitor cell will initiate migratory cell elongation (red), then extend its basal process to the marginal zone (green), and finally translocate the apical compartment to the outer layers (blue) (B, drawing on right). Similar imaging shows migration is initiated in *Dab1* knockout cells, but they fail to advance productive basal extensions, nor do they show translocation (C, drawing on right). (B,C from Franco et al. 2011.)

The same Reelin signal that initiates this migration also triggers a negative feedback such that at the highest levels of Reelin (near the marginal zone), the neurons lose their cell adhesion molecules and integrate into the layers of the cortical plate in a progressive inside-out manner (see Figure 14.13A, cell 2; Feng et al. 2007). Loss of *Reelin*, its receptors, or *Disabled-1* results in an inversion of cortical layering; neurons typically found within the inner layers (layers 4 and 5) are positioned near the marginal zone (layer 1),

Developing Questions

N-cadherin was shown to be important for neural progenitor migration in the neocortex, but how? Cells along the apicobasal axis have increasing levels of cadherin expression that appear to provide a road to the next layer, but cadherins are typically thought of as playing a role in differential adhesion and cell sorting. Could the properties of cell sorting be driving the movement of neural progenitors toward the marginal zone, much like E-cadherin functions in the zebrafish gastrula (see Chapter 4)? In one interesting study, whole-transcriptome analysis of the different proliferative zones of the mouse and human neocortex revealed a high level of expression of cell adhesion and extracellular matrix genes (Fietz et al. 2012), which suggests that the use of cell adhesion is likely both a complex and foundational mechanism for layering the cortex.

Developing Questions

It has been shown that the Hes family of Notch-transcriptional effectors exhibit oscillating periods of gene expression in vRG due to a negative feedback mechanism (see Chapters 4 and 17; Shimojo et al. 2008). It is intriguing to speculate about a model in which the number of Notch-Hes oscillations a vRG cell experiences might regulate its development between renewal, progenitor derivation, and differentiation (Paridaen and Huttner 2014).

and cells of the external layers (layers 2 and 3) are found near the subplate when these genes are lost (**FIGURE 14.13B,C**; Olson et al. 2006; Franco et al. 2011; Sekine et al. 2011).

WATCH DEVELOPMENT 14.3 Observe the morphology of late-migrating neuroblasts as they translocate to the Cajal-Retzius cells.

WEB TOPIC 14.1 NEURONAL VERSUS GLIAL DIFFERENTIATION Explore the role of neurotrophic factors in regulating the differentiation of cortical progenitor cells into neurons versus glia, a mechanism that *maps* to the *stats* you need to know.

WEB TOPIC 14.2 HORIZONTAL AND VERTICAL SPECIFICATION OF THE CEREBRUM Neither the vertical nor the horizontal organization of the cerebral cortex is clonally specified. Cell migration is critical, and paracrine factors from neighboring cells play major roles in migration and specification.

TO BE OR NOT TO BE … A STEM, A PROGENITOR, OR A NEURON? Whether a radial glial cell undergoes symmetrical versus asymmetrical division depends on the plane of division (which in turn depends on the orientation of the mitotic spindle) and is correlated with the type of progeny generated. Cytokinesis that separates the radial glial cell perfectly perpendicular (planar) to the luminal surface (i.e., the mitotic spindle is parallel to the lumen) can result in two radial glial stem cells (Xie et al. 2013). Although such perpendicular cleavages can sometimes yield different progeny—a radial glial cell and a neuron—more often it is oblique division planes that give rise to these two distinct progeny. When the mitotic spindle is altered in such a way that cytokinesis occurs along random axes, it increases early asymmetrical divisions and triggers premature neurogenesis (Xie et al. 2013).

The fate of a daughter cell following cytokinesis has been linked to the centriole it inherits. The two centrioles in a dividing cell are not the same in regard to their age: the parental centriole is "older" than the daughter centriole it creates when it replicates. At each division, the cell receiving the "old" centriole will stay in the ventricular zone as a stem cell, while the cell receiving the "young" centriole leaves and differentiates (Wang et al. 2009). These two centrioles are bound to different proteins and structures, which results in an asymmetrical localization of those factors that influence gene expression and cell fate. Of particular significance is the primary cilium, which is connected to the older centriole and remains with it during cell division. The daughter cell that inherits this older centriole along with the primary cilium can quickly present the primary cilium to the lumen and, consequently, to the cerebrospinal fluid. Cerebrospinal fluid contains factors such as insulin-like growth factors, FGFs, and Sonic hedgehog, which induce proliferation and signal the cell to maintain its radial glial stem cell fate (Lehtinen et al. 2011; Paridaen et al. 2013). The daughter cell that inherits the younger centriole will eventually form a new primary cilium. This cilium, however, will extend from the cell's basal process rather than its apical surface and will experience a different array of signals that will influence its development into a progenitor cell or a neuron (Wilsch-Brauninger et al. 2012).

Another mechanism involved in determining the fate of cells derived from an asymmetrical division of a radial glial stem cell is how the apical protein Par-3 is distributed (**FIGURE 14.14A**). In general, Par-3 maintains the apical-basal polarity of cells. In the developing brain, Par-3 recruits a complex in the apical portion of the cell that can segregate cell-fate-inducing factors such as Notch signaling proteins. In an asymmetrical division, one daughter cell receives more Par-3 protein than the other (**FIGURE 14.14B**). The daughter cell receiving more Par-3 develops high Notch signaling activity and remains a stem cell. The other daughter cell expresses high amounts of the Delta protein (recall that Delta is the Notch receptor) and becomes primed for neuronal differentiation (Bultje et al. 2009).

This separation of high and low Notch is directly due to the co-transport of the Notch-inhibitor Numb with Par-3. It may seem counterintuitive to recruit this inhibitor

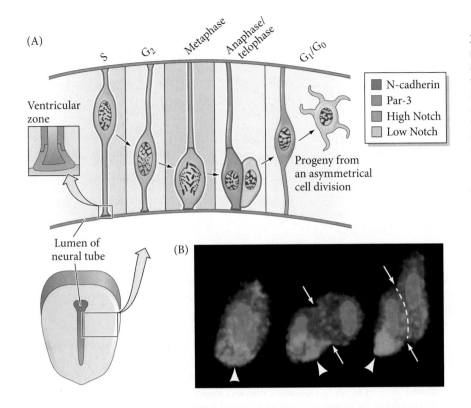

(A)

Ventricular zone

S G_2 Metaphase   Anaphase/telophase G_1/G_0

N-cadherin
Par-3
High Notch
Low Notch

Progeny from an asymmetrical cell division

Lumen of neural tube

(B)

FIGURE 14.14 Asymmetrical division of radial glia mediated by Par3 and Notch. (A) Schematic section of a chick embryo neural tube, showing the position of the nucleus and Par-3 protein in a radial glial cell as a function of the cell cycle. Mitotic cells are found near the inner surface of the neural tube, adjacent to the lumen. The dynamic distribution of Par-3 protein in these luminal stem cells regulates the synthesis of Notch signaling pathway components in the cell membrane of the daughter cells. At mitosis, Par-3 becomes localized primarily to one of the two daughter cells. That daughter cell will express high levels of Notch and remain a stem cell; the cell receiving less Par-3 will express less Notch and become a neural progenitor cell. (B) Fusing the *Par3* gene with GFP enables visualization of Par3 protein movement during division, as seen here in the zebrafish embryonic hindbrain. Par3 (bright green) is isolated primarily to the daughter cell on the left (arrowhead) after an asymmetrical division. As illustrated in (A), this cell will remain a stem cell. (A after Bultje et al. 2009, Lui et al. 2011; B from Alexandre et al. 2010.)

into the stem cell that requires high Notch, but Par-3 actually sequesters and deactivates Numb function. Daughter cells lacking Par3 exhibit freely active Numb, which functions to reduce Notch and thus enable an alternative (Delta-mediated) cell fate (Gaiano et al. 2000; Rasin et al. 2007; Bultje et al. 2009).

Development of the Human Brain

There are many differences between humans and our closest relatives, the chimpanzees and bonobos (Prüfer et al. 2012). These differences include our hairless, sweaty skin and our striding, bipedal posture. Male humans also lack the penile bone and keratinous penile spines that characterize the external genitalia of other primate males. The most striking and significant differences, however, occur in brain development. The enormous growth and asymmetry of the human neocortex and our advanced ability to reason, remember, plan for the future, and learn language and cultural skills make humans unique among the animals (Varki et al. 2008). The development of the human neocortex is strikingly plastic and is an almost constant work in progress. Several developmental phenomena, some of which are shared with other primates, distinguish the development of the human brain from that of other species. These include:

- Cerebral cortical folding
- Activity of human-specific RNA genes
- High levels of transcription
- Human-specific alleles of developmental regulatory genes
- Continuation of brain maturation into adulthood

Fetal neuronal growth rate after birth

If there is one developmental trait that distinguishes humans from the rest of the animal kingdom, it is our retention of the fetal neuronal growth rate. Both human and ape brains have a high growth rate before birth. After birth, however, this rate slows greatly in the apes, whereas human brain growth continues at a rapid rate for about 2 years

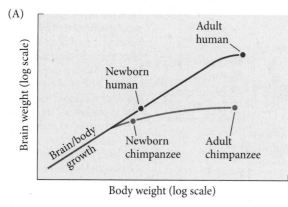

FIGURE 14.15 Brain growth in primates. (A) Whereas other primates (e.g., chimpanzees) attenuate neurogenesis around the time of birth, the generation of neurons in newborn humans occurs at the same rate as in the fetal brain. (B) The brain/body weight ratio (encephalization index) of humans is about 3.5 times higher than that of apes. (After Bogin 1997; see also the more recent quantification done by Herculano-Houzel 2012 and Herculano-Houzel et al. 2015.)

(**FIGURE 14.15A**; Martin 1990; see Leigh 2004). Portmann (1941), Montagu (1962), and Gould (1977) have each made the claim that we are essentially "extrauterine fetuses" for the first year of life.

It has been estimated that during early postnatal development, we add approximately 250,000 neurons per minute (Purves and Lichtman 1985). The ratio of brain weight to body weight at birth is similar for great apes and humans, but by adulthood, the ratio for humans is literally "off the chart" when compared with that of other primates (**FIGURE 14.15B**; Bogin 1997). Indeed, if one follows the charts of ape maturity, human gestation should be 21 months. Our "premature" birth is an evolutionary compromise based on maternal pelvic width, fetal head circumference, and fetal lung maturity. The mechanism for retaining the fetal neuronal growth rate beyond birth has been called **hypermorphosis**, the extension of development beyond its ancestral state (Vrba 1996; Vinicius and Lahr 2003).

In addition to the neurons made after birth, the number of synapses increases by an astronomical number. At the cellular level, no fewer than 30,000 synapses per cm^3 of cortex are formed *every second* during the first few years of human life (Rose 1998; Barinaga 2003). It is speculated that these new neurons and rapidly proliferating neural connections enable plasticity and learning, create an enormous storage potential for memories, and enable us to develop skills such as language, humor, and music—that is, they enable those things that help make us human.

WEB TOPIC 14.3 **NEURONAL GROWTH AND THE INVENTION OF CHILDHOOD** An interesting hypothesis claims that the caloric requirements of early brain growth necessitated a new stage of the human life cycle—childhood—during which the child is actively fed by adults.

Hills raise the horizon for learning

A particularly important feature of the cerebral cortex that is associated with human brain evolution is the number and intricacy of the hills and valleys of the brain—that is, its gyri and sulci (Hofman 1985). There is diversity in the number and complexity of cortical convolutions among mammalian species; for example, the cerebral cortex in humans and elephants is highly folded (**gyrencephalic**), it is only moderately gyrencephalic in ferrets, and it completely lacks folds (is **lissencephalic**) in mice (**FIGURE 14.16**). The amount and complexity of gyrification are usually associated with the level of intelligence[3] and therefore represents a significant adaptation uniquely leveraged in the human brain. What are the mechanisms of cortical folding that may be contributing to the diversity of gyrencephalic brains seen within mammals?

---

[3]This criterion is not exact; dolphins and whales, for example, have a greater amount of folding in the cortex than do humans.

It is not surprising that studying cortical folding is challenging, because it occurs in groups of mammals that are difficult to test in the lab. Recent work on the architecture of the mammalian cortex, however, along with genomic analyses, has begun to unravel the story (reviewed in Lewitus et al. 2013). It is surprising that increased cortical folding is not necessarily associated with an increased number of neurons in the cerebral cortex, although it is correlated with increased surface area of the brain. One study modeled cortical folding in comparison to crumpled paper and demonstrated that when total surface area expands at a faster rate than the thickness of the cortex (or sheets of paper), gyrencephaly will follow (Mota and Herculano-Houzel 2015). In agreement with this finding, larger cerebrums tend to contain more folds than smaller ones. Moreover, in the human disorder pachygyria, the cerebrum has reduced folding and reduced surface area, despite a normal number of neurons (Ross and Walsh 2001).

Cells that could be candidates for providing the mechanical force for creating cerebral folds are the radial glial cells. Remember that in addition to functioning as stem cells, radial glia span the width of the cerebral cortex and provide a structural scaffolding that may generate mechanical forces. Interestingly, there is a greater percentage of proliferative radial glial cells (particularly outer radial glia) in gyrencephalic than in lissencephalic brains. Moreover, in gyrencephalic brains, the distribution and organization of the radial glial cells relative to gyri and sulci are appropriate for providing the tension necessary for folding (**FIGURE 14.17** and Watch Development 14.4; Hansen et al. 2010; Shitamukai et al. 2011; Wang et al. 2011; Pollen et al. 2015). Taken together, the increase of oRG and the biomechanics of their radial fiber arrays provide strong support for the direct involvement of radial glial cells in the evolved mechanisms of cortical folding.

(A) Human

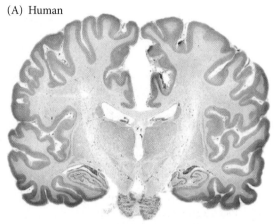

(B) Mouse

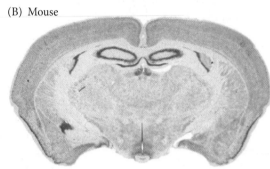

FIGURE 14.16 Transverse sections of the human and mouse brain. Nissl staining marks the nuclei of the gyrencephalic human brain (A) and lissencephalic mouse brain (B). (From Lui et al. 2011.)

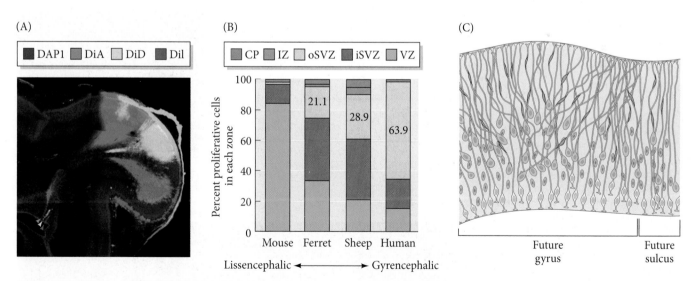

FIGURE 14.17 Characterization of radial glial fanning during cortical folding in the ferret neocortex. (A) Retrograde tracing of radial glial cells in the ferret neocortex over the course of neurogenesis and gyrification. Note the triangle-shaped distribution of the dye-filled cells, indicating a progressive fanning out of the radial fibers along the apicobasal axis. (Arrowheads at the lower left show tight clusters at the luminal surface compared to the extreme width of each dye fill at the pial surface.) (B) Quantification of mitotic cells in the different regions of the cerebrum of lissencephalic and gyrencephalic species. Brains that are more gyrencephalic show higher percentages of proliferative cells in the outer subventricular zone (oSVZ), which is the same region that houses more outer radial glial cells compared to lissencephalic species. (C) The orientation of ventricular (orange) and outer radial glial (brown) fibers in regions that will form a gyrus and sulcus. Outer radial glia are depicted to show more obliquely oriented fibers—a structural organization proposed to support gyrus formation. (A,B after Reillo et al. 2011; C from Lewitus et al. 2013.)

Additional studies looking at whole transcriptomes (total mRNAs expressed by genes in an organism) have found an additional correlation between radial glial cells and cortical folding in humans (Florio et al. 2015; Johnson et al. 2015; Pollen et al. 2015). For instance, a study by Walsh and colleagues (see Johnson et al. 2015) compared the transcriptomes of different radial glia between humans and rodents. They discovered that the outer radial glial cells of humans show differential expression of genes involved in calcium signaling, epithelial-to-mesenchymal transitions, cell migration, and specific activation of the proneural transcriptional regulator neurogenin.

The identification of the *ARHGAP11B* gene focused further attention on the unique role outer radial glial cells may play in human cortex development. This gene is found *only* in humans and is expressed *specifically* in radial glial cells (and not in cortical neurons). When Huttner and colleagues inserted the *ARHGAP11B* gene by electroporation into the developing cortex of the mouse brain (which is normally lissencephalic), the mouse cortex developed folds that resembled gyri (**FIGURE 14.18**). The exact mechanism whereby *ARHGAP11B* expression results in the formation of cortical folds is not clear, but it appears to be connected to a specific and significant increase in the number of outer radial glial cells being produced (Florio et al. 2015). This discovery is of particular significance to our understanding of the evolution of the human brain. The *ARHGAP11B* gene arose in humans from a partial duplication of *ARHGAP11A* (a gene found in animals in general) and arose in the human lineage after early hominids diverged from the chimpanzee lineage (see Figure 14.18A).

Developing Questions

If oRG cells are able to apply tension to the pial surface and promote folding of the cerebral cortex, perhaps something is holding their apical end within the subventricular zone to provide the necessary resistance for tissue folding. What's tethering these oRG cells, given that the hallmark difference between these stem cells and the vRG is the lack of an attachment to the luminal surface? How much pulling force is necessary to fold the cortex?

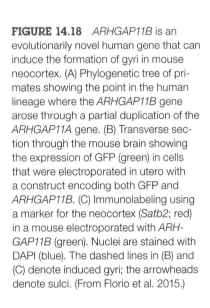

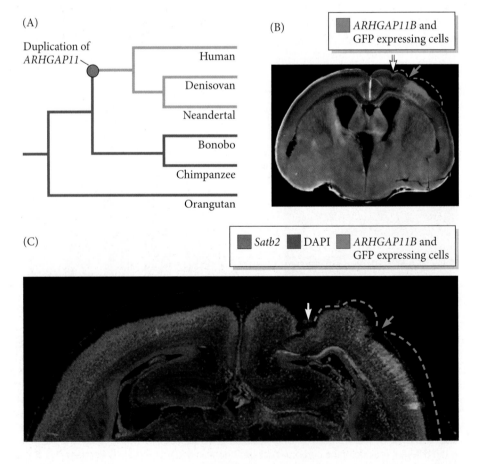

FIGURE 14.18 *ARHGAP11B* is an evolutionarily novel human gene that can induce the formation of gyri in mouse neocortex. (A) Phylogenetic tree of primates showing the point in the human lineage where the *ARHGAP11B* gene arose through a partial duplication of the *ARHGAP11A* gene. (B) Transverse section through the mouse brain showing the expression of GFP (green) in cells that were electroporated in utero with a construct encoding both GFP and *ARHGAP11B*. (C) Immunolabeling using a marker for the neocortex (*Satb2*; red) in a mouse electroporated with *ARHGAP11B* (green). Nuclei are stained with DAPI (blue). The dashed lines in (B) and (C) denote induced gyri; the arrowheads denote sulci. (From Florio et al. 2015.)

WEB TOPIC 14.4 **SPEECH, LANGUAGE, AND THE *FOXP2* GENE** In humans, individuals with mutations of the *FOXP2* gene display severe problems with language. Scientists are studying the possible roles of this gene in the language acquisition abilities of humans versus other primates.

Genes for neuronal growth

What other genes besides *ARHGAP11B* distinguish us from our closest relatives, the chimpanzees and bonobos? Humans and these two non-human primates have remarkably similar genomes. When protein-encoding DNAs are compared, the three genomes are around 99% identical. Protein-coding regions, however, comprise only around 2% of these genomes. When the total genomes are compared, humans and chimpanzees differ at about 4% of their nucleotide sequences, most of the differences occurring in noncoding regions (see Varki et al. 2008). King and Wilson (1975) concluded from their studies of human and chimpanzee proteins that "the organismal differences between chimpanzees and humans would then result chiefly from genetic changes in a few regulatory systems, while amino acid substitutions in general would rarely be a key factor in major adaptive shifts." Theirs was one of the first suggestions that evolution can occur through changes in developmental regulatory genes.

Although there are some brain growth genes (e.g., *ASPM*, also called *microcephalin-5* and *microcephalin-1*) whose DNA sequences differ between humans and apes, these differences have not been correlated with the huge growth of human brains. Rather, the critical differences appear to reside in the sequences that control these genes. These sequences could be in enhancer regions of the DNA or in the DNA that produces noncoding RNAs. Noncoding RNAs are highly expressed in the developing brain and, although not producing any protein products themselves, they may regulate the transcription or translation of neuronal transcription factors. Computer analysis comparing different mammalian genomes may have found such noncoding RNAs to be an important factor in human brain evolution (Pollard et al. 2006a,b; Prabhakar et al. 2006). First, these studies identified a relatively small group of noncoding DNA regions where the sequences were conserved among the non-human mammals studied. This group represents about 2% of the genome, and it was assumed that if these regions have been conserved throughout mammalian evolution, they must be important.

The studies then compared these sequences to their human homologues to see if any of these regions differ between humans and other mammals. About 50 regions were found where the sequence is highly conserved among mammals but has diverged rapidly between humans and chimpanzees. The most rapid divergence is seen in the sequence *HAR1* (*human accelerated region-1*), where 18 sequence changes were seen between chimpanzees and humans. *HAR1* is expressed in the developing brains of humans and apes, especially in the *Reelin*-expressing Cajal-Retzius neurons that are known to be responsible for directing neuronal migration during the formation of the six-layered neocortex (see Figure 14.13). Research is ongoing to discover the function of *HAR1* and the other HAR genes that are in the conserved noncoding region of the genome.

 SCIENTISTS SPEAK 14.2 See a web conference with Dr. Sofie Salama on the *HAR1* gene and human brain development.

A similar search for human-specific DNA deletions in primate genomes found some fascinating candidates. Recalling that the loss of an inhibitor is equivalent to the gain of an activator (think about the Wnt pathway or the double-negative gate in sea urchin blastomeres), McLean and colleagues (2011) uncovered 510 sequences that are present in the genomes of chimpanzees and other mammals but not in humans. One of these deletions is in the forebrain enhancer of the *GADD45G* gene. This gene encodes a growth suppressor that is normally expressed in the ventral forebrain region of chimpanzees and mice, but not humans. When a reporter gene is joined to a chimpanzee *GADD45G* enhancer and inserted into a mouse embryo, *GADD45G* is expressed in the mouse brain. However, when joined to a human *GADD45G* enhancer, it is not expressed in the human brain, evidence that the human *GADD45G* enhancer *does* act to suppress a suppressor (the *GADD45G* gene).

High transcriptional activity

In the 1970s, A. C. Wilson suggested that the difference between humans and chimpanzees might reside in the *amount* of proteins made from their genes (see Gibbons 1998). Today, there is evidence that supports this hypothesis. Using microarrays to study global patterns of gene expression, several recent investigations have found that, although the quantities and types of genes expressed in human and chimpanzee liver and blood are indeed extremely similar, human *brains* produce more than 5 times more mRNA than chimpanzee brains (Enard et al. 2002a; Preuss et al. 2004). In humans, transcription of some genes (such as *SPTLC1*, a gene whose defect causes sensory nerve damage) is elevated 18-fold over the same genes' expression in the chimpanzee cortex. Other genes (such as *DDX17*, whose product is involved in RNA processing) are expressed 10 times *less* in human than in chimpanzee cortices.

Teenage brains: Wired and unchained

Until recently, most scientists thought that in humans, after the initial growth of neurons during fetal development and early childhood, rapid brain growth ceased. However, magnetic resonance imaging (MRI) studies have shown that the brain keeps developing until around puberty and that not all areas of the brain mature simultaneously (Giedd et al. 1999; Sowell et al. 1999). Soon after puberty, brain growth ceases, and pruning of some neuronal synapses occurs. The time of this pruning correlates with the time when language acquisition becomes difficult (which may be why children learn language more readily than adults). There is also a wave of myelin production ("white matter" from the glial cells that surround neuronal axons) in certain areas of the brain at this time. Myelination is critical for proper neural functioning, and although myelination continues throughout adulthood (Lebel and Beaulieu 2011), the greatest differences between brains in early puberty and those in early adulthood involve the frontal cortex (**FIGURE 14.19**; Sowell et al. 1999; Gogtay et al. 2004). These differences in brain development may explain the extreme responses teenagers have to certain stimuli, as well as their ability (or inability) to learn certain tasks.

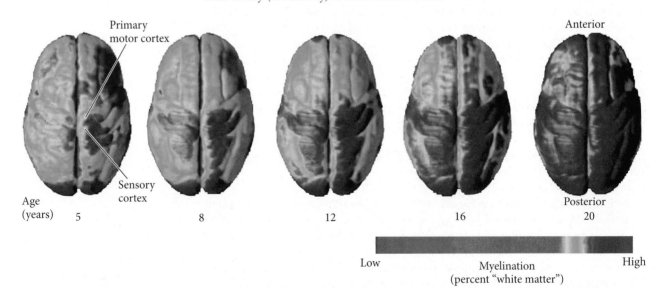

FIGURE 14.19 Dorsal view of the human brain showing the progression of myelination ("white matter") over the cortical surface during adolescence. (Courtesy of N. Gogtay.)

In tests using functional MRI to scan subjects' brains while emotion-charged pictures flashed on a computer screen, the brains of young teenagers showed activity in the amygdala, which mediates fear and strong emotions. When older teens were shown the same pictures, most of their brain activity was centered in the frontal lobe, an area involved in more reasoned perceptions (Baird et al. 1999; Luna et al. 2001). These data come primarily from studies comparing different groups of individuals. Improving technology, however, is beginning to allow assessment of brain maturation of a single individual over time (Dosenbach et al. 2010). The teenage brain is a complicated and dynamic entity that is not easily understood (as any parent knows). Once through the teen years, however, the resulting adult brain is usually capable of making reasoned decisions, even in the onslaught of emotional situations.

Next Step Investigation

Our understanding of the embryonic neural stem cell niche is still quite basic. Unlike the spinal cord, the complexity of the brain has made it more challenging to formulate a comprehensive model that integrates all the components, the cell types, modes of cell division, myriad of molecular regulators, physical forces, gene regulatory networks, and epigenetic modifiers. This challenge increases when adding in the components of time and cell movement. Insight can be gained by studying simpler model systems of brain development, both in invertebrates such as *C. elegans* and *Drosophila*, and in vertebrates such as *Xenopus* and zebrafish.

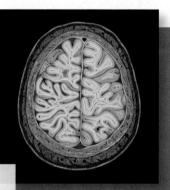

Closing Thoughts on the Opening Photo

The complexities of being human: How deep do they fold? Our big, gyrified brains are part of what makes us human. This photograph is of a sculpture made by Lisa Nilsson, who intricately *folds* colored paper to make anatomical cross-sections of human anatomy—in this case the *folds* of the brain (http://lisanilssonart.com/home.html). Aside from its beauty, this is an artistic representation of the study that modeled cortical folding in comparison to crumpled paper that was discussed in this chapter. Moreover, you have learned that the increase in the number of stem cell types in the cortex and the changes in unique gene expression (e.g. *HAR1* and *ARHGAP11B*) are some of the major contributors to the complexity of the human brain and underlie its evolution. Lastly, throughout adulthood, the human brain continues to grow and develop into the highly myelinated structure that is so remarkable it can create its own structure through art. (Photograph courtesy of Lisa Nilsson.)

14 Snapshot Summary
Brain Growth

1. Dendrites receive signals from other neurons, and axons transmit signals to other neurons. The gap between cells where signals are transferred from one neuron to another (through the release of neurotransmitters) is called a synapse. There is an enormous array of different neuronal morphologies throughout the CNS.

2. Macroglial cell types in the CNS are the astroglia (astrocytes), oligodendrocytes (myelinating cells), and microglia (immune cells of the nervous system).

3. Radial glial cells serve as the neural stem cells in the embryonic and fetal brain. Humans continue making neurons throughout life, although at nowhere near the fetal rate.

4. The neurons of the brain are organized into laminae (layers) and nuclei (clusters).

5. New neurons are formed by the division of neural stem cells (neuroepithelial cells, radial glial cells) in the wall of the neural tube (called the ventricular zone). The resulting newborn neurons can migrate away from the ventricular zone and form a new layer, called the mantle zone (gray matter). Neurons forming later have to migrate through the existing layers. This process forms the cortical layers.

6. Newborn neurons and progenitor cells migrate out of the ventricular zone on the processes of radial glial cells.

7. In the cerebellum, migrating neurons form a second germinal zone, called the external granular layer.

8. The cerebral cortex in mammals, called the neocortex, has six layers. Each layer differs in function and in the type of neurons located there.

9. Ventricular radial glia can give rise to outer radial glial cells that populate the subventricular zone. Both stem cells can also generate intermediate progenitors that are themselves capable of further symmetrical and asymmetrical divisions.

10. In both the cerebellum and cerebrum, secreted Reelin guides migrating neurons to the correct superficial layer in an "inside-out" growth progression; it does so through the regulation of N-cadherin and integrin expression.

11. Asymmetrical partitioning of Par-3 during radial glial cell division restricts active Notch signaling within ventricular radial glia to promote stemness in one daughter cell, while Delta activity in the other daughter supports differentiation.

12. The number and complexity of gyri and sulci (folds) of the neocortex are correlated with level of intelligence. Humans have a highly folded (gyrencephalic) neocortex. Radial glial cells are likely to play a major role in the development of these folds.

13. Human brains appear to differ from those of other primates by their retention of the fetal neuronal growth rate during early childhood, high transcriptional activity of certain genes, the presence of human-specific alleles of developmental regulatory genes, and the loss of transcriptional regulators.

Further Reading

Alexandre, P., A. M. Reugels, D. Barker, E. Blanc and J. D. Clarke. 2010. Neurons derive from the more apical daughter in asymmetric divisions in the zebrafish neural tube. *Nature Neurosci.* 13: 673–679.

Azevedo, F. A. and 8 others. 2009. Equal numbers of neuronal and nonneuronal cells make the human brain an isometrically scaled-up primate brain. *J. Comp. Neurol.* 513: 532–541.

Bentivoglio, M. and P. Mazzarello. 1999. The history of radial glia. *Brain Res. Bull.* 49: 305–315.

Bultje, R. S., D. R. Castaneda-Castellanos, L. Y. Jan, Y. N. Jan, A. R. Kriegstein and S. H. Shi. 2009. Mammalian Par3 regulates progenitor cell asymmetric division via Notch signaling in the developing neocortex. *Neuron* 63: 189–202.

Casper, K. B. and K. D. McCarthy. 2006. GFAP-positive progenitor cells produce neurons and oligodendrocytes throughout the CNS. *Mol. Cell. Neurosci.* 31: 676–684.

Erikksson, P. S., E. Perfiliea, T. Björn-Erikksson, A.-M. Alborn, C. Nordberg, D. A. Peterson and F. H. Gage. 1998. Neurogenesis in the adult human hippocampus. *Nature Med.* 4: 1313–1317.

Fuentealba, L. C., S. B. Rompani, J. I. Parraguez, K. Obernier, R. Romero, C. L. Cepko and A. Alvarez-Buylla. 2015. Embryonic origin of postnatal neural stem cells. *Cell* 161: 1644–1655.

Hatten, M. E. 1990. Riding the glial monorail: A common mechanism for glial-guided neuronal migration in different regions of the mammalian brain. *Trends Neurosci.* 13: 179–184.

Kwan, K. Y., N. Sestan and E. S. Anton. 2012. Transcriptional co-regulation of neuronal migration and laminar identity in the neocortex. *Development* 139: 1535–1546.

Lewitus, E., I. Kelava and W. B. Huttner. 2013. Conical expansion of the outer subventricular zone and the role of neocortical folding in evolution and development. *Front. Hum. Neurosci.* 7: 424.

Mota, B. and S. Herculano-Houzel. 2015. Cortical folding scales universally with surface area and thickness, not number of neurons. *Science* 349: 74–77.

McLean, C. Y. and 12 others. 2011. Human-specific loss of regulatory DNA and the evolution of human-specific traits. *Nature* 471: 216–219.

Paridaen, J. T., M. Wilsch-Brauninger and W. B. Huttner. 2013. Asymmetric inheritance of centrosome-associated primary cilium membrane directs ciliogenesis after cell division. *Cell* 155: 333–344.

Paridaen, J. T. and W. B. Huttner. 2014. Neurogenesis during development of the vertebrate central nervous system. *EMBO Rep.* 15: 351–364.

Pollen, A. A. and 13 others. 2015. Molecular identity of human outer radial glia during cortical development. *Cell* 163: 55–67.

Sekine, K. and 7 others. 2012. Reelin controls neuronal positioning by promoting cell-matrix adhesion via inside-out activation of integrin α5β1. *Neuron* 76: 353–369.

Trommsdorff, M. and 8 others. 1999. Reeler/Disabled-like disruption of neuronal migration in knockout mice lacking the VLDL receptor and ApoE receptor 2. *Cell* 97: 689–701.

Varki, A., D. H. Geschwind and E. E. Eichler. 2008. Explaining human uniqueness: Genome interactions with environment, behaviour, and culture. *Nature Rev. Genet.* 9: 749–763.

GO TO WWW.DEVBIO.COM...

...for Web Topics, Scientists Speak interviews, Watch Development videos, Dev Tutorials, and complete bibliographic information for all literature cited in this chapter.

Neural Crest Cells and Axonal Specificity

CONTINUING THE DISCUSSION OF ECTODERMAL DEVELOPMENT, this chapter focuses on two remarkable entities: (1) the neural crest, whose cells generate the facial skeleton, pigment cells, and peripheral nervous system; and (2) nerve axons, whose growth cones guide them to their destinations. Neural crest cells and axon growth cones share at least two core features: both are motile, and both invade tissues that are external to the nervous system.

The Neural Crest

Although it is derived from the ectoderm, the neural crest is so important that it has sometimes been called the "fourth germ layer" (see Hall 2009). It has even been said—somewhat hyperbolically—that "the only interesting thing about vertebrates is the neural crest" (Thorogood 1989). Certainly the emergence of the neural crest is one of the pivotal events of animal evolution, as it led to the jaws, face, skull, and bilateral sensory ganglia of the vertebrates (Northcutt and Gans 1983).

The neural crest is a transient structure. Adults do not have a neural crest, nor do late-stage vertebrate embryos. Rather, the cells of the neural crest undergo an epithelial-to-

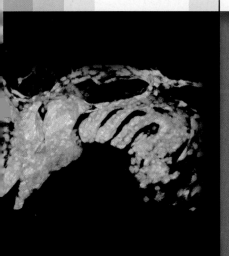

Destined to be a face?

The Punchline

Neural crest cells and axon growth cones both migrate far from their source of origin to specific places in the embryo. In the process, they must recognize and respond to signals that guide them along specific routes to their final destination. Neural crest cells arise from the crest of the neural tube, leaving this position either as a community or as individual cells. These multipotent stem cells navigate through pathways in the trunk and head, differentiating into such diverse cell types as neurons, smooth muscle, pigment, and cartilage. Newly born neurons extend a growing axonal process led by a motile growth cone tip, which navigates through the embryonic environment to form a synapse with its target cell. Both neural crest cells and axonal growth cones use transmembrane receptors to interpret short- and long-range guidance cues. These cues produce cytoskeletal changes that result in the attraction or repulsion of the cell as it travels. Such guidance molecules include members of the stromal-derived factor, ephrin, Slit, and semaphorin families. Equally important are the noncanonical repurposing of common morphogens as guidance cues, and the use of neurotrophins for neuronal survival.

FIGURE 15.1 Neural crest cell migration. (A) The neural crest is a transient structure dorsal to the neural tube. Neural crest cells (stained blue in this micrograph) undergo an epithelial-to-mesenchymal transition from the dorsalmost portion of the neural tube (the top of this micrograph). (B) When the skin has been removed from the dorsal surface of a vertebrate embryo, the neural crest cells (here computer-colored gold against the purple somites) can be seen as a collection of mesenchymal cells above the neural tube. (C) Illustration of the sequential steps of neural crest development, starting with their specification at the border of the neural plate (1) and subsequent location at the apex of neural folds (2), followed by their delamination at the point of neural tube closure (3), and final migration out of ectodermal tissues (4). (A courtesy of J. Briscoe; B courtesy of D. Raible.)

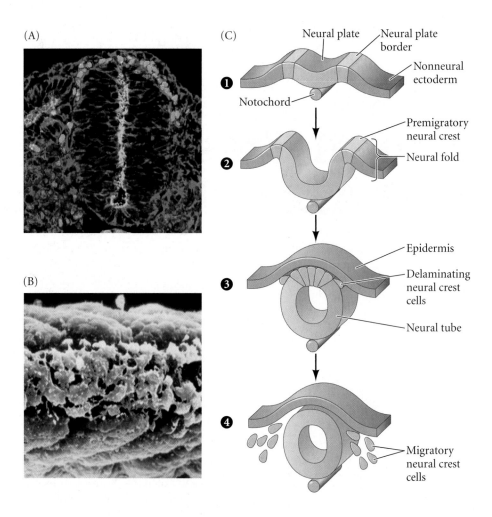

mesenchymal transition from the dorsal neural tube, after which they migrate extensively along the anterior-posterior axis and generate a prodigious number of differentiated cell types (**FIGURE 15.1**; **TABLE 15.1**).

TABLE 15.1 Some derivatives of the neural crest	
Derivative	**Cell type or structure derived**
Peripheral nervous system (PNS)	Neurons, including sensory ganglia, sympathetic and parasympathetic ganglia, and plexuses Neuroglial cells Schwann cells and other glial cells
Endocrine and paraendocrine derivatives	Adrenal medulla Calcitonin-secreting cells Carotid body type I cells
Pigment cells	Epidermal pigment cells
Facial cartilage and bones	Facial and anterior ventral skull cartilage and bones
Connective tissue	Corneal endothelium and stroma Tooth papillae Dermis, smooth muscle, and adipose tissue of skin, head, and neck Connective tissue of salivary, lachrymal, thymus, thyroid, and pituitary glands Connective tissue and smooth muscle in arteries of aortic arch origin

Source: After Jacobson 1991, based on multiple sources.

Regionalization of the Neural Crest

The **neural crest** is a population of cells that can produce tissues as diverse as (1) the neurons and glial cells of the sensory, sympathetic, and parasympathetic nervous systems; (2) the epinephrine-producing (medulla) cells of the adrenal gland; (3) the pigment-containing cells of the epidermis; and (4) many of the skeletal and connective tissue components of the head. The crest can be divided into four main (but overlapping) anatomical regions, each with characteristic derivatives and functions (**FIGURE 15.2**):

1. **Cranial**, or **cephalic**, **neural crest** cells migrate to produce the craniofacial mesenchyme, which differentiates into the cartilage, bone, cranial neurons, glia, pigment cells, and connective tissues of the face. These cells also enter the pharyngeal arches[1] and pouches to give rise to thymic cells, the odontoblasts of the tooth primordia, and the bones of the middle ear and jaw.

2. The **cardiac neural crest** is a subregion of the cranial neural crest and extends from the otic (ear) placodes to the third somites (Kirby 1987; Kirby and Waldo 1990). Cardiac neural crest cells develop into melanocytes, neurons, cartilage, and connective tissue (of the third, fourth, and sixth pharyngeal arches). This region of the neural crest also produces the entire muscular-connective tissue wall of the large arteries (the "outflow tracts") as they arise from the heart; it also contributes to the septum that separates pulmonary circulation from the aorta (Le Lièvre and Le Douarin 1975; Sizarov et al. 2012).

3. **Trunk neural crest** cells take one of two major pathways. One migratory pathway takes trunk neural crest cells ventrolaterally through the anterior half of each somitic sclerotome. **Sclerotomes**, derived from the somites, are blocks of mesodermal cells that will differentiate into the vertebral cartilage of the spine (see Chapter 17). Trunk neural crest cells that remain in the sclerotomes form the **dorsal root ganglia**[2] containing the sensory neurons. Cells that continue traveling more ventrally form the sympathetic ganglia, the adrenal medulla, and the nerve clusters surrounding the aorta. The second major migratory path for trunk neural crest cells proceeds dorsolaterally, allowing the precursors of melanocytes to move through the dermis of the skin from the dorsum to the belly (Harris and Erickson 2007).

4. The **vagal** and **sacral neural crest** cells generate the **parasympathetic (enteric) ganglia** of the gut (Le Douarin and Teillet 1973; Pomeranz et al. 1991). The vagal (neck) neural crest overlaps the cranial/trunk crest boundary, lying opposite chick somites 1–7, while the sacral neural crest lies posterior to somite 28. Failure of neural crest cell migration from these regions to the colon results in the absence of enteric ganglia and thus to the absence of peristaltic movement in the bowels (Hirschprung disease; see pp. 477–479).

Trunk neural crest cells and cranial neural crest cells are not equivalent. Cranial crest cells can form cartilage, muscle, and bone as well connective tissue of the cornea, whereas trunk crest cells cannot. When trunk neural crest cells are transplanted into the head region, they can migrate to the sites of cartilage and cornea formation, but they make neither

---

[1] The pharyngeal (branchial) arches (see Figure 1.8) are outpocketings of the head and neck region into which cranial neural crest cells migrate. The pharyngeal pouches form between these arches and become the thyroid, parathyroid, and thymus.

[2] Recall from Chapter 13 that ganglia are clusters of neurons whose axons form a nerve.

FIGURE 15.2 Regions of the chick neural crest. The cranial neural crest migrates into the pharyngeal arches and the face to form the bones and cartilage of the face and neck. It also contributes to forming the cranial nerves. The vagal neural crest (near somites 1–7) and the sacral neural crest (posterior to somite 28) form the parasympathetic nerves of the gut. The cardiac neural crest cells arise near somites 1 through 3; they are critical in making the division between the aorta and the pulmonary artery. Neural crest cells of the trunk (from about somite 6 through the tail) make sympathetic neurons and pigment cells (melanocytes), and a subset of them (at the level of somites 18–24) forms the medulla portion of the adrenal gland. (After Le Douarin 1982.)

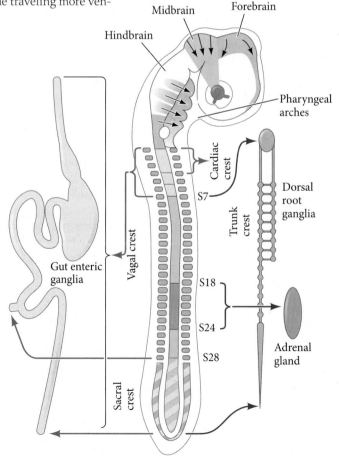

cartilage nor cornea (Noden 1978; Nakamura and Ayer-Le Lievre 1982; Lwigale et al. 2004). However, both cranial and trunk neural crest cells can generate neurons, melanocytes, and glia. The cranial neural crest cells that normally migrate into the eye region to become cartilage cells can form sensory ganglion neurons, adrenomedullary cells, glia, and Schwann cells if the cranial region is transplanted into the trunk region (Noden 1978; Schweizer et al. 1983).

The inability of the trunk neural crest to form skeleton is most likely due to the expression of Hox genes in the trunk neural crest. If Hox genes are expressed in the cranial neural crest, these cells fail to make skeletal tissue; if trunk crest cells lose Hox gene expression, they can form skeleton. Moreover, if transplanted into the trunk region, cranial crest cells participate in forming trunk cartilage that normally does not arise from neural crest components. This ability to form bone may have been a primitive property of the neural crest and may have been critical for forming the bony armor found in several extinct fish species (Smith and Hall 1993). In other words, rather than the cranial crest *acquiring* the ability to form bone, the trunk crest has apparently *lost* this ability. McGonnell and Graham (2002) showed that bone-forming capacity may still be latent in the trunk neural crest: if cultured with certain hormones and vitamins, trunk crest cells become capable of forming bone and cartilage when placed into the head region. Moreover, Abzhanov and colleagues (2003) have shown that the trunk crest cells can act like cranial crest cells (and make skeletal tissue) if the trunk cells are cultured in conditions that cause them to lose the ability to express their Hox genes.

So even though the cells of the cranial neural crest and trunk neural crest are multipotent (a cranial crest cell can form neurons, cartilage, bone, and muscles; a trunk neural crest cell can form glia, pigment cells, and neurons), they have different repertoires of cell types that they can generate under normal conditions.[3]

 DEV TUTORIAL *Neural crest cell development*   Dr. Michael J.F. Barresi describes the remarkable journey that neural crest cells undertake to find their destined tissues outside the CNS and differentiate into a variety of distinct cell types.

Neural Crest: Multipotent Stem Cells?

There has long been controversy as to whether the majority of the individual cells leaving the neural crest are multipotent or whether most are already restricted to certain fates. Bronner-Fraser and Fraser (1988, 1989) provided the first evidence that many individual trunk neural crest cells are multipotent as they leave the crest. They injected fluorescent dextran molecules into individual chick neural crest cells while the cells were still within the neural tube, then recorded what types of cells their descendants became after migration. The progeny of a single neural crest cell could become sensory neurons, melanocytes (pigment-forming cells), glia (including Schwann cells), and cells of the adrenal medulla. Other studies suggested that the initial avian trunk neural crest population was a heterogeneous mixture of precursor cells, and that nearly half of the cells that emerge from the neural crest only generate a single cell type (Henion and Weston 1997; Harris and Erickson 2007).

With the advent of better lineage-tracing methods, this controversy may now be over. Researchers in the Sommers' lab have used the "confetti" mouse model[4] to trace the journey of individual trunk neural crest cells and their progeny from both the premigratory and migratory stages (Baggiolini et al. 2015). Tracings of nearly 100 cell clones of premigratory and migratory neural crest cells showed that approximately

---

[3]To learn the role that neural crest plays in tooth, hair, and cranial nerve development, seek out Chapter 16.

[4]We introduced Brainbow-based fate mapping in Chapter 2 and described the utility of the "confetti" mouse model for lineage tracing in Chapter 5.

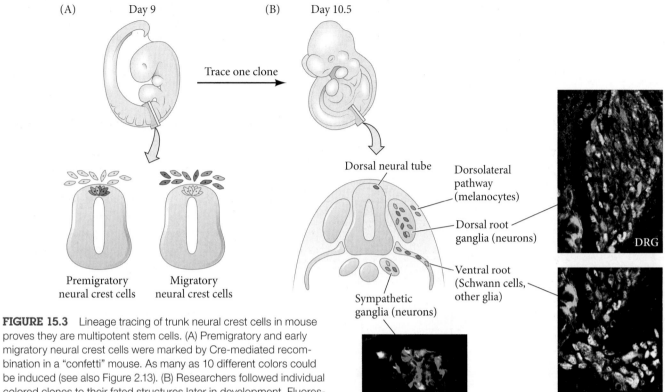

FIGURE 15.3 Lineage tracing of trunk neural crest cells in mouse proves they are multipotent stem cells. (A) Premigratory and early migratory neural crest cells were marked by Cre-mediated recombination in a "confetti" mouse. As many as 10 different colors could be induced (see also Figure 2.13). (B) Researchers followed individual colored clones to their fated structures later in development. Fluorescent cells were seen in the dorsolateral pathway where melanocytes differentiate; in the dorsal root ganglia (DRG), as part of the Schwann cell population on the ventral root, and in the sympathetic ganglia (SG). Micrographs show the tracing of premigratory cells labeled by the Wnt1-CreERT driver, which shows a unique YFP/RFP fluorescent combination in three different peripheral structures visualized with cell type-specific markers. (After Baggiolini et al. 2015.)

75% of them proliferated, and their progeny showed multiple types of lineages that differentiated into different cell types: dorsal root ganglia, sympathetic ganglia, Schwann cells wrapping the ventral nerve root, and melanocytes (**FIGURE 15.3**). Although a small population of these mapped neural crest cells appeared to be unipotent, the large majority displayed multipotency throughout their migration—a finding strongly suggesting that mouse embryonic trunk neural crest cells are multipotent stem cells, an important step forward in the field of neural crest research.

The next logical question is whether cephalic neural crest cells are equally multipotent. It appears that a majority of early migrating cranial neural crest cells from the chick can generate multiple cell types (Calloni et al. 2009), but it is not yet certain whether they are in fact multipotent stem cells.

Nicole Le Douarin and others proposed a model of neural crest development that likely still holds true, even in light of the new information about multipotency. In this model, an original multipotent neural crest cell divides and progressively refines its developmental potentials (**FIGURE 15.4**; see Creuzet et al. 2004; Martinez-Morales et al. 2007; Le Douarin et al. 2008). To test this model directly, an individual neural crest cell would need to be exposed to different environments to determine the array of different cell types it could generate.

Developing Questions

It seems we now can be confident that trunk neural crest cells are multipotent stem cells. Is it possible that some small population of these cells, dispersed widely like novel developmental seeds, could be retained as adult stem cells in each of their final destinations? The ontogeny of adult stem cells is unknown for many tissues. The migratory and multipotent nature of neural crest cells suggests the hypothesis that they may be capable of seeding these adult tissues.

VADE MECUM

The segment on Dr. Nicole Le Douarin's work shows original footage of the experimental techniques and results of her work on neural crest cell reorganization, migration, and differentiation.

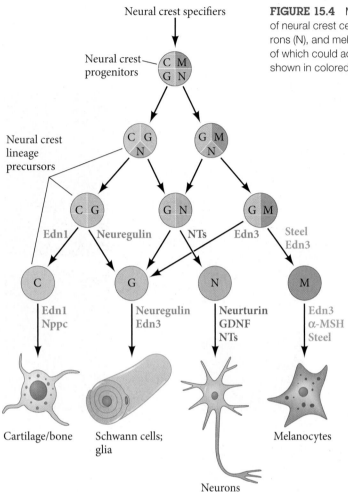

Neural crest specifiers

Neural crest progenitors

Neural crest lineage precursors

Edn1   Neuregulin   NTs   Edn3   Steel
Edn3

C
Edn1
Nppc

G
Neuregulin
Edn3

N
Neurturin
GDNF
NTs

M
Edn3
α-MSH
Steel

Cartilage/bone

Schwann cells; glia

Neurons

Melanocytes

FIGURE 15.4 Model for neural crest lineage segregation and the heterogeneity of neural crest cells. The committed precursors of cartilage/bone (C), glia (G), neurons (N), and melanocytes (M) are derived from intermediate progenitor cells, some of which could act as stem cells. The paracrine factors regulating these steps are shown in colored type. NTs, neurotropin(s). (After Martinez-Morales et al. 2007.)

Specification of Neural Crest Cells

Although neural crest cells are not identifiable until they emigrate from the neural tube, induction of these cells first occurs during early gastrulation, at the border between the presumptive epidermis and the presumptive neural plate (the region that will form the central nervous system). Specification of the neural crest at the neural plate-epidermis boundary is a multistep process (see Huang and Saint-Jeannet 2004; Meulemans and Bronner-Fraser 2004). The first step appears to be the specification of the neural plate border. The cells in this border between the neural plate and the epidermis will become the neural crest and (in the anterior region) the **placodes**—thickenings in the surface ectoderm that will generate the eye lens, inner ear, olfactory epithelium, and other sensory structures (see Chapter 16). In amphibians, the border appears to be specified by the interplay between a number of **neural plate inductive signals**, including BMPs, Wnts, and FGFs. Indeed, in the 1940s, Raven and Kloos (1945) showed that although the presumptive notochord could induce both the amphibian neural plate and neural crest tissue (presumably blocking nearly all BMPs), the somite and lateral plate mesoderm could induce only the neural crest. In chick embryos, neural crest specification occurs during gastrulation, when the borders between the neural and nonneural ectoderm are still forming (Basch et al. 2006; Schmidt et al. 2007; Ezin et al. 2009). Here, the neural plate inductive signals (especially BMPs and Wnts) secreted from the ventral ectoderm and paraxial mesoderm interact to specify the boundaries.

In the anterior region, the timing of BMP and Wnt expression is critical for discriminating between neural plate, epidermis, placode, and neural crest tissues (**FIGURE 15.5**). As we saw in Chapters 11 and 12, if both BMP and Wnt signaling are continuous, the fate of the ectoderm is epidermal; but if BMP antagonists (e.g., Noggin or FGFs) block BMP signaling, the ectoderm becomes neural. Studies by Patthey and colleagues (2008, 2009) have shown that if Wnts induce BMPs and then Wnt signaling is turned *off*, the cells become committed to be anterior placodes, whereas if the Wnt signaling induces BMPs but stays *on*, the cells become capable of becoming neural crest.

The last few decades of research on neural crest specification has helped to assemble the gene regulatory network (GRN) involved in the maturation of neural crest cells (**FIGURE 15.6**). This GRN begins with Wnt and BMP inducing the expression of a set of transcription factors in the ectoderm (including Gbx2, Zic1, Msx1, and Tfap2), which in turn regulate the **neural plate border specifiers**. These specifiers, including Pax3/7 and dlx5/6, collectively confer upon the border region the ability to form neural crest as well as dorsal neural tube cell types. The border-specifying transcription factors then induce a second set of more specific transcription factors, the **neural crest specifiers**, in those cells that are to become the neural crest. These neural crest specifiers include genes encoding the transcription factors FoxD3, Sox9, Snail (premigratory), and Sox10 (migratory) (Simões-Costa and Bronner 2015).

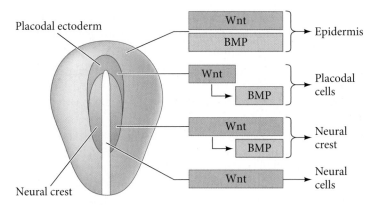

FIGURE 15.5 Specification of neural crest cells. The neural plate is bordered by neural crest anteriorly and caudally and by placodal ectoderm anteriorly. If the ectodermal cells receive both BMP and Wnt for an extended period of time, they become epidermis. If Wnt induces BMPs and is then down-regulated, the cells become placodal cells (expressing the placode specifier genes *Six1*, *Six4*, and *Eya2*). If Wnt induces BMP but remains active, these border cells between the neural plate and epidermis become neural crest (expressing neural crest specifier genes *Pax7*, *Snail2*, and *Sox9*). If they receive Wnt only (because the BMP signal is blocked by Noggin or FGF), the ectodermal cells become neural cells. (After Patthey et al. 2009.)

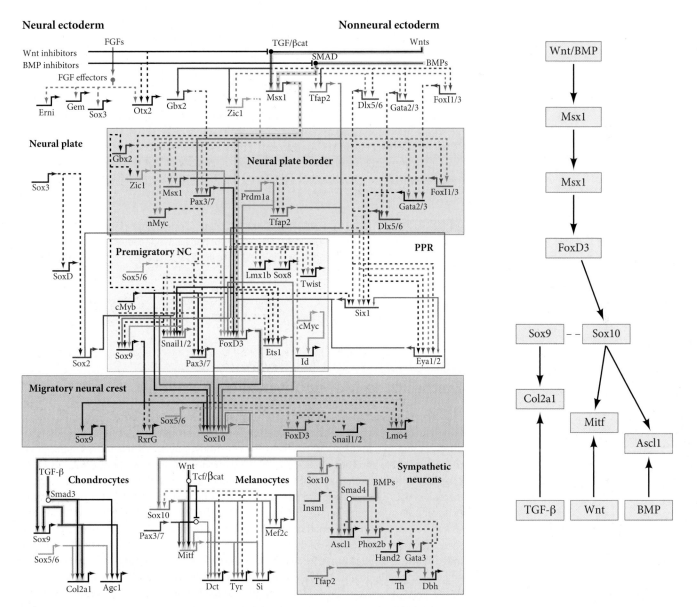

FIGURE 15.6 The gene regulatory network for neural crest development. This GRN is a compilation of data from a variety of vertebrate organisms. One of the most significant circuits (highlighted in yellow) shows gene expression in cells from early ectoderm (top) to the derived cells types (bottom). This circuit is replicated in a simpler linear flow chart to the right. (Note: Not all derived cell types are illustrated.) (After Simões-Costa and Bronner 2015.)

When Foxd3, Snail, Sox9, and Sox10 are experimentally expressed in the lateral neural tube, the lateral neuroepithelial cells become neural crest-like, undergo epithelial-to-mesenchymal transition (EMT) and delaminate from the neuroepithelium:

- Sox9 and Snail together are sufficient to induce EMT in neuroepithelial cells. Sox9 is also required for the survival of trunk neural crest cells after delamination (in the absence of Sox9, neural crest cells undergo apoptosis as soon as they delaminate).

- Foxd3 may play many roles. It is needed for the expression of the cell surface proteins needed for cell migration, and it also appears to be critical for the specification of ectodermal cells as neural crest. Inhibiting expression of the *Foxd3* gene inhibits neural crest differentiation. Conversely, when *Foxd3* is expressed ectopically by electroporating the active gene into neural plate cells, those neural plate cells express proteins characteristic of the neural crest (Nieto et al. 1994; Taneyhill et al. 2007; Teng et al. 2008).

- The *Sox10* gene appears to be one of the most critical regulators of neural crest specification. It is crucial not only for the delamination of neural crest cells from the neural tube, but also for the differentiation of the numerous neural crest lineages (Kelsh 2006; Betancur et al. 2010). Sox10 protein binds to the enhancers of numerous target genes that encode the neural crest effectors; these include the genes for some small G proteins, such as Rho GTPases, that allow cells to change shape and migrate; cell surface receptors, such as receptor tyrosine kinases and endothelin receptor (e.g., ENDRB2), that allow the neural crest cells to respond to patterning and chemotactic proteins in their environments; and transcription factors, such as MITF in the melanocyte lineage that forms pigment cells (see Figure 15.6; Simões-Costa and Bronner 2015).

WEB TOPIC 15.1 **INDUCED NEURAL CREST CELLS** Understanding the gene regulatory network controlling neural crest development has allowed researchers to reprogram human fibroblasts into induced neural crest cells (iNCCs) to study their development and role in disease.

 SCIENTISTS SPEAK 15.1 Dr. Marianne Bronner talks about the gene regulatory network of neural crest development in regard to evolution.

Neural Crest Cell Migration: Epithelial to Mesenchymal and Beyond

The environment through which neural crest cells migrate differs along the anterior to posterior axis, which results in different types of journeys for neural crest cells from different regions. Like cars in traffic, these cells must navigate their routes using environmental cues, and their passage is affected by the cells around them (**FIGURE 15.7**). Like cars using their engines and wheels to move on the road, the cells use the protrusive forces of their cytoskeletons to extend lamellipodia, reaching out to grab hold of the extracellular matrix in front of them while releasing the brakes from behind. A car can scoot freely down an open road or may have to travel collectively in traffic, being responsive to the pace and distance of neighboring cars. Similarly, neural crest cells can migrate individually as well as move as a collective cluster of cells responsive to the distance of neighboring cells. Just as traffic officers, structured barriers, and street signs guide car traffic, local adhesive cues and long-range secreted factors displayed in gradients guide migrating cells through an embryonic environment. And, like drivers who cannot see ahead to the final turn into their parking spot, migrating cells must make decisions in an incremental way, moving from one turn to the next to reach their final destination. Consider this analogy as you continue reading about neural crest cell migration.

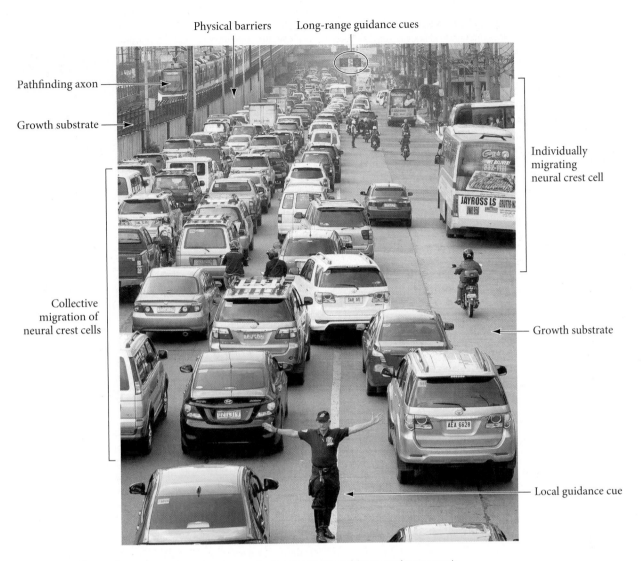

Physical barriers   Long-range guidance cues

Pathfinding axon

Growth substrate

Individually
migrating
neural crest cell

Collective
migration of
neural crest cells

Growth substrate

Local guidance cue

FIGURE 15.7 Analogy of neural crest and axonal migration to the guidance and movement of traffic. See text for narrative. (Photo © Alexis Corpuz.)

Delamination

After cell specification, the first visible indication of neural crest cells is their epithelial-to-mesenchyme transition (EMT) in preparation for leaving the neural tube. Neural crest cells lose their adhesive junctions and separate from the epithelium in a process known as **delamination** (**FIGURE 15.8**). The timing of neural crest delamination is controlled by the neural tube's environment. The trigger for the EMT appears to be the activation of the Wnt genes by BMPs. The BMPs (which can be produced by the dorsal region of the neural tube; see Chapters 13 and 14) are held in check by Noggin produced by the notochord and somites. When Noggin expression is reduced, BMPs can function and activate EMT in the neural crest cells (Burstyn-Cohen et al. 2004).

Prior to delamination, the different regions of ectoderm in the area of the neural crest can be identified by their expression of different cell-cell adhesion molecules: surface

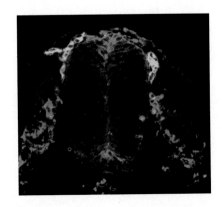

FIGURE 15.8 Migrating neural crest cells are stained red by HNK-1 antibody, which recognizes a cell surface carbohydrate involved in neural crest cell migration. RhoB protein (green stain) is expressed in cells as they delaminate. Cells expressing both HNK-1 and RhoB appear yellow. (From Liu and Jessell 1998, courtesy of T. M. Jessell.)

FIGURE 15.9 Neural crest delamination and migration by contact inhibition. The process of neural crest delamination is shown here at the time when the neural and surface ectoderms have separated and are both in the process of fusing at the midline into the neural tube and epidermis respectively. BMP and Wnt signals specify the three major regions of the neuroepithelium, which are distinguished by their expression of unique adhesion proteins: surface ectoderm (E-cadherin), neural tube (N-cadherin), and the premigratory neural crest (cadherin-6B). In the premigratory domain, BMP levels are the highest, with Wnt at intermediate amounts; this situation supports the upregulation of *Snail-2* (and *Zeb-2*) in these cells. Snail-2 proteins repress N-cadherin and E-cadherin in this domain. Cadherin-6B is upregulated only in the apical half of premigratory neural crest cells, and functions to activate RhoA and actomyosin contractile fibers for apical constriction and the initiation of delamination. Noncanonical Wnt signaling (not shown) establishes the polar activity of RhoA (red) and Rac1 (yellow) along the migratory axis of migrating neural crest cells. When neural crest cells contact one another, they experience contact inhibition, during which they will stop, turn, and migrate away in the opposite direction.

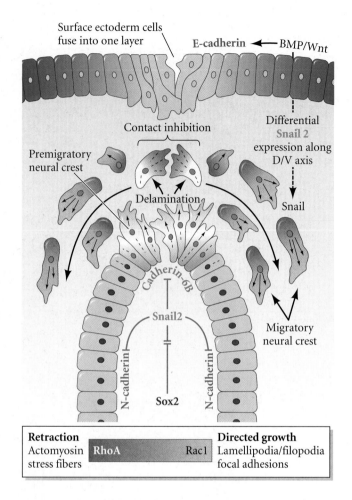

ectoderm expresses E-cadherin, premigratory neural crest expresses cadherin 6B, and the neural tube expresses N-cadherin (**FIGURE 15.9**). The Wnt and BMP signals lead to the expression of the "core EMT regulatory factors" (e.g., Snail-2, Zeb-2, Foxd3, and Twist) in the delaminating premigratory neural crest. Sox2 is expressed by the cells of the neural ectoderm (plate/fold/tube) and functions in part to transcriptionally repress *Snail-2* expression, whereas the more dorsally expressed *Snail-2* in the premigratory neural crest region cross-represses *Sox2* expression (reviewed in Duband et al. 2015). As seen in the patterning of the ventral neural tube (see Chapter 14), this cross-transcriptional repression helps to refine the boundaries between the neural tube epithelium (N-cadherin), premigratory neural crest (cadherin 6B), and surface ectoderm (E-cadherin); see Figure 15.9.

Noncanonical Wnt signaling is critical for activating the small Rho GTPases in premigratory neural crest cells. These Rho GTPases function to (1) facilitate the expression of *Foxd3* and the Snail family genes and (2) establish the cytoskeletal conditions for migration by promoting actin polymerization into microfilaments and the attachment of these microfilaments to focal adhesions in the cell membrane (Hall 1998; De Calisto et al. 2005).

The crest cells cannot leave the neural tube as long as they are tightly connected to one another. Snail has been shown to directly downregulate the expression of cadherin-6B and the tight junction proteins that bind epithelial cells together (see Figure 15.9). In zebrafish, cadherin-6 is transiently maintained only at the apical end of the delaminating cell, which allows RhoA to construct the actomyosin contractile fibers for apical constriction and the commencement of delamination and migration (Clay and Halloran 2014). After migration, neural crest cells re-express cadherins, as demonstrated where they aggregate to form the dorsal root and sympathetic ganglia (Takeichi 1988; Akitaya and Bronner-Fraser 1992; Coles et al. 2007).

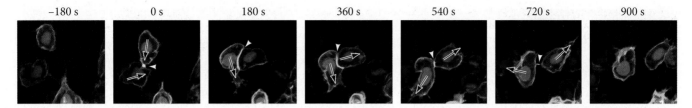

−180 s   0 s   180 s   360 s   540 s   720 s   900 s

The driving force of contact inhibition

The pushing out of neural crest cells from the dorsal neural tube appears to be facilitated by fellow neural crest cells (Abercrombie 1970; Carmona-Fontaine et al. 2008). The phenomenon known as **contact inhibition of locomotion** occurs when two migrating cells make contact. Resulting depolymerizing changes in each cell's cytoskeleton halt the protrusive activity along the contacting cell surfaces, and new protrusive extensions form away from the point of contact (**FIGURE 15.10**; Carmona-Fontaine et al. 2008; Scarpa et al. 2015). As you might imagine, this behavior can cause cells to disperse, a behavior exhibited by other cells as well, such as Cajal-Retzius cells in the cortex of the brain (see Chapter 14; Villar-Cerviño et al. 2013). However, wherever neural crest cells are in close contact with each other as they stream away from the dorsal neural tube, contact inhibition will repress protrusive activity on all sides of the cells, except for those at the leading edge of the stream (Roycroft and Mayor 2016). The mechanism for this contact inhibition involves Wnt and RhoA. On the sides where neural crest cells make contact with each other (but not with another cell type), proteins of the noncanonical Wnt-PCP pathway assemble and activate RhoA (see Figure 4.34), which in turn disaggregates the cytoskeletons of the lamellipodia responsible for migration (see Figure 15.9). Thus, polarization of Wnt-mediated activity resulting in contact inhibition leads to the directed migration of neural crest cells both as individual cells streaming in the trunk regions and as a collective group, a behavior most often seen in cranial neural crest cells (Mayor and Theveneau 2014).

Collective migration

Traveling as part of a group with multiple cars destined for a similar location is a different experience than traveling alone on the open road. Being part of the group requires cooperation and sticking together during the journey. A similar pattern of cell migration in the embryo is called **collective migration** (**FIGURE 15.11**).

Both epithelial and mesenchymal cells can migrate collectively, with the cells on the leading edge guiding and driving the movement of the cluster (Scarpa and Mayor 2016). In neural crest cell migration, cranial neural crest cells typically undergo collective migration and can do so even in culture (Alfandari et al. 2003; Theveneau et al. 2010). This ability suggests that external factors such as

FIGURE 15.10 Migrating neural crest cells demonstrate contact inhibition of locomotion in a live zebrafish embryo. A time series of neural cells that have been made to express *mCherry* in the nucleus (red) and green fluorescent protein in the cell membrane (blue). After membrane contact (yellow; arrowhead), the two cells send protrusions away from the location of contact. Arrows denote the direction of cell movement. (From Scarpa et al. 2015.)

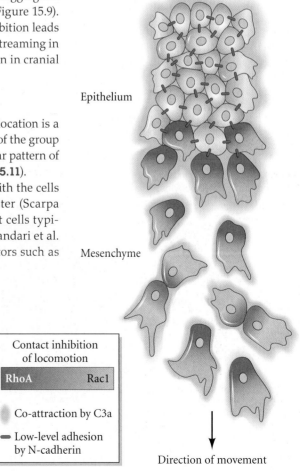

Epithelium

Mesenchyme

Contact inhibition
of locomotion

RhoA   Rac1

Co-attraction by C3a

Low-level adhesion
by N-cadherin

Direction of movement

FIGURE 15.11 Model of collective migration of neural crest cells. Some populations of neural crest cells are known to migrate collectively as a large group of cells. This "collective migration" requires some amount of cell-to-cell adhesion, which is mediated by a low level N-cadherin expression (blue receptors). Additionally, collectively migrating neural crest cells will also secrete an attractive signal (Complement 3a, C3a) to ensure that neural crest cells continually grow towards each other. The pattern of migration by the group has a *collective direction* due to ongoing contact inhibition among the cells at the leading edge. Contact of inhibition is represented by the differential activation of Rho GTPases (red to yellow). (Data from Scarpa and Mayor 2016.)

chemotaxis are not required for collective migration, but that properties intrinsic to the cells are sufficient to maintain cluster integrity and directional movement. Simulations modeling collective migration predict that both contact inhibition of locomotion and co-attraction between cells are necessary for efficient collective migration, which matches what is seen in vivo and in vitro (Carmona-Fontaine et al. 2011; Woods et al. 2014). Indeed, cranial neural crest cells in *Xenopus* show not only a Wnt/PCP-mediated RhoA mechanism of contact inhibition of locomotion but also secretion of Complement 3a (C3a), which attracts neural crest cells expressing the C3a receptor. These same cranial neural crest cells also express low levels of N-cadherin (see Figure 15.11). Experimentally increasing N-cadherin results in a more tightly adherent population of neural crest unable to invade spaces with the same migratory speed, suggesting that optimal levels of N-cadherin may be necessary for normal collective migration of these cells and for their invasion of tissues (Theveneau et al. 2010; Kuriyama et al. 2014).

WATCH DEVELOPMENT 15.1 This time-lapse movie demonstrates the changes in collective migration following manipulation of N-cadherin in neural crest cells.

Following early specification and delamination, neural crest cells migrate along different paths to their specific locations for final differentiation. How do these cells "know" where to go? Does the substrate for travel differ along different paths? What are the "traffic signs" in the environment that provide guidance cues for neural crest cells to colonize their target tissues? Next we will highlight mechanisms behind the guidance of neural crest cell migration into the trunk, head, and cardiac regions of the embryo.

Migration Pathways of Trunk Neural Crest Cells

Neural crest cells migrating from the neural tube at the axial level of the trunk follow one of two major pathways (**FIGURE 15.12**). Many cells that leave early follow a **ventral pathway** away from the neural tube. Fate-mapping experiments show that these cells become sensory (dorsal root) and autonomic neurons, adrenomedullary cells, and Schwann cells and other glial cells (Weston 1963; Le Douarin and Teillet 1974). In birds and mammals (but not in fish and amphibians), these cells migrate ventrally through

FIGURE 15.12 Neural crest cell migration in the trunk of the chick embryo. Schematic diagram of trunk neural crest cell migration. Cells taking the ventral pathway (1) travel through the anterior of the sclerotome (that portion of the somite that generates vertebral cartilage). Those cells initially opposite the posterior portion of a sclerotome migrate along the neural tube until they come to an anterior region. These cells contribute to the sympathetic and parasympathetic ganglia as well as to the adrenomedullary cells and dorsal root ganglia. Somewhat later, other trunk neural crest cells enter the dorsolateral pathway (2) at all axial positions of the somite. These cells travel beneath the ectoderm and become pigment-producing melanocytes. (Migration pathways are shown on only one side of the embryo.)

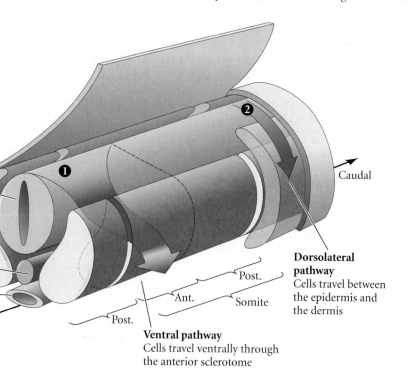

Epidermis
Neural tube
Dermomyotome
Sclerotome
Notochord
Aorta
Rostral
Caudal

Dorsolateral pathway
Cells travel between the epidermis and the dermis

Post.
Ant.
Somite
Post.

Ventral pathway
Cells travel ventrally through the anterior sclerotome

the anterior, but not the posterior, section of the sclerotomes[5] (Rickmann et al. 1985; Bronner-Fraser 1986; Loring and Erickson 1987; Teillet et al. 1987).

Trunk crest cells that emigrate via the second pathway—the **dorsolateral pathway**—become melanocytes, the melanin-forming pigment cells. These cells travel between the epidermis and the dermis, entering the ectoderm through minute holes in the basal lamina (which they themselves may create). Once in the ectoderm, they colonize the skin and hair follicles (Mayer 1973; Erickson et al. 1992). The dorsolateral pathway was demonstrated in a series of classic experiments by Mary Rawles (1948), who transplanted the neural tube and neural crest from a pigmented strain of chickens into the neural tube of an albino chick embryo and saw pigmented feathers on otherwise white wings (see Figure 1.15C).

By transplanting quail neural tubes or neural folds into chick embryos, Teillet and colleagues (1987) were able to mark neural crest cells both genetically and immunologically. The antibody marker recognized and labeled neural crest cells of both species; the genetic marker enabled the investigators to distinguish between quail and chick cells (see Figure 1.15A,B). These studies showed that neural crest cells initially located opposite the posterior region of a somite migrate anteriorly or posteriorly along the neural tube, and then enter the anterior region of their own or an adjacent somite. These cells join with the neural crest cells that initially were opposite the anterior portion of the somite, and they form the same structures. Thus, each dorsal root ganglion comprises neural crest cell populations forming adjacent to three somites: one from the neural crest opposite the anterior portion of the somite, and one each from the two neural crest regions opposite the posterior portions of its own and the neighboring somites.

The ventral pathway

The choice between the dorsolateral versus the ventral trunk pathway is made at the dorsal neural tube shortly after neural crest cell specification (Harris and Erickson 2007). The earliest migrating cells are inhibited from entering the dorsolateral pathway by chondroitin sulfate proteoglycans, ephrins, Slit proteins, and probably several other molecules. Because they are so inhibited, these cells turn around and migrate ventrally, and there they give rise to the neurons and glial cells of the peripheral nervous system.

The next choice concerns whether these ventrally migrating cells migrate *between* the somites (to form the sympathetic ganglia of the aorta) or *through* the somites (Schwarz et al. 2009). In the mouse embryo, the first few neural crest cells that form go between the somites, but this pathway is soon blocked by **semaphorin-3F**, a protein that repels neural crest cells; thus, most neural crest cells traveling ventrally migrate through the somites. These cells migrate through the *anterior* portion of each sclerotome and associate with proteins of the extracellular matrix, such as fibronectin and laminin, that are permissive for migration (Newgreen and Gooday 1985; Newgreen et al. 1986).

The extracellular matrices of the sclerotome differ in the anterior and posterior regions of each somite, and only the extracellular matrix of the *anterior* sclerotome allows neural crest cell migration (**FIGURE 15.13A**). Like the extracellular matrix molecules that prevented neural crest cells from migrating dorsolaterally, the extracellular matrix of the *posterior* portion of each sclerotome contains proteins that actively exclude neural crest cells (**FIGURE 15.13B**). Besides semaphorin-3F, these proteins include the **ephrins**. The ephrin on the posterior sclerotome is recognized by its receptor, Eph, on the neural crest cells. Similarly, semaphorin-3F on the posterior sclerotome cells is recognized by its receptor, neuropilin-2, on the migrating neural crest cells. When neural crest cells are plated on a culture dish containing stripes of immobilized cell membrane proteins alternately with and without ephrins, the cells leave the ephrin-containing

---

[5] Recall that the sclerotome is the portion of the somite that gives rise to the cartilage of the spine. In the migration of fish neural crest cells, the sclerotome is less important; rather, the myotome appears to guide the migration of the crest cells ventrally (Morin-Kensicki and Eisen 1997).

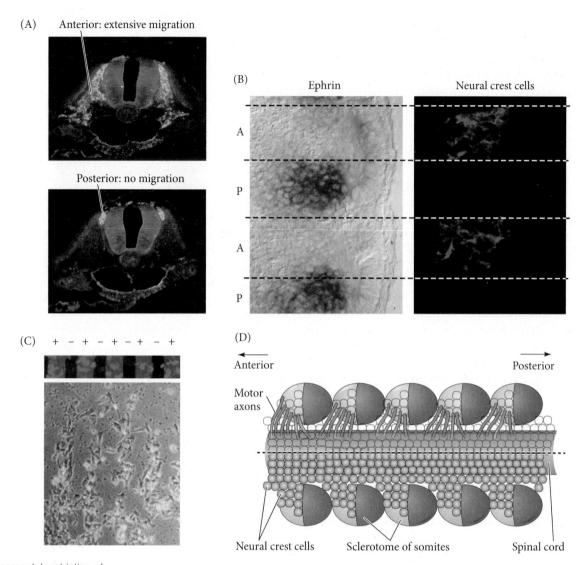

(A) Anterior: extensive migration

Posterior: no migration

(B) Ephrin | Neural crest cells

A

P

A

P

(C) + − + − + − + − +

(D) Anterior ← Posterior →

Motor axons

Neural crest cells | Sclerotome of somites | Spinal cord

FIGURE 15.13 Segmental restriction of neural crest cells and motor neurons by the ephrin proteins of the sclerotome. (A) Cross sections through these areas, showing extensive migration through the anterior portion of the sclerotome (top), but no migration through the posterior portion (bottom). Antibodies to HNK-1 are stained green. (B) Negative correlation between regions of ephrin in the sclerotome (dark blue stain, left) and the presence of neural crest cells (green HNK-1 stain, right). (C) When neural crest cells are plated on fibronectin-containing matrices with alternating stripes of ephrin, they bind to those regions lacking ephrin. (D) Composite scheme showing the migration of spinal cord neural crest cells and motor neurons through the ephrin-deficient anterior regions of the sclerotomes. (For clarity, the neural crest cells and motor neurons are each depicted on only one side of the spinal cord.) (A from Bronner-Fraser 1986, courtesy of the author; B from Krull et al. 1997; C after O'Leary and Wilkinson 1999.)

stripes and move along the stripes that lack ephrin (**FIGURE 15.13C**; Krull et al. 1997; Wang and Anderson 1997; Davy and Soriano 2007). Similarly, neural crest cells fail to migrate on substrates containing semaphorin-3F; mutant mice deficient in either semaphorin-3F or neuropilin-2 have severe migration abnormalities throughout the trunk, with neural crest cells migrating through both anterior and posterior halves of the somite. This patterning of neural crest cell migration generates the overall segmental character of the peripheral nervous system, reflected in the positioning of the dorsal root ganglia and other neural crest-derived structures (**FIGURE 15.13D**).

CELL DIFFERENTIATION IN THE VENTRAL PATHWAY The neural crest cells entering the somites differentiate to become two major types of neurons, depending on their location. Those cells that differentiate within the sclerotome give rise to the dorsal root ganglia. These neural crest cells contain the sensory neurons that relay information regarding touch, pain, and temperature back to the CNS.[6] As they begin to migrate ventrally, it is likely that the neural crest cells produce progeny that express

---

[6]These sensory neurons are *afferent neurons* because they carry information from sensory cells to the central nervous system (i.e., the brain and spinal cord). *Efferent neurons* carry information away from the CNS; these are the motor neurons generated in the ventral region of the neural tube (as discussed in Chapter 14).

FIGURE 15.14 Entry of neural crest cells into the gut and adrenal gland. (A) Migrating neural crest cells (stained red for the Sox8 transcription factor) migrating toward the adrenal cortical cells (stained green for SF1). The limits of the adrenal gland are circled; the dorsal aorta boundary is shown by a dotted line. (B) Neural crest cells form the enteric (gut) ganglia necessary for peristalsis. Confocal image (200× magnification) of an 11.5-day mouse gut showing the migration of neural crest cells (stained for Phox2b) through the foregut and into the cecal bulge of the intestine. (A from Reiprich et al. 2008; B from Corpening et al. 2008.)

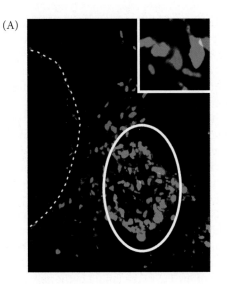

different receptors. Migrating cells that have receptors for neurotrophin and Wnt respond to those proteins (which are produced by the dorsal neural tube) and differentiate close to the neural tube into the glia and neurons of the dorsal root ganglia (Weston 1963). Within the dorsal root ganglia, those cells having more Notch become the glia, while those cells having more Delta (the Notch ligand) become the neurons (Wakamatsu et al. 2000; Harris and Erickson 2007).

Neural crest cells that lack receptors for Wnt and neurotrophin continue migrating. They migrate through the anterior portion of the sclerotome and continue ventrally until they reach the dorsal aorta (but stop before they enter the gut) and become the sympathetic ganglia (Vogel and Weston 1990). At the trunk level, they contribute to the epinephrine-secreting sympathetic (adrenergic, "flight or fight") neurons of the autonomic nervous system, as well as to the adrenal medulla (**FIGURE 15.14A**). At the cardiac and vagal axial levels, they become acetylcholine-secreting parasympathetic (cholinergic, "rest and digest") neurons, including the enteric neurons of the gut (**FIGURE 15.14B**). These cell lineages may each arise from a multipotent neural crest progenitor cell, and the restriction of fate into these three lineages may come relatively late (Sieber-Blum 1989). BMPs from the aorta appear to convert neural crest cells into the sympathetic and adrenal lineages, whereas glucocorticoids from the adrenal cortex block neuron formation, directing the neural crest cells near them to become adrenomedullary cells (Unsicker et al. 1978; Doupe et al. 1985; Anderson and Axel 1986; Vogel and Weston 1990). The cells destined to become the epinephrine-secreting cells of the adrenal medulla retain their responsiveness to BMP and migrate toward the BMP4-secreting adrenal cortical cells. The remaining cells become the sympathetic ganglia surrounding the aorta (Saito et al. 2012).

Moreover, when chick vagal and thoracic neural crests are reciprocally transplanted, the former thoracic crest produces the cholinergic neurons of the parasympathetic ganglia, and the former vagal crest forms adrenergic neurons in the sympathetic ganglia (Le Douarin et al. 1975). Kahn and colleagues (1980) found that premigratory neural crest cells from both the thoracic and the vagal regions contain enzymes for synthesizing both acetylcholine and norepinephrine. Thus, there is good evidence that although some neural crest cells are committed soon after their formation, the differentiation of the ventrally migrating neural crest cells depends on the pathway they follow and their final location.

GOING FOR THE GUT Which neural crest cells colonize the gut and which do not? This distinction involves both extracellular matrix components and soluble paracrine factors. Neural crest cells from the vagal and sacral regions form the enteric ganglia of the gut tube and control intestinal peristalsis. Cells from the vagal neural crest, once past the somites, enter into the foregut and spread to most of the digestive tube, while the sacral neural crest cells colonize the hindgut (see Figure 15.14B). Various inhibitory extracellular matrix proteins (including the Slit proteins) block the more ventral migration of trunk neural crest cells into the gut, but these inhibitory proteins are absent around the vagal and sacral crest, allowing these neural crest cells to reach the gut tissue. Once in the vicinity of the developing gut, these crest cells are attracted to the digestive tube by **glial-derived neurotrophic factor (GDNF)**, a paracrine factor produced by the gut mesenchyme (Young et al. 2001; Natarajan et al. 2002). GDNF from the gut mesenchyme binds to its receptor, Ret, on the neural crest cells. The vagal neural crest cells have more Ret in their cell membranes than do the sacral cells, which makes the vagal cells more invasive (Delalande et al. 2008).

GDNF activates cell division, directs cell migration into the gut mesoderm, and induces neural differentiation (Mwizerwa et al. 2011). If either GDNF or Ret is deficient in mice or humans, the pup or child suffers from Hirschsprung disease, a syndrome wherein the intestine cannot properly void solid wastes. In humans, this condition is most often due to the failure of the vagal neural crest cells to complete their colonization of the hindgut, thus leaving a section of the lower intestine without the ability to undergo peristalsis. By combining experimental analysis of crest cell migration with mathematical modeling, Landman and colleagues (2007) modeled the migration of the vagal crest cells and explained the genetic deficiencies that cause Hirschsprung disease. In their model, the vagal crest cells normally do not migrate in a directed manner once they are in the anterior portion of the gut. Rather, they proliferate until all the niches in that region of the intestine are saturated, after which the migrating front moves posteriorly (Simpson et al. 2007). Meanwhile, the gut itself continues to elongate. Whether or not the colonization is complete depends on the initial number of vagal crest cells entering the anterior gut and the ratio of cell motility to gut growth. These results were not intuitively obvious just from physical observation, and the study shows the power of combining experimental and mathematical approaches to development.

The enteric neural crest cells perform one of the longest migratory journeys because they are chasing a moving target—the caudalmost, or distal extent, of the growing gut. This caudal migration of enteric neural crest cells has been likened to a wave that has a leading (caudal) "crest" of cells (Druckenbrod and Epstein 2007). As the wave progresses down the developing gut, neural crest cells must spread evenly throughout the tissue to ensure complete innervation and function. The process by which enteric neural crest cells are deposited in the gut has been called "directional dispersal" (Theveneau and Mayor 2012), but little has been discovered about the cell behaviors mediating this process. Enteric neural crest do not migrate collectively as clusters but rather in long chains (Corpening et al. 2011; Zhang et al. 2012). Moreover, as the wave progresses, enteric neural crest cells migrate in seemingly random directions and explore domains along all axes; overall dispersal, however, preferentially occurs in the posterior direction (**FIGURE 15.15A**; Young et al. 2014). Enteric neural crest cells are differentiating into neurons along this journey, with both the soma (cell body) and projecting axon still quite mobile during gut development. It is interesting that enteric neural crest cells can be found migrating all along the growing axon as well as just ahead of the leading tip of the growing axon (**FIGURE 15.15B**). These findings suggest a mutualistic relationship between migrating enteric neural crest cells and pathfinding enteric neurons such that the neural crest cells use the nerve axons as a substrate for migration and the neurons' axons follow "trail-blazing" neural crest cells (Young et al. 2014).

Developing Questions

Enteric neural crest cells are positioned at the right time and place to influence the trajectory of pathfinding enteric neurons. Does that in fact occur? Are these leading neural crest cells expressing membrane receptors or secreting diffusible proteins that instruct the growing neuron to extend its axon in the direction of the neural crest cell? If the enteric neural crest cells are leading the way for the axons, what is guiding the neural crest cells toward the caudal end of the gut while also establishing a homogeneously dispersed array of neurons?

WATCH DEVELOPMENT 15.2   The first of two movies features the dispersal of enteric neural crest cells. The second shows a close-up of individual enteric neural crest cells migrating along actively growing axons in the developing gut.

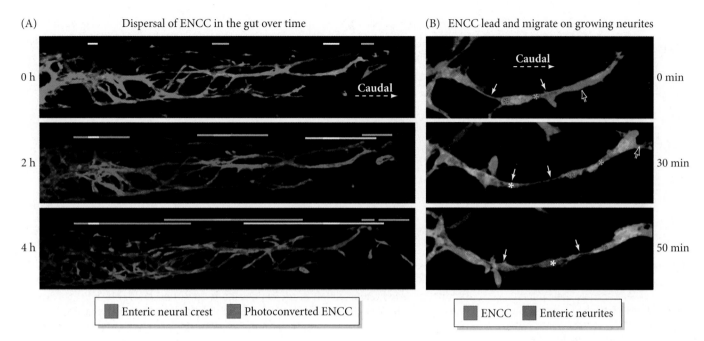

(A) Dispersal of ENCC in the gut over time

0 h

2 h

4 h

Caudal

Enteric neural crest   Photoconverted ENCC

(B) ENCC lead and migrate on growing neurites

Caudal

0 min

30 min

50 min

ENCC   Enteric neurites

FIGURE 15.15 Following the movement of individual enteric neural crest cells (ENCCs) in the developing gut. Transgenic *Ednrb-hKikGR* mice were used to fluorescently label enteric neural crest (green). KikGR is a photoconvertable protein that changes its emission when exposed to ultraviolet light from green to red. (A) Four separate foci were photoconverted in the developing gut (red; bars at top). The caudalmost tip of the moving ENCC wave is located to the far right of the 0-hour time point. Early chains of ENCCs can be seen sparsely spread throughout the gut (green, 0 h). Photoconverted cells actively migrate and disperse with a caudal preference over time (2 h, 4 h; width of bar at top). (B) ENCCs (green) can be seen at the growth tip of differentiating neurites (red, arrows). ENCCs are also seen to use neurites as a substrate for migration (note movement of asterisks over time; different colored asterisks denote different cells). (Data adapted from Young et al. 2014.)

The dorsolateral pathway

In vertebrates, all pigment cells except those of the pigmented retina are derived from the neural crest. It appears that the cells that take the dorsolateral pathway have already become specified as melanoblasts—pigment cell progenitors—and that they are led along the dorsolateral route by chemotactic factors and cell matrix glycoproteins (**FIGURE 15.16**). In the chick (but not in the mouse), the first neural crest cells to migrate enter the ventral pathway, whereas cells that migrate later enter the dorsolateral pathway (see Harris and Erickson 2007). These late-migrating cells remain above the neural tube in what is often called the "staging area," and it is these cells that become specified as melanoblasts (Weston and Butler 1966; Tosney 2004). The switch between glial/neural precursor and melanoblast precursor seems to be controlled by the Foxd3 transcription factor. If Foxd3 is present, it represses expression of the gene for **MITF**,[7] a transcription factor necessary for melanoblast specification and pigment

(A)

[7] *MITF* stands for *m*icrophthalmia-associated *t*ranscription *f*actor, so named because one result of mutations of the gene, as described in mice, is small eyes (i.e., microphthalmia). The effects of MITF are widespread, however, as we will see in this chapter.

(B)

FIGURE 15.16 Neural crest cell migration in the dorsolateral pathway through the skin. (A) Whole mount in situ hybridization of day 11 mouse embryo stained for neural crest-derived melanoblasts (purple). (B) Stage 18 chick embryo seen in cross section through the trunk. Melanoblasts (arrows) can be seen moving through the dermis, from the neural crest region toward the periphery. (A from Baxter and Pavan 2003; B from Santiago and Erickson 2002, courtesy of C. Erickson.)

(A)   (B)   (C)   (D)

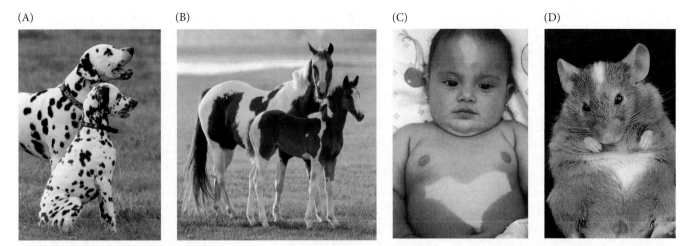

FIGURE 15.17 Variable melanoblast migration, caused by different mutations. (A,B) In several animals, the random death of melanoblasts provides spotted pigmentation. Migrating melanoblasts induce the blood vessels to form in the inner ear, and without these vessels, the cochlea degenerates, and the animal cannot hear in that ear. That is often the case with Dalmation dogs (A), which are heterozygous for *Mitf*, and American Paint horses (B), which are thought to be heterozygous for endothelin receptor B. (C) Piebaldism in a human infant. Pigment fails to form in regions of the body, the result of a mutation in the *KIT* gene. Kit protein is essential for the proliferation and migration of neural crest cells, germ cell precursors, and blood cell precursors. (D) Mice can also have a *Kit* mutation, and they provide important models for piebaldism and melanoblast migration. (A © Robert Pickett/Getty Images; B © M. J. Barrett/Alamy; C,D courtesy of R. A. Fleischman.)

production (see Figure 15.6). If Foxd3 expression is downregulated, MITF is expressed, and the cells become melanoblasts. MITF is involved in three signaling cascades. The first cascade activates those genes responsible for pigment production; the second allows these neural crest cells to travel along the dorsolateral pathway into the skin; and the third prevents apoptosis in the migrating cells (Kos et al. 2001; McGill et al. 2002; Thomas and Erickson 2009). In humans heterozygous for *MITF,* fewer pigment cells reach the center of the body, resulting in a hypopigmented (white) streak through the hair. In some animals, including certain breeds of dogs and horses, heterozygosity for *Mitf* causes a random death of melanoblasts (**FIGURE 15.17**).

Once specified, the melanoblasts in the staging area upregulate the ephrin receptor (Eph B2) and the endothelin receptor (EDNRB2). Doing so allows melanoblasts to migrate along extracellular matrices that contain ephrin and endothelin-3 (see Figure 15.16B; Harris et al. 2008). Indeed, the melanocyte lineage migrates on exactly those same molecules that *repelled* the glial/neural lineage of crest cells. Ephrin expressed along the dorsolateral migration pathway stimulates the migration of melanocyte precursors. Ephrin activates its own receptor, Eph B2, on the neural crest cell membrane, and this Eph signaling appears to be critical for promoting neural crest migration to sites of melanocyte differentiation. Disruption of Eph signaling in late-migrating neural crest cells prevents their dorsolateral migration (Santiago and Erickson 2002; Harris et al. 2008). An interesting recent discovery is that certain breeds of chickens displaying white plumage are the result of naturally occurring mutations in the *Ednrb2* gene (Kinoshita et al. 2014).

In mammals (but not in chicks), the Kit receptor protein is critical in causing the committed melanoblast precursors to migrate on the dorsolateral pathway. This protein is found on those mouse neural crest cells that also express MITF—that is, the presumptive melanoblasts. Kit protein binds **stem cell factor (SCF)**, which is made by the dermal cells. When bound to SCF, Kit prevents apoptosis and stimulates cell division among the melanoblast precursors. If mice or humans do not make sufficient amounts of Kit, the neural crest cells do not proliferate enough to cover the entire skin (see Figure 15.17C,D; Spritz et al. 1992). Moreover, SCF is critical for dorsolateral migration. If SCF is experimentally secreted from tissues (such as cheek epithelium or the footpads) that do not usually synthesize this protein (and do not usually have melanocytes), neural crest cells will enter those regions and become melanocytes (Kunisada et al. 1998; Wilson et al. 2004).

Thus, the differentiation of the trunk neural crest is accomplished by (1) autonomous factors (such as the Hox genes distinguishing trunk and cranial neural crest cells, or MITF committing cells to a melanocyte lineage), (2) specific conditions of the environment (such as the adrenal cortex inducing adjacent neural crest cells into adrenomedullary cells), or (3) a combination of the two (as when cells migrating through the sclerotome respond to Wnt signals depending on their types of receptors). The fate of an individual neural crest cell is determined both by its starting position (anterior-posterior along the neural tube) and by its migratory path.

 SCIENTISTS SPEAK 15.2 Dr. Melissa Harris and Dr. Carol Erickson talk about EphB2/EDNRB2 and migration along the dorsolateral pathway.

Cranial Neural Crest

The head, comprising the face and the skull, is the most anatomically sophisticated portion of the vertebrate body (Northcutt and Gans 1983; Wilkie and Morriss-Kay 2001). The head is largely the product of the cranial neural crest, and the evolution of jaws, teeth, and facial cartilage occurs through changes in the placement of these cells (see Chapter 20).

Like the trunk neural crest, the cranial crest can form pigment cells, glial cells, and peripheral neurons; but in addition, it can generate bone, cartilage, and connective tissue. The cranial neural crest is a mixed population of cells in different stages of commitment, and about 10% of the population is made up of multipotent progenitor cells that can differentiate to become neurons, glia, melanocytes, muscle cells, cartilage, and bone (Calloni et al. 2009). In mice and humans, the cranial neural crest cells migrate from the neural folds even before they have fused together (Nichols 1981; Betters et al. 2010). Subsequent migration of these cells is directed by an underlying segmentation of the hindbrain. As mentioned in Chapter 13, the hindbrain is segmented along the anterior-posterior axis into compartments called rhombomeres. The cranial neural crest cells migrate ventrally from those regions anterior to rhombomere 8 into the pharyngeal arches and the frontonasal process that forms the face (**FIGURE 15.18**). The final destination of these crest cells determines their eventual fate (**TABLE 15.2**).

TABLE 15.2	Some derivatives of the pharyngeal arches in humans			
Pharyngeal arch	Skeletal elements (neural crest plus mesoderm)	Arches, arteries (mesoderm)	Muscles (mesoderm)	Cranial nerves (neural tube)
1	Incus and malleus (from neural crest); mandible, maxilla, and temporal bone regions (from neural crest)	Maxillary branch of the carotid artery (to the ear, nose, and jaw)	Jaw muscles; floor of mouth; muscles of the ear and soft palate	Maxillary and mandibular divisions of trigeminal nerve (V)
2	Stapes bone of the middle ear; styloid process of temporal bone; part of hyoid bone of neck (all from neural crest cartilage)	Arteries to the ear region: corticotympanic artery (adult); stapedial artery (embryo)	Muscles of facial expression; jaw and upper neck muscles	Facial nerve (VII)
3	Lower rim and greater horns of hyoid bone (from neural crest)	Common carotid artery; root of internal carotid	Stylopharyngeus (to elevate the pharynx)	Glossopharyngeal nerve (IX)
4	Laryngeal cartilages (from lateral plate mesoderm)	Arch of aorta; right subclavian artery; original spouts of pulmonary arteries	Constrictors of pharynx and vocal cords	Superior laryngeal branch of vagus nerve (X)
6[a]	Laryngeal cartilages (from lateral plate mesoderm)	Ductus arteriosus; roots of definitive pulmonary arteries	Intrinsic muscles of larynx	Recurrent laryngeal branch of vagus nerve (X)

Source: After Larsen 1993.

[a] The fifth arch degenerates in humans.

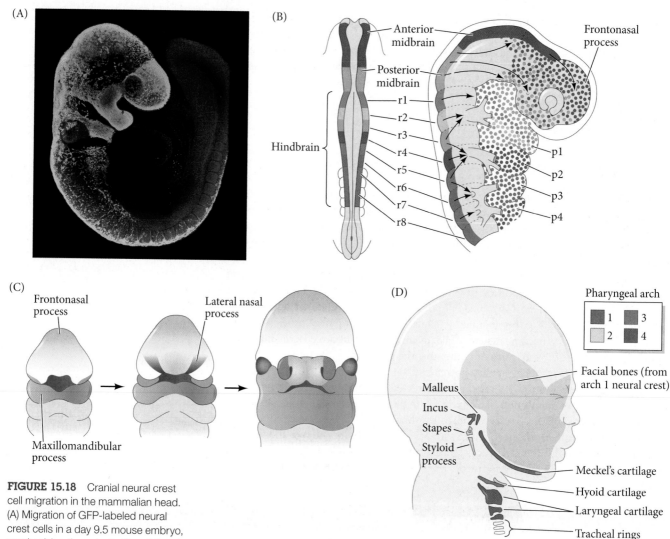

FIGURE 15.18 Cranial neural crest cell migration in the mammalian head. (A) Migration of GFP-labeled neural crest cells in a day 9.5 mouse embryo, emphasizing the colonization of the pharyngeal arches and frontonasal process. (B) Migrational pathways from the cranial neural crest into the pharyngeal arches (p1–p4) and frontonasal process. (C) Continued migration of the cranial neural crest to produce the human face. The frontonasal process contributes to the forehead, nose, philtrum of the upper lip (the area between the lip and nose), and primary palate. The lateral nasal process generates the sides of the nose. The maxillomandibular process gives rise to the lower jaw, much of the upper jaw, and the sides of the middle and lower regions of the face. (D) Structures formed in the human face by the mesenchymal cells of the neural crest. The cartilaginous elements of the pharyngeal arches are indicated by colors, and the darker pink region indicates the facial skeleton produced by anterior regions of the cranial neural crest. (A courtesy of P. Trainor and A. Barlow; B after Le Douarin 2004; C after Helms et al. 2005; D after Carlson 1999.)

The cranial crest cells follow one of three major streams:

1. Neural crest cells from the midbrain and rhombomeres 1 and 2 of the hindbrain migrate to the first pharyngeal arch (the mandibular arch), forming the jawbones as well as the incus and malleus bones of the middle ear. These cells will also differentiate into neurons of the trigeminal ganglion—the cranial nerve that innervates the teeth and jaw—and will contribute to the ciliary ganglion that innervates the ciliary muscle of the eye. These neural crest cells are also pulled by the expanding epidermis to generate the **frontonasal process**, the bone-forming region that becomes the forehead, the middle of the nose, and the primary palate. Thus, the cranial neural crest cells generate much of the facial skeleton (see Figure 15.18B,C; Le Douarin and Kalcheim 1999; Wada et al. 2011).

2. Neural crest cells from rhombomere 4 populate the second pharyngeal arch, forming the upper portion of the hyoid cartilage of the neck as well as the stapes bone of the middle ear (see Figure 15.18B,D). These cells will also contribute neurons of the facial nerve. The hyoid cartilage eventually ossifies to provide the bone in the neck that attaches the muscles of the larynx and tongue.

3. Neural crest cells from rhombomeres 6–8 migrate into the third and fourth pharyngeal arches and pouches to form the lower portion of the hyoid cartilage as well as contributing cells to the thymus, parathyroid, and thyroid glands (see Figure 15.18B; Serbedzija et al. 1992; Creuzet et al. 2005). These neural crest cells also go

to the region of the developing heart, where they help construct the outflow tracts (i.e., the aorta and pulmonary artery). If the neural crest is removed from those regions, these structures fail to form (Bockman and Kirby 1984). Some of these cells migrate caudally to the clavicle (collarbone), where they settle at the sites that will be used for the attachment of certain neck muscles (McGonnell et al. 2001).

The "Chase and Run" Model

Due to the many cell types that the cranial neural crest generates in the anterior embryo, much importance has been placed upon understanding the molecular and cellular mechanisms governing cranial neural crest migration in this region. Recall that streams of cranial neural crest cells collectively migrate through the autonomous mechanisms of contact inhibition of locomotion, co-attraction, and low-level adhesion (see Figure 15.11). But how is each stream kept separate to move collectively in the correct direction? The three streams of cranial neural crest cells are kept from dispersing through interactions of the cells with their environment and with one another. Observations of the migration patterns produced by rhombomeres 3 and 5 in the chick hindbrain revealed that they do not migrate laterally, but rather join the even-numbered streams anterior and posterior to these odd-numbered rhombomeres. The migration of individually marked cranial neural crest cells in these regions can be monitored with cameras focused through a Teflon membrane window in the egg, and such experiments have shown that the migrating cells were "kept in line" not only by restrictions provided by neighboring cells, but also by the lead cells passing material to those cells behind them. It appears that cranial neural crest cells extend long, slender bridges that temporarily connect cells and influence the migration of the later cells to "follow the leader"[8] (Kulesa and Fraser 2000; McKinney et al. 2011).

Recently, analysis of frog and fish cranial neural crest migratory streams revealed that the separate streams seem to be kept apart by the chemorepellent properties of ephrins and semaphorins (**FIGURE 15.19A**; also see Watch Development 15.3). Blocking the activity of the Eph receptors causes cells from the different streams to mix (Smith et al. 1997; Helbling et al. 1998; reviewed in Scarpa and Mayor 2016). In addition, it appears that the ventrally directed stream is guided by placode cells (Theveneau et al. 2013). If an explant of cranial neural crest cells is placed next to an explant of a placode, the neural crest cells appear to "chase" the placodal explant, a behavior that can be abolished by knocking down CXCR4. These and other data led Theveneau and colleagues (2013) to propose the "chase and run" model to explain how this relationship results in directed collective migration.

WATCH DEVELOPMENT 15.3   Watch cranial neural crest cells chase the placode in the frog embryo. In a second movie, cranial neural crest cells chase placodal cells even in isolated of culture conditions.

It was found that placodal cells secrete the chemoattractant **stromal-derived factor-1** (**SDF1**), thereby setting up a gradient of the factor that is highest at the placode. The cranial neural crest cells express the receptor for this ligand, CXCR4, which allows them to sense the attraction of the SDF1 gradient and directs the migration of neural crest cells up the gradient toward the placode (the "chase"). Once the neural crest cells reach the placode, however, contact inhibition of locomotion between the neural crest and placode cells causes the placode cells to migrate away from the site of contact (the "run"). The chemoattractive force of SDF1, however, will start the chase again in the

---

[8]We have seen similar phenomena before, as in the cells of the chick neural folds (some of which probably become neural crest cells) and limb buds, early zebrafish blastomeres, and the extensions of sea urchin micromeres.

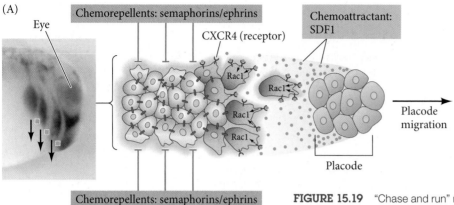

(A)

Eye

Chemorepellents: semaphorins/ephrins

CXCR4 (receptor)

Chemoattractant: SDF1

Rac1

Rac1

Rac1

Rac1

Placode migration

Placode

Chemorepellents: semaphorins/ephrins

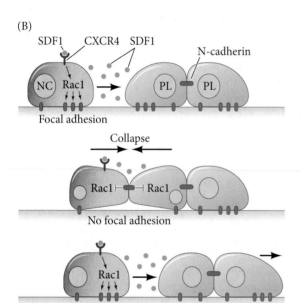

(B)

SDF1   CXCR4   SDF1

N-cadherin

NC   Rac1

PL   PL

Focal adhesion

Collapse

Rac1 ⊢ ⊣ Rac1

No focal adhesion

Rac1

FIGURE 15.19   "Chase and run" model for chemotactic cell migration. (A) The micrograph is a lateral view of streams of cranial neural crest cells in the *Xenopus* head, visualized by in situ hybridization for *FoxD3* and *Dlx2* for both premigratory and migratory cranial neural crest populations (dark purple). The illustration depicts the collective migration of cranial neural crest cells using the autonomous mechanisms described in Figure 15.11. Here, a ventrally positioned placode (blue cells) attracts the leading edge of the cranial neural crest stream via SDF1-CXCR4 signaling (the "chase"). The cranial neural crest contacts the placode, triggering contact inhibition that pushes the placode forward (the "run"). Chemorepellents restrict crest cells from lateral wandering outside of the cluster. (B) Internal molecular events and resulting cellular behaviors that yield forward migratory movement of both placodal (blue) and cranial neural crest (red/yellow) cells. (After Theveneau et al. 2013; Scarpa and Mayor 2016. Photograph from Kuriyama et al. 2014.)

ventral direction toward the "running" placode (**FIGURE 15.19B**; Theveneau et al. 2013; Scarpa and Mayor 2016).

WEB TOPIC 15.2   **INTRAMEMBRANOUS BONE AND THE ROLE OF NEURAL CREST IN BUILDING THE HEAD SKELETON**   Further your understanding of the mechanisms of ossification carried out by both ectodermal and mesodermal cells necessary to build the skull.

Neural Crest-Derived Head Skeleton

The vertebrate skull, or **cranium**, is composed of the **neurocranium** (skull vault and base) and the **viscerocranium** (jaws and other pharyngeal arch derivatives). The bones of the cranium are called **intramembranous bones** because they are created by laying down calcified spicules directly in connective tissue without a cartilaginous precursor. Skull bones are derived from both the neural crest and the head mesoderm (Le Lièvre 1978; Noden 1978; Evans and Noden 2006). Although the neural crest origin of the viscerocranium is well documented, the contributions of cranial neural crest cells to the skull vault are more controversial. In 2002, Jiang and colleagues constructed transgenic mice that expressed β-galactosidase only in their cranial neural crest cells.[9] When the embryonic mice were stained for β-galactosidase, the cells forming the anterior portion of the head—the nasal, frontal, alisphenoid, and squamosal bones—stained blue; the parietal bone of the skull did not (**FIGURE 15.20A,B**). The boundary between neural crest-derived head

Developing Questions

We discussed the progressive maturation of trunk neural crest cells over the course of their migration. Cranial neural crest cells are more tightly adherent to one another and undergo collective migration, implying that a different kind of mechanism might regulate the progressive differentiation of cranial neural crest cells. Do the cells located more centrally within the collective stream get patterned differently from those at the periphery, or is the spatiotemporal positioning of the cells along their migration route correlated with their specification?

(A) *Wnt1-Cre*: Neural crest-derived bone

(B) *Mesp-Cre*: Mesoderm-derived bone

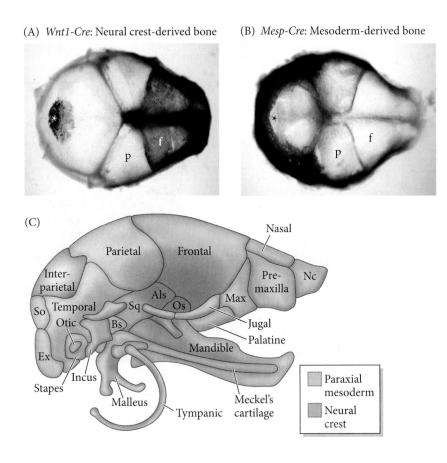

(C)

FIGURE 15.20 Cranial neural crest cells in embryonic mice, stained for β-galactosidase expression. (A) In the *Wnt1-Cre* strain, β-galactosidase is expressed wherever Wnt1 (a neural crest marker) would be expressed. This dorsal view of a 17.5-day embryonic mouse shows staining in the frontal bone (f) and interparietal bone (asterisk) but not in the parietal bone (p). (B) The *Mesp-Cre* strain of mice expresses β-galactosidease in those cells derived from the mesoderm. Here, a reciprocal pattern of staining is seen, and the parietal bone is blue. (C) Summary diagram of results from mapping with Sox9 and Wnt1 markers (Als, alisphenoid; Bs, basisphenoid; Ex, exoccipital; Max, maxilla; Nc, nasal capsule; Os, orbitosphenoid; So, supraoccipital; Sq, squamosal). (A,B from Yoshida 2008, courtesy of G. Morriss-Kay; C from several sources, including Noden and Schneider 2006 and Lee and Saint-Jeannet 2011.)

bone and mesoderm-derived head bone is between the frontal and parietal bones (**FIGURE 15.20C**; Yoshida et al. 2008). Although the specifics may vary among the vertebrate groups, in general the front of the head is derived from the neural crest while the back of the skull is derived from a combination of neural crest-derived and mesodermal bones. The neural crest contribution to facial muscle mixes with the cells of the cranial mesoderm such that facial muscles probably also have dual origins (Grenier et al. 2009).

Given that the neural crest forms our facial skeleton, it follows that even small variations in the rate and direction of cranial neural crest cell divisions will determine what we look like. Moreover, because we look more like our biological parents than our friends do (at least, we hope this is true), such small variations must be hereditary. The regulation of our facial features is probably coordinated in large part by numerous paracrine growth factors. BMPs (especially BMP3) and Wnt signaling cause the protrusion of the frontonasal and maxillary processes, giving shape to the face (Brugmann et al. 2006; Schoenebeck et al. 2012). FGFs from the pharyngeal endoderm are responsible for the attraction of the cranial neural crest cells into the arches as well as for patterning the skeletal elements within the arches. Fgf8 is both a survival factor for the cranial crest cells and is critical for the proliferation of cells forming the facial skeleton (Trocovic et al. 2003, 2005; Creuzet et al. 2004, 2005). The FGFs work in concert with BMPs, sometimes activating them and sometimes repressing them (Lee et al. 2001; Holleville et al. 2003; Das and Crump 2012).

Coordination of face and brain growth

It is a generalization in clinical genetics that "the face reflects the brain." Although this is not always the case, physicians are aware that children with facial anomalies may have brain malformations as well. The coordination between facial form and brain growth was highlighted in studies by Le Douarin and colleagues (2007). First they found that the region of cranial neural crest that forms the facial skeleton is also critical

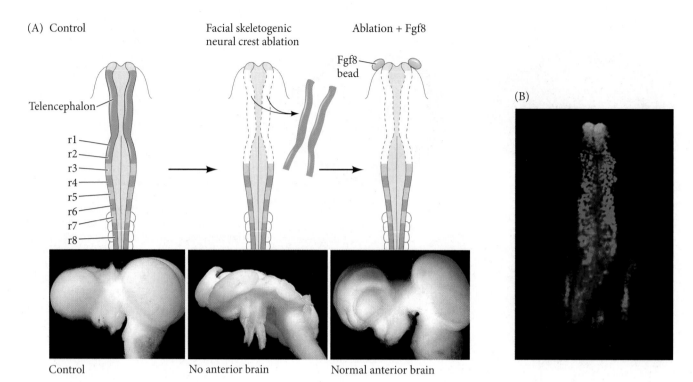

(A) Control

Facial skeletogenic
neural crest ablation

Ablation + Fgf8

Fgf8
bead

Telencephalon

r1
r2
r3
r4
r5
r6
r7
r8

Control

No anterior brain

Normal anterior brain

(B)

FIGURE 15.21 The cranial neural crest that forms the facial skeleton is also critical for the growth of the anterior region of the brain. (A) Removing the facial skeleton-forming neural crest cells from a 6-somite-stage chick embryo stops the telencephalon from forming, as well as inhibiting formation of the facial skeleton. Telencephalon development can be rescued by adding Fgf8-containing beads to the anterior neural ridge. (B) Embryo stained with HNK-1 (which labels neural crest cells green). Fgf8 appears pink in this micrograph. (After Creuzet et al. 2006, 2009; photographs courtesy of N. Le Douarin.)

for the growth of the anterior brain (**FIGURE 15.21**). When that region of chick neural crest was removed, not only did the bird's face fail to form, but the telencephalon failed to grow as well. Next they found that forebrain development could be rescued by adding Fgf8-containing beads to the anterior neural ridge (the neural folds of the anterior neuropore). This finding was strange, however, because cranial neural crest cells do not make or secrete Fgf8; the anterior neural ridge usually does. It seemed that removing the cranial neural crest cells prevented the anterior neural ridge from making the Fgf8 necessary for forebrain proliferation.

Looking at the effects of activated genes added to the anterior neural ridge region, Le Douarin and colleagues hypothesized that the BMP4 from the surface ectoderm was capable of blocking Fgf8. The cranial neural crest cells secreted Noggin and Gremlin, two extracellular proteins that bind to and inactivate BMP4, allowing for the synthesis of Fgf8 in the anterior neural ridge and the development of the forebrain structures. Thus, not only do the cranial neural crest cells provide the cells that build the facial skeleton and connective tissues, they also regulate the production of Fgf8 in the anterior neural ridge, thereby allowing development of the midbrain and forebrain.

WEB TOPIC 15.3 WHY BIRDS DON'T HAVE TEETH The formation of the face and jaw is coordinated by a series of cranial neural crest cell migrations and by the interactions of these neural crest cells with surrounding tissues.

Cardiac Neural Crest

The heart originally forms in the neck region, directly beneath the pharyngeal arches, so it should not be surprising that it acquires cells from the neural crest. The pharyngeal ectoderm and endoderm both secrete Fgf8, which acts as a chemotactic factor to draw neural crest cells into the area. Indeed, if beads containing large amounts of Fgf8 are placed dorsal to the chick pharynx, the cardiac neural crest cells will migrate there instead (Sato et al. 2011). The caudal region of the cranial neural crest is called the

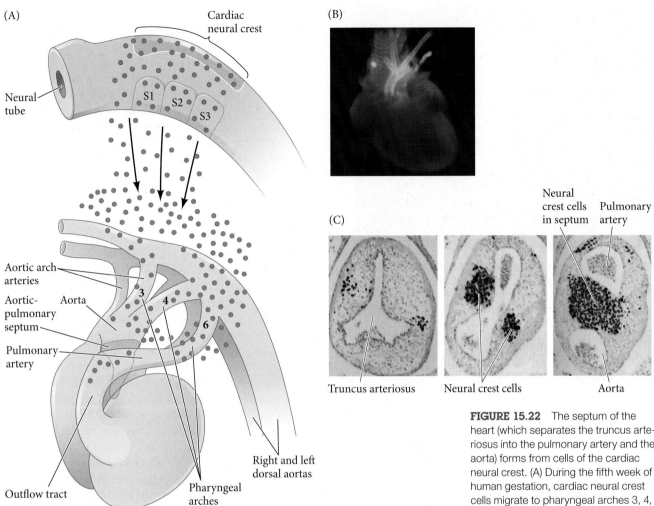

FIGURE 15.22 The septum of the heart (which separates the truncus arteriosus into the pulmonary artery and the aorta) forms from cells of the cardiac neural crest. (A) During the fifth week of human gestation, cardiac neural crest cells migrate to pharyngeal arches 3, 4, and 6 and enter the truncus arteriosus to generate the septum. (B) In a transgenic mouse where the fluorescent green protein is expressed only in cells having the Pax3 cardiac neural crest marker, the outflow regions of the heart become labeled. (C) Quail cardiac neural crest cells were transplanted into the analogous region of a chick embryo, and the embryos were allowed to develop. Quail cardiac neural crest cells are visualized by a quail-specific antibody (dark stain). In the heart, these cells can be seen separating the truncus arteriosus into the pulmonary artery and the aorta. (A after Hutson and Kirby 2007; B from Stoller and Epstein 2005; C from Waldo et al. 1998, courtesy of K. Waldo and M. Kirby.)

cardiac neural crest because its cells (and only those particular neural crest cells) generate the endothelium of the aortic arch arteries and the septum between the aorta and the pulmonary artery (**FIGURE 15.22**; Kirby 1989; Waldo et al. 1998). Cardiac crest cells also enter pharyngeal arches 3, 4, and 6 to become portions of other neck structures such as the thyroid, parathyroid, and thymus glands. These cells are often referred to as the circumpharyngeal crest (Kuratani and Kirby 1991, 1992). In the thymus, neural crest-derived cells are especially important in one of the most critical functions of adaptive immunity: regulating the exit of mature T cells from the thymus and into the circulation (Zachariah and Cyster 2010). It is also likely that the carotid body, which monitors oxygen in the blood and regulates respiration accordingly, is derived from the cardiac neural crest (see Pardal et al. 2007).

In mice, cardiac neural crest cells are peculiar in that they express the transcription factor Pax3. Mutations of the *Pax3* gene result in fewer cardiac neural crest cells, which in turn leads to persistent truncus arteriosus (failure of the aorta and pulmonary artery to separate) as well as to defects in the thymus, thyroid, and parathyroid glands (Conway et al. 1997, 2000). The path from the dorsal neural tube to the heart appears to involve the coordination between the attractive cues provided by semaphorin-3C and the repulsive signals provided by semaphorin-6 (Toyofuku et al. 2008). Congenital heart defects in humans and mice often occur along with defects in the parathyroid, thyroid, or thymus glands. It would not be surprising to find that all these problems are linked to defects in the migration of cells from the neural crest (Hutson and Kirby 2007).

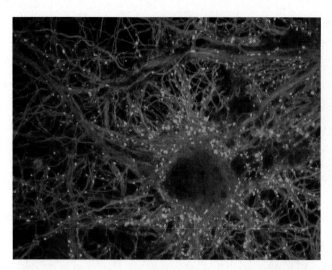

FIGURE 15.23 Connections of axons to a cultured rat hippocampal neuron. The neuron has been stained red with fluorescent antibodies to tubulin. The neuron appears to be outlined by the synaptic protein synapsin (green), which is present in the terminals of axons that contact it. (Photograph courtesy of R. Fitzsimmons and PerkinElmer Life Sciences.)

Establishing Axonal Pathways in the Nervous System

At the beginning of the twentieth century, there were many competing theories on how axons formed. Theodor Schwann (yes, he discovered Schwann cells) believed that numerous neural cells linked themselves together in a chain to form an axon. Viktor Hensen, the discoverer of the embryonic node in birds, thought that axons formed around preexisting cytoplasmic threads between the cells. Wilhelm His (1886) and Santiago Ramón y Cajal (1890) postulated that the axon was an outgrowth (albeit an extremely large one) of the neuron's soma.

In 1907, Ross Granville Harrison demonstrated the validity of the outgrowth theory in an elegant experiment that birthed the science of developmental neurobiology and the technique of tissue culture. Harrison isolated a portion of the neural tube from a 3-mm frog tadpole. (At this stage, shortly after the closure of the neural tube, there is no visible differentiation of axons.) He placed this neuroblast-containing tissue in a drop of frog lymph on a coverslip and inverted the coverslip over a depression slide so that he could watch what was happening within this "hanging drop." What Harrison saw was the emergence of axons as outgrowths from the neuroblasts, elongating at about 56 mm per hour.

Unlike most cells, neurons are not confined to their immediate space; instead, they can produce axons that may extend for meters. Each of the 86 billion neurons in the human brain has the potential to interact in specific ways with thousands of other neurons (**FIGURE 15.23**; Azevedo et al. 2009). A large neuron (such as a Purkinje cell or motor neuron) can receive input from more than 10^5 other neurons (Gershon et al. 1985). Understanding the generation of this stunningly ordered complexity is one of the greatest challenges of modern science. How is this complex circuitry established? We will explore this question as we discuss how neurons extend their axons, how axons are guided to their target cells, how synapses are formed, and what determines whether a neuron lives or dies.

WEB TOPIC 15.4 **THE EVOLUTION OF DEVELOPMENTAL NEUROBIOLOGY**   Santiago Ramón y Cajal, Viktor Hamburger, and Rita Levi-Montalcini helped bring order to the study of neural development by identifying some of the important questions that still puzzle us today.

The Growth Cone: Driver and Engine of Axon Pathfinding

Earlier in this chapter we presented an analogy comparing cell migration to the navigation of cars in traffic (see Figure 15.7). A similar analogy can be used for axon pathfinding by comparing it to a moving train. A neuron needs to build an axonal connection with a target cell that may lie a great distance away. Like the engine of a train, the *locomotory* apparatus of an axon—the **growth cone**—is at the front, and like new carriages added behind the engine, the axon grows through polymerization of microtubules (**FIGURE 15.24A**). The growth cone has been called a "neural crest cell on a leash" because, like neural crest cells, it migrates and senses the environment. Moreover, it can respond to the same types of signals that migrating cells sense.

The growth cone does not move forward in a straight line but rather "feels" its way along the substrate. The growth cone moves by the elongation and contraction

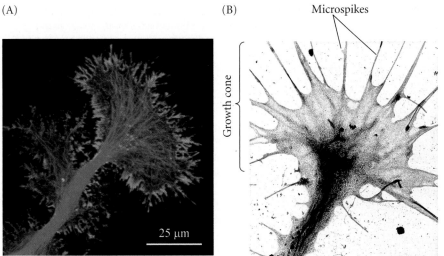

(A)

(B) Microspikes

25 µm

(C)

Encounters substrate

Repulsive signals

Microtubules

F-actin arc

Axon

CLASP

APC

Dynamic MT

Lamellipodium

Filopodium

Attractive signals

Directional turn

Protrusion

Filopodium and Lamellipodium extend forward

Engorgement

C-domain moves forward

Consolidation

New axon shaft forms

FIGURE 15.24 Axon growth cones. (A) Growth cone of the hawkmoth *Manduca sexta* during axon extension and pathfinding. The actin in the filopodia is stained green with fluorescent phalloidin, and the microtubules are stained red with a fluorescent antibody to tubulin. (B) Actin microspikes in an axon growth cone, seen by transmission electron microscopy. (C) The periphery of the growth cone contains lamellipodia and filopodia. The lamellipodia are the major motile apparatus and are seen in the regions that are turning toward a stimulus. The filopodia are sensory. Both structures contain actin microfilaments. There is also a central region of microtubules, some of which extend outward into the filopodia. The microtubules entering the peripheral area can be lengthened or shortened by proteins activated by attractive or repulsive stimuli. During attraction, regulatory proteins bind to the plus ends of the microtubules, stabilizing and lengthening them. On the side opposite the attractive cue, microtubules are removed from the periphery. (A courtesy of R. B. Levin and R. Luedemanan; B from Letourneau 1979; C after Lowery and Van Vactor 2009; Bearce et al. 2015; Cammarata et al. 2016.)

of pointed filopodia called **microspikes** (**FIGURE 15.24B**). These microspikes contain microfilaments, which are oriented parallel to the long axis of the axon. (This mechanism is similar to that seen in the filopodial microfilaments of secondary mesenchyme cells in echinoderms; see Chapter 10). Within the axon itself, structural support is provided by microtubules. If the neuron is placed in a solution of colchicine (an inhibitor of microtubule polymerization), microspikes will be destroyed, and axons retract (Yamada et al. 1971; Forscher and Smith 1988).

As in most migrating cells, the exploratory microspikes of the growth cone attach to the substrate and exert a force that pulls the rest of the cell forward. Axons will not grow if the growth cone fails to advance (Lamoureux et al. 1989). In addition to their structural role in axonal migration, microspikes also have a sensory function. Fanning out in front of the growth cone, each microspike samples the microenvironment and sends signals back to the cell body (Davenport et al. 1993). The navigation of axons to their appropriate targets depends on guidance molecules in the extracellular environment, and it is the growth cone that turns or does not turn in response to guidance cues as the axon seeks to make appropriate synaptic connections. Such differential responsiveness is due to disparities in the expression of receptors on the growth cone cell membrane. Growth cones have the ability to sense the environment and translate the extracellular signals into a directed movement (**FIGURE 15.24C**). This use of directional cues to facilitate specific migration is accomplished by altering the cytoskeleton, changing membrane growth, and coordinating cell adhesion and cell movement (Vitriol and Zheng 2012).

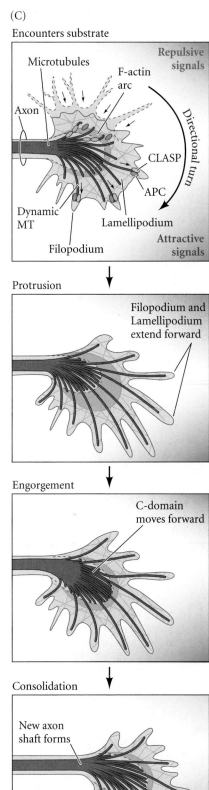

FIGURE 15.25 APC localizes tubulin mRNA at the plus end of microtubules for spatially ▶ targeted translation to fuel growth cone expansion. (A) Schematic showing the localized *Tubb2b* transcript and targeted translation for immediate microtubule elongation at the plus end. (B) Growth cone of a neuronal cell line (white outline) expressing *β2B-tubulin* mRNA (green) as seen by fluorescence in situ hybridization. (C) APC and β2B tubulin (Tubb2b) immunolocalization in the peripheral domain of a neuron from a dissociated rat dorsal root ganglion. APC is co-localized to the tips of Tubb2b-containing microtubules (arrows), seen close up in the inset. (A after Coles and Bradke 2014; B,C from Preitner et al. 2014.)

The growth cone was found to have two major compartments. The central domain of the growth cone contains microtubules that extend the axon shaft and support mitochondria and other organelles (see Figure 15.24C). The peripheral domain contains two types of actin-associated membrane protrusions: the lamellipodia, broad membranous sheets containing short, branched actin networks, that act as the migratory network of the growth cone; and the filopodia, membranes extended by long bundles of filamentous actin, that act as the sensory network. A transition zone between the central and peripheral regions may coordinate actin and tubulin growth (Rodriguez et al. 2003; Lowery and Van Vactor 2009). Lone "pioneer" microtubules from the central zone's core of microtubule bundles protrude through the actin arc that defines the transition zone and extend into the periphery of the growth cone. These pioneer microtubules dynamically associate with the actin microfilaments, and together the microtubules and microfilaments grow and contract to perform the fingerlike movements characteristic of filopodia (Mitchison and Kirschner 1988; Sabry et al. 1991; Tanaka and Kirschner 1991, 1995; Schaefer et al. 2002). The microtubule- and actin-based membrane protrusions, coupled with selective adhesion and membrane recycling, provide the force that drives axon movement and directionality.

"Plus tips" and actin-microtubule interactions

Regulation of actin filaments and microtubules in the peripheral domain plays an important role in the movement of the growth cone. A complex of proteins that interacts with the distal tip or plus end of the microtubules has been identified and is logically called "microtubule plus-end tracking proteins" or simply "plus-tip" proteins (+TIPs). Examples are CLASP (*c*ytoplasmic *l*inker *as*sociated *p*rotein) and APC (*a*denomatous *p*olyposis *c*oli), which, based on their level of phosphorylation, can either stabilize and foster extension of the microtubules or disassociate from the microtubules and inhibit axon outgrowth. Plus-tip proteins bind directly to end-binding proteins (EB1/3) at the distal end of the microtubule. End-binding proteins stay in place whether the microtubule is growing or shrinking (**FIGURE 15.25A**, and see Figure 15.24C; Lowery et al. 2010; reviewed in Lowery and Van Vactor 2009; Bearce et al. 2015; Cammarata et al. 2016).

When considering the incredible journey that awaits the growth cone of a young neuron, a certain "supply and demand" problem presents itself. The demand for a ready supply of proteins at the growth cone, such as tubulin and actin monomers, must be paramount to accomplish the long extension of the axon during pathfinding. As extension proceeds, the growth cone gets farther and farther from the soma and hub of protein generation, suggesting that meeting this demand for protein subunits may get progressively more challenging. However, recent work (Preitner et al. 2014) has shown that transcripts can be housed in the growth cone for onsite translation, and that APC functions to bind and keep transcripts, such as β2b-Tubulin (*Tubb2b*), in position at the distal end of the microtubules. APC also co-localizes with factors that facilitate translation of Tubb2b for rapid microtubule elongation (**FIGURE 15.25B,C**).

((• **SCIENTISTS SPEAK 15.3** Dr. Laura Anne Lowery and Dr. David Van Vactor talk about the identification of CLASP and its role in axon guidance in *Drosophila*.

Developing Questions

Having APC localize mRNA to the growing end of microtubules is like having a constant supply of fuel being pumped into the gas tank of the motorized growth cone. What about CLASP and some of the other seven or more families of +TIPs? Can they also sequester transcripts? Only recently we have realized that regionalized protein synthesis within a cell can be a mechanism for developmental change. How many other events in developmental biology might involve spatially restricted pockets of mRNA and translation machinery?

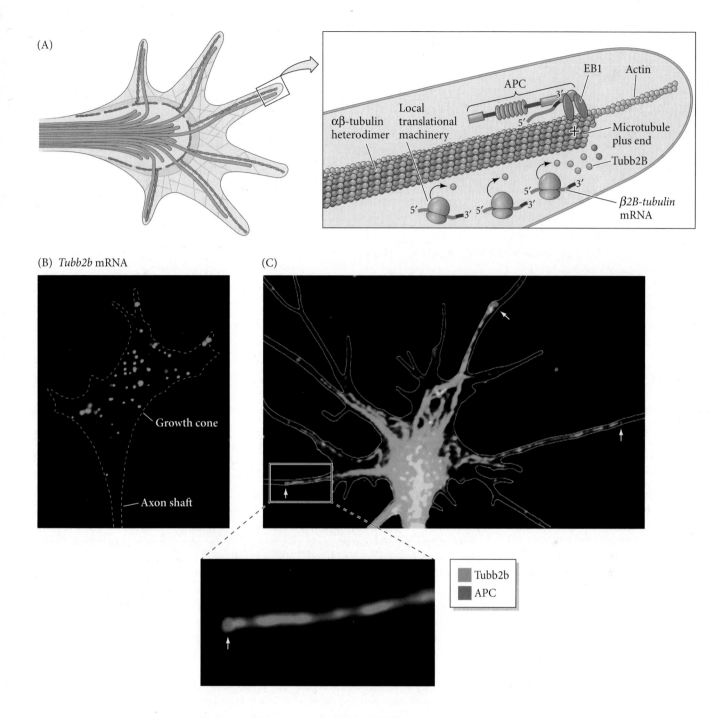

(A)

αβ-tubulin heterodimer

Local translational machinery

APC

EB1

Actin

Microtubule plus end

Tubb2B

β2B-tubulin mRNA

(B) *Tubb2b* mRNA

Growth cone

Axon shaft

(C)

Tubb2b
APC

Rho, Rho, Rho your actin filaments down the signaling stream

The regulation of actin polymerization drives growth cone movement and thus is the target of many molecular guidance pathways. **Rho GTPases** regulate the growth of actin microfilaments. These GTPases can be activated or repressed by receptors binding ephrins, netrins, Slit proteins, or semaphorins (**FIGURE 15.26A**). Similarly, the regulation of tubulin polymerization into microtubules is important because tubulin is encouraged to polymerize on the side of the growth cone receiving attractant stimuli, and it is inhibited from polymerizing (indeed, the tubulin is depolymerized and recycled) on the side opposite the attractive stimuli (Vitriol and Zheng 2012).

Adhesion is thought to provide the "clutch" for directional movement. Visualize actin as being linked to the cell membrane. Now contemplate that the dynamics of

(A)

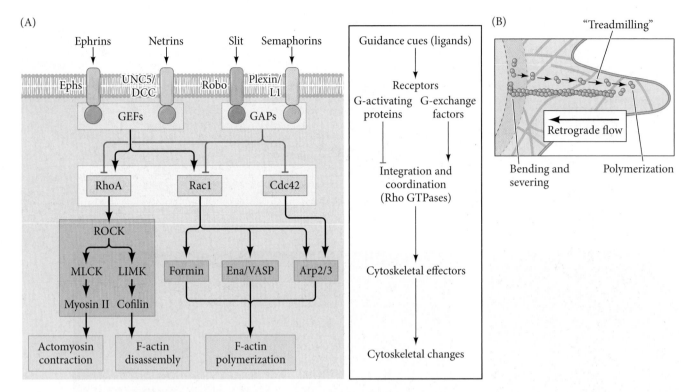

FIGURE 15.26 Rho GTPases interpret and relay external guidance signals to the actin cytoskeleton. The four major ligands providing cues to the growth cone (ephrins, netrins, Slit, and semaphorins) bind to receptors that stabilize or destabilize actin microfilaments. The Rho family of GTPases (RhoA, Rac1, and Cdc42) act as mediators between the receptors and the agents carrying out the cytoskeletal changes. (B) Diagram depicting actin "treadmilling," during which the position of actin monomers flows from the plus end to the minus end, powered by polarized polymerization and depolymerization. (A after Lowery and Van Vactor 2009; B after Cammarata et al. 2016.)

actin assembly and disassembly cause a retrograde flow of actin—that is, picture actin moving *away from* the tip of the growth cone and *toward* the cell body ("treadmilling"; **FIGURE 15.26B**). If the cell membrane is anchored to an external adhesion molecule (through its integrins or cadherins), however, the membrane is propelled forward (Bard et al. 2008; Chan and Odde 2008). If there is no such anchoring adhesion, there is no net movement. But if the adhesion is too stable, the growth cone also stops moving. Thus, adhesions have to be made and broken for the growth cone to progress. These transitory adhesive complexes are referred to as focal adhesions, and they bind actin internally and the extracellular environment externally. Focal adhesions may have as many as 100 different protein components (Geiger and Yamada 2011). One of these components, focal adhesion kinase (FAK), appears to be critical for the assembly, stabilization, and degradation of the focal adhesions (Mitra et al. 2005; Chacon and Fazzari 2011) and appears to be able to recognize both attractive and repulsive stimuli. The investigation of the focal adhesion components is just beginning to delineate the mechanisms by which traction is coordinated with cytoskeletal growth and membrane turnover. As we discussed earlier in this chapter, Rho GTPases and focal adhesions also play essential roles in the migration of neural crest cells.

Because the membranes turn over, the growth cone grows by the exocytosis of vesicles and the incorporation of their membranes into the cell membrane. These vesicles (sometimes called "enlargeosomes") are constructed in the neuron cell body and travel on microtubules to the center of the growth cone (Pfenniger et al. 2003; Rachetti et al. 2010). Most of these vesicles are involved in the constitutive growth of the axon, not the directional growth of the growth cone. Some of these vesicles, however, are transported from the central region into the periphery in response to Ca^{2+} signals coming from membrane receptors. The vesicles then integrate into the tip of the growth cone and function (perhaps exclusively) in turning the growth cone toward attractive stimuli (Tojima et al. 2007). Repulsive cues have been found to initiate endocytosis (the formation of vesicles from cell membranes) in those areas they contact, which would have the effect of both removing the receptor and of diminishing the amount of cell membrane in that area (Hines et al. 2010; Tojima et al. 2010). Thus, by cytoskeleton assembly, cell adhesion, and membrane turnover, the growth cone mechanically transports the axon toward its appropriate target.

Axon Guidance

How does the growth cone "know" to traverse numerous potential target cells and make a specific connection? Harrison (1910) first suggested that the specificity of axonal growth is due to **pioneer nerve fibers**, axons that go ahead of other axons and serve as guides for them.[10] This observation simplified but did not solve the problem of how neurons form appropriate patterns of interconnection. Harrison also noted, however, that axons must grow on a solid substrate, and he speculated that differences among embryonic surfaces might allow axons to travel in certain specified directions. The final connections would occur by complementary interactions on the target cell surface:

> *That it must be a sort of a surface reaction between each kind of nerve fiber and the particular structure to be innervated seems clear from the fact that sensory and motor fibers, though running close together in the same bundle, nevertheless form proper peripheral connections, the one with the epidermis and the other with the muscle. . . . The foregoing facts suggest that there may be a certain analogy here with the union of egg and sperm cell.*

Research on the specificity of neuronal connections has focused on three major systems: (1) motor neurons, whose axons travel from the spinal cord to a specific muscle; (2) commissural neurons, whose axons must cross the midline plane of the embryo to innervate targets on the opposite side of the central nervous system; and (3) the optic system, where axons originating in the retina must find their way back into the brain. In all cases, the specificity of axonal connections unfolds in three steps (Goodman and Shatz 1993):

1. *Pathway selection.* The axons travel along a route that leads them to a particular region of the embryo.

2. *Target selection.* The axons, once they reach the correct area, recognize and bind to a set of cells with which they may form stable connections.

3. *Address selection.* The initial patterns are refined such that each axon binds to a small subset (sometimes only one) of its possible targets.

The first two processes are independent of neuronal activity. The third process involves interactions between several active neurons and converts the overlapping projections into a fine-tuned pattern of connections.

We have known since the 1930s that the axons of motor neurons can find their appropriate muscles even when the neural activity of the axons is blocked. Twitty (who was Harrison's student) and his colleagues found that embryos of the newt *Taricha torosa* secrete tetrodotoxin (TTX), a toxin that blocks neural transmission in other species. By grafting pieces of *T. torosa* embryos onto embryos of other salamander species, they were able to paralyze the host embryos for days while development occurred. Normal neuronal connections were made even though no neural activity could occur. At about the time that the larvae were ready to feed, the neurotoxin wore off, and the young salamanders swam and fed normally (Twitty and Johnson 1934; Twitty 1937). More recent experiments using zebrafish mutants with nonfunctional neurotransmitter receptors similarly demonstrated that motor neurons can establish their normal patterns of innervation in the absence of neuronal activity (Westerfield et al. 1990). But the question remains: *How are the neurons' axons instructed where to go?*

---

[10]The growth cones of pioneer neurons migrate to their target tissue while embryonic distances are still short and the intervening embryonic tissue is still relatively uncomplicated. Later in development, other neurons bind to pioneer neurons and thereby enter the target tissue. Klose and Bentley (1989) have shown that in some cases, pioneer neurons die after the "follow-up" neurons reach their destination. Yet if the pioneer neurons are prevented from differentiating, the other axons do not reach their target tissue.

The Intrinsic Navigational Programming of Motor Neurons

Neurons at the ventrolateral margin of the vertebrate neural tube become motor neurons, and one of their first steps toward maturation involves target specificity (Dasen et al. 2008). The cell bodies of the motor neurons projecting to a single muscle are pooled in a longitudinal column of the spinal cord (**FIGURE 15.27A**; Landmesser 1978; Hollyday 1980; Price et al. 2002). The pools are grouped into the columns of Terni and the lateral and medial motor columns (LMC and MMC, respectively), and neurons in similar places have similar targets (see Figure 13.18C). For instance, in the chick hindlimb, LMC motor neurons innervate the dorsal musculature, whereas the motor neurons of the MMC innervate ventral limb musculature (Tosney et al. 1995; Polleux et al. 2007). This arrangement of motor neurons is consistent throughout the vertebrates.

The targets of motor neurons are specified before their axons extend into the periphery. This was shown by Lance-Jones and Landmesser (1980), who reversed segments of the chick spinal cord so that the motor neurons were placed in new locations. The axons went to their original targets, not to the ones expected from their new positions (**FIGURE 15.27B–D**). The molecular basis for this target specificity resides in the members of the Hox and Lim protein families that are induced during neuronal specification (Tsushida et al. 1994; Sharma et al. 2000; Price and Briscoe 2004; Bonanomi and Pfaff 2010). For instance, all motor neurons express the Lim proteins Islet1 and (slightly later) Islet2. If no other Lim protein is expressed, the neurons project to the ventral limb muscles (**FIGURE 15.28**) because the axons (just like the trunk neural crest cells) synthesize neuropilin-2, the receptor for the chemorepellant semaphorin-3F, which is made in the dorsal part of the limb bud. If Lim1 protein is also synthesized, however, the motor neurons project dorsally to the dorsal limb muscles. This axonal growth toward the dorsal muscle is because Lim1 induces the expression of Eph A4, which is the receptor for the chemorepellent protein ephrin A5 that is made in the ventral part of the limb bud. Thus, the innervation of the limb by motor neurons depends on repulsive signals. The motor neurons entering the axial muscles of the body wall, however, are brought there by chemoattraction—indeed, these axons make an abrupt turn to get to the developing musculature—because these motor neurons express Lhx3, which induces the expression of a receptor for FGFs such as those secreted by the dermomyotome (the somitic region

FIGURE 15.27 Compensation for small dislocations of axonal initiation position in the chick embryo. (A) Axons from motor neurons and sensory neurons group together (fasciculate) before finding their muscle targets. Here, motor nerves (stained green with GFP) and sensory neurons (stained red with antibodies) fasciculate before entering the limb bud of a 10.5-day mouse embryo. (B) A length of spinal cord comprising segments T7–LS3 (seventh thoracic through third lumbosacral segments) is reversed in a 2.5-day embryo. (C) Normal pattern of axon projection to the forelimb muscles at 6 days. (D) Projection of axons from the reversed segment at 6 days. The ectopically placed neurons eventually found their proper neural pathways and innervated the appropriate muscles. (A from Huettl et al. 2011, courtesy of A. Huber-Brösamle; B–D after Lance-Jones and Landmesser 1980.)

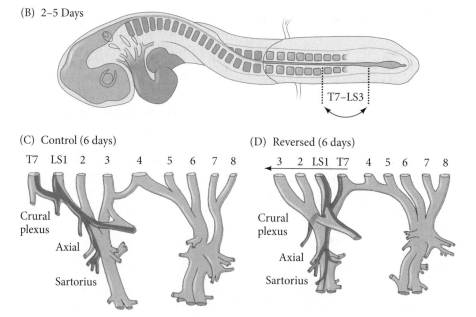

(A)

(B) 2–5 Days

T7–LS3

(C) Control (6 days)

T7 LS1 2 3 4 5 6 7 8

Crural plexus

Axial

Sartorius

(D) Reversed (6 days)

3 2 LS1 T7 4 5 6 7 8

Crural plexus

Axial

Sartorius

FIGURE 15.28 Motor neuron organization and Lim specification in the spinal cord innervating the chick hindlimb. Neurons in each of three different columns express specific sets of Lim family genes (including *Isl1* and *Isl2*), and neurons within each column make similar pathfinding decisions. Neurons of the medial motor column are attracted to the axial muscles by FGFs secreted by the dermomyotome. Neurons of the lateral motor column send axons to the limb musculature. Where these columns are subdivided, medial subdivisions project to ventral positions because they are repelled by semaphorin-3F in the dorsal limb bud, and lateral subdivisions send axons to dorsal regions of the limb bud because they are repelled by the ephrin A5 synthesized in the ventral half. (After Polleux et al. 2007.)

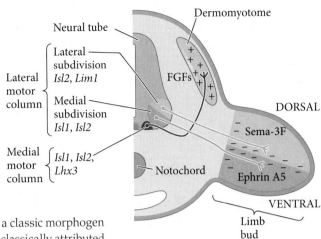

that contains muscle precursor cells). This process is an example of a classic morphogen (FGFs) also functioning to directly guide axons in a manner that is classically attributed to guidance molecules. We will revisit this point again later in this chapter. In summary, motor neurons seek their targets through intrinsic "programs" that assign different motor neurons different cell surface molecules that determine the responsiveness of the axon growth cones to guidance cues in their path and on their targets.

Cell adhesion: A mechanism to grab the road

The initial pathway an axonal growth cone follows is determined by the environment the growth cone experiences. The polarity of a neuron—that is, which part of the cell will extend the axon—is determined largely by the neuron's response to cell adhesive cues in its immediate environment. Integrins and N-cadherins serve as receptors to orient the neuron in accordance with cues from the extracellular matrices and membranes of surrounding cells (Myers et al. 2011; Randlett et al. 2011; Gärtner et al. 2012). These receptors recruit actin, which forms microfilaments in the specified area. The microfilaments transport the motor protein dynein, which in turn recruits microtubules, which can extend the axon (Ligon et al. 2001).

Once an axon begins to form, its growth cone encounters different substrates. The growth cone adheres to certain substrates and moves in their direction. Other substrates cause the growth cone to retract, preventing its axon from growing in that direction. Growth cones prefer to migrate on surfaces that are more adhesive than their surroundings, and a track of adhesive molecules (such as laminin) can direct them to their targets (Letourneau 1979; Akers et al. 1981; Gundersen 1987).

In addition to general extracellular matrix cues, there are important cell-to-cell adhesion contacts that provide permissive substrates for growth cone walking. For instance, the most common railway that growth cones will follow is on the backs of previously laid axons. Whereas motor neurons appear to be intrinsically locked into the location of their final targets (see Figure 15.27B–D), it has long been known that sensory neurons need motor neurons to make appropriate connections (Hamburger 1929; Landmesser et al. 1983; Honig et al. 1986). It appears that the subtypes of motor neurons produce specific compounds (such as Ephs) that cause the sensory neurons to adhere to the motor axon and tract along them (Huettl et al. 2011; Wang et al. 2011). The process of one axon adhering to and using another axon for growth is called **fasciculation** (see Figure 15.27A). It is interesting that axons of the spinal nerve use NCAM to fasciculate together during their shared outgrowth; the dorsal divergence of axons to innervate the epaxial (back) musculature, however, requires the modification of NCAM with polysialic acid (PSA). Modification with PSA transiently breaks the homophilic NCAM interactions, which facilitates defasciculation and the exploration of different pathways in response to such cues as FGFs mentioned above (Tang et al. 1992; Allan and Greer 1998). These are examples of *local contact-mediated guidance cues* (Ephs/NCAMs) that regulate a close adhesive connection between motor neurons and their associated sensory neurons (see Figure 15.28).

Local and long-range guidance molecules: The street signs of the embryo

Navigation through the embryonic environment is literally guided by molecules that function much like the traffic signs, lights, and other directional cues that we use to find our way around our environment. Scientists have interpreted many of the decisions a growth cone makes during its journey as equivalent to choices between being attracted to or repelled from a particular region of the embryo (see Figure 15.24). The signals that elicit attractive and repulsive responses in growing axons fall into four protein families: ephrins, semaphorins, netrins, and the Slit proteins; they are some of the same proteins that we saw regulating neural crest cell migration (see Kolodkin and Tessier-Lavigne 2011). We have already seen that neural crest cells are patterned by their recognition of ephrin and that what is an attractive cue to one set of cells (such as the presumptive melanocytes going through the dermis) can be a repulsive signal to other cells (such as the presumptive sympathetic ganglia). Whether a guidance signal is attractive or repulsive can depend on (1) the type of cell receiving that signal and (2) the time when a cell receives the signal. Most intriguing is that neural development has employed dynamic mechanisms to alter the responsiveness of a growth cone, allowing growth cones to become repelled by cues they either ignored or were actively attracted to previously.

Repulsion patterns: Ephrins and semaphorins

Members of two membrane protein families, the ephrins and the semaphorins, are involved in neural patterning. Just as neural crest cells are inhibited from migrating across the posterior portion of a sclerotome, the axons from the dorsal root ganglia and motor neurons also pass only through the anterior portion of each sclerotome and avoid migrating through the posterior portion (**FIGURE 15.29A**; also see Figure 15.13). Davies and colleagues (1990) showed that membranes isolated from the posterior portion of a somite cause the growth cones of these neurons to collapse (**FIGURE 15.29B,C**). These growth cones contain Eph receptors (which bind ephrins) and neuropilin receptors (which bind semaphorins) and are thus responsive to ephrins and semaphorins on the posterior sclerotome cells (Wang and Anderson 1997; Krull et al. 1999; Kuan et al. 2004). In this way, the same signals that pattern neural crest cell migration also pattern the spinal neuronal outgrowths.

Found throughout the animal kingdom, the semaphorins usually guide growth cones by selective repulsion. They are especially important in forcing "turns" when an axon must change direction. Semaphorin-1, for example, is a transmembrane protein that is expressed in a band of epithelial cells in the developing insect limb. This protein appears to inhibit the growth cones of the Ti1 sensory neurons from moving forward,

FIGURE 15.29 Repulsion of dorsal root ganglion growth cones. (A) Motor axons migrating through the rostral (anterior), but not the caudal (posterior), compartments of each sclerotome. (B) In vitro assay, wherein ephrin stripes were placed on a background surface of laminin. Motor axons grew only where the ephrin was absent. (C) Inhibition of growth cones by ephrin after 10 minutes of incubation. The left-hand photograph shows a control axon exposed to a similar (but not inhibitory) compound; the axon on the right was exposed to an ephrin found in the posterior somite. (From Wang and Anderson 1997, courtesy of the authors.)

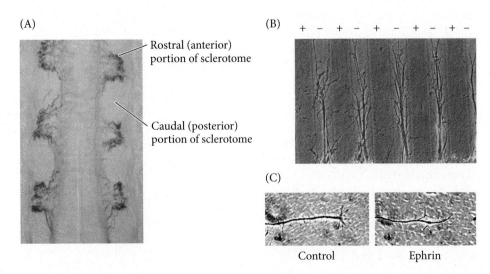

(A)

Rostral (anterior) portion of sclerotome

Caudal (posterior) portion of sclerotome

(B) + − + − + − + − + −

(C)

Control   Ephrin

FIGURE 15.30 Action of semaphorin-1 in the developing grasshopper limb. The axon of sensory neuron Ti1 projects toward the central nervous system. (The arrows represent sequential steps en route.) When it reaches a band of semaphorin-1-expressing epithelial cells, the axon reorients its growth cone and extends ventrally along the distal boundary of the semaphorin-1-expressing cells. When its filopodia connect to the Cx1 pair of cells, the growth cone crosses the boundary and projects into the central nervous system. When semaphorin-1 is blocked by antibodies, the growth cone searches randomly for the Cx1 cells. (After Kolodkin et al. 1993.)

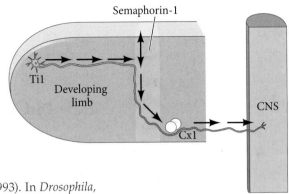

thus causing them to turn (**FIGURE 15.30**; Kolodkin et al. 1992, 1993). In *Drosophila*, semaphorin-2 is secreted by a single large thoracic muscle. In this way, the thoracic muscle prevents itself from being innervated by inappropriate axons (Matthes et al. 1995).

The proteins of the semaphorin-3 family, also known as collapsins, are found in mammals and birds. These secreted proteins collapse the growth cones of axons originating in the dorsal root ganglia (Luo et al. 1993). There are several types of neurons in the dorsal root ganglia whose axons enter the dorsal spinal cord. Most of these axons are prevented from traveling farther and entering the ventral spinal cord; however, a subset of them does travel ventrally through the other neural cells (**FIGURE 15.31**). These particular axons are not inhibited by semaphorin-3, whereas those of the other neurons are (Messersmith et al. 1995). This finding suggests that semaphorin-3 patterns sensory projections from the dorsal root ganglia by selectively repelling certain axons so that they terminate dorsally. A similar scheme is seen in the brain, where semaphorin made in one region of the brain prevents the entry of neurons that originated in another region (Marín et al. 2001).

(A)

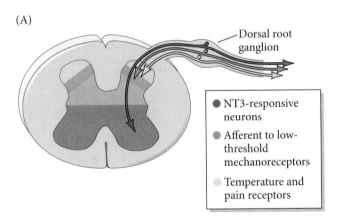

FIGURE 15.31 Semaphorin-3 as a selective inhibitor of axonal projections into the ventral spinal cord. (A) Trajectory of axons in relation to semaphorin-3 expression in the spinal cord of a 14-day embryonic rat. Neurons that are responsive to neurotrophin 3 (NT3) can travel to the ventral region of the spinal cord, but the afferent axons for the mechanoreceptors and for temperature and pain receptor neurons terminate dorsally. (B) Transgenic chick fibroblast cells that secrete semaphorin-3 inhibit the outgrowth of mechanoreceptor axons. These axons are growing in medium treated with nerve growth factor (NGF), which stimulates their growth, but they are still inhibited from growing toward the source of semaphorin-3. (C) Neurons that are responsive to NT3 for growth are not inhibited from extending toward the source of semaphorin-3 when grown with NT3. (A after Marx 1995; B,C from Messersmith et al. 1995, courtesy of A. Kolodkin.)

(B)

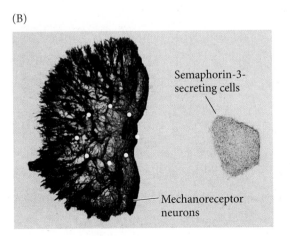

(C)

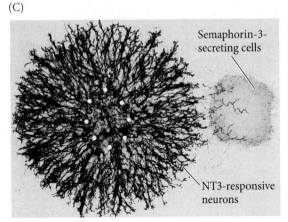

In some circumstances, ephrins and semaphorins can also be attractants. For instance, semaphorin-3A is a classic chemorepellant for *axons* coming from pyramidal neurons in the mammalian cortex; however, it is a chemoattractant for *dendrites* of these same cells. In this way, a target can "reach out" to the dendrites of these cells without also attracting their axons (Polleux et al. 2000).

How Did the Axon Cross the Road?

The idea that chemotactic cues guide axons in the developing nervous system was first proposed by Santiago Ramón y Cajal (1892). He suggested that diffusible molecules might signal the commissural neurons of the spinal cord to send axons from their dorsal positions in the neural tube to the ventral floor plate. Commissural neurons coordinate right and left motor activities. To accomplish this, they somehow must migrate to (and through) the ventral midline. The axons of commissural neurons begin growing ventrally down the side of the neural tube. About two-thirds of the way down, however, the axons change direction and project through the ventrolateral (motor) neuron area of the neural tube toward the floor plate cells (**FIGURE 15.32**).

There appear to be two systems involved in attracting the axons of dorsal commissural neurons to the ventral midline. The first is the Sonic hedgehog protein (Shh), which starts the commissural neurons on their ventral migrations. (Recall from Chapter 13 the importance of Shh as a morphogen for patterning cell fate; see Figures 13.19 and 4.30.) Shh is made in and secreted from the floor plate and is distributed in a concentration gradient that is high ventrally and low dorsally. If Shh signaling is inhibited by cyclopamine (an inhibitor of Smoothened, the main transducer of Shh) or if Smoothened is conditionally knocked out in commissural neurons, the commissural axons have difficulty getting to the ventral midline and turning toward the midline (Charron et al. 2003). It appears, however, that commissural axon guidance by Shh signaling operates in a *noncanonical* manner through the alternative receptor, Brother of Cdo (Boc), and is independent of Gli-mediated transcriptional regulation (see Figure 4.30; Okada et al. 2006; Yam et al. 2009). In addition, loss of the Shh gradient does not eliminate all midline crossing of commissural axons, which suggests that some other factor is also involved.

NETRIN In 1994, Serafini and colleagues developed an assay that allowed them to screen for the presence of a presumptive diffusible molecule that might be guiding the commissural neurons. When dorsal spinal cord explants from chick embryos were explanted onto collagen gels, the presence of floor plate cells near them promoted the outgrowth of commissural axons. Serafini and colleagues took fractions of embryonic chick brain homogenate and tested them to see if any of the proteins therein mimicked explant activity. This research resulted in the identification of two proteins, **netrin-1** and **netrin-2**. Like Shh,

FIGURE 15.32 Trajectory of commissural axons in the rat spinal cord. (A) Schematic drawing of a model wherein commissural neurons first experience a gradient of Sonic hedgehog and netrin-2 and then a steeper gradient of netrin-1. The commissural axons are chemotactically guided ventrally down the lateral margin of the spinal cord toward the floor plate. Upon reaching the floor plate, contact guidance from the floor plate cells causes the axons to change direction. (B) Autoradiographic localization of *netrin-1* mRNA by in situ hybridization of antisense RNA to the hindbrain of a young rat embryo. *netrin-1* mRNA (dark area) is concentrated in the floor plate neurons. (B from Kennedy et al. 1994, courtesy of M. Tessier-Lavigne.)

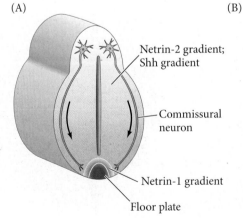

(A)

Netrin-2 gradient; Shh gradient

Commissural neuron

Netrin-1 gradient

Floor plate

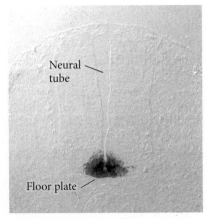

(B)

Neural tube

Floor plate

netrin-1 is made by and secreted from the floor plate cells, whereas netrin-2 is synthesized in the lower region of the spinal cord but not in the floor plate (see Figure 15.32B). It is possible that the commissural neurons first encounter a gradient of netrin-2 and Shh, which brings them into the domain of the steeper netrin-1 gradient. The netrins are recognized by the receptors DCC and DSCAM, which are found in commissural axon growth cones (Liu et al. 2009).

Although they are soluble molecules, both netrins become associated with the extracellular matrix.[11] Such associations can play important roles and may change the effect of netrin from attractive to repulsive, as seen in *Xenopus* retinal neurons (Höpker et al. 1999). The structures of the netrin proteins have numerous regions of homology with UNC-6, a protein implicated in directing the migration of axons around the body wall of *C. elegans*. In the wild-type nematode, UNC-6 induces axons from certain centrally located sensory neurons to move ventrally while inducing ventrally placed motor neurons to extend axons dorsally. In *unc-6* loss-of-function mutations, neither of these migrations occurs (Hedgecock et al. 1990; Ishii et al. 1992; Hamelin et al. 1993). Mutations of the *unc-40* gene disrupt ventral (but not dorsal) axon migration, whereas mutations of the *unc-5* gene prevent only dorsal migration (**FIGURE 15.33**). Genetic and biochemical evidence suggests that UNC-5 and UNC-40 are portions of the UNC-6 receptor complex, and that UNC-5 can convert a UNC-40-mediated attraction into a repulsion (Leonardo et al. 1997; Hong et al. 1999; Chang et al. 2004).

There is reciprocity in science. Just as research on vertebrate netrin genes led to the discovery of their *C. elegans* homologues, research on the nematode *unc-5* gene led to the discovery of the gene encoding the mammalian netrin receptor. This gene turns out to be one whose mutation in mice causes a disease called rostral cerebellar malformation (Ackerman et al. 1997; Leonardo et al. 1997). Similarly, the "netrin DCC" receptor received its name from analysis of mutated genes associated with cancer and garnered its acronym from "*d*eleted in *c*olorectal *c*ancer."

Recently, **Vegf** has been identified as a third midline attractant that cooperates with Shh and Netrin to guide commissural axons ventromedially to the floor plate. Moreover, in vitro studies indicate that all three of these attractants may be involved in signaling **Src family kinases (SFKs)** to mediate growth cone responses (Li et al. 2004; Meriane et al. 2004; Yam et al. 2009; Ruiz de Almodovar et al. 2011). It will be exciting to see if future in vivo studies reveal possible spatiotemporal roles for SFKs in the midline crossing of commissural axons.

SLIT AND ROBO It seems that to cross the midline and grow away from it on the *contralateral side* (opposite to the side of the CNS in which the soma is residing), repulsive cues are needed as a driving force. One important chemorepulsive group of molecules is the **Slit** proteins, which are expressed and secreted by midline cells (recently

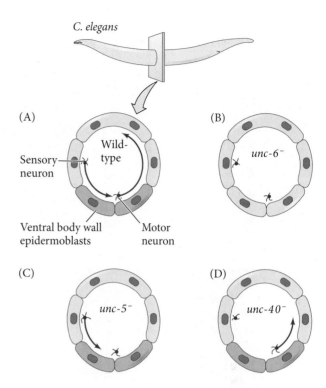

FIGURE 15.33 UNC expression and function in axonal guidance. (A) In the body of the wild-type *C. elegans* embryo, sensory neurons project ventrally, and motor neurons project dorsally. The ventral body wall epidermoblasts expressing UNC-6 are darkly shaded. (B) In the *unc-6* mutant embryo, neither of these migrations occurs. (C) The *unc-5* loss-of-function mutation affects only the dorsal movements of the motor neurons. (D) The *unc-40* loss-of-function mutation affects only the ventral migration of the sensory growth cones. (After Goodman 1994.)

Developing Questions

You have learned that Shh is in part responsible for attracting commissural axons to the floor plate. What about other classical morphogen signals? Can BMPs from the dorsal neural tube affect dorsoventral pathfinding, or can Wnts expressed in an anterior-to-posterior gradient in the neural tube influence longitudinal axon guidance?

---

[11] Nature does not necessarily conform to the categories we humans create. The binding of a soluble factor to the extracellular matrix makes for an interesting ambiguity between *chemotaxis* (movement toward a specific chemical substance) and *haptotaxis* (migration along a preferred substrate). There is also some confusion between the terms *neurotropic* and *neurotrophic*. Neurotropic (Latin, *tropicus*, "a turning movement") means that something attracts the neuron. Neurotrophic (Greek, *trophikos*, "nursing" or "nourishing") refers to a factor's ability to keep the neuron alive, usually by supplying growth factors. Because many agents have both properties, they are alternatively called *neurotropins* and *neurotrophins*. In the recent literature, *neurotrophin* appears to be more widely used.

(A) Slit protein

(B) Robo protein

(C) Wild-type

(D) *Slit$^{-/-}$*

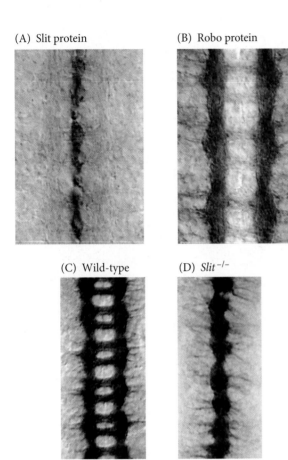

FIGURE 15.34 Robo/Slit regulation of midline crossing by neurons. Robo and Slit in the *Drosophila* central nervous system. (A) Antibody staining reveals Slit protein in the midline glial cells. (B) Robo protein appears along the neurons of the longitudinal tracts of the CNS axon scaffold. (C) Wild-type CNS axon scaffold shows the ladderlike arrangement of neurons crossing the midline. (D) Staining of the CNS axon scaffold with antibodies to all CNS neurons in a *Slit* loss-of-function mutant shows axons entering but failing to leave the midline (instead of running alongside it). (From Kidd et al. 1999, courtesy of C. S. Goodman.)

reviewed in Neuhaus-Follini and Bashaw 2015; Martinez and Tran 2015). In *Drosophila*, Slit is secreted by the glial cells at the midline of the nerve cord, and it acts to prevent most axons from crossing the midline from either side. **Roundabout (Robo)** proteins (Robo1[12], Robo2, and Robo3) are the receptors for Slit (Rothberg et al. 1990; Kidd et al. 1998; Kidd et al. 1999). In *Drosophila,* Robo receptors in the growth cones of pathfinding neurons function to prevent migration across the midline and, depending on the different combinations of expressed Robo receptors, instruct the lateral positioning of longitudinal tracts[13] relative to the midline (Rajagopalan et al. 2000; Simpson et al. 2000; Bhat 2005; Spitzweck et al. 2010). Loss of Slit or the combined loss of Robo1 and Robo2 results in a failure of proper axon crossing of the midline such that the axons grow to the midline and then cross, re-cross, and continue to extend in parallel along the midline (**FIGURE 15.34**). These and other results have led to a model in which the commissural neurons that cross from one side of the midline to the other temporarily avoid this repulsion by downregulating Robo1/2 proteins as they approach the midline. Once the growth cone is across the middle of the embryo, the neurons re-express Robos at the growth cone and once again become sensitive to the midline inhibitory actions of Slit (Brose et al. 1999; Kidd et al. 1999; Orgogozo et al. 2004).

In *Drosophila*, this mechanistic change in responsiveness is controlled by an endosomal protein called Commissureless (Comm), which is expressed only in precrossing axons and functions to misroute Robo proteins to the lysosome instead of permitting their expression at the membrane. In *Drosophila*, this endosomal trafficking mechanism for protein enables a faster change in the responsiveness of midline-crossing commissural axons than would be possible through regulation of *robo* gene expression (**FIGURE 15.35**; Keleman et al. 2002; 2005; Yang et al. 2009).

Vertebrates also use Slit and Robo signaling for midline repulsion, but the switch in growth cone responsiveness between pre- and postcrossing axons is somewhat different. In vertebrates there are several Slit (1–3) and Robo (1–4) proteins with Robo3 having two isoforms, Robo3.1 and Robo3.2, which are expressed in pre- and postcrossing commissural axons, respectively (Mambetisaeva et al. 2005). As neurons extend axons toward the midline, those axons destined to remain ipsilateral (i.e., same side of the CNS on which the soma resides) are expressing Robo1 and Robo2, so they are repelled from crossing the midline by Slit (see Figure 15.38). Axons expressing Robo3.1, however, are able to cross the midline. Although the precise mechanism is unclear, Robo3.1 is suggested to actively promote midline crossing, since knockdown of *Robo3.1* results in the failure of commissural axons to cross the midline. Once the commissural neuron's growth cone has crossed the midline, it downregulates Robo3.1 and upregulates Robo1, 2, and 3.2, which prevents the growth cone from re-crossing the midline and allows the full force of Slit to act as a chemorepellent, thus forcing the growth cone away from the midline (see Figure 15.35, mouse; Long et al. 2004; Sabatier et al. 2004; Woods 2004; Chen et al. 2008).

In addition to the downregulation of Robo3.1, the change in growth cone responsiveness at the midline is reinforced by altering how the neuron interprets gradients

---

[12]Although historically Roundabout1 in *Drosophila* has been referred to simply as Robo (without the number 1), to avoid confusion we call it Robo1 throughout this text.

[13]Axon pathways in the central nervous system are called *tracts*, while axon pathways in the peripheral nervous system are called *nerves*.

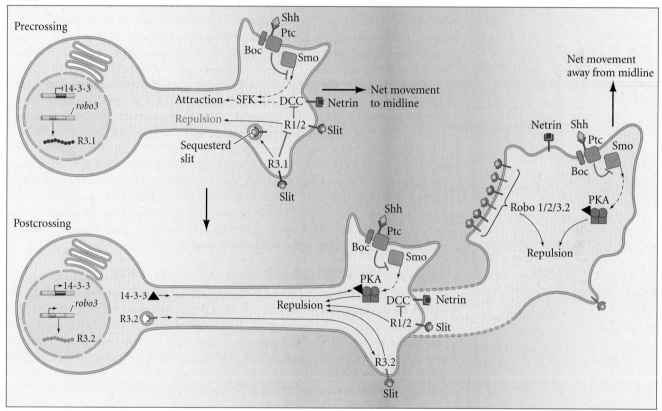

FIGURE 15.35 Model of axon guidance of commissural neurons crossing the midline in the fly and the mouse. Illustration shows a single commissural neuron residing in the left hemisphere of the fly ventral nerve cord (top) or mouse neural tube (bottom). Slit-Robo signaling mediates repulsion. Netrin and Shh elicit attraction to the midline in precrossing commissural axons through their Frazzled/DCC and the Ptc-Boc-Smo receptor complex, respectively. In postcrossing axons, 14-3-3 protein becomes upregulated, which shifts the responsiveness to Shh by influencing PKA downstream of Shh signaling. In flies, precrossing axons are essentially blind to Slit repulsive cues due to the redirecting of Robo receptors to the lysosome by Commissureless (Comm) for degradation. Once at the midline, however, increased signaling through the Netrin receptor Frazzled triggers the downregulation of Comm. Robo is then returned to the growth cone, and Slit-mediated repulsion ensues so that the axon will not recross. In vertebrates, Robo1/2 (R1/2) receptors are capable of inhibiting Netrin-DCC binding; therefore, in precrossing axons, this repression and Slit-Robo repulsion in general needs to be attenuated. The Robo3.1 (R3.1) isoform *may* function to inhibit Robo1/2, thus permitting Netrin-mediated attraction. Moreover, Robo3.1 may also competitively sequester Slit, with no direct downstream guidance outcome, which would serve to reduce available Slit to bind Robo1/2. In postcrossing axons, however, the Robo3.2 (R3.2) isoform is upregulated, and the 3.1 isoform is lost. Robo3.2 seems to function similarly as a canonical Slit repellent.

of Shh. Postcrossing axons upregulate 14-3-3 proteins that function through protein kinase A (PKA) to alter the growth cone's interpretation of the Shh from attractive to repulsive (see Figure 15.35; Yam et al. 2012). Therefore, the precise spatial and temporal regulation of Slit-Robo and Shh signaling in pre- and postcrossing axons enables commissure formation. Mutations in the human *ROBO3* gene disrupt the normal crossing of axons from one side of the brain's medulla to the other (Jen et al. 2004). Among other problems, people with this mutation are unable to coordinate their eye movements.

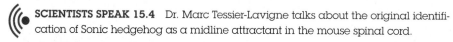 **SCIENTISTS SPEAK 15.4** Dr. Marc Tessier-Lavigne talks about the original identification of Sonic hedgehog as a midline attractant in the mouse spinal cord.

WEB TOPIC 15.5 **THE EARLY EVIDENCE FOR CHEMOTAXIS** Before molecular techniques, investigators used transplantation experiments and ingenuity to reveal evidence that chemotactic molecules were being released by target tissues.

The Travels of Retinal Ganglion Axons

Nearly all the mechanisms for neuronal specification and axon specificity mentioned in this chapter can be seen in the ways individual retinal neurons send axons to the visual processing areas of the brain. Despite some differences, retinal development and axon guidance are largely conserved across vertebrates. Even the strategy for setting up the retinal ganglion layer of neurons has significant similarity to the specification of neuroepithelia throughout the brain. For instance, retinal ganglion cells (RGCs) are initially patterned by the spatiotemporal actions of canonical Sonic hedgehog signaling, which primarily seems to regulate RGC number (Neumann and Nuesslein-Volhard 2000; Zhang and Yang 2001; Dakubo et al. 2003; Wang et al. 2005; Sánchez-Arrones et al. 2013). Also, regulation of cell fate choices between neuron and glia seem to be controlled by Notch-Delta function in the retina such that Notch promotes glial cell fates and represses neuronal cell identities (Austin et al. 1995; Dorsky et al. 1995; Ahmad et al. 1997; Dorsky et al. 1997; Furukawa et al. 2000; Jadhav et al. 2006; Yaron et al. 2006; Nelson et al. 2007; Luo et al. 2012). Last, as with motor neurons, the LIM family of transcription factors (Islet-2) is differentially expressed in the developing retinal ganglion layer specifying cell fate, which ultimately determines the repertoire of growth cone receptors to guide axon pathfinding to the tectum (reviewed in Bejarano-Escobar et al. 2015).

Growth of the retinal ganglion axon to the optic nerve

The first steps in getting retinal ganglion cell axons to their specific regions of the optic tectum take place within the retina (the neural retina of the optic cup). As the RGCs differentiate, their position in the inner margin of the retina is determined by cadherin molecules (N-cadherin and retina-specific R-cadherin) on their cell membranes (Matsunaga et al. 1988; van Horck et al. 2004). The RGC axons grow along the inner surface of the retina toward the optic disc (the head of the optic nerve). The mature human optic nerve will contain more than a million retinal ganglion axons.

INTRARETINAL GUIDANCE The adhesion and growth of retinal ganglion axons along the inner surface of the retina may be governed by the retina's laminin-containing basal lamina. The embryonic lens and the periphery of the retina secrete inhibitory factors (probably chondroitin sulfate proteoglycans) that repel the RGC axons, preventing them from traveling in the wrong direction (**FIGURE 15.36**; Hynes and Lander 1992; Ohta et al. 1999). NCAM may also be especially important here because the directional migration of the retinal ganglion growth cones depends on the N-CAM-expressing glial endfeet at the inner retinal surface (Stier and Schlosshauer 1995). In the mouse retina, RGCs express Robo1 and Robo2 receptors, and Slits are expressed in both the ganglion layer and lens epithelium. Functional analysis of Slit and Robo during *intraretinal* pathfinding of RGC axons suggests that Slits and Robo2 play a role in repelling RGC

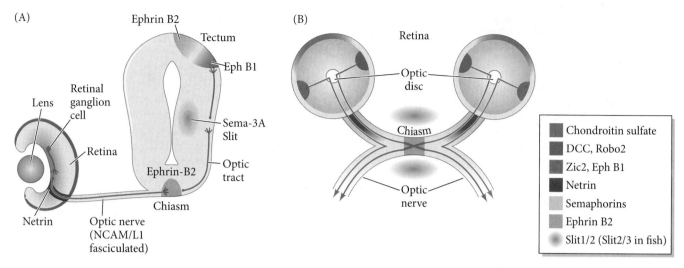

(A)

Ephrin B2
Tectum
Eph B1
Lens
Retinal ganglion cell
Retina
Sema-3A Slit
Optic tract
Ephrin-B2
Chiasm
Netrin
Optic nerve (NCAM/L1 fasciculated)

(B)

Retina
Optic disc
Chiasm
Optic nerve

Chondroitin sulfate
DCC, Robo2
Zic2, Eph B1
Netrin
Semaphorins
Ephrin B2
Slit1/2 (Slit2/3 in fish)

axons out of the retina (Niclou et al. 2000; Thompson et al. 2006, 2009). The secretion of netrin-1 by the cells of the optic disc (where the axons are assembled to form the optic nerve) plays a role in this migration as well. Mice lacking the genes for either netrin-1 or for the netrin receptor (found in the retinal ganglion axons) have poorly formed optic nerves because many of the axons fail to leave the eye and instead grow randomly around the disc (Deiner et al. 1997). The role of netrin may change in different parts of the eye. At the entrance to the optic nerve, netrin-1 is co-expressed with laminin on the surface of the retina. Laminin converts netrin from having an attractive signal to having a repulsive signal. This repulsion might "push" the growth cone away from the retinal surface and into the head of the optic nerve, where netrin is expressed without laminin (Mann et al. 2004; see Figure 15.36A).

Upon their arrival at the optic nerve, the migrating axons fasciculate (form a bundle) with axons already present there. N-CAM and L1 cell adhesion molecules are critical to this fasciculation, and antibodies against L1 or N-CAM cause the axons to enter the optic nerve in a disorderly fashion, which in turn causes them to emerge into the tectum at the wrong positions (Thanos et al. 1984; Brittis et al. 1995; Yin et al. 1995).

Growth of the retinal ganglion axon through the optic chiasm

In non-mammalian vertebrates, the final destination for RGC axons is a portion of the brain called the optic tectum, while mammalian RGC axons go to the lateral geniculate nuclei. At many points, the journey of RGC axons within the brain occurs on an astroglial substrate. When the axons enter the optic nerve, they grow on astroglial cells toward the midbrain (Bovolenta et al. 1987; Marcus and Easter 1995; Barresi et al. 2005). Laminin appears to promote crossing of the optic chiasm. On their way to the optic tectum, the axons of non-mammalian vertebrates travel on a pathway (the optic tract) over glial cells whose surfaces are coated with laminin. Very few areas of the brain contain laminin, and the laminin in this pathway exists only when the optic nerve fibers are growing on it (Cohen et al. 1987).

After leaving the eye, the RGC axons appear to grow on surfaces of netrin, surrounded on all sides by semaphorins keeping them on track by providing repulsive cues (see Harada et al. 2007). Upon entering the brain, mammalian RGC axons reach the optic chiasm, where they have to "decide" if they are to continue straight or if they are to turn 90° and enter the other side of the brain. In the optic chiasm area, semaphorins are no longer present, but Slit proteins take over their function by establishing a corridor (perpendicular to the midline) through which the axons must travel, preventing inappropriate exploration outside of the chiasm location. (Thus, although the optic chiasm occurs at the midline, it is using a very different strategy of Slit-mediated repulsion from that used in the midline of the ventral spinal cord; see Figure 15.35.) As

FIGURE 15.36 Multiple guidance cues direct the movement of retinal ganglion cell (RGC) axons to the optic tectum. Guidance molecules belonging to the netrin, Slit, semaphorin, and ephrin families are expressed in discrete regions at several sites along the pathway to direct the RGC growth cones. RGC axons are repelled from the retinal periphery, probably by chondroitin sulfate. At the optic disc, the axons exit the retina and enter the optic nerve, guided by netrin/DCC-mediated attraction. Once in the optic nerve, the axons are kept within the pathway by inhibitory interactions. Slit proteins in the optic chiasm create zones of inhibition. Zic2-expressing ganglia in the ventrotemporal retina project Eph B1-expressing axons, which are repelled at the chiasm by ephrin B2, thus terminating at ipsilateral (same-side) targets. Neurons from the medial portions of the retina do not express Eph B1 and proceed to the opposite (contralateral) side. (A) Cross section. (B) Dorsal view. Not all cues are shown. (A after van Horck et al. 2004; B after Harada et al. 2007.)

in the retina, Robo2 appears to be the major mediator of RGC guidance at the ventral forebrain where the optic chiasm is forming (see Figure 15.36B).

In fish, all RGC axons cross at the optic chiasm to the contralateral side, but in mammals, some portion of RGC axons remain on the ipsilateral side. It appears that those axons not destined to cross to the opposite side of the brain are repulsed from doing so when they enter the optic chiasm (Godement et al. 1990). This repulsion appears to be influenced by the synthesis of ephrin and Shh in the cells occupying the chiasm. These midline cues are interpreted by the receptors Eph and Boc that are uniquely expressed on ipsilaterally projecting RGCs (Cheng et al. 1995; Marcus et al. 2000; Fabre et al. 2010).

In the mouse eye, the Eph B1 receptor is expressed on those temporal axons that are repelled by the optic chiasm's ephrin B2, and those axons project to the side of the tectum on the same side as their eye; Eph B1 is nearly absent on axons that are allowed to cross over. Mice lacking the *Eph B1* gene show hardly any ipsilateral projections. This pattern of *Eph B1* expression appears to be regulated by the Zic2 transcription factor found in those retinal axons that do form ipsilateral projections (Herrera et al. 2003; Williams et al. 2003; Pak et al. 2004). In addition, loss of the alternative Shh receptor Boc results in the aberrant contralateral pathfinding of specific temporal RGCs that typically remain on the ipsilateral side. Furthermore, ectopic expression of Boc in typically contralateral axons will cause them to now project ipsilaterally (Fabre et al. 2010). These results suggest that both Eph B1/ephrin B2 and Boc/Shh are the major regulators of the decision to cross or not to cross the forebrain midline.

Ephrin appears to play a similar role in the retinotectal mapping in the frog. In the developing frog, the ventral axons express the Eph B receptor, whereas the dorsal axons do not. Before metamorphosis, both axons cross the optic chiasm. When the frog's nervous system is being remodeled during metamorphosis, however, the chiasm expresses ephrin B, which causes a subpopulation of ventral cells to be repulsed and project to the same side rather than cross the chiasm (Mann et al. 2002). This arrangement allows the frog to have binocular vision, which is very good if one is trying to catch flies with one's tongue.

Target Selection: "Are We There Yet?"

In some cases, nerves in the same ganglion can have several different targets. How does a neuron know which cell to form a synapse with? The general mechanisms of ligand-receptor specificity that lead a growth cone to the target tissue in the first place take on a refining role at the final destination. Different neurons in the same ganglion can have different receptors that can respond to certain cues and not to others. Once a neuron reaches a group of cells in which lie its potential targets, it is responsive to various proteins produced by the target cells.[14] Both attractive and repulsive forces steer the axon into its appropriate final "parking spot." As we will see in the RGC axons, the amount of repulsive proteins (such as ephrins) can be critical in getting particular neurons to particular targets (Gosse et al. 2008). What are the important signals that direct axons to the right addresses?

Chemotactic proteins

ENDOTHELINS Some neurons in the superior cervical ganglia (the largest ganglia in the neck) go toward the carotid artery, whereas other neurons from these same ganglia do not. It appears that those axons extending from the superior cervical ganglia to the carotid artery follow blood vessels that also lead there. These blood vessels secrete small peptides called **endothelins**. In addition to their adult roles constricting blood vessels,

---

[14]As is seen throughout developmental biology, the metaphor of a "target" is problematic. Here, the target is not a passive entity, but an importantly active one.

endothelins appear to have an embryonic role, as they are able to direct the migration of certain neural crest cells (such as those entering the gut) and of certain sympathetic axons that have endothelin receptors on their membranes (Makita et al. 2008).

WEB TOPIC 15.6 **BMP4 AND TRIGEMINAL GANGLION NEURONS** Axon bundles from neurons in the trigeminal ganglion innervate the eye regions and the upper and lower jaws. BMP4 from their target organs guides these axons.

NEUROTROPHINS Some target cells produce a set of chemotactic proteins collectively called **neurotrophins.** The neurotrophins include **nerve growth factor** (**NGF**), **brain-derived neurotrophic factor** (**BDNF**), **conserved dopamine neurotrophic factor** (**CDNF**), and **neurotrophins 3 and 4/5** (**NT3**, **NT4/5**). These proteins are released from potential target tissues and work at short ranges as either chemotactic factors or chemorepulsive factors (Paves and Saarma 1997). Each can promote and attract the growth of some axons to its source while inhibiting other axons. For instance, sensory neurons from the rat dorsal root ganglia are attracted to sources of NT3 (**FIGURE 15.37**) but are inhibited by BDNF. Neurotrophins are probably transported from the axon growth cone to the soma of the neuron. For instance, NGF derived from the hippocampus of the brain binds to receptors on the axons of basal forebrain neurons and is endocytosed into these neurons. It is then transported back to the nerve cell body, where it stimulates gene expression. Increased expression of *App* (a gene on chromosome 21 that encodes ameloid precursor protein) is seen in people with Down syndrome and Alzheimer's disease. Increased App protein blocks the retrograde transport of NGF from the axon to the cell body and affects the sensitivity and localization of the receptor for NGF on the cell membrane (Salehi et al. 2006; Matrone et al. 2011). Thus, the NGF pathway is being studied for possible roles in the treatment of impaired cognition.

CHEMOTROPHINS: QUALITY AND QUANTITY The attachment of an axon to its target can be either "digital" or "analogue." In "analogue" mode, different axons recognize the same molecule on the target, but the *amount* of the molecule on the target appears to be critical to the connections that form; this may be the case in the attachment of retinal neurons to the tectum in the fish brain (Gosse et al. 2008). In other situations, there may be extremely molecule-specific qualitative ("digital") binding such that certain connections are neuron-specific. This may be the case for retinal neurons in *Drosophila*. Dscam protein has several thousand splicing isoforms (see Chapter 3), and this variety may enable highly specific recognition of a given neuron with its target neurons (Millard et al. 2010; Zipursky and Sanes 2010). Given the complexity of neural connections, it is probable that both qualitative and quantitative cues are used. Growth cones do not rely on a single type of molecule to recognize their target, but integrate the simultaneously presented attractive and repulsive cues, selecting their targets based on the combined input of multiple signals (Winberg et al. 1998).

Target selection by retinal axons: "Seeing is believing"

When the retinal axons come to the end of the laminin-lined optic tract, they spread out and find their specific targets in the optic tectum. Studies on frogs and fish (in which retinal neurons from each eye project to the opposite side of the brain) have indicated that each retinal ganglion axon sends its impulse to one specific site (a cell or small group of cells) within the optic tectum (**FIGURE 15.38A**; Sperry 1951). There are two optic tecta in the frog brain. The axons from the right eye form synapses with the left optic tectum, and those from the left eye form synapses in the right optic tectum.

The map of retinal connections to the frog optic tectum (the **retinotectal projection**) was detailed by Marcus Jacobson (1967). Jacobson created this map by shining a narrow beam of light on a small, limited region of the retina and noting, by means of a recording electrode in the tectum, which tectal cells were being stimulated. The retinotectal projection of *Xenopus laevis* is shown in **FIGURE 15.38B**. Light illuminating the ventral part of

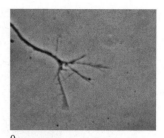

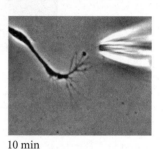

FIGURE 15.37 Embryonic axon from a rat dorsal root ganglion turning in response to a source of NT3. The photographs document the growth cone's turn over a 10-minute period. The same growth cone was insensitive to other neurotrophins. (From Paves and Saarma 1997, courtesy of M. Saarma.)

(A)

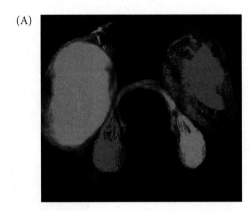

(B)

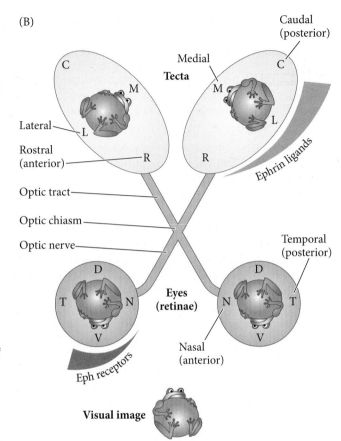

FIGURE 15.38 Retinotectal projections. (A) Confocal micrograph of axons entering the tecta of a 5-day zebrafish embryo. Fluorescent dyes were injected into the eyes of zebrafish embryos mounted in agarose. The dyes diffused down the axons and into each tectum, showing the retinal axons from the right eye going to the left tectum and vice versa. (B) Map of the normal retinotectal projection in adult *Xenopus*. The right eye innervates the left tectum, and the left eye innervates the right tectum. The dorsal (D) portion of the retina innervates the lateral (L) regions of the tectum. The nasal (anterior) region of the retina projects to the caudal (C) region of the tectum. (A courtesy of M. Wilson; B after Holt 2002, courtesy of C. Holt.)

the retina stimulates cells on the lateral surface of the tectum. Similarly, light focused on the temporal (posterior) part of the retina stimulates cells in the caudal portion of the tectum. These studies demonstrate a point-for-point correspondence between the cells of the retina and the cells of the tectum. When a group of retinal cells is activated, a very small and specific group of tectal cells is stimulated. Furthermore, the points form a continuum; in other words, adjacent points on the retina project onto adjacent points on the tectum. This arrangement enables the frog to see an unbroken image. This intricate specificity caused Sperry (1965) to put forward the **chemoaffinity hypothesis**:

> *The complicated nerve fiber circuits of the brain grow, assemble, and organize themselves through the use of intricate chemical codes under genetic control. Early in development, the nerve cells, numbering in the millions, acquire and retain thereafter, individual identification tags, chemical in nature, by which they can be distinguished and recognized from one another.*

Current theories do not propose a point-for-point specificity between each axon and the neuron that it contacts. Rather, evidence now demonstrates that gradients of adhesivity (especially those involving repulsion) play a role in defining the territories that the axons enter, and that activity-driven competition between these neurons determines the final connection of each axon.

Adhesive specificities in different regions of the optic tectum: Ephrins and Ephs

There is good evidence that retinal ganglion cells can distinguish between regions of the optic tectum. Cells taken from the ventral half of the chick neural retina preferentially adhere to dorsal (medial) halves of the tectum and vice versa (Gottlieb et al. 1976; Roth and Marchase 1976; Halfter et al. 1981). Retinal ganglion cells are specified along the dorsal-ventral axis by a gradient of transcription factors. Dorsal retinal cells are characterized by high concentrations of Tbx5 transcription factor, whereas ventral cells have high levels of Pax2. These transcription factors are induced by paracrine factors (BMP4 and retinoic acid, respectively) from nearby tissues (Koshiba-Takeuchi et al. 2000). Misexpression of Tbx5 in the early chick retina results in marked abnormalities of the retinotectal projection. Therefore, the retinal ganglion cells are specified according to their location.

One gradient that has been identified and functionally characterized is a gradient of repulsion, which is highest in the posterior tectum and weakest in the anterior tectum. Bonhoeffer and colleagues (Walter et al. 1987; Baier and Bonhoeffer 1992) prepared a "carpet" of tectal membranes with alternating "stripes" of membrane derived from the posterior and the anterior tecta. They then let cells from the nasal (anterior) or temporal (posterior) regions of the retina extend axons into the carpet. The nasal ganglion cells extended axons equally well on both the anterior and posterior tectal membranes. The neurons from the temporal side of the retina, however, extended axons only on the anterior tectal membranes. When the growth cone of a temporal retinal ganglion axon contacted a posterior tectal cell membrane, the growth cone's filopodia withdrew, and the cone collapsed and retracted (Cox et al. 1990).

The basis for this specificity appears to be two sets of gradients along the tectum and retina. The first gradient set consists of ephrin proteins and their receptors. In the optic tectum, ephrin proteins (especially ephrins A2 and A5) are found in gradients that are highest in the posterior (caudal) tectum and decline anteriorly (rostrally) (**FIGURE 15.39A**). Moreover, cloned ephrin proteins have the ability to repulse axons, and ectopically expressed ephrin will prohibit axons from the temporal (but not from the nasal) regions of the retina from projecting to where it is expressed (Drescher et al. 1995; Nakamoto et al. 1996). The complementary Eph receptors have been found on chick retinal ganglion cells, expressed in a temporal-to-nasal gradient along the retinal ganglion axons (Cheng et al. 1995). This gradient appears to be due to a spatially and temporally regulated expression of retinoic acid (Sen et al. 2005).

Ephrins appear to be remarkably pliable molecules. Concentration differences in ephrin A in the tectum can account for the smooth topographic map (wherein the position of neurons in the retina maps continuously onto the targets). Hansen and colleagues (2004) have shown that ephrin A can be an attractive as well as a repulsive signal for retinal axons. Moreover, their quantitative assay for axon growth showed that the origin of the axon determined whether it was attracted or repulsed by ephrins. Axon growth is promoted by low ephrin A concentrations that are anterior to the proper target and is inhibited by higher concentrations posterior to the correct target (**FIGURE 15.39B**). Each axon is thus led to the appropriate place and then told to go no farther. At

FIGURE 15.39 Differential retinotectal adhesion is guided by gradients of Eph receptors and their ligands. (A) Representation of the dual gradients of Eph receptor tyrosine kinase in the retina and its ligands (ephrin A2 and A5) in the tectum. (B) Experiment showing that temporal, but not nasal, retinal ganglion axons respond to a gradient of ephrin ligand in tectal membranes by turning away or slowing down. An equilibrium of attractive and repulsive forces inherent in the gradient may lead specific axons to their targets. (After Barinaga 1995; Hansen et al. 2004.)

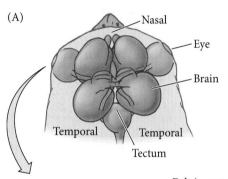

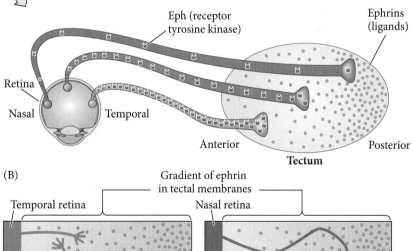

that equilibrium point, there would be no growth and no inhibition, and the synapses with the target tectal neurons could be made.

The second set of gradients parallels the ephrins and Ephs. The tectum has a gradient of Wnt3 that is highest at the medial region and lowest laterally (like the ephrin gradient). In the retina, a gradient of Wnt receptor is highest ventrally (like the Eph proteins). The two sets of gradients are both required to specify the coordinates of axon-to-tectum targets (Schmitt et al. 2006).

Synapse Formation

When an axon contacts its target (usually either a muscle cell or another neuron), it forms a specialized junction called a **synapse**. The axon terminal of the **presynaptic neuron** (i.e., the neuron transmitting the signal) releases chemical neurotransmitters that depolarize or hyperpolarize the membrane of the target cell (the **postsynaptic cell**). The neurotransmitters are released into the synaptic cleft between the two cells, where they bind to receptors in the target cell.

The construction of a synapse involves several steps (Burden 1998). When motor neurons in the spinal cord extend axons to muscles, the growth cones that contact newly formed muscle cells migrate over their surfaces. When a growth cone first adheres to

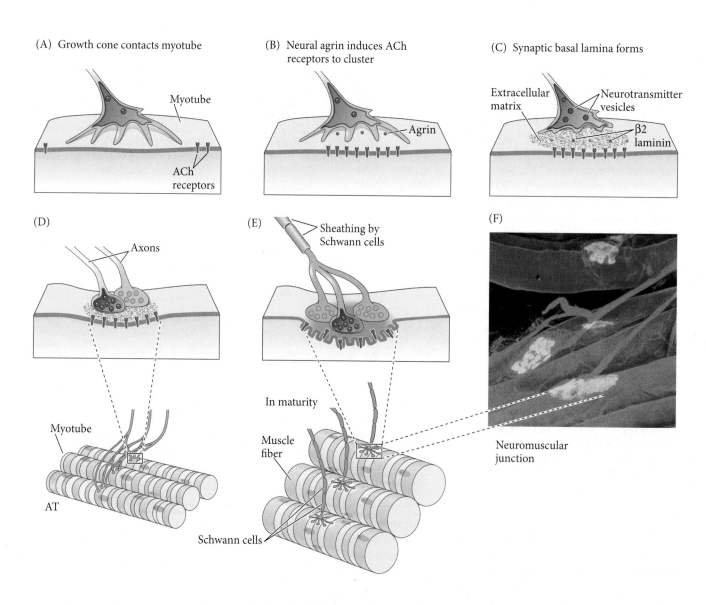

(A) Growth cone contacts myotube

Myotube

ACh receptors

(B) Neural agrin induces ACh receptors to cluster

Agrin

(C) Synaptic basal lamina forms

Extracellular matrix

Neurotransmitter vesicles

β2 laminin

(D)

Axons

Myotube

AT

(E)

Sheathing by Schwann cells

In maturity

Muscle fiber

Schwann cells

(F)

Neuromuscular junction

the cell membrane of a muscle fiber, no specializations can be seen in either membrane. However, the axon terminal soon begins to accumulate neurotransmitter-containing synaptic vesicles, the membranes of both cells thicken at the region of contact, and the synaptic cleft between the cells fills with an extracellular matrix that includes a specific form of laminin (**FIGURE 15.40A–C**). This muscle-derived laminin specifically binds the growth cones of motor neurons and may act as a "stop signal" for axonal growth (Martin et al. 1995; Noakes et al. 1995). In at least some neuron-to-neuron synapses, the synapse is stabilized by N-cadherin. The activity of the synapse releases N-cadherin from storage vesicles in the growth cone (Tanaka et al. 2000).

In muscles, after the first axon makes contact, growth cones from other axons converge at the site to form additional synapses. During mammalian development, all muscles that have been studied are innervated by at least two axons. This *polyneuronal innervation* is transient, however; soon after birth, all but one of the axon branches are retracted (**FIGURE 15.40D–F**). This "address selection" is based on competition between the axons (Purves and Lichtman 1980; Thompson 1983; Colman et al. 1997). When one of the motor neurons is active, it suppresses the synapses of the other neurons, possibly through a nitric oxide-dependent mechanism (Dan and Poo 1992; Wang et al. 1995). Eventually, the less active synapses are eliminated. The remaining axon terminal expands and is ensheathed by a Schwann cell (see Figure 15.40E).

A Program of Cell Death

"To be, or not to be: that is the question." One of the most puzzling phenomena in the development of the nervous system is neuronal cell death. In many parts of the vertebrate central and peripheral nervous systems, more than half the neurons die during the normal course of development. Moreover, there does not seem to be much conservation of apoptosis patterns across species. For example, about 80% of a cat's retinal ganglion cells die, whereas in the chick retina this figure is only 40%. In fish and amphibian retinas, no ganglion cells appear to die (Patterson 1992). *What causes this programmed cell death?*

Although we all are constantly poised over life-or-death decisions, this existential dichotomy is exceptionally stark for embryonic cells. Programmed cell death, or **apoptosis** (both *p*s are pronounced) is a normal part of development (see Fuchs and Steller 2011); the term comes from the Greek word for the natural process of leaves falling from trees or petals falling from flowers. Apoptosis is an active process and is subject to evolutionary selection. (A second type of cell death, **necrosis**, is a pathological death caused by external factors such as inflammation or toxic injury.)

In the nematode *C. elegans*, in which we can count the number of cells as the animal develops, exactly 131 cells die under the normal developmental pattern. All the cells of *C. elegans* are programmed to die unless they are actively told not to undergo apoptosis. In an adult human, as many as 10^{11} cells die each day and are replaced by other cells.

Developing Questions

There is significant synaptic plasticity in the brain over one's life, and problems with forming new synapses may underlie many disorders such as autism spectrum disorder. What role do the guidance and target-specifying cues play in synapse remodeling later in life?

FIGURE 15.40 Differentiation of a motor neuron synapse with a muscle in mammals. (A) A growth cone approaches a developing muscle cell. (B) The axon stops and forms an unspecialized contact on the muscle surface. Agrin, a protein released by the neuron, causes acetylcholine (ACh) receptors to cluster near the axon. (C) Neurotransmitter vesicles enter the axon terminal (AT), and an extracellular matrix connects the axon terminal to the muscle cell as the synapse widens. This matrix contains a nerve-specific laminin. (D) Other axons converge on the same synaptic site. The wider view (below) shows muscle innervation by several axons (seen in mammals at birth). (E) All but one of the axons are eliminated. The remaining axon can branch to form a complex neuromuscular junction with the muscle fiber. Each axon terminal is sheathed by a Schwann cell process, and folds form in the muscle cell membrane. The overview shows muscle innervation several weeks after birth. (F) Whole mount view of mature neuromuscular junction in a mouse. (A–E after Hall and Sanes 1993, Purves 1994, Hall 1995, F courtesy of M. A. Ruegg.)

(Indeed, the mass of cells we lose each year through normal cell death is close to our entire body weight!) During embryonic development, we were constantly making and destroying cells, and we generated about three times as many neurons as we eventually ended up with when we were born. Lewis Thomas (1992) wisely noted:

> *By the time I was born, more of me had died than survived. It was no wonder I cannot remember; during that time I went through brain after brain for nine months, finally contriving the one model that could be human, equipped for language.*

Apoptosis is necessary not only for the proper spacing and orientation of neurons, but also for generating the middle ear space, the vaginal opening in females, and the spaces between our fingers and toes (Saunders and Fallon 1966; Rodriguez et al. 1997; Roberts and Miller 1998). Apoptosis prunes unneeded structures (e.g., frog tails, male mammary tissue), controls the number of cells in particular tissues (neurons in vertebrates and flies), and sculpts complex organs (palate, retina, digits, heart).

The pathways for apoptosis were delineated primarily through genetic studies of *C. elegans*. Indeed, the importance of these pathways was recognized by awarding a Nobel Prize in Physiology or Medicine to Sydney Brenner, H. Robert Horvitz, and John E. Sulston in 2002. Through these studies, it was found that the proteins encoded by the *ced-3* and *ced-4* genes were essential for apoptosis and that, in the cells that did not undergo apoptosis, those genes were turned off by the product of the *ced-9* gene (**FIGURE 15.41A**; Hengartner et al. 1992). The CED-4 protein is a protease-activating factor that activates the gene for CED-3, a protease that initiates destruction of the cell. CED-9 can bind to and inactivate CED-4. Mutations that inactivate the gene for CED-9 cause numerous cells that would normally survive to activate their *ced-3* and *ced-4* genes and die, leading to the death of the entire embryo. Conversely, gain-of-function mutations in the *ced-9* gene cause its protein to be made in cells that would normally die, resulting in those cells surviving. Thus, the *ced-9* gene appears to be a binary switch that regulates the choice between life and death on the cellular level. It is possible that every cell in the nematode embryo is poised to die, with those cells that survive being rescued by the activation of the *ced-9* gene.

(A) *C. elegans*

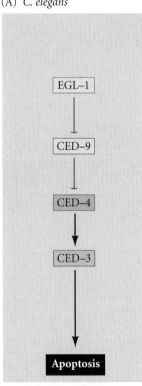

(B) Mammalian neurons

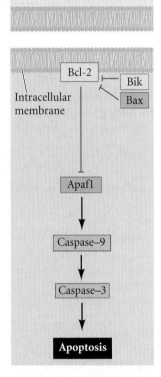

(C) *caspase-9*⁺/⁺ (wild-type)

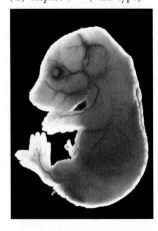

(D) *caspase-9*⁻/⁻ (knockout)

FIGURE 15.41 The loss of apoptosis can disrupt normal brain development. (A) In *C. elegans*, the CED-4 protein is a protease-activating factor that can activate CED-3. The CED-3 protease initiates the cell destruction events. CED-9 can inhibit CED-4 (and CED-9 can be inhibited upstream by EGL-1). (B) A similar pathway exists in mammals and appears to function in a similar manner. In this hypothetical scheme for the regulation of apoptosis in mammalian neurons, Bcl-X_L (a member of the Bcl-2 family) binds Apaf1 and prevents it from activating the precursor of caspase-9. The signal for apoptosis allows another protein (here, Bik) to inhibit the binding of Apaf1 to Bcl-X_L. Apaf1 is now able to bind to the caspase-9 precursor and cleave it. Caspase-9 dimerizes and activates caspase-3, initiating apoptosis. The same colors are used to represent homologous proteins. (C,D) In mice in which the genes for caspase-9 have been knocked out, normal neural apoptosis fails to occur, and the overproliferation of brain neurons is obvious. (C) A 6-day embryonic wild-type mouse. (D) A *caspase-9* knockout mouse of the same age. The enlarged brain protrudes above the face, and the limbs are still webbed. (A,B after Adams and Cory 1998; C,D from Kuida et al. 1998.)

The CED-3 and CED-4 proteins are at the center of the apoptosis pathway that is common to all animals studied. The trigger for apoptosis can be a developmental cue such as a particular molecule (e.g., BMP4 or glucocorticoids), the loss of adhesion to a matrix, or the lack of sufficient neurotrophic signals. Either type of cue can activate CED-3 or CED-4 proteins or inactivate CED-9 molecules. In mammals, the homologues of the CED-9 protein are members of the Bcl-2 family (which includes Bcl-2, Bcl-X, and similar proteins; **FIGURE 15.41B**). The functional similarities are so strong that if an active human *BCL-2* gene is placed in *C. elegans* embryos, it prevents normally occurring cell death (Vaux et al. 1992).

The mammalian homologue of CED-4 is Apaf1 (*apoptotic protease activating factor 1*). Apaf1 participates in the cytochrome *c*-dependent activation of the mammalian CED-3 homologues, the proteases caspase-9 and caspase-3 (see Figure 15.41; Shaham and Horvitz 1996; Cecconi et al. 1998; Yoshida et al. 1998). Activation of the caspase proteins results in a cascade of autodigestion—caspases are strong proteases that digest the cell from within, cleaving cellular proteins and fragmenting DNA.

Although apoptosis-deficient nematodes (i.e., worms deficient for CED-4) are viable despite having 15% more cells than wild-type worms, mice with loss-of-function mutations for either caspase-3 or caspase-9 die around the time of birth from massive cell overgrowth in the nervous system (**FIGURE 15.41C,D**; Jacobson et al. 1997; Kuida et al. 1996, 1998). Similarly, mice homozygous for targeted deletions of *Apaf1* have severe craniofacial abnormalities, brain overgrowth, and webbing between their toes.

WEB TOPIC 15.7 **THE USES OF APOPTOSIS** Apoptosis is used for numerous processes throughout development. This website explores the role of apoptosis in such phenomena as *Drosophila* germ cell development and the eyes of blind cave fish.

Activity-dependent neuronal survival

The apoptotic death of a neuron is not caused by any obvious defect in the neuron itself. Indeed, before dying, these neurons have differentiated and successfully extended axons to their targets. Rather, it appears that the target tissue regulates the number of axons innervating it by selectively supporting the survival of certain synapses. A recent study of motor neuron pathfinding in the limb exemplifies how neuronal survival is dependent on neuron activity (Hua et al. 2013). When manipulations of guidance systems cause motor axons to be misrouted to the incorrect muscle cells in the limb, the motor neurons survive despite the errors in target selection because they form successful synapses. In contrast, when motor neurons are unable to find their target muscle cells in the limb of mice in which the gene for Frizzled-3 (a Wnt/PCP receptor) is knocked out, the neurons never form synapses and consequently undergo apoptosis (**FIGURE 15.42**; Hua et al. 2013). These results strongly support the requirement of successful synapse formation for neuronal survival, and also indicate that the target cell (muscle in this case) must supply a signal to the presynaptic cell that promotes the neuron's survival. *What is this survival signal?*

Differential survival after innervation: The role of neurotrophins

The target tissue regulates the number of axons innervating it by limiting the supply of neurotrophins. In addition to their roles as chemotrophic factors described previously, neurotrophins regulate the survival of different subsets of neurons (**FIGURE 15.43**). NGF, for example, is necessary for the survival of sympathetic and sensory neurons. Treating mouse embryos with anti-NGF antibodies reduces the number of trigeminal sympathetic and dorsal root ganglion neurons to 20% of their control numbers (Levi-Montalcini and Booker 1960; Pearson et al. 1983). Furthermore, removing these neurons' target tissues results in the death of the neurons that would have innervated them, and there is a good correlation between the amount of NGF secreted and the survival of the neurons that innervate these tissues (Korsching and Thoenen 1983; Harper and

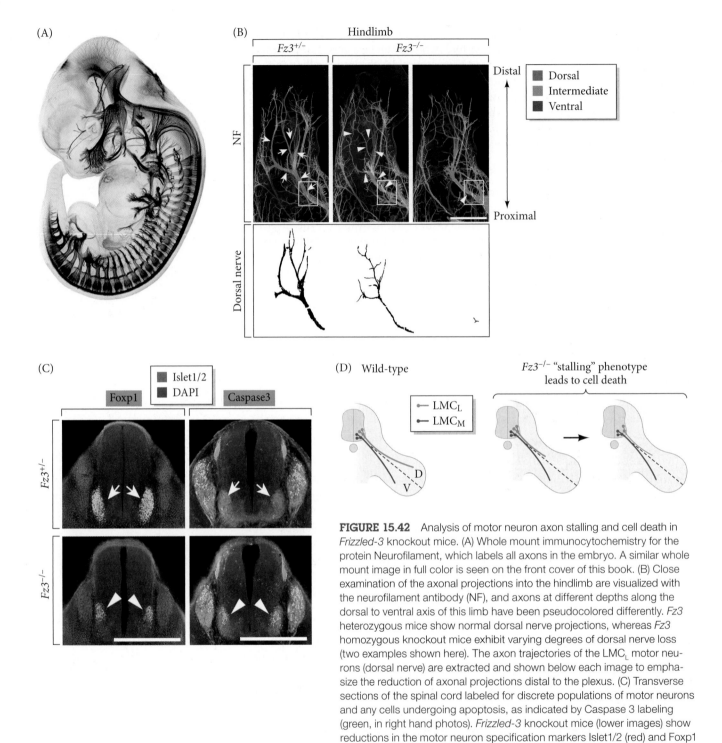

FIGURE 15.42 Analysis of motor neuron axon stalling and cell death in *Frizzled-3* knockout mice. (A) Whole mount immunocytochemistry for the protein Neurofilament, which labels all axons in the embryo. A similar whole mount image in full color is seen on the front cover of this book. (B) Close examination of the axonal projections into the hindlimb are visualized with the neurofilament antibody (NF), and axons at different depths along the dorsal to ventral axis of this limb have been pseudocolored differently. *Fz3* heterozygous mice show normal dorsal nerve projections, whereas *Fz3* homozygous knockout mice exhibit varying degrees of dorsal nerve loss (two examples shown here). The axon trajectories of the LMC$_L$ motor neurons (dorsal nerve) are extracted and shown below each image to emphasize the reduction of axonal projections distal to the plexus. (C) Transverse sections of the spinal cord labeled for discrete populations of motor neurons and any cells undergoing apoptosis, as indicated by Caspase 3 labeling (green, in right hand photos). *Frizzled-3* knockout mice (lower images) show reductions in the motor neuron specification markers Islet1/2 (red) and Foxp1 (green, in left hand photos), along with increased cell death specifically in the motor columns. (D) Schematics describe the phenotypes associated with a loss of *Frizzled-3/PCP* signaling. First, a stalling defect in motor neurons destined for the dorsal limb is followed by cell death. ([A–C] from Hua et al. 2013; D from Yung and Goodrich 2013.)

Davies 1990). In contrast, another neurotrophin, BDNF, does not affect sympathetic or sensory neurons, but it can rescue fetal motor neurons in vivo from normally occurring cell death and from induced cell death following the removal of their target tissue. The results of these in vitro studies have been corroborated by gene knockout experiments

(A) Sympathetic   (B) Dorsal root   (C) Nodose (taste)

NGF

BDNF

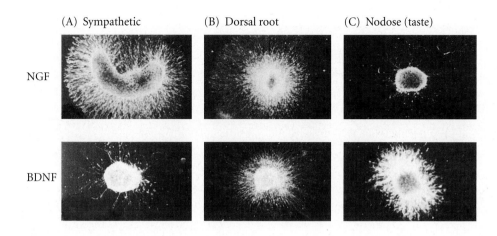

FIGURE 15.43 Effects of nerve growth factor (NGF; top row) and brain-derived neurotrophic factor (BDNF; bottom row) on axonal outgrowths from (A) sympathetic ganglia, (B) dorsal root ganglia, and (C) nodose (taste perception) ganglia. Although both NGF and BDNF had a mild stimulatory effect on dorsal root ganglia axonal outgrowth, the sympathetic ganglia responded dramatically to NGF and hardly at all to BDNF; the converse was true of the nodose ganglia. (From Ibáñez et al. 1991.)

in which the deletion of particular neurotrophic factors results in the loss of only certain subsets of neurons (Crowley et al. 1994; Jones et al. 1994).

WEB TOPIC 15.8 **THE DEVELOPMENT OF BEHAVIORS: CONSTANCY AND PLASTICITY** The correlation of certain neuronal connections with specific behaviors is one of the fascinating aspects of developmental neurobiology.

Neurotrophic factors are produced continuously in adults, and their loss may produce debilitating diseases. BDNF is required for the survival of a particular subset of neurons in the striatum (a region of the brain involved in modulating the intensity of coordinated muscle activity such as movement, balance, and walking), and enables these neurons to differentiate and synthesize the receptor for dopamine. BDNF in this region of the brain is upregulated by huntingtin, a protein that is mutated in Huntington disease. Patients with Huntington disease have decreased production of BDNF, which leads to the death of striatal neurons (Guillin et al. 2001; Zuccato et al. 2001). The result is a series of cognitive abnormalities, involuntary muscle movements, and eventual death. Two other neurotrophins—glial-derived neurotrophic factor (GDNF, discussed earlier in terms of neural crest migration) and conserved dopamine neurotrophic factor (CDNF)—enhance the survival of the midbrain dopaminergic neurons, whose destruction characterizes Parkinson disease (Lin et al. 1993; Lindholm et al. 2007). The midbrain dopaminergic neurons send axons to the cells of the striatum, whose ability to respond to dopamine signals is dependent on BDNF. Drugs that activate the neurotrophic factors are being tested for the ability to cure Parkinson and Alzheimer diseases (Youdim 2013).

Next Step Investigation

The field of neurodevelopment has made great strides in identifying many of the essential factors that establish neural connectivity. From the key transcription factors specifying neural cell fates and the axon guidance receptors required for target specificity, to the mechanical machinery housed in the growth cone and the secreted guidance cues and survival factors, we are beginning to truly understand how a circuit is formed. Scientists are just starting to piece circuits together for a systems-level view of neural development, however. Each neuron sends its axon from one intermediate target to another and then on to the final destination. It is currently rare for us to know the full sequence of guidance mechanisms underlying the entire journey of a particular neuron, never mind linking multiple connections together. With recent advances of live cell imaging in vivo, however, studies in this new frontier of visualizing cell-cell dynamics during axon pathfinding are poised to solve some of the mysteries of connectivity. Go to Eyewire to help connect up a brain (http://eyewire.org/explore).

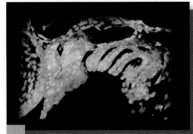

Closing Thoughts on the Opening Photo

This image shows the cranial neural crest cells populating the pharyngeal arches of a 42-hour zebrafish embryo expressing GFP driven by the *fli1a* promoter. It is a lateral view with anterior to the left. The first two major streams contribute neural crest cells to the major jaw cartilages, and the more posterior streams contribute to the "branchial arches" and gill structures. Cranial neural crest cells play a major role in building the craniofacial skeleton—the "face"—in fish and humans alike. We have only recently come to understand that cranial neural crest cells operate through the collective migration mechanism described in this chapter. The pattern of cranial neural crest-built pharyngeal arches may appear complicated, yet there is a way you can grasp these structures—literally. This image was collected with a laser-scanning confocal microscope and as such possesses three-dimensional data. You can go to the National Institutes of Health three-dimensional print exchange website (http://3dprint.nih.gov/discover/3dpx-001506), download a file of the pharyngeal arches of this 42-hour-postfertilization transgenic zebrafish embryo [*tg(fli1a:EGFP)*], and use this file to print a three-dimensional model that you can hold in your hand. (Image and three-dimensional modeling generated and provided by the Barresi lab; Barresi et al. 2015.)

Snapshot Summary
Neural Crest Cells and Axonal Specificity

1. The neural crest is a transitory structure. Its cells migrate to become numerous different cell types. The path a neural crest cell takes depends on the extracellular environment it meets.

   • Trunk neural crest cells can migrate dorsolaterally to become melanocytes and dorsal root ganglia cells. They can also migrate ventrally to become sympathetic and parasympathetic neurons and adrenomedullary cells.

   • Cranial neural crest cells enter the pharyngeal arches to become the cartilage of the jaw and the bones of the middle ear. They also form the bones of the frontonasal process, the papillae of the teeth, and cranial nerves.

   • Cardiac neural crest cells enter the heart and form the septum (separating wall) between the pulmonary artery and aorta.

2. The formation of the neural crest depends on interactions between the prospective epidermis and the neural plate. Paracrine factors from these regions induce the formation of transcription factors that enable neural crest cells to emigrate.

3. Collective migration of neural crest cells is powered by contact inhibition of locomotion and a co-attraction to leading cells, which are cellular behaviors mediated by a combination of low N-cadherin, bipolar Rho-GTPase activity, and attraction to Sdf1 secretion.

4. Trunk neural crest cells will migrate through the anterior portion of each sclerotome, but not through the posterior portion of a sclerotome. Semaphorin and ephrin proteins expressed in the posterior portion of each sclerotome can prevent neural crest cell migration.

5. Some neural crest cells appear to be capable of forming a large repertoire of cell types. Other neural crest cells may be restricted even before they migrate. The final destination of the neural crest cell can sometimes change its specification.

6. The fates of the cranial neural crest cells are influenced by Hox genes. They can acquire their Hox gene expression pattern through interaction with neighboring cells.

7. Motor neurons are specified according to their position in the neural tube. The Lim family of transcription factors plays an important role in this specification before their axons have even extended into the periphery.

8. The growth cone is the locomotory organelle of the neuron and rearranges its cytoskeletal architecture in response to environmental cues. Axons can find their targets without neuronal activity.

9. Some proteins are generally permissive to neuron adhesion and provide substrates on which axons can migrate. Other substances prohibit migration.

10. Some growth cones recognize molecules that are present in very specific areas and are guided by these molecules to their respective targets.

11. Some neurons are "kept in line" by repulsive molecules. If the neurons wander off the path to their target, these molecules send them back. Some molecules, such as the semaphorins and Slits, are selectively repulsive to particular sets of neurons.

12. Some neurons sense gradients of a protein and are brought to their target by following these gradients. The netrins and Shh may work in this fashion.

13. Changes in growth cone responsiveness to the secreted attractive and repulsive cues secreted from the midline enable commissural axons to cross the midline and connect to two sides of the central nervous system.

14. Target selection can be brought about by neurotrophins, proteins that are made by the target tissue and that stimulate the particular set of axons able to innervate it. In some cases, the target makes only enough of these factors to support a single axon. Neurotrophins also play a role in the apoptosis of many neurons.

15. Retinal ganglion cells in frogs and chicks send axons that bind to specific regions of the optic tectum. This process is mediated by numerous interactions, and target selection appears to be mediated through ephrins.

16. Synapse formation has an activity-dependent component. An active neuron can suppress synapse formation by other neurons on the same target.

17. Lack of synapse formation and neuronal activity can lead to the induction of programmed cell death or apoptosis, which unleashes a cascade of caspase enzymes that result in cell death.

Further Reading

Baggiolini, A. and 10 others. 2015. Premigratory and migratory neural crest cells are multipotent in vivo. *Cell Stem Cell* 16: 314–322.

Bard, L., C. Boscher, M. Lambert, R. M. Mège, D. Choquet and O. Thoumine. 2008. A molecular clutch between the actin flow and N-cadherin adhesions drives growth cone migration. *J. Neurosci.* 28: 5879–5890.

Cammarata, G. M., E. A. Bearce and L. A. Lowery. 2016. Cytoskeletal social networking in the growth cone: How +TIPs mediate microtubule-actin cross-linking to drive axon outgrowth and guidance. *Cytoskeleton* doi: 10.1002/cm.21272.

Clay, M. R. and M. C. Halloran. 2014. Cadherin 6 promotes neural crest cell detachment via F-actin regulation and influences active Rho distribution during epithelial-to-mesenchymal transition. *Development* 141: 2506–2515.

Duband, J. L., A. Dady and V. Fleury. 2015. Resolving time and space constraints during neural crest formation and delamination. *Curr. Top. Dev. Biol.* 111: 27–67.

Harada, T., C. Harada and L. F. Parada. 2007. Molecular regulation of visual system development: More than meets the eye. *Genes Dev.* 21: 367–378.

Keleman, K. and 7 others. 2002. Comm sorts Robo to control axon guidance at the *Drosophila* midline. *Cell* 110: 415–427.

Kolodkin, A. L. and M. Tessier-Lavigne. 2011. Mechanisms and molecules of neuronal wiring: A primer. *Cold Spring Harbor Persp. Biol.* 3(6): pii: a001727.

Martinez, E. and T. S. Tran. 2015. Vertebrate spinal commissural neurons: a model system for studying axon guidance beyond the midline. *Wiley Interdiscip. Rev. Dev. Biol.* 4: 283–297.

Okada, A. and 7 others. 2006. Boc is a receptor for Sonic hedgehog in the guidance of commissural axons. *Nature* 444: 369–373.

Preitner, N. and 7 others. 2014. APC is an RNA-binding protein, and its interactome provides a link to neural development and microtubule assembly. *Cell* 158: 368–382.

Sabatier, C. and 7 others. 2004. The divergent Robo family protein rig-1/Robo3 is a negative regulator of slit responsiveness required for midline crossing by commissural axons. *Cell* 117: 157–169.

Sánchez-Arrones, L., F. Nieto-Lopez, C. Sánchez-Camacho, M. I. Carreres, E. Herrera, A. Okada and P. Bovolenta. 2013. Shh/Boc signaling is required for sustained generation of ipsilateral projecting ganglion cells in the mouse retina. *J. Neurosci.* 33: 8596–8607.

Scarpa, E. and R. Mayor. 2016. Collective cell migration in development. *J. Cell Biol* 212: 143–155.

Scarpa, E., A. Szabó, A. Bibonne, E. Theveneau, M. Parsons and R. Mayor. 2015. Cadherin switch during EMT in neural crest cells leads to contact inhibition of locomotion via repolarization of forces. *Dev. Cell* 34: 421–434.

Simões-Costa, M. and M. E. Bronner. 2015. Establishing neural crest identity: A gene regulatory recipe. *Development* 142: 242–257.

Simpson, J. H., K. S. Bland, R. D. Fetter and C. S. Goodman. 2000. Short-range and long-range guidance by Slit and its Robo receptors: A combinatorial code of Robo receptors controls lateral position. *Cell* 103: 1019–1032.

Teillet, M.-A., C. Kalcheim and N. M. Le Douarin. 1987. Formation of the dorsal root ganglia in the avian embryo: Segmental origin and migratory behavior of neural crest progenitor cells. *Dev. Biol.* 120: 329–347.

Theveneau E., B. Steventon, E. Scarpa, S. Garcia, X. Trepat, A. Streit and R. Mayor. 2013. Chase-and-run between adjacent cell populations promotes directional collective migration. *Nature Cell Biol.* 15: 763–772.

Tosney, K. W. 2004. Long-distance cue from emerging dermis stimulates neural crest melanoblast migration. *Dev. Dynam.* 229: 99–108.

Waldo, K., S. Miyagawa-Tomita, D. Kumiski and M. L. Kirby. 1998. Cardiac neural crest cells provide new insight into septation of the cardiac outflow tract: Aortic sac to ventricular septal closure. *Dev. Biol.* 196: 129–144.

Walter, J., S. Henke-Fahle and F. Bonhoeffer. 1987. Avoidance of posterior tectal membranes by temporal retinal axons. *Development* 101: 909–913.

Yoshida, T., P. Vivatbutsiri, G. Morriss-Kay, Y. Saga and S. Iseki. 2008. Cell lineage in mammalian craniofacial mesenchyme. *Mech. Dev.* 125: 797–808.

GO TO WWW.DEVBIO.COM...

...for Web Topics, Scientists Speak interviews, Watch Development videos, Dev Tutorials, and complete bibliographic information for all literature cited in this chapter.

Ectodermal Placodes and the Epidermis

EPITHELIAL SHEETS CAN BE FOLDED into intricate three-dimensional structures as a result of coordinated changes in cell shapes, cell division, and cell movements (Montell 2008; St. Johnston and Sanson 2011). One important type of epithelial reorganization is the formation of **ectodermal placodes**, thickenings of the surface ectoderm that become the rudiment of numerous organs. Ectodermal placodes include the cranial sensory placodes such as olfactory (nasal), auditory (ear), and lens (eye) placodes, as well as placodes that give rise to non-sensory cutaneous structures such as hair, teeth, feathers, mammary, and sweat glands (Pispa and Thesleff 2003).

Cranial Placodes: The Senses of Our Heads

The vertebrate head has a concentration of neurons critical for sensation and perception. In addition to the brain, the eyes, nose, ears, and taste buds are all in the head. The head also has its own highly integrated nervous system for sensing pain (think of the trigeminal nerve that innervates the teeth) and pleasure (think of the receptors on our lips and tongues). The elements of this nervous system arise from the **cranial sensory placodes**—local and transient thickenings of the ectoderm in the head and neck between the prospective neural

What controls hair growth on different parts of the body?

The Punchline

Placodes are thickenings of the surface ectoderm. Cranial placodes form the sensory neurons of our face, becoming the auditory, nasal, gustatory, and optical organs. The lens placode is induced by a bulge from the brain. As the lens placode becomes the lens, it instructs the brain bulge to become the retina. Pax6 appears to be important in the ability of the anterior ectoderm to respond to signals from the presumptive retina. The more posterior ectodermal placodes generate the cutaneous appendages, including hair, feathers, scales, sweat glands, mammary glands, and teeth. These inductions involve the interactions of several paracrine pathways. The epidermis retains epidermal stem cells that allow it to be regenerated constantly, and the Notch pathway is critical in this maintenance. Interactions between the epidermis and the underlying mesenchyme enable the production and shaping of the placodes for these ectodermal appendages. The ability of these cutaneous tissues to grow and regenerate depends on the ability to maintain an epidermal stem cell niche, and different species have different abilities in this respect.

FIGURE 16.1 Cranial placodes form sensory neurons. Fate map of the cranial placodes in the developing chick embryo at the neural plate (left) and 8-somite (right) stages. (After Schlosser 2010.)

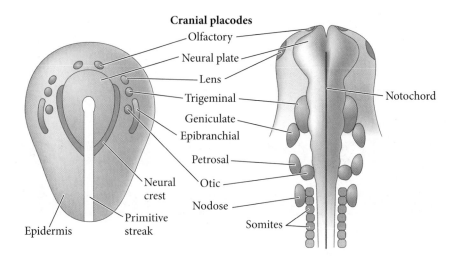

Cranial placodes

Olfactory
Neural plate
Lens
Trigeminal
Geniculate
Epibranchial
Petrosal
Otic
Nodose
Somites
Notochord
Neural crest
Primitive streak
Epidermis

tube and epidermis (**FIGURE 16.1**). With some contributions from the cranial neural crest, the cranial placodes generate most of the peripheral neurons of the head associated with hearing, balance, smell, and taste; the cranial neural crest contributes all of the glia. The olfactory placode gives rise to the sensory neurons involved in smell, as well as to migratory neurons that will travel into the brain and secrete gonadotropin-releasing hormone. The otic placode gives rise to the sensory epithelium of the ear and to neurons that help form the cochlear-vestibular ganglion (see Steit 2008). In the case of the trigeminal ganglion, the proximal neurons are formed from neural crest cells (Baker and Bronner-Fraser 2001) and the distal ones from the trigeminal placode (Hamburger 1961). The lens placode is the only cranial sensory placode that does not form neurons.

The specification of the entire placodal domain and of each region of the specific placodes is defined by the placement and timing of paracrine factor expression by neighboring cells. Both the paracrine factors and their antagonists (such as Noggin and Dickkopf) are critically important (**FIGURE 16.2**). Interestingly, the entire pre-placodal region is originally specified as lens tissue. This propensity has to be suppressed locally by FGFs and by neural crest cells in order for the other cell types to arise (Bailey et al. 2006; Streit 2008).

In addition to these anterior placodes that give rise to specific senses, other placodes provide sensory neurons for the face. These are the **epibranchial placodes**, and they form dorsally to the point at which the pharyngeal pouches contact the epidermis. The epibranchial placodes give rise to the sensory neurons of the facial, glossopharyngeal, and vagal nerves (which relay sensory information about the organs to the brain). The connections made by these placodal neurons are critical in that they enable taste and other facial sensations to be appreciated. But how do these neurons find their way into the hindbrain? Late-migrating cranial neural crest cells do not travel ventrally to enter the pharyngeal arches; rather, they migrate dorsally to generate glial cells (Weston and Butler 1966; Baker et al. 1997). These glial cells form tracks that guide neurons from the epibranchial placodes to the hindbrain (Begbie and Graham 2001). Thus, glial cells from the late-migrating cranial neural crest cells are critical in organizing the innervation of the hindbrain.

The placodes are induced by their neighboring tissues, and there is evidence that the different placodes are each a small portion of what had earlier been a common **pan-placodal field** that is competent to form the placodes if induced (see Figure 16.2B; Streit 2004; Schlosser 2005). The cranial placodes may be induced from a common set of inductive signals arising from the pharyngeal endoderm and head mesoderm (Platt 1896; Jacobson 1966). Jacobson (1963) also showed that the presumptive placodal cells adjacent to the anterior neural tube are competent to give rise to any placode. The anterior-posterior and lateral boundaries of the pan-placodal field are set by retinoic acid, working through Fgf8 (Janesick et al. 2012). Detailed fate mapping studies have

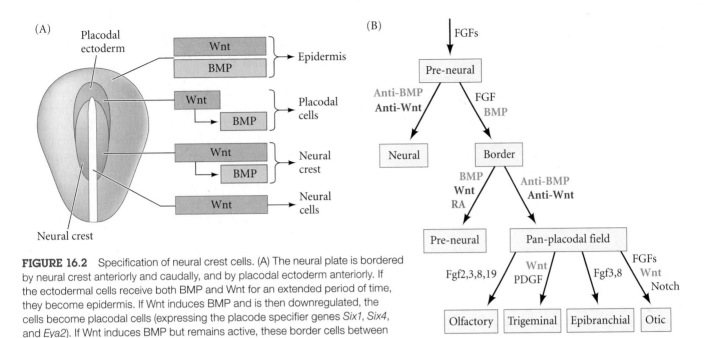

FIGURE 16.2 Specification of neural crest cells. (A) The neural plate is bordered by neural crest anteriorly and caudally, and by placodal ectoderm anteriorly. If the ectodermal cells receive both BMP and Wnt for an extended period of time, they become epidermis. If Wnt induces BMP and is then downregulated, the cells become placodal cells (expressing the placode specifier genes *Six1*, *Six4*, and *Eya2*). If Wnt induces BMP but remains active, these border cells between the neural plate and epidermis become neural crest (expressing the neural crest-specifier genes *Pax7*, *Snail2*, and *Sox9*). If they receive Wnt only (because the BMP signal is blocked by Noggin or FGFs), the ectodermal cells become neural cells. (B) Diagrammatic summary of some of the signaling pathways involved in the specification of the pan-placodal field and the induction of cranial placodes from this region. (A after Patthey et al. 2009; B after Steit 2008.)

confirmed that during the neurula stages, all the placodal precursors are located in a horseshoe-shaped domain that surrounds the anterior neural plate and cranial neural folds (Pieper et al. 2011). This columnar pan-placodal epithelium contains the transcription factors Six1, Six4, and Eya2. These proteins are maintained in all the placodes and are downregulated in the interplacodal regions (Bhattacharyya et al. 2004; Schlosser and Ahrens 2004). Later, the pan-placodal field is separated into discrete placodes; the mechanism for this separation is not known (Breau and Schneider-Maunoury 2014). Different sets of paracrine factors now induce each discrete placode toward its respective fate, such that each placode expresses its own unique set of transcription factors (Groves and LaBonne 2014; Moody and LaMantia 2015). For instance, the chick otic placode, which develops into the sensory cells of the inner ear, is induced by a combination of FGF and Wnt signaling (Ladher et al. 2000, 2005). Fgf19 from the underlying cranial paraxial mesoderm is received by both the presumptive otic vesicle and the adjacent neural plate. Fgf19 induces the neural plate to secrete Wnt8c and Fgf3, which in turn act synergistically to induce formation of the otic placode. The localization of Fgf19 to the specific region of the mesoderm is controlled by Fgf8 secreted in the endodermal region beneath it (**FIGURE 16.3**).

Throughout the head, the neural crest and the sensory placodes—those structures between the epidermis and neural plate—provide the sensory neurons of our ears, nose, tongue, facial skin, and balance system.

WEB TOPIC 16.1 **KALLMANN SYNDROME** Some infertile men have no sense of smell. The relationship between these two conditions was elusive until the gene for Kallmann syndrome was identified.

WEB TOPIC 16.2 **THE HUMAN CRANIAL NERVES** The 12 cranial nerves mediate much of our perception of the outside world.

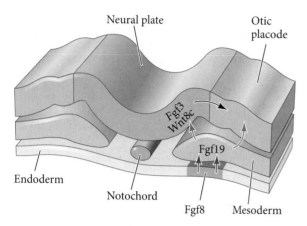

FIGURE 16.3 Induction of the otic (inner ear) placode in the chick embryo. A portion of the pharyngeal endoderm secretes Fgf8, which induces the mesoderm overlying it to secrete Fgf19. Fgf19 is received by both the prospective otic placode and the adjacent neural plate. Fgf19 instructs the neural plate to secrete Wnt8c and Fgf3, two paracrine factors that work synergistically to activate *Pax2* and other genes that allow the cells to produce the otic placode and become sensory cells. (After Schlosser 2010.)

The Dynamics of Optic Development: The Vertebrate Eye

The lens placode does not form neurons. Rather, it forms the transparent lens that allows light to impinge on the retina. The retina develops from a bulge in the forebrain. The interactions between the cells of the lens placode and the presumptive retina structure the eye via a cascade of reciprocal changes that enable the construction of an intricately complex organ. At gastrulation, the head ectoderm is made competent to respond to the signals of the brain bulges. The involuting prechordal plate and foregut endoderm interact with the adjacent prospective head ectoderm to give the head ectoderm a lens-forming bias (Saha et al. 1989). In mammals, this induces the Pax6 transcription factor in the ectoderm, which is critical in the competence of the ectoderm to respond to subsequent signals. But not all parts of the head ectoderm eventually form lenses, and the lens must have a precise spatial relationship with the retina. The activation of the head ectoderm's latent lens-forming ability and the positioning of the lens in relation to the retina are accomplished by **optic vesicles** that extend from the diencephalon of the forebrain.

FIGURE 16.4 shows the development of the vertebrate eye. Where the optic vesicle contacts the head ectoderm, it induces the ectoderm to lengthen, forming the **lens placode**. The optic vesicle then bends to form the two-layered **optic cup**, and in so doing draws the developing lens into the embryo. This invagination is accomplished by the cells of the lens placode extending adhesive filopodia to contact the optic vesicle (Chauhan et al. 2009). As the optic vesicle becomes the optic cup, its two layers differentiate. The cells of the outer layer produce melanin pigment (being one of the few tissues other than the neural crest cells that can form this pigment) and ultimately become the **pigmented retina**. The cells of the inner layer proliferate rapidly and generate a variety of glial cells, ganglion cells, interneurons, and light-sensitive photoreceptor neurons that collectively constitute the **neural retina** (see Figure 16.10). The retinal ganglion cells are neurons whose axons send electric impulses to the brain. Their axons meet at the base of the eye and travel down the optic stalk, which is then called the **optic nerve**. The inner cells of the optic cup (which will become the neural retina) induce the lens placode to become the **lens vesicle**, which will eventually differentiate into the lens cells of the eye.

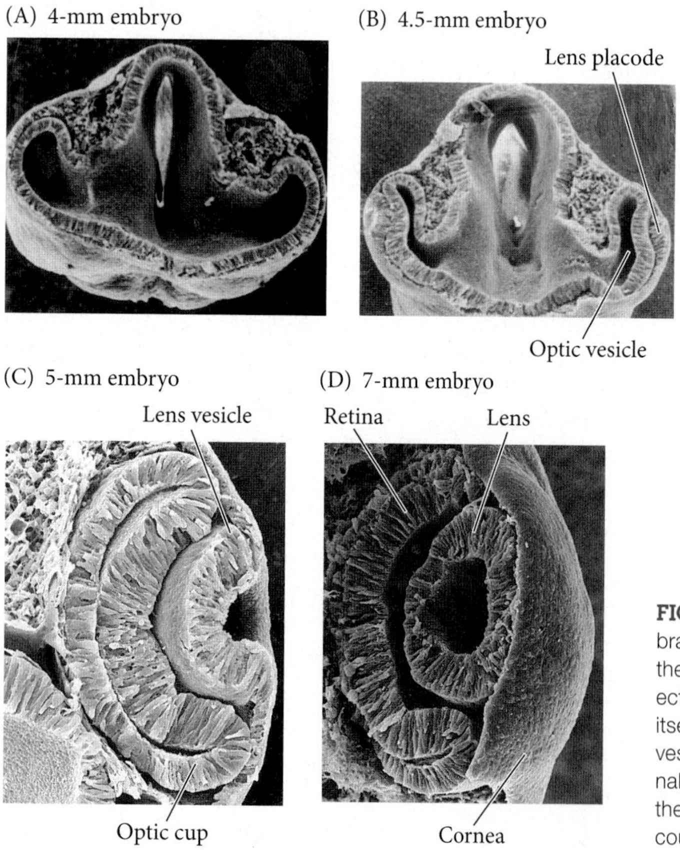

(A) 4-mm embryo

(B) 4.5-mm embryo

(C) 5-mm embryo

(D) 7-mm embryo

FIGURE 16.4 Development and reciprocal induction of the vertebrate eye. (A) The optic vesicle evaginates from the brain and contacts the overlying ectoderm, inducing a lens placode. (B,C) The overlying ectoderm differentiates into lens cells as the optic vesicle folds in on itself, and the lens placode becomes the lens vesicle. (C) The optic vesicle becomes the neural and pigmented retina as the lens is internalized. (D) The lens vesicle induces the overlying ectoderm to become the cornea. (A–C from Hilfer and Yang 1980, courtesy of S. R. Hilfer; D courtesy of K. Tosney.)

Formation of the Eye Field: The Beginnings of the Retina

The details of eye development tell us how eyes come to be made only in the head and why only two eyes normally form. These details show that the precise arrangement of the eye is the result of multiple layers of inductive events involving gene expression differences in both time and place. The story begins with formation of the **eye field** in the anterior of the neural tube. The anterior portion of the neural tube, where both BMP and Wnt pathways are inhibited, is specified by *Otx2* gene expression. Noggin is especially important, as it not only blocks BMPs (thus allowing *Otx2* expression), but also inhibits expression of the transcription factor ET, one of the first proteins expressed in the eye field. However, once Otx2 protein accumulates in the ventral head region, it blocks Noggin's ability to inhibit the *ET* gene, and ET protein is produced.

One of the genes controlled by ET is *Rx*, whose product helps specify the retina. Rx (for "retinal homeobox") is a transcription factor that acts first by inhibiting *Otx2* (since Otx2 has finished its jobs and now might interfere), and second by activating *Pax6*, the major gene in forming the eye field in the anterior neural plate (**FIGURE 16.5A–C**; Zuber et al. 2003; Zuber 2010). Pax6 protein is especially important in specifying the lens and retina; indeed, it appears to be a common denominator for specifying photoreceptive cells in all phyla, vertebrate and invertebrate (Halder et al. 1995).

Humans and mice heterozygous for loss-of-function mutations in *Pax6* have small eyes, whereas homozygotic mice and humans (and *Drosophila*) lack eyes altogether, as do *Rx* mutant mice (**FIGURE 16.5D**; Jordan et al. 1992; Glaser et al. 1994; Quiring et al. 1994). In both flies and vertebrates, Pax6 protein initiates a cascade of transcription

FIGURE 16.5 Dynamic formation of the eye field in the anterior neural plate. (A) Formation of the eye field. Light blue represents the neural plate; moderate blue indicates the area of *Otx2* expression (forebrain); and dark blue indicates the region of the eye field as it forms in the forebrain. (B) Dynamic expression of transcription factors leading to specification of the eye field. Prior to stage 10, Noggin inhibits *ET* expression but promotes *Otx2* expression. Otx2 protein then blocks the inhibition of *ET* by Noggin signaling. The resulting ET transcription factor activates the *Rx* gene, which encodes a transcription factor that blocks *Otx2* and promotes *Pax6* expression. Pax6 protein initiates the cascade of gene expression constituting the eye field (at right). (C) Location of the transcription factors in the nascent eye field of stage 12.5 (early neurula) and stage 15 (mid-neurula) *Xenopus* embryos, showing a concentric organization of transcription factors having domains of decreasing size: Six3 > Pax6 > Rx > Lhx2 > ET. (D) Eye development in a normal mouse embryo (left) and lack of eyes in a mouse embryo whose *Rx* gene has been knocked out (right). (E) Expression pattern of the *Xenopus Xrx1* gene in the single eye field of the early neurula (left) and in the two developing retinas (as well as in the pineal, an organ that has a presumptive retina-like set of photoreceptors) of a newly hatched tadpole (right). (A–C after Zuber et al. 2003; D,E after Bailey et al. 2004, courtesy of M. Jamrich.)

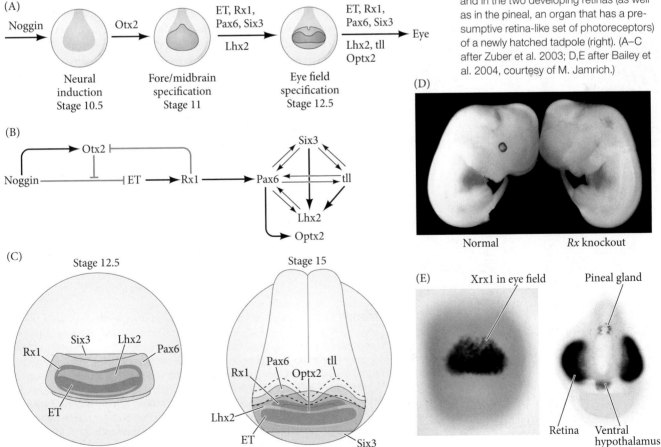

(A)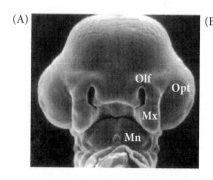

(B)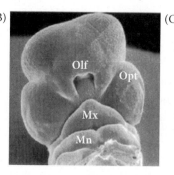

(C)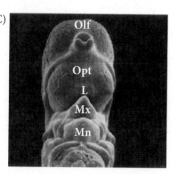

(D)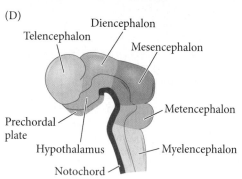

FIGURE 16.6 Sonic hedgehog separates the eye field into bilateral fields. Jervine, an alkaloid found in certain plants, inhibits endogenous Shh signaling. (A) Scanning electron micrograph showing the external facial features of a normal mouse embryo. (B) Mouse embryos exposed to 10 μM jervine had variable loss of midline tissue and resulting fusion of the paired, lateral olfactory processes (Olf), optic vesicles (Opt), and maxillary (Mx) and mandibular (Mn) processes of the jaw. (C) Complete fusion of the mouse optic vesicles and lenses (L) resulted in cyclopia. (D) Drawing showing the location of the prechordal plate (the source of Shh) in the 12-day mouse embryo. (A–C from Cooper et al. 1998, courtesy of P. A. Beachy.)

factors (such as Six3, Rx, and Sox2) with overlapping functions. These factors mutually activate one another to generate a single eye-forming field in the center of the ventral forebrain (**FIGURE 16.5E**; Tétreault et al. 2009; Fuhrmann 2010). The final result, however, is two eyes that are more lateral in the head. The main player in separating the single vertebrate eye field into two bilateral fields is our old friend Sonic hedgehog (Shh).

Shh from the prechordal plate suppresses *Pax6* expression in the center of the neural tube, dividing the field in two (**FIGURE 16.6**). If the mouse *Shh* gene is mutated, or if the processing of this protein is inhibited, the single median eye field does not split. The result is **cyclopia**—a single eye in the center of the face, usually below the nose (see Figure 16.6C; Chiang et al. 1996; Kelley et al. 1996; Roessler et al. 1996). Conversely, if too much Shh is synthesized by the prechordal plate, the *Pax6* gene is suppressed in too large an area and the eyes fail to form at all. This phenomenon may explain why cave-dwelling fish are blind. Yamamoto and colleagues (2004) demonstrated that the difference between surface populations of the Mexican tetra fish (*Astyanax mexicanus*) and eyeless cave-dwelling populations of the same species is the amount of Shh secreted from the prechordal plate. Elevated Shh was probably selected in cave-dwelling species because it resulted in heightened oral sensing and larger jaws (Yamamoto et al. 2009). However, Shh also downregulates *Pax6*, resulting in the disruption of optic cup development, apoptosis of lens cells, and arrested eye development (**FIGURE 16.7**).

(A) Surface-dwelling populations

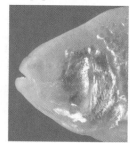

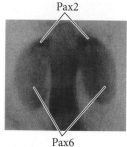

(B) Cave-dwelling populations

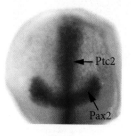

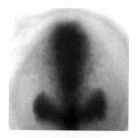

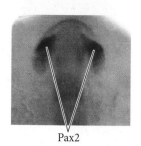

FIGURE 16.7 Surface-dwelling (A) and cave-dwelling (B) Mexican tetras (*Astyanax mexicanus*). The eye fails to form in the population that has lived in caves for more than 10,000 years (top right). Two genes, *Ptc2* and *Pax2*, respond to Shh and are expressed in broader domains in the cavefish embryos than in those of surface dwellers (center). The embryonic optic vesicles (bottom) of surface-dwelling fish are normal size and have small domains of *Pax2* expression (specifying the optic stalk). The optic vesicles of the cave-dwelling fishes' embryos (where *Pax6* is usually expressed) are much smaller, and the *Pax2*-expressing region has grown at the expense of the *Pax6* region. (From Yamamoto et al. 2004; photographs courtesy of W. Jeffery.)

The Lens-Retina Induction Cascade

Once the eye field split is accomplished, how do the two fields form the eyes? Modern studies of vertebrate eye formation were initiated by Hans Spemann (1901), who found that when he destroyed the anterior neural plate on one side of the embryo, no lens formed on the affected side. Something in the neural plate was necessary for the lens to form. Soon afterward, Warren Lewis (1904) found that when he placed anterior neural tube under a different part of the head epidermis, the neural tube became retina and the skin became lens. More recent studies (see Grainger 1992; Ogino et al. 2012) have shown that, although the story is more complicated, the eye field of the anterior neural plate induces the epidermis above it to become the lens, and as each lens begins to form, it induces the eye field to become the retina. The development of the eye is a beautiful example of reciprocal embryonic induction (**FIGURE 16.8**).

As mentioned earlier, the epidermal tissue has to become competent to respond to the signals sent by the presumptive retinal cells. Jacobson (1963, 1966) showed that the epidermis that can respond to the eye field is first conditioned by passing over the pharyngeal endoderm and heart-forming mesoderm during gastrulation. These developing organs probably supply the area with antagonists that block the BMP and Wnt pathways, and these immature organs may be critical for inducing *Pax6* and other genes specific for the anterior ectoderm (Donner et al. 2006). Meanwhile, in the brain, the bilateral ventral forebrain eye fields evaginate as the Rx protein activates *Nlcam*, a gene whose cell-surface product regulates the evagination of the retinal precursor cells from the ventral forebrain (Brown et al. 2010). These evaginations become the optic vesicles. When the cells of the optic vesicles touch surface ectoderm, both tissues are changed. The optic vesicle cells flatten against the surface ectoderm and produce BMP4, Fgf8, and Delta (Ogino et al. 2012). These inducers instruct the cells of the surface ectoderm to elongate and become lens placode cells. As these surface ectoderm cells become the lens placode, they secrete FGFs that instruct the adjacent cells of the optic vesicle to activate the *Vsx2* gene that characterizes the neural retina. The dermal mesenchyme surrounding the optic vesicle instructs most of the outer optic vesicle cells to activate the *Mitf* gene, which will activate the production of melanin pigment (Burmeister et al. 1996; Nguyen and Arnheiter 2000). Thus, the most distal part of the optic vesicle (those cells touching the surface ectoderm) is instructed to become neural retina, while the cells adjacent to this region are instructed to become pigmented retina (see Fuhrmann 2010).

Once the eye fields are specified and divided, much of the development into the optic cup is remarkably autonomous. In mice, a single homogeneous embryonic stem cell population, when placed on a three-dimensional extracellular matrix in the presence of appropriate paracrine factors, can generate an optic vesicle. It will first create an ectodermal sphere, which then produces a "bud" with an inner and an outer wall. These interact such that the outer wall secretes Wnts and become characterized by the MITF transcription factor and melanin pigment. That is, it will become pigmented retina cells. Simultaneously, the inner layer becomes specified as neural retina, characterized by transcription factors such as Six3 and Chx10. Moreover, this optic vesicle, without any external pressure, will invaginate to become an optic cup, and the inner portion will differentiate into a retinalike structure that contains each of the major types of retinal neuron, including the photoreceptors. This indicates that once the eye field is formed, the neural retina and the pigmented retina will segregate from one another (patterning), the folding will occur by intrinsic cell shape changes (morphogenesis), and the neural retina cells will differentiate into different neuronal types (differentiation) in the correct spatial arrangement (Eiraku et al. 2011; Sasai et al. 2013).

The optic vesicle then adheres to the lens placode and changes its shape to form the optic cup. The presumptive neural retina adheres to the lens placode, drawing it into the embryo, while the outer wall of the optic cup becomes the pigmented retina. FGFs from the optic cup activate a new set of genes in the lens placode, transforming the placode into the lens vesicle, which will form the cells of the lens.

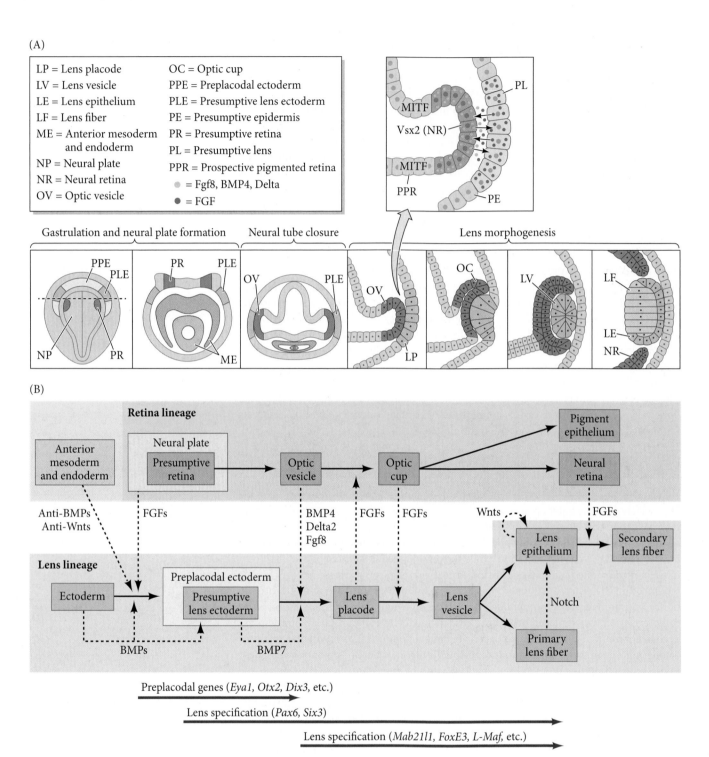

FIGURE 16.8 Reciprocal embryonic interactions between the developing lens placode and the optic vesicle from the brain. (A) Diagrammatic rendering of the major anatomical changes from gastrulation through lens morphogenesis. These interactions start with the presumptive lens ectoderm being influenced by the neural plate, cardiac mesoderm, and pharyngeal mesoderm. Later, the optic vesicle—a bulge from the diencephalon—touches the presumptive lens ectoderm, triggering a series of interactions that turns the optic vesicle into a two-layered optic cup, converts the inner layer of the optic cup into the neural retina, and causes the lens placode to involute and form the lens vesicle. The expanded inset (upper right) shows the critical interactions between the optic vesicle and the presumptive lens cells of the lens placode. (B) Some of the paracrine factors involved in lens development. At different stages and in different tissues, the FGFs and their receptors may be different The three arrows below show some of the genes that become expressed in the presumptive lens during the indicated timeframes. (After Ogino et al. 2012.)

Lens and cornea differentiation

The differentiation of lens tissue into a transparent membrane capable of directing light onto the retina involves changes in cell structure and shape as well as the synthesis of transparent, lens-specific proteins called **crystallins**. Crystallins account for up to 90% of the lens-soluble proteins. The lens cells must curve properly, and this curvature is caused by balancing the Rho-generated apical constriction of microfilaments with Rac-generated actin polymerization that extends the microfilaments along the apical-basal axis (Chauhan et al. 2011).[1]

The crystallin-containing posterior primary fiber cells eventually elongate and fill the lumen of the lens vesicle (**FIGURE 16.9A,B**; Piatigorsky 1981). The anterior cells of the lens vesicle constitute a germinal epithelium, which continues dividing. These dividing cells move toward the equator of the vesicle, and as they pass through the equatorial region, they, too, begin to elongate into secondary cellular fibers (**FIGURE 16.9C,D**). With maturation, these fiber cells lose their cellular organelles and their nuclei are degraded. Thus, the lens contains three regions: an anterior zone of dividing epithelial cells, an equatorial zone of cellular elongation, and a posterior and central

---

[1] As you may recall from Chapter 4, Rac and Rho are two Rho-family GTPases that regulate cell shape and motility by reorganizing the cytoskeletal subunits. Rho is often involved in contractility, while Rac specializes in growth and spreading. We'll see these antagonistic cytoskeletal masons numerous in this book.

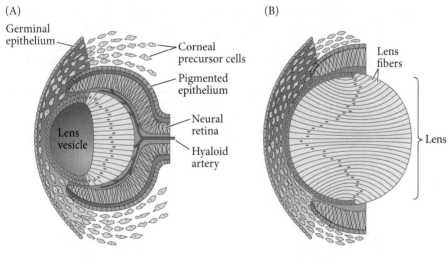

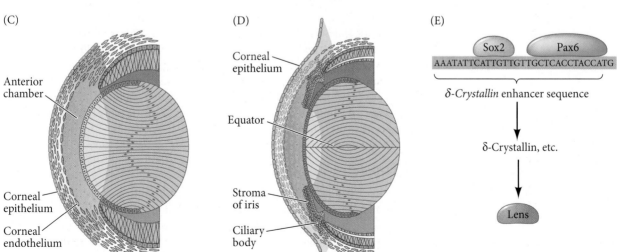

FIGURE 16.9 Differentiation of the lens and anterior portion of the mouse eye. (A) At embryonic day 13, the lens vesicle detaches from the surface ectoderm and invaginates into the optic cup. Corneal precursors (mesenchymal cells) from the neural crest migrate into this space. The elongation of the interior lens cells begins, producing primary lens fibers. (B) At day 14, the lens is filled with crystallin-synthesizing fibers. The neural crest-derived mesenchyme cells between the lens and surface condense to form several layers. (C) At day 15, the lens detaches from the corneal layers, generating an anterior cavity. (D) The surface ectoderm at the anterior side becomes the corneal epithelium, and at day 15.5, corneal layers differentiate and begin to become transparent. The anterior edge of the optic cup enlarges to form a nonneural region containing the iris muscles and the ciliary body. New lens cells are derived from the anterior lens epithelium. As the lens grows, the nuclei of the primary lens cells degenerate and new lens fibers grow from the epithelium on the lateral sides. (E) Close binding of the Sox2 and Pax6 transcription factors on a small region of the δ-*crystallin* enhancer. (A–D after Cvekl and Tamm 2004; E after Kondoh et al. 2004.)

zone of crystallin-containing fiber cells. This arrangement persists throughout the lifetime of the animal as as epithelial cells at the lens equator differentiate into new secondary fibers that are continuously added to the lens mass (Papaconstantinou 1967).

The initial differentiation of lens-forming tissues requires contact between the optic vesicle and the presumptive lens ectoderm. In addition to preventing neural crest cells from inhibiting the intrinsic lens-specifying bias of the anterior ectoderm, this contact appears to permit Delta proteins on the optic vesicle to activate Notch receptors on the presumptive lens ectoderm (Ogino et al. 2008). The Notch intracellular domain binds to an enhancer element of the *Lens1* gene, and in the presence of the Otx2 transcription factor (which is expressed throughout the entire head region), *Lens1* is activated. The Lens1 protein is itself a transcription factor that is essential for epithelial cell proliferation (making and growing the lens placode) and eventually for closing the lens vesicle. In this interaction we see a principle that is observed throughout development—namely, that some transcription factors (such as Otx2) specify a particular field and provide competence for cells to respond to a more specific induction (such as Notch) within that field.

Paracrine factors from the optic vesicle also induce lens-specific transcription factors. Regulation of the crystallin genes is under the control of Pax6, Sox2, and L-Maf (**FIGURE 16.9E**). Like Otx2, Pax6 appears in the head ectoderm before the lens is formed, and Sox2 is induced in the lens placode by BMP4 secreted from the optic vesicle. Coexpression of *Pax6* and *Sox2* in the same cells initiates lens differentiation and activates crystallin genes. Appearing later than Sox2, L-Maf is induced by Fgf8 secreted by the optic vesicle and is needed for the maintenance of crystallin gene expression and the completion of lens fiber differentiation (Kondoh et al. 2004; Reza et al. 2007).

Shortly after the lens vesicle has detached from the surface ectoderm, the lens vesicle stimulates the overlying ectoderm to become cornea. The molecules necessary for this transformation are probably Dickkopf proteins that inhibit Wnts and the Wnt-induced β-catenins. In mice with loss-of-function mutations in their *Dickkopf-2* genes, the corneal epithelium becomes head epidermal tissue (Mukhopadhyay et al. 2006). The presumptive corneal cells secrete layers of collagen into which neural crest cells migrate and make new cell layers while secreting a corneal-specific extracellular matrix (Meier and Hay 1974; Johnston et al. 1979; Kanakubo et al. 2006). These cells condense to form several flat layers of cells, eventually becoming the corneal precursor cells (see Figure 16.9A; Cvekl and Tamm 2004). As these cells mature, they dehydrate and form tight junctions among the cells, uniting with the surface ectoderm (Kurpakus et al. 1994; Gage et al. 2005) to become the cornea. Intraocular fluid pressure (from the aqueous humor) is necessary for the correct curvature of the cornea, allowing light to be focused on the retina (Coulombre 1956, 1965).

Repair and regeneration are critical to the cornea since, like the epidermis, it is exposed to the outside world. The main problem for the cornea is reactive oxygen species (ROS; see Chapter 23) that damage DNA and proteins. The major sources of ROS are the amniotic fluid (as an embryo) and ultraviolet light (as an adult). One protective mechanism is the production of the iron-binding protein ferritin (Linsenmeyer et al. 2005; Beazley et al. 2009). The second mode of protection is a layer of basal cells that continually renew the corneal epithelial cells throughout the life of the individual. Long-lived stem cells found at the edge of the cornea contribute to corneal repair and can regenerate the cornea in humans (Cotsarelis et al. 1989; Tsai et al. 2000; Majo et al. 2008). The cells in this region also secrete Dickkopf to prevent their becoming epidermal cells (Mukhopadhyay et al. 2006).

Neural retina differentiation

Like the cerebral and cerebellar cortices, the neural retina develops into a layered array of different neuronal types (**FIGURE 16.10A**). In mammals, these layers include the light- and color-sensitive photoreceptor cells (rods and cones); the cell bodies of the ganglion cells; and bipolar neurons that transmit electric stimuli from the rods and cones to the ganglion cells (**FIGURE 16.10B**). In addition, the retina contains numerous

(A)

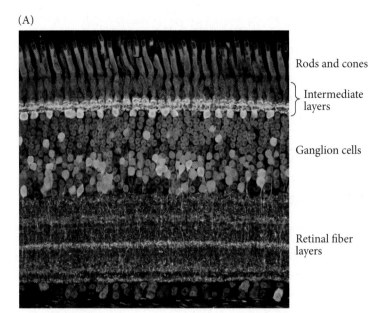

Rods and cones

} Intermediate layers

Ganglion cells

Retinal fiber layers

(B)

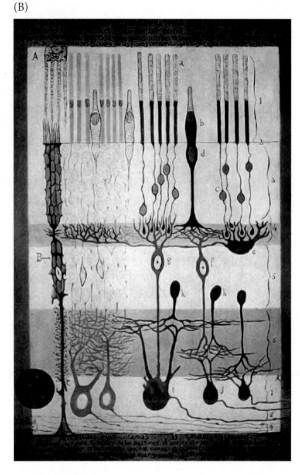

FIGURE 16.10   Retinal neurons sort out into functional layers during development. (A) Fluorescent labeled confocal micrograph of chick retina, showing the three layers of neurons and the synapses between them. (B) The mammalian retina as drawn and labeled by Santiago Ramón y Cajal in 1900. *Numerical labels:* 1, photoreceptor layer; 2, external limiting membrane; 3, outer granular layer; 4, outer plexiform layer; 5, inner granular layer; 6, inner plexiform layer; 7, ganglion cell layer; 8, optic nerve fiber layer; 9, internal limiting membrane. *Alphabetical labels:* A, pigmented cells; B, epithelial cells; a, rods; b, cones; c, rod nucleus; d, cone nucleus; e, large horizontal cell; f, cone-associated bipolar cell; g, rod-associated bipolar cell; h, amacrine cell; i, giant ganglion cell; j, small ganglion cell. (A courtesy of A. J. Fischer; B courtesy of the Instituto Cajal, CSIC, Madrid.)

Müller glial cells that maintain its integrity, amacrine neurons (which lack large axons), and horizontal neurons that transmit electric impulses in the plane of the retina.

The neuroblasts of the retina appear to be competent to generate all of the retinal cell types (Turner and Cepko 1987; Yang 2004). In amphibians, the type of neuron produced from a multipotent retinal stem cell appears to depend on the *timing* of gene translation. Photoreceptor neurons, for instance, are specified through expression of the *Xotx5b* gene, while *Xotx2* and *Xvsx1* expression is critical for specifying the bipolar neurons. Interestingly, these three genes are transcribed in all retinal cells, but they are translated differently. Those neurons whose birthday is at stage 30 translate *Xotx5b* mRNA and become photoreceptors, while those neurons forming later (birthdays at stage 35) translate the *Xotx2* and *Xvsx1* messages and become bipolar interneurons (Decembrini et al. 2006, 2009). This time-dependent regulation of translation is mediated by microRNAs.

Not all the cells of the optic cup become neural tissue. The tips of the optic cup on either side of the lens develop into a pigmented ring of muscular tissue called the **iris**. The iris muscles control the size of the pupil (and give an individual his or her characteristic eye color). At the junction between the neural retina and the iris, the optic cup forms the **ciliary body**. This tissue secretes the **aqueous humor**, a fluid needed for the nutrition of the lens and for forming the pressure needed to stabilize the curvature of the eye and the constant distance between the lens and the cornea.

WEB TOPIC 16.3   **WHY BABIES DON'T SEE WELL**   The retinal photoreceptors are not fully developed at birth but increase in density and discriminatory ability as a child grows.

The Epidermis and Its Cutaneous Appendages

The skin—a tough, elastic, water-impermeable membrane—is the largest organ in our bodies. Mammalian skin has three major components: (1) a stratified epidermis; (2) an underlying dermis composed of loosely packed fibroblasts; and (3) neural crest–derived melanocytes that reside in the basal epidermis and hair follicles. It is the melanocytes (discussed in Chapter 15) that provide the skin's pigmentation. In addition, a subcutaneous ("below the skin") fat layer is present beneath the dermis. Moreover, skin is constantly being renewed. This regenerative ability is possible due to a population of epidermal stem cells that last a lifetime.

Origin of the Epidermis

The epidermis originates from the ectodermal[2] cells covering the embryo after neurulation. As detailed in Chapter 13, this surface ectoderm is induced to form epidermis rather than neural tissue by the actions of BMPs. The BMPs promote epidermal specification and at the same time induce transcription factors that block the neural pathway (see Bakkers et al. 2002). Once again we see the principle that the specification of one tissue also involves blocking the specification of an alternative tissue.

The epidermis is only one cell layer thick to start with, but in most vertebrates it soon becomes a two-layered structure. The outer layer gives rise to the **periderm**, a temporary covering that is shed once the inner layer differentiates to form a true epidermis. The inner layer, called the **basal layer** or **stratum germinativum**, contains epidermal stem cells attached to a basal lamina that the stem cells themselves help to make (**FIGURE 16.11**). Just as in neural stem cells, this differentiation is positively regulated by the Notch pathway (Nguyen et al. 2006; Aguirre et al. 2010). In the absence of Notch signaling, there is hyperproliferation of the dividing cells (Ezratty et al. 2011). The Notch signal promotes the synthesis of the keratins characteristic of skin and joins them into dense intermediate filaments (Lechler and Fuchs 2005; Williams et al. 2011). There is some evidence that, like the neural stem cells of the ependymal layer, epidermal stem cells divide asymmetrically. The daughter cell that remains attached to the basal lamina remains a stem cell, while the cell that leaves the basal layer migrates outward and starts differentiating. However, it is

[2]To review the vocabulary, *epidermis* is the outer layer of skin. The *ectoderm* is the germ layer that forms the epidermis, neural tube, placodes, and neural crest. *Epithelial* refers to a sheet of cells that are held together tightly (as opposed to the loosely connected *mesenchymal* cells; see Chapter 4). Epithelia can be produced by any germ layer. The epidermis and the neural tube both happen to be ectodermal epithelia; the lining of the gut is an *endodermal* epithelium.

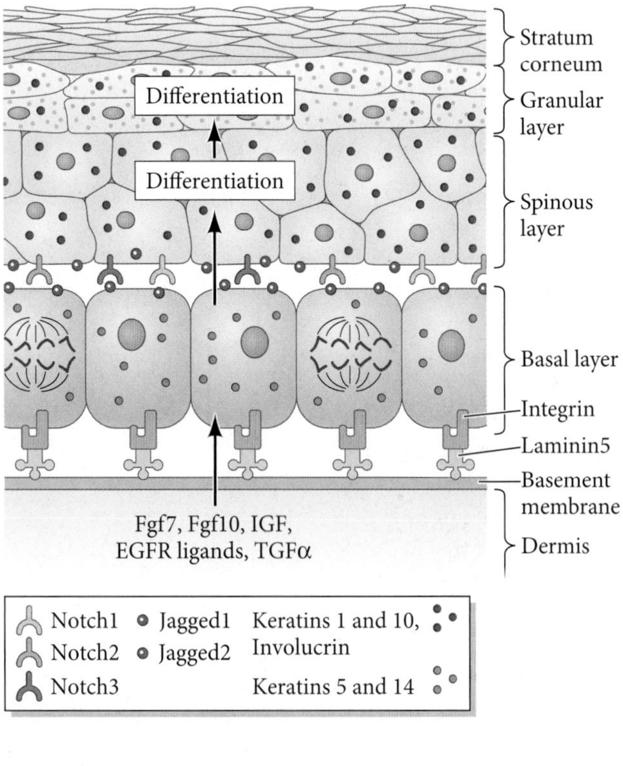

FIGURE 16.11 Layers of the human epidermis and the signals that enable the continued regeneration of mammalian skin. The basal cells are mitotically active, whereas the fully keratinized cells characteristic of external skin are dead and are shed continually. Self-renewing stem cells reside within the basal layer, which adheres through integrins to an underlying laminin-rich basement membrane that separates the epidermis from the underlying dermis. The dermal fibroblasts secrete factors such as Fgf7, Fgf10, IGF, EGF ligands, and TGF-α to promote the proliferation of basal epidermal cells. Proliferative basal progenitor cells generate columns of Notch-activated terminally differentiating cells that pass through three stages, each expressing particular keratins: spinous layers, granular layers, and finally dead stratum corneum layers that are shed from the surface. (After Hsu et al. 2014.)

also possible that both asymmetrical and symmetrical division play important roles in forming and sustaining the epidermis (Hsu et al. 2014; Yang et al. 2015). Moreover, it is still not determined whether there is a discrete population of long-lived stem cells in the basal layer (Mascré et al. 2012), or whether all basal cells have stem cell-like properties (Clevers et al. 2015).

Cell division from the basal layer produces younger cells and pushes the older cells to the surface of the skin. This is unlike the "inside-out" patterning of the neural tube, where newly generated neurons migrate through layers of older cells on their way to the periphery. (Neural cells, however, don't form anew each day like the epidermal cells.) After the synthesis of the differentiated products, the cells cease transcriptional and metabolic activities. These differentiated epidermal cells, the **keratinocytes**, are bound tightly together and produce a water-impermeable seal of lipid and protein.

As they reach the surface, keratinocytes are dead, flattened sacs of keratin protein, and their nuclei are pushed to one edge of the cell. These cells constitute the **cornified layer**, or **stratum corneum**. Throughout life, the dead keratinocytes of the cornified layer are shed[3] and are replaced by new cells. In mice, the journey from the basal layer to the sloughed cell takes about two weeks. Human epidermis turns over a bit more slowly; the proliferative ability of the basal layer is remarkable in that it can supply the cellular material to continuously replace 1–2 m^2 of skin for several decades throughout adult life.

Several factors stimulate development of the epidermis (see Figure 16.11). Dermal fibroblasts activate epidermal stem cell division through the production of FGFs, insulin-like growth factor, and the appropriately named epidermal growth factor (Hsu et al. 2014). BMPs help initiate epidermal production by inducing the p63 transcription factor in the basal layer. This transcription factor's multiple roles may depend in part on different splicing isoforms of p63 that are expressed in the epidermis. The p63 protein is required for keratinocyte proliferation and differentiation (Truong and Khavari 2007); it also appears to stimulate the production of the Notch ligand Jagged. Jagged is a juxtacrine protein in the basal cells that activates the Notch protein on the cells above them, activating the keratinocyte differentiation pathway and preventing further cell divisions (see Mack et al. 2005; Blanpain and Fuchs 2009). Thus, Notch signaling is necessary for the transition from the basal layer to the spinous layer.

The Ectodermal Appendages

The ectodermal epidermis and the mesenchymal dermis interact inductively at specific sites to create the **ectodermal appendages**: hairs, scales, scutes (e.g., the coverings of turtle shells), teeth, sweat glands, mammary glands, or feathers, depending on the species and type of mesenchyme. The formation of these appendages requires a series of reciprocal inductive interactions between the mesenchyme and the ectodermal epithelium, resulting in the formation of **epidermal placodes** that are the precursors of the epithelia of these structures. Remarkably, the early development of structures as different as hair, teeth, and mammary glands follow the same patterns and appear to be governed by reciprocal induction using same paracrine factors.

In all these ectodermal appendages, the first obvious sign of morphogenesis is a local epithelial thickening, the placode. In many regions of the trunk and abdominal ectoderm, thousands of individual hair placodes develop independently. In each jaw, there is a broad epidermal thickening called the **dental lamina**, which (like the panplacodal stage of cranial sensory placodes) later resolves into separate placodes, each of which becomes a tooth (see Figure 16.13). In the ventral ectoderm, two mammary ridges (or "milk lines") extend from the forelimbs to hindlimbs. In the mouse, five pairs of mammary placodes typically survive on each side, each becoming a mammary gland. In humans, usually only one pair survives, although sometimes a third or fourth

---

[3] Humans lose about 1.5 grams of these cells every day. Most of this skin becomes "house dust." Deficient Notch signaling has been implicated in psoriasis (Kim et al. 2016).

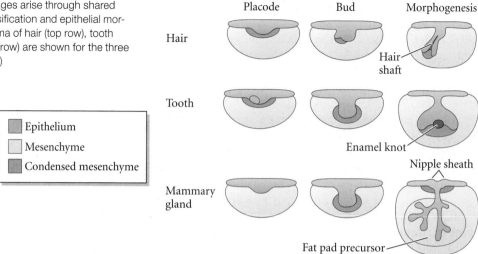

FIGURE 16.12 Ectodermal appendages arise through shared placode and bud stages prior to diversification and epithelial morphogenesis. The developmental schema of hair (top row), tooth (center), and mammary gland (bottom row) are shown for the three stages. (After Biggs and Mikkola 2014.)

Legend:
- Epithelium
- Mesenchyme
- Condensed mesenchyme

Labels in figure: Placode, Bud, Morphogenesis; Hair, Hair shaft; Tooth, Enamel knot; Nipple sheath; Mammary gland, Fat pad precursor

Developing Questions

How is the avian feather formed? What patterns the placement of the feather bud placodes?

placode remains, forming supernumerary nipples. Mammary placodes form in both sexes but are fully developed only in females (Biggs and Mikkola 2014).

After the placode stage, there is a bud stage during which the ectoderm grows into the mesenchyme. In the placode-forming regions, superficial cells of the placode contract and intercalate toward the common center. This causes them to pucker inwardly and drive the epithelium into the underlying mesenchyme. Such contractile forces have been seen in teeth, mammary, and hair placodes (Panoutsopoulou and Green 2016). The initial bud looks very much the same for all the epidermal appendages. However, as the bud continues to interact with the underlying mesenchyme, differences begin to be observed. The hair follicle elongates, grows inward, and grows around the condensed inductive mesenchyme. The tooth epithelium likewise grows into the mesenchyme, and in the center of the epithelium generates an **enamel knot**. This signaling center will control the proliferation and differentiation of the surrounding cells (Jernvall et al. 1996). The mammary epithelium grows through the inductive mesenchymal cells and into the developing fat pad, where it undergoes extensive branching (**FIGURE 16.12**).

Recombination experiments: The roles of epithelium and mesenchyme

The inductive interactions between epithelium and mesenchyme are very specific. By separating epithelial and mesenchymal components and then recombining them, twentieth century developmental biologists were able to discern which part held the specificity. For instance, dental *epithelium* from the jaw of a day-10 mouse embryo caused tooth formation when combined with non-dental jaw mesenchyme of the same stage embryo. By embryonic day 12, however, the dental epithelium lost this ability, while the newly condensed dental mesenchyme gained it (Mina and Kollar 1987). The expression pattern of Bmp4 shifts from the epithelium to the mesenchyme concomitant with this switch in odontogenic ability (Vainio et al. 1993). Moreover, after this transition, the site of the epidermis became less important. Dental mesenchyme could interact with foot epidermis to generate teeth (Kollar and Baird 1970). The reverse combination did not work. In the mouse jaw, the mesenchyme is given tooth-forming ability through Fgf8 and is prevented from tooth formation by BMPs (**FIGURE 16.13**; Neubüser et al. 1997).

Similarly, condensed dermal cells can induce hair follicles even in epithelia (such as the soles of the feet) that usually do not produce hair (Kollar 1970). The hair placode epithelium cannot induce new hairs. Early mouse mammary gland mesenchyme (but not epithelium) will induce the early stages of mammary formation in epidermis derived from head and neck (Propper and Gomot 1967; Kratochwil 1985). Thus, in the hair, mammary gland, and tooth follicles, the condensed mesenchyme appears to have the specific inductive capacity. Interestingly, this capacity to induce placodes is absent

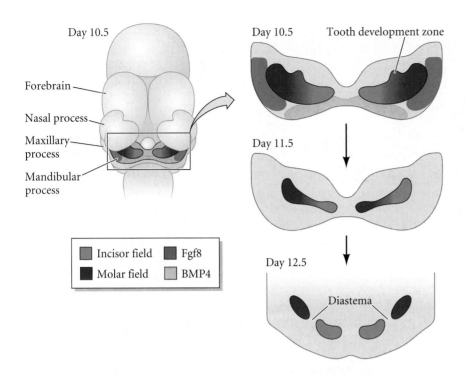

FIGURE 16.13 Division of the dental lamina into incisor and molar fields. (A) Schematic diagram showing the mandible of a 10.5-day mouse embryo. A mutually antagonistic interaction between Bmp4 and Fgf8 is thought to define the tooth field in the oral ectoderm. Each tooth fields is continuous until day 11, but subsequently divides into the incisor and molar fields in the distal and proximal areas of the jaw, respectively. The two fields are separated by a toothless region called diastema. (After Ahn 2015.)

in certain skin mesenchymes, namely, those mesenchyme associated with palms, soles, and external genitalia. In these locations, *HoxA13* gene expression appears to promote the synthesis of paracrine factors that induce the ectodermal epithelium to express specific keratin proteins rather than hair (Rinn et al. 2008; Johansson and Headon 2014).

Recombination experiments with mouse mammary glands have revealed remarkable sex-related differences in development. In female mice, the mammary gland completes the first part of its development in the embryo. Like the human mammary gland, it completes the second part of its development in response to estrogenic hormones made at puberty. Final development, including ductal growth and the differentiation of the milk-producing cells on the branches and tips, is accomplished during pregnancy. In male mice, however, mammary development is literally nipped in the bud. Shortly after placode epidermis has ingressed, the mesenchyme cells condense around it. The mesenchymal cells appear to stretch the epithelial rudiment and separate the male mammary epithelium from the epidermis. The epithelial portion of the male mammary gland then dies.

The condensation of the mesenchyme around the stalk of the male mammary bud is due to testosterone. The addition of testosterone to female cultured mammary likewise arrested their development. Moreover, recombination experiments with mice lacking the testosterone receptor show that testosterone acts only on the mesenchyme cells, instructing them to destroy the male mammary rudiment (**FIGURE 16.14**; Kratochwil and Schwartz 1976; Dürnberger et al. 1978). In humans, however, the duct in males is not severed, and male mammary development follows the female pattern until puberty. At that time, increased estrogen levels permit expanded breast development in women, while male estrogens remain at prepubertal levels. If men are exposed to estrogenic compounds (such as some endocrine disruptors; see Chapter 24), their breasts will enlarge, a condition called gynecomastia. However, since males do not secrete prolactin, gynecomastic breasts do not usually produce milk.

Signaling pathways

Just about all of the major signaling pathways are involved in the formation of the ectodermal appendages (Biggs and Mikkola 2014; Ahn et al. 2015). In some instances, such as the enamel knot of mammalian teeth, the same signaling center sends out paracrine factors belonging to nearly every family (**FIGURE 16.15**). The canonical

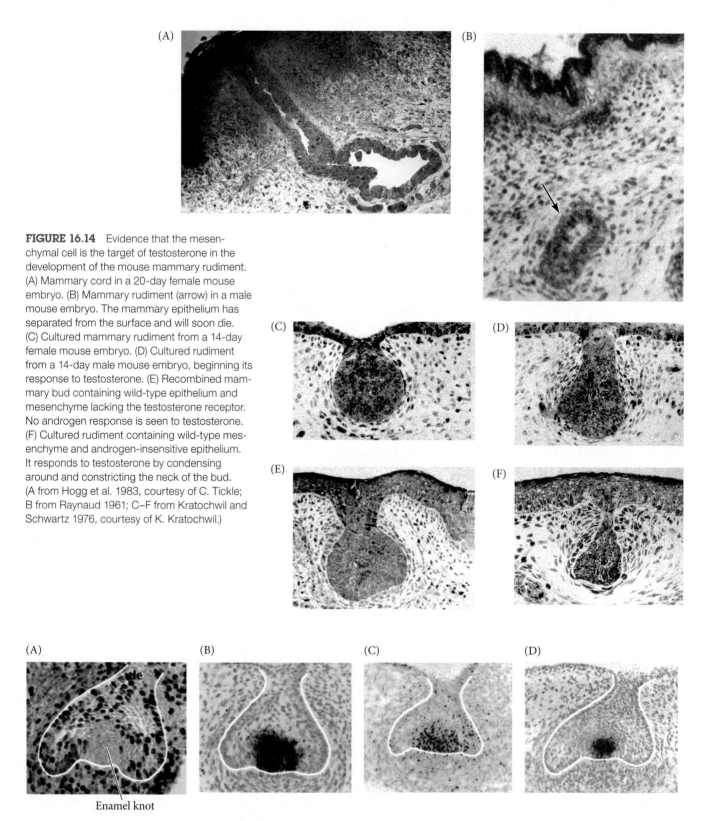

FIGURE 16.14 Evidence that the mesenchymal cell is the target of testosterone in the development of the mouse mammary rudiment. (A) Mammary cord in a 20-day female mouse embryo. (B) Mammary rudiment (arrow) in a male mouse embryo. The mammary epithelium has separated from the surface and will soon die. (C) Cultured mammary rudiment from a 14-day female mouse embryo. (D) Cultured rudiment from a 14-day male mouse embryo, beginning its response to testosterone. (E) Recombined mammary bud containing wild-type epithelium and mesenchyme lacking the testosterone receptor. No androgen response is seen to testosterone. (F) Cultured rudiment containing wild-type mesenchyme and androgen-insensitive epithelium. It responds to testosterone by condensing around and constricting the neck of the bud. (A from Hogg et al. 1983, courtesy of C. Tickle; B from Raynaud 1961; C–F from Kratochwil and Schwartz 1976, courtesy of K. Kratochwil.)

Enamel knot

FIGURE 16.15 Tooth generation in mammals. The enamel knot is the signaling center directing tooth morphogenesis. These photographs show the cap stage of development where the epithelium is growing into the mesenchyme. (A) Staining for cell division with radioactive BrdU indicates a region of non-dividing cells, the enamel knot. (B–D) In situ hybridization reveals that the enamel knot transcribes the genes for paracrine factors initiating several signaling cascades. These genes include *Sonic hedgehog* (B), *Bmp7* (C), and *Fgf4* (D). (From Vaahtokari et al 1996, courtesy of I. Thesleff.)

Wnt/β-catenin pathway was implicated by the loss of properly formed hair, teeth, and mammary glands in mice deficient in components of this pathway (van Genderen et al. 1994). Manipulating embryos to have β-catenin expression throughout the epidermis eventually transforms the entire epidermis to a hair follicle fate (Närhi et al. 2008; Zhang et al. 2008), and it creates supernumerary teeth in the jaw (Järvinen et al. 2006; Liu et al. 2008). Indeed, Wnt signaling may help induce the enamel knot to form and may be critical in the ability (lost in mammals) to regenerate teeth (Järvinen et al. 2006). Mouse mutants lacking negative regulators of the Wnt pathway display more and larger mammary placodes (Närhi et al. 2012; Ahn et al. 2013).

FGFs probably play multiple roles in ectodermal appendage development. One of these roles is to regulate the migration of mesenchymal cells to become the condensates beneath the placode. In teeth, Fgf8 from the placode appears to attract and sustain mesenchyme cells to the tooth placode (Trumpp et al. 1999; Mammoto et al. 2011). In hair, the placodal Wnt activates the secretion of Fgf20, which may stimulate mesenchymal cell migration to the placode (Huh et al. 2014). In the mammary gland, Fgf10 from the somites (and possibly the limb buds) is thought to induce the placode formation (Mailleux et al. 2002; Veltmaat et al. 2006). Mice lacking the genes for Fgf10 or its receptor lack mammary placodes 1,2, 3, and 5. (For some unknown reason, mammary placode 4 survives).

TGF-β family members, especially the BMPs, also play important roles in ectodermal appendage formation. Indeed, the shifting expression pattern of BMP4 from the epithelium to the mesenchyme coordinates the shifting of tooth-generating potential and is critical for the bud-to-cap transition. BMPs are known to induce several genes involved in tooth development (Vainio et al. 1993; Jussila and Thesleff 2012) and BMPs and Wnts most likely regulate each other to control tooth shape (Munne et al. 2009; O'Connell et al. 2012). While BMP expression is necessary for tooth formation, BMP activity must be *suppressed* for the induction of hair placodes (Jussila and Thesleff 2012; Sennett and Rendl 2012).

Other signaling pathways, such as those initiated by hedgehog and ectodysplasin, are also important to varying degrees (Biggs and Mikkola; Ahn 2014). The ectodysplasin pathway (activating the NF-κB transcription factor) is active in every cutaneous appendage. People (and other animals) with anhidrotic ectodermal dysplasia thus have defective growth of hair, teeth, and sweat glands (Mikkola et al. 2008).

> **WEB TOPIC 16.4**   **THE ECTODYSPLASIN PATHWAY AND MUTATIONS OF HAIR DEVELOPMENT**   Genetic conditions can give us insight into the mechanisms of normal hair growth.

Developing Questions

Turtle scutes (the keratinous external plates covering the back and belly) all display the same pattern, whether they are on marine turtles or desert tortoises. The mathematically inclined might want to ask: How is this pattern generated?

Ectodermal appendage stem cells

In many cases, the epidermal appendages generate or retain adult stem cells that allow the regrowth of these structures at particular times. Whether or not such stem cells are present differs importantly between species. Fish and reptiles can regenerate their teeth but mammals cannot. Most mammals have two sets of teeth, one for children ("milk" teeth) and a "permanent" set for adults. Both sets of teeth have started developing before birth. Once we grow adult teeth, the dental lamina decays and we cannot regenerate lost or damaged teeth. While human teeth are set once and for all (one must remember that throughout much of human history most people died before the age of 40), other mammals, including rodents and elephants, have teeth that grow continually (Thesleff and Tummers 2009). In the ever-growing mouse incisors, there is a stem cell niche that retains the epithelial cells that constantly generate the enamel-forming ameloblast cells. In some reptiles, such as alligators, part of the dental lamina is retained, and it contains epithelial stem cells capable of regenerating lost teeth. When teeth are lost, β-catenin accumulates in these cells while Wnt inhibitors are lost (Wu et al. 2013).

(A)

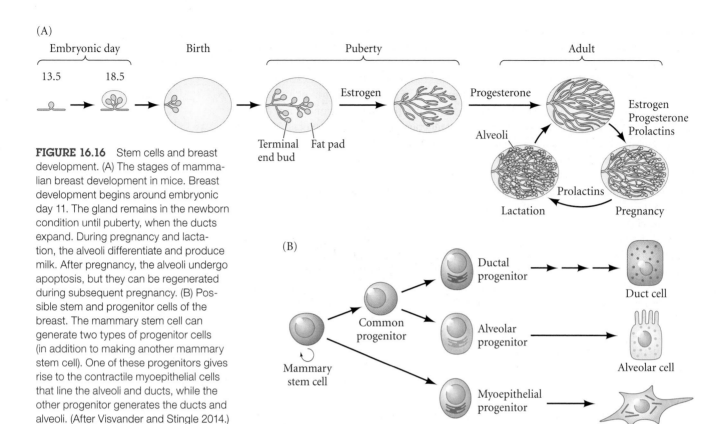

FIGURE 16.16 Stem cells and breast development. (A) The stages of mammalian breast development in mice. Breast development begins around embryonic day 11. The gland remains in the newborn condition until puberty, when the ducts expand. During pregnancy and lactation, the alveoli differentiate and produce milk. After pregnancy, the alveoli undergo apoptosis, but they can be regenerated during subsequent pregnancy. (B) Possible stem and progenitor cells of the breast. The mammary stem cell can generate two types of progenitor cells (in addition to making another mammary stem cell). One of these progenitors gives rise to the contractile myoepithelial cells that line the alveoli and ducts, while the other progenitor generates the ducts and alveoli. (After Visvander and Stingle 2014.)

Developing Questions

Humans have sweat glands throughout their skin, whereas most mammals have them only in their palms and soles. With increasing brain size came the necessity for a very efficient cooling system; thus hairless skin with its many sweat glands. How might sweat glands arise as an evolutionary modification of ectodermal appendages? What is the evidence that sweat glands may be a modification of hair that develop at the expense of hair follicles, thereby explaining both our large number of sweat glands and our bodies' relative hairlessness?

The mammary gland (after which our vertebrate class was named) contains stem cells for the reactivation of its growth at puberty and pregnancy (**FIGURE 16.16A**). During puberty, estrogens cause extensive branching of the ducts and elongation of the **terminal end buds**. During pregnancy, the ducts, stimulated by progesterone and prolactin, form tertiary side branches that differentiate into the milk-producing alveoli (Oakes et al. 2006; Sternlicht et al. 2006). The mammalian mammary gland probably contains a stem cell that is able to generate all the lineages of the gland. Data from genetically labeled mammary cells (Rios et al. 2014; Wang et al. 2015) suggest that there may be a single stem cell type that can make the two major progenitor cells of the breasts: one that generates the ducts and aveoli, and one that generates the myoepithelial cells that contract to push the milk out of the alveoli and toward the nipple (**FIGURE 16.16B**).

The best studied of the epidermal appendage stem cells is the hair stem cell. There appear to be three stem cell populations involved in producing epidermal structures. One, discussed earlier, is found in the germinal layer of the epidermis and generates the keratinocytes that characterize the interfollicular epidermis. A second group of stem cells is critical for forming the sebaceous gland of each hair shaft, and a third group is critical for regenerating the hair shaft itself. Interestingly, it seems that there is also a primitive stem cell that can form all the others (Snippert et al. 2010), and members of each stem cell group can be recruited to any of the other three pools if needed, as when the skin repairs itself when wounded (Levy et al. 2007; Fuchs et al. 2008).

Hair is one structure that mammals are able to regenerate. Throughout life, hair follicles undergo cycles of growth (**anagen**), regression (**catagen**), rest (**telogen**), and regrowth. Hair length is determined by the amount of time the hair follicle spends in the anagen phase. Human scalp hair can spend several years in anagen, whereas arm hair grows for only 6–12 weeks in each cycle. The ability of hair follicles to regenerate depends on the existence of a population of epithelial stem cells that forms in the permanent **bulge** region of the follicle late in embryogenesis. When Philipp Stöhr drew the histology of the human hair for his 1903 textbook, he showed this bulge ("Wulst") as the attachment site for the arrector pili muscles (which give a person "goosebumps" when they contract). Research carried out during the 1990s suggested that the bulge houses populations of at least two adult stem cell types: the **hair follicle stem cells** (HFSCs), which gives rise to the hair shaft and sheath (Cotsarelis et al. 1990; Morris and Potten 1999; Taylor et al. 2000); and the **melanocyte stem cells**, which give rise to the pigment of the skin and hair (Nishimura et al. 2002). The bulge appears to be an important niche that allows adult cells to retain the quality of "stemness." Follicle stem cells in the bulge can regenerate all the epithelial cell types of the hair, and without the stem cells, there is no new follicle. However, if stem cells are selectively ablated by laser, certain bulge epithelial cells (which normally are not used for hair growth) repopulate the stem cell population and can sustain hair follicle regeneration (Rompolas et al. 2013).

There appear to be two populations of HFSCs: a quiescent population in the bulge, and a population primed for cell division just below the bulge. The entire skin organ seems to be involved in hair cycling (**FIGURE 16.17**). The HFSCs reside in the outer layer of the bulge. The inner bulge cells are the progeny of the HFSCs, and they secrete

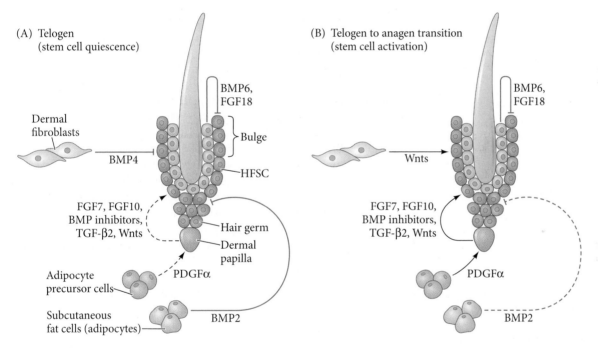

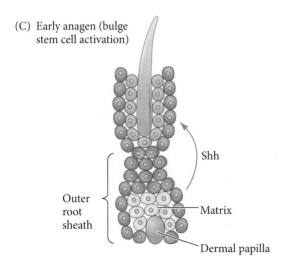

FIGURE 16.17   Regeneration of the hair shaft by the bulge hair follicle stem cells (HFSCs). (A) During quiescence (telogen), the condensed mesenchyme of the dermal papilla contacts the stem cells (blue) in the outer layer of the bulge. The HFSCs are quiescent because of BMP6 and Fgf18 produced by the inner bulge layer, whose cells descend from the HFSC of the outer bulge, and because of other BMPs produced by dermal mesoderm (fibroblasts) and fat cells (adipocytes). (B) At the transition from telogen to anagen (growth), the dermal papilla is induced by the mesenchyme cells to produce activators of hair growth (FGFs and Wnts) as well as antagonists of BMP signals. This causes the proliferation and differentiation of the hair follicle. (C) Contacting the dermal papilla at the base, the cells divide rapidly to generate the hair shaft and its channel. The cells close to the earlier bulge become the outer layer of the bulge and have stem cell properties. An inner layer, several cell layers thick, is also derived from the HFSCs; this layer eventually inhibits HFSC proliferation. At catagen (not shown), most cells undergo apoptosis, but the remaining stem cells survive in the bulge region. An epithelial strand then brings the dermal papilla up to the region of the bulge, and the interaction between them appears to generate the hair germ of the next generation of hair. (After Hsu and Fuchs 2012, Hsu et al. 2014.)

BMP6 and Fgf18, two repressors of HFSC proliferation. In addition, the dermal fibroblasts and the subcutaneous fat cells also make growth-suppressive BMPs. The HFSCs are activated at the beginning of the anagen phase by signals from the dermal papilla of condensed mesenchyme. These signals are FGFs, Wnts, and BMP antagonists, and they direct the epidermal stem cells to migrate out of the bulge. There, the epidermal stem cells produce progenitor cells that proliferate downward and generate the seven concentric columns of cells that form the outer root sheath from bulge to matrix.

This activation of the dermal papilla is regulated by the microenvironment of the dermis. The underlying dermis appears to make more Wnts and less BMPs, while the adipocyte precursor cells make more of the paracrine factor PDGF, which stimulates the dermal papilla (Plikus et al. 2008; Rendl et al. 2008; Hsu and Fuchs 2012). As the dermal papilla is moved farther away by the downward growth of the epithelial cells, its signals are not received by the stem cells, and the bulge return to quiescence. During the latter part of anagen, prostaglandin PGD_2 appears to prevent the production of the progenitor cells. During catagen, most of the basal (outer root sheath) epithelial cells undergo apoptosis. Upper follicular stem cells remain, however. The outer root sheath cells close to the old bulge contain HFSCs and become the outer layer, while those nearer to the matrix differentiate to become the inner layer of the bulge. Apoptosis causes the outer cells to be in contact with the dermal papilla in readiness for the next cycle (Hsu et al. 2011; Mesa et al. 2015).

Remarkably, activation of the bulge stem cells is regulated in part by the progenitor cells they produced. The progenitor cells secrete Sonic hedgehog, which is essential for the division of the bulge HFSCs. It is possible that the BMP6 and FGF signals from these cells inhibit the stem cells beneath the bulge, while their Shh activates the bulge stem cells. This means that the progenitor cells are not merely passive cells on their way to differentiate, but they constitute a signaling center that can activate the quiescent bulge stem cells (**FIGURE 16.18**; see also Figure 16.17C). The dermal papilla can thereby initiate hair regeneration by stimulating the *primed* (sub-bulge) HFSCs to establish a population of transit-amplifying cells, and then this progenitor cell population serves as a signaling center to sustain the dermal papilla signaling needed to expand the transit amplifying cells. These transit-amplifying cells also stimulate proliferation of quiescent stem cells. Thus, the transit-amplifying cells regulate the proliferation of themselves, the primed HFSCs, and the quiescent HFSCs, thereby coordinating the regeneration of the hair follicle (Hsu et al. 2014).

((• **SCIENTISTS SPEAK 16.1** Dr. Elaine Fuchs discusses stem cells in general and the remarkable abilities of hair follicle stem cells in particular.

FIGURE 16.18 During hair follicle regeneration, emerging progenitor cells constitute a signaling center that orchestrates tissue growth. The primed stem cells (HFSCs) generate progenitor cells. The quiescent stem cells divide only after the progenitor cells emerge and begin secreting Sonic hedgehog (Shh). Progenitor cell generation diminishes if the progenitors can't produce Shh. Shh both promotes quiescent stem cell proliferation and regulates the dermal factors that promote progenitor cell expansion. Without input from quiescent HFSCs, replenishment of primed HFSCs for the next hair cycle is reduced. This reduction delays regeneration and can lead to the failure to regenerate the hair follicle. (After Hsu et al. 2014.)

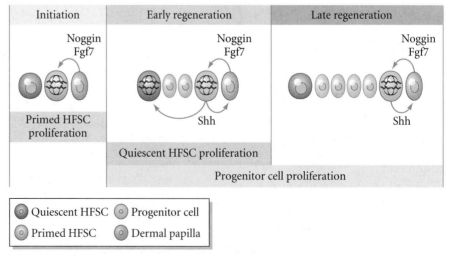

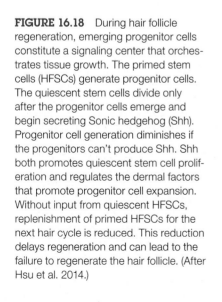

These findings about stem cell activation and quiescence has helped explain two human variations: male pattern baldness and long eyelashes. Male pattern baldness, which is characterized by a decrease in hair follicle size, appears to be caused by the progressive inability of the HFSCs to generate progenitor cells. Indeed, in aged mammals, inhibitory compounds from outside the niche may keep the HFSCs in a dormant state. This cessation of progenitor cell production appears to be due to the prolonged synthesis of prostaglandin PGD_2, which is normally used to stop hair growth at the end of the anagen phase. Bald men have higher levels of this factor, and transgenic mice that overexpress the enzymes leading to prostaglandin PGD_2 synthesis have hair loss. Moreover, the genes encoding the enzymes producing this prostaglandin are upregulated by testosterone[4] (Garza et al. 2011, 2012) and repressed by the Wnt pathway—the same pathway implicated in tooth regeneration. Epithelial Wnt secretion is necessary for adult hair follicle growth and regeneration, and older mice have much higher levels of Wnt inhibitors (such as Dickkopf) than younger mice (Myung et al. 2013; Chen et al. 2014). Normally, PGD_2 probably acts as a counterbalance to the positive growth effects of related prostaglandins PGE_2 and $PGF_{2\alpha}$. These latter two prostaglandins appear to stimulate the growth of hair by prolonging the anagen phase (Johnstone and Albert 2002; Sasaki et al. 2005). Indeed, solutions containing these prostaglandins or their analogues have been approved for the cosmetic use of lengthening eyelashes.

WEB TOPIC 16.5 **NORMAL VARIATION IN HUMAN HAIR PRODUCTION**   The pattern of hair size, length, and thickness (or lack thereof) is determined by paracrine and endocrine factors.

Coda

In 1882, Thomas Huxley, one of Britain's foremost naturalists and one of Charles Darwin's most ardent supporters, reflected, "I remember … the intense satisfaction and delight which I had in listening, by the hour together, to Bach's fugues. … The source of pleasure is exactly the same as in most of my problems in morphology—that you have the theme in one of the old master's works followed out in all its endless variations, always appearing and always reminding you of unity in variety." In the epidermal placodes, one sees amazing variations on the theme of epithelial-mesenchymal interaction. The placodes use similar pathways and similar forces to produce different organs—feathers, hair, scales, mammary glands, and teeth. The epidermis and the eye use similar paracrine factors and signaling pathways to ensure their coordinated development and growth. As Huxley himself noted (see p. 785), few things are "invented" *de novo*. It is the different combinations at different times that makes the difference.

[4]Testosterone and (more importantly, as we saw in Chapter 6) its derivative hydrotestosterone are critical in producing male pattern baldness. Ancient civilizations noted that eunuchs (castrated males) did not become naturally bald. Testosterone does not appear to play a role in female hair thinning (Kaufman 2002).

Next Step Investigation

The ability of hair to regenerate through its stem cells provides a remarkable lesson that we are just beginning to appreciate. Indeed, regenerating skin reactivates some of the pathways seen in normal development. The Foxc1 transcription factor appears to be critically important in retaining the hair follicle niche and in maintaining a population of quiescent stem cells (Lay et al. 2016; Wang et al. 2016). Recent work in has also shown that damaged skin releases double-stranded RNAs that are sensed by the Toll-like receptor-3 protein (TLR3). In addition to activating the cells of the immune system, activated TLR3 protein activates the ectodysplasin pathway (see Nelson et al. 2015).

Closing Thoughts on the Opening Photo

What controls hair length? Our hair usually grows long on our scalps but is much shorter on our torsos, limbs, armpits, and pubic regions. One possibility is local control of the *Fgf5* gene. Long-haired dachshunds differ from their wild-type short-haired sibs by being homozygous for a loss-of-function mutation in the *Fgf5* gene. There are different mutations of this gene in different breeds of dogs (Cadieu et al. 2009; Dierks et al. 2013), and *Fgf5* mutations are seen in long-haired cats and other mammals. Fgf5 is involved in initiating the catagen (regression) stage of the hair growth cycle, so mutations would allow more time in anagen (the growth period). (Photograph © Bigandt Photography/Getty Images.)

16 Snapshot Summary
Ectodermal Placodes and the Epidermis

1. Ectodermal placodes are thickenings of cells, usually in the ectoderm. The anterior placodes form the sensory neurons of the eye, ear, and nose, as well as the lens of the eye. The more posterior placodes form the hair, teeth, feather, scutes, and scales that cover the epidermis.

2. The cranial placodes form in the ectoderm in the border area between the neural tube and ectoderm. They are distinguished from neural crest cells by signals that inhibit BMPs and Wnts.

3. The cranial sensory placodes originate in a pan-placodal region that is separated into individual placodes.

4. The vertebrate retina forms from an optic vesicle that extends from the brain. Pax6 plays a major role in eye formation, and the downregulation of Pax6 by Sonic hedgehog in the center of the brain splits the eye-forming region of the brain in half. If Shh is not expressed there, a single medial eye results.

5. The photoreceptor cells of the retina gather light and transmit an electric impulse to the retinal ganglion cells.

The axons of the retinal ganglion cells form the optic nerve. The lens and cornea form from the surface ectoderm. Both must become transparent.

6. Reciprocal induction is critical in the specification and differentiation of the retina and lens. The cells that form the organs have two "lives." In the embryonic life, they construct the organs; in the adult life, they function as part of an organ. The body is constructed by cells that are not performing their adult roles.

7. The basal layer of the surface ectoderm becomes the germinal layer of the skin. Epidermal stem cells divide to produce differentiated keratinocytes and more stem cells.

8. The enamel knot is the signaling center for tooth shape and development.

9. The hair follicular stem cells, which regenerate hair follicles during periods of cyclical growth, reside in the bulge of the hair follicle. Male baldness appears to result from the inhibition of the Wnt pathway, preventing stem cell division.

Further Reading

Ahn, Y. 2015. Signaling in tooth, hair, and mammary placodes. *Curr. Top. Dev. Biol.* 111: 421–452.

Ahtiainen, L. and 7 others. 2014. Directional cell migration, but not proliferation, drives hair placode morphogenesis. *Dev. Cell* 28: 588–602.

Biggs, L. C. and M. L. Mikkola. 2014. Early inductive events in ectodermal appendage morphogenesis. *Semin. Cell Dev. Biol.* 26: 11–21.

Hsu, Y. C., L. Li and E. Fuchs. 2014. Emerging interactions between skin stem cells and their niches. *Nature Med.* 20: 847-856.

Jernvall, J., T. Aberg, P. Kettunen, S. Keränen and I. Thesleff. 1998. The life history of an embryonic signaling center: BMP4 induces *p21* and is associated with apoptosis in the mouse tooth enamel knot. *Development* 125: 161–169.

Ogino, H., H. Ochi, H. M. Reza and K. Yasuda. 2012. Transcription factors involved in lens development from the preplacodal ectoderm. *Dev. Biol.* 363: 333–347.

Sick, S., S. Reinker, J. Timmer and T. Schlake. 2006. WNT and DKK determine hair follicle spacing through a reaction-diffusion mechanism. *Science* 314: 1447–1450.

GO TO WWW.DEVBIO.COM...

...for Web Topics, Scientists Speak interviews, Watch Development videos, Dev Tutorials, and complete bibliographic information for all literature cited in this chapter.

Paraxial Mesoderm
The Somites and Their Derivatives

SEGMENTATION OF THE BODY PLAN is a highly conserved physical feature across all vertebrate species. Repetition of form through segmentation has provided a developmental mechanism for the evolution of increasingly sophisticated functions. For instance, although humans and giraffes have the same number of cervical vertebrae, the sizes of these segments are profoundly different and adapted to their environmental pressures. Thoracic vertebrae are the only segments to have ribs, which function in part to provide protection to organs. The number of thoracic vertebrae differs wildly among a human, mouse, and snake. The number and size of segments and their bone and muscle derivatives are decided by modifications in the fission of mesoderm along the anterior-posterior axis. How can a tissue be developmentally cut up into precisely sized segments? How can snakes have some 300 segments while humans have only about 35?

One of the major tasks of gastrulation is to create a mesodermal layer between the endoderm and the ectoderm. As seen in **FIGURE 17.1**, the formation of mesodermal tissues in the vertebrate embryo is not subsequent to neural tube formation, but occurs synchronously. The notochord extends beneath the neural tube, from the posterior region of the forebrain into the tail. On either side of the neural tube lie thick bands of mesodermal

What, when, where, and how many?

The Punchline

The paraxial mesoderm lies adjacent to the notochord and neural tube. It gives rise to the vertebrae, the skeletal muscles, and much of the connective tissue of the skin. Anterior paraxial mesoderm is formed during gastrulation; in the posterior, it is generated by multipotent neuromesodermal progenitors in the tailbud. Paraxial mesoderm begins unsegmented, but as the axis elongates, an anterior-to-posterior wave of boundary formation divides it with clocklike precision into similar-sized somites. Somitogenesis is controlled by opposing gradients of FGF/Wnt from the tailbud and retinoic acid from the anterior regions, which maintain progenitor cells or encourage somite differentiation, respectively. Periodic oscillations of Notch-Delta signaling provide the "clock" of somitogenesis, influencing the spatiotemporal collinearity of Hox gene activation along the trunk. Hox expression is also epigenetically regulated and is essential for proper axial identities. Physical boundaries are established through an Eph-EphrinB2 mechanism of cell repulsion and extracellular matrix deposition. Somite differentiation begins with the induction of the sclerotome and dermomyotome, which give rise to bone and muscle, respectively. Signals from the notochord, neural tube, lateral plate mesoderm, surface ectoderm, and migrating neural crest cells contribute to the regulation of chondrogenesis and myogenesis.

FIGURE 17.1 Major lineages of the mesoderm are shown in this schematic of the mesodermal compartments of the amniote embryo.

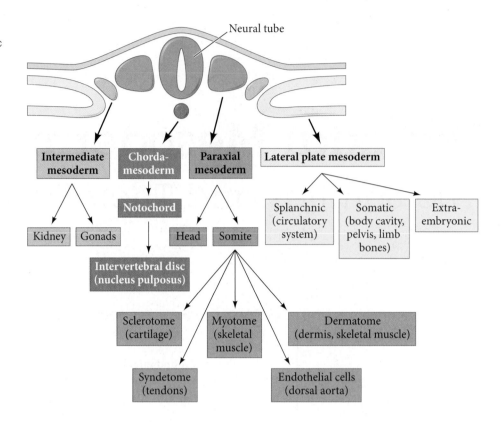

cells divided into the paraxial mesoderm, intermediate mesoderm, and lateral plate mesoderm. The early paraxial mesoderm directly adjacent to the notochord lacks somites; it takes the form of bilateral streaks of continuous mesenchymal cells, referred to as the **presomitic mesoderm** (**PSM**), also known as the **segmental plate**. As studied extensively in amniotes, when the primitive streak is regressing and the neural folds begin to gather at the center of the embryo, the cells of the presomitic mesoderm will form somites. **Somites** are epithelial, "block-like" clusters of cells that are bilaterally positioned adjacent to the neural tube.

The regions of trunk and head mesoderm and their derivatives can be summarized as follows (see Figure 17.1):

1. The central region of trunk mesoderm is the **chordamesoderm** (generally referred to as the axial mesoderm). This tissue forms the notochord, a transient tissue whose major functions include inducing and patterning the neural tube and establishing the anterior-posterior body axis. Notochord cells are hydrostatically pressurized with large vacuoles to provide a rigid rod-like structure for the developing embryo. Despite many notochord cells succumbing to apoptotic clearance, the jelly-like core of the intervertebral disc, called the nucleus pulposus, is derived from notochord cells (**FIGURE 17.2**; Choi et al. 2008; McCann et al. 2011).

2. Flanking the notochord on both sides is the **paraxial**, or **somitic**, **mesoderm**. The tissues developing from this region will be located in the back of the embryo, surrounding the spinal cord and, for some muscle descendants, in the limb and ventral (abdominal wall) regions. Before those regions can be populated, the cells of the paraxial mesoderm will form somites—transitory epithelial blocks of mesodermal cells on either side of the neural tube—which will produce muscle and many of the connective tissues of the back (dermis, muscle, and skeletal elements such as the vertebrae and ribs; see Figure 17.2E,F). The anteriormost paraxial mesoderm does not segment; it becomes the **head mesoderm**, which (along with the neural crest) forms the skeleton, muscles, and connective tissue of the face and skull.

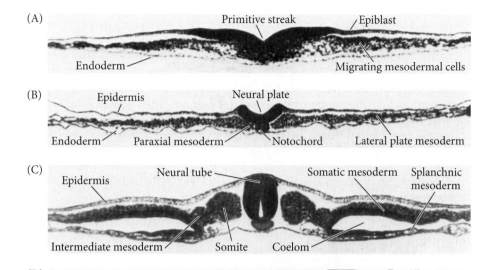

FIGURE 17.2 Gastrulation and neurulation in the chick embryo, focusing on the mesodermal component. (A) Primitive streak region, showing migrating mesodermal and endodermal precursors. (B) Formation of the notochord and paraxial mesoderm. (C,D) Differentiation of the somites, the coelom, and the two dorsal aortae (which will eventually fuse). (A–C) 24-hour embryos. (D) 48-hour embryo. (E,F) Color-coded schematic of one half of a somite from a 48-hour embryo in cross section (E) with the derivative structures those somitic cells contribute to in the adult (F). (E,F adapted from Lawson and Harfe 2015; Scaal 2015.)

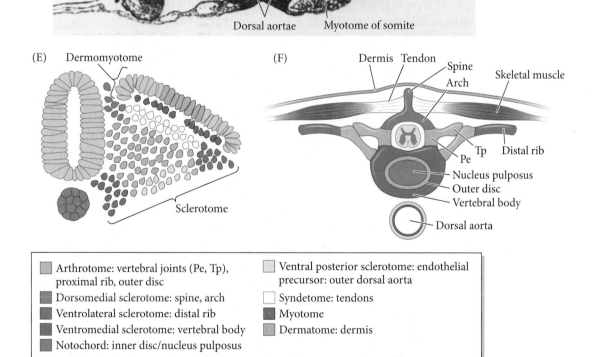

3. The **intermediate mesoderm** is positioned directly lateral to the paraxial mesoderm and forms the urogenital system, consisting of the kidneys, the gonads, and their associated ducts. The outer (cortical) portion of the adrenal gland also derives from this region (see Figure 17.2C).

4. Farthest away from the notochord, the **lateral plate mesoderm** gives rise to the heart, blood vessels, and blood cells of the circulatory system as well as to the

lining of the body cavities. It also gives rise to the pelvic and limb skeleton (but not the limb muscles, which are somitic in origin). Lateral plate mesoderm also helps form a series of extraembryonic membranes that are important for transporting nutrients to the embryo (see Figure 17.2B,C).

5. Anterior to the trunk mesoderm is the head mesoderm, consisting of the unsegmented paraxial mesoderm and prechordal mesoderm. This region provides the head mesenchyme that forms much of the connective tissues and musculature of the head (Evans and Noden 2006). The muscles derived from the head mesoderm form differently than those formed from the somites. Not only do they have their own set of transcription factors, but the head and trunk muscles are affected by different types of muscular dystrophies (Emery 2002; Bothe and Dietrich 2006; Harel et al. 2009).

Cell Types of the Somite

When a somite first forms, it is shaped by epithelial cell contacts that create a block-like mass with mesenchymal cells at its core. The cells within a somite become committed toward a particular cell fate relatively late, after the somite has already formed. When a somite is first separated from the presomitic mesoderm, all its cells are capable of becoming any of the somite-derived structures. As the somite matures, portions of it undergo epithelial-to-mesenchymal transition (EMT), and its various regions become committed to forming only certain cell types (see Figure 17.2A–D). These mature somites contain two major compartments: the sclerotome and the dermomyotome (see Figure 17.2D,E). The sclerotome is formed from the ventromedial cells of the somite (those cells closest to the neural tube and notochord), where they undergo mitosis, lose their round epithelial characteristics, and become mesenchymal cells again. The dermomyotome is formed from the remaining epithelial portion of the somite, which gives rise to the muscle-forming myotome and the dermis-forming dermatome (see Figure 17.2E,F).

The **sclerotome** gives rise to the vertebrae with associated tendons and rib cartilage (see Figure 17.2E). It can be further subdivided into the progenitor zones for specific cell lineages. The **syndetome** arises from the most dorsal sclerotome cells and generates the tendons, while the most internal cells of the sclerotome (sometimes called the **arthrotome**) become the vertebral joints, the outer portion of the intervertebral discs, and the proximal portion of the ribs (Mittapalli et al. 2005; Christ et al. 2007). The lateralmost mesenchyme will build the distalmost aspects of the ribs. Cells occupying the ventromedial sclerotome will migrate to the notochord and form the vertebral body, while additional cells from the arthrotome will combine with the notochord cells to form the intervertebral discs. The most dorsomedial sclerotome cells will establish the spine and arch of the vertebrae. Finally, an as-yet unnamed group of *endothelial precursor cells* in the ventral-posterior[1] sclerotome generates differentiated vascular cells of the dorsal aorta and intervertebral blood vessels (**TABLE 17.1**; Pardanaud et al. 1996; Sato et al. 2008; Ohata et al. 2009).

The **dermomyotome** contains progenitor cells for making skeletal muscle and the dermis of the back (see Figure 17.2E). The ventralmost portion of the dermomyotome is partitioned off as the **myotome**, which forms the musculature of the back, rib cage, and ventral body wall. Additional muscle progenitors are provided by cells that detach from the lateral edge of the dermomyotome that migrate into the limbs to generate the musculature of the forelimbs and hindlimbs. The dorsalmost surface of the dermomyotome develops into the **dermatome**, which gives rise to the dermis of the back.

---

[1] Later in the chapter we will discuss how the somite is polarized along the anterior-to-posterior axis. In addition, as was described in Chapter 16, cells may show a preference to interact with only one half of the somite. In this example, endothelial cell precursors only migrate out of the posterior somite.

TABLE 17.1   Derivatives of the somite

Traditional view	Current view
DERMOMYOTOME	
Myotome forms skeletal muscles	Lateral edges generate primary myotome that forms muscle
Dermatome forms back dermis	Central region forms muscle, muscle stem cells, dermis, brown fat cells
SCLEROTOME	
Forms vertebral and rib cartilage	Forms vertebral and rib cartilage
	Dorsal region forms tendons (syndetome)
	Medial region forms blood vessels and meninges
	Central mesenchymal region forms joints (arthrotome)
	Forms smooth muscle cells of dorsal aorta

Thus, the somite contains a population of multipotent cells whose specification is correlated with (and depends on) their location within the somite. What is it about their position that influences their differentiation? Consider the structures neighboring each of the progenitor domains within the somite. How might they influence the development of sclerotomal or dermomyotomal cells? At the midline are the notochord and neural tube, laterally are the other mesodermal derivatives, and above them all is the epidermis. Later in this chapter, we will discuss how paracrine signals from these surrounding tissues pattern the cell fates of the somite. First, however, we need to understand how the paraxial mesoderm and somites are specified.

Establishing the Paraxial Mesoderm and Cell Fates Along the Anterior-Posterior Axis

Specification of the paraxial mesoderm

The mesodermal subtypes (chordamesoderm, paraxial, intermediate, and lateral plate) are specified along the mediolateral (center-to-side) axis by increasing amounts of BMPs (**FIGURE 17.3A**; Pourquié et al. 1996; Tonegawa et al. 1997). The more lateral mesoderm of the chick embryo expresses higher levels of BMP4 than do the midline areas, and one can change the identity of the mesodermal tissue by altering BMP expression. One mechanism to shape the BMP gradient uses Noggin, an inhibitor of BMP, which is expressed first in the notochord and then in the somitic mesoderm (Tonegawa and Takahashi 1998). If Noggin-expressing cells are placed into presumptive lateral plate mesoderm, the lateral plate tissue will be respecified into somite-forming paraxial mesoderm (**FIGURE 17.3B**; Tonegawa and Takahashi 1998; Gerhart et al. 2011).

Although some differences exist between species, it appears that the different BMP concentrations may result in differential expression of the Forkhead (Fox) family of transcription factors, with *Foxf1* being transcribed in regions that will become the lateral plate mesoderm, and *Foxc1* and *Foxc2* transcribed in the regions that will form the somites (Wilm et al. 2004). Loss of *Foxc1* and *Foxc2* in mouse results in the respecification of paraxial mesoderm into intermediate mesoderm.

Additional transcriptional regulators play conserved roles in the early specification of the presomitic mesoderm, in particular Brachyury (T), Tbx6, and Mesogenin (Van Eeden et al. 1998; Nikaido et al. 2002; Windner et al. 2012). In zebrafish, presomitic mesoderm development requires both *Tbx6* and *Tbx16* (*spadetail*); in the mouse, however, *Tbx6* appears to serve the function of both those genes. Loss of *Tbx6* in mouse converts the presumptive PSM into neural tissue. These *Tbx6* knockout mice express the neural

(A)

(B)

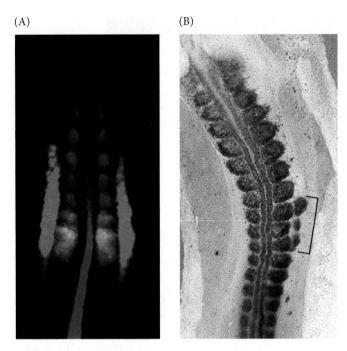

FIGURE 17.3 (A) Staining for the medial mesodermal compartments in the trunk of a 12-somite chick embryo (about 33 hours). In situ hybridization was performed with probes binding to *Chordin* mRNA (blue) in the notochord, *Paraxis* mRNA (green) in the somites, and *Pax2* mRNA (red) in the intermediate mesoderm. (B) Specification of somites. Placing Noggin-secreting cells into a prospective region of chick lateral plate mesoderm will respecify that mesoderm into somite-forming paraxial mesoderm. Induced somites (bracket) were detected by in situ hybridization with a *Pax3* probe. (A from Denkers et al. 2004, courtesy of T. J. Mauch; B from Tonegawa and Takahashi 1998, courtesy of Y. Takahashi.)

progenitor determination factor *Sox2* (among other neural genes) in the presumptive PSM, and, strikingly, this generates ectopic neural tubes in place of the PSM (Chapman and Papaioannou 1998; Takemoto et al. 2011; Nowotschin et al. 2012). These embryos actually have three neural tubes (**FIGURE 17.4**)! These results suggest that Tbx6 normally promotes PSM fates, in part by repressing *Sox2* and neural fates.

Tbx6 is not the only PSM determination factor. Another transcription factor, mesogenin 1, may function as a "master regulator" of PSM fates by acting upstream of *Tbx6* (Yabe and Takada 2012; Chalamalasetty et al. 2014). Gain- and loss-of-function analyses of mouse *mesogenin 1* have demonstrated that it is both sufficient and necessary for *Tbx6* expression in the PSM (**FIGURE 17.5**).

Taken together, these findings posit that a bipotential stem cell population is retained in the posterior region of the embryo that maintains the extreme plasticity needed to give rise to cell fates spanning the mesodermal and ectodermal lineages (reviewed in Kimelman and Martin 2012; Neijts et al. 2014; Beck 2015; Carron and Shi 2015; Henrique et al. 2015). Although we have identified the transcriptional regulators of these stem cells, what signaling systems are in place to induce cell-type maturation from this **caudal progenitor zone**?

Several opposing morphogens are displayed along the anterior-posterior axis of the paraxial mesoderm. Specifically, Fgf8 and Wnt3a are highly expressed in the vertebrate tailbud, whereas an anterior originating gradient of retinoic acid (RA) is produced from the somites and neural plate. RA directly represses *Fgf8* and *Tbx6* expression enough to foster neural cell fate maturation through upregulation of *Sox2* (**FIGURE 17.6A**; Kumar and Duester 2014; Cunningham et al. 2015; Garriock et al. 2015). Importantly, Fgf8 activates Cyp26b (a direct inhibitor of RA synthesis), which promotes mesodermal cell fate development (**FIGURE 17.6B**). Therefore, a balance of signaling between these opposing and antagonistic morphogens patterns cell migration, proliferation, and differentiation into the

Figure 17.4 Three neural tubes: loss of the *Tbx6* gene transforms the paraxial mesoderm into neural tubes. In situ hybridization for the mRNA expression (blue) of the neural specification markers *Sox2* and *Pax6* in wild-type mice (A) and *Tbx6* knockout mice (B). *Sox2* is ectopically expressed throughout the presumptive paraxial mesoderm in the *Tbx6*−/− embryo, which has also taken on a neural tube-like morphology, even displaying a central lumen (arrows). Similarly, the dorsal neural tube marker *Pax6* shows regional cell specification within these ectopic neural tubes (arrowheads). (From Takemoto et al. 2011.)

(A) Wild-type   (B) *Tbx6*−/−

Sox2 expression

Pax6 expression

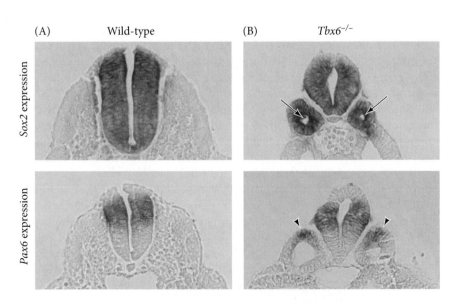

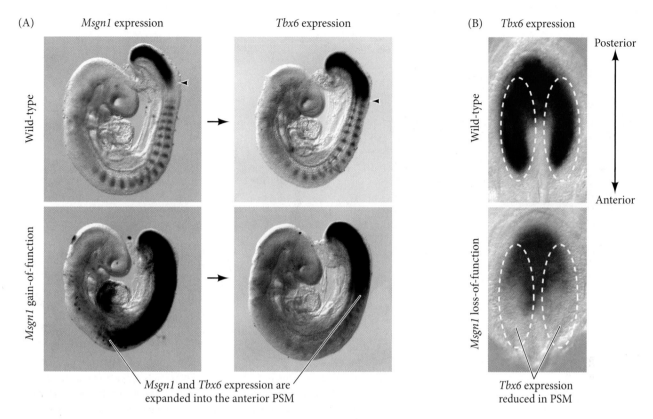

(A) *Msgn1* expression   *Tbx6* expression

Wild-type

Msgn1 gain-of-function

Msgn1 and *Tbx6* expression are
expanded into the anterior PSM

(B) *Tbx6* expression

Posterior

Anterior

Wild-type

Msgn1 loss-of-function

Tbx6 expression
reduced in PSM

FIGURE 17.5 Mesogenin 1 (Msgn1) is both sufficient and necessary for Tbx6 expression. (A) *Msgn1* gain-of-function. As visualized by in situ hybridization, misexpression of *Msgn1* in the presomitic mesoderm (lower two photos) is expanded throughout the trunk (left), which causes an expansion of *Tbx6* expression (right). (B) In contrast, loss of *Msgn1* function results in the reduction of *Tbx6* expression (blue, circled domains). (A, lateral views; B, dorsal views.) (Based on Chalamalasetty et al. 2014; photos courtesy of Terry Yamaguchi and Ravi Chalamalasetty.)

appropriate neural or mesodermal cell fates (Cunningham and Duester 2015; Henrique et al. 2015). However, this model must still enable the development of distinct somite identities along the anterior-posterior axis.

Spatiotemporal collinearity of Hox genes determine identity along the trunk

All somites may resemble one another, but they form different structures along the rostral-to-caudal (anterior-posterior) axis. For instance, the somites that form the cervical vertebrae of the neck and lumbar vertebrae of the abdomen are not capable of forming ribs; ribs are generated only by the somites that form the thoracic vertebrae, and this specification of thoracic vertebrae occurs very early in development. The different regions of presomitic mesoderm are determined by their position along the anterior-posterior axis before somitogenesis. If presomitic mesoderm from the thoracic region of a chick embryo is transplanted to the cervical (neck) region of a younger embryo, the host embryo will develop ribs in its neck on the side of the transplant (**FIGURE 17.7A**; Kieny et al. 1972; Nowicki and Burke 2000).

The anterior-posterior specification of the somites is determined by Hox genes (see Chapter 12). Hox gene expression is spatially presented in a collinear fashion. Thus, the Hox genes organized in more 3′ positions of their cluster in the genome are expressed in more anterior regions of the paraxial mesoderm; conversely, the Hox genes at more 5′ positions are expressed in more posterior regions of the embryo (see Figure 12.23; Wellik and Capecchi 2003). If this pattern of Hox gene expression is altered, so is the

(A)

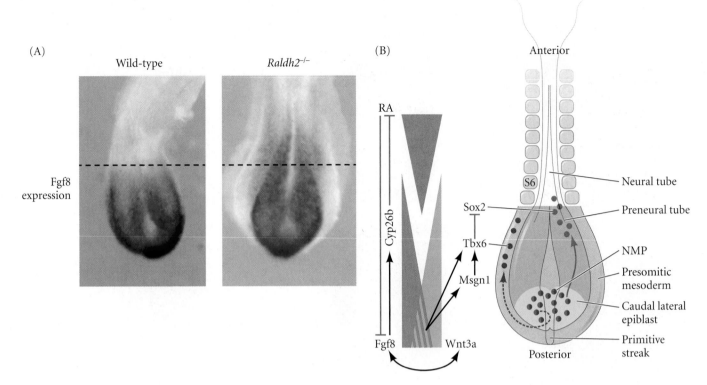

FIGURE 17.6 Antagonistic signals along the anterior-posterior axis pattern neuromesodermal progenitors (NMP) during paraxial mesoderm development. (A) Loss of retinoic acid synthesis in the *Raldh2* knockout mouse shows an anterior expansion of Fgf8 expression (blue past the dashed line). (B) Model of signaling systems regulating the NMP from the caudal progenitor zone (the caudal lateral epiblast) into the preneural tube or presomitic mesoderm. Posterior signals of FGF and Wnt antagonize anterior retinoic acid signaling. Fgf8 and Wnt3a upregulate Mesogenin 1 (Msgn1) and Tbx6 expression to promote presomitic progenitor specification and repress neural cell fate specification (Sox2). (A from Cunningham et al. 2015; B after Henrique et al. 2015.)

specification of the mesoderm. For instance, if the entire presomitic mesoderm ectopically expresses *Hoxa10*, ribs are completely lost due to the replacement of thoracic vertebrae with lumbar vertebra (**FIGURE 17.7B**). If *Hoxb6* is misexpressed throughout the PSM, however, all vertebrae form ribs (**FIGURE 17.7C**; Carapuço et al. 2005; Guerreiro et al. 2013). In both cases, these transformations of vertebral identity were induced by the ectopic expression within the PSM, not in the somites, which suggests that cells of the PSM receive instructions for axial-level specification, and these instructions are then implemented later during somite differentiation. In the chick, prior to cell migration through the primitive streak, Hox gene expression can be labial; once the paraxial mesoderm cells have taken up position in the presomitic mesoderm, however, Hox gene expression appears more fixed. Indeed, once established, each somite retains its pattern of Hox gene expression, even if that somite is transplanted into another region of the embryo (Nowicki and Burke 2000; Iimura and Pourquié 2006; McGrew 2008).

The **temporal collinearity** of Hox genes refers to a temporal control mechanism of Hox gene activation that sets up the spatial collinearity of Hox gene expression along the anterior-posterior axis of the embryo, such that this expression corresponds to the genomic organization of Hox genes along the 3′-to-5′ orientation. In other words, the earlier expressed Hox genes are both expressed by cells positioned more anteriorly in the embryo and found in more 3′ chromosomal locations. In fact, it is the dynamic temporal pairing of 3′-to-5′ Hox gene activation with the timing of cell ingression/migration into the paraxial mesoderm that then lays out the spatial pattern of Hox gene expression along the trunk (**FIGURE 17.8A**; Izpisúa-Belmonte et al. 1991). Initially, a progress zone model was proposed in which the progressive activation of 3′-to-5′ Hox genes occurs,

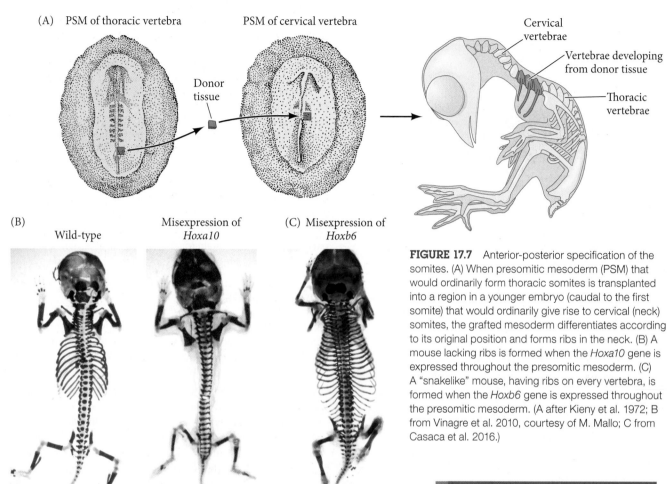

FIGURE 17.7 Anterior-posterior specification of the somites. (A) When presomitic mesoderm (PSM) that would ordinarily form thoracic somites is transplanted into a region in a younger embryo (caudal to the first somite) that would ordinarily give rise to cervical (neck) somites, the grafted mesoderm differentiates according to its original position and forms ribs in the neck. (B) A mouse lacking ribs is formed when the *Hoxa10* gene is expressed throughout the presomitic mesoderm. (C) A "snakelike" mouse, having ribs on every vertebra, is formed when the *Hoxb6* gene is expressed throughout the presomitic mesoderm. (A after Kieny et al. 1972; B from Vinagre et al. 2010, courtesy of M. Mallo; C from Casaca et al. 2016.)

respectively, in earlier-to-later cells of the PSM (Kondo and Duboule 1999; Kmita and Duboule 2003). More recent investigations, however, have favored a model in which Hox genes control the timing of cell ingression in the streak; hence, cells that express anterior Hox genes will ingress early, whereas cells that express more posterior Hox genes ingress later (Iimura and Pourquié 2006; Denans et al. 2015). During this period of paraxial mesoderm development, the axis is elongating as new progenitor cells enter the PSM from the caudal progenitor zone and occupy sequentially more posterior positions over time. Thus, more-anterior PSM cells will express more 3′ Hox genes, while the later-incorporated PSM cells will be located in the posterior and express more 5′ Hox paralogues—all of which results in the final identity of somites along the rostral-caudal axis (see Figure 17.8A; reviewed in Casaca et al. 2014).

This temporal activation of Hox genes has been called the Hox clock (Duboule and Morata 1994). How is such a linear activation of Hox genes carried out on the cellular level? Research has shown that the Hox genes change from a tightly packed to an unpacked architecture[2] in a chronological order that matches the order of their expression in the PSM: first the 3′ Hox genes (*Hoxd4*) show signs of unpacked structure, followed by the *Hoxd8-9* clusters, then the *Hoxd10*, and finally

---

[2] How does one examine chromatin states? The Duboule lab used a new technique called circular chromosome conformation capture (or 4C-seq) to observe the three-dimensional genomic organization of Hox gene clusters and identify which Hox genes are in tightly packed or loosely packed chromatin states during paraxial mesoderm development.

Developing Questions

With the first investigations of the *Drosophila* Hox genes by Morgan (1915) and Lewis (1978) being so long ago, one might think we would understand by now all there is to know about Hox genes, but a variety of questions regarding the regulatory mechanisms of Hox genes persist. You just learned that there is a progressive 3′-to-5′ loosening in chromatin states of Hox clusters over the course of paraxial mesoderm development. What triggers this early epigenetic modification in the chromatin of 3′ Hox genes? Once initiated, is the progressive transformation across the cluster autonomously driven, or are additional regulators required to propel this transition to the 5′ end of the cluster? Last, what are the mechanisms that stabilize the inheritance of the specific epigenetic modifications of Hox genes within a given segment? As you can see, *lots of questions still remain!*

(A)

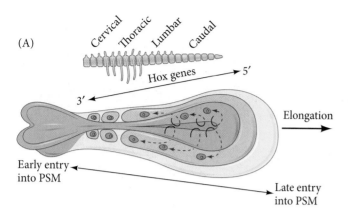

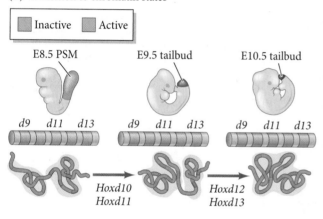

FIGURE 17.8 The spatiotemporal collinearity of Hox gene expression in the presomitic mesoderm is correlated with chromatin remodeling. (A) Illustration showing the successive migration of cells into the PSM as the tailbud elongates, which correlates with the onset of progressively more 5' Hox gene expression and the development of different vertebral identities. (B) Changes in chromatin structure permit progressive access for differential Hox gene expression over the course of PSM elongation. Colored PSM and tailbuds represent tissues used to analyze *Hoxd* chromatin structure at the different embryonic stages (E8.5–E10.5) (B after Noordermeer et al. 2014.)

the more 5' *Hoxd11-12* (**FIGURE 17.8B**; Montavon and Duboule 2013; Noordermeer et al. 2014). Moreover, it appears that once a cell has adopted its position in the PSM, it has also adopted its particular chromatin state for its Hox genes, and all its daughter cells retain a fixed memory of that state.

Somitogenesis

How does the segmental plate become partitioned into the correct number of appropriately sized and bilaterally symmetrical somites? As the mesenchymal cells of the presomitic mesoderm mature, they become organized into "whorls" of cells that constitute the somite precursors and are sometimes called **somitomeres** (Meier 1979). These somite precursors undergo cell organizational changes such that the outer cells adhere together to form an epithelium while the inner cells remain mesenchymal. The first somites appear just posterior to the region of the otic vesicle, and new somites "bud off" from the rostral end of the presomitic mesoderm at regular intervals (**FIGURE 17.9**). The formation of somites is called **somitogenesis**, and this process involves the periodic creation of an epithelial fissure by the mesenchymal cells of the PSM. These medial-to-lateral partitions establish an epithelial boundary between the caudal half of the next anterior somite and the rostral-most extent of the PSM. Therefore, whole somites with anterior and posterior borders are not formed simultaneously; rather, one boundary at a time is created at regular intervals. When a new boundary is formed, it creates the caudal half of a somite (thus completing a fully formed somite) as well as establishing the rostral half of the next somite to be formed.

To accurately describe the somites formed and the position of somitomeres in the presegmental plate, a numbering scheme was established using roman numerals (Pourquié and Tam 2001). The most recently formed somite is always position I, and each additionally older somite is numbered counting up: II, III, IV, and so on. Moving posteriorly into regions where somites have not yet formed, positions of future somites are numbered counting backwards, 0, –I, –II, –III, and so on. Somitomere 0 (zero) shares a boundary with somite I and is always the next somite to be formed.

Because individual embryos in any species can develop at slightly different rates (as when chick embryos are incubated at slightly different temperatures), the number of somites formed is usually the best indicator of how far development has proceeded.

FIGURE 17.9 Neural tube and somites seen by scanning electron microscopy. When the surface ectoderm is peeled away, well-formed somites are revealed, along with paraxial mesoderm (red) that has not yet separated into distinct somites. A rounding of the paraxial mesoderm into a somitomere (area inside brackets) is seen at the lower left, and neural crest cells can be seen migrating ventrally from the roof of the neural tube (yellow). (Courtesy of K. W. Tosney.)

The number of somites in an adult individual is species-specific. Chicks have about 50 pairs of somites, mice have 65 pairs (many of them in the tail), zebrafish have 33 pairs, and humans generally have between 38 and 45 pairs (Muller and O'Rahilly 1986). Some snakes have as many as 500 pairs of somites!

 DEV TUTORIAL *Somite formation* **Dr. Michael Barresi steps through somitogenesis, from segment creation to somite differentiation.**

Axis elongation: A caudal progenitor zone and tissue-to-tissue forces

In comparison to most other vertebrates, the snake clearly has a longer length to its anterior-posterior axis relative to its other axes; thus, one important factor influencing somitogenesis might be the process by which this axis is elongated. Previously, we alluded to the origin of cells that make up the PSM, from which cells in the anterior paraxial mesoderm arise during gastrulation by ingression through the primitive streak in amniotes or via convergence upon the midline in fish and amphibians. As we discussed previously in this chapter, however, the posteriormost region of the tail contains a population of multipotent progenitor cells that have the potential to contribute to both the neural tube (*Sox2*-expressing) and paraxial mesoderm (*Tbx6*-expressing); hence, these cells are called **neuromesoderm progenitors**, or NMPs (Tzouanacou et al. 2009). The nascent paraxial mesoderm cells are released from this pool and become positioned in the tail end of the rods of presomitic mesoderm.

Although the mechanisms vary somewhat among species, the three most significant factors driving axis elongation from the caudal progenitor zone are *cell proliferation, cell migration,*[3] and *intertissue adhesion*. To exemplify the contributions of these three mechanisms, we will examine their roles in the zebrafish tailbud. The zebrafish tailbud can be divided into four regions that reflect different cell behaviors: the dorsal medial zone, the progenitor zone, the maturation zone, and a region occupied by the emerging PSM (**FIGURE 17.10A**). By using a nuclear-localized transgenic reporter, the directional movements of cells within the tailbud have been traced (Lawton et al. 2013). This analysis demonstrated that the bipotential **neuromesodermal** stem cells reside in the dorsal medial zone (DMZ) (see Figure 17.10A), which is positioned just dorsal to the neural tube and the axial and paraxial mesoderm within the tailbud (Martin and Kimelman 2012). These NMP cells first rapidly move posteriorly by **collective migration**[4] into the progenitor zone (tip of the tailbud; see Figure 17.10A). In the progenitor zone, cell velocities slow due to a reduction in "coherence" and concomitant cell mixing. The authors of this study appropriately compare this effect to the flow of traffic. Automobiles all moving in the same direction can achieve high speeds, but when vehicles change direction, swerve in and out of lanes—or even turn around entirely—it triggers a dramatic reduction in speed in the group of vehicles. In the context of the movement of NMP cells, it is thought that this change in community behavior might enable these cells to both change directions and begin to synchronize their developmental trajectories. These progenitor cells turn anteriorly to migrate bilaterally into the maturation zones on either side of the posteriormost axial mesoderm, and finally into the PSM region (see Figure 17.10A).

When NMP cells migrate through the maturation zone, as the term *maturation* implies, they begin to express mesodermal markers (*Msgn1* and *Tbx6* genes). However, they also transiently express Cdc25a, which promotes *one* cycle of division in these

---

[3] In amniotes, cell migration from the caudal progenitor zone has been interpreted to be more a result of tissue deformation as opposed to individual cell migration (see Bénazéraf et al. 2010; Bénazéraf and Pourquié 2013).

[4] *Collective migration* is when self-propelled migrating cells exert directionally coordinated forces upon one another, as opposed to individually migrating cells with loose contacts or the movement of a group of cells by tissue pushing due to proliferation or intercalation. Other cells, such as neural crest and even metastasizing cancer cells, have been suggested to use collective migration.

FIGURE 17.10 A model of axis elongation: notochord vacuole inflation, extracellular matrix deposition, and an active caudal progenitor zone together drive axis elongation in zebrafish. (A) Bipotential neuromesodermal progenitor (NMP) cells from the dorsal medial zone (DMZ) collectively migrate to the progenitor zone, where they diverge either toward a neural lineage and into the neural tube or bilaterally migrate toward the maturation zones and on into the PSM. Progenitor cells transiently express Cdc25a in the maturation zone to trigger a round of division. Chordamesoderm cells inflate vacuoles that exert pressure on surrounding tissues that results in a posterior extension of the notochord (NC). Fibronectin-integrin interactions tether the PSM to the notochord, resulting in a posterior pulling of the PSM during notochord elongation. These three processes—cell migration, cell division, and PSM-NC-coupled tissue shifting—together drive axis elongation. (B) High magnification of notochord cells in the zebrafish embryo shows vacuole (V) filling over time, which contributes to notochord extension (red arrows point out nuclei). (C) Double knockdown of *integrin* α5 and *integrin* αV by morpholino prevents adhesion of notochord cells with the extracellular matrix; as a result, the notochord buckles as it attempts to elongate (red line and arrowheads). (B from Ellis et al. 2013; C from Dray et al. 2013.)

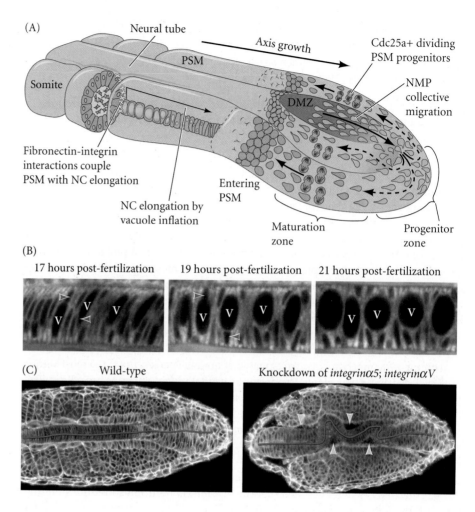

In addition to cell migration and proliferation, intertissue adhesive forces also contribute to the extension of the anterior-posterior axis. The principal tissues at play are the paraxial mesoderm and adjacent notochord. As maturing NMP cells move into the PSM, a fibronectin matrix becomes progressively deposited on the surface of the PSM and the interfaces between the paraxial mesoderm (somites and PSM) and notochord (see Figure 17.10A). During this period of axis elongation, chordamesodermal cells are undergoing significant cellular changes that result in a stiffening and directed extension of the notochord (Ellis et al. 2013a). Specifically, chordamesodermal cells use endosomal trafficking to inflate large vacuoles (yes, that is very cool), which leads to an increase in cell size that exerts pressure on the surrounding tissues (**FIGURE 17.10B**). Chordamesoderm cells also secrete a sheath of extracellular matrix components (collagen, laminin) that surrounds the notochord and resists expansion from the internal pressure of the cells. As a result of this architecture and inflating chordamesoderm, the notochord elongates in the direction of least resistance—toward the tail (see Figure 17.10A; Ellis et al. 2013b). At least in zebrafish, the paraxial mesoderm essentially "hitches a ride" on the notochord by using the fibronectin matrix and integrin receptors to mechanically couple

cells before they move into the PSM and differentiate (see Watch Development 17.1; Bouldin et al. 2014). At least in the zebrafish tailbud, although cells are proliferating, cell *migration* appears to be the more significant contributor to axis elongation (reviewed in McMillen and Holley 2015).

WATCH DEVELOPMENT 17.1 Observe dividing NMPs in the maturation zone of the zebrafish tailbud during somitogenesis.

the posterior extension of the PSM with the elongation of the notochord (**FIGURE 17.10C**; Dray et al. 2013; McMillen and Holley 2015).

Once the PSM is growing, how does it get chopped up into repeated segments? A critical insight into the process was revealed when *Xenopus* and mouse embryos were experimentally reduced in size: the resulting number of somites generated was normal, but each somite was smaller than normal (Tam 1981). This result suggested that a regulatory mechanism controls the number of somites independent of the size of the segmenting tissue. We can therefore refine our questions to ask, What mediates the epithelialization of PSM cells to physically create a boundary and consequently a somite? What mechanism(s) defines the position and timing of this boundary formation?

HOW A SOMITE FORMS: A MESENCHYMAL-TO-EPITHELIAL TRANSITION Somite architecture is built of epithelial blocks, although the PSM only supplies mesenchymal cells. Therefore, the embryo must transform the mesenchymal cells into epithelial cells in a **mesenchymal-to-epithelial transition (MET)**. This process involves the upregulation of the transcription factor *Mesp* (*Mesodermal posterior*), which regulates the onset of the MET. As a somite is forming, *Mesp* expression quickly becomes restricted to the rostral half of the somite (**FIGURE 17.11A**). A principal function of Mesp is to upregulate *Eph* in the anterior portion of somitomeres (**FIGURE 17.11B,C**). Eph activity at the presumptive anterior border of a somitomere (S-I) triggers the upregulation of its own ligands, the ephrins, in the opposing posterior half of the more anterior somitomere (S0; see Figure 17.11B,C), which is a pattern that is sequentially repeated over the course of somitogenesis (**FIGURE 17.12**; Watanabe and Takahashi 2010; Fagotto et al. 2014; Cayuso et al. 2015; Liang et al. 2015).

We saw in Chapter 15 that the Eph tyrosine kinase receptors and their ephrin ligands are able to elicit cell-to-cell repulsion between the posterior region of a somite and migrating neural crest cells. Similarly, separation of the somite from the anterior end of the presomitic mesoderm occurs at the border between ephrin- and Eph-expressing cells (see Figure 17.11C; Durbin et al. 1998). Interfering with this signaling (by injecting embryos with mRNA encoding dominant negative Ephs) leads to the formation of abnormal somite boundaries. Moreover, in *fused somites* (*tbx6*) zebrafish mutants, *eph A4* is lost, and *ephrin B2* is ubiquitously expressed throughout the paraxial mesoderm, and consequently no somite boundaries are formed (Barrios et al. 2003). The Eph A4-Ephrin B2 signaling leads to epithelialization immediately after somite fission by regulating two downstream factors, Rho GTPases and integrin/fibronectin interactions.

A mesenchymal-to-epithelial transition requires significant cytoskeletal rearrangements, which are typically governed by modulating the

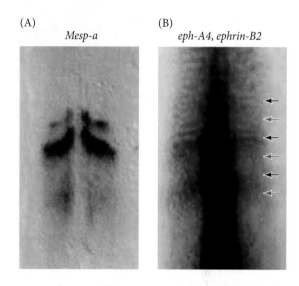

(A) *Mesp-a*

(B) *eph-A4, ephrin-B2*

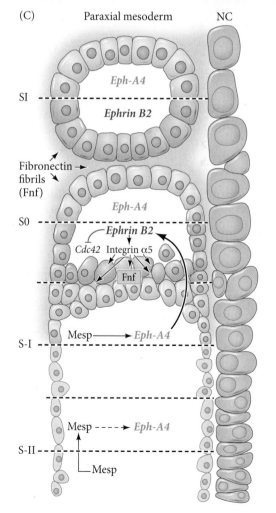

(C)   Paraxial mesoderm   NC

FIGURE 17.11 Eph-Ephrin signaling regulates epithelialization during somite boundary formation. Expression of (A) Mesodermal posterior-a (*Mesp-a*; dark purple) and (B) *Eph-A4* (black arrows) and *ephrin-B2* (red arrows) in the paraxial mesoderm of zebrafish embryos (dorsal views). (C) Model of Mesp-a and Eph-Ephrin signaling fostering mesenchymal-to-epithelial transitions that define the apposing cells of a somite boundary. *Mesp-a* becomes restricted to the anterior half of the S-I somitomere, which upregulates *Eph-A4* within this domain. In turn, *Eph-A4* upregulates its binding partner ephrin-B2 in the cells of the presumptive posterior S-0 somitomere, which triggers epithelialization and formation of a boundary. Fissure formation is facilitated by repression of *Cdc42* and activation of integrin α5-fibronectin interactions downstream of *ephrin-B2*. (A,B from Durbin et al. 2000.)

FIGURE 17.12 Ephrin and its receptor constitute a possible fissure site for somite formation. In situ hybridization shows that Eph A4 (dark blue; arrows) is expressed as new somites form in the chick embryo. (Courtesy of J. Kastner.)

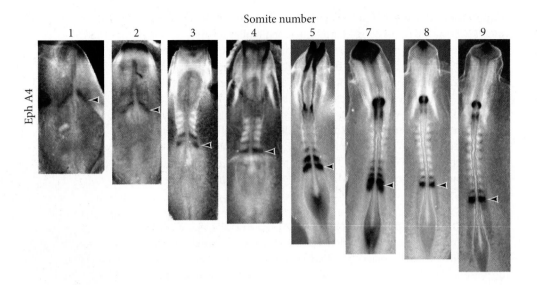

Somite number

Eph A4

Rho family of small GTPases, such as Cdc42. Activated Ephrin B2 signaling leads to the suppression of Cdc42 activity, which establishes a lower level of Cdc42 function in the anterior half of somitomere zero (S0) compared to the posterior half of S0. A forced reduction in Cdc42 activity leads to hyperepithelialized somites, whereas induced Cdc42 activity inhibits epithelialization (Nakaya et al. 2004;Watanabe et al. 2009). Therefore, the progressive formation of boundaries by MET in the PSM is fulfilled by reducing Cdc42 levels only in the peripheral cells of the somitomere, which creates a box around the remaining mesenchymal cells which exhibit a higher level of Cdc42 until the next subsequent fissure event. Although reductions in Cdc42 are required for epithelialization, differential regulation of other RhoGTPases, such as Rac1, are also involved in epithelialization of somites (Burgess et al. 1995; Barnes et al. 1997; Nakaya et al. 2004).

In contrast to the repressive outcome of Eph A4-ephrin B2 signaling on Cdc42, this signaling enhances the activity of integrin α5, which then serves to promote fibronectin assemblies in the extracellular matrix surrounding the immature somite (see Figure 17.11C; Lash and Yamada 1986; Hatta et al. 1987; Saga et al. 1997; Durbin et al. 1998; Linask et al. 1998; Barrios et al. 2003; Koshida et al. 2005; Jülich et al. 2009; Watanabe et al. 2009). This assembly of fibronectin reinforces epithelial cell separation and completes somite boundary formation (Jülich et al. 2015; reviewed by McMillen and Holley 2015).

The clock-wavefront model

Somites appear on both sides of the embryo at exactly the same time. Even when isolated from the rest of the body, presomitic mesoderm will segment at the appropriate time and in the correct direction (Palmeirim et al. 1997). The current predominant model to explain synchronized somite formation is the clock-wavefront model proposed by Cooke and Zeeman (1976; reviewed by Hubaud and Pourquié 2014). In this model, two converging systems interact to regulate (1) *where* a boundary will be capable of forming (the wavefront) and (2) *when* epithelial boundary formation should occur (the clock).

The wavefront, more aptly called the **determination front**, is established by a caudal[HIGH]-to-rostral[LOW] gradient of FGF activity within the PSM, which is inverse to the rostral-[HIGH]-to-caudal[LOW] gradient of retinoic acid (**FIGURE 17.13**). FGF signaling maintains PSM cells within an immature state; therefore, cells that become positioned at lower threshold concentrations of FGF activity will become competent to form a boundary. However, only those cells that are both competent and receiving the temporal instructions to form a boundary will actually undergo epithelialization (*activation of the Mesp-to-Eph cascade*).

The instructions for when to create a segmentation furrow are largely controlled by the clock-like oscillating signals of the Notch pathway. Each oscillation of Notch-Delta organizes groups of presomitic cells that will then segment together at the appropriate threshold of Fgf signaling (see Maroto et al. 2012). In the chick embryo, a new somite is formed about every 90 minutes. In mouse embryos, this time frame is more variable but is closer to every 2 hours, whereas in zebrafish, a somite is formed approximately every 30 minutes (Tam 1981; Kimmel et al. 1995).

WHERE A SOMITE BOUNDARY FORMS: THE DETERMINATION FRONT Recall the maturation of a newly born NMP cell from the tailbud; as it enters the PSM, the axis continues to grow, and this cell will eventually find itself part of a somite. Interestingly, the periodicity at which a group of cells contributes to a newly formed somite generally occurs at the same distance from the tailbud, although some exceptions exist. This finding suggests that the posterior extension of the tailbud strongly influences the location of boundary formation, and this is indeed the case. We discussed above the role of the opposing anterior-to-posterior gradients of RA and Fgf8/Wnt3a signaling in mediating NMP cell specification (see Figure 17.6). This robust morphogenic mechanism is similarly used to influence the maturation of PSM cells to become competent to form boundaries and as such is called the "wavefront," "wave," or "determination front" (see Figure 17.13B). We will use determination front throughout the rest of this chapter to reduce potential confusion with discussions of oscillating waves of gene expression covered later in this section.

Elegant somite inversions were performed in the chick embryo to identify the location of the determination front in the PSM (**FIGURE 17.14A**; Dubrulle et al. 2001). As you are now aware, a rostral-caudal prepattern first develops in the somitomeres of the PSM. Inverting somitomere 0 resulted in an unchanged, committed gene expression pattern as illustrated by the retention of a reversed caudal-to-rostral pattern in this inverted tissue (this somitomere region was "determined"). However, similar inversions at somitomeres –III and –VI caused variable patterning changes ("labial") and a complete reassignment of polarity ("undetermined"), respectively (**FIGURE 17.14B–D**). From this work, the determination front was identified to reside at S-IV.

This determination front has been defined as the trailing edge of an Fgf8 gradient originating from the tailbud and node of the primitive streak. Most interesting is how the Fgf8 gradient is shaped in the PSM. *Fgf8* is transcribed only in the tailbud and *not* in the PSM (**FIGURE 17.14E,F**); therefore, as the tailbud grows caudally, so does the source of cells actively transcribing Fgf8. It is accepted that one mechanism playing a major role in shaping the Fgf8 gradient is RNA decay (Dubrulle and Pourquié 2004). The amount of *Fgf8* transcript in a PSM cell will steadily decline over time due to standard mechanisms of RNA decay, which will by default build a caudal to rostral gradient of Fgf8 activity (**FIGURE 17.14G**). In this way, sloping gradients of Fgf8/Fgf4 and (likely also) Wnt3a provide different concentration thresholds for these morphogens across the PSM. In addition, the lack of *Fgf* transcription in the PSM is further maintained by the increasing concentration of the repressive retinoic acid from the somites and anterior PSM. What is the cellular outcome of these opposing morphogens?

Additional experiments manipulating the axial reach of the Fgf8 gradient by either implanting beads coated with Fgf8 into the PSM (for a gain of function) or using drugs that inhibit Fgf receptors such as SU5402 (for a loss of function) created smaller or larger somites, respectively (Dubrulle et al. 2001). These findings can be interpreted to suggest that the Fgf8 morphogen *is* the molecular determination front for epithelialization, and at somitomere –IV the threshold drops low enough to permit those cells at that axial location to be competent to form a boundary. More specifically, the cells at the Fgf8 determination front become competent to respond to the "molecular clock" whose alarm goes off when a boundary should be created.

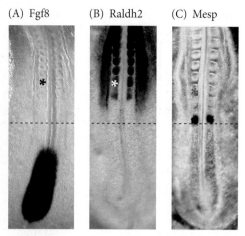

(A) Fgf8 (B) Raldh2 (C) Mesp

FIGURE 17.13 Somites form at the junction of retinoic acid (anterior) and FGF (posterior) domains. (A) Fgf8 expression (purple) in the posterior part of the embryo. (B) RNA for Raldh2 (retinoic acid-synthesizing enzyme) in the central part of the embryo. (C) Mesp stain shows the region where the somite formation will occur later. Asterisks show the last formed somite. The dashed line approximates the boundary region where somites are being determined. (From Pourquié 2011.)

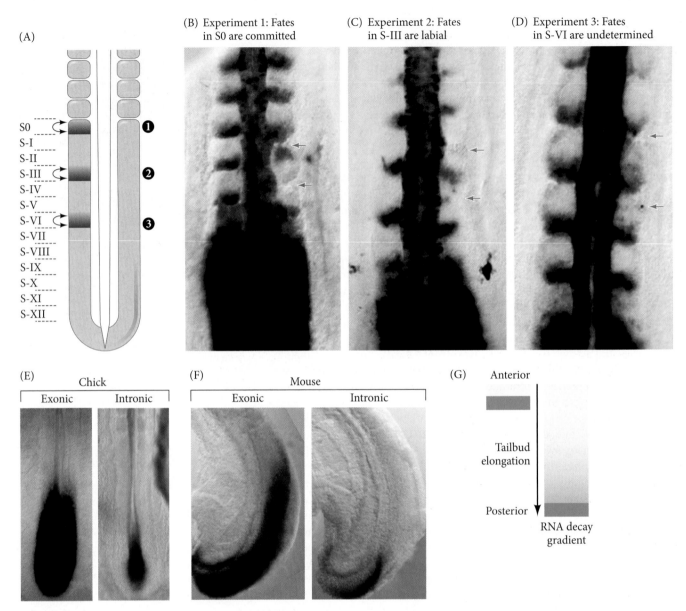

FIGURE 17.14 A caudal Fgf8 gradient establishes the "determination front." (A) Schematic of a series of somitomere tissue reversals in which presomitic tissue at different axial locations was flipped along the anterior-posterior axis. (B–D) *c-delta1* gene expression in chick embryos that had somitomere reversal at the axial level denoted in the schematic. For each example, the control side is on the left and the experimental side is on the right. Flipping somitomere S0 showed complete commitment to maintain the positional expression of *c-delta1*; thus, it was already determined. In contrast, flipping somitomeres at positions S-VI and S-III showed, respectively, normal posterior expression of *c-delta1* and disorganized expression. These data suggested that cells become determined to form boundary locations at S-IV. Red arrows indicated points of surgical inversion. (E,F) *Fgf8* expression in the chick and mouse. Exonic and intronic in situ probes to reveal, respectively, any cell with *Fgf8* mRNA or nuclear RNA (pre-RNA). Careful examination of these results demonstrates two important properties of the *Fgf8* gradient. The first is that the *Fgf8* gene is actively transcribed only in the tailbud (intronic probe). (G) The second property is that a gradient of Fgf8 in the presomitic mesoderm is established by a mechanism of RNA decays. Green bar represents the caudal movement of cells actively transcribing Fgf8 and the trailing gradient is made by RNA decay over time in cells no longer transcribing Fgf8. (A–D from Dubrelle et al. 2001; E–G from Dubrelle and Pourquié 2004.)

WHEN A SOMITE BOUNDARY FORMS: THE CLOCK Scientists use the analogy of a clock to describe the controls behind the periodicity of boundary (and somite) formation. What would a molecular clock look like in an embryo? Does it have hands that physically tell a cell what time it is? What duration constitutes a period for this clock? In the context of a cell, a clock can simply be the regular fluctuation of a protein's activity on and off, provided this change in activity is in fact repeated and rhythmic in nature. In the context of a tissue, however, this fluctuating protein timepiece needs to somehow be communicated across a field of cells. Thus, one model for a molecular clock of somitogenesis would posit that the activity of a protein that regulates METs might become functional in one cell of the PSM and as part of its function relay this event to its neighboring cells through cell-to-cell interactions until its activity is cyclically inhibited. In this way, each cell across a tissue could experience "on" and "off" states of protein activity, or *the ticking of a clock*.

One of the key "clock" components that maintains the pace of somitogenesis is the Notch signaling pathway (see Wahi et al. 2014). When a small group of cells from a region constituting the posterior border at the presumptive somite boundary is transplanted into a region of presomitic mesoderm that would not ordinarily be part of the boundary area, a new boundary is created. The transplanted boundary cells instruct the cells anterior to them to epithelialize and separate. Nonboundary cells will not induce border formation when transplanted to a nonborder area. However, these cells can acquire boundary-forming ability if an activated Notch protein is electroporated into them, which demonstrates that Notch signaling can induce the METs that underlie somite formation (see Web Topic 17.1; Sato et al. 2002).

WEB TOPIC 17.1   **NOTCH SIGNALING AND SOMITE FORMATION**   The Notch pathway is one of the key agents in signaling where somites are to form.

Promoting METs is the behavioral output of this particular molecular clock, but for Notch to represent this timekeeper, it would have to also satisfy the on-off rhythmicity and transfer across cells. The endogenous level of Notch activity in the mouse PSM has been visualized and shown to oscillate in a segmentally defined pattern that correlates with boundary formation (Morimoto et al. 2005; Aulehla et al. 2008). Like a wave of gene expression across the PSM, cells experience Notch upregulation and then downregulation from the caudal to rostral extents of the PSM. This wave of Notch expression crashes on somite 0, where a somite boundary forms at the interface between the Notch-expressing and Notch-nonexpressing areas.

In addition, Notch signaling provides a mechanism to transfer this signal from cell-to-cell across the PSM. As discussed in Chapter 4, full-length Notch forms a transmembrane protein that binds to its receptor, Delta, in adjacent cells. Delta is also a transmembrane protein, and initial upregulation and presentation of Notch will trigger a concomitant upregulation of Delta in adjacent cells, which in turn will reinforce Notch in the other cells surrounding it. This is a pattern generating mechanism by Notch-Delta signaling known as **lateral inhibition**. This receptor-to-receptor pairing provides a mechanism for signal transfer throughout the PSM; however, this mechanism predicts that the expression of Notch and Delta should display a mosaic pattern over the entire PSM, yet it doesn't. Like Notch, Delta also displays oscillatory expression patterns with a posterior to anterior progression across the PSM—a key feature of the clock mechanism. How is this possible?

Although there are interspecies differences in exactly which gene products oscillate, in all vertebrate species the on-off ticking of the clock involves a negative feedback loop of the Notch signaling pathway (Krol et al. 2011; Eckalbar et al. 2012). Thus, in all vertebrates, at least one of the Notch target genes having dynamic oscillating expression in the presomitic mesoderm is also able to *inhibit* the gene for Notch, which establishes a negative feedback loop. These inhibitory proteins are unstable, and when the inhibitor is degraded, Notch becomes active again. Such feedback creates a cycle (the "clock") in which the gene for Notch is turned on and off by the absence or presence of a protein that Notch itself induces. These on-off oscillations could provide the molecular basis for the periodicity of somite segmentation (Holley and Nüsslein-Volhard 2000; Jiang et al. 2000; Dale et al. 2003). Such oscillating Notch targets include *Hairy1*, *Hairy/Enhancer of split-related proteins* (*Her*), and *lunatic fringe*, which are all activated by Notch, expressed in a similar oscillating pattern throughout the PSM from the tailbud to the last formed

FIGURE 17.15 Somite formation correlates with wavelike expression of the *Hairy1* gene in the chick. (A) In the posterior portion of a chick embryo, somite SI has just budded off the presomitic mesoderm. Expression of the *Hairy1* gene (purple) is seen in the caudal half of this somite as well as in the posterior portion of the presomitic mesoderm and in a thin band that will form the caudal half of the next somite (S0). (B) A caudal fissure (small arrow) begins to separate the new somite from the presomitic mesoderm. The posterior-most region of *Hairy1* expression shifts anteriorly. (C) The newly formed somite is now referred to as SI, it retains the expression of *Hairy1* in its caudal half. Again, the posterior-most region of *Hairy1* expression in the PSM continues to shift anteriorly as well as shortens. The former SI somite, now called SII, undergoes differentiation. (D) Creation of this newly formed somite SI is complete, and a new cycle of *Hairy1* expression begins again. In the chick, formation of each somite and the wave of *Hairy1* expression through the PSM takes about 90 minutes.

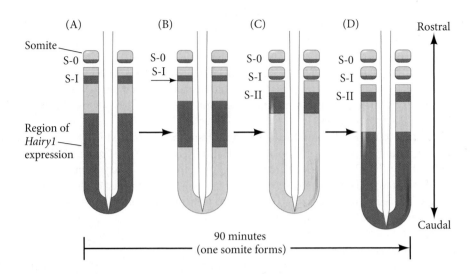

somite, and which all negatively feed back to repress Notch signaling (Chipman and Akam 2008; Pueyo et al. 2008). For instance, the *Hairy1* gene was the first Notch target found to show a rhythmic pattern of expression (**FIGURE 17.15**). The *Hairy1* gene is expressed first in a broad domain at the caudal end of the presomitic mesoderm. This expression domain moves anteriorly while narrowing until it reaches the rostral end of the PSM, at which time a new wave of expression begins in the caudal end. Aulehla and colleagues observed the waves of transcriptional activity of *lunatic fringe* in live mice embryos, which mimics the pattern seen of *Hairy1*. The time it takes for a wave of expression to cross the presomitic mesoderm is 90 minutes in the chick, which is—not coincidentally—the time it takes to form a pair of somites in this species. This dynamic expression is not due to cell movement but to cells turning the gene on and off in different regions of the tissue through loops of negative feedback (Johnston et al. 1997; Palmeirim et al. 1997; Jouve et al. 2000, 2002; Dale et al. 2003). Of relevance, loss of function of Notch or its downstream cycling target genes in mice and humans leads to major segmentation defects of the vertebrae, such as the spinal deformities in scoliosis and spondylocostal dysostosis (**FIGURE 17.16**; Zhang et al. 2002; Sparrow et al. 2006).

> **WATCH DEVELOPMENT 17.2** Conceptualizing the periodicity of genes being turn on and off in a given tissue can be abstract until you actually see the oscillations occurring. Using a fluorescent reporter mouse, watch the gene expression of *lunatic fringe* cycle through the PSM.

TERMINATING THE NOTCH CLOCK WITH EPITHELIALIZATION As we discussed above, Mesp is a global regulator of the Eph-Ephrin cascade that triggers MET and boundary formation. Mesp is activated by Notch and, as a transcription factor, Mesp contributes to the suppression of Notch (Morimoto et al. 2005). This cycle of alternating activation and suppression causes Mesp expression to oscillate in time and space as well. Mesp is initially expressed in a somite-wide domain; it is then repressed in the posterior half of this domain but maintained in the anterior half, where it in turn represses Notch activity. Wherever Mesp expression is maintained, that group of cells becomes the most anterior in the next somite; Eph A4 is then induced, and the boundary forms immediately anterior to those cells (see Figure 17.11C; Saga et al. 1997). In the posterior half of the prospective somite, where Mesp is not expressed, Notch activity induces the expression of the transcription factor Uncx4.1, which contributes to the specification of the somite's posterior identity (Takahashi et al. 2000; Saga 2007). In this way, the somite boundary is determined, and the somite is given anterior/posterior polarity at the same time.

(A) Mouse

Wild-type

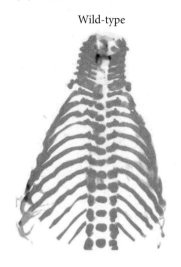

Lfng⁻/⁻

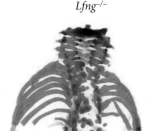

Dll3⁻/⁻

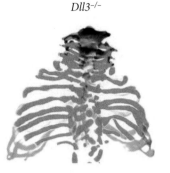

(B) Human

Lfng 564 C-to-A missense
mutation (inactive enzyme)

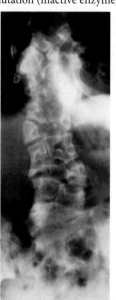

FIGURE 17.16 Notch-Delta signaling is essential for proper somitogenesis in the mouse and in humans. In the mouse, loss of either the Notch target *Lunatic fringe* (*Lfng*) or its binding partner *distaless3* (*Dll3*) result in severe vertebral malformations (A), which are particularly similar to the phenotype caused by known mutations in the human *Lunatic fringe* (B). (A from Fisher et al. 2012; B from Sparrow et al. 2006.)

BEING AT THE RIGHT PLACE AT THE RIGHT TIME: CONNECTING THE CLOCK AND THE DETERMINATION FRONT More posteriorly positioned PSM cells experiencing waves of Notch signaling do not prematurely epithelialize because these cells are not competent to sufficiently respond to Notch signaling due to the influence of FGFs. As long as the presomitic mesenchyme is in a region of relatively high Fgf8 concentration, the clock will not function. At least in zebrafish, this lack of function appears to be due to the repression of Delta, the major ligand of Notch. The binding of Fgf8 to its receptor enables the expression of the Her13.2 protein, which is necessary to inhibit transcription of Delta (see Dequéant and Pourquié 2008). FGF signals are needed to get cells to migrate anteriorly out of the tailbud, but as long as FGFs activate the ERK transcription factors, the cells remain unresponsive to Notch ligands. It has recently been proposed that Fgf8 synthesis also cycles, but at a different frequency than the Notch ligands (Niwa et al. 2011; Pourquié 2011). Thus, through a combination of a sloping concentration gradient of Fgf8 and its own unique pattern of cycling (probably by synthesizing its own inhibitors and by inhibition by retinoic acid), FGF signaling is downregulated in certain areas of the paraxial mesoderm, and cells in these regions become progressively more competent to respond to Notch signals (**FIGURE 17.17**). FGFs therefore establish the *placement* (i.e., determination front) of cells that are competent to respond to oscillating Notch signals (the clock), which can induce Mesp2 to initiate METs and Eph-mediated segment formation.[5]

HOW MANY SOMITES TO FORM? THE RATE OF OSCILLATIONS TO THE RATE OF AXIS ELONGATION A consequence of epithelialization depending on both receiving a Notch-mediated wave of gene expression (as an "okay to segment") and becoming competent based on an Fgf8 concentration threshold means that the size and number of somites is based on two factors: the rate of segmenting oscillations and the rate of axis elongation. In fact, it is the ratio of these two rates that sets the parameters for somite number and size. For instance, let us represent the rate of the clock as τ and the rate of axis elongation as α. If α is both sustained and in balance with τ, an infinite number of identically sized somites are theoretically possible. Alternatively, if τ is faster than

[5] Pourquié (2011) has noted that this situation seems a lot like the one that exists in the limb bud (see Chapter 19), wherein a pool of newly formed cells that has been maintained by FGFs in a relatively undifferentiated and migratory state becomes differentiated into periodic elements (limb cartilage) by interacting gradients of FGFs and retinoic acid.

Developing Questions

Is the clock analogy sufficient? It certainly helps us understand the critical mechanisms influencing somitogenesis. Like all clock analogies, however, it makes the assumption that the rate of time stays constant. Work in the lab of Andrew Oates has demonstrated that, due to the gradual change in the rate of axis elongation over time in zebrafish, the waves of oscillating segmentation genes (the period from onset to arrest) are also shortening, a phenomenon the researchers compare to the Doppler effect (Soroldoni et al. 2014). How are the different facets of somitogenesis (axis elongation, the determination front, epithelialization, and the clock) intricately coordinated to respond to change? If time is not steady, what new model can you come up with to integrate all these mechanisms?

FIGURE 17.17 Possible model for "clock-wave-front" somite specification. In each panel, the FGF-induced transcription factor (pERK) and the somite-specifying transcription factor (Mesp) are shown on the left, and the Notch-generated transcription factor (NICD) and the inhibitory protein Hes7 induced by NICD are shown on the right. (After Niwa et al. 2011.)

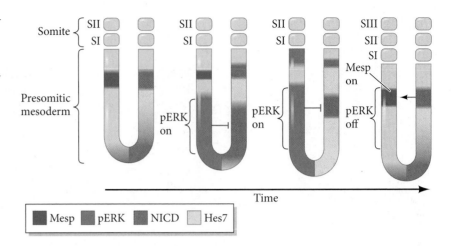

α somite formation will eventually catch up to the tailbud and terminate somitogenesis. The current model for the termination of somitogenesis is that axis elongation (α) slows, and as new somites become progressively closer to the tailbud, they bring with them the inhibitory actions of retinoic acid that arrest tailbud development (Gomez and Pourquié 2009).

Somite size could also be altered in predictable ways by manipulating these rates. Comparative analysis of somitogenesis across species has suggested that modulation of the rate of the molecular clock has been a major mechanism for adaptation of somite number. As an example, snakes can have several hundred somites, compared to the approximately 60 found in the mouse and chick and approximately 30 found in zebrafish. Olivier Pourquié and colleagues compared the rates of PSM elongation and the segmentation clock exhibited by corn snake, mouse, and chick (among other species). The snake's somites are about three times smaller than the somites of the mouse or chick (Gomez et al. 2008). Although the PSM of snakes is modestly larger than that of the chick or mouse, the most significant contribution to the greater number of smaller somites in snakes is a highly accelerated clock. Genes such as *lunatic fringe* were found to show upwards of 9 bands of oscillations in snakes, compared to 1–3 bands found in the chick and mouse (**FIGURE 17.18**; Gomez et al. 2008). Therefore, a greater number of Notch-mediated oscillations over the course of axis elongation will divide the PSM more times, creating more somites and more vertebrae.

Linking the clock-wavefront to Hox-mediated axial identity and the end of somitogenesis

Somitogenesis cannot continue indefinitely; rather, it must end and reach this end with the appropriate rostral-to-caudal identities laid down. As mentioned above, Hox genes play master regulatory roles in the specification of axial identity from head to tail with spatial and temporal collinearity. How are the clock and the determination front connected to the role of Hox genes?

If Fgf8 protein levels are manipulated to create extra (albeit smaller) somites, the appropriate Hox gene expression will be activated in the appropriately numbered somite, even if it is in a different position along the anterior-posterior axis, which suggests that the determination front (FGF gradient) primarily influences somite size as opposed to Hox gene expression. Mutations that affect the autonomous segmentation clock, however, do affect activation of the appropriate Hox genes (Dubrulle et al. 2001; Zákány et al. 2001). The regulation of Hox genes by the segmentation clock presumably allows coordination between the formation and the specification of the new segments. How the mechanisms of somitogenesis feed into Hox gene regulation is not completely known. In *Xenopus*, research has shown that *XDelta2*, an oscillating gene and receptor

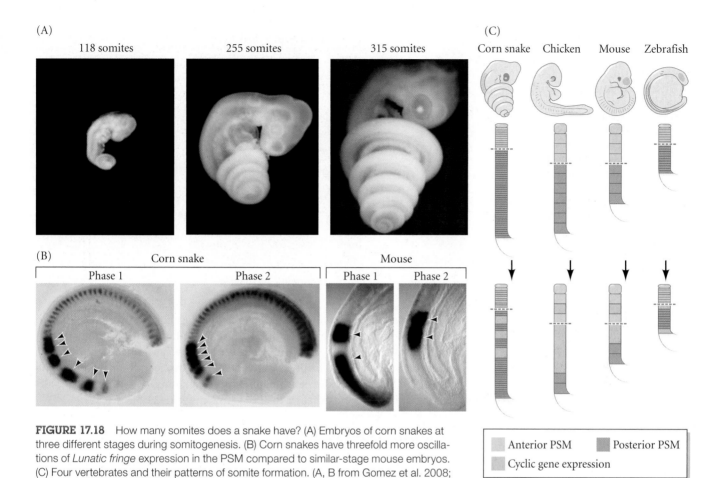

FIGURE 17.18 How many somites does a snake have? (A) Embryos of corn snakes at three different stages during somitogenesis. (B) Corn snakes have threefold more oscillations of *Lunatic fringe* expression in the PSM compared to similar-stage mouse embryos. (C) Four vertebrates and their patterns of somite formation. (A, B from Gomez et al. 2008; C after Gomez and Pourquié 2009.)

to Notch, can upregulate at least three different Hox paralogue groups and initiate a positive feedback loop with these Hox proteins (Peres et al. 2006). Based on this association, it is tempting to speculate that the segmentation clock might directly trigger the timed activation of Hox genes; it is unknown, however, whether this influence occurs in a collinear fashion or involves chromatin modifications as described above for Hox gene expression.

Once Hox gene expression along the trunk is initiated, it will feed back upon the determination front to terminate axis elongation and end somitogenesis. Specifically, the collinear activation of the expression of Hox genes closer to the 5′ end of the clusters results in a progressively greater repression of Wnt signaling in the tailbud (**FIGURE 17.19**; Denans et al. 2015). Recall that Wnt3a is secreted from the tailbud and displayed in a caudal-to-rostral gradient similar to Fgf8. Wnt3a functions to promote the migration of neuromesoderm progenitors into the PSM, and in doing so fuels PSM growth and axis elongation (Dunty et al. 2008). Thus, as new cells move into the PSM over time and start to express increasingly more 5′ Hox genes with temporal collinearity, Wnt signaling is progressively inhibited, and tailbud growth slows. In amniotes, the segmentation clock does not significantly alter its rate during this time; therefore, the rate of somite formation will outpace tailbud growth and exhaust the PSM, leading to the end of somitogenesis (Denans et al. 2015). Moreover, Hox gene-mediated slowing of tailbud growth is reinforced by the indirect repression of FGF signaling in two ways. First, the encroaching reach of retinoic acid from the somites will inhibit FGF expression with greater effectiveness; second, Wnt and FGF signaling will function in a positive feedback loop, maintaining each other's expression in the tailbud (Aulehla et al. 2003; Young et al. 2009; Naiche et al. 2011). Due to the progressive repression

FIGURE 17.19 Model of the regulatory mechanisms governing somitogenesis. The molecular segmentation clock through Notch-Delta dictates the order of Hox gene expression, which functions in part to repress Wnt signaling and, indirectly, Fgf8 expression. Thus, the levels of the posterior-originating Wnt/Fgf8 determination front are influenced by Hox genes as well as by the development of anterior structures through retinoic acid signaling. Retinoic acid inhibits Fgf8 and Wnt3a, whereas Fgf8 is capable of repressing retinoic acid through Cyp26A1. Via this balance of signaling, anterior somites form before posterior somites.

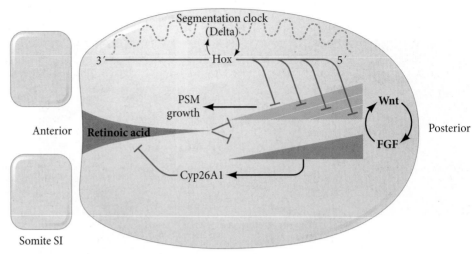

Presomitic mesoderm

of Wnt signaling, FGF will indirectly also become downregulated over time. That, in turn, will lead to the downregulation of Cyp26A1, providing further redundancy for the hyperactivation of retinoic acid (Iulianella et al. 1999). Taken together, we have a model in which the temporal collinear activation of 5′ Hox genes puts the brakes on axis elongation through the direct inhibition of Wnt signaling, which indirectly halts the determination front and exhausts the PSM (Denans et al. 2015).

Sclerotome Development

As a somite matures, it becomes divided into two major compartments, the sclerotome and the dermomyotome. These compartments are found in all vertebrates, and similar structures are found in the cephalochordate amphioxus (lancelets), our closest invertebrate relatives, indicating that they are ancient embryonic structures (Devoto et al. 2006; Mansfield et al. 2015). How the sclerotome and dermomyotome develop is a complex story involving epithelial-to-mesenchymal transitions (EMTs) and signaling cascades. Here we turn to the development of the sclerotome, with a discussion of dermomyotome formation to follow.

Shortly after somite formation, the outer epithelial cells and inner core of mesenchymal cells begin to show signs of differentiation (**FIGURE 17.20A,E**). The first visible indicator occurs in the ventromedial portion of the somite, where an EMT occurs to form the **sclerotome** (**FIGURE 17.20B**). This EMT is important to establish a migratory population of cells capable of moving into position around the midline axial structures and building the vertebral column. Just prior to the transition, sclerotome progenitor cells express the transcription factor Pax1, which is required for their transition into a mesenchyme and subsequent differentiation into cartilage (Smith and Tuan 1996). In this transition, the epithelial cells lose N-cadherin expression and become motile mesenchyme (**FIGURE 17.20C,D,F**; Sosic et al. 1997). Sclerotome cells also express inhibitors of the muscle-forming transcription factors—myogenic regulatory factors, or MRFs—which we will discuss soon (Chen et al. 1996).

As mentioned near the start of this chapter, the mesenchymal cells that make up the sclerotome can be subdivided into several regions (see Figure 17.2E). Although most sclerotome cells become precursors of the vertebral and rib cartilage, the dorsal sclerotome forms the **syndetome**, giving rise to tendons, and the medial sclerotome cells closest to the neural tube generate the meninges (coverings) of the spinal cord as well as giving rise to blood vessels that will provide the spinal cord with nutrients and oxygen (Halata et al. 1990; Nimmagadda et al. 2007). The cells in the center of the somite (which remain mesenchymal) also contribute to the sclerotome, becoming the

(A) 2-day embryo

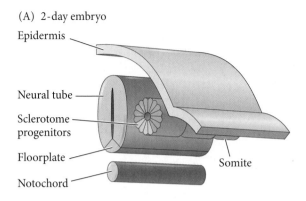

Epidermis

Neural tube

Sclerotome progenitors

Floorplate

Somite

Notochord

(B) 3-day embryo

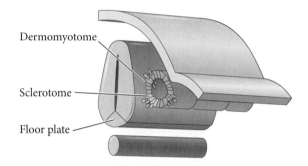

Dermomyotome

Sclerotome

Floor plate

(C) 4-day embryo

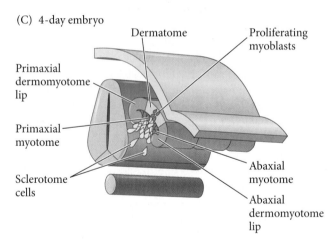

Dermatome

Proliferating myoblasts

Primaxial dermomyotome lip

Primaxial myotome

Sclerotome cells

Abaxial myotome

Abaxial dermomyotome lip

(D) Late 4-day embryo

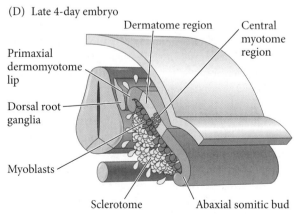

Dermatome region

Central myotome region

Primaxial dermomyotome lip

Dorsal root ganglia

Myoblasts

Sclerotome

Abaxial somitic bud

(E)

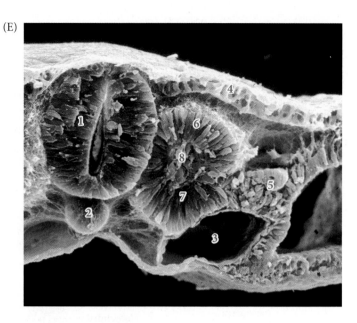

FIGURE 17.20 Transverse section through the trunk of a chick embryo on days 2–4. (A) In the 2-day somite, the sclerotome cells can be distinguished from the rest of the somite. (B) On day 3, the sclerotome cells lose their adhesion to one another and migrate toward the neural tube. (C) On day 4, the remaining cells divide. The medial cells form a primaxial myotome beneath the dermomyotome, and the lateral cells form an abaxial myotome. (D) A layer of muscle cell precursors (the myotome) forms beneath the epithelial dermomyotome. (E,F) Scanning electron micrographs correspond to (A) and (D), respectively; 1, neural tube; 2, notochord; 3, dorsal aorta; 4, surface ectoderm; 5, intermediate mesoderm; 6, dorsal half of somite; 7, ventral half of somite; 8, somito-coel/arthrotome; 9, central sclerotome; 10, ventral sclerotome; 11, lateral sclerotome; 12, dorsal sclerotome; 13, dermomyotome. (A,B after Langman 1981; C,D after Ordahl 1993; E,F from Christ et al. 2007, courtesy of H. J. Jacob and B. Christ.)

(F)

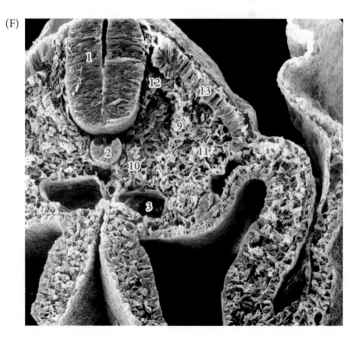

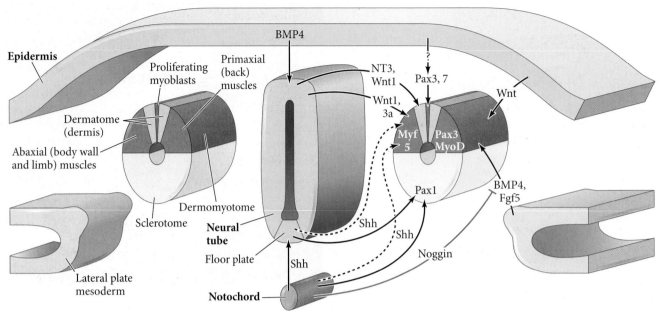

FIGURE 17.21 Model of major postulated interactions in the patterning of the somite. The sclerotome is white; dermomyotome regions are red and pink. A combination of Wnts (probably Wnt1 and Wnt3a) is induced by BMP4 in the dorsal neural tube. These Wnt proteins, in combination with low concentrations of Sonic hedgehog from the notochord and floor plate, induce the primaxial myotome, which synthesizes the myogenic transcription factor Myf5. High concentrations of Shh from the notochord and neural tube floor plate induce Pax1 expression in those cells fated to become the sclerotome. Certain concentrations of neurotrophin-3 (NT3) from the dorsal neural tube appear to specify the dermatome, while Wnt proteins from the epidermis, in conjunction with BMP4 and Fgf5 from the lateral plate mesoderm, are thought to induce the primaxial myotome. The proliferating myoblasts are characterized by Pax3 and Pax7 and are induced by Wnts in the epidermis. (After Cossu et al. 1996b.)

vertebral joints, the cartilagenous discs between the vertebrae (intervertebral discs), and the portions of the ribs closest to the vertebrae (Mittapalli et al. 2005; Christ et al. 2007; Scaal 2015). This region of the somite has been called the **arthrotome**.

Like the proverbial piece of real estate, the destiny of a particular region of the somite depends on three things: location, location, location. As shown in **FIGURE 17.21**, the locations of the somitic regions place them close to different signaling centers, such as the notochord and floor plate (sources of Sonic hedgehog and Noggin), the neural tube (source of Wnts and BMPs), and the surface epithelium (also a source of Wnts and BMPs). Sclerotome precursors reside in the ventromedial portion of the somite and therefore are in closest proximity to the notochord. These cells are induced to become the sclerotome by notochord-derived paracrine factors, especially Sonic hedgehog (Fan and Tessier-Lavigne 1994; Johnson et al. 1994). If portions of the chick notochord are transplanted next to other regions of the somite, those regions will also become sclerotome cells. The notochord and somites also secrete Noggin and Gremlin, two BMP antagonists. The absence of BMPs is critical in permitting Sonic hedgehog to induce cartilage expression, and if either of these inhibitors is deficient, the sclerotome fails to form, and the chicks lack normal vertebrae.

Vertebrae formation

Sonic hedgehog is required for specification of sclerotome fates, but what directs sclerotome cell migration toward and around the notochord and neural tube to form the vertebrae? The notochord appears to induce its surrounding mesenchyme cells to secrete epimorphin. Epimorphin then attracts sclerotome cells to the region around the notochord and neural tube, where they begin to condense and differentiate into cartilage (**FIGURE 17.22A**). In addition, the more dorsal migration of sclerotome cells over the top of the neural tube to form the spinous process of the vertebra appears to be induced by the secretion of platelet-derived growth factor (PDGF) from the sclerotome cells immediately below them. The migrating cells are able to respond to these PDGF signals by expressing TGF-β type II receptors (Wang and Serra 2012).

Before the sclerotome-derived cells form a vertebra, each sclerotome must split into a rostral (anterior) and a caudal segment (**FIGURE 17.22B**). As motor neurons from the neural tube grow laterally to innervate the newly forming muscles, the rostral segment of each sclerotome recombines with the caudal segment of the next anterior sclerotome

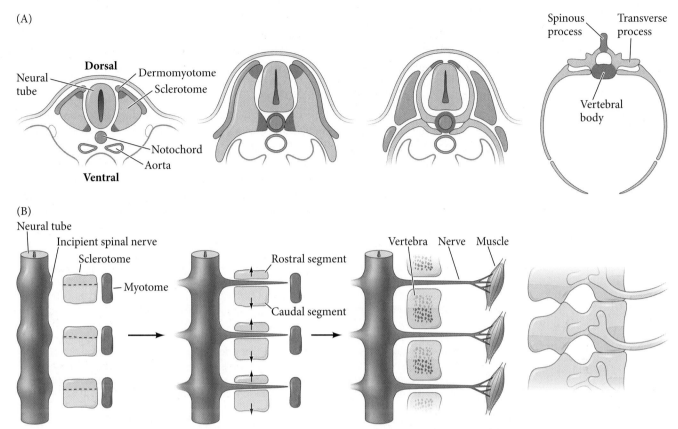

FIGURE 17.22 Resegmentation of the sclerotome to form vertebrae. (A) Illustration of the sequential development of sclerotome into the vertebrae. (B) Each sclerotome splits into a rostral and a caudal segment. As the spinal neurons grow outward to innervate the muscles from the myotome, the rostral segment of each sclerotome combines with the caudal segment of the next anterior sclerotome to form a vertebral rudiment. (A, after Christ et al. 2000; B, after Larson 1998 and Aoyama and Asamoto 2000.)

to form the vertebral rudiment in a process known as **resegmentation** (Remak 1850). This division of neighboring somitic contributions was confirmed through quail-chick chimeras in which the rostral or caudal portion of quail somites was transplanted into a chick somite at the identical location. Quail-specific antigens are easily identified on cells, enabling the differentiated structures to be traced back to the donor tissue (Aoyama and Asamoto 2000; Huang et al. 2000). Although these experiments in the chick supported distinct resegmentation of sclerotome halves, in zebrafish there may be more mixing of contributions from both halves of the sclerotome to vertebrae (Morin-Kensicki et al. 2002). This resegmentation involving the sclerotome but not the myotome enables the muscles to coordinate the movement of the skeleton and permits the body to move laterally, a movement that is reminiscent of the strategy used by insects when constructing segments out of parasegments (see Chapter 9). The bending and twisting movements of the spine are permitted by the intervertebral (synovial) joints that form from the arthrotome region of the sclerotome. Removal of sclerotome cells from the arthrotome leads to the failure of synovial joints to form and to the fusion of adjacent vertebrae (Mittapalli et al. 2005).

THE NOTOCHORD SUPPORTS VERTEBRAL MORPHOGENESIS AND BECOMES PART OF THE INTERVERTEBRAL DISC   The notochord, with its secretion of Sonic hedgehog, is critical for sclerotome development. As we discussed earlier, the notochord is also important for elongation of the axis, which has ramifications for spine morphogenesis. In zebrafish, improper vacuole filling of notochordal cells results in buckling of the notochord and subsequent vertebral fusions and associated spine defects (Ellis et al. 2013). Additional evidence that proper notochord formation is necessary for proper spine formation comes from experiments that destroy the integrity of the extracellular matrix sheath surrounding the notochord. When one of the collagens in this sheath is inhibited from forming in zebrafish embryos, the notochord bends, which leads to irregular bone deposition, vertebral fusion and curvature of the spine akin to scoliosis in humans (**FIGURE 17.23A,B**; Gray et al. 2014).

What happens to the notochord in the adult? A common misconception is that after the notochord has provided inductions and axial support, it completely degenerates. There is some truth to this idea in that some of the notochordal cells appear to die by apoptosis once the vertebrae have formed, likely signaled by mechanical forces. It is interesting, however, that those same tensile forces from invading vertebrae also segment the notochord into smaller units, which are retained and develop into the **nuclei pulposi** (Aszódi et al. 1998; Choi et al. 2008; Guehring et al. 2009; McCann et al. 2011; Risbud and Shapiro 2011; reviewed in Chan et al. 2014; Lawson and Harfe 2015). Most important is that the notochordal origin of nuclei pulposi has been demonstrated through lineage tracing experiments in the mouse (Choi et al. 2008; McCann et al. 2011). The nuclei pulposi form a gel-like mass in the center of the intervertebral discs, which

FIGURE 17.23 Development of the spinal column and intervertebral discs. (A) Collagen 8a1a is normally expressed throughout the spinal column, as revealed by this *Col2a1* transgenic zebrafish reporter (green). (B) Loss of *Col8a1a* clearly results in a failure to properly form a straight spine and the presence of fused vertebrae, as visualized with Alizarin red staining (magenta) for bone. (C) A vertebra with associated nucleus pulposus (NP) in an E15.5 mouse. Vertebra (V); annulus fibrosis (AF). (D) A model of how the notochordal sheath functions to maintain small portions of the notochord as they develop into the nuclei pulposi. Loss of proper Sonic hedgehog signaling (in *smoothened* mutants) results in varying degrees of notochord sheath reductions and consequently a failure to form nuclei pulposi. (dpf, days post fertilization.) (A,B from Gray et al. 2014; C from Lawson and Harfe 2015; D after Choi and Harfe 2011.)

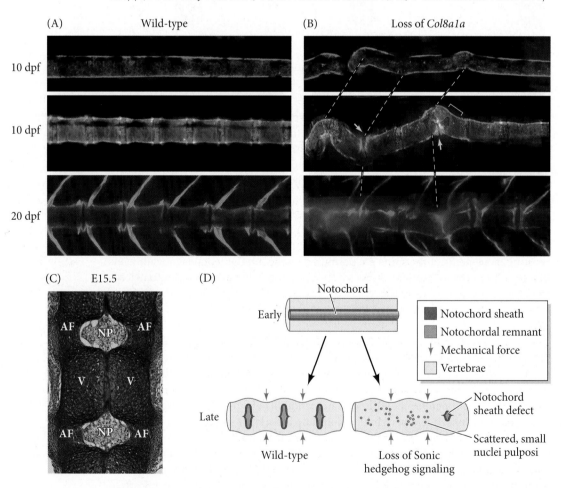

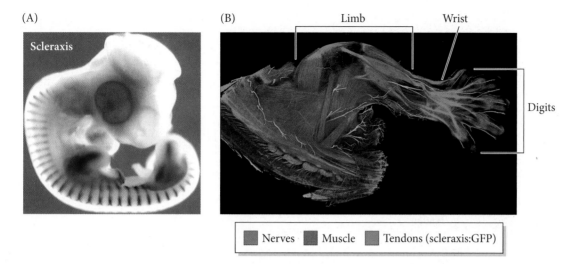

(A)

Scleraxis

(B)

Limb   Wrist

Digits

Nerves   Muscle   Tendons (scleraxis:GFP)

FIGURE 17.24 Scleraxis is expressed in the progenitors of the tendons. (A) In situ hybridiza-tion showing the pattern of scleraxis expression in the developing chick embryo. (B) Limb, wrist, and digits of a newborn mouse, showing scleraxis (fused to GFP) in the tendons (green) connecting muscles (stained red with antibodies to myosin). The neurons have been stained blue by antibodies to neurofilament proteins. (A from Schweitzer et al. 2001, courtesy of R. Schweitzer; B courtesy of A. K. Lewis and G. Kardon.)

are surrounded by the annulus fibrosus, a connective tissue derived from sclerotome (**FIGURE 17.23C**). Those are the spinal discs that "slip" in certain back injuries.

Little is known about the mechanisms regulating intervertebral disc formation, but it appears that the notochordal sheath is essential to the development of the nuclei pulpo-si (Choi and Harfe 2011; Choi 2012). In experiments that cause a weakened extracellular sheath to form, pressure from the forming vertebral bodies disperse the notochord cells, and the nuclei pulposi fail to form (**FIGURE 17.23D**; Choi and Harfe 2011).

Tendon formation: The syndetome

The most dorsal part of the sclerotome will become the fourth compartment of the somite, the syndetome. The tendon-forming cells of the syndetome can be visualized by their expression of the *scleraxis* gene (**FIGURE 17.24**; Schweitzer et al. 2001; Brent et al. 2003). Because there is no obvious morphological distinction between the sclero-tome and syndetome cells (they are both mesenchymal), our knowledge of this somitic compartment had to wait until we had molecular markers (Pax1 for the sclerotome, scleraxis for the syndetome) that could distinguish between them and allow researchers to follow the cells' fates.

Because the tendons connect muscles to bones, it is not surprising that the syndetome (Greek *syn*, "connected") is derived from the most dorsal portion of the sclerotome—that is, it is derived from sclerotome cells adjacent to the muscle-forming myotome (**FIG-URE 17.25A**). The syndetome is made from the myotome's secretion of Fgf8 onto the immediately subjacent row of sclerotome cells (Brent et al. 2003; Brent and Tabin 2004). Other transcription factors limit the expression of scleraxis to the anterior and posterior portions of the syndetome, causing two stripes of scleraxis expression (**FIGURE 17.25B**). Meanwhile, the developing cartilage cells, under the influence of Sonic hedgehog from the notochord and floor plate, synthesize the Sox5 and Sox6 transcription factors that block scleraxis transcription while activating the cartilage-promoting factor Sox9 (Yamashita et al. 2012). In this way, the cartilage protects itself from any spread of the Fgf8 signal. The tendons then associate with the muscles directly above them and with the skeleton (including the ribs) on either side of them (**FIGURE 17.25C**; Brent et al. 2005).

Developing Questions

Some species do not develop nuclei pulposi; the chick is one such species. What becomes of notochord cells in these animals? Do they all degenerate, or do some contribute to other structures of the spine? In addition, although mechanical forces seem to contribute to the development of nuclei pulposi, do any molecular cues also guide their formation? Consider the array of guidance cues expressed in midline structures (Eph-ephrins, netrins, slits); could they be instructing the coalescence of notochord cells into segmented foci? Amazingly, despite the "central" role the notochord plays in development, fundamental questions about the development of its own cell lineage remain.

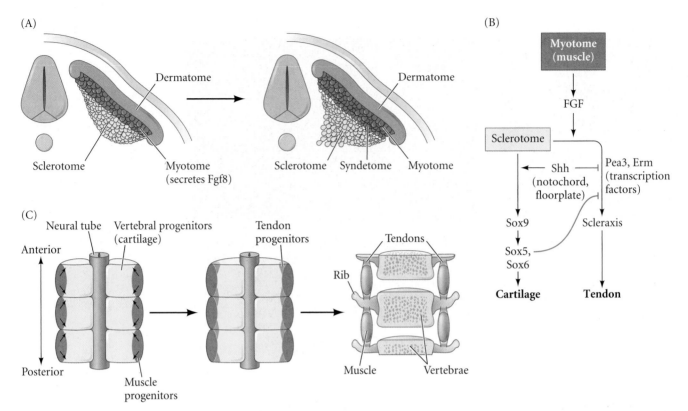

FIGURE 17.25 Induction of scleraxis in the chick sclerotome by Fgf8 from the myotome. (A) The dermatome, myotome, and sclerotome are established before the tendon precursors are specified. Tendon precursors (syndetome) are specified in the dorsalmost tier of sclerotome cells by Fgf8 received from the myotome. (B) Pathway by which Fgf8 signals from the muscle precursor cells induce the subjacent sclerotome cells to become tendons. (C) Syndetome cells migrate (arrows) along the developing vertebrae. They differentiate into tendons that connect the ribs to the intercostal muscles beloved by devotees of spareribs. (A,C after Brent et al. 2003.)

Formation of the dorsal aorta

Most of the circulatory system of an early-stage amniote embryo is directed outside the embryo, its job being to obtain nutrients from the yolk or placenta. The *intraembryonic* circulatory system begins with the formation of the dorsal aorta. The dorsal aorta is composed of two cell layers: an internal lining of endothelial cells that is surrounded concentrically by a layer of smooth muscle cells. Elsewhere in the body, these two layers of blood vessels are usually derived from the lateral plate mesoderm, as will be detailed in Chapter 18. The posterior sclerotome, however, provides the endothelial cells and smooth muscle cells for the dorsal aorta and intervertebral blood vessels (see Figure 17.2E,F; Pardanaud et al. 1996; Wiegreffe et al. 2007). The presumptive endothelial cells are induced by Notch signaling in an ephrin B2-dependent manner. These sclerotomal cells are instructed to migrate ventrally by a presumed chemoattractant made by the primary dorsal aorta, a transitory structure made by the lateral plate mesoderm. Eventually, the endothelial cells from the sclerotome replace the cells of the primary dorsal aorta, which will become part of the blood stem cell population (Pouget et al. 2008; Sato et al. 2008; Ohata et al. 2009).

Dermomyotome Development

The **dermomyotome** occupies the dorsolateral half of the somite, and in contrast to the complete epithelial-mesenchymal transition (EMT) exhibited by the sclerotome, it maintains much of its epithelial structure. Through a variety of analyses, including fate mapping with chick-quail chimeras, the dermomyotome can be subdivided into three functionally distinct regions: the dermatome, the myotome, and migratory myoblasts (see Figure 17.2; Ordahl and Le Douarin 1992; Brand-Saberi et al. 1996; Kato and Aoyama 1998). The cells in the two lateralmost portions of this epithelium are called the dorsomedial and ventrolateral lips (closest to and farthest from the neural tube, respectively), which together function as progenitor zones that generate the myotome for the formation of skeletal muscle cells of the body and limbs. Muscle precursor cells—**myoblasts**—from the dorsomedial and ventrolateral lips will migrate beneath

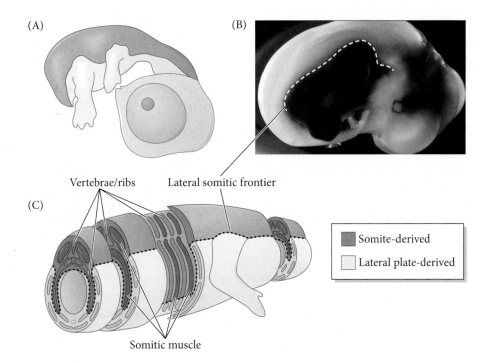

(A)

(B)

Vertebrae/ribs   Lateral somitic frontier

(C)

Somite-derived

Lateral plate-derived

Somitic muscle

FIGURE 17.26 Primaxial and abaxial domains of vertebrate mesoderm. (A) Mesoderm (red) differentiation in an early-stage chick embryo. (B) Day-9 chick embryo in which *Prox1* gene expression is revealed by a dark stain. *Prox1* is expressed in the abaxial region of the chick trunk. The boundary between the stained and unstained regions is the lateral somitic frontier (dotted line). (C) Day-13 chick; regionalization of the mesoderm is apparent. (A,C after Winslow et al. 2007; B from Durland et al. 2008, courtesy of A. C. Burke.)

the dermomyotome to produce the myotome (see Figure 17.25A). Those myoblasts in the myotome closest to the neural tube form the centrally located **primaxial muscles**,[6] which include the intercostal musculature between the ribs and the deep muscles of the back; those myoblasts farthest from the neural tube produce the **abaxial muscles** of the body wall, limbs, and tongue. The **dermatome** is located in the centralmost region of the dermomyotome and will form back dermis and several other derivatives. The boundary between the primaxial and abaxial muscles and between the somite-derived and lateral plate-derived dermis is called the **lateral somitic frontier** (**FIGURE 17.26**; Christ and Ordahl 1995; Burke and Nowicki 2003; Nowicki et al. 2003). Various transcription factors distinguish the primaxial and abaxial muscles.

The dermis of the ventral portion of the body is derived from the lateral plate, and the dermis of the head and neck comes, at least in part, from the cranial neural crest. However, in the trunk the dermatome's major product is the precursors of the dermis of the back. In addition, recent studies have shown that this central region of the dermomyotome also gives rise to a population of muscle cells (Gros et al. 2005; Relaix et al. 2005). Therefore, some researchers (Christ and Ordahl 1995; Christ et al. 2007) prefer to retain the term *dermomyotome* (or *central dermomyotome*) for this epithelial region. Soon, this part of the somite also undergoes an epithelial-to-mesenchymal transition. FGF signals from the myotome activate the transcription of the *Snail2* gene in the central dermomyotome cells, and the Snail2 protein is a well-known regulator of EMT (see Figure 15.9; Delfini et al. 2009). During EMT, the mitotic spindles of the epithelial cells are realigned so that cell division takes place along the dorsal-ventral axis. The ventral daughter cell joins the other myoblasts from the myotomes, while the other daughter cell locates dorsally, becoming a precursor of the dermis. Reminiscent of sclerotome EMT progression, the N-cadherin holding these cells together is downregulated, and the two daughter cells go their separate ways, with the remaining N-cadherin found only on those cells entering the myotome (Ben-Yair and Kalcheim 2005).

---

[6] As used here, the terms *primaxial* and *abaxial* designate the muscles from the medial and lateral portions of the somite, respectively. The terms *epaxial* and *hypaxial* are commonly used, but these terms are derived from secondary modifications of the adult anatomy (the hypaxial muscles being innervated by the ventral regions of the spinal cord) rather than from the somitic myotome lineages (see Nowicki 2003).

The muscle precursor cells that delaminate from the epithelial plate to join the primary myotome cells remain undifferentiated, and they proliferate rapidly to account for most of the myoblast cells. While most of these progenitor cells differentiate to form muscles, some remain undifferentiated and surround the mature muscle cells. These undifferentiated cells become skeletal muscle stem cells called **satellite cells**, and they are responsible for postnatal muscle growth and muscle repair.

Determination of the central dermomyotome

The central dermomyotome generates muscle precursors as well as the dermal cells that constitute the dermis of the dorsal skin. The dermis of the ventral and lateral sides of the body is derived from the lateral plate mesoderm that forms the body wall. The maintenance of the central dermomyotome depends on Wnt6 from the epidermis (Christ et al. 2007), and its EMT appears to be regulated by neurotrophin 3 (NT3) and Wnt1, two factors secreted by the neural tube (see Figure 17.21). Antibodies that block NT3 activity prevent the conversion of epithelial dermatome into the loose dermal mesenchyme that migrates beneath the epidermis (Brill et al. 1995). Removing or rotating the neural tube prevents this dermis from forming (Takahashi et al. 1992; Olivera-Martinez et al. 2002). The Wnt signals from the epidermis promote the differentiation of the dorsally migrating central dermomyotome cells into dermis (Atit et al. 2006).

Muscle precursor cells and dermal cells are not the only derivatives of the central dermomyotome, however. Atit and colleagues (2006) have shown that **brown adipose cells** ("brown fat") are also somite-derived and appear to come from the central dermomyotome. Brown fat plays an active role in energy utilization by burning fat (unlike the better-known white adipose tissue, or "white fat," which stores fat). Tseng and colleagues (2008) have found that skeletal muscle and brown fat cells share the same somitic precursor that originally expresses myogenic regulatory factors. In brown fat precursor cells, the transcription factor PRDM16 is induced (probably by BMP7); PRDM16 appears to be critical for the conversion of myoblasts to brown fat cells because it activates a battery of genes that are specific for the fat-burning metabolism of brown adipocytes (Kajimura et al. 2009).

Determination of the myotome

All the skeletal musculature in the vertebrate body with the exception of the head muscles comes from the dermomyotome of the somite. The myotome forms from the lateral edges or "lips" of the dermomyotome and folds to form a layer between the more peripheral dermomyotome and the more medial sclerotome. The major transcription factors associated with (and causing) muscle development are the **myogenic regulatory factors** (**MRFs**, sometimes called the myogenic bHLH proteins). This family of transcription factors includes MyoD, Myf5, myogenin, and MRF4 (**FIGURE 17.27**). Each member of this family can activate the genes of the other family members, leading to positive feedback regulation so powerful that the activation of an MRF in nearly any cell in the body converts that cell into muscle.[7]

The MRFs bind to and activate genes that are critical for muscle function. For instance, the MyoD protein appears to directly activate the muscle-specific creatine phosphokinase gene by binding to the DNA immediately upstream from it (Lassar et al. 1989). There are also two MyoD-binding sites on the DNA adjacent to the genes encoding a subunit of the chicken muscle acetylcholine receptor (Piette et al. 1990). MyoD also directly activates its own gene. Therefore, once the *myoD* gene is activated, its protein product binds to the DNA immediately upstream of *myoD* and keeps this gene active. Many of the MRFs are active only if they associate with a muscle-specific cofactor from

---

[7]A general rule of development, as in the U.S. Constitution, is that powerful entities must be powerfully regulated. As a result of their power to convert any cell into muscle, the MRFs are among the most powerfully controlled entities of the genome. They are controlled at several points in transcription as well as in processing, translation, and posttranslational modification (see Sartorelli and Juan 2011; Ling et al. 2012).

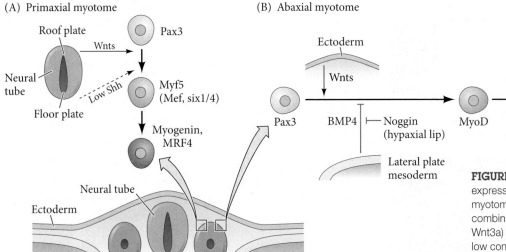

FIGURE 17.27 Differential gene expression in myotome. (A) The primaxial myotome is thought to be specified by a combination of Wnts (probably Wnt1 and Wnt3a) from the dorsal neural tube and low concentrations of Sonic hedgehog from the floor plate of the neural tube. Pax3 in somitic cells allows expression of *Myf5* in response to paracrine factors, which allows the cells to synthesize the myogenic transcription factor Myf5. In combination with a Six protein and Mef2, Myf5 activates the genes responsible for activating myogenin and MRF4. (B) BMP4 is inhibited by Noggin produced by cells that migrate specifically to the lips of the somite. In the absence of BMP4, Wnt proteins from the epidermis are thought to induce the abaxial myotome. (After Punch et al. 2009.)

the Mef2 (myocyte enhancer factor-2) family of proteins. MyoD can activate the *Mef2* gene and thereby regulate the differential timing of muscle gene expression.

As discussed above, the myotome is induced in the somite at two different places by at least two distinct signals (see Punch et al. 2009). Studies using transplantation and knockout mice indicate that the *primaxial* myoblasts from the medial portion of the somite are induced by factors from the neural tube—probably Wnt1 and Wnt3a from the dorsal region and low levels of Sonic hedgehog from the floor plate of the neural tube (see Figure 17.21; Münsterberg et al. 1995; Stern et al. 1995; Borycki et al. 2000). These factors induce the Pax3-containing cells of the somite to activate the *Myf5* gene in the primaxial myotome. Myf5 (in concert with Mef2 and either Six1 or Six4) activates the myogenin and MRF4 genes, whose proteins activate the muscle-specific gene regulatory network (see Figure 17.27A; Buckingham et al. 2006). The cells of the primaxial myotome appear to be originally confined by the laminin extracellular matrix that outlines the dermomyotome and myotome. As the myoblasts mature, however, this matrix dissolves, and the primaxial myoblasts migrate along fibronectin cables. Eventually, they align, fuse, and elongate to become the deep muscles of the back, connecting to the developing vertebrae and ribs (Deries et al. 2010, 2012).

The abaxial myoblasts that form the limb and ventral body wall musculature arise from the lateral edge of the somite. Two conditions appear necessary to produce these muscle precursors: (1) the presence of Wnt signals and (2) the absence of BMPs (see Figure 17.27B; Marcelle et al. 1997; Reshef et al. 1998). Wnt proteins (especially Wnt7a) are made in the epidermis (see Figure 17.23; Cossu et al. 1996a; Pourquié et al. 1996; Dietrich et al. 1998), but the BMP4 made by the adjacent lateral plate mesoderm would normally prevent muscles from forming.

What, then, is inhibiting BMP activity? Several studies on chick embryos have found that the dorsomedial and ventrolateral lips of the dermomyotome have attached at their tips a population of cells that secrete the BMP inhibitor Noggin (Gerhart et al. 2006, 2011). These Noggin-secreting cells arise in the blastocyst, become part of the epiblast, and distinguish themselves by expressing the mRNA for MyoD but not translating this mRNA into protein. These particular cells migrate to become paraxial mesoderm, specifically sorting out to the dorsomedial and ventrolateral lips of the dermomyotome. There they synthesize and secrete Noggin, thus promoting differentiation of myoblasts. If Noggin-secreting cells are removed from the epiblast, there is a decrease in the skeletal musculature throughout the body, and the ventral body wall is so weak that the heart

(A) Control

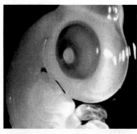

(B) Ablated

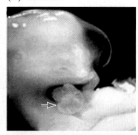

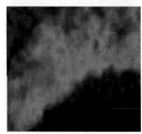

FIGURE 17.28 Ablating Noggin-secreting epiblast cells results in severe muscle defects. Noggin-secreting epiblast cells were ablated in stage 2 chick embryos using antibodies against G8. (A) The control embryo has normal morphology (upper photograph) and abundant staining of myosin (lower photograph, red) in the muscles. (B) Embryos whose Noggin-secreting epiblast cells are ablated have severe eye defects, severely reduced somatic musculature, and herniation of abdominal organs through the thin abdominal wall (upper photograph, arrow). Severely reduced musculature (sparse myosin in lower photograph) is characteristic of these embryos. (From Gerhart et al. 2006, courtesy of J. Gerhart and M. George-Weinstein.)

and abdominal organs often are herniated through it (**FIGURE 17.28**). This defect can be prevented by implanting Noggin-releasing beads into somites lacking noggin-secreting cells. Once BMP is inhibited, Wnt7 can induce MyoD in the competent dermomyotome cells, which activates the battery of MRF proteins that generate the muscle precursor cells.

An emerging model of neural crest-regulated myogenesis

Both the dorsomedial and ventrolateral lips (DML and VLL) of the dermomyotome are considered to function as self-perpetuating "cellular growth engines" capable of self-renewal and creating differentiating myocytes (Denetclaw and Ordahl 2000; Ordahl et al. 2001). What regulates which cells within the DML will adopt a renewal fate and which will mature into muscle? We have described an array of paracrine factors from the neural tube, surface ectoderm, and notochord that influence myogenesis; however, matriculation of cells into muscle development from the DML appears to occur in a highly mosaic and random fashion (Hirst and Marcelle 2015). Although transient, another potential source of signals could be the migrating population of neural crest cells that passes directly adjacent to the DML (see also Chapter 15).

In agreement with the mosaic maturation of muscle from the DML is the correlation that the Notch signaling pathway is activated in muscle progenitors. Within the tip of the DML, some cells begin to express *Notch1*, *Hes1*, and *lunatic fringe* (among others), and these cells develop into myofibers of the myotome. Christophe Marcelle's research group has shown that a sporadic portion of migrating neural crest cells express Delta1 and come in direct contact with the Notch-containing membranes at the DML (often through outstretched filopodia) (**FIGURE 17.29A,B**). Removal of neural crest cells or loss of Delta1 function in neural crest greatly reduces the myotome, whereas increased expression of Delta1 only in neural crest is sufficient to induce greater *Myf5* expression in the dermomyotome and expanded myogenesis (Rios et al. 2011).

WATCH DEVELOPMENT 17.3 Video from the laboratory of Dr. Christophe Marcelle and colleagues shows the amazingly dynamic extension of filopodia coming from the DML as neural crest cells migrate between the DML and neural tube, which suggests direct but transient cell-to-cell contact between these two cell types.

Rios and colleagues (2011) have named this neural crest transported mode of signaling "kiss and run" because it may represent a more widespread mechanism of signal dispersal. In fact, ventrally migrating neural crest cells also carry Wnt1 (**FIGURE 17.29C**), the same protein we previously described as being important in myogenesis and supplied by the dorsal neural tube. These neural crest cells require the GPC4 heparin sulfate proteoglycan to bind Wnt1 and present it to the DML as they pass by, which consequently establishes a gradient of Wnt1 protein based on the rate of migration (**FIGURE 17.29D**). Neural crest-derived Wnt1 is required for the upregulation of Wnt11 within the dermomyotome and the proper organization of the myotome (Serralbo and Marcelle 2014). Finally, as neural crest cells migrate through the sclerotome, they secrete neuregulin-1, a paracrine factor that prevents the premature differentiation of

Developing Questions

Why not recruit a courier for important signals to be delivered over long distances? What are the molecular mechanisms that can decide which neural crest cells come to express these signals at the correct time and in the correct form of presentation? The answers to these questions are as yet unknown. It is likely that neural crest cells are not the only migratory cells transporting "hot" goods across the borders that divide up the embryo. The inner somite is a chaotic intersection of neural crest cells, sclerotome, pathfinding axons, and other mesenchyme. We can assume that they all may transiently express important messages to adjacent tissues as well as to other migratory passersby.

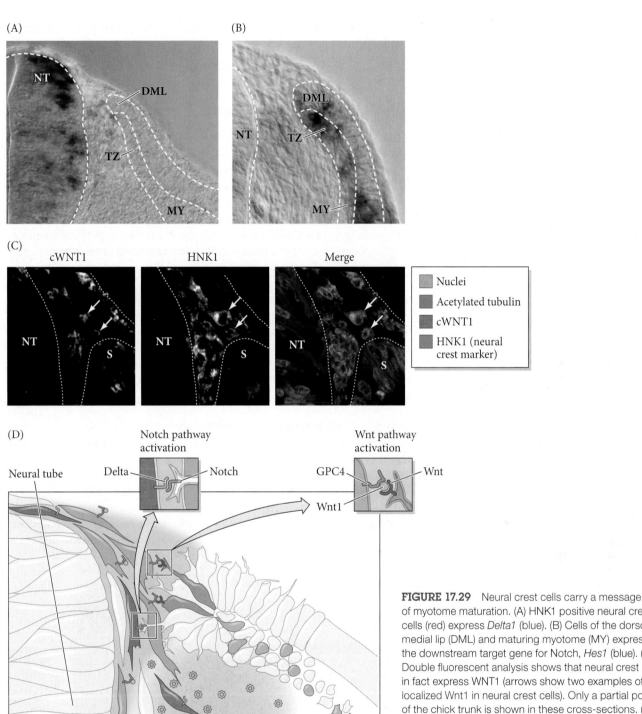

FIGURE 17.29 Neural crest cells carry a message of myotome maturation. (A) HNK1 positive neural crest cells (red) express *Delta1* (blue). (B) Cells of the dorso-medial lip (DML) and maturing myotome (MY) express the downstream target gene for Notch, *Hes1* (blue). (C) Double fluorescent analysis shows that neural crest cells in fact express WNT1 (arrows show two examples of co-localized Wnt1 in neural crest cells). Only a partial portion of the chick trunk is shown in these cross-sections. (NT, neural tube; S, somite; TZ, transition zone.) (D) Model of the transient interactions between neural crest cells and the DML myogenic precursors through Notch-Delta and Wnt signaling. DML cells extend long filopodia into the periphery of the dermomyotome through the path of migrating neural crest cells. Different neural crest cells are illustrated with different colors to identify Delta expressing (red) from Wnt1 expressing (blue) neural crest cells. In addition, neural crest cells also secrete neuregulin-1 (gold), which prevents precocious differentiation of myoblast (green cells) from DML progenitors (purple cells). (A,B from Rios et al. 2011; C courtesy of Olivier Serralbo and Christophe Marcelle.)

myoblasts into muscle cells, which helps to maintain the pool of myogenic progenitors (Ho et al. 2011). Thus, like a bee unknowingly carrying pollen to flowers along its journey, a neural crest cell delivers morphogenic signals to cells throughout the somite that influence their differentiation and growth.

Osteogenesis: The Development of Bones

Three distinct lineages generate the skeleton. The paraxial mesoderm generates the vertebral and craniofacial bones, the lateral plate mesoderm generates the limb skeleton, and the cranial neural crest gives rise to some of the craniofacial bones and cartilage. There are two major modes of bone formation, or **osteogenesis**, and both involve the transformation of preexisting mesenchymal tissue into bone tissue. The direct conversion of mesenchyme into bone is called intramembranous ossification and was discussed in Chapter 11. In other cases, the mesenchymal cells differentiate into cartilage, which is later replaced by bone in a process called **endochondral ossification**.

Endochondral ossification

Endochondral ossification involves the formation of cartilage tissue from aggregated mesenchymal cells and the subsequent replacement of cartilage tissue by bone (Horton 1990). This type of bone formation is characteristic of the vertebrae, ribs, and limbs. The vertebrae and ribs form from the somites, while the limb bones form from the lateral plate mesoderm (see Chapter 19). Endochondral ossification can be divided into five stages: commitment, compaction, proliferation, growth, and, finally, chondrocyte death and the generation of new bone.

PHASES 1 AND 2: COMMITMENT AND COMPACTION   First, the mesenchymal cells commit to becoming cartilage (**FIGURE 17.30A**). This commitment is stimulated by Sonic hedgehog, which induces nearby sclerotome cells to express the Pax1 transcription factor (Johnson et al. 1994; Teissier-Lavigne 1994). Pax1 initiates a cascade that is dependent on external paracrine factors and internal transcription factors.

During the second phase of endochondral ossification, the committed mesenchyme cells condense into compact nodules (**FIGURE 17.30B**). These inner cells become committed to generating cartilage, and the outer cells become committed to becoming bone.

FIGURE 17.30 Schematic diagram of endochondral ossification. (A) Mesenchymal cells commit to becoming cartilage cells (chondrocytes). (B) Committed mesenchyme condenses into compact nodules. (C) Nodules differentiate into chondrocytes and proliferate to form the cartilage model of bone. (D) Chondrocytes undergo hypertrophy and apoptosis while they change and mineralize their extracellular matrix. (E) Apoptosis of chondrocytes allows blood vessels to enter. (F) Blood vessels bring in osteoblasts, which bind to the degenerating cartilaginous matrix and deposit bone matrix. (G) Bone formation and growth consist of ordered arrays of proliferating, hypertrophic, and mineralizing chondrocytes. Secondary ossification centers also form as blood vessels enter near the tips of the bone. (After Horton 1990.)

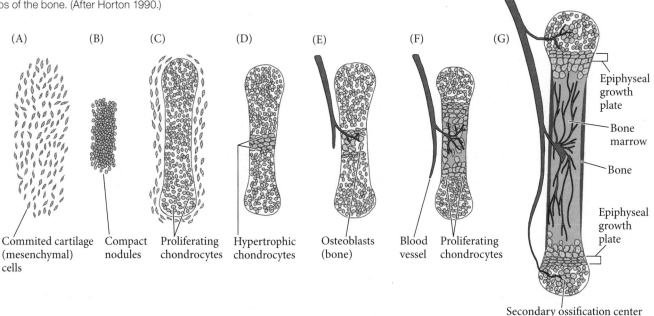

(A)   (B)   (C)   (D)   (E)   (F)   (G)

Commited cartilage (mesenchymal) cells

Compact nodules

Proliferating chondrocytes

Hypertrophic chondrocytes

Osteoblasts (bone)

Blood vessel

Proliferating chondrocytes

Epiphyseal growth plate

Bone marrow

Bone

Epiphyseal growth plate

Secondary ossification center

BMPs appear to be critical in this stage. They are responsible for inducing the expression of the adhesion molecules N-cadherin and N-CAM and the transcription factor Sox9. N-cadherin appears to be important in the initiation of these condensations, and N-CAM critical for maintaining them (Oberlender and Tuan 1994; Hall and Miyake 1995). Sox9 activates other transcription factors as well as a suite of genes, including those encoding collagen II and aggrecan, which are required in cartilage function. In humans, mutations of the *SOX9* gene cause camptomelic dysplasia, a rare disorder of skeletal development that results in deformities of most of the bones of the body. Most affected babies die from respiratory failure due to poorly formed tracheal and rib cartilage (Wright et al. 1995).

PHASES 3 AND 4: PROLIFERATION AND GROWTH   During the third phase of endochondral ossification, the chondrocytes proliferate rapidly to form a cartilaginous model for the bone (**FIGURE 17.30C**). As they divide, the chondrocytes secrete a cartilage-specific extracellular matrix. The outermost cells become the **perichondrium** that ensheaths the cartilage.

In the fourth phase, the chondrocytes stop dividing and increase their volume dramatically, becoming **hypertrophic chondrocytes** (**FIGURES 17.30D** and **17.31**). This step appears to be mediated by the transcription factor Runx2 (also called CBFα1), which

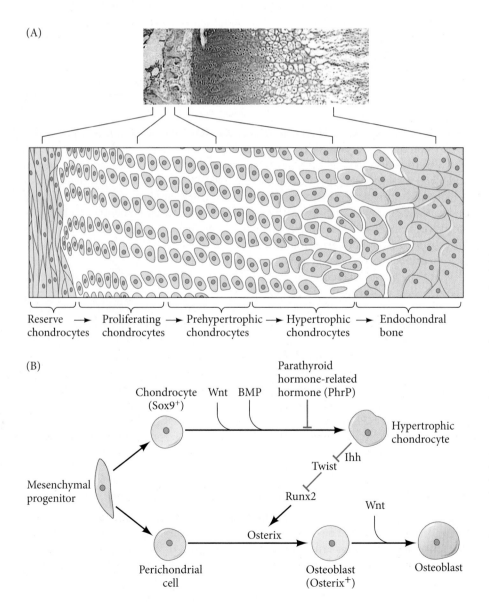

(A)

Reserve → Proliferating → Prehypertrophic → Hypertrophic → Endochondral
chondrocytes   chondrocytes   chondrocytes   chondrocytes   bone

(B)

FIGURE 17.31 Endochondral ossification. (A) Long bone undergoing endochondral ossification. The cartilage is stained with Alcian blue, and the bone is stained with alizarin red. Below is a diagram of the transition zone wherein the chondrocytes (cartilage cells) divide, enlarge, die, and are replaced by osteocytes (mature bone cells). (B) Paracrine and transcription factors active in the transition from cartilage to bone. The mesenchymal sclerotome cell can become a chondrocyte (characterized by the Sox9 transcription factor) or an osteocyte (characterized by the Osterix transcription factor), depending on the types of paracrine factors it experiences. The paracrine factor Indian hedgehog (Ihh), secreted by the growing chondrocytes, appears to repress Twist, an inhibitor of Runx2. Runx2 is critical for directing cell fate into the bone pathway; it activates Osterix, which in turn activates bone-specific proteins. (After Long 2012.)

is necessary for the development of both intramembranous and endochondral bone. *Runx2* expression is regulated by histone deacetylase-4 (HDAC4), a form of chromatin-restructuring enzyme that is expressed solely in the prehypertrophic cartilage. If HDAC4 is overexpressed in the cartilaginous ribs or limbs, ossification is seriously delayed; if the *Hdac4* gene is knocked out of the mouse genome, the limbs and ribs ossify prematurely (Vega et al. 2004). Hypertrophic cartilage is exceptionally important in regulating the final size of the long bone. Indeed, the greatest contribution to the growth rate in mammals is the relative size of the hypertrophic cartilage (Cooper et al. 2013). The swelling of this cartilage determines the elongation rate of each skeletal element and is responsible for the differences in the growth rates between different skeletal elements both within an organism (hands vs. legs, for instance) and between related organisms (the legs of a mouse vs. the legs of a jerboa).

These large chondrocytes alter the matrix they produce (by adding collagen X and more fibronectin) to enable it to become mineralized (calcified) by calcium phosphate. These hypertrophic cartilage cells also secrete two factors that will be critical for the transformation of cartilage into bone. First, they secrete the angiogenesis factor **VEGF (vascular endothelial growth factor)**, which can transform mesodermal mesenchyme cells into blood vessels (see Chapter 18; Gerber et al. 1999; Haigh et al. 2000). Second, they secrete Indian hedgehog, a member of the Hedgehog family and a close cousin of Sonic hedgehog, which activates *Runx2* transcription in the perichondrial cells surrounding the cartilage primordium. This step initiates the differentiation of those cells into bone-forming osteoblasts. Mice lacking the *Indian hedgehog* gene completely lack the osteoblasts of the endochondral skeleton (trunk and limbs), although the osteoblasts formed in the head and face by intramembranous ossification form normally (St-Jacques et al. 1999).

PHASE 5: CHONDROCYTE DEATH AND BONE CELL GENERATION In the fifth phase, the hypertrophic chondrocytes die by apoptosis (Hatori et al. 1995; Rajpurohit et al. 1999). The hypertrophic cartilage is replaced by bone cells both on the outside and inside, and blood vessels invade the cartilage model (**FIGURE 17.30E–G**). On the *outside*, the osteoblasts begin forming bone matrix, constructing a bone collar around the calcified and partially degraded cartilage matrix (Bruder and Caplan 1989; Hatori et al. 1995; St-Jacques et al. 1999). The osteoblasts become responsive to Wnt signals that upregulate Osterix, a transcription factor that instructs the osteoblasts to become mature bone cells, or **osteocytes** (Nakashima et al. 2002; Hu et al. 2005).

New bone material is added peripherally from the *internal surface* of the **periosteum**, a fibrous sheath covering the developing bone. The periosteum contains connective tissue, capillaries, and bone progenitor cells (Long et al. 2012). At the same time, there is a hollowing out of the internal region of the bone to form the bone marrow cavity. As cartilage cells die, they alter the extracellular matrix, releasing VEGF, which stimulates the formation of blood vessels around the dying cartilage. If the blood vessels are inhibited from forming, bone development is significantly delayed (Karsenty and Wagner 2002; Yin et al. 2002). The blood vessels bring in both osteoblasts and **osteoclasts**, multinucleated cells that eat the debris of the apoptotic chondrocytes and thus create the hollow bone marrow cavity (Kahn and Simmons 1975; Manolagas and Jilka 1995). Osteoclasts are not derived from the somite; rather, they are derived from a blood cell lineage (in the lateral plate mesoderm) and come from the same precursors as macrophage blood cells (Ash et al. 1980; Blair et al. 1986).

Mechanotransduction and vertebrate bone development

The ability of cells to sense their environment and convert mechanical forces into molecular signals is called mechanotransduction, and the importance of mechanotransduction to development is just beginning to be recognized. We saw this importance in the discussion of how extracellular mechanical signals changed the differentiation of stem cells, and mechanical forces appear to be significant in the formation of bones, muscles, and tendons, and perhaps also for their repair and regeneration in the adult. However, very little is known about how mechanical stress is sensed, quantified, and transmitted as a change in cytoplasmic chemicals.

Vertebrate skeletal bone development shows some dependence on mechanotransduction. Tension and stress forces activate the gene for Indian hedgehog (Ihh), a paracrine factor that activates the bone morphogenetic proteins (BMPs; Wu et al. 2001). In the chick, several bones do not form if embryonic movement in the egg is suppressed. One of these bones is the fibular crest, which connects the tibia directly to the fibula (**FIGURE 17.32A,B**). This direct connection is believed to be important in the evolution of birds, and the fibular crest is a universal feature of the bird hindlimb (Müller and Steicher 1989; Müller 2003).

The jaws of cichlid fish differ enormously, depending on the food they eat (**FIGURE 17.32C**; Meyer 1987). Similarly, normal primate jaw development may be predicated on how much tension is produced by grinding food: mechanical tension appears to stimulate *Indian hedgehog* expression in mammalian mandibular cartilage (Tang et al. 2004). If an infant monkey is given soft food, its lower jaw is smaller than normal. Corruccini and Beecher (1982, 1984) and Varrela (1992) have shown that people in cultures where infants are fed hard food have jaws that "fit" better, and these researchers speculate that soft baby food may explain why so many children in Western societies need braces. Indeed, the notion that mechanical tension can change jaw size and shape is the basis of the functional hypothesis of modern orthodontics (Moss 1962, 1997).

In mammals, muscle force within the embryo is critical for the normal shaping of bone and the development of load-bearing capacity (Sharir et al. 2011). After birth, the patella (kneecap) is formed by pressure on the skeleton, and it is thought that the aberrant skeletal development seen in persons with cerebral palsy is caused by the absence of pressure on these bones (Shefelbine and Carter 2004; Ward et al. 2006).

WEB TOPIC 17.2 **PARACRINE FACTORS, THEIR RECEPTORS, AND HUMAN BONE GROWTH** Mutations in the genes encoding paracrine factors and their receptors cause numerous skeletal anomalies in humans and mice.

Maturation of Muscle

Myoblasts and myofibers

The cells producing the myogenic regulatory factors are the myoblasts—committed muscle cell precursors—but unlike most cells of the body, muscle cells do not function as "individuals." Rather, several myoblasts align together and fuse their cell membranes to form a **myofiber**, a single large cell with several nuclei that is characteristic of muscle tissue (Konigsberg 1963; Mintz and Baker 1967; Richardson et al. 2008). Myofibers in the adult can be the result of thousands of fusion events involving mononucleated cells. Studies on mouse embryos show that by the time a mouse is born, it has the adult number of myofibers and that these multinucleated myofibers grow during the first week after birth by the fusion of singly nucleated myoblasts (Ontell et al. 1988; Abmayr and Pavlath 2012). After the first week, muscle cells can continue to grow by the fusion of muscle stem cells (satellite cells, discussed later) into existing myofibers and by an increase in contractile proteins within the myofibers.

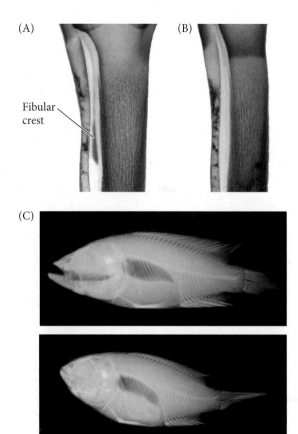

FIGURE 17.32 Phenotypes can be produced by stress force on muscular and skeletal tissues. (A,B) The avian fibular crest (syndesmosis tibiofibularis) connects the tibia directly to the fibula. The fibular crest is formed when the movement of the active embryo inside the egg puts physical stress on the tibia. (A) Fibular crest forming in the connective tissue of a 13-day chick embryo. (B) Absence of fibular crest in the connective tissue of a 13-day embryo whose movement was inhibited. The blue dye stains cartilage; the red dye stains the bone elements. (C) The jaws of cichlid fish are shaped by the hardness of the food they eat. Different diets give different jaw structures. (A,B from Müller 2003, courtesy of G. Müller; C from Meyer 1987, courtesy of A. Meyer.)

MYOBLAST FUSION   The first step in fusion requires the myoblasts to exit the cell cycle, which involves expression of cyclin D3 (Gurung and Pamaik 2012). Next, the myoblasts secrete fibronectin and other proteins onto their extracellular matrices and bind to it through α5β1 integrin, a major receptor for these extracellular matrix components (Menko and Boettiger 1987; Boettiger et al. 1995; Sunadome et al. 2011). If this adhesion is experimentally blocked, no further muscle development ensues; thus, it appears that the signal from the integrin-fibronectin attachment is critical for instructing myoblasts to differentiate into muscle cells (**FIGURE 17.33**).

The third step is the alignment of the myoblasts into chains. This step is mediated by cell membrane glycoproteins, including several cadherins (Knudsen 1985; Knudsen et al. 1990). Recognition and alignment between cells take place only if the cells are myoblasts. Fusion can occur even between chick and rat myoblasts in culture (Yaffe and Feldman 1965); the identity of the species is not critical. The internal cytoplasm is rearranged in preparation for the fusion, with actin regulating the regions of contact between the cells (Duan and Gallagher 2009).

The fourth step is the cell fusion event itself. As in most membrane fusions, calcium ions are critical, and fusion can be activated by calcium transporters, such as A23187, that carry Ca^{2+} across cell membranes (Shainberg et al. 1969; David et al. 1981). Fusion appears to be mediated by a set of metalloproteinases called **meltrins**. Meltrins were discovered during a search for myoblast proteins that would be homologous to fertilin, a protein implicated in sperm-egg membrane fusion. Yagami-Hiromasa and colleagues (1995) found that one of these proteins, meltrin-α, is expressed in myoblasts at about the same time that fusion begins, and that antisense RNA to the meltrin-α message inhibited fusion when added to myoblasts. As the myoblasts become capable of fusing, another myogenic regulatory factor—**myogenin**—becomes active. Myogenin binds to the regulatory region of several muscle-specific genes and activates their expression. Thus, whereas MyoD and Myf5 are active in the lineage specification of muscle cells,

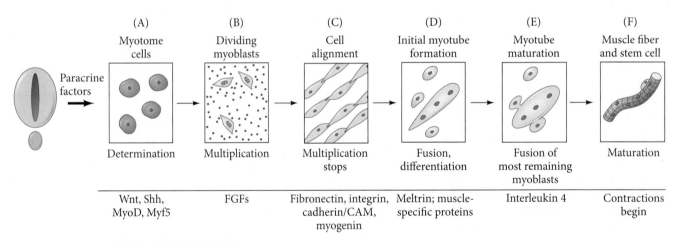

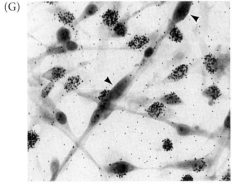

FIGURE 17.33 Conversion of myoblasts into muscles in culture. (A) Determination of myotome cells by paracrine factors. (B) Committed myoblasts divide in the presence of growth factors (primarily FGFs) but show no obvious muscle-specific proteins. (C–E) When the growth factors are used up, the myoblasts cease dividing, align, and fuse into myotubes. (F) The myotubes become organized into muscle fibers that spontaneously contract. (G) Autoradiograph showing DNA synthesis in myoblasts and the exit of fusing cells from the cell cycle. Phospholipase C can "freeze" myoblasts after they have aligned with other myoblasts but before their membranes fuse. Cultured myoblasts were treated with phospholipase C and then exposed to radioactive thymidine. Unattached myoblasts continued to divide and thus incorporated the radioactive thymidine into their DNA. Aligned (but not yet fused) cells (arrowheads) did not incorporate the label. (A–F after Wolpert 1998; G from Nameroff and Munar 1976, courtesy of M. Nameroff.)

myogenin appears to mediate muscle cell differentiation (Bergstrom and Tapscott 2001).

Cell fusion ends with the re-sealing ("healing") of the newly apposed membranes. This step is accomplished by proteins such as myoferlin and dysferlin, which stabilize the membrane phospholipids (Doherty et al. 2005). These proteins are similar to those that re-seal the membranes at axon nerve synapses after membrane vesicle fusion releases neurotransmitters.

MYOFIBER GROWTH After the original fusion of myoblasts into a myofiber, the myofiber secretes the paracrine factor interleukin 4 (IL4). Although IL4 was originally believed to work exclusively in the adult immune system, Horsely and colleagues (2003) found that IL4 secreted by new myofibers recruits other myoblasts to fuse with the myotube, thereby forming the mature myofiber (see Figure 17.3).

The number of muscle fibers in the embryo and the growth of these fibers after birth appear to be negatively regulated by **myostatin**, a member of the TGF-β family (McPherron et al. 1997; Lee 2004). Myostatin is made by developing and adult skeletal muscle and most probably works in an autocrine fashion. As mentioned in Chapter 3, *myostatin* loss-of-function mutations allow both hyperplasia (more fibers) and hypertrophy (larger fibers) of the muscle (see Figure 3.26). These changes give rise to Herculean phenotypes in dogs, cattle, mice, and humans (**FIGURE 17.34**).

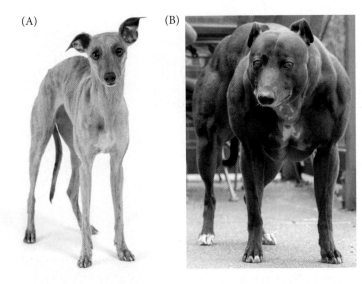

(A) (B)

FIGURE 17.34 A loss-of-function mutation in the *myostatin* gene of whippets. (A) Whippets are a typically slender breed, bred for speed and dog racing. (B) Although the homozygous loss-of-function condition is not advantageous, heterozygotes have more muscle power and are significantly overrepresented among the top racers (see Mosher et al. 2007). (A © kustudio/Shutterstock; B © Bruce Stotesbury/PostMedia News/Zuma Press.)

WEB TOPIC 17.3 **MUSCLE FORMATION** Research on chimeric mice has shown that skeletal muscle becomes multinucleate by the fusion of cells, whereas heart muscle becomes multinucleate by nuclear divisions within a cell.

Satellite cells: Unfused muscle progenitor cells

Any dancer, athlete, or sports fan knows that (1) adult muscles grow larger when they are exercised, and (2) muscles are capable of limited regeneration following injury. The growth and regeneration of muscles both arise from **satellite cells**, populations of stem cells and progenitor cells that reside alongside the adult muscle fibers. Satellite cells respond to injury or exercise by proliferating into myogenic cells that fuse and form new muscle fibers. Lineage tracing using chick-quail chimeras indicates that satellite cells are somite-derived myoblasts that have not fused and that remain potentially available throughout adult life (Armand et al. 1983).

The source of mouse and chick satellite cells appears to be the central part of the dermomyotome (Ben-Yair and Kalcheim 2005; Gros et al. 2005; Kassar-Duchossoy et al. 2005; Relaix et al. 2005). Although the myoblast-forming cells of the dermomyotome form at the lips and express Myf5 and MyoD, the cells that enter the myotome from the central region usually express Pax3 and Pax7 as well as microRNAs miRNA-489 and miRNA-31. The combination of Pax3 and Pax7 appears to inhibit MyoD expression (and thus muscle differentiation) in these cells; Pax7 also protects the satellite cells against apoptosis (Olguin and Olwin 2004; Kassar-Duchossoy et al. 2005; Buckingham et al. 2006). The two microRNAs appear to prevent the translation of factors such as Myf5 that would promote muscle cell differentiation (Cheung et al. 2012; Crist et al. 2012).

Satellite cells are not a homogeneous population; rather, they contain both stem cells and progenitor cells. The stem cells represent only about 10% of satellite cells and are

(A)

(B)

(C)

(D)

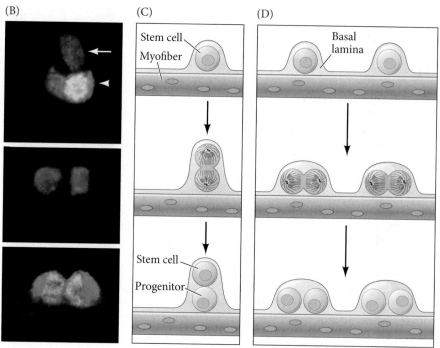

FIGURE 17.35 Satellite cells and muscle growth. (A) Satellite cells (stained with antibodies to the Pax7 protein) reside between the myofiber cell membrane and the basal lamina. (B) The top photograph shows asymmetric cell division of a satellite stem cell and the distinction between the daughter cell keeping Pax7 (stem cell; red) and the daughter cell downregulating Pax7 and expressing Myf5 (progenitor cell; green). This corresponds to the diagrammatic representation in (C). The bottom two photographs in (B) show symmetric division, where stem cells and progenitor cells make more stem and progenitor cells, respectively, as panel (D) shows in diagrammatic form. (After Bentzinger et al. 2012; photographs courtesy of F. Bentzinger and M. A. Rudnicki.)

found, with the other satellite cells, between the cell membrane and the extracellular basal lamina of mature myofibers. Satellite stem cells express Pax7 but not Myf5 (this is designated Pax7$^+$/Myf5$^-$) and can divide asynchronously to produce two types of cells: another Pax7$^+$/Myf5$^-$ stem cell and a Pax7$^+$/Myf5$^+$ satellite progenitor cell that differentiates into muscle (**FIGURE 17.35**). The Pax7$^+$/Myf5$^-$ stem cells, when transplanted into other muscles, contribute to the stem cell population there (Kuang et al. 2007).

The factor responsible for the asymmetry of this division appears to be miRNA-489, which is found in quiescent stem cells. Upon division, miRNA-489 remains in the daughter that remains a stem cell but is absent in the cell that becomes part of the muscle. MiRNA-489 inhibits the translation of the message for the Dek protein, which becomes translated in the daughter cell that differentiates. Dek is a chromatin protein that promotes the transient proliferation of progenitor cells (Cheung et al. 2012). Thus, miRNA-489 maintains the quiescent state of an adult muscle stem cell population.

Mechanotransduction in the musculoskeletal system

We know that physical forces generated by exercise cause muscles to enlarge. Exercise stimulates protein synthesis in the muscle cells, and each nucleus in the multinucleate fiber appears to have a region around it where protein synthesis is regulated (Lai et al. 2004; Quaisar et al. 2012). If physical stress continues, the force appears to cause the muscle satellite cells to proliferate and fuse with the existing muscle fibers. Indeed, endurance exercise has been shown to increase the number of satellite cells in the elderly (Shefer et al. 2010). Insulin-like growth factor, acting as an autocrine secretion from muscle cells, is a candidate for causing such muscle growth (Yang 1996; Goldspink 2004; Sculthorpe et al. 2012), but how this factor or any other is induced by stress remains unknown.

Also, in a way that is not yet understood, the tension produced by weight-bearing loads activates production of TGF-β2 and 3 in the tendon cells (Maeda et al. 2011). Indeed, mice lacking these genes completely lack tendons. The TGF-β pathway (through the Smad2/3 transcription factors) continues to activate the gene for the transcription factor scleraxis after the initial FGF signaling; in turn, scleraxis activates the genes responsible for forming the extracellular matrix. Moreover, TGF-β produced by the developing tendon may recruit cells from the cartilage and muscle to make the bridge between these three tissues (Blitz et al. 2009; Pryce et al. 2009).

Next Step Investigation

Development of the axial and paraxial mesoderm and their derivatives involves a highly complex integration of multiple pathways, epigenetic regulation, cell shape changes, migrating cells, and the ever-present mechanical forces. The most challenging objective for a developmental biologist today might be to design experiments that address how these varied processes are integrated. For instance, we are starting to gain insight into how the segmentation clock, the determination front, and Hox gene regulation interact to achieve both somite formation and axial identity. One approach that has provided new perspectives has been mathematical modeling, which enables a researcher to theoretically manipulate a seemingly infinite array of parameters to help identify predictable outcomes to complex processes. You can start by picking a paraxial mesoderm event and identifying all the parameters that define this event: the cell(s); the changes in size, shape, number, and position over time; and how the event changes when a particular aspect is altered. Now you can begin playing with these parameters to "model" development in silico.

Closing Thoughts on the Opening Photo

The formation of segments or somites is a highly regulated process that determines "what, when, where, and how many" somites an organism makes. This beautiful image of a garter snake embryo stained with Alcian blue was produced by Anne C. Burke, and illustrates the grand nature of somitogenesis. Segments of paraxial mesoderm are carved up into sequential blocks, through the orchestration of an Fgf8 determination front, a Notch-Delta molecular clock, and an Eph-Ephrin mediated boundary formation. (Courtesy of Anne C. Burke.)

17 Snapshot Summary
Paraxial Mesoderm

1. The paraxial mesoderm forms blocks of tissue called somites. Somites give rise to three major divisions: the sclerotome, the myotome, and the central dermomyotome.

2. The spatiotemporal expression of 3′-to-5′ Hox genes along the paraxial mesoderm correlates with a progressive loosening of chromatin structures via epigenetic regulation as well as with the timing of ingression into the paraxial mesoderm along the anterior-posterior axis. Caudal gradient signals of FGFs and Wnts maintain NMP cells in the progenitor state, whereas opposing gradients of retinoic acid promote differentiation of these cells. These antagonistic signals establish where a new somitic boundary will form in the segmental plate.

3. Cyclic activation of Notch-Delta signaling throughout the presomitic mesoderm establishes the timing of segment formation, and Eph receptor systems are involved in the physical formation of boundaries. Moreover, N-cadherin, fibronectin, and Rac1 also appear to be important in causing presomitic mesodermal cells to become epithelial.

4. The sclerotome forms the vertebral cartilage. In thoracic vertebrae, the sclerotome cells also form the ribs. The intervertebral joints as well as the meninges and dorsal aortic cells also come from the sclerotome.

5. The primaxial myotome forms the back musculature. The abaxial myotome forms the muscles of the body wall, limb, diaphragm, and tongue.

6. The central dermomyotome forms the dermis of the back as well as precursors of muscle and brown fat cells.

7. The somite regions are specified by paracrine factors secreted by neighboring tissues. The sclerotome is specified to a large degree by Sonic hedgehog, which is secreted by the notochord and floor plate cells. The two myotome regions are specified by different factors, and in both instances, myogenic regulatory factors are induced in the cells that will become muscles.

8. To form muscles, the myoblasts stop dividing, align themselves into myotubes, and fuse. Further myofiber growth is facilitated by stem cells on the myotube periphery called satellite cells.

9. The major lineages that form the skeleton are the somites (axial skeleton), lateral plate mesoderm (appendages), and neural crest and head mesoderm (skull and face).

10. There are two major types of osteogenesis. In intramembranous ossification, which occurs primarily in the skull

and facial bones, neural crest and head mesenchyme are converted directly into bone. In endochondral ossification, mesenchyme cells become cartilage. These cartilaginous models are later replaced by bone cells.

11. Hypertrophic cartilage cells make Indian hedgehog (Ihh) and vascular endothelial growth factor (VEGF). Ihh initiates osteoblast differentiation into bone, and VEGF induces the construction of capillaries that enable bone cells to be brought into the degenerating cartilage.

12. Osteoclasts continually remodel bone throughout a person's lifetime. The hollowing out of bone for the bone marrow is accomplished by osteoclasts.

13. Tendons are formed through the conversion of the dorsalmost layer of sclerotome cells into syndetome cells by FGFs secreted by the myotome.

Further Reading

Barrios, A., R. J. Poole, L. Durbin, C. Brennan, N. Holder and S. W. Wilson. 2003. Eph/Ephrin signaling regulates the mesenchymal-to-epithelial transition of the paraxial mesoderm during somite morphogenesis. *Curr. Biol.* 13: 1571–1582.

Bouldin, C. M., C. D. Snelson, G. H. Farr and D. Kimelman. 2014. Restricted expression of cdc25a in the tailbud is essential for formation of the zebrafish posterior body. *Genes Dev.* 28: 384–395.

Brent, A. E., R. Schweitzer and C. J. Tabin. 2003. A somitic compartment of tendon precursors. *Cell* 113: 235–248.

Cayuso, J., Q. Xu and D. G. Wilkinson. 2015. Mechanisms of boundary formation by Eph receptor and ephrin signaling. *Dev Biol.* 40: 122–131.

Chalamalasetty, R. B. and 5 others. 2014. Mesogenin 1 is a master regulator of paraxial presomitic mesoderm differentiation. *Development* 141: 4285–4297.

Choi, K. S. and B. D. Harfe. 2011. Hedgehog signaling is required for formation of the notochord sheath and patterning of nuclei pulposi within the intervertebral discs. *Proc. Natl. Acad. Sci. USA* 108: 9484–9489.

Christ, B., R. Huang and M. Scaal. 2007. Amniote somite derivatives. *Dev. Dyn.* 236: 2382–2396.

Denans, N., T. Iimura and O. Pourquié. 2015. Hox genes control vertebrate body elongation by collinear Wnt repression. *Elife* 26: 4.

Dubrulle, J. and O. Pourquié. 2004. fgf8 mRNA decay establishes a gradient that couples axial elongation to patterning in the vertebrate embryo. *Nature* 427: 419–422.

Ellis, K., J. Bagwell and M. Bagnat. 2013. Notochord vacuoles are lysosome-related organelles that function in axis and spine morphogenesis. *J. Cell Biol.* 200: 667–679.

Gomez, C. and 5 others. 2008. Control of segment number in vertebrate embryos. *Nature* 454: 335–339.

Henrique D., E. Abranches, L. Verrier and K. G. Storey. 2015. Neuromesodermal progenitors and the making of the spinal cord. *Development* 142: 2864–2875.

Hubaud, A. and O. Pourquié. 2014. Signalling dynamics in vertebrate segmentation. *Nat. Rev. Mol. Cell Biol.* 15: 709–721.

Jülich, D. and 7 others. 2015. Cross-scale integrin regulation organizes ECM and tissue topology. *Dev. Cell* 34: 33–44.

Kumar, S. and G. Duester. 2014. Retinoic acid controls body axis extension by directly repressing Fgf8 transcription. *Development* 141: 2972–2977.

Mansfield, J. H., E. Haller, N. D. Holland, and A. E. Brent. 2015. Development of somites and their derivatives in amphioxus, and implications for the evolution of vertebrate somites. *Evodevo* May 14: 21.

Noordermeer, D. and 5 others. 2014. Temporal dynamics and developmental memory of 3D chromatin architecture at Hox gene loci. *Elife* Apr 29; 3:e02557. doi: 10.7554/eLife.02557.

Nowotschin, S., A. Ferrer-Vaquer, D. Concepcion, V. E. Papaioannou and A. K. Hadjantonakis. 2012. Interaction of Wnt3a, Msgn1 and Tbx6 in neural versus paraxial mesoderm lineage commitment and paraxial mesoderm differentiation in the mouse embryo. *Dev. Biol.* 367: 1–14.

Ordahl, C. P., E. Berdougo, S. J. Venters and W. F. J. Denetclaw. 2001. The dermomyotome dorsomedial lip drives growth and morphogenesis of both the primary myotome and dermomyotome epithelium. *Development* 128:1731–1744.

Rios, A. C., O. Serralbo, D. Salgado and C. Marcelle. 2011. Neural crest regulates myogenesis through the transient activation of NOTCH. *Nature* 473: 532–535.

Serralbo, O. and C. Marcelle. 2014. Migrating cells mediate long-range WNT signaling. *Development* 141: 2057–2063.

Wahi, K., M. S. Bochter and S. E. Cole. 2014. The many roles of Notch signaling during vertebrate somitogenesis. *Semin. Cell Dev. Biol.* Dec 4. pii: S1084-9521(14)00320-6.

GO TO WWW.DEVBIO.COM...

...for Web Topics, Scientists Speak interviews, Watch Development videos, Dev Tutorials, and complete bibliographic information for all literature cited in this chapter.

Intermediate and Lateral Plate Mesoderm

Heart, Blood, and Kidneys

How is the heart formed, and how does it connect to the arteries and veins?

WHILE THE AXIAL AND PARAXIAL MESODERM form the notochord and somites of the dorsum, the intermediate and lateral plate mesoderms extend around the sides and front of the body. The **intermediate mesoderm** forms the urogenital system, consisting of the kidneys, the gonads, and their associated ducts. The outer (cortical) portion of the adrenal gland also derives from this region. Farthest away from the notochord, the **lateral plate mesoderm** gives rise to the heart, blood vessels, and blood cells of the circulatory system, as well as to the lining of the body cavities. It gives rise to the pelvic and limb skeleton (but not the limb muscles, which are somitic in origin). Lateral plate mesoderm also helps form a series of extraembryonic membranes that are important for transporting nutrients to the embryo.

These four subdivisions are thought to be specified along the mediolateral (center-to-side) axes by increasing amounts of BMPs (Pourquié et al. 1996; Tonegawa et al. 1997). The more lateral mesoderm of the chick embryo expresses higher levels of BMP4 than do the midline areas, and one can change the identity of the mesodermal tissue by altering BMP expression. While it is not known how this patterning is accomplished, it is thought that the different BMP concentrations may cause differential expression of the Forkhead (Fox) family of transcription factors. The *Foxf1* gene is transcribed in those regions that

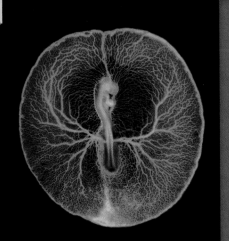

The Punchline

The heart, blood vessels, and kidneys are mesodermal organs involved in the transport of another mesodermal tissue, blood cells, throughout the body. The vertebrate kidney develops from the interactions of two regions of intermediate mesoderm, the nephric duct and the metanephric mesenchyme. Their interactions cause the mesenchyme to form the nephron that filters the blood, and the collecting ducts and ureter that delivers the filtrate to the bladder. Mesodermal progenitor cells form the primary heart field and migrate ventromedially to build a linear heart tube. A secondary heart field, derived from pharyngeal mesoderm, adds another large number of cells to this tube. As a result of uneven cell proliferation and physical forces, the heart tube loops rightward and begins forming ventricles and atria. Vasculogenesis involves the condensing of splanchnic mesoderm cells to form blood islands, the outer cells of which become endothelial (blood vessel) cells. Angiogenesis involves the remodeling of existing blood vessels. Hematopoiesis—the formation of blood cells—involves a stem cell population capable of generating numerous cell types.

(A)

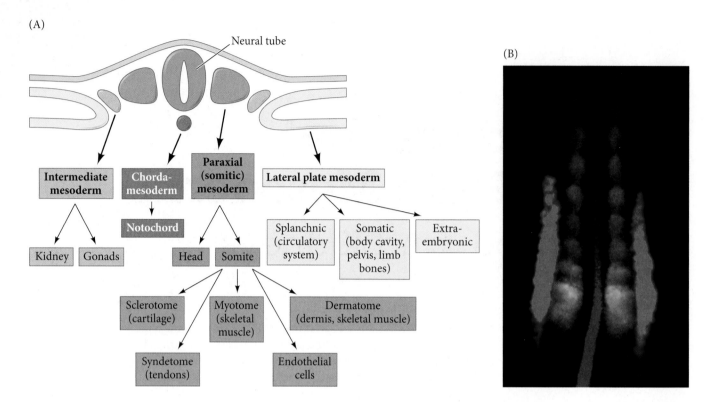

FIGURE 18.1 Major lineages of the amniote mesoderm. (A) Schematic of the mesodermal compartments of the amniote embryo. (B) Staining for the medial mesodermal compartments in the trunk of a 12-somite chick embryo (about 33 hours). In situ hybridization was performed with probes binding to *Chordin* mRNA (blue) in the notochord, *Paraxis* mRNA (green) in the somites, and *Pax2* mRNA (red) in the intermediate mesoderm. (B from Denkers et al. 2004, courtesy of T. J. Mauch.)

will become the lateral plate and extraembryonic mesoderm, whereas *Foxc1* and *Foxc2* are expressed in the paraxial mesoderm that will form the somites (Wilm et al. 2004). If *Foxc1* and *Foxc2* are both deleted from the mouse genome, the paraxial mesoderm is respecified as intermediate mesoderm and initiates expression of the *Pax2* gene, which encodes a major transcription factor of the intermediate mesoderm (**FIGURE 18.1**).

In this chapter, we will focus on those organs that make and circulate the blood. The blood cells are made by the lateral plate mesoderm, as is the heart and most of the blood vessels that circulate the blood. The kidney, from the intermediate mesoderm, filters wastes out of the blood, and it also has a major influence on blood pressure, composition, and volume.

Intermediate Mesoderm: The Kidney

The physiologist and philosopher of science Homer Smith noted in 1953 that "our kidneys constitute the major foundation of our philosophical freedom. Only because they work the way they do has it become possible for us to have bone, muscles, glands, and brains." While this statement may smack of hyperbole, the human kidney is a remarkably intricate organ whose importance cannot be overestimated. Its functional unit, the **nephron**, contains more than 10,000 cells and at least 12 different cell types, each cell type having a specific function and being located in a particular place in relation to the others along the length of the nephron.

Mammalian kidney development progresses through three major stages. The first two stages are transient; only the third and final stage persists as a functional kidney. Early in development (day 22 in humans; day 8 in mice), the **pronephric duct** arises in the intermediate mesoderm just ventral to the anterior somites. The cells of this duct migrate caudally, and the anterior region of the duct induces the adjacent mesenchyme to form the **pronephros**, or tubules of the initial kidney (**FIGURE 18.2A**). The pronephric tubules form functioning kidneys in fish and in amphibian larvae, but they are not believed to be active in amniotes. In mammals, the pronephric tubules and the anterior portion of the pronephric duct degenerate, but the more caudal portions of the

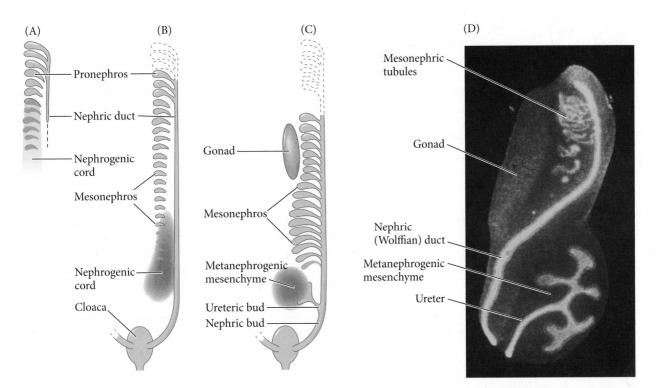

(A)

Pronephros

Nephric duct

Nephrogenic cord

Mesonephros

Nephrogenic cord

Cloaca

(B)

(C)

Gonad

Mesonephros

Metanephrogenic mesenchyme

Ureteric bud
Nephric bud

(D)

Mesonephric tubules

Gonad

Nephric (Wolffian) duct

Metanephrogenic mesenchyme

Ureter

FIGURE 18.2 General scheme of development in the vertebrate kidney. (A) The original tubules, constituting the pronephros, are induced from the nephrogenic mesenchyme by the pronephric duct as it migrates caudally. (B) As the pronephros degenerates, the mesonephric tubules form. (C) The final mammalian kidney, the metanephros, is induced by the ureteric bud, which branches from the nephric duct. (D) Intermediate mesoderm of a 13-day mouse embryo showing initiation of the metanephric kidney (bottom) while the mesonephros is still apparent. The duct tissue is stained with a fluorescent antibody to a cytokeratin found in the pronephric duct and its derivatives. (A–C after Saxén 1987; D courtesy of S. Vainio.)

pronephric duct and its derivatives persist and serve as the central component of the excretory system throughout development (Toivonen 1945; Saxén 1987). This remaining duct is often referred to as the **nephric**, or **Wolffian, duct**.

As the pronephric tubules degenerate, the middle portion of the nephric duct induces a new set of kidney tubules in the adjacent mesenchyme. This set of tubules constitutes the **mesonephros**, sometimes called the mesonephric kidney (**FIGURE 18.2B**; Sainio and Raatikainen-Ahokas 1999). In some mammalian species, the mesonephros functions briefly in urine filtration, but in mice and rats it does not function as a working kidney. In humans, about 30 mesonephric tubules form, beginning around day 25. As more tubules are induced caudally, the anterior mesonephric tubules begin to regress through apoptosis (although in mice, the anterior tubules remain while the posterior ones regress; **FIGURE 18.2C,D**). Although it remains unknown whether the human mesonephros actually filters blood and makes urine, it definitely provides important developmental functions during its brief existence. First, it is one of the main sources of the hematopoietic stem cells necessary for blood cell development (Medvinsky and Dzierzak 1996; Wintour et al. 1996). Second, in male mammals, some of the mesonephric tubules persist to become the tubes that transport the sperm from the testes to the urethra (the epididymis and vas deferens; see Chapter 6).

The permanent kidney of amniotes, the **metanephros**, originates through a complex set of interactions between epithelial and mesenchymal components of the intermediate mesoderm (reviewed in Costantini and Kopan 2010; McMahon 2016). In the first steps, the kidney-forming **metanephric mesenchyme** (also called the metanephrogenic mesenchyme) becomes committed in the posterior regions of the intermediate mesoderm, where it induces the formation of a branch from each of the paired nephric ducts. These epithelial branches are called the **ureteric buds**. These buds eventually grow out from the nephric duct to become the collecting ducts and ureters that take the urine to the bladder. When the ureteric buds emerge from the nephric duct, they enter the metanephric mesenchyme. The ureteric buds induce this mesenchymal tissue to condense around them and to differentiate into the nephrons of the mammalian kidney. As this mesenchyme begins to differentiate, it tells the ureteric bud to branch and grow. These reciprocal inductions form the kidneys.

Specification of the Intermediate Mesoderm: Pax2, Pax8, and Lim1

The intermediate mesoderm of the chick embryo acquires its ability to form kidneys through its interactions with the paraxial mesoderm. While its bias to become intermediate mesoderm is probably established through a BMP gradient, specification appears to become stabilized through signals from the paraxial mesoderm. Mauch and her colleagues (2000) showed that signals from the paraxial mesoderm induced primitive kidney formation in the intermediate mesoderm of the chick embryo. They cut developing embryos such that the intermediate mesoderm could not contact the paraxial mesoderm on one side of the body. That side of the body (where contact with the paraxial mesoderm was abolished) did not form kidneys, but the undisturbed side was able to form kidneys (**FIGURE 18.3A,B**). Thus, paraxial mesoderm appears to be both necessary and sufficient for inducing kidney-forming ability in the intermediate mesoderm. In support of this, paraxial mesoderm can even induce lateral plate mesoderm to generate pronephric tubules when co-cultured together. No other cell type can accomplish this.

These interactions induce the expression of a set of homeodomain transcription factors—including Lim1 (sometimes called Lhx1), Pax2, and Pax8—that cause the intermediate mesoderm to form the kidney (**FIGURE 18.3C**; Karavanov et al. 1998; Kobayashi et al. 2005; Cirio et al. 2011). In the chick embryo, Pax2 and Lim1 are expressed in the intermediate mesoderm, starting at the level of the sixth somite (i.e., only in the trunk, not in the head). If Pax2 is experimentally induced in the presomitic mesoderm, it converts that paraxial mesoderm into intermediate mesoderm, causing it to express Lim1 and form kidneys (Mauch et al. 2000; Suetsugu et al. 2005). Similarly, in mouse embryos with knockouts of both the *Pax2* and *Pax8* genes, the mesenchymal-epithelial transition necessary to form the nephric duct fails, the cells undergo apoptosis, and no kidney structures form (Bouchard et al. 2002). Moreover, in the mouse, Lim1 and Pax2 appear to induce one another.

Lim1 plays several roles in the formation of the mouse kidney. Early in development it is needed for converting the intermediate mesenchyme into the nephric duct (Tsang et al. 2000), and later it is required for the formation of the ureteric bud and the nephrons that form from mesonephric and metanephric mesenchyme (Shawlot and Behringer 1995; Karavanov et al. 1998; Kobayashi et al. 2005).

The anterior border of the Lim1- and Pax2-expressing cells appears to be established by the cells above a certain region losing their competence to respond to activin, a TGF-β family paracrine factor secreted by the neural tube. This competence to respond to activin is established by the transcription factor Hoxb4, which is not expressed in

FIGURE 18.3 Signals from the paraxial mesoderm induce pronephros formation in the intermediate mesoderm of the chick embryo. (A) The paraxial mesoderm was surgically separated from the intermediate mesoderm on the right side of the body. (B) As a result, a pronephric kidney (Pax2-staining duct) developed only on the left side. (C) Lim1 expression in an 8-day mouse embryo, showing the prospective intermediate mesoderm. (A,B after Mauch et al. 2000; B courtesy of T. J. Mauch and G. C. Schoenwolf; C courtesy of K. Sainio and M. Hytönen.)

(A)

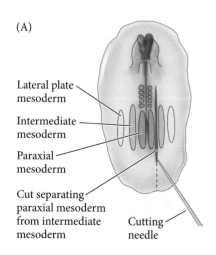

Lateral plate mesoderm

Intermediate mesoderm

Paraxial mesoderm

Cut separating paraxial mesoderm from intermediate mesoderm

Cutting needle

(B)

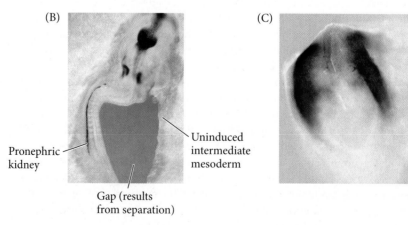

Pronephric kidney

Gap (results from separation)

Uninduced intermediate mesoderm

(C)

the anteriormost region of the intermediate mesoderm. The anterior boundary of *Hoxb4* expression is established by a retinoic acid gradient; adding activin locally will overcome this gradient and allow the kidney to extend anteriorly (Barak et al. 2005; Preger-Ben Noon et al. 2009).

Reciprocal Interactions of Developing Kidney Tissues

The kidney forms from two distinct progenitor cell populations derived from the intermediate mesenchyme—the ureteric bud and the metanephric mesenchyme. The ureteric bud gives rise to all of the cell types that comprise the mature collecting ducts and the ureter, while the metanephric mesenchyme gives rise to all of the cell types that comprise the mature nephron, as well as to some vascular and stromal derivatives. These two groups of cells—the ureteric bud and the metanephric mesenchyme—interact and reciprocally induce each other to form the kidney (**FIGURE 18.4**). The metanephric mesenchyme causes the ureteric bud to elongate and branch. The tips of these branches induce the loose mesenchyme cells to form pretubular renal aggregates. Each aggregated nodule proliferates and differentiates into the intricate structure of a renal nephron. Each pretubular aggregate first undergoes a mesenchymal-epithelial transition, becoming a polarized renal vesicle. Subsequently, this renal vesicle will elongate into a comma shape and then forms a characteristic S-shaped tube. Soon afterward, the cells of this epithelial structure begin to differentiate into regionally specific cell types, including the Bowman's capsule cells, the podocytes, and cells of the proximal and distal renal tubules. While this transformation is happening, the cells of the S-shaped tubule closest to the ureteric bud break down the basal lamina of the ureteric bud epithelium and migrate into the duct region. This creates an open connection between the ureteric bud and the newly formed nephron tubule, allowing material to pass from one into the other (Bard et al. 2001; Kao et al. 2012). These mesenchyme-derived tubules

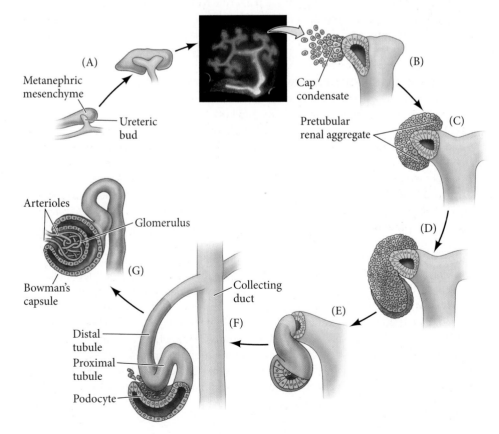

FIGURE 18.4 Reciprocal induction in the development of the mammalian kidney. (A) As the ureteric bud enters the metanephric mesenchyme, the mesenchyme induces the bud to branch. (B–G) At the tips of the branches, the epithelium induces the mesenchyme to aggregate and cavitate to form the renal tubules and glomeruli (where the blood from the arteriole is filtered). When the mesenchyme has condensed into an epithelium, it digests the basal lamina of the ureteric bud cells that induced it and connects to the ureteric bud epithelium. A portion of the aggregated mesenchyme (the pretubular condensate) becomes the nephron (renal tubules and Bowman's capsule), while the ureteric bud becomes the collecting duct for the urine. (After Saxén 1987; Sariola 2002.)

FIGURE 18.5 Kidney branching observed in vitro. (A) A kidney rudiment from an 11.5-day mouse embryo was placed into culture. This transgenic mouse had a GFP gene fused to a Hoxb7 promoter, so it expressed green fluorescent protein in the nephric (Wolffian) duct and in the ureteric buds. Since GFP can be photographed in living tissues, the kidney could be followed as it developed. (Srinivas et al. 1999, courtesy of F. Costantini.)

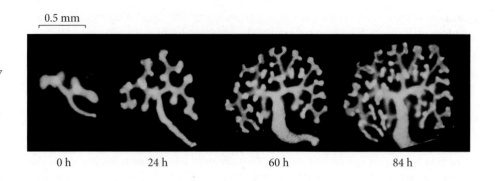

0.5 mm

0 h   24 h   60 h   84 h

form the mature nephrons of the functioning kidney, while the branched ureteric bud gives rise to the collecting ducts and to the ureter, which drains the urine from the kidney.

Clifford Grobstein (1955, 1956) documented this reciprocal induction in vitro. He separated the ureteric bud from the metanephric mesenchyme and cultured them either individually or together. In the absence of mesenchyme, the ureteric bud does not branch. In the absence of the ureteric bud, the mesenchyme soon dies. But when they are placed together, the ureteric bud grows and branches, and nephrons form throughout the mesenchyme. This has been confirmed by experiments using GFP-labeled proteins to monitor cell division and branching (**FIGURE 18.5**; Srinvas et al. 1999).

Mechanisms of reciprocal induction

The induction of the metanephros can be viewed as a dialogue between the ureteric bud and the metanephric mesenchyme. As the dialogue continues, both tissues are altered. We will eavesdrop on this dialogue more intently than we have done for other organs, in part because the kidney has become a model for organogenesis (Costantini 2012; Krause et al. 2015a). Many of the paracrine factors causing the mutual induction of the kidney nephron and its collecting ducts have been identified, and there is the possibility (Krause et al. 2015b) that these proteins are packaged as exosomes whose contents would be concentrated on the neighboring cells.

STEP 1: FORMATION OF METANEPHRIC MESENCHYME AND THE URETERIC BUD   The metanephric mesenchyme and the ureteric bud are more alike than they appear. Both come from the intermediate mesoderm, and both are generated through the actions of Wnt and FGF signaling pathways. The ureteric epithelium comes from early migrating intermediate mesoderm, which is exposed to Wnt signals for only a short time and then is exposed longer to the posterior signals Fgf9 and retinoic acid. The cells that become the metanephric mesenchyme migrate through the primitive streak later and are thus exposed to Wnt signals for a longer period of time. They then experience FGF and retinoic acid signals (**FIGURE 18.6A**; Takasato et al. 2015), which induce a set of transcription factors that enables the metanephric mesenchyme to respond to the ureteric bud. Only metanephric mesenchyme has the competence to respond to the ureteric bud and form kidney tubules (Saxén 1970; Sariola et al. 1982).

The ability of Wnt and FGF signals to generate these two progenitor cell populations has been shown using human induced pluripotential cells. When human iPS cells are cultivated sequentially in activators of the Wnt and FGF pathways, they become either ureteric epithelium or metanephric mesenchye, depending on their length of stay in each factor. Cells exposed briefly to Wnt signals became the epithelium, while cells exposed longer gave rise to the kidney-forming mesenchyme, just as in the embryo. Even more remarkably, when these cell types were cultured together, organoids resembling the kidney were generated (**FIGURE 18.6B**; Takasato and Little 2015). While these organoids did not have the intricate nephron structure, the major cell types of the nephron and collecting ducts were formed.

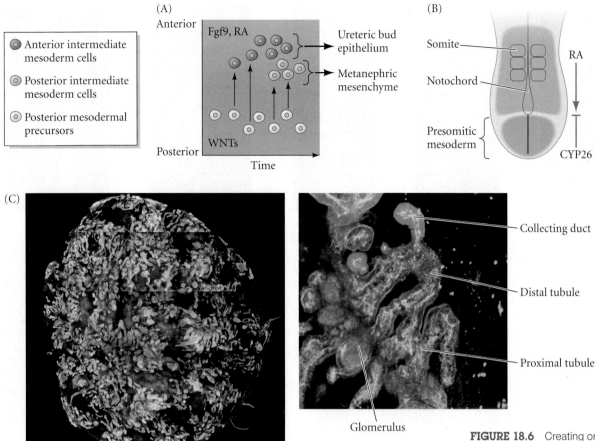

FIGURE 18.6 Creating organoids of mouse kidneys from induced pluripotent stem cells. (A) Schematic mechanism of generating the ureteric bud and the metanephric mesoderm from the posterior mesodermal precursor cells. Those precursor cells migrating from the posterior mesoderm early during gastrulation leave the area of Wnt and proceed toward areas of more FGFs and retinoic acid. These become the progenitors of the ureteric bud epithelium. Those migrating later, after remaining longer in the Wnt-dominated area, become the progenitors of the metanephric mesenchyme. (B) Retinoic acid signaling in the late the primitive streak stage. An RA-degrading enzyme, CYP26, is expressed in the presomitic mesoderm region and shields PMP cells from RA signaling. (C) Immunofluorescence microscope analysis of a kidney organoid formed from human iPS cells that were exposed sequentially to Wnt signal promoters and FGF signals. The insert is a higher-magnification view of a nephron segmented into four compartments, including the collecting duct (green), distal (yellow) and proximal (blue) tubules, and the glomerulus (red). (After Takasato et al. 2015.)

STEP 2: THE METANEPHRIC MESENCHYME SECRETES GDNF TO INDUCE AND DIRECT THE URETERIC BUD The stage is now set for the secretion of paracrine factors that can induce the ureteric buds to emerge. Under the influence of retinoic acid (which is made in many of the surrounding tissues), the nearby nephric duct is instructed to express the Ret receptor on its cell surfaces (Rosselot et al. 2010). Ret is the receptor for **glial-derived neurotrophic factor** (**GDNF**), and GDNF is now secreted from the metanephric mesenchyme. GDNF secreted from the metanephric mesenchyme causes the outgrowth of the ureteric bud from the nephric duct. Indeed, during formation of the ureter, a subset of nephric duct cells (those in which the Ret receptor is most highly active) migrate to positions closest to the source of GDNF and thus form the tip of the emerging urteric bud (**FIGURE 18.7A**; Chi et al. 2009). Mice whose genes for either GDNF or its receptor are knocked out die soon after birth from renal agenesis (lack of kidneys) (**FIGURE 18.7B–D**; Moore et al. 1996; Pichel et al. 1996; Sánchez et al. 1996). The ability of other regions of the nephric duct to proliferate appears to be suppressed by activin, and one of the the major mechanisms of GDNF action may be to locally suppress this inhibitory activin; when activin was experimentally inhibited, numerous ureteric buds formed (Maeshima et al. 2006).

STEP 3: THE URETERIC BUD SECRETES FGF2 AND BMP7 TO PREVENT MESENCHYMAL APOPTOSIS The third signal in kidney development is sent from the ureteric bud to the metanephric mesenchyme. If left uninduced by the ureteric bud, the mesenchyme cells undergo apoptosis (Grobstein 1955; Koseki et al. 1992). However, if induced by the ureteric bud, the mesenchyme cells are rescued from the precipice of death and are converted into proliferating stem cells (Bard and Ross 1991; Bard et al. 1996). The factors secreted

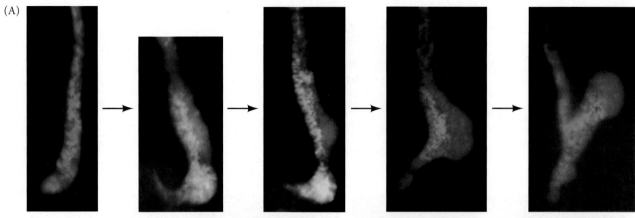

FIGURE 18.7 Ureteric bud growth is dependent on GDNF and its receptors. (A) When mice are constructed from Ret-deficient cells (green) and Ret-expressing cells (blue), the cells expressing Ret migrate to form the tips of the ureteric bud. (B) The ureteric bud from an 11.5-day wild-type mouse kidney cultured for 72 hours has a characteristic branching pattern. (C) In embryonic mice heterozygous for a mutation of the gene encoding GDNF, both the size of the kidney and the number and length of its ureteric bud branches are reduced. (D) In mouse embryos missing both copies of the Gdnf gene, the ureteric bud does not form. (A from Chi et al. 2009, courtesy of F. Costantini; B–D from Pichel et al. 1996, courtesy of J. G. Pichel and H. Sariola.)

from the ureteric bud include Fgf2, Fgf9, and BMP7. FGFs have three modes of action in that they (1) inhibit apoptosis, (2) promote the condensation of mesenchyme cells, and (3) maintain the synthesis of WT1, a transcription factor necessary for bud outgrowth (Perantoni et al. 1995). BMP7 has similar effects (Dudley et al. 1995; Luo et al. 1995).

STEP 4: SIGNALS FROM THE MESENCHYME INDUCE THE BRANCHING OF THE URETERIC BUD Paracrine factors including GDNF, Wnts, FGFs, and BMPs have been implicated in the branching of the ureteric bud, probably working as "pushes" and "pulls" on cell division and on the extracellular matrix (Ritvos et al. 1995; Miyazaki et al. 2000; Lin et al. 2001; Majumdar et al. 2003). The first protein regulating ureteric bud branching is GDNF from the mesenchyme, which not only induces the initial ureteric bud from the nephric duct but can also induce secondary buds from the ureteric bud once the bud enters the mesenchyme (**FIGURE 18.8**; Sainio et al. 1997; Shakya et al. 2005; Chi et al. 2008).

FIGURE 18.8 Effect of GDNF on branching of the ureteric epithelium. The ureteric bud and its branches are stained orange (with antibodies to cytokeratin 18), while the nephrons are stained green (with antibodies to nephron brush border antigens). (A) A 13-day embryonic mouse kidney cultured 2 days with a control bead (circle) has a normal branching pattern. (B) A similar kidney cultured 2 days with a GDNF-soaked bead shows a distorted pattern, as new branches are induced in the vicinity of the bead. (From Sainio et al. 1997, courtesy of K. Sainio.)

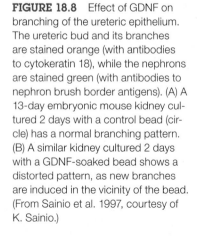

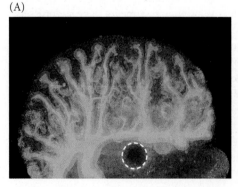

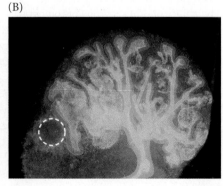

GDNF also appears to induce Wnt11 synthesis in the responsive cells at the tip of the bud (see Figure 18.9A), and Wnt11 reciprocates by regulating GDNF levels (Majumdar et al. 2003; Kuure et al. 2007). The cooperation between the GDNF/Ret pathway and the Wnt pathway appears to coordinate the balance between branching and metanephric mesenchyme proliferation such that continued kidney development is insured. In this way, two groups of stem cells are maintained: the **ureteric bud tip cells** and the **mesenchyme cap cells** (Mugford et al. 2009; Barak et al. 2012).

> **WATCH DEVELOPMENT 18.1** Videos show the progressive branching of the mouse kidney in vitro.

STEP 5: WNT SIGNALS CONVERT THE AGGREGATED MESENCHYME CELLS INTO A NEPHRON Some mesenchyme is not told to remain in a undifferentiated state. These mesenchyme cells become progenitor cells of the nephron, and they respond to Wnt9b and Wnt6 from the sides of the ureteric bud. Wnts 9b and 6 are critical for transforming the metanephric mesenchyme cells into tubular epithelium. The mesenchyme has receptors for these Wnts (Itäranta et al. 2002), which appear to induce Wnt4 in the mesenchyme. (**FIGURE 18.9**). Wnt4 acts in an autocrine fashion to complete the transition from mesenchymal mass to epithelium (Stark et al. 1994; Kispert et al. 1998). In mice lacking the *Wnt4* gene, the mesenchyme becomes condensed but does not form epithelia.

The epithelium hollows out to form the renal vesicle, which immediately becomes polarized in a proximal (near the ureteric bud) to distal direction. A combination of signaling factors (especially Notch proteins) are critical for differential gene expression along the length of the new epithelium. As the epithelium changes shape to form the C-and S-shaped tubules, the regions of the nephron become specified (Georgas et al. 2009). The mechanism by which the nephron connects to the ureteric bud remains elusive.

STEP 6: INSERTING THE URETER INTO THE BLADDER The branching epithelium becomes the collecting system of the kidney. This epithelium collects the filtered urine from the nephron and secretes antidiuretic hormone for the resorption of water (a process that, not so incidentally, makes life on land possible). The original stalk of the ureteric bud, situated above the first branch point, becomes the ureter, the tube that carries urine into the bladder. The junction between the ureter and bladder is extremely important, and hydronephrosis, a birth defect leading to abnormalities of renal filtration, occurs when this junction is not properly placed and urine cannot enter the bladder. The ureter is made into a watertight connecting duct by the condensation of mesenchymal cells around it (but not around the collecting ducts). These mesenchymal

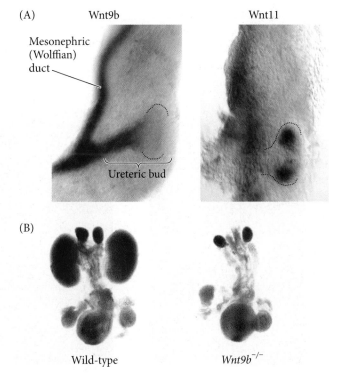

FIGURE 18.9 Wnts are critical for kidney development. (A) In the 11-day mouse kidney, Wnt9b is found on the stalk of the ureteric bud, while Wnt11 is found at the tips. Wnt9b induces the metanephric mesenchyme to condense; Wnt11 will partition the metanephric mesoderm to induce branching of the ureteric bud. Borders of the bud are indicated by a dashed line. (B) A wild-type 18.5-day male mouse (left) has normal kidneys, adrenal glands, and ureters. In a mouse deficient for Wnt9b (right), the kidneys are absent. (From Carroll et al. 2005.)

FIGURE 18.10 Development of the bladder and its connection to the kidney via the ureter. (A) The cloaca originates as an endodermal collecting area that opens into the allantois. (B) The urogenital septum divides the cloaca into the future rectum and the urogenital sinus. The bladder forms from the anterior portion of that sinus, and the urethra develops from the posterior region of the sinus. The space between the rectal opening and the urinary opening is the perineum. (C–F) Insertion of the ureter into the embryonic mouse bladder. (C) Day-10 mouse urogenital tract. The nephric duct is stained with GFP fused to a Hoxb7 promoyter. (D) Urogenital tract from a day-11 embryo, after ureteric bud outgrowth. (E) Whole mount urogenital tract from a day-12 embryo. The ducts are stained green and the urogenital sinus red. (F) The ureter separates from the nephric duct and forms a separate opening into the bladder. (A,B after Cochard 2002; C–F from Batourina et al. 2002, courtesy of C. Mendelsohn.)

cells become smooth muscle cells capable of wavelike contractions (peristalsis) that allow the urine to move into the bladder. These cells also secrete BMP4 (Cebrian et al. 2004), which upregulates genes for uroplakin, a cell membrane protein that causes differentiation of this region of the ureteric bud into the ureter. BMP inhibitors protect the region of the ureteric bud that forms the collecting ducts from this differentiation.

The bladder develops from a portion of the cloaca (**FIGURE 18.10A,B**). The **cloaca**[1] is an endodermally lined chamber at the caudal end of the embryo that will become the waste receptacle for both the intestine and the kidney. Adult amphibians, reptiles, and birds use the cloaca to void both liquid and solid wastes. In mammals, the cloaca becomes divided by a septum into the urogenital sinus and the rectum. Part of the urogenital sinus becomes the bladder, while another part becomes the urethra (which will carry the urine out of the body). The ureteric bud originally empties into the bladder via the nephric (Wolffian) duct, which grows toward the bladder through an ephrin-mediated pathway (Weiss et al. 2014). Once at the bladder, the urogenital sinus cells of the bladder wrap themselves around both the ureter and the nephric duct. Then the nephric ducts migrate ventrally, opening into the

---

[1]The term *cloaca* is Latin for "sewer"—a bad joke on the part of early European anatomists.

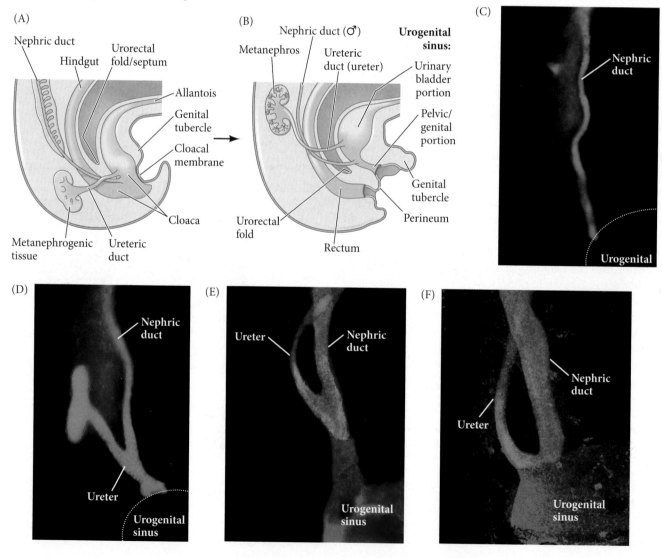

urethra rather than into the bladder and the nephric duct (**FIGURE 18.10C–F**). The caudal end of the nephric duct appears to undergo apoptosis, allowing the ureter to separate from the nephric duct. Expansion of the bladder then moves the ureter to its final position at the neck of the bladder (Batourina et al. 2002; Mendelsohn 2009). In females, the entire nephric duct degenerates, while the Müllerian duct opens into the vagina (see Chapter 6). In males, the nephric duct also forms the sperm outflow track, so males expel sperm and urine through the same opening.

Thus, the blood-filtering kidneys emerge from the mutual induction of two parts of the intermediate mesoderm, the ureteric bud and the metanephric mesenchyme. Now we can focus more laterally, on the lateral plate mesoderm, and discern the genesis of the heart, vessels, and blood.

Lateral Plate Mesoderm: Heart and Circulatory System

In 1651, amid the chaos of the English civil wars, William Harvey, physician to the king, was comforted by viewing the heart as the undisputed ruler of the body, through whose divinely ordained powers the lawful growth of the organism was assured. Later embryologists saw the heart as more of a servant than a ruler, the chamberlain of the household who assured that nutrients reached the apically located brain and the peripherally located muscles. In either metaphor, the heart and its circulation (which Harvey discovered) were seen to be critical for development. As Harvey argued persuasively in 1651, the chick embryo must form its own blood without any help from the hen, and this blood is crucial in embryonic growth. How this happened was a mystery to him.

The heart and circulatory system that were so intriguing to Harvey arise from the vertebrate embryo's lateral plate mesoderm. The lateral plate mesoderm resides on the lateral side of each of the two bands of intermediate mesoderm (see Figure 18.1). Each of these lateral plates splits horizontally into two layers. The dorsal layer is the **somatic (parietal) mesoderm**, which underlies the ectoderm and, together with the ectoderm, forms the **somatopleure**. The ventral layer is the **splanchnic (visceral) mesoderm**, which overlies the endoderm and, together with the endoderm, forms the **splanchnopleure** (**FIGURE 18.11A**). The space between these two layers becomes the body cavity—the **coelom**—which stretches from the future neck region to the posterior of the body.

WEB TOPIC 18.1   **COELOM FORMATION**   An animation illustrates the formation of the coelom and the expansion of the lateral plate mesoderm.

During later development, the right- and left-side coeloms fuse and folds of tissue extend from the somatic mesoderm, dividing the coelom into separate cavities. In mammals, these mesodermal folds subdivide the coelom into the **pleural**, **pericardial**, and **peritoneal cavities**, enveloping the thorax, heart, and abdomen, respectively. The mechanism for creating the linings of these body cavities from the lateral plate mesoderm has changed little throughout vertebrate evolution, and the development of the amniote mesoderm can be compared with similar stages of frog embryos (**FIGURE 18.11B,C**).

Consisting of a heart, blood cells, and an intricate system of blood vessels, the circulatory system is the vertebrate embryo's first functional unit, providing nourishment to the developing organism. Few events in biology are as thought-provoking and accessible as watching the heart beating in a two-day chick embryo, pumping the first blood cells into vessels that have not even formed valves yet. The development of the circulatory system provides excellent examples of induction, specification, cell migration, organ formation, and the role of stem cells in both embryonic development and adult tissue regeneration.

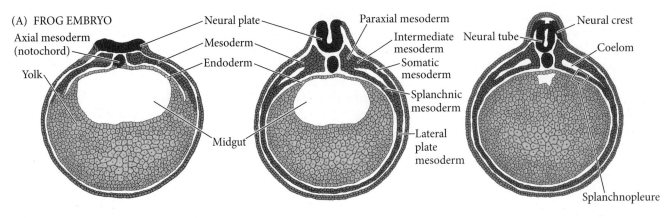

(A) FROG EMBRYO

Neural plate

Axial mesoderm (notochord)

Mesoderm

Yolk

Endoderm

Midgut

Paraxial mesoderm

Intermediate mesoderm

Somatic mesoderm

Splanchnic mesoderm

Lateral plate mesoderm

Neural crest

Neural tube

Coelom

Splanchnopleure

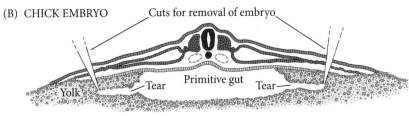

(B) CHICK EMBRYO   Cuts for removal of embryo

Yolk   Tear   Primitive gut   Tear

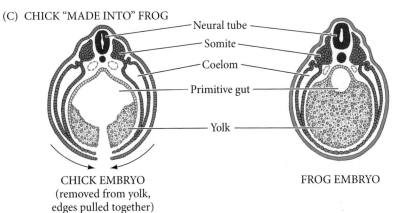

(C) CHICK "MADE INTO" FROG

Neural tube

Somite

Coelom

Primitive gut

Yolk

CHICK EMBRYO (removed from yolk, edges pulled together)

FROG EMBRYO

FIGURE 18.11 Mesodermal development in frog and chick embryos. (A) Neurula-stage frog embryos, showing progressive development of the mesoderm and coelom. (B) Transverse section of a chick embryo. (C) When the chick embryo is separated from its enormous yolk mass, it resembles the amphibian neurula at a similar stage. (A after Rugh 1951; B,C after Patten 1951.)

Heart Development

The circulatory system is the first working unit in the developing embryo, and the heart is the first functional organ. Like other organs, the heart arises through the specification of precursor cells, the migration of these precursor cells to the organ-forming region, the specification of cell types through signaling interactions within and between tissues, and the coordination of morphogenesis, growth, and cell differentiation.

A minimalist heart

Both the chick and the mammalian heart are complex, rather baroque structures. However, the heart evolved from far simpler pumps (Stolfi et al. 2010). The four-chambered masterpiece that is the mammalian heart is a developmental elaboration of the single-chambered tunicate heart that forms from about two dozen cells. In tunicates (the closest invertebrate relative of the vertebrates; see Chapter 10), cardiac precursor cells form bilateral cell clusters that migrate anteriorly and ventrally along the endoderm and fuse at the ventral midline (Davidson et al. 2005). The few cells that form the tunicate heart appear to have the same basic pattern of transcription factors that we see in the chick and mouse heart lineages.

In the gastrulating tunicate embryo, only two pairs of mesodermal cells near the vegetal pole represent the heart lineage. Each side of the embryo contains a pair of B8.9 and B8.10 blastomeres that express the MesP transcription factor, just like the cardiac

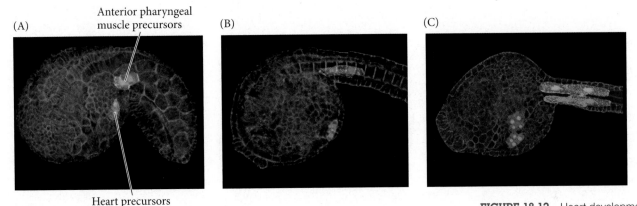

(A) Anterior pharyngeal muscle precursors

Heart precursors

(B)

(C)

FIGURE 18.12   Heart development in the tunicate *Ciona*. (A) In the tailbud-stage embryo, transgenic MesP-GFP glows in regions where MesP is activated by Tbx6 in the B8.9 and B8.10 blastomeres. (B) At a slightly later stage, the heart precursors migrate into the head region. (C) Ventrolateral view in which both left and right heart and muscle precursors can be observed. Cell divisions are forming both the heart (left) and the anterior muscles. (From Davidson et al. 2005.)

precursor cells of vertebrates (**FIGURE 18.12**; Davidson et al. 2006). During neurulation, each of these four cardiac founder cells divides asymmetrically to produce a small cell that generates the cardiac precursors and a larger cell that generates anterior pharyngeal muscle precursors. The anterior tail cells do not migrate, but they express retinaldehyde dehydrogenase and initiate a retinoic acid gradient that specifies the heart cells as in vertebrate embryos.

Moreover, as in the vertebrate cardiac precursors, it appears that FGF signaling is critical for the production of heart cells, with FGF signals combining with MesP to induce the expression of the Nkx2-5, GATA, and Hand-family transcription factors (Davidson and Levine 2003; Simões-Costa et al. 2005). When the cardiopharyngeal precursor cell divides, the cell that remains bound to the extracellular matrix of the epidermis retains the FGF receptors, while the cell that does not adhere to the epidermis internalizes and degrades such receptors. As a result, the cell attached to the epidermis can respond to FGF and become the heart precursor cell, while the cell without FGF receptors cannot respond to the paracrine factor and produces pharyngeal muscle instead (Cota and Davidson 2015).

Formation of the heart fields

While tunicate embryos develop rapidly and from a small number of cells, the vertebrate heart arises from two regions of splanchnic mesoderm—one on each side of the body—that interact with adjacent tissue to become specified for heart development.

In the early amniote gastrula, the heart progenitor cells (about 50 of them in mice) are located in two small patches, one on each side of the epiblast, close to the rostral portion of the primitive streak. These cells migrate together through the streak and form two groups of lateral plate mesoderm cells, positioned anteriorly at the level of the node (Tam et al. 1997; Colas et al. 2000). The general specification of a **heart field**, also known as **cardiogenic mesoderm**, has already started during this cellular migration. Labeling experiments by Stalberg and DeHann (1969) and Abu-Issa and Kirby (2008) have shown that the progenitor cells of the heart field migrate such that the medial-lateral (center to side) arrangement of these early cells will become the anterior-posterior (rostral-caudal) axis of a linear **heart tube**.

The vertebrate **heart field** is divided into at least two regions (**FIGURE 18.13**). The first heart field appears to form the scaffold of the developing heart. The progenitor cells of the first field fuse at the midline to form the primary heart tube that gives rise to the muscular regions of the left and right ventricles (de la Cruz and Sanchez-Gomez 1998). However, these cells have limited proliferative ability and therefore will generate only the major portion of the left ventricle of the adult heart (i.e., the chamber that pumps blood into the aorta). The progenitors of the second heart field add cells to both anterior and posterior ends of the heart tube (Meilhac et al. 2015). On posterior end, these cells will produce the two atria and contribute to the inlet part of the heart. On the anterior end, the second heart field will generate the right ventricle as well as the

(A)

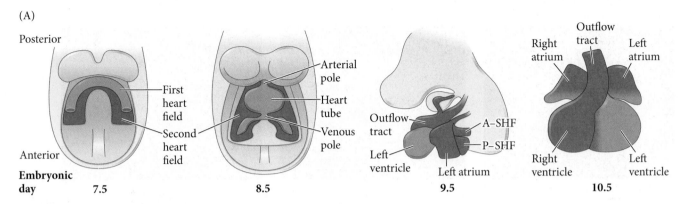

Posterior

First heart field

Second heart field

Anterior

Embryonic day   7.5

Arterial pole

Heart tube

Venous pole

8.5

Outflow tract

Left ventricle

A–SHF

P–SHF

Left atrium

9.5

Outflow tract

Right atrium

Left atrium

Right ventricle

Left ventricle

10.5

FIGURE 18.13 The heart fields in the mouse embryo. (A) On embryonic day 7.5, the heart fields from each side of the body have joined into a common cardiac crescent that contains the first and second heart fields. The first heart field contributes primarily to the left ventricle. By day 10.5, the second heart field contributes to the other three chambers—the right ventricle and the left and right atria—as well as to the outflow tract that originally includes the aorta and the pulmonary artery. (B) A possible lineage tree showing the cooperation of first and second heart fields in forming the heart and also showing the mixture of heart, lung, and pulmonary blood vessel cells existing in the second heart field. The dotted line indicates that the exact location of the separation of lung, pulmonary, facial muscles, and heart cell progenitors is not known. Some of the transcription factors associated with these progenitor cells are listed beneath them. (A after Kelly 2012; B after Diogo et al. 2015 and Peng et al. 2013.)

(B)

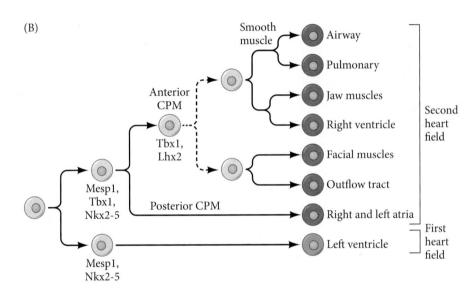

Smooth muscle → Airway, Pulmonary

Anterior CPM
Tbx1, Lhx2

Jaw muscles

Right ventricle

Facial muscles

Outflow tract

Mesp1, Tbx1, Nkx2-5

Posterior CPM

Right and left atria

Mesp1, Nkx2-5

Left ventricle

Second heart field

First heart field

outflow region (the conus arteriosus and the truncus arteriosus), which becomes the base of the aorta and pulmonary arteries (de la Cruz et al. 1989; Kelly 2012). It is only through the process of looping that the atria are brought anterior to the ventricles to form the adult four-chambered heart.

The second heart field is a remarkable group of cells, because it not only contains the progenitor cells for the heart, it also contains the cells that will generate the facial muscles, the pulmonary artery and vein, and the lung mesenchyme (Lescroart et al. 2010, 2015; Peng et al. 2013). Therefore, in a remarkable way, the heart precursor cells are coordinated with the development of the face and lungs. The common precursor of pharyngeal and cardiac mesoderm is thought to be derived from a similar group of pharynx and heart progenitor cells that are found in certain deuterostome invertebrates such as tunicates (Diogo et al. 2015).

All the cell types of the heart—the **cardiomyocytes** that form the muscular layers, the **endocardium** that forms the internal layer, the **endocardial cushions** of the valves, the **epicardium** that forms the coronary blood vessels that feed the heart, and the **Purkinje fibers**[2] that coordinate the heartbeat—are generated from these heart fields[3]

[2] Note that these specialized myocardial nerve fibers are not the same thing as the Purkinje *neurons* of the cerebellum mentioned in Chapter 13. Both were named for the nineteenth-century Czech anatomist and histologist Jan Purkinje.

[3] A third heart field, extending more posteriorly, may exist (Bressan et al. 2013). In chick embryos, this third heart field includes those cells that generate the pacemaker myocytes that stimulate the rhythmic contractions of the heart muscles.

(Mikawa 1999; van Wijk et al. 2009). Moreover, as we will discuss later, it appears that each progenitor cell is capable of becoming any of the differentiated heart cell types. The cardiac precursor cells will be supplemented by cardiac neural crest cells; the latter help make the outflow tract and the septum that separates the aorta from the pulmonary trunk (see Figure 15.22; Porras and Brown 2008).

Specification of the cardiogenic mesoderm

The cardiogenic mesoderm cells are specified by their interactions with the pharyngeal endoderm and notochord. The heart does not form if this anterior endoderm is removed, and posterior endoderm cannot induce heart cells to form (Nascone and Mercola 1995; Schultheiss et al. 1995). BMPs (especially BMP2) from the anterior endoderm promote both heart and blood development. Endodermal BMPs also induce Fgf8 synthesis in the endoderm directly beneath the cardiogenic mesoderm, and Fgf8 appears to be critical for the expression of cardiac proteins (Alsan and Schultheiss 2002).

Inhibitory signals prevent heart structures from forming where they should not be made. First, the notochord secretes Noggin and Chordin, blocking BMP signaling in the center of the embryo, and specific Noggin-secreting cells in the myotome prevent heart cell specification of the somites. Second, Wnt proteins from the neural tube, especially Wnt3a and Wnt8, inhibit heart formation but promote blood formation. The anterior endoderm, moreover, produces Wnt inhibitors such as Cerberus, Dickkopf, and Crescent, which prevent Wnts from binding to their receptors. In this way, cardiac precursor cells are specified in those places where BMPs (from the lateral mesoderm and endoderm) and Wnt antagonists (from the anterior endoderm) coincide (**FIGURE 18.14A**; Marvin et al. 2001; Schneider and Mercola 2001; Tzahor and Lassar 2001; Gerhart et al. 2011).

In the absence of Wnt signals, BMPs activate **Nkx2-5** and **Mesp1**, two genes that are critical in the regulatory network that specifies the heart cells (**FIGURE 18.14B**). The *nkx2-5* gene has functions in heart development that are conserved across species (Komuro and Izumo 1993; Lints et al. 1993; Sugi and Lough 1994; Schultheiss et al. 1995; Andrée et al. 1998). In *Drosophila*, the *nkx2-5* homologue is called *tinman*, as loss-of-function mutants lack a heart. Nkx2-5 can also downregulate BMPs, and in early heart cell development it limits the number of heart cell precursors that can form the heart fields. If the *nkx2-5* gene is specifically knocked out in those cells destined to become ventricles, these chambers express BMP10, resulting in massive overgrowth of the ventricles such that the ventricular chambers fill with muscle cells (Pashmforoush et al. 2004; Prall et al. 2007).

The other gene activated by BMPs is *Mesp1*.[4] Mesp1 and Nkx2-5 proteins cooperate to activate the genes that specify the heart. Mesp1 also acts to prevent heart progenitors from being respecified as some other type of mesoderm. First, it activates the *dickkopf* gene in the heart progenitors (David et al. 2008), thereby preventing Wnts from transforming these cells into vascular cells. Second, Mesp1 represses *brachyury*, *sox17*, and *goosecoid* genes so the cardiac precursor cells will not become endoderm, somite, or notochord (Bondue and Blanpain 2010). Mesp1 also promotes the expression of those genes whose products allow cell migration and, once the cardiogenic precursor cells are committed to become a heart, the cells migrate to the midline to form the heart tube (Lazic and Scott 2011).

---

[4] *Mesp1* is a close relative of *Mesp2*, which directs somitogenesis (see Chapter 17). The tunicate (see Chapter 10) has only one *Mesp* gene, and it specifies heart development through the activation of the *Nkx* and *Hand* genes, just as occurs in vertebrates (Satou et al. 2004). BMPs may activate *Mesp1* indirectly by inducing the expression of the Eomesodermin, a transcription factor that is important for both endoderm and mesodermal lineages. In the early mouse epiblast (which has low amounts of Nodal), Eomesodermin activates *Mesp1*. Later, as the primitive streak elongates, Eomesodermin acts with Nodal to activate genes for the Sox17 and Foxa2 transcription factors that specify the definitive endoderm (Costello et al. 2011).

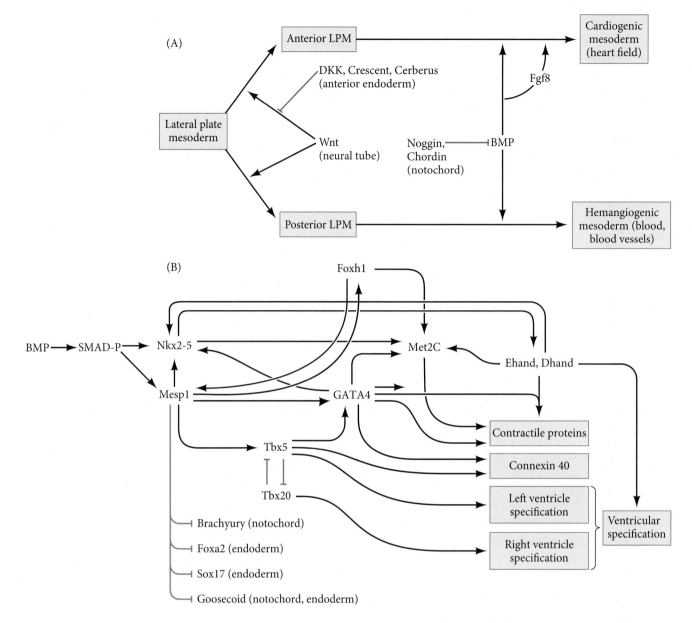

FIGURE 18.14 Model of inductive interactions involving the BMP and Wnt pathways that form the boundaries of the cardiogenic mesoderm. (A) Wnt signals from the neural tube instruct lateral plate mesoderm to become precursors of the blood and blood vessels. In the anterior portion of the body, Wnt inhibitors (Dickkopf, Crescent, Cerberus) from the pharyngeal endoderm prevent Wnt from functioning, allowing later signals (BMP, Fgf8) to convert lateral plate mesoderm into cardiogenic mesoderm. BMP signals will also be important for the differentiation of hemangiogenic (blood, blood vessel) mesoderm. In the center of the embryo, Noggin and Chordin signals from the notochord block BMPs. Thus, the cardiac and blood-forming fields do not form in the center of the embryo. (B) Model gene regulatory network for the vertebrate heart initiated by BMP signals. BMP signaling activates the pivotal switches Nkx2-5 and Mesp1. These transcription factors act in concert to activate numerous heart-forming genes. Mesp1 has also been shown to repress genes that would otherwise specify the cell into other fates. The antagonism between Tbx20 (right side) and Tbx5 (left side) can also be seen. This model is provisional, as new ChIP-Seq techniques have identified thousands of promoters activated at different stages of heart development. (A after Davidson 2006; B after May et al. 2012.)

Migration of the cardiac precursor cells

As the presumptive heart cells move anteriorly between the ectoderm and endoderm toward the middle of the embryo, they remain in close contact with the endoderm surface (Linask and Lash 1986). In the chick, the directionality of this migration appears to be provided by the foregut endoderm. If the cardiac region endoderm is rotated with respect to the rest of the embryo, migration of the cardiogenic mesoderm cells is reversed. It is thought that the endodermal component responsible for this movement is an anterior-to-posterior concentration gradient of fibronectin. Antibodies against fibronectin stop the migration, whereas antibodies against other extracellular matrix components do not (Linask and Lash 1988).

This movement produces two populations of migrating cardiac precursor cells, one on the right side of the embryo and another on the left. Each side has its own first and second heart fields, and each of these populations starts to form its own heart tube. In the chick, the fields are brought together around the 7-somite stage, when the foregut is formed by the inward folding of the splanchnopleure. This movement places the two cardiac tubes together (Varner and Taber 2012). The two endocardial tubes lie within the common tube for a short time, but eventually these two tubes also fuse. The bilateral origin of the heart can be demonstrated by surgically preventing the merger of the lateral plate mesoderm (Gräper 1907; DeHaan 1959). This manipulation results in a condition called **cardia bifida**, in which two separate hearts form, one on each side of the body (**FIGURE 18.15A**). Thus, endoderm specifies heart progenitors, gives directionality to their migration, and mechanically pulls the two heart fields together.

Although the chick is an excellent model for surgical manipulation, mouse and zebrafish embryos have been more tractable genetically. In the zebrafish, heart precursor cells migrate actively from the lateral edges toward the midline. Several mutations affecting endoderm differentiation disrupt this process, indicating that, as in the chick, the endoderm is critical for cardiac precursor specification and migration. The *faust* gene, which encodes the GATA5 protein, is expressed in the endoderm and is required for the migration of cardiac precursor cells to the midline and also for their division and specification. It appears to be important in the pathway leading to activation of the zebrafish *nkx2-5* gene in the cardiac precursor cells (Reiter et al. 1999). Another particularly interesting zebrafish mutation is *miles apart*. Its phenotype is limited to cardiac precursor migration to the midline and resembles the cardia bifida seen in experimentally manipulated chick embryos (**FIGURE 18.15B,C**). Differentiation is not affected; the fish form two normal heart tubes, but the tubes are not connected properly to the blood vessels, and thus cannot support circulation. The *miles apart* gene

(A)
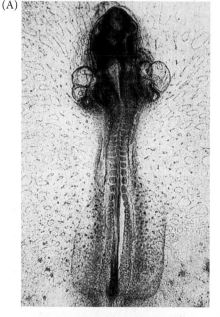

(B)
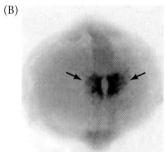

(C)
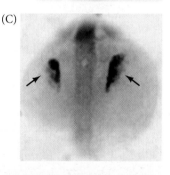

(D)
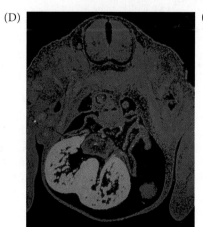

(E)
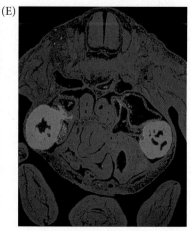

FIGURE 18.15 Migration of heart primordia. (A) Cardia bifida (two hearts) in a chick embryo, induced by surgically cutting the ventral midline, thereby preventing the two heart primordia from fusing. (B) Wild-type zebrafish and (C) miles apart mutant, stained with probes for the cardiac myosin light chain. There is a lack of migration in the miles apart mutant. (D) Mouse heart stained with antisense RNA probe to ventricular myosin shows fusion of the heart primordia in a wild-type 13.5-day embryo. (E) Cardia bifida in a Foxp4-deficient mouse embryo. Interestingly, each of these hearts has ventricles and atria, and they both loop and form all four chambers with normal left-right asymmetry. (A courtesy of R. L. DeHaan; B,C from Kupperman et al. 2000, courtesy of Y. R. Didier; D,E from Li et al. 2004, courtesy of E. E. Morrisey.)

encodes a protein that regulates the interactions of the cardiac cells with fibronectin, and it is expressed in the endoderm on either side of the midline (Kupperman et al. 2000; Matsui et al. 2007).

In mice, cardia bifida can also be produced by mutations of genes that are expressed in the endoderm. One of these genes, *Foxp4*, encodes a transcription factor expressed in the early foregut cells, along the pathway the cardiogenic precursors travel toward the midline. In these mutants, each heart primordia develops separately, and the embryonic mouse contains two hearts, one on each side of the body (**FIGURE 18.15D,E**; Li et al. 2004).

However, as the cells of the first heart field migrate along the endoderm to form the heart tube, the cells of the second heart field remain in contact with the pharyngeal endoderm. Here they are kept in a state of proliferation by a combination of paracrine factors (probably Sonic hedgehog, Fgf8, and Wnts) (Chen et al. 2007; Lin et al. 2007). The cells of the second heart field can be distinguished by their expression of the Islet1 transcription factor. These cells also begin to synthesize and secrete Fgf8, which acts in an autocrine manner to stimulate the cells to migrate and add themselves to the anterior and posterior portions of the heart tube formed by the first heart field progenitor cells (Park et al. 2008). The anterior region of the second heart field contributes to the right ventricle and the outflow tract, whereas its posterior region generates the atria (Zaffran et al. 2004; Verzi et al. 2005; Galli et al. 2008).

As the second heart field precursor cells migrate, the posterior region becomes exposed to increasingly higher concentrations of retinoic acid (RA) produced by the posterior mesoderm. RA is critical in specifying these posterior precursor cells to become the inflow, or "venous," portions of the heart—the sinus venosus and atria. Originally, these fates are not fixed, as transplantation or rotation experiments show that these precursor cells can regulate and differentiate in accordance with a new environment. But once the posterior cardiac precursors enter the realm of active RA synthesis, they express the gene for retinaldehyde dehydrogenase; they then can produce their own RA, and their posterior fate becomes committed (**FIGURE 18.16**; Simões-Costa et al. 2005).

As in kidney development, retinoic acid regulates the expression of Hox genes (especially *Hoxa1*, *Hoxb1*, and *Hoxa3*), which appear to promote different regional identities in the second heart field precursors (Bertrand et al. 2011). In mice, the outflow tract region, as well as the cardiac neural crest cells that enter this region of the second heart field, display differential Hox gene expression based on their exposure to RA (Diman et al. 2011). This ability of RA to specify and commit heart precursor to become atria explains its teratogenic effects on heart development, wherein exposure of vertebrate embryos to RA can cause expansion of atrial tissues at the expense of ventricular tissues (Stainier and Fishman 1992; Hochgreb et al. 2003).

VADE MECUM

The vertebrate heart begins to function early in development, as is seen in movies of the early-stage chick embryo.

(A) Chick, stage 8

(B) Mouse, 8 days

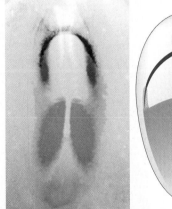

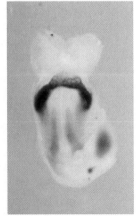

Ventricular
(outflow)

Atrial
(inflow)

FIGURE 18.16 Double in situ hybridization for the expression of RADH2 (orange), which encodes the retinoic acid-synthesizing enzyme retinaldehyde dehydrogenase-2; and Tbx5 (purple), a marker for the early heart fields. In the developmental stages seen here, the heart precursor cells are exposed to progressively increasing amounts of retinoic acid. (A) Chick, stage 8 (26–29 hours). (B) Mouse, 8 days. (From Simões-Costa et al. 2005, courtesy of J. Xavier-Neto.)

WEB TOPIC 18.2 **FUSION OF THE HEART AND THE FIRST HEARTBEATS** Pulsations of the chick heart begin while the paired primordia are still fusing. Isolated embryonic heart cells will beat when placed in petri dishes.

Initial heart cell differentiation

One of the most important discoveries of cardiac development was the demonstration that the different cells of the heart—the ventricular myocytes, the atrial myocytes, the smooth muscles that generate the venous and arterial vasculature, the endothelial lining of the heart and valves, and the epicardium that forms an envelope for the heart—all are derived from the same progenitor cell type (Kattman et al. 2006; Moretti et al. 2006; Wu et al. 2006). The heart fields contain multipotent progenitor cells. Indeed, there appears to be an early progenitor cell population that bears responsibility for forming the entire circulatory system. Under one set of influences, its descendants become **hemangioblasts**, those cells that form blood vessels and blood cells; under the conditions in the heart fields, its descendants form **multipotent cardiac precursor cells** (**FIGURE 18.17**; Anton et al. 2007). Several investigators have proposed slightly different pathways for generating these cells, but the differences may be caused by the ability of the heart precursor cells to differentiate according to their microenvironment (Linask 2003).

Several proteins are expressed very early during heart development (see Figure 18.14B). Nkx2-5 and Mesp1 are also critical in initiating a self-sustaining gene regulatory network. One of the genes active in this network encodes the GATA4 transcription factor, which is first seen in the precardiac cells of chicks and mice when these cells emerge from the primitive streak. GATA4 is necessary for activating numerous heart-specific genes as well as for activating expression of the gene for N-cadherin, a protein that is critical for both the formation of the cardiac epithelium and the fusion of the two heart rudiments into one tube (Linask 1992; Zhang et al. 2003).

In addition to activating a group of core heart-forming genes, Mesp1 also helps activate different patterns of protein synthesis in the heart fields on each side of the embryo. Mesp1 and Nkx2-5 instruct the cells of the second heart field to express the *Foxh1* gene, which commits these heart precursor cells to become the right ventricle and outflow tract (von Both et al. 2004). In the first heart field, Mesp1 activates the *Tbx5* gene, whose product is critical for heart tube and left ventricle development (see Figure 18.16; Koshiba-Takeuchi et al. 2009). In these early cells, Tbx5 acts with GATA4 and Nkx2-5 to activate numerous genes involved in heart specification. Later, Tbx5 becomes restricted to the atria and left ventricle. The ventricular septum (the wall separating the left and right

FIGURE 18.17 Model for early cardiovascular lineages. The splanchnic mesoderm gives rise to two lineages, both of which have Flk1 (a VEGF receptor) on their cell membranes. The earlier population gives rise to the hemangioblasts (precursors to blood cells and blood vessels), whereas the later population gives rise to the cardiac (heart) precursor cells. This latter population in turn gives rise to a variety of cell types whose relationships are still obscure; however, all the cell types of the heart can be traced back to the cardiac precursor cells. (After Anton 2007 and DeLaughter et al. 2011.)

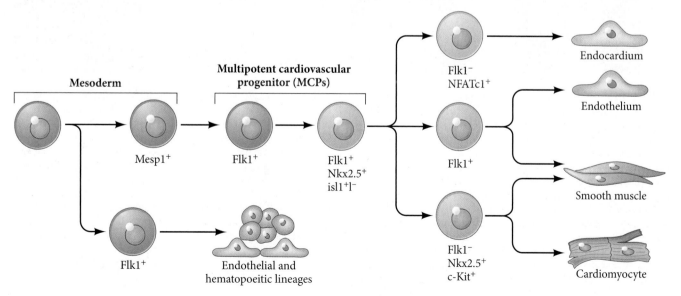

ventricles) is formed at the boundary between those cells that express *Tbx5* and those that do not. Tbx5 protein works antagonistically to Tbx20, which becomes expressed in the right ventricle. When the *Tbx5* expression domain is ectopically expanded, the location of the ventricular septum shifts to this new location. Moreover, a conditional knockout of the mouse *Tbx5* gene—specifically inactivating it during ventricular development—leads to the formation of a lizardlike ventricle that lacks any septum (Takeuchi et al. 2003; Koshiba-Takeuchi et al. 2009). Thus, Tbx5 is extremely important in separating the left and right ventricles. Mutations in the human *TBX5* gene cause Holt-Oram syndrome (Bruneau et al. 1999), characterized by abnormalities of the heart and upper limbs.

In 3-day chick embryos and 4-week human embryos, the heart is a two-chambered tube, with an atrium to receive blood and a ventricle to pump blood out. (In the chick embryo, the unaided eye can see the remarkable cycle of blood entering the lower

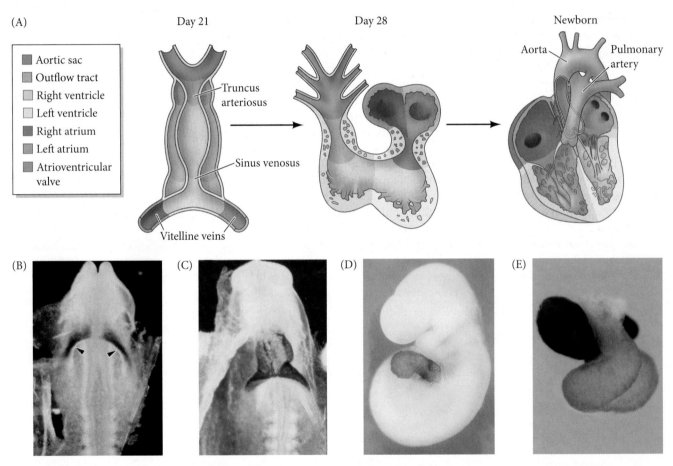

FIGURE 18.18 Cardiac looping and chamber formation. (A) Schematic diagram of cardiac morphogenesis in humans. On day 21, the heart is a single-chambered tube. Regional specification of the tube is shown by the different colors. By day 28, cardiac looping has occurred, placing the presumptive atria anterior to the presumptive ventricles. In the newborn, the valves and chambers establish circulatory routes such that the left ventricle pumps into the aorta and the right ventricle pumps into the pulmonary artery to the lungs. (B,C) *Xin* expression in the fusion of left and right heart primordia of a chick. The cells fated to form the myocardium are stained for the *Xin* message, whose protein product is essential for heart-tube looping. (B) Stage-9 chick neurula, in which Xin protein (purple) is seen in the two symmetrical heart-forming fields (arrows). (C) Stage-10 chick embryo showing fusion of the two heart-forming

regions prior to looping. (D,E) Specification of the atria and ventricles occurs even before heart looping. The atria and ventricles of the mouse embryo express different types of myosin proteins; here, atrial myosin stains blue and ventricular myosin stains orange. (D) In the tubular heart (prior to looping), the two myosins (and their respective stains) overlap at the atrioventricular channel joining the future regions of the heart. (E) After looping, the blue stain is seen in the definitive atria and inflow tract, while the orange stain is seen in the ventricles. The unstained region above the ventricles is the truncus arteriosus. Derived primarily from the neural crest, the truncus arteriosus becomes separated into the aorta and pulmonary arteries. (A after Srivastava and Olson 2000; B,C from Wang et al. 1999, courtesy of J. J.-C. Lin; D,E from Xavier-Neto et al. 1999, courtesy of N. Rosenthal.)

chamber and being pumped out through the aorta.) Looping of the heart converts the original anterior-posterior polarity of the heart tube into the right-left polarity seen in the adult organism. When this looping is complete, the portion of the heart tube destined to become the atria lies anterior to the portion that will become the ventricles (**FIGURE 18.18**).

This critically important process begins with the anterior part of the heart specifying the direction of the bend. Looping begins immediately after the onset of rhythmic heart contractions and the initiation of blood flow; pressure from the blood flow helps drive looping to completion (Groenendijk et al. 2005; Hove et al. 2003). As the bending of the heart tube deepens, an increasing volume of blood enters the heart. The volume differences are thought to be transmitted to the cells through the extracellular matrix and the cytoskeleton (Linask et al. 2005; Garita et al. 2011). Precise chamber alignment is needed for the correct signaling for the formation of the heart valves, the ventricular and atrial septa, and to allow for the heart to become connected to the embryonic vasculature that has been developing concomitantly within the embryo.

SCIENTISTS SPEAK 18.1   Dr. Kersti Linask discusses the looping of the vertebrate heart.

As the heart is looping, changes in the endocardium start forming the valves. The beginning of heart valve development is the formation of endocardial cushions in the canal between the atrium and ventricle and in the outflow tract of the looping heart tube (Armstrong and Bischoff 2004). Cushion development is initiated by the myocardium's signaling to endocardial cells to express the *Twist* gene. Twist protein is a transcription factor that initiates epithelial-mesenchymal transformation and cell migration. And these endocardial cells leave the endocardium and migrate to form the endocardial cushions (Barnett and Desgrosellier 2003; Shelton and Yutey 2008). Twist also activates the gene for Tbx20, and together Twist and Tbx20 activate the proteins that cause the proliferation and strengthening of the valves.

WATCH DEVELOPMENT 18.2   This video provides a medical view of human heart development.

WEB TOPIC 18.3   **CHANGING HEART ANATOMY AT BIRTH**   The process of taking the first breath actually alters the anatomy of the heart, allowing for pulmonary circulation.

Blood Vessel Formation

Although the heart is the first functional organ of the body, it does not begin to pump until the vascular system of the embryo has established its first circulatory loops. Rather than sprouting from the heart, the blood vessels form independently, linking up to the heart soon afterward. Everyone's circulatory system is different, since the genome cannot encode the intricate series of connections between the arteries and veins. Indeed, chance plays a major role in establishing the microanatomy of the circulatory system. However, all circulatory systems in a given species look very much alike because the development of the circulatory system is severely constrained by physiological, evolutionary, and physical parameters.

Vasculogenesis: The initial formation of blood vessels

The development of blood vessels occurs by two temporally separate processes: **vasculogenesis** and **angiogenesis** (**FIGURE 18.19**). During vasculogenesis, a network of blood vessels is created de novo from the lateral plate mesoderm. During angiogenesis, this primary network is remodeled and pruned into a distinct capillary bed, arteries, and veins.

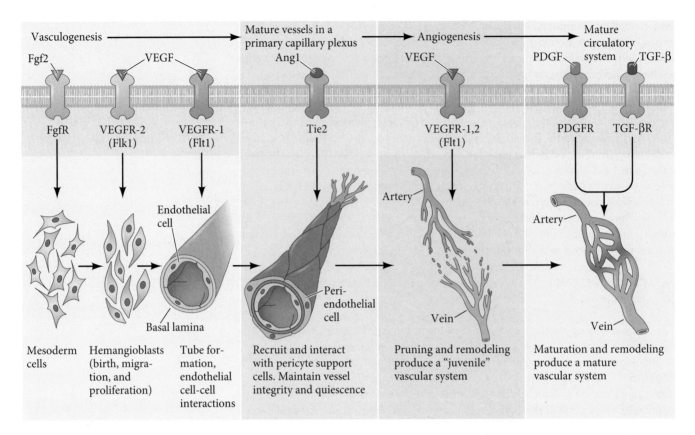

FIGURE 18.19 Vasculogenesis and angiogenesis. Vasculogenesis involves the formation of blood islands and the construction of capillary networks from them. Angiogenesis involves the formation of new blood vessels by remodeling and building on older ones. Angiogenesis finishes the circulatory connections begun by vasculogenesis. The major paracrine factors involved in each step are shown at the top of the diagram, and their receptors (on the vessel-forming cells) are shown beneath them. (After Hanahan 1997 and Risau 1997.)

In the first phase of vasculogenesis, a combination of BMP, Wnt, and Notch signals activates the Etv2 transcription factor in lateral plate mesoderm cells leaving the primitive streak in the posterior of the embryo, converting them into hemangioblasts.[5] Labeling zebrafish embryos with fluorescent probes to make single-cell fate maps confirms that hemangioblasts are the common progenitor for both the hematopoietic (blood cells) and endothelial (blood vessels) lineages in zebrafish (Paik and Zon 2010). This population of bipotential progenitor cells is found only in the ventral portion of the aorta, the region that had been known to produce these two cell types. The pathway permitting such aortic cells to differentiate into hemangioblasts appears to be induced by the *Cdx4* gene, while the determination of whether the hemangioblast becomes a blood cell precursor or a blood vessel precursor is regulated by the Notch signaling pathway. Notch signaling increases the conversion of hemangioblasts into blood cell precursors, whereas reduced amounts of Notch cause hemangioblasts to become endothelial (Vogeli et al. 2006; Hart et al. 2007; Lee et al. 2009). Notch signaling activates the expression of the Runx1 transcription factor, which, as we shall shortly see, appears to be conserved throughout vertebrates in inducing the conversion of endothelial cells to blood stem cells (Burns et al. 2005, 2009).

WEB TOPIC 18.4 **CONSTRAINTS ON THE FORMATION OF BLOOD VESSELS** Not all types of circulatory system can be functional. Physical constraints limit the types of vascular systems.

---

[5]The prefixes *hem-* and *hemato-* refer to blood (as in hemoglobin). Similarly, the prefix *angio-* refers to blood vessels. The suffix *-blast* denotes a rapidly dividing cell, usually a stem cell. The suffixes *-poiesis* and *-poietic* refer to generation or formation (*poeisis* is also the root of the word *poetry*). Thus, hematopoietic stem cells are those cells that generate the different types of blood cells. The Latin suffix *-genesis* (as in *angiogenesis*) means the same as the Greek *-poiesis*. The names of zebrafish hematopoietic mutants can be very poetic. Most are named after wines, and one of the genes producing a bloodless phenotype is named *vlad tepes,* after the historic Vlad Dracula.

SITES OF VASCULOGENESIS In amniotes, formation of the primary vascular networks occurs in two distinct and independent regions. First, **extraembryonic vasculogenesis** occurs in the **blood islands** of the yolk sac. These are the blood islands formed by the hemangioblasts, and they give rise to the early vasculature needed to feed the embryo and also to a red blood cell population that functions in the early embryo (**FIGURE 18.20A**). New studies (Frame et al. 2016) suggest that definitive (adult) blood stem cells also emerge in these yolk sac blood islands. Second, **intraembryonic vasculogenesis** forms the dorsal aorta, and vessels from this large vessel connect with capillary networks that form from mesodermal cells within each organ. The first scaffold of the dorsal aorta came, as we saw in Chapter 17, came from cells of the somite that migrate ventrally.

The aggregation of endothelium-forming cells in the yolk sac is a critical step in amniote development, for the blood islands that line the yolk sac produce the veins that bring nutrients to the embryo and transport gases to and from the sites of respiratory exchange (**FIGURE 18.20B**). In birds, these vessels are called the **vitelline veins**; in mammals, they are called the **omphalomesenteric veins**, or more usually, **umbilical veins**. In the chick, blood islands are first seen in the area opaca, when the primitive streak is at its fullest extent (Pardanaud et al. 1987). They form cords of hemangioblasts, which soon become hollowed out and become the flat endothelial cells lining the vessels (while the central cells give rise to blood cells). As the blood islands grow, they eventually merge to form the capillary network draining into the two vitelline veins, which bring food and blood cells to the newly formed heart.

GROWTH FACTORS AND VASCULOGENESIS Three growth factors are critically responsible for initiating vasculogenesis (see Figure 18.19). One of these, **basic fibroblast growth factor** (**Fgf2**), is required for the generation of hemangioblasts from the splanchnic mesoderm. When cells from quail blastodiscs are dissociated in culture, they do not form blood islands or endothelial cells. However, when these cells are cultured

(A)

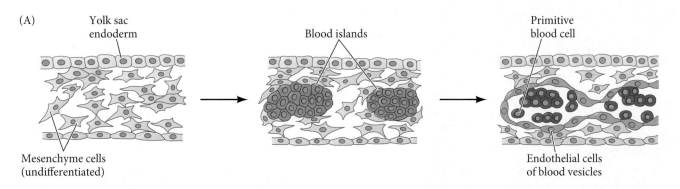

Yolk sac endoderm

Blood islands

Primitive blood cell

Mesenchyme cells (undifferentiated)

Endothelial cells of blood vesicles

(B)

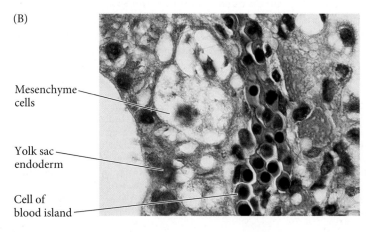

Mesenchyme cells

Yolk sac endoderm

Cell of blood island

FIGURE 18.20 Vasculogenesis. (A) Blood vessel formation is first seen in the wall of the yolk sac, where undifferentiated mesenchyme cells cluster to form blood islands. The centers of these clusters form the blood cells, and the outsides of the clusters develop into blood vessel endothelial cells. (B) A human blood island in the mesoderm surrounding the yolk sac. (The photomicrograph is from a tubal pregnancy; the embryo had to be removed because it implanted in an oviduct rather than in the uterus.) (A after Langman 1981; B from Katayama and Kayano 1999, courtesy of the authors.)

with Fgf2 protein, blood islands emerge and form endothelial cells (Flamme and Risau 1992). Fgf2 is synthesized in the chick embryonic chorioallantoic membrane and is responsible for the vascularization of this tissue (Ribatti et al. 1995).

The second family of proteins involved in vasculogenesis is the **vascular endothelial growth factors** (**VEGFs**). This family includes several VEGFs, as well as placental growth factor (PlGF), which direct the expansive growth of blood vessels in the placenta. Each VEGF appears to enable the differentiation of the angioblasts and their multiplication to form endothelial tubes. The most important VEGF in normal development, VEGF-A, is secreted by the mesenchymal cells near the blood islands, and hemangioblasts and angioblasts have receptors for this VEGF (Millauer et al. 1993). If mouse embryos lack the genes encoding either VEGF-A or its major receptor (the Flk1 receptor tyrosine kinase), yolk sac blood islands fail to appear and vasculogenesis fails to take place (**FIGURE 18.21A**; Ferrara et al. 1996). Mice lacking genes for the Flk1 receptor protein have blood islands and differentiated endothelial cells, but these cells are not organized into blood vessels (Fong et al. 1995; Shalaby et al. 1995). VEGF-A is also important in forming blood vessels to the developing bone and kidney.[6]

A third set of proteins, the **angiopoietins**, mediate the interaction between the endothelial cells and the **pericytes**—smooth-musclelike cells that the endothelial cells recruit to cover them. Mutations of either the angiopoietins or their receptor protein, Tie2, lead to malformed blood vessels deficient in the smooth muscles that normally surround them (Davis et al. 1996; Suri et al. 1996; Vikkula et al. 1996; Moyon et al. 2001).

WEB TOPIC 18.5 **ARTERIAL, VENOUS, AND LYMPHATIC VESSELS** The structure and function of the vascular tissues differs between the arteries, the veins, and the vessels that specialize in transporting lymph throughout the body.

Angiogenesis: Sprouting of blood vessels and remodeling of vascular beds

After an initial phase of vasculogenesis, angiogenesis begins. By this process, the primary capillary networks are remodeled and veins and arteries are made (see Figure 18.19). The critical factor for angiogenesis is VEGF-A (Adams and Alitalo 2007). In many cases, VEGF-A secreted by an organ will induce the migration of endothelial cells from existing blood vessels into that organ, causing the endothelial cells to form capillary networks there. Other factors, including hypoxia (low oxygen levels), can also induce the secretion of VEGF-A and thus induce blood vessel formation.

During angiogenesis, some endothelial cells in the existing blood vessel respond to the VEGF signal and begin "sprouting" to form a new vessel. These are known as the **tip cells**, and they differ from the other vessel cells. (If all the endothelial cells responded equally, the original blood vessel would fall apart.) The tip cells express the Notch ligand Delta-like-4 (Dll4) on their cell surfaces. The Dll4 ligand activates Notch signaling in the adjacent cells, preventing them from responding to VEGF-A (Noguera-Troise et al. 2006; Ridgway et al. 2006; Hellström et al. 2007). If Dll4 expression

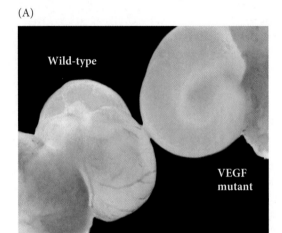

(A)

Wild-type

VEGF mutant

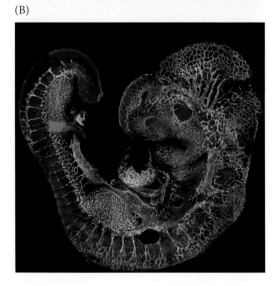

(B)

FIGURE 18.21 VEGF and its receptors in mouse embryos. (A) Yolk sacs of a wild-type mouse and a littermate heterozygous for a loss-of-function mutation of VEGF-A. The mutant embryo lacks blood vessels in its yolk sac and dies. (B) In a 9.5-day mouse embryo, VEGFR-3 (red), a VEGF receptor found on tip cells, is found at the angiogenic front of the capillaries (stained green). (A from Tammela et al. 2008, courtesy of the authors; B from Ferrara and Alitalo 1999, courtesy of K. Alitalo.)

[6]VEGF needs to be regulated very carefully in adults, and studies indicate that it can be affected by diet. The consumption of green tea has been associated with lower incidences of human cancer and the inhibition of tumor cell growth in laboratory animals. Cao and Cao (1999) have shown that green tea and one of its components, epigallocatechin-3-gallate (EGCG), prevent angiogenesis by inhibiting VEGF. Moreover, in mice given green tea instead of water (at levels similar to humans drinking 2–3 cups of green tea), the ability of VEGF to stimulate new blood vessel formation was reduced by more than 50%.

is experimentally reduced, tip cells form along a large portion of the blood vessel in response to VEGF-A.

The tip cells produce filopodia that are densely packed with VEGFR-2 (VEGF receptor-2) on their cell surfaces. They also express another VEGF receptor, VEGFR-3, and blocking VEGFR-3 greatly suppresses sprouting (**FIGURE 18.21B**; Tammela et al. 2008). These receptors enable the tip cell to extend toward the source of VEGF, and when the cell divides, the division is along the gradient of VEGFs. Indeed, the filopodia of the tip cells act just like the filopodia of neural crest cells and neural growth cones, and they respond to similar cues (Carmeliet and Tessier-Lavigne 2005; Eichmann et al. 2005). Semaphorins, netrins, neuropilins, and split proteins all have roles in directing the sprouting tip cells to the source of VEGF.

Anti-angiogenesis in normal and abnormal development

Like any powerful process in development, angiogenesis has to be powerfully regulated. Blood vessel formation must be signaled when to cease, and in some tissues blood vessel formation must be prevented. For example, the cornea of most mammals is avascular.[7] This absence of blood vessels allows the transparency of the cornea and optical acuity. The cornea appears to have two ways of keeping blood vessels out of the cornea. The first mechanism involves preventing the release of VEGF from the extracellular matrix in which it is stored (Seo et al. 2012). In addition, Ambati and colleagues (2006) have shown that the cornea secretes a soluble form of the VEGF receptor that "traps" VEGF and prevents angiogenesis in the cornea.

Soluble VEGF receptor also appears to be one of the normal mechanisms for regulating the increased formation of vasculature in the uterus during pregnancy. However, if too much soluble VEGF receptor is produced during pregnancy, there can be a dramatic reduction of normal angiogenesis. The spiral arteries that supply the fetus with nutrition fail to form and the capillary bed of the kidneys is reduced. These events are thought to be a major cause of **preeclampsia**, a condition of pregnancy characterized by hypertension and poor renal filtration (both of which are kidney problems) and fetal distress. Preeclampsia is the leading cause of premature birth and a major cause of maternal and fetal deaths (Levine et al. 2006; Mutter and Karumanchi 2008).

Too much VEGF can also be dangerous. Abnormal blood vessel formation occurs in solid tumors and in the retina of patients with diabetes. This vascularization results in the growth and spread of tumor cells and blindness, respectively. By targeting the VEGF receptors and the Notch pathway involved in regulating them, researchers are seeking ways to block angiogenesis and prevent cancer cells or the retina from becoming vascularized (Miller et al. 2013; Wilson et al. 2013).

Hematopoiesis: Stem Cells and Long-Lived Progenitor Cells

Each day we lose and replace about 300 *billion* blood cells. As blood cells are destroyed in the spleen, their replacements come from populations of stem cells. As we described in Chapter 5, a stem cell is capable of extensive proliferation, creating both more stem cells (self-renewal) and differentiated cell progeny (see Figures 5.1 and 5.3). In the case of **hematopoiesis**—the generation of blood cells—stem cells divide to produce (1) more stem cells and (2) progenitor cells that can respond to the environment around them to differentiate into about a dozen mature blood cell types (Notta et al. 2016). The critical

---

[7]The manatee is the only mammal known to have a vascularized cornea, and it turns out that this exception proves the rule—the cornea of the manatee does not express the soluble VEGF receptor. The manatee's closest relatives (dugongs and elephants) do express it, and their corneas are avascular (Ambati et al. 2006). This morphological distinction among related taxa provides further evidence of the importance of soluble VEGF in preventing corneal vascularization.

stem cell in hematopoiesis is the **pluripotent hematopoietic stem cell**, or simply the **hematopoietic stem cell** (**HSC**), which is capable of producing all the blood cells and lymphocytes of the body. The HSC can achieve this by generating a series of intermediate progenitor cells whose potency is restricted to certain lineages.

Sites of hematopoiesis

In the early 1900s, numerous investigators (looking at many different vertebrate species, including mongooses, bats, and humans) observed the emergence of blood cells from the ventral endothelium of the aorta (Adamo and Garcia-Cardeña 2012). In the 1960s, however, experiments on mice concluded that all hematopoietic stem cells are derived from cells originating in the extraembryonic blood islands surrounding the yolk sac. The aortic hematopoiesis was thought of as an intermediate stop that the stem cells made on their way to the spleen and bone marrow (the sites of adult hematopoiesis in mice).

However, in 1975 Françoise Dieterlen-Lièvre transplanted early chick yolk sacs onto 2-day (pre-circulation) quail embryos. Chick and quail blood cells can be readily distinguished under the microscope, and the chimeric animal survives. Dieterlen-Lièvre's analysis indicated that all the blood cells of the late quail embryo originated from the quail host and not from the transplanted chick yolk sac. Moreover, hematopoietic activity within the embryo was restricted to one major site: the ventral portion of the aorta (Dieterlen-Lièvre and Martin 1981). The grafting of splanchopleure from this **aorta-gonad-mesonephros** (**AGM**) **region** from one genetically variant mouse to another confirmed that in mammals, too, definitive hematopoiesis takes place from inside the embryo (Godin et al. 1993; Medvinsky et al. 1993). Soon afterward, hematopoietic stem cells were identified in clusters of cells that were observed on the ventral region of the 10.5-day embryonic mouse aorta (Cumano et al. 1996; Medvinsky and Dziermak 1996).

While there is evidence that some yolk sac hematopoietic stem cells persist in the adult mouse (see Samokhvalov et al. 2007; Frame et al. 2016), it is generally thought that the yolk sac hematopoietic stem cells in mammals produce blood cells that allow oxygen to be transported to the early embryo, but that nearly all the stem cells found in the adult are those from the AGM that have migrated to the bone marrow (Jaffredo et al. 2010).

In 2009, several laboratories proposed a new mechanism for blood cell production. This new hypothesis was based on the discovery of a new cell type in the AGM, the **hemogenic endothelial cell**.[8] Recall from the discussion of somites in Chapter 17, that the sclerotome produces angioblasts that migrate to the dorsal aorta and replace most of the primary dorsal aorta cells. Before their replacement, the remaining primary, lateral plate-derived endothelial cells of the dorsal aorta (now in the ventral area of the blood vessel) give rise to blood-forming stem cells. These blood vessel-derived hematopoietic stem cells are the critical source of adult blood stem cells (see Chapter 5). By analyzing the types of cells made by the blood vessel endothelium, researchers were able to isolate the hemogenic endothelial cells and showed that they produce the hematopoietic stem cells that migrate to the liver and bone marrow (Eilken et al. 2009; Lancrin et al. 2009). Furthermore, the transition from endothelial cell to hematopoietic stem cell was mediated by the activation of the Runx1 transcription factor (**FIGURE 18.22**). In mice lacking the *Runx1* gene, blood stem cells failed to form in the yolk sac, umbilical arteries, dorsal aorta, and placental vessels (Chen et al. 2009; Tober et al. 2016).

The *Runx1* gene appears to be regulated by a complex and dynamic circuitry. Moreover, Runx1 protein expression is not initiated until after the heart starts beating. If cardiac mutations prevent fluid flow through the aorta, Runx1 is not expressed. Rather, shear forces (i.e., friction) from the fluid flow are required to activate the *Runx1* gene

---

[8]The relationship of the hemogenic endothelial cell and the hemangioblast is controversial. In general, it is thought that hemangioblasts generate the hemogenic endothelial cells (see Ueno and Weissman 2010) and that the hemangioblast is a precursor to the hemogenic endothelium.

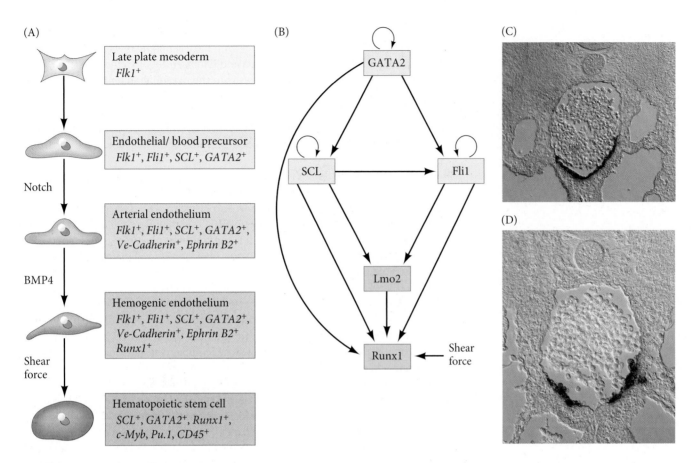

FIGURE 18.22 Pathway for hematopoietic stem cell formation. (A) In the developing mouse, hematopoietic stem cells arise from the hemogenic endothelium of the aorta. Runx1 is critical for this conversion of endothelial cells into blood stem cells. The lateral plate mesodermal heritage of the hemogenic endothelium is seen, as well as the juxtacrine and paracrine factors that brought it to its fate. The transcription factors associated with each stage are shown at the right. (B) A simplified view of the factors activating *Runx1* in the gene regulatory network establishing the hematopoietic stem cell in the mouse. The GATA2, Fli1, and Scl transcription factors bind at adjacent sites on a single enhancer 23 base pairs downstream from the *Runx1* transcription initiation site. Scl is critical for preventing blood and vascular cells from becoming heart muscle. The mechanism by which shear force is mechanotransduced to help activate *Runx1* remains unknown. (C) Runx1 expression (purple) in the stage-19 chick embryo; Runx1-expressing cells have become part of the blood vessel. (D) Runx1-expressing cells at stage 21 in the chick embryo. Hematopoietic colonies are visible. (A after Swiers et al. 2010; B after Pimanda and Göttgens 2010; C,D from Jaffredo et al. 2010.)

in the ventral endothelium of the dorsal aorta (Adamo et al. 2009; North et al. 2009).[9] The shear forces appear to elevate levels of nitric oxide (NO) in the endothelium. NO, in turn, activates (perhaps through cGMP) *Runx1* and other genes known to be critical for blood cell formation. The transition from hemogenic endothelial cell to HSC does not appear to be caused by an asymmetric cell division. Rather, there is a rearrangement of cytoskeleton and tight junctions that resemble an epithelial-mesenchymal transition such as those seen in the sclerotome or dermamyotome (Yue et al. 2012).

In non-amniote vertebrates, the splanchnopleure is also the source of the hematopoietic stem cells, and BMPs are crucial in inducing the blood-forming cells in all vertebrates studied. In *Xenopus*, the ventral mesoderm forms a large blood island that is

---

[9] Mechanotransduction of biophysical forces such as shear stress from blood flow is a major player in cardiovascular development (see Linask and Watanabe 2015). Recall that it is also required for normal heart development (Mironov et al. 2005) and for the correct patterning of blood vessels (Lucitti et al. 2007; Yashiro et al. 2007). It is also needed for the fragmentation of the platelet precursor cell—the megakaryocyte—into platelets. The megakaryocyte in the bone marrow inserts small processes into the blood vessels surrounding the stem cell niche, and the shear force there fragments these processes into platelets (Junt et al. 2007).

the first site of hematopoiesis. Ectopic BMP2 and BMP4 can induce blood cell and blood vessel formation in *Xenopus*, and interference with BMP signaling prevents blood formation (Maéno et al. 1994; Hemmati-Brivanlou and Thomsen 1995). In the zebra-fish, both yolk sac and aortic hematopoiesis are seen. As in the mammalian embryo, the second wave of hematopoiesis is from the aorta. Hematopoietic stem cells can be seen arising from the ventral aortic endothelium (Bertrand et al. 2010; Kissa and Herbomel 2010), and the same genetic pathways leading to Runx1 expression (including the BMPs) regulate this second, definitive wave of hematopoiesis (Mullins et al. 1996; Paik and Zon 2010).

The bone marrow HSC niche

The hematopoietic stem cells of the aorta generate HSCs that come to reside first in the liver and then in the bone marrow (Coskun and Hirschi 2010). In humans, the aorta generates blood cells around days 27 through 40 (Tavian and Péault 2005). The bone marrow HSC is a remarkable cell, in that it is the common precursor of red blood cells (erythrocytes), white blood cells (granulocytes, neutrophils, and platelets), monocytes (macrophages and osteoclasts), and lymphocytes. When transplanted into inbred, irradiated mice (that are genetically identical to the donor cells and whose own stem cells have been eliminated by radiation), HSCs can repopulate the mouse with all the blood and lymphoid cell types. It is estimated that only about 1 in every 10,000 blood cells is a pluripotent HSC (Berardi et al. 1995). In humans, "bone marrow transplants" are used to transfer healthy HSCs into people whose lymphocytes, red blood cells, or white blood cells have been wiped out by disease, drugs, or radiation. In recent years, more than 50,000 such transplantations have been performed annually (Gratwohl et al. 2010).

The maintenance of the HSC depends on the stem cell niche, and especially on the ability of the HSC to receive the paracrine factor **stem cell factor**, or **SCF**. SCF binds to the Kit receptor protein. (This binding is critical for sperm and pigment stem cells as well as HSC.) Because it was important to determine which cells of the stem cell niche were supplying SCF, Ding and colleagues (2012) constructed genetically recombined mice in which the gene for SCF was replaced by the gene for green fluorescent protein in all or in selected cell types. When all the cell types of the niche expressed GFP instead of SCF, the HSCs died. When they deleted SCF production only in certain cell types, they found that replacing SCF with GFP in blood cells, bone cells, or marrow mesenchymal cells did not block HSC maintenance. However, when they got rid of SCF expression in either the endothelial cells or the perivascular cells surrounding the endothelial cells, many fewer HSCs survived. And when SCF synthesis was turned off in both of these cell types (but not in the others), all the HSCs perished. It appears, then, that the SCF needed for HSC survival is made primarily by the perivascular cells, with some contribution from the endothelial cells (**FIGURE 18.23**).

SCF is not the only paracrine factor that HSCs require; there are numerous others, and these probably render the HSCs competent to respond to the paracrine and

FIGURE 18.23 The home for HSCs appears to be a niche where stem cell factor (SCF) can be made by the perivascular (subendothelial) cells as well as by the endothelial cells of the bone marrow sinusoids. (A) Simplified diagram of the sinusoid with its endothelial cells and a surrounding perivascular cell. (B) Development of a stem cell niche when human perivascular cells (stained brown with antibodies to the subendothelial cell marker CD146) implanted into a mouse. At 8 weeks, processes from perivascular cells establish contacts with hematopoietic stem cells (as in human bone marrow). The red arrows show hematopoietic cells between endothelial and perivascular cells. (A after Shestopalov and Zon 2012; B from Sacchetti et al. 2007, courtesy of P. Bianco.)

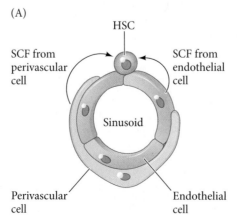

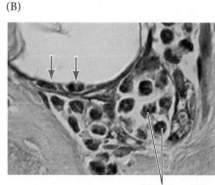

juxtacrine factors that will direct cell differentiation (Morrison and Scadden 2014). Stem cell niches often contain long-term, quiescent HSCs that are used to generate progenitor cells on a continual basis, as well as shorter-acting HSCs that can respond to immediate physiological needs (see Figure 18.24). Wnts that activate the noncanonical pathways are secreted by niche osteoblasts to maintain the quiescent HSCs, whereas the canonical Wnt pathway may be critical for inducing them to become the rapidly proliferating HSCs (Reya et al. 2003; Sugimura et al. 2012). It is probable that in adult mammals, the maintenance of the billions of blood cells is not dependent on a small number of hematopoietic stem cells, but on the steady-state production of numerous long-lived progenitor cells that are specified for either a single lineage or multiple lineages (Sun et al. 2014).

Hematopoietic inductive microenvironments

The major models of blood cell production envision blood differentiation as progressing down a series of less potent precursor cells. At the top are the multipotent hematopoietic stem cells (HSCs) which can give rise to more restricted multipotent stem cells (such as the common myeloid precursor cell, CMP), and finally to lineage-commited progenitor cells. Endocrine, paracrine, and juxtacrine factors are thought to push blood cell differentiation down one path or another (**FIGURE 18.24**).

One of the major endocrine factors (i.e., hormones) is **erythropoietin**, which appears to cause the common myeloid precursor cell (CMP) to make more megakaryocyte/erythroid precursor cells (MEPs) and biases the MEPs to make more erythrocytes (Lu et al. 2008; Klimchenko et al. 2009). The paracrine factors involved in blood cell and lymphocyte formation are the **cytokines**. Cytokines can be made by several cell types, but they are collected and concentrated by the extracellular matrix of the stromal (mesenchymal)

FIGURE 18.24 Hierarchy of hematopoietic lineages. At the top of the hierarchy are the long-term hematopoietic stem cells (LT-HSCs), which give rise to short-term HSCs (ST-HSCs) that retain limited self-renewing capabilities (see Chapter 5). Rapidly dividing multipotent progenitors (MPP) still possess the potential to generate either myeloid (red blood cell types) or lymphoid lineages (white blood cell types), beyond which differentiation becomes increasingly restricted. Progeny of the MPP include the common myeloid progenitors (CMP) and granulocyte–macrophage–lymphocyte progenitors (GMLPs). Further differentiation takes place, producing common lymphoid progenitors (CLPs), granulocyte–macrophage progenitors (GMPs), and megakaryocyte–erythrocyte progenitors (MEPs), These progenitors will further differentiate into the various red and white blood cell types. (After Cullen et al. 2014.)

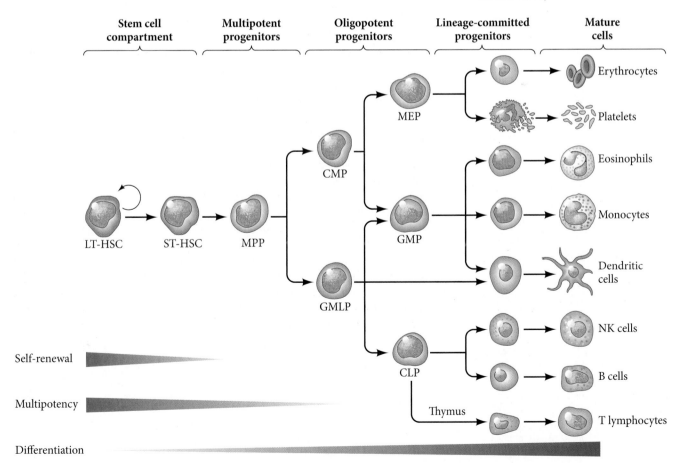

cells at the sites of hematopoiesis (Hunt et al. 1987; Whitlock et al. 1987). For instance, granulocyte-macrophage colony-stimulating factor (GM-CSF) and the multilineage growth factor interleukin 3 (IL3) both bind to the heparan sulfate glycosaminoglycan of the bone marrow stroma (Gordon et al. 1987; Roberts et al. 1988). The extracellular matrix is then able to present these paracrine factors to the stem cells in concentrations high enough to bind to their respective receptors. At different stages of maturation, the stem cells become competent to respond to different factors.

The developmental path taken by a descendant of a pluripotent HSC depends on which growth factors it meets, and is therefore determined by the stromal cells. Wolf and Trentin (1968) demonstrated that short-range interactions between stromal cells and stem cells determine the developmental fates of the stem cells' progeny. These investigators placed plugs of bone marrow in a spleen and then injected stem cells into it. Those CMPs that came to reside in the spleen formed colonies that were predominantly erythroid, whereas those that came to reside in the bone marrow formed colonies that were predominantly granulocytic. Colonies that straddled the borders of the two tissue types were predominantly erythroid in the spleen and granulocytic in the marrow. Such regions of determination are referred to as **hematopoietic inductive microenvironments** (**HIMs**). As expected, the HIMs induce different sets of transcription factors in these cells, and these transcription factors specify the fate of the particular cells (see Kluger et al. 2004).

The transcription factors of the HIM may act by pulling the equilibrium of the stem cells transcription network in different directions (Krumsiek et al. 2011; Wontakal et al. 2012). By decomposing the interactions into negative feedback loops and feedforward loops (of both activation and repression), there appear to be only four stable configurations that this network can have. Such stable configurations are called "attractor states" in systems theory, and these attractor states correspond to four cell types. Moreover, certain mutations will make certain attractor states impossible, and these are the mutations that block the differentiation of certain cell types.

This scheme of blood cell differentiation going through cell types of progressively diminished potency may not work at all stages of life. Notta and colleagues used single-cell transcriptomes to show that the midlevel stem cells (such as the CMP) were not present during later stages of human development. It is possible that later in life, blood cell production follows immediately from the HSC.

Coda

We ask a lot of our circulatory system. We require a precise flow of blood through the valves each second of our lives; we demand fine-tuned coordination between our brain, heart, bone marrow, and hormones such that the cardiac muscle contractions adapt to our physiological needs; and we demand that the production of our blood cells—cells made by precursors that formed in our embryo—be so precise that we get neither cancer nor anemia. Given all this, it is not surprising that blood cell differentiation, heart development, kidney development, and blood vessel formation are now among the most important fields of study in medical science. Congenital heart defects are among the most prevalent types of birth defects, and cardiovascular disease is the most common cause of death in industrialized nations. The questions of cardiogenesis, kidney formation, angiogenesis, and hematopoiesis that engaged Aristotle and Harvey still excite major research programs.

Next Step Investigation

Although only 2% of the human genome encodes proteins, more than 75% of the genome is transcribed. Much of this transcribed but "noncoding RNA" is involved in gene regulation, and research is underway to determine how this ncRNA is integrated into the networks that form organs. For instance, the long, noncoding RNA *Braveheart* is required for the expression of MeSP1, which helps define the cardiovascular lineage (Klattenhoff et al. 2013). MicroRNAs such as

mIR-1 and miR-133 are required for fine-tuning the cardiac muscle division (Philippen et al. 2015). Similarly, microRNAs have been found to be critical in kidney and capillary development (Ho and Kriedberg 2013; Yin et al. 2014). These newly discovered noncoding RNAs may be extremely important in the pathways leading to organ development, and their disruption or absence may explain the basis of several congenital anomalies and adult-onset diseases.

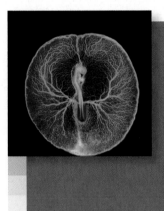

Closing Thoughts on the Opening Photo

The heart and vascular system of the chick have been studied for centuries (see Figure 1.4). This modern image, a fluorescence micrograph, depicts the 2-day chick embryo. It was compiled about 45 hours after the egg was laid, at the point where the heart begins to beat. The vascular system was revealed by injecting fluorescent beads into the circulatory system. The heart is the first functioning organ; yet, as seen here, most of the circulation goes to the extraembryonic region, bringing in nutrients from the yolk and exchanging gases. One of the first events in the formation of the heart is the establishing of polarity. The atria receive the blood, while the ventricles pump it. (Photograph © Vincent Pasque and Wellcome Images.)

18 Snapshot Summary
Intermediate and Lateral Plate Mesoderm

1. The intermediate mesoderm forms the kidneys, adrenal glands, and gonads. It is specified through interactions with the paraxial mesoderm that require Pax2, Pax8, and Lim1.

2. The metanephric kidney of mammals is formed by reciprocal interactions of the metanephric mesenchyme and a branch of the nephric duct called the ureteric bud. The ureteric bud and metanephric mensenchyme are specified depending on the length of time their progenitor cells are exposed to Wnt and FGF signals.

3. The metanephric mesenchyme becomes competent to form nephrons by expressing WT1, and it starts to secrete GDNF. GDNF is secreted by the mesoderm and induces the formation of the ureteric bud.

4. The ureteric bud secretes Fgf2 and BMP7 to prevent apoptosis in the metanephric mesenchyme. Without these factors, this kidney-forming mesenchyme dies.

5. The ureteric bud secretes Wnt9b and Wnt6, which induce the competent metanephric mesenchyme to form epithelial tubules. As they form these tubules, the cells secrete Wnt4, which promotes and maintains their epithelialization.

6. The lateral plate mesoderm splits into two layers. The dorsal layer is the somatic (parietal) mesoderm, which underlies the ectoderm and forms the somatopleure. The

ventral layer is the splanchnic (visceral) mesoderm, which overlies the endoderm and forms the splanchnopleure.

7. The space between the two layers of lateral plate mesoderm forms the body cavity, or coelom.

8. The heart arises from splanchnic mesoderm on both sides of the body. This region of cells is called the heart field, or cardiogenic mesoderm. The cardiogenic mesoderm is specified by BMPs in the absence of Wnt signals.

9. The Nkx2-5, Mesp1, and GATA transcription factors are important in committing the cardiogenic mesoderm to become heart cells. These cardiac precursor cells migrate from the sides to the midline of the embryo, in the neck region.

10. There are two major heart fields one each side of the body. Each heart field has two regions: The first heart field forms the scaffold of the heart tube and will form the left ventricle. The rest of the heart is made largely by the second heart field.

11. A cardiac precursor cell can form each of the major lineages of the heart. The cardiogenic mesoderm forms the endocardium (which is continuous with the blood vessels) and the myocardium (the muscular component of the heart).

12. The endocardial tubes form separately and then fuse. The looping of the heart transforms the original anterior-posterior polarity of the heart tube into a right-left polarity.

13. Retinoic acid is important in determining the anterior-posterior polarity of the heart and kidneys

14. Tbx transcription factors are critical for specifying the heart chambers and for establishing the electrical circuitry of the heart.

15. Coronary arteries and lymphatic vessels both come from the reprogramming of veins.

16. Blood vessels are constructed by two processes, vasculogenesis and angiogenesis. Vasculogenesis involves the condensing of splanchnic mesoderm cells to form blood islands. The outer cells of these islands become endothelial (blood vessel) cells. Angiogenesis involves the remodeling of existing blood vessels.

17. Numerous paracrine factors are essential in blood vessel formation. Fgf2 is needed for specifying the angioblasts. VEGF-A is essential for the differentiation of the angioblasts. Angiopoietins allow the smooth muscle cells (and smooth muscle-like pericytes) to cover the vessels. Ephrin ligands and Eph receptor tyrosine kinases are critical for capillary bed formation.

18. The pluripotent hematopoietic stem cell (HSC) generates other pluripotent stem cells, as well as lineage-restricted stem cells. It gives rise to both blood cells and lymphocytes.

19. In vertebrates, HSCs are thought to originate from hemogenic endothelial cells that characterize the blood islands, the dorsal aorta, and the placental vessels. The definitive HSC appears to be derived from the ventral portion of the aorta.

20. The common myeloid precursor (CMP) is a blood stem cell that can generate the more committed stem cells for the different blood lineages. Hematopoietic inductive microenvironments (HIMs) determine the blood cell differentiation.

21. The HSC depends on stem cell factor, which is provided to it by the perivascular cells of the sinusoids contained in the stem cell niche.

Further Reading

Adamo, L. and G. García-Cardeña. 2012. The vascular origin of hematopoietic cells. *Dev. Biol.* 362: 1–10.

Cooley, J., S. Whitaker, S. Sweeney, S. Fraser and B. Davidson. 2011. Cytoskeletal polarity mediates localized induction of the heart progenitor lineage. *Nat. Cell Biol.* 13: 952–957.

Diogo, R. and 7 others. 2015. A new heart for a new head in vertebrate cardiopharyngeal evolution. *Nature* 520: 466-473.

Ding, L., T. L. Saunders, G. Enikolopov and S. J. Morrison. 2012. Endothelial and perivascular cells maintain haematopoietic stem cells. *Nature* 481: 457–462.

Krause, M., A. Rak-Raszewska, I. Pietilä, S. E. Quaggin and S. Vainio S. 2015. Signaling during kidney development. *Cells* 4:112–132.

Morrison, S. J. and D. T. Scadden. 2014. The bone marrow niche for haematopoietic stem cells. *Nature* 505: 327– 334.

Sizarov, A., J. Ya, B. A. de Boer, W. H. Lamers, V. M. Christoffels and A. F. Moorman. 2011. Formation of the building plan of the human heart: Morphogenesis, growth, and differentiation. *Circulation* 123: 1125–1135.

Takasato, M. and M. H. Little. 2015. The origin of the mammalian kidney: Implications for recreating the kidney in vitro. *Development* 142: 1937–1947.

GO TO WWW.DEVBIO.COM...

...for Web Topics, Scientists Speak interviews, Watch Development videos, Dev Tutorials, and complete bibliographic information for all literature cited in this chapter.

Development of the Tetrapod Limb

CONSIDER YOUR LIMB. It has fingers or toes at one end, a humerus or femur at the other. You won't find anyone with fingers in the middle of their arm. Also consider the subtle but obvious differences between your hands and your feet. If your fingers were replaced by toes, you would certainly know it. Despite these differences, the bones of your feet are similar to the bones of your hand. It's easy to see that they share a common pattern. And finally, consider that both your hands are remarkably similar in size, as are both your feet. These commonplace phenomena present fascinating questions to the developmental biologist. How is it that vertebrates have four limbs and not six or eight? How is it that the little finger develops at one edge of the limb and the thumb at the other? How does the forelimb grow differently than the hindlimb? How can limb size be so precisely regulated? Is there a conserved set of developmental mechanisms that can explain why our hands have five digits, a chick's wing three digits, and the horse's hoof one?

Limb Anatomy

As the name denotes, **tetrapods** are four-limbed vertebrates (amphibians, reptiles, birds, and mammals). The bones of any tetrapod limb—be it arm or leg, wing or flipper—consist

How many fingers am I holding up?

The Punchline

Designed for locomotion on land, tetrapod limbs have joints between the bones and a set of digits at their distal ends. The limb begins as a "bud" of tissue on the sides of the embryo when cells from the developing myotome and lateral plate mesoderm migrate to and proliferate within the presumptive limb field, the limb mesenchyme. This proliferative "progress zone" is covered by ectoderm, with a thickening at the distal tip called the apical ectodermal ridge. Fgf8 signaling from this ridge antagonizes flank-derived retinoic acid, initiating a positive feedback loop with mesenchyme signaling (Fgf10 and Wnt) to promote limb bud outgrowth. Fgf8 specifies the posterior mesenchyme into the "zone of polarizing activity," which secretes Shh to set up the anterior-posterior (thumb-pinkie) axis of the limb. Wnt signaling determines the dorsal-ventral axis (knuckle-palm). Skeletogenesis is controlled by a "Turing-type" model of self-organization through morphogen interactions. In certain animals the early webbed tissue between the digits remains; in others, it dies via BMP-mediated apoptosis. Each limb signaling system influences the differential expression of Hox genes along each axis, modification of which supports the evolution of fins to fingers.

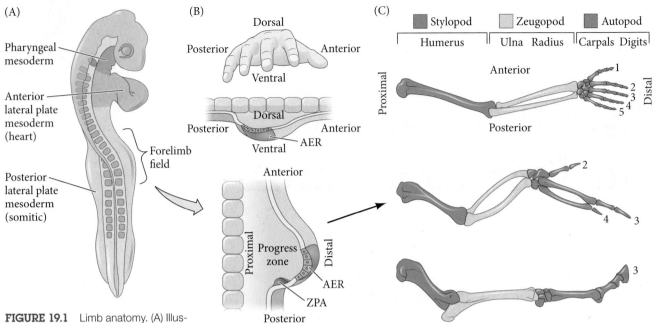

FIGURE 19.1 Limb anatomy. (A) Illustration of a chick embryo just prior to limb growth showing three important mesodermal cell types as well as the emergence of the limb field. (B) Axis orientation and anatomy of the limb bud. Apical ectodermal ridge (AER); zone of polarizing activity (ZPA). (C) Skeletal pattern of the human arm, chick wing, and horse forelimb. (According to convention, the digits of the chick wing are numbered 2, 3, and 4. The cartilage condensations forming the digits appear similar to those forming digits 2, 3, and 4 of mice and humans; however, new evidence suggests that the correct designation may be 1, 2, and 3.) (A after Tanaka 2013; B after Logan 2003.)

of a proximal **stylopod** (humerus/femur) adjacent to the body wall, a **zeugopod** (radius-ulna/tibia-fibula) in the middle region, and a distal **autopod** (carpals-fingers/tarsals-toes)[1] (**FIGURE 19.1**). Fingers and toes can be referred to as phalanges or, more generally, digits. The positional information needed to construct a limb has to function in a three-dimensional[2] coordinate system:

- The first dimension is the *proximal-distal axis* ("close-far"; that is, shoulder-to-finger or hip-to-toe). The bones of the limb are formed by endochondral ossification. They are initially cartilaginous, but eventually most of the cartilage is replaced by bone. Somehow the limb cells develop differently at early stages of limb morphogenesis (when they make the stylopod) than at later stages (when they make the autopod).

- The second dimension is the *anterior-posterior axis* (thumb-to-pinkie). Our little fingers or toes mark the posterior end and our thumbs or big toes are at the anterior end. In humans, it is obvious that each hand develops as a mirror image of the other. One can imagine other arrangements—such as the thumb developing on the left side of both hands—but these patterns do not occur.

- Finally, limbs have a *dorsal-ventral axis*: our palms (ventral) are readily distinguishable from our knuckles (dorsal).

The Limb Bud

The first visible sign of limb development is the formation of bilateral bulges called **limb buds** at the presumptive forelimb and hindlimb locations (**FIGURE 19.2A**). Fate mapping studies on salamanders, pioneered by Ross Granville Harrison's laboratory (see Harrison 1918, 1969), showed that the center of this disc of cells in the somatic region of the lateral plate mesoderm normally gives rise to the limb itself. Adjacent to it are the cells that will form the peribrachial (around the limb) flank tissue and the shoulder

[1] These terms can be difficult to remember, but knowing their word origins can help. *Stylo* = like a pillar; *zeugo* = joining; *auto* = self; *pod* = foot.

[2] Actually, it is a four-dimensional system, in which time is the fourth axis. Developmental biologists get used to seeing nature in four dimensions.

(A)

Somites

Pronephric kidney

Gills

Peribrachial flank tissue

Free limb

Shoulder girdle

(B)

Epaxial myotome bud

Myotome

Spinal cord

Sclerotome

Notochord

Pronephron

Endoderm

Central dermatome

Hypaxial myotome bud

Limb muscle precursors

Limb bud

Limb skeletal precursors

Lateral plate mesoderm

(C)

(D)

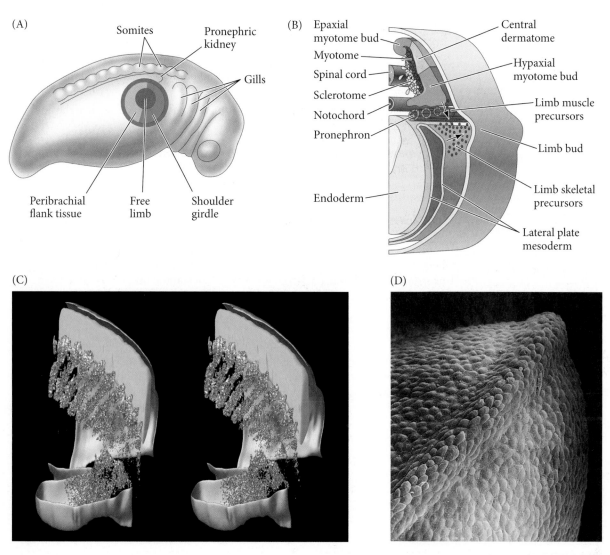

FIGURE 19.2 The limb bud. (A) Prospective forelimb field of the salamander *Ambystoma maculatum*. The central area contains cells destined to form the limb per se (the free limb). The cells surrounding the free limb give rise to the peribrachial flank tissue and the shoulder girdle. The ring of cells outside these regions usually is not included in the limb, but can form a limb if the more central tissues are extirpated. (B) Emergence of the limb bud. Proliferation of mesenchymal cells (arrows) from the somatic region of the lateral plate mesoderm causes the limb bud in the amphibian embryo to bulge outward. These cells generate the skeletal elements of the limb. Contributions of myoblasts from the lateral myotome provide the limb's musculature. (C) Entry of myoblasts (purple) into the limb bud. This computer stereogram was created from sections of an in situ hybridization to the *myf5* mRNA found in developing muscle cells. If you can cross your eyes (or try focusing "past" the page, looking through it to your toes), the three-dimensionality of the stereogram will become apparent. (D) Scanning electron micrograph of an early chick forelimb bud, with the apical ectodermal ridge in the foreground. (A after Stocum and Fallon 1982; C courtesy of J. Streicher and G. Müller; D courtesy of K. W. Tosney.)

girdle. However, if all these cells are extirpated from the embryo, a limb will still form (albeit somewhat later) from an additional ring of cells that surrounds this area but would not normally form a limb. If this surrounding ring of cells is included in the extirpated tissue, no limb will develop. This larger region, representing all the cells in the area capable of forming a limb on their own, is the **limb field**.

The cells that make up the limb bud are derived from the posterior lateral plate mesoderm, adjacent somites, and the bud's overlying ectoderm. Lateral plate mesenchyme cells migrate within the limb fields to form the limb *skeletal* precursor cells, while mesenchymal cells from the somites at the same level migrate in to establish the limb *muscle* precursor cells (**FIGURE 19.2B,C**). This accumulating heterogeneous population of mesenchymal cells proliferates under the ectodermal tissue, creating the limb bud.

Even the early limb bud possesses its own organization such that the main direction of growth occurs along the proximal-to-distal axis (somites-to-ectoderm), with lesser growth occurring along the dorsal-to-ventral and anterior-to-posterior axes (see Figure 19.1B). The limb bud is further regionalized into three functionally distinct domains:

FIGURE 19.3 Deletion of limb bone elements by the deletion of paralogous Hox genes. (A) 5′ Hox gene patterning of the forelimb. *Hox9* and *Hox10* paralogues specify the humerus (stylopod). *Hox10* paralogues are expressed to a lesser extent in the radius and ulna (zeugopod). *Hox11* paralogues are chiefly responsible for patterning the zeugopod. *Hox12* and *Hox13* paralogues function in the autopod, with *Hox12* paralogues functioning primarily in the wrist and to a lesser extent in the digits. (B) A similar but somewhat differing pattern is seen in the hindlimb. (C) Forelimb of a wild-type mouse (left) and of a double-mutant mouse that lacks functional *Hoxa11* and *Hoxd11* genes (right). The ulna and radius are severely reduced or absent in the mutant. (D) Human polysyndactyly ("many fingers joined together") syndrome results from a homozygous mutation at the *HOXD13* loci. This syndrome includes malformations of the urogenital system, which also expresses *HOXD13*. (A,B after Wellik and Capecchi 2003; C from Davis et al. 1995, courtesy of M. Capecchi; D from Muragaki et al. 1996, courtesy of B. Olsen.)

1. The highly proliferative mesenchyme that fuels limb bud growth is known as the **progress zone** (**PZ**) mesenchyme (a.k.a. the undifferentiated zone).

2. The cells found within the most posterior region of the progress zone constitute the **zone of polarizing activity** (**ZPA**), since it patterns cell fates along the anterior-posterior axis.

3. The **apical ectodermal ridge** (**AER**) is a thickening of the ectoderm at the apex of the developing limb bud (**FIGURE 19.2D**).

▶ **DEV TUTORIAL** *Tetrapod Limb Development* Dr. Michael J.F. Barresi describes the basics behind building a limb.

Hox Gene Specification of Limb Skeleton Identity

Homeobox transcription factors, or Hox genes, play an essential role in specifying whether a particular mesenchymal cell will become stylopod, zeugopod, or autopod. Understanding their role has given researchers substantial new insight into the development and evolution of the vertebrate limb.

From proximal to distal: Hox genes in the limb

The 5′ (*AbdB*-like) portions (paralogues 9–13) of the *Hoxa* and *Hoxd* gene complexes appear to be active in the limb buds of mice. Based on the expression patterns of these genes and on naturally occurring and gene knockout mutations, Mario Capecchi's laboratory (Davis et al. 1995) proposed a model wherein these Hox genes specify the identity of a limb region (**FIGURE 19.3A,B**). Here, *Hox9* and *Hox10* paralogues specify the stylopod, *Hox11* paralogues specify the zeugopod, and *Hox12* and *Hox13* paralogues

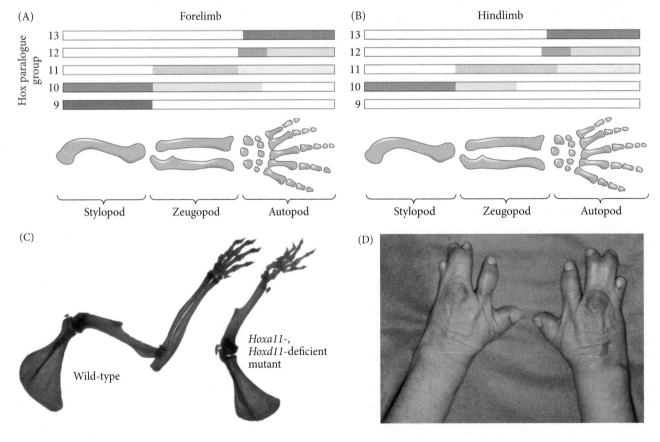

(A) Forelimb

Hox paralogue group: 13, 12, 11, 10, 9

Stylopod Zeugopod Autopod

(B) Hindlimb

Hox paralogue group: 13, 12, 11, 10, 9

Stylopod Zeugopod Autopod

(C) Wild-type   *Hoxa11*-, *Hoxd11*-deficient mutant

(D)

specify the autopod. This scenario has been confirmed by numerous experiments. For instance, when Wellik and Capecchi (2003) knocked out all six alleles of the three *Hox10* paralogues (*Hox10aaccdd*) in mouse embryos, the resulting mice not only had severe axial skeletal defects, they also had no femur or patella (they did have humeruses, however, because the *Hox9* paralogues are expressed in the forelimb stylopod but not in the hindlimb stylopod). When all six alleles of the three *Hox11* paralogues were knocked out, the resulting hindlimbs had femurs but neither tibias nor fibulas (and the forelimbs lacked the ulnas and radii). Thus, the *Hox11* knockout got rid of the zeugopods (**FIGURE 19.3C**). Similarly, knocking out all the paralogous *Hoxa13* and *Hoxd13* loci resulted in loss of the autopod (Fromental-Ramain et al. 1996). Humans homozygous for a *HOXD13* mutation show abnormalities of the hands and feet wherein the digits fuse (**FIGURE 19.3D**), and humans with homozygous mutant alleles of *HOXA13* also have deformities of their autopods (Muragaki et al. 1996; Mortlock and Innis 1997). In both mice and humans, the autopod (the most distal portion of the limb) is affected by the loss of function of the most 5′ Hox genes.

 SCIENTISTS SPEAK 19.1 Watch a Web conference with Dr. Denis Duboule on Hox genes in the limb.

From fins to fingers: Hox genes and limb evolution

How did the vertebrate appendage evolve into the limbs we find so useful today? The fossil record points to an important transition in forelimb morphology from the pectoral fins of ray-finned fish to the digited limbs of the tetrapod, a transition that provided the opportunity for aquatic life to explore terrestrial habitats. Understanding the evolutionary history of the tetrapod limb can help us analyze the developmental mechanisms that are essential for today's limb morphology. The discovery of the Devonian fossil *Tiktaalik roseae*, a "fish with fingers," highlights the importance of joint development in limb evolution. Fish fins, including those of some of the most primitive species, develop using the same three Hox gene expression phases as tetrapods use to form their limbs (Davis et al. 2007; Ahn and Ho 2008). The independent modification of fin bones into limb bones may have been made possible by the joints. The joints of *Tiktaalik*'s pectoral fins are very similar to those of amphibians and indicate that *Tiktaalik* had mobile wrists and a substrate-supported stance in which the elbow and shoulder could flex (**FIGURE 19.4A–C**; Shubin et al. 2006; Shubin 2008). In addition, the presence of wristlike structures and the loss of dermal scales in these regions suggest that this Devonian fish was able to propel itself on moist substrates. Thus, *Tiktaalik* is thought to be a transition between fish and amphibians—a "fishapod" (as one of its discoverers, Neil Shubin, called it) "capable of doing push-ups."

What types of molecular and morphological changes occurred along the different branches that led to the ray-finned fish on the one hand and terrestrial tetrapods on the other? In those fish that are most closely related to tetrapods (lobe-finned fish such as coelocanths and lungfish), the more proximal bones of the pectoral fin are homologous to the stylopod segment of tetrapod forelimbs and are similarly responsible for articulation about the pectoral girdle or shoulder. However, the fin of ray-finned fish diverged in form, and this is most evident in the more distal elements and in particular the autopod (digits). Ray-finned fish lack an endoskeleton associated with the autopod, whereas ancestral fish within the Sarcopterygian clade (lobbed-fin fish) display expanded endochondral skeletons within their fins (as in *Tiktaalik*). Thus, adaptation targeting the developmental mechanisms governing the more distal limb skeleton was the primary basis for limb evolution.

 SCIENTISTS SPEAK 19.2 In this Web conference, Dr. Peter Currie discusses the evolution of limb muscles in fish.

(A)

(B)

(C)

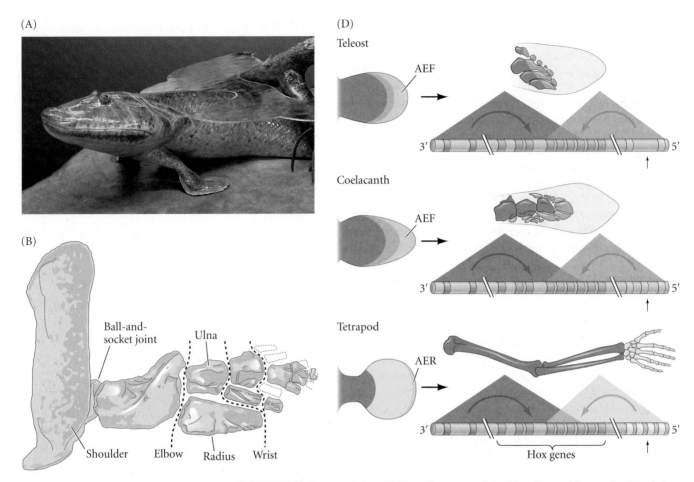

(D)

FIGURE 19.4 Limb evolution. (A) *Tiktaalik roseae*, a fish with wrists and fingers, lived in shallow waters about 375 million years ago. This reconstruction shows *Tiktaalik*'s fishlike gills, fins, scales, and (lack of) neck. The external nostrils on its snout, however, indicate that it could breathe air. (B) Fossilized *Tiktaalik* bones reveal the beginnings of digits, wrists, elbows, and shoulders and suggest that this amphibian-like fish could propel itself on stream bottoms and perhaps live on land for short durations. The joints of the fin included a ball-and-socket joint in the shoulder and a planar joint that allowed the wrist to bend. Other joints allowed the animal to perch on its substrate. (C) Resistant contact with a substrate would allow flexion at the proximal joints (shoulder and elbow) and extension at the distal ones (wrist and digits). (D) Schematic representations of the limb buds, adult limb structures, and the *Hoxd* gene cluster with associated 5′ and 3′ *cis*-regulatory enhancers of a teleost (zebrafish), hypothetical coelacanth, and human. In this illustration of the limb buds, the proximal and distal mesenchyme tissue are red and orange/yellow, respectively, and the apical ectodermal fold and ridge (AEF/AER) are gray. In the adult limbs, the formation of fin radials are seen in both teleost fish and coelacanths (red for proximal, orange for distal, and gray for the dermoskeleton), whereas the digits of the autopod are present in humans (yellow). Adaptation of the 5′ *cis*-regulatory regions of Hox genes may underlie autopod evolution. (A © John Weinstein/Field Museum Library/ Getty Images; B,C after Shubin et al. 2006; D after Schneider and Shubin 2013; Woltering et al. 2014; Zuniga 2015.)

The fish fin bud is homologous to the limb bud and similarly has progress zone mesenchyme and an overlying apical ectodermal ridge (AER). However, after proximal patterning of the stylopod, the AER of the fin bud changes into an **apical ectodermal fold (AEF)** that promotes fin ray development as opposed to digits (**FIGURE 19.4D**). One hypothesis suggests that potential developmental delays in this AER-to-AEF transition would permit longer exposure to distal signals from the AER, enabling the progress zone mesenchyme to become increasingly permissive to autopod fates (digits). Moreover, changes in the spatial and temporal pattern of distal Hox genes may be responsible for the evolution of the tetrapod hand from the distal fin region of ancient lobe-finned fish (Schneider and Shubin 2013; Freitas and Gómez-Skarmeta 2014; Zuniga 2015). Increased numbers of *cis*-regulatory enhancers associated with the *Hoxa/d* clusters could provide one mechanism for heritable adaptation of the autopod (see Figure 19.4D). In further support of this model, researchers have identified both conserved enhancers ("global control region," GCR, and CsB) and tetrapod-specific enhancers (CsS) that are associated with early (proximal) and later (distal) Hox gene expression. In fact, mouse CsS enhancers can functionally drive reporter expression in transgenic zebrafish embryos similarly within the distalmost mesenchyme (**FIGURE 19.5**; Freitas et al. 2012).

In conclusion, this brief examination of the evolution of the tetrapod limb, from fish fins to human hands, has hopefully illuminated the importance of Hox gene regulation during limb development. Hox genes are critical for specifying fates along each axis of the limb, and their expression is under the influence of signals emanating from the flank (proximal) and AER (distal), among others. What are these signals, and how do they function to (1) determine where limbs form, (2) promote limb bud outgrowth and patterning, and (3) specify fates along the anteroposterior and dorsoventral axes?

((• SCIENTISTS SPEAK 19.3 This Web conference with Dr. Sean Carroll covers *cis*-regulatory elements in evolution.

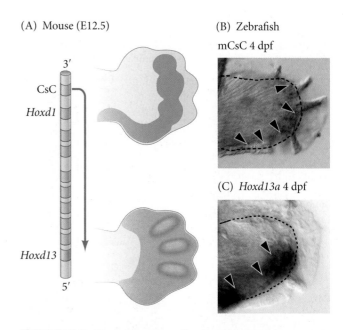

FIGURE 19.5 The tetrapod-specific *cis*-acting regulatory element (Csc) for mouse *Hoxd13* drives similar expression in the zebrafish distal fin bud. (A) *Hoxd13* is positioned downstream and responsive to the actions of the CsC enhancer in the distal mouse limb. (B) In the zebrafish fin, this same mouse CsC enhancer can drive reporter gene expression (mCsC:GFP; blue, arrowheads) in the distal limits of the endoskeletal territory (dashed line). (C) This mCsC:GFP expression pattern in the zebrafish fin is spatially similar to the endogenous expression pattern of *Hoxd13a* (blue, arrowheads) suggesting the CsC could be a conserved enhancer of Hox gene regulation. (After Freitas et al. 2012.)

Determining What Kind of Limb to Form and Where to Put It

Because the limbs, unlike the heart or brain, are not essential for embryonic or fetal life, one can experimentally remove or transplant parts of the developing limb, or create limb-specific mutants, without interfering with the vital processes of the organism. Such experiments have shown that certain basic "morphogenetic rules" for forming a limb appear to be the same in all tetrapods. Grafted pieces of reptile or mammalian limb buds can direct the formation of chick limbs, and regions taken from frog limb buds can direct the patterning of salamander limbs (Fallon and Crosby 1977; Sessions et al. 1989; Hinchliffe 1991). Moreover, *regenerating* salamander limbs appear to follow many of the same rules as developing limbs (see Chapter 22; Muneoka and Bryant 1982). What are these morphogenetic rules?

Specifying the limb fields

Limbs do not form just anywhere along the body axis; rather, they are generated at discrete positions. Early fate-mapping and transplantation studies in the chick demonstrated that there are two specific regions, or fields, of somitic and lateral plate mesoderm that are determined to form limbs long before any visible signs of wings or legs emerge. The mesodermal cells that give rise to a vertebrate limb have been identified by (1) removing certain groups of cells and observing that a limb does not develop in their absence ("lose it"; see Detwiler 1918; Harrison 1918), (2) transplanting groups of cells to a new location and observing that they form a limb in this new place ("move it"; see Hertwig 1925), and (3) marking groups of cells with dyes or radioactive precursors and observing that their descendants partake in limb development ("find it"; Rosenquist 1971).

Vertebrates have no more than four limb buds per embryo, and limb buds are always paired opposite each other with respect to the midline. Although the limbs of different

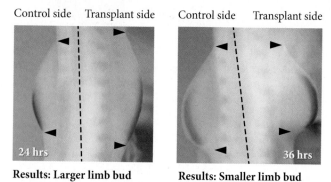

(A)

Transplantation of somites
from limb to flank level

Donor   Recipient

Limb field

Flank (non-limb)

Control side   Transplant side

24 hrs

Results: Larger limb bud

(B)

Transplantation of somites
from flank to limb level

Donor   Recipient

Limb field

Flank (non-limb)

Control side   Transplant side

36 hrs

Results: Smaller limb bud

FIGURE 19.6 Transplantation of different regions of the presomitic mesoderm (PSM) to the limb field causes changes in limb size. (A) Transplanting PSM from the level of the presumptive forelimb to the flank (region between the forelimb and hindlimb) results in a larger forelimb bud (area between arrowheads). (B) Transplantation of PSM from the flank region to the level of the presumptive forelimb results in a smaller forelimb bud. (After Noro et al. 2011.)

vertebrates differ with respect to the somite level at which they arise, their position is constant with respect to the level of Hox gene expression along the anterior-posterior axis (see Chapter 9). For instance, in fish (in which the pectoral and pelvic fins correspond to the anterior and posterior limbs, respectively), amphibians, birds, and mammals, the forelimb buds are found at the most anterior expression region of *Hoxc6*, the position of the first thoracic vertebra[3] (Oliver et al. 1988; Molven et al. 1990; Burke et al. 1995). It is probable that positional information from the Hox gene expression domains causes the paraxial mesoderm in the limb-forming regions to be different from all other paraxial mesoderm. Transplantation experiments in which paraxial mesoderm (somites) from different locations is placed adjacent to the flank lateral plate shows that the paraxial mesoderm from limb-forming regions promotes limb bud formation, whereas the paraxial mesoderm from limbless flank actively represses limb formation (**FIGURE 19.6**; Noro et al. 2011).

When it first forms, the limb field has the ability to regulate for lost or added parts. In the tailbud stage of the yellow-spotted salamander (*Ambystoma maculatum*), any half of the limb disc is able to generate an entire limb when grafted to a new site (Harrison 1918). This potency can also be shown by splitting the limb disc vertically into two or more segments and placing thin barriers between the segments to prevent their reunion. When this is done, each segment develops into a full limb. Thus, like an early sea urchin embryo, the limb field represents a "harmonious equipotent system" wherein a cell can be instructed to form any part of the limb. The regulative ability of the limb bud has recently been highlighted by a remarkable experiment of nature. In numerous ponds in the United States, multilegged frogs and salamanders have been found (**FIGURE 19.7**). The presence of these extra appendages has been linked to the infestation of the larval abdomen by parasitic trematode worms. The eggs of these worms apparently split the developing tadpole limb buds in several places, and the limb bud fragments develop as multiple limbs (Sessions and Ruth 1990; Sessions et al. 1999).

Induction of the early limb bud

Differential Hox gene expression along the anterior-posterior axis in the trunk sets up a prepattern of tissue identities that includes the limb field location, but what mechanisms are then triggered to initiate limb bud formation? The process can be divided into four stages: (1) making mesoderm permissive for limb formation; (2) specifying forelimb and hindlimb; (3) inducing epithelial-to-mesenchymal transitions; and (4) establishing two positive feedback loops for limb bud formation.

1. MAKING MESODERM PERMISSIVE FOR FORELIMB FORMATION BY RETINOIC ACID In Chapter 17, we described the antagonistic relationship of retinoic acid (RA) and Fgf8 during somitogenesis. Recall that Fgf8 is expressed by the caudal progenitor zone located just posterior to the forelimb field (and present in a gradient that is high

---

[3] Interestingly, Hox gene expression in at least some snakes (such as *Python*) creates a pattern in which each somite is specified to become a thoracic (ribbed) vertebra. The patterns of Hox gene expression associated with limb-forming regions are not seen (see Chapter 13; Cohn and Tickle 1999).

FIGURE 19.7 Multilimbed Pacific tree frog (*Hyla regilla*), the result of infestation of the tadpole-stage developing limb buds by trematode cysts. The parasitic cysts apparently split the developing limb buds in several places, resulting in extra limbs. In this adult frog's skeleton, the cartilage is stained blue, and the bones are stained red. (Courtesy of S. Sessions.)

posteriorly along the presomitic mesoderm), whereas RA is generated more anteriorly in somites and anterior presomitic mesoderm; relevant for forelimb development, however, Fgf8 is also expressed within the heart lateral plate mesoderm located just anterior to the forelimb field (**FIGURE 19.8**). Investigations into forelimb development in chick and mouse, as well as the pectoral fin of zebrafish, suggest that the anterior and posterior expression of Fgf8 functions to inhibit forelimb bud initiation (Tanaka 2013; Cunningham and Duester 2015). For instance, gain of Fgf8 function by its direct application or by constitutive activation of FGF receptors (FgfR) results in a loss of the forelimb field (marked by loss of expression of *Tbx5*; see Figure 19.8C) and truncated limbs (Marques et al. 2008; Cunningham et al. 2013).

In contrast, RA is present throughout the somitic regions of the trunk adjacent to the forelimb field, where it is needed to repress trunk Fgf8 expression and thus promote forelimb bud initiation. Targeted loss of RA synthesis (by mutation or pharmacological inhibition) in the vertebrate forelimb results in an expansion of Fgf8 expression toward the presumptive forelimb field, a reduction in *Tbx5* expression, and the failure to form forelimb buds—all of which are consistent with a gain of Fgf8 signaling. Because RA functions as a transcription factor ligand and has been shown to directly repress *Fgf8* gene transcription (Kumar and Duester 2014),

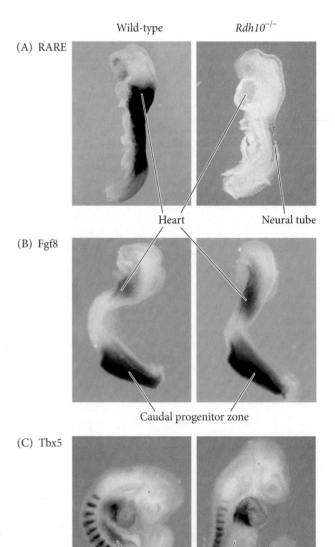

(A) RARE

Wild-type   *Rdh10⁻ᐟ⁻*

Heart   Neural tube

(B) Fgf8

Caudal progenitor zone

(C) Tbx5

Forelimb field

FIGURE 19.8 Antagonism between retinoic acid and Fgf8 determines the pattern of Tbx5 expression in the mouse forelimb field. As seen in control subjects at the left, reporter expression from the RA regulatory element (RARE) falls directly between Fgf8 expression in the heart (anterior) and throughout the caudal progenitor zone. (A) RARE reporter expression is almost completely repressed by the loss of the enzyme retinoid dehydrogenase 10 (*Rdh10* mutant, right), the exception being minimal expression in the neural tube. (B) In contrast, loss of *Rdh10* results in an expansion of Fgf8 expression into the axial region typical occupied by RA. (C) Lack of RA signaling through the loss of *Rdh10* also leads to a reduction of Tbx5 expression in the forelimb field. (From Cunningham et al. 2013.)

FIGURE 19.9 Model for forelimb field initiation. (A) Initially, axial-level Hox genes regulate Fgf8 and retinoic acid expression, which function as antagonistic signals that induce Tbx5 expression for initiation of the forelimb field. (B) Model of the positive feedback loops between signaling factors that promote forelimb development across species. (C) Many of the essential factors patterning hindlimb development between chick and mouse are the same and also generate positive feedback loops of signaling. Some of the initiation factors for hindlimb induction do differ, however; namely, Islet1 is required for the mouse but not chick hindlimb. (A after Cunningham and Duester 2015.)

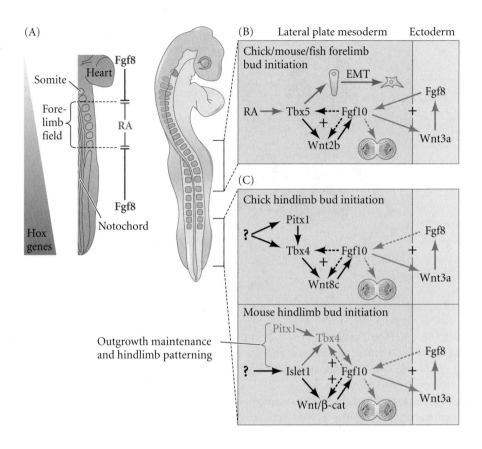

the current model of forelimb bud initiation begins with RA restricting Fgf8 expression from the presumptive limb region. In the absence of Fgf8, the lateral plate mesoderm is permissive for forelimb bud formation and development (**FIGURE 19.9A**; Tanaka 2013; Cunningham and Duester 2015).

Although RA-Fgf8 antagonism plays a role during forelimb initiation and the early stages of somitogenesis, RA is largely dispensable for hindlimb development. Loss of RA synthesis in mice does not affect hindlimb bud formation, and overall hindlimb size and patterning are normal (Cunningham et al. 2013). Thus, it is currently unknown what signaling mechanisms underlie hindlimb bud initiation.

2. FORELIMB AND HINDLIMB SPECIFICATION BY TBX5 AND ISLET1, RESPECTIVELY

Early specification of forelimb and hindlimb identity begins in the limb fields prior to bud formation through the expression of particular transcription factors (Agarwal et al. 2003; Grandel and Brand 2011). In mice, the gene encoding Tbx5 is transcribed in limb fields of the forelimbs, while the genes encoding Islet1, Tbx4, and Pitx1 are expressed in presumptive hindlimbs[4] (Chapman et al. 1996; Gibson-Brown et al. 1996; Takeuchi et al. 1999; Kawakami et al. 2011). Downstream of the regulatory function of these transcription factors is Fgf10, the primary inducer for limb bud formation through the initiation of cell shape changes and proliferation for outgrowth, as discussed below (**FIGURE 19.9B,C**).

Several laboratories (e.g., Logan et al. 1998; Ohuchi et al. 1998; Rodriguez-Esteban et al. 1999; Takeuchi et al. 1999) have provided gain-of-function evidence that Tbx4 and Tbx5 are critical in the specification of hindlimbs and forelimbs, respectively. Before limb formation, there are normally regions of Tbx4 expression in the posterior portion

---

[4] *Tbx* stands for "T-box," a specific DNA-binding domain. The *T* (*Brachyury*) gene and its relatives have a sequence that encodes this domain. We discussed *Tbx5* in the context of heart ventricle development in Chapter 18.

of the lateral plate mesoderm (including the region that will form the hindlimbs) and regions of Tbx5 expression in the anterior portion (including the region that will form the forelimbs). When Fgf10-secreting beads were used to induce an ectopic limb between the chick hindlimb and forelimb buds (**FIGURE 19.10A,B**), the type of limb produced was determined by which Tbx protein was expressed. Limb buds induced by placing FGF beads close to the hindlimb (opposite somite 25) expressed *Tbx4* and became hindlimbs. Limb buds induced close to the forelimb (opposite somite 17) expressed *Tbx5* and developed as forelimbs (wings). Limb buds induced in the center of the flank tissue expressed *Tbx5* in the anterior portion of the limb and *Tbx4* in the posterior portion; these limbs developed as chimeric structures, with the anterior resembling a forelimb and the posterior resembling a hindlimb (**FIGURE 19.10C–E**). Moreover, when a chick embryo was made to express *Tbx4* throughout the flank tissue (by infecting the tissue with a virus that expressed *Tbx4*), limbs induced in the anterior region of the flank often became legs instead of wings (**FIGURE 19.10F,G**).

In further support of Tbx5 being the critical factor for the initiation and specification of the forelimb limb bud, loss of the *Tbx5* gene in chicks, mice, and fish all result in a complete failure of forelimb formation that includes even the most proximal shoulder/girdle structure (Garrity et al. 2002; Agarwal et al. 2003; Rallis et al. 2003). However, the role of Tbx4 in hindlimb specification may differ between chicks and mice. In chicks, loss of Tbx4 function in the hindlimb field completely inhibits leg initiation and growth (Takeuchi et al. 2003); in mice, hindlimb bud growth and initial patterning appear normal

FIGURE 19.10 Fgf10 expression and action in the developing chick limb. (A) Fgf10 becomes expressed in the lateral plate mesoderm in precisely those positions (arrows) where limbs normally form. (B) When transgenic cells that secrete Fgf10 are placed in the flanks of a chick embryo, the Fgf10 can cause the formation of an ectopic limb (arrow). (C) Limb type in the chick is specified by Tbx4 and Tbx5. In situ hybridizations show that during normal chick development, Tbx5 (blue) is found in the anterior lateral plate mesoderm, whereas Tbx4 (red) is found in the posterior lateral plate mesoderm. Tbx5-containing limb buds produce wings, whereas Tbx4-containing limb buds generate legs. If a new limb bud is induced with an FGF-secreting bead, the type of limb formed depends on which *Tbx* gene is expressed in the limb bud. If placed between the regions of *Tbx4* and *Tbx5* expression, the bead will induce the expression of *Tbx4* posteriorly and *Tbx5* anteriorly. The resulting limb bud will also express *Tbx5* anteriorly and *Tbx4* posteriorly and will generate a chimeric limb. (D) Expression of *Tbx5* in the forelimb (w, wing) buds and in the anterior portion of a limb bud induced by an FGF-secreting bead (red arrow). Staining for *Mrf4* mRNA marks the somite positions.) (E) Expression of *Tbx4* in the hindlimb (le, leg) buds and in the posterior portion of an FGF-induced limb bud (red arrow). (F) A chimeric limb (red arrow) induced by an FGF bead. (G) At a later stage of development the chimeric limb contains anterior wing structures (feathers) and posterior leg structures (scales). (After Ohuchi et al. 1998, Ohuchi and Noji 1999; photographs courtesy of S. Noji.)

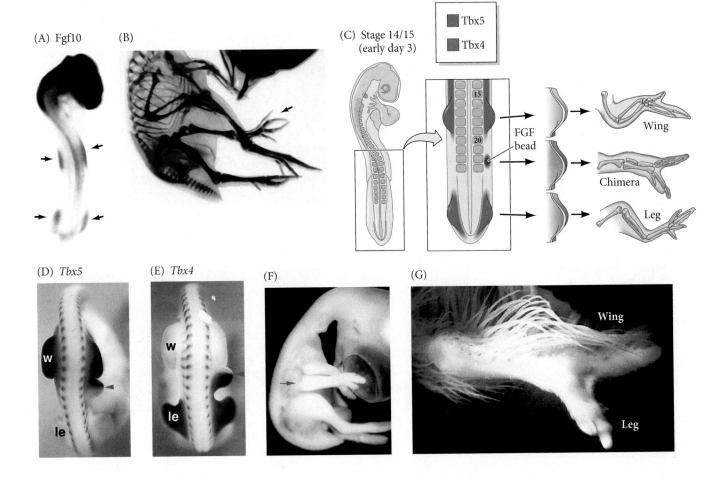

(A) Fgf10

(B)

(C) Stage 14/15 (early day 3)

Tbx5
Tbx4

FGF bead

Wing

Chimera

Leg

(D) *Tbx5*   (E) *Tbx4*   (F)   (G)

w   w   Wing

le   le   Leg

when *Tbx4* is knocked out (Naiche and Papaioannou 2003), although leg development is arrested prematurely. This finding suggests that in mice, Tbx4 normally plays more of a role in the maintenance of hindlimb outgrowth than its initial formation.

More recent investigations have revealed two additional transcription factors involved in the initiation of the hindlimb: Pitx1 and Islet1. Indeed, misexpression of *Pitx1* in the mouse forelimb causes its muscles, bones, and tendons to develop into ones that look like those of a hindlimb (Minguillon et al. 2005; DeLaurier et al. 2006; Ouimette et al. 2009); *Tbx4* expressed in the mouse forelimb will not have this effect. Additionally, Pitx1 protein activates hindlimb-specific genes in the forelimb, including *Hoxc10* and *Tbx4*. Interestingly, a mutation in the human *PITX1* gene that causes a haploinsufficiency in Pitx1 protein results in a bilateral "club foot" phenotype (Alvarado et al. 2011). These results indicate that Pitx1 is sufficient for hindlimb specification; however, the hindlimb is neither completely lost nor severely mispatterned in *Pitx1*-null mice, although some hindlimb structures are malformed (Duboc and Logan 2011). This observation suggests that yet another factor may be involved. *Islet1*, a homeodomain transcription factor, is transiently expressed in the hindlimb field before *Fgf10* expression and leg bud formation in mice (Yang et al. 2006). When *Islet1* is inactivated specifically in the lateral plate mesoderm, hindlimbs do not form, which is consistent with a role in hindlimb initiation (Itou et al. 2012). Transcriptional regulation of *Islet1* and of *Pitx1* are independent of each other's function, as are their roles in hindlimb development. Despite both genes being documented to similarly upregulate *Fgf10* and *Tbx4*, Islet1 functions to induce hindlimb bud initiation (see Figure 19.9C, black arrows), whereas Pitx1 plays a role in hindlimb patterning (see Figure 19.9C, gray arrows).

3. INDUCTION OF EPITHELIAL-TO-MESENCHYMAL TRANSITIONS BY TBX5 Prior to limb bud formation, the lateral plate mesoderm of the somatopleure displays characteristics of a pseudostratified epithelium with apical-to-basal polarity (**FIGURE 19.11**). This tissue architecture is perplexing, since these cells contribute to the progress zone of the limb bud, which is made up of mesenchymal cells. Research by Clifford Tabin's lab has demonstrated that the epithelial cells making up the mesoderm of the early somatopleure undergo an epithelial-to-mesenchymal transition (EMT) specifically in the limb fields before any signs of such cell behavior is observed in the mesoderm of flank regions (Gros and Tabin 2014). Lineage tracing of the somatopleural mesoderm reveals a visible change from epithelial to mesenchymal morphology over the course of 24 hours. At least in the case of the forelimb, *Tbx5* knockout mice show a significant loss of limb bud mesenchyme, suggesting that Tbx5 is a major regulator of EMT in the forelimb field (see Figure 19.9B, green arrows). It is unknown whether Islet1, Fgf10, or other factors (Tbx4, Pitx1) are similarly required for EMT in the hindlimb.

FIGURE 19.11 Epithelial-to-mesenchymal transitions of the epithelial mesoderm of the somatopleure during limb bud formation. (A) The mesoderm (lateral plate mesoderm) of the early somatopleure is epithelial. (B–D) Over a 24-hour period, this mesoderm (labeled with GFP) undergoes an epithelial-to-mesenchymal transition in the areas of the limb fields. (After Gros and Tabin 2014.)

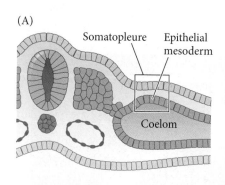

(A)

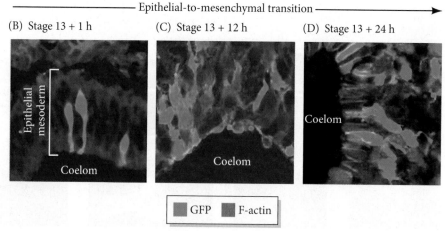

◄──────── Epithelial-to-mesenchymal transition ────────►

(B) Stage 13 + 1 h   (C) Stage 13 + 12 h   (D) Stage 13 + 24 h

GFP   F-actin

4. ESTABLISHMENT OF TWO POSITIVE FEEDBACK LOOPS FOR LIMB BUD FORMATION BY FGF-WNT   Through the upregulation of *Tbx5* in the forelimb and *Islet1* in the hindlimb, the mesenchyme cells commit toward limb bud development and secrete the paracrine factor Fgf10. Fgf10 provides the signal to initiate and propagate the limb-forming interactions between ectoderm and mesoderm, and these signaling interactions directly promote limb bud formation and growth.

Formation of the limb is arguably one of the most remarkable morphological events in embryonic development. Limb development is a prolonged process of outgrowth, and Fgf10 possesses the morphogenetic power to induce limb formation. Remember that a bead containing Fgf10 placed ectopically beneath the flank ectoderm can induce extra limbs to form (see Figure 19.10B,C; Ohuchi et al. 1997; Sekine et al. 1999). If Fgf10 is the signaling factor inducing the growth of the limb bud, how might this growth be maintained?

The answer is, *positive feedback loops*. Among the downstream targets of Fgf10 signaling are Wnt/β-catenin and the transcription factors it initiates, which perpetuate Fgf10 signaling (see Figure 19.9B,C, black dashed arrows). This loop could be thought of as the rate-limiting step of limb induction: once Fgf10 is expressed, limb bud formation and growth will commence. In the presumptive forelimb field, for example, Tbx5 induces Wnt2b, which then upregulates Fgf10, which in turn positively feeds back to maintain activation of both Wnt2b and Tbx5. As a result of maintaining Fgf10 signaling, limb buds will form, and the limb will begin to grow—but why? What else is Fgf10 doing to more directly lead to limb bud outgrowth?

Fgf10 secreted by the limb field mesenchyme induces the overlying ectoderm to form the apical ectodermal ridge (see Figure 19.20; Xu et al. 1998; Yonei-Tamura et al. 1999). The AER runs along the distal margin of the limb bud and will become a major signaling center for the developing limbs (Saunders 1948; Kieny 1960; Saunders and Reuss 1974; Fernandez-Teran and Ros 2008). Fgf10 is capable of inducing the AER in the competent ectoderm between the dorsal and ventral sides of the embryo. The boundary where dorsal and ventral ectoderm meet is critical to the placement of the AER.

In mutants in which the limb bud is dorsalized and there is no dorsal-ventral junction (as in the chick mutant *limbless*), the AER fails to form, and limb development ceases (Carrington and Fallon 1988; Laufer et al. 1997a; Rodriguez-Esteban et al. 1997; Tanaka et al. 1997). Fgf10 stimulates Wnt3a (Wnt3a in chicks; Wnt3 in humans and mice) in the prospective limb bud surface ectoderm. The Wnt protein acts through the canonical β-catenin pathway to induce Fgf8 expression in the ectoderm (Fernandez-Teran and Ros 2008). This relay of signaling represents a foundational stage in the formation of limb buds, because once Fgf8 is made in the surface ectoderm, the surface ectoderm elongates to physically become the AER. One of the main functions of the AER is to tell the mesenchyme cells directly beneath it to continue making Fgf10. In this way, a second *positive feedback loop* is created wherein mesodermal Fgf10 tells the surface ectoderm to continue to make Fgf8, and the surface ectoderm continues to tell the underlying mesoderm to make Fgf10 (see Figure 19.9B,C, red arrows). Each FGF activates the synthesis of the other (Mahmood et al. 1995; Crossley et al. 1996; Vogel et al. 1996; Ohuchi et al. 1997; Kawakami et al. 2001). The continued expression of FGFs maintains mitosis in the mesenchyme beneath the AER, which fuels the outgrowth of the limb.

Outgrowth: Generating the Proximal-Distal Axis of the Limb

The apical ectodermal ridge

The Fgf10-induced apical ectodermal ridge is a multipurpose signaling center that will influence patterning along all axes of limb development (**FIGURE 19.12A,B**). The diverse roles of the AER include (1) maintaining the mesenchyme beneath it in a plastic, proliferating state that enables the linear (proximal-distal, or shoulder-finger) growth

Developing Questions

Autonomous or nonautonomous? Perhaps that should be the question regarding hindlimb bud formation. Retinoic acid antagonism of Fgf8 is an important nonautonomous mechanism required to induce forelimb development, but this "battle of the paracrine factors" does not play out for induction of the hindlimb field. Is there enough prepatterning from Hox and Islet1 gene expression to support an autonomous mechanism of hindlimb bud induction? Moreover, how important is the fourth dimension—that of time—in influencing hindlimb development? How might you experimentally approach these questions?

FIGURE 19.12 Manipulation of the apical ectodermal ridge (AER). (A) In the normal 3-day chick embryo, Fgf8 (dark purple) is expressed in the AER of both the forelimb and hindlimb buds. (B) Expression of *Fgf8* RNA in the AER, the source of mitotic signals to the underlying mesoderm. (C) Summary of experiments demonstrating the effect of the AER on the underlying mesenchyme. (A courtesy of A. López-Martínez and J. F. Fallon; B courtesy of J. C. Izpisúa-Belmonte; C after Wessells 1977.)

(A) (B)

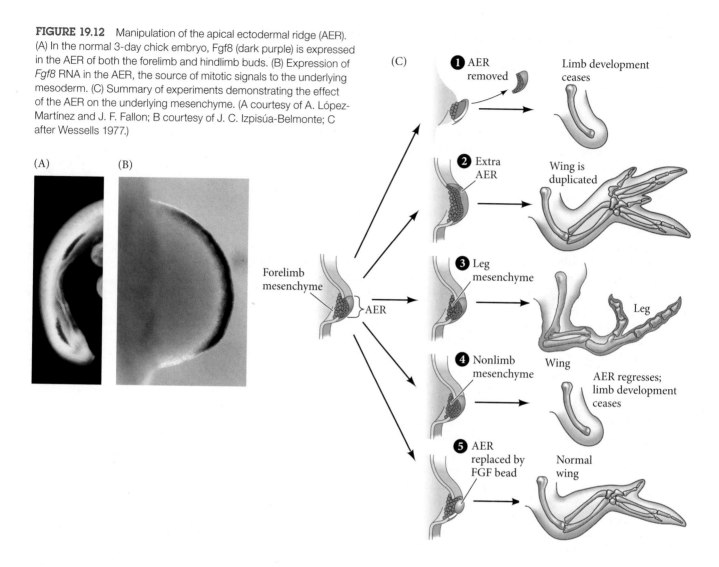

(C)

❶ AER removed — Limb development ceases

❷ Extra AER — Wing is duplicated

Forelimb mesenchyme — AER

❸ Leg mesenchyme — Leg

❹ Nonlimb mesenchyme — Wing / AER regresses; limb development ceases

❺ AER replaced by FGF bead — Normal wing

of the limb; (2) maintaining the expression of those molecules that generate the anterior-posterior (thumb-pinkie) axis; and (3) interacting with the proteins specifying the anterior-posterior and dorsal-ventral (knuckle-palm) axes so that each cell is given instructions on how to differentiate (see Figure 19.1).

The proximal-distal growth and differentiation of the limb bud are made possible by a series of interactions between the AER and the limb bud mesenchyme directly (200 µm) beneath it. As mentioned previously, this distal mesenchyme is called the progress zone (PZ) mesenchyme because its proliferative activity extends the limb bud (Harrison 1918; Saunders 1948; Tabin and Wolpert 2007). These interactions were demonstrated by the results of several experiments on chick embryos (**FIGURE 19.12C**):

1. If the AER is removed at any time during limb development, further development of distal limb skeletal elements ceases.

2. If an extra AER is grafted onto an existing limb bud, supernumerary structures are formed, usually toward the distal end of the limb.

3. If leg mesenchyme is placed directly beneath the wing AER, distal hindlimb structures (toes) develop at the end of the limb. (If this mesenchyme is placed farther from the AER, however, the hindlimb [leg] mesenchyme becomes integrated into wing structures.)

4. If limb mesenchyme is replaced by nonlimb mesenchyme beneath the AER, the AER regresses and limb development ceases.

Thus, although the mesenchyme cells induce and sustain the AER and determine the type of limb to be formed, the AER is responsible for the sustained outgrowth and development of the limb (Zwilling 1955; Saunders et al. 1957; Saunders 1972; Krabbenhoft and Fallon 1989). The AER keeps the mesenchyme cells directly beneath it in a state of mitotic proliferation and prevents them from forming cartilage (see ten Berge et al. 2008).

FGF and Wnt proteins also regulate the shape and growth of the early limb bud. The initial limb mesenchyme cells are not randomly organized. Rather, they show a polarity such that they extend their long axes perpendicular to the ectoderm (Gros et al. 2010; Sato et al. 2010). Wnt signaling appears to determine this orientation and also promotes cell division in the plane of the ectoderm, thereby extending the limb bud outward. FGF signals increase the velocity of cell migration, causing cells to migrate distally. As a result, the limb bud flattens in the anterior-posterior axis while growing distally (i.e., toward the source of Fgf8).

Fgf8 is the major active factor in the AER, and Fgf8-secreting beads can substitute for the AER functions in inducing limb growth (see Figure 19.12C, panel 5). There are other FGFs made by the AER, including Fgf4, Fgf9, and Fgf17 (Lewandoski et al. 2000; Boulet et al. 2004). Loss of any one of these FGFs cause only mild to no defects in skeletal pattern, though, suggesting significant redundancy exists within this family for limb patterning. However, genetic removal of multiple FGF genes demonstrated increasingly severe and specific skeletal malformations with each additional FGF gene removed, which supports the idea that AER-derived FGFs exhibit some control over patterning (see Figure 19.9B,C, red arrows; Mariani et al. 2008).

 SCIENTISTS SPEAK 19.4 A Web conference with Dr. Francesca Mariani on the instructive roles of FGF signaling in proximal-to-distal patterning.

WEB TOPIC 19.1 **INDUCTION OF THE AER** This complex event involves interaction between the dorsal and ventral compartments of the ectoderm. The Notch pathway may be critical; misexpression of the genes in this pathway can result in the absence or duplication of limbs.

Specifying the limb mesoderm: Determining the proximal-distal polarity

THE ROLE OF THE AER In 1948, John Saunders made a simple and profound observation: if the AER is removed from an early-stage wing bud, only a humerus forms. If the AER is removed slightly later, humerus, radius, and ulna form (Saunders 1948; Iten 1982; Rowe et al. 1982). Explaining how this happens has not been easy. First it had to be determined whether the positional information for proximal-distal polarity resided in the AER or in the progress zone mesenchyme. Through a series of reciprocal transplantations, this specificity was found to reside in the mesenchyme. If the AER had provided the positional information—somehow instructing the undifferentiated mesoderm beneath it as to what structures to make—then older AERs combined with younger mesoderm should have produced limbs with deletions in the middle, whereas younger AERs combined with older mesenchyme should have produced duplications of structures. This was not found to be the case; rather, normal limbs formed in both experiments (Rubin and Saunders 1972). But when the entire progress zone (including both the mesoderm and the AER) from an early embryo was placed on the limb bud of a later-stage embryo, new proximal structures were produced beyond those already present (**FIGURE 19.13A**). Conversely, when old progress zones were added to young limb buds, distal structures developed such that digits were seen to emerge from the humerus without an intervening ulna and radius (**FIGURE 19.13B**); Summerbell and Lewis 1975). These experiments demonstrated that the mesenchyme specifies the skeletal identities along the proximal-distal axis, which beckons the next question: "How?"

VADE MECUM

An interview with Dr. John Saunders contains movies of his work on limb development, which identified the AER and the ZPA as two of the major signaling centers in limb formation. His transplantation studies provided the framework for the molecular characterization of the mechanisms of limb formation.

FIGURE 19.13 Control of proximal-distal specification of the limb is correlated with the age of the progress zone (PZ) mesenchyme. (A) An extra set of ulna and radius formed when an early wing-bud progress zone was transplanted to a late wing bud that had already formed ulna and radius. (B) Lack of intermediate structures are seen when a late wing-bud progress zone was transplanted to an early wing bud. (From Summerbell and Lewis 1975, courtesy of D. Summerbell.)

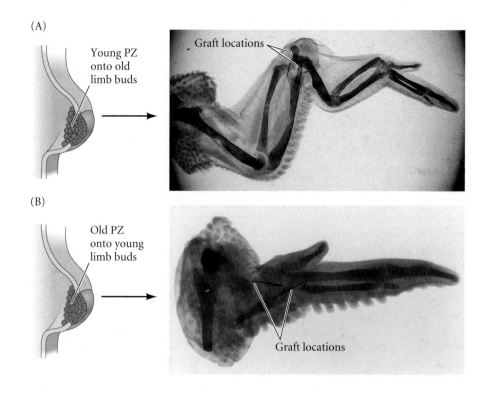

(A) Young PZ onto old limb buds — Graft locations

(B) Old PZ onto young limb buds — Graft locations

GRADIENT MODELS OF LIMB PATTERNING In 2010, evidence on chick limb patterning converged on a model of two opposing gradients: a gradient of FGFs and Wnts from the distal AER, and a second gradient of retinoic acid from the proximal flank tissue (**FIGURE 19.14A**). Such a two-gradient explanation had been proposed earlier for amphibian limb regeneration (see Maden 1985; Crawford and Stocum 1988) and had even been hypothesized for embryonic limb patterning (see Mercader et al. 2000). Actual evidence for the model eventually came from mesenchyme transplantation experiments by Cooper and colleagues in the United States and by Roselló-Díez and colleagues in Spain (Cooper et al. 2011; Roselló-Díez et al. 2011).

Researchers took undifferentiated limb bud mesenchyme cells and "repacked" them into the ectodermal hull of a young limb bud. As expected, the age of the mesenchyme determined the type of bones formed. However, the type of bone formed became more proximal (in the stylopod direction) if young limb bud mesenchyme had been treated with RA in the presence of Wnt and FGF; and it became more distal (toward the autopod) if the mesenchyme had been treated with only FGFs and Wnts (**FIGURE 19.14B,C**). Moreover, if the actions of FGFs were inhibited, the bones became more proximal; and if RA synthesis was inhibited, the bones became more distal. Thus, there appears to be a balance between the proximalizing of the bones by RA from the flank and the distal-izing of bones by the FGFs and Wnts of the AER. The opposing gradients may accomplish this balance by laying down a segmental pattern of different transcription factors in the mesenchyme. Such opposing gradients are probably a common mechanism for cell specification, as we've already seen in the early *Drosophila* embryo (see Chapter 9).

There is mechanistic support for this model based on the functional actions of RA and Fgf8. As we described earlier for forelimb field initiation, RA and Fgf8 exhibit an antagonistic relationship toward one another that is mediated on at least two levels: direct repression and the differential regulation of gene targets (see Figure 19.14A). RA functions as a direct transcriptional repressor of *Fgf8* expression (Kumar and Duester 2014); therefore, as outgrowth of the limb bud progresses, the AER (and source of Fgf8) will move outside the reach of RA, enabling greater *Fgf8* expression over time. In a more direct counter to RA, Fgf8 upregulates cytochrome P450 26 (CYP26) proteins that directly degrade RA (Probst et al. 2011). In addition, RA and Fgf8 differentially regulate

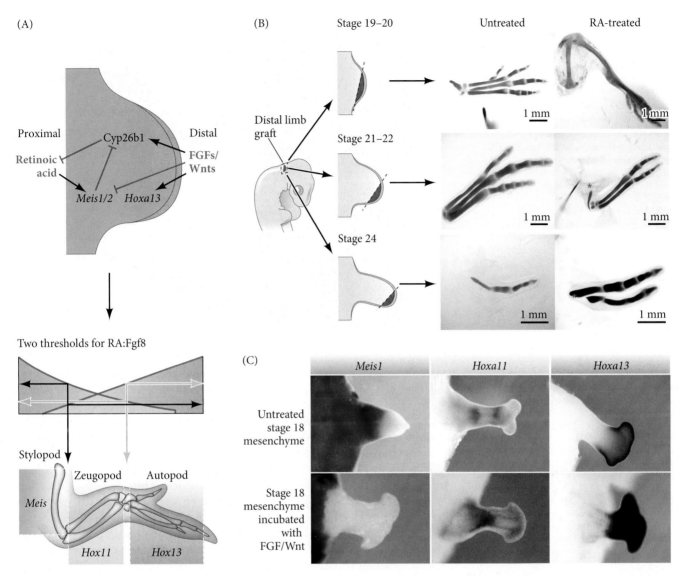

FIGURE 19.14 (A) Model for limb patterning, whereby the proximal-distal axis is generated by opposing gradients of retinoic acid (RA) (blue shading) from the proximal flank and of FGFs and Wnts (pink shading) from the distal AER. (B) Grafting procedure in chick embryos showing transplantation of limb bud tips to the head region of another chick embryo. Bud tips were either untreated or treated by inserting a bead soaked in RA (asterisks mark bead location). Results showed that RA proximalizes the bones forming from the transplanted mesenchyme. Untreated limb bud tips (shaded) generated specific limb cartilage depending on age. However, when the bud tip was treated with 1 mg/ml RA, the skeleton that formed became more proximal. (C) Treatment with FGFs and Wnts changes the expression pattern of specific proximodistal transcription factors in transplanted mesenchyme (dark staining). *Meis1* is specific for the stylopod, *Hoxa11* is specific for the zeugopod, and *Hoxa13* is specific for the autopod. Mesenchyme from the earliest limb bud (stage 18) will form all three types of cartilage. When the mesenchyme is first incubated in Fgf8 and Wnt3a, however, the autopod transcription factor (*Hoxa13*) is greatly expressed, whereas the stylopod marker (*Meis1*) is drastically reduced. The addition of RA in culture is required to maintain competence to express *Meis1* proximally (not shown). (A after Macken and Lewandoski 2011; B after Roselló-Díez et al. 2011; C after Cooper et al. 2011, photographs courtesy of the authors.)

proximal-distal determining genes, such as *Meis1/2*, *Hoxa11*, and *Hoxa13* in the stylopod, zeugopod, and autopod, respectively (Cooper et al. 2011). For instance, RA promotes the expression of the more proximal *Meis1*, whereas Fgf8 inhibits expression of this gene. The opposite relationship holds true for the more distal *Hoxa13* (see Figure 19.14A,C). The upregulation of *Meis1* by RA is a protective one because, aside from promoting extreme proximal cell fates, Meis1 protein also represses *CYP26b1* transcription. In fact, recent refinements of this model suggest that there are two distinct thresholds of RA-to-Fgf8

FIGURE 19.15 Epigenetic regulation of Hox gene expression in the chick limb. (A) *Hoxa13* is normally not expressed during early limb bud outgrowth. (B) However, when histone deacetylation is inhibited with a TSA coated bead (enabling more acetylation and more open states of chromatin) *Hoxa13* expression is upregulated (arrowhead), and the zeugopod is severely reduced (brackets). Asterisks mark the location of beads in the control (A, no TSA) and in the HDAC inhibited (B, +TSA). (From Miguel Torres Sanchez.)

(A) Control

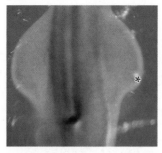

(B) HDAC inhibitor (increased acetylation)

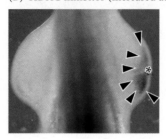

signaling. A relatively high RA-to-Fgf8 threshold determines the stylopod-to-zeugopod transition, while a low RA-to-Fgf8 threshold determines the zeugopod-to-autopod transition (Roselló-Díez et al. 2014). Moreover, there appears to be an *epigenetic hold* on autopod gene (*Hoxa13*) expression to allow the necessary time for zeugopod development. Pharmacological inhibition of histone deacetylases with a TSA-coated bead implanted into the early limb bud results in precocious *Hoxa13* expression and specifically reduced skeletal elements of the zeugopod (**FIGURE 19.15**; Roselló-Díez et al. 2014).

Taken together, these data support a dual gradient model for proximal-to-distal patterning in the chick limb (Roselló-Díez et al. 2014; reviewed by Tanaka 2013; Cunningham and Duester 2015; Zuniga 2015). The patterning stage is set prior to signs of limb formation in the lateral plate mesoderm, where RA is highly expressed and induces *Meis1/2* expression throughout the limb field and early limb bud (thus supporting stylopod specification). Soon after, an opposing gradient of Fgf8 and Wnt from the AER antagonizes RA signaling along the proximal-distal axis. A distancing of the AER from the flank by proliferative outgrowth results in diminished RA signaling and enhancement of Fgf8, until a threshold of RA:Fgf8 signaling triggers *Hoxa11* gene expression and *Meis1/2* downregulation (yielding zeugopod differentiation). As a result of the reduction in *Meis1/2* function, CYP26b1-mediated degradation of RA intensifies to reach the next threshold for autopod development. Although at this point the distal mesenchyme may be permissive for autopod fate specification, it is not until chromatin regulation permits access for the transcription of *Hoxa13* that autopod differentiation commences (see Figure 19.14). This epigenetic regulation of delay in autopod development enables greater cell contributions to the zeugopod lineage, influencing its size, which may represent an important mechanism employed throughout development to shape the embryo.

The dual gradient model is not without its controversies. The model is largely based on data generated solely in the chick, and some inconsistencies exist with comparable data generated in the mouse. Most striking of these inconsistencies is that RA appears to be dispensable for hindlimb development and patterning in mice (Sandell et al. 2007; Zhao et al. 2009; Cunningham et al. 2011, 2013), thus leaving open the question as to whether opposing gradients of RA and FGF/Wnt function in the mouse forelimb as they do in the chick wing. However, several studies examining the loss of RA revealed that although mouse forelimbs were shortened, their patterning was relatively normal, including proximal expression of *Meis1/2* (Sandell et al. 2007; Zhao et al. 2009; Cunningham et al. 2011, 2013). This shortened forelimb phenotype has been interpreted to mean that RA plays a role in establishing the early limb field, as opposed to RA affecting patterning along the proximal-distal axis.

As an alternative to the dual gradient model, Cunningham and colleagues have proposed a single gradient model that focuses on the instructive patterning signals from proteins derived from the AER (Cunningham et al. 2013; reviewed in Cunningham and Duester 2015). Specifically, initial expression of *Meis1/2* throughout the early limb bud specifies stylopod fates; then, distal FGF expression functions both to repress *Meis1/2*, preventing the adoption of more proximal fates by distal mesenchyme, and to repress RA through Cyp26b1 induction, thus preventing RA from interfering with distal patterning. The keys to this proposed single gradient model are, first, that RA does not play an instructive role beyond forelimb field induction and, second, that the autonomously timed collinear expression of Hox genes along the proximal-distal axis (see Tarchini and Duboule 2006) will be permitted to appropriately pattern cell fates in the absence of excessive RA signaling.

Turing's model: A reaction-diffusion mechanism of proximal-distal limb development

Genes and proteins don't produce a skeleton. Cells do. The cell types of the stylopod and the autopod are identical; it is only how they are arranged in space that differs. Amazingly, disassociated limb mesenchyme placed in culture is capable of self-organizing, expressing 5′ Hox genes, and forming limblike structures with rods and nodules of cartilage (Ros et al. 1994), which raises the foundational question, how do these cells "know" to organize appropriately? Applied to the embryonic limb, why is only one cartilage element formed in the stylopod while two are formed in the zeugopod and several in the autopod? How do the gradients surrounding these cells tell them how to create different parts of the skeleton in different places? Why are the fingers and toes always at the distal end of the limb? The answers may come from a model that involves the diffusion of two or more negatively interacting signals. This is known as the reaction-diffusion mechanism for developmental patterning.

THE REACTION-DIFFUSION MODEL   The **reaction-diffusion mechanism** is a mathematical model formulated by Alan Turing (1952) to explain how complex chemical patterns can be generated out of substances that are initially homogenously distributed. Turing was the British mathematician and computer scientist who broke the German "Enigma" code during World War II, as recounted in the 2014 film *The Imitation Game*. Two years before his death, Turing provided biologists with a basic mathematical model for explaining how patterns can be self-organized. Although some scientists began applying his model in the 1970s to pattern chondrogenesis in the limb (Newman and Frisch 1979), it was not until quite recently, with the accumulation of experimental evidence, that the model gained broad acceptance.

The uniqueness of Turing's model lies in the "reaction" portion of his mechanism. There is no dependence on molecular prepatterns; rather, interactions between two molecules can spontaneously produce a nonuniform pattern (reviewed in an approachable manner in Kondo and Miura 2010). Turing realized that generation of such patterns would not occur in the presence of just a single diffusible morphogen, but that it *could* be achieved by two homogeneously distributed substances (that we will call morphogen *A*, for "activator," and morphogen *I*, for "inhibitor") *if the rates of production of each substance depended on the other* (**FIGURE 19.16A,B**).

The Turing model provides a framework for a system of "local autoactivation-lateral inhibition" (LALI) to generate stable patterns that could be used to drive developmental change (Meinhardt 2008). (Other "Turing-type" reaction-diffusion systems are also used by cells, with similar results.) In Turing's model, morphogen *A* promotes the production of more morphogen *A* (autoactivation) as well as production of morphogen *I*. Morphogen *I*, however, inhibits the production of morphogen *A* (lateral inhibition).

Developing Questions

Autonomous or nonautonomous yet again? It seems that the current challenge for the limb field (and you) will be to further investigate the nature of, and interactions between, autonomous and nonautonomous mechanisms. The model should include mechanisms—epigenetic or otherwise—for controlling developmental time. Do you think the evidence is there to support a dual gradient model, or are distal gradients of FGF proteins alone sufficient?

FIGURE 19.16 Reaction-diffusion (Turing) mechanism of pattern generation. (A) The Turing model is based on the interaction of two factors, one that is both auto-activating and able to activate its own inhibitor. These interactions can lead to self-generating patterns of alternative cell fates, which may resemble the stripes of a flag or more labyrinth-like patterns. (B) Generation of periodic spatial heterogeneity can occur spontaneously when two reactants, I and A, are mixed together under the conditions that I inhibits A, A catalyzes production of both I and A, and I diffuses faster than A (Time 1). Time 2 illustrates the conditions of the reaction-diffusion mechanism yielding a peak of A and a lower peak of I at the same place. (C) The distribution of the reactants is initially random, and their concentrations fluctuate over a given average. As A increases locally, it produces more I, which diffuses to inhibit more peaks of A from forming in the vicinity of its production. The result is a series of A peaks ("standing waves") at regular intervals. (D,E) Computer simulations of the limb elements that would result from a Turing mechanism of self-generation. (D) Cross-sectional view of the activator morphogen TGF-β at successive stages of chick limb development (increasing time is shown from bottom to top). The concentration of TGF-β is indicated by color (low = green; high = red). (E) Three-dimensional view of the cells undergoing condensation into bone (gray) as predicted by this computer simulation. Note that the number of "bones" in each region of the limb correlates with the number of TGF-β concentration peaks over developmental time, as shown in (D). (A from Kondo et al. 2010; D,E from Zhang et al. 2013.)

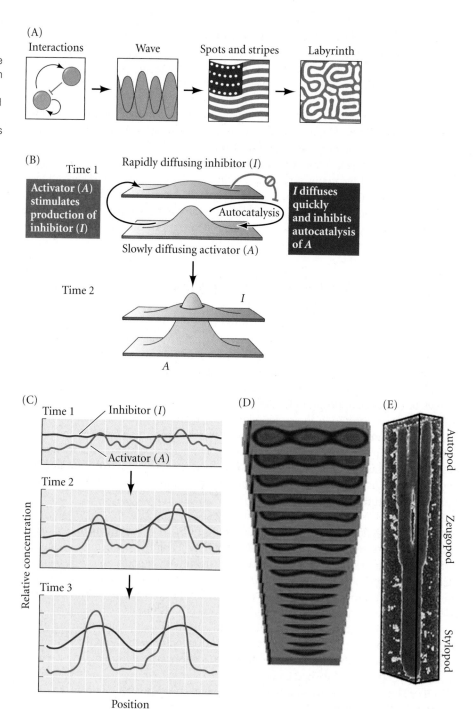

Turing's mathematics show that if I diffuses more readily than A, sharp waves of concentration differences will be generated for morphogen A (**FIGURE 19.16C**).

The diffusion of the interacting signals may initially be random, yet due to the activator-inhibitor dynamics of this LALI-type Turing model, there will be alternating areas of high and low concentrations of a morphogen, which can produce differential cell fates. When the concentration of the activating morphogen is above a certain threshold level, a cell (or group of cells) can be instructed to differentiate in a certain way.

"TURING" ALONG THE LIMB Turing's model has produced fascinating results when applied to limb development (**FIGURE 19.16D,E**). It appears that reaction-diffusion dynamics can tell us how the limb bud acquires its proximal-distal polarity as well as how the number of digits is regulated at the distal tip of the limb (Turing and digits will be covered later in this chapter). The reaction-diffusion system has been proposed to be sufficient for establishing patterns of precartilage and noncartilage tissues (Zhu et al. 2010).

Stuart Newman's laboratory has demonstrated that a reaction-diffusion mechanism can pattern limb mesenchyme, and that size and shape matter (Hentschel et al. 2004; Chaturvedi et al. 2005; Newman and Bhat 2007; Zhu et al. 2010; reviewed in Zhang et al. 2013). To mathematically model limb chondrogenesis within a Turing framework, the key parameters need to be identified. During chondrogenesis along the proximal-distal axis, the AER is seen as dividing the limb into two domains: the *inhibitory domain* (also called the *apical zone*), the most distal mesenchyme subjacent to the AER, in which precartilage condensation is repressed; and the *active zone*, which lies proximally adjacent to the inhibitory domain and is the morphogenetically active domain where cartilage-forming condensations coalesce. A third domain, the "frozen zone" well beyond the influence of the AER, contains the formed cartilage primordia of the skeleton at proximal regions of the developing limb (**FIGURE 19.17**).

As mentioned earlier, Gros and colleagues (2010) found that the Wnt and FGF proteins secreted from the AER induce specific patterns of cell division and growth in the underlying mesenchyme. Moreover, factors secreted by the AER keep the most distal mesenchyme in a plastic, undifferentiated state (Kosher et al. 1979), signaling that is in part mediated via FgfR1. It is within the more proximal active zone of the limb mesenchyme that the Turing parameters apply. The active and frozen zones are further defined (both in tissue and in differential equations of the math model) by their unique expression of FgfR2 and FgfR3, respectively (Szebenyi et al. 1995; Hentschel et al. 2004). The limb mesenchyme cells within the active zone synthesize *activators* of cartilage nodule formation. These activators include TGF-β, BMPs, activins, and certain carbohydrate-binding proteins called galectins. Galectins can induce the formation of certain cell adhesion molecules and extracellular matrix proteins, such as fibronectin, that cause cells to aggregate together to form the cartilaginous skeleton. These same cells, however, also synthesize *inhibitors* of aggregation, such as Noggin and inhibitory galectins. As a result, what were once cartilage-forming aggregates inhibit the areas surrounding them from forming more such aggregates (see Figure 9.17, lower panel). Broader space allocation to the limb allows more aggregates to form.

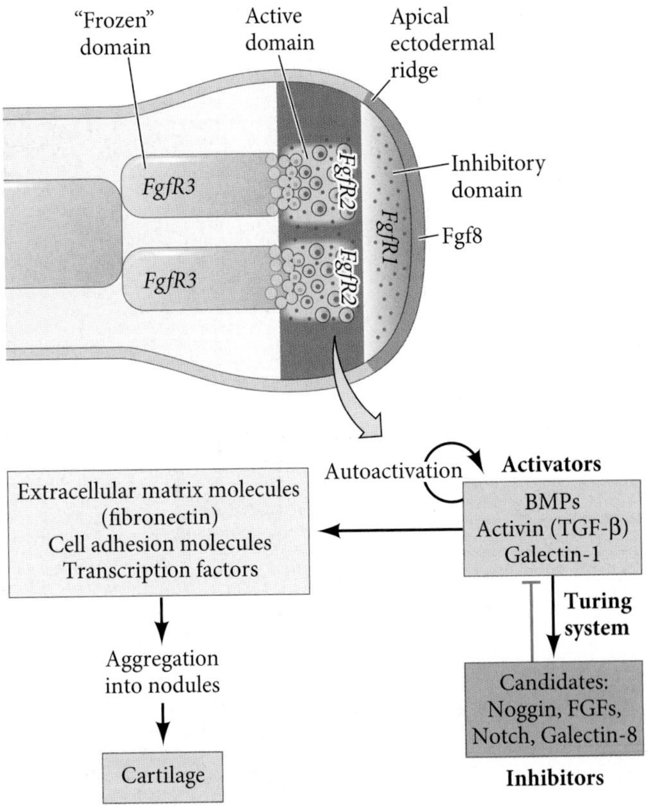

At different sizes of the limb, different numbers of precartilaginous condensations can form. First, a single condensation can fit (humerus), then two (ulna and radius), then several (wrist, digits). In this reaction-diffusion hypothesis, the aggregations of precartilage mesenchyme actively recruit more cells from the surrounding area and laterally inhibit the formation of other foci of condensation. The number of these condensations, then, depends on the geometry of the active zone and the strength of the lateral inhibition. Once formed, the aggregates of mesenchyme interact with one another not only to recruit more cells but also to express the transcription factors (Sox9) and extracellular matrix (collagen 2) characteristic of cartilage (Lorda-Diez et al. 2011).

According to the model, waves of synthesis and inhibition would form the original pattern of the limb. By placing such constraints as geometry, diffusibility, and the rates

FIGURE 19.17 Reaction-diffusion mechanism for proximal-distal limb specification. In the inhibitory domain immediately outside the AER, cells are kept dividing by FGFs and Wnts and are prevented from forming cartilage. Behind this area, in the active domain, cartilaginous nodules actively form according to a reaction-diffusion mechanism. Here each cell secretes and can respond to activating paracrine factors of the TGF-β family (TGF-β, BMPs, activin) and cell adhesion factors such as galectin-1. These factors stimulate their own synthesis as well as that of the extracellular matrix and cell adhesion proteins that promote aggregation. The activating cells also stimulate the synthesis of diffusible inhibitors of aggregation (including Noggin and galectin-8), preventing cell adhesion in neighboring regions. The places where nodules can form are governed by the geometry of the limb bud (i.e., the geometry decides how many "waves" of activator will be allowed). In the "frozen" domain, the aggregated nodules can now differentiate into cartilage, thus "freezing" the configuration. (After Zhu et al. 2010.)

of synthesis and degradation of each activator and inhibitor, Zhu and colleagues have been able to model the types of skeleton formed as the limb bud grows. First, the computer model accurately mimicked the normal patterning of the limb (**FIGURE 19.18A**). Next, it simulated the aberrant skeletons formed as a result of manipulations (**FIGURE**

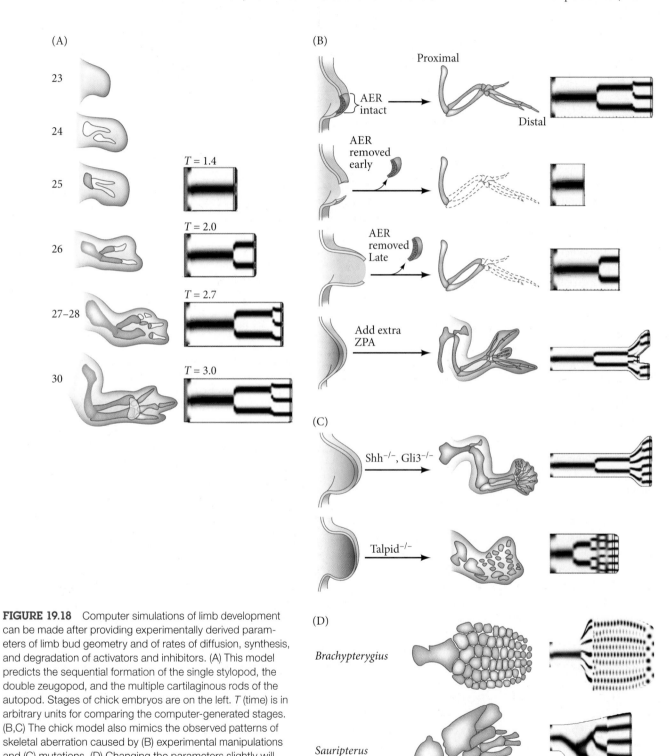

FIGURE 19.18 Computer simulations of limb development can be made after providing experimentally derived parameters of limb bud geometry and of rates of diffusion, synthesis, and degradation of activators and inhibitors. (A) This model predicts the sequential formation of the single stylopod, the double zeugopod, and the multiple cartilaginous rods of the autopod. Stages of chick embryos are on the left. *T* (time) is in arbitrary units for comparing the computer-generated stages. (B,C) The chick model also mimics the observed patterns of skeletal aberration caused by (B) experimental manipulations and (C) mutations. (D) Changing the parameters slightly will also generate the observed limb skeletons found in fossils such as the fishlike aquatic reptile *Brachypterygius* (whose forelimb had a paddlelike shape) and *Sauripterus*, one of the first land-dwelling reptiles. (After Zhu et al. 2010.)

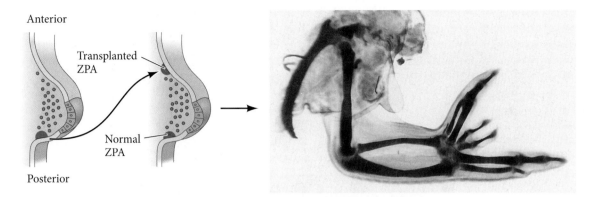

FIGURE 19.19 When a ZPA is grafted to anterior limb bud mesoderm, duplicated digits emerge as a mirror image of the normal digits. (After Honig and Summerbell 1985; photograph courtesy of D. Summerbell.)

19.18B) and mutations (**FIGURE 19.18C**). Altering the geometries also could yield the patterns seen in fossil limbs (**FIGURE 19.18D**).

WATCH DEVELOPMENT 19.1   Witness a computer simulation of the acquisition of skeletal limb patterns over time based upon a reaction-diffusion model.

Specifying the Anterior-Posterior Axis

The specification of the anterior-posterior axis of the limb is the earliest restriction in limb bud cell potency from the pluripotent condition. In the chick, this axis is specified shortly before a limb bud is recognizable.

Sonic hedgehog defines a zone of polarizing activity

Viktor Hamburger (1938) showed that as early as the 16-somite stage, prospective wing mesoderm transplanted to the flank area develops into a limb with the anterior-posterior and dorsal-ventral polarities of the donor graft, not those of the host tissue. Several later experiments (Saunders and Gasseling 1968; Tickle et al. 1975) suggested that the anterior-posterior axis is specified by a small block of mesodermal tissue near the posterior junction of the young limb bud and the body wall. When tissue from this region is taken from a young limb bud and transplanted to a position on the anterior side of another limb bud, the number of digits on the resulting wing is doubled (**FIGURE 19.19**). Moreover, the structures of the extra set of digits are mirror images of the normally produced structures. Polarity is maintained, but the information is now coming from both an anterior and a posterior direction. Thus, this region of the mesoderm has been called the **zone of polarizing activity**, or **ZPA**.

The search for the molecule(s) that confer polarizing activity on the ZPA became one of the most intensive quests in developmental biology. In 1993, Riddle and colleagues showed by in situ hybridization that *Sonic hedgehog* (*Shh*), a vertebrate homologue of the *Drosophila hedgehog* gene, was expressed specifically in that region of the limb bud known to be the ZPA (**FIGURE 19.20A**). As evidence that this association between the ZPA and Sonic hedgehog was more than just a correlation, Riddle and colleagues (1993) demonstrated that the secretion of Shh protein is sufficient for polarizing activity. They transfected embryonic chick fibroblasts (which normally would never synthesize Shh) with a viral vector containing the *Shh* gene (**FIGURE 19.20B**). The gene became expressed, translated, and secreted in these fibroblasts, which were then inserted under the anterior ectoderm of an early chick limb bud. Mirror-image digit duplications like those induced by ZPA transplants were the result. Moreover,

Developing Questions

Mathematical modeling can point the developmental biologist toward new questions and experiments, and Turing's model has certainly done so with regard to pattern formation during organogenesis. For example, which factors in the active zone are the main "reactive" activators and inhibitors? Although TGF-β is a well-supported candidate for the chondrogenic activator, there has been little experimental data to characterize the potential inhibitors that mathematical modeling is predicting. Another parameter to evaluate is cell movements.

Figure 19.20 Sonic hedgehog protein is expressed in the ZPA. (A) In situ hybridization showing the sites of Sonic hedgehog expression (arrows) in the posterior mesoderm of the chick limb buds. These are precisely the regions that transplantation experiments defined as the ZPA. (B) Shh is sufficient to serve the function of the ZPA. When Shh is ectopically produced on the anterior margin of a limb bud of a chick embryo (by grafting recombinant cells expressing Shh), the resulting limbs exhibit a mirror-image digit duplication. (A courtesy of R. D. Riddle; B after Riddle et al. 1993.)

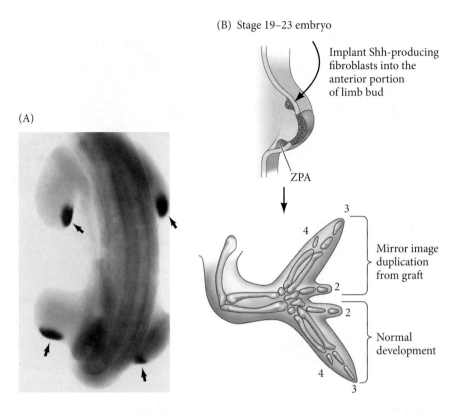

beads containing Sonic hedgehog protein were shown to cause the same duplications (López-Martínez et al. 1995; Yang et al. 1997). Thus, Sonic hedgehog appears to be the active agent of the ZPA.

This fact was confirmed by a remarkable gain-of-function mutation. The *hemimelic extra-toes* (*hx*) mutant of mice has extra digits on the thumb side of the paws (**FIGURE 19.21A,B**). This phenotype is associated with a single base-pair difference in the limb-specific *Shh* enhancer, a highly conserved region located a long distance (about 1 million base pairs) upstream from the *Shh* gene itself (Lettice et al. 2003; Sagai et al. 2005). Maas and Fallon (2005) made a reporter construct by fusing a β-galactosidase gene to this long-range limb enhancer region from both wild-type and *hx*-mutant genes. They injected these reporter constructs into the pronuclei of newly fertilized mouse eggs to obtain transgenic embryos. In the transgenic mouse embryos carrying the reporter gene with wild-type limb enhancer, staining for β-galactosidase activity revealed a single patch of expression in the posterior mesoderm of each limb bud (i.e., in the ZPA; **FIGURE 19.21C**). However, mice carrying the mutant *hx* reporter construct showed β-galactosidase activity in *both* the anterior and posterior regions of the limb bud (**FIGURE 19.21D**). It thus appears that (1) this enhancer has both positive and negative functions, and (2) in the anterior region of the limb bud, some inhibitory factor represses the ability of this enhancer to activate *Shh* transcription. The inhibitor probably cannot bind to the mutated enhancer; thus, in *hx* mutant mice, Shh is expressed in both the anterior and posterior regions of the limb bud, and this anterior Shh expression causes extra digits to develop. Similar mutations in the long-range limb enhancer of Shh produce polydactylous phenotypes in humans and other mammals (**FIGURE 19.21E,F**; Gurnett et al. 2007; Lettice et al. 2008; Sun et al. 2008).

Specifying digit identity by Sonic hedgehog

How does Sonic hedgehog specify the identities of the digits? When scientists were able to perform fine-scale fate mapping experiments on the Shh-secreting cells of the ZPA, they were surprised to find that cells that expressed Shh at any time did not undergo apoptosis (programmed cell death; see Chapter 15) in the way the AER does after it

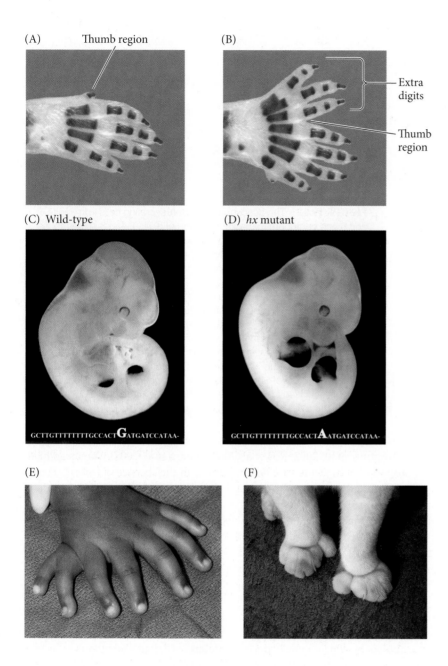

(A) Thumb region

(B) Extra digits

Thumb region

(C) Wild-type

GCTTGTTTTTTTTGCCACT**G**ATGATCCATAA-

(D) *hx* mutant

GCTTGTTTTTTTTGCCACT**A**ATGATCCATAA-

FIGURE 19.21 Ectopic expression of Sonic hedgehog in the anterior limb causes extra digit formation. (A) Wild-type mouse paw. The bones are stained with alizarin red. (B) *Hx* (*hemimelic extra-toes*) mutant mouse paws, showing the extra digits associated with the anterior ("thumb") region. (The small extra nodule of posterior bone is peculiar to the *Hx* phenotype on the genetic background used and is not seen on other genetic backgrounds.) (C) Reporter constructs from wild-type *Shh* limb enhancer direct transcription solely to the posterior part of each mouse limb bud (i.e., in the ZPA). (D) Reporter constructs from the *hx* mutant direct transcription to both the anterior and posterior regions of each mouse limb bud. The wild-type and mutant *Shh* limb-specific enhancer region DNA sequences are shown below and highlight the single G-to-A nucleotide substitution that differentiates the two. (E) A similar mutation in the human long-range enhancer for *SHH* causes similar mirror-image hand duplications. (F) Descendants of Ernest Hemingway's polydactylous pet cats still inhabit the Hemingway home in Key West, Florida, and display a mutation in this long-range enhancer. (A–D from Maas and Fallon 2005, courtesy of B. Robert, Y. Lallemand, S. A. Maas, and J. F. Fallon; E from Yang and Kozin 2009, courtesy of S. Kozin; F, photograph by S. Gilbert.)

(E)

(F)

finishes its job. Rather, the descendants of Shh-secreting cells become the bone and muscle of the posterior limb (Ahn and Joyner 2004; Harfe et al. 2004). Indeed, digits 5 and 4 (and part of digit 3) of the mouse hindlimb are formed from the descendants of Shh-secreting cells (**FIGURE 19.22**).

It seems that specification of the digits is primarily dependent on the amount of time the *Shh* gene is expressed and only a little bit on the concentration of Shh protein that other cells receive (see Tabin and McMahon 2008). The difference between digits 4 and 5 is that the cells of the more posterior digit 5 express *Shh* longer and are exposed to Shh (in an autocrine manner) for a longer time. Digit 3 is made up of some cells that secreted Shh for a shorter period than those of digit 4, and they also depend on Shh diffusion from the ZPA (indicated by digit 4 being lost when Shh is modified such that it cannot diffuse away from cells). Digit 2 is dependent entirely on Shh diffusion for its specification, and digit 1 is specified independently of Shh. Indeed, in a naturally occurring chick mutant that lacks Shh expression in the limb, the only digit that forms is digit 1. Furthermore, when the genes for Shh and Gli3 are conditionally knocked out in the

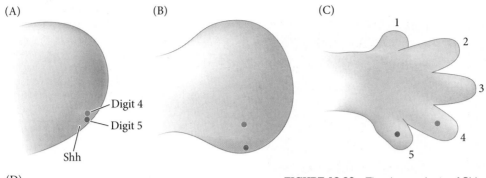

(A)

Digit 4
Digit 5
Shh

(B)

(C)

1
2
3
4
5

(D)

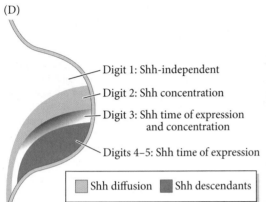

Digit 1: Shh-independent

Digit 2: Shh concentration

Digit 3: Shh time of expression and concentration

Digits 4–5: Shh time of expression

☐ Shh diffusion   ■ Shh descendants

FIGURE 19.22 The descendants of Shh-secreting cells form digits 4 and 5 and contribute to the specification of digits 2 and 3 in the mouse limb. (A) In the early mouse hindlimb bud, the progenitors of digit 4 (green dot) and the progenitors of digit 5 (red dot) are both in the ZPA and express Sonic hedgehog (light green shading). (B) At later stages of limb development, the cells forming digit 5 are still expressing Shh in the ZPA, but the cells that form digit 4 no longer do. (C) When the digits form, the cells in digit 5 will have seen high levels of Shh protein for a longer time than the cells in digit 4. (D) Schematic by which digits 4 and 5 are specified by the amount of time they were exposed to Shh in an autocrine fashion; digit 3 is specified by the amount of time the cells were exposed to Shh both in an autocrine and paracrine fashion. Digit 2 is specified by the concentration of Shh its cells received by paracrine diffusion, and digit 1 is specified independently of Shh. (After Harfe et al. 2004.)

mouse limb, the resulting limbs have numerous digits, but the digits have no obvious specificity (Litingtung et al. 2002; Ros et al. 2003; Scherz et al. 2007). Vargas and Fallon (2005) propose that digit 1 is specified by *Hoxd13* in the absence of *Hoxd12*. Forced expression of *Hoxd12* throughout the digital primordia leads to the transformation of digit 1 into a more posterior digit (Knezevic et al. 1997).

By using conditional knockouts of the mouse *Shh* gene (i.e., researchers could stop Shh expression at different times during mouse development), Zhu and Mackem (2011) found that Sonic hedgehog works by two temporally distinct mechanisms. The first phase involves the specification of digit identity (from the posterior pinky to the anterior thumb). In this phase, Shh acts as a morphogen, with the digit identities being specified first by the concentration of Shh in that region of the limb bud, and then by the duration of exposure to Shh. In the second phase, Shh works as a mitogen to stimulate the proliferation and expansion of the limb bud mesenchyme, thus helping shape the limb bud.

The mechanism by which Sonic hedgehog establishes a digit's identity may involve cell-cycle regulation and the BMP pathway. The time and concentration dependent actions of Shh lead to the graded activation of the downstream transcriptional effector Gli3. These targets include the genes for the BMP antagonist Gremlin, the cell cycle regulator Cdk6, and the genes that synthesize hyaluronic acid (a component of cell adhesion). Shh (through Gli3) restricts the proliferation of cartilage progenitor cells (by downregulating Cdk6) and promotes their BMP-stimulated differentiation into cartilage by inhibiting the BMP antagonist Gremlin and by upregulating hyaluronic acid synthase 2 (Vokes et al. 2008; Liu et al. 2012; Lopez-Rios et al. 2012).

Shh initiates and sustains a gradient of BMP proteins across the limb bud, and this BMP gradient can specify the digits (Laufer et al. 1994; Kawakami et al. 1996; Drossopoulou et al. 2000). Identity is not specified directly in each digit primordium, however. Rather, the identity of each digit is determined by the *interdigital* mesoderm—that is, by the webbing between the digits (the region of mesenchyme that will shortly undergo apoptosis).

The interdigital tissue specifies the identity of the digit forming anteriorly to it (i.e., toward the thumb or big toe). Thus, when Dahn and Fallon (2000) removed the webbing

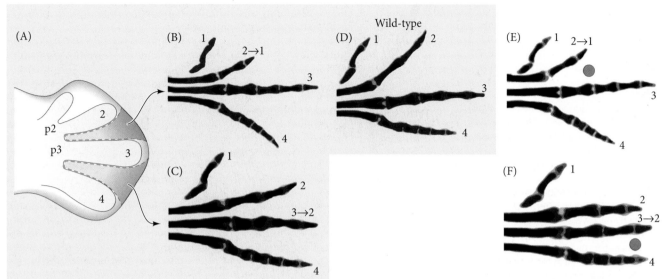

FIGURE 19.23 Regulation of digit identity by BMP concentrations in the interdigital space anterior to the digit and by Gli3. (A) Scheme for removal of interdigital (ID) regions. (B) Removal of ID region 2 between digit primordia 2 (p2) and 3 (p3) causes digit 2 to change to the structure of digit 1. (C) Removing ID region 3 (between digit primordia 3 and 4) causes digit 3 to form the structures of digit 2. (D) Wild-type digits and their ID spaces. (E,F) The same transformations as in (B) and (C) can be obtained by adding beads containing the BMP inhibitor Noggin to the ID regions. (E) When a Noggin-containing bead (green dot) is placed in ID region 2, digit 2 is transformed into a copy of digit 1. (F) When the Noggin bead is placed in ID region 3, digit 3 is transformed into a copy of digit 2. (After Dahn and Fallon 2000; Litingtung et al. 2002; B–F, photographs courtesy of R. D. Dahn and J. F. Fallon.)

between the cartilaginous condensations forming chick hindlimb digits 2 and 3, the second digit was changed into a copy of digit 1 (**FIGURE 19.23A,B**). Similarly, when the webbing on the other side of digit 3 was removed, the third digit formed a copy of digit 2 (**FIGURE 19.23A,C**). Moreover, the positional value of the webbing could be altered by changing the BMP level (**FIGURE 19.23D–F**). Each digit has a characteristic array of nodules that form the digit skeleton, and Suzuki and colleagues (2008) have shown that the different levels of BMP signaling in the interdigital webbing regulate the recruitment of progress zone mesenchymal cells into the nodules that make the digits.

Sonic hedgehog and FGFs: Another positive feedback loop

When the limb bud is relatively small, an initial positive feedback loop is established between Fgf10 produced in the mesoderm and Fgf8 produced in the ectoderm, promoting limb outgrowth (**FIGURE 19.24A**). As the limb bud grows, the ZPA is established, and another regulatory loop is created (**FIGURE 19.24B**). BMPs in the mesoderm would downregulate FGFs in the AER were it not for the Shh-dependent expression of a BMP inhibitor, Gremlin (Niswander et al. 1994; Zúñiga et al. 1999; Scherz et al. 2004; Vokes et al. 2008). Sonic hedgehog in the ZPA activates Gremlin, which inhibits BMPs, thus promoting the maintenance of FGF expression and continued limb bud outgrowth. FGF in turn inhibits repressors of Shh to complete the positive feedback loop. As with most multigene pathways, however, this interactivity is more complicated.

Depending on the levels of FGFs in the apical ectodermal ridge, the zone of polarizing activity can be either activated or shut down; two feedback loops have been demonstrated (**FIGURE 19.24C**; Verheyden and Sun 2008; Bénazet et al. 2009). At first, relatively low levels of AER FGFs activate Shh and keep the ZPA functioning. The FGF signals appear to inhibit the proteins Etv4 and Etv5, which are repressors of Sonic hedgehog transcription (see Figure 21.24B; Mao et al. 2009; Zhang et al. 2009). Thus, the AER and ZPA mutually support each other through the positive loop of Sonic hedgehog and FGFs (Todt and Fallon 1987; Laufer et al. 1994; Niswander et al. 1994). In the more anterior region of the limb bud, Fgf8 positively regulates Etv4/5, which in turn repress Shh in this region, further reinforcing the posterior-to-anterior gradient of Shh from the ZPA (Mao et al. 2009).

As a result of Shh stimulation through FGF signaling, levels of Gremlin (a powerful BMP antagonist) become high, and the positive FGF/Shh loop sustains limb growth (**FIGURE 19.24D**). As long as the Gremlin signal can diffuse to the AER, FGFs will be made and the AER maintained. However, as FGF levels consequently also rise, a

FIGURE 19.24 Early interactions between the AER and limb bud mesenchyme. (A) In the limb bud, Fgf10 from mesenchyme generated by the lateral plate mesoderm activates a Wnt (Wnt3a in chicks; Wnt3 in mice and humans) in the ectoderm. Wnt activates the β-catenin pathway, which induces synthesis of Fgf8 in the region near the AER. Fgf8 activates Fgf10, causing a positive feedback loop. (B) As the limb bud grows, Sonic hedgehog (Shh) in the posterior mesenchyme creates a new signaling center that induces posterior-anterior polarity, and it also activates Gremlin (Grem1) to prevent mesenchymal BMPs from blocking FGF synthesis in the AER. Moreover, Fgf8 operates in part by differentially regulating Etv4/5 (genes within the E-twenty-six superfamily of transcription factors) along the anterior-posterior axis of the limb bud, which in turn reinforces a gradient of Shh expression from the posterior. (C) Two feedback loops link the AER and ZPA. In the positive feedback loop (black arrow; below), FGFs 4, 9, and 17 from the AER activate Shh, stabilizing the ZPA. In the reciprocal inhibitory loop (red; above), Shh from the ZPA activates Gremlin (Grem1), which blocks BMPs, thus preventing BMP-mediated inactivation of FGFs in the AER. (D) The feedback loops create a mutual accelerated synthesis of Shh (ZPA) and FGFs (AER). (E) As FGF concentration climbs, it eventually reaches a threshold where it inhibits Gremlin, thus allowing the BMPs to begin repressing the AER FGFs. As more cells multiply in the area not expressing Gremlin, the Gremlin signal near the AER is too weak to prevent the BMPs from repressing the FGFs. (F) At that point, the AER disappears, removing the signals that stabilize the ZPA. The ZPA then also disappears. (A,B after Fernandez-Teran and Ros 2008; C after Verheyden and Sun 2008.)

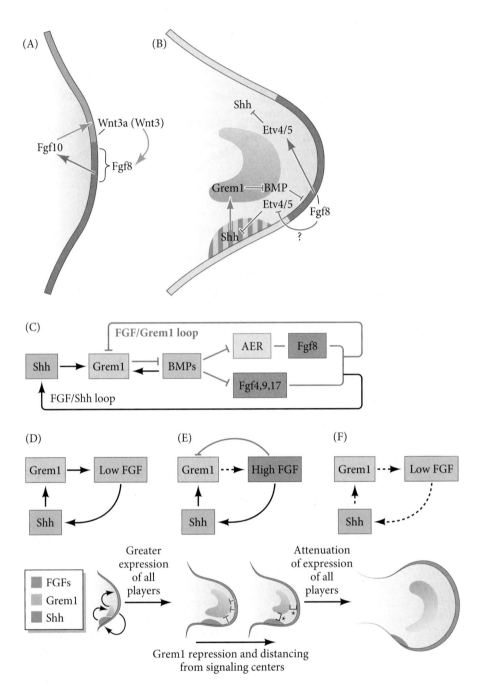

negative feedback loop to *block* Gremlin expression in the distal mesenchyme is triggered (**FIGURE 19.24E**). This repression of Gremlin synthesis paired with the progressive expansion of the limb bud creates increased distance between Gremlin and the signaling centers (AER and ZPA) in the distalmost mesenchyme. At that time, the BMPs abrogate FGF synthesis, the AER collapses, and the ZPA (with no FGFs to support it) is terminated. The embryonic phase of limb development ends (**FIGURE 19.24F**).

Hox specification of the digits

As mentioned early in this chapter, Hox genes are critical for specifying fates along each axis of the limb, and their expression—especially that of the *Hoxd* cluster—functions in two phases (Zakany et al. 2004; Tarchini and Duboule 2006; also see Abbasi 2011). The first phase is important for the specification of the stylopod and zeugopod, as discussed earlier (**FIGURE 19.25**). The later phase of *Hoxd* expression helps specify the autopod.

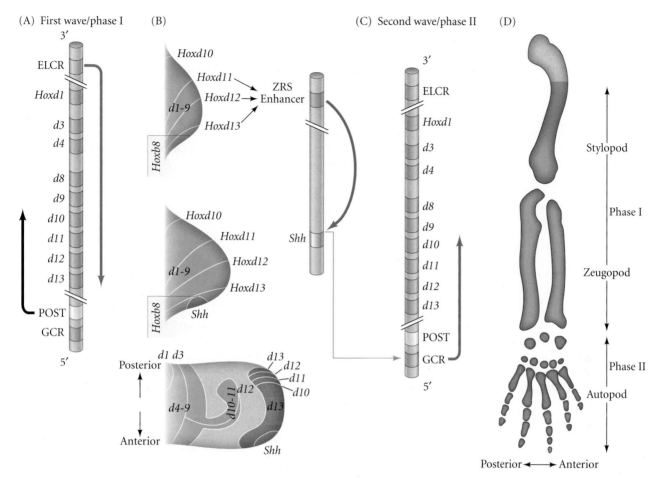

FIGURE 19.25 Changes in *Hoxd* gene expression regulate patterning of the tetrapod limb in two independent phases. (A) The first phase of *Hoxd* expression is initiated as the limb bud forms. The early limb control regulatory (ELCR) element activates those genes closest to it earlier than those genes away from it, whereas the POST regulatory element acts negatively to restrict the anterior expression of these genes in the opposite direction. (B) This results in expression domains such that *Hoxd13* expression is confined to the most posterior region, whereas *Hoxd12* is allowed to expand more anteriorly. The 5′ *Hoxd* genes activate the long-range Shh enhancer (ZRS), thereby creating the ZPA in the posterior limb mesoderm. (C) In the second phase, Shh activates the GCR regulatory locus, which inverts the *Hoxd* expression pattern such that *Hoxd13* is located more anteriorly than the other *Hoxd* genes (B, bottom red). (D) Skeletal elements specified by the early (blue) and later (red) phases. White lines in the limb bud show the boundaries of gene expression. (After Abbasi 2011.)

There are two major "early" *cis*-regulatory regions involved, composed of numerous enhancers that work together to activate the *Hoxd* genes in a specific temporal and spatial array. The major early regulatory, or early limb control regulatory (ELCR), region activates transcription in a time-dependent manner: the closer the gene is to the ELCR, the earlier it is activated. The second early regulatory region, POST ("posterior restriction"), imposes spatial restrictions on the expression of the 5′ *Hoxd* genes (*Hoxd10–13*) such that the genes closest to this region have the most restricted expression domains, starting from the posterior margin of the limb bud (see Figure 19.25A,B). This situation creates a pattern of nested Hoxd proteins essential for activating the long-range enhancer (ZRS enhancer) of the *Sonic hedgehog* gene, thereby activating Shh expression in the posterior limb bud mesoderm and forming the ZPA (Tarchini et al. 2006; Galli et al. 2010). In addition, the presence of Hoxb8 in the mesenchyme appears to help define the posterior boundary of the forelimb bud (see Figure 19.25B). If *Hoxb8* is expressed ectopically in the anterior compartment of the mouse forelimb bud, a ZPA will also form there (Charite et al. 1994; Hornstein et al. 2005).

The ZPA now acts to alter the *Hoxd* gene expression patterns. Sonic hedgehog expressed from the posterior margin activates a second set of enhancers called the global control region, or GCR (see Figure 19.25C,D; Spitz et al. 2003; Montavon et al. 2011.) The Hox genes closest to the GCR are expressed most broadly. This expression inverts the original pattern of *Hoxd10–13* expression such that *Hoxd13* is expressed at the highest level and extends most anteriorly. *Hoxd12*, *Hoxd11*, and *Hoxd10* are expressed in slightly narrower domains, so that the most anterior digit (e.g., the thumb) expresses *Hoxd13* but no other Hox gene (see Figure 19.25B, last diagram in panel; Montavon et al. 2008). Thus, the first phase of *Hoxd* gene expression helps specify the ZPA, while in the second phase of *Hoxd* expression the ZPA instructs the expression patterns, and these patterns define the identities of the digits. Moreover, transplantation of either the ZPA or other Shh-secreting cells to the anterior margin of the limb bud at this stage leads to the formation of mirror-image patterns of *Hoxd* expression and results in mirror-image digits (Izpisúa-Belmonte et al. 1991; Nohno et al. 1991; Riddle et al. 1993).

What genes are these Hox proteins regulating? Some clues come from the analysis of mutations of the *Hox13* series of genes. As mentioned above, people with mutations in the *HOXD13* gene have portions of their autopods that fuse together rather than separate. Ectopic expression of the chicken *Hoxa13* gene (usually expressed in the distal ends of developing chick limbs) appears to make the cells expressing it "stickier," as well as making the "wavelength" between cartilage and web smaller. These properties might cause the cartilaginous nodules to condense in specific ways (Yokouchi et al. 1995; Newman 1996; Sheth et al. 2012).

A Turing model for self-organizing digit skeletogenesis

We have discussed the importance of Shh and Gli3 in regulating digit patterning along the anterior-to-posterior axis. However, we neglected to mention that *Shh* and *Gli3* single- and double-null mutants still form digits; in fact, these mutants form a lot of digits—a polydactylous limb phenotype (see Litingtung et al. 2002; te Welscher et al. 2002). These data imply either that some other inductive system generates digits, or that digit formation is built on an intrinsic molecular prepattern of skeletogenesis. The anterior-to-posterior striped pattern of five digits in the mouse paw is reminiscent of Turing-type patterning (see Figure 19.18). If a Turing-type mechanism is enabling the distal mesenchyme to self-organize during chondrogenesis, what are the core factors representing the activating and inhibiting nodes of this pattern-generating system?

Knowing the essential roles that distal Hox genes play in the gene regulatory network of digit identity and their regulatory interactions with Shh/Gli3, Sheth and colleagues (2012) hypothesized that *Hoxa13/Hoxd11–13* genes function through a Turing mechanism to control digit number. One way to theoretically increase the number of digits would be to narrow the wavelength of patterned chondrogenesis; that is, one could divide the distal mesenchyme into smaller stripes of precartilage development. Remarkably, Sheth and colleagues demonstrated that the progressive loss of distal Hox genes combined with similarly dosed reductions in *Gli3* correlated with gradual increases in the number of digits (**FIGURE 19.26**). This collaborative work among the Ros,

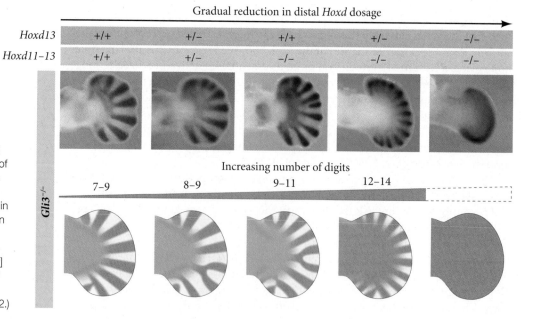

FIGURE 19.26 Gli3 and distal Hox gene expression. The loss of *Gli3* combined with progressive reductions in distal Hox genes causes a concomitant increase in the number of digits. The pattern of supernumerary digits follows a Turing-type mechanism of formation (simulations [bottom row] match the expression pattern of *Sox9* in the mouse forelimb [photos]). (From Sheth et al. 2012.)

Sharpe, and Kmita labs used a "reaction-diffusion" simulation with assumed generic activator and inhibitor morphogens. This simulation showed that distal Hox genes in combination with AER-derived Fgf gradients were sufficient modulators of the *wavelength* of a Turing system to recapitulate the skeletogenic pattern of digits in normal mice and in *Gli3*-null backgrounds (Sheth et al. 2012).

This Turing model predicts that slight size changes of the distal limb bud will alter the number of digits. That was indeed found to be the case, and it may be a simple way of gaining or losing digits during evolution.[5] In fact, comparison of the polydactyly in the combined *Hox/Gli3* mutants with the limbs of early tetrapods and the fins of Sarcopterygian (lobe-finned) and Actinopterygian (ray-finned) fishes strongly suggest that there may be a conserved, self-organizing, reaction-diffusion mechanism for digit skeletogenesis.

The assumptions built into this Turing model predict that morphogen activators and inhibitors should be present in the distal limb mesenchyme at the time of chondrogenesis. But what morphogens? The search was on. How might you identify the activators and inhibitors of this system?

The Sharpe lab approached this problem by first characterizing the temporal and spatial patterns of early precartilage formation in the distal limb, by looking at expression patterns of the precartilage marker Sox9. Meanwhile, Raspopovic and colleagues compared the transcriptomes of limb mesenchyme cells expressing Sox9 with those of cells not expressing Sox9 and found that developmental genes were expressed differently in the two populations. Namely, Wnt- and BMP-related genes were highly upregulated only in cells that were *not* expressing Sox9 ("out of phase") (**FIGURE 19.27A**). In addition, it is known that loss of the *Sox9* gene eliminates any periodic expression pattern of Wnt and BMP in the limb (Akiyama et al. 2002), suggesting that Sox9 is not

---

[5] Thus, dogs with larger limb buds may generate enough cells that an additional cartilage condensation can fit into the autopod (Alberch 1985). Such appears to be the case with the St. Bernard and Great Pyrenees breeds. Fondon and Garner (2004) have shown that one allele of the *Alx-4* gene is homozygous in only one breed of dog, the Great Pyrenees. These dogs are characterized by polydactyly (an extra toe, the dew claw). Here, there is apparently more growth of the limb bud such that another condensation can emerge in the autopod.

Figure 19.27 A BMP-Sox9-Wnt Turing-type mechanism governs digit formation. (A) *Sox9* gene expression is in alternating stripes with BMP and the Wnt pathway genes, *Axin2* and *Lef1*. (B) Proper limb growth and digit formation is correctly simulated through a Turing mechanism of BMP/Wnt and Sox9 interactions (BSW) paired with distal expression of *Hoxd13* and *Fgf8*. (C) These computer simulations of digit development under this model (top panel) match the endogenous in situ pattern of *Sox9* gene expression remarkably well (bottom panel). (D) Illustration of BMP-Sox9-Wnt expression in the limb bud with a representation of the quantitative differences among these genes along the anterior-posterior axis. (A–C from Raspopovic et al. 2014; D after Zuniga and Zeller 2014.)

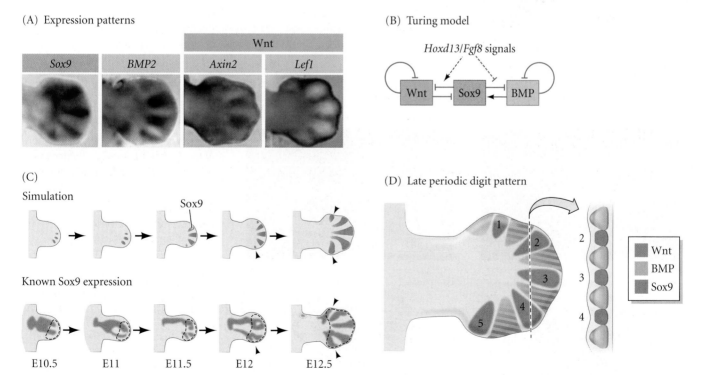

(A) Expression patterns

(B) Turing model

(C) Simulation

Known Sox9 expression

(D) Late periodic digit pattern

only a marker for precartilage but also potentially a direct component of the regulatory gene network. Based on these results, Raspopovic created a "three-node" BMP-Sox9-Wnt (BSW) Turing-type network—in which BMP functions as the activator of Sox9 and Wnt the inhibitor—to simulate the skeletogenesis of digits in the mouse (**FIGURE 19.27B**). It is interesting that only by including the *wavelength* modulating parameters of an FGF gradient for the timing of proximal-distal growth and the spatial restrictions of distal Hox genes did the BSW Turing model simulate accurately the self-organizing nature of digit development (**FIGURE 19.27C**).

> **WATCH DEVELOPMENT 19.2** "Turing" mesenchyme into digits. Watch a computer simulation of how the BSW model builds the digit pattern.

In summary, it appears that the chondrogenic pattern of digit formation is governed by a self-organizing Turing system of molecular interactions (**FIGURE 19.27D**). BMP and Wnt morphogens differentially regulate *Sox9* expression, which function under the tuning control of FGF and distal Hox genes. Lastly, the Sonic hedgehog morphogen provides an earlier polarization of the distal mesenchyme that influences the specification of digit identity along the anterior-posterior axis.

Generating the Dorsal-Ventral Axis

The third axis of the limb distinguishes the dorsal half of the limb (knuckles, nails, claws) from the ventral half (pads, soles). In 1974, MacCabe and colleagues demonstrated that the dorsal-ventral polarity of the limb bud is determined by the ectoderm encasing it. If the ectoderm is rotated 180° with respect to the limb bud mesenchyme, the dorsal-ventral axis is partially reversed—that is, the distal elements (digits) are "upside-down"—which suggests that the late specification of the dorsal-ventral axis of the limb is regulated by its ectodermal component(s).

One molecule that appears to be particularly important in specifying dorsal-ventral polarity is **Wnt7a**. The *Wnt7a* gene is expressed in the dorsal (but not the ventral) ectoderm of chick and mouse limb buds (**FIGURE 19.28A**; Dealy et al. 1993; Parr et al. 1993). When Parr and McMahon (1995) knocked out the *Wnt7a* gene, the resulting mouse embryos had ventral footpads on both surfaces of their paws, showing that Wnt7a is needed for the dorsal patterning of the limb.

Wnt7a is the first known dorsal-ventral axis gene expressed in limb development. It induces activation of the *Lmx1b* (also known as *Lim1*) gene in the dorsal mesenchyme. *Lmx1b* encodes a transcription factor that appears to be essential for specifying dorsal cell fates in the limb. If Lmx1b protein is expressed in

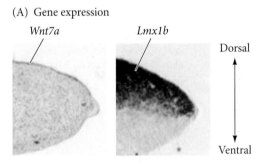

(A) Gene expression

Wnt7a   *Lmx1b*

Dorsal

Ventral

(B) Wild-type

Plane of section

Footpad   Dorsal

Ventral

(C) *Lmx1b* mutant

Plane of section

Footpads   Dorsal (ventralized)

Ventral

FIGURE 19.28 Lmx1b-dependent dorsal/ventral patterning by Wnt7a. (A) *Wnt7a* and *Lmx1b* are both expressed in the dorsal limb bud. *Wnt7a* is restricted to the epidermis, however, whereas *Lmx1b* is present throughout the dorsal mesenchyme. (B,C) Loss of *Lmx1b* ventralizes the forelimb as evidenced by the presence of footpads on both sides of the paw in the mutant phenotype. (Courtesy of Randy Johnson and Kenneth Dunner.)

the ventral mesenchyme cells, those cells develop a dorsal phenotype (Riddle et al. 1995; Vogel et al. 1995; Altabef and Tickle 2002). Human and mouse *lmx1b* mutants also reveal this gene's importance for specifying dorsal limb fates. *Lmx1b* knockouts in mice produce a syndrome in which the dorsal limb phenotype is lacking and those cells have taken on ventral fates, exhibiting footpads, ventral tendons, and sesamoids (all ventral-specific structures; **FIGURE 19.28B,C**). Similarly in humans, loss-of-function mutations in the *LMX1B* gene result in nail-patella syndrome (no nails on the digits, no kneecaps), in which the dorsal sides of the limbs have been ventralized (Chen et al. 1998; Dreyer et al. 1998). The Lim1 protein probably specifies the cells to differentiate in a dorsal manner, which is critical, as we saw in Chapter 15, for the innervation of motor neurons (whose growth cones recognize inhibitory factors made differentially in the dorsal and ventral compartments of the limb bud). Conversely, the transcription factor Engrailed-1 marks the ventral ectoderm of the limb bud and is induced by BMPs in the underlying meso-derm (**FIGURE 19.29**). If BMPs are knocked out in the early limb bud, Engrailed-1 is not expressed, and Wnt7a is expressed in both dorsal and ventral ectoderm. The result is a malformed limb that is dorsal on both sides (Ahn et al. 2001; Pizette et al. 2001).

The dorsal-ventral axis is also coordinated with the other two axes. Indeed, the *Wnt7a*-deficient mice described earlier lacked not only dorsal limb structures but also posterior digits, suggesting that Wnt7a is also needed for the anterior-posterior axis (Parr and McMahon 1995). Yang and Niswander (1995) made a similar set of observations in chick embryos. These investigators removed the dorsal ectoderm from developing limbs and found that the result was the loss of posterior skeletal elements from the limbs. The reason these limbs lacked posterior digits was that *Shh* expression was greatly reduced. Viral-induced expression of *Wnt7a* was able to substitute for the dorsal ectoderm signal and restore *Shh* expression and posterior phenotypes. These findings showed that Sonic hedgehog synthesis is stimulated by the combination of Fgf4 and Wnt7a proteins. Conversely, overactive Wnt signaling in the ventral ectoderm causes an overgrowth of the AER and extra digits, indicating that the proximal-distal patterning is not independent of dorsal-ventral patterning either (Loomis et al. 1998; Adamska et al. 2004).

Thus, at the end of limb patterning, BMPs are responsible for simultaneously shutting down the AER, indirectly shutting down the ZPA, and inhibiting the Wnt7a signal along the dorsal-ventral axis (Pizette et al. 2001). The BMP signal eliminates growth and patterning along all three axes. When exogenous BMP is applied to the AER, the elongated epithelium of the AER reverts to a cuboidal epithelium and ceases to produce FGFs; and when BMPs are inhibited by Noggin, the AER continues to persist days after it would normally have regressed (Gañan et al. 1998; Pizette and Niswander 1999).

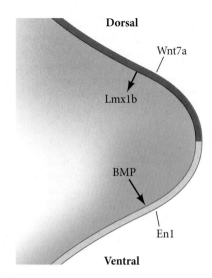

FIGURE 19.29 Model of dorsal-to-ventral patterning in the limb bud by Wnt and BMP signaling. Wnt7a induces dorsal cell fates of the limb bud through Lmx1b, whereas BMP signaling functions through Engrailed-1 (En1) to regulate ventral limb patterning.

Cell Death and the Formation of Digits and Joints

Apoptosis—programmed cell death—plays a role in sculpting the tetrapod limb. Indeed, cell death is essential if our joints are to form and if our fingers are to become separated (Zaleske 1985; Zuzarte-Luis and Hurle 2005). The death (or lack of death) of specific cells in the vertebrate limb is genetically programmed and has been selected for over the course of evolution.

Sculpting the autopod

The difference between a chicken's foot and the webbed foot of a duck is the presence or absence of cell death between the digits (**FIGURE 19.30**). Saunders and colleagues have shown that after a certain stage, chick cells between the digit cartilage are destined to die, and will do so even if transplanted to another region of the embryo or placed in culture (Saunders et al. 1962; Saunders and Fallon 1966). Before that time, however, transplantation to a duck limb will save them. Between the time when the cell's death is determined and when death actually takes place, levels of DNA, RNA, and protein synthesis in the cell decrease dramatically (Pollak and Fallon 1976).

FIGURE 19.30 Patterns of cell death in leg primordia of (A) duck and (B) chick embryos. Blue shading indicates areas of cell death. In the duck, the regions of cell death are very small, whereas there are extensive regions of cell death in the interdigital tissue of the chick leg. (After Saunders and Fallon 1966.)

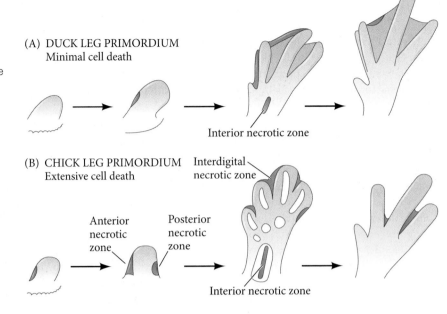

(A) DUCK LEG PRIMORDIUM
Minimal cell death

Interior necrotic zone

(B) CHICK LEG PRIMORDIUM
Extensive cell death

Interdigital necrotic zone

Anterior necrotic zone

Posterior necrotic zone

Interior necrotic zone

In addition to the **interdigital necrotic zone**, three other regions of the limb are "sculpted" by cell death. The ulna and radius are separated from each other by an **interior necrotic zone**, and two other regions, the **anterior** and **posterior necrotic zones**, further shape the end of the limb (see Figure 19.30B; Saunders and Fallon 1966). Although these zones are referred to as "necrotic," this term is a holdover from the days when no distinction was made between necrotic (pathologic or traumatic) cell death and apoptotic cell death. These cells die by apoptosis, and the death of the interdigital tissue is associated with the fragmentation of their DNA (Mori et al. 1995).

The signal for apoptosis in the autopod is provided by the BMP proteins, whose expression, interestingly, is dependent on upregulated synthesis of interdigital RA (Cunningham and Duester 2015). BMP2, BMP4, and BMP7 are each expressed in the interdigital mesenchyme, and blocking BMP signaling (by infecting progress zone cells with retroviruses carrying dominant negative BMP receptors) prevents interdigital apoptosis (Yokouchi et al. 1996; Zou and Niswander 1996; Abara-Buis et al. 2011). Because these BMPs are expressed throughout the progress zone mesenchyme, it is thought that cell death would be the default state unless there were active suppression of the BMPs. This suppression may come from the Noggin protein, which is made in the developing cartilage of the digits and in the perichondrial cells surrounding it (Capdevila and Johnson 1998; Merino et al. 1998). If Noggin is expressed throughout the limb bud, no apoptosis is seen.

Forming the joints

The function first ascribed to BMPs was the formation, not the destruction, of bone and cartilage tissue. In the developing limb, BMPs induce the mesenchymal cells—depending on the stage of development—either to undergo apoptosis or to become cartilage-producing chondrocytes. The same BMPs can induce death or differentiation, depending on the responding cell's history. This **context dependency** of signal action is a critical concept in developmental biology. It is also critical for the formation of joints. Macias and colleagues (1997) have shown that during early limb bud stages (before cartilage condensation), beads secreting BMP2 or BMP7 cause apoptosis. Two days later, the same beads cause the limb bud cells to form cartilage.

In the normally developing limb, BMPs use both of these properties to form joints. Several BMPs are made in the perichondrial cells surrounding the condensing chondrocytes and promote further cartilage formation[6] (**FIGURE 19.31A**). Another BMP, Gdf5, is

---

[6] In bats, the amount of BMPs synthesized is extraordinarily high and recruits more mesenchymal cells into the cartilage, thereby extending the digits to a much greater length than in most other mammals (Cooper et al. 2012).

expressed at the regions between the bones, where joints will form, and appears critical for joint formation (**FIGURE 19.31B**; Macias et al. 1997; Brunet et al. 1998). Mouse mutations of the gene for Gdf5 produce brachypodia, a condition characterized by the lack of limb joints (Storm and Kingsley 1999). In mice homozygous for loss-of-function of the BMP antagonist Noggin, no joints form. Rather, BMP7 in these *Noggin*-defective embryos appears to recruit nearly all the surrounding mesenchyme into the digits (**FIGURE 19.31C**).

Blood vessels and Wnt proteins also appear to be critical in joint formation. The conversion of mesenchyme cells into nodules of cartilage-forming tissue establishes the bone boundaries. The mesenchyme will not form such nodules in the presence of blood vessels, and one of the first indications of cartilage formation is the regression of blood vessels in the region wherein the nodule will form (Yin and Pacifici 2001). Wnt proteins are critical in sustaining transcription of *Gdf5*, and β-catenin produced by the Wnts is able to suppress the *Sox9* and *collagen-2* genes that characterize precartilage cells (Hartmann and Tabin 2001; Tufan and Tuan 2001).

Joints are not just absences of bone. Rather, joints are complex structures that incorporate a lubrication system, an immune system, and a ligament system, all joined to the proper articulation of the skeleton. One critical element of joint formation that allows this differentiation is muscle contraction. In normal joint formation, the cells that will form the joint lose their chondrocyte characteristics (such as expression of collagen-2 and Sox9) and instead begin to express Gdf5, Wnt4, Wnt9a, and Ext1 (a protein necessary for synthesizing heparan sulfate). These cells will form the articulate cartilage and the synovium, which secretes lubricating synovial fluid (Koyama et al. 2008; Mundy et al. 2011). Kahn and colleagues (2009) have shown that the movement of the bones is necessary for maintaining this commitment to form joints. In mutant mice where the muscles do not form or are paralyzed, the joint cells revert back to a cartilaginous phenotype.

Continued limb growth: Epiphyseal plates

The three axes of human limbs are specified in a highly asymmetrical fashion. However, their growth over the next 16 years is so symmetrical that the length of one arm matches the other to within 0.2% (Ballock and O'Keefe 2003; Wolpert 2010). If all our cartilage were turned into bone before birth, we could not grow any larger, and our bones would be only as large as the original cartilaginous model. But growth zones—**epiphyseal plates**—form at the proximal and distal ends of each developing long bone. At the portion of the epiphyseal plate farthest from the new bone is a germinal region of cartilaginous stem cells, followed by regions of proliferating cartilage cells, mature cartilage cells (chondrocytes), and hypertrophic cartilage cells. The hypertrophic cartilage cells (which increase their size five- to tenfold) undergo apoptosis and are replaced by bone cells (osteocytes).

In the long bones of many mammals (including humans), endochondral ossification spreads outward in both directions from the center of the bone (**FIGURE 19.32**). Although more than 10,000 new cartilage cells may be made daily, the number appears to be identical in each arm. Cartilage proliferation in humans is high through about 3 years of age. The major factor of differential growth (between the arms and legs of a

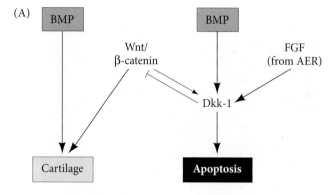

(A)

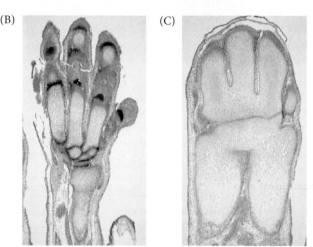

(B)   (C)

FIGURE 19.31 Possible involvement of BMPs in stabilizing cartilage and apoptosis. (A) Model for the dual role of BMP signals in limb mesodermal cells. BMP can be received in the presence of FGFs (to produce apoptosis) or Wnts (to induce bone). When FGFs from the AER are present, Dickkopf (Dkk) is activated. This protein mediates apoptosis and at the same time inhibits Wnt from aiding in skeleton formation. (B,C) Effects of Noggin. (B) A 16.5-day autopod from a wild-type mouse, showing *Gdf5* expression (dark blue) at the joints. (C) A 16.5-day *Noggin*-deficient mutant mouse autopod, showing neither joints nor *Gdf5* expression. Presumably, in the absence of Noggin, BMP7 was able to convert nearly all the mesenchyme into cartilage. (A after Grotewold and Rüther 2002; B,C from Brunet et al. 1998, courtesy of A. P. McMahon.)

(A)

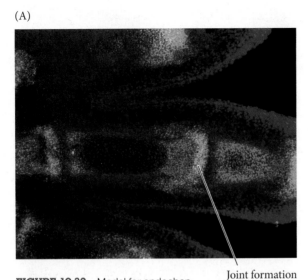

Joint formation

(B)

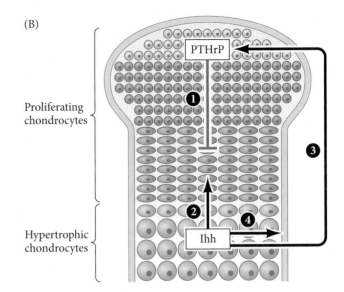

Proliferating chondrocytes

Hypertrophic chondrocytes

PTHrP

Ihh

FIGURE 19.32 Model for endochondral ossification in the limb. (A) In situ hybridization in the limb of a 21.5-day mouse embryo shows *collagen-2* mRNA (green) in the proliferative growth plate cartilage and *Gdf5* mRNA (red) in the region destined to become a joint. Nuclei are stained blue. (B) Regulation of cartilage cell proliferation by PTHrP (parathyroid hormone related protein) and Indian hedgehog (Ihh). (1) PTHrP acts on receptors on proliferating chondrocytes to keep them dividing and thereby delays the production of Ihh. (2) When the source of PTHrP production becomes sufficiently far away, Ihh is synthesized; Ihh acts on its receptor (present on chondrocytes) to increase the rate of chondrocyte proliferation and (3) to stimulate the production of PTHrP at the ends of bones. (4) Ihh also acts on the perichondrial cells enveloping the cartilage, converting them into the osteoblasts of the bone collar. (A courtesy of Dr. P. Tylzanowski; B after Kronenberg 2003.)

human or between the legs of a human versus the legs of a dog), however, is probably the swelling of the hypertrophic cartilage. Not only can different numbers of cartilage cells be made, but those cells can enlarge to different sizes (Cooper et al. 2013). After a growth spurt at puberty, the epiphyseal growth plates fuse, and there are no longer any stem cells for growth. As long as the epiphyseal plates are able to produce chondrocytes, the bone continues to grow.

Fibroblast growth factor receptors: Dwarfism

The rate of growth appears to be intrinsic to each bone. Each growth plate is controlled locally (probably by differences in the sensitivity to growth factors), but the coordinated growth of the entire skeleton is maintained by circulating factors. Thus, when transplantations are made between growth plates of old and young mammals, the growth rate of the transplanted growth plate depends on the age of the donor animal, not that of the host animal (Wolpert 2010). However, discoveries of human and mouse mutations resulting in abnormal skeletal development have provided remarkable insights into how hormones and paracrine factors can control the eventual size of the limbs.

Fibroblast growth factors are critically important in halting the growth of the epiphyseal plates, telling the cells to differentiate rather than divide (Deng et al. 1996; Webster and Donoghue 1996). In humans, mutations of the receptors for FGFs can cause these receptors to become active prior to receiving the normal FGF signal. Such mutations give rise to the major types of human dwarfism. **Achondroplasia** is a dominant condition caused by mutations in the transmembrane region of FGF receptor 3 (FgfR3). Roughly 95% of achondroplastic dwarfs have the same mutation of the *FgfR3* gene: a base-pair substitution that converts glycine to arginine at position 380 in the transmembrane region of the protein. In addition, mutations in the extracellular portion of the FgfR3 protein or in the tyrosine kinase intracellular domain may result in thanatophoric dysplasia, a lethal form of dwarfism that resembles homozygous achondroplasia (Bellus et al. 1995; Tavormina et al. 1995).

As mentioned in Chapter 1, dachshunds have an achondroplastic mutation, but its cause is slightly different than that of the human form. Dachshunds have an extra copy of the *Fgf4* gene, which is also expressed in the developing limb. This extra copy causes excess production of Fgf4, activating FgfR3 and accelerating the pathway that stops the growth of chondroblasts and hastens their differentiation. The same extra copy of *Fgf4* has been found in other short-limbed dogs, such as corgis and basset hounds (Parker et al. 2009).

WEB TOPIC 19.2 **GROWTH HORMONE AND ESTROGEN RECEPTORS**   As its name implies, growth hormone (GH) is a major factor in the regulation of growth, including limb growth. Interactions with the gonadal steroids, especially estrogen, appear to be important for bone growth.

Evolution by Altering Limb Signaling Centers

Charles Darwin wrote in *On the Origin of Species*, "What can be more curious than that the hand of a man, formed for grasping, that of a mole for digging, the leg of a horse, the paddle of the porpoise, and the wing of the bat should all be constructed on the same pattern and should include similar bones, and in the same relative positions?" Darwin recognized that the differences between horse legs, porpoise flippers, and human hands are underlain by a similar pattern of bone formation. He proposed that these bones evolved from a common ancestor, although he did not know how. C. H. Waddington pointed out that such evolution was predicated on changing the development of the limb in ways that could be selected. In other words, he proposed that changes in development caused the arrival of new variations. These variations could then be tested by natural selection. We have now discovered several ways by which "tinkering" with limb signaling molecules can generate new limb morphologies. In these ways, limb evolution can be caused by developmental changes.[7]

WEB TOPIC 19.3 **DINOSAURS AND CHICKEN FINGERS**   One piece of evidence that birds were descended from dinosaurs is that both groups retain three digits. To some researchers, however, it seems that birds have digits 2, 3, and 4, whereas dinosaurs had digits 1, 2, and 3. Recent evidence has uncovered a change in gene transcription that might explain this conundrum.

 SCIENTISTS SPEAK 19.5   Watch three web conferences on limb evolution. Dr. Peter Currie discuses the evolution of limb muscle in fish, Dr. Michael Shapiro describes the evolution of pelvic reduction in sticklebacks, and Dr. James Noonan talks about the evolution of the human thumb.

WEB-FOOTED FRIENDS   We can start with the sculpting of the autopod. The regulation of BMPs is critical in creating the webbed feet of ducks (Laufer et al. 1997b; Merino et al. 1999). The interdigital regions of duck feet exhibit the same pattern of BMP expression as the webbing of chick feet. However, whereas the interdigital regions of the chick feet appear to undergo BMP-mediated apoptosis, developing duck feet synthesize the BMP inhibitor Gremlin and block this regional cell death (**FIGURE 19.33**). Moreover, the webbing of chick feet can be preserved if Gremlin-soaked beads are placed in the interdigital regions. Thus, the evolution of web-footed birds probably involved the inhibition of BMP-mediated apoptosis in the interdigital regions. In Chapter 26, we will find that the bat embryo uses a similar mechanism to acquire its wings.

TINKERING WITH THE SIGNALING CENTERS: MAKING WHALES   Numerous transition fossils attest to the evolution of modern cetaceans (whales, dolphins, porpoises) from hoofed land mammals (Gingrich et al. 1994; Thewissen et al. 2007, 2009). Numerous changes in the anatomy were made, but few are as striking as the conversion of a forelimb into a flipper and the elimination of the hindlimb altogether. These events were accomplished by modifying the signaling centers of the ancestral cetacean limb buds in

---

[7] Earlier in this chapter, we noted that developmental biologists get used to thinking in four dimensions. Evolutionary developmental biologists have to think in *five* dimensions: the three standard dimensions of space, the dimension of developmental time (hours or days), and the dimension of evolutionary time (millions of years).

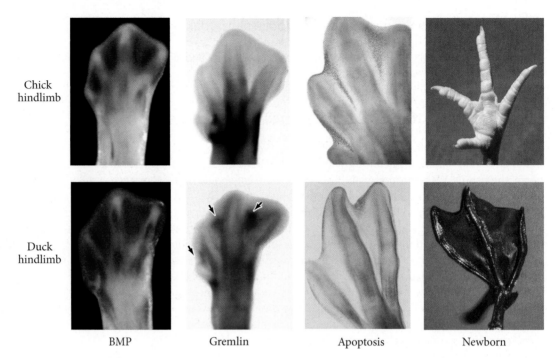

Chick hindlimb

Duck hindlimb

BMP   Gremlin   Apoptosis   Newborn

FIGURE 19.33 Autopods of chick (upper row) and duck (lower row) are shown at similar stages. Both show BMP4 expression (dark blue) in the interdigital webbing; BMP4 induces apoptosis. The duck foot (but not the chicken foot) expresses the BMP4-inhibitory protein Gremlin (dark brown; arrows) in the interdigital webbing. Thus, the chicken foot undergoes interdigital apoptosis (as seen by neutral red dye accumulation in the dying cells), but the duck foot does not. (Courtesy of J. Hurle and E. Laufer.)

three ways. First, the FGF signaling of the forelimb AER was preserved for a much longer duration, which caused the formation of longer fingers by the continual addition of phalanges. Second, interdigital apoptosis was prevented by blocking BMP activity in a manner similar to that described above for duck feet. Third, the Sonic hedgehog signal from the hindlimb ZPA ceased early in development. Once the ZPA signal diminished, the AER could not be sustained, and the hindlimb ceased to develop (Thewissen et al. 2006). **FIGURE 19.34** shows the elongated flipper phalanges and truncated hindlimb of an embryonic dolphin. Thus, despite creationists claiming there is no way whales could have evolved from land mammals (see Gish 1985), in fact the combination of developmental biology and paleontology explains the phenomenon extremely well.

Remnants of pelvic girdle

Hyperphalangy in digits 2 and 3

FIGURE 19.34 A 110-day embryo of a pantropical spotted dolphin (*Stella attenuata*), stained to show bones (red) and cartilage (blue). Hyperphalangy (extremely long fingers) is seen in the forelimb (correlating with continued *Fgf8* expression in the AER), and a rudimentary hindlimb is seen (correlating with the reduction of AER signaling following the elimination of Shh from the ZPA). (From Cooper 2009; courtesy of L. N. Cooper and the Thewissen laboratory.)

Next Step Investigation

All the major families of paracrine factors act in coordination to build the limb. Although many of the "executives" of limb bud formation have been identified, we are just starting to discern how the activity of these paracrine factors influences where the cartilaginous condensations will form, how the skeletal elements are sculpted, how each digit becomes specified, and where the tendons

and muscles will insert onto the skeletal elements. Limb development is therefore a meeting place for developmental biology, evolutionary biology, and medicine. Within the next decade, we may uncover the bases for numerous congenital defects of limb formation, and perhaps we will better understand how limbs are modified into flippers, wings, hands, and legs. Maybe you can "lend a hand."

Closing Thoughts on the Opening Photo

Perhaps the more appropriate question for this image might have been, "Which fingers am I holding up?" The skeletal elements in this chick wing reveal a mirror-image duplication of the digits, which we now know was due to the misexpression of Sonic hedgehog from the anterior side of the limb bud. Gradients of signaling factors along the major axes play essential roles in establishing the correct number and pattern of structures in the arm and hand. Just as important for limb development is the underlying gene regulatory network that Hox genes regulate as well as the self-organizing interactions between the cells that ultimately build the tissues of the limb. (After Honing and Summerbell 1985; photograph courtesy of D. Summerbell.)

19 Snapshot Summary
Development of the Tetrapod Limb

1. The positions where limbs emerge from the body axis depend on Hox gene expression.

2. The proximal-distal axis of the developing limb is initiated by the induction of the ectoderm at the dorsal-ventral boundary by Fgf10 from the mesenchyme. This induction forms the apical ectodermal ridge (AER). The AER secretes Fgf8, which keeps the underlying mesenchyme proliferative and undifferentiated. This area of mesenchyme is called the progress zone.

3. Tbx5 induces the forelimb, whereas Tbx4 (chick) and Islet1 (mouse) induce hindlimb identity.

4. Two opposing gradients—one of FGFs and Wnts from the AER, the other of retinoic acid from the flank—pattern the proximal-distal axis of the chick limb. In mice, however, retinoic acid appears unnecessary for proximal-distal patterning, suggesting a single gradient of FGFs and Wnts in mice.

5. As the limb grows outward, the stylopod forms first, then the zeugopod, and last, the autopod. Each phase of limb development is characterized by a specific pattern of Hox gene expression. The evolution of the autopod involved a duplication and reversal of Hox gene expression that distinguishes fish fins from tetrapod limbs.

6. Turing-type models suggest that a reaction-diffusion mechanism can explain the constant pattern of stylopod-zeugopod-autopod seen in tetrapod limbs.

7. The anterior-posterior axis is defined by the expression of Sonic hedgehog in the zone of polarizing activity, a region in the posterior mesoderm of the limb bud. If ZPA tissue (or Shh-secreting cells or beads) is placed in the anterior margin of a limb bud, a second, mirror-image pattern of Hox gene expression occurs, along with a corresponding mirror-image duplication of the digits.

8. The ZPA is maintained by the interaction of FGFs from the AER with mesenchyme made competent to express Sonic hedgehog by its expression of particular Hox genes. Sonic hedgehog acts in turn, probably in an indirect manner, and probably through the Gli factors, to change the expression of the Hox genes in the limb bud.

9. Sonic hedgehog specifies digits in at least two ways. It works through BMP inhibition in the interdigital mesenchyme, and it also regulates the proliferation of digit cartilage. Mutations in the long-range enhancer for Shh can cause polydactyly by creating a second ZPA in the anterior margin of the limb bud.

10. The dorsal-ventral axis is formed in part by the expression of Wnt7a in the dorsal portion of the limb ectoderm. Wnt7a also maintains the expression level of Sonic hedgehog in the ZPA and of Fgf4 in the posterior AER. Fgf4 and Shh reciprocally maintain each other's expression.

11. Levels of FGFs in the AER can either support or inhibit the production of Shh by the ZPA. As the limb bud grows and more FGFs are produced in the AER, Shh expression is inhibited. This in turn causes the lowering of FGF levels, and eventually proximal-distal outgrowth ceases.

12. A "three-node" BMP-Sox9-Wnt Turing-type network underlies skeletogenesis of digits in the mouse such that cells expressing BMP function to activate Sox9, with Wnt serving as its inhibitor.

13. Cell death in the limb is mediated by BMPs and is necessary for the formation of digits and joints. Differences between the unwebbed chicken foot and the webbed duck foot can be explained by differences in the expression of Gremlin, a protein that antagonizes BMPs.

14. The ends of the long bones of humans and other mammals contain cartilaginous regions called epiphyseal growth plates. The cartilage in these regions proliferates, allowing the resulting bone to grow larger. The cartilage is eventually replaced by bone and growth stops.

15. The differences in bone growth between different parts of the body and between different species is due primarily to the expansion of hypertrophic cartilage (which will later be replaced by bone).

16. By modifying paracrine factor secretion, different limb morphologies can form, initiating the development of webbed feet, flippers, or hands. By eliminating the synthesis of certain paracrine factors, limbs can be prevented from forming (as in whales and snakes).

17. BMPs are involved both in inducing apoptosis and in differentiating the mesenchymal cells into cartilage. The regulation of BMP effects by Noggin and Gremlin proteins is critical in forming the joints between the bones of the limb and in regulating proximal-distal outgrowth.

Further Reading

Cooper, K. L., J. K. Hu, D. ten Berge, M. Fernandez-Teran, M. A. Ros and C. J. Tabin. 2011. Initiation of proximal-distal patterning in the vertebrate limb by signals and growth. *Science* 332: 1083–1086.

Cunningham, T. J. and G. Duester. 2015. Mechanisms of retinoic acid signalling and its roles in organ and limb development. *Nature Rev. Mol. Cell Biol.* 16: 110–123.

Freitas, R., C. Gómez-Marín, J. M. Wilson, F. Casares and J. L. Gómez-Skarmeta. 2012. *Hoxd13* contribution to the evolution of vertebrate appendages. *Dev. Cell* 23: 1219–1229.

Kawakami, Y. and 12 others. 2011. Islet1-mediated activation of the β-catenin pathway is necessary for hindlimb initiation in mice. *Development* 138: 4465–4473.

Mahmood, R. and 9 others. 1995. A role for Fgf8 in the initiation and maintenance of vertebrate limb outgrowth. *Curr. Biol.* 5: 797–806.

Merino, R., J. Rodriguez-Leon, D. Macias, Y. Gañan, A. N. Economides and J. M. Hurle. 1999. The BMP antagonist Gremlin regulates outgrowth, chondrogenesis, and programmed cell death in the developing limb. *Development* 126: 5515–5522.

Niswander, L., S. Jeffrey, G. R. Martin and C. Tickle. 1994. A positive feedback loop coordinates growth and patterning in the vertebrate limb. *Nature* 371: 609–612.

Raspopovic, J., L. Marcon, L. Russo and J. Sharpe. 2014. Digit patterning is controlled by a BMP-Sox9-Wnt Turing network modulated by morphogen gradients. *Science* 345: 566–570.

Riddle, R. D., R. L. Johnson, E. Laufer and C. Tabin. 1993. Sonic hedgehog mediates the polarizing activity of the ZPA. *Cell* 75: 1401–1416.

Roselló-Díez, A., C. G. Arques, I. Delgado, G. Giovinazzo and M. Torres. 2014. Diffusible signals and epigenetic timing cooperate in late proximo-distal limb patterning. *Development* 141: 1534-1543.

Schneider, I. and N. H. Shubin. 2013. The origin of the tetrapod limb: From expeditions to enhancers. *Trends Genet.* 29: 419–426.

Sekine, K. and 10 others. 1999. Fgf10 is essential for limb and lung formation. *Nature Genet.* 21: 138–141.

Todt, W. L. and J. F. Fallon. 1987. Posterior apical ectodermal ridge removal in the chick wing bud triggers a series of events resulting in defective anterior pattern formation. *Development* 101: 501–515.

Verheyden, J. M. and X. Sun. 2008. An FGF-Gremlin inhibitory feedback loop triggers termination of limb bud outgrowth. *Nature* 454: 638–641.

Zhang, Y. T., M. S. Alber and S. A. Newman. 2013. Mathematical modeling of vertebrate limb development. *Math. Biosci.* 243: 1–17.

Zuniga, A. 2015. Next generation limb development and evolution: Old questions, new perspectives. *Development* 142: 3810–3820.

GO TO WWW.DEVBIO.COM...

...for Web Topics, Scientists Speak interviews, Watch Development videos, Dev Tutorials, and complete bibliographic information for all literature cited in this chapter.

The Endoderm
Tubes and Organs for Digestion and Respiration

THE ENDODERM FORMS THE GUT AND RESPIRATORY TUBES of the *adult* amniote, where it is essential for the exchange of gases and foods. In the *embryonic* amniote, whose food and oxygen come from the mother via the placenta, the endoderm's first major function is to induce the formation of several mesodermal organs. As we have seen in earlier chapters, the endoderm is critical for instructing the formation of the notochord, heart, blood vessels, and even the mesodermal germ layer. The endoderm's second embryonic function is to construct the linings of two tubes within the vertebrate body. The **digestive tube** extends the length of the body, and buds from the digestive tube form the liver, gallbladder, and pancreas. The **respiratory tube** forms as an outgrowth of the digestive tube and eventually bifurcates into the two lungs. The region of the digestive tube anterior to the point where the respiratory tube branches off is the **pharynx**. A third embryonic function is to form the epithelium of several glands. Epithelial outpockets of the pharynx give rise to the tonsils and to the thyroid, thymus, and parathyroid glands.

The endoderm arises from two sources. The main source is the set of cells that enters the interior of the embryo through the primitive streak during gastrulation. This is often called the **definitive endoderm**. It replaces the **visceral endoderm** that is primarily forming the yolk sac. However, not all the visceral endoderm is removed. Live-cell imaging studies using fluorescent markers show that definitive endoderm doesn't replace visceral endoderm as a sheet (Kwon et al. 2008; Viotti et al. 2014a). Rather, individual definitive endoderm cells can be seen intercalating into the visceral endoderm layer. The descendants of these epiblast cells remain in the embryonic region while most (but not all) of

How do some gut cells become pancreas cells while the neighboring gut cells become liver or intestine?

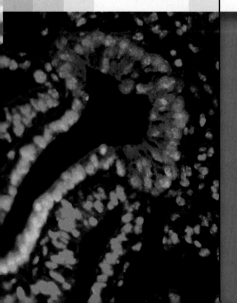

The Punchline

The endoderm consists of the digestive tube, the respiratory tube, and their associated glands. The Sox17 transcription factor plays a major role in specifying the endoderm. The gut tube becomes defined along the anterior-posterior axis. The anterior cells become the pharynx, lung, and thyroid glands; the posterior cells become the intestine; and the cells between them become the precursors of the pancreas, gallbladder, and liver. Splanchnic mesoderm plays important roles in specifying the morphogenesis of the different regions of the digestive tube. Its secretion of Sonic hedgehog induces a nested pattern of Hox genes, which in turn distinguish regions of the mesoderm that interact with the adjacent endoderm. The β cells of the pancreas produce insulin, and the delineation of the pathway leading to their development has allowed the generation of insulin-producing β cells from induced pluripotent stem cells.

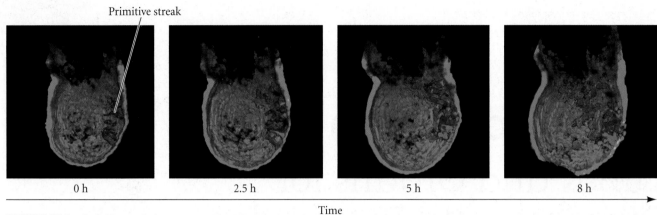

Primitive streak

0 h　　　　　2.5 h　　　　　5 h　　　　　8 h

Time

FIGURE 20.1 Definitive endoderm cells from mouse epiblast replace the cells of the visceral endoderm. Here, the visceral endoderm of the embryo has been genetically marked with green fluorescent protein and can be seen enveloping the epiblast. Epiblast cells were randomly labeled with red dye and filmed as they migrated through the primitive streak and replaced the visceral endoderm cells. (From Viotti et al. 2014b, courtesy of A. K. Hadjantonakis.)

the original visceral endoderm becomes extraembryonic. The Sox17 transcription factor marks the endoderm in many species, and the *Sox17* gene appears to be activated in some cells as they leave the primitive streak. In mutants lacking *Sox17*, the definitive endoderm does not form (Viotti et al. 2014b).

The presence of Sox17 protein makes the definitive endoderm cells different from the mesodermal cells that also travel through the primitive streak; the mesodermal cells express the gene for the Brachyury transcription factor, and Brachyury appears to be critical for the development of the mesoderm. Whether *Sox17* or *Brachyury* is

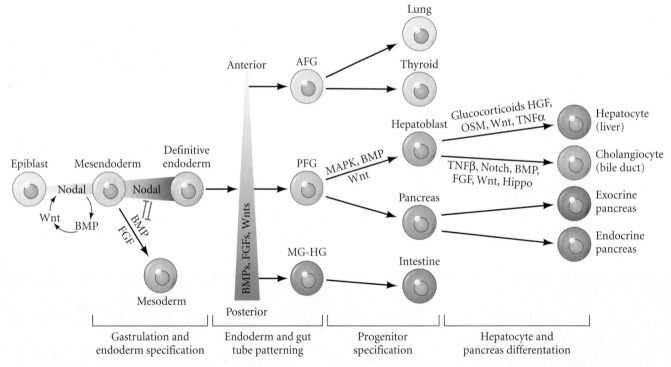

FIGURE 20.2 Signals mediating endoderm progenitor fates. Nodal signaling activates epiblast cells to assume a mesendodermal fate and migrate through the primitive streak. Those cells exposed to high Nodal concentrations then tend to become definitive endoderm cells. Those receiving BMPs and FGFs tend to become mesodermal. The later fate of the definitive endoderm depends to a large degree on where the cell resides along the anterior-posterior axis. Those in the posterior region are exposed to high levels of Wnts, BMPs, and FGFs and become midgut-hindgut (MG-HG) cells that generate the intestines. Cells exposed to somewhat lower levels of these paracrine factors give rise to the posterior foregut cells (PFG), the precursors of the liver (hepatocyte and cholangiocytes) and pancreas (which will divide into exocrine and endocrine progenitors). Those definitive endoderm cells seeing very low levels of these paracrine factors generate the anterior foregut cells (AFG) that become the precursors of the lung and thyroid gland. Many lineages were left out for simplification. (After Gordillo et al. 2015.)

expressed appears to depend on the concentration of Nodal secreted from the visceral endoderm. High levels of Nodal induce *Sox17*, while BMPs and FGFs act against Nodal and specify the migrating cells to become mesoderm (**FIGURE 20.1**; Vincent et al. 2003; Dunn et al. 2004).

The definitive endoderm of the gut is then defined by three regions, each with a distinct embryonic origin and delineated by their location along the anterior-posterior axis (Gordillo et al. 2015). As should be all too apparent by now, the A-P axis of vertebrates is specified by gradients of Wnts, FGFs, and BMPs, each of which is at the highest concentration posteriorly (**FIGURE 20.2**). The endoderm near the head will form the anterior foregut cells, which will generate the precursors of the lung and thyroid glands. The endoderm in the posterior becomes a collection of midgut-hindut precursor cells and will form the intestinal progenitor cells. The region between them—in the area of moderate BMPs, FGFs, and Wnts—become the posterior foregut precursors. These are the cells that give rise to the pancreas and the liver.

The gut cells initially form a flat sheet beneath the embryo (in the chick or human) or around the embryo (in the mouse). From the flat sheet of definitive endoderm, these cells form a tube. Mammalian gut tube development begins at two sites that migrate toward each other and fuse in the center (Lawson et al. 1986; Franklin et al. 2008). In the foregut, cells from the lateral portions of the anterior endoderm move ventrally to form the tube of the **anterior intestinal portal** (**AIP**); the **caudal intestinal portal** (**CIP**) forms from the posterior endoderm. The AIP and CIP migrate toward each other and come together to form the midgut (**FIGURE 20.3**).

The openings of the anterior and posterior ends of the gut tube are unique in that they are the only regions of the embryo where endoderm meets ectoderm. At first, the oral end is blocked by a region of endodermal cells that join to the mouth ectoderm (the **stomodeum**) at the **oral plate**. Eventually (at about 22 days in human embryos), the oral plate breaks, creating the oral opening of the digestive tube. The opening itself is lined by ectodermal cells. This arrangement creates an interesting situation, because the oral plate ectoderm is in contact with the brain ectoderm, which has curved around toward the ventral portion of the embryo. These two ectodermal regions interact with each other, with the roof of the oral region forming Rathke's pouch and becoming the glandular portion of the pituitary gland. The neural tissue on the floor of the diencephalon gives rise to the infundibulum, which becomes the neural portion of the pituitary. Thus, the pituitary gland has a dual origin, which is reflected in its adult functions. There is a similar meeting of endoderm and ectoderm at the anus; this is called the **anorectal junction**.

The Pharynx

The anterior endodermal portion of the digestive and respiratory tubes begins in the pharynx. Using a reporter gene (the above-mentioned *Sox17*) that becomes activated only in the endoderm, Rothova and colleagues (2012) found that there is a dividing line between ectoderm and endoderm in the mammalian mouth. In mammals, the teeth and the major salivary glands are from the ectoderm. The anterior taste buds are ectodermal (generated by cranial placodes; see Chapter 16), but the posterior taste buds, as well as some of the posterior salivary and mucus glands, are derived from the endoderm.

The embryonic pharynx in mammals contains four pairs of endoderm-derived **pharyngeal pouches**. Between these pouches are four **pharyngeal arches** (**FIGURE 20.4**). The first pair of pharyngeal pouches become the auditory cavities of the middle ear and the associated eustachian tubes. The second pair of pouches give rise to the walls of the tonsils. The thymus is derived from the third pair of pharyngeal pouches; the thymus will direct the differentiation of T lymphocytes during later stages of development. One pair of parathyroid glands is also derived from the third pair of pharyngeal pouches, while the other pair is derived from the fourth pair of pouches. In addition to these

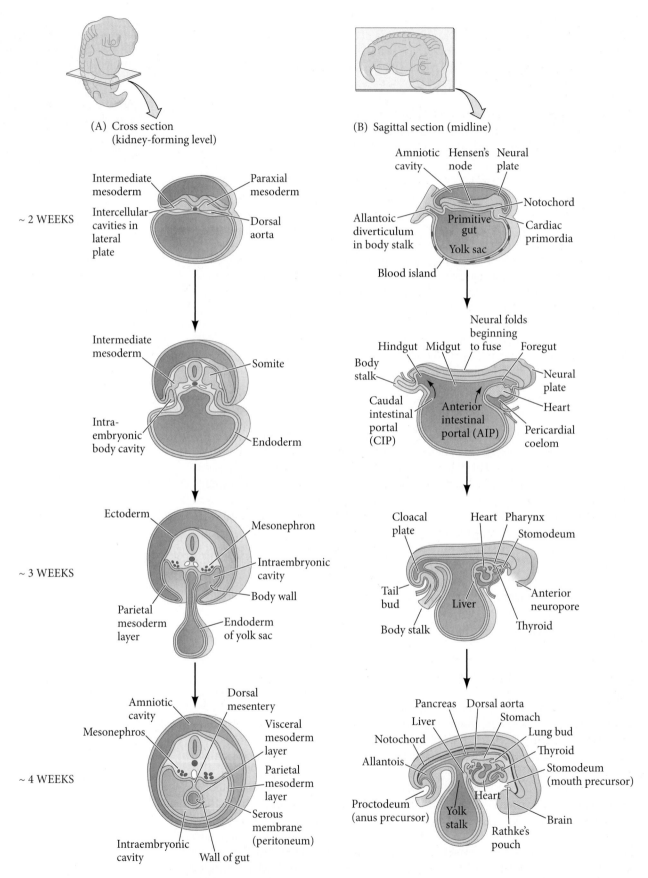

FIGURE 20.3 Endodermal folding during early human development. (A) Cross sections through the kidney-forming region. (B) Sagittal sections through the embryo's midline. (After Sadler 2009.)

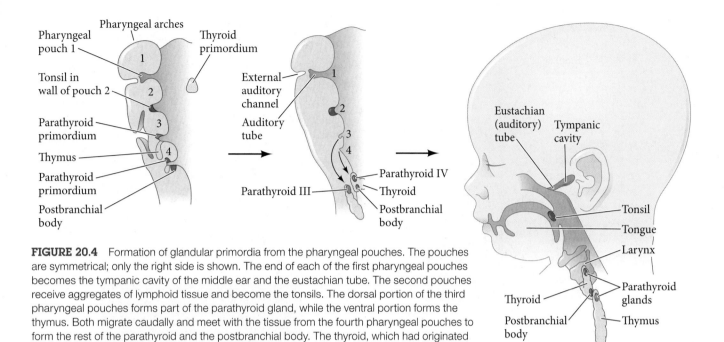

FIGURE 20.4 Formation of glandular primordia from the pharyngeal pouches. The pouches are symmetrical; only the right side is shown. The end of each of the first pharyngeal pouches becomes the tympanic cavity of the middle ear and the eustachian tube. The second pouches receive aggregates of lymphoid tissue and become the tonsils. The dorsal portion of the third pharyngeal pouches forms part of the parathyroid gland, while the ventral portion forms the thymus. Both migrate caudally and meet with the tissue from the fourth pharyngeal pouches to form the rest of the parathyroid and the postbranchial body. The thyroid, which had originated in the midline of the pharynx, also migrates caudally into the neck region. (After Carlson 1981.)

paired pouches, a small, central diverticulum is formed between the second pharyngeal pouches on the floor of the pharynx. This pocket of endoderm and mesenchyme will bud off from the pharynx and migrate down the neck to become the thyroid gland. The respiratory tube sprouts from the pharyngeal floor (between the fourth pair of pharyngeal pouches) to form the lungs, as we will see below.

Many streams of cranial neural crest cells migrate into the pouches that are forming the thyroid, parathyroid, and thymus glands. Sonic hedgehog from the endoderm appears to act as a survival factor, preventing apoptosis of the neural crest cells (Moore-Scott and Manley 2005; see Chapter 15). In addition, genetic analysis combined with transplantation studies of zebrafish has shown that FGFs (mainly Fgf3 and Fgf8) from the ectoderm and mesoderm are important not only for the migration and survival of neural crest cells, but also for the formation of the pouches themselves. Mice deficient in both *Fgf8* and *Fgf3* genes lack all the pharyngeal pouches, even when endoderm is present. Instead of migrating laterally and ventrally to form pouches, the endoderm remains in the anterior pharynx and does not spread out (Crump et al. 2004).

The Digestive Tube and Its Derivatives

During formation of the endodermal tube, mesenchyme cells from the splanchnic portion of the lateral plate mesoderm wrap around the endoderm (see Figure 20.3). The endodermal cells generate only the lining of the digestive tube and its glands, while mesenchyme cells from the splanchnic mesoderm surround the tube and will provide the future smooth muscles that generate peristaltic contraction. Posterior to the pharynx, the digestive tube constricts to form the esophagus, which is followed in sequence by the stomach, small intestine, and large intestine.

As **FIGURE 20.5A** shows, the stomach develops as a dilated region of the gut close to the pharynx. The intestines develop more caudally, and the connection between the intestine and yolk sac is eventually severed. The intestine originally ends in the endodermal cloaca, but after the cloaca separates into the bladder and rectal regions (see Chapter 18), the intestine joins with the rectum. At the caudal end of the rectum, a depression forms where the endoderm meets the overlying ectoderm, and a thin

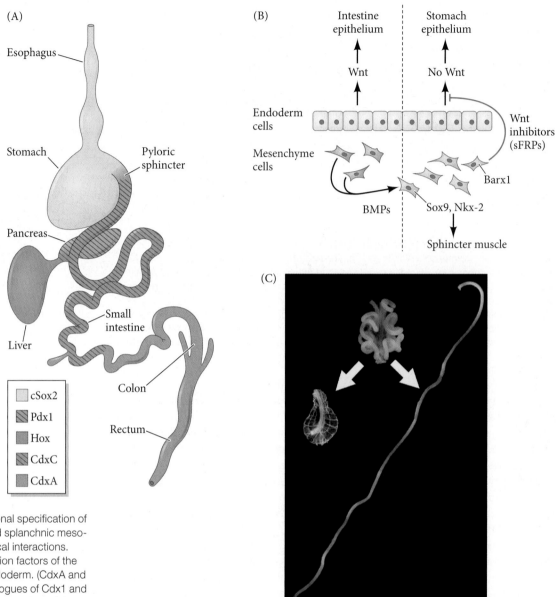

(A)

Esophagus

Stomach

Pyloric
sphincter

Pancreas

Liver

Small
intestine

Colon

Rectum

☐ cSox2
▨ Pdx1
■ Hox
▨ CdxC
■ CdxA

(B)

Intestine
epithelium

Stomach
epithelium

Wnt

No Wnt

Endoderm
cells

Wnt
inhibitors
(sFRPs)

Mesenchyme
cells

Barx1

BMPs

Sox9, Nkx-2

Sphincter muscle

(C)

FIGURE 20.5 Regional specification of the gut endoderm and splanchnic mesoderm through reciprocal interactions. (A) Regional transcription factors of the (mature) chick gut endoderm. (CdxA and C are the avian homologues of Cdx1 and 2.) These factors are seen prior to interactions with the mesoderm, but they are not stabilized. (B) Possible mechanism by which mesenchymal cells may induce endoderm to become either intestine or stomach, depending on the region. (C) Surgically separating the chick endodermal gut tube from the dorsal mesentery at embryonic day 12 causes the mesentery to shrivel up and the gut tube to straighten. The original gut-mesentery association (top) holds the digestive tube in place. When the two parts are separated on embryonic day 20, the mesentary (left) shrivels up, while the gut tube (right) straightens. (A after Grapin-Botton et al. 2001; C after Savin et al. 2011.)

cloacal membrane separates the two tissues. When the cloacal membrane eventually ruptures, the resulting opening becomes the anus.

Specification of the gut tissue

The production of endoderm is one of the first decisions made by the embryo, and the transcription factor **Sox17** is critical in this specification. In amphibian embryos, endoderm is specified autonomously by the presence of Sox17. Dominant negative forms of Sox17 (having repressive instead of activator subunits) block endoderm formation in the vegetal blastomeres of amphibians, while the overexpression of the wild-type form expands the endodermal domain (Hudson et al. 1997; Henry and Melton 1998). Mice and zebrafish deficient in *Sox17* have defective gut endoderm, and when *Sox17* is expressed experimentally in embryonic stem cells these stem cells produce endodermal derivatives (Kanai-Azuma et al. 2002; Takayama et al. 2011).

Although Sox17 helps specify the digestive tube, it does not give the tube its remarkable polarity. The digestive tube proceeds from the pharynx to the anus, differentiating along the way into the esophagus, stomach, duodenum, and intestines, and putting out branches that become (among other things) the thyroid, thymus, pancreas, and liver. What tells the endodermal tube to become these particular tissues at particular places? Why do we never see a mouth opening directly into a stomach? The endoderm and the splanchnic lateral plate mesoderm undergo a complicated set of interactions, and the signals for generating the different gut tissues appear to be conserved throughout the vertebrate classes (see Wallace and Pack 2003). One possible model for the polarity of the gut tube starts off with the specification of the pharynx and then follows with the specification of the remainder of the gut tube. Studies on chick embryos using beads containing either retinoic acid or inhibitors of its synthesis (Bayha et al. 2009) show that the pharynx can develop only in areas containing little or no RA, whereas the RA gradient patterns the pharyngeal arch endoderm in a graded manner. This is probably accomplished by activating and repressing particular sets of transcription factor genes.

The second phase of gut specification is thought to involve signals from the splanchnic mesoderm-derived mesenchyme surrounding the endodermal tube. As the digestive tube encounters different mesenchymes, mesenchyme cells instruct the endoderm to differentiate into esophagus, stomach, small intestine, and colon (Okada 1960; Gumpel-Pinot et al. 1978; Fukumachi and Takayama 1980; Kedinger et al. 1990). Wnt signals are thought to be especially important. The initial ("default") specification of the entire gut tube is thought to be anterior (i.e., stomach/esophagus). However, graded Wnt signaling from the posterior mesoderm (instructed by RA and FGF gradients) provides a signal that induces in the gut endoderm the posteriorizing transcription factors Cdx1 and Cdx2 (see Figure 20.5A), as well as the paracrine factor Indian hedgehog. At high concentrations, the Cdx transcription factors induce formation of the large intestine, whereas at lower concentrations they induce formation of the small intestine. Indeed, when β-catenin is artificially expressed in the *foregut* tissue, the *Cdx2* gene is activated and the anterior endoderm tissue is transformed into the more posterior, intestinal type of tissue (Sherwood et al. 2011; Stringer et al. 2012).

The molecular pathways by which Wnt signals from the mesenchyme influence the gut tube are just becoming known (**FIGURE 20.5B**). Cdx2, for instance, suppresses genes such as *Hhex* and thereby prevents the stomach, liver, and pancreas from forming in the posterior (Bossard and Zaret 2000; McLin et al. 2007). In the anterior regions of the gut tube (which form the thymus, pancreas, stomach, and liver), Wnt signaling is blocked. In the stomach-forming domain, the mesenchyme lining the gut tube expresses the transcription factor Barx1, which activates production of two Frzb-like Wnt antagonists (the proteins sFRP1 and sFRP2) that block Wnt signaling in the vicinity of the stomach but not around the intestine. (Indeed, *Barx1*-deficient mice do not develop stomachs and express intestinal markers in that tissue; Kim et al. 2005.)

Wnt-based polarity may be transient and needs to be strengthened and refined by further interactions between the endoderm and the surrounding mesenchymes. Roberts and colleagues (1995, 1998) have implicated Sonic hedgehog (Shh) in endodermal specification. Shh is thought to be made by the endoderm and secreted in different concentrations at different sites. Its targets appear to be the mesodermal cells surrounding the gut tube. Secretion of Shh by the hindgut endoderm induces a nested pattern of "posterior" Hox gene expression in the mesoderm. As in the vertebrae (see Chapter 12), the anterior borders of Hox gene expression delineate the morphological boundaries of the regions that will form the cloaca, large intestine, cecum, mid-cecum (at the midgut-hindgut border), and posterior portion of the midgut (Roberts et al. 1995; Yokouchi et al. 1995). When experimentally generated Hox-expressing viruses cause misexpression of Hox genes in the mesoderm, the mesodermal cells alter the differentiation of the adjacent endoderm (Roberts et al. 1998). The Hox genes are thought to specify the mesoderm so that it can further interact with the endodermal tube and more finely specify its regions.

Once the boundaries of the transcription factors are established, differentiation can begin. The regional differentiation of the mesoderm (into smooth muscle types) and the regional differentiation of the endoderm (into different functional units such as the stomach, duodenum, and small intestine) are synchronized. For instance, in certain regions the intestinal mesenchyme secretes BMP4, which instructs the mesoderm anterior to it to express the Sox9 and Nkx2-5 transcription factors. Sox9 and Nkx2-5 tell the mesoderm to become the smooth muscles of the pyloric sphincter rather than the smooth muscle that normally lines the stomach and intestine (Theodosiou and Tabin 2005).

The interaction between the splanchnic mesoderm and the endoderm continues long after the specification stage of development. One derivative of the splanchnic mesoderm is the **dorsal mesentery**, a fibrous membrane that connects the endoderm to the body wall. The looping of the intestinal tube is driven by a combination of growth intrinsic to the endoderm coupled with the connection of that tube to the dorsal mesentery (Savin et al. 2011). If the connection is severed, the mesentery shrinks and the gut becomes a long, thin, tube with no folding (**FIGURE 20.5C**).

Interactions of mesoderm and endoderm are also important for the formation of villi in the gut. The differentiation of smooth muscle in the mesoderm constricts the underlying growing endoderm and mesenchyme, thus creating compressive stresses that cause the endoderm to buckle and ultimately lead to villi formation (Shyer et al. 2013). This buckling is critical for localizing the intestinal stem cells at the base of the villi. Originally, all gut tube cells have the potential to become stem cells, but the buckling allows certain tissues to interact more readily than others and causes inhibitory paracrine factors (most importantly BMP4) to restrict stem cell formation to those regions farthest away from the tip of the villi (Shyer et al. 2015).

 **SCIENTISTS SPEAK 20.1**   Dr. Cliff Tabin and Dr. Amy Shyer speak about how the stem cells of the intestine get to the correct places.

The further development of the intestine involves (1) the differentiation of the Paneth cells and the progeny of intestinal stem cells, which was discussed in Chapter 5; and (2) interactions between the gut epithelium and symbiotic bacteria to complete the differentiation of the cell types, which will be discussed in Chapter 25. Interestingly, in at least parts of the endoderm, differentiated cells can revert to becoming stem cells. This may be due to their being exposed to the outside environment as we eat and breathe. When stem cells are removed from the stomach, a differentiated secretory cell (called the chief cell) loses its differentiated properties and becomes a stem cell (Stang et al. 2013). Similarly, the differentiated cells of the trachea can divide to generate lung stem cells when the original stem cells are deleted (Tata et al. 2013). Even the Paneth cell precursor of the intestine, which is usually committed to maturing into differentiated Paneth cells and enteroendocrine cells, can return to a stem cell status if the intestine is injured (Buczaki et al. 2013). So the endoderm appears to be remarkable, if not unique, in having a high degree of plasticity between differentiated cells and stem cells.

Accessory organs: The liver, pancreas, and gallbladder

Endoderm forms the lining of three accessory organs—the liver, pancreas, and gallbladder—that develop immediately caudal to the stomach. The **hepatic diverticulum** buds off endoderm and extends out from the foregut into the surrounding mesenchyme. The endoderm comprising this bud comes from two populations of cells: a lateral group that exclusively forms liver cells; and ventral-medial endoderm cells that form several midgut regions, including the liver (Tremblay and Zaret 2005). The mesenchyme induces this endoderm to proliferate, branch, and form the glandular epithelium of the liver. A portion of the hepatic diverticulum (the region closest to the digestive tube) continues to function as the drainage duct of the liver, and a branch from this duct produces the gallbladder (**FIGURE 20.6**). The pancreas develops from the fusion of distinct dorsal and ventral diverticula. As they grow, they come closer

Developing Questions

One of the major human birth defects is pyloric stenosis, a condition in which the muscles of the pyloric sphincter thicken and prevent food from entering the intestine. How do these muscles arise and function, and what might cause the defect?

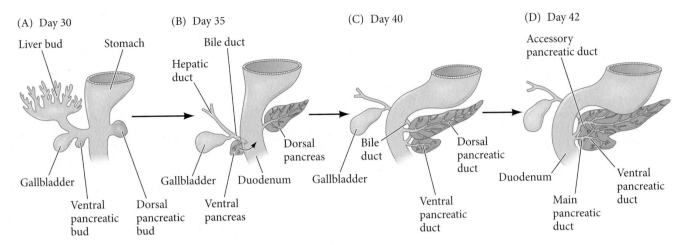

FIGURE 20.6 Pancreatic development in humans. (A) At 30 days, the ventral pancreatic bud is close to the liver primordium. (B) By 35 days, it begins migrating posteriorly, and (C) comes into contact with the dorsal pancreatic bud during the sixth week of development. (D) In most individuals, the dorsal pancreatic bud loses its duct into the duodenum; however, in about 10% of the population, the dual duct system persists.

together and eventually fuse. In humans, only the ventral duct survives to carry digestive enzymes into the intestine. In other species (such as the dog), both dorsal and ventral ducts empty into the intestine.

The posterior foregut endoderm contains progenitor cells that can give rise to pancreas, liver, and gallbladder. There is an intimate relationship between the splanchnic lateral plate mesoderm and the foregut endoderm. Just as the foregut endoderm is critical in specifying the cardiogenic mesoderm, the mesoderm, especially the blood-vessel endothelial cells, induce the endodermal tube to produce the liver primordium and the pancreatic rudiments.

The chromatin of the multipotent stem cells of the ventral foregut endoderm may be primed for their differential activation. The genes involved in forming the liver progenitor cells are silenced in a different manner than those genes involved in forming the pancreatic progenitor cells. Thus, a single signal may be able to de-repress an entire battery of specification genes (Xu et al. 2011; Zaret 2016).

LIVER FORMATION The expression of liver-specific genes (such as those for α-fetoprotein and albumin) can occur in any region of the gut tube that is exposed to cardiogenic mesoderm. However, this induction can occur only if the notochord is removed. If the notochord is placed adjacent to the portion of the endoderm normally induced by cardiogenic mesoderm to become liver, the endoderm will not form liver (hepatic) tissue. Therefore, the developing heart appears to induce the liver to form, while the presence of the notochord inhibits liver formation (**FIGURE 20.7**). This induction is probably due to FGFs secreted by the developing heart and endothelial cells (Le Douarin 1975; Gualdi et al. 1996; Jung et al. 1999; Matsumoto et al. 2001). BMP (and possibly Wnt) signals from the lateral plate mesoderm are also needed for liver formation (Zhang et al. 2004; Ober et al. 2007). Thus, the heart and endothelial cells have a developmental function in addition to their circulatory roles: they help induce the formation of the liver bud by secreting paracrine factors.

But in order to respond to the FGF signal, the endoderm has to become competent. This competence is given to the foregut endoderm by the Forkhead transcription factors. Forkhead transcription factors Foxa1 and Foxa2 are required to open the chromatin surrounding the liver-specific genes. These pioneer transcription factors displace nucleosomes from the regulatory regions surrounding these genes and are required

FIGURE 20.7 Positive and negative signaling in the formation of the mouse hepatic (liver) endoderm. The ectoderm and the notochord block the ability of the endoderm to express liver-specific genes. The cardiogenic mesoderm, probably through Fgf1 or Fgf2, promotes liver-specific gene transcription by blocking the inhibitory factors induced by the surrounding tissue. (After Gualdi et al. 1996.)

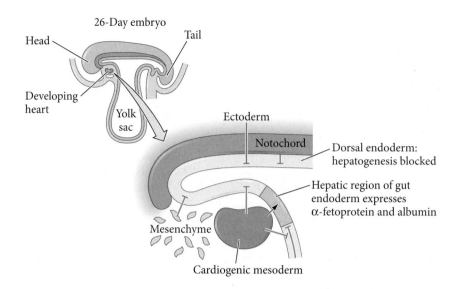

before the FGF signal is given (Lee et al. 2005; Hirai et al. 2010). Mouse embryos lacking *Foxa1* and *Foxa2* gene expression in their endoderm fail to produce a liver bud or to express liver-specific enzymes.

Once the signal is given, other Forkhead transcription factors, such as HNF4α, become critical. HNF4α is essential for the morphological and biochemical differentiation of the hepatic bud into liver tissue (Parviz et al. 2003). When conditional mutants of HNF4α were made such that this factor was absent only in the developing liver, neither tissue architecture, cellular structure, nor liver-specific enzymes were seen in the liver bud cells.

The two major liver cell types are the **hepatocytes** (liver cells) and the **cholangiocytes** that form the cells of the bile ducts (see Figure 20.2). Paracrine factors of the TGF-β family as well as Notch signaling from the blood vessels appear to stimulate the production of the duct cells, while glucocorticoid hormones and other paracrine factors (hepatocyte growth factor and Wnts) help specify the hepatocytes (Schmidt et al. 1995; Clotman et al. 2005). In addition, there are other cells in the liver, including the liver **sinusoidal endothelial cells**. These are specialized endodermal cells that create the blood channels in the liver (Goldman et al. 2014). These cells are critical for liver functioning, as they bring nutrients and poisonous substances into the liver to be metabolized and degraded. Sinusoidal endothelial cells also appear to play a critical role in liver regeneration, since they are the source of two paracrine factors—hepatocyte growth factor and angiopoietin-2—that are critical for the organized division of hepatoblast stem cells (Ding et al. 2010; DeLeve 2013; Hu et al. 2014). The ability of the mammalian liver to regenerate has been a fascinating topic, since it grows back only the same amount of liver tissue as was removed or destroyed. We are just beginning to understand how our body accomplishes this, and we will return to discuss liver regeneration more fully in Chapter 22.

PANCREAS FORMATION The formation of the pancreas may be the flip side of liver formation. Whereas the heart cells promote and the notochord prevents liver formation, the notochord may actively promote pancreas formation, while the heart may block the pancreas from forming. It seems this particular region of the digestive tube has the ability to become either pancreas or liver. One set of conditions (presence of heart, absence of notochord) induces the liver, while the opposite conditions (presence of notochord, absence of heart) cause the pancreas to form.

The notochord activates pancreas development by repressing *Shh* expression in the endoderm (Apelqvist et al. 1997; Hebrok et al. 1998). (This was a surprising finding, since the notochord is a source of Shh protein and an inducer of further *Shh* gene expression in ectodermal tissues.) Sonic hedgehog is expressed throughout the gut endoderm, *except* in the region that will form the pancreas. The notochord in this region secretes Fgf2 and activin, which are able to downregulate *Shh* expression. If *Shh* is experimentally expressed in this region, the tissue reverts to being intestinal (Jonnson et al. 1994; Ahlgren et al. 1996; Offield et al. 1996).

The lack of Shh in the pancreas-forming region of the gut seems to enable this region to respond to signals coming from the blood vessel endothelium. Indeed, pancreatic development is initiated at precisely those three locations where the foregut endoderm contacts the endothelium of the major blood vessels. It is at those points—where the endodermal tube meets the aorta and the vitelline veins—that the transcription factors Pdx1 and Ptf1a are expressed (**FIGURE 20.8A–C**; Lammert et al. 2001; Yoshitomi and Zaret 2004). If the blood vessels are removed from this area, the *Pdx1*- and *Ptf1a*-expressing regions fail to form and the pancreatic endoderm fails to bud. If more blood vessels form in this area, more of the endodermal tube becomes pancreatic tissues.

INSULIN-SECRETING PANCREATIC CELLS The association of the pancreatic tissues with blood vessels is critical in the formation of the insulin-secreting cells of the pancreas. Pdx1 appears to act in concert with other transcription factors to form the endocrine cells of the pancreas, the islets of Langerhans (Odom et al. 2004; Burlison et al. 2008; Dong et al. 2008). The exocrine cells (which produce digestive enzymes such as chymotrypsin) and the endocrine cells (which produce insulin, glucagon, and somatostatin) appear to have the same progenitor (Fishman and Melton 2002), and the level of Ptf1a appears to regulate the proportion of cells in these lineages. The exocrine pancreatic cells have higher amounts of Ptf1a (Dong et al. 2008). The islet cells secrete VEGF to attract blood vessels, and these vessels surround the developing islet (**FIGURE 20.8D**).

The endocrine progenitor cells form two populations. One is the progenitor of the β and δ cells of the islets of Langerhans. The other is the progenitor of the α and pancreatic polypeptide (PP) cells (PP is a hormone that regulates gut endocrine secretion). The progenitor of the β and δ cells expresses the transcription factor Pax4, while the α/PP progenitor cell expresses Arx. These are mutually exclusive states, so a cell becomes one cell type or the other. If the cell expresses Pax4 (becoming a β/δ progenitor), it has a further choice. If it expresses the gene for MafA, it becomes a β cell that can secrete insulin. If it doesn't express *MafA*, it becomes a δ cell (**FIGURE 20.9**).

The hierarchical dichotomous system in Figure 20.9 resembles the scheme of blood production from a single hematopoietic stem cell (see Figure 18.24). Modeling by Zhou and colleagues (2011) suggests that the different cell types can be seen as attractor states that result from the possible interactions of a common set of transcription factors. Indeed, Dhawan and colleagues (2011) have found that when the gene for Dnmt1 methyltransferase is knocked out in insulin-producing β cells, the methylation patterns change such that the *Arx* promoter is de-repressed and the *Arx* gene activated, converting β cells into glucagon-producing δ cells. Pancreatic β cell identity is thereby maintained by the methylation-mediated repression of *Arx*.

These transcription factor networks may enable the reprogramming of one cell type into another. Horb and colleagues (2003) have shown that Pdx1 can re-specify developing liver tissue into pancreas. When *Xenopus* tadpoles were given a *pdx1* gene attached

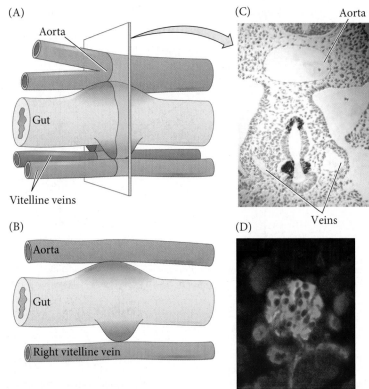

FIGURE 20.8 Induction of *Pdx1* gene expression in the gut epithelium. (A) In the chick embryo, *Pdx1* (purple) is expressed in the gut tube and is induced by contact with the aorta and vitelline veins. The regions of *Pdx1* gene expression create the dorsal and ventral rudiments of the pancreas. (B) In the mouse embryo, only the right vitelline vein survives, and it contacts the gut endothelium. *Pdx1* gene expression is seen only on this side, and only one ventral pancreatic bud emerges. (C) In situ hybridization of *Pdx1* mRNA in a section through the region of contact between the blood vessels and the gut tube of a mouse embryo. The regions of *Pdx1* expression show as deep blue. (D) Blood vessels (stained red) direct islets (stained green with antibodies to insulin) of chick embryo to differentiate. The nuclei are stained deep blue. (After Lammert et al. 2001, courtesy of D. Melton.)

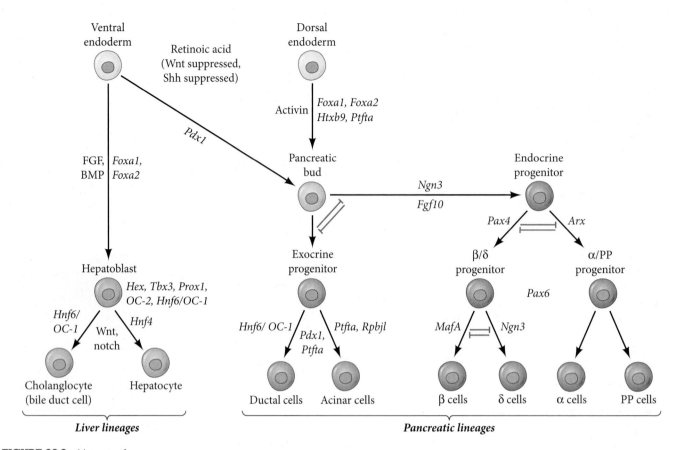

FIGURE 20.9 Lineage of pancreatic and liver cells. All pancreatic cells express *Pdx1*, distinguishing them from the cells that will become the liver. Within the pancreatic lineage, the *Ngn3*-expressing endocrine progenitor cells give rise to the endocrine lineages, while *Ptf1a*-expressing cells give rise to the exocrine progenitor that makes the ducts and the acinar cells (which secrete digestive enzymes). The endocrine progenitor can give rise to two lineages, one that can form β and δ cells and another that can form α and PP cells. (After Zhou et al. 2011.)

to a promoter active in liver cells, Pdx1 was made in the liver, and the liver was converted into a pancreas that contained both exocrine and endocrine cells. Thus, Pdx1 appears to be the critical factor in distinguishing the liver from the pancreatic mode of development. As mentioned in Chapter 5, the expression of Ngn3, Pdx1, and MafA reprograms differentiated exocrine pancreatic cells of adult mice into functional β cells (see Figure 3.15; Zhou et al. 2007).

GENERATING FUNCTIONAL PANCREATIC β CELLS One of the most important potential medical applications of developmental biology is the conversion or replacement of missing or damaged cells with new functional cells. In Chapter 5, we discussed induced pluripotent stem cells (iPSCs). Human skin cells can be transformed into pluripotent stem cells by activating certain transcription factors that revert the adult cell to a condition similar to—perhaps identical with— the embryonic stem cells of the inner cell mass. Paracrine factors added at the right times and in the right amounts can replicate in vitro the conditions of the embryo and cause the cell to differentiate into the particular cell type. In 2014, two laboratories found the right sequence of conditions to sequentially induce the formation of functional, insulin-secreting pancreatic β cells (Pagliuca et al. 2014; Rezania et al. 2014). These β cells were able to cure diabetes in mice.

For this work, knowledge of the paracrine factors involved and their inhibitors were crucial (**FIGURE 20.10**). For instance, TGF-β family members and inhibitors of the canonical Wnt pathway were needed to transform an iPSC into a cell expressing the same transcription factors as the definitive endoderm (i.e., SOX17 and FOXA2). FGF family members could transform definitive endoderm cells into cells having the same transcription factor complex as the cells of the primitive foregut (FOXA2, HNF1B). These cells were then exposed to an inhibitor of the Hedgehog pathway (mimicking the notochord) as well as other factors, to generate the pancreatic endoderm. Other sets of paracrine factors, vitamins, and inhibitors allowed the cells to sequentially become

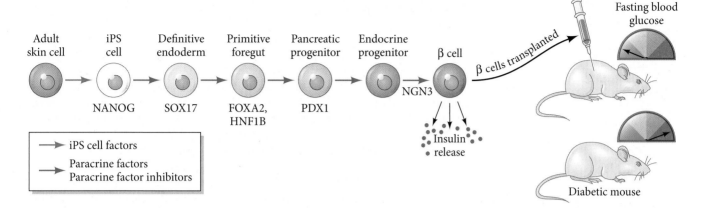

pancreatic endocrine precursor cells and finally mature insulin-secreting pancreatic β cells. Not only did these cells look like normal pancreatic β cells and have the same transcription factor activation pattern, they also secreted insulin upon stimulation with glucose. When injected into mice, they regulated glucose levels just like normal pancreatic β cells (see Figure 20.10).

These results still don't provide a "cure" for human diabetes, however. Human type 1 diabetes is an autoimmune disease wherein the person makes antibodies that destroy his or her own β cells, and the newly generated β cells are still prone to this destruction. However, because induced β cells can now be grown by the millions, they may provide a palliative therapy until we find a way to block the autoimmune destruction of β cells. The goal would be to have the induced β cells remain in the person throughout their life.

SCIENTISTS SPEAK Dr. Doug Melton reflects on a life in science and how knowledge of developmental biology can help cure diseases.

THE GALLBLADDER The origin of the gallbladder is not well characterized. In fact, fate mapping of the endoderm that gives rise to mouse gallbladder was accomplished only in 2015. Using DiI labeling, Uemura and colleagues showed that most of the gallbladder progenitors were located in the lateralmost region of the foregut endoderm, at the level corresponding to the junction of the first and second somites (Uemura et al. 2015).

One interesting finding about the development of the gallbladder is that some mammals (including some human infants) are born with the disease biliary atresia, where the bile ducts of the gallbladder are blocked. No one knows how this happens. However, an outbreak of biliary atresia in Australian livestock focused attention on this disease. An international group of scientists concluded that prolonged drought had resulted in expanded use of the pigweed plant as fodder, and that this plant contained a teratogenic toxin, biliatresone, that specifically interfered with gallbladder development (**FIGURE 20.11**). Screening thousands of compounds in zebrafish (which can be done relatively cheaply compared to other organisms), they found that this particular compound occluded the developing bile ducts. This showed that an environmental compound could cause this disease, and biologists are searching for this compound in other plants.

FIGURE 20.10 Production of functional insulin-secreting human β cells from adult cells. The adult skin cell is converted into an induced pluripotent stem cell (iPSC) by the transcription factors mentioned in Chapter 5. The iPSC can become almost any cell in the embryo. To make the cell into a pancreatic β cell, researchers sequentially mimicked the conditions seen in the embryo. This meant providing it certain paracrine factors and paracrine factor inhibitors. The iPSC first became a cell type having the transcription factor pattern of primitive endoderm. Then it became sequentially, like a foregut cell, a pancreatic cell, a pancreatic endocrine cell, and finally (after some intermediary steps not shown here), a pancreatic β cell. These β cells could be transferred into mice, where they were able to regulate glucose levels and cure the mouse model of diabetes. (After Pagliuca et al. 2014; Rezania et al. 2014.)

FIGURE 20.11 Biliary atresia caused by plant compound. (A) Gallbladder cells in 3D culture form spheres with open lumen, resembling normal bile ducts. (B) In mice treated with the teratogenic compound biliatresone from pigweed, the polarity of the cells is altered and the lumen is occluded. (From Lorent et al. 2015, courtesy of M. Pack.)

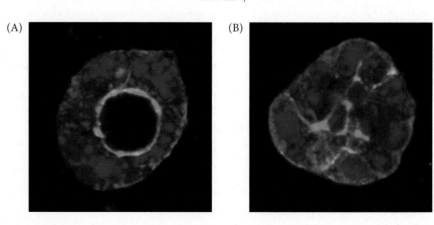

(A)   (B)

The Respiratory Tube

Although they have no role in digestion, the lungs are in fact a derivative of the digestive tube. In the center of the pharyngeal floor, between the fourth pair of pharyngeal pouches, the **laryngotracheal groove** extends ventrally (**FIGURE 20.12A–C**). This groove then bifurcates into the branches that form the paired bronchi and lungs. The laryngotracheal endoderm becomes the lining of the trachea, the two bronchi, and the air sacs (alveoli) of the lungs. Sometimes this separation is not complete and a baby is born with a connection between the gut tube and the respiratory tube. This digestive and respiratory condition is called a **tracheal-esophageal fistula** and must be surgically repaired so the baby can breathe and swallow properly.

The splitting of the trachea from the esophagus is another example of the interactions between the endoderm and specific mesenchyme. At this later point in development, the difference is between dorsal and ventral regions of the body. Wnt signals from the mesenchyme cause β-catenin to accumulate in the region of the gut tube that will become the lung and trachea. Without these signals, the separation of the gut tube

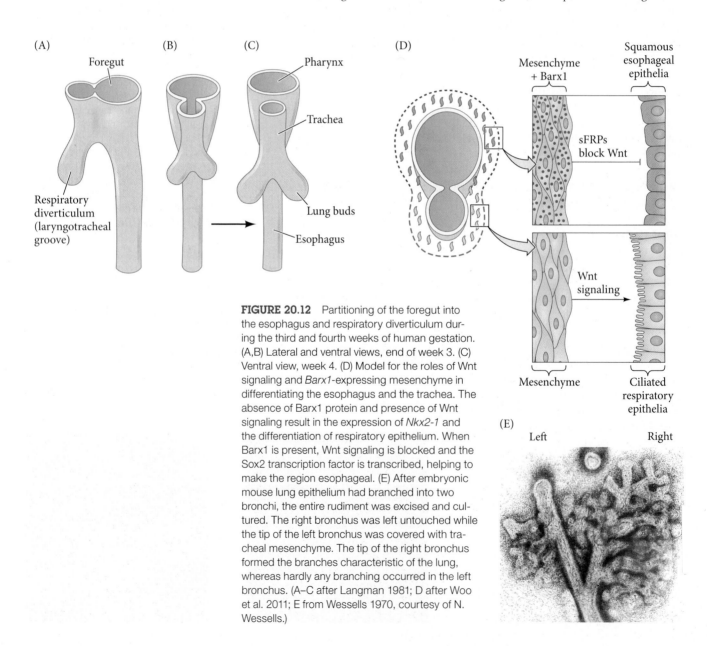

FIGURE 20.12 Partitioning of the foregut into the esophagus and respiratory diverticulum during the third and fourth weeks of human gestation. (A,B) Lateral and ventral views, end of week 3. (C) Ventral view, week 4. (D) Model for the roles of Wnt signaling and *Barx1*-expressing mesenchyme in differentiating the esophagus and the trachea. The absence of Barx1 protein and presence of Wnt signaling result in the expression of *Nkx2-1* and the differentiation of respiratory epithelium. When Barx1 is present, Wnt signaling is blocked and the Sox2 transcription factor is transcribed, helping to make the region esophageal. (E) After embryonic mouse lung epithelium had branched into two bronchi, the entire rudiment was excised and cultured. The right bronchus was left untouched while the tip of the left bronchus was covered with tracheal mesenchyme. The tip of the right bronchus formed the branches characteristic of the lung, whereas hardly any branching occurred in the left bronchus. (A–C after Langman 1981; D after Woo et al. 2011; E from Wessells 1970, courtesy of N. Wessells.)

from the tracheal tube and the trachea's development into the lungs fails to happen (Goss et al. 2009). Conversely, extra lungs can form if β-catenin is expressed ectopically in the gut tube (Harris-Johnson et al. 2009).

The dorsal portion of the respiratory tube remains in contact with mesenchyme that contains the Barx1 transcription factor and is producing the Wnt-blocking sFRPs. The sFRPs are soluble Frizzled-related proteins similar to Frzb (see Chapter 11); sFRPs can bind to Wnts and prevent their reaching their cell membrane receptors, thus blocking Wnt activity and helping to specify the epithelium of the esophagus. The ventral portion of the respiratory tube, however, comes into contact with a mesenchyme that does not produce sFRPs. The Wnt signals, which were blocked earlier, here convert the tube into the ciliated respiratory epithelium of the trachea (**FIGURE 20.12D**; Woo et al. 2011).

As in the digestive tube, the regional specificity of the mesenchyme determines the differentiation of the developing respiratory tube. In the developing mammal, respiratory epithelium responds in two distinct fashions. In the region of the neck, it grows straight, forming the trachea. After entering the thorax, it branches, forming the two bronchi and then the lungs. The respiratory epithelium of an embryonic mouse can be isolated soon after it has split into two bronchi, and the two sides can be treated differently. **FIGURE 20.12E** shows the result when the right bronchial epithelium was allowed to retain its lung mesenchyme while the left bronchus was surrounded with tracheal mesenchyme (Wessells 1970). The right bronchus proliferated and branched under the influence of the lung mesenchyme, whereas the left bronchus continued to grow in an unbranched manner. Moreover, the differentiation of the respiratory epithelia into trachea cells or lung cells depends on the mesenchyme it encounters (Shannon et al. 1998).

The stem cells of the airway passages are fascinating in many ways. As mentioned earlier, the endoderm can replace its stem cells by reverting differentiated cells to a stem cell condition. In the lungs, one group of differentiated cells, the Clara cells, can divide to produce stem cells when the original stem cells are removed. In addition, the parent stem cells provide niches for their daughter cells to develop (Pardo-Saganta et al. 2015). The basal stem cells of the trachea produce progenitor cells that will become the secretory and ciliated cells of the airways. The progenitor cells, however, need the Notch signal from their parent cells in order to remain progenitor cells. Without Notch, they undergo terminal differentiation into ciliated cells. Thus, the parent stem cell here is still "taking care" of its daughters and maintaining their proliferative capacities.

WEB TOPIC 20.1 **INDUCTION OF THE LUNG** The induction of the lung involves interplay between FGFs and Sonic hedgehog. However, it appears to be different from the induction of either the pancreas or the liver.

The lungs are among the last of the mammalian organs to fully differentiate. The lungs must be able to draw in oxygen at the newborn's first breath. To accomplish this, the alveolar cells secrete a surfactant into the fluid within the lungs. This surfactant, consisting of specific proteins and phospholipids such as sphingomyelin and lecithin, is secreted very late in gestation; in humans, surfactant usually reaches physiologically useful levels at about week 34 of gestation. Surfactant enables the alveolar cells to touch one another without sticking together. Thus, infants born prematurely—that is, before their surfactant has reached functional levels—often have difficulty breathing and have to be placed on respirators until their surfactant-producing cells mature.

Mammalian birth occurs very soon after lung maturation. Some evidence suggests that the embryonic lung may actually signal the mother to start delivery. Condon and colleagues (2004) showed that surfactant protein A—one of the final products produced by the embryonic mouse lung—activates macrophages in the amniotic fluid. These macrophages migrate from the amnion to the uterine muscle, where they produce immune system proteins such as interleukin-1β (IL1β). IL1β initiates the contractions of labor, both by activating cyclooxygenase-2 (which stimulates production of the prostaglandins that contract the uterine muscle cells) and by antagonizing the progesterone

Developing Questions

When a baby is born prematurely, its lung cells are often not differentiated. What can physicians do to accelerate lung development?

FIGURE 20.13 The immune system relays a signal from the embryonic lung. Surfactant protein A (SP-A) activates macrophages in the amniotic fluid to migrate into the uterine muscles, where the macrophages secrete IL1β. IL1β stimulates production of cyclooxygenase-2, an enzyme that in turn triggers the production of the prostaglandin hormones responsible for initiating uterine muscle contractions and birth.

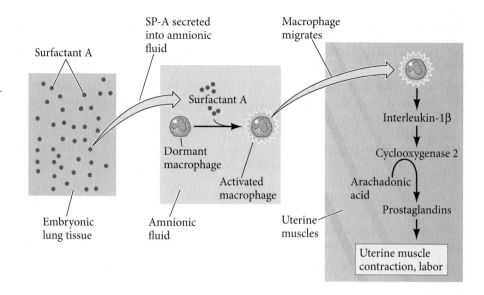

receptor (**FIGURE 20.13**). Mice deficient in surfactant proteins have a significant delay in the onset of labor, while surfactant-stimulated macrophages injected into the uteri of female mice induce early labor (Montalbano et al. 2013). Thus, one of the critical signals initiating birth is given only when the lungs have matured to the point where a newborn can take its first breath, and this signal may be transmitted to the mother via her immune system.

Next Step Investigation

This chapter has mentioned that not only molecular agents (paracrine factors), but also physical agents (such as buckling) can cause cell differentiation. Recently, a third set of factors have been found—symbiotic microbes. In mice, microbes are responsible for activating the expression of several intestine-specific genes to normal levels.

The proteins encoded by these genes are important in the morphogenesis and function of the intestine. In zebrafish, bacterial products induce normal stem cell division. How bacteria are integrated into the normal development of the gut is topic that is just beginning to be explored. We will discuss this in more detail in Chapter 25.

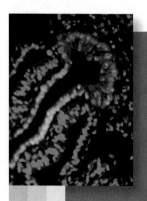

Closing Thoughts on the Opening Photo

The budding of the liver rudiment can be seen here in this 9-day mouse embryo. The nuclei are stained green, and the gut precursor cells are stained orange with an antibody to the FoxA2 transcription factor. The blue-staining cells are hepatoblasts and will become liver; these cells express a homeodomain-containing transcription factor, Hex, that changes the structure of the epithelial cells and enables them to proliferate into the mesenchyme. (Photograph courtesy of Ken S. Zaret.)

20 Snapshot Summary
The Endoderm

1. The Sox17 transcription factor is crucial for the specification of the endoderm. In vertebrates, the endoderm constructs the digestive (gut) tube and the respiratory tube.

2. The gut tube is divided into three regions by the gradient of Wnts, BMPs, and FGFs along its anterior-posterior axis. The endoderm in the posterior becomes a collection of midgut-hindut precursor cells, and forming the intestines. The endoderm near the head will form the anterior foregut cells, which will generate the precursors of the lung and thyroid glands. The region between them becomes the posterior foregut precursor cells, giving rise to the pancreas, gallbladder, and the liver.

3. Four pairs of pharyngeal pouches become the endodermal lining of the eustacian tubes, the tonsils, the thymus, and the parathyroid glands. The thyroid also forms in this region of endoderm.

4. Gut tissues form by reciprocal interactions between the endoderm and the mesoderm. Wnt signals from the mesoderm and Sonic hedgehog from the endoderm appear to play a role in inducing a nested pattern of Hox gene expression in the mesoderm surrounding the gut. The regionalized mesoderm then instructs the endodermal tube to become the different organs of the digestive tract.

5. In some areas of the gut, removal of stem cells results in the de-differentiation of some functional cells to form a new stem cell population.

6. The two major cell types of the liver are the hepatocytes, which regulate body metabolism, and the cholangiocytes that line the ducts.

7. The endoderm helps specify the splanchnic mesoderm; the splanchnic mesoderm, especially the heart and the blood vessels, helps specify the endoderm.

8. The pancreas forms in a region of endoderm that lacks *Shh* expression. The Pdx1 and Ptf1a transcription factors are expressed in this region.

9. The endocrine and exocrine cells of the pancreas have a common origin. The Ngn3 transcription factor probably decides endocrine fate.

10. By mimicking the conditions of embryonic development, human iPSCs can be transformed into precursors of pancreatic β cells and generate cells that secrete insulin.

11. The respiratory tube is derived as an outpocketing of the digestive tube. The regional specificity of the mesenchyme it meets determines whether the tube remains straight (as in the trachea) or branches (as in the bronchi and alveoli).

Further Reading

Clotman, F. and 7 others. 2005. Control of liver cell fate decision by a gradient of TGF-β signaling modulated by Onecut transcription factors. *Genes Dev.* 19: 1849–1854.

Fishman, M. P. and D. A. Melton. 2002. Pancreatic lineage analysis using a retroviral vector in embryonic mice demonstrates a common progenitor for endocrine and exocrine cells. *Int. J. Dev. Biol.* 46: 201–207.

Kim, B.-M., G. Buchner, I. Miletich, P. T. Sharpe and R. A. Shivdasani. 2005. The stomach mesenchymal transcription factor Barx1 specifies gastric epithelial identity through inhibition of transient Wnt signaling. *Dev. Cell* 8: 611–622.

Odom, D. T. and 12 others. 2004. Control of pancreas and liver gene expression by HNF transcription factors. *Science* 303: 1378–1381.

Pagliuca, F. W. and 8 others. 2014. Generation of functional human pancreatic β cells in vitro. *Cell* 159: 428–439.

Rezania, A. and 12 others. 2014. Reversal of diabetes with insulin-producing cells derived in vitro from human pluripotent stem cells. *Nature Biotechnol.* 32: 1121–1133.

Viotti , M., S. Nowotschin and A. K. Hadjantonakis. 2014. Sox17 links gut endoderm morphogenesis and germ layer segregation. *Nature Cell Biol.* 16: 1146–1156.

GO TO WWW.DEVBIO.COM...

...for Web Topics, Scientists Speak interviews, Watch Development videos, Dev Tutorials, and complete bibliographic information for all literature cited in this chapter.

Metamorphosis

The Hormonal Reactivation of Development

FEW EVENTS IN ANIMAL DEVELOPMENT are as spectacular as **metamorphosis**, the hormonal reactivation of developmental phenomena that gives the animal a new form. Animals (including humans) whose young are essentially smaller, less sexually mature, versions of the adult are referred to as **direct developers**. Most animal species, however, are **indirect developers** whose life cycle includes a larval stage with characteristics very different from those of the adult organism, which emerges only after a period of metamorphosis.

Metamorphosis is both a developmental and an ecological transition. Very often, larval forms are specialized for some function such as growth or dispersal, whereas the adult is specialized for reproduction. *Cecropia* moths, for example, hatch from eggs and develop as wingless juveniles—caterpillars—for several months. After metamorphosis, the adult insects spend only a day or so as fully developed winged moths and must mate quickly before they die. The adult moths never eat, and in fact have no mouthparts during this brief reproductive phase of the life cycle. Metamorphosis is initiated by specific hormones that reactivate developmental processes throughout the entire organism, changing it morphologically, physiologically, and behaviorally to prepare itself for a new mode of existence. Ecologically, metamorphosis is associated with changes of habitat, food, and behaviors (Jacobs et al. 2006).

Among indirect developers, there are two major types of larvae. Larvae that represent dramatically different body plans than the adult form and that are morphologically distinct from the adult are called **primary larvae**. Sea urchin larvae, for instance, are bilaterally symmetrical organisms that float among and collect food in the plankton of

How might the larval foods help the survival of the adult form?

The Punchline

Metamorphosis is the suite of dramatic developmental changes whereby an immature organism is given a new, usually sexually mature, form. Changes are initiated by endocrine factors—hormones—that potentially affect every cell in the body. Some cells are told to divide, some to differentiate, and some to die. The transformation from larva (caterpillar, tadpole, pluteus) to adult provides striking examples of rapid developmental change. In insects and amphibians, the axes of the larvae become the axes of the adult, even though the cells forming these axes have changed. In sea urchins, none of the axes of the larva are preserved, as the adult forms from a small pouch on the left side of the larval gut. Metamorphosis in all three groups is accomplished by lipid hormones that activate transcription factors.

the open ocean. The sea urchin *adult*, on the other hand, is pentameral (i.e., has fivefold symmetry) and feeds by scraping algae from rocks on the seafloor. There is no trace of the adult form in the body plan of the larva.[1]

Secondary larvae are found among those animals whose larvae and adults possess the same basic body plan. Thus, despite the obvious differences between the caterpillar and the butterfly, these two life stages retain the same major body axes and develop by deleting and modifying old parts while adding new structures into a preexisting framework. Similarly, the frog tadpole, although specialized for an aquatic environment, is a secondary larva, organized on the same pattern as the adult will be (Jagersten 1972; Raff and Raff 2009).

Metamorphosis is one of the most striking of developmental phenomena, and the extensive *morphological* changes undergone by some species have fascinated developmental anatomists for centuries (Merian 1705; Swammerdam 1737). But we know only an outline of the *molecular* bases of metamorphosis, and only for a handful of species.

Amphibian Metamorphosis

Amphibians are actually named for their ability to undergo metamorphosis, their appellation coming from the Greek *amphi* ("double") and *bios* ("life"). Amphibian metamorphosis is associated with morphological changes that prepare an aquatic organism for a primarily terrestrial existence. In **urodeles** (salamanders), these changes include the resorption of the tail fin, the destruction of the external gills, and a change in skin structure. In **anurans** (frogs and toads), the changes are more dramatic, with almost every organ subject to modification (**TABLE 21.1**; see also Figure 1.3). The changes in amphibian metamorphosis are initiated by thyroid hormones such as **thyroxine** (T_4) and **tri-iodothyronine** (T_3) that travel through the blood to reach all the organs of the larva. When the larval organs encounter these thyroid hormones, they can respond in any of four ways: growth (as in the hindlimbs of the frog), death (as in the tail of the frog), remodeling (as in the frog's intestine), and respecification (as in the liver enzymes of the frog).

---

[1] Although there is controversy on the subject, larvae probably evolved after the adult form had been established. In other words, animals evolved through direct development, and larval forms came about as specializations for feeding or dispersal during the early part of the life cycle (Jenner 2000; Rouse 2000; Raff and Raff 2009). Even so, the biphasic life cycle may be a trait characteristic of metazoans (see Degnan and Degnan 2010).

TABLE 21.1 Some metamorphic changes in anurans

System	Larva	Adult
Locomotory	Aquatic; tail fins	Terrestrial; tailless tetrapod
Respiratory	Gills, skin, lungs; larval hemoglobins	Skin, lungs; adult hemoglobins
Circulatory	Aortic arches; aorta; anterior, posterior, and common jugular veins	Carotid arch; systemic arch; cardinal veins
Nutritional	Herbivorous; long spiral gut; intestinal symbionts; small mouth, horny jaws, labial teeth	Carnivorous; short gut; proteases, large mouth with long tongue
Nervous	Lack of nictitating membrane; porphyropsin, lateral line system, Mauthner neurons	Development of ocular muscles, nictitating membrane, tympanic membrane; rhodopsin; lateral line system lost, Mauthner neurons degenerate
Excretory	Largely ammonia, some urea (ammonotelic)	Largely urea; high activity of enzymes of ornithine-urea cycle (ureotelic)
Integument	Thin, bilayered epidermis with thin dermis; no mucous or granular glands	Stratified squamous epidermis with adult keratins; well-developed dermis contains mucous and granular glands secreting antimicrobial peptides

Source: Data from Turner and Bagnara 1976 and Reilly et al. 1994

Morphological changes associated with amphibian metamorphosis

GROWTH OF NEW STRUCTURES The hormone tri-iodothyronine induces certain adult-specific organs to form. As seen in Chapter 1, the limbs of the adult frog emerge from specific sites on the metamorphosing tadpole. Similarly, in the eye, eyelids and the nictitating membranes (the so-called "third eyelid" in frogs) both emerge. Moreover, T_3 induces the proliferation and differentiation of new neurons to serve these organs. As the limbs grow out from the body axis, new neurons proliferate and differentiate in the spinal cord. These neurons send axons to the newly formed limb musculature (Marsh-Armstrong et al. 2004). Blocking T_3 activity prevents these neurons from forming and causes paralysis of the limbs.

One readily observed consequence of anuran metamorphosis is the movement of the eyes to the front of the head from their originally lateral position (**FIGURE 21.1A,B**).[2] The lateral eyes of the tadpole are typical of preyed-upon herbivores, whereas the frontally located eyes of the frog befit its more predatory lifestyle. To catch its prey, the frog needs to see in three dimensions. That is, it has to acquire a *binocular field of vision*, where inputs from both eyes converge in the brain (see Figure 15.38B). In the tadpole, the right eye innervates the left side of the brain, and vice versa; there are no ipsilateral (same-side) projections of the retinal neurons. During metamorphosis, however, ipsilateral pathways emerge alongside the contralateral (opposite-side) pathways, enabling input from both eyes to reach the same area of the brain (Currie and Cowan 1974; Hoskins and Grobstein 1985a).

In *Xenopus*, these new pathways result not from the remodeling of existing neurons but from the formation of new neurons that differentiate in response to thyroid hormones (Hoskins and Grobstein 1985a,b). The ability of these axons to project ipsilaterally results from the induction of ephrin B in the optic chiasm by the thyroid hormones (Nakagawa et al. 2000). Ephrin B is also found in the optic chiasm of mammals (which have ipsilateral projections throughout life) but not in the chiasm of fish and birds (which have only contralateral projections). As we saw in Chapter 15, ephrins can repel certain neurons, causing them to project in one direction rather than another (**FIGURE 21.1C,D**).

CELL DEATH DURING METAMORPHOSIS The hormone T_3 also induces certain larval-specific structures to die. T_3 causes the degeneration of the paddlelike tail and the oxygen-procuring gills that were important for larval (but not adult) movement and respiration. While it is obvious that the tadpole's tail muscles and skin die, is this death murder or induced suicide? In other words, is T_3 telling the cells to kill themselves, or is T_3 telling something else to kill the cells? Recent evidence suggests that the first part of tail resorption is caused by suicide, but the last remnants of the tadpole tail must be killed off by other means. When tadpole muscle cells were injected with a dominant negative T_3 receptor (and therefore could not respond to T_3), the muscle cells survived, indicating that T_3 told them to kill themselves by apoptosis (Nakajima and Yaoita 2003; Nakajima et al. 2005). This was confirmed

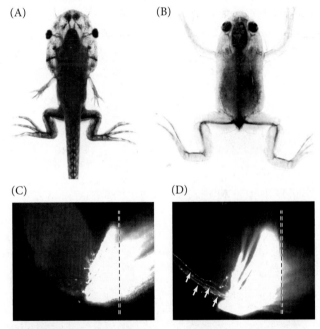

FIGURE 21.1 Eye migration and associated neuronal changes during metamorphosis of the *Xenopus laevis* tadpole. (A) The eyes of the tadpole are laterally placed, so there is relatively little binocular field of vision. (B) The eyes migrate dorsally and rostrally during metamorphosis, creating a large binocular field for the adult frog. (C,D) Retinal projections of metamorphosing tadpole. The dye DiI was placed on a cut stump of the optic nerve to label the retinal projection. (C) In early and middle stages of metamorphosis, axons project across the midline (dashed line) from one side of the brain to the other. (D) In late metamorphosis, ephrin B is produced in the optic chiasm as certain neurons (arrows) are formed that project ipsilaterally. (A,B from Hoskins and Grobstein 1984, courtesy of P. Grobstein; C,D from Nakagawa et al. 2000, courtesy of C. E. Holt.)

---

[2]One of the most spectacular movements of eyes during metamorphosis occurs in flatfish such as flounder. Originally, a flounder's eyes, like the lateral eyes of other fish species, are on opposite sides of its face. However, during metamorphosis, one of the eyes migrates across the head to meet the eye on the other side (Hashimoto et al. 2002; Bao et al. 2005). This allows the flatfish to dwell on the ocean bottom, looking upward.

VADE MECUM

For photographs of amphibian metamorphosis (and for the sounds of the adult frogs), check out the metamorphosis and frog calls sections of the "Amphibians" segment.

by the demonstration that the death of the tadpole muscle cells is prevented by blocking the activity of the apoptosis-inducing enzyme caspase-9 (Rowe et al. 2005). However, later in metamorphosis, the tail muscles are eaten by macrophages, perhaps because the extracellular matrix that supported the muscle cells has been digested by proteases.

Death also comes to the tadpole's red blood cells. During metamorphosis, tadpole hemoglobin is replaced by adult hemoglobin, which binds oxygen more slowly and releases it more rapidly (McCutcheon 1936; Riggs 1951). The red blood cells carrying the tadpole hemoglobin have a different shape than the adult red blood cells, and these larval red blood cells are specifically digested by macrophages in the liver and spleen after the adult red blood cells are made (Hasebe et al. 1999).

REMODELING DURING METAMORPHOSIS   Among frogs and toads, certain larval structures are remodeled for adult needs. The larval intestine, with its numerous coils for digesting plant material, is converted into a shorter intestine for a carnivorous diet. Schrieber and his colleagues (2005) have demonstrated that the new cells of the adult intestine are derived from functioning cells of the larval intestine (instead of there being a subpopulation of stem cells that give rise to the adult intestine). As the extracellular matrix of the old intestine dissolves, most of the intestinal epithelial cells die. Those that survive appear to dedifferentiate and become intestinal stem cells (Stolow and Shi 1995; Ishizuya-Oka et al. 2001; Fu et al. 2005; Hasebe et aal. 2013).

Much of the nervous system is remodeled as neurons grow and innervate new targets. While some neurons (like those in the optic pathway) emerge, other larval neurons, such as certain motor neurons in the tadpole jaw, switch their allegiances from larval muscle to newly formed adult muscle (Alley and Barnes 1983). Still others, such as the cells innervating the tongue muscle (a newly formed muscle not present in the larva), lie dormant during the tadpole stage and form their first synapses during metamorphosis (Grobstein 1987). The lateral line system of the tadpole (which allows the tadpole to sense water movement and helps it hear) degenerates, and the ears undergo further differentiation (see Fritzsch et al. 1988). The middle ear develops, as does the tympanic membrane characteristic of frog and toad outer ears.[3] Thus, the anuran nervous system undergoes enormous restructuring as some neurons die, others are born, and others change their specificity.

The shape of the anuran skull also changes significantly as practically every structural component of the head is remodeled (Trueb and Hanken 1992; Berry et al. 1998). The most obvious change is that new bone is being made. The tadpole skull is primarily neural crest-derived cartilage; the adult skull is primarily neural crest-derived bone (**FIGURE 21.2**; Gross and Hanken 2005). As the lower jaw of the adult forms, Meckel's cartilage elongates to nearly double its original length, and dermal bone forms around it. While Meckel's cartilage is growing, the gills and pharyngeal arch cartilage (which were necessary for aquatic respiration in the tadpole) degenerate. Other cartilage is extensively remodeled. Thus, as in the nervous system, some skeletal elements proliferate, some die, and some are remodeled. The mechanisms by which one hormone signals differential effects in different, and often adjacent tissues, remains unknown.

(A) Tadpole

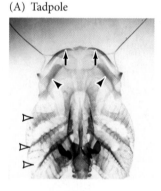

(B) Early metamorphosis

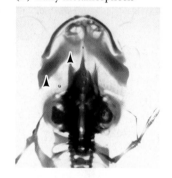

(C) Late metamorphosis

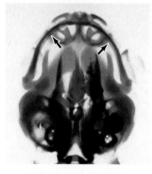

(D) Froglet

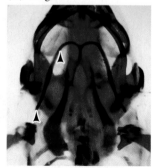

FIGURE 21.2   Changes in the *Xenopus* skull during metamorphosis. Whole mounts were stained with alcian blue to stain cartilage and alizarin red to stain bone. (A) Prior to metamorphosis, the pharyngeal (branchial) arch cartilage (open arrowheads) is prominent, Meckel's cartilage (arrows) is at the tip of the head, and the ceratohyal cartilage (arrowheads) is relatively wide and anteriorly placed. (B–D) As metamorphosis ensues, the pharyngeal arch cartilage disappears, Meckel's cartilage elongates, the mandible (lower jawbone) forms around Meckel's cartilage, and the ceratohyal cartilage narrows and becomes more posteriorly located. (From Berry et al. 1998, courtesy of D. D. Brown.)

---

[3]Tadpoles experience a brief period of deafness as the neurons change targets; see Boatright-Horowitz and Simmons 1997.

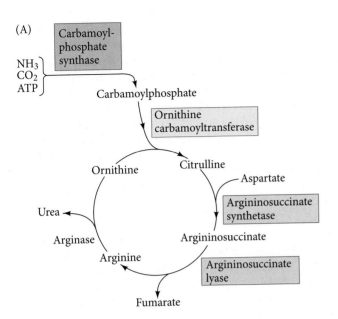

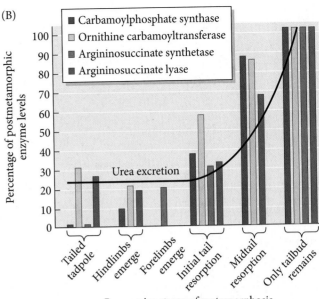

FIGURE 21.3 Development of the urea cycle during anuran metamorphosis. (A) Major features of the urea cycle, by which nitrogenous wastes are detoxified and excreted with minimal water loss. (B) The emergence of urea-cycle enzyme activities correlates with metamorphic changes in the frog *Rana catesbeiana*. (After Cohen 1970.)

BIOCHEMICAL RESPECIFICATION IN THE LIVER   In addition to the obvious morphological changes, important biochemical transformations occur during metamorphosis as T_3 induces a new set of proteins in existing cells. One of the most dramatic biochemical changes occurs in the liver. Tadpoles, like most freshwater fish, are ammonotelic—that is, they excrete ammonia. However, like most *terrestrial* vertebrates, many adult frogs (such as the genus *Rana*, although not the more aquatic *Xenopus*) are ureotelic: they excrete urea, which requires less water than ammonia excretion. During metamorphosis, the liver begins to synthesize the enzymes necessary to create urea from carbon dioxide and ammonia (**FIGURE 21.3**).

T_3 may regulate this change by inducing a set of transcription factors that specifically activates expression of the urea-cycle genes while suppressing the genes responsible for ammonia synthesis (Cohen 1970; Atkinson et al. 1996, 1998). Mukhi and colleagues (2010) showed that T_3 activates adult hepatic genes while repressing larval hepatic genes in the same cell. Moreover, for a brief time during metamorphosis, the same liver cell contains mRNAs for both larval and adult proteins.

WATCH DEVELOPMENT 21.1   Time-lapse cinematography allows us to watch the metamorphosis of frogs and flatfish.

Hormonal control of amphibian metamorphosis

The control of metamorphosis by thyroid hormones was first demonstrated in 1912 by J. F. Gudernatsch, who discovered that tadpoles metamorphosed prematurely when fed powdered horse thyroid glands. In a complementary study, Bennet Allen (1916) found that when he removed or destroyed the thyroid rudiment of early tadpoles, the larvae never metamorphosed but instead grew into giant tadpoles. Subsequent studies showed that the sequential steps of anuran metamorphosis are regulated by increasing amounts of thyroid hormone (see Saxén et al. 1957; Kollros 1961; Hanken and Hall 1988). Some events (such as the development of limbs) occur early, when the concentration of thyroid hormones is low; other events (such as the resorption of the tail and remodeling of the intestine) occur later, after the hormones reach higher concentrations. These observations gave rise to a **threshold model**, wherein the different events of metamorphosis are triggered by different concentrations of thyroid hormones. Although the threshold model remains useful, molecular studies have shown that the timing of the events

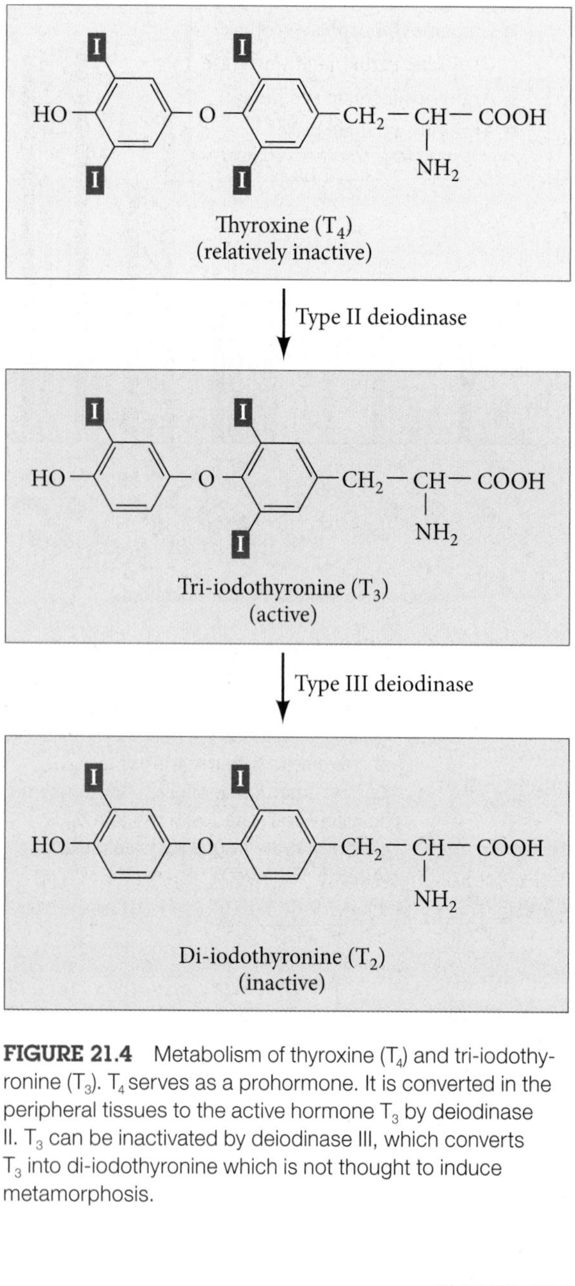

FIGURE 21.4 Metabolism of thyroxine (T_4) and tri-iodothyronine (T_3). T_4 serves as a prohormone. It is converted in the peripheral tissues to the active hormone T_3 by deiodinase II. T_3 can be inactivated by deiodinase III, which converts T_3 into di-iodothyronine which is not thought to induce metamorphosis.

of amphibian metamorphosis is more complex than just increasing hormone concentrations.

The metamorphic changes of frog development are brought about by (1) the secretion of the hormone thyroxine (T_4) into the blood by the thyroid gland; (2) the conversion of T_4 into the more active hormone, tri-iodothyronine (T_3) by the target tissues; and (3) the degradation of T_3 in the target tissues (**FIGURE 21.4**). Once inside the cell, T_3 binds to the nuclear **thyroid hormone receptors** (**TRs**) with much higher affinity than does T_4, and causes these transcription factors to become transcriptional activators of gene expression. Thus, the levels of both T_3 and TRs in the target tissues are essential for producing the metamorphic response in each tissue (Kistler et al. 1977; Robinson et al. 1977; Becker et al. 1997).

The concentration of T_3 in each tissue is regulated by the concentration of T_4 in the blood and by two critical intracellular enzymes that remove iodine atoms from T_4 and T_3. **Type II deiodinase** removes an iodine atom from the outer ring of the precursor hormone (T_4) to convert it into the more active hormone T_3. **Type III deiodinase** removes an iodine atom from the inner ring of T_3 to convert it into an inactive compound (T2) that will eventually be metabolized to tyrosine (Becker et al. 1997). Tadpoles that are genetically modified to overexpress type III deiodinase in their target tissues never complete metamorphosis (Huang et al. 1999); therefore the regulation of metamorphosis involves tissue specific regulation of the form of the hormone that binds most effectively to its receptor.

Thyroid hormone receptors are nuclear proteins, and there are two major types. In *Xenopus*, **thyroid hormone receptor α** (**TRα**) is widely distributed throughout all tissues and is present even before the organism has a thyroid gland. Yet, in an example of a positive feedback loop, the gene encoding **thyroid hormone receptor β** (**TRβ**) is itself directly activated by thyroid hormones. TRβ levels are very low before the advent of metamorphosis; as the levels of thyroid hormone increase during metamorphosis, so do intracellular levels of TRβ (**FIGURE 21.5**; Yaoita and Brown 1990; Eliceiri and Brown 1994). As we will see, this positive regulation of hormone receptor gene expression by its own gene product is a common feature of metamorphosis across animal taxa.

The TRs do not work alone, but form dimers with the retinoid receptor RXR. TR-RXR dimers bind thyroid hormones and can then upregulate transcription (Mangelsdorf and Evans 1995; Wong and Shi 1995; Wolffe and Shi 1999). Importantly, the TR-RXR receptor complex is physically associated with appropriate promoters and enhancers even before it binds T_3 (Grimaldi et al. 2012). In its unbound state, TR-RXR is a transcriptional *repressor*, recruiting histone deacetylases and other co-repressor proteins to its target genes and stabilizing repressive nucleosomes around the promoter. However, when the TR-RXR complex binds T_3, the repressors leave the complex and are replaced by co-activators such as histone acetyltransferase. These co-activators cause the dispersal of the nucleosomes and the activation of those same genes previously inhibited (Sachs et al. 2001; Buchholz et al. 2003; Grimaldi et al. 2013). Thus, the TRs have a dual function: when unliganded, they repress gene expression, preventing early metamorphosis; but when bound to T_3, they activate the expression of these same genes (see Figure 21.5).

Metamorphosis is often divided into stages based on the concentration of thyroid hormones in the blood. During the first stage, **pre-metamorphosis**, the thyroid gland has begun to mature and is secreting low levels of T_4 and very low levels of T_3. The TRα receptor is present but the TRβ receptor is not. T_4 secretion may be initiated by corticotropin-releasing hormone (CRH, which can be activated either developmentally or by external stresses). CRH generates steroids such as cortisosterone. Corticosterone probably has two modes of action (Kulkarni and Bucholz 2014). First, it acts directly on the frog pituitary, instructing it to release thyroid-stimulating hormone (TSH), thereby initiating thyroid hormone synthesis. Second, it may act at the level of the promoter and enhancer of the hormone-responsive genes to make the responding cells more responsive to low amounts of T_3 (Denver 1993, 2003).

The tissues that respond earliest to the thyroid hormones are those that express high levels of deiodinase II and thus can convert T_4 directly into T_3 (Cai and Brown 2004). For instance, the limb rudiments, which have high levels of both deiodinase II and TRα, can convert T_4 into T_3 and use it immediately through the TRα receptor. Thus, during the early stage of metamorphosis, the limb rudiments are able to receive thyroid hormone and use it to initiate leg growth (Becker et al. 1997; Huang et al. 2001; Schreiber et al. 2001).

As the thyroid matures to the stage of **prometamorphosis**, it secretes more hormones. However, many major changes (such as tail resorption, gill resorption, and intestinal remodeling) must wait until the **metamorphic climax** stage. At that time, the concentration of T_4 rises dramatically and TRβ levels peak inside the cells. Since one of the targets of T_3 is the *TRβ* gene, it is thought that TRβ is the principal receptor mediating the metamorphic climax. In the tail, there is only a small amount of TRα during premetamorphosis, and deiodinase II is not detectable. However, during prometamorphosis, rising levels of thyroid hormones induce higher levels of TRβ. At metamorphic climax, deiodinase II is expressed and the tail begins to be resorbed. In this way, the tail undergoes absorption only *after* the legs are functional (otherwise, the poor amphibian would have no means of locomotion). The wisdom of the frog is simple: "Never get rid of your tail before your legs are working."

Some tissues do not seem to be responsive to thyroid hormones. For instance, thyroid hormones instruct the *ventral* retina to express ephrin B and to generate the ipsilateral neurons seen in Figure 21.1D. The *dorsal* retina, however, is not responsive to thyroid hormones and does not generate new neurons. The dorsal retina appears to be insulated from thyroid hormones by expressing deiodinase III, which degrades the T_3 produced by deiodinase II. If deiodinase III is activated in the ventral retina, neurons will not proliferate and no ipsilateral axons will form (Kawahara et al. 1999; Marsh-Armstrong et al. 1999).

The frog brain also undergoes changes during metamorphosis, and one of its functions is to downregulate metamorphosis once metamorphic climax has been reached. Thyroid hormones eventually induce a negative feedback loop, shutting down the pituitary cells that instruct the thyroid to secrete them (Saxén et al. 1957; Kollros 1961; White and Nicoll 1981). Huang and colleagues (2001) have shown that, at the climax of metamorphosis, deiodinase II expression is seen in those cells of the anterior pituitary that secrete thyrotropin, the hormone that activates thyroid hormone expression. The resulting T_3 is thought to activate genes that block secretion of thyrotropin, thereby initiating the negative feedback loop so that less thyroid hormone is made (Sternberg et al. 2011).

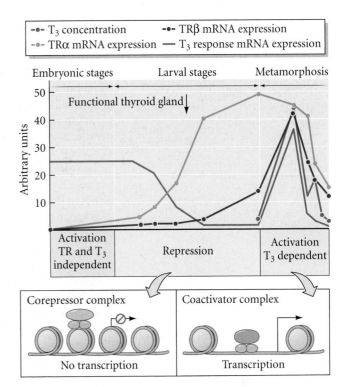

FIGURE 21.5 The hormonal control of *Xenopus* metamorphosis. During premetamorphosis, thyroid hormone titres are low, and unliganded TRα binds to the chromatin and fixes transcriptional repressors that stabilize nucleosomes. During metamorphic climax, blood levels of thyroid hormones increase, and the TRα binds thyroxine. This causes the exchange of transcriptional repressors to transcriptional activators. The nucleosomes disperse, and the T3-sensitive genes are activated. One of these genes encodes TRβ, which further accelerates the metamorphic responses. Eventually, feedback inhibition lowers the amount of circulating thyroid hormones, and metamorphosis ends. (After Grimaldi 2013.)

WEB TOPIC 21.1 VARIATIONS ON THE THEMES OF AMPHIBIAN METAMORPHOSIS
Direct development and paedomorphosis are themes of amphibian development that alter the life cycles.

Regionally specific developmental programs

By regulating the amount of T_3 and TRs in their cells, the different regions of the tadpole body can respond to thyroid hormones at different times. The type of response (proliferation, apoptosis, differentiation, migration) is determined by other factors already present in the different tissues. The same stimulus causes some tissues to degenerate while stimulating others to develop and differentiate, as exemplified by the process of tail degeneration: thyroid hormone instructs the limb bud muscles to grow (they die without thyroxine) while instructing the tail muscles to undergo apoptosis (Cai et al. 2007).

The resorption of the tail structures is relatively rapid, since the bony skeleton does not extend to the tail (Wassersug 1989). After apoptosis has taken place, macrophages collect in the tail region and digest the debris with their enzymes (especially collagenases and metalloproteinases) and the tail becomes a large sac of proteolytic enzymes[4] (Kaltenbach et al. 1979; Oofusa and Yoshizato 1991; Patterson et al. 1995). The tail epidermis acts differently than the head or trunk epidermis. During metamorphic climax, the larval skin is instructed to undergo apoptosis. The tadpole head and body are able to generate new epidermis from epithelial stem cells. The tail epidermis, however, lacks these stem cells and fails to generate new skin (Suzuki et al. 2002).

Organ-specific responses to thyroid hormones have been dramatically demonstrated by transplanting a tail tip to the trunk region and placing an eye cup in the tail (Schwind 1933; Geigy 1941). Tail-tip tissue placed in the trunk is not protected from degeneration, but the eye cup retains its integrity even when it lies within the degenerating tail (**FIGURE 21.6**). Thus, the way a tissue responds to the thyroid hormone is inherent in the tissue, itself. It is not dependent on its position within the larva.

---

[4] Interestingly, the degeneration of the human tail, which takes place during week 4 of gestation, resembles the resorption of the tadpole tail (see Fallon and Simandl 1978).

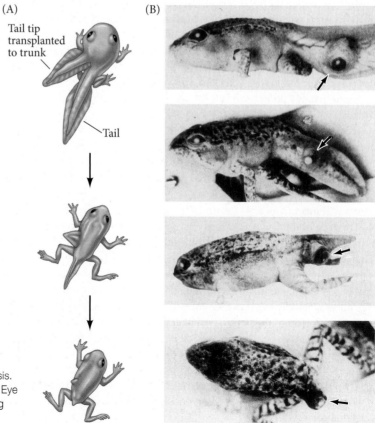

(A) Tail tip transplanted to trunk

Tail

(B)

FIGURE 21.6 Regional specificity during frog metamorphosis. (A) Tail tips regress even when transplanted into the trunk. (B) Eye cups remain intact even when transplanted into the regressing tail. (After Schwind 1933.)

The metamorphosis of tadpoles into frogs is one of the most rapid and accessible examples of development, obvious even to the eyes of children. Yet it still presents an enormous set of enigmas. As Don Brown and Liquan Cai (2007) have asked, "What will encourage the modern generation of scientists to study the wonderful biological problems presented by amphibian metamorphosis?" Recent work has shown the importance of metamorphosis for studying regeneration, and it is also a critical area area where development and ecology have a marked impact on each other.

Metamorphosis in Insects

Insects are the most speciose of Earth's animals, and the diversity of their life cycles makes science fiction pale by comparison. There are three major patterns of insect development. A few insects, such as springtails, have no larval stage and undergo direct, or **ametabolous**, development (**FIGURE 21.7A**). Immediately after hatching, ametabolous insects have a **pronymph** stage bearing the structures that enabled it to get out of the egg. But after this transitory stage, the insect looks like a small adult; it grows larger after each molt with a new cuticle, but is unchanged in form (Truman and Riddiford 1999).

Other insects, notably grasshoppers and bugs, undergo a gradual, or **hemimetabolous**, metamorphosis (**FIGURE 21.7B**). After spending a very brief period of time as a pronymph (whose cuticle is often shed as the insect hatches), the insect looks like an immature adult and is called a **nymph**. The rudiments of the wings, genital organs, and

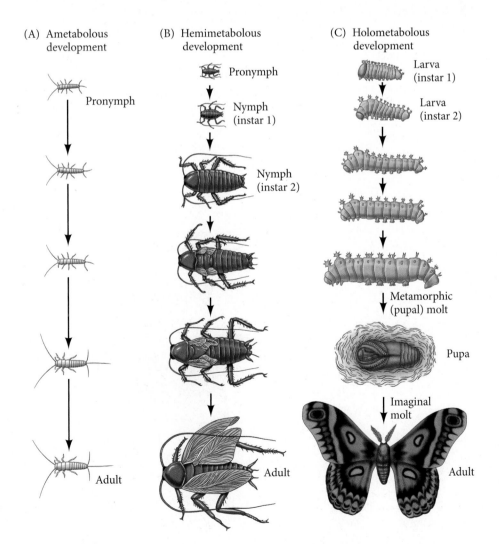

(A) Ametabolous development

Pronymph

Adult

(B) Hemimetabolous development

Pronymph

Nymph (instar 1)

Nymph (instar 2)

Adult

(C) Holometabolous development

Larva (instar 1)

Larva (instar 2)

Metamorphic (pupal) molt

Pupa

Imaginal molt

Adult

FIGURE 21.7 Modes of insect development. Molts are represented as arrows. (A) Ametabolous (direct) development in a silverfish. After a brief pronymph stage, the insect looks like a small adult. (B) Hemimetabolous (gradual) metamorphosis in a cockroach. After a very brief pronymph phase, the insect becomes a nymph. After each molt, the next nymphal instar looks more like an adult, gradually growing wings and genital organs. (C) Holometabolous (complete) metamorphosis in a moth. After hatching as a larva, the insect undergoes successive larval molts until a metamorphic molt causes it to enter the pupal stage. Then an imaginal molt turns it into an adult that ecloses from the pupal case with a new cuticle.

other adult structures are present and become progressively more mature with each molt. At the final molt, the emerging insect is a winged and sexually mature adult, or **imago**.

In the **holometabolous** development of insects such as flies, beetles, moths, and butterflies, there is no pronymph stage (**FIGURE 21.7C**). The juvenile form that hatches from the egg is called a **larva**. The larva (a caterpillar, grub, or maggot) undergoes a series of molts as it becomes larger. The stages between these larval molts are called **instars**. The number of larval molts before becoming an adult is characteristic of a species, although environmental factors can increase or decrease the number. The larval instars grow in a stepwise fashion, each instar being larger than the previous one. Finally, there is a dramatic and sudden transformation between the larval and adult stages: after the final instar, the larva undergoes a **metamorphic molt** to become a **pupa**. The pupa does not feed, and its energy must come from those foods it ingested as a larva. During pupation, adult structures form and replace the larval structures. Eventually, an **imaginal molt** forms the adult (imago) cuticle beneath the pupal cuticle, and then the adult later emerge from the pupal at adult eclosion. While the larva is said to hatch from an egg, the imago is said to eclose from the pupa. Carroll Williams (1958) characterized holometabolous metamorphosis as the switch between foraging and reproduction: "The earth-bound early stages built enormous digestive tracts and hauled them around on caterpillar treads. Later in the life-history these assets could be liquidated and reinvested in the construction of an entirely new organism—a flying-machine devoted to sex."

WATCH DEVELOPMENT 21.2   Two remarkable time-lapse videos document the development and metamorphosis of monarch butterflies and honeybees.

Imaginal discs

In holometabolous insects, the transformation from juvenile into adult occurs within the pupal cuticle. Most of the larval body is systematically destroyed by programmed cell death, while new adult organs develop from relatively undifferentiated nests of

FIGURE 21.8   Locations and developmental fates of imaginal discs and imaginal tissues in the third instar larva (left) of *Drosophila melanogaster*. (After Kalm et al. 1995.)

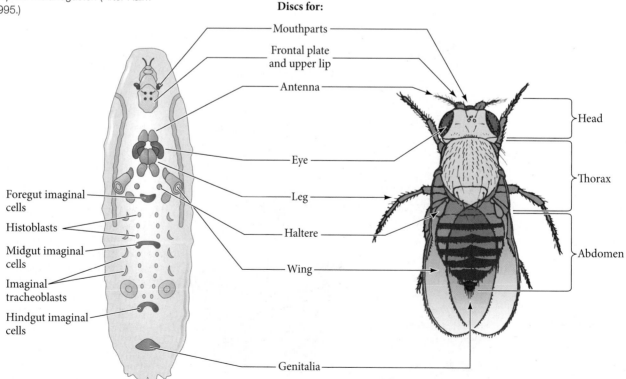

Discs for:

Mouthparts
Frontal plate and upper lip
Antenna
Eye
Leg
Haltere
Wing
Genitalia

Foregut imaginal cells
Histoblasts
Midgut imaginal cells
Imaginal tracheoblasts
Hindgut imaginal cells

Head
Thorax
Abdomen

imaginal cells. Thus, within any larva there are two distinct populations of cells: the larval cells, which are used for the functions of the juvenile insect; and thousands of imaginal cells, which lie within the larva in clusters, awaiting the signal to differentiate.

There are three main types of imaginal cells (**FIGURE 21.8**):

1. The cells of **imaginal discs** will form the cuticular structures of the adult, including the wings, legs, antennae, eyes, head, thorax, and genitalia.

2. **Histoblasts** (tissue-forming cells) are imaginal cells that will form the adult abdomen.

3. There are clusters of imaginal cells within each organ that will proliferate to form the adult organ as the larval organ degenerates.

In the newly hatched larvae, the imaginal discs are visible as local thickenings of the epidermis. Each disc in the early *Drosophila* larva has about 10–50 cells, and there are 19 such discs in these flies. The epidermis of the head, thorax, and limbs comes from nine bilateral pairs of discs, whereas the epidermis of the genitalia is derived from a single disc at the midline.

Whereas most larval cells have a very limited mitotic capacity, imaginal discs divide rapidly at specific characteristic times. As their cells proliferate, the discs form a tubular epithelium that folds in on itself in a compact spiral (**FIGURE 21.9A**). At metamorphosis, these cells proliferate even further as they differentiate and elongate (**FIGURE 21.9B**). The fate map and elongation sequence of one of the six *Drosophila* leg discs is shown in **FIGURE 21.10**. At the end of the third instar, just before pupation, the leg disc is an epithelial sac connected by a thin stalk to the larval epidermis. On one side of the sac, the epithelium is coiled into a series of concentric folds "reminiscent of a Danish pastry" (Kalm et al. 1995). As pupation begins, the cells

(A)

(B)

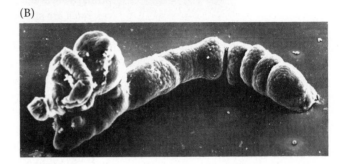

FIGURE 21.9 Imaginal disc elongation. Scanning electron micrograph of *Drosophila* third instar leg disc (A) before and (B) after elongation. (From Fristrom et al. 1977; courtesy of D. Fristrom.)

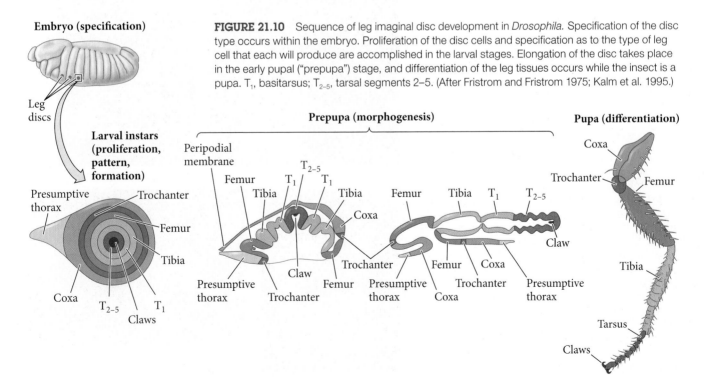

FIGURE 21.10 Sequence of leg imaginal disc development in *Drosophila*. Specification of the disc type occurs within the embryo. Proliferation of the disc cells and specification as to the type of leg cell that each will produce are accomplished in the larval stages. Elongation of the disc takes place in the early pupal ("prepupa") stage, and differentiation of the leg tissues occurs while the insect is a pupa. T_1, basitarsus; T_{2-5}, tarsal segments 2–5. (After Fristrom and Fristrom 1975; Kalm et al. 1995.)

at the center of the disc telescope out to become the most distal portions of the leg—the claws and the tarsus. The outer cells become the proximal structures—the coxa and the adjoining epidermis (Schubiger 1968). After differentiating, the cells of the appendages and epidermis secrete a cuticle appropriate for each specific region. Although the disc is composed primarily of epidermal cells, a small number of **adepithelial cells** migrate into the disc early in development. During the pupal stage, these cells give rise to the muscles and nerves that serve the legs.

SPECIFICATION AND PROLIFERATION   Specification of the general cell fates (i.e., that the disc is to be a leg disc and not a wing disc) occurs in the embryo and is mediated primarily by the Hox genes such as *Ultrabithorax* and *Antennapedia*. Increasingly specific cell fates are specified in the larval stages, as the cells proliferate (Kalm et al. 1995). The type of leg structure (claw, femur, etc.) generated is determined by the interactions between several genes in the imaginal disc. **FIGURE 21.11** shows the expression of three genes involved in determining the proximal-distal axis of the fly leg. In the third instar leg disc, the center of the disc secretes the highest concentration of two morphogens, Wingless (Wg, a Wnt paracrine factor) and Decapentaplegic (Dpp, a BMP paracrine factor). High concentrations of these paracrine factors induce expression of the *Distal-less* gene. Moderate concentrations induce the expression of the *dachshund* gene, and lower concentrations induce the expression of the *homothorax* gene.

Those cells expressing *Distal-less* telescope out to become the most distal structures of the leg—the claw and distal tarsal segments. Those expressing *homothorax* become the most proximal structure, the coxa. Cells expressing *dachshund* become the femur and proximal tibia. Areas where the transcription factors overlap produce the trochanter and distal tibia (Abu-Shaar and Mann 1998). These regions of gene expression are stabilized by inhibitory interactions between the protein products of these genes and of the neighboring genes. In this manner, the gradient of Wg and Dpp proteins is converted into discrete domains of gene expression that specify the different regions of the *Drosophila* leg.

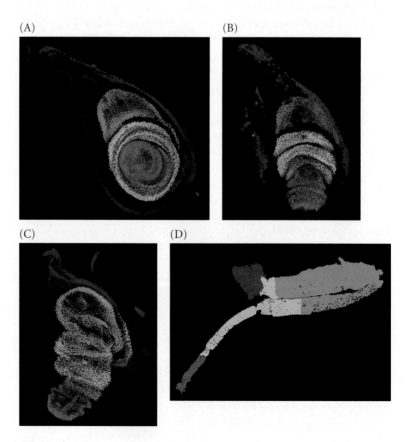

FIGURE 21.11   Expression of transcription factor genes in the *Drosophila* leg disc. At the periphery, the *homothorax* gene (purple) establishes the boundary for the coxa. The expression of the *dachshund* gene (green) locates the femur and proximal tibia. The most distal structures, the claw and distal tarsal segments, arise from the expression domain of *Distal-less* (red) in the center of the imaginal disc. The overlap of *dachshund* and *Distal-less* appears yellow and specifies the distal tibia and trochanter segments. (A–C) Gene expression at successively later stages of pupal development. (D) Localization of expression domains of the genes onto a leg immediately prior to eclosion. The areas where there is overlap between expression domains are shown in yellow, aqua, and orange. (From Abu-Shaar and Mann 1998, courtesy of R. S. Mann.)

EVERSION AND DIFFERENTIATION   The mature leg disc in the third instar of *Drosophila* does not look anything like the adult structure. It is determined but not yet differentiated; its differentiation requires a signal, in the form of a set of pulses of the "molting" hormone 20-hydroxyecdysone (20E; see Figure 21.12A). The first pulse, occurring in the late larval stages, initiates formation of the pupa, arrests cell division in the disc, and initiates the cell shape changes that drive the eversion of the leg. The elongation of imaginal discs occurs is due largely to cell shape changes within the disc epithelium, supplemented by cell division (Condic et al. 1991; Taylor et al. 2008). Using fluorescently labeled phalloidin to stain the peripheral microfilaments of leg disc cells, Condic and coworkers showed that the cells of early third instar discs are tightly arranged along the proximal-distal axis. When the hormonal signal to differentiate is given, the cells change their shape and the leg is everted, the central cells of the disc becoming the most distal (claw) cells of the limb. The leg structures will differentiate within the pupa, so that by the time the adult fly ecloses they are fully formed and functional: in fact they use their legs in their final escape from the pupal case.

WEB TOPIC 21.2   INSECT METAMORPHOSIS   The three links in this web topic discuss (1) the experiments of Wigglesworth and others who identified the hormones of metamorphosis and the glands producing them; (2) the variations that *Drosophila* and other insects play on the general theme of metamorphosis; and (3) the remodeling of the insect nervous system during metamorphosis.

WEB TOPIC 21.3   PARASITOID WASP DEVELOPMENT   Parasitoid wasps helped convince Darwin that God could not be both benevolent and all-powerful. The life cycles of the predatory insects are fascinating examples of one species exploiting the development of another species to its own advantage.

Hormonal control of insect metamorphosis

Although the details of insect metamorphosis differ among species, the general pattern of hormonal action is very similar. Like amphibian metamorphosis, the metamorphosis of insects is regulated by systemic hormonal signals, which are controlled by neurohormones from the brain (for reviews, see Gilbert and Goodman 1981; Riddiford 1996). Insect molting and metamorphosis are controlled by two effector hormones: the steroid **20-hydroxyecdysone** (**20E**) and the lipid **juvenile hormone** (**JH**) (**FIGURE 21.12A**). 20E initiates and coordinates each molt (whether larva-to-larva, larva-to-pupa, or pupa-to-adult) and regulates the changes in gene expression that occur during metamorphosis. High levels of JH prevents the ecdysone-induced changes in gene expression that are necessary for metamorphosis. Thus, its presence during a larval molt ensures that the result of that molt is another larval instar, not a pupa or an adult.

The molting process is initiated in the brain, where neurosecretory cells release **prothoracicotropic hormone** (**PTTH**) in response to neural, hormonal, or environmental signals (**FIGURE 21.12B**). PTTH is a peptide hormone with a molecular weight of approximately 40,000, and it stimulates the production of **ecdysone** by the **prothoracic gland** by activating the RTK (receptor tyrosine kinase) pathway in those cells (Rewitz et al. 2009; Ou et al. 2011). Ecdysone is modified in peripheral tissues to become the active molting hormone 20E. Each molt is initiated by one or more pulses of 20E. For a larval molt, the first pulse produces a small rise in the 20E concentration in the larval hemolymph (blood) and elicits a change in cellular commitment in the epidermis. A second, larger pulse of 20E initiates the differentiation events associated with molting. These pulses of 20E commit and stimulate the epidermal cells to synthesize enzymes that digest the old cuticle and synthesize a new one.

Larval-to-larval molts are produced when there are large circulating titres of juvenile hormone. Juvenile hormone is secreted by the **corpora allata**. The secretory cells of the corpora allata are active during larval molts but inactive during the metamorphic molt and the imaginal molt. As long as JH is present, the 20E-stimulated molts result in a

Developing Questions

How might diseases such as malaria be controlled by altering insect metamorphosis?

(A) Juvenile hormone (JH)

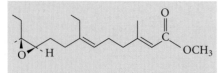

Ecdysone

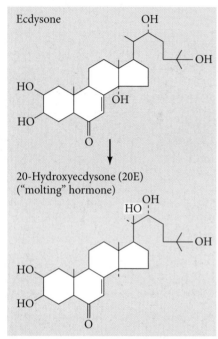

20-Hydroxyecdysone (20E)
("molting" hormone)

FIGURE 21.12 Regulation of insect metamorphosis. (A) Structures of juvenile hormone (JH), ecdysone, and the active molting hormone 20-hydroxyecdysone (20E). (B) General pathway of insect metamorphosis. 20E and JH together cause molts that form the next larval instar. When the concentration of JH becomes low enough, the 20E-induced molt produces a pupa instead of a larva. When 20E acts in the absence of JH, the imaginal discs differentiate and the molt gives rise to an adult (imago). (After Gilbert and Goodman 1981.)

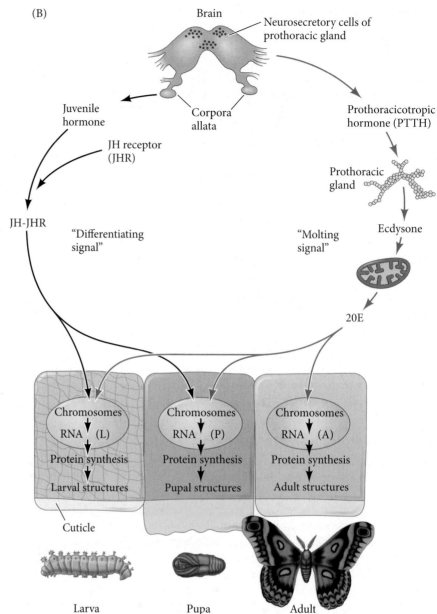

new larval instar. In the last larval instar, however, the medial nerve from the brain to the corpora allata inhibits these glands from producing JH, and there is a simultaneous increase in the body's ability to degrade existing JH (Safranek and Williams 1989). Both these mechanisms cause JH levels to drop below a critical threshold value, triggering the release of PTTH from the brain (Nijhout and Williams 1974; Rountree and Bollenbacher 1986). PTTH, in turn, stimulates the prothoracic gland to secrete a small amount of ecdysone. The resulting pulse of 20E, in the absence of high levels of JH, commits the epidermal cells to pupal development. Larva-specific mRNAs are not replaced, and new mRNAs are synthesized whose protein products inhibit the transcription of the larval messages.

There are two major pulses of 20E during *Drosophila* metamorphosis. The first pulse occurs in the third instar larva and triggers the "prepupal" morphogenesis of the leg and wing imaginal discs, as well as the death of the larval hindgut. The larva stops eating and migrates to find a site to begin pupation. The second 20E pulse occurs 10–12 hours later and tells the prepupa to become a pupa. The head inverts and the salivary glands degenerate (Riddiford 1982; Nijhout 1994). It appears, then, that the first pulse of 20E during the last larval instar triggers the processes that inactivate larva-specific genes and initiates the morphogenesis of imaginal disc structures. The second pulse transcribes pupa-specific genes and initiates the molt (Nijhout 1994). At the imaginal molt, when 20E acts in the absence of juvenile hormone, the imaginal discs fully differentiate and the molt gives rise to an adult.

> **WATCH DEVELOPMENT 21.3**   Juvenile hormone is a versatile hormone that can regulate metamorphosis and can also be used as a potent insecticide.

The molecular biology of 20-hydroxyecdysone activity

ECDYSONE RECEPTORS   Like amphibian thyroid hormones, 20E cannot bind to DNA by itself. It must first bind to nuclear proteins called **ecdysone receptors (EcRs)**. EcRs are evolutionarily related to, and almost identical in structure to, the thyroid hormone receptors of amphibians. An EcR protein forms an active molecule by dimerizing with an Ultraspiracle (Usp) protein. Usp is the homologue of amphibian RXR, which we learned earlier dimerizes with TR to form the active thyroid hormone receptor (Koelle et al. 1991; Yao et al. 1992; Thomas et al. 1993). So the EcR-Usp receptor complex of insects is very similar in structure to the TR-RXR complex of amphibians. The EcR and Usp proteins each attach to the DNA; they then dimerize on the enhancer or promoter element of the ecdysone-responsive genes (Szamborska-Gbur et al. 2014).

It is thought that in the absence of hormone-bound EcR, Usp recruits inhibitors of the transcription of ecdysone-responsive genes (Tsai et al. 1999). This inhibition is converted into activation when ecdysone binds to its receptor. The presence of ecdysone-bound EcR-USp recruits histone methyltransferases that *activate* the ecdysone-responsive genes (Sedkov et al. 2003). However, other mechanisms may also be at work. For instance, Johnston and colleagues (2011) have found that in the absence of hormone-bound EcR, another transcription factor, E75, binds to these ecdysone-responsive sites and inhibits transcription. Thus, ecdysone helps mediate the competition of EcR-Usp with E75 for the same *cis*-regulatory sites.

VADE MECUM

The segment on the fruit fly contains a sequence showing how to perform a chromosome squash using the *Drosophila* larval salivary gland.

BINDING OF 20-HYDROXYECDYSONE TO DNA   The early events of *Drosophila* metamorphosis were first elucidated after seeing the effects of ecdysone on polytene chromosomes. During *Drosophila* molting and metamorphosis, certain regions of these chromosomes "puff out" in the cells of certain organs at particular times (**FIGURE 21.13**; Clever 1966; Ashburner 1972; Ashburner and Berondes 1978). These chromosome puffs are areas where DNA is being actively transcribed. When 20E is added to larval salivary glands, certain puffs are produced and others regress. Fluorescent antibodies against 20E demonstrated that the hormone localizes at the puff sites, further confirming the involvement of 20E in transcription of target genes at the puff sites (Gronemeyer and Pongs 1980). At these sites, the

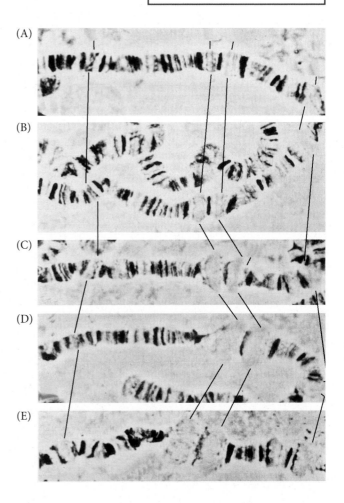

(A)

(B)

(C)

(D)

(E)

FIGURE 21.13   20E-induced puffs in cultured salivary gland cells of *D. melanogaster*. (A) Uninduced control. (B–E) 20E-stimulated chromosomes at (B) 25 minutes, (C) 1 hour, (D) 2 hours, and (E) 4 hours. (Courtesy of M. Ashburner.)

ecdysone-bound receptor complex recruits a histone methyltransferase that methylates lysine-4 of histone H3, thereby loosening the nucleosomes in that area (Sedkov et al. 2003). Remarkably, after the initial puffs had formed, other puffs were seen to regress. And hours later, more puffs formed. Ashburner (1974, 1990) hypothesized that the "early puff" genes make a protein product that is essential for the activation of the "late puff" genes and that, moreover, the early regulatory protein itself turns off the transcription of the early puff genes.[5] These insights have been confirmed by molecular analyses.

FIGURE 21.14A shows a simplified schematic for the framework of metamorphosis in *Drosophila*. First, 20E binds to the EcR/Usp receptor complex. It activates the "early response genes," including *E74* and *E75* (the puffs in Figure 21.18), as well as *Broad* and the *EcR* gene itself. The transcription factors encoded by these genes activate a second series of genes, such as *E75, DHR4,* and *DHR3*. The products of these genes are transcription factors that work together to form the pupa. Second, the products of the second-wave genes shut off the early response genes so that they do not interfere with this second burst of 20E. Third, 20E activates the genes whose products inactivate and degrade ecdysone itself. In this way, the nucleus is cleared of the hormone so that it can respond to a second pulse. Moreover, 20E usually inhibits the gene encoding βFTZ-F1. Now this transcription factor can be synthesized, and it enables a new set of genes to respond to the second burst of 20E (Rewitz et al. 2009). Moreover, DHR4 coordinates growth and behavior in the larva. It allows the larva to stop feeding once it reaches a certain weight and to begin searching for a place to glue itself to and form a pupa (Urness and Thummel 1995; Crossgrove et al. 1996; King-Jones et al. 2005).

The effects of these two 20E pulses can be extremely different. One example of this is the ecdysone-mediated changes in the larval salivary gland. The early pulse of 20E activates the *Broad* gene, which encodes a family of transcription factors through differential RNA splicing. The targets of the Broad complex proteins include genes that encode the salivary gland "glue proteins"—proteins that allow the larva to adhere to a solid surface, where it will become a pupa (Guay and Guild 1991). At this time, 20E binds to the EcR-A isoform of the ecdysone receptor (**FIGURE 21.14B**). When complexed with Usp, it activates the transcription of early response genes *E74, E75,* and *Broad*. But now a different set of targets is activated. The transcription factors encoded by the early genes activate the genes encoding the apoptosis-promoting proteins Hid and Reaper, as well as blocking the expression of the *diap2* gene (which would otherwise repress apoptosis). Thus, the first 20E pulse stimulates the function of the larval salivary gland, whereas the second pulse of 20E calls for the destruction of this larval organ (Buszczak and Segraves 2000; Jiang et al. 2000).

Like the ecdysone receptor gene, the *Broad* gene can generate several different transcription factor proteins through differentially initiated and spliced messages. Moreover, the variants of the ecdysone receptor (EcR-A, EcR-B1, and EcR-B2), when partnered with Usp, may induce the synthesis of particular variants of the Broad proteins. Organs such as the larval salivary gland that are destined for death during metamorphosis express the Broad Z1 isoform; imaginal discs destined for differentiation express the Z2 isoform; and the central nervous system (which undergoes marked remodeling during metamorphosis) expresses all isoforms, with Z3 predominating (Emery et al. 1994; Crossgrove et al. 1996).

When juvenile hormone is present, however, the *Broad* gene is repressed, and metamorphosis does not take place (Riddiford 1972; Zhou and Riddiford 2002; Hiruma and Kaneko 2013). JH maintains the status quo of larval-to-larval molts by binding to its

---

[5]The observation that 20E controlled the transcriptional units of chromosomes was an extremely important and exciting discovery. One could see transcription occurring, using only a light microscope. This was our first real glimpse of gene regulation in eukaryotic organisms. At the time when this discovery was made, the only examples of transcriptional gene regulation were in bacteria.

nuclear receptor—the Met protein[6]—and converting this receptor into a transcription factor. The JH-bound Met protein activates the *Kr-h1* gene, whose product, a repressive transcription factor, blocks the activation of the *Broad* gene (**FIGURE 21.14C**; Minakuchi et al. 2008; Charles et al. 2011; Li et al. 2011). Thus, in the presence of JH, the *Broad* gene is not activated and metamorphosis is blocked.

[6]Not to be confused with the unrelated Met receptor in vertebrates (which is a cell membrane receptor for hepatocyte growth factor), the Met receptor for JH was identified by its ability to bind methoprene, an insecticide that works by mimicking JH, thus preventing metamorphosis (Konopova and Jindra 2007; Charles et al. 2011).

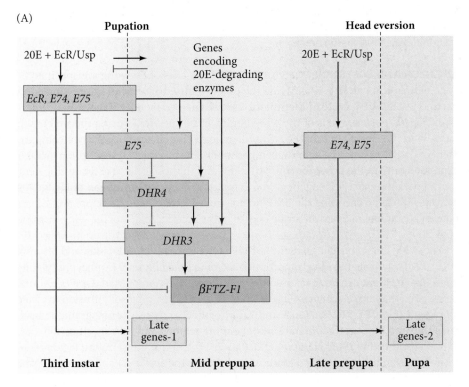

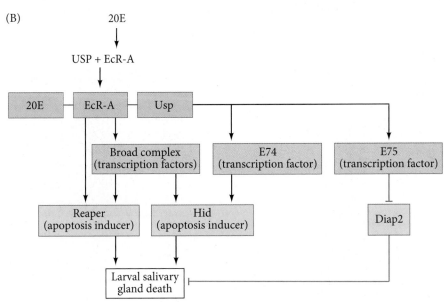

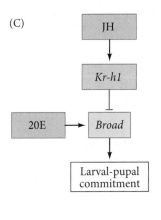

FIGURE 21.14 20-hydroxyecdysone initiates developmental cascades. (A) Schematic of the major gene expression cascade in *Drosophila* metamorphosis. When 20E binds to the EcR/Usp receptor complex, it activates the early response genes, including *E74*, *E75*, and *Broad*. Their products activate the "late genes." The activated EcR/Usp complex also activates a series of genes whose products are transcription factors and which activate the *βFTZ-F1* gene. The βFTZ-F1 protein modifies the chromatin so that the next 20E pulse activates a different set of late genes. The products of these genes also inhibit the early-expressed genes, including those for the EcR receptor. (B) Postulated cascade leading from ecdysone reception to death of the larval salivary gland. The 20E binds to the EcR-A isoform of the ecdysone receptor. After complexing with Usp, the activated transcription factor complex stimulates transcription of the early response genes *E74A*, *E75B*, and the *Broad* complex. These promote apoptosis in the salivary gland cells. (C) When juvenile hormone binds to its receptor, Met, it activates the *Kr-h1* gene. Kr-h1 protein is a repressive transcription factor that blocks activation of the *Broad* gene by 20E. (A after King-Jones et al. 2005, Rewitz et al. 2010; B after Buszczak and Segraves 2000; C after Hiruma and Kaneko 2013.)

WEB TOPIC 21.4 **PRECOCENES AND SYNTHETIC JH** Given the voracity of insect larvae, it is amazing that any plants survive. However, many plants get revenge on their predators by making compounds that alter insect metamorphosis, thus preventing the animals from developing or reproducing.

Determination of the wing imaginal discs

When ecdysone signaling is unaffected by juvenile hormone, it activates the growth and differentiation of imaginal discs that have already been determined. For example, the largest of *Drosophila*'s imaginal discs is that of the wing, containing some 60,000 cells. (In contrast, the leg and haltere discs contain about 10,000 cells each; Fristrom 1972.) The wing discs are distinguished from the other imaginal discs by the expression of the *vestigial* gene (Kim et al. 1996). When this gene is expressed in any other imaginal disc, wing tissue emerges.

ANTERIOR AND POSTERIOR COMPARTMENTS   The axes of the wing are specified by gene expression patterns that divide the embryo into discrete but interacting compartments (**FIGURE 21.15A**; Meinhardt 1980; Causo et al. 1993; Tabata et al. 1995). The anterior-posterior axis of the wing begins to be specified during the first instar larva. Here, the expression of the *engrailed* gene distinguishes the posterior compartment of the wing from the anterior compartment. The Engrailed transcription factor is expressed only in the posterior compartment, and in those cells, it activates the gene for the paracrine factor Hedgehog. In a complex manner, the diffusion of Hedgehog activates the gene encoding the BMP homologues Decapentaplegic (Dpp) and Glass-bottom boat (Gbb) in a narrow stripe of cells in the anterior region of the wing disc (Ho et al. 2005).

These BMPs establish a gradient of BMP signaling activity (Matsuda and Shimmi 2012). BMPs activate the Mad transcription factor (a Smad protein) by phosphorylating it, so this gradient can be measured by the phosphorylation of Mad. Dpp is a short-range paracrine factor, whereas Gbb exhibits a much longer range of diffusion to create a gradient (**FIGURE 21.15B**; Bangi and Wharton 2006). This signaling gradient regulates the amount of cell proliferation in the wing regions and also specifies cell fates (Rogulja and Irvine 2005; Hamaratoglu et al. 2014). Several transcription factor genes respond differently to activated Mad. At high levels, the *spalt* (*sal*) and *optomotor blind* (*omb*) genes are activated, whereas at low levels (where Gbb provides the primary signal), only *omb* is activated. Below a particular level of phosphorylated Mad activity, the

FIGURE 21.15 Compartmentalization and anterior-posterior patterning in the wing imaginal disc. (A) In the first instar larva, the anterior-posterior axis has been formed and can be recognized by the expression of the *engrailed* gene in the posterior compartment. Engrailed, a transcription factor, activates the *hedgehog* gene. Hedgehog acts as a short-range paracrine factor to activate *decapentaplegic* (*dpp*) in the anterior cells adjacent to the posterior compartment, where Dpp and a related protein, Glass-bottom boat (Gbb), act over a longer range. (B) Dpp and Gbb proteins create a concentration gradient of BMP-like signaling, measured by the phosphorylation of Mad (pMad). High concentrations of Dpp plus Gbb near the source activate both the *spalt* (*sal*) and *optomotor blind* (*omb*) genes. Lower concentrations (near the periphery) activate *omb* but not *sal*. When Dpp plus Gbb levels drop below a certain threshold, *brinker* (*brk*) is no longer repressed. L2–L5 mark the longitudinal wing veins, with L2 being the most anterior. (After Bangi and Wharton 2006.)

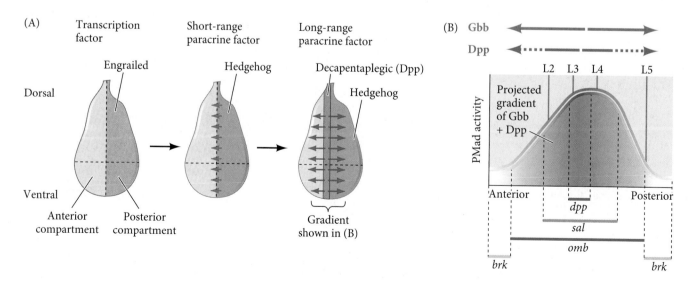

brinker (*brk*) gene is no longer inhibited; thus, *brk* is expressed outside the signaling domain. Specific cell fates of the wing are specified in response to the action of these transcription factors. (For example, the fifth longitudinal vein of the wing is formed at the border of *optomotor blind* and *brinker*; see Figure 21.15B). Although experimental evidence shows that Dpp regulates wing growth, the mechanisms by which it does this remain unknown (Hamaratoglu et al. 2014; Hariharan 2015).

DORSAL-VENTRAL AND PROXIMAL-DISTAL AXES The dorsal-ventral axis of the wing is formed at the second instar stage by the expression of the *apterous* gene in the prospective dorsal cells of the wing disc (Blair 1993; Diaz-Benjumea and Cohen 1993). Here, the upper layer of the wing is distinguished from the lower layer of the wing blade (Bryant 1970; Garcia-Bellido et al. 1973). The *vestigial* gene remains active in the ventral portion of the wing disc (**FIGURE 21.16A**). The dorsal portion of the wing synthesizes transmembrane proteins that prevent the intermixing of the dorsal and ventral cells (Milán et al. 2005). At the boundary between the dorsal and ventral compartments, the Apterous and Vestigial transcription factors interact to activate the gene encoding the Wnt paracrine factor Wingless (**FIGURE 21.16B**). Neumann and Cohen (1996) showed that Wingless protein acts as a growth factor to promote the cell proliferation that extends the wing. Wingless also helps establish the proximal-distal axis of the wing: high levels of Wingless activate the *Distal-less* gene, which specifies the most distal regions of the wing (Neumann and Cohen 1996, 1997; Zecca et al. 1996). This occurs in the central region of the disc and "telescopes" outward as the distal margin of the wing blade (**FIGURE 21.16C**). Thus, a battery of paracrine factors patterns the wing disc, giving each cell an identity along the dorsal-ventral, proximal-distal, and anterior-posterior axes. In metamorphosis, we see a reprising of the developmental phenomena that generated the larva, itself.

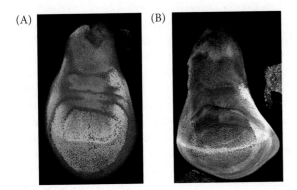

(A)   (B)

FIGURE 21.16 Determining the dorsal-ventral axis. (A) The prospective ventral surface of the wing is stained by antibodies to Vestigial protein (green), while the prospective dorsal surface is stained by antibodies to Apterous protein (red). The region of yellow illustrates where the two proteins overlap in the margin. (B) Wingless protein (purple) synthesized at the marginal juncture organizes the wing disc along the dorsal-ventral axis. The expression of Vestigial (green) is seen in cells close to those expressing Wingless. (C) The dorsal and ventral portions of the wing disc telescope out to form the two-layered wing. Gene expression patterns are indicated on the double-layered wing. (A,B courtesy of S. Carroll and S. Paddock.)

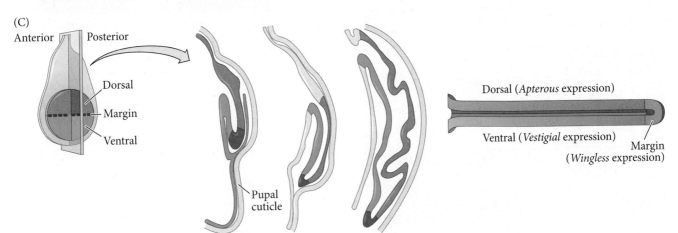

(C)

WEB TOPIC 21.5 **HOMOLOGOUS SPECIFICATION**   If a group of cells in one imaginal disc are mutated such that they give rise to a structure characteristic of another imaginal disc (for instance, cells from a leg disc giving rise to antennal structures), the regional specification of those structures will be in accordance with their position in the original disc.

Metamorphosis of the Pluteus Larva

Sea urchins undergo complete metamorphosis, forming a "primary larva"—the pluteus—which is later jettisoned pretty much in its entirety. Almost the entire body of the adult sea urchin comes from the left coelomic sac of the pluteus archenteron. The larval skeleton is abandoned, while the rest of the larval body undergoes programmed cell death, providing raw materials for juvenile growth. As the pluteus forms, the top of the archenteron meets the blastocoel wall. Here, the secondary mesenchyme cells form the right and left coelomic pouches (see Figure 10.10, 13.5 hours). Under the influence of Nodal protein, the right coelomic sac remains rudimentary while the left coelomic sac undergoes extensive development to form the structures of the adult sea urchin. This growth involves activating BMP signaling on the left side. The small micromeres, which give rise to the germ cells, are attracted to the coelomic sacs by chemoattractant factors synthesized by means of gene regulatory network similar to that of the *Drosophila* eye (Yajima and Wessels 2012; Campanale et al. 2014; Martik and McClay 2015). These primordial germ cells are preferentially retained by the left coelomic pouch (Luo and Su 2012; Warner et al. 2012).

The left sac eventually splits into three smaller sacs. An invagination from the ectoderm fuses with the middle sac to form the **imaginal rudiment**, and only the aboral epidermis is derived from the larval dermis. This rudiment develops a fivefold (pentaradial) symmetry (**FIGURE 21.17**), and skeletogenic mesenchyme cells enter the rudiment to synthesize the first skeletal plates of the shell. The left side of the pluteus in effect becomes the future oral surface of the adult sea urchin (Bury 1895; Aihara and Amemiya 2001; Minsuk et al. 2009). During metamorphosis, the larva settles on the sea floor and the imaginal rudiment separates from the larva, which then degenerates. As we will see in Chapter 25, the cues for larval settlement often include a mixture of environmental factors such as photoperiod, turbulence, and chemicals released by potential food sources. In some sea urchin species, the shear force generated by turbulence enables the larvae to sense and respond to chemicals emanating from algae and bacteria (indicating that food for the adults is abundant) (Rowley 1989; Gaylord et al. 2013; Nielsen et al. 2015). While the imaginal rudiment (now called a juvenile) is re-forming its digestive tract, it is dependent on the nutrition it receives from the disintegrating larval structures.

The pentaradial symmetry of adult echinoderms is unique and distinguishes them from the many bilaterally symmetrical animals. Note, however, that pluteus larvae *are* bilaterally symmetrical—evidence that echinoderms share a common ancestor with the bilaterally symmetrical chordates (Zamora et al. 2012, 2015). The sea urchin larvae and adult differ dramatically. Indeed, the pluteus larva is a free-swimming planktonic dispersal stage with no sexual reproductive capacity, whereas the adult grazes on algae on the sea floor and produces thousands of gametes. It is also interesting that, like amphibians, echinoderms such as sea urchins use thyroid hormones to cue metamorphosis (Chino et al. 1994; Heyland and Hodin 2004).

Developing Questions

The pluteus larva makes no gametes and thus cannot reproduce sexually. However, it has been found to reproduce asexually. Under what conditions might asexual larvae replicate by budding off parts of themselves?

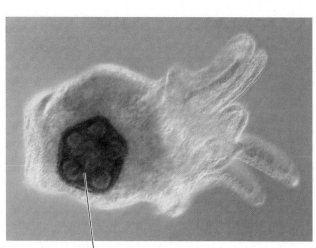

Imaginal rudiment

FIGURE 21.17   The imaginal rudiment growing in the left side of the pluteus larva of a sea urchin. The rudiment will become the adult sea urchin, while most of the larval stage will be dissolved. The fivefold symmetry of the rudiment is obvious. (Courtesy of G. Wray.)

Next Step Investigation

Our ignorance about metamorphosis is stunning. Metamorphosis is understood in its broad outlines, but only in very few species. Moreover, the mechanisms by which genetic instructions are transformed into bodily changes is just beginning to be understood. For instance, experimental evidence shows that Dpp controls the growth of the *Drosophila* imaginal discs, but we don't yet know how. We're not even certain how paracrine factors such as Dpp

and Hedgehog are transported from their sites of synthesis to the responsive cells. One of the enormous unanswered questions about metamorphosis concerns the different response to the same stimulus. The same hormone can instruct one group of cells to proliferate and an adjacent group of cells to die. We still do not understand what mediates these different responses.

Closing Thoughts on the Opening Photo

As Alfred Lord Tennyson (1886) intuited, "The old order changeth, yielding place to the new." Metamorphosis separates an individual into two distinct life cycle stages, with different anatomy, different physiology, and different ecological niches. The insect life cycle was discovered and documented in the early eighteenth century by Maria Merian, an artist who, among other things, painted the butterflies of Surinam in South America. This portion of a lithograph by Merian (1705) shows the larval, pupal, and adult forms of *Morpho deidamia*. The caterpillar is eating the leaves of the Barbados cherry tree; the pupa of this species resembles that tree's leaves. Merian also noticed that the larvae of different species need different plants than the adult butterfly. In many instances, the larval food plant contains noxious chemicals that the adult absorbs. Monarch butterfly caterpillars, for instance, obtain toxic alkaloids from plants; these toxins render the metamorphosed adult very unpalatable to birds (and the birds learn not to eat a monarch butterfly).

21 Snapshot Summary
Metamorphosis

1. Amphibian metamorphosis includes both morphological and biochemical changes. Some structures are remodeled, some are replaced, and some new structures are formed.

2. The hormone responsible for amphibian metamorphosis is tri-iodothyronine (T_3). The synthesis of T_3 from thyroxine (T_4) and the degradation of T_3 by deiodinases can regulate metamorphosis in different tissues. T_3 binds to thyroid hormone receptors and acts predominantly at the transcriptional level.

3. Many changes during amphibian metamorphosis are regionally specific. The tail muscles degenerate; the trunk muscles persist. An eye will persist even if transplanted into a degenerating tail.

4. Metamorphic change in amphibians can be brought about by cell death, cell differentiation, or by cell-type switching.

5. The specific timing of metamorphic events can be orchestrated by the different events occurring at different levels of thyroid hormones.

6. Animals with direct development do not have a larval stage. Primary larvae (such as those of sea urchins)

specify their body axes differently than the adult, whereas secondary larvae (such as those of insects and amphibians) have body axes that are the same as adults of the species.

7. Ametabolous insects undergo direct development. Hemimetabolous insects pass through nymph stages wherein the immature organism is usually a smaller version of the adult.

8. In holometabolous insects, there is a dramatic metamorphosis from larva to pupa to sexually mature adult. In the stages between larval molts, the larva is called an instar. After the last instar, the larva undergoes a metamorphic molt to become a pupa. The pupa undergoes an imaginal molt to become an adult.

9. During the pupal stage, the imaginal discs and histoblasts grow and differentiate to produce the structures of the adult body.

10. The anterior-posterior, dorsal-ventral, and proximal-distal axes are sequentially specified by interactions between different compartments in the imaginal discs. The disc "telescopes out" during development, its central regions becoming distal.

11. Molting is caused by the hormone 20-hydroxyecdysone (20E). In the presence of high levels of juvenile hormone, the molt gives rise to another larval instar. In low concentrations of juvenile hormone, the molt produces a pupa; if no juvenile hormone is present, the molt is an imaginal molt.

12. The ecdysone receptor gene produces a nuclear RNA that can form at least three different proteins. The types of ecdysone receptors in a cell may influence the response of that cell to 20E. The ecdysone receptors bind to DNA to activate or repress transcription.

13. Sea urchins undergo complete metamorphosis, forming a primary larva, the pluteus. Almost the entire body of the adult sea urchin comes from the left coelomic sac of the pluteus archenteron, known as the imaginal rudiment.

Further Reading

Cai, L. and D. D. Brown. 2004. Expression of type II iodothyronine deiodinase marks the time that a tissue responds to thyroid hormone-induced metamorphosis in *Xenopus laevis. Dev. Biol.* 266: 87–95.

Grimaldi, A., N. Buisien, T. Miller, Y.-B. Shi and L. M. Sachs. 2013. Mechanisms of thyroid hormone receptor action during development: Lessons from amphibian studies. *Bioch. Biophys. Acta* 1830: 3882–3892.

Hamaratoglu, F., M. Affolter and G. Pyrowolakis. 2014. Dpp/BMP signaling in flies: From molecules to biology. *Sem. Cell Dev. Biol.* 32: 128–136.

Hiruma. K. and Y. Kaneko. 2013. Hormonal regulation of insect metamorphosis with special reference to juvenile hormone biosynthesis. *Curr. Top. Dev. Biol.* 103: 73–100.

Jiang, C., A. F. Lamblin, H. Steller and C. S. Thummel. 2000. A steroid-triggered transcriptional hierarchy controls salivary gland cell death during *Drosophila* metamorphosis. *Mol. Cell* 5: 445–455.

GO TO WWW.DEVBIO.COM...

...for Web Topics, Scientists Speak interviews, Watch Development videos, Dev Tutorials, and complete bibliographic information for all literature cited in this chapter.

Regeneration

DEVELOPMENT NEVER CEASES. Throughout life, we continuously generate new blood cells, epidermal cells, and digestive tract epithelium from stem cells. A more obvious recurrence of embryonic-like development is regeneration, the replacement of a body part by an adult animal after the original one has been removed. Whether it's Ponce de Leon's Fountain of Youth or Marvel's superhero Wolverine, regeneration is a process that has captivated the imagination of writers, artists, and Hollywood alike. Fortunately, it is not all science fiction. More important is that regeneration has caught the wonder of scientists who have made great strides in dissecting the developmental mechanisms underlying the ability of some species to exhibit a fantastic potential for regeneration. Some adult salamanders, for instance, can regrow limbs and tails after these appendages have been amputated (even Wolverine has not been depicted to do that—at least in this universe![1]).

[1] For the comic book and X-Men aficionado, different universe dimensions have been described, and our Earth's universe is 616. Technically in a different universe (295), Wolverine's hand was severed off during the "Age of Apocalypse." Only when his regenerative powers were restored did it grow back. So it is plausible that Wolverine's powers of regeneration may in fact mimic those of a salamander as opposed to those of *Hydra* (and no, we don't mean Captain America's arch enemy here).

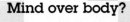

Mind over body?

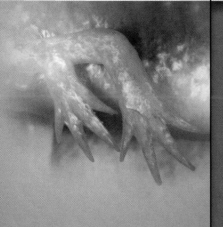

The Punchline

Regeneration is a postembryonic event that operates under familiar embryonic developmental programs. The ability to regenerate tissues varies across species, from a capacity for near-total regeneration in hydras and planarians, to the ability to replace complex structures in salamanders and fish, to the limited ability to add and replace cells for growth and maintenance in mammals. During regeneration, focal cell proliferation or cell migration to the wound location creates an undifferentiated blastema whose outgrowth and differentiation replace the damaged tissues. A blastema can be created by pluripotent stem cells, as in the planarian, or lineage-restricted progenitor cells derived by dedifferentiation, as in the salamander limb. Regeneration can also occur through a remodeling of existing tissues through compensatory proliferation or the transdifferentiation of differentiated cells, as exemplified during regeneration of the zebrafish heart. The mammalian liver cannot regenerate completely, but when a lobe of this organ is damaged or removed, the remaining liver mass can grow to compensate for the loss. Virtually every developmental pathway discussed in this book plays a role in regeneration. Among these recurring players is the Wnt/β-catenin pathway.

It is difficult to behold the phenomenon of limb regeneration in salamanders without wondering why we humans cannot grow back our arms and legs. What gives these animals an ability we so sorely lack? Experimental biology was born of the efforts of eighteenth-century naturalists to answer this question (see Morgan 1901). The regeneration experiments of Abraham Tremblay (using *Hydra*, a cnidarian), René Antoine Ferchault de Réaumur (crustaceans), and Lazzaro Spallanzani (salamanders) set the standard for experimental research and for the intelligent discussion of one's data (see Dinsmore 1991). More than two centuries later, we are beginning to find answers to the great questions of regeneration, and at some point we may be able to alter the human body so as to permit our own limbs to regenerate.

Many Ways to Rebuild

"I'd give my right arm to know the secret of regeneration." This quote from Oscar E. Schotté (1950) captures the fascination science has had with the remarkable ability of some organisms to rebuild themselves. **Regeneration** is the reactivation of development in postembryonic life to restore missing or damaged tissues. The potential benefit of harnessing the powers of regeneration in humans would mean that severed limbs could be restored; diseased organs could be removed and then regrown; and nerve cells altered by age, disease, or trauma could once again function normally. Before modern medicine can succeed in coaxing human bone or neural tissue to regenerate, we must first understand how regeneration occurs in those species that already have this ability. Our knowledge of the roles of paracrine factors in organ formation and our ability to clone the genes that produce those factors have propelled what Susan Bryant (1999) has called "a regeneration renaissance." *Renaissance* literally means "rebirth," and because regeneration can involve a return to the embryonic state, the term is apt in many ways.

Although regeneration takes place in nearly all species, several organisms have emerged as particularly fruitful models for the study of regeneration (**FIGURE 22.1**). The near totality of hydra and planarian regeneration is unmatched. They are able to regenerate complete organisms following amputation or even complete individuals from very small fragments. Certain salamanders are unique among tetrapods in being able to regenerate whole limbs, and frog larvae are often used to study the regeneration of the tail and the lens of the eye. Zebrafish have recently proved advantageous for investigating the mechanisms of central nervous system, retina, heart, liver, and fin regeneration. And, although mammals are unable to rebuild whole appendages, individual tissues and organs do possess variable regenerative capabilities; most notable are the antlers of deer.

Aside from the differences in regenerative potential, each of these model systems exemplifies one or more of the four modes of regeneration (**FIGURE 22.2**):

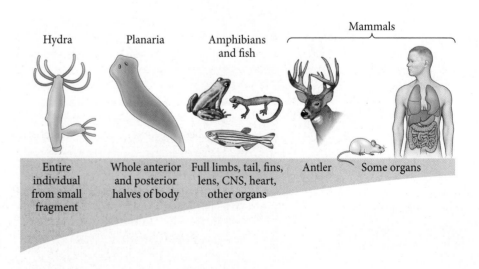

FIGURE 22.1 Representative organisms and their comparative regenerative capabilities.

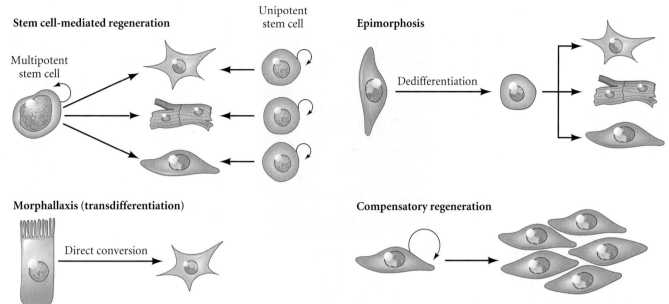

Stem cell-mediated regeneration

Multipotent stem cell

Unipotent stem cell

Epimorphosis

Dedifferentiation

Morphallaxis (transdifferentiation)

Direct conversion

Compensatory regeneration

FIGURE 22.2 Four different modes of regeneration.

1. *Stem-cell mediated regeneration.* Stem cells allow an organism to regrow certain organs or tissues that have been lost; examples include the regrowth of hair shafts from follicular stem cells in the hair bulge and the continual replacement of blood cells from the hematopoietic stem cells in the bone marrow.

2. *Epimorphosis.* In some species, adult structures can undergo *de*differentiation to form a relatively undifferentiated mass of cells (a blastema) that then redifferentiates to form the new structure. Such regeneration is characteristic of regenerating amphibian limbs.

3. *Morphallaxis.* Here, regeneration occurs through the repatterning of existing tissues (transdifferentiation), and there is little new growth. Such regeneration is seen in the hydra.

4. *Compensatory regeneration.* Here the differentiated cells divide but maintain their differentiated functions. The new cells do not come from stem cells, nor do they come from the dedifferentiation of the adult cells. Each cell produces cells similar to itself; no mass of undifferentiated tissue forms. This type of regeneration is characteristic of the mammalian liver.

Hydra: Stem Cell-Mediated Regeneration, Morphallaxis, and Epimorphosis

Hydra is a genus of freshwater cnidarians.[2] Most hydras are tiny—about 0.5 cm long. A hydra has a tubular, radially symmetric body with a "head" at its distal end and a "foot" at its proximal end. The "foot," or **basal disc**, enables the animal to stick to rocks or the undersides of pond plants. The "head" consists of a conical **hypostome** region that contains the mouth and a ring of tentacles (which catch food) beneath it. Hydras are diploblastic animals, having only ectoderm and endoderm (**FIGURE 22.3A**). Their two epithelial layers are referred to as **myoepithelia** because they possess characteristics

---

[2] *Hydra* is both the genus name and the common name for these animals. For simplicity, we will use the common (unitalicized) form in this discussion. The animal is named after the Hydra, the many-headed serpent from Greek mythology. Whenever one of the Hydra's heads was chopped off, it regenerated two new ones. Hercules finally defeated the monster by cauterizing the stumps of its heads with fire. Hercules seems to have had a significant interest in regeneration: he also freed the bound Prometheus, thereby stopping his daily series of partial hepatectomies (see p. 718).

(A)

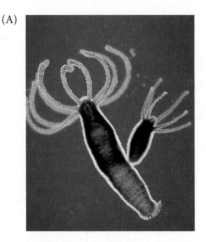

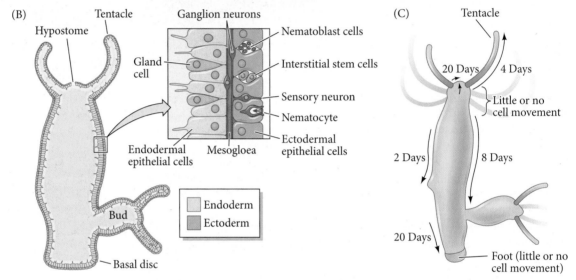

of both epithelial and muscle cells. Although hydras lack a true mesoderm, they do contain secretory cells, gametes, stinging cells (nematocytes), and neurons that are not part of the two epithelial layers (**FIGURE 22.3B**; Li et al. 2015). Hydras can reproduce sexually, but they do so only under adverse conditions (such as crowding or cold temperatures). They usually multiply asexually, by budding off a new individual. The buds form about two-thirds of the way down the animal's body axis.

Routine cell replacement by three types of stem cells

A hydra's body is not particularly stable. In humans and flies, for instance, a skin cell in the body's trunk is not expected to migrate and eventually be sloughed off from the face or foot—but that is exactly what happens in hydra. The cells of the body column are constantly undergoing mitosis and are eventually displaced to the extremities of the column, from which they are shed (**FIGURE 22.3C**; Campbell 1967a,b). Thus, each cell plays several roles, depending on how old it is, and the signals specifying cell fate must be active all the time. In a sense, a hydra's body is always regenerating.

This cellular replacement is generated from three cell types. Endodermal and ectodermal cells are unipotent progenitor cells that divide continuously, producing more lineage-restricted epithelia. The third cell type is a multipotent **interstitial stem cell** found within the ectodermal layer (see Figure 22.3B). This stem cell generates neurons, secretory cells, nematocytes, and gametes. The most significant cell proliferation by each of these three stem cells occurs primarily within the central region of the body, after which displaced myoepithelia and migrating interstitial progeny move to and differentiate within the apical and basal extremities (Buzgariu et al. 2015). Compared to myoepithelial stem cells, interstitial stem cells are paused in the G2 phase of the cell cycle for a longer period and cycle at a faster rate (Buzgariu et al. 2014), suggesting that the interstitial stem cells are poised to immediately respond to a need for cell replacement through rapid proliferation. These three cell types are all that are needed to form a hydra, and if hydra cells are separated and reaggregated, a new hydra will form (Gierer et al. 1972; Technau 2000; Bode 2011).

The head activator

Experimental embryology—indeed, experimental biology—can be said to have started with Tremblay's studies of hydra regeneration.[3] In 1741, Tremblay reported that "the story of the Phoenix who is reborn from his own ashes, fabulous as it is, offers nothing more marvelous than the discovery of which we are going to speak." He found that when he cut a hydra into as many as 40 pieces, "there are reborn as many complete animals similar to the first." Each piece would regenerate a head at its original apical end and a foot at its original basal end. (Imagine if a person could be generated from a piece as small as a kneecap!)

Every portion of the hydra's body column along the apical-basal axis is potentially able to form both a head and a foot. The animal's polarity, however, is coordinated by a series of morphogenetic gradients that permit the head to form only at one place and the basal disc to form only at another. Evidence for such gradients was first obtained from grafting experiments begun by Ethel Browne in the early 1900s. When hypostome tissue from one hydra is transplanted into the middle of another hydra, the transplanted tissue forms a new apical-basal axis, with the hypostome extending outward (**FIGURE 22.4A**). When a basal disc is grafted to the middle of a host hydra, a new axis also forms, but with the opposite polarity, extending a basal disc (**FIGURE 22.4B**). When tissues from both ends are transplanted simultaneously into the middle of a host, either no new axis is formed or the new axis has little polarity (**FIGURE 22.4C**; Browne 1909; Newman 1974). These experiments have been interpreted to indicate the existence of a **head activation gradient** (highest at the hypostome) and a **foot activation gradient** (highest at the basal disc). The head activation gradient can be measured by implanting rings of tissue from various levels of a donor hydra into a particular region of the host trunk (Wilby and Webster 1970; Herlands and Bode 1974; Mac-Williams 1983b). The higher the level of head activator in the donor tissue, the greater the percentage of implants that will induce the formation of new heads. The head activation factor is concentrated in the hypostome and decreases linearly toward the basal disc.

THE HYPOSTOME AS ORGANIZER Ethel Browne (1909; also see Lenhoff 1991) noted that the hypostome acted as an "organizer" of the hydra. This notion has been confirmed by Broun and Bode (2002), who demonstrated that (1) when transplanted, the

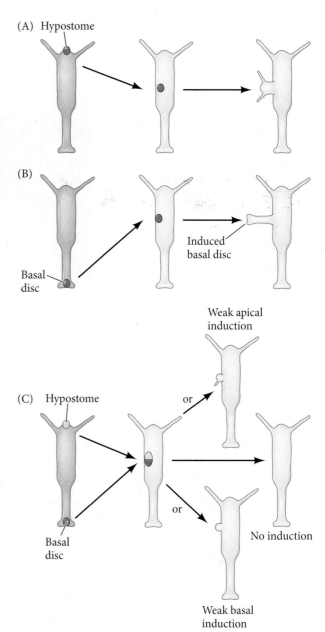

FIGURE 22.4 Grafting experiments demonstrating different morphogenetic capabilities in different regions of the hydra apical-basal axis. (A) Hypostome tissue grafted onto a host trunk induces a secondary axis with an extended hypostome. (B) Basal disc tissue grafted onto a host trunk induces a secondary axis with an extended basal disc. (C) If hypostome and basal disc tissues are transplanted together, only weak (if any) inductions are seen. (After Newman 1974.)

---

[3] Tremblay's advice to researchers is pertinent even today: He advises us to go directly to nature and to avoid the prejudices that our education has given us. Moreover, "one should not become disheartened by want of success, but should try anew whatever has failed. It is even good to repeat successful experiments a number of times. All that is possible to see is not discovered, and often cannot be discovered, the first time." (Quoted in Dinsmore 1991.)

FIGURE 22.5 Formation of secondary axes following transplantation of head regions into the trunk of a hydra. The host endoderm was stained with India ink. (A) Hypostome tissue grafted onto the trunk induces the host's own trunk tissue to become tentacles and head. (B) Sub-hypostomal donor tissue placed on the host trunk self-differentiates into a head and upper trunk. (From Broun and Bode 2002, courtesy of H. R. Bode.)

(A)　　Host tissue　Donor hypostome

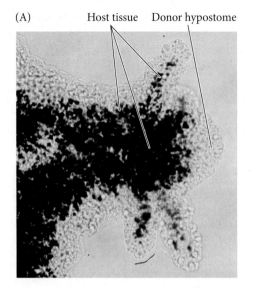

(B)　　Host tissue

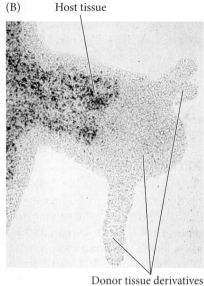

Donor tissue derivatives

hypostome can induce host tissue to form a second body axis; (2) the hypostome produces the head activation signal; (3) the hypostome is the only "self-differentiating" region of the hydra; and (4) the hypostome also produces a "head inhibition signal" that suppresses the formation of new organizing centers.

By inserting small pieces of hypostome tissue into a host hydra whose cells were labeled with India ink (colloidal carbon), Broun and Bode found that the hypostome induced a new body axis and that almost all of the resulting head tissue came from *host* tissue, not from the differentiation of donor tissue (**FIGURE 22.5A**). In contrast, when tissues from other regions (such as the subhypostomal region) were grafted onto a host trunk, a head and apical trunk of a new hydra formed from the grafted *donor* tissue (**FIGURE 22.5B**). In other words, only the hypostome region could alter the fates of the trunk cells and cause them to become head cells. Broun and Bode also found that the signal did not have to emanate from a permanent graft. Even transient contact with the hypostome region was sufficient to induce a new axis from a host hydra. In these cases, *all* the tissue of the new axis came from the host.

WEB TOPIC 22.1 **ETHEL BROWNE AND THE ORGANIZER** As detailed in Chapter 11, Spemann and Mangold's work with amphibians introduced the concept of "the organizer" into embryology, and Spemann's laboratory helped make the idea a unifying principle of embryology. It has been argued, however, that the concept actually had its origins in Browne's experiments on hydra.

A GRADIENT OF WNT3 IS THE INDUCER The major head inducer of the hypostome organizer is a set of Wnt proteins acting through the canonical β-catenin pathway (Hobmayer et al. 2000; Broun et al. 2005; Lengfeld et al. 2009; also see Bode 2009). These Wnt proteins are seen in the apical end of the early bud, defining the hypostome region as the bud elongates (**FIGURE 22.6A**). If the Wnt signaling inhibitor GSK3 is itself inhibited throughout the body axis, ectopic tentacles form at all levels, and each piece of the trunk has the ability to stimulate the outgrowth of new buds. Similarly, transgenic hydra made to globally misexpress the downstream Wnt effector β-catenin form ectopic buds all along the body axis and even on top of newly formed ectopic buds (**FIGURE 22.6B**; Gee et al. 2010). When the hypostome is brought into contact with the trunk of an adult hydra, it induces expression of the *Brachyury* gene in a Wnt-dependent manner—just as vertebrate organizers do—even though hydras lack mesoderm (Broun

(A) *Wnt3* mRNA expression

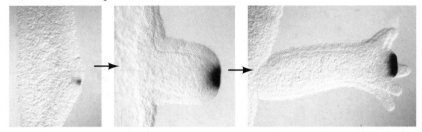

(B) Misexpression of β-catenin

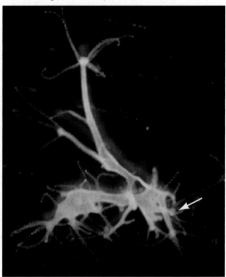

Figure 22.6 Wnt/β-catenin signaling during hydra budding. (A) *Wnt3* mRNA expression (purple) in the hypostome during early bud (left), midstage bud (center), and a bud with early tentacles (right). (B) Transgenic hydra made to misexpress β-catenin (the downstream Wnt effector) have numerous ectopic buds (including buds formed on top of other buds, such as the example marked with an arrow). (A from Hobmayer et al. 2000, courtesy of T. W. Holstein and B. Hobmeyer; B from Gee et al. 2010.)

et al. 1999; Broun and Bode 2002). These results strongly indicate that Wnt proteins (in particular Wnt3) function as the head organizer during normal hydra development, but do they function similarly during regeneration?

MORPHALLAXIS AND EPIMORPHOSIS IN HYDRA When a hydra is decapitated, the Wnt pathway is activated in the apical portion that will form a new head. If the cut is made just below the hypostome, Wnt3 is upregulated in the epithelial cells near the cut surface, which causes the remodeling of existing cells to form the head. No proliferation is seen in this case; hence, it is **morphallactic regeneration** (regeneration by cell trans-differentiation). If the hydra is cut at its midsection, however, the cells derived from the interstitial stem cell (neurons, nematocytes, secretory cells, and gametes) undergo apoptosis immediately below the cut site. Before dying, however, these cells produce a burst of Wnt3, which activates β-catenin in the interstitial cells beneath them. This β-catenin surge causes a wave of proliferation in the interstitial cells as well as remodeling in the epithelial cells. Here we have **epimorphic regeneration** (regeneration by cell dedifferentiation), or **epimorphosis** (Chera et al. 2009). Canonical Wnt signaling is thus important both in normal budding and in head regeneration.

The head inhibition gradients

If any region of the hydra body column is capable of forming a head, how is head formation restricted to a specific location? In 1926, Rand and colleagues showed that normal regeneration of the hypostome is inhibited when an intact hypostome is grafted adjacent to the amputation site (**FIGURE 22.7A**). Moreover, if a graft of subhypostomal tissue (from the region just below the hypostome, where there is a relatively high concentration of head activator) is placed in the same region of a host hydra, no secondary axis forms (**FIGURE 22.7B**). The host head appears to make an inhibitor that prevents the grafted tissue from forming a head and secondary axis. Supporting this hypothesis is the fact that if subhypostomal tissue is grafted onto a decapitated host hydra, a second axis does form (**FIGURE 22.7C**). A gradient of this inhibitor appears to extend from the head down the body column and can be measured by grafting subhypostomal tissue into various regions along the trunks of host hydras. This tissue will not produce a head when implanted into the apical area of an intact host hydra (see Figure 22.7B), but it will form a head if placed lower on the host (**FIGURE 22.7D**). The head inhibitor remains unknown, but it appears to be labile, with a half-life of only 2–3 hours (Wilby

Developing Questions

In hydra, what triggers Wnt3 upregulation upon amputation? Moreover, if a midsection cut and a more apical cut both result in Wnt activation, how is apoptosis triggered in the first scenario and not in the other?

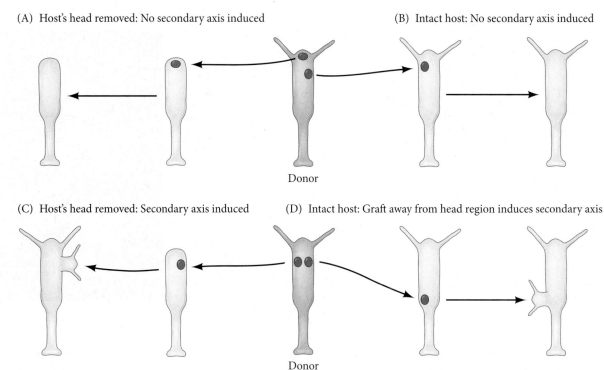

(A) Host's head removed: No secondary axis induced

(B) Intact host: No secondary axis induced

Donor

(C) Host's head removed: Secondary axis induced

(D) Intact host: Graft away from head region induces secondary axis

Donor

FIGURE 22.7 Grafting experiments provide evidence for a head inhibition gradient. (A) Hypostomal tissue grafted to the amputated region inhibits regeneration of the head. (B) Subhypostomal tissue does not generate a new head when placed close to an existing host head. (C) Subhypostomal tissue generates a head if the existing host head is removed. A head also forms at the site where the host's head was amputated. (D) Subhypostomal tissue generates a new head when placed far away from an existing host head. (After Newman 1974.)

and Webster 1970; MacWilliams 1983a). It is thought that the head inhibitor and the head activator (Wnts) are both made in the hypostome, but that the head inhibition gradient falls off more rapidly than the head activator gradient (see Bode 2011, 2012). The place where the head activator is uninhibited by the head inhibitor becomes the budding zone.

But that does not account for the bottom third of the column. What prevents cells there from becoming heads? Head formation at the base appears to be prevented by the production of another substance, a foot activator (MacWilliams et al. 1970; Hicklin and Wolpert 1973; Schmidt and Schaller 1976; Meinhardt 1993; Grens et al. 1999). The inhibition gradients for the head and the foot may be important in determining where and when a bud can form. In young adult hydras, the gradients of head and foot inhibitors appear to block bud formation. However, as the hydra grows, the sources of these labile substances grow farther apart, creating a region of tissue about two-thirds down the trunk where levels of both inhibitors are minimal. This region is where the bud forms (**FIGURE 22.8A**; Shostak 1974; Bode and Bode 1984; Schiliro et al. 1999).

Certain hydra mutants have defects in their ability to form buds, and these defects can be explained by alterations of the inhibition gradients. The *L4* mutant of *Hydra magnipapillata*, for instance, forms buds very slowly, and does so only after reaching a size about twice as long as wild-type individuals. The amount of head inhibitor in these mutants was found to be much greater than in wild-type individuals (Takano and Sugiyama 1983).

Several small peptides have been found to activate foot formation, and researchers are beginning to sort out the mechanisms by which these proteins arise and function (see Harafuji et al. 2001; Siebert et al. 2005). The specification of cells from the basal region through the body column may be mediated by a gradient of tyrosine kinase, however. The product of the *shinguard* gene is a tyrosine kinase that extends in a gradient from the ectoderm just above the basal disc through the lower region of the trunk. Buds appear to form where this gradient fades (**FIGURE 22.8B**). The *shinguard* gene appears to be activated through the product of the *manacle* gene, a putative transcription factor that is expressed earlier in the basal disc ectoderm (Bridge et al. 2000).

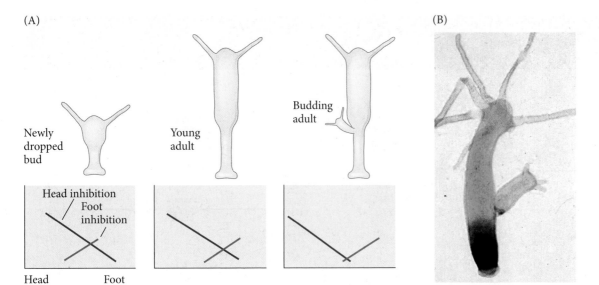

(A)

Newly
dropped
bud

Young
adult

Budding
adult

Head inhibition
Foot
inhibition

Head   Foot

(B)

FIGURE 22.8   Bud location as a function of head and foot inhibition gradients. (A) Head inhibition (blue) and foot inhibition (red) gradients in newly dropped buds, young adults, and budding adults. (B) Expression of the Shinguard protein in a graded fashion in a budding hydra. (A after Bode and Bode 1984; B from Bridge et al. 2000.)

Stem Cell-Mediated Regeneration in Flatworms

Planarian flatworms can reproduce asexually by binary fission, during which they split themselves in half, separating their posterior end from the anterior end, and each segment regenerates the lost parts. During regeneration, each piece re-creates all the appropriate cell types that make up the planarian, such as photoreceptors, nervous system, epithelium, muscle, intestines, pharynx, and gonads (see Roberts-Galbraith and Newmark 2015). Only recently has it been shown that the cells capable of this regeneration are the same pluripotent stem cells that repair and replace body parts. It has been known since the 1700s that when planarians are cut in half, just as occurs in asexual reproduction, the head half will regenerate a tail from the wound site while the tail half will regenerate a head (**FIGURE 22.9A,B**; Pallas 1766). Not until 1905, however, did Thomas Hunt Morgan and C. M. Child realize that such polarity indicated an important principle of development[4] (see Sunderland 2010). Morgan pointed out that if both the head and the tail were cut off a flatworm, thus trisecting the animal, the medial segment would regenerate a head from the former anterior end and a tail from the former posterior end, but never the reverse (**FIGURE 22.9C**). Furthermore, if the medial segment were sufficiently thin, the regenerating portions would be abnormal (**FIGURE 22.9D**). Both Morgan (1905) and Child (1905) postulated a gradient of anterior-producing materials concentrated in the head region. The middle segment would be told what to regenerate at both ends by the concentration gradient of these materials. If the piece were too narrow, however, the gradient would not be sensed within the segment.

THE BLASTEMA AND ADULT PLURIPOTENT STEM CELLS   A major question to ask about planarian regeneration is, What cells form the new head or tail? It is now known that immediately following amputation, a wound response is initiated, eliciting from all cells in the vicinity of the wound the same "generic" transcriptional response no matter where the amputation occurred. However, after this initial wound response, the transcriptional profile becomes regionally distinct as regeneration proceeds (Wurtzel et al. 2015). For decades it was believed that the old cells *dedifferentiated* at the cut ends of the planarian to form a **regeneration blastema**, a collection of relatively undifferentiated cells

---

[4] Before 1910, "fly lab" maestro Morgan was well known for his research into flatworm regeneration. Indeed, it was only in 1900 that Morgan first mentioned *Drosophila*—as food for his flatworms! He was even able to "stain" the flatworms' digestive tubes by feeding them pigmented *Drosophila* eyes. Later, when he founded modern genetics, Morgan renounced flatworms as a model for heredity in favor of *Drosophila* (see Mittman and Fausto-Sterling 1992).

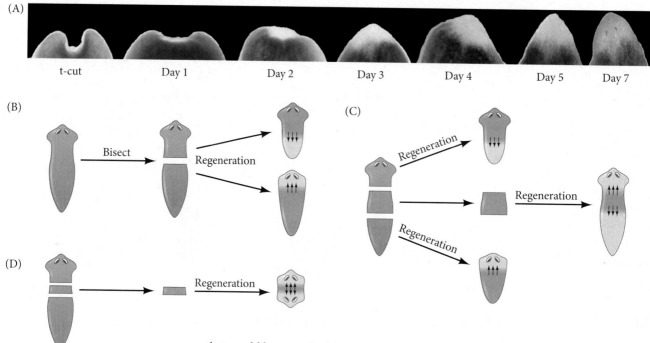

(A) t-cut   Day 1   Day 2   Day 3   Day 4   Day 5   Day 7

FIGURE 22.9 Flatworm regeneration and its limits. (A) Time course of a planarian flatworm regenerating a new head following head amputation. (B) If the planarian is cut in half, the anterior portion of the lower half regenerates a head while the posterior of the upper half regenerates a tail. The same tissue can generate a head (if it is at the anterior portion of the tail piece) or a tail (if it is at the posterior portion of the head piece). (C) If a flatworm is cut into three pieces, the middle piece will regenerate a head from its anterior end and a tail from its posterior end. (D) If the middle slice is too narrow, there is no discernible morphogen gradient within it, and regeneration is abnormal. (A from Gentile et al. 2011; B–D after Gosse 1969.)

that would be organized into new structures by paracrine factors located at the wound surface (see Baguña 2012). In 2011, however, a series of experiments by Wagner and colleagues provided substantial evidence that dedifferentiation does *not* occur. Rather, the regeneration blastema forms from pluripotent stem cells called **clonogenic neoblasts (cNeoblast)**, a set of pluripotent cells in flatworms that serve as stem cells to replace the aging cells of the adult body (**FIGURE 22.10A**; Newmark and Sánchez Alvarado 2000; Pellettieri and Sánchez Alvarado 2007; reviewed in Adler and Sánchez Alvarado 2015).

Clonogenic neoblasts can migrate to a wound site and regenerate the tissue. Wagner and colleagues were able to show that if planaria are irradiated at a dosage that destroys nearly all neoblasts (dividing cells are killed more readily by radiation than nondividing cells, which is the basis for irradiating cancer sites), there would be some individual animals in which a single cNeoblast survived. From this neoblast, dividing progenitor cells formed, ultimately producing cell types of all germ layer origins. This single-cell response demonstrated that neoblasts are pluripotent cells residing in the adult body, capable of regenerating all tissues of the planarian (**FIGURE 22.10B**).

If cNeoblasts are essential for regeneration, their total loss should prevent regeneration. Next the researchers irradiated planaria so that all dividing cells were destroyed (**FIGURE 22.10C**). These planaria died because of failed tissue replacement. However, transplantation of a single clonogenic neoblast into such an irradiated flatworm could, in some cases, restore all the cells of the organism. Not only did the flatworm survive, it split into more planaria, and all cells of these new planaria had the same genotype as the single donor neoblast. These results conclusively demonstrated that the regeneration of the flatworm was the result of the production of new cells from adult pluripotent stem cells (Wagner et al. 2011).

NEOBLAST SPECIALIZATION The use of an adult population of pluripotent stem cells to fuel planarian regeneration presents several important questions. Is this population heterogeneous, or is it derived from a single pluripotent population, as the above clonal studies suggest is possible? Moreover, what is the process whereby a neoblast can create the 30 or so cell types of the adult planarian? When does cell specification take place? Could it be that lineage-restricted multipotent stem cells are derived from a cNeoblast and seeded throughout the flatworm prior to any injury? Or are differentiated postmitotic cells produced directly from a pluripotent neoblast at the time of injury (**FIGURE 22.11**)?

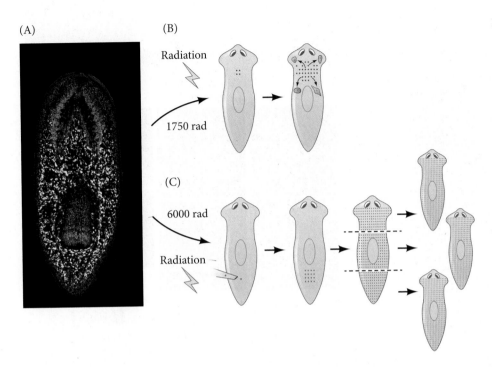

FIGURE 22.10 Planarian regeneration is accomplished by a pluripotent stem cell population of neoblasts. (A) Neoblasts in the planarian flatworm *Schmidtea mediterranea*, stained with antibodies to phosphorylated histone 3. Each pluripotent neoblast generates a colony of neoblast cells (red; nuclei are stained blue) in the flatworm. These clonogenic neoblast cells produce the differentiating cells of the regenerating flatworm. Neoblasts are scattered throughout the body posterior to the eyes (although they are not present in the centrally located pharynx). (B) Irradiation with 1750 rad kills almost all neoblasts. If even one survives, a single clonogenic neoblast can divide to generate a colony of dividing cells that will ultimately produce the differentiated cells of the organs. (C) Irradiation with 6000 rad eliminates all dividing cells. Transplanting a single clonogenic neoblast from a donor strain (red) results not only in the production of all the cell types in the organism but also restores the organism's capacity for regeneration. (A courtesy of P. W. Reddien; B,C after Tanaka and Reddien 2011.)

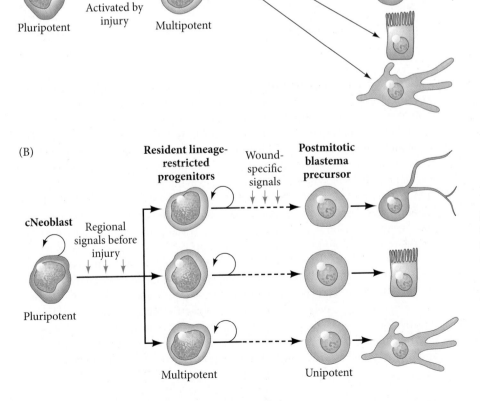

FIGURE 22.11 Two possible mechanisms for neoblast specification during regeneration. (A) One proposed mechanism is that a single pluripotent neoblast is responsible for generating new cell progenitors that form the multipotent cells of the blastema. (B) Alternatively, multipotent progenitor cells are positioned throughout the planarian with their lineages specified based on their position. These divide and produce postmitotic blastema cells that differentiate into their specified fates.

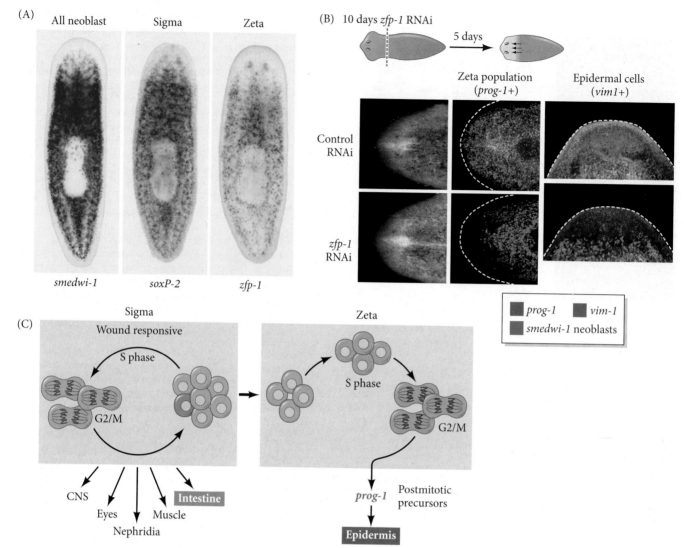

FIGURE 22.12 Two distinct populations of neoblasts exist with one of them responsive to injury. (A) In situ hybridizations marking (blue) all neoblast cells (*smedwi-1*; left); the sigma neoblast population (*soxP-2*; center); and the zeta neoblast population (*zfp-1*; right). (B) Planarian heads largely regenerate even in the absence of zeta neoblasts (leftmost, brightfield photos). Loss of the zeta population by RNAi is indicated by a lack of *prog-1* expression (center photos) as seen with fluorescence in situ hybridization (FISH). Although, epidermal cell types (right) do not fully regenerate in the absence of zeta neoblasts (*vim-1*, purple), the remaining neoblasts (green; likely sigma) and their derived cell types are unaffected. (C) Model of sigma neoblast-led development and regeneration. Sigma neoblasts direct the generation of numerous cell types as well as the zeta progenitor cells. Zeta neoblasts give rise to epidermal cell fates through a *prog-1*-positive postmitotic precursor state. (A,B from van Wolfswinkel et al. 2014; C after van Wolfswinkel et al. 2014.)

Genetic studies are beginning to unravel some of the answers. The gene *smedwi-1* is expressed by all neoblasts and has become the most common marker for their identification across planarian species (Reddien et al. 2005; Reddien 2013). Several studies have revealed that some neoblasts express different sets of transcription factors that correlate with specific cell fates in the planarian adult, which suggests that lineage specification of stem cell populations may exist in the organism for normal development and in response to injury (Hayashi et al. 2010; Pearson et al. 2010; Shibata et al. 2012). To further investigate this possibility, the Reddien lab performed transcriptomic analysis of individual neoblasts during homeostasis and regeneration (van Wolfswinkel et al. 2014). Through this comprehensive analysis, the researchers found that two distinct populations of neoblasts exist, which they named zeta and sigma (**FIGURE 22.12A**). Although zeta and sigma neoblasts are morphologically indistinguishable, they have several defining characteristics: they express different gene regulatory networks, and zeta neoblasts are postmitotic, whereas sigma neoblasts are highly proliferative and are the only stem cells directly responsive to injury. Upon amputation, sigma neoblasts (*soxP-2*-expressing) generate a huge variety of cell types (brain, intestine, muscle, excretory, pharynx, and eyes), as well as the progenitor population for zeta neoblasts. Zeta neoblasts (*zfp-1*-expressing) are then directly responsible for creating the remaining epidermal cell types. Planaria lacking a zeta population can be made through *zfp-1* RNAi knockdown. When the head of these planarian were amputated, the sigma neoblasts were able to fuel the regeneration of all the cells of a new head, except for the epidermal lineages (**FIGURE 22.12B,C**; van Wolfswinkel et al. 2014).

These data all support a vital role for stem cells in regeneration. But we are still left with the question, How are the specific cell types patterned correctly? How does the flatworm tell the posterior blastema to become tail and the anterior blastema to become head?

HEAD-TO-TAIL POLARITY As we saw with hydra, Wnt signaling appears to play a major role in establishing a polarity of differential cell fates. In hydra, this polarity was positively regulated by Wnt/β-catenin signaling along the apical-basal axis. In planaria, Wnt/β-catenin functions to establish anterior-posterior polarity of the regenerating flatworm; here, though, Wnts functioning through β-catenin promotes tail development while repressing head regeneration (Gurley et al. 2008; Petersen and Reddien 2008, 2011). In fact, *Wnt* expression is excluded from the head, and functional proteins are presumed to be present in a tail-to-head gradient.[5]

Several labs took a comparative approach to understanding the control over polarity during head regeneration in planaria. It is interesting that some species of planaria are incapable of regeneration. When planaria of the species *Procotyla fluviatilis* and *Dendrocoelum lacteum* are decapitated, they are unable to regenerate their heads. Researchers saw this distinction as an opportunity to identify the genes that are essential for the regeneration capabilities of planarian species such as *Dugesia japonica*. Comparing the transcriptomes of anterior-facing blastema cells from regeneration-competent species with regeneration-deficient species revealed an upregulation of genes indicative of highly active Wnt/β-catenin signaling only in the blastemas of regeneration-deficient planaria (Sikes and Newmark 2013). Most remarkable is that inhibition of Wnt/β-catenin signaling in regeneration-deficient species yields fully functional regenerated heads (**FIGURE 22.13**; Liu et al. 2013; Sikes and Newmark 2013; Umesono et al. 2013). These results demonstrate the inhibitory function of Wnt signaling on head specification during regeneration. In regenerating planaria, β-catenin is activated (via Wnts) in the posterior-facing blastema, which generates a tail. As in vertebrate development, an anterior polarity in planarian regeneration is dependent on repression of Wnt signaling, which prevents β-catenin accumulation and allows head formation. If β-catenin is eliminated from the posterior (tail-forming) blastema by RNA interference, that blastema will form a head (**FIGURE 22.14A–C**). Indeed, when RNAi completely eliminates β-catenin from non-regenerating flatworms, the entire organism becomes a head, with eyes all around the periphery (Gurley et al. 2008; Iglesias et al. 2008).

What are the inhibitory influences that Wnt imposes on anterior cell specification? Alternatively, what do anterior cells do to protect themselves from Wnt signaling? In Chapter 4, we discussed one such Wnt inhibitor called Notum (see Figure 4.33), which in planaria is specifically expressed at the apex of the head in opposition to the posterior Wnt expression. Notum is upregulated in the anterior-facing blastema

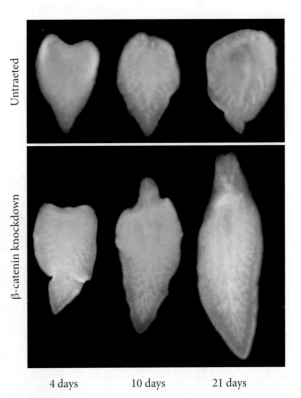

FIGURE 22.13 Restoration of head regeneration in *Dendrocoelum lacteum* by knockdown of β-catenin. This regeneration-incompetent flatworm cannot regrow a head following amputation (top row). If, however, the gene for β-catenin is knocked down with RNA interference (bottom row), this species is capable of regenerating its head following amputation over a 21-day period. (From Liu et al. 2013.)

---

[5] It is interesting that although Wnt signaling is used in both flatworm and hydra regeneration, its effects are different. In flatworms, Wnt signals the formation of the tail, and its inhibition is needed to form heads. In hydra, Wnt signaling appears to establish heads, or at least the part that has a mouth and resembles a head. (Just because we call it a head does not make it homologous to the face of bilaterians. There is still controversy about this point.)

FIGURE 22.14 Polarity in planarian regeneration. (A) Normally, Wnts are produced in the posterior blastema, and the result is a tail. If the Wnt pathway is blocked by using RNA interference against either β-catenin (B) or Wnt1 messages (C), though, the posterior blastema regenerates a head, thereby forming a worm with heads at both ends. (D) Proposed model of anterior-posterior polarity through the interactions of three signals: Erk, Notum, and Wnt. Posterior Wnt promotes tail specification while repressing head specification. Wnt inhibits the anteriorly expressed head inducer Erk. Wnt is restricted from the most anterior head regions through antagonism by Notum, however. (A–C from Reddien 2011, photographs courtesy of D. Reddien.)

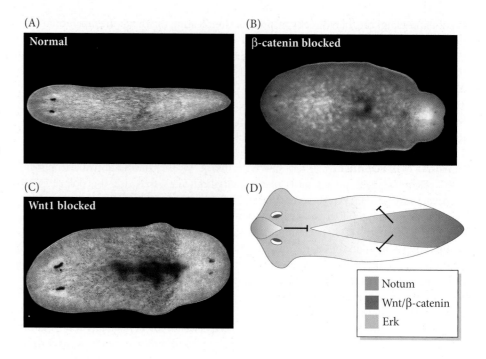

(**FIGURE 22.15**). Not surprisingly, it was found that *Notum* expression is downregulated in regeneration-deficient species (Lui et al. 2013). Moreover, transcriptome analysis of blastemas in regenerating planaria found that only *one* out of 4,401 genes examined was differentially expressed between anterior- and posterior-facing blastemas! That gene was *Notum*, expressed only in the anterior blastema (Wurtzel et al. 2015). If *Notum* expression

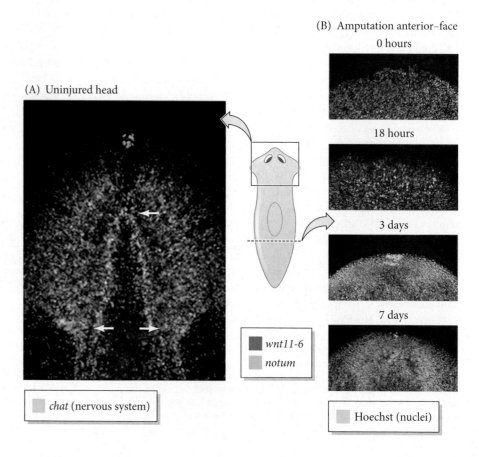

FIGURE 22.15 Wnt and Notum expression in the planarian head (A) and in the anterior-facing blastema (B). In both the uninjured head and blastema, Notum is expressed by midline cells in the most anterior region. The illustration in the middle denotes the area of the flatworm image in (A) and the tail cut that (B) is based on. *Chat* is a gene marker for nervous system cells; Hoechst labels all nuclei. (From Hill and Petersen 2015.)

is knocked down, causing too much Wnt to become functionally active, the anterior-facing blastema will form a tail (Petersen and Reddien 2011). These results strongly suggest that the anterior expression of *Notum* functions to antagonize the posteriorly produced Wnt, leading to head specification. It has also been proposed that regulation of the balance between Wnt and Notum signals may underlie not only head-tail specification but also regulation of organ size (Hill and Petersen 2015). In addition, there appears to be an anterior-to-posterior gradient of Erk signaling that functions as a positive inducer of head specification. Wnt signaling achieves its repression of head regeneration by inhibiting Erk. Therefore, only in those more anterior regions lacking Wnt (due to Notum's repression of Wnt) can Erk induce head regeneration (**FIGURE 22.14D**; Umesono et al. 2013).

 SCIENTISTS SPEAK 22.1 Dr. Alejandro Sánchez Alvarado describes the role of stem cells during planarian regeneration in this short documentary.

Salamanders: Epimorphic Limb Regeneration

When an adult salamander limb is amputated, the remaining limb cells reconstruct a new limb, complete with all its differentiated cells arranged in the proper order. Remarkably, the limb regenerates only the missing structures and no more. For example, when the limb is amputated at the wrist, the salamander forms a new wrist and foot, but not a new elbow. In some way, the salamander limb "knows" where the proximal-distal axis has been severed and is able to regenerate from that point on (**FIGURE 22.16**).

Salamanders accomplish epimorphic regeneration by cell dedifferentiation to form a **regeneration blastema**, which in this case is an aggregation of relatively undifferentiated cells derived from the originally differentiated tissue that then proliferates and redifferentiates into the new limb parts (**FIGURE 22.17**; see Brockes and Kumar 2002; Gardiner et al. 2002; Simon and Tanaka 2013). Bone, dermis, and cartilage just beneath the site of amputation contribute to the regeneration blastema, as do satellite cells from nearby muscles (Morrison et al. 2006). Hence, unlike the flatworm, which formed its regeneration blastema from adult pluripotent stem cells, much of the salamander limb's

Developing Questions

Many details of the interactions between anterior and posterior signaling systems still need to be determined. For instance, does Erk signaling positively regulate *Notum* gene expression? Moreover, although much of the current work has focused on the role of canonical Wnt/β-catenin signaling, noncanonical Wnt signaling has long been known to influence cell and tissue polarity. What role, if any, do the noncanonical Wnt signaling pathways have in planarian regeneration?

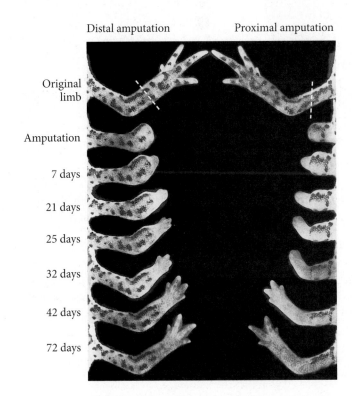

FIGURE 22.16 Regeneration of a salamander forelimb. The amputation shown on the left was made below the elbow; the amputation shown on the right cut through the humerus. In both instances, the correct positional information was respecified, and a normal limb was regenerated within 72 days. (From Goss 1969, courtesy of R. J. Goss.)

(A)

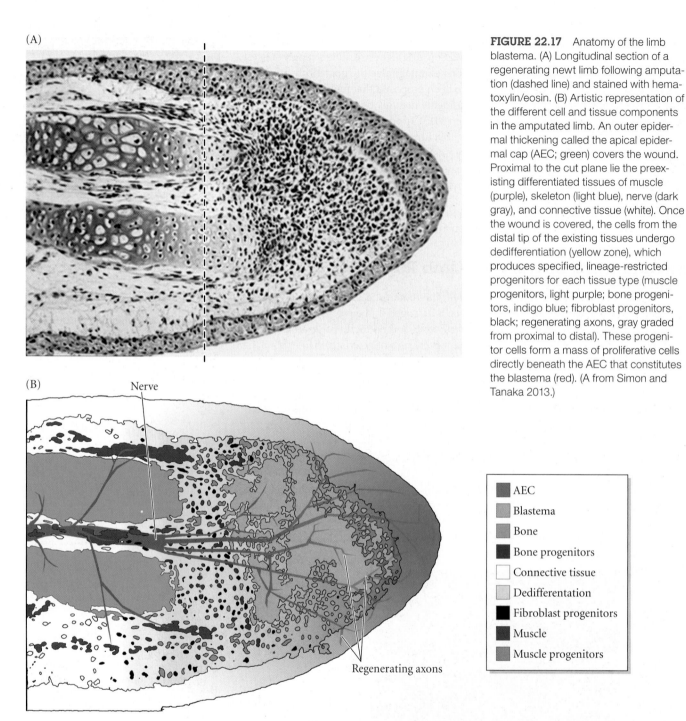

(B)

Nerve

Regenerating axons

■	AEC
■	Blastema
■	Bone
■	Bone progenitors
□	Connective tissue
■	Dedifferentation
■	Fibroblast progenitors
■	Muscle
■	Muscle progenitors

FIGURE 22.17 Anatomy of the limb blastema. (A) Longitudinal section of a regenerating newt limb following amputation (dashed line) and stained with hematoxylin/eosin. (B) Artistic representation of the different cell and tissue components in the amputated limb. An outer epidermal thickening called the apical epidermal cap (AEC; green) covers the wound. Proximal to the cut plane lie the preexisting differentiated tissues of muscle (purple), skeleton (light blue), nerve (dark gray), and connective tissue (white). Once the wound is covered, the cells from the distal tip of the existing tissues undergo dedifferentiation (yellow zone), which produces specified, lineage-restricted progenitors for each tissue type (muscle progenitors, light purple; bone progenitors, indigo blue; fibroblast progenitors, black; regenerating axons, gray graded from proximal to distal). These progenitor cells form a mass of proliferative cells directly beneath the AEC that constitutes the blastema (red). (A from Simon and Tanaka 2013.)

regeneration blastema appears to arise from the dedifferentiation of adult cells followed by cell division and the redifferentiation of those cells back into their original cell types.[6]

Formation of the apical epidermal cap and regeneration blastema

When a salamander limb is amputated, a plasma clot forms. Within 6–12 hours, epidermal cells from the remaining stump migrate to cover the wound surface, forming the **wound epidermis**. In contrast to wound healing in mammals, no scar forms, and the dermis does not move with the epidermis to cover the site of amputation. The nerves

---

[6] There is still controversy over whether this is "true" dedifferentiation of normally postmitotic cells or whether much of the limb blastema is formed from the activation of uncommitted stem cells (see Nacu and Tanaka 2011).

innervating the limb degenerate for a short distance proximal to the plane of amputation (see Chernoff and Stocum 1995).

During the next four days, the extracellular matrices of the tissues beneath the wound epidermis are degraded by proteases, liberating single cells that undergo dramatic dedifferentiation: bone cells, cartilage cells, fibroblasts, and myocytes all lose their differentiated characteristics. Genes that are expressed in differentiated tissues (such as the *mrf4* and *myf5* genes expressed in muscle cells) are downregulated, while there is a dramatic increase in the expression of genes such as *msx1* that are associated with the proliferating progress zone mesenchyme of the embryonic limb (Simon et al. 1995). This cell mass is the regeneration blastema, and these cells are the ones that will continue to proliferate and that will eventually redifferentiate to form the new structures of the limb (**FIGURE 22.18**; Butler 1935). Moreover, during this time,

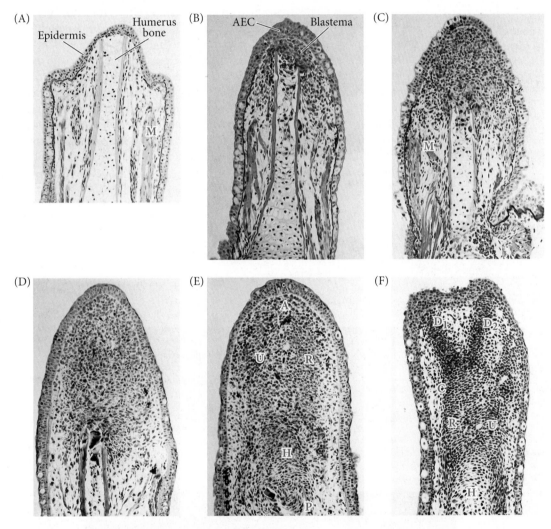

FIGURE 22.18 Regeneration in the larval forelimb of the spotted salamander *Ambystoma maculatum*. (A) Longitudinal section of the upper arm, 2 days after amputation. The skin and muscle (M) have retracted from the tip of the humerus. (B) At 5 days after amputation, a thin accumulation of blastema cells is seen beneath the thickened epidermis, where the apical epidermal cap (AEC) forms. (C) At 7 days, a large population of mitotically active blastema cells lies distal to the humerus. (D) At 8 days, the blastema elongates by mitotic activity; much dedifferentiation has occurred.

(E) At 9 days, early redifferentiation can be seen. Chondrogenesis has begun in the proximal part of the regenerating humerus (H). The letter A marks the apical mesenchyme of the blastema, and U and R are the precartilaginous condensations that will form the ulna and radius, respectively. P represents the stump where the amputation was made. (F) At 10 days after amputation, the precartilaginous condensations for the carpal bones (wrist, C) and the first two digits (D1, D2) can also be seen. (From Stocum 1979, courtesy of D. L. Stocum.)

FIGURE 22.19 Blastema cells retain their specification even when they dedifferentiate. (A,B) Schematic representation of the procedure wherein a particular tissue (in this case, cartilage) is transplanted from a salamander expressing a *GFP* transgene into a wild-type salamander limb. Later, the limb is amputated through the region of the limb containing GFP expression, and a blastema is formed containing GFP-expressing cells that had been cartilage precursors. The regenerated limb is then studied to see if GFP is found only in the regenerated cartilage tissues or in other tissues. The dashed lines in (B) mark the position of the amputation. (C) Longitudinal section of a regenerated limb 30 days after amputation. Muscle cells are stained red; nuclei are stained blue. The majority of GFP-expressing cells (green) were found in regenerated cartilage; no GFP was seen in the muscle. (After Kragl et al. 2009, courtesy of E. Tanaka.)

the wound epidermis thickens to form the **apical epidermal cap** (**AEC**) (see Figure 22.17B), which acts similarly to the apical ectodermal ridge during normal limb development (see Chapter 19; Han et al. 2001).

Thus, the previously well-structured limb region at the cut edge of the stump forms a proliferating mass of indistinguishable cells just beneath the apical epidermal cap. One of the major questions of regeneration is whether the cells keep a "memory" of what they had been. In other words, do new muscles arise from old muscle cells that dedifferentiated, or can any cell of the blastema become a muscle cell? Kragl and colleagues (2009) found that the blastema is not a collection of homogeneous, fully dedifferentiated cells. Rather, in the regenerating limbs of the axolotl salamander, muscle cells arise only from old muscle cells, dermal cells come only from old dermal cells, and cartilage can arise only from old cartilage or old dermal cells. Thus, the blastema is not a collection of unspecified multipotent progenitor cells. Rather, the cells retain their specification, and the blastema is a heterogeneous assortment of *restricted* progenitor cells.

Kragl and colleagues (2009) performed an experiment in which they transplanted limb tissue from a salamander whose cells expressed green fluorescent protein (GFP) into different regions of limbs of normal salamanders that did not have the *GFP* transgene (**FIGURE 22.19**). If they transplanted the GFP-expressing limb cartilage into a salamander limb that did not contain the *GFP* transgene, the GFP-expressing cartilage would integrate normally into the limb skeleton. They later amputated the limb through the region containing GFP-marked cartilage cells. The blastema was found to contain GFP-expressing cells, and when the blastema differentiated, the only GFP-expressing cells found were in the limb cartilage. Similarly, GFP-marked muscle cells gave rise only to muscle, and GFP-marked epidermal cells only produced the epidermis of the regenerated limb.

(A)

Cartilage transplanted

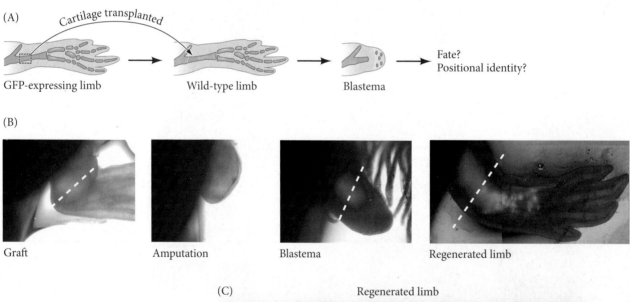

GFP-expressing limb → Wild-type limb → Blastema → Fate? Positional identity?

(B)

Graft   Amputation   Blastema   Regenerated limb

(C)   Regenerated limb

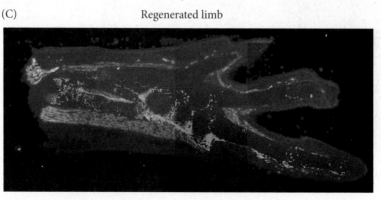

Proliferation of the blastema cells:
The requirement for nerves and the apical epidermal cap

The growth of the regeneration blastema depends on the presence of both the apical epidermal cap and nerves. The AEC stimulates the growth of the blastema by secreting Fgf8 (just as the apical ectodermal ridge does in normal limb development), but the effect of the AEC is only possible if nerves are present (Mullen et al. 1996). Both sensory and motor axons innervate the blastema such that sensory axons make direct contact with the AEC and motor axons terminate in the blastema mesenchyme (see Figure 22.17B). Singer and others demonstrated that a minimum number of nerve fibers of either sensory or motor type must be present for regeneration to take place (Todd 1823; Sidman and Singer 1960; Singer 1946, 1952, 1954; Singer and Craven 1948). Most important is that regenerating nerves are necessary for the proliferation and outgrowth of the blastema. If the limb is first denervated and then amputated, no regeneration will occur (**FIGURE 22.20**). If a wound is made in the epidermis of the proximal limb and a nerve is then diverted to the wound area, a blastema-like bud will form, but not a fully regenerated limb. To induce a complete ectopic limb, not only does a nerve need to be diverted to the wound site, but an epidermal graft from the opposite side of the limb (from a posterior to an anterior location) needs to be placed near the wound (**FIGURE 22.21**; Endo et al. 2004). These results suggest that during normal limb regeneration, the regenerating nerves deliver important signals to the AEC. They also suggest, however, that signals from nerves are not sufficient for ectopic limb growth; for that growth, positional cues from an epidermis that are different from the positional cues at the wound site itself are also needed (see Yin and Poss 2008; McCusker and Gardiner 2011, 2014).

ANTERIOR GRADIENT PROTEIN   As we discussed in Chapter 15, neural activity with target cells is required for synapse maturation and neuronal survival. Thus, it was initially postulated that neural activity might be the necessary stimulus for limb

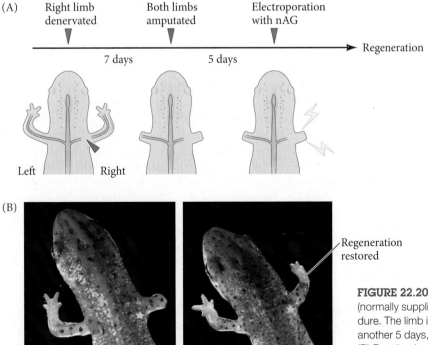

FIGURE 22.20 Regeneration of newt limbs depends on nAG (normally supplied by the limb nerves). (A) Schematic of the procedure. The limb is denervated and 7 days later is amputated. After another 5 days, nAG is electroporated into the limb blastema. (B) Results show that in the denervated control (not given nAG), the amputated limb (yellow star) remains a stump. The limb that is given nAG regenerates tissues with appropriate proximal-distal polarity. (After Yin and Poss 2008, courtesy of K. Poss.)

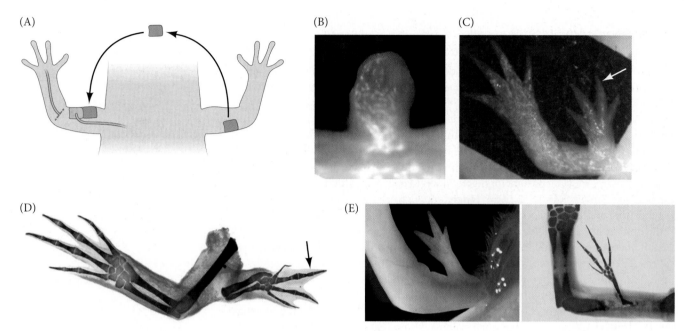

(A)

(B)

(C)

(D)

(E)

FIGURE 22.21 Induction of ectopic limbs in salamanders. (A) Schematic showing experiment in which a nerve is diverted to a wound site in the limb epidermis (gray square) and an epidermal skin graft from the posterior portion of the contralateral limb (blue square) is grafted next to the wound site. (B,C) Results of this experiment show that a limb blastema is induced (B), which develops into a full limb (C; arrow). (D) The regenerated accessory limb (arrow) is correctly patterned, as seen by the Alcian blue staining of cartilaginous elements. (E) Accessory limbs can be induced solely through the application of beads coated with BMP2 (or BMP7) and Fgf2/8 to a wound location in the limb. (A–C from Endo et al. 2004; D from McCusker et al. 2014; E from Makanae et al. 2014.)

regeneration. However, experiments have shown that neural conductance (action potentials and release of acetylcholine) is not required to promote limb regeneration (Sidman and Singer 1951; Drachman and Singer 1971; Stocum 2011). If neural activity is not required, then what are the regenerating axons providing to the limb blastema? These neurons are believed to release factors necessary for the proliferation of the blastema cells (Singer and Caston 1972; Mescher and Tassava 1975). There have been many candidates for such a nerve-derived blastema mitogen, but probably the best candidate is **newt anterior gradient protein (nAG)**. This protein can cause blastema cells to proliferate in culture, and it permits normal regeneration in limbs that have been denervated (see Figure 22.20; Kumar et al. 2007a). If activated nAG genes are electroporated into the dedifferentiating tissues of amputated limbs that have been denervated, the limbs are able to regenerate. If nAG is not administered, the limbs remain stumps. Moreover, nAG is only minimally expressed in normal limbs, but it is induced in the Schwann cells that surround the regenerating axons within 5 days of amputation.

Further support for the stimulatory role of nAG comes from the study of aneurogenic limbs. In this experiment, two embryonic salamanders were joined together by parabiosis, ensuring similar conditions and survival of both. In one of the salamanders, the neural tube was removed. Both salamanders survived this procedure, and all limbs grew successfully, but the limbs of the salamander lacking a neural tube were aneurogenic, completely devoid of any neural innervation. Based on the previous findings that denervated limbs are unable to regenerate, one would hypothesize that these aneurogenic limbs would likewise be unable to regenerate following amputation. Strikingly, these aneurogenic limbs *did* regenerate. When researchers compared the expression of nAG of normal limbs with that of both denervated limbs and these aneurogenic limbs, they discovered that the aneurogenic limbs had a uniquely high level of nAG expression in the epidermis (**FIGURE 22.22A**; Kumar et al. 2011). Moreover, upon limb amputation, nAG increased first in the nerve sheath of normal limbs, but was present throughout the blastema in the aneurogenic limbs (**FIGURE 22.22B**). These results suggest that nAG alone is the primary mitogen responsible for nerve-dependent regeneration. The receptor for nAG, Prod1, has since been discovered and found to be expressed in a proximal-to-distal gradient in the salamander limb (Morais et al. 2002; Kumar et al. 2007a,b). This ligand-receptor relationship seems to be conserved across regeneration-competent salamander species (Geng et al. 2015).

(A) Limb development

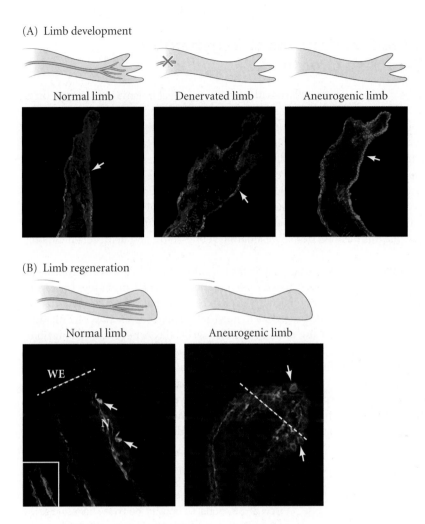

Normal limb   Denervated limb   Aneurogenic limb

(B) Limb regeneration

Normal limb   Aneurogenic limb

FIGURE 22.22   Differential expression of nAG in the aneurogenic salamander limb. (A) Expression of nAG protein (green) is localized to a small subset of epidermal cells in the normal and denervated limbs during development but is highly upregulated throughout the epidermis of aneurogenic limbs. Arrows indicate the epidermis. (B) Schematic representation of limbs prior to amputation. Lower images show the regenerated limbs, with a white dotted line indicating the plane of amputation; the wound epidermis (WE) is positioned distal to this line. In normal limb regeneration, nAG is upregulated only in the sheath cells of regenerating nerves (N); the inset is a low magnification view of neurofilaments (red) in same image. In the regenerating aneurogenic limb, nAG is upregulated throughout the limb blastema. (From Kumar et al. 2011.)

As mentioned above, the regeneration of the salamander limb along the anterior-posterior axis appears to follow rules similar to those that generate the developing limb. Indeed, the opposing retinoic acid-FGF gradients postulated for the development of the limb were first hypothesized for the regeneration of limb structures along this axis (Crawford and Stocum 1988). It is known that the size and pattern of the regenerated limb depends on the proximal-to-distal position of the amputation such that the limb only generates tissues that had been distal to the cut, replacing them in the appropriate pattern. What factors beyond retinoic acid and FGFs are controlling this pattern formation? Many different signaling molecules have been implicated, including Wnts, BMPs, Hedgehogs, and Notch (reviewed in Satoh et al. 2015; Singh et al. 2015). But the story is not yet clear, and odd patterns of regeneration can result from applying some of these molecules. For example, if a regenerating salamander limb is exposed to retinoic acid, the blastema is reprogrammed to produce a full limb with all proximodistal structures, regardless of the position of the amputation (Maden 1983; Niazi et al. 1985; McCusker et al. 2014). In addition, if beads coated with BMP proteins or a combination of BMPs and FGFs are implanted in a wound site on a salamander limb, an accessory limb will form (see Figure 22.21E; Makanae et al. 2014).

WEB TOPIC 22.2   AXES OF REGENERATION   The phenomena of epimorphic regeneration can be seen formally as events that reestablish continuity among tissues that the amputation has severed. How does the amputated stump know to start at the cut site and not start making a whole new arm from the shoulder? How does it build the same dorsal-ventral axis as the stump?

Developing Questions

The limb will regenerate only those distal tissues removed, which suggests that key positional information is resident in the salamander limb to provide instructions for cell fate patterning during regeneration. Treatment with retinoic acid, use of an epidermal graft from a contralateral position, or BMP and FGF signals can induce different types of growth where it normally would not occur, however. What are the key components of positional information? How is this positional information able to be maintained and accessed during regeneration?

Luring the Mechanisms of Regeneration from Zebrafish Organs

So far in this chapter, we have discussed such diverse mechanisms of regeneration as morphallaxis in hydra, the planarian's deployment of pluripotent stem cells, and the elegant epimorphosis exhibited by the salamander's blastema. Next we will expand our understanding of these regenerative processes by looking at organ regeneration in zebrafish (*Danio rerio*). The zebrafish has increasingly been employed to study organ regeneration due to its regenerative competencies and its genetic and technical advantages. Most notable has been its use in studying the molecular regulation of regeneration in the fin, heart, central nervous system, eye, liver, pancreas, kidney, bone, and sensory hair cells (reviewed in Shi et al. 2015; Zhong et al. 2015). Below we highlight some of the insights gleaned from investigating zebrafish fin and heart regeneration.

WNT UPON A FIN The zebrafish caudal fin fans out along the dorsal-to-ventral axis with 16 to 18 segmented boney rays separated from one another by inter-ray tissue. The regenerative capacity of the fin is so robust and easily accessible that clipping caudal fins for molecular analysis is a routine procedure in most zebrafish research labs. The fin ray is primarily made of bone, but it also includes a diverse set of other cell types, including fibroblasts, blood vessels, nerves, and pigment cells. Like the salamander limb, upon amputation the zebrafish fin rays first close the wound with epidermal cells that form an apical epidermal cap, and most tissue types undergo dedifferentiation, proliferation, and distal migration to establish a blastema (Knopf et al. 2011; Stewart and Stankunas 2012). Some contributions may occur from as yet unidentified resident stem/progenitor cells capable of generating osteoblasts in the absence of any committed bone cells (Singh et al. 2012).

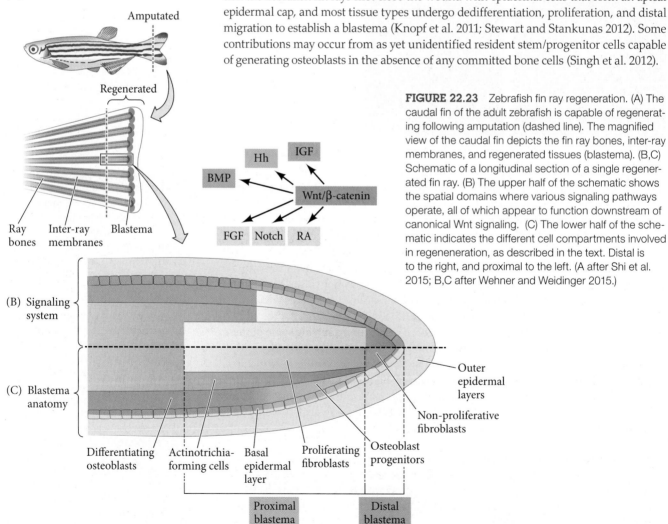

FIGURE 22.23 Zebrafish fin ray regeneration. (A) The caudal fin of the adult zebrafish is capable of regenerating following amputation (dashed line). The magnified view of the caudal fin depicts the fin ray bones, inter-ray membranes, and regenerated tissues (blastema). (B,C) Schematic of a longitudinal section of a single regenerated fin ray. (B) The upper half of the schematic shows the spatial domains where various signaling pathways operate, all of which appear to function downstream of canonical Wnt signaling. (C) The lower half of the schematic indicates the different cell compartments involved in regeneneration, as described in the text. Distal is to the right, and proximal to the left. (A after Shi et al. 2015; B,C after Wehner and Weidinger 2015.)

The regenerated fin can be divided into four basic sections (**FIGURE 22.23**): (1) the *distal blastema* made up of nonproliferative fibroblasts, (2) the *proliferative proximal blastema* that constitutes the major portion of the dedifferentiated mesenchyme, (3) the *differentiating proximal blastema* that functions to add differentiated cells to existing and newly formed tissues during outgrowth, and (4) the lateral-encompassing *epidermal layers* that serve as complex signaling centers throughout regeneration. Despite the seemingly simple distal outgrowth that typifies fin regeneration, the molecular regulation of this process is highly complex and involves all major signaling pathways known to play roles during embryonic development. We will focus on just one here, the Wnt/β-catenin signaling pathway, which appears to be the first "domino" that begins to reveal the regenerative pattern of fin rays (comprehensively reviewed in Wehner and Weidinger 2015). The Wnt/β-catenin pathway is active in the distal blastema and the lateralmost regions of the proximal proliferative blastema constituting progenitors of osteoblasts and actinotrichia (fibrous elements of the fin; see Figure 22.23). Loss and gain of function of Wnt/β-catenin signaling results in the decrease and increase of blastema proliferation and regeneration rate, respectively (Kawakami et al. 2006; Stoick-Cooper et al. 2007; Huang et al. 2009; Wehner et al. 2014). The Weidinger lab demonstrated that misexpression of the β-catenin inhibitor Axin1 throughout the fin blastema or just in the lateral progenitor zones resulted in significantly reduced fin rays as well as a failure of fin rays to ossify (**FIGURE 22.24**; Wehner et al. 2014). It appears, however, that Wnt/β-catenin functions indirectly through the modulation of other mitogenic regulators, such as Hedgehog proteins (Sonic hedgehog and Indian hedgehog), Fgf8, retinoic acid, and insulin-like growth factor (see Figure 22.23B).

EPIMORPHOSIS, COMPENSATION, AND TRANSDIFFERENTIATION: THE HEART OF THE MATTER The zebrafish has a relatively simple tubular heart. Venous blood enters the sinus venosus, passes into the single atrium, is pumped into the single ventricle, and leaves the heart through the bulbus arteriosis. Several different injury models—involving surgical removal of pieces of the myocardium, cryo-injury, and genetically induced tissue-specific ablation—have been adopted to study regeneration of heart tissues in zebrafish (reviewed in Shi et al. 2015). The zebrafish heart retains the ability to regenerate throughout the life of the fish, which is in part due to the sustained mitotic capacity of the cardiac myocytes (muscle cells) that constitute a majority of the heart tissue (Poss et al. 2002). It is clear that a major contribution to regeneration of the adult heart comes directly from preexisting cardiac myocytes. Indeed, use of the "Zebrabow" transgenic

FIGURE 22.24 Testing the spatiotemporal requirements of Wnt/β-catenin signaling during fin regeneration. (A) Experimental design using the Tet/On system for inducible gene expression. One transgenic fish possesses a tissue specific promoter (*ubiquitin* or *her4.3*), driving the expression of tetracycline and cyan fluorescent protein. A different transgenic fish has a transgene with the *tet* promoter driving Axin1-YFP expression; functional transcription of this transgene will only occur in combination with doxycycline (DOX). Crossing the two fish generates double transgenic individuals (represented by dots and stripes) that enable the spatial (tissue-specific promoter) and temporal (DOX) expression of Axin1. (B,C) Misexpression of Axin1 disrupts regeneration. The photos at left show the pattern of Axin1 misexpression (yellow) driven by the *ubiquitin* (B) or *her4.3* (C) promoter. The photos at right show the tail fin 12 days after amputation, with (+DOX) or without (–DOX) Axin1 misexpression. (B) When Wnt/β-catenin is inhibited by misexpressing Axin1 throughout the fin (*ubiquitin* promoter), there is a reduced rate of fin regeneration. (C) When Axin1 is misexpressed only in the osteoblast progenitor cells of the fin (*her4.3* promoter), ossification of the fin ray bones during regeneration is severely impaired (red stain, brackets). (A after Wehner et al. 2014; B,C from Wehner et al. 2014.)

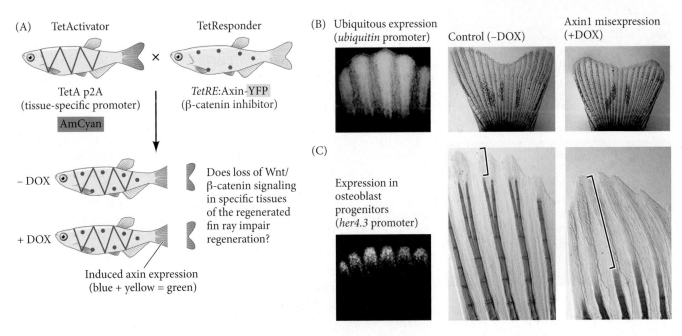

(A)

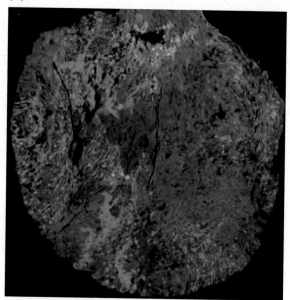

(B)

Control (uninjured)

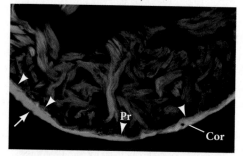

14 days postablation

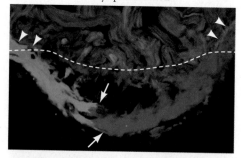

60 days postablation

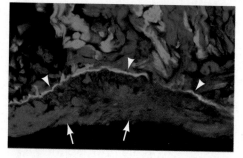

FIGURE 22.25 Preexisting cardiomyocytes contribute to ventricular regeneration in zebrafish. (A) Double-transgenic zebrafish were used to produce multicolor clonal labeling only in cardiomyocytes, as controlled by the *cmlc2* promoter for Cre expression. Recall that the "ER" in CreER denotes its estrogen-responsive control, which enables researchers to use the drug Tamoxifen to induce recombination at any time they desire. This image is of a 6-week-postfertilization heart ventricle following recombination at 4 days postfertilization. Patches of distinct colors can be seen, which indicates that the heart is derived from only several dozen cardiac progenitors. (B) These preexisting clonally labeled cardiac myocytes are seen contributing to a majority of the regenerated ventricular tissue. Arrowheads and arrows indicate the primordial (Pr) and cortical layers (Cor), respectively. (From Gupta and Poss 2012.)

system (see Chapter 2) for tracing the lineage of cells under the control of heart-specific promoters has demonstrated that previously differentiated cardiac myocytes give rise to clones of regenerated cells (**FIGURE 22.25**; Gupta and Poss 2012).

Work from many labs has shown that heart regeneration in the adult zebrafish is carried out primarily through the dedifferentiation of preexisting cardiomyocytes (epimorphosis), establishment of a blastema at the wound site through local proliferation and migration, and, finally, redifferentiation of blastema cells to repair the heart (Curado and Stainier 2006; Lepilina et al. 2006; Kikuchi et al. 2010; Zhang et al. 2013). In addition, it has been shown that healthy ventricular tissue far from the acute injury also responds by increasing proliferation (hyperplasia), which is a *compensatory* mechanism of regeneration (**FIGURE 22.26**; Poss et al. 2002; Sallin et al. 2015). Compensatory regeneration is often accompanied by hypertrophic growth (i.e., increase in cell size), but this has yet to be shown to occur in zebrafish heart regeneration

The developing zebrafish heart also appears to be capable of morphallaxis or transdifferentiation during regeneration. Researchers working with zebrafish larvae induced severe injury to the ventricular tissue of the larval heart by causing apoptosis of the ventricular cardiomyocytes (Zhang et al. 2013). They did so by targeting the expression of nitroreductase (NTR) in ventricular cardiomyocytes using the *ventricular myosin heavy chain* (*vmhc*) promoter and inducing cell death by administering an NTR-reactive cytotoxic prodrug. This procedure severely ablated ventricular tissue in this larval heart. What happened next was remarkable. Neighboring differentiated *atrial* cardiomyocytes responded to the injury by migrating into the damaged ventricular tissue and upregulating ventricle-specific genes such as *vmhc* (**FIGURE 22.27A**). Months later, fate mapping of these migrating atrial cardiomyocytes revealed that they remained in the ventricular wall, contributing to a fully regenerated and functioning ventricle and heart (**FIGURE 22.27B**). Zhang and colleagues went on to show that Notch-Delta signaling is highly upregulated in the atrial myocardium and is required for this atrial-mediated repair of ventricular damage (**FIGURE 22.27C**). Pharmacological inhibition of Notch signaling with DAPT exposure during ablation of ventricle tissue severely impairs heart regeneration (**FIGURE 22.27D**; Zhang et al. 2013). These results suggest that, at least during larval development, cardiac myocytes are capable of undergoing transdifferentiation to support regeneration of the heart. Thus, it appears the cells of the zebrafish heart employ multiple mechanisms to power its regeneration: blastema formation through epimorphosis, compensatory proliferation, and Notch-mediated transdifferentiation.

WATCH DEVELOPMENT 22.1   Videos from the laboratory of Dr. Neil Chi document the fascinating work being done on zebrafish heart regeneration.

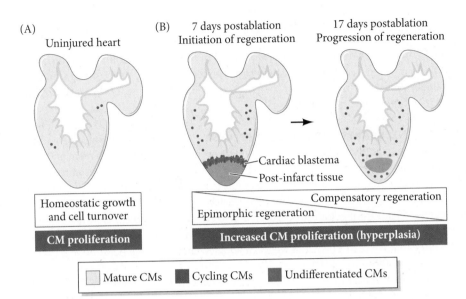

FIGURE 22.26 Epimorphosis and compensation in adult zebrafish heart regeneration. (A) Normal heart hemostasis with respect to the levels of growth and proliferation. (B) Regenerative response 7 and 17 days following ventricular injury. The regenerating heart first produces a focal blastema at the wound surface, fueled by epimorphic processes during which cardiomyocytes (CMs) dedifferentiate, proliferate, and redifferentiate into new ventricular cardiomyoctyes. Compensatory regeneration through increased levels of proliferation is triggered throughout the ventricular tissue, even in areas devoid of any injury. (After Sallin et al. 2015.)

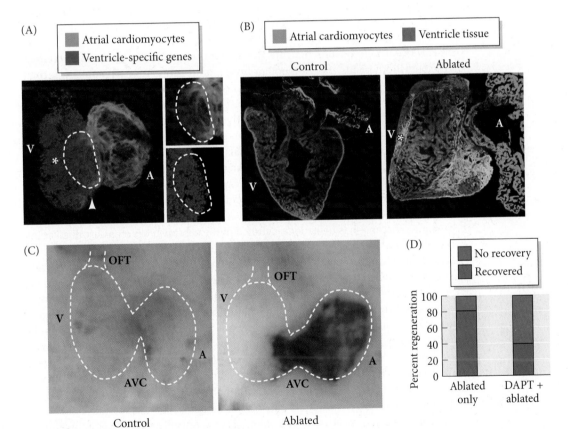

FIGURE 22.27 Transdifferentiation of atrial cardiomyocytes into ventricular cardiomyocytes during larval heart regeneration. (A) After researchers induced apoptosis of ventricular cardiomyocytes, fate mapping of differentiated atrial cardiomyocytes (green) shows migration from the atrium to the ablated ventricle tissue. At the transition point during this migration, the atrial cells start expressing ventricular markers, suggesting transdifferentiation (see Watch Development 22.1). (B) Twelve months following ablation of ventricular tissue, infiltrating atrial cardiomyocytes (green) fully differentiate into the ventricular tissue (red) and contribute to a functional adult heart. (C) Notch-Delta signaling is required for the successful regenerative contributions from atrial cardiomyocytes. Twenty-four hours after ventricular cell ablation (right photograph), *deltaD* along with other related genes (blue stain) are highly upregulated in the atrial cardiomyocytes, particularly in those migrating toward the ventricular tissue. (D) Pharmacological inhibition of Notch signaling with DAPT exposure during ablation of ventricle tissue severely impairs heart regeneration. A, atrium; V, ventricle; AVC, atrioventricular canal; OFT, outflow tract. (From Zhang et al. 2013.)

Regeneration in Mammals

Although mammals do not have the same level of regenerative abilities as the other organisms covered in this chapter, they can regenerate certain structures. Mammals operate less on the principle of "start again from scratch" than from the premise that "if you can't remake it, make it bigger."

FINGERS AND HEARTS ARE NOT SO DIFFERENT Mammals such as rodents and humans have been shown to regenerate the tips of their digits if the organism is young enough. First, a regeneration blastema composed of progenitor cells forms at the tip of the digit (Fernando et al. 2011). As in salamander limb regeneration, respecification does not occur (Lehoczky et al. 2011; Rinkevich et al. 2011). Instead, the new epidermis is derived from ectoderm-restricted progenitor cells, and the new bone comes from osteoblast progenitor cells. This similarity to salamander limb regeneration provides hope that we can apply what we learn about salamander regeneration to enhance regenerative abilities in humans.

It is interesting that heart tissue can also regenerate in mice, but only within the first week of neonatal life. After that, the ability is lost, presumably because the cardiomyocytes withdraw from the cell cycle (Porrello et al. 2011). It is known, though, that cardiomyocytes in adult mammals (as in adult zebrafish) will respond to a heart attack by re-entering the cell cycle, thus contributing to injury repair (Senyo et al. 2013).

COMPENSATORY REGENERATION IN THE MAMMALIAN LIVER According to Greek mythology, Prometheus' punishment for bringing the gift of fire to humans was to be chained to a rock and have an eagle tear out and eat a portion of his liver each day. His liver then regenerated each night, providing a continuous food supply for the eagle and eternal punishment for Prometheus. Today the standard assay for liver regeneration is a partial hepatectomy, where specific lobes of the liver are removed (*after* anesthesia is administered, unlike the fate of Prometheus), leaving the other hepatic lobes intact. Although the removed lobe does not grow back, the remaining lobes enlarge to compensate for the loss of the missing tissue (Higgins and Anderson 1931). The amount of liver regenerated is equivalent to the amount of liver removed. Such **compensatory regeneration**—the division of differentiated cells to recover the structure and function of an injured organ—has been demonstrated in the zebrafish heart as described above and in the mammalian liver.

The human liver regenerates by the proliferation of existing tissue. Surprisingly, the regenerating liver cells do not fully dedifferentiate when they reenter the cell cycle. No regeneration blastema is formed. Rather, mammalian liver regeneration appears to have two other lines of defense, the first of which consists of normal, mature, adult hepatocytes. These mature cells, which are usually not dividing, are instructed to rejoin the cell cycle and proliferate until they have compensated for the missing part. The second line of defense, discussed below, is a population of hepatic progenitor cells that are normally quiescent but that are activated when the injury is severe and adult hepatocytes cannot regenerate well due to senescence, alcohol abuse, or disease.

In normal liver regeneration, the five types of liver cells—hepatocytes, duct cells, fat-storing (Ito) cells, endothelial cells, and Kupffer macrophages—all begin dividing to produce more of themselves. Each type retains its cellular identity, and the liver retains its ability to synthesize the liver-specific enzymes necessary for glucose regulation, toxin degradation, bile synthesis, albumin production, and other hepatic functions even as it regenerates itself (Michalopoulos and DeFrances 1997).

There are probably several redundant pathways that initiate liver cell proliferation and regeneration (**FIGURE 22.28**; Riehle et al. 2011). Global gene profiling indicates that the end result of these pathways is to downregulate (but not totally suppress) the genes involved in the differentiated functions of liver cells while activating those genes committing the cell to mitosis (White et al. 2005). The removal or injury of the liver is sensed through the bloodstream: some liver-specific factors are lost while others (such as bile acids and gut lipopolysaccharides) increase. These lipopolysaccharides activate some non-hepatocytes to secrete paracrine factors that allow the remaining hepatocytes to re-enter the cell cycle. The Kupffer cell secretes interleukin 6 (IL6) and tumor necrosis factor-α (which are usually involved in activating the adult immune system), and the stellate cells secrete the paracrine factors **hepatocyte growth factor** (**HGF**, or **scatter factor**) and TGF-β. The specialized blood vessels of the liver also produce HGF as well as Wnt2 (Ding et al. 2010).

Hepatocytes that are still connected to one another in an epithelium cannot respond to HGF, however. Hepatocytes activate cMet (the receptor for HGF) within an hour of partial hepatectomy, and the blocking of cMet (by RNA interference or knockout) blocks liver regeneration (Borowiak et al. 2004; Huh et al. 2004; Paranjpe et al. 2007). The trauma of partial hepatectomy may activate metalloproteinases that digest the extracellular matrix and permit the hepatocytes to separate and proliferate. These enzymes also may cleave HGF to its active form (Mars et al. 1995). Together, the factors produced by the endothelial cells, Kupffer cells, and stellate cells allow the hepatocytes to divide by preventing apoptosis, activating cyclins D and E, and repressing cyclin inhibitors such as p27 (see Taub 2004).

The liver stops growing when it reaches the appropriate size; the mechanism for how this is achieved is not yet known. One clue, though, comes from parabiosis experiments, in which the circulatory systems of two rats are surgically joined together. Partial hepatectomy in one parabiosed rat causes the other rat's liver to enlarge (Moolten and Bucher 1967). Therefore, some factor or factors in the blood appear to be establishing the size of the liver. Huang and colleagues (2006) have proposed that these factors are bile acids that are secreted by the liver and positively regulate hepatocyte growth. Partial hepatectomy stimulates the release of bile acids into the blood. These bile acids are received by the hepatocytes and activate the Fxr transcription factor, which promotes cell division. Mice without functional Fxr protein cannot regenerate their livers. Therefore, bile acids (a relatively small percentage of the products secreted by the liver) appear to regulate the size of the liver, keeping it at a particular volume of cells. The molecular mechanisms by which these factors interact and by which the liver is first told to begin regenerating and then to stop regenerating after reaching the appropriate size remain to be discovered.

Because human livers have the power to regenerate, a patient's diseased liver can be replaced by compatible liver tissue from a living donor (usually a genetically close relative, whose own liver grows back). Human livers regenerate more slowly than those of mice, but function is restored quickly (Pascher et al. 2002; Olthoff 2003). In addition, mammalian livers possess a "second line" of regenerative ability. If the hepatocytes are unable to regenerate the liver sufficiently within a certain amount of time, the **oval cells** divide to form new hepatocytes. Oval cells are a small progenitor cell population that can produce hepatocytes and bile duct cells. They appear to be kept in reserve and are used only *after* the hepatocytes have attempted to heal the liver (Fausto and Campbell 2005; Knight et al. 2005).

(A) DNA synthesis

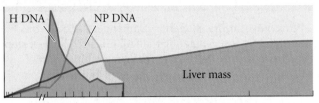

(B) Growth-regulated genes

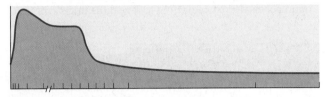

(C) Cell cycle-regulated genes

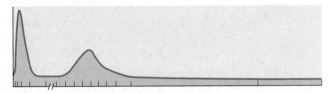

(D) Gene expression after growth phase

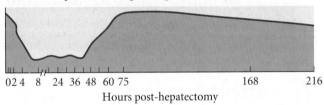

FIGURE 22.28 Correlation of changes in gene expression with increases in liver mass following partial hepatectomy in mammals. (A) Initial peaks in DNA synthesis are seen in both hepatocytes (H DNA; green) and thereafter in nonparenchymal cells (NP DNA; yellow). This burst in DNA synthesis corresponds with the upregulation of growth-regulated (B) and cell cycle-regulated (C) gene expression, both of which taper off as the liver mass (brown shading in panel A) reaches its normal volume. (D) Overall gene expression remains high after the growth phase, reflecting the functionality of the regenerated liver tissue. (After Taub 2004.)

Next Step Investigation

What makes hydra, planaria, salamanders, and fish so much better at regenerating than mammals? The next step in investigating regeneration could be nothing short of harnessing those unique capabilities to induce regeneration in mammalian tissues. In this chapter, we highlighted several studies taking a whole-transcriptome approach to identifying genes essential for regeneration. Those approaches are just beginning to reveal the important players. It is also clear that regeneration is often a localized redeployment of embryonic development. Why is this return to developmental states more challenging in mammals? Continuing to identify what makes regeneration possible in these different model organisms will help us gain a better understanding of what is possible as well as the evolutionary advantages and disadvantages to regeneration.

Closing Thoughts on the Opening Photo

After reading this chapter, you should be able to identify this image as an ectopic or accessory salamander limb. This image was produced by the Gardiner lab after researchers conducted the same type of experiment described in the chapter, in which a nerve is diverted to a wounded area in the anterior epidermis and posterior epidermis is grafted to the same area. "Mind over body?" Although this regeneration was obviously not a brain-originating thought, it did take "some nerve." More specifically, it took regenerating nerves with affiliated Schwann cells that expressed the newt anterior growth factor required to initiate limb regeneration. Even in the absence of nerves, this factor can promote epimorphic regeneration. It is currently unknown whether this or other similar factors can function in mammals to stimulate limb regeneration, but the excitement to learn about such potential regeneration-promoting factors is most definitely real. (Photograph from C. McCusker and D. M. Gardiner 2011.)

22 Snapshot Summary
Regeneration

1. There are four major types of regeneration. In stem-cell mediated regeneration (such as planarian regeneration), new cells are routinely produced to replace the ones that die. In epimorphosis (such as regenerating salamander limbs and fish fins), tissues form into a regeneration blastema, divide, and redifferentiate into the new structure. In morphallaxis (characteristic of hydra), there is a repatterning of existing tissue with little or no growth. In compensatory regeneration (such as in the mammalian liver), cells divide but retain their differentiated state.

2. Hydra appears to have a head activation gradient, a head inhibition gradient, a foot activation gradient, and a foot inhibition gradient. Budding occurs where these gradients are minimal.

3. The hypostome region of hydra appears to be an organizer region that secretes paracrine factors (Wnt3) to alter the fates of surrounding tissue through an epimorphic mechanism.

4. In planarian flatworms, regeneration occurs by forming a regeneration blastema produced by pluripotent clonogenic neoblasts. Wnt signaling gradients appear to direct the anterior-posterior differentiation of these cells in a pattern regulated by the head-expressed Wnt inhibitor Notum.

5. In the regenerating limb blastemas of amphibians, cells do not become multipotent. Rather, cells retain their specifications, with bone arising from preexisting cartilage, neurons coming from preexisting neurons, and muscles coming either from preexisting muscle cells or from muscle stem cells.

6. Mitogens such as nAG are provided by the apical epidermal cap and by the glia surrounding the limb axons, which are capable of inducing regeneration of limbs even in the absence of nerves. Salamander limb regeneration appears to use the same pattern formation system as the developing limb.

7. Multiple modes of regeneration have been discovered to operate in zebrafish. Distal regeneration of the zebrafish fin occurs largely through the dedifferentiation of existing cell types followed by the active proliferation of a blastema-like outgrowth. Zebrafish heart tissue also employs an initial mode of epimorphsis followed by a compensatory regenerative period of proliferation.

8. In the mammalian liver, no regenerating blastema is formed, and the liver regenerates the same volume as it lost. Each cell appears to generate its own cell type. A reserve population of multipotent progenitor cells divides when these tissues cannot regenerate the missing portions.

Further Reading

Bode, H. R. 2012. The head organizer in *Hydra*. *Int. J. Dev. Biol.* 56: 473–478.

Gee, L., J. Hartig, L. Law, J. Wittlie, K. Khalturin, T. C. Bosch and H. R. Bode. 2010. β-Catenin plays a central role in setting up the head organizer in hydra. *Dev. Biol.* 340: 116–124.

Gupta, V. and K. D. Poss. 2012. Clonally dominant cardiomyocytes direct heart morphogenesis. *Nature* 484: 479–84.

Hill, E. M. and C. P. Petersen. 2015. Wnt/Notum spatial feedback inhibition controls neoblast differentiation to regulate reversible growth of the planarian brain. *Development* 142: 4217–4229.

Kragl, M., D. Knapp, E. Nacu, S. Khattak, M. Maden, H. H. Epperlein and E. M. Tanaka. 2009. Cells keep a memory of their tissue origin during axolotl limb regeneration. *Nature* 460: 60–65.

Kumar, A., J. W. Godwin, P. B. Gates, A. A. Garza-Garcia and J. P. Brockes. 2007. Molecular basis for the nerve dependence of limb regeneration in an adult vertebrate. *Science* 318: 772–777.

Kumar, A., J. P. Delgado, P. B. Gates, G. Neville, A. Forge and J. P. Brockes. 2011. The aneurogenic limb identifies developmental cell interactions underlying vertebrate limb regeneration. *Proc. Natl. Acad. Sci. USA* 108: 13588–13593.

Li, Q., H. Yang and T. P. Zhong. 2015. Regeneration across metazoan phylogeny: Lessons from model organisms. *J. Genet. Genomics* 42: 57–70.

Liu, S. Y. and 9 others. 2013. Reactivating head regrowth in a regeneration-deficient planarian species. *Nature* 500: 81–84.

McCusker, C. D. and D. M. Gardiner. 2014. Understanding positional cues in salamander limb regeneration: Implications for optimizing cell-based regenerative therapies. *Dis. Model Mech.* 7: 593–599.

McCusker, C., J. Lehrberg and D. M. Gardiner. 2015. Position-specific induction of ectopic limbs in non-regenerating blastemas on axolotl forelimbs. *Regeneration* 1: 27–34.

Sikes, J. M. and P. A. Newmark. 2013. Restoration of anterior regeneration in a planarian with limited regenerative ability. *Nature* 500: 77–80.

Taub, R. 2004. Liver regeneration: From myth to mechanism. *Nature Rev. Mol. Cell Biol.* 5: 836–847.

Umesono, Y. and 9 others. 2013. The molecular logic for planarian regeneration along the anterior-posterior axis. *Nature* 500: 73–76.

van Wolfswinkel, J. C., D. E. Wagner and P. W. Reddien. 2014. Single-cell analysis reveals functionally distinct classes within the planarian stem cell compartment. *Cell Stem Cell* 15: 326–339.

Wagner, D. E., I. E. Wang and P. W. Reddien. 2011. Clonogenic neoblasts are pluripotent adult stem cells that underlie planarian regeneration. *Science* 332: 811e816.

Wagner, D. E., J. J. Ho and P. W. Reddien. 2012. Genetic regulators of a pluripotent adult stem cell system in planarians identified by RNAi and clonal analysis. *Cell Stem Cell* 10: 299–311.

Wehner, D. and 11 others. 2014. Wnt/β-catenin signaling defines organizing centers that orchestrate growth and differentiation of the regenerating zebrafish caudal fin. *Cell. Rep.* 6: 467–481.

Wurtzel, O., L. E. Cote, A. Poirier, R. Satija, A. Regev and P. W. Reddien. 2015. A generic and cell-type-specific wound response precedes regeneration in planarians. *Dev. Cell* 35: 632–645.

Zhang, R. and 11 others. 2013. In vivo cardiac reprogramming contributes to zebrafish heart regeneration. *Nature* 498: 497–501.

GO TO WWW.DEVBIO.COM...

...for Web Topics, Scientists Speak interviews, Watch Development videos, Dev Tutorials, and complete bibliographic information for all literature cited in this chapter.

Aging and Senescence

ENTROPY ALWAYS WINS. A multicellular organism is able to develop and maintain its identity for only so long before deterioration prevails over synthesis and the organism ages. **Aging** can be defined as the time-related deterioration of the physiological functions necessary for survival and fertility. The characteristics of aging—as distinguished from diseases of aging, such as cancer and heart disease—affect all the individuals of a species. The aging process has two major facets. The first is simply how long an organism lives; the second concerns the physiological deterioration, or **senescence**, that characterizes old age. These topics are often viewed as being interrelated. Both aging and senescence have genetic and environmental components, and so far there is no unified theory of aging that puts them all together. As one recent review (Underwood 2015) noted, "In the race to find a biological clock, there are plenty of contenders."

Genes and Aging

"Death," according to Steve Jobs (2005), "is very likely the single best invention of life." But it is obvious that death occurs differently in different organisms and that these differences are inherited. A mouse can live for 3 years; humans can live for decades. The **maximum lifespan** is the maximum number of years an individual of a given species has been known to survive and is characteristic of that species (Coles 2004). As of 2013, the maximum verified human life span stood at 122.5 years. The life spans of some tortoises and lake trout are uncertain but are estimated to extend beyond 150 years. The maximum life span of a domestic dog is about 20 years, and that of a laboratory mouse is 4.5 years; most mice in the wild do not live to celebrate even their first birthday. If a fruit fly survives to eclose (in the wild, more than 90% die as larvae), it has a maximum life span of 3 months.

How might this jellyfish circumvent death and become potentially immortal?

The Punchline

There is no single hypothesis that explains aging and senescence. Mutations in genes encoding DNA repair enzymes and factors regulating metabolic rates may be crucial in aging, and the insulin signaling cascade (involving the interplay of metabolism and diet) mediates between fertility and aging. In addition, random epigenetic changes in DNA and chromatin can inactivate genes as one ages. New research suggests that stem cells and their niches are critical for normal organismal function and that as mutations and epigenetic changes accumulate, these stem cells may die or become nonfunctional. As cells become senescent, they have been seen to secrete paracrine factors that mimic inflammatory responses and suppress organ function.

Genetic factors play roles in determining longevity both between and within species (Wilson et al. 2007). Most people cannot expect to live 122 years. **Life expectancy**—the average length of time a given individual of a given species can expect to live—is not characteristic of species but of populations. It is sometimes defined as the age at which half the population still survives. A baby born in England during the 1780s could expect to live to be 35 years old. In Massachusetts during that same time, life expectancy was 28 years. These ages represent the normal range of human life expectancy for most of the human race throughout recorded history (Arking 1998). Even today, in some countries (Angola, Chad, Lesetho, and several others) life expectancy is only around 45 years. Males in the United States today have a life expectancy of about 76 years, and females can expect to live around 81 years.[1]

Given that in most times and places people did not live much past the age of 40, our awareness of human aging is relatively new. In 1900, 50% of Americans were dead before the age of 60; a 70-year-old person was exceptional in 1900 but is commonplace today. People in 1900 did not have the "luxury" of dying from heart attacks or cancers, because these conditions are most likely to affect people over 50. Rather, many people died (as they are still dying in large parts of the world) from microbial and viral infections. Until recently, relatively few people exhibited the general human senescent phenotype: gray hair, sagging and wrinkling skin, arthritic joints, osteoporosis (loss of bone calcium), loss of muscle fibers and muscular strength, memory loss, eyesight deterioration, and slowed sexual responsiveness. As the melancholy Jacques notes in Shakespeare's *As You Like It*, those who did survive to senescence left the world "*sans* teeth, *sans* eyes, *sans* taste, *sans* everything."

Species-specific life spans appear to be determined by genes that effect a trade-off between the energy used for early growth and reproduction (which results in somatic damage) versus the energy allocated for maintenance and repair. In other words, aging results from natural selection operating more strongly on early survival and reproduction than on having a vigorous postreproductive life. Molecular evidence indicates that certain genetic components of longevity are conserved between species—flies, worms, mammals, and even yeast all appear to use the same set of genes to promote survival and longevity (see Vijg and Campisi 2008; Kenyon 2010). Four sets of genes are well known to be involved in aging and its prevention, and each set appears to be conserved between phyla and even kingdoms. These are the genes encoding (1) DNA repair enzymes, (2) proteins of the insulin signaling pathway, (3) proteins in the mTORC1 signaling pathway (a cascade that regulates translation), and (4) chromatin remodeling enzymes.

DNA repair enzymes

DNA repair enzymes appear be critically important in preventing senescence (Gorbunova et al. 2007). Individuals of species whose cells have more efficient DNA repair enzymes live longer (**FIGURE 23.1**; Hart and Setlow 1974). Certain premature aging syndromes, called **progerias**, in humans and mice appear to be caused by mutations that prevent the functioning of DNA repair enzymes (**FIGURE 23.1B**; Sun et al. 1998; Shen and Loeb 2001; de Boer et al. 2002).

"Wear-and-tear" theories of aging are among the oldest hypotheses proposed to account for the human senescent phenotype (Weismann 1891; Medawar 1952). As one gets older, small traumas to the body and its genome build up. At the molecular level, the number of point mutations increases with age, and the efficiency of the enzymes encoded by our genes decreases (Singh et al. 2001; Bailey et al. 2004; Rossi et al. 2007). Moreover, if mutations occur in the genes encoding transcriptional or translational

---

[1] When Social Security was enacted in the United States in 1935, the average working citizen died before age 65. Thus, he (and it usually was a he) was not expected to get back as much as he had paid into the system. Similarly, marriage "until death do us part" was easier to achieve when death occurred in the third or fourth decade of life. Before antibiotics, the death rate of young women due to infections associated with childbirth was high throughout the world.

(A)

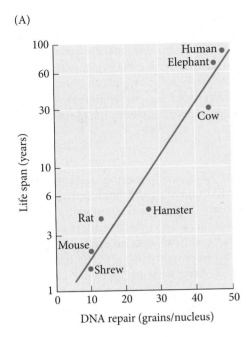

(B)

FIGURE 23.1 Life span and the aging phenotype. (A) Correlation between life span and the ability of fibroblasts to repair DNA in various mammalian species. Repair capacity is represented in autoradiography by the number of grains from radioactive thymidine per cell nucleus. Note that the *y* axis (life span) is logarithmic. (B) Hutchinson-Gilford progeria. Although they are not yet 8 years old, these children have a phenotype similar to that of an aged person. The hair loss, fat distribution, and skin transparency are characteristic of the normal aging pattern as seen in elderly adults. The mutation causes an aberrant nuclear envelope protein that appears to prevent DNA repair (Coppedè and Migliore 2010). (A after Hart and Setlow 1974; B © Associated Press.)

proteins, the cell may make an even greater number of faulty proteins (Orgel 1963; Murray and Holliday 1981; Kamileri et al. 2012).

REACTIVE OXYGEN SPECIES Two major sources of mutation are radiation and **reactive oxygen species (ROS)**. The ROS produced by normal metabolism can oxidize and damage cell membranes, proteins, and nucleic acids. Some 2–3% of the oxygen atoms taken up by our mitochondria are reduced insufficiently and form ROS: superoxide ions, hydroxyl ("free") radicals, and hydrogen peroxide. Evidence that ROS molecules are critical in the aging process includes observations that fruit flies and nematodes overexpressing the enzymes that destroy ROS (catalase and superoxide dismutase) live significantly longer than do control animals (Orr and Sohal 1994; Parkes et al. 1998; Sun and Tower 1999; Feng et al. 2001). However, these correlations have not held up in some other studies, so the genetic ability to destroy free oxygen radicals may not be essential for a long life (Pérez et al. 2009; van Raamsdonk and Hekimi 2012).

TELOMERASE AND P53 The transcription factor **p53** is one of the most important regulators of cell division. This factor can stop the cell cycle, cause cellular senescence in rapidly dividing cells, instruct genes to initiate cellular apoptosis, and activate DNA repair enzymes. In most cells, p53 is bound to a repressor protein that keeps p53 inactive. However, ultraviolet radiation, oxidative stress, and other factors that cause DNA damage will separate p53 from its repressor, allowing it to function. The induction of apoptosis or cellular senescence by p53 can be beneficial (when destroying cancer cells) or deleterious (when destroying, say, neurons or stem cells).

One of the chief ways of activating p53 (and related proteins such as p63) is to damage the **telomeres**, the protective nucleoprotein caps on the tips of the chromosomes (similar to the way aglets on the tips of shoelaces keep them from unwinding). When p53 is activated by damaged telomeres, DNA replication halts, and if the repair doesn't work, apoptosis is initiated. If the cell is a stem cell or some other rapidly replicating cell, this will reduce the numbers of cells produced, and the lack of stem cells will produce an "aged" phenotype. The relationship between shortened telomeres and stem cell depletion has been seen in degenerative diseases such as mouse muscular dystrophy (Sacco et al. 2010).

There is a positive correlation between telomere length and longevity in humans (Atzmon et al. 2010), and telomeres appear to shorten with age in the stem cell compartments of mice and humans (Zhang and Ju 2010). The enzyme complex that maintains telomere integrity is **telomerase**, which acts as an antisenescence complex. Mice and humans with telomerase deficiencies age prematurely (Mitchell et al. 1999). Overexpressing telomerase or reactivating it in senescent cells extends longevity in mice without increasing cancer (Tomás-Loba et al. 2008; Jaskelioff et al. 2011; Bernardes de Jesus et al. 2012). However, except for the rare genetic telomere syndromes, telomere length only gives a statistical probability of age; it does not predict the lifespan of an individual (Blackburn et al. 2015).

Aging and the insulin signaling cascade

One criticism of the idea that there are genetic "programs" for aging asks how evolution could have selected for them. Once an organism has passed reproductive age and raised its offspring to sexual maturity, it becomes "an excrescence on the tree of life" (Rostand 1962); natural selection presumably cannot act on traits that affect an organism only after it has reproduced. But "How can evolution select for a way and time to degenerate?" may be the wrong question. Evolution probably can't select for such traits. The right question may be, "How can evolution select for phenotypes that postpone reproduction or sexual maturity?" There is often a trade-off between reproduction and maintenance, and in many species reproduction and senescence are closely linked.

Recent studies of mice, *Caenorhabditis elegans*, and *Drosophila* suggest that there is a conserved genetic pathway that regulates aging, and that it can indeed be selected for. This pathway involves the response to insulin and insulin-like growth factors. In *C. elegans*, a larva proceeds through four larval stages, after which it becomes an adult. If the nematodes are overcrowded or if there is insufficient food, the larva can enter a metabolically dormant **dauer larva** stage, a nonfeeding state of **diapause**, a condition in which development and aging are suspended. The nematode can remain in the dauer stage for up to 6 months (rather than becoming an adult that lives only a few weeks). In this state it has increased resistance to oxygen radicals that can crosslink proteins and destroy DNA. The pathway that regulates both dauer larva formation and longevity has been identified as the **insulin signaling pathway** (Kimura et al. 1997; Guarente and Kenyon 2000; Gerisch et al. 2001; Pierce et al. 2001).

In *C. elegans*, favorable environments signal activation of the insulin receptor homologue DAF-2, and this receptor stimulates the onset of adulthood (**FIGURE 23.2A**). Poor environments fail to activate the DAF-2 receptor, and dauer formation ensues. While severe loss-of-function alleles in the insulin signaling pathway cause the formation of dauer larvae in any environment, weak mutations in the pathway enable the animals to reach adulthood and live four times longer than wild-type animals.

Downregulation of the insulin signaling pathway has several other functions. First, it appears to influence metabolism, decreasing mitochondrial electron transport. Second, when the DAF-2 receptor is not active, cells increase the production of enzymes that prevent oxidative damage, as well as DNA repair enzymes (Honda and Honda 1999; Tran et al. 2002). Third, this lack of insulin signaling decreases fertility (Gems et al. 1998). This increase in DNA synthetic enzymes and in enzymes that protect against ROS is due to the Foxo/DAF-16 transcription factor. This Forkhead-type transcription factor is inhibited by the insulin receptor (DAF-2) signal. When that signal is absent, Foxo/DAF-16 can function, and this factor promotes longevity in ways not yet deciphered. It is possible that Foxo/DAF-16 activates the expression of genes involved in producing anti-stress proteins within the cell as well as lipid signals that help extend life to those cells nearby (Zhang et al. 2013). The Foxo transcription factor has been associated with longevity throughout the animal kingdom. Indeed, it has recently been shown to be one of the major drivers of stem cell renewal in potentially immortal hydras (Boehm et al. 2012).

It is possible that this system also operates in mammals, but the mammalian insulin and insulin-like growth factor pathways are so integrated with embryonic development and adult metabolism that mutations often have numerous and deleterious effects (such as diabetes or Donahue syndrome). However, there is evidence that the insulin

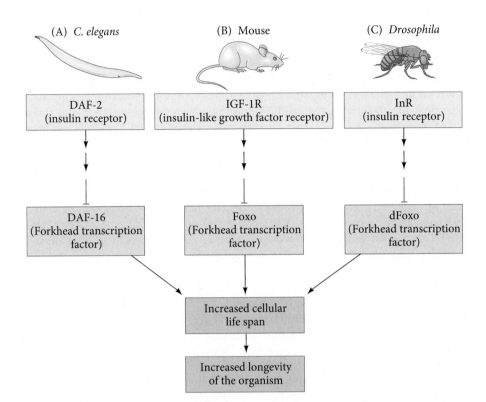

(A) *C. elegans*

(B) Mouse

(C) *Drosophila*

| DAF-2 (insulin receptor) | IGF-1R (insulin-like growth factor receptor) | InR (insulin receptor) |

DAF-16 (Forkhead transcription factor)

Foxo (Forkhead transcription factor)

dFoxo (Forkhead transcription factor)

Increased cellular life span

Increased longevity of the organism

FIGURE 23.2 A possible pathway for regulating longevity. In each case, the insulin signaling pathway inhibits the synthesis of the Foxo transcription factor proteins that would otherwise increase cellular longevity.

signaling pathway does affect life span in mammals (**FIGURE 23.2B**). Dog breeds with low levels of insulin-like growth factor 1 (IGF-1) live longer than breeds with higher levels of this factor. Mice with loss-of-function mutations of the insulin signaling pathway live longer than their wild-type littermates (see Partridge and Gems 2002; Blüher et al. 2003; Kurosu et al. 2005). Holzenberger and colleagues (2003) found that mice heterozygous for the insulin receptor IGF-1R not only lived about 30% longer than their wild-type littermates, they also had greater resistance to oxidative stress. In addition, mice lacking one copy of their IGF-1R gene lived about 25% longer than wild-type mice.

The insulin signaling pathway also appears to regulate life span in *Drosophila* (**FIGURE 23.2C**). Flies with weak loss-of-function mutations of the insulin receptor gene or genes in the insulin signaling pathway live nearly 85% longer than wild-type flies (Clancy et al. 2001; Tatar et al. 2001). These long-lived mutants are sterile, and their metabolism resembles that of flies that are in diapause (Kenyon 2001). The insulin receptor in *Drosophila* is thought to regulate a Forkhead transcription factor (dFoxo) similar to the Foxo/DAF-16 protein of *C. elegans*. When the *Drosophila dFoxo* gene is activated in the fat body, it can lengthen the fly's life span (Giannakou et al. 2004; Hwang-bo et al. 2004). From an evolutionary point of view, the insulin pathway may mediate a trade-off between reproduction and survival/maintenance. Many (although not all) of the long-lived mutants have reduced fertility. Thus, it is interesting that another longevity signal originates in the gonad. When the germline cells are removed from *C. elegans*, the worms live longer. The germline stem cells produce a substance that blocks the effects of a longevity-inducing steroid hormone (Hsin and Kenyon 1999; Gerisch et al. 2001; Shen et al. 2012).

Calorie restriction is another way of downregulating the insulin pathway (Kenyon 2001; Roth et al. 2002; Holzenberger et al. 2003). Calorie restriction may reduce levels of IGF-1 (the main ligand of IGF-1R) and of circulating insulin, although other mechanisms are also being explored (e.g., Selman et al. 2009). Studies in primates (including humans) have not concluded that low calorie intake extends their longevity, although it does appear to retard the age-associated decline of heartbeat variability and motor coordination (see Colman et al. 2009; Mattison et al. 2012; Stein et al. 2012).

The mTORC1 pathway

One of the main ways by which the insulin signaling pathway might function to lower longevity is to activate mTORC1, a protein kinase complex that promotes the translation of mRNA into proteins in response to nutrients and hormones (Lamming et al. 2012; Johnson et al. 2013). Thus, the insulin signaling pathway depresses Foxo and at the same time activates mTORC1. Dietary restriction reduces mTORC1 activity, and mice with reduced mTORC1 levels had longer lives, better protection against age-related cognitive dysfunction, and more functional stem cells than control mice (Chen et al. 2009; Harrison 2009; Halloran et al. 2012; Majumder et al. 2012; Yilmaz et al. 2012). Reducing mTORC1 also increases the amount of **autophagy**, the removal and replacement of damaged organelles and senescent cells. Many of the maladies associated with old age appear to be the result of failed autophagy and replacement (Baker et al. 2011). The mechanisms by which reduced mTORC1 accomplishes these feats are still unknown, and this pathway is an area of active study.

Chromatin modification

Chromatin modification may be very important in aging. The **sirtuin** genes, which encode histone deacetylation (chromatin-silencing) enzymes, have been found to prevent aging throughout the eukaryotic kingdoms, including in yeasts and mammals (Howitz et al. 2003; Oberdoerffer et al. 2008). Sirtuins prevent genes from being expressed at the wrong times and places, and they help repair chromatin breaks. When DNA strands break (as inevitably happens as the body ages), sirtuin proteins are called on to fix them and cannot attend to their usual functions. Thus, genes that are usually silenced become active as the cells age.

Alternatively, there are other areas of the body, such as the brain, where histone deacetylases can generate an aging phenotype. Cognitive decline, especially in the ability to recall past experiences, is a normal part of the mammalian aging syndrome. Long-term memories are stabilized by chromatin remodeling in the hippocampus and frontal lobes of the brain, a process involving DNA methylation and histone modifications (Swank et al. 2001; Korzus 2004; Miller et al. 2008; Penner et al. 2011). Peleg and colleagues (2010) have shown that the normal transcription associated with long-term memory stabilization is disrupted as mice age, and that this lack of transcription is associated with lessened H4K12 acetylation. Indeed, this ability to store memory can be retrieved by infusing into the hippocampus an inhibitor of histone deacetylase (**FIGURE 23.3**).

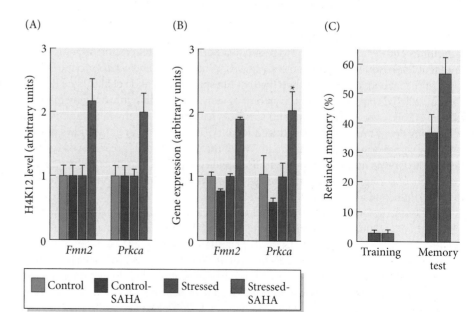

FIGURE 23.3 Age-related memory decline in mice can be reversed by inhibitors of histone deacetylases. Mice were either unstressed (control) or stressed to form a new memory. (A) H4K12 identified by chromatin immunoprecipitation (ChIP) assays in the coding regions of three genes. The stressed mice treated with the inhibitor of histone deacetylase (SAHA) had the highest level of H4K12. (B) The stressed mice treated with SAHA also had the highest levels of expression of *Fmn2* and *Prkca*, two genes that have been associated with memory formation. (C) Mice that were stressed stabilized a memory of this stress better if they had been treated with the inhibitor of histone deacetylases. (After Peleg et al. 2010.)

Random Epigenetic Drift

The idea that random epigenetic drift inactivates important genes without any particular environmental cue gives rise to an entirely new hypothesis of aging. Instead of randomly accumulated mutations—which might be due to specific mutagens—we are at the mercy of chance accumulations of errors made by the DNA-methylating and DNA-demethylating enzymes. Indeed, unlike the DNA polymerases, our DNA-methylating enzymes are prone to errors. At each round of DNA replication, DNA methyltransferases must methylate the appropriate cytosines while leaving other cytosines unmethylated, and they are not the most fastidious of enzymes, making errors at the rate of about 2–4% (Ushijima et al. 2005). Within certain genetic parameters (which may affect the speed at which methylation changes occur, and which may differ between species and individuals), our cells may accumulate errors of gene expression throughout our lives.

Random epigenetic drift may have profound effects on our physiology. For instance, methylation of the promoter regions of the α and β estrogen receptors is known to increase linearly with age (**FIGURE 23.4**; Issa et al. 1994), and such methylation is thought to bring about the inactivation of these genes in the smooth muscle cells of blood vessels. This decline in estrogen receptors would prevent estrogen from maintaining the elasticity of these muscles, thereby leading to "hardening of the arteries." Increased methylation of the estrogen receptor genes is even more prominent in the atherosclerotic plaques that occlude the blood vessels (**FIGURE 23.5**); these plaques show more methylation of estrogen receptor genes than do the surrounding tissues (Post et al. 1999; Kim et al. 2007). Thus, methylation-associated inactivation of the estrogen receptor genes in these cells may play a role in the age-related deterioration of the vascular system. This potentially reversible defect may provide a new target for intervention in heart disease. Several neurological diseases, including bipolar disorder, depression, and stress responses, have been linked to DNA methylation and/or histone modifications (Sweatt 2013). Recent studies of Alzheimer's disease (and its mouse model) showed epigenetic signals (chromatin methylation and acetylation) indicating a loss of synaptic plasticity functions and a gain of immune function in the hippocampus. These strongly implicate the immune system (and inflammation) in predisposing people to Alzheimer's dementia (Gjoneska et al. 2015).

Horvath (2013) extended and refined the epigenetic aging clock by wide-scale genomic investigations of over 50 healthy individuals at different ages. He was able to analyze over 350 sites of possible DNA methylation, showing that as people age, these sites become progressively more methylated. Cells removed from early embryos have hardly any methylation at these sites, while cells taken from centenarians are heavily methylated. Using cells from a person's saliva, Horvath's analysis of DNA methylation allows prediction of a person's age to within 2 years (**FIGURE 23.6**). Moreover, Hannum and colleagues (2013) showed that tumors of breast, kidney, and lung tissues had more heavily methylated DNA than surrounding nontumor tissues, causing them to appear about 40% "older" than the patients from whom they were removed. This could be due to the methylation of a gene involved with the chromatin remodeling processes.

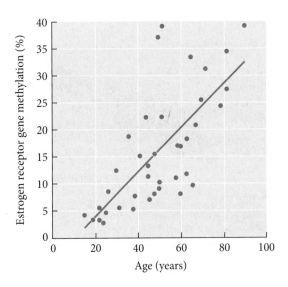

FIGURE 23.4 Methylation of a β estrogen receptor gene occurs as a function of normal physiological aging. (After Issa et al. 1994.)

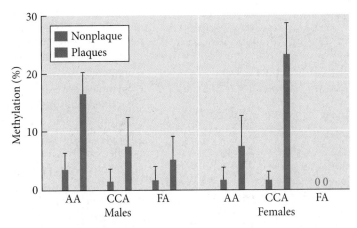

FIGURE 23.5 Methylation of the estrogen receptor gene in atherosclerotic plaques and adjacent nonplaque blood vessel tissue in the ascending aorta (AA), common carotid artery (CCA), and femoral artery (FA). (After Kim et al. 2007.)

(A) Saliva

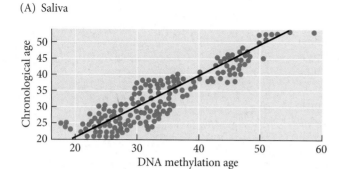

(B) Blood

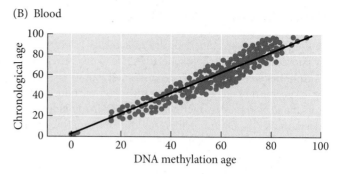

FIGURE 23.6 Chronological age (*y*-axis) of a person versus his or her DNA methylation "age" (*x*-axis). Each point corresponds to a DNA methylation sample taken from cells in the saliva (A) or cells in the blood (B). (After Horvath 2013.)

Developing Questions

Is aging really part of normal development? Or is aging just random degeneration?

So in this new hypothesis for aging, there appears to be random epigenetic drift that is not determined by the type of allele or any specific environmental factor. Random epigenetic drift may be the cause of the various phenotypes associated with aging as different genes are randomly repressed or ectopically activated. Mistakes in the DNA methylation process accumulate with age and may be responsible for the deterioration of our physiology and anatomy. If this is so, some genes may be more important targets than others. (The above-mentioned estrogen receptors, for instance, are critical not only in the vascular system but also for skeletal and muscular health.) It is not known what mechanisms set the rate of random DNA methylation, but these may be critically important for regulating aging both within and between species.

SCIENTISTS SPEAK 23.1 Dr. Silvia Gravina discusses her research on the epigenetic regulation of aging.

Stem Cells and Aging

One of the hallmarks of aging is the declining ability of stem cells and progenitor cells to restore damaged or nonfunctioning tissues. A decline in muscle progenitor (satellite) cell activity is seen when Notch signaling is lost, resulting in a significant decrease in the ability to maintain muscle function. Similarly, an age-dependent decline in liver progenitor cell division impairs liver regeneration due to a decline in transcription factor cEBPα. And the age-associated graying of mammalian hair appears to be due to the apoptosis of melanocyte stem cells in the hair bulge niche (Nishimura et al. 2005; Robinson and Fisher 2009). One of the questions, then, becomes: Is this part of the aging syndrome caused by the declining function of stem cells or by a declining ability of the stem cell niche to support them?

One way to test this is by "fusing" an old mouse to a young mouse. This can be done by a technique called parabiosis, wherein the animals' circulatory systems are surgically joined so that the two mice share one blood supply. If an aged and a young mouse are parabiosed—a technique called **heterochronic parabiosis**—the stem cells of the old mouse are exposed to factors in young blood serum (and vice versa). Heterochronic parabiosis has been seen to restore the activity of old stem cells. Notch signaling of the muscle stem cells regained its youthful levels, and muscle cell regeneration was restored. Similarly, liver progenitor cells regained "young" levels of cEBPα—and with it their ability to regenerate (Conboy et al. 2005; Conboy and Rando 2012). Young blood promoted the repair of aged spinal cords, reversed the thickening (hypertrophy) of the heart walls, and stimulated the formation of new neurons in aged mice (**FIGURE 23.7**; Villeda et al. 2011, 2014; Ruckh et al. 2012).

Loffredo and colleagues (2013) identified the "rejuvenating" blood-borne agent as paracrine factor GDF11, an extracellular signaling protein. GDF11 circulates through the blood of young mice, and its levels decline with age. When GDF11 was transfused into older mice, the youthful levels reversed age-related hypertrophy. In the brain, GDF11 appeared to counteract some of the deterioration of aging; injecting GDF11

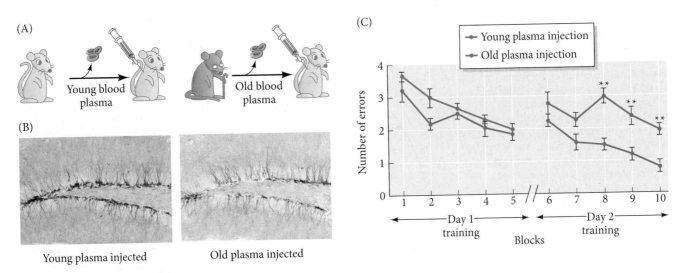

(A)

Young blood plasma

Old blood plasma

(B)

Young plasma injected

Old plasma injected

(C)

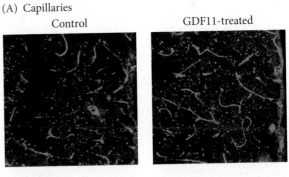

FIGURE 23.7 Factors in plasma (the liquid portion of blood) of old mice alter the development of new neurons and behaviors in young mice. (A) Protocol whereby plasma from young mice is injected into other young mice (left), or plasma from old mice is injected into young mice (right). (B) Young mice receiving young plasma continue to manufacture new neurons (dark stain), whereas the number of new neurons decreases in young mice injected with old plasma. (C) In training to do a particular task, mice receiving young or old plasma initially had the same number of errors. One day later, mice that received young plasma remembered their former mistakes and made fewer errors than mice that received old plasma. (After Villeda et al. 2011.)

into older mice increased brain capillary production, neuron formation, and olfactory discrimination (**FIGURE 23.8**; Katsimpardi et al. 2014; Poggioli et al. 2015).

Since the stem cells are not transfused from one animal to the other, it appears that GDF11 helps the function of the stem cell niche. It is not yet known whether GDF11 (or young blood plasma in general) extends the lifespan or improves the health of mice or humans (see Scudellari 2015). There is also the danger that, when working with stem cells, there is a fine balance between underproliferation (leading to aging) and overproliferation (leading to cancer).

WEB TOPIC 23.1 **CELLULAR SENESCENCE AND AGING** There may be a cell state state opposite that of the stem cell. "Senescent cells" produce paracrine factors that appear to cause many of the symptoms of aging.

SCIENTISTS SPEAK 23.2 Dr. Nadia Rosenthal discusses the ways by which stem cells might be used to stop or reverse the aging process.

Exceptions to the Aging Rule

There are a few species in which aging seems to be optional, and these may hold some important clues to how animals can live longer and retain their health. Turtles, for instance, are a symbol of longevity in many cultures. Many turtle species not only live a long time,

Developing Questions

How might induced pluripotent stem cells (iPSCs) link aging, regeneration, and development?

(A) Capillaries

Control   GDF11-treated

(B) Neural stem cells

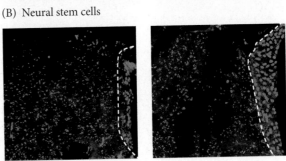

FIGURE 23.8 A possible "rejuvenating" agent. The blood-borne paracrine factor GDF11 promotes vascular remodeling (A) and neurogenesis (B) in the mouse brain. Micrographs show a region of the dentate gyrus of the brain stained for new capillaries or neural stem cells. The GDF11-treated brains were from 22-month old (elderly) mice injected with GDF11 for 4 weeks. (From Katsimpardi et al. 2014, courtesy of L. Katsimpardi and L. Rubin.)

FIGURE 23.9 Life cycle of *Turritopsis dohrnii* and *Hydractinia carnea*. In the normal life cycle of cnidarians, the colonies of polyps bud off medusa (jellyfish) into the sea water. After a period of planktonic life, the mature medusa release their gametes. Fertilization occurs and the mature medusa dies. The embryo forms a larva (planula), which then transforms itself into a ball-like stage from which a new polyp emerges. In *T. dohrnii* and *H. carnea*, the medusa can dedifferentiate into a ball-like stage, which can generate a polyp and start the life cycle again ("reverse development"; red arrows). (After Schmich et al. 2007.)

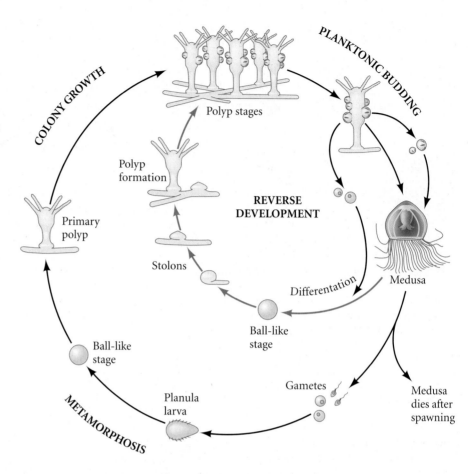

they don't seem to undergo a typical aging syndrome. Turtles seem to have "negligible senescence," in that their mortality rate does not increase with age, nor does their reproductive rate decrease with age. In these species, older females lay as many eggs (if not more) as their younger counterparts. Miller (2001) showed that a 60-year-old female three-toed box turtle (*Terrapene carolina triunguis*) lays as many eggs annually as she ever did. If turtle telomeres shorten with age, it happens (like so many turtle things) extremely slowly (Girondot and Garcia 1998; Hatase et al. 2008). Interestingly, turtles have special adaptations against oxygen deprivation, and these enzymes also protect against ROS (Congdon et al. 2003; Lutz et al. 2003; Krivoruchko and Storey 2010).

In monarch butterflies (*Danaus plexippus*), adults that migrate to wintering grounds in the mountains of central Mexico live several months (August–March), whereas their summer counterparts live only about 2 months (May–July). The regulation of this difference appears to be juvenile hormone (JH; Herman and Tatar 2001). The migrating butterflies are sterile because of suppressed synthesis of JH. If migrants are given JH in the laboratory, they regain fertility but lose their longevity. Conversely, when summer monarchs have their corpora allata removed so they can no longer make JH (see Chapter 21), their longevity increases 100%. Mutations in the insulin signaling pathway of *Drosophila* likewise decrease JH synthesis (Tu et al. 2005). This decrease in JH makes the flies small, sterile, and long-lived, adding to whatever longevity-producing effect protection against ROS might have.

Finally, there may be organisms, especially among the cnidarians, that can actually cheat death. As we saw in Chapter 22, hydra appear to be immortal, retaining their stem cell populations. The hydrozoans *Turritopsis dohrnii* and *Hydractinia carnea* are two cnidarian species known to have evolved a remarkable variation on this theme. These organisms have both a polyp stage (similar to the hydra) and a medusa ("jellyfish") stage in their life cycle. In most hydrozoan species, the polyp stage gives rise the sexual medusa stage. The medusae then produce gametes that spawn the next generation and, like most adults, they senesce and die. However, the medusae of *T. dohrnii* and *H. carnea* can revert to the polyp stage *after* becoming sexually mature (Bavestrello et al. 1992; Piraino et al. 1996), a feat called **reverse development** (Schmich et al. 2007). The multilayered medusa essentially dedifferentiates to become a two-layered ball-like stage similar to that of a larva before becoming a polyp, and then develops into a polyp itself (**FIGURE 23.9**).

Next Step Investigation

Several interacting agents may promote longevity. These include calorie restriction, protection against oxidative stress, factors activated by a suppressed insulin pathway, and factors affected by decreased mTORC1 signaling (e.g., decreased protein translation and augmented autophagy). Stem cells and their niches may play critical roles, as well.

The next step is to integrate these into pathways that might predict the lifespan of organisms and to aid individuals to live a healthy life to the very end. Unless attention is paid to this general aging syndrome, we risk ending up like Tithonos, the miserable wretch of Greek mythology to whom the gods awarded eternal life, but not eternal youth.

Closing Thoughts on the Opening Photo

The hydrozoan cnidarian *Turritopsis dohrnii* appears to be able to cheat death, earning it the nickname "the immortal jellyfish." It is one of a handful of cnidarian species known to be capable of reverting to an essentially embryonic state after having reached its adult form (the sexual medusa; see Figure 23.9). What roles stem cells may play in this reverse development and whether such processes might be applicable to human organs remains to be seen, but it is interesting to know that such organisms exist. (Photograph © Yiming Chen/Getty Images.)

24 Snapshot Summary
Aging and Senescence

1. The maximum life span of a species is the longest time an individual of that species has been observed to survive. Life expectancy is usually defined as the age at which approximately 50% of the members of a given population still survive.

2. Aging is the time-related deterioration of the physiological functions necessary for survival and reproduction. The phenotypic changes of senescence (which affect all members of a species) are not to be confused with diseases of senescence, such as cancer and heart disease (which affect some individuals but not others).

3. Reactive oxygen species (ROS) can damage cell membranes, inactivate proteins, and mutate DNA. Mutations that alter the ability to make or degrade ROS can change the life span.

4. Proteins that regulate DNA repair and cell division (such as p53 and telomerase) may be important regulators of aging.

5. An insulin signaling pathway, involving a receptor for insulin and insulin-like proteins, may be an important component of genetically limited life spans. It may upregulate mTORC1 and downregulate Foxo transcription factors.

6. Random DNA methylation appears to repress gene expression as a cell ages. Enzymes involved with chromatin modification may be important mediators of such aging events.

7. In many cases, the aging phenotype is the result of apoptosis of stem cells or progenitor cells.

8. A few animal species (such as turtles) display "negligible senescence" in that their mortality rates do not increase nor do reproductive rates decrease with age. Some cnidarian species appear to be potentially immortal.

Further Reading

Blackburn, E. H., E. S. Epel and J. Lin. 2015. Human telomere biology: A contributory and interactive factor in aging, disease risks, and protection. *Science* 350: 1193–1198. (*Note:* The December 4, 2015, issue of *Science* focuses on aging and contains several relevant articles.)

Fraga, M. F. and 20 others. 2005. Epigenetic differences arise during the lifetime of monozygotic twins. *Proc. Natl. Acad. Sci. USA* 102: 10604–10609.

Johnson, S. C., P. S. Rabinovitch and M. Kaeberlein. 2013. mTOR is a key modulator of ageing and age-related disease. *Nature* 493: 338–345.

Kenyon, C. J. 2010. The genetics of ageing. *Nature* 464: 504–512.

Underwood, E. 2015. The final countdown. *Science* 350: 1188–1190.

GO TO WWW.DEVBIO.COM...

...for Web Topics, Scientists Speak interviews, Watch Development videos, Dev Tutorials, and complete bibliographic information for all literature cited in this chapter.

Development in Health and Disease

Birth Defects, Endocrine Disruptors, and Cancer

"THE AMAZING THING ABOUT MAMMALIAN DEVELOPMENT," says British medical geneticist Veronica van Heyningen (2000), "is not that it sometimes goes wrong, but that it ever succeeds." It is indeed amazing that any of us is here, because relatively few human conceptions develop successfully to birth. Recent data (Mantzouratou and Delhanty 2011; Chavez et al. 2012) suggest that only 20–50% of human cleavage-stage embryos successfully implant in the uterus. Many human embryos have chromosomal anomalies that are expressed so early the embryo fails to implant and is spontaneously aborted (miscarried), usually before a woman realizes she has conceived. Of the embryos that *do* implant successfully, studies from the 1980s suggest that only about 40% survive to term (Edmonds et al. 1982; Boué et al. 1985). Further studies (Winter 1996; Epstein 2008) estimate that some 2.5% of babies who do come to term have a recognizable birth defect.

With so many genes, cells, and tissues becoming organized simultaneously and changing together, it is not surprising that some developmental events do not happen properly. Although the body has remarkable back-up pathways and redundancies that permit a great deal of flexibility, if developmental phenomena are absent when they should be activate or activated when they should be repressed, abnormal phenotypes emerge. There are three major pathways to abnormal development:

1. *Genetic mechanisms.* Mutations in genes or changes in the number of chromosomes can alter development.

2. *Environmental mechanisms.* Agents (usually chemicals) from outside the body cause deleterious phenotypic changes by inhibiting or enhancing developmental signals.

A cat with six digits. Can such anomalies ever be useful?

The Punchline

Development sometimes (even often) goes wrong. Genetic and chromosomal mutations can delete genes, destroy gene function, or render genes or proteins active in the wrong times or places, all of which may lead to congenital anomalies (birth defects). Teratogens are chemicals in the environment (such as alcohol and retinoic acid) that can prevent particular genes or proteins from functioning or can activate processes at the wrong places and times. Endocrine disruptors are environmental chemicals (such as BPA and DES) that activate or suppress hormone function, thus altering normal development. Cancer can result when developmental pathways are either reactivated or suppressed in the adult, leading to abnormal cell proliferation. Like birth defects, cancers can be caused by genetic alterations, environmental agents or chance circumstances.

3. *Stochastic (random) events.* Chance plays a role in determining the phenotype, and some developmental anomalies are just "bad luck" (Molenaar et al. 1993; Holliday 2005; Smith 2011).

Most of this chapter will deal with genetic and environmental effects.[1] We will start, however, by briefly examining the role of random events.

The Role of Chance

Although physicians and researchers often parse developmental anomalies into those caused by internal (genetic) means versus those caused by external (environmental) agents, more consideration is now being given to the role of stochastic factors—randomness—in developmental defects. Even an embryo with wild-type genes and a favorable environment may develop an abnormal phenotype as the result of "bad luck." Developmental outcomes are probabilistic rather than predetermined (Wright 1920; Gottlieb 2003; Kilfoil et al. 2009). Consider for example, X-chromosome inactivation in females. If a woman carries one normal and one mutant allele for an X-linked blood clotting factor, statistically we would expect the wild-type allele to be inactivated in about 50% of her cells. If the wild-type allele is inactivated in 50% of the liver cells that produce clotting factor, she is phenotypically normal. But what would happen if, just by chance, 95% of the wild-type X chromosomes were inactivated in those liver cells? Only 5% of her X chromosomes would express the wild-type allele, and she would have an abnormality. Indeed, there have been cases of female identical twins where, in one twin, chance resulted in the inactivation of a large percentage of her X chromosomes carrying the normal clotting factor allele; that twin had severe hemophilia (inability of the blood to clot). The other twin, with a lower percentage of her normal X chromosomes inactivated, was not affected (Tiberio 1994; Valleix et al. 2002).

Such variability is not limited to genes on the X chromosome. Measurements of gene expression in individual cells show that protein synthesis is a stochastic process, with random fluctuations in both transcription and translation leading to variations in the levels of proteins produced at any given time (Raj and van Oudenaarden 2008; Stockholm et al. 2010). Cell specification, developmental signaling, and cell migration are thought to be influenced by chance fluctuations in the amounts of transcription factors, paracrine factors, and receptors produced at a particular moment. Thus, genetically identical animals raised in precisely the same environments can have vastly different phenotypes (Gilbert and Jorgensen 1998; Vogt et al. 2008; Ruvinsky 2009), and random fluctuations of gene expression can produce developmental anomalies. Mathematical modeling has permitted scientists to study these stochastic events, enabling researchers to "quantitatively demonstrate that development represents a combination of stochastic and deterministic events, offering insight into how chance influences normal development and may give rise to birth defects" (Zhou et al. 2013).

Genetic Errors of Human Development

Congenital ("present at birth") abnormalities and losses of the fetus prior to birth have both intrinsic and extrinsic causes. Those abnormalities caused by genetic events may result from mutations, aneuploidies (improper chromosome number), and translocations (Opitz 1987).

The nature of human syndromes

Human birth defects, which range from life-threatening to relatively benign, are often linked into **syndromes** (Greek, "running together"), with several abnormalities occurring together. Genetically based syndromes are caused either by (1) a chromosomal

---

[1]This chapter focuses on human health and provides general information about a variety of medical topics. It is not intended to provide medical advice for specific persons or disorders. For a more thorough account of development and disease, please see Gilbert and Epel (2015).

(A) Mosaic pleiotropy

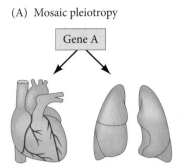

(B) Relational pleiotropy

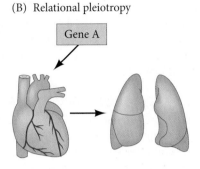

FIGURE 24.1 Mosaic and relational pleiotropy. (A) In mosaic pleiotropy, a gene is independently expressed in several tissues. Each tissue needs the gene product and develops abnormally in its absence. (B) In relational pleiotropy, a gene product is needed by only one particular tissue. However, a second tissue needs a signal from the first tissue in order to develop properly. If the first tissue develops abnormally, the signal is not given, so the second tissue also develops abnormally.

event (such as an aneuploidy) in which several genes are deleted or added, or (2) **pleiotropy**—the production of several effects by a single gene or pair of genes (see Grüneberg 1938; Hadorn 1955). Syndromes are said to have **mosaic pleiotropy** when the effects are produced independently as a result of the gene being critical in different parts of the body (**FIGURE 24.1A**). For instance, the *KIT* gene is expressed in blood stem cells, pigment stem cells, and germ stem cells, where it is needed for their proliferation. When this gene is defective, the resulting syndrome of anemia (lack of red blood cells), albinism (lack of pigment cells), and sterility (lack of germ cells) is evidence of mosaic pleiotropy. Syndromes are said to have **relational pleiotropy** when a defective gene in one part of the embryo causes a defect in another part, even though the gene is not expressed in the second tissue (**FIGURE 24.1B**). For example, failure of *MITF* expression in the pigmented retina prevents this structure from fully differentiating. This failure of pigmented retina growth in turn causes a malformation of the choroid fissure of the eye, resulting in the drainage of vitreous humor. Without this fluid, the eye fails to enlarge (hence microphthalmia, "small eye"). The lenses and corneas are therefore smaller, even though they themselves do not express *MITF*.

Mosaic syndromes can be the result of **aneuploidies**—errors in the number of particular chromosomes. Even an extra copy of the tiny chromosome 21 disrupts numerous developmental functions. This **trisomy 21** causes a set of anomalies—among them facial muscle changes, heart and gut abnormalities, and cognitive problems—collectively known as **Down syndrome** (**FIGURE 24.2**). Certain genes on chromosome 21 are thought to encode transcription factors and regulatory microRNAs, and the extra copy of chromosome 21 probably causes an overproduction of these regulatory proteins. Such overproduction results in the misregulation of genes necessary for heart, muscle, and nerve formation (Chang and Min 2009; Korbel et al. 2009). One such regulatory

(A)

(B)

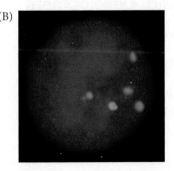

FIGURE 24.2 Down syndrome. (A) Down syndrome, caused by a third copy of chromosome 21, is characterized by a particular facial pattern, cognitive deficiencies, the absence of a nasal bone, and often heart and gastrointestinal defects. (B) The procedure shown here tests for chromosome number using fluorescently labeled probes that bind to DNA on chromosomes 21 (pink) and 13 (blue). This person has Down syndrome (trisomy 21) but has the normal two copies of chromosome 13. (A © MoodBoard/Alamy; B courtesy of Vysis, Inc.)

microRNA encoded on chromosome 21, miRNA-155, is found throughout the developing human fetus. This miRNA downregulates translation of the messages for certain transcription factors necessary for normal neural and heart development and is highly elevated in the brains and hearts of people with Down syndrome (Elton et al. 2010; Wang et al. 2013).

Genetic and phenotypic heterogeneity

In pleiotropy, the same gene can produce different effects in different tissues. However, the opposite phenomenon is an equally important feature of genetic syndromes: mutations in different genes can produce the same phenotype. If several genes are part of the same signal transduction pathway, a mutation in any of them often produces a similar phenotypic result. This production of similar phenotypes by mutations in different genes is called **genetic heterogeneity**. The syndrome of sterility, anemia, and albinism caused by the absence of Kit protein (discussed above) can also be caused by the absence of its paracrine ligand, stem cell factor (SCF). Another example is cyclopia (see Figure 4.31B), a phenotype that can be produced either by mutations in the gene for Sonic hedgehog, *or* by mutations in the genes activated by Shh, *or* in the genes controlling cholesterol synthesis (since cholesterol is essential for Shh signaling).

Not only can different mutations produce the same phenotype, but the same mutation can produce a different phenotype in different individuals, a phenomenon known as **phenotypic heterogeneity** (Wolf 1995, 1997; Nijhout and Paulsen 1997). Phenotypic heterogeneity comes about because genes are not autonomous agents. Rather, they interact with other genes and gene products, becoming integrated into complex pathways and networks. Bellus and colleagues (1996) analyzed the phenotypes derived from the same mutation in the *FGFR3* gene in 10 unrelated families. These phenotypes ranged from relatively mild anomalies to potentially lethal malformations. Similarly, Freire-Maia (1975) reported that within one family, the homozygous state of a mutant gene affecting limb development caused phenotypes ranging from severe phocomelia (lack of limb development) to a mild abnormality of the thumb. The severity of a mutant gene's effect often depends on the other genes in the pathway, as well as on environmental and stochastic (random) factors.

WEB TOPIC 24.1 **PREIMPLANTATION GENETICS** The ability to identify allelic variants in a single cell has enabled medical professionals to determine whether an embryo carries deleterious genes. It also enables people to determine the embryo's sex, allowing a parent to implant into the uterus only embryos of the desired sex.

Teratogenesis: Environmental Assaults on Animal Development

The summer of 1962 brought two portentous events. The first was the publication of the landmark book *Silent Spring* by Rachel Carson, in which she documented that the pesticide DDT was destroying bird eggs and preventing reproduction in several species. Her work is credited with spurring the modern environmental movement. The second event was the discovery that thalidomide, a sedative used to help manage pregnancies, could cause limb and ear abnormalities in the human fetus (Lenz 1962; see Chapter 1). These two revelations showed that the embryo was vulnerable to environmental agents. Indeed, Rachel Carson made the connection, commenting, "It is all of a piece, thalidomide and pesticides. They represent our willingness to rush ahead and use something without knowing what the results will be" (Carson 1962).

WEB TOPIC 24.2 **THALIDOMIDE AS A TERATOGEN** The drug thalidomide caused thousands of babies to be born with malformed arms and legs. It provided the first major evidence that drugs could induce congenital anomalies. The mechanism of its action is still hotly debated.

TABLE 24.1 Some agents thought to cause disruptions in human fetal development[a]	
DRUGS AND CHEMICALS	Valproic acid
Alcohol	Warfarin
Aminoglycosides (Gentamycin)	**IONIZING RADIATION (X-RAYS)**
Aminopterin	**HYPERTHERMIA (FEVER)**
Antithyroid agents (PTU)	**INFECTIOUS MICROORGANISMS**
Bromine	Coxsackie virus
Cortisone	Cytomegalovirus
Diethylstilbesterol (DES)	Herpes simplex
Diphenylhydantoin	Parvovirus
Heroin	Rubella (German measles)
Lead	*Toxoplasma gondii* (toxoplasmosis)
Methylmercury	*Treponema pallidum* (syphilis)
Penicillamine	Zika virus
Retinoic acid (Isotretinoin, Accutane)	**METABOLIC CONDITIONS IN THE MOTHER**
Streptomycin	Autoimmune disease (including Rh incompatibility)
Tetracycline	Diabetes
Thalidomide	Dietary deficiencies, malnutrition
Trimethadione	Phenylketonuria

[a] This list includes known and possible teratogenic agents and is not exhaustive.

Exogenous agents that cause birth defects are called **teratogens** (**TABLE 24.1**). Most teratogens produce their effects during certain critical timeframes. Human development is usually divided into an **embryonic period** (to the end of week 8) and a **fetal period** (the remaining time in utero). Most organ systems form during the embryonic period; the fetal period is generally one of growth and modeling. Thus, maximum fetal susceptibility to teratogens is between weeks 3 and 8 (**FIGURE 24.3**). The nervous system, however, is constantly forming and remains susceptible throughout development. Prior to week 3, exposure does not usually produce congenital anomalies because a teratogen encountered at this time either damages most or all of the cells of an embryo, resulting in its death, or it kills only a few cells, allowing the embryo to fully recover.

 SCIENTISTS SPEAK 24.1 A video created by students at Smith College explains the bases of teratology and endocrine disruption.

The largest class of teratogens includes drugs and chemicals. Viruses, radiation, high body temperature, and metabolic conditions in the mother can also act as teratogens. Some chemicals found naturally in the environment can cause birth defects. For example, jervine and cyclopamine are products of the plant *Veratrum californicum* that block Sonic hedgehog signaling and lead to cyclopia (see Figure 4.31B). Nicotine, a natural product concentrated in tobacco smoke, is associated with impaired lung and brain development (Dwyer et al. 2008; Maritz and Harding 2011). Some viruses can cause congenital anomalies. The mosquito-borne Zika virus has been implicated in microcephaly, a birth defect characterized by small brains and heads (CDC 2016; Mlakar et al. 2016). Evidence indicates that, in pregnant women, Zika virus directly infects the neural progenitor cells of the fetal cortex, resulting in the death of these cells, which could result in the smaller brain and head of the newborn (Tang et al. 2016).

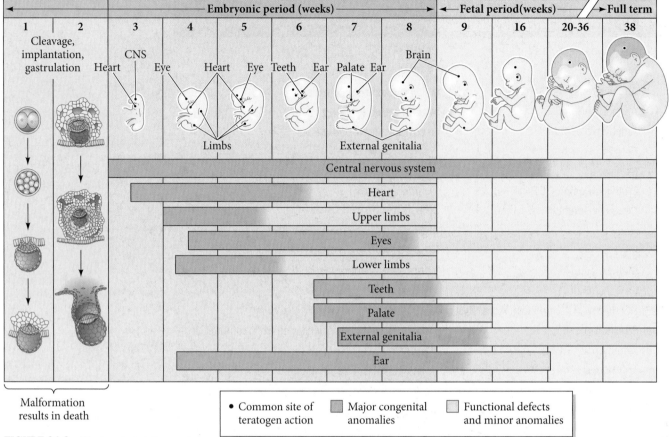

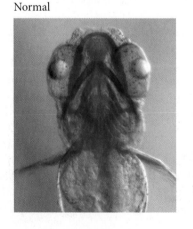

FIGURE 24.3 Weeks of gestation and sensitivity of embryonic organs to teratogens. (After Moore and Persaud 1993.)

Although different agents are teratogenic in different organisms (see Gilbert and Epel 2015), animals have been used to screen compounds that have a high probability of being hazardous. *Xenopus* and zebrafish, as we saw in Chapters 11 and 12, undergo early development using the same basic paracrine factors and transcription factor pathways that we use. These model organisms have been especially important in identifying teratogenic molecules in the environment. Studies on zebrafish found, for instance, that water-soluble components from the 2010 *Deepwater Horizon* oil spill in the Gulf of Mexico caused numerous developmental anomalies traceable to neural crest cell migration (**FIGURE 24.4**; de Soysa et al. 2012).

Developing Questions

In the United States, pregnant women are warned not to drink water from lakes that lie near abandoned mines. Do you think this warning is warranted? Why?

Normal   Affected

FIGURE 24.4 Water-soluble crude oil components from the *Deepwater Horizon* oil spill were teratogenic in zebrafish. Compared to normal zebrafish of the same age, zebrafish embryos exposed to oil spill components produced larvae with severe developmental anomalies, including reduction in the size of head, gill, and thoracic cartilages (blue staining) associated with cranial neural crest migration. (From de Soysa et al. 2012.)

SCIENTISTS SPEAK 24.2   Dr. Daniel Gorelick discusses using transgenic zebrafish as organisms to detect BPA and other developmentally disruptive compounds in water.

Alcohol as a teratogen

In terms of the frequency of its effects and its cost to society, the most devastating teratogen is undoubtedly alcohol (ethanol). Babies born with **fetal alcohol syndrome (FAS)** are characterized by small head size, an indistinct philtrum (the pair of ridges that runs between the nose and mouth above the center of the upper lip), a narrow vermillion border on the upper lip, and a low nose bridge (Lemoine et al. 1968; Jones and Smith 1973). The brains of such children may be dramatically smaller than normal and often show poor development, the results of deficiencies of neuronal and glial migration (**FIGURE 24.5A,B**; Clarren 1986). FAS is the most prevalent type of congenital mental retardation syndrome, occurring in approximately 1 out of every 650 children born in the United States (May and Gossage 2001). Although the IQs of children with FAS vary substantially, the mean is about 68 (Streissguth and LaDue 1987). Most adults and adolescents with FAS cannot handle money and have difficulty learning from past experiences (see Dorris 1989; Kulp and Kulp 2000).

Fetal alcohol syndrome represents only a portion of a range of defects caused by prenatal alcohol exposure. The term **fetal alcohol spectrum disorder (FASD)** has been coined to encompass all of the alcohol-induced malformations and functional deficits that occur. In many FASD children, behavioral abnormalities exist without any gross physical changes in head size or notable reductions in IQ (NCBDD 2009). However,

FIGURE 24.5   Effects of alcohol on fetal brains. (A,B) Comparison of a brain from an infant with fetal alcohol syndrome (A) with a brain from a normal infant of the same age (B). The brain from the infant with FAS is smaller, and the pattern of convolutions is obscured by glial cells that have migrated over the top of the brain. (C,D) Regionally specific abnormalities of the corpus callosum seen by diffusion tensor imaging of myelinated neurons. The difference in fiber tracks in a child with FASD (C) compared with those of a same-age unaffected child (D) suggests that there are significant abnormalities in neurons that would normally project through the posterior regions of the brain into the cortex of the parietal and temporal lobes. (A,B courtesy of S. Clarren; C,D from Wozniak and Muetzel 2011, courtesy of the authors.)

(A)   (C)

(B)   (D)

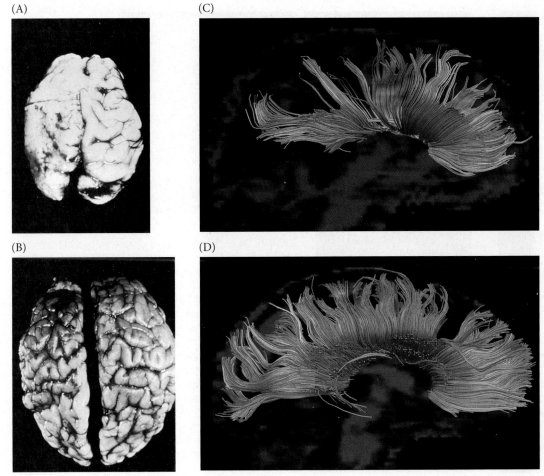

recent techniques that can identify neural tracts in the brain have found subtle abnormalities that correlate with altered mental processing speed and executive functioning (such as planning, memorizing, and retaining information) (**FIGURE 24.5C,D**; Wozniak and Muetzel 2011).

As with other teratogens, the amount and timing of fetal exposure to alcohol, as well as the genetic background of the fetus, contribute to the developmental outcome. Variability in the mother's ability to metabolize alcohol also may account for some outcome differences (Warren and Li 2005). While FASD is most strongly associated with high levels of alcohol consumption, the results of animal studies suggest that even a single episode of consuming the equivalent of two alcoholic drinks during pregnancy may lead to loss of fetal brain cells. ("One drink" is defined as 12 oz. of beer, 5 oz. of wine, or 1.5 oz. of hard liquor.) It is important to note that alcohol can cause permanent damage to a fetus at a time before most women even realize they are pregnant.

HOW ALCOHOL AFFECTS DEVELOPMENT: LESSONS FROM THE MOUSE   When mice are exposed to alcohol at the time of gastrulation, ethanol induces defects of the face and brain that are comparable to those in humans with FAS (**FIGURE 24.6**; Sulik 2005). As in human fetuses, the nose and upper lip of the ethanol-exposed pups are poorly developed, and nervous system problems involve failure to close the neural tube and incomplete development of the forebrain (see Chapters 13 and 14). This mouse model of FAS can be used to study the ways by which ethanol causes its effects on the embryo.

It appears that ethanol works on several processes, and can interfere with cell migration, proliferation, adhesion, and survival. Hoffman and Kulyk (1999) showed that

Normal

Alcohol-exposed

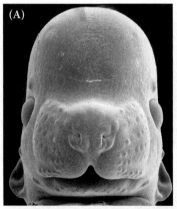
(A)

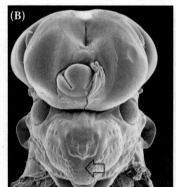
(B)

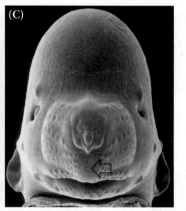
(C)

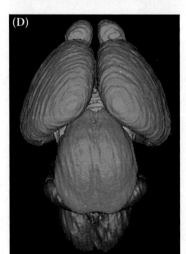
(D)

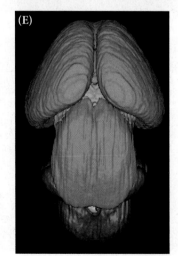
(E)

FIGURE 24.6   Alcohol-induced craniofacial and brain abnormalities in mice. (A–C) Normal (A) and abnormal (B,C) day-14 embryonic mice. In (B), the anterior neural tube failed to close, resulting in exencephaly, a condition in which the brain tissue is exposed to the exterior. Later in development, the exposed brain tissue will erode away, resulting in anencephaly. (B,C) Prenatal alcohol exposure can also affect facial development, resulting in a small nose and an abnormal upper lip (open arrow). These facial features are present in fetal alcohol syndrome. (D,E) Three-dimensional reconstructions prepared from magnetic resonance images of the brains of normal (D) and alcohol-exposed (E) 17-day embryonic mice. In the alcohol-exposed specimen, the olfactory bulbs (pink) are absent and the cerebral hemispheres (red) are abnormally united in the midline. Light green, diencephalon; magenta, mesencephalon; teal, cerebellum; dark green, pons and medulla. (Courtesy of K. Sulik.)

instead of migrating and dividing, neural crest cells of alcohol-exposed fetuses prematurely initiate their differentiation into facial cartilage. Among the numerous genes that are misregulated following maternal alcohol exposure in mice are several involved in the cytoskeletal reorganization that enables cell movement (Green et al. 2007). In addition, cell death is apparent shortly after alcohol exposure. In later-stage mouse embryos exposed to ethanol, the death of neural crest-derived cells is seen as early as 12 hours following the exposure. When the time of alcohol exposure corresponds to the third and fourth weeks of human development, cells that should form the median portion of the forebrain, upper midface, and cranial nerves are killed. This has been confirmed in early chick embryos, where transient ethanol exposure at environmentally relevant doses (about 25 mM) decimates migrating cranial neural crest cells, causing cell death throughout the head region (Flentke et al. 2011).

One reason for this cell death in mouse embryos is that alcohol treatment results in the generation of superoxide radicals that can damage cell membranes (**FIGURE 24.7A–C**; Davis et al. 1990; Kotch et al. 1995; Sulik 2005). In model systems, antioxidants have been effective in reducing both the cell death and the malformations caused by alcohol (Chen et al. 2004).

Abnormal signaling may also underlie excessive cell death. In alcohol-exposed embryos, expression of Sonic hedgehog (which is important in establishing the facial

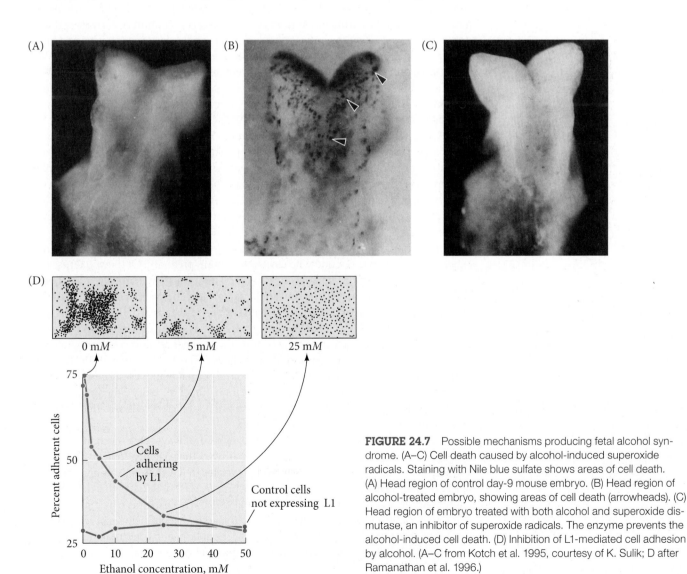

FIGURE 24.7 Possible mechanisms producing fetal alcohol syndrome. (A–C) Cell death caused by alcohol-induced superoxide radicals. Staining with Nile blue sulfate shows areas of cell death. (A) Head region of control day-9 mouse embryo. (B) Head region of alcohol-treated embryo, showing areas of cell death (arrowheads). (C) Head region of embryo treated with both alcohol and superoxide dismutase, an inhibitor of superoxide radicals. The enzyme prevents the alcohol-induced cell death. (D) Inhibition of L1-mediated cell adhesion by alcohol. (A–C from Kotch et al. 1995, courtesy of K. Sulik; D after Ramanathan et al. 1996.)

midline structures) is downregulated. While the mechanism for this downregulation remains incompletely understood, the finding that Shh-secreting cells placed into the head mesenchyme can prevent the alcohol-induced death of cranial neural crest cells highlights the importance of the Shh pathway as a target for alcohol's teratogenesis (Ahlgren et al. 2002; Chrisman et al. 2004).

Another mechanism that may be involved in alcohol's teratogenesis is its interference with the ability of the cell adhesion molecule L1 to hold cells together. Ramanathan and colleagues (1996) have shown that at levels as low as 7 mM, an alcohol concentration produced in the blood or brain with a single drink, alcohol can block the adhesive function of the L1 protein in vitro (**FIGURE 24.7D**). Moreover, mutations in the human L1 gene cause a syndrome of mental retardation and malformations similar to that seen in severe FAS cases. Thus, alcohol can cross the placenta, enter the fetus, and block several critical functions in brain and facial development.

 SCIENTISTS SPEAK 24.3   Dr. Kathy Sulik discusses the biological bases of fetal alcohol syndrome.

Retinoic acid as a teratogen

In some instances, even a compound involved in normal development can have deleterious effects if it is present in large enough amounts or at particular times. As we have seen throughout this book, retinoic acid (RA) is a vitamin A derivative that is important in specifying the anterior-posterior axis and in forming the jaws and heart of the mammalian embryo. In its pharmaceutical form, 13-*cis*-retinoic acid (also called isotretinoin and sold under the trademark Accutane) has been useful in treating severe cystic acne and has been available for this purpose since 1982. The deleterious effects of administering large amounts of RA (or its vitamin A precursor) to pregnant animals have been known since the 1950s (Cohlan 1953; Giroud and Martinet 1959; Kochhar et al. 1984). However, about 160,000 women of childbearing age (15–45 years) have taken isotretinoin since it was introduced, and some have used it during pregnancy. Isotretinoin-containing drugs now carry a strong warning against their use by pregnant women. In the United States, retinoic acid exposure is a critical public health concern because there is significant overlap between the population using acne medicine and the population of women of childbearing age—and because an estimated 50% of pregnancies in the U.S. are unplanned (Finer and Zolna 2011).

Lammer and co-workers (1985) studied a group of women who inadvertently exposed themselves to RA and who elected to remain pregnant. Of their 59 fetuses, 26 were born without any noticeable anomalies, 12 aborted spontaneously, and 21 were born with obvious anomalies. The affected infants had a characteristic pattern of anomalies, including absent or defective ears, absent or small jaws, cleft palate, aortic arch abnormalities, thymus deficiencies, and abnormalities of the central nervous system. These anomalies are largely due to the failure of cranial neural crest cells to migrate into the pharyngeal arches of the face to form the jaws and ear (Moroni et al. 1994; Studer et al. 1994). Radioactively labeled RA binds to the cranial neural crest cells and arrests both their proliferation and their migration (Johnston et al. 1985; Goulding and Pratt 1986). The teratogenic period during which cranial neural crest cells are affected occurs on days 20–35 in humans (days 8–10 in mice).

Retinoic acid probably disrupts these cells in several ways. One mechanism is that excess RA activates the negative feedback pathway that usually ensures the proper amount of this compound. Transient large increases in RA thus activate the synthesis of RA-degrading enzymes, causing a long-lasting *decrease* of RA. It is this deficiency in RA that results in the malformations (Lee et al. 2012). This explains why high amounts of retinoic acid produce phenotypes similar to those seen in deficiencies of retinoic acid.

Interference with RA signaling may be a wider public health concern for another reason. Glyphosate-based herbicides (such as Roundup) have been reported to upregulate

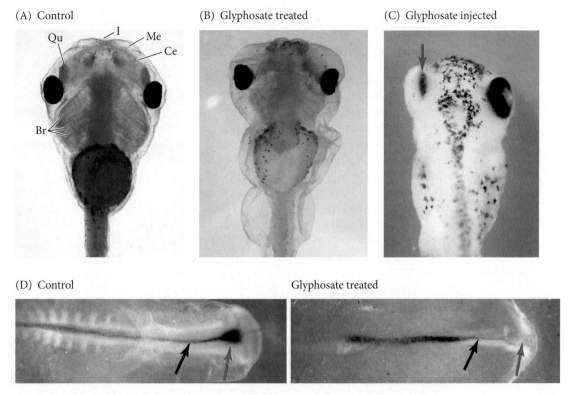

(A) Control   (B) Glyphosate treated   (C) Glyphosate injected

(D) Control   Glyphosate treated

FIGURE 24.8 Glyphosate herbicide teratogenicity. (A) *Xenopus* tadpole raised under control conditions, stained with Alcian blue to show facial cartilages. Br, branchial; Ce, ceratohyal; I, infrarostral; Me, Meckel; Qu, quadrate. (B) *Xenopus* tadpole raised in environmentally relevant concentrations of glyphosate and similarly stained. Its branchial arches and midline facial cartilage (cranial neural crest derivatives) failed to develop properly. (C) If an embryo is injected such that only one side (arrow) is exposed to glyphosate, that side shows cranial neural crest anomalies. (D) Control chick embryos show sonic hedgehog gene expression in the notochord (black arrow) and prechordal mesoderm (red arrow). Chick embryos grown in glyphosate show a severe reduction of sonic hedgehog expression in the prechordal (craniofacial) mesoderm. (Photographs courtesy of A. Carrasco.)

the activity of endogenous RA (Paganelli et al. 2010). When *Xenopus* embryos were incubated in solutions containing ecologically relevant concentrations of these herbicides, RA-responsive reporter gene activation was dramatically altered, and the embryos exhibited cranial neural crest defects and facial disorders similar to those seen in RA teratogenesis (**FIGURE 24.8**). Glyphosate is the most widely used (and profitable) herbicide in North America, where over 180 tons of it have been applied. It acts by blocking a plant enzyme that is critical for the synthesis of certain amino acids. One of the powerful abilities of genetic engineering has been to manufacture wide-spectrum herbicides such as Roundup and then breed crop plants that are resistant to this herbicide. This means that if you spray a large area, all the weeds will be killed and the only plants remaining will be those that are glyphosate-resistant. In 2010, 70% of the corn and 93% of the soybeans grown in the United States were from genetically modified herbicide-resistant seeds (Hamer 2010).

Developing Questions

In the northern United States, certain ponds contain a high proportion of frogs that have six or more limbs. What are the possible causes of these malformations?

WEB TOPIC 24.3 **THE DEVELOPMENTAL ORIGINS OF ADULT HUMAN DISEASE** Can food be considered a teratogen? In certain instances, the foods a mammal eats during pregnancy may activate certain genes in the fetus that can either help or predispose it to certain illnesses later in life.

Endocrine Disruptors:
The Embryonic Origins of Adult Disease

A specialized area of teratogenesis involves the misregulation of the endocrine system during development. Endocrine disruptors are exogenous (coming from outside the body) chemicals that disrupt development by interfering with the normal functions of hormones (Colborn et al. 1993, 1997). The phenotypic changes produced by endocrine disruptors are not the obvious anatomical birth defects produced by classic teratogens. Rather, the anatomical alterations induced by endocrine disruptors are often seen only microscopically; the major changes are physiological. These functional changes are more subtle than those produced by other teratogens, but they can be extremely important phenotypic alterations. Their effects often manifest later in adult life and may persist for generations after exposure to the disruptor.

Endocrine disruptors can interfere with hormone function in many ways:

- They can mimic the effect of a natural hormone. A paradigmatic example is the endocrine disruptor diethylstilbestrol (DES), which binds to the estrogen receptor and mimics estradiol, a hormone that is very active in building the tissues of the female reproductive tract.

- They can act as antagonists and inhibit the binding of a hormone to its receptor or block the synthesis of a hormone. DDE, a metabolic product of the insecticide DDT, can act as an anti-testosterone, binding to the androgen receptor and preventing normal testosterone from functioning properly.

- They can affect the synthesis, elimination, or transportation of a hormone in the body. The herbicide atrazine, for example, elevates the synthesis of estrogen and can convert testes into ovaries in frogs. One of the ways that polychlorinated biphenyls (PCBs) disrupt the endocrine system is by interfering with the elimination and degradation of thyroid hormones.

- Some endocrine disruptors can "prime" the organism to be more sensitive to hormones later in life. As we will see, bisphenol A exposure during fetal development makes breast tissue more responsive to steroid hormones during puberty.

It was long thought that there were only a few teratogenic agents, and that the only ones in danger were the fetuses of women who were inadvertently exposed to high doses of these chemicals during pregnancy. We now recognize that endocrine disruptors are everywhere in our technological society (including rural areas, where pesticides and herbicides are abundant), and that low-dose exposure to endocrine disruptors can be sufficient to produce significant disabilities later in life. Endocrine disruptors include chemicals in the materials that line baby bottles and the brightly colored plastic containers from which we drink our water; chemicals used in cosmetics, sunblocks, and hair dyes; and chemicals that prevent clothing from being highly flammable. As expected when so many chemicals are involved, we are exposed to not just one but to multiple endocrine disruptors, simultaneously and continuously. And, in another important difference between endocrine disruptors and classic teratogens, more damage may be done by a "moderate" dose of endocrine disruptor than by higher doses, since higher concentrations may activate negative feedback processes that detoxify or eliminate the chemical (see Myers et al. 2009; Belcher et al. 2012; Vandenberg et al. 2012).

There are numerous endocrine disruptors, including diethylstilbestrol, bisphenol A, phthalates, and tributyltin. Other compounds having endocrine-disrupting capabilities include the herbicide atrazine, and compounds released as a result of hydraulic fracturing ("fracking") (see Gilbert and Epel 2015; Kabir et al. 2015). We encounter many of these substances on a daily basis and we are exposed to them prenatally. The developmental effects of this level of exposure are just beginning to be studied (Wild 2005; Rappaport and Smith 2010).

WEB TOPIC 24.4 **DDT AS AN ENDOCRINE DISRUPTOR**   DDT was first found to be damaging to the development of birds. Since then it has been implicated in disrupting human development as well.

SCIENTISTS SPEAK 24.4   In an online documentary, Stéphane Horel interviews some of the leading scientists involved in identifying endocrine disruptors.

Diethylstilbestrol (DES)

One of the first endocrine disruptors to be identified was the potent environmental estrogen **diethylstilbestrol**, or **DES**. This drug was thought to ease pregnancy and prevent miscarriages, and it is estimated that in the United States more than 1 million pregnant women and their fetuses were exposed to DES between 1947 and 1971. (This is probably a small fraction of exposures worldwide.) Although research from the 1950s showed that in fact DES had no beneficial effects on pregnancy, it continued to be prescribed until the FDA banned it in 1971. The ban was imposed when a specific type of tumor (clear-cell adenocarcinoma) was discovered in the reproductive tracts of some of the women whose mothers took DES during pregnancy (**FIGURE 24.9**).

DES interferes with sexual and gonadal development by causing cell type changes in the female reproductive tract (the derivatives of the Müllerian duct, which forms the upper portion of the vagina, cervix, uterus, and oviducts; see Figure 6.1). In many cases, DES causes the boundary between the oviduct and the uterus (the uterotubal junction) to be lost, resulting in infertility, low fertility, and a high risk for other reproductive health problems (Robboy et al. 1982; Newbold et al. 1983; Hoover et al. 2011).

Symptoms similar to human DES syndrome occur in mice exposed to DES in utero, allowing the mechanisms of this endocrine disruptor to be uncovered. Normally, the regions of the female reproductive tract are specified by the *Hoxa* genes, which are expressed in a nested fashion throughout the Müllerian duct (**FIGURE 24.10**). Ma

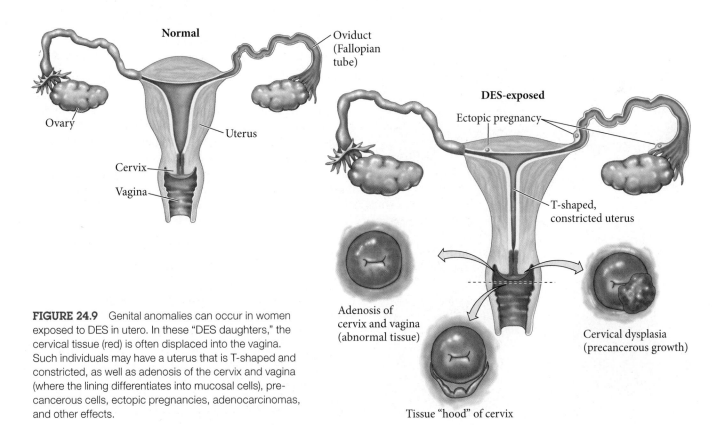

FIGURE 24.9   Genital anomalies can occur in women exposed to DES in utero. In these "DES daughters," the cervical tissue (red) is often displaced into the vagina. Such individuals may have a uterus that is T-shaped and constricted, as well as adenosis of the cervix and vagina (where the lining differentiates into mucosal cells), precancerous cells, ectopic pregnancies, adenocarcinomas, and other effects.

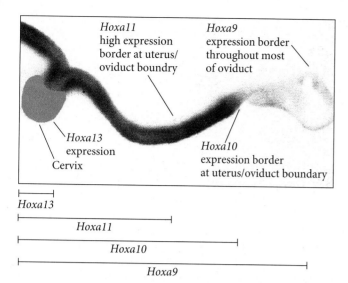

Hoxa11
high expression
border at uterus/
oviduct boundry

Hoxa9
expression border
throughout most
of oviduct

Hoxa13
expression
Cervix

Hoxa10
expression border
at uterus/oviduct boundary

Hoxa13
Hoxa11
Hoxa10
Hoxa9

FIGURE 24.10 Effects of DES exposure on the female reproductive system. (A) The chemical structure of DES. (B) *Hoxa* gene expression in the reproductive system of a normal 16.5-day embryonic female mouse. A whole mount in situ hybridization of the *Hoxa13* probe is shown (red) along with a probe for *Hoxa10* (purple). *Hoxa9* expression extends throughout the uterus and through much of the presumptive oviduct. *Hoxa10* expression has a sharp anterior border at the transition between the presumptive uterus and the oviduct. *Hoxa11* has the same anterior border as *Hoxa10*, but its expression diminishes closer to the cervix. *Hoxa13* expression is found only in the cervix and upper vagina. (After Ma et al. 1998.)

and colleagues (1998) showed that the effects of DES on the female mouse reproductive tract could be explained as the result of altered *Hoxa10* expression in the Müllerian duct. DES was injected under the skin of pregnant mice and the fetuses were allowed to develop almost to birth. When the fetuses from DES-injected mothers were compared with fetuses from mothers that had not received DES, it was seen that DES almost completely repressed the expression of *Hoxa10* in the Müllerian duct (**FIGURE 24.11**). This repression was most pronounced in the stroma (mesenchyme) of the duct, where experimental embryologists had localized the effect of DES (Boutin and Cunha 1997). The case for DES acting through repression of *Hoxa10* is strengthened by the phenotype of the *Hoxa10* knockout mouse (Benson et al. 1996; Ma et al. 1998), in which there is a transformation of the proximal quarter of the uterus into oviduct tissue, as well as abnormalities of the uterotubal junction.

One link between Hox gene expression and uterine morphology is the Wnt proteins, which are associated with cell proliferation and protection against apoptosis. The Hox and Wnt proteins are both involved in the specification and morphogenesis of the reproductive tissues (**FIGURE 24.12**). The reproductive tracts of DES-exposed female mice resemble those of *Wnt7a* knockout mice. Hox genes and the Wnt genes communicate to keep each other activated; however, DES, acting through the estrogen receptor, represses the *Wnt7a* gene. This repression prevents the maintenance of the Hox gene expression pattern, and also prevents the activation of another Wnt gene, *Wnt5a*, which encodes a protein necessary for cell proliferation (Miller et al. 1998; Carta and Sassoon 2004).

WEB TOPIC 24.5 **DES AS AN OBESOGEN** Some endocrine disruptors, including DES, increase the production of fat cells and the accumulation of fat within these cells.

The effects of DES on fertility is a complex story of public policy, medicine, and developmental biology (Bell 1986; Palmlund 1996), and endocrine disruption by estrogenic compounds is ongoing. The Chapel Hill Consensus of 2007 (stemming from a conference sponsored by the Environmental Protection Agency and the National Institute of Environmental Health Sciences) claimed that some of the major constituents

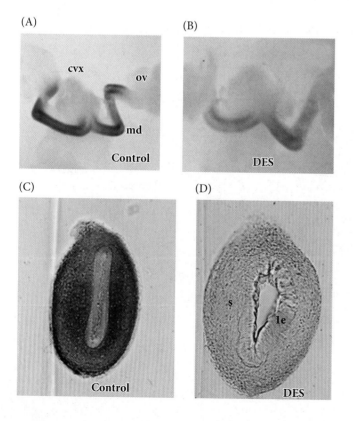

FIGURE 24.11 In situ hybridization of a *Hoxa10* probe shows that DES exposure represses *Hoxa10*. (A) Normal 16.5-day embryonic female mice show *Hoxa10* expression from the boundary of the cervix through the uterus primordium and most of the oviduct (cvx, cervix; md, Müllerian duct; ov, ovary). (B) In mice exposed prenatally to DES, this expression is severely repressed. (C) In control female mice at 5 days after birth (when reproductive tissues are still forming), a section through the uterus shows abundant expression of *Hoxa10* in the uterine mesenchyme. (D) In female mice that are given high doses of DES 5 days after birth, *Hoxa10* expression in the mesenchyme is almost completely suppressed (le, luminal epithelium; s, stroma). (After Ma et al. 1998.)

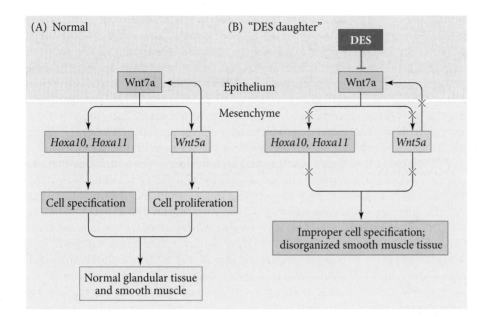

(A) Normal
(B) "DES daughter"

FIGURE 24.12 Misregulation of Müllerian duct morphogenesis by DES. (A) During normal morphogenesis, the *Hoxa10* and *Hoxa11* genes in the mesenchyme are activated and maintained by Wnt7a from the epithelium. Wnt7a also induces *Wnt5a* in the mesenchyme, and Wnt5a protein both maintains *Wnt7a* expression and causes mesenchymal cell proliferation. Together, these factors specify and order the morphogenesis of the uterus. (B) DES, acting through the estrogen receptor, blocks *Wnt7a* expression. Proper activation of the Hox genes and *Wnt5a* in the mesenchyme does not occur, leading to a radically altered morphology of the female genitalia. (After Kitajewsky and Sassoon 2000.)

of plastics were estrogenic compounds, and that they were present in doses large enough to have profound effects on human sexual development (vom Saal et al. 2007). The most important of these compounds is bisphenol A.

Bisphenol A (BPA)

In the early years of hormone research, the steroid hormones were very difficult to isolate, so chemists manufactured synthetic analogues that would accomplish the same tasks. Bisphenol A, one of these analogues, was first synthesized as an estrogenic compound in the 1930s. Later, polymer chemists realized that BPA could be used in plastic production, and today it is one of the top 50 chemicals produced worldwide. Four corporations in the United States make almost 2 billion pounds of it each year for use in the resin lining most cans, as well as the polycarbonate plastic in baby bottles, children's toys, and water bottles. It is also used in dental sealant and (strange as it sounds) in cash register receipts. In its modified form, tetrabromo-bisphenol A, it is the major flame retardant on fabrics.

Human exposure comes primarily from BPA that has leached from food containers (von Goetz et al. 2010). Babies and infants acquire BPA through polycarbonate bottles; teenagers and adults get most of their BPA by consuming canned food that has been stored in containers lined with BPA-containing resins. Since 95% of urine samples taken from people in the U.S. and Japan have measurable BPA levels (Calafat et al. 2005), public health concerns have been raised over the roles BPA might play in causing reproductive failure, cancer, and behavioral anomalies.

WEB TOPIC 24.6 BPA AND ALTERED BEHAVIOR   Fetal exposure of mice to BPA leads to changes in behavior. In humans, prenatal BPA exposure has been associated with aggression and hyperactivity.

BPA AND REPRODUCTIVE HEALTH   BPA does not remain fixed in plastic forever (Krishnan et al. 1993; vom Saal 2000; Howdeshell et al. 2003). If you let water sit in an old polycarbonate rat cage at room temperature for a week, you can measure about 300 μg per liter of BPA in the water. That is a biologically active amount—a concentration that will reverse the sex of a male frog and cause weight changes in the uterus of a young mouse. It also can cause chromosome anomalies. When a laboratory technician mistakenly rinsed some polycarbonate cages in an alkaline detergent, BPA was released

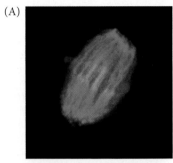

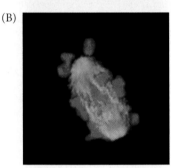

FIGURE 24.13 Bisphenol A causes meiotic defects in maturing mouse oocytes. (A) Chromosomes (red) normally line up at the center of the spindle during first meiotic metaphase. (B) Short exposures to BPA cause chromosomes to align randomly on the spindle. Diffrent numbers of chromosomes then enter the egg and polar body, resulting in aneuploidy and infertility. (From Hunt et al. 2003, courtesy of P. Hunt.)

from the plastic and the female mice housed in the cages showed meiotic abnormalities in 40% of their oocytes (the normal level of such abnormalities is about 1.5%). When BPA was administered to pregnant mice under controlled circumstances, Hunt and her colleagues (2003) showed that short, low-dose exposure to BPA was sufficient to cause meiotic defects in maturing mouse oocytes (**FIGURE 24.13**). This effect was also seen in primates. Exposure of fetal female monkeys to low doses of BPA (at levels comparable to that found in human serum) caused ovarian and meiotic abnormalities similar to those observed in mice. There were several abnormalities of ovarian function, including abnormal meiotic chromosome behavior and aberrant follicle formation (Hunt et al. 2012).

BPA crosses the human placenta and accumulates in concentrations that can alter development in laboratory animals (Ikezuki et al. 2002; Schönfelder et al. 2002). Indeed, women exposed to high levels of BPA during pregnancy had an 83% higher rate of miscarriages than women who had not been so heavily exposed (Lathi et al. 2014). In model organisms, BPA at environmentally relevant concentrations can cause abnormalities in fetal gonads, prostate enlargement, low sperm counts, and behavioral changes when these fetuses become adults (vom Saal et al. 1998, 2005; Palanza et al. 2002; Kubo et al. 2003). When vom Saal and colleagues (1997) gave pregnant mice 2 parts per billion BPA—that is, 2 nanograms per gram of body weight—for the 7 days at the end of pregnancy (equivalent to the period when human reproductive organs are developing), male offspring showed an increase in prostate size of about 30% (Wetherill 2002; Timms et al. 2005). Female mice exposed to low doses of BPA in utero had reduced fertility and fecundity as adults (Cabaton et al. 2007).

This lower fertility may be the result of several actions in addition to the above-mentioned effects on developing eggs. First, BPA and other endocrine disruptors are found to prevent the sex-specific maturation of those parts of the mouse brain regulating ovulation (Ruben et al. 2006; Gore et al. 2011). Second, female mice exposed in utero to low doses of BPA (2000 times lower than the dosage considered safe by the U.S. government) had alterations in the organization of their uterus, vagina, breast tissue, and ovaries, as well as altered estrous cycles as adults (Howdeshell et al. 1999, 2000; Markey et al. 2003). And third, BPA alters the gamete-specific methylation pattern of imprinted genes in mouse embryos and placentas (Susiarjo et al. 2013).

WEB TOPIC 24.7 **TESTICULAR DYSGENESIS** The amount of sperm produced by human males appears to have declined rapidly during the past 50 years. There is evidence that estrogen-enhancing endocrine disruptors are causing this decline.

BPA AND CANCER SUSCEPTIBILITY BPA appears to make breast tissue more sensitive to estrogens, and it is thought that in utero exposure to BPA may predispose women to breast cancer later in life. Fetal exposure to BPA caused the development of early-stage cancer in the mammary glands of one-third of the rats exposed to environmentally relevant doses of BPA later in life (Murray et al. 2006). None of the control rats developed such cancers. Furthermore, daily gestational exposure to as little as 25 ng BPA per kilogram of body weight, followed at puberty by a "subcarcinogenic dose" of a chemical carcinogen, resulted in the formation of tumors *only* in those animals exposed to BPA (Durando et al. 2006). Indeed, altered mammary development had already manifested during fetal life in BPA-exposed mice, and at puberty, the mammary glands produced more terminal buds and were more sensitive to estrogen, which may have predisposed these mice toward breast cancer as adults (Muñoz-de-Toro et al. 2005). Moreover, exposure of female monkey fetuses to low doses of BPA (at levels comparable to those found in human blood serum) caused changes in mammary development similar to those seen in BPA-exposed mice (**FIGURE 24.14**). In the above experiments, BPA was shown to be a factor that predisposed the rats to develop a cancer when they encountered estrogenic chemicals later in life. However, new studies with a different strain of rats have shown that when rat embryos are exposed to relatively small doses of BPA (levels considered safe by the EPA), they can develop palpable postnatal tumors without

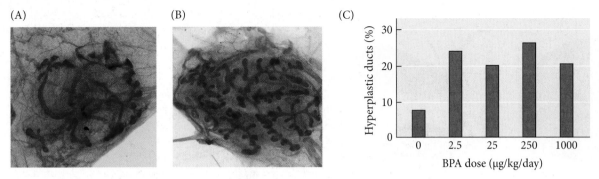

(A) (B) (C)

FIGURE 24.14 Bisphenol A induces altered mammary gland development. (A,B) Whole mount stained preparation of mammary glands from newborn female rhesus monkeys. (A) Control mammary gland. (B) Mammary gland from a fetus exposed in utero to BPA. Twice as many buds (incipient branches) are seen in the BPA-exposed tissue. (C) The percentage of mouse mammary glands showing intraductal hyperplasia (a cancer-prone state) is significantly increased at postnatal day 50 in BPA-exposed animals. (A,B from Tharp et al. 2012; C from Murray et al. 2007.)

having to have a second BPA experience later in life. Acevedo and colleagues (2013) conclude that "BPA may act as a complete mammary gland carcinogen."

The plastics industry claims that BPA is safe and that rodents exposed to BPA in utero do not show developmental anomalies (see Cagen et al. 1999; Lamb 2002). However, reviews of the industry's studies points out that these experiments were improperly done (the positive controls did not show the expected effects) and concluded that BPA is one of the most dangerous chemicals known and that governments should consider banning its use in products containing liquids that humans and animals might drink (vom Saal and Hughes 2005; Chapel Hill Consensus 2007; Myers et al. 2009; Gioiosa et al. 2015).

Indeed, when confronted with one study (Lernath et al. 2008) showing that, at concentrations *lower* than what the U.S. EPA considers safe, BPA disrupted monkey brain development, the American Chemical Council replied that "there is no direct evidence that exposure to bisphenol A adversely affects human reproduction or development" (see Layton 2008; Gilbert and Epel 2015). The catch is that "direct evidence" would mean testing the drugs in known concentrations on human fetuses, which cannot morally be done. In the absence of government regulation, Nalgene and Wal-Mart voluntarily stopped making and selling BPA-containing bottles.

 SCIENTISTS SPEAK 24.5 Dr. Frederick vom Saal discusses the endocrine disruptive effects of BPA and the problems of regulating this compound.

Atrazine: Endocrine disruption through hormone synthesis

The enzyme aromatase can convert testosterone into estrogen, and this estrogen is able to induce female sex determination in many vertebrates. In turtles, for instance, estrogen downregulates testis-forming genes and upregulates those genes producing ovaries (Valenzuela et al. 2013; Bieser and Wibbels 2014). PCBs and BPA can also reverse the sex of turtles raised at "male" temperatures (Jandegian et al. 2015). This and other studies (e.g., see Bergeron et al. 1994, 1999) have important consequences for environmental conservation efforts to protect endangered species (including turtles, amphibians, and crocodiles) in which hormones can effect changes in primary sex determination.

The survival of some amphibian species may be at risk from herbicides that promote estrogens at the expense of testosterone, severely depleting the number, function, and fertility of the males. One such case involves the development of hermaphroditic and demasculinized frogs after exposure to extremely low doses of the weed killer atrazine, one of the most widely used herbicides in the world, and one that is found in streams and ponds throughout the United States (**FIGURE 24.15**). Atrazine induces aromatase, which as mentioned, above converts testosterone into estrogen. Hayes and colleagues (2002a) found that exposing tadpoles to atrazine concentrations as low as 0.1 part per billion (ppb) produced gonadal and other sexual anomalies in male frogs. At 0.1 ppb and higher, many male tadpoles developed ovaries in addition to testes. At 1 ppb atrazine, the vocal sacs (which a male frog must have in order to signal and obtain a potential mate) failed to develop properly. Similar experiments in outdoor environments more

(A)

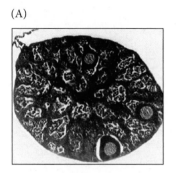

(B)

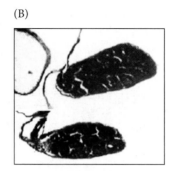

(C)

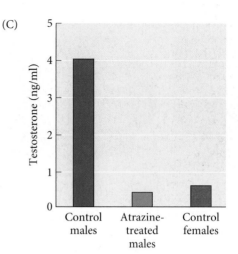

FIGURE 24.15 Demasculinization of frogs by low amounts of atrazine. (A) Testis of a frog from a natural site having 0.5 parts per billion (ppb) atrazine. The testis contains three lobules that are developing both sperm and an oocyte. (B) Two testes of a frog from a natural site containing 0.8 ppb atrazine. These organs show severe testicular dysgenesis, which characterized 28% of the frogs found at that site. (C) Effect of a 46-day exposure to 25 ppb atrazine on testosterone levels in the blood plasma of sexually mature male *Xenopus*. Levels in control males were some tenfold higher than in control females; atrazine-treated males had plasma testosterone levels at or below those of control females. (A,B after Hayes et al. 2003, photographs courtesy of T. Hayes; C after Hayes et al. 2002a.)

similar to natural conditions (Langlois et al. 2010) also showed that male frogs (*Rana pipiens*) had been transformed into females by atrazine.

In laboratory experiments, the testosterone levels of adult male frogs were reduced by 90% (to levels of control females) when they were exposed, as sexually mature adults, to 25 ppb atrazine (Hayes et al. 2002a). This is an ecologically relevant dose, since the allowable amount of atrazine in U.S. drinking water is 3 ppb, and atrazine levels can reach 224 ppb in streams of the midwestern United States (Battaglin et al. 2000; Barbash et al. 2001). Even at doses as low as 2.5 ppb, the sexual behavior of the male frogs was severely diminished. Matings became relatively rare, and in 10% of the cases, males exposed to atrazine became functional egg-laying females (Hayes et al. 2010).

In a field study, Hayes and his colleagues collected leopard frogs and water samples from eight sites across the central United States (Hayes et al. 2002b, 2003). They sent the water samples to two separate laboratories to determine their atrazine levels, and they coded the frog specimens so that the technicians dissecting the gonads did not know which site the animals came from. The results showed that the water from all but one site contained atrazine—and this was the only site from which the frogs had no gonadal abnormalities. At concentrations as low as 0.1 ppb, leopard frogs displayed testicular dysgenesis (stunted growth of the testes) or conversion to ovaries. In many examples, oocytes were found in the testes (see Figure 24.15).

Atrazine's ability to feminize male gonads has been seen in all classes of vertebrates. Indeed, low sperm count, poor semen quality and decreased fertility have been observed in men who are routinely exposed to atrazine (Swan et al. 2003; Hayes et al. 2011). Concern over atrazine's apparent ability to disrupt sex hormones in both wildlife and humans has resulted in bans on the use of this herbicide by France, Germany, Italy, Norway, Sweden, and Switzerland (Dalton 2002). However, drug companies in the United States have lobbied successfully to keep atrazine in the American markets (see Blumenstyk 2003; Aviv 2014).

 SCIENTISTS SPEAK 24.6 Dr. Tyrone Hayes describes his studies documenting that a leading herbicide turns male frogs into females and is associated with genital abnormalities and cancers in humans.

Fracking: A potential new source of endocrine disruption

U. S. government regulations do not cover the compounds added to the environment by the hydraulic fracturing ("fracking") procedures used to extract methane (natural gas) from shale. A total of 632 chemicals have been identified as being used in this procedure. About 25% of them are known to cause tumors, and more than 35% of them are known affect the endocrine system (Colburn et al. 2011). It is estimated that about

50% of the fluid used in fracking returns to the surface (DOE 2009). Water samples taken from both standing water and groundwater at fracking sites contained estrogenic compounds, antiestrogenic compounds, and antiandrogenic (antitestosterone) compounds (Kassotis et al. 2014; Webb et al. 2014). Some compounds in the water activated enhancers of estrogen-responsive genes and others prevented the activation of testosterone-responsive genes. One of the sites where the water was tested had been a ranch prior to drilling and fracking, but ranching had to be discontinued because the animals were no longer producing offspring. A recent study concerning fracking in rural Colorado documented an increased incidence of congenital heart disease in children born in families residing close to the fracking wells (McKenzie et al. 2014).

 SCIENTISTS SPEAK 24.7 Dr. Susan Nagel discusses her identification of endocrine-disrupting chemicals in water produced by hydraulic fracturing (fracking) activities.

Transgenerational Inheritance of Developmental Disorders

A lifetime of chopping wood will not give your offspring bulging biceps, nor would the loss of your arms in an accident cause your offspring to be born limbless. This is because the environmental agents—exercise or trauma in these cases—do not cause mutations in the DNA. In order to be transmitted, mutations must not only be somatic, they must enter the germ line. Thus, genetic mutations acquired in skin cells that are overexposed to sunlight will not be transmitted. However, one of the most surprising results of contemporary developmental genetics has been the discovery that certain environmentally induced phenotypes *can* be transmitted from generation to generation. DNA methylation seems to be a mechanism that can circumvent the mutational block to the transmission of acquired traits.

Certain agents can cause the same alterations of DNA methylation throughout the body, and these alterations in methylation can be transmitted by the sperm and egg. Jablonka and Raz (2009) have documented dozens of cases where different "epialleles"—DNA containing different methylation patterns—can be stably transmitted from generation to generation. In mammals, epiallelic inheritance was first documented by studies of the endocrine disruptor vinclozolin, a fungicide widely used on grapes. When injected into pregnant rats during particular days of gestation, vinclozolin caused testicular dysgenesis in the male offspring. The testes started forming normally, but as the mouse got older, its testes degenerated and quit producing sperm. What's more interesting is that the male mice fathered by mice with this induced testicular dysgenesis (often by artificial means) also get testicular dysgenesis. So do their male offspring and the subsequent generation's male offspring (Anway et al. 2005, 2006; Guerrero-Bosagna et al. 2010). Thus, when a pregnant rat is given vinclozolin, even her great-grandsons are affected (**FIGURE 24.16**).

The mechanism for this inheritance in rats appears to be DNA methylation. The promoters of more than 100 genes in the Sertoli cells (see Chapter 6) have their methylation patterns changed by vinclozolin, and altered promoter methylation can be seen in the sperm DNA for at least three subsequent generations (Guerrero-Bosagna et al. 2010; Stouder and Paolini-Giacobino 2010). These genes include those whose products are necessary for cell proliferation, G proteins, ion channels, and receptors. It is important to note that by the third (F_3) generation, there could have been no direct exposure to vinclozolin. The fetus is inside the treated mother; and the fetus has germ cells (of the F_2 generation) inside itself. But even though the offspring of the F_3 and F_4 generations have never been exposed to vinclozolin, their phenotype is changed by the initial injection to their great-grandmother.

Similar studies have indicated that other endocrine disruptors—DES, bisphenol A, and PCBs—also have transgenerational effects (Skinner et al. 2010; Walker and Gore

FIGURE 24.16 Epigenetic transmission of endocrine disruption. (A) Transmission of testicular dysgenesis syndrome (red circles) is shown through four generations of mice. The only mice exposed in utero were the F_1 generation. (B-C) Cross section of the seminiferous tubules from the testes of (A) a control male rat and (B) a male rat whose grandfather was born from a mother injected with vinclozolin. The arrow in (B) shows the tails of normal sperm. The arrow in (C) shows the lack of germ cells in the much smaller tubule of the rat descended from the vinclozolin-injected female; this rat was infertile under normal conditions. (After Anway et al. 2005; Anway and Skinner 2006; courtesy of M. K. Skinner.)

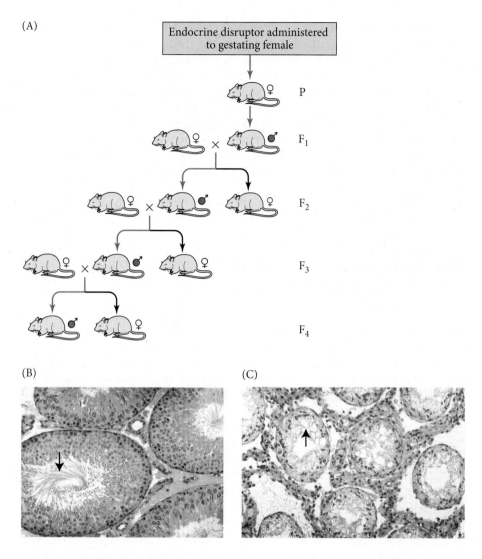

2011). Indeed, the behavioral changes induced by BPA in mice may last at least four generations (Wolstenhome et al. 2012). The public health ramifications of this type of inheritance are just beginning to be explored.

SCIENTISTS SPEAK 24.8 Dr. Michael Skinner discusses his research on transgenerational epigenetic diseases.

Cancer as a Disease of Development

Because endocrine disruptors are known to cause tumors as well as developmental abnormalities, cancer is increasingly being studied as a disease of development. However, the idea that cancer is a disease of development has been recognized for many years (see, e.g., Stevens 1953; Auerbach 1961; Pierce et al. 1978). Carcinogenesis is more than just genetic changes in the cells giving rise to the tumor (see Hanahan and Weinberg 2000). Rather, carcinogenesis can be viewed as aberrations of the very processes that underlie differentiation and morphogenesis. Indeed, a recent study of melanomas in zebrafish (Kaufman 2016) showed that when the melanocytes (pigment cells) had two of the mutations found in numerous cancer cells, they formed pigmented nodules but not tumors. Melanomas (cancerous growth of melanocytes) occurred only when the pigment cells also expressed markers of the neural crest precursor cells that had given rise to the melanocytes in the embryo.

It once was thought that carcinogensis and metastasis were caused by the proliferation of a cell that had acquired mutations enabling it to become "autonomous," thus defining cancer by intracellular mechanisms that enable a cell to become independent of its environment. But this turns out to be only part of the explanation. We now know that the initial cancer cells modify their environment, turning it into a cancer-promoting niche. Cancer is being recast as the result of a stepwise progression of conditions that depends on reciprocal interactions between incipient cancer cells and the supporting cells of their tissue environment. The progressive alteration of cell-cell interactions leads to aberrant tissue architecture, and possibly to the formation of niches that generate cancer cells. Indeed, cancer cells appear to proceed by recapitulating steps of normal development, including the formation of a niche in which to proliferate. Thus, both carcinogenesis and congenital anomalies can be seen as diseases of tissue organization, differentiation, and intercellular communication. As we will see, they are often caused by defects in the same pathways. There are many reasons to view malignancy and metastasis in terms of development, four of which will be discussed here:

1. Context-dependent tumor formation
2. Defects in cell-cell communication as the initiator of cancers
3. Cancer stem cells
4. Epigenetic reprogramming of cancer cells

CONTEXT-DEPENDENT TUMORS Many tumor cells have normal genomes, and whether or not these tumors become malignant depends on their environment (Pierce et al. 1974; Mack et al. 2014). The most remarkable of these cases is the **teratocarcinoma**, a tumor of germ cells or stem cells (Illmensee and Mintz 1976; Stewart and Mintz 1981). Teratocarcinomas are malignant growths of cells that resemble the inner cell mass of the mammalian blastocyst, and they can kill the organism. However, if a teratocarcinoma cell is placed on the inner cell mass of a mouse blastocyst, it will integrate into the blastocyst, lose its malignancy, and divide normally. Its cellular progeny can become part of numerous embryonic organs. Should its progeny form part of the germ line, the sperm or egg cells formed from the tumor cell will transmit the tumor genome to the next generation. Thus, whether the cell becomes a tumor or becomes part of the embryo can depend on its surrounding cells.

It is possible that the stem cell environment suppresses tumor formation by its secretion of inhibitors of the paracrine pathways. For instance, many tumor cells, such as melanomas, secrete the paracrine factor Nodal. This aids their proliferation and also helps supply them with blood vessels. When placed in an environment of embryonic stem cells (which secrete Nodal inhibitors), aggressive melanoma tumors (which are derived from neural crest cells) become normal pigment cells (Hendrix et al. 2007; Postovit et al. 2008). Remarkably, such malignant melanoma cells, when transplanted into early chick embryos, downregulate their Nodal expression and migrate as nonmalignant cells along the neural crest cell pathways (**FIGURE 24.17**; Kasemeier-Kulesa et al. 2008).

DEFECTS IN CELL-CELL COMMUNICATION In many cases, tissue interactions are required to prevent cells from dividing, leading to the premise that cancer can be caused by miscommunication between cells. Thus, tumors can arise through defects in tissue architecture, and the surroundings of a cell are critical in determining malignancy (Sonnenschein and Soto 1999, 2000; Bissell et al. 2002). Studies have shown that tumors can be caused by altering the structure of the tissue, and that these tumors can be suppressed by restoring an appropriate tissue environment (Coleman et al. 1997; Weaver et al. 1997; Booth et al. 2010). In particular, although 80% of human tumors are from epithelial cells, these cells do not always appear to be the site of the cancer-causing lesion. Rather, epithelial cell cancers are often caused by defects in the mesenchymal stromal cells that surround and sustain the epithelia. When Maffini and colleagues

FIGURE 24.17 When aggressively metastatic human melanoma cells are injected into a 2-day chick embryo dorsal neural tube (A), they form normal migratory chains (B) and follow the neural crest migration roots to integrate into facial cartilage (C) and sympathetic ganglia (D). There, they form non-malignant melanocytes. (From Kasemeier-Kulesa et al. 2008.)

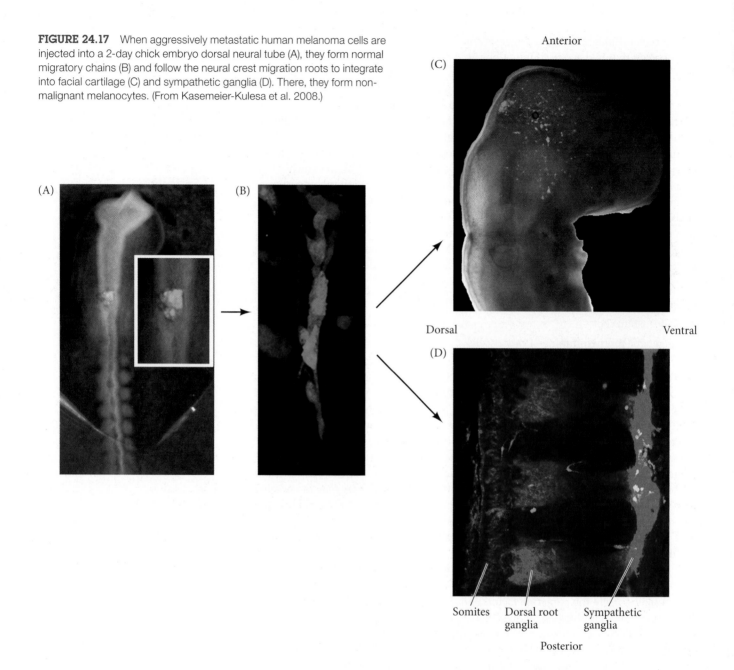

(A)   (B)   (C) Anterior

Dorsal   Ventral

(D)

Somites   Dorsal root   Sympathetic
           ganglia   ganglia

Posterior

(2004) recombined normal and carcinogen-treated epithelia and mesenchyme in rat mammary glands, tumorous growth of mammary epithelial cells occurred not in carcinogen-treated epithelia, but only in epithelia placed in combination with mammary mesenchyme that had been exposed to the carcinogen. Thus, the carcinogen caused defects in the mesenchymal stroma of the mammary gland, and apparently the treated mammary stroma could no longer provide the epithelial cells the instructions to form normal structures. In turn, these abnormal structures exhibited a loose control of cell proliferation (**FIGURE 24.18**). These findings have led to a new appreciation of the ways in which stromal cells can regulate cancer initiation in the adjacent epithelium (see Wagner et al. 2016).

DEFECTS IN PARACRINE PATHWAYS This brings us to the next notion: that tumors can occur by disruptions of paracrine signaling between cells. Rubin and de Sauvage (2006) concluded that "several key signaling pathways, such as Hedgehog, Notch, Wnt and BMP/TGF-β/Activin, are involved in most processes essential to the proper development of an

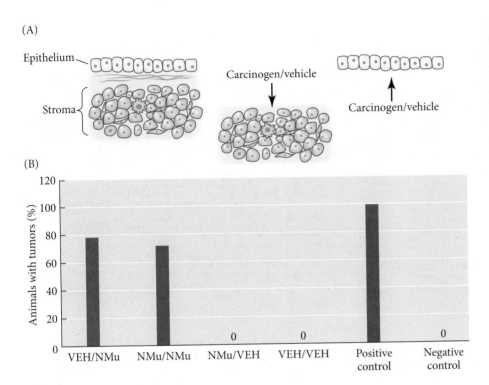

FIGURE 24.18 Evidence that the stroma regulates the production of epithelial (parenchymal) tumors. (A) Schematic drawing of the experimental protocol. The mammary gland tissue contains both epithelium and stroma (mesenchymal cells). The two groups of cells can be isolated and then recombined. One can add a cancer-forming substance (carcinogen) to the epithelium and not the stroma, or to the stroma but not the epithelium. Then one can combine them so that the cancer-causing substance has been experienced by the epithelium (but not the stroma), by the stroma (but not the epithelium), or by both stroma and epithelium, or by neither. (B) Results when the cancer-causing mutagen Nmethylnitrosourea (NMu) or just the control vehicle (VEH) was applied to either the stroma or epithelium and transplanted back into the rat mammary gland. On the horizontal axis, the upper denominator refers to the epithelium and the lower denominator refers to the stroma. Only animals whose stroma was treated with NMu developed tumors, regardless of whether the epithelium was exposed to NMu or not. NMu-treated, intact animals (positive controls) developed tumors, and none of the rats receiving control solutions (negative controls) had tumors. (After Maffini et al. 2004.)

embryo. It is also becoming increasingly clear that these pathways can have a crucial role in tumorigenesis when reactivated in adult tissues through sporadic mutations or other mechanisms." We saw this above, in the discussion of Nodal secretion by melanoma cells. These findings demonstrate the importance of stromal tissue just mentioned. Many tumors, for instance, secrete the paracrine factor Sonic hedgehog (Shh), which can act in one of two ways. First, it can act in an autocrine fashion, stimulating the cells that produce it to grow. Autocrine Shh is normally required for maintenance of cerebellar granule neuron progenitor cells and hematopoietic stem cells; Shh pathway inhibitors can reverse certain medulloblastomas and leukemias, which are tumors of these cell types (**FIGURE 24.19A,B**; Rubin and de Sauvage et al. 2006; Zhao et al. 2009). An autocrine requirement for hedgehog has also been reported for small-cell lung carcinoma, pancreatic adenocarcinoma, prostate cancer, breast cancer, colon cancer, and liver cancer. Second, in some instances, the Shh produced by the tumor cells may not act on the tumor cells themselves but on the stromal cells, causing the stromal cells to produce factors (such as insulin-like growth factor, IGF) that support the tumor cells (**FIGURE 24.19C**). If the Shh pathway is blocked, the tumor regresses (Yauch et al. 2008, 2009; Tian et al. 2009). Cyclopamine, a teratogen that blocks Shh signaling, can prevent certain of these tumors from growing (Berman et al. 2002, 2003; Thayer et al. 2003; Song et al. 2011).

Thus, the same chemicals that can cause teratogenesis by blocking a pathway in embryonic development may be useful in blocking the activation of cancer stem cells. Cyclopamine and other antagonists of the Shh pathway, for instance, can cause malformation in embryos, but they appear to be useful in preventing the generation and proliferation of medulloblastoma stem cells (Berman et al. 2002; De Smale et al. 2010). Even the classic teratogen thalidomide is being "rehabilitated" for use in the fight against cancer.

THE CANCER STEM CELL HYPOTHESIS Another aspect of viewing cancers as diseases of development is that the properties of tumors may emerge because of a population of cells that are analogous to adult stem cells. The idea that cancers had stem cells was one of the first links connecting cancer research and developmental biology. Pierce and Johnson (1971) reported that "malignant tissue, like normal tissue, maintains itself by proliferation and differentiation of its stem cells." That same year, Pierce and

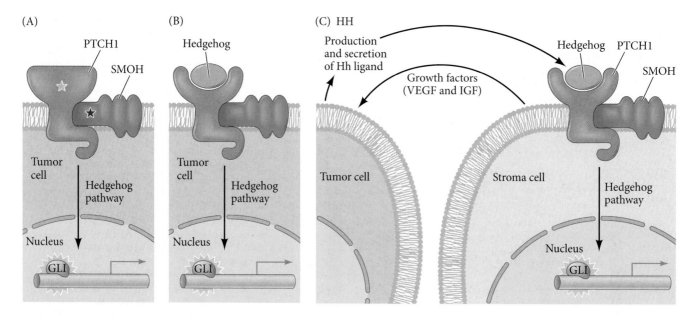

FIGURE 24.19 Mechanisms by which the Hedgehog pathway (see Figure 4.30) can lead to cancer. (A) When Shh is a mitogen (as it is for cerebellar granule neuron progenitor cells or hematopoietic stem cells), loss-of-function mutations in the Hh ligand Patched (PTCH1; yellow star) or gain-of-function mutations in the Patched inhibitor Smoothened (SMOH; blue star) activate the Hedgehog pathway, even in the absence of Shh or another Hedgehog protein. (B) In the autocrine model, tumor cells both produce and respond to the Hh ligand. (C) In the paracrine model, tumor cells produce and secrete the Hh ligand, and the surrounding stromal cells receive the Hh protein. The stromal cells respond by producing growth factors such as VEGF or IGF that support tumor growth or survival. (After Rubin and de Sauvage 2006.)

Wallace (1971) demonstrated the presence of stem cells in rat carcinomas. The similarities between normal stem cells and cancer stem cells was highlighted when lineage tracing revealed that the stem cells of intestinal adenomas (the precursor of intestinal cancer) are Lgr5+ and have the same relationship to the Paneth cells (see Chapter 5) as do normal intestinal stem cells (Schepers et al. 2013).

In numerous cancers, including glioblastomas (the most common brain tumor), prostate cancer, melanomas, and myeloid leukemias, a rapidly dividing cancer stem cell (CSC) population gives rise to more cancer stem cells and to populations of relatively slowly dividing differentiated cells (Lapidot et al. 1994; Chen et al. 2012; Driessens et al. 2012; Schepers et al. 2012). These CSCs can self-renew as well as generating the non-stem-cell populations of the tumor. Indeed, when tumor cells are transplanted from one animal to another, only the CSCs can give rise to new heterogeneous tumors (Gupta et al. 2009; Singh and Settleman 2010). The origins of CSCs remains uncertain and may be different for different tumor types. Most researchers feel that the CSC comes from either a normal adult stem cell or a progenitor (transit-amplifying) cell.

While the tumor is forming, CSCs produce more cancer stem cells as well as the bulk of the more differentiated tumor cells. Remarkably, it appears that the CSCs of aggressive glioblastomas not only make immature glia-like cells (glioblasts), they also make blood vessel endothelial cells. In this way, the tumor can create its own vasculature (El Hallani et al. 2010; Ricci-Vitiani et al. 2010; Wang et al. 2010).

Developmental therapies for cancer

Cancer is not so much the result of a cell gone bad as it is of cell relationships gone awry. Cancers are often diseases of developmental signaling, and several types of cancer cells can be normalized when placed back into regions of embryos that express certain paracrine factors or their inhibitors. This developmental view of cancer allows us to explore new avenues for cancer treatment. One such mode of treatment, **differentiation therapy**, was considered possible as long as 30 years ago but was not feasible at the time.

In 1978, Pierce and his colleagues noted that cancer cells were in many ways reversions to embryonic cells, and they hypothesized that cancer cells should revert to normalcy if they were made to differentiate. Also in 1978, Sachs discovered that certain leukemias could be controlled by making the leukemic cells differentiate rather than proliferate. One of these leukemias, acute promyelocytic leukemia (APL), is caused by a somatic recombination creating a "new" transcription factor, one of whose subunits is a retinoic acid receptor. This receptor, even in the absence of retinoic acid, binds to

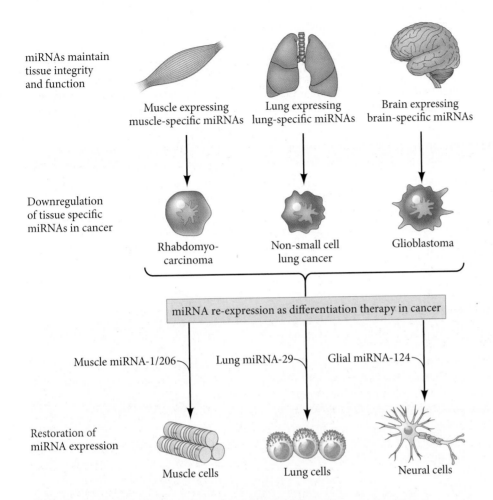

FIGURE 24.20 MicroRNA insertion as a possible means of differentiation therapy. Many cell types (such as those in muscle, lung, and brain) dedifferentiate when forming a tumor. This dedifferentiaton is accompanied by loss of specific microRNAs that maintain the differentiated cell's DNA methylation patterns. Restoring those microRNAs to tumor cells reestablished the differentiated pattern of DNA methylation. (After Mishra and Merlino 2009.)

the RA binding sites in DNA, where it represses RA-responsive genes as well as creates a larger condensed chromatin structure (Nowak et al. 2009). Expression of this "new" transcription factor in neutrophil progenitors causes the cell to become malignant (Miller et al. 1992; Grignani et al. 1998). Treatment of APL patients with all-*trans* retinoic acid results in remission in more than 90% of cases because the additional RA is able to effect the differentiation of the leukemic cells into normal neutrophils (Hansen et al. 2000; Fontana and Rishi 2002).

Recently, microRNAs have begun to be tested in differentiation therapy. In many tumors, there are specific microRNAs that are downregulated (Berdasco 2009; Mishra and Merlino 2009). These downregulated microRNAs are usually tumor suppressors that prevent changes in DNA methylation. Taulli and colleagues (2009) have shown that the microRNA miRNA-206, which is normally present in skeletal muscle cells, is downregulated in muscle cell tumors. Adding miRNA-206 to muscle tumor cells restores their differentiated phenotype and blocks cancer formation. This suggests a tissue-specific mechanism for stopping cancers by causing them to differentiate (**FIGURE 24.20**).

Coda

Developmental biology is increasingly important in modern medicine. Preventive medicine, public health, and conservation biology demand that we learn more about the mechanisms by which industrial chemicals and drugs can damage embryos. The ability to effectively and inexpensively assay compounds for potential harm is critical. Developmental biology also provides new ways of understanding carcinogenesis and new approaches to preventing and curing cancers. And finally, developmental biology is providing the explanations for how mutated genes and aneuploidies cause their aberrant phenotypes.

It is critical to realize that the agents we put into the environment, the cosmetics we put on our skin, and the substances we eat and drink can reach developing embryos, fetuses, and larvae. Developing organisms have different physiologies as they construct, rather than merely sustain, their phenotypes, and chemicals that appear harmless to adults may disrupt the development of embryos. It takes a community to raise an embryo.

Next Step Investigation

One of the most important aspects of current investigation concerns the effects of endocrine disruptors on behavior, the immune system, and cancer. It is critically important to know whether teratogens and endocrine disruptors experienced in utero predispose us to cancers, asthma, obesity, or cognitive problems later in life. Some endocrine disruptors appear to affect the immune systems of children and can nullify the expected immunity of vaccination. It will also be important to know whether such environmental chemicals (or agents such as stress) can affect the chromatin in ways that cause transgenerational inheritance of the phenotype.

Closing Thoughts on the Opening Photo

Defects in development can be due to either genetic or environmental causes. In the polydactylous cat seen here, the cause is genetic, and it involves an enhancer for the *sonic hedgehog* gene (see Chapter 19). Polydactyly is not uncommon among mammals, including humans. (Robert Chambers, an early evolutionist, was a complete hexadactyl, having twelve toes and twelve fingers.) Such relatively minor "glitches" in development add diversity to a population; you can probably envision environments where large numbers of digits might be useful. It is possible that many evolutionary adaptations start off as congenital anomalies. (Photograph © Jane Burton/Dorling Kindersley/Corbis.)

24 Snapshot Summary
Development in Health and Disease

1. Developmental anomalies due to genetic errors and environmental influences result in a relatively low rate of survival of all human conceptions.

2. Chance plays a role in developmental outcomes. There is large variation in the amounts of transcription and translation, such that at different times, cells are making more or less developmentally important proteins.

3. Pleiotropy occurs when several different effects are produced by a single gene. In mosaic pleiotropy, each effect is caused independently by the expression of the same gene in different tissues. In relational pleiotropy, abnormal gene expression in one tissue influences other tissues, even though those other tissues do not express that gene.

4. Genetic heterogeneity occurs when mutations in more than one gene can produce the same phenotype. Phenotypic heterogeneity arises when the same gene can produce different defects (or differing severities of the same defect) in different individuals.

5. Teratogenic agents include chemicals such as alcohol and retinoic acid, as well as heavy metals, certain pathogens, and ionizing radiation. These agents adversely affect normal development and can result in malformations and functional deficits.

   - There may be multiple effects of alcohol on cells and tissues that result in this syndrome of cognitive and physical abnormalities.

   - The compound retinoic acid is active in development, and too much or too little of it can cause congenital anomalies.

6. Endocrine disruptors can bind to or block hormone receptors or block the synthesis, transport, or excretion of hormones. Presently, bisphenol A and other endocrine disruptive compounds are being considered as possible agents of low sperm counts in men and a predisposition to breast cancer in women.

   - Environmental estrogens can cause reproductive system anomalies by suppressing Hox gene expression and Wnt pathways. These substances can also cause obesity, and in some cases they activate the transcription factors predisposing mesenchymal stem cells to differentiate into adipose tissue.

- In some instances, endocrine disruptors methylate DNA, and these patterns of methylation can be inherited from one generation to the next. Such methylation can alter metabolism and development by turning genes off.

7. Cancer can be seen as a disease of altered development. Cancers metastasize in manners similar to embryonic cell movement, and some tumors revert to nonmalignancy when placed in environments that support normal morphogenesis and curtail excessive cell proliferation.

- Cancers can arise from errors in cell-cell communication. These errors include alterations of paracrine factor synthesis.

- In many instances, tumors have a rapidly dividing cancer stem cell population which produces more cancer stem cells as well as more quiescent and differentiated cells.

Further Reading

Anway, M. D., A. S. Cupp, M. Uzumcu and M. K. Skipper. 2005. Epigenetic transgeneration effects of endocrine disruptors and male fertility. *Science* 308: 1466–1469.

Bissell, M. J., D. C. Radisky, A. Rizki, V. M. Weaver and O. W. Petersen. 2002. The organizing principle: Microenvironmental influences in the normal and malignant breast. *Differentiation* 70: 537–546.

Gilbert, S. F. and D. Epel. 2015. *Ecological Developmental Biology: The Environmental Regulation of Development, Health, and Evolution,* 2nd edition. Sinauer Associates. Sunderland, MA.

Guerrero-Bosagna, C., M. Settles, B. Lucker and M. K. Skinner. 2010. Epigenetic transgenerational actions of vinclozolin on promoter regions of the sperm epigenome. *PLoS One* Sep 30;5(9). pii: e13100.

Hanahan, D., and R. A. Weinberg. 2011. Hallmarks of cancer: The next generation. *Cell* 144: 646–674.

Hayes, T. B. and 21 others. 2011. Demasculinization and feminization of male gonads by atrazine: Consistent effects across vertebrate classes. *J. Steroid Biochem. Mol. Biol.* 127: 64–73.

Howdeshell, K. L., A. K. Hotchkiss, K. A. Thayer, J. G. Vandenbergh and F. S. vom Saal. 1999. Plastic bisphenol A speeds growth and puberty. *Nature* 401: 762–764.

Lammer, E. J. and 11 others. 1985. Retinoic acid embryo-pathy. *New Engl. J. Med.* 313: 837–841.

Maffini, M. V., A. M. Soto, J. M. Calabro, A. A. Ucci and C. Sonnenschein. 2004. The stroma as a crucial target in mammary gland carcinogenesis. *J. Cell Sci.* 117: 1495–1502.

Steingraber, S. 2003. *Having Faith: An Ecologist's Journey to Motherhood.* New York: The Berkley Publishing Group.

Sulik, K. K. 2005. Genesis of alcohol-induced craniofacial dysmorphism. *Exp. Biol. Med.* 230: 366–375.

Tang, H. and 14 others. 2016. Zika virus infects human cortical neural progenitors and attenuates their growth. *Cell Stem Cell.* doi: org/10/1016/j.stem.2016.02.016.

GO TO WWW.DEVBIO.COM...

...for Web Topics, Scientists Speak interviews, Watch Development videos, Dev Tutorials, and complete bibliographic information for all literature cited in this chapter.

Development and the Environment

Biotic, Abiotic, and Symbiotic Regulation of Development

Why should microbes be considered an important part of normal development?

IT WAS LONG THOUGHT THAT THE ENVIRONMENT played only a minor role in development. Nearly all developmental phenomena were believed to be a "readout" of nuclear genes, and those organisms whose development *was* significantly controlled by the environment were considered interesting oddities. When environmental agents played roles in development, they appeared to be destructive, such as the roles played by teratogens and endocrine disruptors (see Chapter 24). However, recent studies have shown that *the environmental context plays significant roles in the normal development of almost all species, and that animal genomes have evolved to respond to environmental conditions.* Moreover, there are symbiotic associations wherein the development of one organism is regulated by the molecular products of organisms of other species. In fact, such cases appear to be the rule rather than the exception.

One reason developmental biologists have largely ignored the environment's effects is that most animals studied in developmental biology—*C. elegans*, *Drosophila*, zebrafish, *Xenopus*, chicks, and laboratory mice—have been selected for their lack of such effects (Bolker 2012). These model organisms make it easier to study the genes that regulate development, but they can leave us with the erroneous impression that everything needed to form the embryo is present in the fertilized egg. With new concerns about the loss of organismal diversity and the effects of environmental pollutants, there is renewed interest in the regulation of development by the environment (see van der Weele 1999; Bateson and Gluckman 2011; Gilbert and Epel 2015).

The Punchline

The material within a fertilized egg does not fully determine the organism's phenotype. Rather, the inherited genome can respond to numerous environmental factors. Chemical signals from symbionts, usually bacteria, are needed for normal development. In mammals, the gut microbiome is acquired at birth and is critical for the development of the gut, the capillary network, and the immune system. It is possible that it is also needed for normal development of the brain. In addition, abiotic conditions such as temperature can be critical for normal development. In many non-mammalian vertebrates, the sex of an organism depends on the temperature it experiences during development. And finally, biotic conditions such as diet, conspecific crowding, or the presence of predators can alter development in ways that enable the phenotype to be more adaptive. Responding to environmental cues can help the organism to integrate into its habitat.

The Environment as a Normal Agent in Producing Phenotypes

Phenotypic plasticity is the ability of an organism to react to an environmental input with a change in form, state, movement, or rate of activity (West-Eberhard 2003; Beldade et al. 2011). When the difference occurs in the embryonic or larval stages of animals or plants, this ability to change phenotype is often called **developmental plasticity**. We have already encountered several examples of developmental plasticity. When we discussed environmental sex determination in turtles (see Chapter 6, pp. 201–202), we were aware that the sexual phenotype was being instructed not by the genome but by the environment. When we mentioned in Chapter 18 the ability of shear stress to activate gene expression in capillary, heart, and bone tissue, we similarly were noting the effect of an environmental agent on phenotype. Although studies of phenotypic plasticity have played a central role in plant developmental biology, the mechanisms of plasticity have only recently been studied in animals. These studies now show that developmental plasticity is a critical means for integrating animals into their ecological communities.

Two main types of phenotypic plasticity are currently recognized: reaction norms and polyphenisms (Woltereck 1909; Schmalhausen 1949; Stearns et al. 1991). In a **reaction norm**, the genome encodes the potential for a *continuous range* of potential phenotypes, and the environment the individual encounters determines the phenotype (often the most adaptive one) that emerges. For instance, human muscle phenotype is determined by the amount of exercise the body is exposed to over time (even though there is a genetically defined limit to how much muscular hypertrophy is possible). The upper and lower limits of a reaction norm, as well as the kinetics of how rapidly the trait changes in response to the environment, are properties of the genome that can be selected. The different phenotypes produced by environmental conditions are called **morphs** (or occasionally **ecomorphs**).

The second type of phenotypic plasticity, **polyphenism**, refers to *discontinuous* (either/or) phenotypes elicited by the environment. One obvious example is sex determination in turtles, where one range of temperatures induces female development in the embryo and another set of temperatures induces male development. Between these sets of temperatures is a small band of temperatures that will produce different proportions of males and females, but these intermediate temperatures do not induce intersexual animals. Another important example of polyphenism is found in the migratory locust *Schistocerca gregaria*. These grasshoppers exist either as a short-winged, green, solitary morph or as a long-winged, brown, gregarious morph (**FIGURE 25.1A,B**). Cues in the environment determine which morphology a grasshopper nymph will develop upon molting (Rogers et al. 2003; Simpson and Sword 2008).

Diet-induced polyphenisms

Diet can play major roles in determining a developing animal's phenotype. The effects of diet in development are seen in the caterpillar of *Nemoria arizonaria*. When the caterpillar hatches on oak trees in the spring, it has a form that blends remarkably with the young oak flowers (catkins). But those larvae hatching from their eggs in the summer would be very obvious if they still looked like oak flowers. Instead, they resemble newly formed twigs. Here, it is the diet (young versus old oak leaves) that determines the phenotype (**FIGURE 25.1C,D**; Greene 1989).

Diet is also largely responsible for the formation of fertile "queens" in ant, wasp, and bee colonies (**FIGURE 25.1E,F**). In honeybees, adult females are either workers or queens. The queen is the only reproductive member of the hive, laying up to 2,000 eggs per day. Queens also live 10 times longer than the average worker. The larvae are fed by the workers, and only those larvae fed adequately become queens. The protein inducing these queen-forming activities is called **royalactin**. Royalactin binds to the

(A)

(B)

(C)

(D)

(E)

(F)

FIGURE 25.1   Developmental plasticity in insects. (A,B) Density-induced polyphenism in the desert ("plague") locust *Schistocerca gregaria*. (A) The low-density morph has green pigmentation and miniature wings. (B) The high-density morph has deep pigmen-tation and wings and legs suitable for migration. (C,D) *Nemoria arizonaria* caterpillars. (C) Caterpillars that hatch in the spring eat young oak leaves and develop a cuticle that resembles the oak's flowers (catkins). (D) Caterpillars that hatch in the summer, after the catkins are gone, eat mature oak leaves and develop a cuticle that resembles a young twig. (E) Gyne (reproductive queen) and worker of the ant *Pheidologeton*. This picture shows the remarkable dimorphism between the large, fertile queen and the small, sterile worker (seen near the queen's antennae). The difference between these two sisters is the result of larval feeding. (F) Nutrition-induced size difference in a queen honeybee *Apis mellifera* compared with her sister workers. (A,B from Tawfik et al. 1999, courtesy of S. Tanaka; C,D courtesy of E. Greene; E © Mark W. Moffett/Getty Images; F courtesy of D. McIntyre.)

EGF receptor in the fat body of the honeybee larvae and stimulates the production of juvenile hormone, which elevates the levels of yolk proteins that are necessary for egg production (**FIGURE 25.2**; Kamakura 2011). RNAi against either the EGF receptor or its downstream targets abolishes the effects of royalactin.

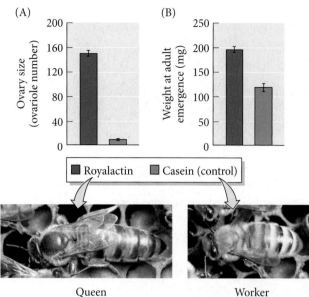

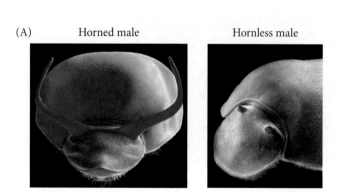

FIGURE 25.2 Diet-induced developmental changes can produce either reproductively competent queens or sterile workers. Royalactin induces functional ovaries (A) and increased body weight (B) in the honeybee *Apis mellifera*. (After Kamakura 2011.)

WEB TOPIC 25.1 **INDUCIBLE CASTE DETERMINATION IN ANT COLONIES** In some species of ants, the loss of soldier ants creates conditions that induce more workers to become soldiers.

WHEN DUNG REALLY MATTERS For a male dung beetle (*Onthophagus*), what really matters in life is the amount and quality of the dung he eats as a larva. The hornless female dung beetle digs tunnels, then gathers balls of dung and buries them in these tunnels. She then lays a single egg on each dung ball; when the larvae hatch, they eat the dung. Metamorphosis occurs when the dung ball is finished, and the anatomical and behavioral phenotypes of the males are determined by the quality and quantity of this maternally provided food (Emlen 1997; Moczek and Emlen 2000). The amount and quality of food determines the titer of juvenile hormone during the larva's last molt. This, in turn, determines the size of the larva at metamorphosis and positively regulates the growth of the imaginal discs that make the horns (**FIGURE 25.3A**; Emlen and Nijhout 1999; Moczek 2005).

If juvenile hormone is added to tiny *O. taurus* males during the sensitive period of their last molt, the cuticle in their heads expands to produce horns. Thus, whether a male is horned or hornless depends not on the male's genes but on the food his mother left for him. Horns do not grow until the male beetle larva reaches a certain size. After this threshold body size, horn growth is very rapid.[1] Thus, although body size has a normal distribution, there is a bimodal distribution of horn sizes: about half the males have no horns, whereas the other half have horns of considerable length (**FIGURE 25.3B**).

---

[1] Developmentally, there is a trade-off between primary and secondary male sexual characters here. Making a large horn appears to take away resources from making the penis. The growth rates of both the horn and the penis may be regulated through the Foxo transcription factor, which is upregulated by diet (Parzer and Moczek 2008; Snell-Rood and Moczek 2012).

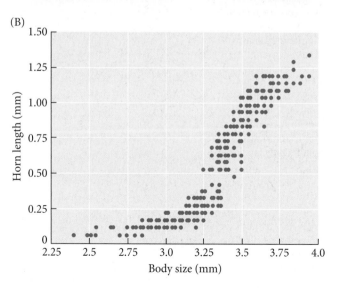

FIGURE 25.3 Diet and *Onthophagus* horn phenotype. (A) Horned and hornless males of the dung beetle *Onthophagus acuminatus* (horns have been artificially colored). Whether a male is horned or hornless is determined by the titer of juvenile hormone at the last molt, which in turn depends on the size of the larva. (B) There is a sharp threshold of body size under which horns fail to form and above which horn growth is linear with the size of the beetle. This threshold effect produces males with no horns and males with large horns, but very few with horns of intermediate size. (After Emlen 2000; photographs courtesy of D. Emlen.)

Horned males guard the females' tunnels and use their horns to prevent other males from mating. The size of the horns determines a male's behavior and chances for reproductive success; the male with the biggest horns wins such contests. But what about the males with no horns? Hornless males do not fight with the horned males for mates. Since they, like the females, lack horns, they are able to dig their own tunnels. These "sneaker males" dig tunnels that intersect those of the females and mate with the females while the horned male stands guard at the tunnel entrance (**FIGURE 25.4**; Emlen 2000; Moczek and Emlen 2000). Indeed, about half the fertilized eggs in most populations are from hornless males. In short, then, the *ability* to produce a horn is inherited, but *whether* a horn is produced and how big the horn becomes is regulated by the environment.

DIET AND DNA METHYLATION Dietary alterations can produce changes in DNA methylation, and these methylation changes can affect the phenotype. In the above-mentioned case of honeybee phenotype, larvae can also be transformed from workers into queens by preventing DNA methylation (Kucharski et al. 2008; Lyko et al. 2010).

Such diet-induced changes in DNA methylation can also affect mammalian phenotypes. Waterland and Jirtle (2003) demonstrated this by using mice containing the *viable-yellow* allele of *Agouti*. *Agouti* is a dominant gene that gives mice yellowish hair color; it also affects lipid metabolism such that the mice become fatter. The *viable-yellow* allele has a transposable element inserted into the *Agouti* gene. This transposable element contains a *cis*-regulatory element that allows *Agouti* to be expressed throughout the skin. Moreover, whereas most regions of the adult genome have hardly any intraspecies variation in CpG methylation, there are large DNA methylation differences between individuals at these transposon sites. Such CpG methylation can block transcription of the gene. When the *Agouti* promoter is methylated, the gene is not transcribed. The mouse's fur remains black, and lipid metabolism is not altered.

Waterland and Jirtle fed pregnant *viable-yellow* Agouti mice methyl donor supplements, including folate, choline, and betain. They found that the more methyl supplementation, the greater the methylation of the transposon insertion site in their fetuses' genomes, and the darker the pigmentation of the offspring. Although the mice in **FIGURE 25.5** are genetically identical, their mothers were fed different diets during pregnancy. The mouse whose mother did not receive methyl donor supplementation is fat and yellow—the *Agouti* promoter was unmethylated, so the gene was active. The mouse born to the mother that was given folate supplements is sleek and dark; the methylated *Agouti* gene was not transcribed.

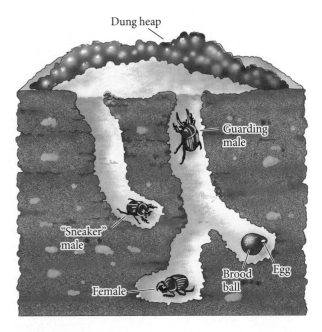

FIGURE 25.4 The presence or absence of horns determines the male reproductive strategy in some dung beetle species. Females dig tunnels in the soil beneath a pile of dung and bring dung fragments into the tunnels. These will be the food supply of the larvae. Horned males guard the tunnel entrances and mate repeatedly with the females. They fight to prevent other males from entering the tunnels, and the males with long horns usually win such contests. Smaller, hornless males do not guard tunnels, but dig their own to connect with those of females. They can then mate and exit, unchallenged by the guarding male. (After Emlen 2000.)

FIGURE 25.5 Maternal diet can affect phenotype. These two mice are genetically identical; both contain the *viable-yellow* allele of the *Agouti* gene, whose protein product converts brown pigment to yellow and accelerates fat storage. The obese yellow mouse is the offspring of a mother whose diet was not supplemented with methyl donors (e.g., folate) during her pregnancy. The embryo's *Agouti* gene was not methylated, and Agouti protein was made. The sleek brown mouse was born of a mother whose prenatal diet was supplemented with methyl donors. The *Agouti* gene was turned off, and no Agouti protein was made. (After Waterland and Jirtle 2003, photograph courtesy of R. L. Jirtle.)

Differential gene methylation has been linked to human health problems. Dietary restrictions during a woman's pregnancy may show up as heart or kidney problems in her adult children. Moreover, studies in rats showed that differences in protein and methyl donor concentration in the mother's prenatal diet affected gene expression and subsequent metabolism in the pups' livers (Lillycrop et al. 2005; Gilbert and Epel 2015). This has led to a new area of preventative medicine, the field of developmental origin of health and disease (see Web Topic 24.3).

Predator-induced polyphenisms

Imagine a species whose larvae are frequently confronted by a particular predator in their pond or tidepool. One could then imagine an individual that could recognize soluble molecules secreted by that predator and could use those molecules to activate the development of structures that would make this individual less palatable to the predator. This ability to modulate development in the presence of predators is called predator-induced defense, or **predator-induced polyphenism**.

To demonstrate predator-induced polyphenism, one has to show that the phenotypic modification is caused by the presence of the predator, and that the modification increases the fitness of its bearers when the predator is present (Adler and Harvell 1990; Tollrian and Harvell 1999). **FIGURE 25.6A** shows the typical and predator-induced morphs for several species. Two things characterize each case: (1) the induced morph is more successful at surviving the predator, and (2) soluble filtrate from water surrounding the predator is able to induce the changes. Chemicals that are released by a predator and can induce defenses in its prey are called **kairomones**.

Several rotifer species will alter their morphology when they develop in pond water in which their predators were cultured (Dodson 1989; Adler and Harvell 1990). The predatory rotifer *Asplanchna* releases a soluble compound that induces the eggs of a prey rotifer species, *Keratella slacki*, to develop into individuals with slightly larger bodies and anterior spines 130% longer than they otherwise would be, making the prey more difficult to eat. When exposed to the effluent of the crab species that preys on it, the snail *Thais lamellosa* develops a thickened shell and a "tooth" in its aperture. In a mixed snail population, crabs will not attack the thicker-shelled snails until more than half of the typical-morph snails are devoured (Palmer 1985).

One of the more interesting mechanisms of predator-induced polyphenism is that of certain echinoderm larvae. When exposed to the mucus of their fish predator, sand dollar plutei larvae clone themselves, budding off small groups of cells that quickly become larvae themselves. The tiny plutei are too small to be seen by the fish, and thereby escape being eaten (Vaughn and Strathmann 2008; Vaughn 2009).

DAPHNIA AND THEIR KIN   The predator-induced polyphenism of the parthenogenetic water flea *Daphnia* is beneficial not only to itself but also to its offspring (Harris et al. 2012). When juveniles of *D. cucullata* encounter the predatory larvae of the fly *Chaeoborus*, their "helmets" grow to twice the normal size (**FIGURE 25.6B**). This increase lessens the chances that *Daphnia* will be eaten by the fly larvae. This same helmet induction occurs if the juvenile *Daphnia* are exposed to extracts of water in which the fly larvae had been swimming. Agrawal and colleagues (1999) have shown that the offspring of such an induced *Daphnia* are born with this same altered head morphology in the absence of a predator. It is possible that the *Chaeoborus* kairomone regulates gene expression both in the adult and in the developing embryo. Although we still do not know the identity of the kairomone, the receptor may be a particular set of neurons and, in some species, the antennae are critical in perceiving the kairomone (Weiss et al. 2012, 2015). The effect does appear to work through the endocrine pathways. The kairomone upregulates the juvenile hormone and the insulin signaling pathways, activating the transcription of several transcription factor genes (**FIGURE 25.6C**; Miyakawa et al. 2010). As in the dung beetles, there are trade-offs: the induced *Daphnia*, having put resources into making protective structures, produce fewer eggs (Tollrian 1995; Imai et al. 2009).

(A)

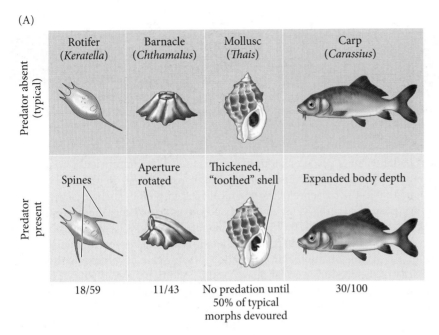

(B)

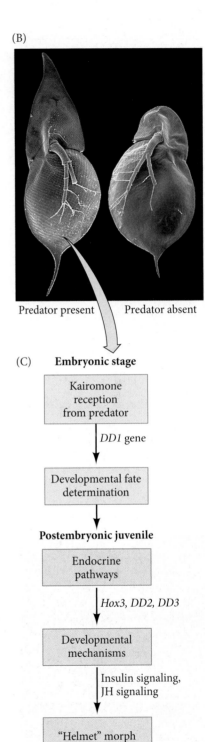

Predator present   Predator absent

FIGURE 25.6 Predator-induced defenses. (A) Typical (upper row) and predator-induced (lower row) morphs of various organisms. The numbers beneath each column represent the percentages of organisms surviving predation when both induced and uninduced individuals were presented with predators (in various assays). (B) Scanning electron micrographs show predator-induced (left) and typical (right) morphs of genetically identical individuals of the water flea *Daphnia*. In the presence of chemical signals from a predator, *Daphnia* grows a protective "helmet." (C) Possible pathway for the development of *Daphnia's* defensive phenotype through the endocrine system. *DD1* is thought to be involved in kairomone reception and/or fate determination during the embryonic stage. It may play a role in the neural reception of the signal. The other genes are thought to play roles in the morphogenesis of postembryonic juveniles. (A after Adler and Harvell 1990 and references cited therein; B courtesy of A. A. Agrawal; C after Miyakawa et al. 2010.)

(C)   **Embryonic stage**

Kairomone reception from predator

↓ *DD1* gene

Developmental fate determination

Postembryonic juvenile

Endocrine pathways

↓ *Hox3, DD2, DD3*

Developmental mechanisms

↓ Insulin signaling, JH signaling

"Helmet" morph

AMPHIBIAN PHENOTYPES INDUCED BY PREDATORS   Predator-induced polyphenism is not limited to invertebrates.[2] Among amphibians, tadpoles found in ponds or reared in the presence of other species may differ significantly from tadpoles reared by themselves in aquaria. For instance, newly hatched wood frog tadpoles (*Rana sylvatica*) reared in tanks containing the predatory larval dragonfly *Anax* (confined in mesh cages so they cannot eat the tadpoles) grow smaller than those reared in similar tanks without predators. Moreover, their tail musculature deepens, allowing faster turning and swimming speeds (Van Buskirk and Relyea 1998). The addition of more predators to the tank causes a continuously deeper tail fin and tail musculature, and in fact what initially appeared to be a polyphenism may be a reaction norm that can assess the number (and type) of predators.

McCollum and Van Buskirk (1996) have shown that in the presence of its predators, the tail fin of the tadpole of the tree frog *Hyla chrysoscelis* grows larger and becomes bright red (**FIGURE 25.7**). This phenotype allows the tadpole to swim away faster and to deflect predator strikes toward the tail region. The trade-off is that non-induced tadpoles grow more slowly and survive better in predator-free environments. In some species, phenotypic plasticity is reversible, and removing the predators can restore the non-induced phenotype (Relyea 2003a).

---

[2] Indeed, the vertebrate immune system is a wonderful example of predator-induced polyphenism. Here, our immune cells use chemicals from our predators (viruses and bacteria) to change our phenotype so that we can better resist them (see Frost 1999).

(A) Predator present

(B) Predator absent

FIGURE 25.7 Predator-induced polyphenism in frog tadpoles. (A) Tadpoles of the tree frog *Hyla chrysoscelis* developing in the presence of cues from a predator's larvae develop strong trunk muscles and a red coloration. (B) When predator cues are absent, the tadpoles grow sleeker, which helps them compete for food. (Photographs courtesy of T. Johnson/USGS.)

The metabolism of predator-induced morphs may differ significantly from that of the uninduced morphs, and this has important consequences. Relyea (2003b, 2004) has found that in the presence of the chemical cues emitted by predators, the toxicity of pesticides such as carbaryl (Sevin™) can become up to 46 times more lethal than it is without the predator cues. Bullfrog and green frog tadpoles were especially sensitive to carbaryl when exposed to predator chemicals. Relyea has related these findings to the global decline of amphibian populations, saying that governments should test the toxicity of the chemicals under more natural conditions, including that of predator stress. He concludes (Relyea 2003b) that "ignoring the relevant ecology can cause incorrect estimates of a pesticide's lethality in nature, yet it is the lethality of pesticides under natural conditions that is of utmost interest. The accumulated evidence strongly suggests that pesticides in nature could be playing a role in the decline of amphibians."

VIBRATIONAL CUES ALTER DEVELOPMENTAL TIMING The phenotypic changes induced by environmental cues are not confined to anatomical structures. They can also include the timing of developmental processes. Embryos of the Costa Rican red-eyed tree frog (*Agalychnis callidryas*) use vibrations transmitted through their egg masses to escape egg-eating snakes. These egg masses are laid on leaves that overhang ponds. Usually, the embryos develop into tadpoles within 7 days, and these tadpoles wiggle out of the egg mass and fall into the pond water. However, when snakes feed on the eggs, the vibrations they produce cue the remaining embryos inside the egg mass to begin the twitching movements that initiate their hatching (within seconds!) and dropping into the pond. The embryos are competent to begin these hatching movements at day 5 (**FIGURE 25.8**). Interestingly, the embryos have evolved to respond this way only to vibrations given at a certain frequency and interval (Warkentin et al. 2005, 2006; Caldwell et al. 2009). Up to 80% of the remaining embryos can escape snake predation in this way, and research has shown that these vibrations alone (and not smell or sight) cue these hatching movements in the embryos. There is a trade-off here, too. Although these embryos have escaped their snake predators, they are now at greater risk from waterborne predators than are fully developed embryos because the musculature of the early hatchers is underdeveloped.

(A)

(B)

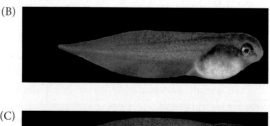

(C)

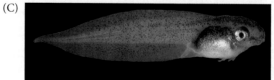

FIGURE 25.8 Predator-induced polyphenism in the red-eyed tree frog (*Agalychnis callidryas*). (A) When a snake eats a clutch of *Agalychnis* eggs, most of the remaining embryos inside the egg mass respond to the vibrations by hatching prematurely (arrow) and falling into the water. (B) Immature tadpole, induced to hatch at day 5. (C) A normal tadpole hatches at day 7 and has better-developed musculature. (Courtesy of K. Warkentin.)

WATCH DEVELOPMENT 25.1   See an experiment performed in Dr. Karen Warkentin's laboratory where a snake tries to eat a clutch of red-eyed tree frog embryos, only to see his potential food hatching as it swallows its earlier meal.

 SCIENTISTS SPEAK 25.1   Dr. Karen Warkentin discusses her work on the plasticity of tadpole hatching.

Temperature as an environmental agent

TEMPERATURE AND SEX   In many species, temperature controls whether testes or ovaries develop; such temperature-dependent sex determination is described briefly in Chapter 6. The mechanisms for temperature-dependent sex determination may differ widely between species, but in many instances, certain transcription factors appear to be induced by the temperature, and these transcription factors are probably promoting the formation of either ovaries or testes. This type of determination is not uncommon among "cold-blooded" vertebrates such as fish, turtles, and alligators (Crews and Bull 2009). Temperature-dependent sex determination has advantages and disadvantages. One probable advantage is that it can give the species the benefits of sexual reproduction without tying the species to a 1:1 sex ratio. In crocodiles, in which extreme temperatures produce females and moderate temperatures produce males, the sex ratio may be as great as 10 females to each male (Woodward and Murray 1993). In such species, where the number of females limits the population size, this ratio is better for survival than the 1:1 ratio demanded by genotypic sex determination.

The major disadvantage of temperature-dependent sex determination may be its narrowing of the temperature limits within which a species can persist. Thus, thermal pollution (either locally or because of global warming) could conceivably eliminate a species from a given area (Janzen and Paukstis 1991). Among marine turtles, females are usually produced at higher temperatures (29°C being the temperature that produces an even sex ratio), and these animals may be particularly vulnerable if the temperature rises for an extended period of time[3] (Hawkes et al. 2009; Fuentes et al. 2010).

Charnov and Bull (1977) argued that environmental sex determination would be adaptive in habitats characterized by patchiness—that is, a habitat having some regions where it is advantageous to be male and other regions where it is advantageous to be female. Conover and Heins (1987) provided evidence for this hypothesis. In certain fish species, females benefit from being larger because larger size translates into higher fecundity. If you are a female Atlantic silverside (*Menidia menidia*), it can be advantageous to be born early in the breeding season, because you have a longer feeding season and thus can grow larger. The size of male fish, however, doesn't influence mating success or outcomes. In the southern range of *Menidia*, females are indeed born early in the breeding season, and temperature appears to play a major role in this pattern. However, in the northern reaches of its range, the species shows no environmental sex determination and a 1:1 sex ratio is generated at all temperatures. Conover and Heins speculate that the more northern populations have a very short feeding season, so there is no advantage for females in being born earlier. Thus, this fish species displays environmental sex determination in those regions where it is adaptive and genotypic sex determination in those regions where it is not.

---

[3] Researchers have speculated that some dinosaurs may have had temperature-dependent sex determination and that their sudden demise may have been caused by a slight change in temperature, creating conditions where only males or only females hatched (see Ferguson and Joanen 1982; Miller et al. 2004). Unlike many turtle species, whose members have long reproductive lives, can hibernate for years, and whose females can store sperm, dinosaurs may have had relatively brief reproductive lives and no ability to hibernate through prolonged bad times.

WEB TOPIC 25.2 **WHEN ADVERSITY CHANGES DEVELOPMENT**   This topic describes some of the organisms that alter their development when faced with environmental stress. In *Volvox*, rising temperatures of the spring bring about sexuality. In *Dictyostelium*, lack of food leads to multicellularity. Other organisms would rather skip development and wait for better times.

 SCIENTISTS SPEAK 25.2   Dr. John Tyler Bonner discusses his pioneering work demonstrating how the environment alter development to turn a single cell organism multicellular.

BUTTERFLY WINGS   Tropical regions of the world often have a hot wet season and a cooler dry season. In Africa, a polyphenism of the dimorphic Malawian butterfly (*Bicyclus anynana*) is adaptive to these seasonal changes. The dry (cool) season morph is a mottled brown butterfly that survives by hiding in dead leaves on the forest floor. In contrast, the wet (hot) season morph, which routinely flies, has prominent ventral eyespots that deflect attacks from predatory birds and lizards (**FIGURE 25.9**; Brakefield and Frankino 2009; Olofsson et al. 2010; Prudic et al. 2015).

The factor determining the seasonal pigmentation of *B. anynana* is not diet, but the temperature during its pupation. Low temperatures produce the dry-season morph; higher temperatures produce the wet-season morph (Brakefield and Reitsma 1991), and the molecular mechanisms by which temperatures regulate phenotype are known. In the late larval stages, transcription of the *Distal-less* gene in the wing imaginal discs is restricted to a set of cells that will become the signaling center of each eyespot. In the early pupa, higher temperatures elevate the formation of 20-hydroxyecdysone (20E; see Chapter 21). This hormone sustains and expands the expression of *Distal-less* in those regions of the wing imaginal disc, resulting in prominent eyespots. In the dry season, the cooler temperatures prevent the accumulation of 20E in the pupa, and the foci of Distal-less signaling are not sustained. In the absence of the Distal-less signal, eyespots do not form (Brakefield et al. 1996; Oostra et al. 2014). Distal-less protein is believed to be the transcription factor that determines the size of the eyespot (see Figure 25.9). In *Bicyclus*, we see the adaptive significance of polyphenism and how this type of developmental plasticity integrates an organism into its environment.

The importance of hormones such as 20E for mediating environmental signals controlling wing phenotypes has been documented in the *Araschnia* butterfly (**FIGURE 25.10**). *Araschnia* develops alternative phenotypes depending on whether the fourth and fifth instars experience a photoperiod (hours of daylight) that is longer or shorter than a particular critical day length. Below this critical day length, ecdysone levels are

FIGURE 25.9   Phenotypic plasticity in *Bicyclus anynana* is regulated by temperature during pupation. High temperature (either in the wild or in controlled laboratory conditions) allows the accumulation of 20-hydroxyecdysone (20E), a hormone that is able to sustain *Distal-less* expression in the pupal imaginal disc. The region of *Distal-less* expression becomes the focus of each eyespot. In cooler weather, 20E is not formed, *Distal-less* expression in the imaginal disc begins but is not sustained, and eyespots fail to form. (Courtesy of S. Carroll and P. Brakefield.)

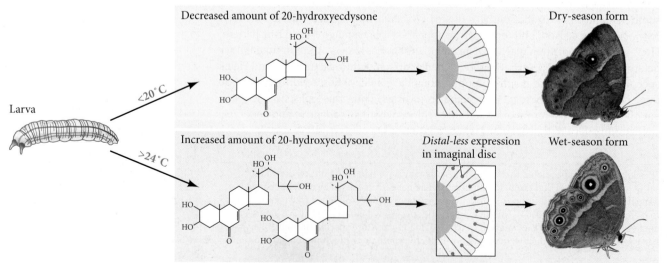

low and the butterfly has the orange wings characteristic of spring butterflies. Above the critical point, ecdysone is made and the summer pigmentation forms. The summer form can be induced in spring pupae by injecting 20E into the pupae. Moreover, by altering the timing of 20E injections, one can generate a series of intermediate forms not seen in the wild (Koch and Bückmann 1987; Nijhout 2003).

WEB TOPIC 25.3 **THE ENVIRONMENTAL INDUCTION OF BEHAVIORAL PHENOTYPES** Several behaviors, including learning and the propensity for anxiety, may be induced by the environment through developmental pathways.

Polyphenic Life Cycles

Larval settlement

Free-swimming marine larvae often need to settle near a source of food or on a firm substrate on which they can undergo metamorphosis. The ability of marine larvae to suspend development until they experience a particular environmental cue is called **larval settlement**. Particularly among the mollusks, there are often very specific cues for settlement (Hadfield 1977; Hadfield and Paul 2001; Zardus et al. 2008). In some cases, the mollusk's prey supply the cues, while in other cases the substrate itself gives off molecules used by the larvae to initiate settlement. These cues may not be constant, but they need to be part of the environment if further development is to occur[4] (Pechenik et al. 1998).

In many marine invertebrate species, larval settlement and the subsequent distribution of invertebrate populations are regulated by mats of bacteria called **biofilms** (Hadfield 2011). Chemicals from these potential food sources are used by the larvae as signals to settle and undergo metamorphosis. Humans are affecting population distributions with our desire to place large objects in the oceans. Such objects readily acquire biofilms and the marine fauna that attach to them. As early as 1854, Charles Darwin speculated that barnacles were transported to new locales when their larvae settled on the hulls of ships. Indeed, the ability of biofilms to aid invertebrate larval settlement and colony formation explains the ability of barnacles and tubeworms ("biofouling invertebrates") to accumulate on ship keels, clog sewer pipes, and deteriorate underwater structures (Zardus et al. 2008).

UNDER THE SEA Most of the cues known for larval settlement and metamorphosis involve chemicals emanating from the substrate; these chemicals can signal the presence of a food source or potentially induce larval metamorphosis. However, in at least one case, vibrational cues appear to direct marine larvae to coral reefs. Coral reefs are the largest biological structures on Earth, and they grow by recruiting planktonic coral (cnidarian) larvae. While chemical cues work within a small distance of the reef, it is the "noise of the reef"—the snapping of shrimp claws and the noises made by thousands

FIGURE 25.10 Environmentally induced morphs of the European map butterfly (*Araschnia levana*). The orange morph (bottom) forms in the spring, when levels of ecdysone in the larva are low. The dark morph with a white stripe (top) forms in summer, when higher temperatures and longer photoperiods induce greater ecdysone production in the larva. Linnaeus classified the two morphs as different species. (Courtesy of H. F. Nijhout.)

---

[4] The importance of substrates for larval settlement and metamorphosis was first demonstrated in 1880, when William Keith Brooks, an embryologist at Johns Hopkins University, was asked to help the ailing oyster industry of Chesapeake Bay. For decades, oysters had been dredged from the bay, and there had always been a new crop to take their place. But by 1880, each year brought fewer oysters. What was responsible for the decline? Experimenting with larval oysters, Brooks discovered that the American oyster (*Crassostrea virginica*), unlike its better-studied European relative *Ostrea edulis*, needs a hard substrate on which to metamorphose. For years, oystermen had thrown the shells back into the sea, but with the advent of paved sidewalks, they started selling the shells to cement factories. Brooks's solution: throw the shells back into the bay. The oyster population responded, and the Baltimore wharves still sell their descendants. Biofilms on the oyster shells appear to be critical (Turner et al. 1994).

of reef fish—that attract coral larvae from long distances. Vermeij and colleagues (2010) made recordings of Caribbean reefs and found that the larvae swam to the source of the sound, even in the laboratory. These findings mean that coral reefs face danger from noise pollution as well as from thermal and chemical pollution. Steve Simpson (2010), who headed the study, has warned, "Anthropogenic noise has increased dramatically in recent years, with small boats, shipping, drilling, pile driving and seismic testing now sometimes drowning out the natural sounds of fish and snapping shrimps."

The hard life of spadefoot toads

Spadefoot toads (*Scaphiopus couchii, Spea multiplicata,* and their relatives) have a remarkable strategy for coping with a harsh environment (Ledón-Rettig and Pfennig 2011). They can let the environment trigger different morphs. In *Scaphiopus*, the toads are called out from hibernation by the thunder that accompanies the first spring rainstorms in the Sonoran desert.[5] They breed in temporary ponds formed by the rain, and the embryos develop quickly into larvae. After the larvae metamorphose, the young toads return to the desert, burrowing into the sand until the next year's storms bring them out.

Desert ponds are ephemeral pools that can either dry up quickly or persist, depending on their initial depth and the frequency of rainfall. One might envision two alternative scenarios confronting a tadpole in such a pond: either (1) the pond persists until you have time to fully metamorphose, and you live; or (2) the pond dries up before your metamorphosis is complete, and you die. In some spadefoot toad species, however, a third alternative has evolved. The timing of their metamorphosis is controlled by the pond. In several *Scaphiopus* species, development continues at its normal rate if the pond persists at a viable level, and the algae-eating tadpoles develop into juvenile toads. However, if the pond is drying out and getting smaller, some of the tadpoles embark on an alternative developmental pathway. They develop a wider mouth and powerful jaw muscles, which enables them to eat (among other things) other *Scaphiopus* tadpoles (**FIGURE 25.11**). These carnivorous tadpoles metamorphose quickly, albeit into a smaller version of a juvenile spadefoot toad. But they survive while other *Scaphiopus* tadpoles perish from desiccation (Newman 1989, 1992).

The signal for accelerated metamorphosis in *Scaphiopus* appears to be the change in water volume. In the laboratory, *Scaphiopus hammondi* tadpoles are able to sense the removal of water from aquaria, and their acceleration of metamorphosis depends on the rate at which the water is removed. The stress-induced corticotropin-releasing hormone signaling system appears to modulate this effect (Denver et al. 1998; Middlemis Maher 2013). This increase in brain corticotropin-releasing hormone is thought

---

[5] Like coral larvae, the toads are sensitive to vibration, and noise pollution may affect their survival. Motorcycles produce the same sounds as thunder, causing the toads to come out of hibernation only to die in the scorching desert sun.

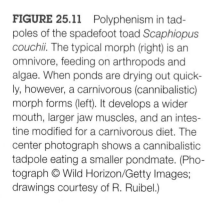

FIGURE 25.11 Polyphenism in tadpoles of the spadefoot toad *Scaphiopus couchii*. The typical morph (right) is an omnivore, feeding on arthropods and algae. When ponds are drying out quickly, however, a carnivorous (cannibalistic) morph forms (left). It develops a wider mouth, larger jaw muscles, and an intestine modified for a carnivorous diet. The center photograph shows a cannibalistic tadpole eating a smaller pondmate. (Photograph © Wild Horizon/Getty Images; drawings courtesy of R. Ruibel.)

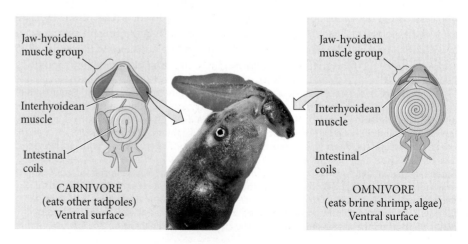

to be responsible for the subsequent elevation of the thyroid hormones that initiate metamorphosis (Boorse and Denver 2003). As in many other cases of polyphenism, the developmental changes are mediated through the endocrine system. Sensory organs send a neural signal to regulate hormone release. The hormones then can alter gene expression in a coordinated and relatively rapid fashion.

In other species of spadefoot toads, the polyphenism is based on diet. In *Spea multiplicata*, the carnivorous morph appears to be induced by the eating of shrimp by the young larva. This carnivorous morph can then eat larger prey, including the tadpoles of *Scaphiopus*, while the other larvae primarily eat organic detritus (Levis et al. 2015). The carnivores grow faster than the detritus-eaters and are more likely to survive if the ponds dry up quickly.

WEB TOPIC 25.4 **PRESSURE AS AN AGENT OF DEVELOPMENT**   Mechanical stress is critical for gene expression in numerous tissues, including bone, heart, and muscle. Without physical force, we would not develop our kneecaps.

Developmental Symbioses

In addition to the above-mentioned abiotic and biotic relationships in which the environment regulates development, there is a special type of biotic relationship called symbiosis. Contrary to the popular use of the term to mean a mutually beneficial relationship, the word **symbiosis** (Greek, *sym*, "together"; *bios*, "life") can refer to any close association between organisms of different species (see Sapp 1994). In many symbiotic relationships, one of the organisms involved is much larger than the other, and the smaller organism may live on the surface or inside the body of the larger. In such relationships, the larger organism is referred to as the **host** and the smaller as the **symbiont**. There are two important categories of symbiosis:[6]

- **Parasitism** occurs when one partner benefits at the expense of the other. An example of a parasitic relationship is that of a tapeworm living in the human digestive tract, wherein the tapeworm steals nutrients from its host.

- **Mutualism** is a relationship that benefits both partners. A striking example of this type of symbiosis can be found in the partnership between the Egyptian plover (*Pluvianus aegyptius*) and the Nile crocodile (*Crocodylus niloticus*). Although it regards most birds as lunch, the crocodile allows the plover to roam its body, feeding on the harmful parasites there. In this mutually beneficial relationship, the bird obtains food while the crocodile is rid of parasites.

In addition, the term **endosymbiosis** ("living inside") is widely used to describe the situation in which one cell lives inside another cell, a circumstance thought to account for the evolution of the organelles of the eukaryotic cell (see Margulis 1971), and one that describes the *Wolbachia* developmental symbioses discussed at length later in this chapter.

Symbiosis, and especially mutualism, is the basis for life on Earth. The symbiosis between *Rhyzobium* bacteria and the roots of legume plants is responsible for converting atmospheric nitrogen into a usable form for generating amino acids, and is therefore essential for life. Symbioses between fungi and plants are ubiquitous and are often necessary for plant development (see Gilbert and Epel 2015; Pringle 2009). Orchid seeds, for example, contain no energy reserves, so a developing orchid plant must acquire carbon from mycorrhizal fungi. (This is why orchids grow best in moist tropical environments, where fungi are plentiful.) The coastal zone ecosystem throughout the world

[6]Commensalism—defined as a relationship that is beneficial to one partner and neither beneficial nor harmful to the other partner—is sometimes thought of as a third category of symbiosis. Although many symbioses appear on the surface to be commensal, recent studies suggest that very few symbiotic relationships are truly neutral with respect to either party.

is sustained by a triple symbiosis among seagrass, clams, and the sulfide-oxidizing bacteria living inside the clam's gills (van der Heide et al. 2012).

In some cases, the development of an organ is brought about by signals from organisms of a different species. In some organisms, this relationship has become obligatory—the symbionts have become so tightly integrated into the host organism that the host cannot develop without them (Sapp 1994). Indeed, recent evidence indicates that developmental symbioses are the rule rather than the exception (McFall-Ngai 2002; McFall-Ngai et al. 2013, 2014). The term for the composite organism of a host and its persistent symbionts is **holobiont** (Rosenberg et al. 2007; Gilbert and Epel 2015).

 DEV TUTORIAL *Developmental symbiosis* Scott Gilbert summarizes some fascinating cases of developmental symbiosis, where development needs two or more species to be complete.

Mechanisms of developmental symbiosis: Getting the partners together

All symbiotic associations must meet the challenge of maintaining their partnerships over successive generations. In the partnerships that are the main subject here, in which microbes are crucial to the development of their animal hosts, the task of transmission is usually accomplished by either vertical or horizontal transmission.

VERTICAL TRANSMISSION **Vertical transmission** refers to the transfer of symbionts from one generation to the next through the germ cells, usually the eggs (Krueger et al. 1996).

Bacteria of the genus *Wolbachia* reside in the egg cytoplasm of invertebrates and provide important signals for the development of the individuals produced by those eggs. As we shall see, many species of invertebrates have "outsourced" important developmental signals to *Wolbachia* bacteria, which are transmitted like mitochondria—that is, in the oocyte cytoplasm. In numerous *Drosophila* species, *Wolbachia* provide resistance against viruses (Teixeira et al. 2008; Osborne et al. 2009). Ferree and colleagues (2005) have shown that in *Drosophila* development, *Wolbachia* use the host's nurse cell microtubule system and dynein motors to travel from the nurse cells into the developing oocyte (**FIGURE 25.12A**). In other words, the bacteria use the same cytoskeletal pathway as mitochondria, ribosomes, and *bicoid* mRNA (see Chapter 9). Once in the oocyte, the bacteria enter every cell, becoming endosymbionts. The *Wolbachia* appear to aid in their propagation by entering the stem cell niches that make ovaries and oocytes (Fast et al. 2011). The females infected with *Wolbachia* make four times more eggs than their uninfected sisters, thereby furthering the spread of *Wolbachia*.

WEB TOPIC 25.5 **DEVELOPMENTAL SYMBIOSIS AND PARASITISM** Some embryos acquire protection and nutrients by forming symbiotic associations with other organisms. The mechanisms by which these associations form are now being elucidated. In other situations, one species uses material from another to support its development. Blood-sucking mosquitoes are examples of such parasites.

HORIZONTAL TRANSMISSION *Wolbachia* can be transmitted horizontally as well as vertically. In **horizontal transmission**, the metazoan host is born free of symbionts but subsequently becomes infected, either by its environment or by other members of the species. In pill bugs such as *Amadillidium vulgare*, genetically male insects infected with *Wolbachia* are transformed by the bacteria into females (**FIGURE 25.12B**). As females, the pill bugs can then transmit the *Wolbachia* symbionts to the next generation (Cordaux et al. 2004).

A different type of horizontal transmission involves aquatic eggs that attract photosynthetic algae. Clutches of both amphibian and snail eggs, for example, are packed together in tight masses. The supply of oxygen limits the rate of their development, and

(A)

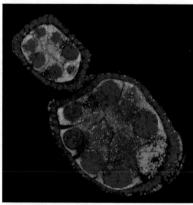

(B)

FIGURE 25.12 Vertical and horizontal transmission of *Wolbachia* bacteria. (A) In *Drosophila*, *Wolbachia* are transmitted vertically through the female germ cells. In the germinarium, 15 nurse cells transport proteins, RNAs, and organelles to the distalmost oocyte cell. The symbiotic bacterium (stained red) is also transported by these microtubules into the oocyte. Cytoplasm of the ovary is green, and blue indicates DNA. (B) Male and female *Armadillidium vulgare*. Genetically male pill bugs (right) can be transformed into phenotypic egg-producing females (left) by infection (i.e., horizontal transmission) of *Wolbachia* bacteria. (A from Ferree et al. 2005, courtesy of H. M. Frydman and E. Wieschaus; B courtesy of D. McIntyre.)

there is a steep gradient of oxygen from the outside of the cluster to deep within it; thus embryos on the inside of the cluster develop more slowly than those near the surface (Strathmann and Strathmann 1995). The embryos seem to get around this problem by coating themselves with a thin film of photosynthetic algae, which they obtain from the pond water. In clutches of amphibian and snail eggs, photosynthesis from this algal "fouling" enables net oxygen production in the light, whereas respiration exceeds photosynthesis in the dark (Bachmann et al. 1986; Pinder and Friet 1994; Cohen and Strathmann 1996). Thus, the symbiotic algae "rescue" the eggs by their photosynthesis.

Horizontal transmission is crucial for the symbiotic gut bacteria found in many animals, including humans. As we will see later in this chapter, mammalian gut bacteria are critical in forming the blood vessels of the intestine, and possibly in regulating stem cell proliferation (Pull et al. 2005; Liu et al. 2010). Human infants usually acquire these symbionts as they travel through the birth canal. Once the amnion breaks, the microbiota of the mother's reproductive tract can colonize the infant's skin and gut. This is supplemented by bacteria from the parents' skin, especially the mother's skin during nursing. The colonization of the infant by the microbes is a critically important event, and the mammalian immune system appears to encourage certain bacteria to enter the body, while discouraging others (see Gilbert et al. 2012). Indeed, some of the complex sugars found in human mothers' milk are not digestible by the infant. Rather, they serve as food for certain bacterial symbionts that help the infants' bodies develop (Zivkovic et al. 2011). Although each baby starts with a unique bacterial profile, within a year the types and proportions of bacteria have converged to the adult human profile that characterizes the human digestive tract (Palmer et al. 2007).

The Euprymna-Vibrio symbiosis

Horizontal transmission plays a major role in one of the best-studied examples of developmental symbiosis: that between the squid *Euprymna scolopes* and the luminescent bacterium *Vibrio fischeri* (McFall-Ngai and Ruby 1991; Montgomery and McFall-Ngai 1995). The adult *Euprymna* is equipped with a light organ composed of sacs filled with these bacteria (**FIGURE 25.13A**). The newly hatched squid, however, does not contain these light-emitting symbionts, nor does it have the light organ to house them. Rather, the symbiotic bacteria interact with the larval squid to build the light organ together. The juvenile squid acquires *V. fischeri* from seawater pumping through its mantle cavity (Nyholm et al. 2000). The bacteria bind to a ciliated epithelium in this cavity; the epithelium binds *only V. fischeri*, allowing other bacteria to pass through (**FIGURE 25.13B**). The bacteria then induce hundreds of genes in the epithelium, leading to the apoptotic death of the epithelial cells, their replacement by a nonciliated epithelium, the differentiation of the surrounding cells into storage sacs for the bacteria, and the expression of genes encoding opsins and other visual proteins in the light organ (**FIGURE 25.13C**; Chun et al. 2008; McFall-Ngai 2008b; Tong et al. 2009).

FIGURE 25.13 The *Euprymna scolopes-Vibrio fischeri* symbiosis. (A) An adult Hawaiian bobtail squid (*E. scolopes*) is about 2 inches long. The symbionts are housed in a two-lobed light organ on the squid's underside. (B) The light organ of a juvenile squid is poised to receive *V. fischeri*. Ciliary currents and mucus secretions create an environment (diffuse yellow stain) that attracts seaborne Gram-negative bacteria, including *V. fischeri*, to the organ. Over time all bacteria except *V. fischeri* will be eliminated by mechanisms yet to be exactly elucidated. (C) Once *V. fischeri* are established in the crypts of the light organ, they induce apoptosis of the epithelial cells (yellow dots) and shut down production of the mucosal secretions that attracted other bacteria. (Courtesy of M. McFall-Ngai.)

(A)

(B)

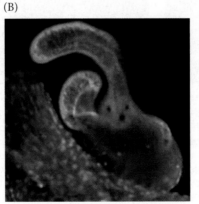

(C)

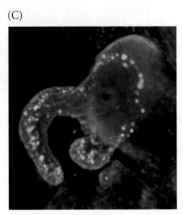

The substance *V. fischeri* secretes to effect these changes turns out to be fragments of the bacterial cell wall, and the active agents are tracheal cytotoxin and lipopolysaccharide (Koropatnick et al. 2004). This finding was surprising, because these two agents are known to cause inflammation and disease that endanger the survival of the host (and therefore the bacteria). Indeed, tracheal cytotoxin is responsible for the tissue damage in both whooping cough and gonorrheal infections. The destruction and replacement of ciliated tissue in the respiratory tract and oviduct are due to these bacterial compounds. After the bacteria have induced the morphological changes in the host, the host secretes a peptide into the *Vibrio*-containing crypts which neutralizes the bacterial toxin (Troll et al. 2010). Both organisms change their gene expression patterns, and both benefit from their association: The bacteria get a home and express their light-generating enzymes, and the squid develops a light organ that allows it to swim at night in shallow waters without casting a shadow.

Obligate developmental mutualism

The species involved in an **obligate mutualism** are interdependent to such an extent that neither could survive without the other. The most common example of obligate mutualism is the lichens, in which fungal and algal species are joined in a relationship that results in an essentially new species. More and more examples of obligate mutualism are being described, and most of these have important consequences for medicine and conservation biology.

One example of obligate developmental mutualism has been described in the parasitic wasp *Asobara tabida*. In these insects, symbiotic bacteria are found in the egg cytoplasm and are vertically transferred through the female germ plasm. In *Asobara*, the *Wolbachia* bacteria enable the wasp to complete yolk production and egg maturation (Dedeine et al. 2001; Pannebakker et al. 2007). If the symbionts are removed, the ovaries undergo apoptosis and no eggs are produced (**FIGURE 25.14**). Another example is the nematode *Brugia malayi*. Here, *Wolbachia* bacteria ride to the posterior pole on the microtubules that form the cellular mitotic spindle. Once in the posterior pole, they become essential for regulating the cell divisions that create the anterior-posterior boundary critical to the early development of the nematode. If they are removed from the egg before first cell division, that division is often abnormal, and a proper anterior-posterior polarity fails to form (Landmann et al. 2014). Here we see very starkly that the developing organism is a holobiont.

In obligate developmental mutualisms, the death of the host can result from killing the symbiont. In Chapter 24 we described atrazine and its ability to induce aromatase

(A) Control Antibiotic-treated

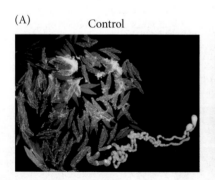
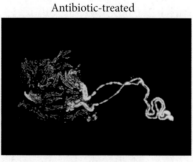

(B) Control Antibiotic-treated

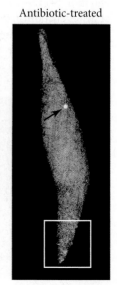
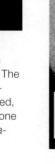

FIGURE 25.14 Comparison of ovaries and oocytes of the wasp *Asobara tabida* from control females and females treated with rifampicin antibiotic to remove *Wolbachia*. (A) The ovaries of control females had an average of 228 oocytes, whereas those of rifampicin-treated females had an average of 36 oocytes. (B) When DNA in the oocytes was stained, oocytes from control females had a nucleus (arrow) as well as a mass of *Wolbachia* at one end (boxed area). Oocytes from rifampicin-treated females had a nucleus but no *Wolbachia*; these eggs were sterile. (From Dedeine et al. 2001.)

FIGURE 25.15 Obligate developmental symbionts. Spotted salamander (*Ambystoma maculatum*) eggs at the center of the cluster cannot survive the lack of oxygen when their algal symbiont is eliminated by herbicides. (Photograph © Gustav Verderber/OSF/Visuals Unlimited.)

and cause sex-determination anomalies in amphibians. But the major use and effect of atrazine is to kill plant life; it is a potent nonspecific herbicide. Once applied, atrazine can remain active in the soil for more than 6 months and can be carried by wind and rainwater to new sites. However, the egg masses of many amphibian and snail species depend on algal symbionts to provide oxygen to the eggs deepest in the clutch. The spotted salamander (*Ambystoma maculatum*) lays eggs that recruit a green algal symbiont so specific that its name is *Oophilia amblystomatis* ("lover of *Ambystoma* eggs"). The algae is actually stored in the mother's body and appears to be deposited along with the eggs (Kerney et al. 2011). Concentrations of atrazine as low as 50 μg/L completely eliminate this algae from the eggs, and the amphibian's hatching success is greatly lowered (**FIGURE 25.15**; Gilbert 1944; Mills and Barnhart 1999; Olivier and Moon 2010).

Developmental symbiosis in the mammalian intestine

Even mammals maintain developmental symbioses with bacteria. Using the polymerase chain reaction and high-speed sequencing techniques, researchers have recently been able to identify many anaerobic bacterial species present in the human gut (see Qin et al. 2010). Their presence was not realized earlier because these species cannot yet be cultured in the laboratory.

These studies have revealed particular distributions of the bacterial symbionts in our bodies. The hundreds of different bacterial species of the human colon are stratified into specific regions along the length and diameter of the gut tube, where they can attain densities of 10^{11} cells per milliliter (Hooper et al. 1998; Xu and Gordon 2003). Indeed, more than half the cells in our body are microbial. We never lack these microbial components; we pick them up from the reproductive tract of our mother as soon as the amnion bursts. We have coevolved to share our space with them, and we have even codeveloped such that our cells are primed to bind to them, and the bacteria induce gene expression in the intestinal epithelial cells (Bry et al. 1996; Hooper et al. 2001).

BACTERIA HELP REGULATE GUT DEVELOPMENT   Bacteria-induced expression of mammalian genes was first demonstrated in the mouse gut. Umesaki (1984) noticed that a particular fucosyl transferase enzyme characteristic of mouse intestinal villi was induced by bacteria. Further studies (Hooper et al. 1998) have shown that the intestines of germ-free mice can initiate, but not complete, their differentiation. For complete development, the microbial symbionts of the gut are needed. Normally occurring gut bacteria can upregulate the transcription of several mouse genes, including those encoding colipase, which is important in nutrient absorption; angiogenin-4, which helps form blood vessels; and Sprr2a, a small, proline-rich protein thought to fortify the extracellular matrices that line the intestine (**FIGURE 25.16**; Hooper et al.

FIGURE 25.16 Induction of mammalian genes by symbiotic microbes. Mice raised in germ-free environments were either left alone or inoculated with one or more types of bacteria. After 10 days, their intestinal mRNAs were isolated and tested on microarrays. Mice grown in germ-free conditions had very little expression of the genes encoding colipase, angiogenin-4, or Sprr2a. Several different bacteria—*Bacteroides thetaiotaomicron, Escherichia coli, Bifidobacterium infantis,* and an assortment of gut bacteria harvested from conventionally raised mice—induced the genes for colipase and angiogenin-4. *B. thetaiotaomicron* appeared to be totally responsible for the 50-fold increase in *Sprr2a* expression over that of germ-free animals. This ecological relationship between the gut microbes and the host cells could not have been discovered without the molecular biological techniques of polymerase chain reaction and microarray analysis. (After Hooper et al. 2001.)

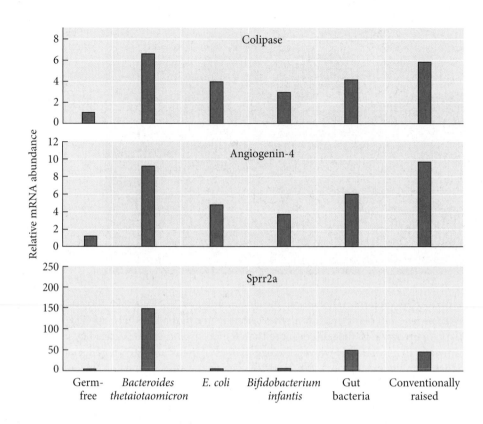

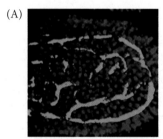

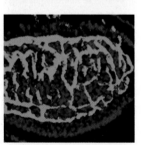

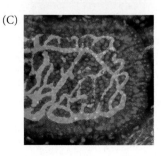

2001). Stappenbeck and colleagues (2002) have demonstrated that in the absence of particular intestinal microbes, the capillaries of the small intestinal villi fail to develop their complete vascular networks (**FIGURE 25.17**). In zebrafish, microbes regulate (through the canonical Wnt pathway) the normal proliferation of the intestinal stem cells. Without these microbes, the intestinal epithelium has fewer cells, and it lacks goblet cells, entroendocrine cells, and the characteristic intestinal brush border enzymes (**FIGURE 25.18**; Rawls et al. 2004, 2006; Bates et al. 2006).

BACTERIA HELP REGULATE DEVELOPMENT OF THE IMMUNE AND NERVOUS SYSTEMS Intestinal microbes also appear to be critical for the maturation of the **mammalian gut-associated lymphoid tissue (GALT)**. The GALT mediates mucosal immunity and oral immune tolerance, allowing us to eat food without making an immune response to it (see Rook and Stanford 1998; Cebra 1999; Steidler 2001). When introduced into germ-free rabbit appendices, neither *Bacillus fragilis* nor *B. subtilis* alone was capable of consistently inducing the proper formation of GALT. However, the combination of these two common mammalian gut bacteria consistently induced GALT (Rhee et al. 2004). The major inducer appears to be the protein bacterial polysaccharide A (PSA), especially that encoded by the genome of *B. fragilis*. The PSA-deficient mutant of *B. fragilis* is not able to restore normal immune function to germ-free mice (Mazmanian et al. 2005). Thus, a bacterial compound appears to play a major role in inducing the host's immune system. Exposure to microbes early in life prevents the development of the T lymphocytes associated with allergies and inflammatory bowel disease, while it enhances the helper T-cell repertoire. The T-lymphocytes associated with protection

FIGURE 25.17 Gut microbes are necessary for mammalian capillary development. (A) The capillary network (green) of germ-free mice is severely reduced compared with (B) the capillary network in those same mice 10 days after inoculation with normal gut bacteria. (C) The addition of *Bacteroides thetaiotaomicron* alone is sufficient to complete capillary formation. (From Stappenbeck et al. 2002.)

FIGURE 25.18 Bacteria stimulate stem cell division and cell differentiation in the zebrafish gut. (A) Quantitation of S-phase (dividing) intestinal epithelial cells in conventionally raised (control), germ-free, and germ-free plus added bacteria specimens. (B) Germ-free zebrafish given bacteria have normal amounts of stem cell division and epithelial cell differentiation after 6 days. Here and in (C), nondividing cells are stained blue and dividing cells are stained magenta. The inner cells are intestinal epithelia; cells in the white outline are mesenchyme and muscle. (C) The intestines of germ-free zebrafish are smaller and contain fewer dividing stem cells. (After Rawls et al. 2004.)

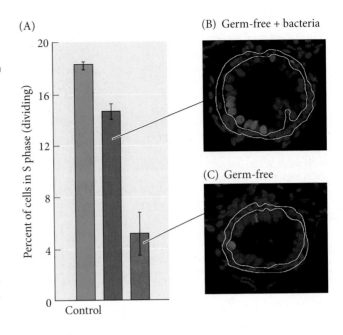

against allergies are induced by certain bacteria (Ohnmacht et al. 2015), and these bacteria may be encouraged to grow by sugars found in mother's milk (Ardeshir et al. 2014). Germ-free mice have an immunodeficiency syndrome, and the full complement of T lymphocytes is made possible only with the host species-specifc microbes (Niess et al. 2008; Duan et al. 2010; Chung et al. 2012; Olszak et al. 2012). So symbiotic bacteria are critically important in the differentiation of the lymphocytes of the mammalian immune system.

Although it may sound like science fiction, there is now evidence that symbiotic bacteria stimulate the postnatal development of the mammalian brain. Germ-free mice have lower levels of the transcription factor Egr1 and the paracrine factor BDNF in relevant portions of their brains than do conventionally raised mice, while having elevated levels of the neural hormone serotonin (**FIGURE 25.19**; Diaz Heijtz et al. 2011; Clarke et al. 2013). This correlates with behavioral differences between groups of mice, leading Diaz Heijtz and colleagues (2011) to conclude that "during evolution, the colonization of gut microbiota has become integrated into the programming of brain development, affecting motor control and anxiety-like behavior." In another investigation, a particular *Lactobacillus* strain has been reported to help regulate emotional behavior through a vagus nerve-dependent regulation of GABA receptors (Bravo et al. 2011). Thus, there may be pathways wherein products made by bacteria can enter the blood and help regulate the development of the brain (Grenham et al. 2011; McLean et al. 2012).

The gut bacteria change dramatically during human pregnancy. Indeed, they appear to respond to the hormonal status and help a pregnant woman adapt to the physiological stresses of carrying a fetus. When transferred into a germ-free mouse, bacteria from women in the early stages of pregnancy cause a normal phenotype to develop in the hosts. When bacteria from women late in their pregnancy are transferred into germ-free mice, the mice get fatter and display some of the metabolic changes (such as insulin desensitization) associated with pregnant women (**FIGURE 25.20**; Koren et al. 2012).

In short, mammals have coevolved with bacteria to the point that our bodily phenotypes do not fully develop without them. The microbial community of our gut can be viewed as an "organ" that provides us with certain functions that we haven't evolved (such as the ability to process plant polysaccharides). And, like our developing organs, microbes induce changes in neighboring tissues. As Mazmanian and colleagues (2005) have

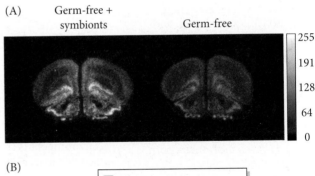

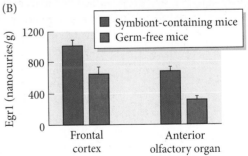

FIGURE 25.19 *Egr1* gene expression in mice depends on symbiotic microbes. (A) In situ hybridization of *Egr1* mRNA in a section through the frontal cortex of the brain, showing high levels of Egr1 protein in a mouse that has conventional microbes compared with a mouse remaining germ-free. (B) Quantitation using radioactive probes showed that symbiont-containing mice had significantly higher levels of Egr1 in the frontal cortex and anterior olfactory region than germ-free mice did. (After Diaz Heijtz et al. 2011.)

FIGURE 25.20 The composition of a woman's gut microbe population changes dramatically during pregnancy. This is associated with weight gain and with the progressive insensitivity to insulin characteristic of human pregnancy. When transplanted into the guts of germ-free mice, the bacteria from women early in their pregnancy (first trimester, roughly weeks 1–12) gave a normal phenotype to the mice. When bacteria from women late in their pregnancy (third trimester, roughly weeks 27–40) were transplanted into the germ-free mouse gut, the bacteria induced pregnancy-like metabolism, including weight gain and insulin resistance, in the mice. (After Koren et al. 2012.)

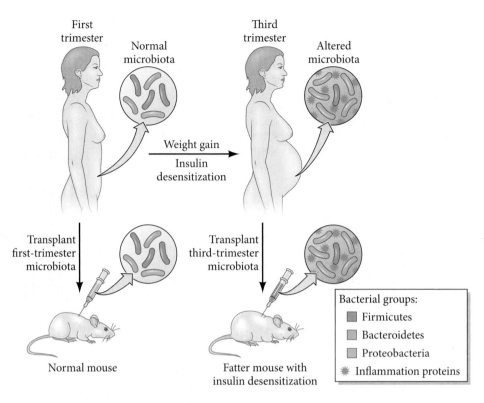

Developing Questions

Given that mammals need microbes to help construct their guts and immune systems, is there any evidence that children born by Cesarean delivery (surgical "C-section"; i.e., without passing through the birth canal) have different gut bacteria and may be more prone to certain diseases than those born without surgery?

concluded, "The most impressive feature of this relationship may be that the host not only tolerates but has evolved to require colonization by commensal microorganisms for its own development and health."

Coda

Phenotype is not merely the expression of one's inherited genome. Rather, there are interactions between an organism's genotype and environment that elicit a particular phenotype from a genetically controlled repertoire of possible phenotypes. Environmental factors such as temperature, diet, physical stress, the presence of predators, and crowding can generate a phenotype that is suited for that particular environment. Environment is therefore considered to play a role in the *generation* of phenotypes in addition to its well-established role in the *selection* of phenotypes. The fact that we codevelop with other organisms is an important concept for developmental biology and for evolutionary biology. This may be extremely important for medicine, especially if brain development can be affected by bacteria. Indeed, research is ongoing as to whether bacteria may be responsible for autism spectrum disorders (see Gonzalez et al. 2011).

Ecological developmental biology is providing the molecular basis for the genotype-by-environment interactions that are increasingly attracting the interests of evolutionary biologists (Schlichting and Pigliucci 1998). Different genomes alter their expression patterns differently in response to the same environmental change. The sensitivity of dung beetle larvae to changes in juvenile hormone, the threshold for sex specification changes in turtles, and numerous other developmental responses to the environment are selectable phenotypes (see Moczek and Nijhout 2002; McGaugh and Janzen 2011; Moczek et al. 2011). As the genes for these environmentally induced developmental pathways are becoming known (Matsumoto and Crews 2012; Snell-Rood and Moczek 2012), molecular mechanisms can be proposed for the genome-environment interactions.

Ecological developmental biology also calls into question the notions of autonomy and independent development. Moreover, if we are not truly "individuals" but have a phenotype based on community interactions, what exactly is natural selection selecting? Can natural selection select teams or relationships? The ramifications of developmental plasticity and developmental symbiosis on the rest of biology are just beginning to be appreciated (see Bateson and Gluckman 2011; Gilbert et al. 2012; McFall-Ngai et al. 2013; Gilbert and Epel 2015).

Next Step Investigation

The ability of genomes to respond to the outside environment by changing the developmental trajectories of organisms opens up entire worlds of research. Is evolution selecting "teams" of organisms? Are microbes responsible for normal brain development? How frequently are environmentally induced changes in chromatin transmitted from one generation to the next? Can behavioral traits be so transmitted? And what signals are the microbes using to effect normal signaling during development? The field of ecological developmental biology is being organized around such questions.

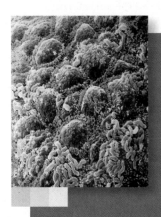

Closing Thoughts on the Opening Photo

"Honor thy symbionts," exhorted Jian Xu and Jeffrey Gordon (2003). Few people 25 years ago would have imagined that normal human development would depend on the interactions of several species. It now appears, however, that the gut microbiome, acquired during the first few days after birth, is crucial not only to normal development but also to our adult physiology. As people in the industrial world are exposed to fewer bacteria and fewer bacterial types, there are concerns that we might be missing some of the bacteria that are essential to healthy development. (Computer-colored SEM of symbiotic bacterial rods within the human colon © P. M. Motta and F. Carpino/Univ. "La Sapienza"/Science Source.)

25 Snapshot Summary
Development and the Environment

1. The environment plays critical roles during normal development. These agents include temperature, diet, crowding, and the presence of predators.

2. Developmental plasticity makes it possible for environmental circumstances to elicit different phenotypes from the same genotype. The genome encodes a repertoire of possible phenotypes. The environment often selects which of those phenotypes will become expressed.

3. Reaction norms are phenotypes that quantitatively respond to environmental conditions, such that the phenotype reflects small differences in the environmental conditions.

4. Polyphenisms represent "either/or" phenotypes; that is, one set of conditions elicits one phenotype, while another set of conditions elicits another.

5. Seasonal cues such as photoperiod, temperature, or type of food can alter development in ways that make the organism more fit under the conditions it encounters. Changes in temperature also are responsible for determining sex in several organisms, including many reptiles and fish.

6. Predator-induced polyphenisms have evolved such that prey species can respond morphologically to the presence of a specific predator. In some instances, this induced adaptation can be transmitted to the progeny of the prey.

7. There are several routes through which gene expression can be influenced by the environment. Environmental factors can methylate genes differentially; they can induce gene expression in surrounding cells; and they can be monitored by the nervous system, which then produces hormones that affect gene expression.

8. Behavioral phenotypes can also be induced by the environment. Conditions experienced as the brain matures after birth can alter patterns of DNA methylation and thereby change hormone reception and behaviors.

9. Organisms usually develop with symbiotic organisms, and signals from the symbionts can be critical for normal development.

10. Symbionts can be acquired horizontally (through infection) or vertically (through the oocyte).

11. In an obligate mutualism, both partners are needed for the survival of the other; in an obligate developmental mutualism, at least one partner is needed for the proper development of another.

12. The mammalian gut contains symbionts that actively regulate intestinal gene expression to generate proteins that are normal physiological components of intestinal development and function. Without these symbionts, the intestinal blood vessels and gut-associated lymphoid tissue of some mammalian species fail to form properly.

13. Symbionts can induce normal gene expression in hosts; and the host phenotype is deficient without the bacterial-induced patterns of gene expression. The differentiation of certain immune cells, gut cells, and neural cells may depend on symbiont-induced gene expression.

14. In vertebrates, gut symbionts may be important for the development of the gut, the immune system, and perhaps even portions of the nervous system.

Further Reading

Agrawal, A. A., C. Laforsch and R. Tollrian. 1999. Transgenerational induction of defenses in animals and plants. *Nature* 401: 60–63.

Brakefield, P. M. and N. Reitsma. 1991. Phenotypic plasticity, seasonal climate, and the population biology of *Bicyclus* butterflies (Satyridae) in Malawi. *Ecol. Entomol.* 16: 291–303.

Caldji, C., I. C. Hellstrom, T. Y. Zhang, J. Diorio and M. J. Meaney. 2011. Environmental regulation of the neural epigenome. *FEBS Lett.* 585: 2049–2058.

Gilbert, S. F. and D. Epel. 2015. *Ecological Developmental Biology: Integrating Epigenetics, Medicine, and Evolution*, 2nd Ed. Sinauer Associates, Sunderland, MA.

Gilbert, S. F., T. C. Bosch and C. Ledón-Rettig. 2015. Eco-Evo Devo: Developmental symbiosis and developmental plasticity as evolutionary agents. *Nature Rev. Genet.* 16: 611–622.

Hooper, L. V., M. H. Wong, A. Thelin, L. Hansson, P. G. Falk and J. I. Gordon. 2001. Molecular analysis of commensal host-microbial relationships in the intestine. *Science* 291: 881–884.

McFall-Ngai, M. J. 2002. Unseen forces: The influence of bacteria on animal development. *Dev. Biol.* 242: 1–14.

McFall-Ngai, M. J. 2014. The importance of microbes in animal development: Lessons from the squid-*Vibrio* symbiosis. *Annu. Rev. Microbiol.* 68: 177–194.

Moczek, A. P. 2005. The evolution of development of novel traits, or how beetles got their horns. *BioScience* 55: 937–951.

Relyea, R. A. and N. Mills. 2001. Predator-induced stress makes the pesticide carbaryl more deadly to grey treefrog tadpoles (*Hyla versicolor*). *Proc. Natl. Acad. Sci. USA* 2491–2496.

Stappenbeck, T. S., L. V. Hooper and J. I. Gordon. 2002. Developmental regulation of intestinal angiogenesis by indigenous microbes via Paneth cells. *Proc. Natl. Acad. Sci. USA* 99: 15451–15455.

Waterland, R. A. and R. L. Jirtle. 2003. Transposable elements: Targets for early nutritional effects of epigenetic gene regulation. *Mol. Cell. Biol.* 23: 5293–5300.

GO TO WWW.DEVBIO.COM...

...for Web Topics, Scientists Speak interviews, Watch Development videos, Dev Tutorials, and complete bibliographic information for all literature cited in this chapter.

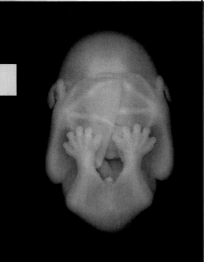

Development and Evolution

Developmental Mechanisms of Evolutionary Change

WHILE HE WAS WRITING *ON THE ORIGIN OF SPECIES*, Charles Darwin consulted his friend Thomas Huxley concerning the origins of variation. In his response, Huxley noted that many differences between organisms could be traced to differences in their development, and that these differences "result not so much of the development of new parts as of the modification of parts already existing and common to both the divergent types" (Huxley 1857).

Huxley's response expresses a major tenet of **evolutionary developmental biology**, a relatively new science that views evolution as the result of changes in development. If development is the change of gene expression and cell position over time, then evolution is the change of development over time. This new field—colloquially referred to as **evo-devo**—is producing a new model of evolution that integrates developmental biology, paleontology, and population genetics to explain and define the diversity of life (Raff 1996; Hall 1999; Arthur 2004; Carroll et al. 2005; Kirschner and Gerhart 2005). In other words, evolutionary developmental biology links genetics with evolution through the agencies of development. As Thomas's grandson, Julian Huxley, observed in 1942, "A study of the effects of genes during development is as essential for an understanding of evolution as are the study of mutation and that of selection." Contemporary evolutionary developmental biology is analyzing how changes in development can create the diverse variation that natural selection can act on. Rather than concentrating on the "survival of the fittest," evolutionary developmental biology gives us new insights into the "arrival of the fittest" (Carroll et al. 2005; Gilbert and Epel 2015).

What changes in development might be needed for the evolution of a nonflying mammal into a bat?

The Punchline

Changes in anatomy come about through changes in development. These changes in development form the bases of morphological variation needed for evolution. Much of the ability to alter development comes from the flexibility of enhancers. Development can be altered by changing the sequence of enhancer elements, which can change the cell type a gene is expressed in, the time at which a gene is expressed, or the amount of gene expression. Altering enhancers can also lead to the recruitment of a battery of genes or the formation of a new cell type. Evolution can also occur by changing the protein-encoding region of genes producing transcription factors. In addition, developmental plasticity may accelerate the evolutionary process and bias it toward certain phenotypes.

Descent with Modification: Why Animals Are Alike and Different

In the nineteenth century, debates over the origin of species pitted two views of nature against each other. One view, championed by Georges Cuvier and Charles Bell, focused on the *differences* between species that allowed each to adapt to its environment. Thus, they believed, the hand of the human, the flipper of the seal, and the wings of birds and bats were marvelous contrivances, each fashioned by the Creator to adapt these animals to their "conditions of existence." The other view, championed by Étienne Geoffroy Saint-Hilaire and Richard Owen, was that "unity of type" (the *similarities* among organisms, which Owen called "homologies") was critical. The human hand, the seal flipper, and the wings of bats and birds, said Owen, were all modifications of the same basic plan (see Figure 1.18). In discovering that plan, one could find the form upon which the Creator designed these animals. The adaptations were secondary.

Darwin acknowledged his debt to these earlier debates when he wrote in 1859, "It is generally acknowledged that all organic beings have been formed on two great laws—Unity of Type, and Conditions of Existence." Darwin went on to explain that his theory would explain unity of type by descent from a common ancestor, while the adaptations to the conditions of existence could be explained by natural selection. Darwin called this concept **descent with modification**. Darwin noted that the homologies between the embryonic and larval structures of different phyla provided excellent evidence for descent with modification. He was thrilled that the larval anatomy of barnacles demonstrated them to be crustaceans, and he was especially pleased by Kowalevsky's demonstration that the larvae of tunicates had both a notochord and pharyngeal pouches (1871). This showed them to be chordates, thereby uniting the invertebrates and vertebrates into a coherent animal kingdom. In the late 1800s, developmental change was seen as being the motor of evolution (Gould 1977). Or, as Thomas Huxley aptly remarked, "Evolution is not a speculation but a fact; and it takes place by epigenesis" (Huxley 1893, p. 202).

WEB TOPIC 26.1   **RELATING EVOLUTION TO DEVELOPMENT IN THE NINETEENTH CENTURY**   Attempts to relate evolution to changes in development began almost immediately after the publication of *On the Origin of Species*. This topic highlights the attempts of three scientists—Frank Lillie, Edmund B. Wilson, and Ernst Haeckel—to expand on the connection between evolution and development.

Preconditions for Evolution: The Developmental Structure of the Genome

If natural selection can only operate on existing variants, where does all that variation come from? If, as Darwin (1868) and Huxley concluded, variation arose from changes in development, then how could the development of an embryo change when development is so finely tuned and complex? How could such change occur without destroying the entire organism?[1] Even after the molecular biology of protein synthesis became understood, the problem did not go away. If a protein-encoding gene were mutated, the abnormal protein would be made everywhere the protein was normally expressed. There was no way a mutation could cause the protein to be made in one place and not another. The matter remained a mystery until evolutionary developmental biologists

---

[1]Darwin's German contemporary Ernst Haeckel proposed that most organisms evolved by adding a step to the end of embryonic development. But there turned out to be so many exceptions to that rule that it fell into disrepute. Two of Darwin's British contemporaries, Herbert Spencer and Robert Chambers, also saw development as the motor of evolution; they used von Baer's laws (see Chapter 1) as its mechanism (see Gould 1977; Friedman and Diggle 2011).

demonstrated that large morphological changes could arise during development because of two conditions that underlie the development of all multicellular organisms: **modularity** and **molecular parsimony**.

 DEV TUTORIAL *EvoDevo* **In two lectures, Scott Gilbert summarizes some of the basic principles of evolutionary developmental biology.**

Modularity: Divergence through dissociation

We now know that even early stages of development can be altered to produce evolutionary novelties. Such changes can occur because development occurs through a series of discrete and interacting modules (Riedl 1978; Bonner 1988; Kuratani 2009). Examples of developmental modules include morphogenetic fields (for example, those for the heart, limb, or eye), signal transduction pathways (such as the Wnt or BMP cascades), imaginal discs, cell lineages (such as inner cell mass or the trophoblast), insect parasegments, and vertebrate organ rudiments (Gilbert et al. 1996; Raff 1996; Wagner 1996; Schlosser and Wagner 2004). The ability of one module to develop differently from other modules (a phenomenon sometimes called **dissociation**) was well known to early experimental embryologists. For instance, when Victor Twitty grafted a limb bud from the early larva of a large salamander onto the trunk of a small salamander larva, the limb grew to its normal large dimensions within the small larva, indicating that the limb field module was independent from the global growth patterning of the embryo (Twitty and Schwind 1931; Twitty and Elliott 1934). The same independence was seen for the eye field. Modular units allow certain parts of the body to change without interfering with the functions of other parts.

One of the most important discoveries of evolutionary developmental biology is that not only are anatomical units modular (such that one part of the body can develop differently than the others), but the DNA regions that form the enhancers of genes are also modular. This genetic modularity—i.e., that there can be multiple enhancers for each gene and that each enhancer region can have binding sites for multiple transcription factors—was illustrated in Figure 3.11. The modularity of enhancer elements allows particular sets of genes to be activated together and permits a particular gene to become expressed in several discrete places. Thus, if by mutation a particular gene loses or gains a modular enhancer element, the organism containing that particular allele will express that gene in different places or at different times than organisms retaining the original allele. This mutability can result in the development of different anatomical and physiological morphologies (Sucena and Stern 2000; Shapiro et al. 2004), and major morphological changes can proceed through a mutation in a DNA regulatory region. Thus, the modularity of enhancers can be critical in providing selectable variation. Indeed, mutations affecting enhancer sequences are now thought to be the most important cause of morphological divergence between groups of animals (Carroll 2008; Stern and Orgogozo 2008).

PITX1 AND STICKLEBACK EVOLUTION The importance of enhancer modularity has been dramatically demonstrated by the analysis of evolution in threespine stickleback fish (*Gasterosteus aculeatus*). Freshwater sticklebacks evolved from marine sticklebacks about 12,000 years ago, when marine populations colonized the newly formed freshwater lakes at the end of the last ice age. Marine sticklebacks (**FIGURE 26.1A**) have pelvic spines that serve as protection against predation, lacerating the mouths of predatory fish that try to eat the stickleback. (Indeed, the scientific name of the fish translates as "bony stomach, with spines.") Freshwater sticklebacks do not have pelvic spines (**FIGURE 26.1B**). This may be because the freshwater fish lack the piscine predators that the marine fish face, but instead must deal with invertebrate predators that can easily capture them by grasping onto such spines. Thus, a pelvis without spines was selected in freshwater populations of this species.

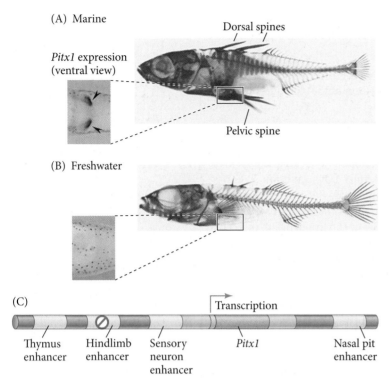

(A) Marine

Pitx1 expression
(ventral view)

Dorsal spines

Pelvic spine

(B) Freshwater

(C)

Transcription

Thymus
enhancer

Hindlimb
enhancer

Sensory
neuron
enhancer

Pitx1

Nasal pit
enhancer

FIGURE 26.1 Modularity of development: enhancers. Loss of *Pitx1* gene expression in the pelvic region of freshwater populations of the threespine stickleback (*Gasterosteus aculeatus*). Bony plates and pelvic spines characterize marine populations of this species (A). In freshwater populations (B), the pelvic spines are absent, as is much of the bony armor. In magnified ventral views of embryos (inset photos), in situ hybridization reveals *Pitx1* expression (purple) in the pelvic area (as well as in sensory neurons, thymic cells, and nasal regions) of the marine population. The staining in the pelvic region is absent in freshwater populations, although it is still seen in the other areas. The arrowheads point to *Pitx1* expression in the ventral region that forms the pelvic spines of the marine populations. (C) Model for the evolution of pelvic spine loss. Four enhancer regions are postulated to reside near the *Pitx1* coding region. These enhancers direct the expression of this gene in the thymus, pelvic spines, sensory neurons, and nasal pit, respectively. In freshwater populations of threespine sticklebacks, the pelvic spine (hindlimb) enhancer module has been mutated and the *Pitx1* gene fails to function there. (After Shapiro et al. 2004; photographs courtesy of D. M. Kingsley.)

To determine which genes might be involved in these pelvic differences, researchers mated individuals from marine (spined) and freshwater (spineless) stickleback populations. The resulting offspring were bred to each other and produced numerous progeny, some of which had pelvic spines and some of which didn't. Using molecular markers to identify specific regions of the parental chromosomes, Shapiro and co-workers (2004) found that the major gene for pelvic spine development mapped to the distal end of chromosome 7. That is to say, nearly all the fish with pelvic spines inherited a "pelvic appendage-encoding" chromosomal region from the marine parent, whereas fish lacking pelvic spines inherited this region from the freshwater parent. The researchers then tested numerous candidate genes (e.g., genes known to be active in the pelvic and hindlimb structures of mice) and found that the gene encoding the transcription factor Pitx1 was located on this region of chromosome 7.

When Shapiro and colleagues compared the amino acid sequences of the Pitx1 proteins of marine and freshwater sticklebacks, there were no differences. However, there was a critically important difference when they compared the expression patterns of the *Pitx1* gene. In both populations, *Pitx1* was expressed in the precursors of the thymus, nose, and sensory neurons. In the marine populations, *Pitx1* was also expressed in the pelvic region. But in the freshwater populations, the pelvic expression of *Pitx1* was absent or severely reduced (**FIGURE 26.1C**). Since the coding region of *Pitx1* was not mutated (and since the gene involved in the pelvic spine differences maps to the site of the *Pitx1* gene, and the difference between the freshwater and marine populations involves the expression of this gene at a particular site), it was reasonable to conclude that the enhancer region allowing expression of *Pitx1* in the pelvic area (i.e., the pelvic spine enhancer) does not function in the freshwater populations.

This conclusion was confirmed when high-resolution genetic mapping showed that the DNA of the "hindlimb" enhancer of *Pitx1* differed between sticklebacks with pelvic spines and those without pelvic spines[2] (Chan et al. 2010). When this 2.5-kb DNA fragment from marine (spined) fish was fused to a gene for green fluorescent protein and inserted into fertilized freshwater stickleback eggs, GFP was expressed in the pelvis. Moreover, when this same fragment taken from marine sticklebacks was placed next to the *Pitx1*-coding sequence of freshwater (spine-deficient) fish and then injected into fertilized eggs of the spine-deficient fish, pelvic spines formed in the freshwater fish.

---

[2] Interestingly, the loss of the pelvic spines in several stickleback populations appears to have been the result of independent losses of this *Pitx1* expression domain. This finding suggests that if the loss of *Pitx1* expression in the pelvis occurs, this trait can be readily selected (Colosimo et al. 2004). Here we see that by combining the approaches of population genetics and developmental genetics one can determine the mechanisms by which evolution can occur.

(A)

(B)

RECRUITMENT   Modularity allows the recruitment (or "co-option") of entire suites of characters into new places. In Chapter 10, we discussed the recruitment of the skeleton-forming genes (the skeletonogenic "subroutine") into the developmental repertoire of the sea urchin micromeres. In most echinoderm groups, the skeletogenic genes are activated only in the adult and are used to form the hard exoskeletal plates. However, in sea urchins (and not in any other echinoderm group) this set of genes has come under the control of the micromere double-negative gate because of changes in the enhancer of one of these genes. Thus, the skeleton is made by larval mesenchymal cells (Gao and Davidson 2008).

Another example of recruitment is seen in insects, in the wing structure that defines the beetles. Beetles are the most successful animal group on the planet, accounting for more than 20% of extant animal species (Hunt et al. 2007). They differ from other insects in forming an elytron, a forewing encased in a hard exoskeleton. This makes them the "living jewels" so beloved of naturalists (**FIGURE 26.2**).[3] In beetles, as in *Drosophila*, the *Apterous* gene is expressed in the dorsal compartment of the wing imaginal discs, and the Apterous transcription factor organizes the tissue to differentiate dorsal wing structures. However, in beetles (and in no other known insect), Apterous protein also activates the exoskeleton genes in the forewing while repressing them in the hindwing (Tomoyasu et al. 2009). Thus, a new type of wing emerges from the recruitment of one module (the subroutine of exoskeletal development) into another (the subroutine of dorsal forewing development).

FIGURE 26.2   Elytra are the hardened forewings that are characteristic of Coleoptera, the beetles. Elytra are formed through the recruitment of the genetic module for exoskeleton development into the module for dorsal forewing development. (A) The elytra of a "ladybug" beetle. Its forewings are ornamented with exoskeleton, and its hindwings are extended. (B) These "living jewels" from the Oxford Museum of Natural History illustrate some of the diversity of beetle elytra. (A © F1online digitale Bildagentur GmbH/Alamy; B © Jochen Tack/Alamy.)

WEB TOPIC 26.2   **CORRELATED PROGRESSION**   In many cases, modules must coevolve. The upper and lower jaws, for instance, have to fit together properly; if one changes, so must the other. If the sperm-binding proteins on the egg change, then so must the egg-binding proteins on the sperm. This site looks at correlated changes during evolution.

Molecular parsimony: Gene duplication and divergence

The second precondition for macroevolution through developmental change is molecular parsimony, sometimes called the "small toolkit." In other words, although development differs enormously from lineage to lineage, development within all lineages uses the same types of molecules. The transcription factors, paracrine factors, adhesion

---

[3]Both Darwin and Wallace were avid beetle collectors, but it was the geneticist J. B. S. Haldane whose remark may best reflect the prominence of these insects. When asked by a cleric what the study of nature could tell us about God, Haldane is said to have replied, "He has an inordinate fondness for beetles."

(A)

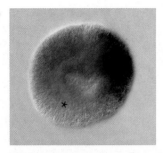

(B)

(C)

FIGURE 26.3 Evidence of the evolutionary conservation of regulatory genes. (A) The cnidarian homologue of the vertebrate *Bmp4* and *Drosophila Decapentaplegic* genes is expressed asymmetrically at the edge of the blastopore (marked with an asterisk) in the embryo of the sea anemone *Nematostella*. This gene represents an ancestral form of the protostome and deuterostome forms of the gene. (B) The Hox gene *Anthox6*, a cnidarian member of the paralogue 1 group of Hox genes, is expressed at the blastopore side (asterisk) of the larval sea anemone. (C) The *Pax6* gene for eye development is an example of a gene ancestral to both protostomes and deuterostomes. The micrograph shows ommatidia of the compound insect eye emerging in the leg of a fruit fly (a protostome) in which mouse (deuterostome) *Pax6* cDNA was expressed in the leg disc. (A,B from Finnerty et al. 2004, courtesy of M. Martindale; C from Halder et al. 1995, courtesy of W. J. Gehring and G. Halder.)

molecules, and signal transduction cascades are remarkably similar from one phylum to another. Indeed, it appears that the development of jellyfish and flatworms uses the same major kit of transcription factors and paracrine factors as flies and vertebrates (Finnerty et al. 2004; Carroll et al. 2005; Putnam et al. 2007; Ryan et al. 2007; Hejnol et al. 2009).

THE SMALL TOOLKIT Certain transcription factors (such as those of the BMP, Hox, and Pax groups) are found in all animal phyla. In fact, some "toolkit genes" appear to play the same *roles* in multiple animal lineages. The BMP levels appear to be used throughout the animal kingdom to specify the dorsal-ventral axis (**FIGURE 26.3A**); the Wnt and Hox genes appear to specify the anterior-posterior axis in all the bilaterians (**FIGURE 26.3B**); and the *Pax6* gene appears to be involved in specifying light-sensing organs, irrespective of whether the eye is that of a mollusk, an insect, or a primate[4] (**FIGURE 26.3C**). Similarly, homologues of *Otx* specify head formation in both vertebrates and invertebrates; and though insect and vertebrate hearts are very different, both are formed using *tinman/ Nkx2-5* (see Erwin 1999). Certain microRNAs appear to be found in all animals, and these appear to play the same or very similar developmental roles in whatever phylum they are found (Christodoulou et al. 2010). These include miRNA-124, which is found in the central nervous systems of protostomes and deuterostomes; miRNA-12, which is found in guts throughout the animal kingdom; and miRNA-92, which helps specify ciliated locomotor cells in deuterostome and protostome larvae. Discovering that the same set of transcription factors and microRNAs causes the specification of the same types of cells throughout the animal kingdom is a very powerful argument that the protostomes and deuterostomes are derived from a common ancestor that used these factors in similar ways to specify its organs (Davidson and Erwin 2010).

DUPLICATION AND DIVERGENCE One theme that resounds through studies of paracrine and transcription factors is that these proteins (and the genes that encode them) come in families. How do gene families come into existence? The answer is through duplication of an original gene and the subsequent independent mutation of the original duplicates (**FIGURE 26.4**). This creates a family of genes that are related by common descent (and which are often still adjacent to each other.) This scenario of **duplication and divergence** is seen in the Hox genes, the globin genes, the collagen genes, the *Distal-less* genes, and in many paracrine factor families (e.g., the Wnt genes). Each member of such a gene family is homologous to the others (that is, their sequence similarities are due to descent from a common ancestor and are not the result of convergence for a particular function), and they are called **paralogues**. Susumu Ohno (1970), one of the founders of the gene family concept, likened gene duplication to a

---

[4]This doesn't mean that the eye is the only thing that is specified by Pax6, or that Pax6 hasn't become regulated by different proteins in different phyla (Lynch and Wagner 2010).

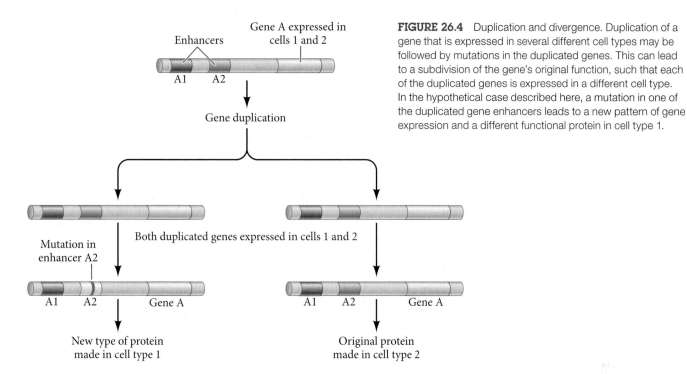

Enhancers

Gene A expressed in cells 1 and 2

A1 A2

Gene duplication

FIGURE 26.4 Duplication and divergence. Duplication of a gene that is expressed in several different cell types may be followed by mutations in the duplicated genes. This can lead to a subdivision of the gene's original function, such that each of the duplicated genes is expressed in a different cell type. In the hypothetical case described here, a mutation in one of the duplicated gene enhancers leads to a new pattern of gene expression and a different functional protein in cell type 1.

Both duplicated genes expressed in cells 1 and 2

Mutation in enhancer A2

A1   A2   Gene A

A1   A2   Gene A

New type of protein made in cell type 1

Original protein made in cell type 2

sneaky criminal circumventing surveillance. While the "police force" of natural selection makes certain that there is a "good" gene properly performing its function, that gene's duplicate, unencumbered by the constraints of selection, can mutate and undertake new functions.

Such "subfunctionalization" has since been shown to be the case in many gene families, including the Hox genes. Hox genes represent an especially complex and important case of duplication and divergence. We find that (1) there are related Hox genes in each animal group (such as *Deformed, Ultrabithorax,* and *Antennapedia* in *Drosophila,* or the 39 Hox genes in mammals); and (2) there are several clusters of Hox genes in vertebrates (the 39 mammalian Hox genes, for example, are clustered on four different chromosomes). The similarity of all the Hox genes is best explained by descent from a common ancestral gene, probably in the single-celled protozoa or sponges. This would mean that in *Drosophila,* the *Deformed, Ultrabithorax,* and *Antennapedia* genes all emerged as duplications of an original gene. The sequence patterns of these three genes (especially in the homeodomain region) are extremely well conserved. Such tandem gene duplications are thought to be the result of errors in DNA replication, and such errors are not uncommon. Once replicated, the gene copies can diverge by random mutations in their coding sequences and enhancers, developing different expression patterns and new functions (Lynch and Conery 2000; Damen 2002; Locascio et al. 2002).

Thus, every *Drosophila* Hox gene has a homologue (and sometimes several) in vertebrates. In some cases, the homologies go very deep and can also be seen in the gene's functions. Not only is the vertebrate *Hoxb4* gene similar in sequence to its *Drosophila* homologue, *Deformed* (*Dfd*), but human *HOXB4* can perform the functions of *Dfd* when introduced into *Dfd*-deficient *Drosophila* embryos (Malicki et al. 1992). As mentioned in Chapter 12, the Hox genes of insects and humans are not just homologous—they occur in the same order on their respective chromosomes. Their expression patterns are also remarkably similar: the more 3' Hox genes have more anterior expression boundaries[5]

---

[5]The conservation of Hox genes and their colinearity demands an explanation. One possibility (Kmita et al. 2000, 2002) is that the Hox genes "compete" for a remote enhancer that recognizes the Hox genes in a polar fashion. This enhancer most efficiently activates Hox genes at the 5' end. Another proposal (Gaunt 2015) suggests that spatial colinearity evolved as a mechanism to physically separate genes, thus avoiding accidents of transcriptional activation.

FIGURE 26.5 Duplication and divergence of human *SRGAP2*. (A) The *SRGAP2* gene is found as a single copy in the genomes of all mammals except humans. In the lineage giving rise to humans, duplication events gave rise to four similar versions of the gene, designated A–D. (B) The "ancestral" gene, *SRGAP2A*, with minor contributions from *SRGAP2B* and *D*, enables the maturation of dendritic spines (protuberances) on the surfaces of neurons. *SGRAP2C* is a partial duplication, and its product inhibits *SRGAP2A*, slowing dendritic spine maturation and promoting neuronal migration. This partial duplication may have allowed for the evolution of longer maturation time and greater flexibility in the human brain. (After Geschwind and Konopka 2012.)

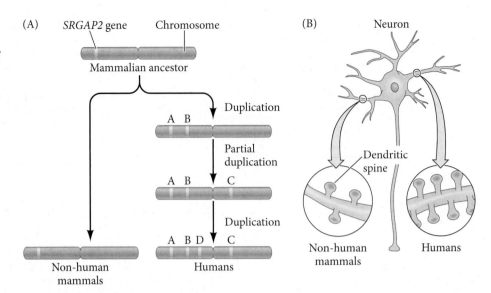

(see Figure 12.23). Thus, these genes are homologous between species (as opposed to members of a gene family being homologous within a species). Genes that are homologous between species are called **orthologues**.

One of the most important gene duplication events in human evolution may have been the duplication of *SRGAP2*, a gene that may have enabled the expansion of the human cerebral cortex. The protein encoded by this gene is expressed in the mammalian brain cortex and appears to *slow down* cell division and decrease the length and density of dendritic processes. However, humans differ from all other animals (including chimpanzees) by having duplicated this gene twice. Moreover, the second duplication event was not complete, so one of the newly formed genes is only a partial duplicate. This partial gene produces a truncated SRGAP2 protein, SRGAP2C, that is also made in the cerebral cortex and which *inhibits* the activity of normal SRGAP2 made from the complete genes. As a result, cell division in the cerebral cortex continues for longer periods of time, and the dendrites are larger with more connections (**FIGURE 26.5**; Charrier et al. 2012; Dennis et al. 2012). Based on genomic evidence, these gene duplication events are calculated to have taken place about 2.4 million years ago. This would be about the time of *Australopithecus*, the increase in primate brain size, and the first known use of tools (Tyler-Smith and Xue 2012).

Deep Homology

One of the most exciting contributions of evolutionary developmental biology has been the discovery not only of homologous regulatory genes, but also of homologous signal transduction pathways, many of which have been mentioned earlier in this book. In different organisms, these pathways are composed of homologous proteins arranged in a homologous manner (Zuckerkandl 1994; Gilbert 1996; Gilbert et al. 1996). This shows a level of parsimony even deeper than that of the individual genes.

In some instances, homologous pathways made of homologous components are used for the same function in both protostomes and deuterostomes. This has been called **deep homology** (Shubin et al. 1997, 2009). Conserved similarities in both the pathway and its function over millions of years of phylogenetic divergence are considered to be evidence of deep homology between these modules (Shubin et al. 1997). One example is the Chordin/BMP4 interaction discussed in Chapter 11. In both vertebrates and invertebrates, Chordin/Short-gastrulation (Sog) inhibits the lateralizing effects of BMP4/Decapentaplegic (Dpp), thereby allowing the ectoderm protected by

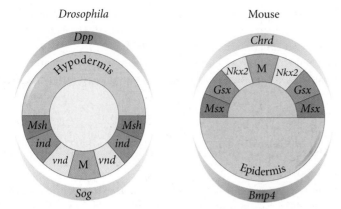

Drosophila   Mouse

FIGURE 26.6 The same set of instructions forms the nervous systems of both protostomes and deuterostomes. In the fruit fly (a protostome), the TGF-β family member *Dpp* (*Decapentaplegic*) is expressed dorsally and is opposed by *Sog* ventrally. In the mouse (a deuterostome), the TGF-β family member *Bmp4* is expressed ventrally and is countered dorsally by *Chordin* (*Chrd*). The highest concentration of Chordin/Sog becomes the midline (M). The midline is dorsal in vertebrates and ventral in insects, and the concentration gradient of the TGF-β family protein (BMP4 or Dpp) activates genes specifying the regions of the nervous system in the same order in both groups: *vnd/Nkx2*, followed by *ind/Gsx*, and finally *Msh/Msx*. These genes have been seen to be expressed in a similar fashion in cnidarians. (After Ball et al. 2004.)

Chordin/Sog to become the neurogenic ectoderm.[6] These reactions are so similar that *Drosophila* Dpp protein can induce ventral fates in *Xenopus* and can substitute for Sog (**FIGURE 26.6**; Holley et al. 1995).

According to this scheme, the central nervous system of the bilaterian animals originated only once, and the BMP-Chordin mechanism was already being used in the bilaterian ancestor of protostomes and deuterostomes. The positioning of BMP signaling at the ventral (vertebrate) or dorsal (invertebrate) location was a later occurrence (see Mizutani and Bier 2008). This idea has been supported by evidence that annelid worms and cephalochordates also use inhibition of the BMP pathway in forming their central nervous systems (Danes et al. 2007; Yu et al. 2007). Thus, despite their obvious differences, the protostome and deuterostome nervous systems seem to be formed by the same set of instructions. Indeed, deep homology has also been proposed for the formation of certain parts of the vertebrate and invertebrate brains (Strausfeld and Hirth 2013).

 SCIENTISTS SPEAK 26.1 Dr. Sean Carroll, a pioneer in the field, talks about the questions underlying evolutionary developmental biology.

Mechanisms of Evolutionary Change

In 1975, Mary-Claire King and Alan Wilson published a paper titled "Evolution at Two Levels in Humans and Chimpanzees." This study showed that despite the large anatomical differences between chimpanzees and humans, their DNA was almost identical. The differences were to be found in the regulatory genes that acted during development:

> *The organismal differences between chimpanzees and humans would … result chiefly from genetic changes in a few regulatory systems, while amino acid substitutions in general would rarely be a key factor in major adaptive shifts.*

In other words, the allelic substitutions of the genes that encode protein sequences—which seem to be pretty much the same for chimpanzees and humans—were not seen as being important. The important differences are where, when, and how much the genes are activated. In 1977, the idea that change within regulatory genes is critical to evolution was extended by François Jacob, the Nobel laureate who helped establish the operon model of gene regulation. First, Jacob said, evolution works with what it has: it

---

[6] In addition to this central inhibitory reaction, there are other reactions that add to the deep homology of the instructions for forming the protostome and deuterostome neural tube. The proteins involved in the diffusion and stability of BMPs and Chordin also are conserved between insects and vertebrates (Larrain et al. 2001).

combines existing parts in new ways rather than creating new parts. Second, he predicted that such "tinkering" would be most likely to occur in those genes that construct the embryo, not in the genes that function in adults (Jacob 1977).

Wallace Arthur (2004) catalogued four ways in which Jacob's "tinkering" can take place at the level of gene expression to generate phenotypic variation available for natural selection:

1. Heterotopy (change in location)
2. Heterochrony (change in time)
3. Heterometry (change in amount)
4. Heterotypy (change in kind)

These changes can only be accomplished if the gene expression patterns are modular—that is, if they are controlled by different enhancer elements. The modularity of development allows one part of the organism to change without necessarily affecting the other parts.[7]

Heterotopy

One important way of creating new structures is to alter the *location* where a transcription factor or paracrine factor is expressed. This spatial alteration of gene expression is called **heterotopy** (Greek, "different place"). Heterotopy allows different cells to take on a new identity (as sea urchin micromeres did when they recruited the genes for the skeleton formation; see Chapter 10) or to activate or inhibit a paracrine factor-mediated process in a new area of the body (as when Gremlin inhibits BMP-mediated apoptosis in the webbing between digits; see Figure 19.33). There are many other examples, some of which we will describe next.

HOW THE BAT GOT ITS WINGS AND THE TURTLE GOT ITS SHELL In Chapter 1, we mentioned that the bat evolved its wing by changing the development of the forelimb such that the cells in the interdigital webbing did not die. It turns out that the bat retains its forelimb webbing in a manner very similar to how the duck embryo retains its hindlimb webbing—by blocking the BMPs that would otherwise cause the interdigital cells to undergo apoptosis (see Figure 1.19). Both Gremlin and FGF signaling appears to block BMP functions in the bat wing. Unlike other mammals, bats express Fgf8 in their interdigital webbing, and this protein is critical for maintaining the cells there. If FGF signaling is inhibited (by drugs such as SU5402), BMPs can induce apoptosis of the forelimb webbing, just as in other mammals (Laufer et al. 1997; Weatherbee et al. 2006). The Fgf8 in the webbing also appears to be responsible for providing the mitotic signal that extends the digits of the bat, thereby expanding its wing (Hockman et al. 2008; Sears 2008).

The formation of the turtle shell also uses BMPs and FGFs, but in different ways. What distinguishes turtles from other vertebrates are their ribs—they migrate laterally into the dermis instead of forming a rib cage (**FIGURE 26.7**). Certain regions of the turtle dermis attract rib precursor cells, and these dermal regions differ from those of other vertebrates because they synthesize Fgf10. Fgf10 seems to attract the ribs, since the ribs do not enter the dermis if the Fgf10 signal is blocked (Burke 1989; Cebra-Thomas et al. 2005). The lateral growth of the ribs causes some muscles to establish new attachment sites and causes the scapula (shoulder blades) to reside inside the ribs. This phenomenon is seen only in the turtles (Nagashima et al. 2009). Once inside the dermis, the rib cells do what rib cells are expected to do—they undergo endochondral ossification wherein the cartilage cells are replaced by bone. To do this, BMPs are made.

---

[7]This chapter concentrates on transcriptional-level changes that can generate new morphological forms, but morphological changes can be instigated at these levels as well. Abzhanov and Kaufman (1999), for instance, have shown that post-transcriptional regulation of the *Sex combs reduced* gene is critical in converting legs into maxillipeds in the terrestrial crustacean sowbug *Porcellio scaber*.

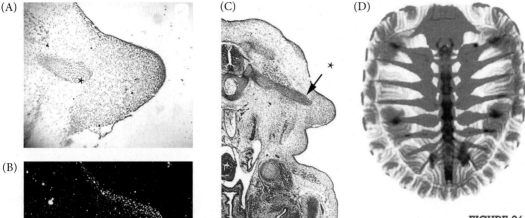

FIGURE 26.7 Heterotopy on several levels in turtle development. The carapace (dorsal shell) of the turtle is formed through sequential layers of heterotopies. *Fgf10* expression in certain regions of the dermis impels rib precursor cells to migrate laterally into the dermis instead of forming a rib cage. (A,B) Cross section of early turtle embryo as the rib enters the dermis (A, brightfield; B, autoradiograph staining for *Fgf10*). (C) Half cross section of a slightly later turtle embryo, showing a rib (arrow) extending from the vertebra into the region of the dermis that will expand to form the shell. (D) Hatchling turtle stained with alizarin to show bones. Bones can be seen in the dermis around the ribs that entered into it. Heterotopies include *Fgf10* expression, rib placement, and bone location. (After Loredo et al. 2001.)

But the rib is embedded in dermis, and the dermal cells can also respond to the BMPs by becoming bone (Cebra-Thomas et al. 2005; Rice et al. 2015). In this way, each of the newly positioned ribs instructs the dermis around it to become bone, and thus the turtle gets its shell. These conclusions about turtle development have facilitated new paleontological theories of the turtle's evolutionary origins (Lyson et al. 2013).

Heterochrony

Heterochrony (Greek, "different time") is a shift in the relative order or timing of two developmental processes. Heterochrony can be seen at any level of development, from gene regulation to adult animal behaviors (West-Eberhard 2003). In heterochrony, one module changes its time of expression or growth rate relative to the other modules of the embryo. One sees heterochronic changes in development throughout the animal kingdom. As Darwin (1859, p. 209) noted, "we may confidently believe that many modifications, wholly due to the laws of growth, and at first in no way advantageous to a species, have been subsequently taken advantage of by the still further modified descendants of this species."

Heterochronies are quite common in vertebrate evolution. We have already discussed the extended growth of the human brain and the heterochronies of amphibian metamorphosis. Another example is found in marsupials, whose jaws and forelimbs develop at a faster rate than do those of placental mammals, allowing the marsupial newborn to climb into the maternal pouch and suckle (Smith 2003; Sears 2004). Birds are thought to have arisen, in part, through heterochronic growth of the dinosaur skeleton (McNamara and Long 2012; Bhullar et al. 2012). The enormous number of vertebrae and ribs formed in embryonic snakes (more than 500 in some species) is likewise due to heterochrony (as well as to changes in the *Hox13* paralogue group). The segmentation reactions cycle nearly four times faster relative to tissue growth in snake embryos than they do in related vertebrate embryos (Gomez et al. 2008).

In some instances we can determine the heterochronic changes in expression of certain genes. The elongated fingers in the dolphin flipper appear to be the result of the heterochronic expression of *Fgf8*, which as we saw in Chapter 19 encodes a major paracrine factor for limb outgrowth (Richardson and Oelschläger 2002; Cooper 2010). Another "digital" example of molecular heterochrony occurs in the lizard genus *Hemiergis*, which includes species with three, four, or five digits per limb. The number of digits is regulated by the length of time the *Sonic hedgehog* gene remains active in the limb bud's zone of polarizing activity. The shorter the duration of *Shh* expression, the fewer the number of digits (Shapiro et al. 2003). In primates, there is a heterochronic shift in the transcription of a set of cerebral mRNAs, such that the expression pattern in adult humans resembles that seen in juvenile chimpanzees (Somel et al. 2009).

Heterometry

Heterometry is a change in the *amount* of a gene product or structure. We mentioned such heterometric changes in Chapter 16 when we discussed the evolution of the blind Mexican cavefish (see Figure 16.7). We saw that overproduction of Sonic hedgehog protein (Shh) in the midline prechordal plate downregulates the *Pax6* gene, preventing eye formation. But overexpression of *Shh* has other consequences as well. Not only does it cause the degeneration of the eyes, it also causes the jaw size and number of taste buds to increase (Franz-Odendaal and Hall 2006; Yamamoto et al. 2009). Since cavefish live in complete darkness, the expansion of their jaw size and gustatory sense at the expense of sight can be selected. Heterometry can also be seen in the human response to parasitic worms: a mutation causing overproduction of interleukin 4 has been (and is being) selected in populations where such parasites are endemic (Rockman et al. 2003).

WEB TOPIC 26.3 **ALLOMETRY**   Modularity and heterochrony combine in allometry: when different parts of an organism grow at different rates than their ancestor. The skulls of different dog breeds is a good example of this, as are the skulls of whales. But the human brain allometry is especially spectacular.

DARWIN'S FINCHES   One of the best examples of heterometry involves Darwin's celebrated finches, a set of 15 closely related birds collected by Charles Darwin and his shipmates during their visit to the Galápagos and Cocos islands in 1835. These birds helped Darwin frame his evolutionary theory of descent with modification, and they still serve as one of the best examples of adaptive radiation and natural selection (see Weiner 1994; Grant and Grant 2008). Systematists have shown that these finch species evolved in a particular manner, with a major speciation event being the split between the cactus finches and the ground finches. The ground finches evolved deep, broad beaks that enable them to crack seeds open, whereas the cactus finches evolved narrow, pointed beaks that allow them to probe cactus flowers and fruits for insects and flower parts. Earlier research (Schneider and Helms 2003) had shown that species differences in the beak pattern were caused by changes in the growth of the neural crest-derived mesenchyme of the frontonasal process (i.e., those cells that form the facial bones). Abzhanov and his colleagues (2004) found a remarkable correlation between the beak shape of the finches and the timing and amount of *Bmp4* expression (**FIGURE 26.8**). No other paracrine factor showed such differences. The expression of *Bmp4* in ground finches started earlier and was much greater than

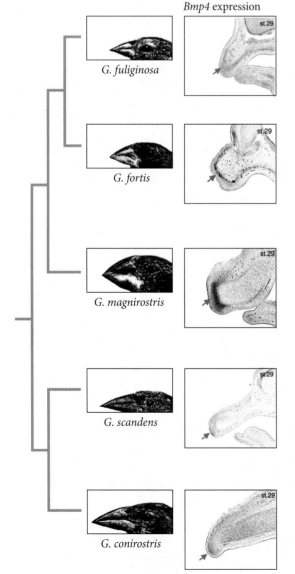

Bmp4 expression

G. fuliginosa

G. fortis

G. magnirostris

G. scandens

G. conirostris

FIGURE 26.8   Correlation between beak shape and the expression of *Bmp4* in five species of Darwin's finches. In the genus *Geospiza*, the ground finches (represented by *G. fuliginosa*, *G. fortis*, and *G. magnirostris*) diverged from the cactus finches (represented by *G. scandens* and *G. conirostris*). The differences in beak morphology correlate to heterochronic and heterometric changes in *Bmp4* expression in the beak. BMP4 (red arrow) is expressed earlier and at higher levels in the seed-crushing ground finches. The photographs of the embryonic beaks were taken at the same stage (stage 29) of development. This gene expression difference provides one explanation for the role of natural selection on these birds. (After Abzhanov et al. 2004.)

Bmp4 expression in cactus finches. In all cases, the *Bmp4* expression pattern correlated with the breadth and depth of the beak.

The importance of these expression differences was confirmed experimentally by changing the *Bmp4* expression pattern in chick embryos to mimic the heterometric and heterochronic changes in the ground finches (Abzhanov et al. 2004; Wu et al. 2004). When *Bmp4* expression was enhanced in the frontonasal process mesenchyme, the chick developed a broad beak reminiscent of the beaks of the ground finches. Conversely, when BMP signaling was inhibited in this region (by Noggin, a BMP inhibitor), the beak lost depth and width.

But this was only the beginning of the story. Gene chip technology showed that the level of *Calmodulin* gene expression in the beak primordia of the sharp-beaked cactus finches was 15-fold greater than in the beak primordia of the blunt-beaked ground finches. Calmodulin is a protein that combines with many enzymes to make their activity dependent on calcium ions. In situ hybridization and other techniques demonstrated that the *Calmodulin* gene is expressed at higher levels in the embryonic beaks of cactus finches than in the embryonic beaks of ground finches (**FIGURE 26.9**). When Calmodulin was upregulated in the embryonic chicken beak to mimic the finchlike expression domain, the chick beak too became long and pointed.

The frontonasal mesenchyme gives rise to two modules that form the adult beak: the premaxillary bone and the prenasal cartilage. The prenasal cartilage develops earlier in beak development and establishes the species-specific beak morphology. The morphology of prenasal cartilage is coordinately regulated by BMP and Calmodulin signals, and these signals correlate well with the exact scaling parameters of the evolving beak

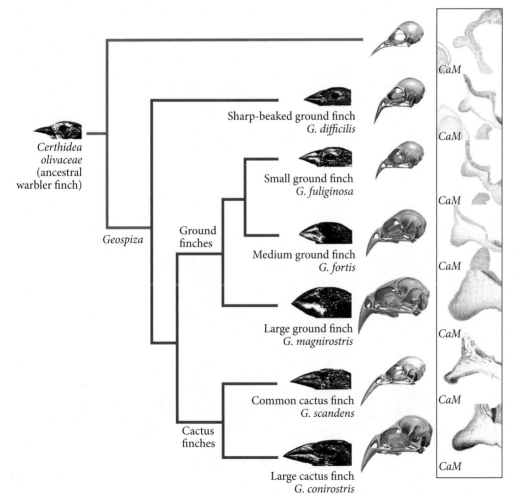

FIGURE 26.9 Correlation between beak length and the amount of *Calmodulin* (*CaM*) gene expression in six species of Darwin's finches. The *Geospiza* species displaying distinct beak morphologies are a monophyletic group, and the differences in beak morphology can be seen skeletally. *CaM* is expressed in a strong distal-ventral domain in the mesenchyme of the upper beak prominence of the large cactus finch (*G. conirostris*), in somewhat lower levels in the common cactus finch (*G. scandens*), and in very low levels in the large ground finch (*G. magnirostris*) and medium ground finch (*G. fortis*). Very low levels of *CaM* expression were also detected in the mesenchyme of *G. difficilis*, *G. fuliginosa*, and the basal warbler finch *Certhidea olivacea*. (After Abzhanov et al. 2006.)

shapes (Campàs et al. 2011; Mallarino et al. 2011, 2012). Thus, enhancers controlling the amount of beak-specific BMP4 and Calmodulin synthesis may have been critically important in the evolution of Darwin's finches. BMP4 and Calmodulin represent two targets for natural selection, and together they explain the shape variations of Darwin's finches (Abzhanov et al. 2006; Campàs et al. 2011).

Heterotypy

In heterochrony, heterotopy, and heterometry, mutations affect the regulatory regions of the gene. The gene's product—the protein—remains the same, although it may be synthesized in a new place, at a different time, or in different amounts. The changes of **heterotypy** affect the actual coding region of the gene, and thus can change the functional properties of the protein being synthesized. These changes in the protein-encoding regions of the gene are usually seen in genes that are expressed in only one or a few tissues, suggesting that pleiotropy (see below) constrains such changes in broadly expressed genes (Haygood et al. 2010; Wu et al. 2011). However, changes in the coding sequence of transcription factors can have profound consequences in animal and plant evolution (Wang et al. 2005).

HOW PREGNANCY MAY HAVE EVOLVED IN MAMMALS   One of the most amazing features of mammals is the female uterus, a structure that can hold, nourish, and protect a developing fetus within its mother's body. One of the key proteins enabling this internal gestation is prolactin. Prolactin promotes differentiation of the uterine epithelial cells, regulates trophoblast growth, allows blood vessels to spread toward the embryo, and helps downregulate the immune and inflammatory responses so the mother's body does not perceive the embryo as a "foreign body" and reject it.

At about the same time the mammalian uterus and pregnancy evolved, one of the mammalian Hox genes—*Hoxa11*—appears to have undergone intensive mutation and selection in the lineage that gave rise to placental mammals. Analysis shows that the sequence of the Hoxa11 protein changed in mammals such that it associates and interacts with another transcription factor, Foxo1a (**FIGURE 26.10**; Lynch et al. 2004, 2008). Association with Foxo1a enables Hoxa11 to upregulate prolactin expression from the enhancer used in uterine epithelial cells. Hoxa11 from non-eutherian mammals (i.e., opossum and platypus) and from chickens does not upregulate prolactin. If the *Hoxa11* mRNA in mouse uterine cells is experimentally knocked out, no prolactin is expressed.

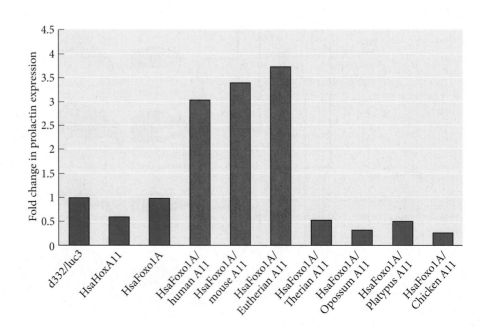

FIGURE 26.10 Ability of the mammalian Hoxa11 protein, in combination with Foxo1a, to promote expression of the uterine *Prolactin* enhancer. The activated luciferinase reporter gene (*d332/luc3*), activated human *Hoxa11* gene (HsaHoxa11), and activated human *FOXO1A* gene (HsaFoxo1a) each failed to activate the *Prolactin* gene from the enhancer transcription. Mammalian (but not opossum, platypus, or chicken) Hoxa11 increased transcription from this enhancer, but only in the presence of Foxo1a. "Eutherian A11" indicates generalized Hoxa11 from placental mammals. "Therian A11" indicates the consensus Hoxa11 sequence from all mammals. (After Lynch et al. 2008.)

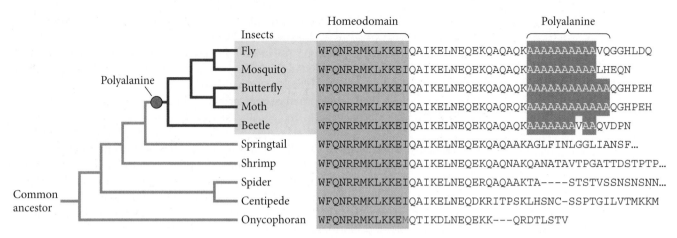

FIGURE 26.11 Changes in Ubx protein associated with the insect clade in the evolution of arthropods. Of all arthropods, only the insects have Ubx protein that is able to repress *Distal-less* gene expression and thereby inhibit abdominal legs. This ability to repress *Distal-less* is due to a mutation that is seen only in the insect *Ubx* gene. (After Galant and Carroll 2002; Ronshaugen et al. 2002.)

Therefore, it appears that one of the most important evolutionary changes in the lineage leading to mammals involved the heterotypic alteration of the Hoxa11 sequence.

WEB TOPIC 26.4 **TRANSPOSABLE ELEMENTS AND THE ORIGIN OF PREGNANCY**
The uterine decidual cell is the progesterone-responsive cell type that allows the fetus to reside within the uterus. This cell type may have emerged through the re-patterning of gene expression due to retroviruses.

WHY INSECTS HAVE SIX LEGS Insects have only six legs while most other arthropod groups (think of spiders, millipedes, centipedes, lobsters, and shrimp) have many more. How is it that the insects form legs only in their three thoracic segments and have no legs on their abdominal segments? The answer seems to be found in the relationship between Ultrabithorax protein (Ubx) and the *Distal-less* gene. In most of the arthropod groups, Ubx does not inhibit *Distal-less*. However, in the insect lineage, a mutation occurred in the *Ubx* gene wherein the original 3′ end of the protein-coding region was replaced by a group of nucleotides encoding a stretch of about 10 alanine residues at the C-terminus (**FIGURE 26.11**; Galant and Carroll 2002; Ronshaugen et al. 2002). This polyalanine region represses *Distal-less* transcription in the abdominal segments.

When a brine shrimp *Ubx* gene is experimentally modified to encode the insect polyalanine region, the shrimp embryo represses the *Distal-less* gene. The ability of Ubx to inhibit *Distal-less* thus appears to be the result of a gain-of-function mutation that characterizes the insect lineage.

WEB TOPIC 26.5 **HOW THE CHORDATES GOT A HEAD** The neural crest is responsible for forming the heads of chordates. But how did the neural crest come into existence? It is probable that ancestral deuterostomes had all the requisite genes, but only in the chordates did these genes become linked together into the network that became the neural crest cell.

Developmental Constraints on Evolution

There are only about three dozen major animal lineages, and they encompass all the different body plans seen in the animal kingdom. One can easily envision other body plans by imagining animals that do not exist; science fiction writers do it all the time. So why don't we see more body plans among the living animals? To answer this, we have to consider the constraints imposed on evolution. This notion of constraint is used differently by different groups of scientists. While many population biologists see constraints as limiting "ideal" adaptations (such as constraints on optimal foraging), developmental biologists see constraints as limiting the possibility of certain phenotypes even existing (see Amundson 1994, 2005).

The number and forms of possible phenotypes are limited by the interactions that are possible among molecules and between modules. These interactions also allow change to occur in certain directions more easily than in others. Collectively, these restraints are called **developmental constraints**, and they fall into three major categories: physical, morphogenetic, and phyletic (see Richardson and Chipman 2003).

PHYSICAL CONSTRAINTS The laws of diffusion, hydraulics, and physical support are immutable and will permit only certain physical phenotypes to arise. For example, blood cannot circulate to a rotating organ; thus a vertebrate on wheeled appendages (of the sort that Dorothy saw in Oz) cannot exist, and this entire evolutionary avenue is closed off. Similarly, structural parameters and fluid dynamics would prohibit the existence of 6-foot-tall mosquitoes or 25-foot-long leeches.

MORPHOGENETIC CONSTRAINTS Bateson (1894) and Alberch (1989) noted that when organisms depart from their normal development, they do so in only a limited number of ways. For instance, although there have been many modifications of the vertebrate limb over 300 million years, some modifications (such as a middle digit shorter than its surrounding digits, or a zeugopod more proximal than the stylopod; see Chapter 19) are never seen (Holder 1983; Wake and Larson 1987). These observations suggest a limb construction scheme that follows certain rules (Oster et al. 1988; Newman and Müller 2005).

One of the major sources of morphogenetic constraints lies in the limited ways that differentiated patterns can arise from homogeneity. Chief among these patterning mechanisms is the **reaction-diffusion mechanism**. This mechanism for developmental patterning, formulated by Alan Turing (1952), is a way of generating complex chemical patterns out of substances that are initially homogenously distributed. Turing realized that this patterning would not occur in the presence of a single morphogen, but that it could be achieved by two homogeneously distributed substances ("substance P" and "substance S") if the rates of production of each substance depended on the other. He went on to show that the dynamics of such a network could produce stable patterns that could be used to drive developmental change.

FIGURE 26.12 Reaction-diffusion (Turing) mechanism of pattern generation. Generation of periodic spatial heterogeneity can come about spontaneously when two reactants, S and P, are mixed together under the conditions that S inhibits P, P catalyzes production of both S and P, and S diffuses faster than P. (A) The conditions of the reaction-diffusion mechanism yielding a peak of P and a lower peak of S at the same place. (B) The distribution of the reactants is initially random, and their concentrations fluctuate over a given average. As P increases locally, it produces more S, which diffuses to inhibit more peaks of P from forming in the vicinity of its production. The result is a series of P peaks ("standing waves") at regular intervals.

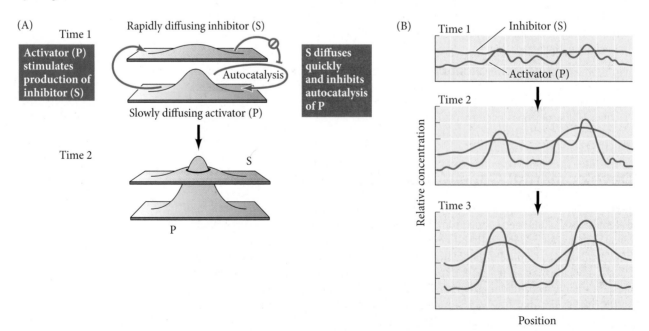

In Turing's model, substance P promotes the production of more substance P as well as substance S. Substance S, however, inhibits the production of substance P. Turing's mathematics show that if S diffuses more readily than P, sharp waves of concentration differences will be generated for substance P (**FIGURE 26.12**). The reaction-diffusion mechanism predicts alternating areas of high and low concentrations of some substance. When the concentration of the substance is above a certain threshold level, a cell (or group of cells) can be instructed to differentiate in a certain way.

An important feature of Turing's model is that particular chemical wavelengths will be amplified while all others will be suppressed. As local concentrations of P increase, the values of S form a peak centering on the P peak, but becoming broader and shallower because substance S diffuses more rapidly. These S peaks inhibit other P peaks from forming. But which of the many P peaks will survive? That depends on the size and shape of the tissues in which the oscillating reaction is occurring. This pattern is analogous to the harmonics of vibrating strings, as in a guitar: only certain resonance vibrations are permitted. The wavelength comes from the constants, particularly the ratio of the diffusion constants. The mathematics describing which particular wavelengths are selected consist of complex polynomial equations (which are now solved computationally).

Turing's model has been used to explain the formation of the limbs and digits of tetrapods (see pp. 631–633 and 642–644), the stripes of zebras and angelfish, and the formation of tooth cusps.

WEB TOPIC 26.6 **HOW DO ZEBRAS (AND ANGELFISH) GET THEIR STRIPES?** The reaction-diffusion mechanism appears to play critically important roles in generating stripes and spots on the skin of animals. "How the zebra got its stripes" may be predicated on such mechanisms, and different species of zebras might form their stripes by modifying diffusion.

WEB TOPIC 26.7 **HOW DO THE CORRECT NUMBER OF CUSPS FORM IN A TOOTH?** The vertebrate tooth has evolved according to what foods the animal can eat, and the different tooth shapes reflect different times and amounts of paracrine factor expression.

PLEIOTROPIC CONSTRAINTS As genes acquire new functions, they may become active in more than one module, making evolutionary change more difficult. **Pleiotropy**, the ability of a gene to play different roles in different cells, is the "opposite" of modularity, involving the connections between parts rather than their independence. Pleiotropies may underlie the constraints seen in mammalian development. Galis speculates that mammals have only seven cervical vertebrae (whereas birds may have dozens) because the Hox genes that specify these vertebrae have become linked to stem cell proliferation in mammals (Galis 1999; Galis and Metz 2001; Abramovich et al. 2005; Schiedlmeier et al. 2007). Thus, changes in Hox gene expression that might facilitate evolutionary changes in the skeleton might also *mis*regulate cell proliferation and lead to cancers. Galis supports this speculation with epidemiological evidence showing that changes in skeletal morphology correlate with childhood cancer. The intraembryonic selection against having more or fewer than seven cervical ribs appears to be remarkably strong. At least 78% of human embryos with an extra anterior rib (i.e., six cervical vertebrae) die before birth, and 83% die by the end of the first year. These deaths appear to be caused by multiple congenital anomalies or cancers (**FIGURE 26.13**; Galis et al. 2006).

Selectable Epigenetic Variation

Changes in development provide the raw material of variation. But we have seen earlier in the book (especially in Chapter 25) that developmental signals can come from the environment as well as from the nuclei and cytoplasm. Might this environmentally induced variation be inherited and selectable? This idea smacks of Lamarckism,

(A)

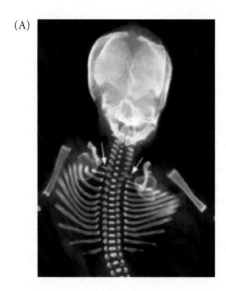

(B)

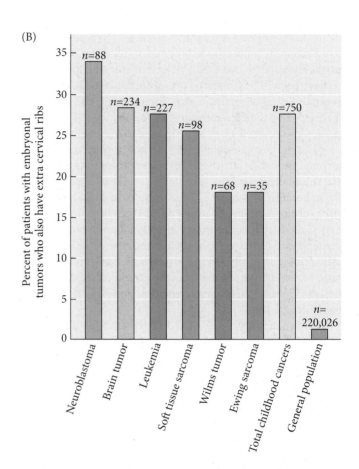

FIGURE 26.13 Extra cervical ribs are associated with childhood cancers. (A) Radiogram showing an extra cervical rib. (B) Nearly 80% of fetuses with extra cervical ribs die before birth. Those surviving often develop cancers very early in life. This indicates strong selection against changes in the number of mammalian cervical ribs. (After Galis et al. 2006, courtesy of F. Galis.)

wherein environmentally induced traits could be inherited through the germ line. We now know that Lamarck was wrong in thinking that phenotypes acquired by use or disuse could be transmitted. Children of weightlifters don't inherit their parents' physiques, and accident victims who have lost limbs can rest assured that their children will be born with normal arms and legs. If the DNA of the germ cells is not altered, environmentally induced variation will not be transmitted from one generation to the next.

But what if an environmental agent were to cause changes not only in the somatic DNA but also in the germline DNA? Then the effect might be able to be transmitted from one generation to the next. There are two known major "epigenetic inheritance systems"—epialleles and symbionts—that allow environmentally induced changes to be transmitted from generation to generation. A third process, genetic assimilation, shows that some environmentally induced traits, when continually selected, are stabilized genetically so that the trait is inherited without having to be induced in each generation.

EPIALLELES While the alleles that are the basis of the genetic inheritance system are variants of the DNA sequence, the **epialleles** of epigenetic inheritance systems are variants of chromatin structure that can be inherited between generations. In most known cases, epialleles are differences in DNA methylation patterning that are able to affect the germ line and thereby be transmitted to offspring. The asymmetrical *peloria* variant of the toadflax plant (*Linaria vulgaris*; **FIGURE 26.14**) was first described by Linnaeus in 1742 as a stably inherited form. In 1999, Coen showed that this variant was due not to a distinctive allele, but rather to a stable epiallele. Instead of carrying a mutation in the *cycloidea* gene, the *peloria* form of this gene was hypermethylated. It does not matter to the developing system whether a gene has been inactivated by a mutation or by altered chromatin configuration (Cubas and Coen 1999). The effect is the same.

(A)

FIGURE 26.14 Epigenetic forms of toadflax. (A) Typical *Linaria*, with a relatively unmethylated *cycloidea* gene. (B) The *cycloidea* gene of the *peloria* variant is relatively heavily methylated. The epialleles that create the different phenotypes of this species are stably inherited. (Courtesy of R. Grant-Downton.)

(B)

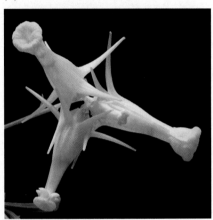

There are dozens of examples of epiallelic inheritance (Jablonka and Raz 2009; Gilbert and Epel 2015). These include:

- **Diet-induced DNA methylation**. In the viable *Agouti* phenotype in mice, methylation differences affect coat color and obesity. When a pregnant female is fed a diet high in methyl donors, the specific methylation pattern at the *Agouti* locus is transmitted not only to the progeny developing in utero, but also to the progeny of those mice and to their progeny (Jirtle and Skinner 2007). Similarly, enzymatic and metabolic phenotypes are established in utero by protein-restricted diets in rats when protein restriction during a grandmother rat's pregnancy leads to a specific methylation pattern in her pups and grandpups (Burdge et al. 2007).

- **Endocrine disruptor-induced DNA methylation**. The endocrine disruptors vinclozolin, methoxychlor, and bisphenol A have the ability to alter DNA methylation patterns in the germ line, thereby causing developmental anomalies and predispositions to diseases in the grandpups of mice exposed to these chemicals in utero (see Figure 24.16; Anway et al. 2005, 2006a,b; Newbold et al. 2006; Crews et al. 2007, 2012).

- **Behavior-induced DNA methylation**. Stress-resistant behavior of rats was shown to be due to methylation patterns, induced by maternal care, in the glucocorticoid receptor genes. Meaney (2001) found that rats that received extensive maternal care had less stress-induced anxiety and, if female, developed into mothers that gave their offspring similar levels of maternal care.

SYMBIONT VARIATION As we explored in Chapter 25, one important aspect of phenotypic plasticity involves interactions with an expected population of symbionts. When symbionts are transmitted through the germ line (as *Wolbachia* bacteria are in many insects), the symbionts provide a second system of inheritance (Gilbert and Epel 2009).

Most symbiotic relationships involve microorganisms that have fast growth rates and can thus change more rapidly under environmental stresses than multicellular organisms can. Rosenberg et al. (2007) describe four mechanisms by which microorganisms may confer greater adaptive potential to the whole organism than can the host genome alone. First, the relative abundance of microorganisms associated with the host can be changed efficiently when environmental pressures shift. Second, adaptive

(A) Without *Rickettsiella*

4 days old   8 days old   12 days old

(B) With *Rickettsiella*

4 days old   8 days old   12 days old

FIGURE 26.15 The color of adult pea aphids depends on whether or not their cells contain *Rickettsiella* bacterial symbionts. (A) Without *Rickettsiella*, red aphid newborns become red adults. (B) With *Rickettsiella*, red aphid newborns become green adults. (From Tsuchida et al. 2010, photographs courtesy of T. Tsuchida.)

Developing Questions

Recent evidence suggests that symbionts may be pivotal in evolutionary events. How might bacteria be involved with reproductive isolation and with the origins of multicellular animals?

variation can result from the introduction of a new symbiont to the community. Third, changes that occur through recombination or random mutation accumulate more rapidly in a microbial symbiont than in the host. And fourth, there is the possibility of horizontal gene transfer between members of the symbiotic community.

Symbionts can be a source of selectable variation. The pea aphid *Acrythosiphon pisum*, for example, has numerous species of symbionts living within most of it cells. One species of symbiotic bacteria, *Buchnera aphidicola*, can provide the aphid with either higher fecundity or greater heat tolerance, depending on which allele of a heat-shock protein the bacteria produces. Another symbiotic bacterium, a species of *Rickettsiella*, contains alleles that can alter the aphid's color (**FIGURE 26.15**). A third bacterial symbiont, *Hamiltonella defensa* can (if it is the appropriate strain) provide proteins that defend the host aphid against parasitoid wasps (Dunbar et al. 2007; Oliver et al. 2009; Tsuchida et al. 2010). These symbionts are usually inherited through the aphid's egg cytoplasm (see Chapter 25). Thus, selectable epigenetic variation may be acquired through the egg, but using a different set of genes.

Genetic assimilation

In the early 1900s, some evolutionary biologists speculated that the environment could select one of a variety of environmentally induced phenotypes and this phenotype would then become "fixed"—i.e., dominant for the species. In other words, the environment could both induce and select for a phenotype. But at that time scientists had no theory of development or genetics to provide mechanisms for their hypotheses. When the idea was revisited in the middle of the twentieth century, several models

(A)

(B)

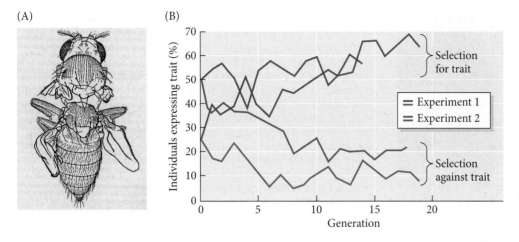

FIGURE 26.16 Phenocopy of the *bithorax* mutation. (A) A *bithorax* (four-winged) phenotype produced after treatment of the embryo with ether. The forewings have been removed to show the aberrant metathorax. This particular individual is actually from the "assimilated" stock that produced this phenotype without being exposed to ether. (B) Selection experiments for or against the *bithorax*-like response to ether treatment. Two experiments are shown (red and blue lines). In both cases, one group was selected for the trait and the other group was selected against the trait. (After Waddington 1956.)

were proposed to explain how constant selection could fix a particular environmentally induced phenotype in a population.

One of the most important hypotheses of such plasticity-driven adaptation schemes is the concept of **genetic assimilation**, defined as the process by which a phenotypic character initially produced only in response to some environmental influence becomes, through a process of selection, taken over by the genotype so that the phenotype forms even in the absence of the environmental influences that gave rise to it (King and Stanfield 1985). The idea of genetic assimilation was introduced independently by Waddington (1942, 1953, 1961) and Schmalhausen (1949) to explain the remarkable outcomes of artificial selection experiments in which an environmentally induced phenotype became expressed even in the *absence* of the external stimulus that was initially necessary to induce it.

GENETIC ASSIMILATION IN THE LABORATORY   Genetic assimilation is readily demonstrated in the laboratory. For example, Waddington showed that his laboratory strains of *Drosophila* had a particular reaction norm in their response to ether. Embryos exposed to ether at a particular stage developed a phenotype similar to the *bithorax* mutation and had four wings instead of two. The flies' halteres—balancing structures on the third thoracic segment—were transformed into wings (see Chapter 9). Generation after generation was exposed to ether, and individuals showing the four-winged state were selectively bred each time. After 20 generations, the mated *Drosophila* produced the mutant phenotype even when no ether was applied (**FIGURE 26.16**; Waddington 1953, 1956).

In 1996, Gibson and Hogness repeated Waddington's *bithorax* experiments and got similar results. Moreover, they found four distinct alleles of the *Ultrabithorax* (*Ubx*) gene existing in the population. Ubx is the homeotic gene whose loss-of-function mutations are responsible for the genetically inherited four-winged fly phenotype (see Figure 9.25). "Waddington's experiment showed some fruit flies were more sensitive to ether-induced phenocopies than others, but he had no idea why," Gibson said. "In our experiment, we show that differences in the *Ubx* gene are the cause of these morphological changes."

Genetic assimilation also has also been demonstrated in Lepidoptera (butterflies and moths). Brakefield and colleagues (1996) were able to genetically assimilate the different morphs of the adaptive polyphenism in *Bicyclus* butterflies (see Figure 25.9), and Suzuki and Nijhout (2006) have shown genetic assimilation in the larvae of the tobacco hornworm moth *Manduca sexta* (**FIGURE 26.17**). By judicious selection protocols, Suzuki and Nijhout were able to breed lines in which the environmentally induced phenotype (larval color) was selected for and was eventually produced without the environmental agent (temperature shock). The underlying genetic differences concerned the ability of heat stress to raise juvenile hormone titres in the larvae. Therefore, at least in the laboratory, genetic assimilation can be shown to work.

(A)

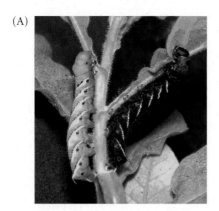

FIGURE 26.17 Effect of selection on temperature-mediated larval color change in the black mutant of the moth *Manduca sexta*. (A) The two color morphs of *Manduca sexta* larvae. (B) Changes in the coloration of heat-shocked larvae in response to selection. One group was selected for increased greenness upon heat treatment (polyphenic; green line), with the "greenest" larvae being bred for the next generation, another for decreased color change (i.e., remaining black) upon heat treatment (monophenic; red line). The remainder of the larvae was not selected (blue line). The color score (0 for completely black, 4 for completely green) indicates the relative amount of colored regions in the larvae. The monophonic line lost its plasticity after the seventh generation. (C) Reaction norm for generation-13 flies reared at constant temperatures between 20°C and 33°C, and heat shocked at 42°C. Note the steep polyphenism at about 28°C. (After Suzuki and Nijhout 2006; photograph courtesy of Fred Nijhout.)

(B)

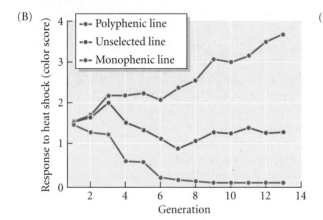

(C)

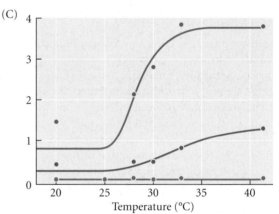

GENETIC ASSIMILATION IN NATURAL ENVIRONMENTS   We know of several instances where it appears that phenotypic variation due to developmental plasticity was later fixed by genes. The first involves pigment variations in butterflies (Hiyama et al. 2012). As early as the 1890s (Standfuss 1896; Goldschmidt 1938), scientists used heat shock to disrupt the pattern of butterfly wing pigmentation. In some instances, the color patterns that develop after temperature shock mimic the normal genetically controlled patterns of races (ecotypes) living at different temperatures. Further observations on the mourning cloak butterfly (*Nymphalis antiopa*; Shapiro 1976), the buckeye butterfly (*Precis coenia*; Nijhout 1984), and the lycaenid butterfly (*Zizeeria maha*; Otaki et al. 2010) have confirmed the view that temperature variation can induce phenotypes that mimic genetically controlled patterns of related races or species existing in colder or warmer conditions. Even "instinctive" behavioral phenotypes associated with these color changes (such as mating and flying) are phenocopied (see Burnet et al. 1973; Chow and Chan 1999). Thus, an *environmentally* induced phenotype might become the standard *genetically* induced phenotype in one part of the range of that organism.

Yet another case of genetic assimilation concerns the tiger snake (*Notechis scutatus*), which, like many fish, has a head structure that can be altered by diet. The tiger snake can develop a bigger head to ingest bigger prey. This plasticity is seen when its diet includes both large and small mice. However, on some islands the diet contains only large mice, and here the snakes are born with large heads, and there is no plasticity (**FIGURE 26.18**). Thus, Aubret and Shine (2009) claim to show "clear empirical evidence of genetic assimilation, with the elaboration of an adaptive trait shifting from phenotypically plastic expression through to canalization within a few thousand years."

Fixation of environmentally induced phenotypes

There are at least two important evolutionary advantages to the fixation of environmentally induced phenotypes (West-Eberhard 1989, 2003):

FIGURE 26.18 Genetic assimilation proposed in tiger snakes. The tiger snakes on the right-hand side of the cage are from mainland populations. They are born with small heads, and they achieve large heads through their plasticity, eating larger prey items (such as rodents and birds). The snakes on the left are from populations that have emigrated to islands where there are no small prey species. These island snakes are born with larger heads. (From Aubret and Shine 2009, courtesy of F. Aubret.)

1. *The phenotype is not random.* The environment elicited the novel phenotype, and the phenotype has already been tested by natural selection. This would eliminate a long period of testing phenotypes derived by random mutations. As Garson and colleagues (2003) note, although mutation is random, developmental parameters may account for some of the directionality in morphological evolution.

2. *The phenotype already exists in a large portion of the population.* One of the problems of explaining new phenotypes is that the bearers of such phenotypes are "monsters" compared with the wild-type. How would such mutations, perhaps present only in one individual or one family, become established and eventually take over a population? The developmental model solves this problem: this phenotype has been around for a long while, and the capacity to express it is widespread in the population; it merely needs to be genetically stabilized by modifier genes that already exist in the population.

Given these two strong advantages, the genetic assimilation of morphs originally produced through developmental plasticity may contribute significantly to the origin of new species. Ecologist Mary Jane West-Eberhard has noted that "contrary to popular belief, environmentally initiated novelties may have greater evolutionary potential than mutationally induced ones. Therefore, the genetics of speciation can profit from studies in changes of gene expression as well as changes in gene frequency and genetic isolation." Evolutionary developmental biology is a young science, and the relative importance of environmentally induced novelties is just beginning to be explored.

Coda

In the late 1800s, experimental embryology separated itself from evolutionary biology to mature on its own. However, one of those pioneers, Wilhelm Roux, promised that once it did mature, embryology would return to evolutionary biology with powerful mechanisms to help explain how evolution takes place. Evolution is a theory of change, and population genetics can identify and quantify the dynamics of such change. However, Roux realized that evolutionary biology needed a theory of body construction that would provide a means by which a specific mutation becomes manifest as a selectable phenotype.

When confronted with the question of how the arthropod body plan arose, Hughes and Kaufman (2002) begin their study by saying:

> To answer this question by invoking natural selection is correct—but insufficient. The fangs of a centipede … and the claws of a lobster accord these organisms a fitness advantage. However, the crux of the mystery is this: From what developmental genetic changes did these novelties arise in the first place?

This is exactly the question for which modern developmental biology has been able to provide some answers, and it continues to do so.

1. Developmental biology has established how the underpinnings of variation—modularity, molecular parsimony, and gene duplication—enable extensive changes in development to occur without destroying the organism.

2. Developmental biology has explained how four modes of genetic change—heterotopy, heterochrony, heterometry, and heterotypy—can act during development to produce new and large variations in morphology.

3. Finally, developmental biology has shown that epigenetic inheritance—epialleles, symbionts, and genetic assimilation—can provide selectable variations and aid their propagation through a population.

In 1922, Walter Garstang declared that ontogeny (an individual's development) does not recapitulate phylogeny (evolutionary history). Rather, ontogeny *creates* phylogeny, and evolution is generated by heritable changes in development. "The first bird," said Garstang, "was hatched from a reptile's egg." The developmental model has been formulated to account for both the homologies and the differences seen in evolution. Indeed, we are still approaching evolution in the two ways that Darwin recognized, and descent with modification remains central. We are now at the point, however, where we can answer evolutionary questions using both population genetics and developmental biology. By integrating population genetics with developmental biology, we can begin to explain the construction and evolution of biodiversity.

WEB TOPIC 26.8 **"INTELLIGENT DESIGN" AND EVOLUTIONARY DEVELOPMENTAL BIOLOGY**   Evolutionary developmental biology explains many of the "problems" (such as the evolution of the vertebrate eye and the evolution of turtle shells) that proponents of "intelligent design" and other creationists once claimed were impossible to explain by evolution.

Next Step Investigation

There are different types of novelty. One well documented type is the introduction of changes in existing cell types. Here, evolution tinkers with what is already there. The turtle alters the direction of its ribs' growth, the beetle takes the program for making exoskeleton and uses it on its wing, etc. However, another type of newness is the origin of new cell types. Research is now ongoing to look at the interactions of enhancers and transcription factors to find out if such changes are responsible for the origins of neural crest cells in vertebrates (see Web Topic 26.5) and nematocysts in cnidarians. One exceptionally interesting series of investigations concerns the origin of the enhancers that enabled the evolution of uterine decidual cells—those cells that enable the uterus to house a pregnancy. These enhancers may have been introduced into the mammalian lineage by viruses. If so, this would link evolutionary developmental biology to the acquisition of new genomes through symbiosis and viral transfection.

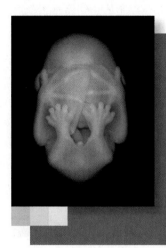

Closing Thoughts on the Opening Photo

How does something new enter into the world? Evolutionary changes in anatomy take place through changes in development. Bats provide us with excellent examples of characteristic anatomical features that can be linked to changes in the expression of developmental regulatory genes. In the evolution of a flying mammal, small changes in the expression of developmental regulatory genes have driven large changes in forelimb morphology. Developmental biologists have identified molecular changes that are critically important in retaining the bat forelimb webbing (such as the expression of BMP inhibitors and FGFs in the webbing), elongating the fingers, and reducing the ulna (see Sears 2008; Behringer et al. 2009). This embryo of a fruit bat (*Carollia perspicillata*) shows forelimb webbing and extension of its digits. (Photograph courtesy of R. R. Behringer.)

26 Snapshot Summary
Development and Evolution

1. Evolution is the result of inherited changes in development. Modifications of embryonic or larval development can create new phenotypes that can then be selected.

2. Darwin's concept of "descent with modification" explained both homologies and adaptations. The similarities of structure are due to common ancestry (homology), while the modifications are due to natural selection (adaptation to the environmental circumstances).

3. Homology means that similarity between organisms or genes can be traced to descent from a common ancestor. In some instances, certain genes specify the same traits throughout the animal phyla.

4. Evolution can occur through the "tinkering" of existing genes. The ways of effecting evolutionary change through development at the level of gene expression are: change in location (heterotopy), change in timing (heterochrony), change in amount (heterometry), and change in kind (heterotypy).

5. Changes in gene sequence can give Hox genes new properties that may have significant developmental effects. The constraint on insect anatomy of having only six legs is one example; the evolution of the uterus is another.

6. Changes in the location of gene expression during development appear to account for the evolution of the turtle shell, the loss of limbs in snakes, the emergence of feathers, and the evolution of differently shaped molars.

7. Changes in the timing of gene expression have been important in the formation of limbs throughout the animal kingdom.

8. Changes in the amount and timing of gene expression can account for the development of beak phenotypes in Darwin's finches and the size of the human brain.

9. Changes in Hox gene number may allow these genes to take on new functions. Large changes in the numbers of Hox genes correlate with major transitions in evolution.

10. The formation of new cell types may result from duplicated genes whose regulation has diverged. The Hox genes and many other gene families started as single genes that were duplicated.

11. Like structures and genes, signal transduction pathways can be homologous, with homologous proteins organized in homologous ways. These pathways can be used for different developmental processes both in different organisms and within the same organism.

12. The modularity of development allows parts of the embryo to change without affecting other parts. This modularity of development is due in large part of the modularity of enhancers.

13. Co-option (recruitment) of existing genes and pathways for new functions is a fundamental mechanism for creating new phenotypes. Such instances include the use of the limb-development signaling pathway to form eye spot pigmentation in fly wings, the formation of beetle elytra, and the production of the larval skeleton of sea urchins.

14. Developmental constraints prevent certain phenotypes from arising. Such constraints may be physical (no rotating limbs), morphogenetic (no middle digit smaller than its neighbors), or phyletic (no neural tube without a notochord).

15. New gene transcription can be caused by modifying existing DNA elements to become enhancers, mutating the DNA sequences bound by transcription factors to eliminate an enhancer, or by having a transposable element add an enhancer sequence or mutate an existing one.

16. Epigenetic inheritance systems include epialleles, wherein inherited patterns of DNA methylation can regulate gene expression. A heavily methylated gene can be as nonfunctional as a genetically mutant allele.

17. Symbiotic organisms are often needed for development to occur, and variants of these organisms may cause different modes of development.

18. Genetic assimilation, wherein a phenotypic character initially induced by the environment becomes, though a process of selection, produced by the genotype in all permissive environments, has been well documented in the laboratory.

19. Evolutionary developmental biology is able to show how small genetic or epigenetic changes can generate large phenotypic changes and enable the production of new anatomical structures. The merging of the population genetic model with the developmental genetic model of evolution is creating a new evolutionary synthesis that can account for macroevolutionary as well as microevolutionary phenomena.

Further Reading

Abzhanov, A., W. P. Kuo, C. Hartmann, P. R. Grant, R. Grant and C. Tabin. 2006. The calmodulin pathway and evolution of beak morphology in Darwin's finches. *Nature* 442: 563–567.

Amundson, R. 2005. *The Changing Role of the Embryo in Evolutionary Thought: Roots of Evo-Devo.* Cambridge University Press, New York.

Carroll, S. B. 2006. *Endless Forms Most Beautiful: The New Science of Evo-Devo.* Norton, New York.

Cohn, M. J. and C. Tickle. 1999. Developmental basis of limblessness and axial patterning in snakes. *Nature* 399: 474–479.

Davidson, E. H. and D. H. Erwin. 2009. An integrated view of Precambrian eumetazoan evolution. *Cold Spring Harb. Symp. Quant. Biol.* 74: 65–80.

Gilbert, S. F. and D. Epel. 2015. *Ecological Developmental Biology: Integrating Epigenetics, Medicine, and Evolution*, 2nd Ed. Sinauer Associates, Sunderland, MA.

Kuratani, S. 2009. Modularity, comparative embryology and evo-devo: Developmental dissection of body plans. *Dev. Biol.* 332: 61–69.

Lynch, V. J., A. Tanzer, Y. Wang, F. C. Leung, B. Gelelrsen, D. Emera and G. P. Wagner. 2008. Adaptive changes in the transcription factor HoxA-11 are essential for the evolution of pregnancy in mammals. *Proc. Natl. Acad. Sci. USA.* 105: 14928–14933.

Merino, R., J. Rodríguez-Leon, D. Macias, Y. Ganan, A. N. Economides and J. M. Hurle. 1999. The BMP antagonist Gremlin regulates outgrowth, chondrogenesis and programmed cell death in the developing limb. *Development* 126: 5515–5522.

Rockman, M. V., M. W. Hahn, N. Soranzo, D. B. Goldstein and G. A. Wray. 2003. Positive selection on a human-specific transcription factor binding site regulating IL4 expression. *Curr. Biol.* 13: 2118–2123.

Shapiro, M. D., and 7 others. 2004. Genetic and developmental basis of evolutionary pelvic reduction in three-spine sticklebacks. *Nature* 428: 717–723.

Shubin, N., C. Tabin and S. B. Carroll. 2009. Deep homology and the origins of evolutionary novelty. *Nature* 457: 818–823.

Smith, K. 2003. Time's arrow: Heterochrony and the evolution of development. *Int. J. Dev. Biol.* 47: 613–621.

GO TO WWW.DEVBIO.COM...

...for Web Topics, Scientists Speak interviews, Watch Development videos, Dev Tutorials, and complete bibliographic information for all literature cited in this chapter.

Glossary

A

Abaxial muscles Muscles derived from the lateral portions of the myotome.

Acetylation See **Histone acetylation**.

Achondroplasia Condition wherein chondrocytes stop proliferating earlier than usual, resulting in short limbs (achondroplasic dwarfism). Often caused by mutations that activate the *FgfR3* gene prematurely.

Acron Anterior region of the body of an arthropod (including insects); in front of the mouth and includes the brain.

Acrosomal process A fingerlike process in the sperm head extended by the polymerization of actin filaments during the early stages of fertilization in sea urchins and many other species. It contains surface molecules for species-specific recognition between sperm and egg.

Acrosome (acrosomal vesicle) Cap-like organelle that, together with the sperm nucleus, forms the sperm head. Contains proteolytic enzymes that can digest the extracellular coats surrounding the egg, allowing the sperm to gain access to the egg cell membrane, to which it fuses.

Acrosome reaction The Ca^{2+}-dependent fusion of the acrosome with the sperm cell membrane, resulting in exocytosis and release of proteolytic enzymes that allow the sperm to penetrate the egg extracellular matrix and fertilize the egg.

Actinomyosin contractions Contractile forces within a cell caused by myosin attaching to and moving along filamentous actin. Examples: contractions of muscle cells; apical constrictions of neural plate cells at the hinge points.

Actinopterygian fishes Ray-finned fish; includes teleosts.

Activins Members of the TGF-β superfamily of proteins; with Nodal, important in specifying the different regions of the mesoderm and for distinguishing the left and right body axes of vertebrates.

Adepithelial cells Cells that migrate into the imaginal discs early in the development of the holometabolous insect larva; these cells give rise to muscles and nerves in the pupal stage.

Adhesion Attachment between cells or between a cell and its extracellular substrate. The latter provides a surface for migrating cells to travel along.

Adult pluripotent stem cells Stem cells in an adult organism capable of regenerating all cell types of the adult. Example: neoblasts of planarian flatworms.

Adult stem cell niche Niche that houses adult stem cells and regulates stem cell self-renewal, survival, and differentiation of the progeny that leave the niche.

Adult stem cells Stem cells found in the tissues of organs after the organ has matured. Adult stem cells are usually involved in replacing and repairing tissues of that particular organ, forming a subset of cell types. Compare with **Adult pluripotent stem cells; Embryonic stem cells**.

Afferent Carrying to, as in neurons that carry information to the central nervous system (spinal cord and brain), from sensory receptor cells (e.g., sound waves from the ear, light signals from the retina, touch sensations from the skin); or vessels that carry fluid (e.g., blood) to a structure.

Aging The time-related deterioration of the physiological functions necessary for survival and reproduction.

Allantois In amniote species, extraembryonic membrane that stores urinary wastes and helps mediate gas exchange. It is derived from splanchnopleure at the caudal end of the primitive streak. In mammals, the size of the allantois depends on how well nitrogenous wastes can be removed by the chorionic placenta. In reptiles and birds, the allantois becomes a large sac, as there is no other way to keep the toxic by-products of metabolism away from the developing embryo.

Allometry Developmental changes that occur when different parts of an organism grow at different rates.

Alternative nRNA splicing A means of producing multiple different proteins from a single gene by splicing together different sets of exons to generate different types of mRNAs.

Amacrine neurons Neurons of the vertebrate neural retina that lack large axons. Most are inhibitory. See also **Neural retina**.

Ametabolous A pattern of insect development in which there is no larval stage and the insect undergoes direct development to a small adult form following a transitory pronymph stage.

Amnion "Water sac." A membrane enclosing and protecting the embryo and its surrounding amniotic fluid. Derived from the two layers of the somatopleure: ectoderm that supplies epithelial cells, and mesoderm that generates the connective tissue.

Amniote egg Egg that develops extraembryonic membranes (the amnion, chorion, allantois, and yolk sac) that provide nourishment and other environmental needs to the developing embryo. Characteristic of the amniote vertebrates: the reptiles and birds, in which the egg typically develops in a shell outside the mother's body; and the mammals, where the egg has become modified to develop inside the mother.

Amniotes The groups of vertebrates in which the embryo develops an amnion (water sac) that surrounds the body of the embryo. Includes reptiles, birds, and mammals. Compare with **Anamniote**.

Amniotic fluid A secretion that serves as a "shock absorber" for the developing embryo while preventing it from drying out.

Ampulla Latin, "flask." The segment of the mammalian oviduct, distal to the uterus and near the ovary, where fertilization takes place.

Anagen The growth phase of a hair follicle, during which the hair grows in length.

Analogous Structures and/or their respective components whose similarity arises from their performing a similar function rather than their arising from a common ancestor (e.g., the wing of a butterfly vs. the wing of a bird). Compare with **Homologous**.

Anamniotes The fish and amphibians; i.e., the vertebrate groups that do not form an amnion during embryonic development. Compare with **Amniotes**.

Anchor cell The cell connecting the overlying gonad to the vulval precursor cells in *C. elegans*. If the anchor cell is destroyed, the VPCs will not form a vulva, but instead become part of the hypodermis.

Androgen Masculinizing substance, usually a steroid hormone such as testosterone.

Androgen insensitivity syndrome Intersex condition in which an XY individual has a mutation in the gene encoding the androgen receptor protein that binds testosterone. This results in a female external phenotype, lack of a uterus and oviducts and presence of abdominal testes.

Anencephaly   A congenital defect (almost always lethal) resulting from failure to close the anterior neuropore. The forebrain remains in contact with the amniotic fluid and subsequently degenerates, so the vault of the skull fails to form.

Aneuploidy   Condition in which one or more chromosome(s) is either lacking or present in multiple copies.

Aneurogenic   Devoid of any neural innervation.

Angioblasts   From *angio*, blood vessel; and *blast*, a rapidly dividing cell (usually a stem cell). The progenitor cells of blood vessels.

Angiogenesis   Process by which the primary network of blood vessels created by vasculogenesis is remodeled and pruned into a distinct capillary bed, arteries, and veins.

Angiopoietins   Paracrine factors that mediate the interaction between endothelial cells and pericytes.

Animal cap   In amphibians, the roof of the blastocoel (in the animal hemisphere).

Animal hemisphere   The upper half of an egg containing the animal pole. In the amphibian embryo, the cells in the animal hemisphere having little yolk divide rapidly and become actively mobile ("animated").

Animal pole   The pole of the egg or embryo where the concentration of yolk is relatively low; opposite end of the egg from the vegetal pole.

Anoikis   Rapid apoptosis that occurs when epithelial cells lose their attachment to the extracellular matrix.

Anorectal junction   The meeting of endoderm and ectoderm at the anus in vertebrate embryos.

Antennapedia complex   A region of *Drosophila* chromosome 3 containing the homeotic genes *labial* (*lab*), *Antennapedia* (*Antp*), *sex combs reduced* (*scr*), *deformed* (*dfd*), and *proboscipedia* (*pb*), which specify head and thoracic segment identities.

Anterior heart field   Cells of the heart field forming the outflow tract (conus and truncus arteriosus, right ventricle).

Anterior intestinal portal (AIP)   The posterior opening of the developing foregut region of the primitive gut tube; it opens into the future midgut region which is contiguous with the yolk sac at this stage.

Anterior necrotic zone   A zone of programmed cell death on the anterior side of the developing tetrapod limb that helps shape the limb.

Anterior neuropore   See **Neuropore**.

Anterior visceral endoderm (AVE)   Mammalian equivalent to the chick hypoblast and similar to the head portion of the amphibian organizer, it creates an anterior region by secreting antagonists of Nodal.

Anterior-posterior (anteroposterior or AP) axis   The body axis defining the head versus the tail (or mouth versus anus). When referring to the limb, this refers to the thumb (anterior)-pinkie (posterior) axis.

Anti-Müllerian hormone (AMH)   TGF-β family paracrine factor secreted by the embryonic testes that induces apoptosis of the epithelium and destruction of the basal lamina of the Müllerian duct, preventing formation of the uterus and oviducts. Also known as anti-Müllerian factor, or AMF. Sometimes called Müllerian-inhibiting factor (MIF).

Anurans   Frogs and toads. Compare with **Urodeles**.

Aorta-gonad-mesonephros region (AGM)   A mesenchymal area in the lateral plate splanchnopleure near the ventral aorta that produces the hematopoietic stem cells.

Aortic arches   These begin as symmetrically arranged, paired vessels that develop within the paired pharyngeal arches and link the ascending/ventral and descending/dorsal paired aortae. Some of the aortic arches degenerate.

Apical constriction   Constriction of the apical end of a cell, caused by localized contraction of actinomyosin complexes at the apical border.

Apical ectodermal fold (AEF)   The ectoderm overlying the mesenchyme of the developing fish fin that promotes fin ray development; derived from the original apical ectodermal ridge, which becomes the AEF in ray-finned fish after the proximal patterning of the stylopod.

Apical ectodermal ridge (AER)   A ridge along the distal margin of the limb bud that will become a major signaling center for the developing limb. Its roles include (1) maintaining the mesenchyme beneath it in a plastic, proliferating state that enables the linear (proximal-distal) growth of the limb; (2) maintaining the expression of those molecules that generate the anterior-posterior axis; and (3) interacting with the proteins specifying the anterior-posterior and dorsal-ventral axes so that each cell is given instructions on how to differentiate.

Apical epidermal cap (AEC)   Forms in the wound epidermis of an amputated

salamander limb and acts similarly to the apical ectodermal ridge during normal limb development.

Apicobasal axis   Apical-to-basal axis.

Apoptosis   Programmed cell death. Apoptosis is an active process that prunes unneeded structures (e.g., frog tails, male mammary tissue), controls the number of cells in particular tissues, and sculpts complex organs (e.g., palate, retina, digits, and heart). See also **Anoikis**; **Necrotic zones**.

Aqueous humor   Nourishing fluid that bathes the lens of the vertebrate eye and supplies pressure needed to stabilize the curvature of the eye.

Archenteron   The primitive gut of an embryo. In the sea urchin, it is formed by invagination of the vegetal plate into the blastocoel.

Area opaca   The peripheral ring of avian blastoderm cells that have not shed their deep cells.

Area pellucida   A 1-cell-thick area in the center of the avian blastoderm (following shedding of most of the deep cells) that forms most of the actual embryo.

Aromatase   Enzyme that converts testosterone to estradiol (a form of estrogen). Excess aromatase in the environment is linked to herbicides and other chemicals and is believed to contribute to reproductive disorders (demasculinization and feminization, particularly in male amphibians).

Arthrotome   Mesenchymal cells in the center of the somite that contribute to the sclerotome, becoming the vertebral joints, the intervertebral discs, and those portions of the ribs closest to the vertebrae.

Astrocytes   See **Astroglial cells**.

Astroglial cells (astrocytes)   A diverse class of star (astro) shaped glial cells that carry out an array of functions, including establishing the blood-brain barrier, responding to inflammation in the CNS, and supporting synapse homeostasis and neural transmission.

Autocrine interaction   The same cells that secrete paracrine factors also respond to them.

Autonomous specification   A mode of cell commitment in which the blastomere inherits a determinant, usually a set of transcription factors from the egg cytoplasm, and these transcription factors regulate gene expression to direct the cell into a particular path of development.

Autophagy   Intracellular system that removes and replaces damaged organelles and senescent cells.

Autopod   The distal bones of a vertebrate limb: carpals and metacarpals (forelimb), tarsals and metatarsals (hindlimb), and phalanges ("digits"; the fingers and toes).

Axial protocadherin   A type of proto-cadherin expressed in the presumptive notochord cells that allow them to separate from the paraxial (somite-forming) mesoderm to form the notochord. Found in amphibian embryos. See **Protocadherin**.

Axon   Thin extension of the nerve cell body. Transmits signals (action potentials) to targets in the central and peripheral nervous systems. Axonal migration is crucial to development of the vertebrate nervous system.

Axoneme   The portion of a cilium or flagellum consisting of two central microtubules surrounded by a row of 9 doublet microtubules. The motor protein dynein attached to the doublet microtubules provides the force for ciliary and flagellar function.

B

Basal disc   The "foot" of a hydra; enables the animal to stick to rocks or the undersides of pond plants.

Basal lamina   Specialized, closely knit sheets of extracellular matrix that underlie epithelia, composed largely of laminin and type IV collagen. Epithelial cells adhere to the basal lamina in part via binding between integrins and laminin. Sometimes called the basement membrane.

Basal layer (stratum germinativum)   The inner layer of both the embryonic and adult epidermis. This layer contains epidermal stem cells attached to a basement membrane.

Basal transcription factors   Transcription factors that specifically bind to the CpG-rich sites, forming a "saddle" that can recruit RNA polymerase II and position it for transcription.

Basic fibroblast growth factor (Fgf2)   One of three growth factors required for the generation of hemangioblasts from the splanchnic mesoderm. See also **Angiopoietins**; **Vascular endothelial growth factors (VEGFs)**.

β-Catenin   A protein that can act as an anchor for cadherins or as a transcription factor (induced by the Wnt pathway). It is important in the specification of germ layers throughout the animal phyla.

Bergmann glia   Type of glial cell; extends a thin process throughout the developing neuroepithelium of the cerebellum.

bHLH proteins   The basic helix-loop-helix family of transcription factors, including such proteins as scleraxis, the MRFs (MyoD, Myf5, and myogenin), and c-Myc.

Bicoid   Anterior morphogen critical for establishing anterior-posterior polarity in the *Drosophila* embryo. Functions as a transcription factor to activate anterior-specific gap genes and as a translational repressor to suppress posterior-specific gap genes.

Bilaminar germ disc   An amniote embryo prior to gastrulation; consists of epiblast and hypoblast layers.

Bilateral holoblastic cleavage   Cleavage pattern, found primarily in tunicates, in which the first cleavage plane establishes the right-left axis of symmetry in the embryo and each successive division orients itself to this plane of symmetry. Thus the half-embryo formed on one side of the first cleavage plane is the mirror image of the other side.

Bilaterians (triploblasts)   Those animals characterized by bilaterian body symmetry and the presence of three germ layers (endoderm, ectoderm, and mesoderm). Includes all animal groups except the sponges, cnidarians, ctenophores, and placozoans.

Bindin   A 30,500-Da protein on the acrosomal process of sea urchin sperm that mediates the species-specific recognition between the sperm and the egg vitelline envelope during fertilization.

Bindin receptors   Species-specific receptors on the vitelline envelope of sea urchin eggs that bind to the bindin on the acrosomal process of sperm during fertilization.

Biofilm   Mats of microorganisms, such as bacteria, that generate an extracellular matrix. These regulate larval settlement of many marine invertebrate species.

Bipolar interneurons   Neurons in the neural retina, positioned between the photoreceptors (rods and cones) and ganglion cells for transmitting signals from the photoreceptors to the ganglion cells.

Bipotential (indifferent) gonad   Common precursor tissue derived from the genital ridge in mammals, from which the male and female gonads diverge.

Bisphenol A (BPA)   Synthetic estrogenic chemical compound used in plastics and flame retardants. BPA has been associated with meiotic defects, reproductive abnormalities, and precancerous conditions in rodents.

Bithorax complex   The region of *Drosophila* chromosome 3 containing the homeotic gene *Ultrabithorax* (*Ubx*), which is required for the identity of the third thoracic segment, and the *abdominal A* (*abdA*) and *Abdominal B* (*AbdB*) genes, which are responsible for the segmental identities of the abdominal segments.

Bivalent   The four chromatids of a homologous pair of chromosomes and

their synaptonemal complex during prophase of the first meiotic division. Also called a tetrad.

Blastocoel   A fluid-filled cavity of the blastula stage of an embryo.

Blastocyst   A mammalian blastula. The blastocoel is expanded and the inner cell mass is positioned on one side of the ring of trophoblast cells.

Blastoderm   The layer of cells formed during cleavage at the animal pole in telolecithal eggs, as in fish, reptiles, and birds. Because the yolk concentrated in the vegetal region of the egg impedes cleavage, only the small amount of yolk-free cytoplasm at the animal pole is able to divide in these eggs. During development, the blastoderm spreads around the yolk as it forms the embryo.

Blastodisc   Small region at the animal pole of the telolecithal eggs of fish, birds, and reptiles, containing the yolk-free cytoplasm where cleavage can occur and that gives rise to the embryo. Following cleavage, the blastodisc becomes the blastoderm.

Blastomere   A cleavage-stage cell resulting from mitosis.

Blastopore   The invagination point where gastrulation begins. In deuterostomes, this marks the site of the anus. In protostomes, this marks the site of the mouth.

Blastula   Early-stage embryo consisting of a sphere of cells surrounding an inner fluid-filled cavity, the blastocoel.

Blood islands   Aggregations of hemangioblasts in the splanchnic mesoderm. It is generally thought that the inner cells of these blood islands become blood progenitor cells, while the outer cells become angioblasts.

BMP family   See **Bone morphogenetic proteins**.

BMP4   A protein in the BMP family, used extensively in neural development; for example, BMP4 is produced by target organs innervated by the trigeminal nerve and causes differential growth and differentiation of the neurons. Also involved in bone differentiation. See **Bone morphogenetic proteins**.

Bone marrow-derived stem cells (BMDCs)   See **Mesenchymal stem cells**.

Bone morphogenetic proteins (BMPs)   Members of the TGF-β superfamily of proteins. Originally identified by their ability to induce bone formation, BMPs are extremely multifunctional, having been found to regulate cell division, apoptosis, cell migration, and differentiation.

Bottle cells   Invaginating cells during amphibian gastrulation, in which the main body of each cell is displaced toward the inside of the embryo while maintaining contact with the outside surface by way of a slender neck.

Brainbow   Genetic method used to trigger the expression of different combinations and amounts of different fluorescent proteins within cells, labeling them with a seeming "rainbow" of possible colors that can be used to identify each individual cell in a tissue, organ, or whole embryo.

Brain-derived neurotrophic factor (BDNF)   A paracrine factor that regulates neural activity and appears to be critical for synapse formation by inducing local translation of neural messages in the dendrites. BDNF is required for the survival of a particular subset of neurons in the striatum (a region of the brain involved in movement).

Branchial arches   See **Pharyngeal arches**.

Brown adipose cells (brown fat)   Adipose cells derived from the central dermomyotome. Brown fat cells produce heat, as opposed to white adipose cells that store lipids. Brown fat cells contain numerous mitochondria and dissipate energy as heat instead of synthesizing ATP.

Bulge   A region of the hair follicle that is a niche for adult stem cells.

C

Cadherins   **Ca**lcium-**d**ependent ad**he**sion molecules. Transmembrane proteins that interact with other cadherins on adjacent cells and are critical for establishing and maintaining intercellular connections, spatial segregation of cell types, and the organization of animal form.

Cajal-Retzius cells   Reelin-secreting cells in the neocortex just under the pial surface. Reelin directs the migration of newly born neurons toward the pial surface.

Calorie restriction   Dietary restriction as a means of extending mammalian longevity (at the expense of fertility).

Canalization   See **Robustness**.

Cancer stem cell hypothesis   The hypothesis that the malignant part of a tumor is either an adult stem cell that has escaped the control of its niche or a more differentiated cell that has regained stem cell properties.

Cap sequence   See **Transcription initiation site**.

Capacitation   The set of physiological changes by which mammalian sperm become capable of fertilizing an egg.

Cardia bifida   A condition in which two separate hearts form, resulting from manipulation of the embryo or genetic defects that prevent fusion of the two endocardial tubes.

Cardiac neural crest   Subregion of the cranial neural crest that extends from the otic (ear) placodes to the third somites. Cardiac neural crest cells develop into melanocytes, neurons, cartilage, and connective tissue. Cardiac neural crest also contributes to the muscular-connective tissue wall of the large arteries (the "outflow tracts") of the heart, as well as contributing to the septum that separates pulmonary circulation from the aorta.

Cardiogenic mesoderm   See **Heart fields**.

Cardiomyocytes   Cardiac cells derived from heart field tissue that form the muscular layers of the heart and its inflow and outflow tracts.

Catagen   Regression phase of the hair follicle regeneration cycle.

Catenins   A complex of proteins that anchor cadherins inside the cell. The cadherin-catenin complex forms the classic adherens junctions that help hold epithelial cells together and, by binding to the actin (microfilament) cytoskeleton of the cell, integrate the epithelial cells into a mechanical unit. One of them, β-catenin, can also be a transcription factor.

Caudal   Referring to the tail.

Caudal intestinal portal (CIP)   The anterior opening of the developing hindgut region of the primitive gut tube; it opens into the future midgut region, which is contiguous with the yolk sac at this stage.

Caudal progenitor zone   A region in the tailbud of vertebrate embryos that is made up of multipotent neuromesoderm progenitor cells. See also **Neuromesoderm progenitors**.

Cavitation   In mammalian embryos, a process whereby the trophoblast cells secrete fluid into the morula to create a blastocoel. The membranes of trophoblast cells pump sodium ions (Na^+) into the central cavity, drawing in water osmotically and thus creating and enlarging the blastocoel.

Cell adhesion molecules   Adhesion molecules that hold cells together. The major group of these is the cadherins. See also **Cadherins**.

Cell lineage   The series of cell types starting from an undifferentiated, pluripotent stem cell through stages of increasing differentiation to the terminally differentiated cell type.

Cellular blastoderm   Stage of *Drosophila* development in which all the cells are arranged in a single-layered jacket around the yolky core of the egg.

Central dogma   The explanation of the transfer of information coded in DNA to make proteins: DNA is transcribed into RNA, which is then translated into proteins.

Central nervous system   The brain and spinal cord of vertebrates.

Centrolecithal   Type of egg, such as those of insects, that has yolk in the center and undergoes superficial cleavage.

Centromere   A region of a chromosome where sister chromatids are attached to each other by the kinetochore.

Centrosome-attracting body (CAB)   Cellular structure that, in some invertebrate blastomeres, positions the centrosomes asymmetrically and recruits particular mRNAs so that the resulting daughter cells are different sizes and have different properties.

Cephalic   Referring to the head.

Cephalic furrow   A transverse furrow formed during gastrulation in *Drosophila* that separates the future head region (procephalon) from the germ band, which will form the thorax and abdomen.

Cephalic neural crest   See **Cranial neural crest**.

Chemoaffinity hypothesis   Hypothesis put forth by Sperry in 1965 suggesting that nerve cells in the brain acquire individual chemical tags that distinguish them from one another and that these guide the assembly and organization of the neural circuits in the brain.

Chemoattractant   A biochemical that causes cells to move toward it.

Chemotaxis   Movement of a cell down a chemical gradient, such as sperm following a chemical (chemoattractant) secreted by the egg.

Chiasmata   Points of attachment between homologous chromosomes during meiosis that are thought to represent regions where crossing-over is occurring.

Chimera   An organism consisting of a mixture of cells from two individuals.

Chimeric embryo   Embryo made from tissues of more than one genetic source.

ChIP-Seq   **Ch**romatin **i**mmuno**p**recipitation **seq**uencing. A lab protocol used to identify the precise DNA sequences bound by particular transcription factors or nucleosomes containing specific modified histones.

Chondrocyte-like osteoblasts   Cranial neural crest cells undergoing early stages of intramembranous ossification. These cells downregulate Runx2 and begin expressing the *osteopontin* gene, giving them a phenotype similar to a developing chondrocyte.

Chondrocytes   Cartilage cells.

Chondrogenesis   Formation of cartilage, in which chondrocytes differentiate from condensed mesenchyme.

Chordamesoderm Axial mesoderm in a chordate embryo that produces the notochord.

Chordate An animal that has, at some stage of its life cycle, a notochord and a dorsal nerve cord or neural tube.

Chordin A paracrine factor with organizer activity. Chordin binds directly to BMP4 and BMP2 and prevents their complexing with their receptors, thus inducing dorsal ectoderm to form neural tissue.

Chorioallantoic membrane Forms in some amniote species, such as birds, by fusion of the mesodermal layer of the allantoic membrane with the mesodermal layer of the chorion. This extremely vascular envelope is crucial for bird development and is responsible for transporting calcium from the eggshell into the embryo for bone production.

Chorion An extraembryonic membrane essential for gas exchange in amniote embryos. It is generated from the extraembryonic somatopleure. The chorion adheres to the shell in birds and reptiles, allowing the exchange of gases between the egg and the environment. It forms the embryonic/fetal portion of the placenta in mammals.

Chorionic villus sampling Taking a sample from the placenta at 8–10 weeks of gestation to grow fetal cells to be analyzed for the presence or absence of certain chromosomes, genes, or enzymes.

Chromatid Half of a mitotic prophase chromosome, which consists of duplicate "sister" chromatids that are attached to each other by the kinetochore.

Chromatin The complex of DNA and protein in which eukaryotic genes are contained.

Chromosome diminution The fragmentation of chromosomes just prior to cell division, resulting in cells in which only a portion of the original chromosome survives. Chromosome diminution occurs during cleavage in *Parascaris aequorum* in the cells that will generate the somatic cells while the future germ cells are protected from this phenomenon and maintain an intact genome.

Ciliary body A vascular structure at the junction between the neural retina and the iris that secretes the aqueous humor.

***Cis*-regulatory elements** Regulatory elements (promoters and enhancers) that reside on the same stretch of DNA as the gene they regulate.

Cleavage A series of rapid mitotic cell divisions following fertilization in many early embryos; cleavage divides the embryo without increasing its mass.

Cleavage furrow A groove formed in the cell membrane in a dividing cell due to tightening of the microfilamentous ring.

Cloaca Latin, "sewer." An endodermally lined chamber at the caudal end of the embryo that will become the receptacle for waste from the intestine and the kidneys and products from the gonads. Amphibians, reptiles, and birds retain this organ and use it to void gametes and both liquid and solid wastes. In mammals, the cloaca becomes divided by a septum into the urogenital sinus and the rectum.

Cloacal membrane At the caudal end of the hindgut formed by closely apposed endoderm and ectoderm; future site of the anus.

Cloning See **Somatic cell nuclear transfer**.

Clonogenic neoblasts (cNeoblasts) Pluripotent stem cells in flatworms that migrate to a wound site and regenerate the tissue; form the regeneration blastema of flatworms.

cNeoblasts See **Clonogenic neoblasts**.

Coelom Space between the somatic mesoderm and splanchnic mesoderm that becomes the body cavity. In mammals, the coelom becomes subdivided into the pleural, pericardial, and peritoneal cavities, enveloping the thorax, heart, and abdomen, respectively.

Coherence Scientific evidence that fits into a system of other findings and is therefore more readily accepted.

Cohesin proteins Protein rings that encircle the sister chromatids during meiosis, provide a scaffold for the assembly of the meiotic recombination complex, resist the pulling forces of the spindle microtubules, and thereby keep the sister chromatids attached during the first meiotic division and promote pairing of homologous chromosomes, allowing recombination.

Collective migration Migration of self-propelled cells exerting directionally coordinated forces upon one another, as opposed to individually migrating cells or the movement of a group of cells caused by tissue pushing due to proliferation or intercalation.

Combinatorial association In developmental genetics, the principle that enhancers contain regions of DNA that bind transcription factors, and it is this combination of transcription factors that activates the gene.

Commensalism A symbiotic relationship that is beneficial to one partner and neither beneficial nor harmful to the other partner.

Commissureless (Comm) An endosomal protein in *Drosophila*, expressed in axons prior to their crossing the midline; functions to route Robo proteins to the lysosome instead of permitting their expression in the cell membrane.

Commitment Describes a state in which a cell's developmental fate has become restricted even though it is not yet displaying overt changes in cellular biochemistry and function.

Committed stem cells Includes multipotent and unipotent stem cells that have the potential to become any of a relatively few cell types (multipotent) or only one cell type (unipotent).

Compaction A unique feature of mammalian cleavage, mediated by the cell adhesion molecule E-cadherin. The cells in the early (around eight-cell) embryo change their adhesive properties and become tightly attached to each other.

Comparative embryology Study of how anatomy changes during the development of different organisms.

Compensatory regeneration Form of regeneration in which the differentiated cells divide but maintain their differentiated functions (e.g., mammalian liver).

Competence The ability of cells or tissues to respond to a specific inductive signal.

Conditional specification The ability of cells to achieve their respective fates by interactions with other cells. What a cell becomes is in large measure specified by paracrine factors secreted by its neighbors.

Cones Color-sensitive photoreceptor cells in the neural retina. See **Neural retina**.

Congenital adrenal hyperplasia A condition causing female pseudohermaphroditism due to the presence of excess testosterone.

Congenital defect Any defect that an individual is born with. Congenital defects can be hereditary. or they can have an environmental cause (e.g., exposure to teratogenic plants, drugs, chemicals, radiation, etc.). They can also be idiopathic (i.e., cause is unknown).

Conjoined twins Monozygotic twins that share some part of their bodies; they may even share a vital organ, such as a heart or liver.

Consensus sequence When referring to an intron, these are located at the 5' and 3' ends of the introns that signal the "splice sites" of the intron.

Conserved dopamine neurotrophic factor (CDNF) A neurotrophin that enhances the survival of the midbrain dopaminergic neurons. See also **Neurotrophin**.

Contact inhibition of locomotion The mechanism whereby cells are prohibited from forming locomotory pseudodopia along contact surfaces with other cells. These interactions with the cell membranes of other cells prevent "backward" migration over other cells and result in "forward" migration of the leading edge of cells.

Context dependency　The meaning or role of an individual component of a system (such as a transcription factor) is dependent on its context. For example, in the formation of tetrapod limb joints, the same BMPs can induce either cell death or cell differentiation, depending on the stage of the responding cell.

Conus arteriosus　Cardiac outflow tract; along with the truncus arteriosus will become the base of the aorta and pulmonary arteries.

Convergent extension　A phenomenon wherein cells intercalate to narrow the tissue and at the same time move it forward. Mechanism used for elongation of the archenteron in the sea urchin embryo, notochord of the tunicate embryo, and involuting mesoderm of the amphibian. This movement is reminiscent of traffic on a highway when several lanes must merge to form a single lane.

Coordinated gene expression　The simultaneous expression of many different genes in a specific cell type. Its basis is often a single transcription factor (e.g., Pax6) that is crucial to several different enhancer sequences; the different enhancers are differentially "primed," and the binding of the same factor to all of them activates all the genes at once.

Cornified layer (stratum corneum)　The outer layer of the epidermis, consisting of keratinocytes that are now dead, flattened sacs of keratin protein with their nuclei pushed to one edge of the cell. These cells are continually shed throughout life and are replaced by new cells.

Corona radiata　The innermost layer of cumulus cells around a mammalian egg, immediately adjacent to the zona pellucida.

Corpora allata　Insect glands that secrete juvenile hormone (JH) during larval molts.

Correlative evidence　Evidence based on the association of events. The "find it" of "find it, lose it, move it." See also **Gain-of-function evidence; Loss-of-function evidence**.

Cortex　An outer structure (in contrast with medulla, an inner structure).

Cortical cytoplasm　A thin layer of gel-like cytoplasm lying immediately beneath the cell membrane of a cell. In an egg, the cortex contains high concentrations of globular actin molecules that will polymerize to form microfilaments and microvilli during fertilization.

Cortical granule reaction　The basis of the slow block to polyspermy in many animal species, including sea urchins

and most mammals. A mechanical block to polyspermy that in sea urchins becomes complete about a minute after successful sperm-egg fusion, in which enzymes from the egg's cortical granules contribute to the formation of a fertilization envelope that blocks further sperm entry.

Cortical granules　Membrane-bound, Golgi-derived structures located in the egg cortex; contain enzymes and other components. The exocytosis of these granules at fertilization is homologous to the exocytosis of the acrosome in sperm in the acrosome reaction.

Cortical plate　The layer of cells in the developing cerebrum of mammals formed by neurons in the ventricular zone migrating outward along radial glial processes to a position near the outer surface of the brain, where they will set up the six layers of the neocortex.

CpG islands　Regions of DNA rich in the CpG sequence: a cytosine and a guanosine connected by a normal phosphate bond. Promoters often contain such islands, and transcription is often initiated nearby, possibly because they bind the basal transcription factors that recruit RNA polymerase II.

Cranial (cephalic) neural crest　Neural crest cells in the future head region that migrate to produce the craniofacial mesenchyme, which differentiates into the cartilage, bone, cranial neurons, glia, and connective tissues of the face. These cells also enter the pharyngeal arches and pouches to give rise to thymic cells, the odontoblasts of the tooth primordia, and the bones of the middle ear and jaw.

Cranial sensory placodes　Ectodermal thickenings that form in the cranial region of the vertebrate embryo; includes the olfactory (nasal), otic (ear), and lens (eye) placodes, and placodes that give rise to sensory neurons of various cranial nerves. Also called cranial ectodermal placodes.

Craniorachischisis　The failure of the entire neural tube to close.

Cranium　The vertebrate skull, composed of the neurocranium (skull vault and base) and the viscerocranium (jaws and other pharyngeal arch derivatives).

Cre-lox　A site-specific recombinase technology that allows control over the spatial and temporal pattern of a gene knockout or gene misexpression.

CRISPR　Clustered Regularly Interspaced Short Palindromic Repeat. A stretch of DNA in prokaryotes that when transcribed into RNA serve as guides for recognizing segments of viral DNA. Used in association with Cas9 (CRISPR-associated enzyme 9) in a method for gene editing that is relatively fast and inexpensive.

Crossing over　The exchange of genetic material during meiosis, whereby genes from one chromatid are exchanged with homologous genes from another.

Crown cells　Cells neighboring the nodal cells, critical to setting up the left-right axis in the mammalian embryo. Crown cells each have a single immobile cilium that senses the left-to-right movement of fluids caused by the motile cilia on node cells. This sets up a cascade of events within the crown cells that serve to maintain Nodal expression on the left side, where it can activate *Pitx1* genes, which determine the left- and right-sidedness.

Crypt　A deep tubular recess or pit. Example: intestinal crypts between the intestinal villi.

Crystallins　Transparent, lens-specific proteins.

Cumulus　A layer of cells surrounding the mammalian egg, made up of ovarian follicular (granulosa) cells that nurture the egg until it is released from the ovary. The innermost layer of cumulus cells, the corona radiate, is released with the egg at ovulation.

Cutaneous appendages　Species-specific epidermal modifications that include hairs, scales, scutes, feathers, hooves, claws, and horns.

CXCR4　The receptor for stromal-derived factor 1 (SDF1). See also **Stromal-derived factor 1**.

Cyclic adenosine 3′,5′-monophosphate (cAMP)　An important component of several intracellular signaling cascades and the soluble chemotactic substance that directs the aggregation of the myxamoebae of *Dictyostelium* to form a grex.

Cyclin B　The larger subunit of mitosis-promoting factor, shows the cyclical behavior that is key to mitotic regulation, accumulating during S and being degraded after the cells have reached M. Cyclin B regulates the small subunit of MPF, the cyclin-dependent kinase.

Cyclin-dependent kinase　Small subunit of MPF, activates mitosis by phosphorylating several target proteins, including histones, the nuclear envelope lamin proteins, and the regulatory subunit of cytoplasmic myosin, resulting in chromatin condensation, nuclear envelope depolymerization, and the organization of the mitotic spindle. Requires cyclin B to function.

Cyclooxygenase-2 (COX2)　An enzyme that generates prostaglandins from the fatty acid arachidonic acid.

Cyclopia　Congenital defect characterized by a single eye, caused by mutations in genes that encode either Sonic hedgehog or the enzymes that synthesize cholesterol and can be induced by certain chemicals that interfere with the cholesterol biosynthetic enzymes.

Cystoblasts/cystocytes Derived from the asymmetric division of the germline stem cells of *Drosophila*, a cystoblast undergoes four mitotic divisions with incomplete cytokinesis to form a cluster of 16 cystocytes (one ovum and 15 nurse cells) interconnected by ring canals.

Cytokines Paracrine factors important in cell signaling and the immune response. During blood formation, they are collected and concentrated by the extracellular matrix of the stromal (mesenchymal) cells at the sites of hematopoiesis and are involved in blood cell and lymphocyte formation.

Cytokinesis The division of the cell cytoplasm into two daughter cells. The mechanical agent of cytokinesis is a contractile ring of microfilaments made of actin and the motor protein myosin. Each daughter cell receives one of the nuclei produced by nuclear division (karyokinesis).

Cytonemes Specialized filopodial projections that extend out from a cell (sometimes more than 100 μm) to make contact with another cell producing a paracrine factor. Paracrine factors may be delivered to target cells by attaching to receptors on the tips of cytonemes and traveling down the length of the cytonemes to the body of the target cells. A cytoneme can also extend from a cell producing a paracrine factor to make contact with a target cell.

Cytoplasmic bridges Continuity between adjacent cells that results from incomplete cytokinesis, e.g., during gametogenesis.

Cytoplasmic determination factors Factors found in the cytoplasm of a cell that determine cell fate. Example: gradients of different cytoplasmic determination factors that dictate cell fate along the anterior-posterior axis are found in the syncytial blastoderm of the *Drosophila* embryo.

Cytoplasmic polyadenylation-element-binding-protein (CPEB) Protein that binds to mRNA in the 3′ UTR and helps to control translation. When phosphorylated, it allows for the elongation of the polyadenine (polyA) tail on the mRNA.

Cytotrophoblast Mammalian extraembryonic epithelium composed of the original trophoblast cells, it adheres to the endometrium through adhesion molecules and, in species with invasive placentation such as the mouse and human, secretes proteolytic enzymes that enable the cytotrophoblast to enter the uterine wall and remodel the uterine blood vessels so that the maternal blood bathes fetal blood vessels.

D

Dauer larva A metabolically dormant larval stage in *C. elegans*. See also **Diapause**.

Decidua The maternal portion of the placenta, made from the endometrium of the uterus.

Deep cells A population of cells in the zebrafish blastula between the enveloping layer (EVL) and the yolk syncytial layer (YSL) that give rise to the embryo proper.

Deep homology Signal transduction pathways composed of homologous proteins arranged in a homologous manner that are used for the same function in both protostomes and deuterostomes.

Definitive endoderm The endoderm that enters the interior of the amniote embryo through the primitive streak during gastrulation, replacing the visceral endoderm, which is primarily forming the yolk sac and allantois along with the splanchnic lateral plate mesoderm.

Delamination The splitting of one cellular sheet into two more or less parallel sheets.

Delta protein Cell surface ligand for Notch; participates in juxtacrine interaction and activation of the Notch pathway.

Dendrites The fine, branching extensions (dendritic arbor) emanating from neurons; dendrites pick up electric impulses from other cells.

Dendritic arbor Extensive branching found in the dendrites of some neurons, such as Purkinje neurons.

Dental lamina A broad epidermal thickening in the jaw that later resolves into separate placodes, which together with the underlying mesenchyme form teeth.

Dermal bone Bone that forms in the dermis of the skin, such as most of the bones of the skull and face. They can be derived from head mesoderm or cranial neural crest-derived mesenchymal cells.

Dermal papilla A component of mesenchymal-epithelial induction during hair formation; a small node formed by dermal fibroblasts beneath the epidermal hair germ that stimulates proliferation of the overlying epidermal basal stem cells, which will give rise to the hair shaft.

Dermatome The central portion of the dermomyotome that produces the precursors of the dermis of the back and a population of muscle cells.

Dermomyotome Dorsolateral portion of the somite that contains skeletal muscle progenitor cells (including those that migrate into the limbs) and the cells that generate the dermis of the back.

Descent with modification Darwin's theory to explain unity of type by descent from a common ancestor and to explain adaptations to the conditions of particular environments by natural selection.

Determination The stage of commitment following specification; the determined stage, assumed irreversible, is when a cell or tissue is capable of differentiating autonomously even when placed into a non-neutral environment.

Determination front Equivalent to the "wavefront" of the "clock-wavefront" model for somite formation; where boundaries of somites form, determined by a caudal^{HIGH}-to-rostral^{LOW} gradient of FGF in the presomitic mesoderm.

Determined Committed to a given fate. A cell is determined if it maintains its developmental maturation toward its fate even when placed in a new environment. See also **Determination**.

Deuterostomes In the deuterostome animal groups (including echinoderms, tunicates, cephalochordates, and vertebrates), during embryonic development the first opening (i.e., the blastopore) becomes the anus while the second opening becomes the mouth (hence, *deutero stoma*, "mouth second"). Compare with **Protostomes**.

Development The process of progressive and continuous change that generates a complex multicellular organism from a single cell. Development occurs throughout embryogenesis, maturation to the adult form, and continues into senescence.

Developmental biology The discipline that studies embryonic and other developmental processes, such as replacement of old cells by new, regeneration, metamorphosis, aging, and development of disease states such as cancer.

Developmental constraints In evolution, the limitation of the number and forms of possible phenotypes that can be created by the interactions that are possible among molecules and between modules in the developing organism.

Developmental plasticity The ability of an embryo or larva to react to an environmental input with a change in form, state, movement, or rate of activity (i.e., phenotypic change).

Dextral coiling Right-coiling. In a snail, having its coils open to the right of its shell. See also **Sinistral coiling**.

Diacylglycerol (DAG) Second messenger generated in the IP$_3$ pathway from membrane phospholipid phosphatidylinositol 4,5-bisphosphate (PIP2), along with IP$_3$. DAG activates protein kinase C, which in turn activates a protein that exchanges sodium ions for hydrogen ions, raising the pH within the cell.

Diakinesis   Greek, "moving apart." In the first meiotic division, the stage that marks the end of prophase I, when the nuclear envelope breaks down and chromosomes migrate to the metaphase plate.

Diapause   A metabolically dormant, nonfeeding stage of an organism during which development and aging are suspended; can occur at the embryonic, larval, pupal, or adult stage.

Dickkopf   German, "thick head," "stubborn." A protein that interacts directly with the Wnt receptors, preventing Wnt signaling.

Dictyate resting stage   The prolonged diplotene stage of the first meiotic division in mammalian primary oocytes. They remain in this stage until just prior to ovulation, when they complete meiosis I and are ovulated as secondary oocytes.

Diencephalon   The caudal subdivision of the prosencephalon that will form the optic vesicles, retinas, pineal gland, and the thalamic and hypothalamic brain regions, which receive neural input from the retina.

Diethylstilbestrol (DES)   A potent environmental estrogen. DES administration to pregnant women interferes with sexual and gonadal development in their female offspring resulting in infertility, subfertility, ectopic pregnancies, adenocarcinomas, and other effects.

Differential adhesion hypothesis   A model explaining patterns of cell sorting based on thermodynamic principles. Cells interact so as to form an aggregate with the smallest interfacial free energy and therefore, the most thermodynamically stable pattern.

Differential gene expression   A basic principle of developmental genetics: In spite of the fact that all the cells of an individual body contain the same genome, the specific proteins expressed by the different cell types are widely diverse. Differential gene expression, differential nRNA processing, differential mRNA translation, and differential protein modification all work to allow the extensive differentiation of cell types.

Differential RNA processing   The splicing of mRNA precursors into messages that specify different proteins by using different combinations of potential exons.

Differentiation   The process by which an unspecialized cell becomes specialized into one of the many cell types that make up the body.

Differentiation therapy   Treatments for cancer that use transcription factors and other molecules to "normalize" cancers—that is, to cause cancerous cells

to revert to differentiation rather than continued proliferation.

Digestive tube   The primitive gut of the embryo, which extends the length of the body from the pharynx to the cloaca. Buds from the digestive tube form the thyroid, thymus, and parathyroid glands, lungs, liver, gallbladder, and pancreas.

5α-Dihydrotestosterone (DHT)   A steroid hormone derived from testosterone by the action of the enzyme 5α-ketosteroid reductase 2. DHT is required for masculinization of the male urethra, prostate, penis, and scrotum.

Diploblasts   "Two-layer" animals; they posses endoderm and ectoderm but most species lack true mesoderm. Includes the ctenophores (comb jellies) and cnidarians (jellyfish, corals, hydra, sea anemones). Compare with **Bilaterians**.

Diplotene   Greek, "double threads." In the first meiotic division, the fourth and last stage of prophase I, when the synaptonemal complex breaks down and the two homologous chromosomes start to separate but remain attached at points of chiasmata where crossing-over is occurring. Follows pachytene stage.

Direct development   Embryogenesis characterized by the lack of a larval stage, where the embryo proceeds to construct a small adult.

Discoidal cleavage   Meroblastic cleavage pattern for telolecithal eggs, in which the cell divisions occur only in the small blastodisc, as in birds, reptiles, and fish.

Disruption   Abnormality or congenital defect caused by exogenous agents (teratogens) such as plants, chemicals, viruses, radiation, or hyperthermia.

Dissociation   The ability of one module to develop differently from other modules.

Distal tip cell   A single nondividing cell located at the end of each gonad in *C. elegans* that maintains the nearest germ cells in mitosis by inhibiting their going into meiosis.

Dizygotic twins   "Two eggs." Twins that result from two separate but approximately simultaneous fertilization events. Genetically such "fraternal" (Latin *frater*, "brother") twins are full siblings. Compare with **Monozygotic twins**.

Dmrt1   Protein that in birds, frogs, and fish appears to activate *Sox9*, the central male-determining gene in vertebrates. Dmrt1 is also needed for maintaining testicular structures in mammals.

DNA methylation   A method of controlling the level of gene transcription in vertebrates by the enzymatic methylation of the promoters of inactive genes.

Certain cytosine residues that are followed by guanosine residues are methylated and the resulting methylcytosine stabilizes nucleosomes and prevents transcription factors from binding. Important in X chromosome inactivation and DNA imprinting.

DNA-binding domain   Transcription factor domain that recognizes a particular DNA sequence.

Dorsal blastopore lip   Location of the involuting marginal zone cells of amphibian gastrulation. Migrating marginal cells sequentially become the dorsal lip of the blastopore, turn inward and travel along the inner surface of the outer animal cap cells (i.e., the blastocoel roof).

Dorsal closure   A process that brings together the two sides of the epidermis of the *Drosophila* embryo at the dorsal surface.

Dorsal mesentery   A derivative of the splanchnic mesoderm, this fibrous membrane connects the endoderm to the body wall. Involved in the looping of the developing intestines.

Dorsal root ganglia (DRG)   Sensory spinal ganglia derived from the trunk neural crest that migrate along the ventral pathway and stay in the sclerotome. Sensory neurons of the DRG connect centrally with neurons in the dorsal horn of the spinal cord.

Dorsal-ventral (dorsoventral or DV) axis   The plane defining the back (dorsum) versus the belly (ventrum). When referring to the limb, this axis refers to the knuckles (dorsal) and palms (ventral).

Dorsolateral hinge points (DLHPs)   In the formation of the avian and mammalian neural tube, two hinge regions in the lateral sides of the neural plate that bend the two sides of the plate inward toward each other after the medial hinge point (MHP) has bent the plate along its midline.

Dorsolateral pathway   Pathway taken by trunk neural crest cells traveling dorsolaterally beneath the ectoderm to become melanocytes.

Dosage compensation   Equalization of expression of X chromosome-encoded gene products in male and female cells. Can be achieved by (1) doubling the transcription rate of the male X chromosomes (*Drosophila*), (2) partially repressing both X chromosomes (*C. elegans*), or (3) inactivating one X chromosome in each female cell (mammals).

Double-negative gate   A mechanism whereby a repressor locks the genes of specification, and these genes can be unlocked by the repressor of that repressor. (In other words, activation by the repression of a repressor.)

Doublesex (*Dsx*) A *Drosophila* gene active in both males and females, but whose RNA transcript is spliced in a sex-specific manner to produce sex-specific transcription factors: the female-specific transcription factor activates female-specific genes and inhibits male development; the male-specific transcription factor inhibits female traits and promotes male traits.

Down syndrome Syndrome caused by having an extra copy of chromosome 21 in humans; includes anomalies such as facial muscle changes, heart and gut abnormalities, and cognitive problems.

Ductus arteriosus A vessel that forms from left aortic arch VI, serves as a shunt between the embryonic/fetal pulmonary artery and the descending aorta in mammals. It normally closes at birth (if not, a pathological condition results called patent ductus arteriosus).

Duplication and divergence Tandem gene duplications resulting from replication errors. Once replicated, the gene copies can diverge by random mutations developing different expression patterns and new functions.

Dynein A motor protein that travels along microtubules. It is an ATPase, an enzyme that hydrolyzes ATP, converting the released chemical energy into mechanical energy. In cilia and flagella, dynein is attached to the axoneme microtubules that provides the force for propulsion by allowing the active sliding of the outer doublet microtubules, causing the flagellum or cilium to bend.

Dysgenesis Greek, "bad beginning." Refers to defective development.

E

20E See **20-Hydroxyecdysone**.

Early allocation and progenitor expansion model An alternative to the progress zone model of proximal-distal specification of the limb, wherein the cells of the entire early limb bud are already specified; subsequent cell divisions simply expand these cell populations.

E-cadherin A type of cadherin expressed in epithelial tissues as well as all early mammalian embryonic cells (the E stands for epithelial). See **Cadherins**.

Ecdysone Insect steroid hormone, secreted by the prothoracic glands, that is modified in peripheral tissues to become the active molting hormone 20-hydroxyecdysone. Crucial to insect metamorphosis.

Ecdysone receptor Nuclear protein that binds to ecdysone in insects; when bound it forms an active complex with another protein that binds to DNA, inducing transcription of ecdysone-responsive genes. Evolutionarily related to, and almost identical in structure to, thyroid hormone receptors.

Ecdysozoans One of the two major protostome groups; characterized by exoskeletons that periodically molt. The arthropods (including insects and crustaceans) and the nematodes (roundworms, including the model organism *C. elegans*) are two prominent groups. See also **Lophotrochozoans**.

Ecomorph See **Morph**.

Ectoderm Greek *ektos*, "outside." The cells that remain on either the outside (amphibian) or dorsal (avian, mammalian) surface of the embryo following gastrulation. Of the three germ layers, the ectoderm forms the nervous system from the neural tube and neural crest; also generates the epidermis covering the embryo.

Ectodermal appendage placodes Thickenings of the epidermal ectoderm involved in forming non-sensory structures such as hair, teeth, feathers, mammary and sweat glands.

Ectodermal appendages Structures that form from specific regions of epidermal ectoderm and underlying mesenchyme through a series of interactive inductions; includes hairs, scales, scutes (such as the coverings of turtle shells), teeth, sweat glands, mammary glands, and feathers.

Ectodermal placodes Thickenings of the surface ectoderm in embryos that become the rudiment of numerous organs. Includes cranial placodes, ectodermal appendage placodes.

Ectodysplasin (EDA) cascade A gene cascade specific for cutaneous appendage formation. Vertebrates with dysfunctional EDA proteins exhibit a syndrome called anhidrotic ectodermal dysplasia characterized by absent or malformed cutaneous appendages (hair, teeth, and sweat glands).

Efferent Carried away from. Often used in reference to neurons that carry information away from the central nervous system (brain and spinal cord) to be acted on by the peripheral nervous system (muscles), or a vessel that carries fluid away from a structure. Compare with **Afferent**.

Efferent ducts Ducts that link the rete testis to the Wolffian duct, formed from remodeled tubules of the mesonephric kidney.

Egg chamber An ovariole or egg tube (over a dozen per ovary) in which the *Drosophila* oocyte will develop, containing 15 interconnected nurse cells and a single oocyte.

Egg jelly A glycoprotein meshwork outside the vitelline envelope in many species, most commonly it is used to attract and/or to activate sperm.

Embryo A developing organism prior to birth or hatching. In humans, the term embryo generally refers to the early stages of development, starting with the fertilized egg until the end of organogenesis (first 8 weeks of gestation). After this, the developing human is called a fetus until its birth.

Embryogenesis The stages of development between fertilization and hatching (or birth).

Embryology The study of animal development from fertilization to hatching or birth.

Embryonic axis Any of the positional axes in an embryo; includes anterior-posterior (head-tail), dorsal-ventral (back-belly), and right-left.

Embryonic epiblast In mammals, the epiblast cells that contribute to the embryo proper that split off from the epiblast cells that line the amniotic cavity.

Embryonic germ cells (EGCs) Pluripotent embryonic cells with characteristics of the inner cell mass derived from PGCs that have been treated particular paracrine factors to maintain cell proliferation.

Embryonic period In human development, the first 8 weeks in utero prior to the fetal period; the time during which most organ systems form.

Embryonic shield A localized thickening on the future dorsal side of the fish embryo; functionally equivalent to the dorsal blastopore lip of amphibians.

Embryonic stem cells (ESCs) Pluripotent stem cells of the mammalian inner cell mass blastomeres that are capable of generating all the cell types of the body.

Emergent properties See **Level-specific properties and emergence**.

EMT See **Epithelial-to-mesenchymal transition**.

Enamel knot The signaling center for tooth development, a group of cells induced in the epithelium by the neural crest-derived mesenchyme that secretes paracrine factors that pattern the cusp of the tooth.

Endoblast See **Secondary hypoblast**.

Endocardial cushions Tissue in the developing vertebrate heart derived from the endocardium. It forms the septa that divide the atrioventricular area of the originally tubular heart into left and right atria and ventricles in amniotes; in amphibians, it separates the two atria (the ventricle remains undivided; in fish, all chambers remain undivided). The endocardial cushions also form the atrioventricular valves.

Endocardium The internal lining of the heart chambers, derived from the heart fields.

Endochondral ossification Bone formation in which mesodermal mesenchyme becomes cartilage and the cartilage is replaced by bone. It characterizes the bones of the trunk and limbs.

Endocrine disruptors Hormonally active compounds in the environment (e.g., DES; BPS; aromatase) that can have major detrimental effects on development, particularly of the gonads. Many endocrine disruptors are also obesogens (cause increased production of fat cells and fat accumulation).

Endocrine factors Hormones that travel through the blood to their target cells and tissues to exert their effects.

Endoderm Greek *endon,* "within." The innermost germ layer; forms the epithelial lining of the respiratory tract, the gastrointestinal tract, and the accessory organs (e.g., liver, pancreas) of the digestive tract. In the amphibian embryo, the yolk-containing cells of the vegetal hemisphere become endoderm. In amniote embryos, the endoderm is the most ventral of the three germ layers, and also forms the epithelium of the yolk sack and allantois.

Endometrium The epithelial lining of the uterus.

Endosome A membrane-bound vesicle that is internalized by a cell through endocytosis. Internalizing ligand-receptor complexes in endosomes is a common mechanism in paracrine signaling.

Endosteal osteoblasts Osteoblasts that line the bone marrow and are responsible for providing the niche that attracts hematopoietic stem cells (HSCs), prevents apoptosis, and keeps the HSCs in a state of plasticity.

Endosymbiosis Greek, "living within." Describes the situation in which one cell lives inside another cell or one organism lives within another.

Endothelins Small peptides secreted by blood vessels that have a role in vasoconstriction and can direct the migration of certain neural crest cells as well as the extension of certain sympathetic axons that have endothelin receptors, e.g. targeting of neurons from the superior cervical ganglia to the carotid artery.

Endothelium The single-layer sheet of epithelial cells lining of the blood vessels.

Energids In *Drosophila,* the nuclei at the periphery of the syncytial blastoderm and their associated cytoplasmic islands of cytoskeletal proteins.

Enhancer A DNA sequence that controls the efficiency and rate of transcription from a specific promoter. Enhancers bind specific transcription factors that activate the gene by (1) recruiting enzymes (such as histone acetyltransferases) that break up the nucleosomes in the area or (2) stabilizing the transcription initiation complex.

Enhancer modularity The principle that having multiple enhancers allows a protein to be expressed in several different tissues while not being expressed at all in others, according to the combination of transcription factor proteins the enhancers bind.

Enteric ganglia See **Parasympathetic ganglia**.

Enterocoely The embryonic process of forming the coelom by extending mesodermal pouches from the gut. Typical of most deuterostomes. See also **Schizocoely**.

Enveloping layer (EVL) A cell population in the zebrafish embryo at the midblastula transition made up of the most superficial cells from the blastoderm, which form an epithelial sheet a single cell layer thick. The EVL is an extraembryonic protective covering that is sloughed off during later development.

Environmental integration Describes the influence of cues from the environment surrounding the embryo, fetus, or larva on their development.

Ependyma Epithelial lining of the spinal cord canal and the ventricles of the brain.

Ependymal cells Epithelial cells that line the ventricles of the brain and canal of the spinal cord; they secrete cerebrospinal fluid.

Eph receptors Receptor for ephrin ligands.

Ephrins Juxtacrine ligands. Binding between an ephrin ligand on one cell and an Eph receptor on an adjacent cell results in signals being sent to both cells. These signals are often those of either attraction or repulsion, and ephrins are often seen directing cell migration and defining where cell boundaries are to form. As well as directing neural crest cell migration, ephrins and Eph receptors function in the formation of blood vessels, neurons, and somites.

Epialleles Variants of chromatin structure that can be inherited between generations. In most known cases, epialleles are differences in DNA methylation patterning that are able to affect the germ line and thereby be transmitted to offspring.

Epiblast The outer layer of the thickened margin of the epibolizing blastoderm in the gastrulating fish embryo or the upper layer of the bilaminar gastrulating embryonic disc in amniotes (reptiles, birds and mammals. The epiblast contains ectoderm precursors in fish and all three germ layer precursors of the embryo proper (plus the amnion) in amniotes. It also forms the avian chorion and allantois.

Epiboly The movement of epithelial sheets (usually of ectodermal cells) that spread as a unit (rather than individually) to enclose the deeper layers of the embryo. Epiboly can occur by the cells dividing, by the cells changing their shape, or by several layers of cells intercalating into fewer layers. Often, all three mechanisms are used.

Epibranchial placodes A subgroup of cranial placodes that form in the pharyngeal region of the vertebrate embryo. Gives rise to the sensory neurons of three cranial nerves: facial (VII), glossopharyngeal (IX), and vagus (X).

Epicardium The outer surface of the heart that forms the coronary blood vessels that feed the heart, derived from the heart fields.

Epidermal placodes The thickenings of epidermal ectoderm associated with ectodermal appendages. See **Ectodermal appendage placodes**.

Epidermis Outer layer of the skin, derived from ectoderm.

Epididymis Derived from the Wolffian duct, the tube adjacent to the testis that links the efferent tubules to the ductus deferens.

Epigenesis The view supported by Aristotle and William Harvey that the organs of the embryo are formed de novo ("from scratch") at each generation.

Epigenetics The study of mechanisms that act on the phenotype without changing the nucleotide sequence of the DNA. Specifically, these changes work "outside the gene" (i.e., epigenetically) by altering gene *expression* rather than by altering the gene sequence as mutation does. Epigenetic changes can sometimes be transmitted to future generations, a phenomenon referred to as **epigenetic inheritance**.

Epimorphic regeneration See **Epimorphosis**.

Epimorphin A multifunctional protein: in the membranes of mesenchymal cells, it directs epithelial morphogenesis; when expressed by sclerotome cells, it acts to attract further prechondrogenic sclerotome cells to the region of the notochord and neural tube, where the cells form vertebrae.

Epimorphosis Form of regeneration observed when adult structures undergo *de*differentiation to form a relatively undifferentiated mass of cells that then redifferentiates to form the new structure (e.g., amphibian limb regeneration).

Epiphyseal plates Cartilaginous growth zones at the proximal and distal ends of the long bones that allow continued bone growth.

Episomal vectors   Vehicles for gene delivery usually derived from viruses that do not insert themselves into host DNA.

Epithelial-mesenchymal interactions   Induction involving interactions of sheets of epithelial cells with adjacent mesenchymal cells. Properties of these interactions include regional specificity (when placed together, the same epithelium develops different structures according to the region from which the mesenchyme was taken), genetic specificity (the genome of the epithelium limits its ability to respond to signals from the mesenchyme, i.e., the response is species-specific).

Epithelial-to-mesenchymal transition (EMT)   An orderly series of events whereby epithelial cells are transformed into mesenchymal cells. In this transition, a polarized stationary epithelial cell, which normally interacts with basement membrane through its basal surface, becomes a migratory mesenchymal cell that can invade tissues and form organs in new places. See also **Mesenchymal-to-epithelial transition**.

Epithelium   Epithelial cells tightly linked together on a basement membrane to form a sheet or tube with little extracellular matrix.

Equatorial region   The junction between the inner acrosomal membrane and the sperm cell membrane in mammals. It is exposed by the acrosomal reaction and is where membrane fusion between sperm and egg begins.

Equivalence group   In the development of *C. elegans*, the group of six vulval precursor cells, each of which is competent to become induced by the anchor cell.

Erythroblast   Cell that matures from the proerythroblast and synthesizes enormous amounts of hemoglobin.

Erythrocyte   The mature red blood cell that enters the circulation where it delivers oxygen to the tissues. It is incapable of division, RNA synthesis, or protein synthesis. Amphibians, fish, and birds retain the functionless nucleus; mammals extrude it from the cell.

Erythroid progenitor cell   A committed stem cell that can form only red blood cells.

Erythropoietin   A hormone that acts on erythroid progenitor cells to produce proerythroblasts, which will generate red blood cells.

ESCs   See **Embryonic stem cells**.

Estrogen   A group of steroid hormones (including **estradiol**) needed for complete postnatal development of the Müllerian ducts (in females) and the Wolffian ducts (in males). Necessary for fertility in both sexes.

Estrus   Greek *oistros*, "frenzy." The estrogen-dominated stage of the ovarian cycle in female non-human mammals that are spontaneous or periodic ovulators, characterized by the display of behaviors consistent with receptivity to mating. Also called "heat."

Euchromatin   The comparatively open state of chromatin that contains most of the organism's genes, most of which are capable of being transcribed. Compare with **Heterochromatin**.

Eukaryotic initiation factor-4 (eIF4E)   A protein that is important for the initiation of translation. Binds to the 5' cap of mRNAs and contributes to the protein complex that mediates RNA unwinding; brings the 3' end of the message next to the 5' end, allowing the mRNA to bind to and be recognized by the ribosome. Interacts with eukaryotic initiation factor-4G (eIF4G), a scaffold protein that allows the mRNA to bind to the ribosome.

Eukaryotic organisms   Organisms whose cells contain membrane-bound organelles, including a nucleus with chromosomes that undergo mitosis. Can be single-celled or multicellular.

Evolutionary developmental biology ("evo-devo")   A model of evolution that integrates developmental genetics and population genetics to explain and the origin of biodiversity.

Exon   In a gene, the region or regions of DNA that encode the protein. Compare with **Intron**.

External granular layer   A germinal zone of cerebellar neuroblasts that migrate from the germinal neuroepithelium to the outer surface of the developing cerebellum.

External YSL (eYSL)   A region of the yolk syncytial layer (YSL) in fish embryos. The eYSL forms from yolk syncytial nuclei that move further vegetally, staying ahead of the blastoderm margin as the blastoderm expands to surround the yolk cell. See **Yolk syncytial layer**.

Extracellular matrix (ECM)   Macromolecules secreted by cells into their immediate environment, forming a region of noncellular material in the interstices between the cells. Extracellular matrices are made up of collagen, proteoglycans, and a variety of specialized glycoprotein molecules such as fibronectin and laminin.

Extraembryonic endoderm   Formed by delamination of the hypoblast cells from the avian epiblast or mammalian inner cell mass to line the yolk sac.

Extraembryonic vasculogenesis   The formation of blood islands in the yolk sac (i.e., outside the embryo).

Eye field   Region in the anterior portion of the neural tube that will develop into the neural and pigmented retinas.

F

Fasciculation   In neural development, the process of one axon adhering to and using another axon for growth.

Fast block to polyspermy   Mechanism by which additional sperm are prevented from fusing with a fertilized sea urchin egg by changing the electric potential to a more positive level. Has not been demonstrated in mammals.

Fate map   Diagrams based on having followed cell lineages from specific regions of the embryo in order to "map" larval or adult structures onto the region of the embryo from which they arose. The superimposition of a map of "what is to be" onto a structure that has yet to develop into these organs.

Female pronucleus   The haploid nucleus of the egg.

Fertilization   Fusion of male and female gametes followed by fusion of the haploid gamete nuclei to restore the full complement of chromosomes characteristic of the species and initiation in the egg cytoplasm of those reactions that permit development to proceed.

Fertilization cone   An extension from the surface of the egg where the egg and sperm have fused during fertilization. Caused by polymerization of actin, it provides a connection that widens the cytoplasmic bridge between egg and sperm, allowing the sperm nucleus and proximal centriole to enter the egg.

Fertilization envelope   Forms from the vitelline envelope of the sea urchin egg following cortical granule release. Glycosaminoglycans released by the cortical granules absorb water to expand the space between the cell membrane and fertilization envelope.

Fetal alcohol syndrome (FAS)   Condition of babies born to alcoholic mothers, characterized by small head size, specific facial features, and small brain that often shows defects in neuronal and glial migration. FAS is the most prevalent congenital mental retardation syndrome. Another term, **fetal alcohol spectrum disorder** (**FASD**) has been coined to encompass the less visible behavioral effects on children exposed prenatally to alcohol.

Fetal period   In human development, the period following the embryonic period, from the end of 8 weeks to birth; the period after organ systems have mostly formed and generally growth and modeling is occurring.

α-Fetoprotein   In mammals, a protein that binds and inactivates fetal estrogen, but not testosterone, in both male and female fetuses and shown in rodents to be critical for normal sexual differentiation of the brain.

Fetus   The stage in mammalian development between the embryonic stage and birth, characterized by growth and modeling. In humans, from the ninth week of gestation to birth.

Fibroblast growth factor 9 (Fgf9)   A growth factor involved in testes development in mammals, stimulating proliferation and differentiation in Sertoli cells and their maintenance of *Sox9* expression. Suppresses Wnt4 signaling, which would otherwise direct ovarian development. Also involved in metanephric kidney development, promoting development of a population of stem cell for nephrons. See also **Fibroblast growth factors**.

Fibroblast growth factor receptors (FGFRs)   A set of receptor tyrosine kinases that are activated by FGFs, resulting in activation of the dormant kinase and phosphorylation of certain proteins (including other FGF receptors) within the responding cell.

Fibroblast growth factors (FGFs)   A family of paracrine factors that regulate cell proliferation and differentiation.

Fibronectin   A very large (460 kDa) glycoprotein dimer synthesized by numerous cell types and secreted into the extracellular matrix. Functions as a general adhesive molecule, linking cells to one another and to other substrates such as collagen and proteoglycans, and provides a substrate for cell migration.

Filopodia   Long, thin processes containing microfilaments; cells can move by extending, attaching, and then contracting filopodia. Produced e.g. by migrating mesenchyme cells in sea urchin embryos, growth cones for nerve outgrowth, tip cells in blood vessel formation.

Fin bud   Bud of tissue in fish embryos that gives rise to a fin; homologous to the limb bud of tetrapods.

First polar body   The smaller cell produced when a primary oocyte goes through its first meiotic division, producing one large cell, the secondary oocyte, which retains most of the cytoplasm, and a tiny cell, the first polar body, which is ultimately lost. Both cells are haploid.

Flagellum   A long, motile extension of a cell containing a central axoneme of microtubules in a 9+2 arrangement (9 outer doublets and 2 central singlets). Its whipping action ("flagellation") functions for propulsion, as in the tail of a sperm.

Floor plate   Ventral region of the neural tube important in establishing dorsal-ventral polarity. Induced to form by Sonic hedgehog secreted from the adjacent notochord. It becomes a secondary signaling center that also secretes Sonic hedgehog, establishing a gradient that is highest ventrally.

Fluorescent dye   Compounds, such as fluorescein and green fluorescent protein (GFP), that emit bright light at a specific wavelength when excited with ultraviolet light.

Focal adhesions   Where the cell membrane adheres to the extracellular matrix in migrating cells, mediated by connections between actin, integrin and the extracellular matrix.

Follicle   A small group of cells around a cavity. E.g., mammalian ovarian follicle, composed of a single ovum surrounded by granulosa cells and thecal cells; hair follicle, feather follicle, where a hair or feather is produced.

Follicle-stimulating hormone (FSH)   A peptide hormone secreted by the mammalian pituitary that promotes ovarian follicle development and spermatogenesis.

Follicular stem cells   Multipotent adult stem cells that reside in the bulge niche of the hair follicle. They give rise to the hair shaft and sheath.

Follistatin   A paracrine factor with organizer activity, an inhibitor of both activin and BMPs, causes ectoderm to become neural tissue.

Foot activation gradient   A gradient, highest at the basal disc, that appears to be present in *Hydra* that permits the basal disc to form only in one place.

Foramen ovale   In the fetal mammalian heart, an opening in the septum separating the right and left atria.

Forkhead transcription factors   Transcription factors (e.g., Fox proteins, HNF4α) that are especially important in the endoderm that will form liver, where they help activate the regulatory regions surrounding liver-specific genes.

Forward genetics   Genetic technique of exposing an organism to an agent that causes random mutations and screening for particular phenotypes. Compare with **Reverse genetics**.

Frizzled   Transmembrane receptor for Wnt family of paracrine factors.

Frontonasal process   Cranial prominence formed by neural crest cells from the midbrain and rhombomeres 1 and 2 of the hindbrain that forms the forehead, the middle of the nose, and the primary palate.

G

G protein   A protein that binds GTP and is activated or inactivated by GTP modifying enzymes (such as GTPases). They play important roles in the RTK pathway and in cytoskeletal maintenance.

Gain-of-function evidence   A strong type of evidence, wherein the initiation of the first event causes the second event to happen even in instances where or when neither event usually occurs. The "move it" of "Find it, lose it, move it." See also **Correlative evidence**; **Loss-of-function evidence**.

Gamete   A specialized reproductive cell through which sexually reproducing parents pass chromosomes to their offspring; an egg or a sperm.

Gametogenesis   The production of gametes.

Ganglia   Clusters of neuronal cell bodies whose axons form a nerve. Singular, ganglion.

Gap genes   *Drosophila* zygotic genes expressed in broad (about three segments wide), partially overlapping domains. Gap mutants lacked large regions of the body (several contiguous segments).

Gastrula   A stage of the embryo following gastrulation that contains the three germ layers that will interact to generate the organs of the body.

Gastrulation   A process involving movement of the blastomeres of the embryo relative to one another resulting in the formation of the three germ layers of the embryo.

GDNF   See **Glial-derived neurotrophic factor**.

GEF   See **GTP exchange factor**; **RTK pathway**.

Gene regulatory networks (GRNs)   Patterns generated by the interactions among transcription factors and their enhancers that help define the course that development follows.

Genetic assimilation   The process by which a phenotypic character initially produced only in response to some environmental influence becomes, through a process of selection, taken over by the genotype so that it is formed even in the absence of the environmental influence that had first been necessary.

Genetic heterogeneity   The production of similar phenotypes by mutations in different genes.

Genital disc   Region of the *Drosophila* larva that will generate male or female genitalia. Male and female genitalia are derived from separate cell populations of the genital disc, as induced by paracrine factors.

Genital ridge A thickening of the splanchnic mesoderm and of the underlying intermediate mesodermal mesenchyme on the medial edge of the mesonephros; it forms the testis or ovary. Also called the germinal ridge. See also **germinal epithelium**.

Genital tubercle A structure cranial to the cloacal membrane during the indifferent stage of differentiation of the mammalian external genitalia. It will form either the clitoris in the female fetus or the penis in the male.

Genome The complete DNA sequence of an individual organism.

Genomic equivalence The theory that every cell of an organism has the same genome as every other cell.

Genomic imprinting A phenomenon in mammals whereby only the sperm-derived or only the egg-derived allele of the gene is expressed, sometimes due to inactivation of one allele by DNA methylation during spermatogenesis or oogenesis.

Germ band A collection of cells along the ventral midline of the *Drosophila* embryo that forms during gastrulation by convergence and extension that includes all the cells that will form the trunk of the embryo and the thorax and abdomen of the adult.

Germ cells A group of cells set aside for reproductive function; germ cells become the cells of the gonads (ovary and testis) that undergo meiotic cell divisions to generate the gametes. Compare with **Somatic cells**.

Germ layer One of the three layers of the embryo, ectoderm, mesoderm, and endoderm, in triploblastic organisms, or of the two layers, ectoderm and endoderm, in diploblastic organisms, generated by the process of gastrulation, that will form all of the tissues of the body except for the germ cells.

Germ line The line of cells that become germ cells, separate from the somatic cells, found in many animals, including insects, roundworms, and vertebrates. Specification of the germ line can occur autonomously from determinants found in cytoplasmic regions of the egg, or can occur later through induction by neighboring cells.

Germ plasm Cytoplasmic determinants (mRNA and proteins) in the eggs of some species, including frogs, nematodes, and flies, that autonomously specify the primordial germ cells.

Germ plasm theory The first testable model of cell specification, proposed by Weismann in 1888, in which each cell of the embryo would develop autonomously. Instead of dividing equally, the chromosomes were hypothesized to divide in such a way that different chromosomal determinants entered different cells. Only the nuclei in those cells destined to become germ cells (gametes) were postulated to contain all the different types of determinants. The nuclei of all other cells would have only a subset of the original determinants.

Germ ring A thickened ring of cells in the margin of the deep cells that appears in a fish embryo once the blastoderm has covered about half of the yolk cell. Composed of a superficial layer, the epiblast, and an inner layer, the hypoblast.

Germ stem cell (GSC) In female *Drosophila*, the stem cell that gives rise to the oocyte.

Germarium In female *Drosophila*, niche in the anterior region of an ovariole containing germ stem cells and several somatic cell types.

Germinal crescent A region in the anterior portion of the avian and reptilian blastoderm area pellucida containing the hypoblasts displaced by migrating endodermal cells. It contains the primordial germ cells (precursors of the germ cells), which later migrate through the blood vessels to the gonads.

Germinal epithelium Epithelium of the bipotential gonad, derived from splanchnic mesoderm, that will form the somatic (i.e., non-germ cell) component of the gonads.

Germinal neuroepithelium A layer of rapidly dividing neural stem cells one cell layer thick that constitute the original neural tube.

Germinal ridge See **Genital ridge**.

Germinal vesicle breakdown (GVBD) Disintegration of the primary oocyte nuclear membrane (germinal vesicle) upon resumption of meiosis during oogenesis.

Germline stem cells In *Drosophila*, pole cell (primordial germ cells) derivatives that divide asymmetrically to produce another stem cell and a differentiated daughter cell called a cystoblast, which in turn produces a single ovum and 15 nurse cells.

GFP See **Green fluorescent protein**.

Glia Supportive cells of the central nervous system, derived from the neural tube, and of the peripheral nervous system, derived from neural crest.

Glial guidance A mechanism important for positioning young neurons in the developing mammalian brain (e.g., the granule neuron precursors travel on the long processes of the Bergmann glia in the cerebellum).

Glial-derived neurotrophic factor (GDNF) A paracrine factor that binds to the Ret receptor tyrosine kinase. It is produced by the gut mesenchyme that attracts vagal and sacral neural crest cells, and it is produced by the metanephrogenic mesenchyme to induce the formation and branching of the ureteric buds.

Glochidium The larva of some freshwater bivalve molluscs, such as uniod clams; has a shell resembling a tiny bear trap, used to attach to the gills or fins of fish. It feeds on the fish's body fluids until it drops off to metamorphose into an adult clam.

Glycogen synthase kinase 3 (GSK3) Targets β-catenin for destruction.

Glycosaminoglycans (GAGs) Complex acidic polysaccharides consisting of unbranched chains assembled from many repeats of a two-sugar unit. The carbohydrate component of proteoglycans.

Gonadotropin-releasing hormone (GRH; GnRH) Peptide hormone released from the hypothalamus that stimulates the pituitary to release the gonadotropins follicle-stimulating hormone and luteinizing hormone, which are required for mammalian gametogenesis and steroidogenesis.

Gonialblast In male *Drosophila*, a committed progenitor cell that divides to become the precursors of the sperm.

Gonocytes Mammalian primordial germ cells (PGCs) that have arrived at the genital ridge of a male embryo and have become incorporated into the sex cords.

Granule cells Derived from neuroblasts of the external granule layer of the developing cerebellum. Granule neurons migrate back toward the ventricular (ependymal) zone, where they produce a region called the internal granule layer.

Granulosa cells Cortical epithelial cells of the fetal ovary, granulosa cells surround individual germ cells that will become the ova and will form, with thecal cells, the follicles that envelop the germ cells and secrete steroid hormones. The number of granulosa cells increase and form concentric layers around the oocyte as the oocyte matures prior to ovulation.

Gray crescent A band of inner gray cytoplasm that appears following a rotation of the cortical cytoplasm with respect to the internal cytoplasm in the marginal region of the 1-cell amphibian embryo. Gastrulation starts in this location.

Gray matter Regions of the brain and spinal cord rich in neuronal cell bodies. Compare with **White matter**.

Green fluorescent protein (GFP) A protein that occurs naturally in certain jellyfish. It emits bright green fluorescence when exposed to ultraviolet light. The *GFP* gene is widely used as a transgenic label for cells in developmental and other research, since cells that express GFP are easily identified by a bright green glow.

GRNs   See **Gene regulatory networks**

Growth and differentiation factors (GDFs)   See **Paracrine factors**.

Growth cone   The motile tip of a neuronal axon; leads nerve outgrowth.

Growth factor   A secreted protein that binds to a receptor and initiates signals to promote cell division and growth.

Growth plate closure   Causes the cessation of bone growth at the end of puberty. High levels of estrogen induce apoptosis in the hypertrophic chondrocytes and stimulate the invasion of bone-forming osteoblasts into the growth plate.

GTP exchange factor (GEF)   In the RTK (receptor tyrosine kinase) pathway, this factor exchanges a phosphate that transforms a bound GDP on a G protein into a bound GTP, activating the G protein. See also **RTK pathway**.

GTPase-activating protein (GAP)   The protein that enables the G protein Ras to quickly return to its inactive state. This is done by hydrolyzing Ras's bound GTP back to GDP. Ras becomes activated through the RTK pathway.

Gurken   A protein coded for by the *gurken* gene. The *gurken* message is synthesized in the nurse cells of the *Drosophila* ovary and transported into the oocyte, where it is translated into the protein along an anterior-posterior gradient. It signals follicle cells closest to the oocyte nucleus to posteriorize; part of the process that will set up the anterior-posterior axis of the egg and future embryo.

Gynandromorph   Greek *gynos*, "female"; *andros*, "male." An animal in which some body parts are male and others are female. Compare with **Hermaphrodite**.

Gyrencephalic   Having numerous folds in the cerebral cortex, as in humans and cetaceans. Compare with **Lissencephalic**.

H

Hair follicle stem cells   See **Follicular stem cells**.

Halteres   A pair of balancers on the third thoracic segment of two-winged flies, such as *Drosophila*.

Haptotaxis   Directional migration of cells on a substrate, up a gradient of adhesiveness.

Hatched blastula   Free-swimming sea urchin embryo, after the cells of the animal hemisphere synthesize and secrete a hatching enzyme that digests the fertilization envelope.

Head activation gradient   A morphogenetic gradient in *Hydra* that is highest at the hypostome and permits the head to form.

Head mesoderm   Mesoderm located anterior to the trunk mesoderm, consisting of the unsegmented paraxial mesoderm and prechordal mesoderm. This region provides the head mesenchyme that forms much of the connective tissues and musculature of the head.

Head process   In avian embryos, the anterior portion of the chordamesoderm that passes through Hensen's node and migrates anteriorly, ahead of the notochordal mesoderm, to come to lie underneath cells that will form forebrain and midbrain.

Heart fields (cardiogenic mesoderm)   In vertebrates, two regions of splanchnic mesoderm, one on each side of the body, that become specified for heart development. In amniotes, the cardiac cells of the heart field migrate through the primitive streak during gastrulation such that the medial-lateral arrangement of these early cells will become the anterior-posterior (rostral-caudal) axis of the developing heart tube.

Heart tube   Linear (anterior-to-posterior) structure formed at the midline of the heart fields; will become the atria, ventricles, and the base of the aorta and pulmonary arteries.

Hedgehog   A family of paracrine factors used by the embryo to induce particular cell types and to create boundaries between tissues. Hedgehog proteins must become complexed with a molecule of cholesterol in order to function. Vertebrates have at least three homologues of the *Drosophila hedgehog* gene: *sonic hedgehog* (*shh*), *desert hedgehog* (*dhh*), and *indian hedgehog* (*ihh*).

Hedgehog pathway   Proteins activated by the binding of a Hedgehog protein to the Patched receptor. When Hedgehog binds to Patched, the Patched protein's shape is altered such that it no longer inhibits Smoothened. Smoothened acts to release the Ci protein from the microtubules and to prevent its being cleaved. The intact Ci protein can now enter the nucleus, where it acts as a transcriptional *activator* of the same genes it once repressed.

Hemangioblasts   Rapidly dividing cells, usually stem cells, that form blood vessels and blood cells.

Hematopoiesis   The generation of blood cells.

Hematopoietic inductive microenvironments (HIMs)   Cell regions that induce different sets of transcription factors in multipotent hematopoietic stem cells; these transcription factors specify the developmental path taken by descendants of those cells.

Hematopoietic stem cell (HSC)   A pluripotent stem cell type that generates a series of intermediate progenitor cells whose potency is restricted to certain blood cell lineages. These lineages are then capable of producing all the blood cells and lymphocytes of the body.

Hemimetabolous   A form of insect metamorphosis that includes pronymph, nymph, and imago (adult) stages.

Hemogenic endothelial cell   Primary endothelial cells of the dorsal aorta, in the ventral area, derived from the lateral plate. They give rise to the hematopoietic stem cells (HSCs) that migrate to the liver and bone marrow and become the adult hematopoietic stem cells.

Hensen's node   In avian embryos, a regional thickening of cells at the anterior end of the primitive streak. The center of Hensen's node contains a funnel-shaped depression (sometimes called the primitive pit) through which cells can enter the embryo to form the notochord and prechordal plate. Hensen's node is the functional equivalent of the dorsal lip of the amphibian blastopore (i.e., the organizer) and the fish embryonic shield. Also known as the **primitive knot**.

Heparan sulfate proteoglycans (HSPGs)   One of the most widespread proteoglycans in the extracellular matrix, it can bind many members of different paracrine families and present them in high concentrations to their receptors.

Hepatectomy   Surgical removal of part of the liver.

Hepatic diverticulum   The liver precursor, a bud of endoderm that extends out from the foregut into the surrounding mesenchyme.

Hepatocyte growth factor (HGF)   Paracrine factor secreted by the stellate cells of the liver that allows the hepatocytes to re-enter the cell cycle during compensatory regeneration. Also called scatter factor.

Hermaphrodite   An individual in which both ovarian and testicular tissues exists, having either ovotestes (gonads containing both ovarian and testicular tissue) or an ovary on one side and a testis on the other. Compare with **Gynandromorph**.

Heterochromatin   Chromatin that remains condensed throughout most of the cell cycle and replicates later than most of the other chromatin. Usually transcriptionally inactive. Compare with **Euchromatin**.

Heterochronic parabiosis Surgically joining the circulatory systems of two animals of different ages; has been used to study the effects of aging on stem cells in mice.

Heterochrony Greek, "different time." A shift in the relative timing of two developmental processes as a mechanism to generate phenotypic variation available for natural selection. One module changes its time of expression or growth rate relative to the other modules of the embryo.

Heterometry Greek, "different measure." A change in the amount of a gene product as a mechanism to generate phenotypic variation available for natural selection.

Heterophilic binding Binding between different molecules, as when a receptor in the membrane of one cell binds to a different type of receptor in the membrane of another cell.

Heterotopy Greek, "different place." The spatial alteration of gene expression (e.g., transcription factors or paracrine factors) as a mechanism to generate phenotypic variation available for natural selection.

Heterotypy Greek, "different kind." The alteration of the actual coding region of the gene, changing the functional properties of the protein being synthesized, as a mechanism to generate phenotypic variation available for natural selection.

High CpG-content promoters (HCPs) Promoters with many CpG islands; these promoters often regulate developmental genes required for the construction of the organism; their default state is "on." See also **CpG islands**.

Histoblast nests Clusters of imaginal cells that will form the adult abdomen in holometabolous insects.

Histone Positively charged proteins that are the major protein component of chromatin. See also **Nucleosome**.

Histone acetylation The addition of negatively charged acetyl groups to histones, which neutralizes the basic charge of lysine and loosens the histones, and thus activates transcription.

Histone acetyltransferases Enzymes that place acetyl groups on histones (especially on lysines in histones H3 and H4). Acetyltransferases destabilize the nucleosomes so that they come apart easily, thus facilitating transcription.

Histone deacetylases Enzymes that remove acetyl groups from histones, stabilizing the nucleosomes and preventing transcription.

Histone methylation The addition of methyl groups to histones. Can either activate or further repress transcription, depending on the amino acid that is methylated and the presence of other methyl or acetyl groups in the vicinity.

Histone methyltransferases Enzymes that add methyl groups to histones and either activate or repress transcription.

Holobiont Term for the composite organism of a host and its persistent symbionts.

Holoblastic cleavage Greek *holos*, "complete." Refers to a cell division (cleavage) pattern in the embryo in which the entire egg is divided into smaller cells, as it is in echinoderms, amphibians and mammals.

Holometabolous The type of insect metamorphosis found in flies, beetles, moths, and butterflies. There is no pronymph stage. The insect hatches as a larva (a caterpillar, grub, or maggot) and progresses through instar stages as it gets bigger between larval molts, a metamorphic molt to become a pupa, an imaginal molt and finally the emergence (eclose) of the adult (imago).

Homeobox A 180-base pair DNA sequence that characterizes genes that code for homeodomain proteins, including Hox genes.

Homeorhesis How the organism stabilizes its different cell lineages while it is still constructing itself.

Homeostasis Maintenance of a stable physiological state by means of feedback responses.

Homeotic complex (Hom-C) The region of *Drosophila* chromosome 3 containing both the Antennapedia complex and the bithorax complex.

Homeotic mutants Result from mutations of homeotic selector genes, in which one structure is replaced by another (as where an antenna is replaced by a leg).

Homeotic selector genes A class of *Drosophila* genes regulated by the protein products of the gap, pair-rule, and segment polarity genes whose transcription determines the developmental fate of each segment.

Homing The ability of a cell to migrate and find its tissue specific destination.

Homodimer Two identical protein molecules bound together.

Homologous Structures and/or their respective components whose similarity arises from their being derived from a common ancestral structure. For example, the wing of a bird and the forelimb of a human. Compare with **Analogous**.

Homologue (1) One of a pair (or larger set) of chromosomes with the same overall genetic composition. For example, diploid organisms have two copies (homologues) of each chromosome, one inherited from each parent. (2) Evolutionary features in different species that are similar by reason of descent from a common ancestor.

Homophylic binding Binding between like molecules, as when a receptor in the membrane of one cell binds to the same type of receptor in the cell membrane of another cell.

Horizontal neurons Neurons in the neural retina that transmit electrical impulses in the plane of the retina; help to integrate sensory signals coming from many photoreceptor cells.

Horizontal transmission When a host that is born free of symbionts but subsequently becomes infected, either by its environment or by other members of the species. Can also refer to the transfer of genes from one organism to another without involving reproduction, as can occur in bacteria. Compare with **Vertical transmission**.

Host The larger organism in a symbiotic relationship in which one of the organisms involved is much larger than the other, and the smaller organism may live on the surface or inside the body of the larger. Also refers to the organism receiving a graft from a donor in a tissue transplant.

Hox genes Large family of related genes that dictate (at least in part) regional identity in the embryo, particularly along the anterior-posterior axis. Hox genes encode transcription factors that regulate the expression of other genes. All known mammalian genomes contain four copies of the Hox complex per haploid set, located on four different chromosomes (*Hoxa* through *Hoxd* in the mouse, *HOXA* through *HOXD* in humans). The mammalian Hox/HOX genes are numbered from 1 to 13, starting from that end of each complex that is expressed most anteriorly.

Hu proteins RNA binding proteins that stabilize mRNAs involved in neuronal development, preventing them from being quickly degraded. Examples: HuA, HuB, HuC, and HuD.

Hub A regulatory microenvironment in *Drosophila* testes where the stem cells for sperm reside.

Hyaline layer A coating around the sea urchin egg formed by the cortical granule protein hyalin. The hyaline layer provides support for the blastomeres during cleavage.

Hydatidiform mole A human tumor which resembles placental tissue, arise when a haploid sperm fertilizes an egg in which the female pronucleus is absent and the entire genome is derived from the sperm, which precludes normal development and is cited as evidence for genomic imprinting.

20-Hydroxyecdysone (20E)   An insect hormone, the active form of ecdysone, that initiates and coordinates each molt, regulates the changes in gene expression that occur during metamorphosis, and signals imaginal disc differentiation.

Hyperactivation   The increased and more forceful motility displayed by capacitated sperm of some mammalian species. Hyperactivation has been proposed to help detach capacitated sperm from the oviductal epithelium, allow sperm to travel more effectively through viscous oviductal fluids, and facilitate penetration of the extracellular matrix of the cumulus cells.

Hypertrophic chondrocytes   Formed during the fourth phase of endochondral ossification, when the chondrocytes, under the influence of the transcription factor Runx2, stop dividing and increase their volume dramatically.

Hypoblast   The inner layer of the thickened margin of the epibolizing blastoderm in the gastrulating fish embryo or the lower layer of the bilaminar embryonic blastoderm in birds and mammals. The hypoblast in fish (but not in birds and mammals) contains the precursors of both the endoderm and mesoderm. In birds and mammals, it contains precursors to the extraembryonic endoderm of the yolk sac.

Hypoblast islands (primary hypoblast)   Derived from area pellucida cells of the avian blastoderm that migrate individually into the subgerminal cavity to form individual disconnected clusters containing 5–20 cells each. Does not contribute to the embryo proper.

Hypomorphic mutations   Mutations that reduce gene function, as opposed to a "null" mutation that results in the loss of a protein's function.

Hypostome   A conical region of the "head" of a hydra that contains the mouth.

I

Imaginal cells   Cells carried around in the larva of the holometabolous insect that will form the structures of the adult. During the larval stages, these cells increase in number, but do not differentiate until the pupal stage; include imaginal discs, histoblasts, and clusters of imaginal cells within each larval organ.

Imaginal discs   Clusters of relatively undifferentiated cells set aside to produce adult structures. Imaginal discs will form the cuticular structures of the adult, including the wings, legs,

antennae, halters, eyes, head, thorax, and genitalia in holometabolous insects.

Imaginal molt   Final molt in a holometabolous insect when the adult (imago) cuticle forms beneath the pupal cuticle, and the adult later emerges from the pupal case at adult eclosion.

Imaginal rudiment   Develops from the left coelomic sac of the pluteus larva and will form many of the structures of the adult sea urchin.

Imago   A winged and sexually mature adult insect.

In situ   Latin, "on site." In its natural position or environment.

In situ probe   Complementary DNA or RNA used to localize a specific DNA or RNA sequence in a tissue.

Indel   An insertion or deletion of DNA bases.

Indifferent gonad   See **Bipotential gonad**.

Indirect developers   Animals for which embryonic development includes a larval stage with characteristics very different from those of the adult organism, which emerges only after a period of metamorphosis.

Induced pluripotent stem (iPS) cells   Adult cells that have been converted to cells with the pluripotency of embryonic stem cells. Usually accomplished by the activation of certain transcription factors.

Inducer   Tissue that produces a signal (or signals) that induces a cellular behavior in some other tissue.

Induction   The process by which one cell population influences the development of neighboring cells via interactions at close range.

Ingression   Migration of individual cells from the surface layer into the interior of the embryo. The cells become mesenchymal (i.e., they separate from one another) and migrate independently.

Inner cell mass (ICM)   A small group of internal cells within a mammalian blastocyst that will eventually develop into the embryo proper and its associated yolk sac, allantois, and amnion.

Inositol 1,4,5-trisphosphate (IP$_3$)   A second messenger generated by the phospholipase C enzyme that releases intracellular Ca^{2+} stores. Important in the initiation of both cortical granule release and sea urchin development.

Inside-out gradient of development   The developmental process in the neocortex and the rest of the brain in which the neurons with the earliest birthdays form the layer closest to the ventricle and subsequent neurons travel greater distances to form the more superficial layers.

Instar   The stages between larval molts in holometabolous insects. During these stages, the larva (caterpillar, grub, or maggot) feeds and grows larger between each molt, until the end of the final instar stage, when the larva is transformed into a pupa.

Instructive interaction   A mode of inductive interaction in which a signal from the inducing cell is necessary for initiating new gene expression in the responding cell.

Insulator   DNA sequence that limits the range within which an enhancer can activate a given gene's expression (thereby "insulating" a promoter from being activated by another gene's enhancers).

Insulin signaling pathway   Pathway involving a receptor for insulin and insulin-like proteins; may be an important component of genetically limited life spans, wherein downregulation of the pathway can correspond to an increased lifespan.

Insulin-like growth factors (IGFs)   Growth factors that initiate an FGF-like signal transduction cascade that interferes with the signal transduction pathways of both BMPs and Wnts. IGFs are required for the formation of the anterior neural tube, including the brain and sensory placodes of amphibians.

Integration   A principle of the theoretical systems approach: How the parts are put together, and how they interact to form the whole.

Integrins   A family of receptor proteins, named for the fact that they *integrate* extracellular and intracellular scaffolds, allowing them to work together. On the extracellular side, integrins bind to sequences found in several adhesive proteins in extracellular matrix, including fibronectin, vitronectin (in the basal lamina of the eye), and laminin. On the cytoplasmic side, integrins bind to talin and α-actinin, two proteins that connect to actin microfilaments. This dual binding enables the cell to move by using myosin to contract the actin microfilaments against the fixed extracellular matrix.

Interdigital necrotic zone   A zone of programmed cell death in the developing tetrapod limb that separates the digits from one another; when the cells in this zone do not die, webbing between the digits remains, as in the duck foot.

Interior necrotic zone   A zone of programmed cell death in the developing tetrapod limb that separates the radius from the ulna.

Interkinesis   The brief period between the end of meiosis I and the beginning of meiosis II.

Interkinetic nuclear migration The movement of nuclei within certain cells as they go through the cell cycle; seen in the germinal neuroepithelium in which nuclei translocate from the basal end to the apical end of the cells near the ventricular surface, where they undergo mitosis, after which they slowly migrate basally again.

Intermediate mesoderm Mesoderm immediately lateral to the paraxial mesoderm. It forms the outer (cortical) portion of the adrenal gland and the urogenital system, consisting of the kidneys, gonads, and their associated ducts.

Intermediate progenitor cells (IP cells) Neuron precursor cells of the subventricular zone; derived from radial glial cells.

Intermediate spermatogonia The first committed stem cell type of the mammalian testis, they are committed to becoming spermatozoa.

Intermediate zone See **Mantle zone**.

Internal granular layer A layer in the cerebellum that is formed by the migration of granule cells from the external granular layer back toward the ventricular zone.

Internal YSL (iYSL) A region of the yolk syncytial layer (YSL) in fish embryos. The iYSL forms from yolk syncytial nuclei that move under the blastoderm as it expands to surround the yolk cell. See **Yolk syncytial layer**.

Intersex A condition in which male and female traits are observed in the same individual.

Interstitial stem cell A type of stem cell found within the ectodermal layer of *Hydra* that generates neurons, secretory cells, nematocytes, and gametes.

Intraembryonic vasculogenesis The formation of blood vessels during embryonic organogenesis. Compare with **Extraembryonic vasculogenesis**.

Intramembranous bone Bone formed by intramembranous ossification

Intramembranous ossification Formation of bone directly in mesenchyme without a cartilaginous precursor. There are three main types of intramembranous bone: sesamoid bone and periosteal bone, which come from mesoderm, and dermal bone which originate from cranial neural crest-derived mesenchymal cells.

Introns Non-protein-coding regions of DNA within a gene. Compare with **Exon**.

Invagination The infolding of a region of cells, much like the indenting of a soft rubber ball when it is poked.

Involuting marginal zone (IMZ) Cells that involute during *Xenopus* gastrulation, includes precursors of the pharyngeal endoderm, head mesoderm, notochord, somites, and heart, kidney, and ventral mesoderm.

Involution Inturning or inward movement of an expanding outer layer so that it spreads over the internal surface of the remaining external cells.

Ionophore A compound that allows the diffusion of ions such as Ca^{2+} across lipid membranes, permitting them to traverse otherwise impermeable barriers.

Iris A pigmented ring of muscular tissue in the eye that controls the size of the pupil and determines eye color.

Isolecithal Greek, "equal yolk." Describes eggs with sparse, equally distributed yolk particles, as in sea urchins, mammals, and snails.

Isthmus The narrow segment of the mammalian oviduct adjacent to the uterus.

J

Jagged protein Ligand for Notch, participates in juxtacrine interaction and activation of the Notch pathway.

JAK **Ja**nus **k**inase proteins. Linked to FGF receptors in the JAK-STAT cascade.

JAK-STAT cascade A pathway activated by paracrine factors binding to receptors that span the cell membrane and are linked on the cytoplasmic side to JAK (**Ja**nus **k**inase) proteins. The binding of ligand to the receptor phosphorylates the STAT (**s**ignal *t*ransducers and *a*ctivators of *t*ranscription) family of transcription factors.

Junctional neurulation Formation of the neural tube in the transition zone between the primary neural tube (which lies anterior to the hindlimbs) and the secondary neural tube (which extends posteriorly from the sacral region in mammals or is just in the tail region in fish and amphibians).

Juvenile hormone (JH) A lipid hormone in insects that prevents the ecdysone-induced changes in gene expression that are necessary for metamorphosis. Thus, its presence during a molt ensures that the result of that molt is another larval instar, not a pupa or an adult.

Juxtacrine interactions When cell membrane proteins on one cell surface interact with receptor proteins on adjacent (juxtaposed) cell surfaces.

Juxtacrine signaling Signaling between cells that are juxtaposed, i.e., in direct contact with one another.

K

Kairomones Chemicals that are released by a predator and can induce defenses in its prey.

Karyokinesis The mitotic division of the cell's nucleus. The mechanical agent of karyokinesis is the mitotic spindle.

Keratinocytes Differentiated epidermal cells that are bound tightly together and produce a water-impermeable seal of lipid and protein.

Koller's sickle See **Primitive streak**.

Kupffer's vesicle Transient fluid-filled organ housing the cilia that control left-right asymmetry in zebrafish.

L

Labioscrotal folds Folds surrounding the cloacal membrane in the indifferent stage of differentiation of mammalian external genitalia. They will form the labia majora in the female and the scrotum in the male. Also called urethral folds or genital swellings.

lacZ **gene** The *E. coli* gene for β-galactosidase; commonly used as a reporter gene.

Lamellipodia Broad locomotory pseudopods containing actin networks; found on migrating cells, also found on growth cones of neurons.

Laminae Layers. In the brain, neurons are organized into laminae and clusters (nuclei).

Laminin A large glycoprotein and major component of the basal lamina, plays a role in assembling the extracellular matrix, promoting cell adhesion and growth, changing cell shape, and permitting cell migration.

Lampbrush chromosomes Chromosomes in an amphibian primary oocyte during the diplotene stage of the first meiotic prophase that stretch out large loops of DNA, representing sites of upregulated RNA synthesis.

Lanugo The first hairs of human embryos, usually shed before birth.

Large micromeres A tier of cells produced by the fifth cleavage in the sea urchin embryo when the micromeres divide. Become the primary mesenchyme cells, which form the skeletal spicules of the larva.

Larva The sexually immature stage of an organism, often of significantly different appearance than the adult and frequently the stage that lives the longest and is used for feeding or dispersal.

Larval settlement Ability of marine larvae to suspend development until they experience a particular environmental cue for settlement.

Laryngotracheal groove An outpouching of endodermal epithelium in the center of the pharyngeal floor, between the fourth pair of pharyngeal pouches, that extends ventrally. The laryngotracheal groove then bifurcates into the branches that form the paired bronchi and lungs.

Lateral inhibition   The inhibition of a cell by the activity of a neighboring cell.

Lateral plate mesoderm   Mesodermal sheet lateral to the intermediate mesoderm. Gives rise to appendicular bones, connective tissues of the limb buds, circulatory system (heart, blood vessels, and blood cells), muscles and connective tissues of the digestive and respiratory tracts, and lining of coelom and its derivatives. It also helps form a series of extraembryonic membranes that are important for transporting nutrients to the embryo.

Lateral somitic frontier   The boundary between the primaxial and abaxial muscles and between the somite-derived and lateral plate-derived dermis.

Leader sequence   See **5′ Untranslated region**.

Lens placode   Paired epidermal thickenings induced by the underlying optic cups to invaginate to form the lens vesicles, which differentiate into the adult transparent eye lenses that allow light to impinge on the retinas.

Lens vesicle   The vesicle that forms from the lens placode. It differentiates into the lens. It also induces the overlying ectoderm to become the cornea, and induces the inner side of the optic cup to differentiate into the neural retina.

Leptotene   Greek, "thin thread." In the first meiotic division, the first stage of prophase I, when the chromatin is stretched out thinly such that one cannot identify individual chromosomes. DNA replication has already occurred, and each chromosome consists of two parallel chromatids.

Level-specific properties and emergence   A principle of the theoretical systems approach: The properties of a system at any given level of organization cannot be totally explained by those of levels "below" it.

Leydig cells   Testis cells derived from the interstitial mesenchyme cells surrounding the testis cords that make the testosterone required for secondary sex determination and, in the adult, required to support spermatogenesis.

Life expectancy   The length of time an average individual of a given species can expect to live; it is characteristic of populations, not of species.

Ligand   A molecule secreted by a cell that elicits a response in another cell by binding to a receptor on that cell.

Lim genes   Genes coding for transcription factors that are structurally related to proteins encoded by Hox genes.

Limb bud   A circular bulge that will form the future limb. The limb bud is formed by the proliferation of mesenchyme cells

from the somatic layer of the limb field lateral plate mesoderm (the limb *skeletal* precursor cells) and from the somites (the limb *muscle* precursor cells).

Limb field   An area of the embryo containing all of the cells capable of forming a limb.

Lineage-restricted stem cells   Stem cells derived from multipotent stem cells, and which can now generate only a particular cell type or set of cell types.

Lissencephalic   Having a cerebral cortex that lacks folds, as in mice. Compare with **Gyrencephalic**.

Long noncoding RNAs (lncRNAs)   Transcriptional regulators that inactivate genes on one of the two chromosomes of a diploid organism. For example, Xist is a lncRNA involved in the inactivation of genes on the second X chromosome of females. Some lncRNAs appear to be specific for either the maternal or paternal copy of a gene.

Lophotrochozoans   One of two major protostome groups, many of which are characterized spiral cleavage and by the larval form known as the trochophore. A diverse group that includes the annelids (segmented worms such as earthworms), molluscs (e.g., snails), and flatworms (e.g., *Planaria*). See also **Ecdysozoans**.

Loss-of-function evidence   The absence of the postulated cause is associated with the absence of the postulated effect. The "lose it" of "find it, lose it, move it." See also **Correlative evidence**; **Gain-of-function evidence**.

Low CpG-content promoters (LCPs)   These promoters are usually found in the genes whose products characterize mature, fully differentiated cells. The CpG sites are usually methylated and their default state is "off," although they can be activated by specific transcription factors. See also **CpG islands**.

Lumen   The hollow space within any tubular or globular structure or organ.

Luteinizing hormone (LH)   A hormone secreted by the mammalian pituitary that stimulates the production of steroid hormones, such as estrogen from the ovarian follicle cells and testosterone from the testicular Leydig cells. A surge in LH levels causes the primary oocyte to complete meiosis I and prepares the follicle for ovulation.

Lymphatic vasculature   The vessels of the circulatory system that transport lymph (as opposed to the blood vessels of the circulatory system).

M

Macromeres   Larger cells generated by asymmetrical cleavage, e.g., the four large cells generated by the fourth cleavage when the vegetal tier of the sea

urchin embryo undergoes an unequal equatorial cleavage.

Male pronucleus   The haploid nucleus of the sperm.

Malformation   Abnormalities caused by genetic events such as gene mutations, chromosomal aneuploidies, and translocations.

Mammalian gut-associated lymphoid tissue (GALT)   Lymphoid tissue that mediates mucosal immunity and oral immune tolerance, allowing mammals to eat food without creating an immune response to it. Intestinal microbes are critical for the maturation of GALT.

Mantle zone (intermediate zone)   Second layer of the developing spinal cord and medulla that forms around the original neural tube. Because it contains neuronal cell bodies and has a grayish appearance grossly, it will form the gray matter.

Marginal zone   (1) The third and outer zone of the developing spinal cord and medulla composed of a cell-poor region composed of axons extending from neurons residing in the mantle zone. Will form the white matter as glial cells cover the axons with myelin sheaths, which have a whitish appearance. (2) In amphibian gastrula, where gastrulation begins, the region surrounding the equator of the blastula, where the animal and vegetal hemispheres meet. (3) In bird and reptile gastrulae (= marginal belt), a thin layer of cells between the area pellucida and the area opaca, important in determining cell fate during early development.

Maskin   Protein in amphibian oocytes that creates a repressive loop structure in messenger RNA, preventing its translation. It creates the loop by binding to two other proteins, cytoplasmic polyadenylation-element-binding protein (CPEB) and eIF4E factor, which are bound to opposite ends of the mRNA.

Master regulator   Transcription factors that can control cell differentiation by (1) being expressed when the specification of a cell type begins, (2) regulating the expression of genes specific to that cell type, and (3) being able to redirect a cell's fate to this cell type.

Maternal contributions   The stored mRNAs and proteins within the cytoplasm of the egg, produced from the maternal genome during the primary oocyte stage. See also **Maternal message**.

Maternal effect genes   Encode messenger RNAs that are localized to different regions of the *Drosophila* egg.

Maternal message   Messenger RNA that is made in the egg and stored in the egg's cytoplasm while the egg is a primary oocyte. At this point, the egg is still diploid within the ovary. The mRNA, therefore, is being made from the maternal genome.

Maternal-to-zygote transition The embryonic stage when maternally provided mRNAs are degraded and control of development is handed over to the zygote's own genome; often occurs in the mid-blastula stage. Seen in many different animal groups.

Maximum lifespan Maximum number of years an individual of a given species has been known to survive and is characteristic of that species.

Medial hinge point (MHP) In birds and mammals, formed by the cells at the midline of the neural plate. MHP cells become anchored to the notochord beneath them and form a hinge, which forms a furrow at the dorsal midline and helps bend the neural plate as it forms a neural tube.

Mediator A large, multimeric complex of nearly 30 protein subunits that in many genes is the link that connects RNA polymerase II (bound to the promoter) to an enhancer sequence, thus forming a pre-initiation complex at the promoter.

Medullary cord Forms by condensation of mesenchyme cells and then mesenchymal-to-epithelial transition in the caudal region of the avian embryo during the process of secondary neurulation. It will then cavitate to form the caudal section of the neural tube.

Meiosis A unique division process that in animals occurs only in germ cells, to reduce the number of chromosomes to a haploid complement. All other cells divide by mitosis. Meiosis differs from mitosis in that (1) meiotic cells undergo two cell divisions without an intervening period of DNA replication, and (2) homologous chromosomes (each consisting of two sister chromatids joined at a kinetochore) pair together and recombine genetic material.

Melanoblasts Pigment progenitor cells.

Melanocyte stem cells Adult stem cells derived from trunk neural crest cells that form melanoblasts and come to reside in the bulge niche of the hair or feather follicle and which give rise to the pigment of the skin, hair, and feathers.

Melanocytes Cells containing the pigment melanin. Derived from neural crest cells that undergo extensive migration to all regions of the epidermis.

Meltrins Set of metalloproteinases involved in cell fusion events, such as the fusion of myoblasts to form a myofiber and of macrophages to form osteoclasts. See also **Metalloproteinases**.

Meroblastic cleavage Greek *meros*, "part." Refers to the cell division (cleavage) pattern in zygotes containing large amounts of yolk, wherein only a portion of the cytoplasm is cleaved. The cleavage furrow does not penetrate the yolky portion of the cytoplasm because the yolk platelets impede membrane formation there. Only part of the egg is destined to become the embryo, while the other portion—the yolk—serves as nutrition for the embryo, as in insects, fish, reptiles, and birds.

Meroistic oogenesis Type of oogenesis found in certain insects (including *Drosophila* and moths), in which cytoplasmic connections remain between the cells produced by the oogonium.

Mesencephalon The midbrain, the middle vesicle of the developing vertebrate brain; major derivatives include optic tectum and tegmentum. Its lumen becomes the cerebral aqueduct.

Mesenchymal stem cells (MSCs) Also called bone marrow-derived stem cells, or BMDCs. Multipotent stem cells that originate in the bone marrow, MSCs are able to give rise to numerous bone, cartilage, muscle, and fat lineages.

Mesenchymal-to-epithelial transition (MET) Transformation of mesenchymal cells into epithelial cells. Occurs, for example, during somite formation, when somites form from presomitic mesoderm. See also **Epithelial-to-mesenchymal transition**.

Mesenchyme Loosely organized embryonic connective tissue consisting of scattered fibroblast-like and sometimes migratory mesenchymal cells separated by large amounts of extracellular matrix.

Mesenchyme cap cells A population of multipotent stem cells, derived from metanephric kidney mesenchyme, that cover the tips of the ureteric bud branches and can form all the cell types of the nephron.

Mesentoblast In snail embryos, the 4d blastomere whose progeny give rise to most of the mesodermal (heart, kidney and muscles) and endodermal (gut tube) structures.

Mesoderm Greek *mesos*, "between." The middle of the three embryonic germ layers, lying between the ectoderm and the endoderm. The mesoderm gives rise to muscles and skeleton, connective tissue, the urogenital system (kidneys, gonads, and ducts), blood and blood vessels, and most of the heart.

Mesodermal mantle The cells that involute through the ventral and lateral blastopore lips during amphibian gastrulation and will form the heart, kidneys, bones, and parts of several other organs.

Mesomeres The eight cells generated in the sea urchin embryo by the fourth cleavage when the four cells of the animal tier divide meridionally into eight blastomeres, each with the same volume.

Mesonephric duct See **Wolffian duct**.

Mesonephros The second kidney of the amniote embryo, induced in the adjacent mesenchyme by the middle portion of the nephric (Wolffian) duct. It functions briefly in urine filtration in some mammalian species and mesonephric tubules form the tubes that transport the sperm from the testes to the urethra (the epididymis and vas deferens). Forms the adult kidney of anamniotes (fish and amphibians).

Messenger RNA (mRNA) RNA that codes for a protein and leaves the nucleus after being processed from nuclear RNA in a manner that excises noncoding domains and protects the ends of the strand.

MET See **Mesenchymal-to-epithelial transition**.

Metalloproteinases Matrix metalloproteinases (MMP). Enzymes that digest extracellular matrices and are important in many types of tissue remodeling in disease and development, including metastasis, branching morphogenesis of epithelial organs, placental detachment at birth, and arthritis.

Metamorphic climax When the major metamorphic changes, such as tail and gill resorption, and intestinal remodeling, occur in the amphibian. The concentration of T_4 rises dramatically and TRβ levels peak.

Metamorphic molt Pupal molt; in holometabolous insects, the molt at the end of the final instar stage, when the larva becomes a pupa.

Metamorphosis Changing from one form to another, such as the transformation of an insect larva to a sexually mature adult or a tadpole to a frog.

Metanephric mesenchyme An area of mesenchyme, derived from posterior regions of the intermediate mesoderm, involved in mesenchymal-epithelial interactions that generate the metanephric kidney and will form the secretory nephrons. Also called metanephrogenic mesenchyme.

Metanephros/metanephric kidney The third kidney of the embryo and the permanent kidney of amniotes.

Metaphase plate A structure present during mitosis or meiosis in which the chromosomes are attached via their kinetochores to the microtubule spindle and are lined up between the two poles of the cell. If the metaphase plate forms midway between the two poles the division will be symmetrical; if it is closer to one pole, it will be asymmetrical, producing one larger cell and one smaller.

Metastasis   The invasion of cancerous cells into other tissues.

Metazoa   Animals.

Metencephalon   The anterior subdivision of the rhombencephalon; gives rise to the cerebellum (coordinates movements, posture, and balance) and pons (fiber tracts for communication between brain regions).

Methylation   See **Histone methylation**.

5-Methylcytosine   A "fifth" base in DNA, made enzymatically after DNA is replicated by converting cytosine to 5-methycytosine—only cytosines followed by a guanosine can be converted. In mammals, about 5% of cytosines in DNA are converted to 5-methylcytosine.

Microfilaments   Long cables of polymerized actin and a major component of the cytoskeleton. In combination with myosin, forms contractile forces necessary for cytokinesis; formed during fertilization in the egg's cortex to extend microvilli; attached indirectly by other molecules to transmembrane adhesion molecules, such as cadherins and integrins.

Microglia   Small glial cells of the central nervous system that carry out an immune function by engulfing dying and dysfunctional neurons and glia.

Micromeres   Small cells created by asymmetrical cleavage, e.g., four small cells generated by the fourth cleavage at the vegetal pole when the vegetal tier of the sea urchin embryo undergoes an unequal equatorial cleavage.

Micropyle   The only place where *Drosophila* sperm can enter the egg, at the future dorsal anterior region of the embryo, a tunnel in the chorion (eggshell) that allows sperm to pass through it one at a time.

MicroRNA (miRNA)   Small (about 22 nucleotide) RNAs complementary to a portion of a particular mRNA that regulates translation of a specific message. MicroRNAs uusally bind to the 3′ UTR of mRNAs and inhibit their translation.

Microspikes   Essential for neuronal pathfinding, microfilament-containing pointed filopodia of the growth cone that elongate and contract to allow axonal migration. Microspikes also sample the microenvironment and send signals back to the soma.

Microvilli   Small microfilament-containing projections that extend from the surface of cells; e.g., on the egg surface during fertilization where they may aid sperm entry into the cell.

Mid-blastula transition   The transition from the early rapid biphasic (only M and S phases) mitoses of the embryo to a stage characterized by (1) mitoses that include the "gap" stages (G1 and G2) of the cell cycle, (2) loss of synchronicity of cell division, and (3) transcription of new (zygotic) mRNAs needed for gastrulation and cell specification.

Midpiece   Section of sperm flagellum near the head that contains rings of mitochondria that provide the ATP needed to fuel the dynein ATPases and support sperm motility.

miRNA   See **MicroRNA**.

MITF   **Mi**crophthalmia-associated **t**ranscription **f**actor. A transcription factor necessary for melanoblast specification and pigment production. Its name comes from the fact that a mutation in the gene for this transcription factor causes small eyes (microphthalmia) in mice.

Mitosis-promoting factor (MPF)   Consists of cyclin B and a cyclin-dependent kinase (CDK), required to initiate entry into the mitotic (M) phase of the cell cycle in both meiosis and mitosis.

Model systems   Species that are easily studied in the laboratory and have special properties that allow their mechanisms of development to be readily observed (e.g., sea urchins, *Drosophila*, *C. elegans*, zebrafish, and mouse).

Modularity   A principle of the theoretical systems approach. The organism develops as a system of discrete and interacting modules.

Module   A discrete unit of growth, characterized by more internal than external integration.

Molecular parsimony   The principle that development in all lineages uses the same types of molecules (the "small toolkit"). The "toolkit" includes transcription factors, paracrine factors, adhesion molecules, and signal transduction cascades that are remarkably similar from one phylum to another.

Monospermy   Only one sperm enters the egg, and a haploid sperm nucleus and a haploid egg nucleus combine to form the diploid nucleus of the fertilized egg (zygote), thus restoring the chromosome number appropriate for the species.

Monozygotic twins   Greek, "one-egg." Genetically "identical" twins; form when the cells of a single early-cleavage embryo become dissociated from one another, either by the separation of early blastomeres or by the separation of the inner cell mass into two regions within the same blastocyst. Compare with **Dizygotic twins**.

Morph   One of several different potential phenotypes that result from environmental conditions. Also called an ecomorph.

Morphallactic regeneration   See **Morphallaxis**.

Morphallaxis   Type of regeneration that occurs through the repatterning of existing tissues with little new growth (e.g., *Hydra*).

Morphogenesis   The organization of the cells of the body into functional structures via coordinated cell growth, cell migration, and cell death.

Morphogenetic determinants   Transcription factors or their mRNAs that will influence the cell's development.

Morphogens   Greek, "form-givers." Diffusible biochemical molecules that can determine the fate of a cell by their concentrations, in that cells exposed to high levels of a morphogen will activate different genes than those exposed to lower levels.

Morpholino   An antisense oligonucleotide against an mRNA; used to experimentally inhibit protein expression.

Morula   Latin, "mulberry." Vertebrate embryo of 16–64 cells; precedes the blastula or blastocyst stage. Mammalian morula occurs at the 16-cell stage, consists of a small group of internal cells (that will form the inner cell mass) surrounded by a larger group of external cells (that will form the trophoblast).

Mosaic embryos   Embryos in which most of the cells are determined by autonomous specification, with each cell receiving its instructions independently and without cell-cell interaction.

Mosaic pleiotropy   A gene is independently expressed in several tissues. Each tissue needs the gene product and develops abnormally in its absence.

mRNA cytoplasmic localization   The spatial regulation of mRNA translation, mediated by (1) diffusion and local anchoring, (2) localized protection, and (3) active transport along the cytoskeleton.

Müller glial cells   Cells of the neural retina that support and maintain the neurons therein.

Müllerian duct (paramesonephric duct)   Duct running lateral to the mesonephric (Wolffian) duct in both male and female mammalian embryos. These ducts regress in the male fetus, but form the oviducts, uterus, cervix, and upper part of the vagina in the female fetus. Compare with **Wolffian duct**.

Müllerian-inhibiting factor (MIF)   See **Anti-Müllerian hormone**.

Multicellular eukaryotic organism   A eukaryotic organism with multiple cells that remain together as a functional whole; subsequent generations form the same coherent individuals composed of multiple cells. (Includes plants, fungi, and animals.)

Multipotent cardiac progenitor cells   Progenitor cells of the heart field that form cardiomyocytes, endocardium, epicardium, and the Purkinje fibers of the heart.

Multipotent stem cells   Adult stem cells whose commitment is limited to a relatively small subset of all the possible cells of the body.

Mutualism   A form of symbiosis in which the relationship benefits both partners.

Myelencephalon   The posterior subdivision of the rhombencephalon; becomes the medulla oblongata.

Myelin sheath   Modified oligodendrocyte (in CNS) or Schwann cell (in peripheral NS) plasma membrane that surrounds nerve cell axons, providing insulation that confines and speeds electrical impulses transmitted along axons.

Myoblast   Muscle precursor cell.

Myocardium   Heart muscle.

Myoepithelia   Epithelia whose cells possess characteristics of both epithelial and muscle cells, e.g., the two epithelial layers of *Hydra*.

Myofiber   Muscle cell that is multinucleate and forms from the fusion of myoblasts; skeletal muscle cell.

Myogenic regulatory factors (MRFs)   Basic helix-loop-helix transcription factors (such as MyoD, Myf5 and myogenin) that are critical regulators of muscle development.

Myogenin   A myogenic regulatory factor that regulates several genes involved in the differentiation and repair of skeletal muscle cells. See also **Myogenic regulatory factors**.

Myostatin   Greek, "muscle stopper." A member of the TGF-β family, it negatively regulates muscle development. Genetic defects in the gene or its negative regulatory miRNA cause huge muscles to develop in some mammals, including humans.

Myotome   Portion of the somite that gives rise to all skeletal muscles of the vertebrate body except for those in the head. The myotome has two components: the primaxial component, closest to the neural tube, which forms the musculature of the back and rib cage, and the abaxial component, away from the neural tube, which forms the muscles of the limbs and ventral body wall.

N

N-cadherin   A type of cadherin that is highly expressed on cells of the developing central nervous system (the N stands for Neural). May play roles in mediating neural signals. See also **Cadherins**.

NAD⁺ kinase   Activated during the early response of the sea urchin egg to the sperm, converts NAD⁺ to NADP⁺, which can be used as a coenzyme for lipid biosynthesis and may be important in the construction of the many new cell membranes required during cleavage. NADP⁺ is also used to make NAADP.

Naïve pluripotent state   The most immature, undifferentiated state of embryonic stem cells with the greatest potential for pluripotency.

Nanos   Protein critical for the establishment of anterior-posterior polarity of the *Drosophila* embryo. *Nanos* mRNA is tethered to the posterior end of the oocyte and is translated after ovulation and fertilization. Nanos protein diffuses from the posterior end, while Bicoid diffuses from the anterior end, setting up two opposing gradients that establish the anterior-posterior polarity of the embryo.

Necrosis   Pathologic cell death caused by factors such as inflammation of toxic injury. Compare with **Apoptosis**.

Necrotic zones   Regions of the tetrapod limb "sculpted" by apoptotic (programmed) cell death; the term "necrotic" zone is a holdover from a time when no distinction was made between necrosis and apoptosis. The four necrotice regions are interdigital, anterior, posterior, and interior.

Negative feedback loop   A process in which the product of the process inhibits an earlier step in the process.

Neocortex   A layer of gray matter in the cerebrum that is a distinguishing feature of the mammalian brain; it stratifies into six layers of neuronal cell bodies, each with different functional properties.

Neoteny   Retention of the juvenile body form throughout life while the germ cells and reproductive system mature (e.g., the Mexican axolotl). See also **Progenesis**.

Nephric duct   See **Wolffian duct**.

Nephron   Functional unit of the kidney.

Nerve growth factor (NGF)   Neurotrophin involved primarily in the growth of nerve cells. Released from potential target tissues, it works at short ranges as either a chemotactic factor or chemorepulsive factor for axonal guidance. Also important in the selective survival of different subsets of neurons.

Netrins   Paracrine factors found in a gradient that guide axonal growth cones. They are important in commissural axon migration and retinal axon migration. Netrin-1 is secreted by the floor plate; netrin-2 is secreted by the lower region of the spinal cord.

Neural crest   A transient band of cells, arising from the lateral edges of the neural plate, that joins the neural tube to the epidermis. It gives rise to a cell population—the neural crest cells—that detach during formation of the neural tube and migrate to form a variety of cell types and structures, including sensory neurons, enteric neurons, glia, pigment cells, and (in the head) bone and cartilage.

Neural crest effectors   Transcription factors (e.g., MITF and Rho GTPase) activated by neural crest specifiers that give the neural crest cells their migratory properties and some of their differentiated properties.

Neural crest specifiers   A set of transcription factors (e.g., FoxD3, Sox9, Id, Twist, and Snail) induced by the neural plate border-specifying transcription factors, that specify the cells that are to become the neural crest.

Neural folds   Thickened edges of the neural plate that move upward during neurulation and migrate toward the midline and eventually fuse to form the neural tube.

Neural groove   U-shaped groove that forms in the center of the neural plate during primary neurulation.

Neural keel   A band of neural precursor cells that are brought into the dorsal midline during convergence and extension movements in the epiblast of the fish embryo. It extends over the axial and paraxial mesoderm and eventually forms a rod of tissue that separates from the epidermal ectoderm and develops a slit-like lumen to become the neural tube.

Neural plate   The region of the dorsal ectoderm that is specified to be neural ectoderm. It later folds upward to become the neural tube.

Neural plate border   The border between the neural plate and the epidermis.

Neural plate border specifiers   A set of transcription factors (e.g., Distalless-5, Pax3, and Pax7), induced by the neural plate inductive signals, that collectively confer upon the border region the ability to form neural crest and dorsal neural tube cell types. Induce expression of neural crest specifiers.

Neural plate inductive signals   Paracrine factors (e.g., BMPs, Wnts, FGFs and Notch) that interact to specify the boundaries between neural and non-neural ectoderm during gastrulation. In amphibians, inductive signals secreted by the notochord are sufficient to specify neural plate; in chick, signals secreted by the ventral ectoderm and paraxial mesoderm specify the boundaries.

Neural restrictive silencer element (NRSE)   A regulatory DNA sequence found in several mouse genes that prevents a promoter's activation in any tissue except neurons, limiting the expression of these genes to the nervous system.

Neural restrictive silencer factor (NRSF) A zinc finger transcription factor that binds the NRSE and is expressed in every cell that is *not* a mature neuron.

Neural retina Derived from the inner layer of the optic cup, composed of a layered array of cells that include the light- and color-sensitive photoreceptor cells (rods and cones), the cell bodies of the ganglion cells, bipolar interneurons that transmit electric stimuli from the rods and cones to the ganglion cells, Müller glial cells that maintain its integrity, amacrine neurons (which lack large axons), and horizontal neurons that transmit electric impulses in the plane of the retina.

Neural stem cells (NSCs) Stem cells of the central nervous system capable of neurogenesis throughout life. In vertebrates, NSCs retain much of the characteristics of their embryonic progenitor cell, the radial glial cell.

Neural tube The embryonic precursor to the central nervous system (brain and spinal cord).

Neuroblast An immature dividing precursor cell that can differentiate into the cells of the nervous system.

Neurocranium The vault and base of the skull.

Neuromesoderm progenitors (NMPs) A population of multipotent progenitor cells with the potential to contribute to both the neural tube and paraxial mesoderm, found in the posteriormost region of the tail.

Neurons Nerve cells; cells specialized for the conduction and transmission of information via electrical and chemical signals.

Neuropore The two open ends (anterior neuropore and posterior neuropore) of the neural tube that later close.

Neurotransmitters Molecules (e.g., acetylcholine, GABA, serotonin) secreted at the ends of axons. These molecules cross the synaptic cleft and are received by the adjacent neuron, thus relaying the neural signal. See also **Synapse**.

Neurotrophin/neurotropin Neuro*trophic* (Greek, "nourishing") refers to a factor's ability to keep the neuron alive, usually by supplying growth factors. Neuro*tropic* (Latin, "turning") refers to a substance that attracts or repulses neurons. Because many factors have both properties, both terms are used; in the recent literature, neurotrophin appears to be preferred. See also **Nerve growth factor**; **Brain-derived neurotrophic factor**; **Conserved dopamine neurotrophic factor (CDNF)**; **Neurotrophins 3 and 4/5**.

Neurotrophins 3 and 4/5 (NT3, NT4/5) Neurotrophin 3 attracts sensory neurons from the dorsal root ganglia; NT4/5 attracts facial motor neurons and cerebellar granule cells.

Neurula Refers to an embryo during neurulation (i.e., while the neural tube is forming).

Neurulation Process of folding of the neural plate and closing of the cranial and caudal neuropores to form the neural tube.

Newt anterior gradient protein (nAG) Factor released by neurons in the blastema of a regenerating salamander limb that is thought to be the nerve-derived factor necessary for proliferation of the blastema cells.

Nieuwkoop center The dorsalmost vegetal blastomeres of the amphibian blastula, formed as a consequence of the cortical rotation initiated by the sperm entry; an important signaling center on the dorsal side of the embryo. One of its main functions is to induce the Organizer.

NMPs See **Neuromesoderm progenitors**.

Nodal A paracrine factor and member of the TGF-β family involved in establishing left-right asymmetry in vertebrates and invertebrates.

Node The mammalian homologue of Hensen's node.

Noggin A BMP antagonist (i.e., blocks BMP signaling).

Noninvoluting marginal zone (NIMZ) Region of cells on the exterior of the gastrulating amphibian embryo that do not involute. They expand by epiboly along with the animal cap cells to cover the entire embryo, eventually forming the surface ectoderm.

Non-skeletogenic mesenchyme Formed from the veg2 layer of the 60-cell sea urchin embryo, it generates pigment cells, immunocytes, and muscle cells. Also called secondary mesenchyme.

Notch protein Transmembrane protein that is a receptor for Delta, Jagged, or Serrate, participants in juxtacrine interactions. Ligand binding causes Notch to undergo a conformational change that enables a part of its cytoplasmic domain to be cut off by the presenilin-1 protease. The cleaved portion enters the nucleus and binds to a dormant transcription factor of the CSL family. When bound to the Notch protein, the CSL transcription factors activate their target genes.

Notochord A transient mesodermal rod in the most dorsal portion of the embryo that plays an important role in inducing and patterning the nervous system. Characteristic feature of chordates.

Nuclear RNA (nRNA) The original transcription product. Sometimes called heterogeneous nuclear RNA (hnRNA) or pre-messenger RNA (pre-mRNA); contains the cap sequence, the 5' UTR, exons, introns, and the 3' UTR.

Nuclear RNA selection Means of controlling gene expression by processing specific subsets of the nRNA population into mRNA in different types of cells.

Nuclei pulposi A gel-like mass in the center of the intervertebral discs derived from notochordal cells.

Nucleosome The basic unit of chromatin structure, composed of an octamer of histone proteins (two molecules each of histones H2A, H2B, H3, and H4) wrapped with two loops containing approximately 147 base pairs of DNA.

Nucleus (1) The membrane-enclosed organelle housing the eukaryotic chromosomes. (2) An organized cluster of the cell bodies of neurons in the brain with specific functions and connections.

Nurse cells Cells that provides nourishment to a developing egg. In *Drosophila* ovarioles, fifteen interconnected nurse cells generate mRNAs and proteins that are transported to a single developing oocyte.

Nymph Insect larval stage that resembles an immature adult of the species. Becomes progressively more mature though a series of molts.

O

Obesogens Substances the increase the production and accumulation of fat and adipose (fat) cells in the body. Several endocrine disruptors, including DES and BPA, have been shown to be obesogens.

Obligate mutualism Symbiosis in which the species involved are interdependent with one another to such an extent that neither partner could survive without the other.

Olfactory placodes Paired epidermal thickenings that form the nasal epithelium (smell receptors) as well as the ganglia for the olfactory nerves.

Oligodendrocytes A type of glial cell within the central nervous system that wrap themselves around axons to produce a myelin sheath. Also called oligodendroglia.

Omphalomesenteric (umbilical) veins The veins that form from yolk sac blood islands. These veins bring nutrients to the mammalian embryo and transport gases to and from sites of respiratory exchange with the mother.

Oncogenes Regulatory genes that promote cell division, reduce cell adhesion, and prevent cell death. Can promote tumor formation and metastasis. Proto-oncogenes are the normal version of these genes, which, when overexpressed or misexpressed through mutations or inappropriate methylations, are called oncogenes and can result in cancer.

Oocyte A developing egg. A **primary oocyte** is in a stage of growth, has not gone through meiosis, and has a diploid nucleus. A **secondary oocyte** has completed its first meiotic division but not the second, and is haploid.

Oogonium A single female germ cell that is mitotic. When it leaves this stage, it becomes a primary oocyte. Plural, oogonia.

Optic cups Double-walled chambers formed by the invagination of the optic vesicles.

Optic nerve Cranial nerve (CN II) that forms from axons of the neural retina that grow back to the brain by traveling down the optic stalk.

Optic vesicle Extend from the diencephalon and activate the head ectoderm's latent lens-forming ability.

Oral plate A region where the ectoderm of the stomodeum meets the endoderm of the primitive gut. It later breaks open to form the oral opening.

Organization/activation hypothesis The theory that sex hormones act during the fetal or neonatal stage of a mammal's life to organize the nervous system in a sex-specific manner, and that during adult life, the same hormones may have transitory motivational (or "activational") effects.

Organizer In amphibians, the dorsal lip of the blastopore and their derivatives (notochord and head endomesoderm). Functionally equivalent to Hensen's node in chick, the node in mammals, and the shield in fish. Organizer action establishes the basic body plan of the early embryo. Also known as the Spemann Organizer or (more correctly) the Spemann-Mangold organizer.

Organogenesis Interactions between, and rearrangement of, cells of the three germ layers to produce tissues and organs.

Organoids Rudimentary organs, usually the size of a pea, grown in culture from pluripotent stem cells.

Orthologues Genes from different species that are similar in DNA sequence because those genes were inherited from a common ancestor. Compare with **Paralogues**.

Oskar A protein involved in setting up the anterior-posterior axis of the *Drosophila* egg and future embryo by binding *nanos* mRNA in the posterior region of the egg, which will establish the posterior end of the future embryo.

Ossification See **Osteogenesis**.

Osteoblast A committed bone precursor cell.

Osteoclasts Multinucleated cells derived from a blood cell lineage that enter the bone through the blood vessels and destroy bone tissue during remodeling.

Osteocytes Bone cells. Derived from osteoblasts that become embedded in the calcified osteoid matrix.

Osteogenesis Bone formation; the transformation of mesenchyme into bone tissue through a progression from osteoclast to osteoblast to osteocyte. See **Endochondral ossification**; **Intramembranous ossification**.

Osteoid matrix A collagen-proteoglycan secreted by osteoblasts that is able to bind calcium.

Otic placodes Paired epidermal thickenings that invaginate to form the inner ear labyrinth, whose neurons form the acoustic ganglia that enable us to hear.

Outer radial glia (oRG) Progenitor cells that reside in the subventricular zone of the cerebrum and give rise to intermediate progenitor (IP) cells.

Outflow tract In the developing heart, made up of the conus arteriosus and truncus arteriosus; becomes the base of the aorta and the pulmonary arteries.

Oval cells A population of progenitor cells in the liver that divide and form new hepatocytes and bile duct cells when hepatocytes themselves are unable to regenerate the liver sufficiently.

Ovariole The *Drosophila* egg chamber.

Oviparity Young hatch from eggs ejected by the mother, as in birds, amphibians, and most invertebrates.

Ovoviviparity Young hatch from eggs held within the mother's body where they continue to develop for a period of time, as in certain reptiles and sharks. Compare with **Viviparity**.

Ovulation Release of the egg from the ovary.

Ovum The mature egg (at the stage of meiosis at which it is fertilized). Plural, ova.

P

p53 A transcription factor that can stop the cell cycle, cause cellular senescence in rapidly dividing cells, instruct the initiation of apoptosis, and activate DNA repair enzymes. One of the most important regulators of cell division.

Pachytene Greek, "thick thread." In the first meiotic division, the third stage of prophase I during which the chromatids thicken and shorten and can be seen by light microscopy as individual chromatids. Crossing over occurs during this stage.

Pair-rule genes *Drosophila* zygotic genes, regulated by gap gene proteins. Pair-rule genes are each expressed in seven stripes that divide the embryo into transverse bands perpendicular to the anterior-posterior axis. Pair-rule mutants lack portions of every other segment.

PAL-1 A maternally expressed transcription factor in the oocyte of the nematode *C. elegans* that is required for the differentiation of the P1 lineage of cells. P1 is one of the cells of the two-cell embryo.

PAR proteins Found in the cytoplasm of oocytes of the nematode *C. elegans*; involved in determining the anterior-posterior axis of the embryo following fertilization.

Paracrine factor A secreted, diffusible protein that provides a signal that interacts with and changes the cellular behavior of neighboring cells and tissues.

Paracrine signaling Signaling between cells that occurs across long distances through the secretion of **paracrine factors** into the extracellular matrix.

Paralogues Genes that are similar in sequence because they are the result of gene duplication events in an ancestral species. Compare with **Orthologues**.

Parasegment A "transegmental" unit in *Drosophila* that includes the posterior compartment of one segment and the anterior compartment of the immediately posterior segment; appears to be the fundamental unit of embryonic gene expression.

Parasitism Type of symbiosis in which one partner benefits at the expense of the other.

Parasympathetic (enteric) ganglia Ganglia of the parasympathetic ("rest and digest") nervous system derived from vagal and sacral neural crest cells.

Paraxial (somitic) mesoderm Thick bands of embryonic mesoderm immediately adjacent to the neural tube and notochord. In the trunk, paraxial mesoderm gives rise to somites, in the head it (along with the neural crest) gives rise to the skeleton, connective tissues and musculature of the face and skull.

Paraxial protocadherin Adhesion protein expressed specifically in the paraxial (somite-forming) mesoderm during amphibian gastrulation; essential for convergent extension.

Parietal endoderm Cells of the primitive endoderm that contact the trophoblast of the mammalian embryo. See **Primitive endoderm**.

Parietal mesoderm See **Somatic mesoderm**.

Parthenogenesis   Greek, "virgin birth." When an ovum is activated in the absence of sperm. Normal development can proceed in many invertebrates and some vertebrates.

Pathway selection   The first step in the specification of axonal connection, wherein the axons travel along a route that leads them to a particular region of the embryo.

Pattern formation   The set of processes by which embryonic cells form ordered spatial arrangements of differentiated tissues.

P-cadherin   A type of cadherin found predominantly on the placenta, where it helps the placenta stick to the uterus (the P stands for placenta). See also **Cadherins**.

Pericardial cavity   The division of the coelom that surrounds the heart. Compare with **Peritoneal cavity**; **Pleural cavity**.

Perichondrium   Connective tissue that surrounds most cartilage, except at joints.

Pericytes   Smooth-musclelike cells recruited to cover endothelial cells during vasculogenesis.

Periderm   A temporary epidermis-like covering in the embryo that is shed once the inner layer differentiates to form a true epidermis.

Periosteal bone   Bone that adds thickness to long bones and is derived from mesoderm via intramembranous ossification.

Periosteum   A fibrous sheath containing connective tissue, capillaries, and bone progenitor cells and that covers the developing and adult bone.

Peritoneal cavity   The division of the coelom that encloses the abdominal organs. Compare with **Pericardial cavity**; **Pleural cavity**.

Permissive interaction   Inductive interaction in which the responding tissue has already been specified, and needs only an environment that allows the expression of these traits.

P-granules   The germ plasm in *C. elegans*. Isolated to a single germline precursor cell (P4 blastomere) early in cleavage.

Pharyngeal arches   Paired bars of mesenchymal tissue (derived from paraxial mesoderm, lateral plate mesoderm, and neural crest cells), covered by endoderm internally and ectoderm externally. Found near the pharynx of the vertebrate embryo, the arches form gill supports in fish, and many skeletal and connective tissue structures in the face, jaw, mouth, and larynx in other vertebrates. Also called branchial arches.

Pharyngeal clefts   Clefts (invaginations) of external ectoderm that separate the pharyngeal arches. In amniotes, there are four pharyngeal clefts in the early embryo, but only the first becomes a structure (the external auditory meatus).

Pharyngeal pouches   Inside the pharynx, these are where the pharyngeal epithelium (endoderm) pushes out laterally to form pairs of pouches between the pharyngeal arches. These give rise to the auditory tube, wall of the tonsil, thymus gland, parathyroids and thyroid.

Pharyngula   Term often applied to the late neurula stage of vertebrate embryos.

Pharynx   The region of the digestive tube anterior to the point at which the respiratory tube branches off.

Phenotypic heterogeneity   Refers to the same mutation producing different phenotypes in different individuals.

Phenotypic plasticity   The ability of an organism to react to an environmental input with a change in form, state, movement, or rate of activity.

Pheromones   Vaporized chemicals emitted by an individual that results in communication with another individual. Pheromones are recognized by the vomeronasal organ of many mammalian species and play a major role in sexual behavior.

Phosphatidylinositol 4,5-bisphosphate (PIP$_2$)   A membrane phospholipid that during the IP$_3$ pathway is split by the enzyme phospholipase C (PLC) to yield two active compounds: IP$_3$ and diacylglycerol (DAG). The IP$_3$ pathway is activated during fertilization, launching the slow block to polyspermy and activating the egg to start developing.

Phospholipase C (PLC)   Enzyme in the IP$_3$ pathway that splits membrane phospholipid phosphatidylinositol 4,5-bisphosphate (PIP$_2$) to yield IP$_3$ and diacylglycerol (DAG).

Phylotypic stage   The stage that typifies a phylum, such as the late neurula or pharyngula of vertebrates, and which appears to be relatively invariant and to constrain its evolution.

Pial surface   The outer surface of the brain; "pial" refers to its being next to the pia mater, one of the meninges of the brain.

PIE-1   A maternally expressed transcription factor in the oocyte of the nematode *C. elegans* that is necessary for germline cell fate.

Pigmented retina   The melanin-containing layer of the vertebrate eye that lies behind the neural retina. It forms from the outer layer of the optic cup. The black melanin pigment absorbs light coming through the neural retina, preventing it from bouncing back through the neural retina, which would distort the image perceived.

Pioneer nerve fibers   Axons that go ahead of other axons and serve as guides for them.

Pioneer transcription factors   Transcription factors (e.g., Fox A1 and Pax7) that can penetrate repressed chromatin and bind to their enhancer DNA sequences, a step critical to establishing certain cell lineages.

Piwi   One of the proteins, along with Tudor, Vasa, and Nanos, expressed in germ cells to suppress gene expression.

Placenta   The organ in placental mammals that serves as the interface between fetal and maternal circulations and has endocrine, immune, nutritive and respiratory functions. It consists of a maternal portion (the uterine endometrium, or decidua, which is modified during pregnancy) and a fetal component (the chorion).

Placodes   An area of ectodermal thickening. These include the cranial placodes (e.g., the olfactory, lens, and otic placodes); and the epidermal placodes of cutaneous appendages such as hair and feathers, which are formed via inductive interactions between the dermal mesenchyme and the ectodermal epithelium.

PLC   See **Phospholipase C**.

PLCζ (Phospholipase C zeta)   A soluble form of phospholipase C found in the head of mammalian sperm that is released during gamete fusion in fertilization. It sets off the IP$_3$ pathway in the egg that results in Ca^{2+} release and activation of the egg.

Pleiotropy   The production of several effects by one gene or pair of genes.

Pleural cavity   The division of the coelom that surrounds the lungs. Compare with **Pericardial cavity**; **Peritoneal cavity**.

Pluripotent   Latin, "capable of many things." A single pluripotent stem cell has the ability to give rise to different types of cells that develop from the three germ layers (mesoderm, endoderm, ectoderm) from which all the cells of the body arise. The cells of the mammalian inner cell mass (ICM) are pluripotent, as are embryonic stem cells. Each of these cells can generate any cell type in the body, but because the distinction between ICM and trophoblast has been established, it is thought that ICM cells are not able to form the trophoblast. Germ cells and germ cell tumors (such as teratocarcinomas) can also form pluripotent stem cells. Compare with **Totipotent**.

Pluripotent hematopoietic stem cell   See **Hematopoietic stem cell (HSC)**.

Pluteus larva   Type of larva found in sea urchins and brittle stars; a planktonic larva that is bilaterally symmetrical, ciliated, and has long arms supported by skeletal spicules.

Polar body The smaller cell, containing hardly any cytoplasm, generated during the asymmetrical meiotic division of the oocyte. The first polar body is haploid and results from the first meiotic division and the secondary polar body is also haploid and results from the second meiotic division.

Polar granule component (PGC) A protein important for germ line specification and localized to *Drosophila* polar granules. PGC inhibits transcription of somatic cell-determining genes by preventing the phosphorylation of RNA polymerase II.

Polar granules Particles containing factors important for germ line specification that are localized to the pole plasm and pole cells of *Drosophila*.

Polar lobe An anucleate bulb of cytoplasm extruded immediately before first cleavage, and sometimes before the second cleavage, in certain spirally cleaving embryos (mostly in the mollusc and annelid phyla). It contains the determinants for the proper cleavage rhythm and the cleavage orientation of the D blastomere.

Polarization The first stage of cell migration, wherein a cell defines its front and its back ends, directed by diffusing signals (such as a chemotactic protein) or by signals from the extracellular matrix. These signals will reorganize the cytoskeleton such that the front part of the cell will form lamellipodia (or filopodia) with newly polymerized actin.

Pole cells About five nuclei in the *Drosophila* embryo that reach the surface of the posterior pole during the ninth division cycle and become enclosed by cell membranes. The pole cells give rise to the gametes of the adult.

Pole plasm Cytoplasm at the posterior pole of the *Drosophila* oocyte that contains the determinants for producing the abdomen and the germ cells.

PolyA tail A series of adenine (A) residues that are added by enzymes to the 3' terminus of the mRNA transcript in the nucleus. The polyA tail confers stability on the mRNA, allows the mRNA to exit the nucleus, and permits the mRNA to be translated into protein.

Polyadenylation The insertion of a "tail" of some 200–300 adenylate residues on the RNA transcript, about 20 bases downstream of the AAUAAA sequence. This polyA tail (1) confers stability on the mRNA, (2) allows the mRNA to exit the nucleus, and (3) permits the mRNA to be translated into protein.

Polycomb proteins Family of proteins that bind to condensed nucleosomes, keeping the genes in an inactive state.

Polydactyly The presence of extra (supernumerary) digits, such as the dew claw on Great Pyrenees dogs.

Polyphenism A type of phenotypic plasticity, refers to discontinuous ("either/or") phenotypes elicited by the environment. Compare with **Reaction norm**.

Polyspermy The entrance of more than one sperm during fertilization resulting in aneuploidy (abnormal chromosome number) and either death or abnormal development. An exception, called physiological polyspermy, occurs in some organisms such as *Drosophila* and birds, where multiple sperm enter the egg but only one sperm pronucleus fuses with the egg pronucleus.

Polytene chromosomes Chromosomes in the larval cells of *Drosophila* (but not the imaginal cells that give rise to the adult) in which the DNA undergoes many rounds of replication without separation, forming large "puffs" that are easily visible and indicate active gene transcription.

Population asymmetry Mode of maintaining homeostasis in a population of stem cells in which some of the cells are more prone to producing differentiated progeny, while others divide to maintain the stem cell pool.

Posterior marginal zone (PMZ) The end of the chick blastoderm where primitive streak formation begins and acts as the equivalent of the amphibian Nieuwkoop center. The cells of the PMZ initiate gastrulation and prevent other regions of the margin from forming their own primitive streaks.

Posterior necrotic zone A zone of programmed cell death on the posterior side of the developing tetrapod limb that helps shape the limb.

Posterior neuropore See **Neuropore**.

Postsynaptic cell The target cell that receives chemical neurotransmitters from a presynaptic neuron, causing depolarization or hyperpolarization of the target cell's membrane.

Posttranslational regulation Modifications that determine whether the translated protein will be active. These modifications can include cleaving an inhibitory peptide sequence; sequestration and targeting to specific cell regions; assembly with other proteins to form a functional unit; binding an ion (such as Ca^{2+}); or modification by the covalent addition of a phosphate or acetate group.

Potency In referring to stem cells, the power to produce different types of differentiated cells.

Prechordal plate mesoderm Precursor of the head mesoderm. The mesoderm cells that move inward during gastrulation ahead of the chordamesoderm.

Precursor cells (precursors) Widely used term to denote any ancestral cell type (stem or progenitor cells) of a particular lineage (e.g., neuronal precursors; blood cell precursors).

Predator-induced polyphenism The ability to modulate development in the presence of predators in order to express a more defensive phenotype.

Preeclampsia Medical condition of pregnant women characterized by hypertension, poor renal filtration, and fetal distress. A leading cause of premature birth and both fetal and maternal deaths.

Preformationism The view, supported by the early microscopist Marcello Malpighi, that the organs of the embryo are already present, in miniature form, within the egg (or sperm). A corollary, *embôitment* (encapsulation), stated that the next generation already existed in a prefigured state within the germ cells of the first prefigured generation, thus ensuring that the species would remain constant.

Preimplantation genetics Testing for genetic diseases using blastomeres from embryos produced by in vitro fertilization before implanting the embryo in the uterus.

Pre-initiation complex The complex of RNA polymerase II at the promoter with transcription factors on the enhancer, as brought together by the Mediator molecules. See also **Mediator**.

Pre-metamorphosis The first stage in amphibian metamorphosis; the thyroid gland has begun to mature and is secreting low levels of T_4 and very low levels of T_3. The TRα receptor is present, but the TRβ receptor is not.

Prenatal diagnosis The use of chorionic villus sampling or amniocentesis to diagnose many genetic diseases before a baby is born.

Presomitic mesoderm (PSM) The mesoderm that will form the somites. Also known as the segmental plate.

Presynaptic neuron Neuron that transmits chemical neurotransmitters to a target cell, causing the depolarization or hyperpolarization of the target cell's membrane.

Primary capillary plexus A network of capillaries formed by endothelial cells during vasculogenesis.

Primary cilium A single, non-motile cilium found on most cells; lacks a central pair of microtubules and is involved in part of the hedgehog signaling pathway by transporting signaling molecules on its microtubules using motor proteins.

Primary embryonic induction The process whereby the dorsal axis and central nervous system forms through interactions with the underlying mesoderm, derived from the dorsal lip of the blasto-pore in amphibian embryos.

Primary larvae   Larvae that represent dramatically different body plans than the adult form and that are morphologi-cally distinct from the adult; the plutei of sea urchins are such larvae. Compare with **Secondary larvae**.

Primary neurulation   The process that forms the anterior portion of the neural tube. The cells surrounding the neural plate direct the neural plate cells to pro-liferate, invaginate, and pinch off from the surface to form a hollow tube.

Primary oocytes   A developing egg that has passed through the oogonial stage and is in a stage of growth prior to any meiotic division. Contains a large nucleus called a germinal vesicle. At this stage, mRNA (maternal mRNA) is being made and stored in the egg. In mammals, a primary oocyte is arrested in first meiotic prophase until just prior to ovulation, when the first meiotic divi-sion is completed and the egg becomes a secondary oocyte. The second meiotic division is then arrested and is not com-pleted until after fertilization.

Primary sex determination   The deter-mination of the gonads to form either the egg-forming ovaries or sperm-form-ing testes. Primary sex determination is chromosomal and is not usually influ-enced by the environment in mammals, but can be affected by the environment in other vertebrates.

Primary spermatocytes   Derived from mitotic division of the type B spermato-gonia, these are the cells that first go through a period of growth and then enter meiosis.

Primaxial muscles   The intercostal mus-culature between the ribs and the deep muscles of the back, formed from those myoblasts in the myotome closest to the neural tube.

Primed pluripotent state   The state of an embryonic stem cell that has undergone some maturation toward the epiblast lineage.

Primitive endoderm   The layer of endo-derm cells created during early mam-malian development when the inner cell mass splits into two layers. The lower layer, in contact with the blas-tocoel, is the primitive endoderm, and is homologous to the hypoblast of the avian embryo. It will form the inner lining of the yolk sac, and will be used for positioning the site of gastrulation, regulating the movements of cells in the epiblast, and promoting the maturation of blood cells. It is an extraembryonic layer that does not provide cells to the body of the embryo.

Primitive groove   A depression that forms within the primitive streak that serves as an opening through which migrating cells pass into the deep layers of the embryo.

Primitive knot/pit   See **Hensen's node**.

Primitive streak   The first morphologi-cal sign of gastrulation in amniotes, it first arises from a local thickening of the epiblast at the posterior edge of the area pellucida, called Koller's sickle. Homol-ogous to the amphibian blastopore.

Primordial germ cells (PGCs)   Gamete progenitor cells, which typically arise elsewhere and migrate into the develop-ing gonads.

Proacrosin   The inactive form of a mam-malian sperm proteinase that is stored in the acrosome and released during the acrosomal reaction and helps the sperm move through the zona pellucida of the egg.

Proerythroblast   A red blood cell precursor.

Progenesis   Condition in which the gonads and germ cells develop at a fast-er rate than the rest of the body, becom-ing sexually mature while the rest of the body is still in a juvenile phase. Com-pare with **Neoteny**.

Progenitor cells   Relatively undiffer-entiated cells that have the capacity to divide a few times before differentiating and, unlike stem cells, are not capable of unlimited self-renewal. They are some-times called **transit amplifying cells** because they divide while migrating.

Progerias   Premature aging syndromes; in humans and mice, appear to be caused by mutations that prevent the functioning of DNA repair enzymes.

Progesterone   A steroid hormone important in the maintenance of preg-nancy in mammals. Progesterone secreted from the cumulus cells may act as a chemotactic factor for sperm.

Programmed cell death   See **Apoptosis**.

Progress zone (PZ)   Highly proliferative limb bud mesenchyme directly beneath the apical ectodermal ridge (AER). The proximal-distal growth and differentia-tion of the limb bud are made possible by a series of interactions between the AER and the progress zone. Also called the undifferentiated zone.

Progress zone model   Model for speci-fication of proximal-distal specifica-tion of the limb that postulates that each mesoderm cell is specified by the amount of time it spends dividing in the progress zone. The longer a cell spends in the progress zone, the more mitoses it achieves and the more distal its specifi-cation becomes.

Prometamorphosis   The second stage in amphibian metamorphosis, during which the thyroid matures and secretes more thyroid hormones.

Promoter   Region of a gene containing the DNA sequence to which RNA poly-merase II binds to initiate transcription. See also **CpG islands**; **Enhancer**.

Pronephric duct   Arises in the interme-diate mesoderm, migrates caudally, and induces the adjacent mesenchyme to form the pronephros, or tubules of the initial kidney of the embryo. The pro-nephric tubules form functioning kid-neys in fish and in amphibian larvae but are not believed to be active in amni-otes. As the duct continues growing downward it induces the mesonephric mesenchyme to form tubules, at which point it is called the mesonephric duct. Also called Wolffian duct and nephric duct.

Pronephros   The first region of kidney mesenchyme to form kidney tubules in vertebrates. The pronephros is a func-tioning kidney in fish and amphibian larvae, but is not believed to be active in amniotes, and degenerates after other regions of the kidney develop.

Pronuclei   The male and female haploid nuclei within a fertilized egg that fuse to form the diploid nucleus of the zygote.

Pronymph   The stage immediately after hatching in ametabolous insects, when the organism bears the structures that enabled it to get out of the egg; after this stage, the insect looks like a small adult.

Prosencephalon   The forebrain; the most anterior vesicle of the developing vertebrate brain. Will form two second-ary brain vesicles: the telencephalon and the diencephalon.

Protamines   Basic proteins, tightly com-pacted through disulfide bonds, that package the DNA of the sperm nucleus.

Protein-protein interaction domain   A domain of a transcription factor that enables it to interact with other proteins on the enhancer or promoter.

Proteoglycans   Large extracellular matrix molecules consisting of core proteins (such as syndecan) with cova-lently attached glycosaminoglycan polysaccharide side chains. Two of the most widespread are heparan sulfate proteoglycan and chondroitin sulfate proteoglycan.

Proteome   The number and type of pro-teins encoded by the genome.

Prothoracic gland   In insects, a gland that secretes ecdysone, a molting hormone; production of ecdysone is stimulated by the prothoracicotropic hormone.

Prothoracicotropic hormone (PTTH)   A peptide hormone that initiates the molt-ing process in insects when it is released by neurosecretory cells in the brain in response to neural, hormonal, or envi-ronmental signals. PTTH stimulates the production of ecdysone by the prothoracic gland.

Protocadherins   A class of cadherins that lack the attachment to the actin skeleton through catenins. They are an important means of keeping migrating epithelia together, and they are important in separating the notochord from surrounding mesoderm during its formation.

Protostomes   Greek, "mouth first." Animals that form their mouth regions from the blastopore, such as molluscs. Compare with **Deuterostomes.**

Proximal-distal axis   The close-far axis, e.g., shoulder-finger or hip-toe (in relation to the body's center).

Pseudohermaphroditism   Intersex conditions in which the secondary sex characteristics differ from what would be expected from the gonadal sex. Male pseudohermaphroditism (e.g., androgen insensitivity syndrome) describes conditions wherein the gonadal sex is male and the secondary sex characteristics are female, while female pseudohermaphroditism describes the reverse situation (e.g., congenital adrenal hyperplasia).

Pupa   A non-feeding stage of a holometabolous insect following the last instar when the organism is going through metamorphosis, being transformed from a larva into an adult (imago).

Purkinje fibers   Modified heart muscle cells in the inner walls of the ventricles, specialized for rapid conduction of the contractile signal. Essential for synchronizing the contractions of the ventricles in amniotes.

Purkinje neurons   Large, mutibranched neurons that are the major cell type of the cerebellum.

R

R-cadherin   A type of cadherin critical in forming the retina (the R stands for retina). See **Cadherins.**

RA   See **Retinoic acid.**

Radial glial cells (radial glia)   Neural progenitor cells found in the ventricular zone (VZ) of the developing brain. At each division, they generate another VZ cell and a more committed cell type that leaves the VZ to differentiate.

Radial holoblastic cleavage   Cleavage pattern in echinoderms. The cleavage planes, which divide the egg completely into separate cells (holoblastic), are parallel or perpendicular to the animal-vegetal axis of the egg.

Radial intercalation   In fish embryos, the movement of deep epiblast cells into the more superficial epiblast layer, helping to power epiboly during gastrulation.

Random epigenetic drift   The hypothesis that the chance accumulation of inappropriate epigenetic methylation due to errors made by the DNA methylating and demethylating enzymes could be the critical factor in aging and cancers.

Ras   A G-protein in the RTK pathway. Mutations in the *RAS* gene account for a large proportion of cancerous human tumors.

Rathke's pouch   An outpocketing of the ectoderm in the roof of the oral region that forms the glandular portion of the pituitary gland in vertebrates. It meets the infundibulum, an outpocketing of the floor of the diencephalon, which forms the neural portion of the pituitary gland.

Reaction norm   A type of phenotypic plasticity in which the genome encodes the potential for a continuous range of potential phenotypes; the environment the individual encounters determines which of the potential phenotypes develops. Compare with **Polyphenism**.

Reaction-diffusion model   Model for developmental patterning, especially that of the limb, wherein two homogeneously distributed substances (an activator, substance *A*, that activates itself as well as forming its own, faster-diffusing inhibitor, substance *I*) interact to produce stable complex patterns during morphogenesis. According to this model, set forth in the early 1950s by mathematician Alan Turing, the patterns generated by this reaction-diffusion mechanism represent regional differences in the concentrations of the two substances.

Reactive oxygen species (ROS)   Metabolic by-products that can damage cell membranes and proteins and destroy DNA. ROS are generated by mitochondria due to insufficient reduction of oxygen atoms and include superoxide ions, hydroxyl ("free") radicals, and hydrogen peroxide.

Receptor   A protein that functions to bind a ligand. See also **Ligand.**

Receptor tyrosine kinase (RTK)   A receptor that spans the cell membrane and has an extracellular region, a transmembrane region, and a cytoplasmic region. Ligand (paracrine factor) binding to the extracellular domain causes a conformational change in the receptor's cytoplasmic domains, activating kinase activity that uses ATP to phosphorylate specific tyrosine residues of particular proteins.

Reciprocal inductions   A common sequential feature of induction: One tissue induces another, and that tissue then acts back on the original inducing tissue and induces it, thus the inducer becomes the induced.

Reelin   An extracellular matrix protein found in the developing cerebellum and cerebrum. In the cerebellum it permits neurons to bind to glial cells as neurons migrate and form layers; in the cerebrum it directs migration of neurons toward the pial surface.

Regeneration   The ability to reform body structure or organ that has been damaged or destroyed by trauma or disease.

Regeneration blastema   A collection of relatively undifferentiated cells that are organized into new structures by paracrine factors located at the cut surface. The collection of cells may be derived from differentiated tissue near the site of amputation that dedifferentiate, go through a period of mitosis, and then redifferentiate into the lost structures, as in the regenerating salamander limb, or may be from pluripotent stem cells that migrate to the cut surface, as in flatworm regeneration.

Regenerative medicine   The therapeutic use of stem cells to correct genetic pathologies (e.g., sickle-cell anemia) or repair damaged organs.

Regulation   The ability to respecify cells so that the removal of cells destined to become a particular structure can be compensated for by other cells producing that structure. This is seen when an entire embryo is produced by cells that would have contributed only certain parts to the original embryo. It is also seen in the ability of two or more early embryos to form one chimeric individual rather than twins, triplets, or a multiheaded individual.

Relational pleiotropy   The action of a gene in one part of the embryo that affects other parts, not by being expressed in these other parts but by having initiated a cascade of events that affect these other parts.

Reporter gene   A gene with a product that is readily identifiable and not usually made in the cells of interest. Can be fused to regulatory elements from a gene of interest, inserted into embryos, and then monitored for reporter gene expression. If the sequence contains an enhancer, the reporter gene should become active at particular times and places. The genes for green fluorescent protein (*GFP*) and β-galactosidase (*lacZ*) are common examples.

Resact   A 14-amino-acid peptide that has been isolated from the egg jelly of the sea urchin *Arbacia punctulata* that acts as a chemotactic factor and sperm-activating peptide for sperm of the same species, i.e., it is species-specific and is thereby a mechanism to ensure that fertilization is also species-specific. See also **Sperm-activating peptide.**

Resegmentation   Occurs during formation of the vertebrae from sclerotomes;

the rostral segment of each sclerotome recombines with the caudal segment of the next anterior sclerotome to form the vertebral rudiment and this enables the muscles of the vertebral column derived from the myotomes to coordinate the movement of the skeleton, permitting the body to move laterally.

Respiratory tube   The future respiratory tract, which forms as an epithelial out-pocketing of the pharynx, and eventually bifurcates into the two lungs.

Responder   During induction, the tissue being induced. Cells of the responding tissue must have receptors for the inducing molecules and be competent to respond to the inducer.

Resting membrane potential   The membrane potential (membrane voltage) normally maintained by a cell, determined by the concentration of ions on either side of the membrane. Generally this is –70mV, where the inside of the cell is negatively charged with respect to the exterior.

Rete testis   A network of thin canals that convey sperm from the seminiferous tubules to the efferent ducts.

Reticulocyte   Cell derived from the mammalian erythroblast that has expelled its nucleus. Although reticulocytes, lacking a nucleus, can no longer synthesize globin mRNA, they can translate existing messages into globins. A reticulocyte differentiates into a mature red blood cell (erythrocyte), in which even translation of mRNA doesn't take place.

Retina   See **Neural retina**.

Retinal ganglion cells (RGCs)   Neurons in the retina of the eye whose axons are guided to the optic tectum of the brain. Guidance cues come from netrin, slit, semaphorin, and ephrin families of molecules.

Retinal homeobox (Rx)   A transcription factor coded for by the *Rx* gene. Produced in the eye field and helps specify the retina.

Retinoic acid (RA)   A derivative of vitamin A and morphogen involved in anterior-posterior axis formation. Cells receiving high levels of RA express posterior genes.

Retinoic acid-4-hydroxylase   An enzyme that degrades retinoic acid.

Retinotectal projection   The map of retinal connections to the optic tectum. Point-for-point correspondence between the cells of the retina and the cells of the tectum that enables the animal to see an unbroken image.

Reverse development   The transformation of a mature stage of an organism to a more juvenile stage of its life cycle. Seen in certain hydrozoan species where

the sexually mature adult-stage medusa is able to revert to the polyp stage.

Reverse genetics   Genetic technique of knocking out or knocking down the expression of a gene in an organism and then studying the phenotype that results. Compare with **Forward genetics**.

Rho GTPases   A family of molecules including RhoA, Rac1, and Cdc42 that convert soluble actin into fibrous actin cables that anchor at the cadherins. These help mediate cell migration by lamellipodia and filopodia and the cadherin-dependent remodeling of the cytoskeleton.

Rhombencephalon   The hindbrain, the most caudal vesicle of the developing vertebrate brain; will form two secondary brain vesicles, the metencephalon and myelencephalon.

Rhombomeres   Periodic swellings that divide the rhombencephalon into smaller compartments, each with a different fate and different associated nerve ganglia.

Right-left axis   Specification of the two lateral sides of the body.

Ring canals   The cytoplasmic interconnections between the cystocytes that become the ovum and nurse cells in an ovariole of *Drosophila*.

RNA interference   Process by which miRNAs inhibit expression of specific genes by degrading their mRNAs.

RNA polymerase II   An enzyme that binds to a promoter on DNA and, when activated, catalyzes the transcription of an RNA template from the DNA.

RNA-induced silencing complex (RISC)   A complex containing several proteins and a microRNA, which can then bind to the 3′ UTR of messages and inhibit their translation.

RNA-Seq (RNA sequencing)   Using next-generation sequencing technology to sequence and quantify the RNA present in a biological sample.

Robo proteins   See **Roundabout proteins**.

Robustness (canalization)   The ability of an organism to develop the same phenotype despite perturbations from the environment or from mutations. It is a function of interactions within and between developmental modules.

Rods   Photoreceptors in the neural retina of the vertebrate eye that are more sensitive to low light than cones. They contain only one light-sensitive pigment and therefore do not transmit information about color.

Roof plate   Dorsal region of the neural tube important in the establishment of dorsal-ventral polarity. The adjacent epidermis induces expression of BMP4

in the roof plate cells, which in turn induces a cascade of TGF-β proteins in adjacent cells of the neural tube.

Rosettes   Pinwheel-like structures, such as the structures made up of small clusters of neural stem cells surrounded by ciliated ependymal cells found in the V-SVZ of the mammalian cerebrum.

Rostral-caudal   Latin, "beak-tail." An anterior-posterior positional axis; often used when referring to vertebrate embryos or brains.

Rotational cleavage   The cleavage pattern for mammalian and nematode embryos. In mammals, the first cleavage is a normal meridional division while in the second cleavage, one of the two blastomeres divides meridionally and the other divides equatorially. In *C. elegans*, each asymmetrical division produces one founder cell that produces differentiated descendants; and one stem cell. The stem cell lineage always undergoes meridional division to produce (1) an anterior founder cell and (2) a posterior cell that will continue the stem cell lineage.

Roundabout proteins (Robo)   Proteins that are receptors for slit proteins, involved in controlling the crossing of the midline of commissural axons.

Royalactin   Protein that induces a honeybee larva to become a queen. Fed to the larva by worker bees, the protein binds to EGF receptors in the larva fat body and stimulates the production of juvenile hormone, which elevates the levels of yolk proteins necessary for egg production.

R-spondin1 (Rspo1)   Small, soluble protein that upregulates the Wnt pathway and is critical for ovary formation in mammals.

RTK pathway   The receptor tyrosine kinase (RTK) is dimerized by ligand, which causes autophosphorylation of the receptor. An adaptor protein recognizes the phosphorylated tyrosines on the RTK and activates an intermediate protein, GEF, which activates the Ras G protein by allowing the phosphorylation of the GDP-bound Ras. At the same time, the GAP protein stimulates the hydrolysis of this phosphate bond, returning Ras to its inactive state. The active Ras activates the Raf protein kinase C (PKC), which in turn phosphorylates a series of kinases. Eventually, an activated kinase alters gene expression in the nucleus of the responding cell by phosphorylating certain transcription factors (which can then enter the nucleus to change the types of genes transcribed) and certain translation factors (which alter the level of protein synthesis). In many cases, this pathway is reinforced by the release of Ca^{2+}.

S

Sacral neural crest   Neural crest cells that lie posterior to the trunk neural crest and along with the vagal neural crest generate the parasympathetic (enteric) ganglia of the gut that are required for peristaltic movement in the bowels.

Sarcopterygian fishes   Lobe-finned fish, including coelacanths and lungfish. Tetrapods evolved from sarcopterygian ancestors.

Satellite cells   Populations of muscle stem cells and progenitor cells that reside alongside adult muscle fibers and can respond to injury or exercise by proliferating into myogenic cells that fuse and form new muscle fibers.

Scatter factor   See **Hepatocyte growth factor**.

Schizocoely   The embryonic process of forming the coelom by hollowing out a previously solid cord of mesodermal cells. Typical of protostomes. See also **Enterocoely**.

Schwann cell   Type of glial cell of the peripheral nervous system that generates a myelin sheath, allowing rapid transmission of electrical signals along an axon.

Sclerotomes   Blocks of mesodermal cells in the ventromedial half of each somite that will differentiate into the vertebrae, intervertebral discs (except for the nuclei pulposi) and ribs, in addition to the meninges of the spinal cord and the blood vessels that serve the spinal cord. They are also critical in patterning the neural crest and motor neurons.

Sebaceous gland   Glands that are associated with hair follicles and produce an oily substance, **sebum**, that serves to lubricate the hair and skin.

Secondary hypoblast   Underlies the epiblast in the bilaminar avian blastoderm. A sheet of cells derived from deep yolky cells at the posterior margin of the blastoderm that migrates anteriorly, displacing the hypoblast islands (primary hypoblast). Hypoblast cells do not contribute to the avian embryo proper, but instead form portions of the external membranes, especially the yolk sac, and provide chemical signals that specify the migration of epiblast cells. Also called endoblast.

Secondary larvae   Larvae that possess the same basic body plan as the adult; caterpillars and tadpoles are examples. Compare with **Primary larvae**.

Secondary neurulation   The process that forms the posterior portion of the neural tube by the coalescence of mesenchyme cells into a solid cord that subsequently forms cavities that coalesce to create a hollow tube.

Secondary oocyte   The haploid oocyte following the first meiotic division (this division also generates the first polar body).

Secondary sex determination   Developmental events, directed by hormones produced by the gonads that affect the phenotype outside the gonads. This includes the male or female duct systems and external genitalia, and, in many species, sex-specific body size, vocal cartilage, and musculature.

Secondary spermatocytes   A pair of haploid cells derived from the first meiotic division of a primary spermatocyte, which then complete the second division of meiosis to generate the four haploid spermatids.

Segment polarity genes   *Drosophila* zygotic genes, activated by the proteins encoded by the pair-rule genes, whose mRNA and protein products divide the embryo into segment-sized units, establishing the periodicity of the embryo. Segment polarity mutants showed defects (deletions, duplications, polarity reversals) in every segment.

Segmentation genes   Genes whose products divide the early *Drosophila* embryo into a repeating series of segmental primordia along the anterior-posterior axis. Include gap genes, pair-rule genes, and segment polarity genes.

Selective affinity   Principle that explains why disaggregated cells reaggregate to reflect their embryonic positions. Specifically, the inner surface of the ectoderm has a positive affinity for mesodermal cells and a negative affinity for the endoderm, while the mesoderm has positive affinities for both ectodermal and endodermal cells.

Self-renewal   The ability of a cell to divide and produce a replica of itself.

Semaphorins   Extracellular matrix proteins that repel migrating neural crest cells and axonal growth cones.

Seminiferous tubules   In male mammals, form in the gonad from the testis cords. They contain Sertoli cells (nurse cells) and spermatogonia (sperm stem cells).

Senescence   The physiological deterioration that characterizes old age.

Septum   A partition that divides a chamber, such as the atrial septa that split the developing atrium into left and right atria. Plural, septa.

Sertoli cells   Large secretory support cells in the seminiferous tubules of the testes involved in spermatogenesis in the adult through their role in nourishing and maintaining the developing sperm cells. They secrete AMH in the fetus and provide a niche for the incoming germ cells. They are derived from

somatic cells, which are in turn derived from the genital ridge epithelium.

Sesamoid bone   Small bones at joints that form as a result of mechanical stress (such as the patella). They are derived from mesoderm via intramembranous ossification.

Sex-lethal (Sxl)   An autosomal gene in *Drosophila* involved in sex determination. It codes for a splicing factor that initiates a cascade of RNA processing events, which eventually lead to male-specific and female-specific transcription factors, the Doublesex proteins. See ***Doublesex***.

Shh   See **Sonic hedgehog**.

Shield   See **Embryonic shield**.

Signal transduction cascades   Pathways of response whereby paracrine factors bind to a receptor that initiates a series of enzymatic reactions within the cell that in turn have often several responses as their end point, such as the regulation of transcription factors (such that different genes are expressed in the cells reacting to these paracrine factors) and/or the regulation of the cytoskeleton (such that the cells responding to the paracrine factors alter their shape or are permitted to migrate).

Silencer   A DNA regulatory element that binds transcription factors that actively repress the transcription of a particular gene.

Single stem cell asymmetry   A mode of stem cell division in which two types of cells are produced at each division, a stem cell and a developmentally committed cell.

Sinistral coiling   Left-coiling. In a snail, having its coils open to the left of its shells. See also **Dextral coiling**.

Sinus venosus   The posterior region of the developing heart, where the two major vitelline veins bringing blood to the heart fuse. Inflow tract to the atrial area of the heart.

Sinusoidal endothelial cells   Cells that line the large blood channels (sinusoids) of the liver and critical to liver function. Also provides paracrine factors needed for division of hepatoblast stem cells during liver regeneration. Long considered mesodermal in origin, they now are known to be derived at least in part by specialized endodermal cells.

Sirtuin genes   Encode histone deacetylation (chromatin-silencing) enzymes that guard the genome, preventing genes from being expressed at the wrong times and places, and may help repair chromosomal breaks. They may be important defenses against premature aging.

Skeletogenic mesenchyme   Also called primary mesenchyme, formed from the first tier of micromeres (the large micromeres) of the 60-cell sea urchin embryo. They ingress, moving into the blastocoel, and form the larval skeleton.

SKN-1   A maternally expressed transcription factor in the oocyte of the nematode *C. elegans* that controls the fate of the EMS cell, one of the cells of the 4-cell stage that marks the ventral region of the developing embryo.

Slit proteins   Proteins of the extracellular matrix that are chemorepulsive; involved in inhibiting migration of neural crest cells and in controlling growth of commissural axons.

Slow block to polyspermy   See **Cortical granule reaction**.

Smad family   Transcription factors activated by members of the TGF-β superfamily that function in the SMAD pathway. See also **SMAD pathway**.

SMAD pathway   The pathway activated by members of the TGF-β superfamily. The TGF-β ligand binds to a type II TGF-β receptor, which allows that receptor to bind to a type I TGF-β receptor. Once the two receptors are in close contact, the type II receptor phosphorylates a serine or threonine on the type I receptor, thereby activating it. The activated type I receptor can now phosphorylate the Smad proteins. Smads 1 and 5 are activated by the BMP family of TGF-β factors, while the receptors binding activin, Nodal, and the TGF-β family phosphorylate Smads 2 and 3. These phosphorylated Smads bind to Smad4 and form the transcription factor complex that will enter the nucleus.

Small micromeres   A cluster of cells produced by the fifth cleavage at the vegetal pole in the sea urchin embryo when the micromeres divide.

Solenoids   Structures, created from tightly wound nucleosomes stabilized by histone H1, that inhibit transcription of genes by preventing transcription factors and RNA polymerases from gaining access to the genes.

Soma   Greek, "body." Can refer to the cell body (particularly of neurons) or to the cells that form an organism's body (as distinct from the germ cells).

Somatic cell nuclear transfer (SCNT)   Less accurately known as "cloning," the procedure by which a cell nucleus is transferred into an activated enucleated egg and directs the development of a complete organism with the same genome as the donor cell.

Somatic cells   Cells that make up the body—i.e., all cells in the organism that are not germ cells. Compare with **Germ cells**.

Somatic (parietal) mesoderm   Derived from lateral mesoderm closest to the ectoderm (dorsal) and separated from

other components of lateral mesoderm (splanchnic, near endoderm, ventral) by the intraembryonic coelom. Together with the overlying ectoderm, the somatic mesoderm comprises the somatopleure, which will form the body wall. The somatic mesoderm also forms part of the lining of the coelom. Not to be confused with somitic (paraxial) mesoderm.

Somatopleure   Made up of somatic lateral plate mesoderm and overlying ectoderm.

Somites   Segmental blocks of mesoderm formed from paraxial mesoderm adjacent to the notochord (the axial mesoderm). Each contain major compartments: the sclerotome, which forms the axial skeleton (vertebrae and ribs), and the dermomyotome, which goes on to form dermatome and myotome. The dermatome forms the dermis of the back; the myotome forms musculature of the back, rib cage, and ventral body. Additional muscle progenitors detach from the lateral edge of the dermomyotome and migrate into the limbs to form the muscles of the fore- and hindlimbs.

Somitic mesoderm   See **Paraxial mesoderm**. Not to be confused with **somatic mesoderm**.

Somitogenesis   The process of segmentation of the paraxial mesoderm to form somites, beginning cranially and extending caudally. Its components are (1) periodicity, (2) fissure formation (to separate the somites), (3) epithelialization, (4) specification, and (5) differentiation.

Somitomeres   Early pre-somites, consisting of paraxial mesoderm cells organized into whorls of cells.

Sonic hedgehog (Shh)   The major hedgehog family paracrine factor. Shh has distinct functions in different tissues of the embryo. For example, it is secreted by the notochord inducing the ventral region of the neural tube to form the floor plate. It is also involved in the establishment of left-right asymmetry, primitive gut tube differentiation, proper feather formation in birds, differentiation of the sclerotome, and patterning the anterior-posterior axis of limb buds.

Sox9   An autosomal gene involved in several developmental processes, most notably bone formation. In the genital ridge of mammals, it induces testis formation, and XX humans with an extra copy of *SOX9* develop as males.

Specification   The first stage of commitment of cell or tissue fate during which the cell or tissue is capable of differentiating autonomously (i.e., by itself) when placed in an environment that is neutral with respect to the developmental pathway. At the stage of specification,

cell commitment is still capable of being reversed.

Spemann's Organizer   See **Organizer**.

Sperm head   Consists of the nucleus, acrosome, and minimal cytoplasm.

Sperm-activating peptides (SAPs)   Small chemotactic peptides found in the jelly of echinoderm eggs. They diffuse away from the egg jelly in seawater and are species specific, only attracting sperm of the same species. Resact, found in the sea urchin *Arbacia punctulata*, is an example.

Spermatids   Haploid sperm cells, the stage following the second meiotic division. In mammals, spermatids are still connected to one another by cytoplasmic bridges, allowing for diffusion of gene products across the cytoplasmic bridges.

Spermatogenesis   The production of sperm.

Spermatogonia   Sperm stem cells. When a spermatogonium stops undergoing mitosis, it becomes a primary spermatocyte and increases in size prior to meiosis.

Spermatozoa   The male gamete or mature sperm cell.

Spermiogenesis   The differentiation of the mature spermatozoa from the haploid round spermatid.

Spina bifida   A congenital defect resulting from incomplete closure of the spine around the spinal cord, usually in the lower back. There are differing degrees of severity, the most severe being when the neural folds also fail to close.

Spiral holoblastic cleavage   Characteristic of several animal groups, including annelid worms, some flatworms, and most molluscs. Cleavage is at oblique angles to the animal-vegetal axis, forming a "spiral" arrangement of daughter blastomeres. The cells touch one another at more places than do those of radially cleaving embryos, assuming the most thermodynamically stable packing orientation.

Splanchnic (visceral) mesoderm   Also called the visceral mesoderm and splanchnic lateral plate mesoderm; derived from lateral mesoderm closest to the endoderm (ventral) and separated from other component of lateral mesoderm (somatic, near ectoderm, dorsal) by the intraembryonic coelom. Together with the underlying endoderm, it forms the splanchnopleure. The splanchnic mesoderm will form the heart, capillaries, gonads, the visceral peritoneum and serous membranes that cover the organs, the mesenteries, and blood cells.

Splanchnopleure   Made up of splanchnic lateral plate mesoderm and underlying endoderm. See **Splanchnic mesoderm**.

Spliceosome A complex made up of small nuclear RNAs (snRNAs) and splicing factors, that binds to splice sites and mediates the splicing of nRNA.

Splicing enhancer A *cis*-acting sequence on nRNA that promotes the assembly of spliceosomes at RNA cleavage sites.

Splicing factors Proteins that bind to splice sites or to the areas adjacent to them.

Splicing isoforms Different proteins encoded by the same gene and generated by alternative splicing.

Splicing silencer A *cis*-acting sequence on nRNA that acts to exclude exons from an mRNA sequence.

Src family kinases (SFK) Family of enzymes that phosphorylate tyrosine residues; involved in many signaling events, including the responses of growth cones to chemoattractants.

Sry **S**ex-determining **r**egion of the **Y** chromosome. The *Sry* gene encodes the mammalian testis-determining factor. It is probably active for only a few hours in the genital ridge, during which time it synthesizes the Sry transcription factor, whose primary role is to activate the *Sox9* gene required for testis formation.

STAT **S**ignal **t**ransducers and **a**ctivators of **t**ranscription. A family of transcription factors, part of the JAK-STAT pathway. Important in the regulation of human fetal bone growth.

Stem cell A relatively undifferentiated cell from the embryo, fetus, or adult that, that divides and when it does so, produces (1) one cell that retains its undifferentiated character and remains in the stem cell niche; and (2) a second cell that leaves the niche and can undergo one or more paths of differentiation. See also **Adult stem cell; Embryonic stem cell.**

Stem cell factor (SCF) Paracrine factor important for maintaining certain stem cells, including hematopoietic, sperm, and pigment stem cells. Binds to the Kit receptor protein.

Stem cell mediated regeneration Process by which stem cells allow an organism to regrow certain organs or tissues (e.g., hair, blood cells) that have been lost.

Stem cell niche An environment (regulatory microenvironment) that provides a milieu of extracellular matrices and paracrine factors that allows cells residing within it to remain relatively undifferentiated. Regulates stem cell proliferation and differentiation.

Stereoblastulae Blastulae that have no blastocoel, e.g., blastulae produced by spiral cleavage.

Steroidogenic factor 1 (Sf1) A transcription factor that in mammals is necessary for creating the bipotential gonad. It declines in the developing ovary but remains at high levels in the developing testis, masculinizing both Leydig and Sertoli cells.

Stochastic Pertaining to a random process that provides a set of random variables that can be analyzed statistically, but not necessarily predicted.

Stomodeum An ectoderm-lined invagination in the oral region of the embryo that meets the endoderm of the closed gut tube to form the oral plate.

Stratum germinativum See **Basal layer**.

Stromal derived factor 1 (SDF1) A chemoattractant. SDF1is secreted, for example, by ectodermal placodes, thereby attracting cranial neural crest cells toward the placode.

Stylopod The proximal bones of a vertebrate limb, adjacent to the body wall; either the humerus (forelimb) or the femur (hindlimb).

Subgerminal cavity A space between the blastoderm and the yolk of avian eggs which is created when the blastoderm cells absorb water from the albumen ("egg white") and secrete fluid between themselves and the yolk.

Subgranular zone (SGZ) A region of the hippocampus in the cerebrum that contains neural stem cells, allowing for adult neurogenesis in this region.

Subventricular zone A region in the vertebrate cerebrum that is formed as progenitor cells migrate away from the ventricular zone.

Sulcus limitans A longitudinal groove that divides the developing spinal cord and medulla into dorsal (receives sensory input) and ventral (initiates motor functions) halves.

Superficial cleavage The divisions of the cytoplasm of centrolecithal zygotes that occur only in the rim of cytoplasm around the periphery of the cell due to the presence of a large amount of centrally-located yolk, as in insects.

Surfactant A secretion of specific proteins and phospholipids such as sphingomyelin and lecithin produced by the type II alveolar cells of the lungs very late in gestation. The surfactant enables the alveolar cells to touch one another without sticking together.

Symbiont The smaller organism in a symbiotic relationship in which the other organism is much larger and serves as the host, while the smaller organism may live on the surface or inside the body of the larger.

Symbiosis Greek, "living together." Refers to any close association between organisms of different species.

Synapse Junction at which a neuron contacts its target cell (which can be another neuron or another type of cell) and information in the form of neurotransmitter molecules (e.g., acetylcholine, GABA, serotonin) is exchanged across the synaptic cleft between the two cells.

Synapsis The highly specific parallel alignment (pairing) of homologous chromosomes during the first meiotic division.

Synaptic cleft The small cleft that separates the axon of a signaling neuron from the dendrite or soma of its target cell.

Synaptonemal complex The proteinaceous ribbon that forms during synapsis between homologous chromosomes, holding them together. A ladderlike structure with a central element and two lateral bars that are associated with the homologous chromosomes. See **Synapsis**.

Syncytial blastoderm Describes the *Drosophila* embryo during cleavage when nuclei have divided, but no cell membranes have yet formed to separate the nuclei into individual cells.

Syncytial specification The interactions of nuclei and transcription factors, which eventually result in cell specification, that take place in a common cytoplasm, as in the early *Drosophila* embryo.

Syncytiotrophoblast A population of cells from the mammalian trophoblast that undergoes mitosis without cytokinesis resulting in multinucleate cells. The syncytiotrophoblast tissue is thought to further the progression of the embryo into the uterine wall by digesting uterine tissue.

Syncytium Many nuclei residing in a common cytoplasm, results either from karyokinesis without cytokinesis or from cell fusion.

Syndetome Greek *syn*, "connected." Derived from the most dorsal sclerotome cells, which express the *scleraxis* gene and generate the tendons.

Syndrome Greek, "happening together." Several malformations or pathologies that occur concurrently. Genetically based syndromes are caused either by (1) a chromosomal event (such as trisomy 21, or Down syndrome) where several genes are deleted or added, or (2) by one gene having many effects.

Systems theory In development, refers to an approach that views the organism as coming together through the interactions of its component processes. Although the emphasis applied to each varies, the theoretical systems approach can be characterized by six principles: (1) context-dependent properties; (2) level-specific properties and emergence; (D) heterogeneous causation; (4)

integration; (5) modularity and robustness; and (6) homeorhesis (stability while undergoing change).

T

T-box (Tbx)   A specific DNA-binding domain found in certain transcription factors, including the *T* (*Brachyury*) gene, *Tbx4* and *Tbx5*. Tbx4 and Tbx5 help specify hindlimbs and forelimbs, respectively.

Target selection   The second step in the specification of axonal connection, wherein the axons, once they reach the correct area, recognize and bind to a set of cells with which they may form stable connections.

Telencephalon   The anterior subdivision of the prosencephalon; will eventually form the cerebral hemispheres.

Telogen   The resting phase of the hair follicle regeneration cycle.

Telolecithal   Describes the eggs of birds and fish which have only one small area at the animal pole of the egg that is free of yolk.

Telomerase   Enzyme complex that can extend the telomeres to their full length and maintains telomere integrity.

Telomeres   Repeated DNA sequences at the ends of chromosomes that provide a protective cap to the chromosomes.

Telson   A tail-like structure; the posterior most segment of certain arthropods. Seen in insect larvae such as *Drosophila*.

Temporal colinearity   The mechanism that controls the timing of Hox gene activation, which occurs anteriorly first and progressively more posteriorly; sets up spatial colinearity of Hox gene expression relative to their 3′-to-5′ genomic organization.

Teratocarcinoma   A tumor derived from malignant primordial germ cells and containing an undifferentiated stem cell population (embryonal carcinoma, or EC cells) that has biochemical and developmental properties similar to those of the inner cell mass. EC cells can differentiate into a wide variety of tissues, including gut and respiratory epithelia, muscle, nerve, cartilage, and bone.

Teratogens   Greek, "monster-formers." Exogenous agents that cause disruptions in development resulting in teratogenesis, the formation of congenital defects. Teratology is the study of birth defects and of how environmental agents disrupt normal development.

Terminal end bulbs   The ends of the extensive branches of ducts in the mammary glands of mammals. Under the influence of estrogens at puberty, the ducts grow by the elongation of these buds.

Testis cords   Loops in the medullary (central) region of the developing testis formed by the developing Sertoli cells and the incoming germ cells. Will become the seminiferous tubules and site of spermatogenesis.

Testis-determining factor   A protein encoded by the *Sry* gene on the mammalian Y chromosome that organizes the gonad into a testis rather than an ovary.

Testosterone   A steroid hormone that is androgenic. In mammals, it is secreted by the fetal testes and masculinizes the fetus, stimulating the formation of the penis, male duct system, scrotum, and other portions of the male anatomy, as well as inhibiting development of the breast primordia.

Tetrad   See **Bivalent**.

Tetrapods   Latin, "four feet." Includes the vertebrates amphibians, reptiles, birds, and mammals. Evolved from lobe-finned fish (sarcopterygian) ancestors.

TGF-β family   **T**ransforming **g**rowth **f**actor-β. A family of growth factors within the TGF-β superfamily.

TGF-β superfamily   More than 30 structurally related members of a group of paracrine factors. The proteins encoded by TGF-β superfamily genes are processed such that the carboxy-terminal region contains the mature peptide. These peptides are dimerized into homodimers (with themselves) or heterodimers (with other TGF-β peptides) and are secreted from the cell. The TGF-β superfamily includes the TGF-β family, activin family, bone morphogenetic proteins (BMPs), Vg1 family, and other proteins, including glial-derived neurotrophic factor (GDNF; necessary for kidney and enteric neuron differentiation) and anti-Müllerian hormone (AMH; involved in mammalian sex determination).

Thecal cells   Steroid hormone-secreting cells of the mammalian ovary that, together with the granulosa cells, form the follicles surrounding the germ cells. They differentiate from mesenchyme cells of the ovary.

Thermotaxis   Migration that is directed by a gradient of temperature, either up or down the gradient.

Threshold model   A model of development wherein biological events are triggered when a specific concentration of a morphogen or hormone is reached.

Thyroid hormone receptors (TRs)   Nuclear receptors that bind the thyroid hormones tri-iodothyronine (T_3), as well as thyroxine (T_4). Once bound to the hormone, the TR becomes a transcriptional activator of gene expression. There are several different TR types, including TRα and TRβ.

Thyroxine (T_4)   Thyroid hormone containing four iodine molecules; is converted to the more active T_3 form through removal of one iodine molecule. Increases basal metabolic rate in cells. Initiates metamorphosis in amphibians.

Tip cells   Certain endothelial cells that can respond to vascular endothelial growth factor (VEGF) and begin "sprouting" to form a new vessel during angiogenesis. See also **Ureteric bud tip cells**.

Tissue engineering   A regenerative medicine approach whereby a scaffold is generated from material that resembles extracellular matrix or decellularized extracellular matrix from a donor, is seeded with stem cells, and is used to replace an organ or part of an organ.

Torpedo   The receptor protein for Gurken. When expressed in terminal follicle cells in a *Drosophila* egg chamber, it binds to Gurken produced by the egg, which signals these follicle cells to differentiate into posterior follicle cells and synthesize a molecule that activates protein kinase A in the egg; part of the process that sets up the anterior-posterior axis of the egg and future embryo.

Totipotent   Latin, "capable of all." Describes the potency of certain stem cells to form all structures of an organism, such as the earliest mammalian blastomeres (up to the 8-cell stage), which can form both trophoblast cells and the embryo precursor cells. Compare with **Pluripotent**.

trans-activating domain   The transcription factor domain that activates or suppresses the transcription of the gene whose promoter or enhancer it has bound, usually by enabling the transcription factor to interact with the proteins involved in binding RNA polymerase or with enzymes that modify histones.

trans-**regulatory elements**   Soluble molecules whose genes are located elsewhere in the genome and which bind to the *cis*-regulatory elements. They are usually transcription factors or microRNAs.

Transcription   The process of copying DNA into RNA.

Transcription elongation complex (TEC)   A complex of several transcription factors that breaks the connection between RNA polymerase II and the Mediator complex, allowing transcription (which has been initiated) to proceed.

Transcription elongation suppressor   A repressive transcription factor that functions to prevent the transcription elongation complex from associating with RNA polymerase II, pausing transcription.

Transcription factor   A protein the binds to DNA with precise sequence recognition for specific promoters, enhancers, or silencers.

Transcription factor domains   The three major domains are a DNA-binding domain, a trans-activating domain and a protein-protein interaction domain.

Transcription initiation site   DNA sequence of a gene that codes for the addition of a modified nucleotide "cap" at the 5′ end of the RNA soon after it is transcribed. Also called the cap sequence.

Transcription termination sequence   DNA sequence of a gene where transcription is terminated. Transcription continues for about 1000 nucleotides beyond the AATAAA site of the 3′ untranslated region of the gene before being terminated.

Transcription-associated factors (TAFs)   Proteins that stabilize RNA polymerase on the promoter of a gene and enable it to initiate transcription.

Transcriptome   Total messenger RNAs (mRNAs) expressed by genes in an organism or a specific type of tissue or cell.

Transdifferentiation   The transformation of one cell type into another.

Transforming growth factor   See **TGF-β superfamily**.

Transgene   Exogenous DNA or gene introduced through experimental manipulation into a cell's genome.

Transit amplifying cells   See **Progenitor cells**.

Transition zone   In neural tube development in vertebrates, the zone between the region that undergoes primary neurulation and the region that undergoes secondary neurulation. The size of this zone varies among different species. See also **Primary neurulation** and **Secondary neurulation**.

Translation   The process in which the codons of a messenger RNA are translated into the amino acid sequence of a polypeptide chain.

Translation initiation site   The ATG codon (becomes AUG in mRNA), which signals the beginning of the first exon (protein-coding region) of a gene.

Translation termination codon   Sequence in a gene, TAA, TAG, or TGA, which is transcribed as a codon in the mRNA—when a ribosome encounters this codon, the ribosome dissociates and the protein is released.

Trefoil stage   A stage in certain spirally cleaving embryos, wherein a particularly large polar lobe is extruded at first cleavage, giving the appearance of a third cell forming before the polar lobe is reabsorbed back into the CD blastomere.

Tri-iodothyronine (T$_3$)   The more active form of thyroid hormone, produced through the removal of an iodine molecular from thyroxine (T$_4$). See **Thyroxine (T$_4$)**.

Triploblasts   See **Bilaterians**.

Trisomy 21   Condition (in humans) of having three copies of chromosome 2 (an example of aneuploidy). Causes Down syndrome.

Trithorax   Family of proteins that are recruited to retain the memory of the transcriptional state of regions of DNA as the cell goes through mitosis; keeps active genes active.

Trophectoderm cells   In the mammalian embryo, the outer layer of cells of the blastocyst that surround the inner cell mass and blastocoel; develop into the embryonic side of the placenta.

Trophoblast   The external cells of the early mammalian embryo (i.e., the morula and the blastocyst) that will bind to the uterus. Trophoblast cells form the chorion (the embryonic portion of the placenta). Also called trophectoderm.

Truncus arteriosus   Cardiac outflow tract precursor that along with the conus arteriosus will form the base of the aorta and pulmonary artery.

Trunk neural crest   Neural crest cells migrating from this region become the dorsal root ganglia containing the sensory neurons, sympathetic ganglia, adrenal medulla, the nerve clusters surrounding the aorta, and Schwann cells if they migrate along a ventral pathway, and they generate melanocytes of the dorsum and belly if they migrate along a dorsolateral pathway.

Tubulin   A dimeric protein that polymerizes to form microtubules. Microtubules are a major component of the cytoskeleton; they are found in centrioles and basal bodies; they also form the mitotic spindle and axoneme of cilia and flagella.

Tudor   One of the proteins, along with Piwi, Vasa, and Nanos, expressed in germ cells to suppress gene expression. Also involved in anterior-posterior polarity in the *Drosophila* embryo by localizing Nanos, a posterior morphogen.

Tumor angiogenesis factors   Factors secreted by microtumors; these factors (including VEGFs, Fgf2, placenta-like growth factor, and others) stimulate mitosis in endothelial cells and direct the cell differentiation into blood vessels in the direction of the tumor.

Tumor suppressor genes   Regulatory genes whose gene products protect against a cell progressing towards cancer. Gene products may inhibit cell division or increase the adhesion between cells; they can also induce apoptosis of rapidly dividing cells. Cancer can result from either mutations or inappropriate methylations that inactivate tumor suppressor genes.

Tunica albuginea   In mammals, a thick, whitish capsule of extracellular matrix that encases the testis.

"Turing-type" model   See **Reaction-diffusion model**.

Type II deiodinase   Intracellular enzyme that removes an iodine atom from the outer ring of thyroxine (T$_4$), converting it into the more active T$_3$ hormone.

Type III deiodinase   Intracellular enzyme that removes an iodine atom from the inner ring of T$_3$ to convert it into the inactive compound T$_2$, which will eventually be metabolized to tyrosine.

Type IV collagen   A type of collagen that forms a fine meshwork; found in the basal lamina, an extracellular matrix that lies underneath epithelia.

Type A spermatogonia   In mammals, sperm stem cells that undergo mitosis and maintain the population of Type A spermatogonia while also generating Type B spermatogonia.

Type B cells   A type of neural stem cell found in the rosettes of the V-SVZ of the cerebrum; fuel the generation of specific types of neurons in the olfactory bulb and striatum.

Type B spermatogonia   In mammals, precursors of the spermatocytes and the last cells of the line that undergo mitosis. They divide once to generate the primary spermatocytes.

U

Umbilical cord   Connecting cord derived from the allantois that brings the embryonic blood circulation to the uterine vessels of the mother in placental mammals.

Umbilical veins   See **Omphalomesenteric veins** and **Vitelline veins**.

Undifferentiated zone   See **Progress zone**.

Unipotent stem cells   Stem cells that generate only one cell type, such as the spermatogonia of the mammalian testes that only generate sperm.

Unsegmented mesoderm   Bands of paraxial mesoderm prior to their segmentation into somites.

3′ Untranslated region (3′ UTR)   A region of a eukaryotic gene and RNA following the translation termination codon that, although transcribed, is not translated into protein. It includes the region needed for insertion of the polyA tail on the transcript that allows the transcript to exit the nucleus.

5′ Untranslated region (5′ UTR) Also called a leader sequence or leader RNA; a region of a eukaryotic gene or RNA. In a gene, it is a sequence of base pairs between the transcription initiation and translation initiation sites; in an RNA, it is its 5′ end. These are not translated into protein, but can determine the rate at which translation is initiated.

Ureteric bud tip cells A population of stem cells that form at the tips of the ureteric bud branches during metanephric kidney formation.

Ureteric buds In amniotes, paired epithelial branches induced by the metanephrogenic mesenchyme to branch from each of the paired nephric ducts. Ureteric buds will form the collecting ducts, renal pelvis, and ureters that take the urine to the bladder.

Urodeles Amphibian group that includes the salamanders. Compare with **Anurans**.

Urogenital sinus In mammals, the region of the cloaca that is separated from the rectum by the urogenital septum. The bladder forms from the anterior portion of the sinus, and the urethra develops from the posterior region. In females, also forms Skene's glands; in males it also forms the prostate gland.

Uterine cycle A component of the menstrual cycle, the function of the uterine cycle is to provide the appropriate environment for the developing blastocyst.

V

Vagal neural crest Neural crest cells from the neck region, which overlaps the cranial/trunk crest boundary. Together with the sacral neural crest, generates the parasympathetic (enteric) ganglia of the gut, which are required for peristaltic movement of the bowels.

Vas (ductus) deferens Derived from the Wolffian duct, the tube through which sperm pass from the epididymis to the urethra.

Vasa One of the proteins, along with Tudor, Piwi, and Nanos, expressed in germ cells to suppress gene expression. Also involved in anterior-posterior polarity in the *Drosophila* embryo by localizing Nanos, a posterior morphogen.

Vascular endothelial growth factors (VEGFs) A family of proteins involved in vasculogenesis that includes several VEGFs, as well as placental growth factor. Each VEGF appears to enable the differentiation of the angioblasts and their multiplication to form endothelial tubes. Also critical for angiogenesis.

Vasculogenesis The de novo creation of a network of blood vessels from the lateral plate mesoderm. See also **Extra-embryonic vasculogenesis**.

Vegetal hemisphere The bottom portion of an ovum, where yolk is more concentrated. The yolk can be an impediment to cleavage, as in the amphibian embryo, causing the yolk-filled cells to divide more slowly and undergo less movement during embryogenesis.

Vegetal plate Area of thickened cells at the vegetal pole of the sea urchin blastula.

Vegetal pole The yolk containing end of the egg or embryo, opposite the animal pole.

Vegetal rotation During frog gastrulation, internal cell rearrangements place the prospective pharyngeal endoderm cells adjacent to the blastocoel and immediately above the involuting mesoderm.

VEGF See **Vascular endothelial growth factors**.

VegT pathway Involved in dorsal-ventral polarity and specification of the organizer cells in the amphibian embryo. The VegT pathway activates the expression of Nodal-related paracrine factors in the cells of the vegetal hemisphere of the embryo, which in turn activate the Smad2 transcription factor in the mesodermal cells above them, activating genes that give these cells their "organizer" properties.

Vellus Short and silky hair of the fetus and neonate that remains on many parts of the human body that are usually considered hairless, such as the forehead and eyelids. In other areas of the body, vellus hair gives way to longer and thicker "terminal" hair.

Ventral furrow Invagination of the prospective mesoderm, about 1000 cells constituting the ventral midline of the embryo, at the onset of gastrulation in *Drosophila*.

Ventral pathway Migration pathway of trunk neural crest cells that travel ventrally through the anterior of the sclerotome and contribute to the sympathetic and parasympathetic ganglia, adrenomedullary cells, and dorsal root ganglia.

Ventricular (ependymal) cells Cells derived from the neuroepithelium that line the ventricles of the brain and secrete cerebrospinal fluid.

Ventricular radial glia (vRG) Progenitor cells that reside in the ventricular zone. They give rise to neurons, outer radial glia (oRG), and intermediate progenitor (IP) cells. See also **Ventricular zone**.

Ventricular zone (VZ) Inner layer of the developing spinal cord and brain. Forms from the germinal neuroepithelium of the original neural tube and contains neural progenitor cells that are a source of neurons and glial cells. Will form the ependyma.

Ventricular-subventricular zone (V-SVZ) Region of the cerebrum that contains neural stem cells and is capable of neurogenesis in the adult.

Vertical transmission In referring to symbiosis, the transfer of symbionts from one generation to the next through the germ cells, usually the eggs.

Vg1 A family of proteins that is part of the TGF-β superfamily. Important in specifying mesoderm in amphibian embryos. See also **TGF-β superfamily**.

Visceral endoderm A region of the primitive endoderm where the cells contact the epiblast in the mammalian embryo. See **Primitive endoderm**.

Visceral mesoderm See **Splanchnic mesoderm**.

Viscerocranium The jaws and other skeletal elements derived from the pharyngeal arches.

Vital dyes Stains used to label living cells without killing them. When applied to embryos, vital dyes have been used to follow cell migration during development and generate fate maps of specific regions of the embryo.

Vitelline envelope In invertebrates, the extracellular matrix that forms a fibrous mat around the egg outside the cell membrane and is often involved in sperm-egg recognition and is essential for the species-specific binding of sperm. The vitelline envelope contains several different glycoproteins. It is supplemented by extensions of membrane glycoproteins from the cell membrane and by proteinaceous "posts" that adhere the vitelline envelope to the membrane.

Vitelline veins The veins, continuous with the endocardium, that carry nutrients from the yolk sac into the sinus venosus of the developing vertebrate heart. In birds, these veins form from yolk sac blood islands, and bring nutrients to the embryo and transport gases to and from the sites of respiratory exchange. In mammals they are called omphalomesenteric veins or umbilical veins.

Vitellogenesis The formation of yolk proteins, which are deposited in the primary oocyte.

Viviparity Young are nourished in and born from the mother's body rather than hatched from an egg, as in placental mammals. Compare with **Oviparity**.

Vulval precursor cells (VPCs) Six cells in the larval stage of *C. elegans* that will form the vulva via inductive signals.

W

White matter The axonal (as opposed to neuronal) region of the brain and spinal cord. Name derives from the fact that myelin sheaths give the axons a whitish appearance. Compare with **Gray matter**.

Wholist organicism Philosophical notion stating that the properties of the whole cannot be predicted solely from the properties of its component parts, and that the properties of the parts are informed by their relationship to the whole. It was very influential in the construction of developmental biology.

Wnt pathways Signal transduction cascades initiated by the binding of a Wnt protein to its receptor Frizzled on the cell membrane. This binding can initiate any of number of different pathways ("canonical" and "noncanonical") to activate Wnt-responsive genes in the nucleus.

Wnt4 A protein in the Wnt family; in mammals it is involved in primary sex determination, kidney development, and the timing of meiosis. It is expressed in the bipotential gonads, but becomes undetectable in XY gonads becoming testes; it is maintained in XX gonads becoming ovaries. See also **Wnts**.

Wnt7a A Wnt protein especially important in specifying dorsal-ventral polarity in the tetrapod limb; expressed in the dorsal, but not ventral, ectoderm of limb buds. If expression in this region is eliminated, both dorsal and ventral sides of the limb form structures appropriate for the ventral surface, such as ventral footpads on both surfaces of a paw. See also **Wnts**.

Wnts A gene family of cysteine-rich glycoprotein paracrine factors. Their name is a fusion of the name of the *Drosophila* segment polarity gene *wingless* with the name of one of its vertebrate homologues, *integrated*. Wnt proteins are critical in establishing the polarity of insect and vertebrate limbs, promoting the proliferation of stem cells, and in several steps of urogenital system development.

Wolffian (nephric) duct In vertebrates, the duct of the developing excretory system that grows down alongside the mesonephric mesoderm and induces it to form kidney tubules. In amniotes, it later degenerates in females, but in males, becomes the epididymis and vas deferens.

Wound epidermis In salamander limb regeneration, the epidermal cells that migrate over the stump amputation to cover the wound surface immediately following amputation; later thickens to form the apical ectodermal cap.

X

X chromosome inactivation In mammals, the irreversible conversion of the chromatin of one X chromosome in each female (XX) cell into highly condensed heterochromatin—a Barr body—thus preventing excess transcription of genes on the X chromosome. See also **Dosage compensation**.

Y

Yellow crescent Region of the tunicate zygote cytoplasm extending from the vegetal pole to the equator that forms after fertilization by the migration of cytoplasm containing yellow lipid inclusions; will become mesoderm. Contains the mRNA for transcription factors that will specify the muscles.

Yolk cell The cell containing the yolk in a fish embryo, once the yolk-free cytoplasm at the animal pole of the egg divides to form individual cells above the yolky cytoplasm. Initially, all the cells maintain a connection with the underlying yolk cell.

Yolk plug The large endodermal cells that remain exposed on the vegetal surface surrounded by the blastopore of the amphibian gastrulating embryo.

Yolk sac The first extraembryonic membrane to form, derived from splanchnopleure that grows over the yolk to enclose it. The yolk sac mediates nutrition in developing birds and reptiles. It is connected to the midgut by the yolk duct (vitelline duct), so that the walls of the yolk sac and the walls of the gut are continuous.

Yolk syncytial layer (YSL) A cell population in the zebrafish cleavage stage embryo formed at the ninth or tenth cell cycle, when the cells at the vegetal edge of the blastoderm fuse with the underlying yolk cell, producing a ring of nuclei in the part of the yolk cell cytoplasm that sits just beneath the blastoderm. Important for directing some of the cell movements of gastrulation.

Z

Zebrabow Transgenic zebrafish used to trigger the expression of different combinations and amounts of different fluorescent proteins within cells, labeling them with a seeming "rainbow" of possible colors that can be used to identify each individual cell in a tissue, organ, or whole embryo.

Zeugopod The middle bones of the vertebrate limb; the radius and ulna (forelimb) or tibia and fibula (hindlimb).

Zona pellucida Glycoprotein coat (extracellular matrix) around the mammalian egg, synthesized and secreted by the growing oocyte.

Zona proteins 1, 2, and 3 (ZP1, ZP2, ZP3) The three major glycoproteins found in the zona pellucida of the mammalian egg; the human zona pellucida also contains ZP4. Involved in binding sperm in a relatively, but not absolutely, species-specific manner.

Zone of polarizing activity (ZPA) A small block of mesodermal tissue in the very posterior of the limb bud progress zone. Specifies the anterior-posterior axis of the developing limb through the action of the paracrine factor Sonic hedgehog.

Zygote A fertilized egg with a diploid chromosomal complement in its zygote nucleus generated by fusion of the haploid male and female pronuclei.

Zygotene Greek, "yoked threads." In the first meiotic division, it is the second stage of prophase I, when homologous chromosomes pair side by side; follows leptotene.

Author Index

Subject Index

About the Book

Editor: Rachel Meyers

Project Editors: Carol Wigg, Sydney Carroll

Copy Editor: Kathleen Lafferty

Production Manager: Christopher Small

Photo Research: David McIntyre

Book Design: Joanne Delphia

Cover Design: Jefferson Johnson

Book Layout: Jefferson Johnson

Illustration Program: Dragonfly Media Group

Indexes: Grant Hackett

Book and Cover Manufacture: RR Donnelley